Sequential Analysis
Hypothesis Testing and Changepoint Detection

MONOGRAPHS ON STATISTICS AND APPLIED PROBABILITY

General Editors

F. Bunea, V. Isham, N. Keiding, T. Louis, R. L. Smith, and H. Tong

1. Stochastic Population Models in Ecology and Epidemiology *M.S. Barlett* (1960)
2. Queues *D.R. Cox and W.L. Smith* (1961)
3. Monte Carlo Methods *J.M. Hammersley and D.C. Handscomb* (1964)
4. The Statistical Analysis of Series of Events *D.R. Cox and P.A.W. Lewis* (1966)
5. Population Genetics *W.J. Ewens* (1969)
6. Probability, Statistics and Time *M.S. Barlett* (1975)
7. Statistical Inference *S.D. Silvey* (1975)
8. The Analysis of Contingency Tables *B.S. Everitt* (1977)
9. Multivariate Analysis in Behavioural Research *A.E. Maxwell* (1977)
10. Stochastic Abundance Models *S. Engen* (1978)
11. Some Basic Theory for Statistical Inference *E.J.G. Pitman* (1979)
12. Point Processes *D.R. Cox and V. Isham* (1980)
13. Identification of Outliers *D.M. Hawkins* (1980)
14. Optimal Design *S.D. Silvey* (1980)
15. Finite Mixture Distributions *B.S. Everitt and D.J. Hand* (1981)
16. Classification *A.D. Gordon* (1981)
17. Distribution-Free Statistical Methods, 2nd edition *J.S. Maritz* (1995)
18. Residuals and Influence in Regression *R.D. Cook and S. Weisberg* (1982)
19. Applications of Queueing Theory, 2nd edition *G.F. Newell* (1982)
20. Risk Theory, 3rd edition *R.E. Beard, T. Pentikäinen and E. Pesonen* (1984)
21. Analysis of Survival Data *D.R. Cox and D. Oakes* (1984)
22. An Introduction to Latent Variable Models *B.S. Everitt* (1984)
23. Bandit Problems *D.A. Berry and B. Fristedt* (1985)
24. Stochastic Modelling and Control *M.H.A. Davis and R. Vinter* (1985)
25. The Statistical Analysis of Composition Data *J. Aitchison* (1986)
26. Density Estimation for Statistics and Data Analysis *B.W. Silverman* (1986)
27. Regression Analysis with Applications *G.B. Wetherill* (1986)
28. Sequential Methods in Statistics, 3rd edition *G.B. Wetherill and K.D. Glazebrook* (1986)
29. Tensor Methods in Statistics *P. McCullagh* (1987)
30. Transformation and Weighting in Regression *R.J. Carroll and D. Ruppert* (1988)
31. Asymptotic Techniques for Use in Statistics *O.E. Bandorff-Nielsen and D.R. Cox* (1989)
32. Analysis of Binary Data, 2nd edition *D.R. Cox and E.J. Snell* (1989)
33. Analysis of Infectious Disease Data *N.G. Becker* (1989)
34. Design and Analysis of Cross-Over Trials *B. Jones and M.G. Kenward* (1989)
35. Empirical Bayes Methods, 2nd edition *J.S. Maritz and T. Lwin* (1989)
36. Symmetric Multivariate and Related Distributions *K.T. Fang, S. Kotz and K.W. Ng* (1990)
37. Generalized Linear Models, 2nd edition *P. McCullagh and J.A. Nelder* (1989)
38. Cyclic and Computer Generated Designs, 2nd edition *J.A. John and E.R. Williams* (1995)
39. Analog Estimation Methods in Econometrics *C.F. Manski* (1988)
40. Subset Selection in Regression *A.J. Miller* (1990)
41. Analysis of Repeated Measures *M.J. Crowder and D.J. Hand* (1990)
42. Statistical Reasoning with Imprecise Probabilities *P. Walley* (1991)
43. Generalized Additive Models *T.J. Hastie and R.J. Tibshirani* (1990)
44. Inspection Errors for Attributes in Quality Control *N.L. Johnson, S. Kotz and X. Wu* (1991)
45. The Analysis of Contingency Tables, 2nd edition *B.S. Everitt* (1992)

46. The Analysis of Quantal Response Data *B.J.T. Morgan* (1992)
47. Longitudinal Data with Serial Correlation—A State-Space Approach *R.H. Jones* (1993)
48. Differential Geometry and Statistics *M.K. Murray and J.W. Rice* (1993)
49. Markov Models and Optimization *M.H.A. Davis* (1993)
50. Networks and Chaos—Statistical and Probabilistic Aspects *O.E. Barndorff-Nielsen, J.L. Jensen and W.S. Kendall* (1993)
51. Number-Theoretic Methods in Statistics *K.-T. Fang and Y. Wang* (1994)
52. Inference and Asymptotics *O.E. Barndorff-Nielsen and D.R. Cox* (1994)
53. Practical Risk Theory for Actuaries *C.D. Daykin, T. Pentikäinen and M. Pesonen* (1994)
54. Biplots *J.C. Gower and D.J. Hand* (1996)
55. Predictive Inference—An Introduction *S. Geisser* (1993)
56. Model-Free Curve Estimation *M.E. Tarter and M.D. Lock* (1993)
57. An Introduction to the Bootstrap *B. Efron and R.J. Tibshirani* (1993)
58. Nonparametric Regression and Generalized Linear Models *P.J. Green and B.W. Silverman* (1994)
59. Multidimensional Scaling *T.F. Cox and M.A.A. Cox* (1994)
60. Kernel Smoothing *M.P. Wand and M.C. Jones* (1995)
61. Statistics for Long Memory Processes *J. Beran* (1995)
62. Nonlinear Models for Repeated Measurement Data *M. Davidian and D.M. Giltinan* (1995)
63. Measurement Error in Nonlinear Models *R.J. Carroll, D. Rupert and L.A. Stefanski* (1995)
64. Analyzing and Modeling Rank Data *J.J. Marden* (1995)
65. Time Series Models—In Econometrics, Finance and Other Fields *D.R. Cox, D.V. Hinkley and O.E. Barndorff-Nielsen* (1996)
66. Local Polynomial Modeling and its Applications *J. Fan and I. Gijbels* (1996)
67. Multivariate Dependencies—Models, Analysis and Interpretation *D.R. Cox and N. Wermuth* (1996)
68. Statistical Inference—Based on the Likelihood *A. Azzalini* (1996)
69. Bayes and Empirical Bayes Methods for Data Analysis *B.P. Carlin and T.A Louis* (1996)
70. Hidden Markov and Other Models for Discrete-Valued Time Series *I.L. MacDonald and W. Zucchini* (1997)
71. Statistical Evidence—A Likelihood Paradigm *R. Royall* (1997)
72. Analysis of Incomplete Multivariate Data *J.L. Schafer* (1997)
73. Multivariate Models and Dependence Concepts *H. Joe* (1997)
74. Theory of Sample Surveys *M.E. Thompson* (1997)
75. Retrial Queues *G. Falin and J.G.C. Templeton* (1997)
76. Theory of Dispersion Models *B. Jørgensen* (1997)
77. Mixed Poisson Processes *J. Grandell* (1997)
78. Variance Components Estimation—Mixed Models, Methodologies and Applications *P.S.R.S. Rao* (1997)
79. Bayesian Methods for Finite Population Sampling *G. Meeden and M. Ghosh* (1997)
80. Stochastic Geometry—Likelihood and computation *O.E. Barndorff-Nielsen, W.S. Kendall and M.N.M. van Lieshout* (1998)
81. Computer-Assisted Analysis of Mixtures and Applications—Meta-Analysis, Disease Mapping and Others *D. Böhning* (1999)
82. Classification, 2nd edition *A.D. Gordon* (1999)
83. Semimartingales and their Statistical Inference *B.L.S. Prakasa Rao* (1999)
84. Statistical Aspects of BSE and vCJD—Models for Epidemics *C.A. Donnelly and N.M. Ferguson* (1999)
85. Set-Indexed Martingales *G. Ivanoff and E. Merzbach* (2000)
86. The Theory of the Design of Experiments *D.R. Cox and N. Reid* (2000)
87. Complex Stochastic Systems *O.E. Barndorff-Nielsen, D.R. Cox and C. Klüppelberg* (2001)
88. Multidimensional Scaling, 2nd edition *T.F. Cox and M.A.A. Cox* (2001)
89. Algebraic Statistics—Computational Commutative Algebra in Statistics *G. Pistone, E. Riccomagno and H.P. Wynn* (2001)
90. Analysis of Time Series Structure—SSA and Related Techniques *N. Golyandina, V. Nekrutkin and A.A. Zhigljavsky* (2001)
91. Subjective Probability Models for Lifetimes *Fabio Spizzichino* (2001)
92. Empirical Likelihood *Art B. Owen* (2001)
93. Statistics in the 21st Century *Adrian E. Raftery, Martin A. Tanner, and Martin T. Wells* (2001)

94. Accelerated Life Models: Modeling and Statistical Analysis
Vilijandas Bagdonavicius and Mikhail Nikulin (2001)
95. Subset Selection in Regression, Second Edition *Alan Miller* (2002)
96. Topics in Modelling of Clustered Data *Marc Aerts, Helena Geys, Geert Molenberghs, and Louise M. Ryan* (2002)
97. Components of Variance *D.R. Cox and P.J. Solomon* (2002)
98. Design and Analysis of Cross-Over Trials, 2nd Edition *Byron Jones and Michael G. Kenward* (2003)
99. Extreme Values in Finance, Telecommunications, and the Environment
Bärbel Finkenstädt and Holger Rootzén (2003)
100. Statistical Inference and Simulation for Spatial Point Processes
Jesper Møller and Rasmus Plenge Waagepetersen (2004)
101. Hierarchical Modeling and Analysis for Spatial Data
Sudipto Banerjee, Bradley P. Carlin, and Alan E. Gelfand (2004)
102. Diagnostic Checks in Time Series *Wai Keung Li* (2004)
103. Stereology for Statisticians *Adrian Baddeley and Eva B. Vedel Jensen* (2004)
104. Gaussian Markov Random Fields: Theory and Applications *Håvard Rue and Leonhard Held* (2005)
105. Measurement Error in Nonlinear Models: A Modern Perspective, Second Edition
Raymond J. Carroll, David Ruppert, Leonard A. Stefanski, and Ciprian M. Crainiceanu (2006)
106. Generalized Linear Models with Random Effects: Unified Analysis via H-likelihood
Youngjo Lee, John A. Nelder, and Yudi Pawitan (2006)
107. Statistical Methods for Spatio-Temporal Systems
Bärbel Finkenstädt, Leonhard Held, and Valerie Isham (2007)
108. Nonlinear Time Series: Semiparametric and Nonparametric Methods *Jiti Gao* (2007)
109. Missing Data in Longitudinal Studies: Strategies for Bayesian Modeling and Sensitivity Analysis
Michael J. Daniels and Joseph W. Hogan (2008)
110. Hidden Markov Models for Time Series: An Introduction Using R
Walter Zucchini and Iain L. MacDonald (2009)
111. ROC Curves for Continuous Data *Wojtek J. Krzanowski and David J. Hand* (2009)
112. Antedependence Models for Longitudinal Data *Dale L. Zimmerman and Vicente A. Núñez-Antón* (2009)
113. Mixed Effects Models for Complex Data *Lang Wu* (2010)
114. Intoduction to Time Series Modeling *Genshiro Kitagawa* (2010)
115. Expansions and Asymptotics for Statistics *Christopher G. Small* (2010)
116. Statistical Inference: An Integrated Bayesian/Likelihood Approach *Murray Aitkin* (2010)
117. Circular and Linear Regression: Fitting Circles and Lines by Least Squares *Nikolai Chernov* (2010)
118. Simultaneous Inference in Regression *Wei Liu* (2010)
119. Robust Nonparametric Statistical Methods, Second Edition
Thomas P. Hettmansperger and Joseph W. McKean (2011)
120. Statistical Inference: The Minimum Distance Approach
Ayanendranath Basu, Hiroyuki Shioya, and Chanseok Park (2011)
121. Smoothing Splines: Methods and Applications *Yuedong Wang* (2011)
122. Extreme Value Methods with Applications to Finance *Serguei Y. Novak* (2012)
123. Dynamic Prediction in Clinical Survival Analysis *Hans C. van Houwelingen and Hein Putter* (2012)
124. Statistical Methods for Stochastic Differential Equations
Mathieu Kessler, Alexander Lindner, and Michael Sørensen (2012)
125. Maximum Likelihood Estimation for Sample Surveys
R. L. Chambers, D. G. Steel, Suojin Wang, and A. H. Welsh (2012)
126. Mean Field Simulation for Monte Carlo Integration *Pierre Del Moral* (2013)
127. Analysis of Variance for Functional Data *Jin-Ting Zhang* (2013)
128. Statistical Analysis of Spatial and Spatio-Temporal Point Patterns, Third Edition *Peter J. Diggle* (2013)
129. Constrained Principal Component Analysis and Related Techniques *Yoshio Takane* (2014)
130. Randomised Response-Adaptive Designs in Clinical Trials *Anthony C. Atkinson and Atanu Biswas* (2014)
131. Theory of Factorial Design: Single- and Multi-Stratum Experiments *Ching-Shui Cheng* (2014)
132. Quasi-Least Squares Regression *Justine Shults and Joseph M. Hilbe* (2014)
133. Data Analysis and Approximate Models: Model Choice, Location-Scale, Analysis of Variance, Nonparametric Regression and Image Analysis *Laurie Davies* (2014)
134. Dependence Modeling with Copulas *Harry Joe* (2014)
135. Hierarchical Modeling and Analysis for Spatial Data, Second Edition *Sudipto Banerjee, Bradley P. Carlin, and Alan E. Gelfand* (2014)
136. Sequential Analysis: Hypothesis Testing and Changepoint Detection *Alexander Tartakovsky, Igor Nikiforov, and Michèle Basseville* (2015)

Monographs on Statistics and Applied Probability 136

Sequential Analysis

Hypothesis Testing and Changepoint Detection

Alexander Tartakovsky
University of Connecticut
Storrs, USA

Igor Nikiforov
Université de Technologie de Troyes
France

Michèle Basseville
CNRS & IRISA
Rennes, France

CRC Press
Taylor & Francis Group
Boca Raton London New York

CRC Press is an imprint of the
Taylor & Francis Group, an **informa** business
A CHAPMAN & HALL BOOK

CRC Press
Taylor & Francis Group
6000 Broken Sound Parkway NW, Suite 300
Boca Raton, FL 33487-2742

First issued in Paperback 2020

ISBN-13: 978-1-4398-3820-4 (hbk)
ISBN-13: 978-0-3677-4004-7 (pbk)
ISBN-13: 978-0-4291-5234-4 (ebk)

Library of Congress Cataloging-in-Publication Data

Tartakovsky, Alexander.
Sequential analysis : hypothesis testing and changepoint detection / Alexander Tartakovsky, Igor Nikiforov, Michele Basseville.
pages cm. -- (Chapman & Hall/CRC monographs on statistics & applied probability ; 136)
"A CRC title."
Includes bibliographical references and index.
ISBN 978-1-4398-3820-4 (hardcover : alk. paper) 1. Sequential analysis. 2. Mathematical statistics. I. Nikiforov, I. V. (Igor? Vladimirovich) II. Basseville, M. (Mich?le), 1952- III. Title.

QA279.7.T37 2015
519.5'4--dc23 2014022603

Visit the Taylor & Francis Web site at
http://www.taylorandfrancis.com

and the CRC Press Web site at
http://www.crcpress.com

Contents

Preface

About seventy years ago Abraham Wald, while treating the problem of testing two simple hypotheses, showed how the fixed sample size likelihood ratio test of Neyman and Pearson can be modified into the more efficient sequential scheme when observations are collected one at a time and processed on-line. This has led to the modern theory of sequential analysis developed due to a practical demand for more efficient sampling policies and summarized by A. Wald in his monograph *Sequential Analysis* published in 1947.

A separate important branch of sequential analysis is on-line surveillance, the so-called changepoint detection, the goal of which is to detect a change in distribution or anomaly quickly. More specifically, sequential changepoint detection (or quickest change/"disorder" detection) is concerned with the design and analysis of techniques for on-line detection of a change in the state of a phenomenon, subject to a tolerable limit on the risk of false alarms. An observed process of interest may unexpectedly undergo an abrupt change-of-state from "normal" to "abnormal" (or anomalous), each defined as deemed appropriate given the physical context. The sequential setting assumes the observations are made successively, and, as long as their behavior suggests that the process is in the normal state, the process is allowed to continue. However, if the state is believed to have altered, one's aim is to detect the change "as soon as possible," so that an appropriate response can be provided in a timely manner.

Historically, the subject of changepoint detection first began to emerge in the 1920s motivated by considerations of industrial quality control due to the work of Walter Shewhart who successfully brought together the disciplines of statistics, engineering, and economics and became the father of modern statistical quality control. Shewhart's work (in particular Shewhart control charts) was highlighted in his books *Economic Control of Quality of Manufactured Product* (1931) [411] and *Statistical Method from the Viewpoint of Quality Control* (1939) [412], for which he gained recognition in the statistical community, but efficient (optimal and quasi-optimal) sequential detection procedures were developed much later in the 1950–1960s after the emergence of Wald's book *Sequential Analysis* (1947) [494]. The ideas set in motion by Shewhart and Wald have formed a platform for extensive research on both theory and practice of sequential changepoint detection, starting with the seminal paper by Page (1954) where the now famous Cumulative Sum (CUSUM) detection procedure was first proposed, and followed by the series of works of Shiryaev (1961–1969) [414, 413, 415, 416, 417, 418, 419] and Lorden (1971) [271] where the first optimality results in Bayesian and non-Bayesian contexts were established.

During the past 20 years, general stochastic models appropriate for many interesting applications have been treated extensively, as theoretical foundation for asymptotic studies of properties of known sequential tests such as Wald's Sequential Probability Ratio Test (SPRT), matrix versions of this test suitable for multiple decision problems, CUSUM and Shiryaev–Roberts change detection procedures, which are known to be optimal or nearly optimal for the models with independent and identically distributed (iid) observations. Asymptotic optimality of these rules has been established under various conditions, including conventional iid and general non-iid scenarios. Novel procedures have also been proposed and studied. Multihypothesis and multichannel change detection–classification (or detection–isolation) rules have been developed and their asymptotic optimality properties have been established for iid and general non-iid models. Even for relatively simple iid models new results have been obtained, in particular toward very precise analysis via solving integral equations numerically and asymptotic analysis using renewal-theoretic and nonlinear renewal-theoretic approaches. These numerical and asymptotic approaches are in fact complementary, since numerical solutions become very time-consuming when dealing with small error probabilities or low false alarm rates, while asymptotic approximations are usually not too accurate for high and moderate false alarm rates.

The main focus of this book is on a systematic development of the theory of sequential hypothesis testing (Part I) and changepoint detection (Part II). In Part III, we briefly describe certain important applications where theoretical results can be used efficiently, perhaps with some reasonable modifications. We review recent accomplishments in hypothesis testing and changepoint detection both in decision-theoretic (Bayesian) and non-decision-theoretic (non-Bayesian) contexts. The emphasis is not only on more traditional binary hypotheses but also on substantially more difficult multiple decision problems. Scenarios with simple hypotheses and more realistic cases of (two and finitely many) composite hypotheses are considered and treated in detail. While our major attention is on more practical discrete-time models, since we strongly believe that "life is discrete in nature" (not only due to measurements obtained from devices and sensors with discrete sample rates), certain continuous-time models are also considered once in a while, especially when general results can be obtained very similarly in both cases. It should be noted that although we have tried to provide rigorous proofs of the most important results, in some cases we included heuristic argument instead of the real proofs as well as gave references to the sources where the proofs can be found.

While there are many other interesting topics in sequential analysis such as point and interval estimation, selection/ranking, and sequential games, these important topics are out of the scope of our book. A detailed treatment of these additional sequential methods can be found, e.g., in [56, 163, 259, 312, 452].

We would like to thank many colleagues who have directly and indirectly contributed to this project. Several students and postdoctoral fellows at the University of Southern California worked on some of the problems considered in the book at different stages. Aleksey Polunchenko contributed to certain theoretical aspects and numerical methods related to the very precise analysis of minimax changepoint detection procedures as well as helped with simulations and processing real and semi-real data for computer network security applications. The joint work with Georgios Fellouris on minimax tests for discrete composite hypotheses became the basis for the corresponding sections in Part I. Greg Sokolov performed useful numerical analysis and Monte Carlo simulations of multichannel change detection procedures. Collaboration with George Moustakides, Moshe Pollak, and Venugopal Veeravalli as well as frequent discussions with them were extremely fruitful. The joint work with Lionel Fillatre on FSS multiple hypothesis testing has been used for writing Subsections 2.9.6 and 10.2.2.

Alexander Tartakovsky is thankful to various U.S. agencies (Department of Defense, Department of Energy, National Science Foundation) for supporting his work under multiple contracts.[1]

Alexander Tartakovsky wants to thank his wife, Marina Blanco, for her patience, help, and inspiration.

Igor Nikiforov is thankful to the University of Technology of Troyes for supporting this work and for the environment in which the book has been written. A preliminary version of some material of the book has been used for Master and PhD courses at the University of Technology of Troyes. The work reported in Section 11.1 has been partly supported by the SERCEL (Société d'Etudes, Recherches et Constructions Électroniques), by the SAGEM (Société d'Applications Générales d'Électricité et de Mécanique, of the SAFRAN group), by the LRBA (Laboratoire de Recherches Balistiques et Aérodynamiques) and by the DGAC/DTI (Direction de la Technique et de l'Innovation, formerly known as STNA).

[1]In particular, the work of Alexander Tartakovsky was partially supported by the U.S. Air Force Office of Scientific Research under MURI grant FA9550-10-1-0569, by the U.S. Defense Threat Reduction Agency under grant HDTRA1-10-1-0086, by the U.S. Defense Advanced Research Projects Agency under grant W911NF-12-1-0034, the U.S. Army Research Office under MURI grant W911NF-06-1-0044 and under grants W911NF-13-1-0073 and W911NF-14-1-0246, and by the U.S. National Science Foundation under grants CCF-0830419, EFRI-1025043, and DMS-1221888 at the University of Southern California, Department of Mathematics and at the University of Connecticut, Department of Statistics.

Igor Nikiforov wants to thank his wife, Tatiana, for her support, understanding, and encouragement during the writing of this book.

Michèle Basseville is thankful to the Centre National de la Recherche Scientifique (CNRS) for its support and to the Institut de Recherche en Informatique et Systèmes Aléatoires (IRISA) for the environment in which the book has been written. The work reported in Section 11.2 has been partly supported by the Eurêka projects no 1562 SINOPSYS, no 2419 FliTE and no 3341 FliTE2.

Finally, we are grateful to two anonymous referees whose comments have improved the presentation.

Los Angeles, California and Storrs, Connecticut USA[2]	Alexander Tartakovsky
Troyes, France	Igor Nikiforov
Rennes, France	Michèle Basseville

[2]The book was completed after Alexander Tartakovsky joined the Department of Statistics, University of Connecticut at Storrs (September 2013) but most of the book was written while he was with the Department of Mathematics, University of Southern California, Los Angeles.

List of Figures

List of Tables

Notation and Symbols

Notation	Meaning
$X_t \xrightarrow[t\to\infty]{\mathsf{P-a.s.}} Y$	Almost sure convergence under P (or with probability 1).
R_n	*A posteriori* risk (APR); also minimum *a posteriori* risk (MAPR).
R_n^{st}	APR associated with stopping.
$\tilde{R}_n$	APR associated with continuation of observations.
$\rho(\delta)$	Average (or integrated) risk (AR).
ADD	Average delay to detection (detection delay).
$\mathbf{C}(\alpha)$	Class of tests with significance level α.
$[a,b]$	Closed interval.
$X_t \xrightarrow[t\to\infty]{\text{completely}} Y$	Complete convergence.
$(\Omega,\mathscr{F},\mathsf{P})$	Complete probability space.
CADD	Conditional average detection delay.
$\mathsf{E}[X \mid \mathscr{B}]$	Conditional expectation of the random variable X given sigma-algebra $\mathscr{B}$.
$X_t \xrightarrow[t\to\infty]{\text{law}} Y$	Convergence in distribution (or in law or weak).
$X_t \xrightarrow[t\to\infty]{\mathsf{P}} Y$	Convergence in probability.
$F(x) = \mathsf{P}(X \le x)$	Cumulative distribution function (cdf) of a random variable X.
$\det A$	Determinant of the matrix A.
δ	Decision rule, procedure, function.
$\lVert\mathbf{X}\rVert_2 = \sqrt{\sum_{i=1}^n x_i^2}$	Euclidean norm.
E	Expectation.
ESS	Expected sample size (or average sample number).
$\mathsf{Expon}(\theta)$	Exponential distribution (or random variable) with the parameter θ.
$\{\mathscr{F}_t\}$	Filtration (a flow of sub-sigma-algebras $\mathscr{F}_t$).
$\dot{g}(x)$	First derivative of the function $x \mapsto g(x)$.
$\mathcal{F}$	Fisher information.
∇	Gradient (vector of first partial derivatives).
∇^2	Hessian (matrix of second partial derivatives).
$\mathbb{I}_n$	Identity matrix of size $n \times n$.
H_i	i^{th} hypothesis, $0 \le i \le M-1$, where M is the total number of hypotheses.
$\mathbb{1}_{\{A\}}$	Indicator of a set A.
A^{-1}	Inverse of the matrix A.
$\ker A$	Kernel of the matrix A.

I	Kullback–Leibler (K–L) information (or distance or divergence).
$\mathbb{R}^\ell$	ℓ-dimensional Euclidean space.
$\Lambda = \frac{d\mathsf{P}}{d\mathsf{Q}}(\omega)$	Likelihood ratio (Radon–Nikodým derivative of measure P with respect to measure Q).
$\varkappa$	Limiting average overshoot.
$L(\theta, d)$	Loss function.
$X_t \xrightarrow[t\to\infty]{L^p} Y$	L^p-convergence (or in the pth mean).
$A = [a_{ij}]$	Matrix A of size $m \times n$ $(1 \le i \le m,\ 1 \le j \le n)$.
$\mathbb{R}_+ = [0, \infty)$	Nonnegative real line.
$X_t, t \ge 0$	Observed process in continuous time.
$X_n, n \ge 1$	Observations in discrete time.
(a, b)	Open interval.
$\mathcal{P} = \{\mathsf{P}_\theta\}_{\theta\in\Theta}$	Parametric family of probability distributions.
$f_\theta(x)$, $p_\theta(x)$	Parametrized probability density, pdf.
P	Probability measure.
$f(x)$, $p(x)$	Probability density function (pdf).
θ	Parameter or vector of parameters.
ν	Point of change (or changepoint).
$\chi^2_m(p)$	p-quantile of the standard chi-squared distribution with m degrees of freedom.
β	Power of test.
α_{ij}	Probability of accepting H_i when the hypothesis H_j is true.
α_i	Probability of rejecting H_i when it is true.
$\operatorname{rank} A$	Rank of the matrix A.
$\mathbb{R} = (-\infty, \infty)$	Real line.
$X_t \xrightarrow[t\to\infty]{r-\text{quickly}} Y$	r-quick convergence.
$\ddot{g}(x)$	Second derivative of the function $x \mapsto g(x)$.
$\delta = (T, d)$	Sequential test (more generally rule).
$\mathbb{Z}_+ = \{0, 1, 2, \dots\}$	Set of nonnegative integers.
Ω	Set of elementary events ω.
$\{t : \dots\}$	Set of t such that $\dots$.
$\mathscr{F}$	Sigma algebra (field).
$\varphi(x)$	Standard normal density function.
$\Phi(x)$	Standard normal distribution function.
$\mathcal{N}(0, 1)$	Standard normal random variable.
STADD	Stationary average detection delay.
$(\Omega, \mathscr{F}_t, \mathscr{F}, \mathsf{P})$	Stochastic basis.
T	Stopping time.
SADD	Supremum average detection delay.
d	Terminal decision.
$\mathbf{X}_0^t = \{X_u, 0 \le u \le t\}$	Trajectory of a random process observed on the interval $[0, t]$.
$A^\top$	Transpose of the matrix A.

$\mathrm{tr}A$ Trace of the matrix A.

$\mathbf{X}_1^n = \begin{pmatrix} X_1 \\ X_2 \\ \vdots \\ X_n \end{pmatrix}$ Vector of observed n random variables.

$\check{\mathbf{X}}_1^n = \begin{pmatrix} X_n \\ X_{n-1} \\ \vdots \\ X_1 \end{pmatrix}$ Vector of observed n random variables in reverse order.

Chapter 1

Motivation for the Sequential Approach and Selected Applications

In this chapter, we describe the theoretical and applied motivations for the sequential approach in general and for change detection in particular, and we describe the positioning of the book as well. We also introduce several typical application examples.

1.1 Motivation

Sequential analysis refers to statistical theory and methods for processing data in which the total number of observations is not fixed in advance but depends somehow on the observed data as they become available. A sequential method is characterized by two components:

1. A stopping rule that decides whether to stop the observation process with $(X_1, X_2, \ldots, X_n)$ or to get an additional observation X_{n+1} for $n \geq 1$;
2. A decision rule that specifies the action to be taken about the considered problem (estimation, detection, classification, *etc.*) after the observation has stopped.

Denoting by T the stopping variable and d the terminal decision, the pair $\delta \triangleq (T, d)$ specifies the sequential decision rule (or procedure). Such a pair may not be unique for a given problem. The objective of sequential analysis is to determine an optimal decision rule δ that satisfies some criteria. Note that if T is fixed with probability 1 the procedure has an *a priori* fixed size of a sample. We will refer to such procedures as *Fixed Sample Size* procedures.

In sequential changepoint detection problems, however, the situation is slightly different. A change detection procedure is identified with a stopping time depending on the observations and the decision on no-change is equivalent to the decision on continuing observation. Furthermore, typically the observation process is not terminated even after deciding that the change is in effect but rather renewed all over again, leading to a multicyclic detection procedure. This is practically always the case in surveillance applications and often in other applications. See Section 6.3 for further details.

Even though most experiments are essentially sequential, many classical statistical methods are fixed sample size. In his history of sequential analysis, B.K. Ghosh distinguishes several practical motivations for sequential analysis [161].

In some applications sequential analysis is nothing but intrinsic: no fixed sample size procedure can be thought of. This is the case of industrial process control [81, 303, 482, 499, 501, 511]. This is also the case in the classical secretary problem [144] and while monitoring some critical health parameters of a patient in clinical trials [502]. Most surveillance problems are also sequential in nature. It should be noted that in the key area of medical and pharmaceutical research the requirement for sequential analysis may also result from ethical grounds.

In some other statistical inference applications, sequential analysis is the most economic solution, in terms of sample size or cost or duration of the experiment. This is the case of the so-called curtailed sampling procedure that ensures the same power while requiring a smaller sample size than the best fixed sample size procedure [132, 189]. This is also the case of the repeated significance test that also maintains the flexibility of deciding sooner than the fixed sample size procedure at the price of some lower power [13, 514]. The sequential probability ratio test (SPRT) and the Kiefer–Weiss procedure also belong to the category of most economic solutions, since they minimize the expected sample size (resp. the maximum expected sample size). These sequential tests are investigated in detail in Chapters 3 and 5, respectively.

Finally, in some parametric sequential point estimation problems, sequential analysis may reinforce a fixed sample size procedure in a somewhat wider context than usual [311].

1.2 Two Theoretical Tracks

In this book we propose to focus on two tracks: Sequential Hypothesis Tests and Sequential (Quickest) Changepoint Detection.

First, classical settings of hypothesis testing and changepoint detection problems operate with the case of independent and identically distributed (iid) observations and two simple hypotheses. These assumptions may be quite restrictive for many contemporary applications. Therefore, generalizations to general non-iid models are under way. However, even in a relatively simple iid setting there are several challenges that have been addressed in the literature during the last decade, including the work by the authors. All these important results are scattered in the literature (conference proceedings as well as in statistical, applied probability, engineering, computer science, and other kinds of journals) and are not easily accessible and understandable for students and even for professionals in the field. Moreover, the practical needs of various applied areas lead the researchers to study more sophisticated statistical models by considering:

- Non-identically distributed and/or dependent observations,
- Multiple hypotheses,
- Composite hypotheses, including nuisance parameters in the statistical model.

Therefore, we believe that a book that would combine all these results in a synergistic way is timely.

Second, the proposed book contains both theoretical concepts and results and a number of application examples. As explained below and detailed in the table of contents, the book covers sequential hypothesis testing and sequential quickest changepoint detection from theoretical developments to applications in a wide range of engineering and environmental domains. It is the intention of the authors to explain how the theoretical aspects influence the problem statement and the design of algorithms when addressing problems in various application areas.

Third, we would like to mention two recent books related to sequential hypothesis tests and quickest change detection: by G. Peskir and A.N. Shiryaev, *Optimal Stopping and Free Boundary Problems* [360] and by H.V. Poor and O. Hadjiliadis, *Quickest Detection* [376]. While these books cover certain interesting aspects of sequential hypothesis testing and changepoint detection, they both focus mainly on continuous-time models, which are restricted for most applications. The present book covers mostly more practical discrete-time models as well as very general cases that include both continuous- and discrete-time models. In addition, we consider multiple decision making problems, including sequential multihypothesis tests and quickest change detection–isolation procedures, that are not presented in the above referenced books.

1.2.1 Track 1: Sequential Hypothesis Testing

The goal of testing statistical hypotheses is to relate an observed stochastic process to one of N ($N \geq 2$) possible classes based on some knowledge about the distributions of the observations under each class or hypothesis. In a sequential setting, the number of observations is allowed to be random, i.e., a function of the observations. The theoretical study of sequential hypothesis testing has been initiated by A. Wald [492]. A sequential procedure or test includes a stopping time and a terminal decision to achieve a tradeoff between the average observation time and the quality of the decision.

Most efforts have been devoted to testing two hypotheses, namely, to developing optimal strategies and obtaining lower bounds for the average number of observations necessary to decide between the two hypotheses with given error probabilities; see Wald [492, 494], Wolfowitz [496, 497], Hoeffding [192, 193], and many others. Also, these bounds have been compared with the sample size of the best non-sequential, fixed sample size test. It has been shown that the sequential procedure performs significantly better than the classical Neyman–Pearson test in the case of two simple hypotheses.

The problem of sequential testing of many hypotheses is substantially more difficult than that of testing two hypotheses. For multiple-decision testing problems, it is usually very difficult, if even possible, to obtain optimal solutions. The first results have been established by Sobel and Wald [435], Armitage [12], and Paulson [350]. The lower bounds for the average sample number has been established by Simons [432].

A substantial part of the development of sequential multihypothesis testing in the last several decades has been directed toward the study of suboptimal procedures, basically multihypothesis modifications of a sequential probability ratio test, for iid data models. See, e.g., Armitage [12], Chernoff [97], Dragalin [123], Dragalin and Novikov [127], Kiefer and Sacks [231], Lorden [269, 275], Pavlov [351, 352]. The generalization to the case of non-stationary processes with independent increments was made by Tartakovsky [449, 452, 457], Golubev and Khas'minskii [168], and Verdenskaya and Tartakovsky [484]. The condition of independence of the log-likelihood ratio increments was crucial in these works. Further generalizations to the case of non-iid stochastic models that may include both nonhomogeneous and correlated processes observed in continuous or in discrete time were made by Lai [248], Tartakovsky [455], and Dragalin *et al.* [128]. The results obtained in these latter works are indeed very general and cover almost any, and perhaps every, model of interest in the applications. Such popular models as Itô processes, state-space models, and hidden Markov models with discrete and continuous space are particular cases.

1.2.2 Track 2: Quickest Changepoint Detection

Changepoint problems deal with detecting changes in the state of a process. In the sequential setting, as long as the behavior of the observations is consistent with the initial or target state, one is content to let the process continue. If the state changes, then one is interested in detecting that a change is in effect, usually as soon as possible after its occurrence. Any detection policy may give rise to false alarms. The desire to detect a change quickly causes one to be trigger-happy, which will bring about many false alarms if there is no change. On the other hand, attempting to avoid false alarms too strenuously will lead to a long delay between the time of occurrence of a real change and its detection . The gist of the changepoint problem is to produce a detection policy that minimizes the average delay to detection subject to a bound on the average frequency of false alarms.

The theoretical study of quickest changepoint detection has been initiated in two different directions: Bayesian and minimax. In the Bayesian case, it is supposed that the changepoint is a random variable independent of the observations with known distribution. On the contrary, in the minimax case it is assumed that the changepoint is an unknown non-random number. The very first study of the Bayesian quickest changepoint detection approach has been done by Girschick and Rubin [165] in the framework of quality control. An optimal solution to this problem has been obtained by Shiryaev [413, 414, 415] who has also performed the comparison between the optimal procedure, the repeated sequential Wald test and the classical Neyman–Pearson test. Independently, another, minimax approach has been adopted by Lorden [271]. In contrast to the Bayesian approach, the minimax criterion is based on the worst-case mean detection delay, characterized by the essential supremum with respect to pre-change observations and by the supremum over all possible changepoints. An optimal solution to the problem and a lower bound in the class of procedures with a given mean time (average run length) to a false alarm has been studied by Lorden [271] in the asymptotic case for large average run length to false alarm. In this work, Lorden established, for the first time, asymptotic minimax optimality of Page's CUSUM procedure [346], a well-known statistical control chart. Later Moustakides [305] showed that the CUSUM procedure is in fact exactly minimax with respect to Lorden's essential supremum detection speed measure.

In 1961, for detecting a change in the drift of a Brownian motion, Shiryaev [413, 414] introduced a change detection procedure, which is now usually referred to as the Shiryaev–Roberts procedure [394]. This procedure has a number of interesting optimality properties. In particular, it minimizes the *integral average detection delay* being Generalized Bayesian for an improper uniform prior distribution of the changepoint. It is also optimal in the sense of minimizing the stationary average detection delay when a change occurs in a distant future and is preceded by a long interval with a stationary flow of false alarms; see Feinberg and Shiryaev [139] and Pollak and Tartakovsky [370]. On the other hand, Pollak [365] introduced a natural worst-case detection delay measure — *maximal conditional average delay to detection*, which is less pessimistic than Lorden's essential supremum measure, and attempted to find an optimal procedure that would minimize this measure over procedures subject to constraint on the average run length to false alarm. Pollak's idea was to modify the Shiryaev–Roberts statistic by randomization of the initial condition in order to make it an equalizer. Pollak's version of the Shiryaev–Roberts procedure starts from a random point sampled from the quasi-stationary distribution of the Shiryaev–Roberts statistic. He proved that, for a large average run length to false alarm, this randomized procedure is asymptotically nearly minimax within an additive vanishing term. Since the Shiryaev–Roberts–Pollak procedure is an equalizer, it is tempting for one to conjecture that it may in fact be *strictly* optimal for any false alarm rate. However, a recent work of Moustakides *et al.* [310] and Polunchenko and Tartakovsky [373] indicates that the Shiryaev–Roberts–Pollak procedure is not exactly minimax and sheds light on this issue by considering a generalization of the Shiryaev–Roberts procedure that starts from a specially designed deterministic point.

As we mentioned above, in the early stages the theoretical development was focused on iid models. However, in practice the iid assumption may be too restrictive. The observations may be either non-identically distributed or correlated or both, i.e., non-iid. An extension of Lorden's results to the case of dependent stationary random processes before and after the change has been done by Bansal and Papantoni-Kazakos [26]. A general theory of changepoint detection is now available both in the Bayesian and minimax settings due to the work of Tartakovsky and Veeravalli [475, 476], Baron and Tartakovsky [28], Lai [251], and Fuh [154, 155]. In particular, for a low false alarm rate the asymptotic minimax optimality of the CUSUM and Shiryaev–Roberts procedures has been established in [154, 155, 251, 475] and the asymptotic optimality of the Bayesian Shiryaev procedure proven in [28, 476]. Moustakides [306] generalized for the Itô processes the CUSUM minimax optimality result with respect to Lorden's essential supremum measure acting on the total expected Kullback–Leibler information.

For iid data and for large thresholds, the suitably standardized distributions of the CUSUM and Shiryaev–Roberts stopping times are asymptotically exponential and fit well into the geometric distribution even for a very moderate false alarm rate [369]. In this case, the mean time to false alarm, the global false alarm rate metric, is obviously appropriate. However, for non-iid models the limiting distribution is not guaranteed to be exponential or even close to it. In general, we cannot even guarantee that large values of the mean time to false alarm will produce small values of the maximal local false alarm probability. Therefore, the mean time to false alarm, a standard and well accepted measure of false alarms, may not be appropriate in general. Instead of global measures of false alarms, it may be more appropriate to use local measures, for example the local false alarm probability, as suggested in [459]. This issue is extremely important for non-iid models as a discussion in [293, 460] and other discussion pieces published in *Sequential Analysis*, Vol. 27, No. 4, 2008 show.

Another challenging extension is a multidecision change *detection–isolation* problem when, along with detecting a change with a given false alarm rate, an identification/isolation of a true post-change hypothesis with a given misidentification rate is required [48, 49]. An optimal solution to the problem of abrupt change detection–isolation and a non-recursive algorithm that asymptotically attains the lower bound were obtained by Nikiforov in [322] by using a minimax approach based on minimizing the Lorden-type worst-case mean detection–isolation delay for a given mean time before a false alarm and for a given probability of false isolation. The comparison between the optimal sequential and repeated fixed sample size approaches and different recursive sequential detection–isolation algorithms have been studied by Dragalin [125], Nikiforov [326, 328, 331], Oskiper and Poor [343], and Tartakovsky [453, 461]. A multiple hypothesis extension of the Shiryaev–Roberts procedure by adopting a dynamic programming approach has been proposed by Malladi and Speyer [287]. Next, Lai [252] generalized the results obtained for the worst-case mean detection–isolation criterion in [322] to the case of dependent observations. Lai also proposed two new optimality criteria: a non-Bayesian one, where the maximum probabilities of false alarm and false isolation within a given time window are constrained; and a Bayesian one, where a weighted sum of the false alarm and false isolation probabilities is used. Finally, Lai designed a window-limited generalized likelihood ratio-based algorithm with reduced computational complexity for on-line processing that asymptotically attains the lower bounds.

1.3 Several Applications

Hypothesis testing and changepoint problems arise across various branches of science and engineering and have an enormous spectrum of important applications, including environment surveillance and monitoring, biomedical signal and image processing, quality control engineering, link failure detection in communication networks, intrusion detection in computer networks and security systems, detection and tracking of covert hostile activities, chemical or biological warfare agent detection systems as a protection tool against terrorist attacks, detection of the onset of an epidemic, failure detection in manufacturing systems and large machines, target detection in surveillance systems, econometrics, financial markets, detection of signals with unknown arrival time in seismology, navigation, radar and sonar signal processing, speech segmentation, and the analysis of historical texts. In all of these applications, sensors take observations that undergo a change in their distribution in response to changes and anomalies in the environment or changes in the patterns of a certain behavior. The observations are obtained sequentially and, as long as their behavior is consistent with the normal state, one is content to let the process continue. If the state changes, then one is interested in detecting the change as soon as possible while minimizing false detections.

During the last years, a number of new application fields have emerged: structural health monitoring of bridges [24, 25, 43], wind turbines [178, 216], and aircraft [41, 102, 186, 188], detecting multiple sensor faults in an unmanned air vehicle (UAV) [403], monitoring railway vehicle dynamics [87], detecting road traffic incidents [521] or changes in highway traffic condition [170], monitoring low consumption components of road vehicles [36], diagnosing automotive antilock braking systems [285], chemical process control [196], physiological data analysis [398], surveillance of daily disease counts [439], nanoscale analysis of soft biomaterials through atomic force microscopy [402], biosurveillance [110, 342, 424], radio-astronomy [152, 438] and interferometry [341], spectrum sensing in cognitive radio systems [201, 263], landmine detection [379], leak detection in water channels [58], monitoring biological waste water treatment plants [19], environmental monitoring [57, 120, 361, 385, 409], hydrology [286], handling climate changes [284, 393, 526], navigation systems monitoring [295, 336, 408], detecting salient motion for dynamic scene modeling [233], human motion analysis [85], video scene analysis [262], sequential steganography [479, 480], biometric identification [7], onset detection in music signals [59], detecting changes in large payment card datasets [107], running consensus in sensor networks [82, 83], and distributed systems monitoring [382, 461, 475].

In particular a number of computer and network problems are now addressed with the aid of sequential hypothesis testing and change detection algorithms: anomaly detection in IP networks [477], secure IP telephony [386], detection of intrusion, viruses, and other denial of service (DoS) attacks [215, 357, 433, 472], including scanning worms infections [397, 406], bioterrorism detection and other aspects of global security, Internet access patterns characterization [208], teletraffic monitoring [2, 3, 211, 313], tracking the preferences of users in recommendation systems [520], network bandwith monitoring [183], active queue management [74], and even cost estimation for software evolution [383] and software quality and performance monitoring [171].

In this section, we describe several typical application examples of sequential hypothesis testing and change detection techniques. For each example, we give a short description of the particular problem and its context. For some of these models, the detailed information about the possibly complex underlying physical models is given in Part III. This selection of examples is not exclusive; it is intended to give only sufficient initial insights into the variety of problems that can be solved within this framework. In Part III, we come back to some application problems, showing results of the processing of real data with the aid of sequential hypothesis testing and change detection algorithms.

In Subsections 1.3.1 and 1.3.2 we start with quality control and target detection, and we continue with integrity monitoring of navigation systems in Subsection 1.3.3. Then in Subsection 1.3.4 we describe a couple of signal processing problems, namely segmentation of signals and seismic signal processing. Mechanical systems integrity monitoring is discussed in Subsection 1.3.5. Finally, we discuss application to finance and economics and to computer network surveillance and security in Subsections 1.3.6 and 1.3.7.

1.3.1 Quality Control

One of the earliest applications of change detection is the problem of quality control, or continuous production monitoring. On-line quality control deals with scenarios where the measurements are taken one at a time and the decisions are to be reached sequentially as the measurements are taken. Consider a production process that can be *in control* and *out of control*. The events associated with the transitions of this process from the in-control state to the out-of-control state are called *disorders*. For many reasons, it is necessary to detect a disorder as quickly as possible after its occurrence as well as to estimate its onset time. It may be a question of safety of the technological process, quality of the production, or classification of output production items. For all these problems, the best solution is the *quickest detection of the disorder with as few false alarms as possible*. This criterion is used because the delay until detection is a time interval during which the technological process is out of control, but there is no action of the monitoring system to this event. From both the safety and quality points of view, this situation is obviously highly undesirable. On the other hand, frequent false alarms are inconvenient because of the cost of stopping production, verifying whether this is a true or false disorder, and searching for the origin of the defect; nor is this situation desirable from a psychological point of view, because the operator will stop using the monitoring system very quickly if it produces too-frequent false alarms. Thus, an optimal solution is based on a *tradeoff* between the speed of detection or detection delay and the false alarm rate, using a comparison of the losses implied by the true and false detections.

We stress that we are interested in solving this problem using a *statistical approach*, that is, assuming that the measurements are a realization of a random process. Because of the random behavior, large fluctuations can occur in the measurements even when the process is in control, and these fluctuations result in false alarms. On the other hand, any (even the best) decision rule cannot detect the change instantaneously, again because of the random fluctuations in the measurements. When the technological process is in control, the measurements have a specific probability distribution. When the process is out of control, this distribution changes. If a parametric approach is used, we speak about changes in the parameters of this probability distribution. A chemical plant where the quality of the output material is characterized by the concentration of some chemical component is a typical example, where the concentration is distributed according to the Gaussian law. Under normal operating conditions, the mean value and standard deviation of this normal distribution are μ_0 and σ_0, respectively. Under abnormal conditions three types of changes can occur in these parameters:

- Deviation from the reference mean value μ_0 toward μ_1 with constant standard deviation, i.e., a systematic error;

- Increase in the standard deviation from σ_0 to σ_1 with constant mean, i.e., a random error;
- Both the mean and the standard deviation change, i.e., systematic and random errors.

The goal is to design a statistical decision rule (detection procedure, algorithm) that can detect these disorders effectively. Typically a decision procedure involves comparing a statistic sensitive to a change with a threshold that controls a false alarm rate.

If a decision statistic is chosen, the tuning of the statistical decision rule is reduced to selecting a threshold that guarantees the tradeoff between the false alarm rate and the mean delay to detection. Several types of decision rules are used in the industry as standards, they are called *control charts*, and each differs by the detection statistic. In the simplest case, the pre-change and post-change parameters are assumed to be known. In this case the decision statistics should be a function of the likelihood ratio for the pre- and post-change parameters.

The main references in the area of quality control and Statistical Process Control (SPC) are the books [80, 81, 114, 130, 153, 184, 288, 303, 340, 348, 434, 482, 499, 500, 501, 515] and the survey papers [65, 106, 443, 447, 509, 510, 511], with special notice for [381] and [67, 185].

1.3.2 Target Detection and Tracking

Surveillance systems, such as those for ballistic and cruise missile defense, deal with the detection and tracking of moving targets. The most challenging problem for such systems is the quick detection of maneuvering targets that appear and disappear at unknown points in time against a strong cluttered background. To illustrate the importance of this task, we remark that under certain conditions *a few seconds decrease* in the time it takes to detect a sea/surface skimming cruise missile can yield a significant increase in the *probability of raid annihilation*. Furthermore, usually detection systems are multichannel, since the target velocity is unknown. Thus, finding an optimal combination of a multihypothesis testing algorithm with changepoint detection methods is a challenge. This challenging applied problem can be effectively solved using the quickest detection–isolation methods developed in this book.

We also note that standard ad hoc methods for target track initiation and termination [27, 68, 69] can be substantially improved by using advanced quickest detection methods that are the subject of this book. Improving the operating characteristics is especially important for Space-Based Infrared and Space Tracking and Surveillance System sensors with chaotically vibrating lines-of-sight that have to provide early detection and tracking of low observable targets in the presence of highly-structured cluttered backgrounds.

1.3.3 Navigation System Integrity Monitoring

For many safety-critical aircraft navigation modes (landing, takeoff, *etc.*), a major problem of existing navigation systems consists in their lack of integrity. The integrity monitoring concept, defined by the International Civil Aviation Organization, requires a navigation system to detect the faults and remove them from the navigation solution before they sufficiently contaminate the output. Recent research shows that the quickest detection–isolation of the navigation message contamination is crucially important for the safety of the radio-navigation system, e.g., GPS, GLONASS, Galileo, *etc.* It is proposed *to encourage all the transportation modes to give attention to autonomous integrity monitoring of GPS signals* [93].

Monitoring the integrity of a navigation system can be reduced to a quickest change detection–isolation problem [21, 324, 325, 332]. The time when the fault occurs and the type of fault are not just unknown but sometimes can be intentionally chosen to maximize their negative impacts on the navigation system. Therefore, the optimality criterion should favor fast detection in the worst case with few false alarms and false isolations. Fast detection is necessary because abnormal measurements are taken in the navigation system between the changepoint (fault onset time) and its detection, which is clearly very undesirable. On the other hand, false alarms/isolations result in lower accuracy of the estimates because incorrect information is used at certain time intervals. An optimal solution involves a tradeoff between these two contradictory requirements. The changepoint detection–isolation techniques developed in this book can be used for obtaining optimal solutions to this challenging problem. This is discussed in Section 11.1. Historical references related to inertial navigation system monitoring are [315, 506]. The integrity monitoring of navigation systems is investigated in [93, 227, 295, 324, 325, 332, 336, 446]. Some challenges are pointed out in [408].

1.3.4 Signal Processing Applications

1.3.4.1 Segmentation of Signals and Images

A first processing step of recognition-oriented signal processing consists in automatic segmentation of a signal. A segmentation algorithm splits the signal into homogeneous segments, with sizes adapted to the local characteristics of the analyzed signal. The homogeneity of a segment can be formulated in terms of the mean level or in terms of the spectral characteristics. The segmentation approach has proved useful for the automatic analysis of various biomedical signals, in particular electroencephalograms [11, 73, 78, 207, 213, 404] and electrocardiograms [172]. Several segmentation algorithms for recognition-oriented geophysical signal processing are discussed in [39]. A changepoint detection based segmentation algorithm has also been introduced as a powerful tool for the automatic analysis of continuous speech signals, both for recognition [10] and for coding [117].

The main desired properties of a segmentation algorithm are *low false alarm and mis-detection rates* and a *small detection delay*, as in the previous examples. However, we have to keep in mind that signal segmentation is usually only the first step of a recognition procedure. From this point of view, it is obvious that the properties of a given segmentation algorithm also depend upon the processing of the segments which is performed at the next stage. For example, it is often the case that, for segmentation algorithms, false alarms (sometimes called oversegmentation) are less critical than for onset detection algorithms. A false alarm for the detection of an imminent tsunami obviously has severe and costly practical consequences. On the other hand, in a recognition system, false alarms at the segmentation stage can often be easily recognized and filtered at the next stage, which means that the loss due to false alarms is small at the first segmentation stage. A segmentation algorithm exhibiting the above-mentioned properties is potentially a powerful tool for a recognition system.

It should be clear that a segmentation algorithm allows us to detect several types of events. Examples of events obtained through a spectral segmentation algorithm and concerning recognition-oriented speech processing are discussed in [10]. Other examples of events in seismology are mentioned in the previous subsection.

Changepoint detection methods are also efficient and useful in image segmentation and boundary tracking problems [96].

1.3.4.2 Seismic Data Processing

In many situations of seismic data processing, it is necessary to estimate *in situ* the geographical coordinates and other parameters of earthquakes.

The standard sensor equipment of a three-component seismic station results in the availability of records of seismograms with three components, namely the east-west, north-south, and vertical components. When an earthquake arises, the sensors begin to record several types of seismic waves (body and surface waves), among which the more important ones are the P-wave and the S-wave. The P-wave is polarized in the source-to-receiver direction, namely from the epicenter of the earthquake to the seismic station. Hence, it is possible to estimate the source-to-receiver azimuth α using the linear polarization of the P-wave in the direction of propagation of the seismic waves. The two main events to be detected are the P-wave and the S-wave; note that the P-wave can be very low-contrast with respect to seismic noise. The processing of these three-dimensional measurements can be split into three tasks:

1. On-line detection and identification of the seismic waves;

2. Off-line estimation of the onset times of these waves;
3. Off-line estimation of the azimuth using the correlation between the components of the P-wave segments.

The P-wave has to be detected *very quickly with a fixed false alarms rate*, so that the S-wave can also be detected on-line. The detection of the P-wave is a difficult problem, because the data contain many nuisance signals (interference) coming from the environment of the seismic station, and discriminating between these events and a true P-wave is not easy. The same is true for the S-wave, which is an even more difficult problem because of a low signal-to-noise ratio and numerous interferences between the P-wave and the S-wave.

After P-wave and S-wave detection, the *off-line accurate estimation of onset times* is required for both types of waves. A possible solution is to use fixed-size samples of the three-dimensional signals centered at a rough estimate of the onset time provided by the detection algorithm. Some references for seismic data processing are [235, 301, 334, 363, 377, 478].

1.3.5 Mechanical Systems Integrity Monitoring

Detecting and localizing damages for monitoring the integrity of structural and mechanical systems is a topic of growing interest, due to the aging of many engineering constructions and machines and to increased safety norms. Many structures to be monitored, e.g., civil engineering structures subject to wind and earthquakes, aircraft subject to turbulence, are subject to both fast and unmeasured variations in their environment and small slow variations in their modal or vibrating properties. While any change in the excitation is meaningless, damages or fatigues on the structure are of interest. But the available measurements do not separate the effects of the external forces from the effect of the structure. Moreover, the changes of interest, that may be as small as 1% in the eigenfrequencies, are visible neither on the signals nor on their spectra. A global health monitoring method must rather rely on a model which will help in discriminating between the two mixed causes of the changes that are contained in the data. This vibration monitoring problem can be stated as the problem of detecting changes in the autoregressive (AR) part of a multivariable autoregressive moving average (ARMA) model having nonstationary MA coefficients. Change detection turns out to be very useful for this monitoring purpose, for example for monitoring the integrity of the civil infrastructure [24, 25, 45].

The improved safety and performance of aerospace structures and reduced aircraft development and operating costs are major concerns. One of the critical objectives is to ensure that the newly designed aircraft is stable throughout its operating range. A critical aircraft instability phenomenon, known as flutter, results from an unfavorable interaction of aerodynamic, elastic, and inertial forces, and may cause major failures. A careful exploration of the dynamical behavior of the structure subject to vibration and aeroservoelastic forces is thus required. A major challenge is the in-flight use of flight test data. The flight flutter monitoring problem can be addressed on-line as the problem of detecting that some instability indicators decrease below some critical value. CUSUM-type change detection algorithms are useful solutions to these problems [41, 46, 296, 531].

These application examples illustrate change detection with estimating functions different from the likelihood [36, 38].

The vibration-based structural health monitoring problem is explored in Section 11.2.

1.3.6 Finance and Economics

Stochastic modeling in finance is a new application area for optimal stopping and quickest change-point detection. For example, in the *Russian option* [410] the fluctuations in the price of an asset are modeled by geometric Brownian motion (the Black–Sholtz model), and the problem consists in finding a stopping time that maximizes a certain gain. In this optimization problem, the option owner is trying to find an exercise strategy that maximizes the expected value of his future reward with a certain interest rate for discounting. This problem can be effectively solved using the optimal stopping theory which is a part of the book. A similar approach can be applied for finding an optimal solution to the *American put option* with infinite horizon [359].

An application of the optimal stopping theory in financial engineering imposes an analysis for the gain process depending on the future and referring to an optimal prediction problem, which falls outside the scope of the classical optimal stopping framework. A typical setting is related to minimizing over a stopping time a functional of a Brownian motion.

These examples show that the optimal stopping theory can be effectively applied to many probabilistic settings of theoretical and practical interest. In addition, we mention the articles [52, 358] and references therein.

We also argue that quickest changepoint detection schemes can be effectively applied to the analysis of financial data. In particular, quickest changepoint detection problems are naturally associated with rapid detection of the appearance of an arbitrage in a market [421].

1.3.7 Computer Network Surveillance and Security

A considerable interest exhibited over the past decade in the field of defense against cyber-terrorism in general, and network security in particular, has been induced by a series of external and internal attacks on public, private corporate, and governmental computer network resources. Malicious intrusion attempts occur every day and have become a common phenomenon in contemporary computer networks. Examples of malicious activities are spam campaigns, phishing, personal data theft, worms, distributed denial-of-service (DDoS) attacks, address resolution protocol man-in-the-middle (ARP MiM) attacks, fast flux, *etc.* These pose an enormous risk to the users for a multitude of reasons such as significant financial damage, or severe threat to the integrity of personal information. It is therefore essential to devise automated techniques to detect such events as quickly as possible so that an appropriate response can be provided and the negative consequences for the user can be eliminated.

The detection of traffic anomalies is done by employing an intrusion detection system (IDS). Such systems in one way or another capitalize on the fact that malicious traffic is noticeably different from legitimate traffic. Depending on the principle of operation there are two categories of IDSs: either signature or anomaly based [113, 224]. A signature-based IDS inspects the passing traffic with the intent to find matches against already known malicious patterns. By contrast, an anomaly-based IDS is first trained to recognized the normal network behavior and then watches for any deviation from the normal profile.

Currently both types of IDSs are plagued by a high rate of false positives and the susceptibility to carefully crafted attacks that blend themselves into normal traffic. These two systems are complementary, and neither alone is sufficient to detect and isolate the myriad of network malicious or legitimate anomalies generated by attacks or other non-malicious events.

Intrusions usually lead to an abrupt change in the statistical characteristics of the observed traffic. For example, DDoS attacks lead to changes in the average number of packets sent through the victim's link per unit time. It is therefore appealing to formulate the problem of detecting computer intrusions as a *quickest changepoint detection problem*: to detect changes in statistical models as rapidly as possible, i.e., with minimal average delays, while maintaining the false alarm rate at a given low level. The feasibility of this approach has been already demonstrated in [472, 473, 474].

To make the detection delay small one has to increase the false alarm rate (FAR), and *vice versa.* As a result, the FAR cannot be made arbitrarily low without sacrificing other important performance metrics such as the detection delay and the probability of detection in a given time interval. Therefore, while attack detection algorithms can run with very low delay, this comes at the expense of high FAR, and thus changepoint detection techniques may not be efficient enough for intrusion detection. The ability of changepoint detection techniques to run at high speeds and with low delay, combined with the generally low frequency of intrusion attempts, presents an interesting opportunity: What if one could combine such techniques with others that offer very low false alarm rates but are too heavy to use at line speeds? Do such synergistic IDSs exist, and how can they be integrated? Such an approach is explored in Section 11.3. Specifically, a novel hybrid approach to network intrusion detection that combines *changepoint detection based anomaly* IDS with a *flow-based signature* IDS is proposed. The proposed hybrid IDS with profiling capability complements existing anomaly- and signature-based systems. In addition to achieving high performance in terms of the tradeoff between delay to detection, correct detection, and false alarms, the system also allows for isolating the anomalies. Therefore, the proposed approach overcomes common drawbacks and technological barriers of existing anomaly and signature IDSs by combining statistical changepoint detection and signal processing methods.

Chapter 2

Background on Probability and Statistics

2.1 Probability, Expectation, Markov Times, and Stochastic Processes

2.1.1 Probability and Expectation

We begin with outlining some standard definitions in probability theory and stochastic processes. Let $(\Omega, \mathscr{F})$ be a measurable space, i.e., Ω is a set of elementary events ω and $\mathscr{F}$ is a sigma-algebra (a system of subsets of Ω satisfying standard conditions). A *probability space* is a triple $(\Omega, \mathscr{F}, \mathsf{P})$, where P is a probability measure (completely additive measure normalized to 1) defined on the sets from the sigma-algebra $\mathscr{F}$. More specifically, by Kolmogorov's axioms, probability P satisfies: (a) $\mathsf{P}(A) \geq 0$ for any $A \in \mathscr{F}$; (b) $\mathsf{P}(\Omega) = 1$; and (c) $\mathsf{P}(\cup_{i=1}^{\infty} A_i) = \sum_{i=1}^{\infty} \mathsf{P}(A_i)$ for $A_i \in \mathscr{F}, A_i \cap A_j = \varnothing$, $i \neq j$, where $\varnothing$ is an empty set.

A function $X = X(\omega)$ defined on the space $(\Omega, \mathscr{F})$ with values in $\mathscr{X}$ is called *random variable* if it is $\mathscr{F}$-measurable, i.e., the set $\{\omega : X(\omega) \in B\}$, $B \in \mathscr{X}$, belongs to the sigma-algebra $\mathscr{F}$. The function $F(x) = \mathsf{P}(\omega : X(\omega) \leq x)$ is the *Distribution Function* of X. It is also called *Cumulative Distribution Function* (cdf). If the cdf $F(x)$ is continuous, then the function $f(x) = \mathrm{d}F(x)/\mathrm{d}x$ is called the *Probability Density Function* (pdf). This is the absolutely continuous case. In the opposite purely discrete case the cdf $F(x)$ is a step function with jumps at the points x_i, $i \geq 1$. In this case $f(x_i) = \mathsf{P}(X = x_i)$, $i \geq 1$ is called the *Probability Mass Function* (pmf). There is also a third mixed case. The three cases may be combined and unified using a measure-theoretic consideration, as shown in Theorem 2.1.1 below.

The events in a sequence $A_1, A_2, \dots$ are *independent* if for every finite set $j_1, \dots, j_n$ of distinct integers

$$\mathsf{P}\left\{\prod_{i=1}^{n} A_{j_i}\right\} = \prod_{i=1}^{n} \mathsf{P}\{A_{j_i}\},$$

so that the real random variables $X_1, X_2, \dots$ are independent if the events $\{X_1 \leq x_1\}, \{X_2 \leq x_2\}, \dots$ are independent for every sequence $x_1, x_2, \dots$ of real numbers.

Write $X^+ = \max(0, X)$ and $X^- = -\min(0, X)$. The *expectation* of a real random variable ($\mathscr{X} = \mathbb{R} = (-\infty, +\infty)$) denoted as $\mathsf{E}X$ is defined as $\mathsf{E}X = \mathsf{E}X^+ - \mathsf{E}X^-$. It exists if one of the expectations $\mathsf{E}X^+$ or $\mathsf{E}X^-$ is finite. If both these expectations are finite, then $\mathsf{E}|X| = \mathsf{E}X^+ + \mathsf{E}X^- < \infty$ and the random variable X is said to be *integrable*.

It is said that the probability measure P_0 is *absolutely continuous* with respect to another probability measure P_1, which is written as $\mathsf{P}_0 \ll \mathsf{P}_1$, if $\mathsf{P}_0(\mathcal{B}) = 0$ implies $\mathsf{P}_1(\mathcal{B}) = 0$. If P_0 and P_1 are mutually absolutely continuous, i.e., $\mathsf{P}_0 \ll \mathsf{P}_1$ and $\mathsf{P}_1 \ll \mathsf{P}_0$, then it is said that P_0 and P_1 are *equivalent*, which is written as $\mathsf{P}_0 \equiv \mathsf{P}_1$.

Theorem 2.1.1. *Let P_0 and P_1 be two probability measures on the measurable space $(\Omega, \mathscr{F})$.*

(i) *Lebesgue Decomposition. There exist a random variable $\Lambda = \Lambda(\omega)$ and a P_0-zero probability event $\mathcal{A}$ (i.e., $\mathsf{P}_0(\mathcal{A}) = 0$), such that*

$$\mathsf{P}_1(\mathcal{B}) = \int_{\mathcal{B}} \Lambda \mathrm{d}\mathsf{P}_0 + \mathsf{P}_1(\mathcal{A} \cap \mathcal{B}) \quad \text{for all } \mathcal{B} \in \mathscr{F}. \tag{2.1}$$

(ii) *Radon–Nikodým Theorem. Let* $\mathsf{P}_1 \ll \mathsf{P}_0$. *There exists a random variable* $\Lambda = \Lambda(\omega)$ *such that*

$$\mathsf{P}_1(\mathcal{B}) = \int_{\mathcal{B}} \Lambda \mathrm{d}\mathsf{P}_0 \quad \text{for all } \mathcal{B} \in \mathscr{F}. \tag{2.2}$$

The function $\Lambda(\omega)$ *is called the Radon–Nikodým derivative of* P_1 *with respect to* P_0, *or density of the measure* P_1 *with respect to the measure* P_0, *and the notation*

$$\Lambda(\omega) = \frac{d\mathsf{P}_1}{d\mathsf{P}_0}(\omega)$$

is usually used.

If the measures P_i, $i = 0, 1$ have probability density functions $p_i = p_i(x)$ with respect to some sigma-finite measure μ, e.g., the Lebesgue measure on $(\mathbb{R}, \mathscr{B})$, where $\mathscr{B}$ is the Borel sigma-algebra, then

$$\Lambda(x) = \frac{p_1(x)}{p_0(x)}.$$

Let $\mathscr{B} \subseteq \mathscr{F}$ be a sub-sigma-algebra of $\mathscr{F}$ and X be a random variable with expectation $\mathsf{E}X$. The *conditional expectation* of X given $\mathscr{B}$, which is denoted by $\mathsf{E}(X \mid \mathscr{B})$, is the random variable $Y = Y(\omega)$ satisfying

$$\int_A X d\mathsf{P} = \int_A Y \mathrm{d}\mathsf{P} \quad \text{for all } A \in \mathscr{B}.$$

Note that by the Radon–Nikodým theorem $Y = \mathsf{E}(X \mid \mathscr{B})$ always exists.

The conditional expectation of X given an event A is

$$\mathsf{E}[X \mid A] = \frac{\int_A X \, \mathrm{d}\mathsf{P}}{\mathsf{P}(A)},$$

assuming that $\mathsf{P}(A) > 0$.

Let $\mathbb{1}_{\{A\}} = \mathbb{1}_{\{A\}}(\omega)$ denote an indicator of a set A. In the particular case where $X = \mathbb{1}_{\{A\}}$, the conditional expectation $\mathsf{E}[\mathbb{1}_{\{A\}}(\omega) \mid \mathscr{B}] = \mathsf{P}(A \mid \mathscr{B})$ is the conditional probability of the event A with respect to $\mathscr{B}$.

Let Y be another random variable. Then the conditional expectation of X given Y is

$$\mathsf{E}[X \mid Y] = \mathsf{E}[X|\sigma(Y)],$$

where $\sigma(Y)$ is the sigma-algebra generated by Y.

2.1.2 *Exponential Family of Distributions*

Let $\theta = (\theta_1, \dots, \theta_\ell)^\top$ be an ℓ-dimensional vector parameter and let $\mathbf{X} = (X_1, \dots, X_m)^\top$ be an m-dimensional random variable associated with the parameter θ. Note that ℓ and m need not be the same. Let $\{\mathsf{P}_\theta, \theta \in \Theta\}$ be a family of probability measures generated by $\mathbf{X}$ with density $p_\theta(\mathbf{x}) = \mathrm{d}\mathsf{P}_\theta / \mathrm{d}\mu$ with respect to some non-degenerate sigma-finite measure μ.

The family $\{\mathsf{P}_\theta, \theta \in \Theta\}$ is said to be a *multivariate exponential family*[1] of probability measures (or distributions) if

$$p_\theta(\mathbf{x}) = \exp\left\{\theta^\top \mathbf{T}(\mathbf{x}) - b(\theta)\right\}, \quad \theta \in \Theta. \tag{2.3}$$

This representation is the so-called *canonical form* or a *natural parametrization*; see (2.6) below for a different "direct" representation. Since $p_\theta(\mathbf{x})$ must be integrable to 1, it follows that

$$b(\theta) = \log\left[\int_{\mathbb{R}^m} \exp\left\{\theta^\top \mathbf{T}(\mathbf{x})\right\} \mathrm{d}\mu(\mathbf{x})\right] = \log\left\{\mathsf{E}_\theta\left[e^{\theta^\top \mathbf{T}(\mathbf{X})}\right]\right\}.$$

[1] In the statistical literature sometimes the exponential family is referred to as the Pitman–Darmois–Koopman family.

There is no guarantee that the function $b(\theta)$ is finite for all $\theta \in \Theta$, which must be the case to make $p_\theta(\mathbf{x})$ a proper probability density. The subset Θ^* of θ for which $b(\theta)$ is finite,

$$\left\{\theta \in \Theta^* \subseteq \Theta : \log\left\{\mathsf{E}_\theta\left[e^{\theta^\top \mathbf{T}(\mathbf{X})}\right]\right\} < \infty\right\},$$

is called the *natural parameter space*. The natural parameter space is always a convex set.

Setting $\mathbf{Y} = \mathbf{T}(\mathbf{X})$, i.e., replacing $\mathbf{T}(\mathbf{X})$ with a new random variable $\mathbf{Y}$, we can write density (2.3) as

$$p_\theta(\mathbf{y}) = \exp\left\{\theta^\top \mathbf{y} - b(\theta)\right\}, \quad \theta \in \Theta. \tag{2.4}$$

In statistical applications, $\mathbf{X}$ represents original observations and $\mathbf{T}(\mathbf{X})$ is a so-called *sufficient statistic* which contains all the necessary information for inference. Pretending that we observe $\mathbf{Y} = \mathbf{T}(\mathbf{X})$ in place of $\mathbf{X}$, we therefore can think about the exponential family (2.4). Having this in mind, in subsequent chapters we will often define the exponential family as

$$p_\theta(\mathbf{x}) = \exp\left\{\theta^\top \mathbf{x} - b(\theta)\right\}, \quad \theta \in \Theta, \tag{2.5}$$

which means that the required transformation of observations has already been made.

Using (2.5) as the basic definition of the exponential family, we now proceed with discussing the properties of the function $b(\theta)$ and its role in finding the moments of $\mathbf{X}$ or the moments of the sufficient statistic $\mathbf{T}$ if (2.3) is used.

Noting that the cumulant-generating function $\mathsf{K}_\theta(\boldsymbol{t}) := \log \mathsf{E}_\theta[e^{\boldsymbol{t}^\top \mathbf{X}}]$ is equal to

$$\mathsf{K}_\theta(\boldsymbol{t}) = b(\boldsymbol{t} + \theta) - b(\theta)$$

and recalling that the gradient of $\mathsf{K}_\theta(\boldsymbol{t})$ at $\boldsymbol{t} = 0$ is the first moment and the Hessian of $\mathsf{K}_\theta(\boldsymbol{t})$ at $\boldsymbol{t} = 0$ is the covariance, we obtain that

$$\mathsf{E}_\theta \mathbf{X} = \nabla[b(\theta)] = \left[\frac{\partial b(\theta)}{\partial \theta_1}, \dots, \frac{\partial b(\theta)}{\partial \theta_\ell}\right], \quad \mathsf{cov}_\theta[\mathbf{X}] = \nabla^2[b(\theta)] = \left[\frac{\partial^2 b(\theta)}{\partial \theta_i \partial \theta_j}\right].$$

Hereafter we use $\nabla(\cdot)$ and $\nabla^2(\cdot)$ to denote the gradient and the Hessian matrix, respectively. In particular, it follows that the function $b(\cdot)$ is strictly convex unless the measure μ is degenerate, since in the nondegenerate case the covariance matrix is positive definite. Thus, in the case of a scalar parameter, $\mathsf{E}_\theta X = \mathrm{d}b(\theta)/\mathrm{d}\theta = \dot{b}(\theta)$ and $\mathsf{var}_\theta(X) = \mathrm{d}^2 b(\theta)/\mathrm{d}\theta = \ddot{b}(\theta)$. Since b is strictly convex, $\mathsf{E}_\theta X$ is a continuous and strictly increasing function of θ.

Note also that by a transformation of the coordinates, namely shifting the origin, we can always suppose that $b(0) = 0$, which is convenient in many statistical applications. In particular, often it is convenient to take $\mu = \mathsf{P}_0$ as the reference probability measure.

Next, if we are interested in considering $p_\theta(\mathbf{x})$ as a density relative to the Lebesgue measure in $\mathbb{R}^m$ in the continuous case or as the pmf in the discrete case, then the exponential family is defined by

$$p_\theta(\mathbf{x}) = h(\mathbf{x}) \exp\left\{c(\theta)^\top \mathbf{T}(\mathbf{x}) - b(\theta)\right\}, \quad \theta \in \Theta. \tag{2.6}$$

In the case of natural parametrization,

$$p_\theta(\mathbf{x}) = h(\mathbf{x}) \exp\left\{\theta^\top \mathbf{T}(\mathbf{x}) - b(\theta)\right\}, \quad \theta \in \Theta, \tag{2.7}$$

so

$$b(\theta) = \log\left[\int_{\mathbb{R}^m} h(\mathbf{x}) \exp\left\{\theta^\top \mathbf{T}(\mathbf{x})\right\} \mathrm{d}\mathbf{x}\right].$$

It is the logarithm of Laplace transform of $h(\mathbf{x})$ when $\mathbf{T}(\mathbf{x}) = \mathbf{x}$.

Examples.

(a) Normal distribution with unknown mean m:

$$\varphi((x-m)/\sigma) = \frac{1}{\sqrt{2\pi\sigma^2}} \exp\left\{-\frac{(x-m)^2}{2\sigma^2}\right\}.$$

Take $\mu(\mathrm{d}x) = \varphi(x/\sigma)e^{1/\sigma^2}\mathrm{d}x$ to obtain that $\theta = m$ and

$$p_\theta(x) = \exp\{\theta x - \theta^2/2\}, \quad \theta \in \Theta = (-\infty, +\infty),$$

so $b(\theta) = \theta^2/2$ and the natural parameter space is $(-\infty, +\infty)$.

(b) Normal distribution with unknown mean m and variance σ^2:

$$\theta = (m/\sigma^2, -1/2\sigma^2)^\top, \quad \mathbf{T}(x) = (x, x^2)^\top, \quad b(\theta) = m^2/2\sigma^2 + \log\sigma.$$

The natural parameter space is $\Theta = (-\infty, +\infty) \times (-\infty, 0]$.

(c) Bernoulli distribution: $\mathsf{P}(X = x) = p^x(1-p)^{1-x}$, $x = 0, 1$. Setting $\theta = \log[p/(1-p)]$ this pmf can be written as

$$\mathsf{P}_\theta(X = x) = \exp\left\{\theta x - \log\left(1 + e^\theta\right)\right\}, \quad \theta \in \Theta = [-\infty, +\infty],$$

so $b(\theta) = \log(1 + e^\theta)$ and the natural parameter space is $\Theta^* = [-\infty, +\infty)$.

2.1.3 Markov Times

Let $\{\mathscr{F}_t\}_{t\in\mathbb{R}_+}$ be a nondecreasing sequence of sub-sigma-algebras of $\mathscr{F}$, i.e., $\mathscr{F}_u \subseteq \mathscr{F}_t$, $u \le t$, $\mathscr{F}_u \in \mathscr{F}$. This nondecreasing sequence is also called *filtration*.

A random variable $T = T(\omega) \in [0, \infty] = \mathbb{R}_+ \cup \infty$ is said to be a *Markov time* with respect to $\{\mathscr{F}_t\}$ if $\{\omega : T \le t\} \in \mathscr{F}_t$ for all $t \in \mathbb{R}_+$. It is easily shown that also $\{\omega : T < t\} \in \mathscr{F}_t$, $t \in \mathbb{R}_+$, so that $\{\omega : T = t\} \in \mathscr{F}_t$, while the converse is not necessarily true unless the family $\{\mathscr{F}_t\}$ is right-continuous. See, e.g., [420].

In the following chapters we shall mostly, if not always, deal with sigma-algebras generated by some (observed) stochastic process $\{X_t\}$ defined on the probability space $(\Omega, \mathscr{F}, \mathsf{P})$, in which case $\mathscr{F}_t = \mathscr{F}_t^X = \sigma(\omega : X_u(\omega), u \le t)$, $\mathscr{F}_0 = (\Omega, \varnothing)$. This flow of (natural) sigma-algebras is obviously nondecreasing.

In the discrete-time case where $t = n \in \mathbb{Z}_+$ and $T \in \mathbb{Z}_+ \cup \infty$ the conditions $\{T \le n\} \in \mathscr{F}_n$ and $\{T = n\} \in \mathscr{F}_n$ are always equivalent, so that any of them can be taken as a definition of the Markov time. Indeed, clearly $\{T \le n\} \in \mathscr{F}_n$ implies $\{T = n\} \in \mathscr{F}_n$, while the converse also holds since $\{T \le n\} = \bigcup_{i=0}^n \{T = i\} \in \mathscr{F}_n$.

Let T_1 and T_2 be Markov times (with respect to the same sigma-algebra). Then $T_1 \wedge T_2 = \min(T_1, T_2)$, $T_1 \vee T_2 = \max(T_1, T_2)$, and $T_1 + T_2$ are Markov times. Certainly, this applies to any finite number of components. If $\{T_k\}_{k\ge1}$ is a sequence of Markov times, then $\sup_k T_k$ is also a Markov time, since $\{\sup_k T_k \le t\} = \bigcap_k \{T_k \le t\}$.

The Markov time T is also called the *stopping time*. This is due to the fact that usually this object is associated with stopping some process based on observations and making a decision. Often stopping times are considered as a sub-class of Markov times finite with probability 1. However, in this book we will not distinguish between Markov and stopping times, so that these notions are equivalent.

Example 2.1.1. Let $\{X_t\}$ be either a discrete- or a continuous-time stochastic process, and let $\mathscr{F}_t = \sigma(X_u, u \le t)$ be a natural sigma-algebra. Clearly, $\{X_t\}$ is adapted to $\{\mathscr{F}_t\}$. Let $\mathcal{A}$ be a set and define the first-exit time

$$T_{\mathcal{A}} = \inf\{t \ge 0 : X_t \notin \mathcal{A}\}, \quad \inf\{\varnothing\} = \infty,$$

which is a Markov time.

Assume $\mathcal{A} = (A_0, A_1)$, where A_0 and A_1 are some numbers. This special case is important in hypothesis testing, as we will see in Chapter 3. In this case, the Markov time $T_{\mathcal{A}}$ can be written as $T_{\mathcal{A}} = T_{A_0} \wedge T_{A_1}$, where

$$T_{A_0} = \inf\{t \geq 0 : X_t \leq A_0\}, \quad T_{A_1} = \inf\{t \geq 0 : X_t \geq A_1\}.$$

Here and in the following we always assume that $\inf\{\varnothing\} = \infty$, i.e., that $T_{A_i} = \infty$ if no such t exists.

2.1.4 Markov Processes

Write $\mathbb{R}_+ = [0, \infty)$ and $\mathbb{Z}_+ = \{0, 1, 2 \dots\}$. A collection of random variables $X = \{X_t(\omega)\}_{t \in \mathbb{R}_+}$ is called a continuous-time random process, and $X = \{X_t(\omega)\}_{t \in \mathbb{Z}_+}$ – a discrete-time random process, which is also called a random sequence. In the latter case, we will usually write $t = n$ for the time index, where $n = 0, 1, \dots$. We will often write $\{X_t\}$ for the random process without specifying the set of the time index, especially when the conclusion/result holds in both discrete-and continuous-time cases.

A random process $\{X_t\}$ is said to be *adapted* to the family of sigma-algebras $\{\mathscr{F}_t\}$ or simply $\{\mathscr{F}_t\}$*-adapted* if $\{\omega : X_t(\omega) \in B\} \in \mathscr{F}_t$ for all t. In the continuous-time case $t \in \mathbb{R}_+$, we will always suppose that the random process $\{X_t\}$ is either continuous (i.e., $X_t(\omega)$ viewed as a function of t is continuous) or, more generally, right-continuous having left-hand limits.

Let $\{X_t\}$ be a random process adapted to $\{\mathscr{F}_t\}$, and let P_x be a probability measure on $\mathscr{F}$ given $X_t = x \in \mathscr{X}$.

A random process $\{X_t\}$ is called a *homogeneous Markov process* if

(i) For all $x \in \mathscr{X}$ and all $u, t \in \mathbb{R}_+(\mathbb{Z}_+)$

$$\mathsf{P}_x(X_{t+u} \in B \mid \mathscr{F}_t) = \mathsf{P}_{X_t}(X_{t+u} \in B), \quad \mathsf{P}_x - \text{a.s.}$$

(ii) $\mathsf{P}_x(X_0 = x) = 1, x \in \mathscr{X}$.

The first condition is the Markov property of dependence of the future on the past via the present state.

The probability $P(u, x, B) = \mathsf{P}_{X_t = x}(X_{t+u} \in B) = \mathsf{P}_{X_0 = x}(X_u \in B)$ is called the *transition* probability. If it depends on t, the Markov process is *nonhomogeneous.*

If the space $\mathscr{X}$ is finite or countable, then the Markov process is called *Markov chain.*

We now consider several important special discrete- and continuous-time cases.

Example 2.1.2 (Random walk). Let $X_n = x + Y_1 + \cdots + Y_n$, $n \geq 1$ be partial sums, where $\{Y_n\}_{n \geq 1}$ is a sequence of iid random variables with a cdf $F(y)$. The discrete-time process $\{X_n\}_{n \in \mathbb{Z}_+}$ is called a *random walk* and the mean of its increment $d = \mathsf{E}Y_k$ is called the *drift* of the random walk. It is a homogeneous Markov process with mean $\mathsf{E}X_n = x + dn$ and transition probability

$$P(x, y) = \mathsf{P}(X_{n+1} \leq y | X_n = x) = \mathsf{P}(X_1 \leq y | X_0 = x) = \mathsf{P}(Y_1 \leq y - x) = F(y - x).$$

If $\mathsf{var}[Y_k] = \sigma^2 < \infty$, then $\mathsf{var}[X_n] = \sigma^2 n$. Usually it is assumed that the starting point $X_0 = x = 0$.

If $x = 0$ and Y_k takes on two values -1 and $+1$ with equal probabilities $\mathsf{P}(Y_k = -1) = \mathsf{P}(Y_k = +1) = 1/2$, then the random walk is called *simple.* The *Bernoulli random walk* is the sum of Bernoulli random variables taking on values 1 and 0 with probabilities p and $1 - p$.

Random walks play an important role in the theory of discrete-time Markov processes, as will become apparent later.

An important Markov process with continuous Gaussian trajectories similar to a Gaussian random walk — a Brownian motion process — is introduced and discussed in detail in the next subsection.

Example 2.1.3 (Reflected random walk). Again, let $\{Y_n\}_{n\geq 1}$ be a sequence of iid random variables with distribution $F(y)$. Define the process X_n recursively

$$X_0 = x, \quad X_n = \max(0, X_{n-1} + Y_n), \quad n = 1, 2, \dots. \tag{2.8}$$

Clearly, in this case the process X_n is a random walk reflected from the zero barrier—each time it hits the zero level it restarts all over again. If it never hits zero before time n, then $X_n = x + Y_1 + \cdots + Y_n$. This is the homogeneous Markov process with transition probability

$$P(x,y) = \mathsf{P}(X_{n+1} \leq y | X_n = x) = \begin{cases} F(y-x) & \text{if } y \geq 0, \\ 0 & \text{if } y < 0. \end{cases}$$

In statistics, the reflected random walk given by the recursion (2.8) is often called the *Cumulative Sum* (CUSUM) statistic. It is the basis of the most famous changepoint detection procedure, the CUSUM procedure, which is considered in Chapter 8 in detail.

Example 2.1.4 (Autoregression). Define the process X_n recursively

$$X_0 = x, \quad X_n = \rho X_{n-1} + Y_n, \quad n = 1, 2, \dots,$$

where ρ is a finite number and $\{Y_n\}_{n\geq 1}$ is as in the previous examples. Note that if $\rho = 1$ this process reduces to a random walk. For $\rho \neq 1$ this process is called an *autoregressive process of* 1st *order* and an abbreviation AR(1) is usually being used. This is the homogeneous Markov process with transition probability

$$P(x,y) = \mathsf{P}(X_{n+1} \leq y | X_n = x) = \mathsf{P}(Y_1 \leq y - \rho x | X_0 = x) = F(y - \rho x).$$

If $\mathsf{E}Y_k = d$ and $\mathsf{var}[Y_k] = \sigma^2$, then assuming that $|\rho| < 1$,

$$\lim_{n\to\infty} \mathsf{E}X_n = d/(1-\rho), \quad \lim_{n\to\infty} \mathsf{var}[X_n] = \sigma^2/(1-\rho^2).$$

Example 2.1.5 (Poisson process). An important continuous-time process is the Poisson process, whose trajectories are not continuous (piecewise constant with unit jumps). The random process $\{X_t\}_{t\in\mathbb{R}_+}$ is called a (homogeneous) *Poisson process* with intensity $\lambda > 0$ if it has iid increments, $X_0 = 0$, and the distribution $\mathsf{P}(X_t = n)$ has the form

$$\mathsf{P}(X_t = n) = \frac{(\lambda t)^n}{n!} e^{-\lambda t}, \quad t \geq 0, \ n = 0, 1, 2, \dots. \tag{2.9}$$

It is easily seen that $\mathsf{E}X_t = \mathsf{var}[X_t] = \lambda t$ and that the transition probability of the Poisson process is

$$P(u,m,n) = \mathsf{P}(X_{t+u} = n | X_t = m) = \frac{(\lambda u)^{m+n}}{(m+n)!} e^{-\lambda u}, \quad m,n = 0, 1, 2, \dots.$$

More generally, the intensity λ may depend on time, $\lambda = \lambda_t$. In this case the Poisson process is nonhomogeneous and has independent but nonidentically distributed increments that have distributions

$$P(X_{t+u} - X_t = n) = \frac{\left(\int_t^{t+u} \lambda_s \, \mathrm{d}s\right)^n}{n!} \exp\left\{-\int_t^{t+u} \lambda_s \, \mathrm{d}s\right\}, \quad t,u \geq 0, \ n = 0, 1, 2, \dots. \tag{2.10}$$

In this case, $\mathsf{E}X_t = \mathsf{var}[X_t] = \int_0^t \lambda_s \, \mathrm{d}s$.

The Poisson process belongs to the class of *counting* or *point* random processes that will be considered in Subsection 2.1.7. For these processes, X_t can be interpreted as a (random) number of certain events occurring in the time interval $[0,t]$. For the homogeneous Poisson process, the intervals between events (or waiting times) $\tau_n - \tau_{n-1}$, $n \geq 1$ ($\tau_0 = 0$) are iid exponentially distributed random variables with density $p(x) = \lambda e^{-\lambda x}$, $x \geq 0$.

Irreducible, Recurrent and Transient Markov Chains. Let $\{X_n\}$ be a homogeneous Markov chain with states $0,1,2,\dots$. It is said that the state j is *reachable* from i if there exists an integer $n \geq 0$ such that $\mathsf{P}_i(X_n = j) = \mathsf{P}(X_n = j|X_0 = i) > 0$, i.e., the probability that the chain will be in state j at time n when it is initialized at i is positive for some $n \geq 0$. If state j is reachable from i and state i is reachable from j, then it is said that states i and j *communicate*. A Markov chain for which *all states* communicate is said to be an *irreducible* Markov chain.

Let $T_{ii} = \inf\{n \geq 1 : X_n = i|X_0 = i\}$ denote the *return time* to state i when the chain starts at i, where $\inf\{\varnothing\} = \infty$, i.e., $T_{ii} = \infty$ if $X_n \neq i$ for all $n \geq 1$. Next, define $\mathsf{P}_i(T_{ii} < \infty)$, the probability of ever returning to state i, given that the chain started in state i. A state i is said to be *recurrent* if $\mathsf{P}_i(T_{ii} < \infty) = 1$ and *transient* if $\mathsf{P}_i(T_{ii} < \infty) < 1$.

By the Markov property, once the chain revisits state i, the future is independent of the past. Hence, after each time state i is visited, it will be revisited with the same probability $p_i = \mathsf{P}_i(T_{ii} < \infty)$ independent of the past. In particular, if $p_i = 1$, the chain will return to state i over and over again (an infinite number of times), and for this reason it is called recurrent. The transient state will only be visited a finite random number of times.

Formally, let N_i denote the total number of visits to state i, given that $X_0 = i$. The distribution of N_i is geometric, $\mathsf{P}_i(N_i = k) = (1-p_i)p_i^{k-1}, k \geq 1$ (counting the initial visit $X_0 = i$ as the first one). Therefore, the expected number of visits is $\mathsf{E}_i N_i = (1-p_i)^{-1}$, and the state is recurrent if $\mathsf{E}_i N_i = \infty$ and transient if $\mathsf{E}_i N_i < \infty$.

If states i and j communicate and i is recurrent, then so is j; if i is transient, then j is also transient.

The Markov chain is said to be *recurrent* if it is irreducible and all states are recurrent; otherwise the irreducible Markov chain is said to be *transient*. An irreducible Markov chain with *finite* state space is always recurrent.

A recurrent state i is said to be *positive recurrent* if $\mathsf{E}_i T_{ii} < \infty$ and *null recurrent* if $\mathsf{E}_i T_{ii} = \infty$. The irreducible Markov chain is said to be *positive recurrent* if all states are positive recurrent and *null recurrent* if all states are null recurrent.

Stationary Distributions. Consider a general discrete-time homogeneous Markov process $\{X_n\}_{n\in\mathbb{Z}_+}$ with state space $\mathscr{X}$ and transition probabilities $P(x,y) = \mathsf{P}(X_{n+1} \leq y|X_n = x)$, $x,y \in \mathscr{X}$. Let P_x denote the probability for the process with initial state $X_0 = x$, i.e., $\mathsf{P}(X_n \in \mathcal{B}|X_0 = x) = \mathsf{P}_x(X_n \in \mathcal{B})$. A *stationary distribution* of the Markov process $\{X_n\}$ is a limit (if it exists)

$$\lim_{n\to\infty} \mathsf{P}_x(X_n \leq y) = \mathsf{Q}_{\mathrm{st}}(y)$$

for every initial state $X_0 = x$ at continuity points of $\mathsf{Q}_{\mathrm{st}}(y)$. This distribution satisfies the following integral equation

$$\mathsf{Q}_{\mathrm{st}}(y) = \int_{\mathscr{X}} P(x,y)\,\mathrm{d}\mathsf{Q}_{\mathrm{st}}(x). \tag{2.11}$$

Clearly, if the initial variable X_0 has the probability distribution $\mathsf{Q}_{\mathrm{st}}(x)$, then all the other variables $X_1, X_2, \dots$ have the same distribution, which explains why $\mathsf{Q}_{\mathrm{st}}(x)$ is called the stationary distribution—the Markov process started from the stationary distribution is not only homogeneous but also stationary. A stationary distribution is also often called an *invariant* distribution.

Assume that $\{X_n\}$ is a continuous process. Then the stationary distribution has density $\mathsf{q}_{\mathrm{st}}(y) = \mathrm{d}\mathsf{Q}_{\mathrm{st}}(y)/\mathrm{d}y$ with respect to the Lebesgue measure, and it follows from (2.11) that this stationary density satisfies the equation

$$\mathsf{q}_{\mathrm{st}}(y) = \int_{\mathscr{X}} \mathsf{q}_{\mathrm{st}}(x)\mathcal{K}(x,y)\,\mathrm{d}x, \tag{2.12}$$

where $\mathcal{K}(x,y) = \frac{\partial}{\partial x}P(x,y)$. Thus, the stationary density $\mathsf{q}_{\mathrm{st}}(y)$ is the (left) eigenfunction corresponding to the unit eigenvalue of the linear operator $\mathcal{K}(x,y)$.

A stationary distribution exists for recurrent (more generally Harris-recurrent [53, 179]) Markov processes at least in a generalized sense, i.e., possibly an improper distribution (see below).

Example 2.1.6. Let $\{X_n\}$ be given recursively

$$X_0 = x \in [0,\infty), \quad X_{n+1} = (1+X_n)\Lambda_{n+1}, \quad n = 0,1,2,\ldots, \tag{2.13}$$

where $\Lambda_n, n \geq 1$ are nonnegative iid random variables having a Beta type II distribution with density

$$p(y) = \frac{y^{\delta-1}(1+y)^{-2\delta-1}}{\mathsf{B}(\delta,\delta+1)} \mathbb{1}_{\{y\geq 0\}}, \quad \mathsf{B}(r,v) = \int_0^1 t^{r-1}(1-t)^{v-1}\mathrm{d}t \ (r,v>0).$$

Using (2.12), we obtain that the stationary pdf is governed by the equation

$$\mathsf{q}_{\mathrm{st}}(y) = \frac{y^{\delta-1}}{\mathsf{B}(\delta+1,\delta)} \int_0^\infty \mathsf{q}_{\mathrm{st}}(x) \frac{(1+x)^{\delta+1}}{(1+x+y)^{1+2\delta}} \, \mathrm{d}x,$$

and the solution is

$$\mathsf{q}_{\mathrm{st}}(y) = \frac{y^{\delta-1}(1+y)^{-1-\delta}}{\mathsf{B}(\delta,1)} \mathbb{1}_{\{y\geq 0\}} = \delta y^{\delta-1}(1+y)^{-1-\delta} \mathbb{1}_{\{y\geq 0\}},$$

which is the pdf of a Beta type II distribution with parameters δ and 1.

Consider now a discrete case of a homogeneous Markov chain $\{X_n\}$ with the state space $\{0,1,2,\ldots\}$ and transition probabilities $P_{ij} = \mathsf{P}(X_{n+1} = j | X_n = i)$, $n,i,j = 0,1,2,\ldots$. A set of the limiting probabilities

$$\lim_{n\to\infty} \mathsf{P}(X_n = j | X_0 = i) = \mathsf{Q}^*_{\mathrm{st}}(j), \quad j = 0,1,2,\ldots$$

(if they exist) is said to be the *stationary distribution* of the Markov chain X_n. It satisfies the equation

$$\mathsf{Q}^*_{\mathrm{st}}(i) = \sum_{j=0}^{\infty} \mathsf{Q}^*_{\mathrm{st}}(j) P_{ji}, \quad i = 0,1,2,\ldots; \ \mathsf{Q}^*_{\mathrm{st}}(i) \geq 0; \ \sum_{i=0}^{\infty} \mathsf{Q}^*_{\mathrm{st}}(i) = 1. \tag{2.14}$$

If $\{X_n\}$ is a positive recurrent Markov chain, then a stationary distribution exists and is given by $\mathsf{Q}^*_{\mathrm{st}}(i) = 1/\mathsf{E}_i T_{ii}$, $i = 0,1,2,\ldots$. Since every irreducible Markov chain with finite state space is positive recurrent, it follows that a (unique) stationary distribution always exists in the finite state space case. If the Markov chain is null recurrent or transient, then a stationary distribution satisfying (2.14) does not exist. Indeed, the stationary distribution may not exist in the class of probability measures, which is often the case if X_n either goes to ∞ or to 0 (typical for null recurrent and transient chains). However, it may still have a stationary "distribution" in the generalized sense, i.e., an *improper stationary* distribution, satisfying (2.14) with an infinite sum, $\sum_{i=0}^{\infty} \mathsf{Q}^*_{\mathrm{st}}(i) = \infty$. In this more general sense, a stationary measure always exists for irreducible (indecomposable) recurrent Markov chains and sometimes for transient chains. For further details, see Harris [180, Sec. I.11] and references therein.

As an example, let X_n be a two-state Markov chain with strictly positive transition probabilities $P_{01} = p$, $P_{10} = q$. Then the unique stationary distribution is $\mathsf{Q}^*_{\mathrm{st}}(0) = q/(p+q)$, $\mathsf{Q}^*_{\mathrm{st}}(1) = p/(p+q)$.

Quasi-stationary Distributions. Quasi-stationary distributions come up naturally in the context of first-exit times of Markov processes. Let $\{X_n\}$ be a Markov process with state space $\mathscr{X}$ and transition probabilities $P(x,y) = \mathsf{P}(X_{n+1} \leq y | X_n = x)$, $x,y \in \mathscr{X}$. If the process is *absorbing*, its *quasi-stationary distribution* is defined to be the limit (if it exists) as $n \to \infty$ of the distribution of X_n, given that absorption has not occurred by the time n,

$$\mathsf{Q}_{\mathcal{A}}(y) = \lim_{n\to\infty} \mathsf{P}_x(X_n \leq y | X_1 \notin \mathcal{A}, \ldots, X_n \notin \mathcal{A}) \quad \text{for every initial state } X_0 = x,$$

where $\mathcal{A}$ is an absorbing set or state. Equivalently the quasi-stationary distribution can be defined as

$$\mathsf{Q}_{\mathcal{A}}(y) = \lim_{n\to\infty} \mathsf{P}_x(X_n \leq y | T_{\mathcal{A}} > n),$$

where $T_{\mathcal{A}} = \inf\{n \geq 1 : X_n \in \mathcal{A}\}$ is the "killing" time. Therefore, the quasi-stationary distribution is nothing but a stationary conditional distribution and also a limiting conditional distribution in that $X_n \to X_\infty$ as $n \to \infty$, and thus it can be used for modeling the long-term behavior of the process (or system).

Of special interest—in particular in certain statistical applications—is the case of a nonnegative Markov process ($\mathscr{X} = [0, \infty)$), where the first time that the process exceeds a fixed level A ($A > 0$) signals that some action is to be taken or a decision should be made. The quasi-stationary distribution is the distribution of the state of the process if a long time has passed and yet no crossover has occurred, i.e.,

$$\mathsf{Q}_A(y) = \lim_{n\to\infty} \mathsf{P}_x(X_n \leq y | T_A > n) \quad \text{for every initial state } X_0 = x, \tag{2.15}$$

where $T_A = \inf\{n \geq 1 : X_n \geq A\}$ is the corresponding stopping time which makes the process X_n absorbing.

Various topics pertaining to quasi-stationary distributions are existence, calculation, and simulation. For an extensive bibliography see Pollett [371].

The quasi-stationary distribution defined in (2.15) satisfies the following integral equation

$$\lambda_A \mathsf{Q}_A(y) = \int_0^A P(x,y)\, \mathrm{d}\mathsf{Q}_A(x), \tag{2.16}$$

where

$$\lambda_A = \int_0^A P(x,A)\, \mathrm{d}\mathsf{Q}_A(x).$$

If the sequence $\{X_n\}_{n\geq 0}$ is initialized from the random point $X_0 \sim \mathsf{Q}_A$ distributed according to a quasi-stationary distribution, then all the other variables $X_1, X_2, \dots$ are also distributed according to Q_A,

$$\mathsf{P}_{\mathsf{Q}_A}(X_n \in \mathcal{B} | T_A > n) = \mathsf{Q}_A(\mathcal{B}), \quad n = 0, 1, 2, \dots .$$

In this case, for every $A > 0$ the distribution of the stopping time T_A is strictly geometric with parameter $1 - \lambda_A$,

$$\mathsf{P}_{\mathsf{Q}_A}(T_A = k) = (1 - \lambda_A)\lambda_A^{k-1}, \quad k = 1, 2, \dots,$$

so $\mathsf{E}_{\mathsf{Q}_A} T_A = (1 - \lambda_A)^{-1}$. Indeed,

$$\begin{aligned}
\mathsf{P}_{\mathsf{Q}_A}(T_A > n) &= \mathsf{P}_{\mathsf{Q}_A}(X_0 < A, \dots, X_n < A) \\
&= \mathsf{P}_{\mathsf{Q}_A}(T_A > n-1)\mathsf{P}_{\mathsf{Q}_A}(X_n < A | X_0 < A, \dots, X_{n-1} < A) \\
&= \mathsf{P}_{\mathsf{Q}_A}(T_A > n-1)\mathsf{P}_{\mathsf{Q}_A}(X_n < A | X_{n-1} < A) \\
&= \mathsf{P}_{\mathsf{Q}_A}(T_A > n-1)\mathsf{P}_{\mathsf{Q}_A}(X_1 < A | X_0 < A) \\
&= \mathsf{P}_{\mathsf{Q}_A}(T_A > n-1)\mathsf{P}_{\mathsf{Q}_A}(T_A > 1).
\end{aligned}$$

Let $\{X_n\}$ be a continuous process. Write $\mathsf{q}_A(y) = \frac{\mathrm{d}\mathsf{Q}_A(y)}{\mathrm{d}y}$ for the quasi-stationary density. By (2.16), it satisfies the integral equation

$$\lambda_A \mathsf{q}_A(y) = \int_{\mathscr{X}} \mathsf{q}_A(x) \mathcal{K}(x,y)\, \mathrm{d}x. \tag{2.17}$$

Therefore, the quasi-stationary density $\mathsf{q}_A(y)$ is the (left) eigenfunction corresponding to the eigenvalue λ_A of the linear operator $\mathcal{K}(x,y) = \frac{\partial}{\partial x} P(x,y)$.

The quasi-stationary distribution may not exist. For example, if $\mathsf{P}(X_1 \geq A | X_0 < A) = 1$, there is no quasi-stationary distribution since $T_A = 1$ almost surely, but it must be geometric.

If the Markov process $\{X_n\}$ is Harris-recurrent and continuous, then the quasi-stationary distribution always exists [180, Theorem III.10.1].

The above results are also valid for continuous-time Markov processes, in which case the distribution of the stopping time $T_A = \inf\{t \ge 0 : X_t \ge A\}$ is exponential for all $A > 0$ if the process is started at a quasi-stationary distribution, $X_0 \sim \mathsf{Q}_A$.

Example 2.1.7. Again, let $\{X_n\}$ be given recursively as in (2.13) with nonnegative iid random variables Λ_n, $n \ge 1$ that have the distribution

$$\mathsf{P}(\Lambda_1 \le t) = \begin{cases} 1 & \text{if } t \ge 2 \\ t/2 & \text{if } 0 < t < 2 \\ 0 & \text{if } t \le 0. \end{cases}$$

Let $A < 2$. Then, by (2.17), the quasi-stationary density $\mathsf{q}_A(y)$ satisfies the integral equation

$$\lambda_A \, \mathsf{q}_A(y) = \frac{1}{2} \int_0^A \mathsf{q}_A(x) \, \frac{dx}{1+x},$$

which yields $\lambda_A = \frac{1}{2}\log(1+A)$ and $\mathsf{q}_A(y) = A^{-1}\mathbb{1}_{\{x \in [0,A)\}}$. Thus, for $A < 2$ the quasi-stationary distribution $\mathsf{Q}_A(y) = y/A$ is uniform. Note that it is attained already for $n = 1$.

An important question is whether the quasi-stationary distribution converges to the stationary distribution as $A \to \infty$, i.e., $\lim_{A\to\infty} \mathsf{Q}_A(y) = \mathsf{Q}_{\mathrm{st}}(y)$ at continuity points of $\mathsf{Q}_{\mathrm{st}}(y)$. We now give sufficient conditions for this to be true. Recall that we consider a time-homogeneous Markov process $\{X_n\}_{n\in\mathbb{Z}_+}$ with state space $[0,\infty)$. Again, let P_x denote the probability for the process with initial state $X_0 = x$.

We call the process $\{X_n\}$ *stochastically monotone* if $\mathsf{P}_x(X_1 \ge y)$ is non-decreasing and right-continuous in x for all $y \in [0,\infty)$.

The following result due to Pollak and Siegmund [368, Theorem 1] establishes convergence of the quasi-stationary distribution to a stationary one when the latter exists.

Theorem 2.1.2. *Let $\{X_n\}_{n\in\mathbb{Z}_+}$ be a nonnegative stochastically monotone Markov process, and let $T_A = \inf\{n : X_n \ge A\}$, $A > 0$.*

(i) *For arbitrary $x, y \ge 0$, $A > 0$ and $n = 1, 2, \ldots$*

$$\mathsf{P}_x(X_n \le y | T_A > n) \ge \mathsf{P}_x(X_n \le y). \tag{2.18}$$

(ii) *If in addition a stationary distribution $\mathsf{Q}_{\mathrm{st}}(y)$ exists, i.e., $\lim_{n\to\infty} \mathsf{P}_x(X_n \le y) = \mathsf{Q}_{\mathrm{st}}(y)$ for every initial point $X_0 = x$ at continuity points y of $\mathsf{Q}_{\mathrm{st}}(y)$, then for arbitrary $x \ge 0$*

$$\mathsf{P}_x(X_n \le y | T_A > n) \to \mathsf{Q}_{\mathrm{st}}(y) \quad \text{as } A, n \to \infty \tag{2.19}$$

at all continuity points y of $\mathsf{Q}_{\mathrm{st}}(y)$.

Similar assertions also hold for the stochastically monotone continuous-time Markov processes [368, Theorem 2]. For example, homogeneous diffusion processes driven by a Brownian motion are stochastically monotone.

Stochastic monotonicity and an additional mild condition of communicability imply the existence of a stationary distribution $\mathsf{Q}_{\mathrm{st}}(y)$, which in general may not be a probability measure [368, Theorem 3].

2.1.5 Brownian Motion and Itô's Stochastic Integral

The random process $\{W_t\}_{t\in\mathbb{R}_+}$ is called a *Brownian motion process* or a *Wiener process* if it is a continuous (homogeneous) Gaussian process with independent and identically distributed increments and $W_0 = 0$ (P-a.s.), $\mathsf{E}W_t = 0$, $\mathrm{var}[W_t] = \sigma^2 t$. If $\sigma^2 = 1$, then W_t is called the *standard Brownian motion.*[2] Clearly, the distribution of $W_t - W_{t+s}$ coincides with the distribution of W_s, which is normal, $\mathsf{P}(W_s \le x) = \Phi(x/\sigma\sqrt{s})$, where $\Phi(y) = (2\pi)^{-1/2}\int_{-\infty}^{y} e^{-u^2/2}\,du$ is the standard normal cdf.

[2] With a certain abuse of terminology we often say simply "Brownian motion" instead of the "Brownian motion process."

The Brownian motion is a homogeneous Markov process with transition density

$$p(u,x,y) = \frac{\partial \mathsf{P}_{W_t=x}(W_{t+u} \le y)}{\partial y} = \frac{1}{\sqrt{2\pi u}} e^{-(y-x)^2/2\sigma^2 u}$$

and covariance $\mathsf{cov}(W_s, W_t) = \sigma^2 \min(s,t)$.

Also, with probability 1, $\mathsf{E}[W_t|\mathscr{F}_s] = W_s$ and $\mathsf{E}[(W_t - W_s)^2|\mathscr{F}_s] = \sigma^2(t-s)$ for all $s \le t$, where $\mathscr{F}_s = \sigma(W_u, 0 \le u \le s)$.

Note that if the Brownian motion is sampled at discrete time moments $t_n = \Delta n$, $n = 1,2,\dots$ and then we form sums $W_n = \Delta W_1 + \cdots + \Delta W_n$, $n \ge 1$, where $\Delta W_k = W_k - W_{k-1}$ ($W_0 = 0$) are increments, then the discrete-time process $\{W_n\}_{n\ge 0}$ is a Gaussian random walk. Thus, the Brownian motion can be regarded as a limit of the random walk as $\Delta \to 0$ (in some sense).

The Brownian motion plays a special role in the theory of continuous-time stochastic processes as well as in various applications, as will become apparent shortly. Specifically, many engineering systems (e.g., radio, acoustic, infra-red, and video sensors) are stochastic in nature and may be adequately described either by the system of recursive (linear or nonlinear) equations in discrete time or by differential equations with stochastic perturbations (due to sensor noise and other random effects) in continuous time. For example, a wide class of problems in remote sensing deal with detection, tracking, and recognition of objects based on signals corrupted by sensor noise and environmental clutter. In this case, an appropriate model for the observations is $X_t = S_t + V_t + \xi_t$, where S_t is a (generally) random process associated with a signal from the object, V_t is a random process associated with clutter, and ξ_t is sensor noise (also random). Usually the sensor noise has a very wide spectrum with constant intensity so that two even very closely spaced values are (at least approximately) independent, and moreover, often Gaussian. For this reason, practitioners model such noise by a delta-correlated Gaussian process, $\mathsf{E}\xi_t = 0$, $\mathsf{E}[\xi(t)\xi(t+u)] = \sigma_\xi^2 \delta(u)$, where $\delta(u)$ is Dirac's delta-function which is equal to infinity at $u = 0$ and zero otherwise, $\int_{-\infty}^{\infty} \delta(x)\mathrm{d}x = 1$. Such a process is usually called *white noise*, since its spectrum is uniform at any frequency (from 0 to ∞). Also, the processes S_t and V_t can be reasonably modeled as an output of some inertial system driven by another white noise η_t, independent of or dependent from ξ_t. For example, V_t may be thought as a solution of the differential equation

$$\dot{V}_t = d_t(\mathbf{V}_0^t) + \sigma_t(\mathbf{V}_0^t)\eta_t,$$

where $\dot{V}_t = \mathrm{d}V_t/\mathrm{d}t$ is time-derivative and d_t and σ_t are some functions. Recall that by $\mathbf{Y}_0^t$ we denote the trajectory $\{Y_u, 0 \le u \le t\}$. This differential equation is not ordinary but rather stochastic.

White Gaussian noise may be regarded as a derivative of the Brownian motion $\dot{W}_t$. Unfortunately, such a process does not exist in the class of ordinary stochastic processes. While formally it can be rigorously introduced as a generalized random process, this is tedious.

Clearly, the process V_t satisfying the previous differential equation can be written as

$$V_t = V_0 + \int_0^t d_u(\mathbf{V}_0^u)\,\mathrm{d}u + \int_0^t \sigma_u(\mathbf{V}_0^u)\,\mathrm{d}W_u. \tag{2.20}$$

The first integral is an ordinary Reimann–Stieltjes integral, but the second integral cannot be justified in an ordinary way since the white Gaussian process η_u does not exist—its trajectories are discontinuous everywhere. It is this integral that we would like to define in a proper way.

Let $S_t(\omega)$ be a random function (process). We now define the stochastic integral $I_t(S) = \int_0^t S_u \mathrm{d}W_u$ over the standard Brownian motion. Note first that stochastic integrals cannot be defined as ordinary Lebesgue–Stieltjes or Riemann integrals, since trajectories of the Brownian motion are not differentiable, as was mentioned above. In fact, W_t has unbounded variation in small intervals

The process S_t is called *non-anticipative* with respect to the filtration $\{\mathscr{F}_t\}$ if it is $\mathscr{F}_t$-measurable for each $t \ge 0$. This means that it does not depend on the future, i.e., it is $\mathscr{F}_t$-adapted. In particular, it may depend on W_0^t, but not on W_u for $u > t$.

In the following we will always assume that S_t is a non-anticipative process and that it is square integrable in a sense that $\int_0^t \mathsf{E}[S_u^2]\,du < \infty$ for $t < \infty$, and the stochastic integral will be defined for such class of processes. The stochastic integral is first constructed for the step function S_t (i.e., $S_u = S_{u_k}$ for $u \in [u_k, u_{k+1})$, $k = 0, 1, \dots, m-1$, $u_0 = 0$, $u_m = t$) as

$$I_t(S) = \sum_{k=0}^{m-1} S_{u_k}[W_{u_{k+1}} - W_{u_k}].$$

Next, assuming that there exists a sequence of step functions $g_n(t)$ such that

$$\lim_{n\to\infty} \mathsf{E}\left\{\int_0^t [S_u - g_n(u)]^2\right\} du = 0,$$

the stochastic integral $I_t(S)$ is defined as the mean-square limit of the sequence of random variables $\int_0^t g_n(u)\,\mathrm{d}W_u$, i.e.,

$$\int_0^t g_n(u)\,\mathrm{d}W_u \xrightarrow[n\to\infty]{L^2} \int_0^t S_u\,\mathrm{d}W_u.$$

See Subsection 2.4.1 for a definition of L^p-convergence.

In a similar way the stochastic integral can be defined for the class of functions satisfying

$$\mathsf{P}\left(\int_0^t S_u^2(\omega)\,\mathrm{d}u < \infty\right) = 1,$$

in which case it is defined as the limit in probability

$$\int_0^t g_n(u)\,\mathrm{d}W_u \xrightarrow[n\to\infty]{\mathsf{P}} \int_0^t S_u\,\mathrm{d}W_u.$$

See Subsection 2.4.1 for a definition of convergence in probability.

For further details of constructing the stochastic integral, see [164, 266].

If the process S_t is continuous (with probability 1), then

$$\int_0^t S_u\,\mathrm{d}W_u = \lim_{\Delta\to 0} \sum_{k=0}^{m-1} S_{u_k}[W_{u_{k+1}} - W_{u_k}],$$

where $0 = u_0 < u_1 < \cdots < u_m = t$, $\Delta = \max_{0\le k\le m-1}(u_{k+1} - u_k)$. This integral is often referred to as the Itô stochastic integral. It is worth noting that the choice of the left point u_k for sampling the process S_u is important in cases where S_u depends on W_u, unlike in the ordinary case of the Lebesgue–Stieltjes integrals. Changing this point changes the integral. In particular, taking the mid point $(u_k + u_{k+1})/2$ leads to a totally different integral, the Stratonovich stochastic integral [380, 445]. To see this, it suffices to consider the example where $S_u = W_u$. Then

$$\mathsf{E}\left(\sum_{k=0}^{m-1} W_{u_k}[W_{u_{k+1}} - W_{u_k}]\right) = \sum_{k=0}^{m-1}(u_k - u_k) = 0$$

while if we take the mid point, then

$$\mathsf{E}\left(\sum_{k=0}^{m-1} W_{(u_k+u_{k+1})/2}[W_{u_{k+1}} - W_{u_k}]\right) = \sum_{k=0}^{m-1}(u_{k+1} - u_k)/2 = t/2.$$

The Itô stochastic integral has the following properties:

$$\mathsf{E}I_t(S) = 0, \quad \mathsf{E}[I_t(S)]^2 = \int_0^t \mathsf{E}S_u^2\,du.$$

Also, the process $\{I_t(S)\}$ is a zero-mean martingale; see Section 2.3 for a definition. Note that the Stratonovich integral is not a martingale.

2.1.6 *Stochastic Differential Equations, Itô Processes, and Diffusion Processes*

Consider a random process $\{X_t\}$ that is given by equation (2.20) where the second integral is understood as the Itô stochastic integral (over a standard Brownian motion W_t), i.e.,

$$X_t = X_0 + \int_0^t d_u(\mathbf{X}_0^u)\,\mathrm{d}u + \int_0^t \sigma_u(\mathbf{X}_0^u)\,\mathrm{d}W_u. \tag{2.21}$$

Note that the functions d_t and σ_t, which we will refer to as the coefficients, are random, but not just arbitrary random — they depend on the trajectories $\mathbf{X}_0^t$ of the process X_t. In other words, the coefficients are $\mathscr{F}_t^X$-measurable. In this case, it is said that the process $\{X_t\}$ is a *diffusion-type process* . In order to avoid complications related to existence of integrals, we assume that the following conditions hold

$$\mathsf{P}\left(\int_0^t |d_u|\,\mathrm{d}u < \infty\right) = 1, \quad \mathsf{P}\left(\int_0^t \sigma_u^2\,\mathrm{d}u < \infty\right) = 1. \tag{2.22}$$

In the particular case where the corresponding coefficients are functions of only the current state, i.e., $d_t = d_t(X_t)$ and $\sigma_t = \sigma_t(X_t)$, the process $\{X_t\}$ is called a *diffusion process*. It is Markovian (but generally nonhomogeneous). It is a homogeneous Markov process if $d_t(X_t) = d(X_t)$ and $\sigma_t(X_t) = \sigma(X_t)$ (do not depend on time t). The coefficient d_t is called the *coefficient of drift* or just drift and σ_t – the *coefficient of diffusion.*

In the most general case the coefficients $d_t = d_t(\omega)$ and $\sigma_t = \sigma_t(\omega)$ in (2.21) are arbitrary $\{\mathscr{F}_t\}$-adapted random processes satisfying the conditions (2.22). In this case, it is said that $\{X_t\}$ is an *Itô process* (with respect to the Brownian motion W_t). As we will see, any Itô process can be represented in an equivalent form of a diffusion-type process.

For short, instead of representing the Itô process in the integral form (2.21), it is written in the differential form

$$\mathrm{d}X_t = d_t(\omega)\,\mathrm{d}t + \sigma_t(\omega)\,\mathrm{d}W_t, \quad t \geq 0, \tag{2.23}$$

and it is said that the corresponding process has the stochastic differential given in (2.23).

The stochastic integral $I_t^X(S) = \int_0^t S_u\,\mathrm{d}X_u$ over the Itô process X_t with the stochastic differential (2.23) is naturally defined as

$$I_t^X(S) = \int_0^t S_u d_u\,\mathrm{d}u + \int_0^t S_u \sigma_u\,\mathrm{d}W_u,$$

assuming that

$$\mathsf{P}\left(\int_0^t |S_u d_u|\,\mathrm{d}u < \infty\right) = 1, \quad \mathsf{P}\left(\int_0^t S_u^2 \sigma_u^2\,\mathrm{d}u < \infty\right) = 1$$

We now provide *Itô's differentiation formula* for functions of Itô processes. Let the function $f(t,x)$ (on $[0,t) \times \mathbb{R}^1$) be twice continuously differentiable in x and have continuous derivative in time t. If the Itô process X_t has stochastic differential (2.23), then the stochastic differential of the process $f(t,X_t)$ is

$$\mathrm{d}f(t,X_t) = \left[\frac{\partial f(t,X_t)}{\partial t} + \frac{f(t,X_t)}{\partial t} d_t + \frac{1}{2}\frac{\partial^2 f(t,X_t)}{\partial^2 x}\sigma_t^2\right]\mathrm{d}t + \frac{\partial f(t,X_t)}{\partial x}\sigma_t \mathrm{d}W_t. \tag{2.24}$$

Note that this formula is different from the ordinary Newton–Leibniz formula, so that Itô's stochastic calculus differs from the conventional calculus. For example, the conventional calculus suggests that $\int_0^t W_u\,\mathrm{d}W_u = W_t^2/2$, but the Itô formula yields $W_t^2/2 - t/2$.

We complete this section with a discussion of the issues related to mutual absolute continuity of probability measures corresponding to Itô and diffusion-type processes and the structure of the Radon–Nikodým derivatives for these processes. These results will be used in the following chapters for constructing likelihood ratio processes when considering hypothesis testing and changepoint detection problems.

Let $(\Omega, \mathscr{F}, \{\mathscr{F}_t\}, \mathsf{P})$ be a filtered probability space and let $(X_t, \mathscr{F}_t)_{t\geq 0}$ be an Itô process with the stochastic differential

$$\mathrm{d}X_t = S_t(\omega)\,\mathrm{d}t + \mathrm{d}W_t, \quad X_0 = 0, \; t \geq 0, \tag{2.25}$$

where W_t is a standard Brownian motion.[3] By P_X we denote the probability measure corresponding to the process $X = \{X_t\}_{t\geq 0}$ and by P_W the Wiener measure corresponding to $W = \{W_t\}_{t\geq 0}$. Also, as usual, we use the notation P^t for the restriction of the measure P to the sigma-algebra $\mathscr{F}_t$. Define the "likelihood ratio" process $\Lambda_t = \mathrm{d}\mathsf{P}^t_X/\mathrm{d}\mathsf{P}^t_W$. Below we determine conditions under which P^t_X is absolutely continuous with respect to the Wiener measure P^t_W (i.e., $\mathsf{P}^t_X \ll \mathsf{P}^t_W$) and give a convenient representation for Λ_t. This is a Girsanov-type theorem for Itô processes. We write $\Lambda_t(X)$ when the measure P_X is a true measure and $\Lambda_t(W)$ when the measure P_W is a true measure.

For example, in signal detection theory one typically deals with a signal-plus-noise model of the form (2.25), where $S_t(\omega)$ is a random signal that either does not depend on white noise $\dot{W}_t$ or depends on its trajectory till time t (i.e., $S_t(\omega)$ is a nonanticipative process). If we are interested in detecting this signal in noisy observations X_t, then we have to construct the likelihood ratio process $\Lambda_t = \mathrm{d}\mathsf{P}^t_X/\mathrm{d}\mathsf{P}^t_W$ for the hypotheses "signal+noise" *versus* "noise only." It turns out that this problem can be reduced to hypothesis testing for diffusion-type processes, as the following lemma shows.

Lemma 2.1.1. *Let $\{X_t\}$ be an Itô process with stochastic differential* (2.25). *Let $\hat{S}_t(\mathbf{X}_0^t) = \mathsf{E}[S_t(\omega)|\mathscr{F}_t^X]$. If*

$$\int_0^t \mathsf{E}|S_u|\,\mathrm{d}u < \infty, \tag{2.26}$$

then $\{X_t\}$ can be represented in the form of the diffusion-type process

$$\mathrm{d}X_t = \hat{S}_t(\mathbf{X}_0^t)\,\mathrm{d}t + \mathrm{d}\widetilde{W}_t, \quad X_0 = 0, \; t \geq 0 \tag{2.27}$$

with respect to the Brownian motion

$$\widetilde{W}_t = X_t - \int_0^t \hat{S}_u(\mathbf{X}_0^u)\,\mathrm{d}u. \tag{2.28}$$

The proof can be found in Liptser and Shiryaev [266, Theorem 7.12].

The representation (2.27) is often called the *minimal representation.* This representation plays a critical role in forming the likelihood ratio process Λ_t, since it can be computed using the functional $\hat{S}_t(\mathbf{X}_0^t)$ that depends on the observed trajectory $\mathbf{X}_0^t$. Note that the representation via the original random process $S_t(\omega)$ is not constructive, since it is not observable.

Observe that if $\mathsf{E}|S_t|^2 < \infty$, then the functional $\hat{S}_t(\mathbf{X}_0^t)$ is the Bayesian estimate of S_t in the mean square error sense, i.e., it minimizes the expected loss $\mathsf{E}(S_t - \hat{S}_t)^2$.

Obviously, by definition of $\widetilde{W}_t$ an inclusion $\mathscr{F}_t^{\widetilde{W}} \subseteq \mathscr{F}^X$ holds for all $0 \leq t < 0$. If the inverse inclusion $\mathscr{F}_t^{\widetilde{W}} \supseteq \mathscr{F}^X$ holds, then $\mathscr{F}_t^{\widetilde{W}} = \mathscr{F}^X$. In this case, the process $\widetilde{W}$ is called the *innovation process*, and it carries the same information as the process X.

Therefore, in order to investigate the general case of Itô processes, we can use the results of absolute continuity of measures for diffusion processes. Specifically, it follows from (2.27) and Theorems 7.5 and 7.6 of Liptser and Shiryaev [266, Sec. 7.2] that if

$$\mathsf{P}\left(\int_0^t \hat{S}_u^2(\mathbf{X}_0^u)\,\mathrm{d}u < \infty\right) = 1 \quad \text{and} \quad \mathsf{P}\left(\int_0^t \hat{S}_u^2(\mathbf{W}_0^u)\,\mathrm{d}u < \infty\right) = 1,$$

[3]Assuming that W_t is a standard process does not lead to any loss of generality, since if W_t has intensity $\sigma_t > 0$ (deterministic) such that $\int_0^t \sigma_u^2\,\mathrm{d}u < \infty$, $t < \infty$, then all the results presented below hold with S_t replaced by S_t/σ_t and $\mathrm{d}X_t$ by $\mathrm{d}X_t/\sigma_t$.

then the measures P_X and P_W are equivalent ($\mathsf{P}_X \sim \mathsf{P}_W$) and almost surely

$$\Lambda_t(X) = \exp\left\{\int_0^t \hat{S}_u(\mathbf{X}_0^u)\,\mathrm{d}X_u - \frac{1}{2}\int_0^t \hat{S}_u^2(\mathbf{X}_0^u)\,\mathrm{d}u\right\},$$
$$\Lambda_t(W) = \exp\left\{\int_0^t \hat{S}_u(\mathbf{W}_0^u)\,\mathrm{d}W_u - \frac{1}{2}\int_0^t \hat{S}_u^2(\mathbf{W}_0^u)\,\mathrm{d}u\right\}. \tag{2.29}$$

The following theorem gives all the required conditions and an exact result that is of central importance for our statistical applications in the following sections.

Theorem 2.1.3. *Let $\{X_t\}$ be an Itô process with stochastic differential* (2.25). *Assume that the conditions*

$$\mathsf{P}\left(\int_0^t S_u^2\,\mathrm{d}u < \infty\right) = 1 \quad \textit{and} \quad \int_0^t \mathsf{E}|S_u|\,\mathrm{d}u < \infty \tag{2.30}$$

are satisfied and that, in addition,

$$\mathsf{E}\left[\exp\left(-\int_0^t S_t\,\mathrm{d}W_t - \frac{1}{2}\int_0^t S_u^2\,\mathrm{d}u\right)\right] = 1. \tag{2.31}$$

Then $\mathsf{P}_X \sim \mathsf{P}_W$ and the equalities (2.29) *hold for the likelihood ratio processes under X and W.*

Remark 2.1.1. Theorem 2.1.3 holds true if the process S_t is independent of W_t. It also holds in an important particular case where S_t depends on W_t and is a continuous Gaussian process [266, Sec. 7.5]. Also, the following Novikov condition is sufficient for (2.31) to hold:

$$\mathsf{E}\left[\exp\left(\frac{1}{2}\int_0^t S_u^2\,\mathrm{d}u\right)\right] < \infty. \tag{2.32}$$

Finally, we provide a theorem that generalizes Theorem 2.1.3 to the case of two Itô processes. Assume that the Itô processes $\{X_t^{(i)}\}$ are given by the stochastic differentials

$$\mathrm{d}X_t^{(i)} = S_t^{(i)}(\omega)\,\mathrm{d}t + \sigma_t\mathrm{d}W_t, \quad X_0^{(1)} = X_0^{(2)}, \quad t \ge 0, \quad i = 0,1, \tag{2.33}$$

where σ_t is a positive deterministic square integrable function, $\int_0^t \sigma_u^2\,\mathrm{d}u < \infty$. Similar to (2.27), the processes $\{X_t^{(i)}\}$ can be represented as diffusion-type processes

$$\mathrm{d}X_t^{(i)} = \hat{S}_t^{(i)}(\mathbf{X}_t^{(i)})\,\mathrm{d}t + \sigma_t\mathrm{d}\widetilde{W}_t^{(i)}, \qquad t \ge 0, \tag{2.34}$$

where $\mathbf{X}_t^{(i)} = \{X_u^{(i)}, 0 \le u \le t\}$, $\hat{S}_t^{(i)}(\mathbf{X}_t^{(i)}) = \mathsf{E}[S_t^{(i)}(\omega)|\mathscr{F}_t^{X^{(i)}}]$, and $\{\widetilde{W}_t^{(i)}\}$ are $\{\mathscr{F}_t^{X^{(i)}}\}$-adapted standard Brownian motions. Therefore, the problem of the absolute continuity of measures for Itô processes reduces to the equivalent problem for diffusion-type processes, and it suffices to consider this problem for the processes given by the differentials

$$\mathrm{d}X_t^{(i)} = d_t^{(i)}(\mathbf{X}_t^{(i)})\,\mathrm{d}t + \sigma_t\mathrm{d}W_t, \qquad t \ge 0, \quad \int_0^t \sigma_u^2\,\mathrm{d}u < \infty. \tag{2.35}$$

The following theorem is the analogue of Theorem 2.1.3 for measures generated by diffusion-type processes. We write $\Lambda_t = \mathrm{d}\mathsf{P}_1^t/\mathrm{d}\mathsf{P}_0^t$, where P_i^t is the restriction of the probability measure generated by the process $\{X_t^{(i)}\}$ to the sigma-algebra $\mathscr{F}_t$.

Theorem 2.1.4. *Let $\{X_t^{(i)}\}$, $i = 0,1$ be diffusion-type processes with stochastic differentials* (2.35). *Let the conditions*

$$\mathsf{P}\left\{\int_0^t \sigma_u^{-2}\left([d_u^{(0)}(\mathbf{X}_u^{(0)})]^2 + [d_u^{(1)}(\mathbf{X}_u^{(0)})]^2\right)\mathrm{d}u < \infty\right\}$$
$$= \mathsf{P}\left\{\int_0^t \sigma_u^{-2}\left([d_u^{(0)}(\mathbf{X}_u^{(1)})]^2 + [d_u^{(1)}(\mathbf{X}_u^{(1)})]^2\right)\mathrm{d}u < \infty\right\} = 1 \tag{2.36}$$

hold. Then $\mathsf{P}_0 \sim \mathsf{P}_1$ *and*

$$\begin{aligned}
\Lambda_t(\mathbf{X}_t^{(0)}) &= \exp\left\{ \int_0^t \sigma_u^{-2} \left(d_u^{(0)}(\mathbf{X}_u^{(0)}) - d_u^{(1)}(\mathbf{X}_u^{(0)}) \right) \mathrm{d}X_u^{(0)} \right. \\
&\quad \left. - \frac{1}{2} \int_0^t \sigma_u^{-2} \left([d_u^{(0)}(\mathbf{X}_u^{(0)})]^2 - [d_u^{(1)}(\mathbf{X}_u^{(0)})]^2 \right) \mathrm{d}u \right\}, \\
\Lambda_t(\mathbf{X}_t^{(1)}) &= \exp\left\{ \int_0^t \sigma_u^{-2} \left(d_u^{(0)}(\mathbf{X}_u^{(1)}) - d_u^{(1)}(\mathbf{X}_u^{(1)}) \right) \mathrm{d}X_u^{(1)} \right. \\
&\quad \left. - \frac{1}{2} \int_0^t \sigma_u^{-2} \left([d_u^{(0)}(\mathbf{X}_u^{(1)})]^2 - [d_u^{(1)}(\mathbf{X}_u^{(1)})]^2 \right) \mathrm{d}u \right\}.
\end{aligned} \tag{2.37}$$

2.1.7 Point Random Processes

Along with Itô and diffusion processes driven by the Brownian motion, the class of the so-called counting (or point) processes that are piecewise continuous and have jumps is of great interest. Let $\{\tau_n\}_{n\geq 1}$ be a sequence of Markov times (with respect to the filtration $\{\mathscr{F}_t\}_{t\in\mathbb{R}_+}$) such that $\tau_1 > 0$ (a.s.), $\tau_n < \tau_{n+1}$ (a.s.) on $\{\tau_n < \infty\}$ and $\tau_n = \tau_{n+1} = \infty$ on $\{\tau_n = \infty\}$. The moments τ_n, $n \geq 1$ are associated with occurrence of some events, so that the n-th event occurs at time τ_n. Hence $\delta_n = \tau_n - \tau_{n-1}$, $n \geq 1$ are random intervals between consecutive events ($\tau_0 = 0$). The sequence of random variables $\{\tau_n\}_{n\geq 1}$ can be completely characterized by a *counting* process

$$N_t = \sum_{n=1}^{\infty} \mathbb{1}_{\{\tau_n \leq n\}}, \quad t \in \mathbb{R}_+,$$

which is also called the *point* process.[4] Thus, studying the sequence $\{\tau_n\}_{n\geq 1}$ is equivalent to studying the point process $\{N_t\}_{t\in\mathbb{R}_+}$, and *vice versa.* Often the sequence $\{\tau_n\}$ is also referred to as a point process. Note that trajectories of the point process $\{N_t\}_{t\in\mathbb{R}_+}$ are nondecreasing, right-continuous and piecewise constant with unit jumps. Obviously, the value of N_t can be interpreted as the number of events occurring in the interval $[0,t]$.

The Poisson process with intensity λ introduced in Subsection 2.1.4 whose increments are iid with distribution

$$\mathsf{P}(N_{t+u} - N_t = n) = \frac{(\lambda u)^n}{n!} e^{-\lambda u}, \quad t,u \geq 0, \;\; n = 0,1,2,\dots \tag{2.38}$$

($N_0 = 0$) is a point process, so is the nonhomogeneous Poisson process; see (2.10). For the homogeneous Poisson process $\{\tau_n - \tau_{n-1}\}_{n\geq 1}$ are iid with density $p(x) = \lambda e^{-\lambda x} \mathbb{1}_{\{x\geq 0\}}$.

Let τ be a Markov time positive w.p. 1 ($\mathsf{P}(\tau > 0) = 1$). Then $N_t = \mathbb{1}_{\{\tau\leq t\}}$, $t \geq 0$ is a point process, in which case $\tau_1 = \tau$ and $\tau_n = \infty$ for $n \geq 2$.

It turns out that a point process N_t allows for the (Doob–Meyer) decomposition $N_t = m_t + A_t$, where m_t is a martingale as defined in Section 2.3 and A_t is a predictable increasing process [267, Theorem 18.1]. The process A_t is called the *compensator* of the point process N_t, since the difference $N_t - A_t$ is a martingale, so that it compensates N_t to a martingale. For example, if $\{N_t\}$ is a Poisson process with intensity λ, then the process $\{N_t - \lambda t\}$ is a martingale, so that the compensator $A_t = \lambda t$ is deterministic.

In the class of point processes there are some processes whose compensator has the form

$$A_t = \int_0^t \lambda_u \, \mathrm{d}b_u,$$

[4] If the intervals between events δ_n are iid, then the processes $\{\tau_n\}$ and $\{N_t\}$ are called renewal processes.

where $\{\lambda_t(\omega)\}$ is a nonnegative predictable process and b_t is a nonnegative right continuous nondecreasing function. Since in the Poisson case $\lambda_t = \lambda$ and $b_t = t$, these kinds of processes are referred to as Poisson-type processes.

Let $A_{t-} = \lim_{u\uparrow t} A_u$ and $\Delta A_t = A_t - A_{t-}$ (jump). If $\Delta A_t = 0$ for all $t \geq 0$, then the compensator is continuous. An important result that we will need in the following chapters is related to the absolute continuity of measures for Poisson-type processes with continuous compensators $A_t(N)$. Specifically, if $N_t^{(1)}$ and $N_t^{(2)}$ are two Poisson-type processes with continuous compensators $A_1(t) = A_1(t,N)$ and $A_2(t) = A_2(t,N)$, then their measures P_1^t and P_t^2 are equivalent and the likelihood ratio process $\Lambda_t = \mathrm{d}\mathsf{P}_1^t/\mathrm{d}\mathsf{P}_2^t$ can be written as

$$\Lambda_t = \exp\left\{\int_0^t \log\left(\frac{\mathrm{d}A_1(u)}{\mathrm{d}A_2(u)}\right)\mathrm{d}N_u - [A_1(t) - A_2(t)]\right\}, \tag{2.39}$$

where the integral $\int_0^t f(u)\,\mathrm{d}N_u$ over the point process N_t is understood as a Stieltjes stochastic integral [267, Sec. 18.4]. This result can be derived from [267, Theorem 19.7].

2.2 Certain Useful Equalities and Inequalities

We write A^c for the event complementary to A, i.e., for such event that $\mathsf{P}(A^c) = 1 - \mathsf{P}(A)$. We use notation AB for the intersection $A \cap B$.

Let A and B be two events. The following equality holds

$$\mathsf{P}(A \cup B) = \mathsf{P}(A) + \mathsf{P}(B) - \mathsf{P}(AB),$$

which implies the inequality

$$\mathsf{P}(A \cup B) \leq \mathsf{P}(A) + \mathsf{P}(B),$$

where the equality holds when A and B are mutually exclusive. In this case (with a certain abuse of terminology) we will say that the events are independent. This inequality can be immediately extended to a finite and even infinite number of events:

$$\mathsf{P}\left(\bigcup_{i=1}^{\infty} A_i\right) \leq \sum_{i=1}^{\infty} \mathsf{P}(A_i)$$

where equality holds if all the events are mutually independent.

Boole's Inequality:

$$\mathsf{P}(AB) \geq \mathsf{P}(A) - \mathsf{P}(B^c) = 1 - \mathsf{P}(A^c) - \mathsf{P}(B^c),$$

and more generally

$$\mathsf{P}\left(\bigcap_{i=1}^{\infty} A_i\right) \geq 1 - \sum_{i=1}^{\infty} \mathsf{P}(A_i^c).$$

C_m−inequality. Let a and b be real numbers. The following inequality holds:

$$|a+b|^m \leq C_m(|a|^m + |b|^m), \tag{2.40}$$

where $C_m = 1$ if $m \leq 1$ and $C_m = 2^{m-1}$ if $m > 1$. The inequality (2.40) will be referred to as the C_m−inequality.

Generalized Chebyshev's Inequality. Let $a > 0$ and let $g(x)$ be an even and nondecreasing function for $x \geq 0$, $g(0) \geq 0$. Let X be a random variable such that $\mathsf{E}[g(X)]$ exists. Then

$$\mathsf{P}(|X| \geq a) \leq \frac{\mathsf{E}[g(X)]}{g(a)}. \tag{2.41}$$

In particular, if $g(x) = |x|^m$, $m > 0$, then it follows from (2.41) that

$$\mathsf{P}(|X| \geq a) \leq \frac{\mathsf{E}|X|^m}{a^m}. \tag{2.42}$$

For $m = 1$, this is *Markov's inequality*.

Taking $X = Y - \mathsf{E}Y$ and $m = 2$, we obtain *Chebyshev's inequality*

$$\mathsf{P}(|Y| \geq a) \leq \frac{\mathsf{var}(Y)}{a^2}.$$

All three inequalities will be referred to as the Chebyshev inequalities.

Hölder's Inequality. Let $p > 1$ and $1/p + 1/q = 1$. Let X and Y be two random variables such that expectations $\mathsf{E}|X|^p$ and $\mathsf{E}|Y|^q$ exist. Then

$$\mathsf{E}|XY| \leq (\mathsf{E}|X|^p)^{1/p}(\mathsf{E}|Y|^q)^{1/q} \tag{2.43}$$

For $p = q = 2$ this is the Cauchy–Schwarz–Bunyakovsky inequality.

Minkowski's Inequality. Let $p \geq 1$ and assume that expectations $\mathsf{E}|X|^p$ and $\mathsf{E}|Y|^q$ exist. Then

$$(\mathsf{E}|X+Y|^p)^{1/p} \leq (\mathsf{E}|X|^p)^{1/p} + (\mathsf{E}|Y|^p)^{1/p}. \tag{2.44}$$

Jensen's Inequality. Let $g(x)$ be a convex function and let X be a random variable such that the expectation $\mathsf{E}[g(X)]$ exists. Then

$$\mathsf{E}[g(X)] \geq g(\mathsf{E}X). \tag{2.45}$$

Remark 2.2.1. Hölder's, Minkowski's, and Jensen's inequalities also hold for the conditional expectation $\mathsf{E}(\cdot \mid \mathscr{B})$, where $\mathscr{B}$ is a sub-sigma-algebra of $\mathscr{F}$.

2.3 Martingales, Optional Stopping, and Wald's Identities

Let $(\Omega, \mathscr{F}, \mathsf{P})$ be a probability space. Let $\{\mathscr{F}_t\}$ be a filtration, and let $\{X_t\}$ be a random process on $(\Omega, \mathscr{F}, \mathsf{P})$ adapted to $\{\mathscr{F}_t\}$. The time index may be either discrete $t = n \in \mathbb{Z}_+$ or continuous $t \in \mathbb{R}_+$. Recall that in the continuous-time case we always assume that $\{X_t\}_{t\in\mathbb{R}_+}$ is right-continuous having left-hand limits.

A process $\{X_t\}$ is said to be a *martingale* with respect to $\{\mathscr{F}_t\}$ if $\mathsf{E}|X_t| < \infty$ for all t (i.e., integrable), and

$$\mathsf{E}[X_t | \mathscr{F}_u] = X_u \quad \mathsf{P}-\text{a.s. for } u \leq t. \tag{2.46}$$

If $\mathsf{E}[X_t|\mathscr{F}_u] \geq X_u$, then $(X_t, \mathscr{F}_t)$ is said to be a *submartingale*; if $\mathsf{E}[X_t|\mathscr{F}_u] \leq X_u$, then $(X_t, \mathscr{F}_t)$ is said to be a *supermartingale*. Note that $-X_t$ is a supermartingale whenever X_t is a submartingale. Clearly, the martingale is simultaneously sub- and supermartingale.

In most cases we will deal with martingales with respect to the natural filtration $\mathscr{F}_t = \sigma(X_u, u \leq t)$ (generated by the process X_t), in which case we will omit mentioning $\mathscr{F}_t$.

One typical example of the martingale is the likelihood ratio process

$$\Lambda_t = \frac{d\mathsf{P}_1^t}{d\mathsf{P}_0^t}, \quad t \in \mathbb{R}_+ \text{ or } t \in \mathbb{Z}_+, \tag{2.47}$$

where $P_i^t = P_i|_{\mathscr{F}_t}$ $(i = 0, 1)$ are restrictions of probability measures P_i to the sigma-algebra $\mathscr{F}_t$. If P_1^t is absolutely continuous with respect to P_0^t, then Λ_t is the P_0-martingale with unit expectation. Another typical example is the sum $S_n = \sum_{i=1}^n X_i$ of zero-mean independent and identically distributed (iid) random variables X_i, $EX_i = 0$. Also, if θ is a random variable with finite expectation, then $\theta_n = E[\theta \mid \mathscr{F}_n]$ is a martingale.

If $(X_t, \mathscr{F}_t)$ is a martingale, then $(|X_t|^\gamma, \mathscr{F}_t)$ is a submartingale for $\gamma \geq 1$ and a supermartingale for $0 < \gamma < 1$, which follows from Jensen's inequality.

One of the important properties of semimartingales is that the martingale structure is preserved by optional sampling.[5]

Theorem 2.3.1 (Optional sampling theorem). *Let $(X_t, \mathscr{F}_t)$ be a P-supermartingale and let T and τ be two stopping times with respect to $\{\mathscr{F}_t\}$ such that $\tau \leq T$ and $P(T < \infty) = 1$. Let*

$$E|X_\tau| < \infty, \quad E|X_T| < \infty, \tag{2.48}$$

$$\liminf_{t\to\infty} \int_{\{T>t\}} |X_t| \, dP = 0. \tag{2.49}$$

Then

$$E[X_T \mid \mathscr{F}_\tau] \leq X_\tau \quad P-a.s., \tag{2.50}$$

and if $(X_t, \mathscr{F}_t)$ is a martingale, then

$$E[X_T \mid \mathscr{F}_\tau] = X_\tau \quad P-a.s. \tag{2.51}$$

The optional sampling theorem is of fundamental importance in sequential analysis and optimal stopping. In particular, taking $\tau = 0$, we immediately obtain the following identity for martingales that generalizes Wald's identity for sums of iid random variables.

Theorem 2.3.2 (Generalized Wald's identity). *Let $(X_t, \mathscr{F}_t)$ be a P-martingale and let T be a stopping time with respect to $\{\mathscr{F}_t\}$ which is finite w.p. 1. If $E|X_T| < \infty$ and (2.49) holds, then*

$$E[X_T] = E[X_0]. \tag{2.52}$$

Corollary 2.3.1 (Wald's Identities). *Let $Y_1, Y_2, \dots$ be iid random variables with finite mean $E[Y_1] = \mu$, and let $S_n = \sum_{i=1}^n Y_i$.*

(i) *If $E[T] < \infty$, then*

$$E[S_T] = \mu E[T]. \tag{2.53}$$

(ii) *If the variance $\sigma^2 = \mathrm{var}[Y_i]$ is finite and $E[T] < \infty$, then*

$$E[S_T - \mu T]^2 = \sigma^2 E[T]. \tag{2.54}$$

Proof. (i) Noting that the sequence $\{S_n - \mu n\}_{n\geq 1}$ is a zero-mean martingale and using (2.52) with $X_T = S_T - \mu T$, we obtain

$$E[S_T - \mu T] = 0,$$

from which (2.53) follows.

(ii) Noting that $X_n = [S_n - \mu n]^2 - \sigma^2 n$, $n \geq 1$ is a martingale with mean zero and applying (2.52), we obtain (2.54). □

Another important identity repeatedly used in the following chapters concerns the likelihood ratio process Λ_t defined in (2.47). Denote by E_i the expectations under the probability measures P_i, $i = 0, 1$.

[5] Often optional sampling is called optional stopping.

Theorem 2.3.3 (Wald's likelihood ratio identity). *For any stopping time T and nonnegative $\mathscr{F}_T$-measurable random variable Y*

$$\mathsf{E}_1[Y \mathbb{1}_{\{T<\infty\}}] = \mathsf{E}_0[Y \Lambda_T \mathbb{1}_{\{T<\infty\}}]. \tag{2.55}$$

In particular, for $Y = \mathbb{1}_{\{A\}}$ with $A \in \mathscr{F}_T$,

$$\mathsf{P}_1(A, T < \infty) = \mathsf{E}_0[\Lambda_T, A, T < \infty]. \tag{2.56}$$

Note that (2.55) can be written as

$$\int_{\{T<\infty\}} Y \, d\mathsf{P}_1 = \int_{\{T<\infty\}} Y \Lambda_T \, d\mathsf{P}_0.$$

Taking $A = \Omega$, we obtain

$$\mathsf{P}_1(T < \infty) = \mathsf{E}_0(\Lambda_T, T < \infty) = \int_{\{T<\infty\}} \Lambda_T \, d\mathsf{P}_0, \tag{2.57}$$

and hence

$$\mathsf{E}_0(\Lambda_T, T < \infty) = \int_{\{T<\infty\}} \Lambda_T \, d\mathsf{P}_0 = 1$$

whenever $\mathsf{P}_1(T < \infty) = 1$.

Theorem 2.3.4 (Doob's maximal inequality). *Let $(X_t, \mathscr{F}_t)$ be a* P*-submartingale and let $M_t = \sup_{u \le t} X_u$. The following inequality holds*

$$\mathsf{P}(M_t \ge a) \le \frac{1}{a} \mathsf{E}[X_t \mathbb{1}_{\{M_t \ge a\}}] \le \frac{1}{a} \mathsf{E}[X_t^+], \quad a > 0. \tag{2.58}$$

Doob's inequality (2.58) can be proved using the optional sampling theorem.

If X_t is a uniformly integrable martingale, then this inequality can be extended to stopping times T. Specifically, let $M_t^* = \sup_{u \le t} |X_u|$. Then

$$\mathsf{P}(M_T^* \ge a) \le \frac{1}{a} \mathsf{E}[|X_T| \mathbb{1}_{\{M_T^* \ge a\}}] \le \frac{1}{a} \mathsf{E}|X_T|, \quad a > 0.$$

Note that if $Y_1, Y_2, \dots$ are iid zero-mean random variables with $\mathsf{E}|Y_1|^m < \infty$, $m \ge 1$, then taking $X_n = |S_n|^m$, $S_n = \sum_{i=1}^n Y_i$, we obtain from Theorem 2.3.4 that

$$\begin{aligned} \mathsf{P}\left(\max_{k \le n} |S_k| > a\right) &\le \frac{1}{a^m} \mathsf{E}\left[|S_n|^m ; \max_{k \le n} |S_k| > a\right] \\ &\le \frac{1}{a^m} \mathsf{E}|S_n|^m, \quad a > 0, \; n \ge 1. \end{aligned} \tag{2.59}$$

For $m = 2$, this gives the Kolmogorov inequality. For this reason, the inequality (2.58) is often referred to as the Doob–Kolmogorov inequality for martingales. Note that the last inequality in (2.59) is a generalization of the Chebyshev inequality (2.42).

A martingale $(X_t, \mathscr{F}_t)$ is said to be *square integrable* if $\sup_t \mathsf{E}[X_t^2] < \infty$.

Theorem 2.3.5 (Moment inequalities). **(i)** *Let $(X_t, \mathscr{F}_t)$ be a nonnegative submartingale and let $M_t = \sup_{u \le t} X_u$. Let $p > 1$. If $\mathsf{E}[X_t^p] < \infty$, then*

$$\mathsf{E}[M_t^p] \le \left(\frac{p}{p-1}\right)^p \mathsf{E}[X_t^p]. \tag{2.60}$$

Therefore, if $(X_t, \mathscr{F}_t)$ is a square integrable martingale, then

$$\mathsf{E}\left[\sup_{u \le t} X_u^2\right] \le 4 \mathsf{E}[X_t^2]. \tag{2.61}$$

(ii) *Let $(X_t, \mathscr{F}_t)$ be a uniformly integrable martingale and let T be a stopping time. If $\mathsf{E}|X_T|^p < \infty$, then*

$$\mathsf{E}\left[\left(\sup_{t\le T}|X_t|\right)^p\right] \le \left(\frac{p}{p-1}\right)^p \mathsf{E}|X_T|^p. \tag{2.62}$$

(iii) *Let $(X_t, \mathscr{F}_t)$ be a square integrable martingale with independent increments, $X_0 = 0$. For any $0 < p < \infty$ and any stopping time T there are positive constants C_p and c_p, independent of T and X, such that*

$$\mathsf{E}\left[\left(\sup_{t\le T}|X_t|\right)^p\right] \le \begin{cases} C_p\,\mathsf{E}[D_T]^{p/2} & \text{if } 0 < p \le 2, \\ C_p\,\mathsf{E}[D_T]^{p/2} + c_p\,\mathsf{E}[\sum_{t\le T}|\Delta X_t|^p] & \text{if } p > 2, \end{cases} \tag{2.63}$$

where $D_t = \mathsf{var}[X_t]$ and $\Delta X_t = X_t - \lim_{s\uparrow t} X_s$ is a jump of the process X at time t.

This theorem is true both for continuous- and discrete-time cases. A proof of (i) and (ii) may be found, e.g., in Liptser and Shiryaev [265]. The inequalities (2.63) are particular cases of the Burkholder–Gundy–Novikov inequalities for martingales [88, 89, 338, 339]. A proof of (iii) may be found in Tartakovsky [457, Lemma 5]. The inequality (2.60) is often referred to as Doob's maximal moment inequality.

Corollary 2.3.2. *Let $S_n = X_1 + \cdots + X_n$, $n \ge 1$ be a zero mean random walk, $\mathsf{E}X_1 = 0$. If $\mathsf{E}|X_1|^p < \infty$, then for any stopping time T*

$$\mathsf{E}|S_T|^p \le \begin{cases} C_p\,\mathsf{E}|X_1|^p\mathsf{E}T & \text{if } 0 < p \le 2, \\ C_p\,\mathsf{E}[X_1^2]^{p/2}\mathsf{E}T^{p/2} + c_p\,\mathsf{E}|X_1|^p\mathsf{E}T \le \tilde{C}_p\mathsf{E}|X_1|^{p/2}\mathsf{E}T^{p/2} & \text{if } p > 2, \end{cases} \tag{2.64}$$

where C_p, c_p, and $\tilde{C}_p$ are universal constants depending only on p.

2.4 Stochastic Convergence

2.4.1 Standard Modes of Convergence

Let $(\Omega, \mathscr{F}, \mathsf{P})$ be a probability space. In the following we will always assume that all random objects are defined on this space. Let Y be a random variable and let $\{X_t\}$ be a continuous- or discrete-time stochastic process, $t \in \mathbb{R}_+$ or $t = n \in \mathbb{Z}_+$.

Convergence in Probability. We say that the process $\{X_t\}$ converges to Y *in probability* as $t \to \infty$ and write $X_t \xrightarrow[t\to\infty]{\mathsf{P}} Y$ if

$$\lim_{t\to\infty} \mathsf{P}(|X_t - Y| > \varepsilon) = 0 \quad \text{for all } \varepsilon > 0.$$

Almost Sure Convergence. We say that the process $\{X_t\}$ converges to Y *almost surely* (a.s.) or *with probability* 1 (w.p. 1) as $t \to \infty$ and write $X_t \xrightarrow[t\to\infty]{\text{a.s.}} Y$ if

$$\mathsf{P}(\omega : \lim_{t\to\infty} X_t = Y) = 1.$$

Convergence in Distribution (Weak Convergence). Let $F_t(x) = \mathsf{P}(\omega : X_t \le x)$ be the cdf of X_t and let $F(x) = \mathsf{P}(\omega : Y \le x)$ be the cdf of Y. We say that the process $\{X_t\}$ converges to Y *in distribution* or *in law* or *weakly* as $t \to \infty$ and write $X_t \xrightarrow[t\to\infty]{\text{law}} Y$ if

$$\lim_{t\to\infty} F_t(x) = F(x)$$

at all continuity points of $F(x)$.

The a.s. convergence implies convergence in probability, and the convergence in probability implies convergence in distribution, while the converse statements are not generally true, in fact usually not true except for some particular cases. For instance, the weak convergence of X_t to a constant c implies the convergence in probability of X_t to c.

Strong Law of Large Numbers. Let $\{Y_n\}_{n\geq 1}$ be a sequence of iid random variables. Write $S_n = \sum_{i=1}^n Y_i$. The Kolmogorov strong law of large numbers (SLLN) states that if $\mathsf{E}[Y_1]$ exists, then the sample mean S_n/n converges to the mean value $\mathsf{E}[Y_1]$ w.p. 1, i.e.,

$$n^{-1}S_n \xrightarrow[n\to\infty]{\text{a.s.}} \mathsf{E}[Y_1].$$

Note that this is true regardless of the finiteness of $\mathsf{E}[Y_1]$; only the existence is required. If the expectation is finite (i.e., $\mathsf{E}[Y_1] = \mu, |\mu| < \infty$), then the limit is finite. Otherwise $S_n/n \to \pm\infty$.

Often in addition to establishing the a.s. convergence or the convergence in probability of X_t to Y, one is interested in the convergence of the moments, i.e., $\mathsf{E}|X_t|^m \to \mathsf{E}|Y|^m$. The corresponding mode of convergence is the L^p-convergence, $p \geq 1$.

L^p-convergence. We say that the process $\{X_t\}$ converges to Y *in L^p* or *in the p-th mean* as $t \to \infty$ and write $X_t \xrightarrow[t\to\infty]{L^p} Y$ if

$$\lim_{t\to\infty} \mathsf{E}|X_t - Y|^p = 0.$$

In general, the a.s. convergence does not guarantee the convergence of the moments. To overcome this difficulty we need an additional uniform integrability condition.

Definition 2.4.1. A process $\{X_t\}$ is said to be *uniformly integrable* if

$$\sup_t \mathsf{E}[|X_t|\,\mathbb{1}_{\{|X_t|>a\}}] \xrightarrow[a\to\infty]{} 0.$$

Passage to the Limit under the Sign of Expectation. A very useful topic related to convergence in the theory of probability and statistics is establishing conditions under which it is possible to exchange the operations of expectation (i.e., integration) and limit. The following theorem combines three famous results – the *monotone convergence theorem*, *Fatou's lemma*, and *Lebesgue's dominated convergence theorem*. We write $X_n \uparrow X$ if $X_n \to X$ and $\{X_n\}$ is non-decreasing ($X_n \leq X_{n+1}$), and $X_n \downarrow X$ if $\{X_n\}$ is non-increasing.

Theorem 2.4.1. *Let $\{X_n\}_{n\in\mathbb{Z}^+}$ be a sequence of random variables and X a random variable.*

(i) *Monotone convergence theorem. If $X_n \uparrow X$ w.p. 1 as $n\to\infty$ and $\mathsf{E}[X_1^-] < \infty$, then $\mathsf{E}[X_n] \uparrow \mathsf{E}[X]$. If $X_n \downarrow X$ w.p. 1 as $n \to \infty$ and $\mathsf{E}[X_1^+] < \infty$, then $\mathsf{E}[X_n] \downarrow \mathsf{E}[X]$.*

(ii) *Fatou's lemma. Let $\{X_n^+\}$ be uniformly integrable. If $\mathsf{E}[\limsup_n X_n]$ exists, then*

$$\mathsf{E}[\limsup_n X_n] \geq \limsup_n \mathsf{E}[X_n].$$

In particular, this inequality holds if $X_n \leq X$, $n \geq 0$, where X is integrable.

(iii) *Lebesgue's dominated convergence theorem. Let $X_n \xrightarrow[n\to\infty]{\mathsf{P}} X$. If there exists a random variable Y such that $|X_n| \leq Y$, $n \geq 0$ and $\mathsf{E}[Y] < \infty$, then*

$$\lim_{n\to\infty} \mathsf{E}|X_n - X| = 0.$$

Fatou's lemma allows us to establish the following useful result.

Theorem 2.4.2. *Let X_n be nonnegative, $X_n \geq 0$, with finite expectations $\mathsf{E}[X_n] < \infty$, $n \geq 0$. Assume that $X_n \xrightarrow[n\to\infty]{\mathsf{P}} X$. Then $\lim_{n\to\infty} \mathsf{E}[X_n] = \mathsf{E}[X] < \infty$ iff $\{X_n\}$ is uniformly integrable.*

Note that the dominated convergence theorem follows immediately from Theorem 2.4.2. Also, all of the above results hold for conditional expectations.

Convergence of Moments. The following theorem is useful for establishing the convergence of moments. See, e.g., Loève [268, pp. 165–166].

Theorem 2.4.3. *Assume that* $X_n \xrightarrow[n\to\infty]{\mathsf{P}} X$ *and* $\mathsf{E}|X_n|^p < \infty$, $p > 0$. *If* $\{|X_n|^p\}$ *is uniformly integrable, then*

$$\mathsf{E}|X_n|^m \xrightarrow[n\to\infty]{} \mathsf{E}|X|^m \quad \textit{for all } 0 < m \le p.$$

Proof. By the uniform integrability condition, $\sup_n \mathsf{E}|X_n|^p < \infty$, so that by Fatou's lemma

$$\mathsf{E}|X|^p \le \liminf_n \mathsf{E}|X_n|^p < \infty.$$

It is easily seen that $\{|X_n - X|^p\}$ is uniformly integrable. Since $|X_n - X|^p \xrightarrow[n\to\infty]{\mathsf{P}} 0$, it follows from Theorem 2.4.2 that $\lim_{n\to\infty} \mathsf{E}|X_n - X|^p = 0$, which completes the proof. □

Note that the converse is also true: If $\mathsf{E}|X_n|^p < \infty$ and $X_n \xrightarrow[n\to\infty]{L^p} X$, then $\{|X_n|^p\}$ is uniformly integrable.

Rates of Convergence. Let $\{X_n\}_{n\ge1}$ be a discrete-time random process. Assume that X_n converges a.s. to 0. The question is what the rate of convergence is? In other words, how fast does the probability $\mathsf{P}(|X_n| > \varepsilon)$ decay to zero? This question can be answered by analyzing the behavior of the sums

$$\Sigma(r,\varepsilon) = \sum_{n=1}^{\infty} n^{r-1}\mathsf{P}(|X_n| > \varepsilon) \quad \text{for some } r > 0 \text{ and every } \varepsilon > 0.$$

More specifically, if $\Sigma(r,\varepsilon)$ is finite for every $\varepsilon > 0$, then probability $\mathsf{P}(|X_n| > \varepsilon)$ decays with the rate faster than $1/n^r$, so that $n^r\mathsf{P}(|X_n| > \varepsilon) \to 0$ for all $\varepsilon > 0$ as $n \to \infty$.

We now consider modes of convergence that strengthen the almost sure convergence and help to answer the above question.

2.4.2 Complete Convergence

Definition 2.4.2. A process $\{X_n\}_{n\ge1}$ is said to converge completely to zero if

$$\lim_{n\to\infty} \sum_{i=n}^{\infty} \mathsf{P}(|X_i| > \varepsilon) = 0 \quad \text{for every } \varepsilon > 0. \tag{2.65}$$

This mode of convergence has been introduced by Hsu and Robbins [195]. We write $X_n \xrightarrow[n\to\infty]{\text{completely}} 0$ for this mode of convergence.

Obviously, the sequence $\{X_n\}$ converges completely to a random variable Y if $\{X_n - Y\}$ converges completely to zero.

Note that the a.s. convergence of $\{X_n\}$ to 0 can be equivalently written as

$$\lim_{n\to\infty} \mathsf{P}\left(\sum_{i=n}^{\infty} |X_i| > \varepsilon\right) = 0 \quad \text{for every } \varepsilon > 0,$$

so that the complete convergence implies the a.s. convergence, but the converse is not true in general. The two modes of convergence are the same in the case of independent random variables.

The complete convergence is equivalent to the requirement of finiteness for $\Sigma(1,\varepsilon)$, i.e.,

$$\sum_{n=1}^{\infty} \mathsf{P}(|X_n| > \varepsilon) < \infty \quad \text{for every } \varepsilon > 0, \tag{2.66}$$

More generally, let for some $r > 0$

$$\lim_{n\to\infty} \sum_{i=n}^{\infty} n^{r-1} \mathsf{P}(|X_i| > \varepsilon) = 0 \quad \text{for every } \varepsilon > 0, \tag{2.67}$$

which is equivalent to the requirement $\Sigma(r,\varepsilon) < \infty$, i.e.,

$$\Sigma(r,\varepsilon) := \sum_{n=1}^{\infty} n^{r-1} \mathsf{P}(|X_n| > \varepsilon) < \infty \quad \text{for every } \varepsilon > 0. \tag{2.68}$$

For $r > 1$ this strengthens the complete convergence.

Definition 2.4.3. We say that the sequence $\{X_n\}$ converges r-completely to 0 and write

$$X_n \xrightarrow[n\to\infty]{r-\text{completely}} 0,$$

if the condition (2.68) or equivalently (2.67) is satisfied.

We will simply use *completely* for 1-completely.

This type of convergence is close but generally not identical to the so-called r-quick convergence which is introduced next.

2.4.3 r-Quick Convergence

Before introducing the r-quick convergence, consider the SLLN for the sample mean $X_n = S_n/n$, $S_n = \sum_{i=1}^{n} Y_i$, where $\{Y_i\}_{i\geq 1}$ is an iid sequence. As we stated above, by the SLLN $n^{-1}S_n \xrightarrow{\text{a.s.}} 0$ whenever $\mathsf{E}Y_1 = 0$. What can we say about the rate of convergence if we impose additional conditions on higher moments, say on the second moment? It turns out that assuming finiteness of the second moment, $\mathsf{E}|Y_1|^2 < \infty$, we can make one more step and conclude that $n^{-1}S_n \xrightarrow{\text{completely}} 0$. Specifically, it follows from Theorem 2.4.4 below that $\mathsf{E}|Y_1|^2 < \infty$ is both necessary and sufficient for S_n/n to converge completely to zero. Furthermore, by Theorem 2.4.4, finiteness of the $(r+1)$-th moment, $\mathsf{E}|Y_1|^{r+1}$, is necessary and sufficient for the r-complete convergence of S_n/n to 0, in which case we can conclude that the rate of convergence in the strong law is $\mathsf{P}(n^{-1}|S_n| > \varepsilon) = o(1/n^r)$, i.e., $n^r\mathsf{P}(n^{-1}|S_n| > \varepsilon) \to 0$ as $n \to \infty$ for all $\varepsilon > 0$. Note that this result by no means follows from the Chebyshev inequality (2.42).

For $\varepsilon > 0$ and $t \in \mathbb{R}_+$ or $t = n \in \mathbb{Z}_+$ define

$$L_\varepsilon = \sup\{t : |X_t| > \varepsilon\}, \quad \sup\{\varnothing\} = 0, \tag{2.69}$$

the last entry time of the process X_t in the region $(\varepsilon,\infty) \cup (-\infty,-\varepsilon)$. In other words, after the time moment L_ε the trajectory $\{X_t\}_{t>L_\varepsilon}$ always stays in the interval $[-\varepsilon,\varepsilon]$. Note that L_ε is not a Markov time.

Definition 2.4.4. Let $r > 0$. We say that the stochastic process $\{X_t\}$ converges to zero *r-quickly* and write

$$X_t \xrightarrow[t\to\infty]{r-\text{quickly}} 0,$$

if

$$\mathsf{E}[L_\varepsilon^r] < \infty \quad \text{for every } \varepsilon > 0. \tag{2.70}$$

Note that replacing X_t with $\tilde{X}_t - X$ covers the case of the corresponding convergence of $\tilde{X}_t$ to a random variable X.

This type of convergence is used in the following chapters for proving asymptotic optimality of certain hypothesis testing and change detection procedures for general stochastic models.

Note that the a.s. convergence of X_t to 0 is equivalent to the finiteness of L_ε w.p. 1: $\mathsf{P}(L_\varepsilon <$

$\infty) = 1$ for all $\varepsilon > 0$. Therefore, the r-quick convergence strengthens the a.s. convergence requiring the finiteness of the r-th moment of L_ε.

The following lemma contains sufficient conditions for the r-quick convergence that are useful for applications. While an immediate application of this lemma is proving the equivalence of the r-quick convergence with the r-complete convergence of the sample mean S_n/n as $n \to \infty$ to 0 under the $(r+1)$-th moment condition in the iid case, its value is far beyond the iid case, as shown in Chapters 3 and 4.

Lemma 2.4.1. *Let X_t, $t \in \mathbb{R}_+$ or $t = n \in \mathbb{Z}_+$, be a random process, $X_0 = 0$. Define $M_u = \sup_{0 \le t \le u} |X_t|$ in the continuous-time case and $M_u = \max_{1 \le t \le \lceil u \rceil} |X_t|$ in the discrete-time case, where $\lceil u \rceil$ is an integer part of u. Let $f(t)$ be a nonnegative increasing function. Define*

$$L_\varepsilon(f) = \sup\left\{t : \tfrac{1}{f(t)} |X_t| > \varepsilon\right\}, \quad \sup\{\varnothing\} = 0;$$

$$\mathcal{J}_1(\varepsilon, r, f) = \int_0^\infty u^{r-1} \mathsf{P}\left\{|X_u| \ge \varepsilon f(u)\right\} \mathrm{d}u;$$

$$\mathcal{J}(\varepsilon, r, f) = \int_0^\infty u^{r-1} \mathsf{P}\left\{M_u \ge \varepsilon f(u)\right\} \mathrm{d}u.$$

(i) *For any positive number r and any nonnegative increasing function $f(t)$, $f(0) = 0$,*

$$r\mathcal{J}_1(\varepsilon, r, f) \le \mathsf{E}\big[L_\varepsilon(f)\big]^r \le r \int_0^\infty t^{r-1} \mathsf{P}\left\{\sup_{t \ge u} \tfrac{1}{f(u)} |X_u| \ge \varepsilon\right\} \mathrm{d}t. \tag{2.71}$$

(ii) *If $f(t) = t^\lambda$, $\lambda > 0$, then for any $r > 0$*

$$\left\{\mathcal{J}(\varepsilon, r, \lambda) < \infty \;\forall\, \varepsilon > 0\right\} \Longrightarrow \left\{\mathsf{E}\big[L_\varepsilon(\lambda)\big]^r < \infty \;\forall\, \varepsilon > 0\right\}. \tag{2.72}$$

Proof. Obviously,

$$\mathsf{P}\{|X_t| \ge \varepsilon f(t)\} \le \mathsf{P}\{L_\varepsilon(f) \ge t\} \le \mathsf{P}\left\{\sup_{u \ge t} \tfrac{1}{f(u)} |X_u| \ge \varepsilon\right\}$$

from which the inequalities (2.71) follow immediately.

To prove (ii), we note that

$$
\begin{aligned}
\mathsf{E}\big[L_{2\varepsilon}(\lambda)\big]^r &\le r \int_0^\infty t^{r-1} \mathsf{P}\left\{\sup_{u \ge t} u^{-\lambda} |X_u| \ge 2\varepsilon\right\} \mathrm{d}t \\
&\le r \int_0^\infty t^{r-1} \mathsf{P}\left\{\sup_{u \ge t} \big[|X_u| - \varepsilon u^\lambda\big] \ge \varepsilon t^\lambda\right\} \mathrm{d}t \\
&\le r \int_0^\infty t^{r-1} \mathsf{P}\left\{\sup_{u > 0} \big[|X_u| - \varepsilon u^\lambda\big] \ge \varepsilon t^\lambda\right\} \mathrm{d}t \\
&\le r \sum_{n=1}^\infty \int_0^\infty t^{r-1} \mathsf{P}\left\{\sup_{(2^{n-1}-1)t^\lambda < u^\lambda \le (2^n-1)t^\lambda} \big[|X_u| - \varepsilon u^\lambda\big] \ge \varepsilon t^\lambda\right\} \mathrm{d}t \\
&\le r \sum_{n=1}^\infty \int_0^\infty t^{r-1} \mathsf{P}\left\{\sup_{u^\lambda \le 2^n t^\lambda} |X_u| \ge 2^{n-1} \varepsilon t^\lambda\right\} \mathrm{d}t \\
&= r \sum_{n=1}^\infty \int_0^\infty t^{r-1} \mathsf{P}\left\{M_{2^{n/\lambda} u} \ge 2^{n-1} \varepsilon t^\lambda\right\} \mathrm{d}t \\
&= r \left[\sum_{n=1}^\infty 2^{-n/\lambda}\right] \int_0^\infty u^{r-1} \mathsf{P}\left\{M_u \ge (\varepsilon/2) u^\lambda\right\} \mathrm{d}u \\
&= r \big(2^{1/\lambda} - 1\big)^{-1} \mathcal{J}(\varepsilon/2, r, \lambda).
\end{aligned}
$$

Now, the implication (2.72) follows in an obvious manner. □

If $X_n = S_n/n$ is the sample mean in the sequence of the random variables $\{Y_n\}$, then finiteness of the $(r+1)$-th moment, $\mathsf{E}|Y_1|^{r+1}$, is both necessary and sufficient condition for S_n/n to converge r-quickly to 0, as the following theorem shows.

Theorem 2.4.4. *Let $X_1, X_2, \dots$ be iid zero-mean random variables and let $S_n = X_1 + \cdots + X_n$. For every $r > 0$,*

$$\mathsf{E}|X_1|^{r+1} < \infty \Longleftrightarrow n^{-1} S_n \xrightarrow[t\to\infty]{r-completely} 0, \tag{2.73}$$

and

$$\mathsf{E}|X_1|^{r+1} < \infty \Longleftrightarrow n^{-1} S_n \xrightarrow[t\to\infty]{r-quickly} 0, \tag{2.74}$$

that is, finiteness of the $(r+1)$-th absolute moment is both necessary and sufficient condition for r-complete and r-quick convergence of the sample mean to 0 in the iid case. Therefore, in the iid case r-complete convergence and r-quick convergence of S_n/n to 0 are equivalent.

Proof. By Theorem 3 of Baum and Katz [51],

$$\mathsf{E}|X_1|^{r+1} < \infty \Longleftrightarrow \sum_{n=1}^{\infty} n^{r-1} \mathsf{P}\left(|S_n| > \varepsilon n\right) < \infty \quad \forall \varepsilon > 0 \tag{2.75}$$

$$\Longleftrightarrow \sum_{n=1}^{\infty} n^{r-1} \mathsf{P}\left(\sup_{k\ge n} |S_k/k| > \varepsilon\right) < \infty \quad \forall \varepsilon > 0, \tag{2.76}$$

so that (2.73) follows from (2.75).

Write

$$\mathcal{J}(\varepsilon, r) = \int_0^\infty t^{r-1} \mathsf{P}\left\{\sup_{k\ge t} (|S_k|/k) \ge \varepsilon\right\} \mathrm{d}t.$$

By Lemma 2.4.1(i),

$$\mathsf{E}[L_\varepsilon^r] \le r\mathcal{J}(\varepsilon, r) \quad \forall \varepsilon > 0, \tag{2.77}$$

which along with (2.76) implies (2.74). □

The r-quick convergence has been addressed by Strassen [444], Lai [247, 248], Chow and Lai [99], and Tartakovsky [455].

2.5 Elements of Renewal Theory for Random Walks

Throughout this section $X_1, X_2, \dots$ are iid random variables with the common cdf $F(x) = P(X_i \le x)$. Let $S_n = X_1 + \cdots + X_n$, $n = 0, 1, 2, \dots$ ($S_0 = 0$) denote partial sums. The discrete-time process $\{S_n\}_{n\in\mathbb{Z}_+}$ is called a *random walk*. However, if the random variables X_i are nonnegative and interpreted as random times between certain events, say repairs of equipment or renewals of inspection of the production process, then S_n is the time when the n-th renewal occurs, and the process $\{S_n\}_{n\in\mathbb{Z}_+}$ is called a *renewal process*.

Any renewal process can be associated with a continuous-time point process $\{N_t\}_{t\in\mathbb{R}_+}$, where $N_t = k$ means that k events have occurred by the time t (or in the interval $[0,t]$). The X_i is then the time interval between the $(i-1)$-th and i-th events. In particular, if X_1 is exponentially distributed, $F(x) = (1 - e^{-\theta x})\mathbb{1}_{\{x\ge 0\}}$, $\theta > 0$, then $\{N_t\}$ is a Poisson process with intensity θ.

Alternatively, N_t can be regarded as the number of renewals in $[0,t]$: $N_t = \max\{n : S_n \le t\}$, so that $\{N_t\}_{t\in\mathbb{R}_+}$ is the *renewal counting process*. This process is the main object of *classical renewal theory*.

Clearly, $\{N_t \geq n\} = \{S_n \leq t\}$, so that

$$\mathsf{E}[N_t] = \sum_{n=1}^{\infty} \mathsf{P}(N_t \geq n) = \sum_{n=1}^{\infty} \mathsf{P}(S_n \leq t) = \sum_{n=1}^{\infty} F_n(t),$$

where $F_n(t) = \mathsf{P}(S_n \leq t)$ is the cdf of S_n.

The function $U(t) = \sum_{n=1}^{\infty} F_n(t)$ is called the *renewal function*, and by the previous equality, $U(t) = \mathsf{E}[N_t]$.

It is easily seen that the renewal function satisfies the integral equation

$$U(t) = F(t) + \int_0^t U(t-s)\,\mathrm{d}F(s).$$

The *elementary renewal theorem* states that if the renewal process is non-degenerate, i.e., $\mu = \mathsf{E}[X_1] > 0$, then

$$\frac{U(t)}{t} \xrightarrow[t\to\infty]{} \frac{1}{\mu}, \tag{2.78}$$

where the right-hand side is 0 if $\mu = \infty$.

While being useful in certain applications, classical renewal theory is somehow limited since it deals only with positive random variables. Modern renewal theory deals with general, not necessarily positive random walks.

In this book, we are interested not in renewal theory itself, but rather in its application to the problem of excess over a boundary or the "overshoot" problem. It turns out that renewal theory allows us to answer several challenging questions, in particular to obtain practically useful corrections for certain elementary approximations such as the expectation of the first exit times over the boundaries and associated probabilities.

2.5.1 *The Overshoot Problem*

Let $\{S_n\}_{n\geq 0}$ be an arbitrary random walk, i.e., X_1 may take positive and negative values. For $a \geq 0$, let

$$T_a = \inf\{n \geq 1 : S_n \geq a\}, \quad \inf\{\varnothing\} = \infty \tag{2.79}$$

be the first time when the random walk exceeds the level (threshold) a.

Define

$$\kappa_a = S_{T_a} - a \quad \text{on } \{T_a < \infty\} \tag{2.80}$$

the excess of the random walk over the level a at the time which it crosses this level. We will refer to κ_a as an *overshoot* (at stopping). When $\{S_n\}$ is a renewal process, κ_t is nothing but the residual waiting time until the next renewal after time t. The evaluation of the average overshoot $\varkappa_a = \mathsf{E}[\kappa_a]$ as well as expectation of certain functions of the overshoot, for instance $\mathsf{E}[e^{-\lambda\kappa_a}]$, $\lambda > 0$ is of great interest for many statistical applications, including hypothesis testing and changepoint problems.

For example, let P_1 and P_0 be two probability measures and p_1 and p_0 the corresponding densities of the observations Y_i, $i \geq 1$. Let $X_i = \log[p_1(Y_i)/p_0(Y_i)]$. Wald's identities (2.53) and (2.55) can be applied to show that

$$\mathsf{E}_1 T_a = (a + \varkappa_a)/\mathsf{E}_1 X_1, \quad \mathsf{P}_0(T_a < \infty) = e^{-a}\mathsf{E}_1[e^{-\kappa_a}],$$

where E_j stands for the expectation under P_j.

Therefore, the primary interest is finding a distribution of the overshoot κ_a or at least an approximation to this distribution for sufficiently large values of a.

The results are somewhat different in the so-called arithmetic (lattice) and nonarithmetic (nonlattice) cases, so that we have to consider these cases separately.

Definition 2.5.1. It is said that a random variable X is *arithmetic* (or *lattice*) if the cdf $F(x) = \mathsf{P}(X \le x)$ is concentrated on $\{0, \pm d, \pm 2d, \dots\}$, i.e., $\mathsf{P}(X \in \{\dots -2d, -d, 0, d, 2d, \dots\}) = 1$. The largest such d is called the *span*. In this case, we say that the corresponding random variable is *d-arithmetic*. If there is no such d, then the random variable is *nonarithmetic* (or *nonlattice*). We shall say that the random walk $\{S_n\}_{n\ge 0}$ is nonarithmetic if X_1 is nonarithmetic, and that it is d-arithmetic if X_1 is arithmetic with span d.

Clearly, in the d-arithmetic case S_n is a multiple of d w.p. 1, so that the renewal function $U(t)$ increases only by jumps at $\{0, \pm d, \pm 2d, \dots\}$.

2.5.2 *Approximating the Distribution of the Overshoot via Renewal Theoretic Considerations*

We start with providing a heuristic argument how to evaluate the distribution of the overshoot using the renewal-theoretic consideration and ladder variables. Consider first the "purely" renewal process and for the sake of simplicity the continuous case assuming that $X_1 > 0$ w.p. 1 and X_1 is continuous. Then we have

$$\begin{aligned}
\mathsf{P}(\kappa_a > y) &= \sum_{n=0}^{\infty} \mathsf{P}(S_{n+1} > a+y, S_n < a) = \mathsf{P}(X_1 > a+y) + \sum_{n=1}^{\infty} \mathsf{P}(S_{n+1} > a+y, S_n < a) \\
&= \mathsf{P}(X_1 > a+y) + \sum_{n=1}^{\infty} \int_0^a \mathsf{P}(S_{n+1} > a+y \mid S_n = s)\mathsf{P}(S_n \in \mathrm{d}s) \\
&= \mathsf{P}(X_1 > a+y) + \sum_{n=1}^{\infty} \int_0^a \mathsf{P}(X_{n+1} > a+y-s)\mathsf{P}(S_n \in \mathrm{d}s) \\
&= 1 - F(a+y) + \int_0^a [1 - F(a+y-s)] \sum_{n=1}^{\infty} F_n(\mathrm{d}s) \\
&= 1 - F(a+y) + \int_0^a [1 - F(a+y-s)] U(\mathrm{d}s),
\end{aligned} \tag{2.81}$$

where $U(t)$ is the renewal function defined above. Now, by (2.78) (Elementary Renewal Theorem), for a large s

$$U(\mathrm{d}s) \approx \mathrm{d}s/\mu,$$

which along with the previous formula implies the approximation (for a large a)

$$\mathsf{P}(\kappa_a > y) \approx \frac{1}{\mu} \int_y^{\infty} [1 - F(x)]\, \mathrm{d}x. \tag{2.82}$$

This approximation is asymptotically exact:

$$\mathsf{P}(\kappa_a > y) \xrightarrow[a\to\infty]{} \frac{1}{\mu} \int_y^{\infty} [1 - F(x)]\, \mathrm{d}x. \tag{2.83}$$

Note first that approximations (2.82) and (2.83) are exact for any $a > 0$ for the exponential model when $1 - F(x) = e^{-\theta x}$, $x \ge 0$, since in this case $U(s) = \theta s$ for all $s \ge 0$. Hence, for the exponential model, the distribution of the overshoot is exactly exponential

$$\mathsf{P}(\kappa_a > y) = e^{-\theta y} \quad \text{for all } a, y \ge 0.$$

The general case where X_i may take negative values with positive probability can be reduced to the former case as follows. Assume that $\mu = \mathsf{E}X_1 > 0$, so that by the SLLN $\mathsf{P}(T_a < \infty) = 1$. Define

$$T_+^{(k)} = \inf\left\{ n > T_+^{(k-1)} : S_n > S_{T_+^{(k-1)}} \right\}, \quad k = 1, 2 \dots, \tag{2.84}$$

where $T_+^{(0)} = 0$, so that

$$T_+^{(1)} = T_+ = \inf\{n \geq 1 : S_n > 0\}. \tag{2.85}$$

The random variable $S_{T_+^{(k)}}$ is called the k-th ladder height, and $T_+^{(k)}$ – the k-th ladder epoch. Clearly, $(T_+^{(k)} - T_+^{(k-1)}, S_{T_+^{(k)}} - S_{T_+^{(k-1)}})$, $k = 1, 2\dots$ are iid. Furthermore, $T_a = T_+^{(k)}$ for some k and hence

$$T_+^{(\tau)} = T_a \quad \text{and} \quad \kappa_a = S_{T_+^{(\tau)}} - a,$$

where $\tau = \inf\{k \geq 1 : S_{T_+^{(k)}} > a\}$. Now, applying (2.83) to distribution $F_+(s) = \mathsf{P}(S_{T_+} \leq s)$ of the positive random variable S_{T_+}, we can conjecture that the limiting (as $a \to \infty$) distribution of the overshoot

$$\lim_{a\to\infty} \mathsf{P}(\kappa_a > y) = \lim_{a\to\infty} \mathsf{P}\left(S_{T_+^{(\tau)}} - a > y\right) = \frac{1}{\mathsf{E}[S_{T_+}]} \int_y^\infty [1 - F_+(s)]\,\mathrm{d}s. \tag{2.86}$$

This distribution can be used to compute the limiting average overshoot

$$\lim_{a\to\infty} \varkappa_a = \frac{1}{\mathsf{E}[S_{T_+}]} \int_0^\infty \left\{\int_y^\infty [1 - F_+(s)]\,\mathrm{d}s\right\} \mathrm{d}y = \frac{\mathsf{E}[S_{T_+}^2]}{2\mathsf{E}[S_{T_+}]}. \tag{2.87}$$

Computing the distribution $F_+(s)$ of S_{T_+} is the subject of renewal theory.

We now present rigorous statements regarding all the necessary distributions and expectations, omitting proofs as a rule, unless elementary. The proofs can be found in Gut [173], Siegmund [428], and Woodroofe [513] .

Let $\mathsf{H}(y) = \lim_{a\to\infty} \mathsf{P}(\kappa_a \leq y)$ denote the limiting distribution of the overshoot κ_a and let $\varkappa = \lim_{a\to\infty} \varkappa_a$ denote the limiting average overshoot. In the d-arithmetic case we always assume that $a \to \infty$ through multiples of span d, i.e., $a = dj$, $j \to \infty$. Recall that $\mu = \mathsf{E}[X_1]$ and $F_+(s) = \mathsf{P}(S_{T_+} \leq s)$.

The following theorem makes the conjectures (2.86) and (2.87) precise.

Theorem 2.5.1. *Assume that* $0 < \mu < \infty$.

(i) *If the random walk* $\{S_n\}$ *is nonarithmetic, then*

$$\mathsf{H}(y) = \frac{1}{\mathsf{E}[S_{T_+}]} \int_0^y [1 - F_+(s)]\,\mathrm{d}s, \tag{2.88}$$

and if in addition $\mathsf{E}(X_1^+)^2 < \infty$, *then*

$$\varkappa = \frac{\mathsf{E}[S_{T_+}^2]}{2\mathsf{E}[S_{T_+}]}. \tag{2.89}$$

(ii) *If the random walk* $\{S_n\}$ *is d-arithmetic, then*

$$\lim_{j\to\infty} \mathsf{P}(\kappa_{a=jd} = id) = \frac{d}{\mathsf{E}[S_{T_+}]} \mathsf{P}\left(S_{T_+} \geq id\right), \quad i \geq 1 \tag{2.90}$$

and if in addition $\mathsf{E}(X_1^+)^2 < \infty$, *then*

$$\varkappa = \frac{\mathsf{E}[S_{T_+}^2]}{2\mathsf{E}[S_{T_+}]} + \frac{d}{2}. \tag{2.91}$$

Note that since

$$\int_0^\infty [1 - F_+(s)]\,ds = \mathsf{E}[S_{T_+}],$$

(2.88) implies

$$1 - \mathsf{H}(y) = \frac{1}{\mathsf{E}[S_{T_+}]}\int_y^\infty [1 - F_+(s)]\,ds, \tag{2.92}$$

which is the same as (2.86).

In statistical applications, it is often important to evaluate the "exponential average overshoot" $\mathsf{E}[e^{-\lambda\kappa_a}]$ for some $\lambda > 0$. It follows from Theorem 2.5.1(i) that in the nonarithmetic case the limiting value is

$$\lim_{a\to\infty} \mathsf{E}[e^{-\lambda\kappa_a}] = \int_0^\infty e^{-\lambda y}\,d\mathsf{H}(y),$$

i.e., it is equal to Laplace's transform.

More generally, if we are interested in higher moments of the overshoot, then the following result holds. See, e.g., Gut [173, Theorem III.10.9].

Theorem 2.5.2. *Let the random walk $\{S_n\}_{n\geq 0}$ be nonarithmetic. If $\mathsf{E}(X_1^+)^{m+1} < \infty$ for some $m > 0$, then*

$$\lim_{a\to\infty} \mathsf{E}[\kappa_a^m] = \frac{\mathsf{E}[S_{T_+}^{m+1}]}{(m+1)\mathsf{E}[S_{T_+}]}.$$

It turns out that the converse also holds in the sense that if the m-th moment of the overshoot is a constant for sufficiently large threshold values, $\mathsf{E}[\kappa_a^m] = O(1)$ $(a \to \infty)$, then $\mathsf{E}(X_1^+)^{m+1} < \infty$.

While Theorems 2.5.1 and 2.5.2 are undoubtedly important, to make them indeed useful we need to find a way of computing the limiting distribution and such quantities as moments of ladder variables. Note that $\{T_+^{(n)}\}_{n\geq 0}$ and $\{S_{T_+^{(n)}} - S_{T_+^{(n-1)}}\}_{n\geq 1}$ are renewal processes.

Before getting into the details of accurate methods for the evaluation of the average overshoot, we provide less accurate but simple upper bounds. Lorden [270] showed that in the nonarithmetic case

$$\varkappa_a \leq \frac{\mathsf{E}[X_1^+]^2}{\mu} \quad \text{for any threshold } a > 0.$$

This bound can be improved for large threshold values:

$$\varkappa_a \leq \frac{\mathsf{E}[X_1^+]^2}{2\mu} + o(1) \quad \text{as } a \to \infty, \tag{2.93}$$

and in the d-arithmetic case

$$\varkappa_{a=jd} \leq \frac{\mathsf{E}[X_1^+]^2}{2\mu} + \frac{d}{2} + o(1) \quad \text{as } j \to \infty.$$

See Gut [173, Theorem 10.6]. Furthermore, using a slight modification of the argument that leads to the above asymptotic inequalities, it can be shown that in the nonarithmetic case

$$\mathsf{E}[\kappa_a^m] \leq \frac{\mathsf{E}[X_1^+]^{m+1}}{(m+1)\mu} + o(1) \quad \text{as } a \to \infty. \tag{2.94}$$

Also, it follows from Lorden [270, Theorem 3] that in the nonarithmetic case

$$\mathsf{E}[\kappa_a^m] \leq \frac{(m+2)\mathsf{E}[X_1^+]^{m+1}}{(m+1)\mu} \quad \text{for any } a > 0.$$

We now proceed with computational aspects related to the overshoot problem. Along with the first ascending ladder variable T_+ in (2.85) defined as the first time the random walk S_n upper-crosses the zero level, we now define the descending ladder variable

$$T_- = \inf\{n \geq 1 : S_n \leq 0\}. \tag{2.95}$$

The following two lemmas allow one to perform computations when explicit forms of the distributions $F_n(s) = \mathsf{P}(S_n \leq s)$, $n = 1, 2, \ldots$ can be obtained.

Lemma 2.5.1. *Let $\{S_n\}_{n\geq 0}$ be an arbitrary random walk. If $0 < \mu \leq \infty$, then*

$$\mathsf{E}[T_+] = \frac{1}{\mathsf{P}(T_- = \infty)} = \exp\left\{\sum_{n=1}^{\infty} \frac{1}{n} \mathsf{P}(S_n \leq 0)\right\},$$

$$\mathsf{E}[T_-] = \frac{1}{\mathsf{P}(T_+ = \infty)} = \exp\left\{\sum_{n=1}^{\infty} \frac{1}{n} \mathsf{P}(S_n > 0)\right\}.$$

Note that if $\mu = 0$, then both T_+ and T_- are finite w.p. 1, but $\mathsf{E}[T_+] = \mathsf{E}[T_-] = \infty$.

By Theorem 2.5.1(i), in the nonarithmetic case the asymptotic distribution of the overshoot (or a residual time) has density

$$h(y) = \frac{1}{\mathsf{E}[S_{T_+}]}[1 - F_+(y)] = \frac{1}{\mathsf{E}[S_{T_+}]} \mathsf{P}\left(S_{T_+} > y\right), \quad y \geq 0.$$

Introducing the Laplace transform of H,

$$\mathscr{H}(\lambda) = \int_0^\infty e^{-\lambda y} \, d\mathsf{H}(y), \quad \lambda \geq 0,$$

and integrating by parts, we obtain

$$\mathscr{H}(\lambda) = \frac{1 - \mathsf{E}[e^{-\lambda S_{T_+}}]}{\lambda \mathsf{E}[S_{T_+}]}, \quad \lambda > 0.$$

Lemma 2.5.1 allows us to obtain the following useful result. Recall that $Y^- = -\min(0, Y)$.

Theorem 2.5.3. *Let $\{S_n\}_{n\geq 0}$ be a nonarithmetic random walk with a positive drift $\mu > 0$.*

(i) *For every $\lambda > 0$*

$$\lim_{a\to\infty} \mathsf{E}\left[e^{-\lambda \kappa_a}\right] \equiv \mathscr{H}(\lambda) = \frac{1}{\lambda \mu} \exp\left\{-\sum_{n=1}^{\infty} \frac{1}{n} \mathsf{E}\left[e^{-\lambda S_n^+}\right]\right\}.$$

(ii) *Assume in addition that $\mathsf{E}|X_1|^2 < \infty$. Then*

$$\varkappa = \frac{\mathsf{E}[S_{T_+}^2]}{2\mathsf{E}[S_{T_+}]} = \frac{\mathsf{E}X_1^2}{2\mathsf{E}X_1} - \sum_{n=1}^{\infty} \frac{1}{n} \mathsf{E}[S_n^-].$$

Remark 2.5.1. There is an alternative useful expression for the limiting density $h(y)$ of the overshoot, which is useful in certain circumstances:

$$h(y) = \frac{1}{\mu} \mathsf{P}\left(\min_{n\geq 1} S_n > y\right), \quad y \geq 0, \tag{2.96}$$

which holds for any random walk with positive mean μ. See, e.g., Port [378, Theorem 3], Woodroofe [513, Theorem 2.7], and Gut [173, Theorem III.10.4].

Remark 2.5.2. A direct argument shows that for any random walk with positive mean and finite variance

$$\mathsf{E}\left[\min_{0\le n\le N} S_n\right] = -\sum_{n=1}^{N}\frac{1}{n}\mathsf{E}[S_n^-],$$

which along with Theorem 2.5.3(ii) yields

$$\varkappa = \frac{\mathsf{E}[S_{T_+}^2]}{2\mathsf{E}[S_{T_+}]} = \frac{\mathsf{E}X_1^2}{2\mathsf{E}X_1} + \mathsf{E}\left[\min_{n\ge 0} S_n\right] \tag{2.97}$$

if $\mu > 0$ and $\mathsf{E}|X_1|^2 < \infty$.

Example 2.5.1. Let $\{S_n\}$ be the Gaussian random walk with mean $\mathsf{E}[S_n] = \mu n$ and variance $\mathsf{var}[S_n] = \sigma^2 n$, where $\mu > 0, \sigma^2 > 0$. Write $\varphi(x) = (2\pi)^{-1/2}e^{-x^2/2}$ and $\Phi(x)$ for the standard normal pdf and cdf, respectively. Let $q = \mu^2/\sigma^2$ and denote $\beta_n = \mathsf{E}[e^{-\lambda S_n^+}]$. Direct computations show that for $\lambda \ge 0$ and $n \ge 1$

$$\beta_n = \Phi(-\sqrt{qn}) + \Phi\left[-(\lambda\sigma - \sqrt{q})\sqrt{n}\right]\exp\left\{\left[\lambda\sigma\left(\frac{\lambda\sigma}{2} - \sqrt{q}\right)\right]n\right\}, \tag{2.98}$$

so

$$\mathscr{H}(\lambda) = \frac{1}{\lambda\mu}\exp\left\{-\sum_{n=1}^{\infty}\frac{1}{n}\beta_n\right\}$$

is easily computed numerically. Computations become especially simple when $\lambda = 2\mu/\sigma^2$, which is the case when X_1 is the log-likelihood ratio in the problem of testing two hypotheses related to the mean of the Gaussian iid sequence.

The limiting average overshoot is also easily computable

$$\varkappa = \mu\frac{1+q}{2q} - \frac{\mu}{\sqrt{q}}\sum_{n=1}^{\infty}\frac{1}{\sqrt{n}}\left[\varphi(\sqrt{qn}) - \Phi(-\sqrt{qn})\sqrt{qn}\right]. \tag{2.99}$$

We now consider another interesting example where all computations can be performed precisely.

Example 2.5.2. Consider now an exponential case assuming that the cdf $F(x)$ of X_1 has an exponential right tail, i.e., $F(x) = 1 - C_0\, e^{-C_1 x}$, $x \ge 0$ for some positive constants C_0 and C_1. In this case the distribution of the overshoot κ_a is exactly exponential with rate C_1 for all $a \ge 0$ assuming that $\mu > 0$. Indeed, for $y \ge 0$ and $a \ge 0$,

$$\begin{aligned}
\mathsf{P}(\kappa_a > y, T_a < \infty) &= \sum_{n=1}^{\infty}\mathsf{P}(S_n > a + y, T_a = n)\\
&= \sum_{n=1}^{\infty}\mathsf{P}(X_n > a + y - S_{n-1}, T_a = n)\\
&= \sum_{n=1}^{\infty}\mathsf{E}\left[C_0 e^{-C_1(a+y-S_{n-1})}\mathbb{1}_{\{T_a=n\}}\right]\\
&= e^{-C_1 y}\sum_{n=1}^{\infty}\mathsf{E}\left[C_0 e^{-C_1(a-S_{n-1})}\mathbb{1}_{\{T_a=n\}}\right].
\end{aligned}$$

Setting $y = 0$, yields

$$\sum_{n=1}^{\infty}\mathsf{E}\left[C_0 e^{-C_1(a-S_{n-1})}\mathbb{1}_{\{T_a=n\}}\right] = \mathsf{P}(T_a < \infty).$$

Hence

$$\mathsf{P}(\kappa_a > y, T_a < \infty) = \mathsf{P}(T_a < \infty)e^{-C_1 y}, \quad y, a \geq 0.$$

If $\mu > 0$, then $\mathsf{P}(T_a < \infty) = 1$ and therefore

$$\mathsf{P}(\kappa_a > y) = e^{-C_1 y} \quad \text{for all } y \geq 0,\ a \geq 0. \tag{2.100}$$

In particular, for all $a \geq 0$

$$\mathsf{E}[\kappa_a] = 1/C_1, \quad \mathsf{E}\left[e^{-\lambda \kappa_a}\right] = C_1/(\lambda + C_1). \tag{2.101}$$

Note once more that explicit calculations of the characteristics of the overshoot exploiting Theorem 2.5.3 are possible only if tractable expressions for the distributions of the random walk S_n are available. Otherwise, one may resort to Fourier analysis and associated Spitzer's formulas. See Siegmund [428] and Woodroofe [513].

2.5.3 Properties of the Stopping Time T_a

The overshoot problem is directly related to the evaluation of moments of the stopping time T_a defined in (2.79). Indeed, since

$$S_{T_a} = a + \kappa_a \quad \text{on } \{T_a < \infty\},$$

assuming that $0 < \mu < \infty$ and using Wald's identity $\mathsf{E}[S_{T_a}] = \mu \mathsf{E} T_a$, we obtain

$$\mathsf{E} T_a = \frac{1}{\mu}(a + \varkappa_a). \tag{2.102}$$

This equality is true if $\mathsf{E} T_a < \infty$, so we need some additional investigation to start making any conclusions.

It turns out that positiveness of the first moment guarantees finiteness of the mean $\mathsf{E} T_a$. Since in general we cannot compute the average overshoot $\varkappa_a = \mathsf{E}\kappa_a$ for every $a \geq 0$, the natural question is "how accurate an approximation

$$\mathsf{E} T_a \approx \frac{1}{\mu}(a + \varkappa)$$

with $\varkappa_a$ replaced by its limiting value $\varkappa = \lim_{a \to \infty} \varkappa_a$ is?" From Theorem 2.5.3 we may expect that the second moment condition (or at least finiteness of the second moment of the positive part) is required to make it work.

Since the one-sided stopping time T_a is extremely important in hypothesis testing and change-point detection problems, we now present some precise answers to the above questions.

Lemma 2.5.2. *Let $\mu > 0$ and $\mathsf{E}|X_1|^p < \infty$, $p \geq 1$. Then $\mathsf{E}[T_a^p] < \infty$ for all $a \geq 0$.*

Proof. Let $p = 1$ and for $A > 0$ define $X_i^A = \min(A, X_1) = X_1 \wedge A$, $\mu^A = \mathsf{E}[X_n^A]$,

$$S_n^A = \sum_{k=1}^n X_k^A, \quad n \geq 1, \quad T_a^A = \inf\{n \geq 1 : S_n^A \geq a\}.$$

Since $\mu^A \to \mu$ as $A \to \infty$, there is an A (maybe large) for which $\mu^A > 0$. Then $S_{T_a^A \wedge n} \leq a + A$ w.p. 1 and, by Wald's identity,

$$\mathsf{E}[S_{T_a^A \wedge n}] = \mu \mathsf{E}[T_a^A \wedge n] \leq a + A,$$

which implies

$$\lim_{n \to \infty} \mathsf{E}[T_a^A \wedge n] = \mathsf{E} T_a^A \leq (a + A)/\mu < \infty. \tag{2.103}$$

Obviously, $S_n^A \le S_n$, $n \ge 1$ and hence $T_a \le T_a^A$. This implies that $\mathsf{E}T_a < \infty$ for every $a \ge 0$.

Let $p > 1$. Obviously, $a \le S_{T_a} \le a + X_{T_a}$. Hence, by Minkowski's inequality,

$$[\mathsf{E}|S_{T_a}|^p]^{1/p} \le [\mathsf{E}(a + X_{T_a})^p]^{1/p} \le a + [\mathsf{E}|X_{T_a}|^p]^{1/p}.$$

Note that $|X_n|^p$, $n \ge 1$ are iid random variables and

$$\mathsf{E}|X_{T_a}|^p \le \mathsf{E}\left(|X_1|^p + \cdots + |X_{T_a}|^p\right) = \mathsf{E}T_a \cdot \mathsf{E}|X_1|^p,$$

where we used Wald's identity. Recall that we proved that $\mathsf{E}T_a < \infty$. It follows that

$$[\mathsf{E}|S_{T_a}|^p]^{1/p} \le a + [\mathsf{E}|X_{T_a}|^p]^{1/p} \le a + (\mathsf{E}T_a \cdot \mathsf{E}|X_1|^p)^{1/p} < \infty.$$

Finally, we use Corollary 2.3.2 in order to show that $\mathsf{E}|X_1|^p < \infty$ and $\mathsf{E}|S_{T_a}|^p < \infty$ imply that $\mathsf{E}T_a^p < \infty$. The details are omitted. See, e.g., Gut [173, Theorem I.5.5]. □

Remark 2.5.3. Lemma 2.5.2 holds under $\mathsf{E}[(X_1^-)^p] < \infty$.

The following theorem establishes the SLLN for $\{T_a, a > 0\}$ and the first order expansion for the expected value under the sole first moment condition.

Theorem 2.5.4. *Let* $0 < \mu < \infty$. *Then*

$$\frac{T_a}{a} \xrightarrow[a\to\infty]{\mathsf{P}-a.s.} \frac{1}{\mu} \tag{2.104}$$

and

$$\frac{\mathsf{E}T_a}{a} \xrightarrow[a\to\infty]{} \frac{1}{\mu}. \tag{2.105}$$

Proof. Let $\tilde{S}_n = \sum_{k=1}^{n} (X_k - \mu)$. By the SLLN, $\tilde{S}_n/n \to 0$ w.p. 1 as $n \to \infty$. Since the drift μ is positive, obviously, $T_a \to \infty$ w.p. 1 as $a \to \infty$. Hence, also

$$\frac{\tilde{S}_{T_a}}{T_a} \xrightarrow[a\to\infty]{\mathsf{P}-\text{a.s.}} 0.$$

By the definition of T_a, we have $S_{T_a-1} < a$ and $S_{T_a} \ge a$. Therefore, we have

$$\mu \xleftarrow[a\to\infty]{\mathsf{P}-\text{a.s.}} \frac{\tilde{S}_{T_a-1} + \mu(T_a - 1)}{T_a} < \frac{a}{T_a} \le \frac{\tilde{S}_{T_a} + \mu T_a}{T_a} \xrightarrow[a\to\infty]{\mathsf{P}-\text{a.s.}} \mu,$$

which implies the a.s. convergence (2.104).

In order to prove (2.105) note that, by Wald's identity and the fact that $S_{T_a} \ge a$, the following lower bound holds

$$\mathsf{E}T_a \ge a/\mu, \quad a \ge 0. \tag{2.106}$$

Consider again the stopping time T_a^A associated with the truncated random walk introduced in the proof of Lemma 2.5.2. Using (2.103) and the fact that $T_a \le T_a^A$, we obtain

$$\limsup_{a\to\infty} \frac{\mathsf{E}T_a}{a} \le \frac{1}{\mu},$$

which along with the lower bound (2.106) proves (2.105). □

The following uniform integrability result is due to Lai [246].

Lemma 2.5.3. *Let* $p \ge 1$ *and assume that* $\mathsf{E}[(X^-)^p] < \infty$. *Then the family* $\{(T_a/a)^p, a \ge 1\}$ *is uniformly integrable.*

The following theorem generalizes the second part of Theorem 2.5.4 (i.e., (2.105)) for arbitrary moments.

Theorem 2.5.5. *Let $p \ge 1$. Assume that $\mu > 0$ and $\mathsf{E}[(X^-)^p] < \infty$. Then*

$$\left(\frac{\mathsf{E}T_a}{a}\right)^m \xrightarrow[a\to\infty]{} \frac{1}{\mu^m} \quad \text{for all } 0 < m \le p. \tag{2.107}$$

Proof. By Theorem 2.4.3, it suffices to show that $\mathsf{E}[T_a^p] < \infty$, T_a/a converges in probability to $1/\mu$ as $a \to \infty$, and $\{(T_a/a)^p\}$ is uniformly integrable. Finiteness of $\mathsf{E}[T_a^p]$ follows from Remark 2.5.3. By the a.s. convergence (2.104), the second condition (convergence in probability) holds. Finally, the uniform integrability of $\{(T_a/a)^p\}$ has been established in Lemma 2.5.3. □

Theorem 2.5.4 provides a first-order expansion for the expected value of the stopping time T_a:

$$\mathsf{E}T_a = \frac{a}{\mu}(1+o(1)) \quad \text{as } a \to \infty$$

(see (2.105)), which cannot be improved as long as only the first moment condition is assumed. However, such an improvement is possible under the second moment condition. The following theorem makes the approximation (2.102) precise up to the second order.

Theorem 2.5.6. *Assume that $\mu > 0$ and $\mathsf{E}(X_1^+)^2 < \infty$.*

(i) *If the random walk $\{S_n\}$ is nonarithmetic, then*

$$\mathsf{E}[T_a] = \frac{1}{\mu}(a+\varkappa) + o(1) \quad \text{as } a \to \infty, \tag{2.108}$$

where

$$\varkappa = \frac{\mathsf{E}[S_{T_+}^2]}{2\mathsf{E}[S_{T_+}]}.$$

(ii) *If the random walk $\{S_n\}$ is d-arithmetic, then*

$$\mathsf{E}[T_{a=jd}] = \frac{1}{\mu}(jd+\varkappa) + o(1) \quad \text{as } j \to \infty, \tag{2.109}$$

where

$$\varkappa = \frac{\mathsf{E}[S_{T_+}^2]}{2\mathsf{E}[S_{T_+}]} + \frac{d}{2}.$$

Proof. By Wald's identity, $E[S_{T_a}] = \mu \mathsf{E}T_a$. Since $S_{T_a} = a + \kappa_a$, taking expectation we obtain

$$a + \mathsf{E}[\kappa_a] = \mu \mathsf{E}T_a,$$

so that

$$\mathsf{E}T_a = (a + \mathsf{E}[\kappa_a])/\mu.$$

By Theorem 2.5.1(i),

$$\mathsf{E}[\kappa_a] = \frac{\mathsf{E}[S_{T_+}^2]}{2\mathsf{E}[S_{T_+}]} + o(1) \quad \text{as } a \to \infty,$$

which completes the proof for the nonarithmetic case.

In the arithmetic case the proof is essentially the same if we note that expectation $\mathsf{E}T_a$ increases only at multiples of d, i.e., when $a = jd$, $j = 1, 2, \dots$, and that the average overshoot is approximated by

$$\mathsf{E}[\kappa_{a=jd}] = \frac{\mathsf{E}[S_{T_+}^2]}{2\mathsf{E}[S_{T_+}]} + \frac{d}{2} + o(1) \quad \text{as } j \to \infty.$$

See Theorem 2.5.1(ii). □

The condition of finiteness of $\mathsf{E}(X_1^+)^2$ cannot be relaxed for the asymptotic approximations (2.108) and (2.109) to be true, since otherwise the ladder heights will have infinite second moment. However, finiteness of the variance $\mathsf{var}[X_1]$ is not required.

The rate of the small term $o(1)$ seems to be difficult to find in general. In the exponential case considered in Example 2.5.2 this formula is exact with $\varkappa = 1/C_1$, i.e.,

$$\mathsf{E}T_a = (a + 1/C_1)/\mu \quad \text{for every } a \geq 0,$$

assuming that $\mu > 0$.

Finally, we note that if $\mu > 0$ and $\mathsf{var}[X_1] = \sigma^2 < \infty$, then

$$\mathsf{E}T_a \sim a/\mu, \quad \mathsf{var}[T_a] \sim a\sigma^2/\mu^3 \quad \text{as } a \to \infty$$

and the asymptotic distribution (as $a \to \infty$) of the properly normalized stopping time T_a, specifically of

$$\frac{T_a - a/\mu}{\sqrt{a\sigma^2/\mu^3}},$$

is standard normal:

$$\lim_{a\to\infty} \mathsf{P}\left(\frac{T_a - a/\mu}{\sqrt{a\sigma^2/\mu^3}} \leq x\right) = \Phi(x) \quad \text{for all } -\infty < x < \infty \tag{2.110}$$

See, e.g., Gut [173, Theorem III.5.1].

2.5.4 *On the Joint Distribution of the Stopping Time and Overshoot*

By (2.110), the Central Limit Theorem holds for the random variable

$$\tilde{T}_a = \frac{T_a - a/\mu}{\sqrt{a\sigma^2/\mu^3}}$$

as long as $\mu > 0$ and $\mathsf{var}[X_1] = \sigma^2 < \infty$:

$$\lim_{a\to\infty} \mathsf{P}(\tilde{T}_a \leq x) = \Phi(x), \quad x \in (-\infty, +\infty),$$

and by Theorem 2.5.1, whenever $0 < \mu < \infty$ the limiting distribution of the overshoot $\mathsf{H}(y) = \lim_{a\to\infty} \mathsf{P}(\kappa_a \leq y)$ is equal to

$$\mathsf{H}(y) = \frac{1}{\mathsf{E}[S_{T_+}]} \int_0^y \mathsf{P}\left(S_{T_+} > s\right) \mathrm{d}s, \quad y \geq 0$$

in the arithmetic case and

$$\mathsf{H}(y) = \frac{d}{\mathsf{E}[S_{T_+}]} \sum_{i=1}^{K} \mathsf{P}\left(S_{T_+} \geq id\right), \quad Kd \leq y < (K+1)d, \ K \geq 1$$

in the d-arithmetic case.

The following theorem shows that T_a and κ_a are asymptotically independent, which of course can be expected. The proof can be found in Siegmund [428, p. 172].

Theorem 2.5.7. *Let* $\mu > 0$ *and* $\mathsf{var}[X_1] = \sigma^2 < \infty$. *Then*

$$\mathsf{P}\left(\tilde{T}_a \leq x, \kappa_a \leq y\right) = \Phi(x) \cdot \mathsf{H}(y) \quad \text{as } a \to \infty \text{ for all } x \in (-\infty, \infty),\ y \geq 0,$$

where in the d-arithmetic case $a = jd$, $j \to \infty$.

Note that it follows from this theorem that the random variables $S_{T_a} - a$ and $(S_{T_a} - \mu T_a)/\sqrt{T_a}$ are asymptotically independent as $a \to \infty$.

2.6 Nonlinear Renewal Theory

2.6.1 Preliminaries

In the previous section, we considered a first passage of a random walk to a constant level. This problem, however, is limited in scope and scale. In this section, we deal with a generalization for the problem of crossing nonlinear, varying in time boundaries. We begin with considering not arbitrary nonlinearities but a class of problems with additive and slowly varying "perturbations." In this case the problem reduces to the first crossing of a constant threshold by a perturbed random walk. This scenario is characteristic for many practical applications that we are concerned with in this book. However, not for all. For this reason, we also consider a more general case where a perturbed random walk crosses a nonlinear threshold.

More specifically, as before, let $X_1, X_2, \dots$ be iid random variables with distribution $F(x) = \mathsf{P}(X_1 \le x)$ and let $\{S_n\}_{n\ge 0}$, $S_0 = 0$ be the corresponding random walk (i.e., $S_n = \sum_{k=1}^{n} X_k$, $n \ge 1$ are partial sums). Next, let $\{\xi_n\}_{n\ge 1}$ be a sequence of random variables independent of the future generated by $\{X_n\}$, i.e., $\{\xi_i\}_{1\le i\le n}$ are independent of $\{X_k\}_{k>n}$ for every $n \ge 1$. In other words, if $\mathscr{F}_n^X = \sigma(X_1, \dots, X_n)$ denote the sigma-algebra generated by $\mathbf{X}_1^n = (X_1, \dots, X_n)$, then ξ_n is adapted to $\mathscr{F}_n^X$. Our goal is to extend the results of renewal theory for random walks to the sequences

$$Z_n = S_n + \xi_n, \quad n \ge 1, \tag{2.111}$$

assuming certain additional smoothness conditions on the sequence of "perturbations" $\{\xi_n\}$.

Before proceeding with formal definitions and assumptions, we give two important statistical examples that provide motivation for this extension. Let $X_1, X_2, \dots$ be iid random variables with positive mean $\mathsf{E}X_1 = \mu$, and let $S_n^k = X_k + \cdots + X_n$, $S_n^k = 0$ for $k > n$. Define the stopping time

$$\tau_a = \inf\left\{n \ge 1 : \max_{1\le k\le n+1} S_n^k \ge a\right\}. \tag{2.112}$$

Since

$$W_n = \max_{1\le k\le n+1} S_n^k = S_n - \min_{0\le k\le n} S_k,$$

it can be written in the form (2.111) with $\xi_n = -\min_{0\le k\le n} S_k$. The statistic W_n is the famous cumulative sum (CUSUM) statistic for detecting changes in distributions, and the stopping time τ_a is the corresponding sequential CUSUM procedure or the time at which a change is declared. Computing of the expected value $\mathsf{E}\tau_a$ under the hypothesis that a change takes place from the very beginning is one of the important tasks in the change detection area. Note that $\xi_n = -\min_{0\le k\le n} S_k$, $n \ge 1$ are slowly changing compared to the random walk S_n since $\mathsf{E}S_n = \mu n$ and $S_n \to \infty$ w.p. 1 as $n \to \infty$, but

$$\xi_n \xrightarrow[n\to\infty]{\text{law}} \min_{k\ge 0} S_k = \xi \quad \text{and} \quad \mathsf{E}\xi_n \xrightarrow[n\to\infty]{} \mathsf{E}\xi = \bar{\xi},$$

where $\bar{\xi}$ is usually a relatively small positive number. Therefore, the behavior of the CUSUM W_n is mostly determined by the random walk S_n, and using the results of the previous section it is reasonable to assume that for a sufficiently large a

$$\mathsf{E}\tau_a \approx \frac{1}{\mu}(a - \bar{\xi} + \varkappa),$$

which turns out to be the case.

Let $\{Y_n\}_{n\ge 1}$ be iid with mean θ and variance σ^2. Write $\overline{Y}_n = n^{-1} \sum_{k=1}^{n} Y_k$ for the sample mean, and let $g(\cdot)$ be a positive and twice continuously differentiable function in a neighborhood of θ.

Define $Z_n = ng(\overline{Y}_n)$ and $\tau_a = \inf\{n : Z_n \geq a\}$. Then expanding $g(\cdot)$ in a Taylor series, we obtain that Z_n can be written in the form (2.111) with

$$S_n = ng(\theta) + g'(\theta) \sum_{k=1}^{n} (Y_k - \theta) \quad \text{and} \quad \xi_n = \frac{1}{2n} \left[\sum_{k=1}^{n} (Y_k - \theta) \right]^2 g''(\theta_n),$$

where θ_n is an intermediate point between $\overline{Y}_n$ and θ: $|\theta_n - \theta| \leq |\overline{Y}_n - \theta|$. Denoting $X_k = g(\theta) + g'(\theta)(Y_k - \theta)$ one sees that $\{S_n\}$ is a random walk with positive drift $\mu = g(\theta)$. Also, it is easily shown that as $n \to \infty$ ξ_n converges weakly to $\sigma^2 g''(\theta)\chi_1^2/2$, where χ_1^2 is the chi-squared random variable with one degree of freedom. Again, it seems obvious that the behavior of the process $\{Z_n\}$ depends mainly on the behavior of the random walk S_n that changes much faster than ξ_n.

Another relevant example is testing multiple hypotheses; see Chapter 4.

We now formally define the notion of "slowly changing" random variables.

Definition 2.6.1. The sequence of random variables $\{\xi_n\}_{n\geq 1}$ is called *slowly changing* if

$$\frac{1}{n} \max_{1\leq k\leq n} |\xi_k| \xrightarrow[n\to\infty]{\mathsf{P}} 0, \tag{2.113}$$

and for every $\varepsilon > 0$ there are $n^* \geq 1$ and $\delta > 0$ such that

$$\mathsf{P}\left(\max_{1\leq k\leq n\delta} |\xi_{n+k} - \xi_n| > \varepsilon \right) < \varepsilon \quad \text{for all } n \geq n^*. \tag{2.114}$$

Obviously, if ξ_n converges w.p. 1 to a random variable ξ with finite expectation, then both conditions (2.113) and (2.114) are satisfied. This is the case in the first example (CUSUM). Note also that (2.113) holds when $\xi_n/n \to 0$ w.p. 1 and condition (2.114) holds if ξ_n converges to a finite limit w.p. 1 as $n \to \infty$. It can be shown that in the second example both conditions hold.

Let Z_n be as in (2.111) and let $\{a_t(\lambda), \lambda \in \Lambda\}$ be a family of boundaries, where Λ is an index set. For each $\lambda \in \Lambda$, define the stopping time

$$\tau_\lambda = \inf\{n \geq 1 : Z_n \geq a_n(\lambda)\}. \tag{2.115}$$

We start with a particular, relatively simple case where $a_n = a \geq 0$ is a constant threshold:

$$\tau_a = \inf\{n \geq 1 : Z_n \geq a\}. \tag{2.116}$$

This case will be referred to as *nonlinear renewal theory for perturbed random walks*.

Let

$$\tilde{\kappa}_a = Z_{\tau_a} - a \quad \text{on } \{\tau_a < \infty\} \tag{2.117}$$

denote the overshoot.

One of the main results of nonlinear renewal theory for perturbed random walks is that the limiting distribution of the overshoot does not change when the random walk is perturbed with an additive slowly changing nonlinear term, i.e., Theorems 2.5.1 and 2.5.3 still hold for $\tilde{\kappa}_a$. Another important result is that the approximation for the expectation of the stopping time remains exactly the same if the original threshold value a is replaced with $a - \bar{\xi}$, where $\bar{\xi} = \lim_{n\to\infty} \mathsf{E}\xi_n$. These statements are almost obvious from the heuristic standpoint. However, the rigorous proofs are quite involved.

In Subsection 2.6.2, we follow the methods proposed by Lai and Siegmund [254, 255], Hagwood and Woodroofe [174] and Woodroofe [512, 513]; see also Siegmund [428]. In Subsection 2.6.3, we present a more general version due to Zhang [523] for the stopping time (2.115) with curved threshold $a_n(\lambda)$. This latter generalization is important for some applications and is used in the following chapters, in particular when considering multihypothesis testing problems.

2.6.2 *Asymptotic Properties of the Stopping Time and Overshoot for Perturbed Random Walks*

We begin with nonlinear renewal theory for perturbed random walks, in which case the stopping time τ_a and the overshoot $\tilde{\kappa}_a$ are defined as in (2.116) and (2.117), respectively. The general case is postponed to Subsection 2.6.3.

The following results are similar to the first part of Theorem 2.5.4. The proof is elementary and is omitted [513, p. 42].

Lemma 2.6.1. *Let* $0 < \mu < \infty$ *and let condition* (2.113) *hold. Then* $\mathsf{P}(\tau_a < \infty) = 1$ *for all* $a \ge 0$ *and*

$$\frac{\tau_a}{a} \xrightarrow[a\to\infty]{\mathsf{P}} \frac{1}{\mu}. \tag{2.118}$$

If in addition $\xi_n/n \to 0$ *w.p. 1 as* $n \to \infty$*, then*

$$\frac{\tau_a}{a} \xrightarrow[a\to\infty]{\mathsf{P}-a.s.} \frac{1}{\mu} \tag{2.119}$$

Recall that

$$T_a = \inf\{n \ge 1 : S_n \ge a\}, \quad T_+ = \inf\{n \ge 1 : S_n > 0\}, \quad \kappa_a = S_{T_a} - a.$$

By Theorem 2.5.1(i), if $0 < \mu < \infty$ and $\{S_n\}$ is nonarithmetic, then

$$\lim_{a\to\infty} \mathsf{P}(\kappa_a \le y) = \mathsf{H}(y) = \frac{1}{\mathsf{E}[S_{T_+}]} \int_0^y \mathsf{P}(S_{T_+} > s)\,\mathrm{d}s. \tag{2.120}$$

The following theorem is an analog of Theorem 2.5.1. It shows that $\tilde{\kappa}_a$ has the same limiting distribution as κ_a, so that adding a slowly changing term to the random walk does not change the distribution for large a. We use the abbreviation NRT for Nonlinear Renewal Theorem. The following theorem is the first main result of nonlinear renewal theory for perturbed random walks.

Theorem 2.6.1 (First NRT). *Let* $\{S_n\}_{n\ge 0}$ *be a nonarithmetic random walk with a positive drift* $\mu > 0$. *Assume that* $\{\xi_n\}_{n\ge 1}$ *is a slowly changing sequence. Then* $\lim_{a\to\infty} \mathsf{P}(\tilde{\kappa}_a \le y) = \mathsf{H}(y)$*, where* $\mathsf{H}(y)$ *is defined in* (2.120).

Proof. We provide only a sketch of the proof, that is based on a simple idea. Let $Z_n^{(k)}$ denote the process Z_n conditioned on $\mathscr{F}_k^X$. By (2.114) Z_n^k and $Z_k + S_n - S_k$ are close within ε with probability at least $1-\varepsilon$, while by Lemma 2.6.1 τ_a belongs to this interval with high probability, which implies that Z_n^k and $Z_k + S_n - S_k$ cross the threshold a at the same time and have almost the same overshoots. Since conditioned on Z_k the process $\{Z_k + S_n - S_k\}_{n\ge k+1}$ is a random walk, Theorem 2.5.1 applies to yield the result. We refer to Siegmund [428] and Woodroofe [513] for a complete proof; see also the original work of Lai and Siegmund [254] for a different point of view. □

Corollary 2.6.1. *In the conditions of Theorem 2.6.1 for every* $\lambda > 0$

$$\lim_{a\to\infty} \mathsf{E}\left[e^{-\lambda\tilde{\kappa}_a}\right] = \frac{1}{\lambda\mu} \exp\left\{-\sum_{n=1}^{\infty} \frac{1}{n}\mathsf{E}\left[e^{-\lambda S_n^+}\right]\right\}. \tag{2.121}$$

If in addition $\mathsf{E}(X_1^+)^2 < \infty$*, then*

$$\lim_{a\to\infty} \mathsf{E}[\tilde{\kappa}_a] = \varkappa = \frac{\mathsf{E}[S_{T_+}^2]}{2\mathsf{E}[S_{T_+}]} = \frac{\mathsf{E}X_1^2}{2\mathsf{E}X_1} - \sum_{n=1}^{\infty} \frac{1}{n}\mathsf{E}[S_n^-]. \tag{2.122}$$

Proof. The corollary follows directly from Theorems 2.5.3 and 2.6.1. □

A result similar to Theorem 2.5.7 on the asymptotic independence of $\tilde{\kappa}_a$ and τ_a also holds under some additional conditions [254]. Let $\tilde{\tau}_a = (\tau_a - a/\mu)\sqrt{a\sigma^2/\mu^3}$.

Theorem 2.6.2. *Let $\mu > 0$ and $\mathsf{var}[X_1] = \sigma^2 < \infty$. Assume that $\{\xi_n\}_{n\geq 1}$ is slowly changing, that $n^{-1/2}\xi_n \to 0$ in probability as $n \to \infty$, and that $\{S_n\}_{n\geq 0}$ is nonarithmetic. Then*

$$\mathsf{P}(\tilde{\tau}_a \leq x, \tilde{\kappa}_a \leq y) = \Phi(x)\cdot \mathsf{H}(y) \quad \text{as } a \to \infty \text{ for all } x \in (-\infty,\infty),\ y \geq 0,$$

i.e., the stopping time $\tilde{\tau}_a$ and the overshoot $\tilde{\kappa}_a$ are asymptotically independent as $a \to \infty$.

While under some additional conditions Theorem 2.6.1 is valid in the arithmetic case, Theorem 2.6.2 fails for arithmetic random walks except for some special cases [258].

We now study the approximations for the expectation of the stopping time τ_a. Note that

$$S_{\tau_a} + \xi_{\tau_a} = a + \tilde{\kappa}_a,$$

so that taking expectations on both sides and using Wald's identity yields

$$\mu \mathsf{E}\tau_a = a + \mathsf{E}\tilde{\kappa}_a - \mathsf{E}\xi_{\tau_a}.$$

By Corollary 2.6.1, $\mathsf{E}\tilde{\kappa}_a = \varkappa + o(1)$ for large a. If we now assume that $\mathsf{E}\xi_n \to \bar{\xi}$ as $n \to \infty$, then we expect that

$$\mathsf{E}\tau_a = \frac{1}{\mu}\left(a + \varkappa - \bar{\xi}\right) + o(1) \quad \text{as } a \to \infty. \tag{2.123}$$

Although the basic ideas are simple, a rigorous treatment is not trivial and even tedious.

We start with a simple first-order approximation.

Theorem 2.6.3. *Let $\mu > 0$. Assume that $\sigma^2 = \mathsf{var}(X_1) < \infty$, that $\{\xi_n\}_{n\geq 1}$ satisfies the condition* (2.113)*, and that*

$$\sum_{n=1}^{\infty} \mathsf{P}(\xi_n \leq -n\varepsilon) < \infty \quad \text{for some } 0 < \varepsilon < \mu. \tag{2.124}$$

Then

$$\mathsf{E}\tau_a = \frac{a}{\mu}(1 + o(1)) \quad \text{as } a \to \infty, \tag{2.125}$$

where $o(1) \to 0$ as $a \to \infty$.

Proof. By Lemma 2.6.1, τ_a/a converges to $1/\mu$ in probability as $a \to \infty$. Hence, by Theorem 2.4.3, in order to prove the convergence of the expectation $\mathsf{E}\tau_a/a \to 1/\mu$, it is sufficient to show that $\{\tau_a, a \geq 1\}$ is uniformly integrable.

Write $N_{a,\varepsilon} = 2a/(\mu - \varepsilon)$. Obviously, for any $n \geq 1$,

$$\mathsf{P}(\tau_a > n) = \mathsf{P}\left(\max_{1\leq i\leq n} Z_n < a\right) \leq \mathsf{P}(Z_n < a) = \mathsf{P}\{S_n + \xi_n < N_{a,\varepsilon}(\mu-\varepsilon)/2\},$$

so that for $n \geq N_{a,\varepsilon}$

$$\begin{aligned}
\mathsf{P}(\tau_a > n) &\leq \mathsf{P}\{S_n + \xi_n < n(\mu-\varepsilon)/2\} \\
&= \mathsf{P}\{(S_n - \mu n) + \xi_n < -n(\mu-\varepsilon)/2 - \varepsilon n\} \\
&\leq \mathsf{P}(\xi_n < -n\varepsilon) + \mathsf{P}\{S_n - \mu n < -n(\mu-\varepsilon)/2\} \\
&= \mathsf{P}(\xi_n < -n\varepsilon) + \mathsf{P}(S_n - \mu n < -n\delta),
\end{aligned}$$

where $\delta = (\mu - \varepsilon)/2$. Note that $0 < \delta < \mu/2$. Let p_n denote the right-hand side. By Theorem 2.4.4, the second moment condition $\mathsf{E}[X_1^2] < \infty$ is both necessary and sufficient for S_n/n to converge completely to μ and therefore

$$\sum_{n=1}^{\infty} \mathsf{P}\{|S_n - \mu n| > \delta n\} < \infty \quad \text{for all } \delta > 0.$$

Hence

$$\sum_{n=1}^{\infty} \mathsf{P}\{S_n - \mu n < -n\delta\} < \infty \quad \text{for all } \delta > 0,$$

which along with (2.124) implies that $\sum_1^\infty p_n < \infty$. Note that p_n does not depend on a. Thus, we have

$$\begin{aligned}\mathsf{E}\left[\tau_a \mathbb{1}_{\{\tau_a > 2N_{a,\varepsilon}\}}\right] &\le 2\mathsf{E}\left[(\tau_a - N_{a,\varepsilon})\mathbb{1}_{\{\tau_a > 2N_{a,\varepsilon}\}}\right] \\ &\le 2\mathsf{E}\left[(\tau_a - N_{a,\varepsilon})\mathbb{1}_{\{\tau_a > N_{a,\varepsilon}\}}\right] \le 2\sum_{n=N_{a,\varepsilon}}^{\infty} p_n \xrightarrow[a\to\infty]{} 0.\end{aligned}$$

This implies uniform integrability of $\{\tau_a, a \ge 1\}$, which completes the proof. □

Remark 2.6.1. The condition of finiteness of the second moment in Theorem 2.6.3 may be relaxed into $\mathsf{E}(X_1^-)^2 < \infty$.

Theorems 2.6.2 and 2.6.3 allow us to conjecture that $\mathsf{var}[\tau_a] \sim \mathsf{E}(\tau_a - a/\mu)^2 \sim a\sigma^2/\mu^3$ as $a \to \infty$. This is indeed true under some additional conditions. In particular, a similar line of argument as in the proof of Theorem 2.6.3 may be applied if the condition (2.124) is strengthened into

$$\sum_{n=1}^{\infty} n\mathsf{P}(\xi_n \le -n\varepsilon) < \infty \quad \text{for some } 0 < \varepsilon < \mu.$$

Establishing the finiteness of $\sum_{n=1}^\infty n\mathsf{P}\{S_n - \mu n < -n\delta\} < \infty$ for all $\delta > 0$ requires an additional moment condition on the random walk. The third moment condition $\mathsf{E}|X_1|^3 < \infty$ is sufficient for this purpose. The condition $\mathsf{E}[(X_1^-)^3] < \infty$ seems to be necessary.

We now proceed with the second main result — a detailed higher-order asymptotic approximation to the expected value $\mathsf{E}\tau_a$ that was conjectured in (2.123) based on a simple heuristic argument. To this end, we need the following seven additional conditions. Assume there exist events $\mathcal{A}_n \in \mathscr{F}_n^X$, $n \ge 1$, constants ℓ_n, $n \ge 1$, $\mathscr{F}_n^X$-measurable random variables η_n, $n \ge 1$, and an integrable random variable η such that

$$\sum_{n=1}^{\infty} \mathsf{P}\left(\bigcup_{k=n}^{\infty} \mathcal{A}_n^c\right) < \infty, \tag{2.126}$$

$$\xi_n = \ell_n + \eta_n \quad \text{on } \mathcal{A}_n, \; n \ge 1, \tag{2.127}$$

$$\sup_{n\ge 1} \max_{0\le k\le n\delta} |\ell_{n+k} - \ell_n| \xrightarrow[\delta\to 0]{} 0, \tag{2.128}$$

$$\max_{0\le k\le n} |\eta_{n+k}|, \; n \ge 1 \quad \text{are uniformly integrable,} \tag{2.129}$$

$$\sum_{n=1}^{\infty} \mathsf{P}(\eta_n \le -n\varepsilon) < \infty \quad \text{for some } 0 < \varepsilon < \mu, \tag{2.130}$$

$$\eta_n \xrightarrow[n\to\infty]{\text{law}} \eta, \tag{2.131}$$

$$a\mathsf{P}(\tau_a \le \varepsilon a/\mu) \xrightarrow[a\to\infty]{} 0 \quad \text{for some } 0 < \varepsilon < 1. \tag{2.132}$$

Note that if $\{\eta_n\}_{n\ge 1}$ is a slowly changing sequence and if the conditions (2.126)–(2.128) are satisfied, then the sequence $\{\xi_n\}_{n\ge 1}$ is also slowly changing.

Theorem 2.6.4 (Second NRT). *Assume that $\mu > 0$, that $\sigma^2 = \mathsf{var}[X_1] < \infty$, that conditions* (2.126)–(2.132) *hold, and that the sequence $\{\eta_n\}_{n\geq 1}$ is slowly changing.*

(i) *If the random walk $\{S_n\}_{n\geq 0}$ is nonarithmetic, then*

$$\mathsf{E}\tau_a = \mu^{-1}(a + \varkappa - \bar{\eta} - \ell_{N_a}) + o(1) \quad \text{as } a \to \infty, \tag{2.133}$$

where $N_a = \lfloor a/\mu \rfloor$, $\bar{\eta} = \mathsf{E}\eta$, and $\varkappa$ is the limiting average overshoot given by (2.122).

(ii) *If the random walk $\{S_n\}_{n\geq 0}$ is d-arithmetic and if in addition $\ell_n = 0$ and the random variable η is continuous, then*

$$\mathsf{E}[\tau_{a=jd}] = \mu^{-1}(jd + \varkappa - \bar{\eta}) + o(1) \quad \text{as } j \to \infty, \tag{2.134}$$

where

$$\varkappa = \frac{\mathsf{E}[S_{T_+}^2]}{2\mathsf{E}[S_{T_+}]} + \frac{d}{2}.$$

The theorem as presented above is due to Hagwood and Woodroofe [174]. The proof may be also found in Woodroofe [513] and Siegmund [428].

Lai and Siegmund [255, Theorem 3] obtained the expansion (2.133) under slightly different conditions. Specifically, they required the finiteness of $\mathsf{E}|X_1|^{2/\alpha}$ for some $1/2 < \alpha \leq 1$ in place of $\mathsf{E}|X_1|^2 < \infty$ and, for all $\varepsilon > 0$ and some $\rho > 0$ and $n^* \geq 1$,

$$\sum_{i=n}^{n+\rho n^\alpha} \mathsf{P}(|\eta_i - \eta_n| > \varepsilon) < \varepsilon \quad \text{for } n \geq n^*, \tag{2.135}$$

$$\sum_{n=1}^{\infty} \mathsf{P}(|\eta_n| > \varepsilon n^\alpha) < \infty \tag{2.136}$$

in place of (2.114) and (2.130), respectively. These conditions turn out to be more demanding in terms of the moments of X_1. In addition, if η_n are nonnegative, then the condition (2.130) is satisfied automatically, but the condition (2.136) still has to be checked. However, Lai & Siegmund's condition on the deterministic sequence ℓ_n,

$$t^{-\alpha}|\ell_t| + \sup_{t\leq y\leq t+t^\alpha} |\ell_y - \ell_t| \to 0 \quad \text{as } t \to \infty, \tag{2.137}$$

is less demanding than (2.128). Indeed, (2.128) implies that ℓ_n cannot increase faster than $\log n$ for large n, while (2.137) assumes that ℓ_n grows more slowly than $\sqrt{n}$.

While at first glance the conditions (2.114), (2.128)–(2.131) look complicated, the following two examples show that they can often be easily checked. In particular, in many statistical problems that we will face the conditions (2.129)–(2.131) and (2.113)–(2.114) may be reduced to a single moment condition on X_1. Usually, the verification of the condition (2.132) causes the main difficulty and often requires special reasoning.

Example 2.6.1. Let $\{\tilde{S}_n\}_{n\geq 1}$ be a random walk with drift $\mathsf{E}\tilde{S}_1 = \theta \neq 0$ and diffusion $\mathsf{var}[\tilde{S}_1] = \sigma^2$, $0 < \sigma^2 < \infty$ and let

$$\tau_a = \inf\left\{n \geq 1 : |\tilde{S}_n| \geq \sqrt{n(\log n + 2a)}\right\}. \tag{2.138}$$

The stopping time (2.138) may be written in the desired form $\tau_a = \inf\{n : S_n + \xi_n \geq a\}$ with $S_n = \theta\tilde{S}_n - \theta^2 n/2$ and $\xi_n = -(\log n)/2 + (\tilde{S}_n - \theta n)^2/2n$, so that $\ell_n = -(\log n)/2$ and $\eta_n = (\tilde{S}_n - \theta n)^2/2n$. The condition (2.128) obviously holds. The condition (2.130) is satisfied since $\eta_n \geq 0$. The condition (2.131) holds with $\eta = \sigma^2\chi_1^2/2$ by the central limit theorem. Since the second moment $\mathsf{E}|\tilde{S}_1|^2$ is finite, the conditions (2.113) and (2.129) can be verified using Doob's maximal submartingale inequality. Finally it can be shown that if $\mathsf{E}|\tilde{S}_1|^{2+\delta} < \infty$ for some $\delta > 0$, then for some constant $C > 0$

$$\mathsf{P}(\tau_a \leq \varepsilon a/\mu) \leq \frac{C[1 + \log_2(\varepsilon a/\mu)]}{a^{1+\delta/2}} = o(1/a) \quad \text{as } a \to \infty,$$

so that the condition (2.132) is satisfied. Hence the condition $\mathsf{E}|\tilde{S}_1|^{2+\delta} < \infty$ for any small $\delta > 0$ is sufficient for the asymptotic approximation (2.134) to hold. On the other hand, the conditions of [255, Theorem 3] would require the finiteness of $\mathsf{E}|\tilde{S}_1|^4$.

Assuming that $S_n = \theta \tilde{S}_n - \theta^2 n/2$, $n \geq 1$ is nonarithmetic and using Theorem 2.6.4(i), we obtain

$$\mathsf{E}\tau_a = \frac{2}{\theta^2}\left[a + \frac{\log(2a)}{2} - \log\theta - \frac{\sigma^2}{2} + \varkappa\right] + o(1) \quad \text{as } a \to \infty. \tag{2.139}$$

Note also that this example is related to a particular interesting statistical problem. Specifically, assume that the observations Y_n, $n \geq 1$ are normal with mean θ and we wish to test the null hypothesis $\theta = 0$ *versus* $\theta \neq 0$. Write $S_n = Y_1 + \cdots + Y_n$. Let $\Pi(\theta)$ be the prior distribution, and let it be the standard normal distribution. Define the mixture (average) likelihood ratio

$$\Lambda_n = \int_{-\infty}^{\infty} \prod_{i=1}^{n} \frac{p_\theta(Y_i)}{p_0(Y_i)} \, \mathrm{d}\Pi(\theta),$$

where p_θ is the normal density with mean θ. The stopping time τ_a defined above is nothing but the one-sided (open-ended) mixture-based sequential likelihood ratio test

$$\tau_a = \inf\{n \geq 1 : \log\Lambda_n \geq a\}.$$

Let

$$\mathsf{P}(\mathcal{A}) = \int_{-\infty}^{\infty} \mathsf{P}_\theta(\mathcal{A})\,\mathrm{d}\Pi(\theta)$$

and $\chi_a = \log\Lambda_{\tau_a} - a$. The objects of primary interest are the probability

$$\mathsf{P}_0(\tau_a < \infty) = \int_{\{\tau_a < \infty\}} \left(\frac{1}{\Lambda_{\tau_a}}\right) \mathrm{d}\mathsf{P} = e^{-a} \int_{\{\tau_a < \infty\}} e^{-\chi_a}\,\mathrm{d}\mathsf{P}$$

and the expected sample size $\mathsf{E}_\theta\tau_a$, which for large a may be approximated using Theorems 2.6.1 and 2.6.4. This problem has been first considered by Pollak and Siegmund [367] for an exponential family.

Example 2.6.2. Consider the CUSUM stopping time (2.112). In this case,

$$Z_n = \max_{1 \leq k \leq n+1} S_n^k = S_n - \min_{0 \leq k \leq n} S_k,$$

so that $\ell_n = 0$ and $\xi_n = \eta_n = -\min_{0 \leq k \leq n} S_k$. The conditions (2.126)–(2.128) hold trivially. The condition (2.131) is satisfied with $\eta = -\min_{k \geq 0} S_k$. Recall that Theorem 2.6.4 assumes the second moment condition $\mathsf{E}[X_1^2] < \infty$, so that the uniform integrability in (2.129) follows. It remains to establish (2.132). If $X_i = \log[d\mathsf{P}_1(Y_i)/d\mathsf{P}_0(Y_i)]$, where P_0 and P_1 are two probability measures (pre- and post-change, respectively) and $\mathsf{P} = \mathsf{P}_1$, then using a change-of-measure argument it can be shown that

$$\begin{aligned}
\mathsf{P}_1\{\tau_a \leq (1-\varepsilon)a/\mu\} &= \mathsf{P}_1\left(\max_{n \leq (1-\varepsilon)a/\mu} Z_n \geq a\right) \\
&\leq e^{(1-\varepsilon^2)a}\mathsf{P}_0\left(\max_{n \leq (1-\varepsilon)a/\mu} R_n \geq e^a\right) + \mathsf{P}_1\left\{\max_{n \leq (1-\varepsilon)a/\mu} S_n > (1-\varepsilon^2)a\right\}
\end{aligned}$$

where R_n is a nonnegative P_0-submartingale with mean $\mathsf{E}_0 R_n = n$; see Section 2.3 for the details. By Doob's submartingale inequality, the first term is bounded by

$$e^{(1-\varepsilon^2)a}(1-\varepsilon)(a/\mu)e^{-a} = e^{-\varepsilon^2 a}(1-\varepsilon)(a/\mu) = o(1/a) \quad \text{as } a \to \infty.$$

Define $N_a = (1-\varepsilon)a/\mu$. We have

$$\begin{aligned}
\mathsf{P}_1\left\{\max_{1\le n\le(1-\varepsilon)a/\mu} S_n > (1-\varepsilon^2)a\right\} &= \mathsf{P}_1\left\{\max_{1\le n\le N_a} S_n > (1+\varepsilon)\mu N_a\right\} \\
&= \mathsf{P}_1\left\{\max_{1\le n\le N_a} (S_n-\mu N_a) > \varepsilon\mu N_a\right\} \\
&\le \mathsf{P}_1\left\{\max_{1\le n\le N_a} (S_n-\mu n) > \varepsilon\mu N_a\right\}.
\end{aligned}$$

By Lemma 2.6.2 proven below, for every $0<\varepsilon<1$

$$\mathsf{P}_1\left\{\max_{1\le n\le N_a} (S_n-\mu n) > \varepsilon\mu N_a\right\} = o(1/N_a) = o(1/a) \quad \text{as } a\to\infty.$$

Therefore, condition (2.132) holds. We note that even in this relatively simple problem establishing this condition is not a trivial task.

The following useful result is repeatedly used in the book.

Lemma 2.6.2. *Let $Y_1, Y_2, \dots$ be iid with $\mathsf{E}Y_1 = 0$ and $\mathsf{E}Y_1^2 = \sigma^2 < \infty$. Let $\tilde{S}_n = Y_1 + \cdots + Y_n$. Then for all $\varepsilon > 0$*

$$N\,\mathsf{P}\left(\max_{1\le n\le N} |\tilde{S}_n| > \varepsilon N\right) \xrightarrow[N\to\infty]{} 0.$$

Proof. Applying Doob's maximal submartingale inequality to the submartingale $\tilde{S}_n^2$, we obtain

$$\mathsf{P}\left(\max_{1\le n\le N} |\tilde{S}_n| \ge \varepsilon N\right) \le \frac{1}{(\varepsilon N)^2}\mathsf{E}\left[\tilde{S}_N^2 \mathbb{1}_{\{\max_{1\le n\le N}\tilde{S}_n \ge \varepsilon N\}}\right] = \frac{1}{\varepsilon^2 N}\mathsf{E}\left[\left(\frac{\tilde{S}_N^2}{N}\right)\mathbb{1}_{\{\max_{1\le n\le N}\tilde{S}_n \ge \varepsilon N\}}\right].$$

First, it follows that

$$\mathsf{P}\left(\max_{1\le n\le N} |\tilde{S}_n| \ge \varepsilon N\right) \le \frac{\sigma^2}{\varepsilon^2 N} \xrightarrow[N\to\infty]{} 0.$$

Now, we show that

$$\mathsf{E}\left[(\tilde{S}_N^2/N)\,\mathbb{1}_{\{\max_{1\le n\le N}\tilde{S}_n \ge \varepsilon N\}}\right] \xrightarrow[N\to\infty]{} 0,$$

which implies that $\mathsf{P}(\max_{1\le n\le N} |\tilde{S}_n| > \varepsilon N) = o(1/N)$ as $N\to\infty$, i.e., the desired result.

By the second moment condition, $\mathsf{E}(\tilde{S}_N^2/N) = \sigma^2 < \infty$. Hence, by the central limit theorem,

$$\tilde{S}_N^2/(N\sigma^2) \xrightarrow[N\to\infty]{\text{law}} \chi_1^2,$$

where χ_1^2 is a standard chi-squared random variable with 1 degree of freedom.

Finally, for any $L<\infty$ we have

$$\begin{aligned}
\mathsf{E}\left(\frac{\tilde{S}_N^2}{N}\mathbb{1}_{\{\max_{1\le n\le N}\tilde{S}_n > \varepsilon N\}}\right) &= \mathsf{E}\left[\left(L\wedge\frac{\tilde{S}_N^2}{N}\right)\mathbb{1}_{\{\max_{1\le n\le N}\tilde{S}_n > \varepsilon N\}}\right] + \mathsf{E}\left[\left(\frac{\tilde{S}_N^2}{N} - L\wedge\frac{\tilde{S}_N^2}{N}\right)\mathbb{1}_{\{\max_{1\le n\le N}\tilde{S}_n > \varepsilon N\}}\right] \\
&\le L\,\mathsf{P}\left(\max_{1\le n\le N}\tilde{S}_n > \varepsilon N\right) + \mathsf{E}\left(\frac{\tilde{S}_N^2}{N} - L\wedge\frac{\tilde{S}_N^2}{N}\right) \\
&\le \frac{L\sigma^2}{\varepsilon^2 N} + \sigma^2 - \mathsf{E}\left(L\wedge\frac{\tilde{S}_N^2}{N}\right) \\
&\xrightarrow[N\to\infty]{} \sigma^2 - \mathsf{E}\left(L\wedge\chi_1^2\sigma^2\right) \xrightarrow[L\to\infty]{} \sigma^2(1-1) = 0.
\end{aligned}$$

The proof is complete. □

Note that under the condition $\mathsf{E}|Y_1|^{2+\delta} < \infty$ for some $\delta > 0$ the proof is immediate since in this case $\{\tilde{S}_N^2/N\}_{N\geq 1}$ is uniformly integrable. However, the result holds under the sole second moment condition.

Note also that Lemma 2.6.2 allows us to immediately conclude that the condition (2.132) is satisfied if $\xi_n = 0$, i.e., for the stopping time $T_a = \inf\{n : S_n \geq a\}$ whenever $\mu > 0$ and $\mathsf{E}[X_1^2] < \infty$. Indeed, writing $N_a = (1-\varepsilon)a/\mu$ we obtain

$$\begin{aligned}
a\mathsf{P}(T_a \leq N_a) &= a\mathsf{P}\left(\max_{1\leq n\leq N_a} S_n \geq a\right) = a\mathsf{P}\left(\max_{1\leq n\leq N_a} S_n - \mu N_a \geq a - \mu N_a\right) \\
&= a\mathsf{P}\left(\max_{1\leq n\leq N_a} (S_n - \mu N_a) \geq \varepsilon a\right) = a\mathsf{P}\left(\max_{1\leq n\leq N_a} (S_n - \mu N_a) \geq \varepsilon\mu N_a\right) \\
&\leq a\mathsf{P}\left(\max_{1\leq n\leq N_a} (S_n - \mu n) \geq \varepsilon\mu N_a\right) \xrightarrow[a\to\infty]{} 0 \quad \text{for all } 0 < \varepsilon < 1.
\end{aligned} \tag{2.140}$$

2.6.3 *The General Case*

The nonlinear renewal methods presented in the previous subsection are based on the expansion of a nonlinear function of the random walk $g_n(S_n)$ and then applying the classical renewal argument to the leading term. Alternatively, a boundary can be expanded around an appropriate point. In this subsection, we consider a more general approach. We provide a heuristic argument and exact statements. The proofs may be found in Zhang [523].

Specifically, let, as before, $Z_n = S_n + \xi_n$, where $\{S_n\}$ is a random walk with positive and finite drift μ and random variables ξ_n, $n \geq 1$ do not depend on $X_{n+1}, X_{n+2}, \dots$. In the general case the stopping time τ_λ is defined as in (2.115), i.e.,

$$\tau_\lambda = \inf\{n \geq 1 : Z_n \geq a_n(\lambda)\}, \tag{2.141}$$

where $a_n(\lambda)$ is a changing in time threshold indexed with $\lambda \in \Lambda$. In most cases we will be interested in, λ is a constant part of the boundary and $\Lambda = [0,\infty)$, i.e., $a_n(\lambda) = \lambda + b_n$. Thus, if $b_n = 0$, then $a_n(\lambda) = \lambda$ and the stopping time $\tau_\lambda = \tau_{a=\lambda}$ is just the one we have already investigated in the previous subsection.

For $c \geq 0$ and $\nu < \mu$, define the stopping time

$$T_c(\nu) = \inf\{n \geq 1 : S_n - \nu n \geq c\}. \tag{2.142}$$

For $\nu = 0$ and $c = a$ this is exactly the stopping time we considered in classical renewal theory. Let $c = c_\lambda$ and $\nu = \nu_\lambda$ be suitable values of c and ν that depend on λ and write $T_\lambda = T_{c_\lambda}(\nu_\lambda)$. Zhang's basic idea is to consider powers of absolute differences between the stopping times, $\Delta_{\lambda,p} = |\tau_\lambda - T_\lambda|^p$, $p \geq 1$, and to establish uniform integrability of $\Delta_{\lambda,p}, \lambda \in \Lambda$. Then nonlinear renewal theorems may be established using classical renewal theorems for T_λ defined by the random walk $S_n - \nu_\lambda n$ with the drift $\mu - \nu_\lambda$.

In the following we assume that $a_t(\lambda)$ is twice differentiable in t. Let N_λ denote the point of intersection of the boundary $a_t(\lambda)$ with the line μt, assuming it is finite, i.e.,

$$N_\lambda = \sup\{t \geq 1 : a_t(\lambda) \geq \mu t\}, \quad \sup\{\varnothing\} = 1.$$

We write $\tilde{S}_n(\nu) = S_n - \nu n$ for compactness, so that $T_c(\nu) = \inf\{n \ge 1 : \tilde{S}_n(\nu) \ge c\}$. Define

$$\begin{aligned}
\nu_\lambda &= \frac{\partial a_t(\lambda)}{\partial t}\bigg|_{t=N_\lambda}, \\
\nu_{\max} &= \sup_{t \ge N_\lambda, \lambda \in \Lambda} \frac{\partial a_t(\lambda)}{\partial t}, \\
\kappa_c(\nu) &= \tilde{S}_{T_c(\nu)}(\nu) - c \quad \text{on } \{T_c(\nu) < \infty\}, \\
\tilde{\kappa}_\lambda &= Z_{\tau_\lambda} - a_{\tau_\lambda}(\lambda) \quad \text{on } \{\tau_\lambda < \infty\}, \\
T_+(\nu) &= \inf\{n \ge 1 : \tilde{S}_n(\nu) > 0\}, && (2.143) \\
\varkappa(\nu) &= \frac{\mathsf{E}[\tilde{S}_{T_+(\nu)}(\nu)]^2}{2\mathsf{E}[\tilde{S}_{T_+(\nu)}(\nu)]}, && (2.144) \\
\mathsf{H}(y, \nu) &= \frac{1}{\mathsf{E}[\tilde{S}_{T_+(\nu)}(\nu)]} \int_0^y \mathsf{P}\left\{\tilde{S}_{T_+(\nu)}(\nu) > s\right\} \mathrm{d}s, \quad y \ge 0. && (2.145)
\end{aligned}$$

Note that $\kappa_c(\nu)$ and $\tilde{\kappa}_\lambda$ are the overshoots in the linear and nonlinear schemes (2.142) and (2.141), respectively. By Theorem 2.5.1, $\varkappa(\nu) = \lim_{c\to\infty} \mathsf{E}\kappa_c(\nu)$ is the limiting average overshoot and $\mathsf{H}(y, \nu) = \lim_{c\to\infty} \mathsf{P}\{\kappa_c(\nu) \le y\}$ is the limiting distribution of the overshoot in the linear case.

Since $a_t(\lambda)$ is twice differentiable, it can be approximated using the two-term Taylor expansion in the neighborhood of N_λ:

$$a_t(\lambda) \approx a_{N_\lambda} + (t - N_\lambda) \frac{\partial a_t(\lambda)}{\partial t}\bigg|_{t=N_\lambda} = \mu N_\lambda + \nu_\lambda (t - N_\lambda).$$

Assume first that $\xi_n = 0$. Then for large N_λ the behavior of the stopping time τ_λ is in some sense similar to that of

$$\inf\{n : S_n \ge \mu N_\lambda + \nu_\lambda (n - N_\lambda)\} = \inf\{n : \tilde{S}_n(\nu_\lambda) \ge N_\lambda(\mu - \nu_\lambda)\} = T_{c_\lambda}(\nu_\lambda)$$

where $c_\lambda = N_\lambda(\mu - \nu_\lambda)$. Therefore, we expect that asymptotically as $N_\lambda \to \infty$ the distribution of the overshoot $\tilde{\kappa}_\lambda$ converges to $\lim_{N_\lambda\to\infty} \mathsf{H}(y, \nu_\lambda) = \mathsf{H}(y, \nu^*)$, assuming that

$$\lim_{N_\lambda \to \infty} \nu_\lambda = \nu^* < \mu. \tag{2.146}$$

Now, if $\mathsf{E}[X_1^2] < \infty$, the absolute difference $\{|\tau_\lambda - T_{c_\lambda}(\nu_\lambda)|, \lambda \in \Lambda\}$ turns out to be uniformly integrable, so that $\mathsf{E}|\tau_\lambda - T_{c_\lambda}(\nu_\lambda)| = O(1)$ for large N_λ. Since $\mathsf{E}[\tilde{S}_n] = (\mu - \nu_\lambda)n$, this implies that

$$\mathsf{E}\tau_\lambda = \frac{1}{\mu - \nu_\lambda} c_\lambda + O(1) = N_\lambda + O(1) \quad \text{as } N_\lambda \to \infty.$$

However, in order to obtain a higher-order expansion up to a vanishing term $o(1)$ an additional effort is needed. A more accurate approximation gives a three-term Taylor expansion

$$a_t(\lambda) \approx a_{N_\lambda} + (t - N_\lambda) \frac{\partial a_t(\lambda)}{\partial t}\bigg|_{t=N_\lambda} + N_\lambda \frac{1}{2} \frac{\partial^2 a_t(\lambda)}{\partial t^2}\bigg|_{t=N_\lambda} \frac{(t - N_\lambda)^2}{N_\lambda}.$$

It is reasonable to assume that, as $N_\lambda \to \infty$,

$$\mathsf{E}\tau_\lambda = N_\lambda + \lim_{N_\lambda \to \infty} \frac{1}{\mu - \nu_\lambda} \left\{ \varkappa(\nu_\lambda) + \frac{N_\lambda}{2} \frac{\partial^2 a_t(\lambda)}{\partial t^2}\bigg|_{t=N_\lambda} \mathsf{E}\left[\frac{(T_{c_\lambda}(\nu_\lambda) - N_\lambda)^2}{N_\lambda} \right] \right\} + o(1).$$

By Theorem 2.5.7,

$$\tilde{T}_{c_\lambda}(\nu_\lambda) = \frac{T_{c_\lambda}(\nu_\lambda) - N_\lambda}{\sqrt{N_\lambda \sigma^2/(\mu - \nu^*)^2}} \xrightarrow[N_\lambda \to \infty]{\text{law}} \Phi(x) \quad \text{for all } -\infty < x < \infty,$$

if $\mathsf{var}[X_1^2] = \sigma^2 < \infty$. Recall that $\lim_{N_\lambda \to \infty} \nu_\lambda = \nu^*$. If we now assume in addition that

$$\frac{N_\lambda}{2} \frac{\partial^2 a_t(\lambda)}{\partial t^2}\Big|_{t=N_\lambda} \to d,$$

then it is reasonable to conjecture that

$$\mathsf{E}\tau_\lambda = N_\lambda + \frac{1}{\mu - \nu^*} \{\varkappa(\nu^*) + d\sigma^2/(\mu - \nu^*)^2\} + o(1) \quad \text{as } N_\lambda \to \infty.$$

The final step is to add a slowly changing sequence $\xi_n = \ell_n + \eta_n$, where ℓ_n is a deterministic part and η_n is a slowly changing random part. By the results of Subsection 2.6.2, the limiting distribution of the overshoot does not change when adding a slowly changing sequence, so that the limiting distribution of the overshoot $\mathsf{H}(y, \nu^*)$ remains unchanged in this more general scenario. As far as the expectation of the stopping time is concerned, as before in Subsection 2.6.2, we need to add the term $-\mathsf{E}\eta - \ell_{N_\lambda}$ to the threshold c_λ, so that we may conjecture that in the most general case

$$\mathsf{E}\tau_\lambda = \frac{1}{\mu - \nu_\lambda}\left[N_\lambda(\mu - \nu_\lambda) - \ell_{N_\lambda}\right] + \frac{1}{\mu - \nu^*}\{\varkappa(\nu^*) + d\sigma^2/(\mu - \nu^*)^2 - \mathsf{E}\eta\} + o(1)$$

as $N_\lambda \to \infty$.

Below we make these conjectures precise.

Suppose there are functions $\rho(\delta) > 0$ and $\sqrt{y} \le \gamma(y) \le y$, $\gamma(y) = o(y)$ as $y \to \infty$ such that

$$\frac{\tau_\lambda - N_\lambda}{\gamma(N_\lambda)} = O(1) \quad \text{as } N_\lambda \to \infty, \tag{2.147}$$

$$\lim_{n\to\infty} \mathsf{P}\left\{\max_{1 \le i \le \rho(\delta)\gamma(n)} |\xi_{n+i} - \xi_n| \ge \delta\right\} = 0 \quad \text{for every } \delta > 0, \tag{2.148}$$

$$\sup_{\{|t - N_\lambda| \le K\gamma(N_\lambda), \lambda \in \Lambda\}} \left|\gamma(N_\lambda)\frac{\partial^2 a_t(\lambda)}{\partial t^2}\right| < \infty \quad \text{for all } K < \infty. \tag{2.149}$$

The following theorem generalizes Theorem 2.6.1 (the first NRT for perturbed random walks).

Theorem 2.6.5 (First general NRT). *Assume that conditions* (2.146)–(2.149) *are satisfied. Let* $\mathsf{H}(y, \nu)$ *be as in* (2.145) *and* $\mathsf{H}^*(y) = \mathsf{H}(y, \nu^*)$. *If the random walk* $\{S_n - \nu^* n\}_{n \ge 1}$ *is nonarithmetic with a positive and finite drift* $\mu - \nu^*$, *then the limiting distribution of the overshoot* $\tilde{\kappa}_\lambda$ *is*

$$\lim_{N_\lambda \to \infty} \mathsf{P}(\tilde{\kappa}_\lambda \le y) = \mathsf{H}^*(y) \quad \textit{for every } y \ge 0. \tag{2.150}$$

Proof. See the proof of the first part of Theorem 1 in Zhang [523]. □

Let $\sigma^2 = \mathsf{var}[X_1] < \infty$. Under certain additional conditions on the behavior of the threshold $a_n(\lambda)$ and the sequence $\{\xi_n\}$, in particular when $\xi_n/\sqrt{n} \to 0$ w.p. 1 as $n \to \infty$,

$$\tilde{\tau}_\lambda = \frac{\tau_\lambda - N_\lambda}{\sqrt{N_\lambda \sigma^2/(\mu - \nu^*)^2}} \xrightarrow[N_\lambda \to \infty]{\text{law}} \Phi(x) \quad \text{for all } -\infty < x < \infty,$$

and

$$\lim_{N_\lambda \to \infty} \mathsf{P}(\tilde{\tau}_\lambda \le x, \tilde{\kappa}_\lambda \le y) = \Phi(x) \cdot \mathsf{H}^*(y) \quad \text{for all } -\infty < x < \infty,\ y \ge 0.$$

See Theorem 1 and Proposition 1 in Zhang [523].

We now turn to the asymptotic approximations for the expected sample size, and we begin with the second order approximation, up to the term $O(1)$.

Definition 2.6.2. The sequence of random variables $\{\xi_n\}_{n\geq 1}$ is called p-regular ($p \geq 0$) if there exist a nonnegative random variable L with finite expectation $\mathsf{E}L^p < \infty$, a deterministic sequence $\{\ell_n\}_{n\geq 1}$, and a random sequence $\{\eta_n\}_{n\geq 1}$ such that the following five conditions hold:

$$\xi_n = \ell_n + \eta_n \quad \text{for } n \geq L, \tag{2.151}$$

$$\max_{1\leq i \leq n^{1/2}} |\ell_{n+i} - \ell_n| \leq K \quad \text{for some } 0 < K < \infty, \tag{2.152}$$

$$\left\{ \max_{1\leq i\leq n} |\eta_{n+i}|^p, n \geq 1 \right\} \quad \text{is uniformly integrable,} \tag{2.153}$$

$$\lim_{n\to\infty} n^p \mathsf{P}\left(\max_{0\leq i\leq n} \eta_{n+i} \geq \varepsilon n \right) = 0 \quad \text{for all } \varepsilon > 0, \tag{2.154}$$

$$\sum_{n=1}^{\infty} \mathsf{P}(\eta_n \leq -\varepsilon n) < \infty \quad \text{for some } 0 < \varepsilon < \mu - \nu_{\max}. \tag{2.155}$$

Note that in many statistical applications the conditions (2.153)–(2.155) may be reduced to a single moment condition. In particular, this is the case if $Y_n, n \geq 1$ is a sequence of iid zero-mean random vectors with $\mathsf{E}\,\|Y_1\|^{2p} < \infty$ for some $p \geq 1$ and $|V_n| \leq n^{-1}\,\|Y_1 + \cdots + Y_n\|^2$.

Theorem 2.6.6. *Assume that the sequence $\{\xi_n\}_{n\geq 1}$ is p-regular with $p \geq 1$, that $\mathsf{E}|X_1|^{p+1} < \infty$, and that there exist constants $0 < \delta < 1$ and $\mu^* < \mu$ such that*

$$\lim_{N_\lambda\to\infty} N_\lambda^p \mathsf{P}(\tau_\lambda \leq \delta N_\lambda) = 0 \tag{2.156}$$

and

$$\frac{\partial a_t(\lambda)}{\partial t} \leq \mu^* \quad \textit{for } t \geq \delta N_\lambda,\ \lambda \in \Lambda. \tag{2.157}$$

(i) *If $\mathsf{E}|X_1|^{2p} < \infty$ and*

$$\sup_{\{1-K\leq t/N_\lambda \leq 1+K, \lambda\in\Lambda\}} \left| N_\lambda \frac{\partial^2 a_t(\lambda)}{\partial t^2} \right| < \infty \quad \textit{for any } K > 0, \tag{2.158}$$

then $\{|\tau_\lambda - T_{c_\lambda}(\nu_\lambda)|, \lambda \in \Lambda\}$ is uniformly integrable and, consequently, if $p = 1$, then

$$\mathsf{E}\tau_\lambda = N_\lambda - \ell_{N_\lambda}/(\mu - \nu_\lambda) + O(1) \quad \textit{as } N_\lambda \to \infty, \tag{2.159}$$

and if $p = 2$, then

$$\mathsf{var}[\tau_\lambda] = \sigma^2 N_\lambda/(\mu - \nu_\lambda)^2 + O\left(\sqrt{N_\lambda}\right) \quad \textit{as } N_\lambda \to \infty. \tag{2.160}$$

(ii) *If $\partial^2 a_t(\lambda)/\partial t^2 = 0$ on $t \geq \delta N_\lambda$, then $\{|\tau_\lambda - T_{c_\lambda}(\nu_\lambda)|, \lambda \in \Lambda\}$ is uniformly integrable without condition $\mathsf{E}|X_1|^{2p} < \infty$ and asymptotic approximations* (2.159) *and* (2.160) *hold.*

Proof. See the proof of Theorem 2 in Zhang [523]. □

In order to obtain a higher-order approximation for $\mathsf{E}\tau_\lambda$ certain additional conditions are needed. The following additional constraints will be imposed:

$$\lim_{n\to\infty} \mathsf{P}\left(\max_{1\leq i\leq \sqrt{n}} |\eta_{n+i} - \eta_n| \geq \delta \right) = 0 \quad \text{for every } \delta > 0, \tag{2.161}$$

i.e., condition (2.148) holds for η_n with $\gamma(n) = \sqrt{n}$ and $\rho(\delta) = 1$;

$$\max_{1 \le i \le \sqrt{n}} |\ell_{n+i} - \ell_n| \xrightarrow[n\to\infty]{} 0 \tag{2.162}$$

(cf. (2.152));

$$\eta_n \xrightarrow[n\to\infty]{\text{law}} \eta, \tag{2.163}$$

where η is an integrable random variable with $\mathsf{E}\eta = \bar{\eta}$;

$$\lim_{N_\lambda\to\infty} \sup_{\{(t-N_\lambda)^2 \le KN_\lambda\}} \left| \frac{N_\lambda}{2} \frac{\partial^2 a_t(\lambda)}{\partial t^2} - d \right| = 0 \quad \text{for any } K > 0 \text{ and some constant } d; \tag{2.164}$$

and

$$\lim_{N_\lambda\to\infty} N_\lambda \mathsf{P}(\tau_\lambda \le \delta N_\lambda) = 0 \quad \text{for some } 0 < \delta < 1. \tag{2.165}$$

See (2.156) with $p = 1$.

The following theorem is the *Second General Nonlinear Renewal Theorem.*

Theorem 2.6.7 (Second general NRT). *Suppose that the sequence $\{\xi_n\}_{n\ge1}$ is 1-regular and the random walk $\{S_n - \nu^* n\}_{n\ge1}$ is nonarithmetic. If* $\mathsf{var}[X_1] = \sigma^2 < \infty$ *and conditions* (2.157), (2.158), *and* (2.161)–(2.165) *hold, then*

$$\mathsf{E}\tau_\lambda = N_\lambda - \frac{\ell_{N_\lambda}}{\mu - \nu_\lambda} + \frac{1}{\mu - \nu^*}\left\{\varkappa(\nu^*) + \frac{d\sigma^2}{(\mu - \nu^*)^2} - \bar{\eta}\right\} + o(1) \quad \text{as } N_\lambda \to \infty. \tag{2.166}$$

Proof. See the proof of Theorem 3(i) in Zhang [523]. □

It is not difficult to verify that Theorem 2.6.7 implies Theorem 2.6.4(i) for perturbed random walks. Indeed, since the threshold $a_t(\lambda) = a$ is constant, $\partial a_t(\lambda)/\partial t = 0$ and $\lambda = a$ and we need to set $\nu_\lambda = \nu^* = 0$, $d = 0$ and $N_\lambda = a/\mu$. Therefore, the following corollary for perturbed random walks holds true.

Corollary 2.6.2. *Let $a_n(\lambda) = a > 0$, let the stopping time τ_a be defined as in* (2.116), *and let $N_\lambda = N_a = a/\mu$. Suppose that the sequence $\{\xi_n\}_{n\ge1}$ is 1-regular and the random walk $\{S_n\}_{n\ge1}$ is nonarithmetic. If* $\mathsf{var}[X_1] = \sigma^2 < \infty$ *and the conditions* (2.161)–(2.163), (2.165) *hold, then*

$$\mathsf{E}\tau_a = \frac{1}{\mu}\{a - \ell_{N_a} - \bar{\eta} + \varkappa\} + o(1) \quad \text{as } a \to \infty, \tag{2.167}$$

where $\varkappa = \varkappa(0) = \mathsf{E}[S_{T_+}^2]/2\mathsf{E}[S_{T_+}]$.

For the sake of convenience, we list all the regularity conditions of Corollary 2.6.2 compactly:

$$\xi_n = \ell_n + \eta_n \quad \text{for } n \ge L,\ \mathsf{E}L < \infty; \tag{2.168}$$

$$\max_{1 \le i \le \sqrt{n}} |\ell_{n+i} - \ell_n| \le K \quad \text{for some } 0 < K < \infty; \tag{2.169}$$

$$\max_{1 \le i \le \sqrt{n}} |\ell_{n+i} - \ell_n| \xrightarrow[n\to\infty]{} 0; \tag{2.170}$$

$$\left\{\max_{1 \le i \le n} |\eta_{n+i}|, n \ge 1\right\} \quad \text{is uniformly integrable}; \tag{2.171}$$

$$\lim_{n\to\infty} n\mathsf{P}\left(\max_{0 \le i \le n} \eta_{n+i} \ge \varepsilon n\right) = 0 \quad \text{for all } \varepsilon > 0; \tag{2.172}$$

$$\lim_{n\to\infty} \mathsf{P}\left(\max_{1\le i\le\sqrt{n}} |\eta_{n+i}-\eta_n| \ge \varepsilon\right) = 0 \quad \text{for all } \varepsilon > 0; \tag{2.173}$$

$$\eta_n \xrightarrow[n\to\infty]{\text{law}} \eta, \text{ integrable random variable with } \mathsf{E}\eta = \bar{\eta}; \tag{2.174}$$

$$\sum_{n=1}^{\infty} \mathsf{P}(\eta_n \le -\varepsilon n) < \infty \quad \text{for some } 0 < \varepsilon < \mu - \nu_{\max}; \tag{2.175}$$

$$\lim_{a\to\infty} a\mathsf{P}(\tau_a \le \varepsilon a/\mu) = 0 \quad \text{for some } 0 < \varepsilon < 1. \tag{2.176}$$

It is seen that condition (2.168) is almost identical and conditions (2.171), (2.174)–(2.176) are identical to the corresponding conditions required in Theorem 2.6.4. However, the conditions (2.169), (2.170), (2.172), and (2.173) required for Corollary 2.6.2 are weaker than those of Theorem 2.6.4. This is particularly true for the deterministic sequence $\{\ell_n\}$, which in Theorem 2.6.4 cannot increase faster than $O(\log n)$, while in Theorem 2.6.7 it can grow at the rate $O(\sqrt{n})$. The conditions (2.169) and (2.170) also beat the corresponding condition of Lai and Siegmund [255] for ℓ_n that allows for an increase rate lower than $\sqrt{n}$.

We now consider two examples that illustrate the usefulness of the general nonlinear renewal theorems.

Example 2.6.3. Let

$$\tau_\lambda = \inf\left\{n \ge 1 : S_n \ge \sqrt{n(\log n + 2\lambda)}\right\}, \quad \lambda > 0,$$

where $S_n = X_1 + \cdots + X_n$ and $X_1, X_2, \ldots$ are iid with mean $\mathsf{E}X_1 = \mu > 0$ and $\mathsf{var}[X_1] = \sigma^2$. Observe that the stopping time τ_λ is a particular, one-sided version of the stopping time (2.138) of Example 2.6.1 if $\lambda = a$. Previously we reduced the problem to stopping of a perturbed random walk $S_n^* + \xi_n$ with a constant threshold $a = \lambda$, where $S_n^* = \mu S_n - \mu^2 n/2$ and $\xi_n = -(\log n)/2 + (S_n - \mu n)^2/2n$. Now we will use a direct approach expanding the nonlinear boundary $a_t(\lambda) = \sqrt{t(\log t + 2\lambda)}$.

It is easy to see that

$$N_\lambda = (\log N_\lambda + 2\lambda)/\mu^2 \quad \text{and} \quad \nu_\lambda = \mu/2 + (2\mu N_\lambda)^{-1},$$

so, as $\lambda \to \infty$,

$$N_\lambda = [2\lambda + \log(2\lambda) - 2\log\mu]/\mu^2 + o(1), \quad \nu_\lambda = \nu^* + o(1), \quad \nu^* = \mu/2 < \mu.$$

Note that $\xi_n = 0$, so that the regularity conditions related to this term are satisfied and the problem is simplified. Theorem 2.6.5 obviously holds. Condition (2.156) holds with $p = 1$ if $\mathsf{E}|X_1|^{2+\delta} < \infty$ and with $p = 2$ if $\mathsf{E}|Y_1|^{3+\delta} < \infty$ for some $\delta > 0$, so that by Theorem 2.6.6, as $\lambda \to \infty$,

$$\mathsf{E}\tau_\lambda = [2\lambda + \log(2\lambda)]/\mu^2 + O(1)$$

if $\mathsf{E}|X_1|^{2+\delta} < \infty$ and

$$\mathsf{var}[\tau_\lambda] = 8\sigma^2\lambda/\mu^4 + O\left(\sqrt{\lambda}\right)$$

if $\mathsf{E}|Y_1|^4 < \infty$.

Finally, if the random walk $\{S_n - \mu n/2\}$ is nonarithmetic, then by Theorems 2.6.5 and 2.6.7

$$\mathsf{E}\tau_\lambda = \frac{1}{\mu^2}[2\lambda + \log(2\lambda) - 2\log\mu - \sigma^2 + 2\mu\varkappa(\mu/2)] + o(1), \tag{2.177}$$

where $\varkappa(\mu/2)$ is given by (2.143) and (2.144) with $\nu = \mu/2$ and $\tilde{S}_n(\mu/2) = S_n - n\mu/2$. This expression is exactly the same as (2.139) if we set $\mu = \theta$. Indeed, the average overshoot $\varkappa$ in (2.139) is associated with the random walk $\mu(S_n - \mu n/2)$ while $\varkappa(\mu/2)$ with $S_n - \mu n/2$, so $\varkappa = \mu\varkappa(\mu/2)$. This is not surprising since the two-sided stopping time τ_a of (2.138) can be represented as $\tau_a = \min\{\tau_a^{(-)}, \tau_a^{(+)}\}$, where

$$\tau_a^{(-)} = \inf\left\{n \ge 1 : S_n \le -\sqrt{n(\log n + 2a)}\right\}, \quad \tau_a^{(+)} = \inf\left\{n \ge 1 : S_n \ge \sqrt{n(\log n + 2a)}\right\}.$$

When $\mu > 0$,

$$\mathsf{E}\tau_a = \mathsf{E}\tau_a^{(+)} + o(1) \quad \text{as } a \to \infty,$$

and hence the approximation (2.177) follows from (2.139).

Certainly, similar results hold for $a_\lambda(n) = \sqrt{2\lambda n}$, the problem that has been extensively studied by Woodroofe [512] and Hagwood and Woodroofe [174].

In the previous example the asymptotic expansion for $\mathsf{E}\tau_\lambda$ has been obtained using both Nonlinear Renewal Theorem and General Nonlinear Renewal Theorem. Now we consider an example where Theorem 2.6.4 cannot be applied but Theorem 2.6.7 is applicable.

Example 2.6.4. Let $\{S_n^{(j)}\}_{n\ge1}$, $j = 0, 1, \dots, M$ be $M+1$ ($M \ge 2$) mutually independent random walks with drifts $\mathsf{E}[S_1^{(0)}] = \mu_0 > 0$ and $\mathsf{E}[S_1^{(j)}] = -\mu_1 \le 0$ for $j = 1, \dots, M$ and variances $\mathsf{var}[S_1^{(0)}] = \sigma_0^2$ and $\mathsf{var}[S_1^{(j)}] = \sigma^2$, $j = 1, \dots, M$, $0 < \sigma_0^2, \sigma^2 < \infty$. Define the stopping time

$$\tau_\lambda = \inf\left\{n \ge 1 : S_n^{(0)} - \max_{1\le j\le M} S_n^{(j)} \ge \lambda\right\}, \quad \lambda > 0. \tag{2.178}$$

This stopping time appears, in particular, when testing sequentially $(M+1)$ hypotheses in the symmetric $(M+1)$-sample slippage problem [129]. Let $S_n = S_n^{(0)} + n\mu_1$, $\tilde{S}_n^{(j)} = S_n^{(j)} + n\mu_1$ for $j = 1, \dots, M$ and

$$\xi_n = -\max_{1\le j\le M}\left[\tilde{S}_n^{(j)}\right] + h_M\sqrt{n},$$

where h_M is the expected value of the M-th order standard normal statistic, i.e., $h_M = \mathsf{E}[\max_{1\le j\le M} \zeta_j]$, where $\zeta_k \sim \mathcal{N}(0,1)$, $k = 1, 2, \dots$ are iid standard normal random variables. Then the stopping time (2.178) can be rewritten as

$$\tau_\lambda = \inf\{n \ge 1 : S_n + \xi_n \ge a_n(\lambda)\}, \quad a_n(\lambda) = \lambda + h_M\sqrt{n}.$$

Note that $\mathsf{E}[\tilde{S}_n^{(j)}] = 0$ and $\mathsf{E}[S_n] = n\mu$, $\mu = \mu_0 + \mu_1$. Assume for simplicity that $S_1^{(j)}$, $j = 1, \dots, M$ are identically distributed (under P). Using an argument similar to (but simpler than) that in the proof of Theorem 3.3 in [129], it can be shown that the sequence $\{\xi_n\}$ is 1-regular. If, in addition, the third moment $\mathsf{E}|\tilde{S}_1^{(1)}|^3$ is finite and the Cramér condition on the characteristic function of $\tilde{S}_1^{(1)}$ holds,

$$\limsup_{t\to\infty} \mathsf{E}\left[\exp\left\{it\tilde{S}_1^{(1)}\right\}\right] < 1,$$

then ξ_n converges weakly as $n \to \infty$ to the random variable ξ with expectation $\mathsf{E}\xi = -C_M\mathsf{E}[\tilde{S}_1^{(1)}]/(6\sigma^2)$ with

$$C_M = M\int_{-\infty}^{\infty} x\varphi(x)\Phi^{M-2}(x)\left[(M-1)\varphi(x)(1-x^2) + (x^3 - 3x)\Phi(x)\right] \mathrm{d}x,$$

where $\varphi(x)$ and $\Phi(x)$ are standard normal pdf and cdf, respectively. The value of C_M is tabulated in [129] for $M = 2 \div 1000$. In order to apply Theorem 2.6.7 it remains to show that

$$\lambda\mathsf{P}\{\tau_\lambda \le (1-\varepsilon)\lambda/\mu\} \xrightarrow[\lambda\to\infty]{} 0 \quad \text{for some } 0 < \varepsilon < 1. \tag{2.179}$$

To this end, we rewrite τ_λ as

$$\tau_\lambda = \inf\{n \ge 1 : S_n + \gamma_n \ge \lambda\}, \quad \gamma_n = -\max_{1\le j\le M}\left[\tilde{S}_n^{(j)}\right]$$

Let $K = (1-\varepsilon)\lambda/\mu$. Clearly,

$$\mathsf{P}(\tau_\lambda \le K) = \mathsf{P}\left\{\max_{n\le K}(S_n+\gamma_n) \ge \lambda\right\} \le \mathsf{P}\left\{\max_{n\le K} S_n \ge (1-\varepsilon/2)\lambda\right\} + \mathsf{P}\left\{\max_{n\le K}\gamma_n \ge \varepsilon\lambda/2\right\}.$$

By Lemma 2.6.2 (similar to (2.140)),

$$\mathsf{P}\left\{\max_{n\le K} S_n \ge (1-\varepsilon/2)\lambda\right\} = o(1/K)$$

and

$$\mathsf{P}\left\{\max_{n\le K}\gamma_n \ge \varepsilon\lambda/2\right\} \le \mathsf{P}\left\{\max_{n\le K}(-\tilde{S}_n^{(1)}) \ge \varepsilon\mu K/2(1-\varepsilon)\right\} = o(1/K).$$

Hence, (2.179) holds.

By Theorem 2.6.7,

$$\mathsf{E}\tau_\lambda = N_\lambda + \frac{1}{\mu}(\varkappa - \mathsf{E}\xi) + o(1) \quad \text{as } \lambda\to\infty,$$

where $\varkappa = \mathsf{E}[S_{T_+}]^2/[2\mathsf{E}S_{T_+}]$ is the limiting average overshoot associated with stopping the random walk $\{S_n\}$ and

$$N_\lambda = \sup\{t\ge 1 : \lambda + h_M\sqrt{t} \ge \mu t\} = \frac{\lambda}{\mu} + \frac{\sigma h_M}{\mu}\sqrt{\frac{\lambda}{\mu} + \frac{\sigma^2 h_M^2}{4\mu^2}} + \frac{\sigma^2 h_M^2}{2\mu^2}.$$

Therefore, as $\lambda\to\infty$,

$$\mathsf{E}\tau_\lambda = \frac{1}{\mu}\left\{\lambda + \sigma h_M\sqrt{\frac{\lambda}{\mu} + \frac{\sigma^2 h_M^2}{4\mu^2}} + \frac{\sigma^2 h_M^2}{2\mu^2} + \frac{C_M}{6\sigma^2}\mathsf{E}\left[\tilde{S}_1^{(1)}\right]^3 + \varkappa\right\} + o(1).$$

2.7 Sequential Decision Rules and Optimal Stopping Theory

Statistical Decision Theory deals with a wide variety of statistical problems, such as parameter estimation or hypothesis testing or a combination thereof, and it follows the principle that any statistical rule/procedure has to be evaluated based on the produced results in various circumstances. This principle was first formulated by Neyman and Pearson in their hypothesis testing theory, and then in 1939 Wald suggested to extend it to all possible statistical problems. He developed, in a series of papers, the general statistical decision theory, which was finally published in his seminal monograph *Statistical Decision Functions* in 1950 [495].

Wald's mathematical model in decision theory is in fact a particular case of the models introduced by Borel [77] in 1921 and von Neumann [488] in 1928 in the game theory, which was later developed by von Neumann and Morgenstern [489] in 1944. Both theories have tight connection, which was demonstrated in the excellent book by Blackwell and Girshick [70] in 1954.

The ultimate goal of a decision-maker is to make a decision on the state of the nature or process based on the observation of a certain stochastic process. As opposed to the classical (non-sequential) decision rules that deal with a batch of observations the size of which is fixed in advance (i.e., before the observations become available), sequential decision rules are multistage in nature, and at each stage along with a terminal decision a decision on whether to continue observations or to stop and make a terminal decision is made. As a result, the whole process becomes a controllable multistage decision-making process with a random number of observations prior to the final decision being made. Since the number of observations (or stages) is random, a sequential decision rule unavoidably involves stopping times: choose a random time depending on the observations when to stop and at this time make a terminal decision. Sequential decision-making methods belong to the field of *Sequential Analysis*, which is an important branch of Statistical Sciences, and this branch also started with seminal works of Wald in the 1940s [492, 494, 496].

As we will see, finding optimal sequential decision-making strategies can be almost always reduced to finding optimal stopping times, which is the goal of *Optimal Stopping Theory*.

2.7.1 Sequential Decision Rules

Since in subsequent chapters we will deal mostly with discrete-time models, we will consider only optimization of decision-making rules in discrete time assuming that both unobservable informative parameters θ_t and observations X_t are sampled at discrete time moments $t = n \in \mathbb{Z}_+$. Consider first the N-truncated multistage decision-making policies, where at the n-th stage one may observe a random variable $X_n \in \mathcal{X}_n$, and based on the sequence of observations $\{X_n, n = 1, 2, \dots, N\}$ one has to make a decision (at one of the stages $n = 1, \dots, N$) $u_n \in \mathcal{D}_n$ on the value of the unobserved informative parameter or situation $\theta_n \in \Theta_n$. For example, if the problem is to estimate the parameter θ_n (that in general may change in time), then u_n is the estimated value of the parameter, which usually belongs to Θ_n, the parameter space. If, however, we are interested in testing two hypotheses $H_0 : \theta_n = 0$ and $H_1 : \theta_n = 1$, then $u_n \in \{0, 1\}$, where $u_n = j$ means that the hypothesis H_j is accepted.

The observation X_n may be a scalar or vector random variable or a result of preliminary processing of some random process, and $\{\theta_n\}_{n \geq 0}$ is in general a random process (while often it is just a random or nonrandom variable). Write $\mathbf{X}_1^n = (X_1, \dots, X_n)$ and $\theta_1^n = (\theta_1, \dots, \theta_n)$ for the corresponding vectors, which are defined on the spaces $\mathcal{X}_1^n = \cup_{i=1}^n \mathcal{X}_i$ and $\Theta_1^n = \cup_{i=1}^n \Theta_i$. Let $\mathscr{F}_n^X = \sigma(\mathbf{X}_1^n)$ and $\mathscr{F}_n^\theta = \sigma(\theta_1^n)$ be sigma-algebras generated by the corresponding vectors. In the following we will use the same notation for random variables and their particular values unless it leads to confusion, in which case we will make remarks and clarifications.

Suppose we are given the joint conditional densities for the observations $\mathbf{X}_1^n$ (with respect to some sigma-finite measures $\mu_n(\mathrm{d}\mathbf{X}_n)$, and for simplicity's sake let it be Lebesgue's measures $\mathrm{d}\mathbf{X}_1^n$), $p_{\theta_1^n}(\mathbf{X}_1^n) = p(\mathbf{X}_1^n \mid \theta_1^n)$, $n = 1, \dots, N$ as well as the prior distributions (densities) for parameters $\pi_n = \pi_n(\theta_1^n)$, $n = 1, \dots, N$. Here we suppose that θ_n does not depend on the observations. A generalization to the case where θ_n may depend on the data is more or less straightforward. Next, suppose that we are given the loss function $L_n(\theta_n, u_n, \mathbf{X}_1^n)$, which determines the losses that occur when at the time moment n (or the n-th stage) we make a terminal decision u_n in the situation where the value of the informative parameter is θ_n and we observed the sample $\mathbf{X}_1^n$. Losses $L_n(\cdot)$ are nonnegative, include the cost of experiment (or the cost of the delay in making a final decision) and the loss related to making a terminal decision. For example, $L_n(\theta_n, u_n, \mathbf{X}_1^n) = L(\theta_n, u_n) + c(n)$, where $c(n)$ is the cost of performing n stages of observations and $L(\theta_n, u_n)$ is the loss associated with making a terminal decision u_n when θ_n is the value of the parameter at the n-th stage. More generally, losses may depend on the whole trajectory θ_1^n.

It is assumed that after making a terminal decision irreversible actions are performed and the observations are terminated. Therefore, if the decision is made at the n-th stage, we have to account for the fact that it was not made at the previous stages $1, 2, \dots, n-1$. Formally, this can be identified with making a decision on continuation of observations, so that the stopping time T generally speaking is a random variable that depends on the previous observations and does not depend on the future observations that are not still available, i.e., $\{T = n\} \in \mathscr{F}_n^X$. Note that the sigma-algebra $\mathscr{F}_n^X$ is nothing but the total information available (through observations $\mathbf{X}_1^n$) at the point in time n. According to our previous definition such stopping times are called Markov times, and they correspond to physically realizable systems.

Let $d_n = d_n(\mathbf{X}_1^n)$, $n = 1, \dots, N$ be the decision functions whose particular values are terminal decisions u_n, so that $d_n(\mathbf{X}_1^n) \in \mathcal{D}_n$, and let T_N be a N-truncated stopping time, i.e., a stopping time such that $T_N \le N$. The decision function d_n determines the terminal decision rule (based on the observation of the sample $\mathbf{X}_1^n$), while the function $T_N(\mathbf{X})$ determines the stopping rule (when to stop and then make a final decision). In the most general case d_n and T_N are randomized decision rules which are defined as probability measures, but we will restrict our attention to nonrandomized rules keeping in mind that all optimal Bayesian rules turn out to be nonrandomized. If randomization is necessary (in variational, conditionally extreme problems) we will consider it on a case-by-case basis. The *nonrandomized truncated sequential decision rule* (or procedure or policy) is a pair $\delta_N(\mathbf{X}) = (\mathbf{d}_N(\mathbf{X}), T_N(\mathbf{X}))$, where $\mathbf{d}_N(\mathbf{X}) = d_n(\mathbf{X}_1^n)$ when $T_N = n$ is the terminal decision rule and T_N is the stopping rule (time). The truncated rules are also referred to as the *finite horizon* rules. If $N = \infty$, then the corresponding sequential rules are called *nontruncated sequential rules*, or just *sequential rules*. In this case we write $\delta(\mathbf{X}) = (\mathbf{d}(\mathbf{X}), T(\mathbf{X}))$ and refer to these rules as the *infinite horizon* rules.

For any sequential decision rule $\delta = \delta(\mathbf{X})$, introduce the *expected loss* $\rho(\delta) = \mathsf{E}[L_T(\theta_T, d_T(\mathbf{X}_1^T), \mathbf{X}_1^T)]$, which will be called the *average risk* (AVR). According to the decision theory, an optimal Bayesian decision rule δ^0 is the rule that minimizes the AVR:

$$\rho(\delta^0) = \rho^0 = \inf_{\delta} \rho(\delta). \tag{2.180}$$

It may happen that an optimal rule does not exist. However, in most particular problems it exists, i.e., $\rho^0 < \infty$ and there is an element for which the infimum in (2.180) is attained (it may not be unique). If such an element does not exist, then we can always find an ε-optimal rule δ_ε^0 for which $\rho(\delta_\varepsilon^0) - \rho^0 \le \varepsilon$ for every $\varepsilon > 0$. In the following we assume that an optimal (i.e., 0-optimal) rule exists. We also consider only procedures that require at least one observation. An option to make a decision without observations (which does not have any practical significance) can be accounted for in an obvious way.

Assume that $T = n$, so that we have made n observations and observed the sample $\mathbf{X}_1^n$. Write

$$p_{\theta_n}(\mathbf{X}_1^n) = \int_{\Theta_1^{n-1}} p_{\theta_1^n}(\mathbf{X}_1^n)\pi(\theta_1^{n-1} \mid \theta_n)\mathrm{d}\theta_1^{n-1}, \quad p(\mathbf{X}_1^n) = \int_{\Theta_1^n} p_{\theta_1^n}(\mathbf{X}_1^n)\pi(\theta_1^n)\mathrm{d}\theta_1^n,$$

where $\pi(\theta_1^{n-1} \mid \theta_n)$ is the conditional prior pdf for θ_1^{n-1} for given θ_n. By the Bayes formula, the posterior density of θ_n given $\mathbf{X}_1^n$ is

$$\pi_n(\theta_n) = \frac{p_{\theta_n}(\mathbf{X}_1^n)\pi_n(\theta_n)}{p(\mathbf{X}_1^n)}.$$

Introduce the posterior loss associated with the decision u_n

$$R_n(\mathbf{X}_1^n, u_n) = \mathsf{E}[L_n(\theta_n, u_n, \mathbf{X}_1^n) | \mathbf{X}_1^n] = \int_{\Theta_n} L_n(\theta_n, u_n, \mathbf{X}_1^n)\pi_n(\theta_n)\,\mathrm{d}\theta_n,$$

which we will call the *a posteriori risk* (APR). Let

$$R_n^0(\mathbf{X}_1^n) = R_n(\mathbf{X}_1^n, u_n^0) = \inf_{u_n \in \mathcal{D}_n} R_n(\mathbf{X}_1^n, u_n) \tag{2.181}$$

be the minimal APR, assuming that such element u_n^0 exists.

It is straightforward to show that

$$\begin{aligned}
\rho(\delta) &= \sum_{n=1}^{\infty} \mathsf{P}(T = n) \int_{\{T=n\}} \left(\int_{\Theta_n} L_n(\theta_n, d_n, \mathbf{X}_1^n)\pi_n(\theta_n)\,\mathrm{d}\theta_n \right) p(\mathbf{X}_1^n)\,\mathrm{d}\mathbf{X}_1^n \\
&= \sum_{n=1}^{\infty} \mathsf{P}(T = n) \int_{\{T=n\}} R_n(\mathbf{X}_1^n, u_n) p(\mathbf{X}_1^n)\,\mathrm{d}\mathbf{X}_1^n \\
&\ge \sum_{n=1}^{\infty} \mathsf{P}(T = n) \int_{\{T=n\}} R_n^0(\mathbf{X}_1^n) p(\mathbf{X}_1^n)\,\mathrm{d}\mathbf{X}_1^n = \rho(T, \mathbf{d}^0),
\end{aligned}$$

where $\mathbf{d}^0 = d_n^0$ when $T = n$ and

$$d_n^0(\mathbf{X}_1^n) = u_n^0 = \arg \inf_{u_n \in \mathcal{D}_n} R_n(\mathbf{X}_1^n, u_n). \tag{2.182}$$

Hence, the optimal terminal rule $\mathbf{d}^0$ does not depend on the stopping rule T and finding an optimal sequential decision rule reduces to finding an optimal stopping rule T^0 that minimizes the AVR

$$\rho(T, \mathbf{d}^0) = \sum_{n=1}^{\infty} \mathsf{P}(T = n) \int_{\{T=n\}} R_n^0(\mathbf{X}_1^n) p(\mathbf{X}_1^n) \, \mathrm{d}\mathbf{X}_1^n = \mathsf{E}[R_T^0(\mathbf{X}_1^T)].$$

This problem is the subject of *Optimal Stopping Theory*. However, usually the theory of optimal stopping deals with rewards and payoffs rather than with losses and risks. Noting that $-L_T$ and $-R_T^0 = \Psi_T$ may be considered as a reward (or gain) and a payoff, we now consider an equivalent optimal stopping problem

$$V^0 = \sup_{T \in \mathcal{M}} \mathsf{E}[\Psi_T], \tag{2.183}$$

where the supremum is taken over the totality of all Markov stopping times $\mathcal{M}$ (or possibly over a certain subset if needed). In the following we will always assume that $\mathcal{M}$ is a set of Markov times finite w.p. 1, i.e., $\mathsf{P}(T < \infty) = 1$, since otherwise an optimal stopping time may not exist. The problem is twofold: we need to find an optimal stopping time T^0 that maximizes the payoff $V(T) = \mathsf{E}[\Psi_T]$ as well as the value of the optimal payoff $V^0 = V(T^0)$.

2.7.2 *Optimal Stopping Rules*

To be more specific, we consider the following general scheme. Let $\Psi = \{\Psi_n\}_{n \geq 1}$ be a sequence of random variables (rewards or gains) on the filtered probability space $(\Omega, \mathscr{F}, \{\mathscr{F}_n\}, \mathsf{P})$ adapted to the filtration $\{\mathscr{F}_n\}_{n \geq 1}$, where $\{\mathscr{F}_n\}$ is an increasing sequence of sigma-algebras. Usually $\mathscr{F}_n$ is a natural sigma-algebra $\mathscr{F}_n^{\Psi} = \sigma(\Psi_1, \dots, \Psi_n)$, but not necessarily in general. The sigma-algebra $\mathscr{F}_n$ is cumulative information that we obtain up to time n.

2.7.2.1 *Truncated Stopping Rules*

We begin with considering the N-truncated (finite horizon) case where maximization is performed over $\mathcal{M}^N = \{T \in \mathcal{M} : T \leq N\}$, for some finite and positive N. Let us embed this problem into a sequence of related optimization problems $\mathcal{M}_n^N = \{T \in \mathcal{M} : n \leq T \leq N\}$, $n = 1, 2, \dots, N$, and let

$$V_N^n = \sup_{T \in \mathcal{M}_n^N} \mathsf{E}[\Psi_T] \tag{2.184}$$

be the maximal payoff in the class of stopping times $\mathcal{M}_n^N$. In the following we write $V_N = V_N^1$, $V^n = V_\infty^n$, and $V^1 = V$, so that V_N and V are optimal (maximal) gains in N-truncated and nontruncated problems, respectively. In other words,

$$V_N = \sup_{T \in \mathcal{M}^N} \mathsf{E}[\Psi_T], \quad V = \sup_{T \in \mathcal{M}} \mathsf{E}[\Psi_T].$$

Bellman's dynamic programming approach plays a central role in solving the optimization problem (2.184). According to this approach we have to perform backward induction, i.e., to consider the multistage process from the end N and move backward to the beginning. Specifically, let W_n^N, $n = 1, \dots, N$ be the optimal rewards when information $\mathscr{F}_n$ is available. If $n = N$, then we must stop w.p. 1, so that $W_N^N = \Psi_N$. If $n = N - 1$, then we can either stop or continue getting new information $\mathscr{F}_N$. If we stop, then the reward is $W_{N-1}^N = \Psi_{N-1}$. If we continue, then our reward is $W_{N-1}^N = \mathsf{E}[W_N^N \mid \mathscr{F}_{N-1}] = \mathsf{E}[\Psi_N \mid \mathscr{F}_{N-1}]$. Clearly, we must stop at $N-1$ if the current reward Ψ_{N-1} exceeds the projected reward $W_{N-1}^N = \mathsf{E}[W_N^N \mid \mathscr{F}_{N-1}]$ associated with continuation, and continue if

$\Psi_{N-1} \le W_{N-1}^N$. Continuing by induction, we conclude that the optimal sequence of rewards satisfies the recursion

$$W_N^N = \Psi_N, \quad W_n^N = \max\left\{\Psi_n, \mathsf{E}[W_{n+1}^N \mid \mathscr{F}_n]\right\}, \quad n = N-1, N-2, \dots, 1, \tag{2.185}$$

and that the optimal stopping rule is

$$T_N^n = \min\left\{n \le k \le N : \Psi_k = W_k^N\right\}, \quad n = 1, \dots N. \tag{2.186}$$

Therefore, the optimal N-truncated stopping time T_N^0 (optimal in the class $\mathcal{M}^N$) has the form

$$T_N^0 = \min\left\{1 \le n \le N : \Psi_n = W_n^N\right\}. \tag{2.187}$$

The following theorem gives the exact result in the finite horizon case.

Theorem 2.7.1 (Finite horizon optimal stopping). *Let the sequence $\{W_n^N\}_{1 \le n \le N}$ be defined by the recursion* (2.185)*, and let the stopping time T_N^0 be defined as in* (2.187)*. Assume that the sequence $\{\Psi_n\}_{n \ge 1}$ is integrable. Then T_N^0 is the optimal stopping rule in the class of N-truncated stopping rules $\mathcal{M}^N$, i.e.,*

$$V_N = \sup_{T \in \mathcal{M}^N} \mathsf{E}[\Psi_T] = \mathsf{E}[\Psi_{T_N^0}] \tag{2.188}$$

and

$$V_N = \mathsf{E}[W_1^N]. \tag{2.189}$$

Proof. Let W_n^N be given by the recursion (2.185) and let T_N^n be the stopping time defined in (2.186). We now prove that for every $n = 1, \dots, N$

$$\mathsf{E}[\Psi_{T_N^n} \mid \mathscr{F}_n] = W_n^N \ge \mathsf{E}[\Psi_T \mid \mathscr{F}_n] \quad \text{for all } T \in \mathcal{M}_n^N, \tag{2.190}$$

which implies that

$$\mathsf{E}[\Psi_{T_N^n}] = \mathsf{E}[W_n^N] \ge \mathsf{E}[\Psi_T] \quad \text{for all } T \in \mathcal{M}_n^N \text{ and all } n = 1, \dots, N. \tag{2.191}$$

Relations (2.191) yield in particular the statement of the theorem setting $n = 1$.

For $n = N$ the result holds trivially. The proof will be conducted by induction. Suppose (2.190) is true for some $n \in \{1, \dots, N\}$. Let $\mathcal{A}_{n-1} \in \mathscr{F}_{n-1}$ be an arbitrary event and $\tau_n = \max(T_N^{n-1}, n)$. Obviously, $\tau_n \in \mathcal{M}_n^N$. We have

$$\begin{aligned}
\mathsf{E}[\Psi_{T_N^{n-1}} 1\!\!1_{\{\mathcal{A}_{n-1}\}}] &= \mathsf{E}[\Psi_{T_N^{n-1}} 1\!\!1_{\{\mathcal{A}_{n-1} \cap \{T_N^{n-1} \ge n\}\}}] + \mathsf{E}[\Psi_{T_N^{n-1}} 1\!\!1_{\{\mathcal{A}_{n-1} \cap \{T_N^{n-1} = n-1\}\}}] \\
&= \mathsf{E}[\Psi_{\tau_n} 1\!\!1_{\{\mathcal{A}_{n-1} \cap \{T_N^{n-1} \ge n\}\}}] + \mathsf{E}[\Psi_{n-1} 1\!\!1_{\{\mathcal{A}_{n-1} \cap \{T_N^{n-1} = n-1\}\}}] \\
&= \mathsf{E}\{\mathsf{E}[\mathsf{E}(\Psi_{\tau_n} \mid \mathscr{F}_n) \mid \mathscr{F}_{n-1}] 1\!\!1_{\{\mathcal{A}_{n-1} \cap \{T_N^{n-1} \ge n\}\}}\} + \mathsf{E}[\Psi_{n-1} 1\!\!1_{\{\mathcal{A}_{n-1} \cap \{T_N^{n-1} = n-1\}\}}] \\
&= \mathsf{E}[\mathsf{E}(W_n^N \mid \mathscr{F}_{n-1}] 1\!\!1_{\{\mathcal{A}_{n-1} \cap \{\Psi_{n-1} < \mathsf{E}[W_n^N \mid \mathscr{F}_{n-1}\}\}}] + \mathsf{E}[\Psi_{n-1} 1\!\!1_{\{\mathcal{A}_{n-1} \cap \{\Psi_{n-1} \ge \mathsf{E}[W_n^N \mid \mathscr{F}_{n-1}\}\}}] \\
&= \mathsf{E}[\max(\Psi_{n-1}, \mathsf{E}(W_n^N \mid \mathscr{F}_{n-1}) 1\!\!1_{\{\mathcal{A}_{n-1}\}}] \\
&= \mathsf{E}[W_{n-1}^N 1\!\!1_{\{\mathcal{A}_{n-1}\}}].
\end{aligned}$$

Since the event $\mathcal{A}_{n-1} \in \mathscr{F}_{n-1}$ is arbitrary, it follows that

$$W_{n-1}^N = \mathsf{E}[\Psi_{T_N^{n-1}} \mid \mathscr{F}_{n-1}],$$

and the proof of (2.190) is complete. □

It is useful to provide an additional characterization of random variables (rewards) W_n^N, $n = 1,\dots,N$ defined by the recursion (2.185). Indeed, W_n^N can be written as

$$W_n^N = \operatorname*{ess\,sup}_{T \in \mathcal{M}_n^N} \mathsf{E}[\Psi_T | \mathscr{F}_n], \quad n = 1,\dots,N, \tag{2.192}$$

where ess sup stands for *essential supremum*. This characterization is especially important in nontruncated (infinite horizon) optimal stopping problems.

We recall the definition of ess sup. Let S be an arbitrary set, and let $\{X_t\}_{t\in S}$ be a family of real random variables defined on the probability space $(\Omega, \mathscr{F}, \mathsf{P})$. A random variable Y is said to be the *essential supremum* of $\{X_t\}_{t\in S}$, which is written $Y = \operatorname{ess\,sup}_{t\in S} X_t$, if

(a) $\mathsf{P}(Y \geq X_t) = 1$ for every $t \in S$; and

(b) if Y^* is a random variable such that $\mathsf{P}(Y^* > X_t)$ for every $t \in S$, then $\mathsf{P}(Y^* \geq Y) = 1$.

It turns out that $Y = \operatorname{ess\,sup}_{t\in S} X_t$ always exists, and that there is a countable subset S_c of S such that $Y = \sup_{t\in S_c} X_t$.

2.7.2.2 *Nontruncated Stopping Rules*

Consider now the nontruncated case where $N = \infty$, i.e., we are interested in the optimization problem

$$V^n = \sup_{T \in \mathcal{M}_n} \mathsf{E}[\Psi_T], \quad n \geq 1, \tag{2.193}$$

where $\mathcal{M}_n = \mathcal{M}_n^\infty = \{T : T \geq n\}$. Recall also that we consider only stopping times finite w.p. 1, $\mathsf{P}(T < \infty) = 1$. In other words, the class $\mathcal{M}_n$ contains only such stopping rules that $\mathsf{P}(n \leq T < \infty) = 1$.

Similar to (2.192) let

$$W_n = \operatorname*{ess\,sup}_{T \in \mathcal{M}_n} \mathsf{E}[\Psi_T | \mathscr{F}_n], \quad n \geq 1, \tag{2.194}$$

and similar to (2.187) let

$$T^0 = \inf\{n \geq 1 : \Psi_n = W_n\}, \quad \inf\{\varnothing\} = \infty. \tag{2.195}$$

Theorem 2.7.2 (Infinite horizon optimal stopping)**.** *Let the sequence $\{W_n\}_{n\geq 1}$ obey the recursion*

$$W_n = \max\{\Psi_n, \mathsf{E}[W_{n+1} \mid \mathscr{F}_n]\}, \quad n \geq 1, \tag{2.196}$$

and let the stopping time T^n be defined as

$$T^n = \inf\{k \geq n : \Psi_k = W_k\}, \quad n \geq 1, \quad \inf\{\varnothing\} = \infty. \tag{2.197}$$

If

$$\mathsf{E}\left[\sup_{n\geq 1} |\Psi_n|\right] < \infty \tag{2.198}$$

and

$$\mathsf{P}(T^n < \infty) = 1, \tag{2.199}$$

then T^n is the optimal stopping rule in the problem (2.193) *and, therefore, the stopping time T^0 defined in* (2.195) *is optimal in the class of nontruncated stopping rules $\mathcal{M}$, i.e.,*

$$V = \sup_{T \in \mathcal{M}} \mathsf{E}[\Psi_T] = \mathsf{E}[\Psi_{T^0}] \tag{2.200}$$

and

$$V = \mathsf{E}[W_1]. \tag{2.201}$$

The sequence $\{W_n\}$ is the essential supremum of the posterior gain as defined in (2.194)*.*

The proof of this theorem is quite technical and for this reason is omitted. The detailed proof may be found in Chow, Robbins, and Siegmund [101] and Peskir and Shiryaev [360].

While Theorem 2.7.2 provides a theoretical characterization of the optimal stopping rule, it is usually useless in particular problems and applications, since the recursion (2.196) cannot be solved explicitly and thus defines W_n non-constructively.

It is intuitively clear that there should be a connection between truncated and nontruncated cases, in particular for sufficiently large N an optimal truncated stopping rule should approximate an optimal nontruncated one fairly well, and in the limit when $N \to \infty$ there should be convergence, that is, $\lim_{N\to\infty} W_n^N = W_n$ and $\lim_{N\to\infty} V_N = V$. This is indeed true under quite general conditions.

Observe that the rewards W_n^N and the stopping times T_N^n defined in (2.185) and (2.186) are increasing in N, so that the limits $W_n^\infty = \lim_{N\to\infty} W_n^N$ and $T_\infty^n = \lim_{N\to\infty} T_N^n$ exist w.p. 1 for every $n \ge 1$. For the same reason the limit $V_\infty^n = \lim_{N\to\infty} V_N^n$ exists for every $n \ge 1$. Letting $N \to \infty$ in (2.185) and using the Lebesgue (conditional) monotone convergence theorem, we obtain the following recursion for $\{W_n^\infty\}$:

$$W_n^\infty = \max\left\{\Psi_n, \mathsf{E}[W_{n+1}^\infty \mid \mathscr{F}_n]\right\}, \quad n \ge 1. \tag{2.202}$$

The corresponding stopping rule is

$$T_\infty^n = \inf\{k \ge n : \Psi_k = W_k^\infty\}, \quad n \ge 1. \tag{2.203}$$

In general,

$$W_n^\infty \le W_n, \quad V_\infty^n \le V^n, \quad n \ge 1, \tag{2.204}$$

and the inequalities may be strict. The following theorem establishes conditions under which the equalities hold in (2.204) and the stopping times T_∞^n, $n \ge 1$ defined by (2.202) and (2.203) are optimal.

Theorem 2.7.3 (From finite to infinite horizon). *Consider the optimal stopping problem* (2.193). *Let the sequences* $\{W_n\}_{n\ge1}$ *and* $\{W_n^\infty\}_{n\ge1}$ *satisfy the recursions* (2.196) *and* (2.202), *respectively. Let* T^n *and* T_∞^n *be defined as in* (2.197) *and* (2.203)*. If the condition* (2.198) *holds, then* $W_n = W_n^\infty$, $T^n = T_\infty^n$, *and* $V^n = V_\infty^n$ *for any* $n \ge 1$. *In particular,* T_∞^1 *is the optimal stopping rule in the problem* (2.200) *and*

$$V = V_\infty^1 = \mathsf{E}[W_1^\infty]. \tag{2.205}$$

Proof. By the recursive relation (2.202), the process $(W_n^\infty, \mathscr{F}_n)_{n\ge1}$ is a supermartingale. By condition (2.198),

$$\mathsf{E}|\Psi_n| < \infty, \quad \mathsf{E}|\Psi_T| < \infty, \quad \liminf_{k\to\infty} \mathsf{E}\left[|\Psi_k| \mathbb{1}_{\{T>k\}}\right] = 0,$$

so that the optional sampling theorem (Theorem 2.3.1) can be applied to conclude that, for every $T \in \mathcal{M}_n$,

$$\mathsf{E}(W_T^\infty \mid \mathscr{F}_n) \ge W_n^\infty \quad \mathsf{P}-\text{a.s.} \quad n = 1, 2, \dots,$$

which implies (since $W_k^\infty \ge \Psi_k$ for all $k \ge n$ by definition) that $W_T^\infty \ge \Psi_T$ P-a.s. for every $T \in \mathcal{M}_n$. Hence,

$$\mathsf{E}(W_T^\infty \mid \mathscr{F}_n) \ge \mathsf{E}(\Psi_T \mid \mathscr{F}_n) \quad \mathsf{P}-\text{a.s.} \quad n = 1, 2, \dots$$

for every $T \in \mathcal{M}_n$, which yields

$$W_n^\infty \ge \mathsf{E}(\Psi_T \mid \mathscr{F}_n) \quad \mathsf{P}-\text{a.s.} \quad n = 1, 2, \dots.$$

It follows that $W_n^\infty \ge W_n$. This completes the proof of the fact that $W_n = W_n^\infty$ P-a.s. for all $n \ge 1$, since the reverse inequality $W_n^\infty \le W_n$ always holds. The equality $T^n = T_\infty^n$ P-a.s. holds trivially. Finally, the equality $V^n = V_\infty^n$ for any $n \ge 1$ follows from the monotone convergence theorem. □

Remark 2.7.1. At a certain additional effort it can be shown that the condition (2.198) in Theorem 2.7.3 can be relaxed into

$$\liminf_{n\to\infty} \mathsf{E}\left[(W_n^\infty)^- 1\!\!1_{\{T>n\}}\right] = 0$$

and/or

$$\mathsf{E}[W_1^\infty] < \infty, \quad \mathsf{E}\left[\sup_{n\ge 1} \Psi_n^-\right] < \infty.$$

Also, Theorem 2.7.2 holds under the weaker condition $\mathsf{E}[\sup_{n\ge 1} \Psi_n^+] < \infty$. See [101, 360].

The following example shows that $W_n \ne W_n^\infty$ in general. Let $X_1, X_2, \dots$ be iid random variables taking values ± 1 w.p. $1/2$, and let $\Psi_n = X_1 + \cdots + X_n$, $\mathscr{F}_n = \mathscr{F}_n^X = \sigma(X_1,\dots,X_n)$. Since $(\Psi_n, \mathscr{F}_n^X)_{n\ge 1}$ is a zero-mean martingale, by the optional sampling theorem $\mathsf{E}(\Psi_T \mid \mathscr{F}_n^X) = X_n$ on $\{T \ge n\}$ for any finite w.p. 1 stopping time T. Hence $W_n^\infty = \Psi_n$, $n \ge 1$, but $W_n = \infty$ for all $n \ge 1$ since $\mathsf{P}(\limsup \Psi_n = +\infty) = 1$.

Theorem 2.7.3 shows that the optimal nontruncated stopping rules can be approximated by the N-truncated stopping rules for sufficiently large values of N.

2.7.3 *Optimal Sequential Decision-Making Rules*

We now return to the initial sequential decision-making problem in a Bayesian context formulated in Subsection 2.7.1. Recall that $\mathscr{F}_n = \mathscr{F}_n^X$ and that we rule out the procedures that do not require making observations so that $n = 0$ is not an option.

2.7.3.1 *Optimal Truncated Rules*

Let $R_n^{\rm st} = R_n^{\rm st}(\mathbf{X}_1^n)$ be the minimal APR associated with making an optimal terminal decision when stopping after n observations defined in (2.181), and let $\rho_N(T) = \mathsf{E}[R_T^{\rm st}]$ be the AVR when an optimal terminal decision d_T^0 is made at a stopping time T. Replacing Ψ_T with $-R_T^{\rm st}$ in (2.184), the N-truncated optimal decision-making problem is reduced to the following optimal stopping problem

$$\rho_N^0 = \inf_{T\in\mathcal{M}^N} \mathsf{E}[R_T^{\rm st}]. \tag{2.206}$$

Note that $\rho_N^0 = \rho_N(\delta_N^0) = \mathsf{E}[R_{T_N^0}^{\rm st}]$, where $\delta_N^0 = (\mathbf{d}_N^0, T_N^0)$ and T_N^0 is an optimal stopping rule to be found.

Define the posterior risk function $R_n^N = R_n^N(\mathbf{X}_1^N)$ recursively

$$R_N^N = R_N^{\rm st}, \quad R_n^N = \min\left\{R_n^{\rm st}, \mathsf{E}[R_{n+1}^N \mid \mathbf{X}_1^n]\right\}, \quad n = N-1, N-2, \dots, 1 \tag{2.207}$$

(cf. (2.185)). Write $\tilde{R}_n^N = \tilde{R}_n^N(\mathbf{X}_1^n) = \mathsf{E}[R_{n+1}^N \mid \mathbf{X}_1^n]$. The risk $R_n^{\rm st}$ determines the losses associated with stopping at the stage n when the sample $\mathbf{X}_1^n$ has been observed and an optimal terminal decision $d_n^0(\mathbf{X}_1^n)$ has been made, while the risk $\tilde{R}_n^N$ determines the losses associated with continuation of observations and making an optimal terminal decision in the future.

Finding an optimal terminal decision rule $\{d_n^0\}_{1\le n\le N}$ is usually not a difficult task. The main problem is finding the APR $\tilde{R}_n^N$ related to the continuation of observations. Therefore, finding the optimal decision rule $\delta_N^0 = (\mathbf{d}_N^0, T_N^0)$ is reduced to finding the optimal stopping rule T_N^0, which is nothing but the optimal stopping problem (2.188).

Applying Theorem 2.7.1, we obtain the following general result that determines the structure of the truncated sequential Bayesian decision-making rule.

Theorem 2.7.4. *Assume that in the sets of terminal decisions $\mathcal{D}_n$, $n = 1,\dots,N$ there exist the elements u_n^0, $n = 1,\dots,N$ for which the infimums in* (2.181) *are attained. Let the sequence $\{R_n^N\}_{1\le n\le N}$ be defined by the recursion* (2.207). *Suppose that $\mathsf{E}[\max_{n\le N} R_n^{\rm st}] < \infty$.*

(i) *Then the optimal Bayesian N-truncated rule exists and has the form* $\delta_N^0 = (\mathbf{d}_N^0, T_N^0)$, *where*

$$\begin{aligned} T_N^0 &= \min\left\{1 \le n \le N : R_n^{\mathrm{st}} = R_n^N\right\} = \min\left\{1 \le n \le N : R_n^{\mathrm{st}} \le \tilde{R}_n^N\right\}, \\ \mathbf{d}_N^0 &= u_n^0 \quad \textit{if } T_N^0 = n. \end{aligned} \tag{2.208}$$

(ii) *The optimal Bayesian risk is*

$$\rho_N^0 = \mathsf{E}[R_1^N]. \tag{2.209}$$

Therefore, Theorem 2.7.4 suggests that the optimal stopping time is the first time when the APR R_n^{st} associated with the best terminal decision is less or equal to the APR $\tilde{R}_n^N$ associated with continuation of observations. In the following the risk function R_n^N defined in (2.207) will be referred to as the *minimal a posteriori risk* (MAPR).

Indeed, alternatively this risk function can be characterized as

$$R_n^N = \operatorname*{ess\,inf}_{\{T \in \mathcal{M} : n \le T \le N\}} \mathsf{E}[L_T(\theta_T, d_T^0(\mathbf{X}_1^T), \mathbf{X}_1^T) \mid \mathbf{X}_1^n].$$

The backward induction (2.207) reflects Bellman's optimality principle according to which optimization of the multistage process starts from the end (i.e., from the stage N). At the N-th stage, we stop w.p. 1 and the MAPR R_N^N is equal to the APR R_N^{st} associated with the best terminal decision $d_N^0(\mathbf{X}_1^N)$. At the $(N-1)$-th stage we have two possibilities – either to stop and make an optimal terminal decision d_{N-1}^0, in which case we lose R_{N-1}^{st}, or to continue the observations and lose $\tilde{R}_{N-1}^N = \mathsf{E}[R_N^N \mid \mathbf{X}_1^{N-1}]$. As a result, we stop if $R_{N-1}^{\mathrm{st}} \le \tilde{R}_{N-1}^N$ and make another observation X_N otherwise. The MAPR is equal to $R_{N-1}^N = \min\{R_{N-1}^{\mathrm{st}}, \tilde{R}_{N-1}^N\}$. The situation is analogous for any $n = N-1, \dots, 1$, and the minimal average risk $\rho_N^0 = \inf_\delta \rho_N(\delta)$ is equal to $\mathsf{E}[R_1^N]$.

2.7.3.2 Optimal Nontruncated Rules

Consider now the nontruncated case where $N = \infty$. Analogously to the truncated case the optimization problem of finding an optimal nontruncated decision rule $\delta^0 = (\mathbf{d}^0, T^0)$ that minimizes the average risk $\rho(\delta)$ is reduced to the optimal stopping problem

$$\rho^0 = \inf_{T \in \mathcal{M}} \mathsf{E}[R_T^{\mathrm{st}}], \tag{2.210}$$

and $\rho^0 = \rho(\delta^0) = \mathsf{E}[R_{T^0}^{\mathrm{st}}]$. We recall that $\mathcal{M}$ stands for the class of Markov stopping times that are finite w.p. 1.

The difference here is that there is no N to start with. However, as Theorem 2.7.3 suggests, under certain conditions an optimal nontruncated decision rule can be obtained by the limiting transition $N \to \infty$ from the optimal N-truncated rule.

To be more specific, since increasing of N expands the set of decisions, $\{\rho_N^0\}_{N \ge m}$ is a (nonnegative) monotone nonincreasing sequence assuming that $\rho_m < \infty$ for some $m < \infty$, so that the limit $\lim_{N \to \infty} \rho_N^0 = \rho_\infty^0$ exists. In general, $\rho_\infty^0 \ge \rho^0$, but Theorem 2.7.3 allows us to establish conditions under which the equality $\rho_\infty^0 = \rho^0$ holds, and hence a reasonable candidate for the optimal nontruncated rule is

$$\begin{aligned} T^0 &= \inf\left\{n \ge 1 : R_n^{\mathrm{st}} = R_n\right\} = \inf\left\{n \ge 1 : R_n^{\mathrm{st}} \le \tilde{R}_n\right\}, \\ \mathbf{d}^0 &= u_n^0 \quad \text{if } T^0 = n, \end{aligned} \tag{2.211}$$

where R_n is given by the recursion

$$R_n = \min\left\{R_n^{\mathrm{st}}, \mathsf{E}[R_{n+1} \mid \mathbf{X}_1^n]\right\}, \quad n \ge 1, \tag{2.212}$$

and $\tilde{R}_n = \mathsf{E}[R_{n+1} \mid \mathbf{X}_1^n]$. As before, R_n^{st} is the APR associated with stopping at the n-th stage and making an optimal terminal decision d_n^0 and $\tilde{R}_n$ is the APR associated with the optimal continuation, so that R_n is the minimal APR (MAPR). In other words, we continue the observation on the n-th stage if $\tilde{R}_n$ is below $R_n^{\rm st}$ and stop otherwise.

Theorem 2.7.3 and the above argument imply the following result.

Theorem 2.7.5. *Let $\{R_n\}_{n\ge1}$ satisfy the recursion* (2.212) *and assume that*

$$\mathsf{E}\left[\sup_{n\ge1} R_n^{\rm st}\right] < \infty. \tag{2.213}$$

Then $R_n = \lim_{N\to\infty} R_n^N$ and $\delta^0 = (\mathbf{d}^0, T^0)$ of (2.211) *is the optimal nontruncated decision rule. In addition,*

$$\rho^0 = \lim_{N\to\infty} \rho_N^0 = \mathsf{E}[R_1]. \tag{2.214}$$

The above statements are true if in place of (2.213) *the following conditions hold*

$$\rho_m^0 < \infty \text{ for some } m < \infty \quad \text{and} \quad \lim_{n\to\infty} \mathsf{E}\left[R_n^{\rm st} 1\!\!1_{\{T^0>n\}}\right] = 0. \tag{2.215}$$

The important conclusions that we can make from Theorem 2.7.5 are: (a) the limiting function $\lim_{N\to\infty} R_n^N$ coincides with the MAPR function

$$R_n = \operatorname*{ess\,inf}_{\{T\in\mathcal{M}:T\ge n\}} \mathsf{E}[L_T(\theta_T, d_T^0(\mathbf{X}_1^T), \mathbf{X}_1^T) \mid \mathbf{X}_1^n];$$

and (b) the optimal nontruncated rule can be approximated by the truncated one for sufficiently large N, which makes it useful for practical purposes.

Note that the limiting APR $R_n^\infty = \lim_{N\to\infty} R_n^N$ always satisfies the recursive equation (2.212) (i.e., regardless of conditions (2.213) or (2.215)), but it is not the MAPR function R_n in general.

2.7.3.3 The Case of a Markov Sufficient Statistic

It follows from Theorems 2.7.4 and 2.7.5 that finding optimal sequential decision rules requires solving recurrent equations (2.207) for the truncated case and (2.212) for the nontruncated case. Consider the N-truncated case. For a large N this problem is extremely difficult in general since the MAPR $R_n^N(\mathbf{X}_1^n)$ depends on the vector of observations which dimensionality changes with n. This complication can be avoided if there is a statistic $\mathcal{S}_n = \mathcal{S}_n(\mathbf{X}_1^n)$ with fixed dimensionality such that $R_n^N(\mathbf{X}_1^n) = R_n^N(\mathcal{S}_n)$ for all $n = 1, \dots, N$. As we will see, the dimensionality of the problem is indeed constant if $R_n^{\rm st}(\mathbf{X}_1^n) = R_n^{\rm st}(\mathcal{S}_n)$ and $\{\mathcal{S}_n\}_{n\ge1}$ is a Markov process. A good example is the problem of testing hypotheses for independent observations when the likelihood ratio is a Markov process; see Section 3.2.2. While the Markov property is not a necessary condition, it is instructive and often the case in particular problems.

We are now in a position to introduce the important notion of *sufficient statistic* in sequential decision problems.

Definition 2.7.1. Assume that there exists a statistic $\mathcal{S}_n = \mathcal{S}_n(\mathbf{X}_1^n)$ such that

$$\rho_N^0 = \inf_{\delta(\mathbf{X})} \rho_N(\delta(\mathbf{X})) = \inf_{\delta(\mathcal{S})} \rho_N(\delta(\mathcal{S})), \tag{2.216}$$

where the first infimum is taken over all N-truncated sequential rules that are functions of the observations $\mathbf{X} = \{X_n\}$ and the second one over all N-truncated sequential rules that are functions of the sequence of statistics $\mathcal{S} = \{\mathcal{S}_n\}$. The sequence of statistics $\{\mathcal{S}_n\}_{1\le n\le N}$ satisfying (2.216) will be called *sufficient* in the N-truncated sequential decision-making problem. In the nontruncated problem the definition is the same with $N = \infty$.

In the rest of this subsection we will consider only the truncated case with the understanding that the nontruncated case can be covered by taking the limit $N \to \infty$.

In the following instead of the "sufficient sequence of statistics" we will use the "sufficient statistic." The meaning is obvious since if there exists $\delta_N^0(\mathcal{S})$ such that $\rho_N^0 = \rho_N(\delta_N^0(\mathcal{S}))$, then by (2.216) one may search an optimal sequential rule among the rules that depend on the observations only via $\mathcal{S}$. This means that the sufficient statistic contains all useful information for making the decision.

Theorem 2.7.4 implies that (2.216) holds, in particular, if

$$R_n^N(\mathbf{X}_1^n) = R_n^N(\mathcal{S}_n), \quad n = 1,\dots,N. \tag{2.217}$$

Often condition (2.217) is taken as the definition of the sufficient statistic in sequential decision problems. However, the condition (2.217) may not be satisfied while the statistic will be sufficient in the sense of the given definition.

Example 2.7.1. Assume that

$$R_n^{\mathrm{st}}(\mathbf{X}_1^n) = \gamma_n(\mathcal{S}_n) + \beta_n(\mathbf{X}_1^n), \quad \tilde{R}_n^N(\mathbf{X}_1^n) = \tilde{\gamma}_n^N(\mathcal{S}_n) + \beta_n(\mathbf{X}_1^n).$$

By Theorem 2.7.5, $T_N^0(\mathbf{X}) = T_N^0(\mathcal{S})$. However, $R_n^N(\mathbf{X}_1^n) = \min\{\tilde{\gamma}_n^N(\mathcal{S}_n), \gamma_n(\mathcal{S}_n)\} + \beta_n(\mathbf{X}_1^n)$ is a function of all observations $\mathbf{X}_1^n$, so that the condition (2.217) does not hold, but nevertheless $\mathcal{S}_n$ is the sufficient statistic. We will deal with this situation in the problem of quickest change detection in Subsection 7.1.1.

We call the sequence of statistics $\{\mathcal{S}_n\}_{n\ge 1}$ *transitive* if there exists a function $\varphi_n(\cdot)$ such that

$$\mathcal{S}_{n+1} = \varphi_n(\mathcal{S}_n, X_{n+1}), \quad n \ge 1 \text{ w.p. } 1. \tag{2.218}$$

The following conditions are sufficient for $\mathcal{S}_n$ to be a sufficient statistic in the sequential decision problem:

(i) The sequence $\{\mathcal{S}_n\}_{n\ge 1}$ is transitive;

(ii) The conditional pdf writes

$$p_{n+1}(X_{n+1} \mid \mathbf{X}_1^n) = p_{n+1}(X_{n+1} \mid \mathcal{S}_n), \quad n \ge 1 \text{ w.p. } 1; \tag{2.219}$$

(iii) The following equality holds:

$$R_n^{\mathrm{st}}(\mathbf{X}_1^n) = R_n^{\mathrm{st}}(\mathcal{S}_n), \quad n \ge 1 \text{ w.p. } 1. \tag{2.220}$$

Indeed, we need to show that the MAPR depends on the observations only via $\mathcal{S}_n$ for every $n \le N$, i.e., that the equalities (2.217) hold. For $n = N$, this is immediate since $R_N^N = R_N^{\mathrm{st}}(\mathcal{S}_N)$ by (2.220). For $n = N-1$ we have

$$R_{N-1}^N(\mathcal{S}_{N-1}) = \min\left\{R_{N-1}^{\mathrm{st}}(\mathcal{S}_{N-1}), \tilde{R}_{N-1}^N(\mathcal{S}_{N-1})\right\},$$

where

$$\tilde{R}_{N-1}^N(\mathcal{S}_{N-1}) = \mathsf{E}[R_N^{\mathrm{st}}(\mathcal{S}_N) \mid \mathbf{X}_1^{N-1}] = \int_{\mathcal{X}_N} R_N^{\mathrm{st}}(\varphi_N(\mathcal{S}_{N-1}, X_N)) p_N(X_N \mid \mathcal{S}_{N-1}) \, \mathrm{d}X_N.$$

Continuing by induction we obtain

$$R_N^N(\mathcal{S}_N) = R_N^{\mathrm{st}}(\mathcal{S}_N), \quad R_n^N(\mathcal{S}_n) = \min\left\{R_n^{\mathrm{st}}(\mathcal{S}_n), \tilde{R}_n^N(\mathcal{S}_n)\right\}, \quad n = 1,\dots,N-1,$$

$$\tilde{R}_n^N(\mathcal{S}_n) = \mathsf{E}[R_{n+1}^N(\mathcal{S}_{n+1}) \mid \mathcal{S}_n] = \int_{\mathcal{X}_{n+1}} R_n^N(\varphi_{n+1}(\mathcal{S}_n, X_{n+1})) p_{n+1}(X_{n+1} \mid \mathcal{S}_n) \, \mathrm{d}X_{n+1}.$$

The optimal stopping time is the first entry time in the stopping region $A_n^N = \{\mathcal{S}_n : R_n^{\mathrm{st}}(\mathcal{S}_n) \le R_n^N(\mathcal{S}_n)\}$:

$$T_N^0(\mathcal{S}) = \min\left\{1 \le n \le N : \mathcal{S}_n \in A_n^N\right\},$$

and if $T_N^0 = n$, then an optimal terminal decision $d_n^0(\mathcal{S}_n)$ is made.

Note that conditions (2.218) and (2.219) imply that

$$p_{n+1}(\mathcal{S}_{n+1} \mid \mathbf{X}_1^n) = p_{n+1}(\mathcal{S}_{n+1} \mid \mathcal{S}_n), \quad n \ge 1. \tag{2.221}$$

The latter property has been used by Bahadur [20] as the definition of the transitive sequence of statistics.

Furthermore, if the family of conditional distributions corresponding to $p(\mathbf{X}_1^n \mid \mathcal{S}_1, \dots, \mathcal{S}_n), n \ge 1$ is complete, then the sequence of statistics $\{\mathcal{S}_n\}$ is a Markov process, i.e.,

$$p_{n+1}(\mathcal{S}_{n+1} \mid \mathcal{S}_1, \dots, \mathcal{S}_n) = p_{n+1}(\mathcal{S}_{n+1} \mid \mathcal{S}_n), \quad n \ge 1.$$

Therefore, the postulated conditions (2.218) and (2.219) are in fact very close to the Markov property of the sufficient statistic $\mathcal{S}_n$. In other words, in most applications an efficient solution of the sequential decision problem can be obtained if there is a Markov sufficient statistic.

2.7.3.4 The Degenerate Case

There are problems where an optimal sequential rule is a fixed-size sample rule, i.e., an optimal stopping time $T^0 = n_0$ does not depend on the observations and can be determined in advance. This is the case when the APR R_n^{st} does not depend on the observations.

Theorem 2.7.6. *Suppose that the APR $R_n^{\mathrm{st}}(\mathbf{X}_1^n) = R_n^{\mathrm{st}}$ is a deterministic function of n. Then the N-truncated and nontruncated decision rules are degenerate and are based on the fixed number of stages n_0^N and n_0, respectively, where n_0^N and n_0 are determined from*

$$\rho_N^0 = R_{n_0^N}^{\mathrm{st}} = \min_{1 \le n \le N} R_n^{\mathrm{st}}, \quad \rho^0 = R_{n_0}^{\mathrm{st}} = \min_{n \ge 1} R_n^{\mathrm{st}}. \tag{2.222}$$

The proof is trivial and follows directly from Theorems 2.7.4 and 2.7.5.

Fixed-size decision rules are a subclass of sequential rules, but it is convenient to consider them separately and call non-sequential decision rules. Usually a sequential rule is optimal and if so, then it is of interest to compare it with the best non-sequential rule. The following example illustrates the situation where the optimal rule is non-sequential.

Example 2.7.2. Assume we are interested in sequential estimation of the Gaussian random parameter $\theta_n \sim \mathcal{N}(\mu_n, \sigma_{\theta n}^2)$ observed in Gaussian noise (or measurement error) $\xi_n \sim \mathcal{N}(\lambda_n, \sigma_{\xi n}^2)$:

$$X_n = \theta_n + \xi_n, \quad n \ge 1.$$

Let the loss function have the form

$$L_n(\theta_n, u_n, \mathbf{X}_1^n) = |\theta_n - u_n|^\alpha + C(n), \quad n \ge 1, \alpha > 0,$$

where $u_n = \hat{\theta}_n$ is an estimate of θ_n and $C(n)$ is the cost of the n stages of the experiment.

It is straightforward to check that the posterior distribution $\mathsf{P}(\theta_n \le y | \mathbf{X}_1^n)$ is normal with mean $\bar{\theta}_n(\mathbf{X}_1^n) = \mathsf{E}(\theta_n \mid \mathbf{X}_1^n)$ and variance $v_n^2 = \mathsf{E}[(\theta_n - \bar{\theta}_n(\mathbf{X}_1^n))^2 \mid \mathbf{X}_1^n]$ that depends on n but does not depend on the observations $\mathbf{X}_1^n$. As a result the posterior risk

$$R_n^{\mathrm{st}} = (2v_n^2)^{\alpha/2} \frac{\Gamma\left(\frac{\alpha+1}{2}\right)}{\sqrt{\pi}} + C(n)$$

is a deterministic function of n, so that the optimal procedure is non-sequential. Here $\Gamma(\cdot)$ is a Gamma-function.

If $\theta_n = \theta \sim \mathcal{N}(\mu, \sigma_\theta^2)$ is the normal random variable and $\{\xi_n\}_{n\ge 1}$ are iid normal $\mathcal{N}(0, \sigma_\xi^2)$, then the posterior variance is equal to

$$v_n^2 = \frac{\sigma_\xi^2}{n + \sigma_\xi^2/\sigma_\theta^2}.$$

In particular, for $\alpha = 2$ and $C(n) = cn$, we obtain that the optimal number of observations is

$$n_0 = \left\lceil \sqrt{\sigma_\xi^2/c} - \sigma_\xi^2/\sigma_\theta^2 \right\rceil,$$

and the optimal estimator is the posterior mean

$$\bar{\theta}_{n_0}(\mathbf{X}_1^{n_0}) = \mu + \frac{1}{n_0 + \sigma_\xi^2/\sigma_\theta^2} \sum_{k=1}^{n_0} (X_k - \mu).$$

In the N-truncated case, $n_0^N = \min(N, n_0)$.

2.7.3.5 *The Monotone Case in Sequential Decision Problems*

It follows from Theorems 2.7.4 and 2.7.5 that finding optimal truncated and nontruncated sequential rules is a fairly difficult problem. For instance, in order to obtain a solution of the N-truncated problem, that is, an explicit form of the N-truncated rule, we have to perform a sequence of minimizations and averaging, starting with the last stage N. In other words, we have to obtain an explicit solution of the recursion (2.207) for the MAPR. For large N, this is a difficult task that usually cannot be solved analytically, and its numerical solution is also time consuming. A one-step ahead approximation, where $\tilde{R}_n^N = \mathsf{E}[R_{n+1}^N \mid \mathbf{X}_1^n]$ is replaced with $R_n^* = \mathsf{E}[R_{n+1}^{\mathrm{st}} \mid \mathbf{X}_1^n]$, is usually not adequate when N is large, and the corresponding rule has a very poor performance. Indeed, in this approximation it is assumed that one stops at the stage $n+1$ w.p. 1, which in general is a lousy assumption.

However, it appears that under certain monotonicity conditions the one-step approximation provides the optimal solution. We now proceed with considering this case.

Let $R_n^*(\mathbf{X}_1^n)$ be defined as above, and let

$$\mathcal{X}_n^* = \{\mathbf{X}_1^n : R_n^{\mathrm{st}}(\mathbf{X}_1^n) \le R_n^*(\mathbf{X}_1^n)\}$$

be the stopping region when a one-step ahead approximation for the MAPR is being used. We say that the *monotone case* takes place if

$$\mathcal{X}_1^* \subseteq \mathcal{X}_2^* \cdots \subseteq \mathcal{X}_n^* \subseteq \cdots; \quad \bigcup_{n=1}^{\infty} \mathcal{X}_n^* = \Omega. \tag{2.223}$$

In the N-truncated case, ∞ is replaced with N in the latter union.

Therefore, in the monotone case,

$$\{\mathbf{X}_1^n \in \mathcal{X}_n^*\} \Rightarrow \{\mathbf{X}_1^{n+1} \in \mathcal{X}_{n+1}^*\}, \quad n \ge 1,$$

which is not true in general.

Recall that

$$\mathcal{X}_n^{\mathrm{st}(N)} = \{\mathbf{X}_1^n : R_n^{\mathrm{st}}(\mathbf{X}_1^n) \le \tilde{R}_n^N(\mathbf{X}_1^n)\} \quad \text{and} \quad \mathcal{X}_n^{\mathrm{st}} = \{\mathbf{X}_1^n : R_n^{\mathrm{st}}(\mathbf{X}_1^n) \le \tilde{R}_n(\mathbf{X}_1^n)\}$$

are the optimal stopping regions for N-truncated and nontruncated rules, respectively.

Clearly, the following inclusions hold in general

$$\mathcal{X}_n^{\mathrm{st}} \subseteq \mathcal{X}_n^*, \quad n \ge 1. \tag{2.224}$$

If we now show that in the monotone case the reverse inclusions hold, that is,

$$\mathcal{X}_n^{\mathrm{st}} \supseteq \mathcal{X}_n^*, \quad n \ge 1, \tag{2.225}$$

then

$$\mathcal{X}_n^{\mathrm{st}} = \mathcal{X}_n^*, \quad n \ge 1.$$

Therefore, the stopping time

$$T^0 = \inf\left\{n \ge 1 : R_n^{\mathrm{st}}(\mathbf{X}_1^n) \le R_n^*(\mathbf{X}_1^n)\right\} \tag{2.226}$$

deserves special attention from the optimality standpoint. However, certain additional conditions that guarantee the existence of $\mathsf{E}[R_{T^0}^{\mathrm{st}}]$ similar to those used in Theorem 2.7.5 are needed for its optimality. See, e.g., a counterexample in Chow et al. [101, Sec. 3.5].

The following is a basic result in the monotone case. Recall that we restrict the attention to the class of stopping times $\mathcal{M}$ that includes only Markov times finite w.p. 1, $\mathsf{P}(T < \infty) = 1$.

Theorem 2.7.7 (Optimality in the monotone case). **(i)** *Consider the nontruncated case. Suppose we are in the monotone case* (2.223) *and that* $\mathsf{E}[\sup_n R_n^{\mathrm{st}}] < \infty$. *Then the sequential decision rule* $\delta^0 = (d^0, T^0)$ *where* T^0 *is defined in* (2.226) *is the optimal rule and*

$$\rho^0 = \inf_\delta \rho(\delta) = \rho(\delta^0) = \mathsf{E}[R_1(X_1)].$$

The condition $\mathsf{E}[\sup_n R_n^{\mathrm{st}}] < \infty$ *may be relaxed into* (2.215).

(ii) *Consider the N-truncated case. Let* $T_N^0 = \min(N, T^0)$. *Suppose that* $\mathsf{E}[\max_{1\le n\le N} R_n^{\mathrm{st}}] < \infty$ *and that*

$$\mathcal{X}_n^* \subseteq \mathcal{X}_{n+1}^*, \quad n = 1, \dots, N-1; \quad \bigcup_{n=1}^N \mathcal{X}_n^* = \Omega.$$

Then the sequential decision rule $\delta_N^0 = (\mathbf{d}_N^0, T_N^0)$ *is optimal and*

$$\rho_N^0 = \inf_\delta \rho(\delta) = \rho_N(\delta_N^0) = \mathsf{E}[R_1^N].$$

Proof. (i) By Theorem 2.7.5, the stopping time

$$\tau_0 = \inf\left\{n \ge 1 : R_n^{\mathrm{st}}(\mathbf{X}_1^n) \le \tilde{R}_n(\mathbf{X}_1^n)\right\} = \inf\left\{n \ge 1 : \mathbf{X}_1^n \in \mathcal{X}_n^{\mathrm{st}}\right\}$$

is optimal whenever $\mathsf{E}[\sup_n R_n^{\mathrm{st}}] < \infty$. Hence, if $\mathcal{X}_n^{\mathrm{st}} = \mathcal{X}_n^*$ for all $n \ge 1$, then $\tau_0 = T^0$ and T^0 is optimal. To prove the former equality we need only to prove that along with the obvious inclusions (2.224) the reverse inclusions (2.225) hold.

If the inclusions (2.223) hold, then for every $n \ge 1$ we have

$$\begin{aligned}\{\mathbf{X}_1^n \in \mathcal{X}_n^*\} &\Rightarrow \{\mathbf{X}_1^{n+1} \in \mathcal{X}_{n+1}^*\} \Rightarrow \{\mathbf{X}_1^{n+1} \in \mathcal{X}_{n+1}^{\mathrm{st}}\} \\ &\Rightarrow R_{n+1}(\mathbf{X}_1^{n+1}) = R_{n+1}^{\mathrm{st}}(\mathbf{X}_1^{n+1}) \Rightarrow \mathsf{E}[R_{n+1}(\mathbf{X}_1^{n+1}) \mid \mathbf{X}_1^n] = \mathsf{E}[R_{n+1}^{\mathrm{st}}(\mathbf{X}_1^{n+1}) \mid \mathbf{X}_1^n],\end{aligned}$$

so that

$$\tilde{R}_n(\mathbf{X}_1^n) = R_n^*(\mathbf{X}_1^n) \quad \text{for } \mathbf{X}_1^n \in \mathcal{X}_n^*,$$

and hence

$$\left\{R_n^*(\mathbf{X}_1^n) = R_n^{\mathrm{st}}(\mathbf{X}_1^n)\right\} \Rightarrow \left\{\tilde{R}_n(\mathbf{X}_1^n) = R_n^{\mathrm{st}}(\mathbf{X}_1^n)\right\}, \quad n \ge 1.$$

The latter implications are equivalent to (2.225).

The fact that $\rho^0 = \mathsf{E}[R_1(X_1)]$ follows from Theorem 2.7.5.

(ii) In the N-truncated case, it is sufficient to apply the same argument and Theorem 2.7.4. □

At the first glance one may think that in the monotone case the minimal average risk is computed as $\rho^0 = \mathsf{E}[\min(R_1^{\mathrm{st}}, R_1^*)]$, since in (2.226) the risk $\tilde{R}_n$ is replaced by R_n^*. However, this is not true. Indeed, the equality $R_n^* = \tilde{R}_n$ holds true only in the stopping region $\mathcal{X}_n^{\mathrm{st}}$ but not in the continuation region $\tilde{R}_n(\mathbf{X}_1^N) < R_n^{\mathrm{st}}(\mathbf{X}_1^n)$, where $\tilde{R}_n(\mathbf{X}_1^N) < R_n^*(\mathbf{X}_1^n)$. For this reason, for the minimal average risk we have the same equality as in the general case, i.e., $\rho^0 = \mathsf{E}[\min(R_1^{\mathrm{st}}, \tilde{R}_1)]$.

Unfortunately, the monotone case is not typical in sequential decision problems of primary interest in this book – hypothesis testing and changepoint detection. Nevertheless, some interesting nontrivial examples may be found in Chow *et al.* [101] and Tartakovsky [452].

Example 2.7.3. Let $\{X_n\}_{n\ge 1}$ be iid having the common uniform distribution on $[0, 1/\theta]$ with the pdf $f_\theta(X_n) = \theta\,\mathbb{1}_{\{0\le X_n \le 1/\theta\}}$, where θ is a positive scale parameter to be estimated. Let $\hat{\theta}_n(\mathbf{X}_1^n)$ be an estimator of θ based on the n observations. Assume that the loss is quadratic in the estimation error and the cost of the n-th observation is c_n, so that the total loss on the n-th stage is

$$L_n(\theta, \hat{\theta}_n, \mathbf{X}_1^n) = (\hat{\theta}_n - \theta)^2 + \sum_{i=1}^n c_i.$$

Next, suppose that θ is random with the uniform distribution on $[0, b]$, $b > 0$, i.e., $\pi_0(\theta) = b^{-1}\mathbb{1}_{\{0\le\theta\le b\}}$. Write $X_{(n)} = \max_{1\le i\le n} X_i$ and $\mathcal{S}_n = \min(b, 1/X_{(n)})$.

The posterior density is

$$\pi_n(\theta) = \frac{(n+1)\theta^n}{\mathcal{S}_n^{n+1}}\,\mathbb{1}_{\{0\le\theta\le\mathcal{S}_n\}},$$

and the optimal estimator if one stops at the n-th stage is the posterior mean

$$\hat{\theta}_n^0 = \mathsf{E}[\theta \mid \mathbf{X}_1^n] = \frac{n+1}{n+2}\mathcal{S}_n.$$

The MAPR associated with stopping at n is equal to

$$R_n^{\mathrm{st}}(\mathcal{S}_n) = \frac{n+1}{(n+2)^2(n+3)}\mathcal{S}_n^2 + \sum_{i=1}^n c_i. \tag{2.227}$$

Further computations yield

$$p_{n+1}(X_{n+1} \mid \mathbf{X}_1^n) = \begin{cases} \frac{n+1}{n+2}\mathcal{S}_n, & X_{n+1} \in [0, 1/\mathcal{S}_n], \\ \frac{n+1}{n+2}\frac{1}{\mathcal{S}_n^{n+1}X_{n+1}^{n+2}}, & X_{n+1} > 1/\mathcal{S}_n, \\ 0 & X_{n+1} < 0. \end{cases} \tag{2.228}$$

Using (2.227) and (2.228), we obtain

$$R_n^*(\mathcal{S}_n) = \frac{n+1}{(n+2)^3(n+3)}[(n+1)\mathcal{S}_n^2 + 1] + \sum_{i=1}^{n+1} c_i. \tag{2.229}$$

Let

$$V_n^* = \{\mathcal{S}_n : R_n^{\mathrm{st}}(\mathcal{S}_n) \le R_n^*(\mathcal{S}_n)\} \quad \text{and} \quad A_n = \left(\frac{c_{n+1}(n+2)^3(n+3)}{n+1} + 1\right)^{1/2}$$

It follows from (2.227) and (2.229) that $V_n^{\mathrm{st}} = \{\mathcal{S}_n \le A_n\}$. Since $\{\mathcal{S}_n\}$ is non-increasing ($\mathcal{S}_{n+1} \le \mathcal{S}_n$), it is easily seen that we are in the monotone case as long as $\{c_n\}$ is non-decreasing, in particular when $c_n = c$, $n \ge 1$. Therefore, when the cost of the observation stage does not decrease with n the optimal nontruncated sequential estimation rule is

$$T^0 = \inf\{n \ge 1 : \mathcal{S}_n \le A_n\}, \quad \hat{\theta}_{T^0}^0 = \frac{T^0+1}{T^0+2}\mathcal{S}_{T^0}.$$

2.7.4 Non-Bayesian Optimality Criteria: Minimax Decision Rules

In most applied statistical problems losses consist of two components — the cost of the experiment $C(n)$ (or of the delay in making a final decision) and the loss related to the accuracy of the terminal decision $L^*(\theta_n, u_n)$:

$$L_n(\theta_n, u_n, \mathbf{X}_1^n) = L^*(\theta_n, u_n) + C(n).$$

Often one is interested in minimizing an average cost under a constraint imposed on the average accuracy, e.g.,

$$\mathsf{E}[C(T^0)] = \inf_{\delta \in \mathbf{C}(G)} \mathsf{E}[C(T)] \quad \text{subject to } \mathsf{E}[L^*(\theta_T, u_T)] \le G,$$

where $\mathbf{C}(G)$ is a class of admissible decision rules satisfying the latter constraint and possibly some other natural constraints.

If the prior distribution is given, this conditional Bayesian problem is solved essentially analogously as the unconditional one, since it can be always reduced to an unconditional one. The problem arises when prior distributions are not known, which is usually the case in practice.

Consider the case where the parameter $\theta_n = \theta \in \Theta$ does not vary in time, where Θ is a parameter space (or space of nature states). If the prior distribution $\pi(\theta)$ is unknown, we may try to find a solution that minimizes the conditional risk (for the fixed θ)

$$r_\theta(\delta) = \mathsf{E}_\theta[L^*(\theta, u_n)] + \mathsf{E}_\theta[C(T)] \quad \text{for all } \theta \in \Theta_0 \subset \Theta,$$

where E_θ is the conditional expectation for the fixed θ. Such a uniformly optimal solution may not exist; in which case we may resort to the minimax criterion $\inf_\delta \sup_{\theta \in \Theta} r_\theta(\delta)$.

In the rest of this subsection, we consider arbitrary decision rules that may be either sequential or non-sequential (fixed sample size). All results are equally true for both cases, while fixed sample size examples are simpler.

A decision rule δ^* is said to be *minimax* if

$$\sup_{\theta \in \Theta} r_\theta(\delta^*) = \inf_\delta \sup_{\theta \in \Theta} r_\theta(\delta), \tag{2.230}$$

where the value on the right side is called the *minimax risk* (or *value* in game theory) or *upper value* of the minimax decision problem. Obviously, the decision rule δ^* is minimax if, and only if,

$$r_{\tilde{\theta}}(\delta^*) \le \sup_{\theta \in \Theta} r_\theta(\delta) \quad \text{for all } \tilde{\theta} \in \Theta \text{ and all } \delta.$$

Often the minimax problem may be reduced to a Bayesian one with a certain specially designed (so called *least favorable*) prior, or more generally to the generalized Bayesian problems with improper priors. Specifically, let

$$\rho(\pi, \delta) = \int_\Theta r_\theta(\delta)\, \pi(\mathrm{d}\theta)$$

be the integrated (average Bayes) risk corresponding to the prior distribution π and rule δ. A prior distribution π^* is said to be *least favorable* if

$$\inf_\delta \rho(\pi^*, \delta) = \sup_\pi \inf_\delta \rho(\pi, \delta), \tag{2.231}$$

where the value on the right side is called the *maximin risk* or *lower value* of the minimax decision problem. Clearly, the prior distribution π^* is least favorable if, and only if,

$$\rho(\pi^*, \tilde{\delta}) \ge \inf_\delta \rho(\pi, \delta) \quad \text{for all } \tilde{\delta} \text{ and all } \pi.$$

However, there may not be a least favorable distribution, especially when we deal with infinite sets Θ. In this case, we need two extensions in the notion of Bayes rules. The first one generalizes the notion of a prior distribution to include nonfinite measures on Θ, so-called *improper* prior distributions. A rule δ_0 is said to be a *generalized Bayes rule* if there is a measure π on Θ such that the average risk $\rho(\pi, \delta)$ takes on a finite minimum value when $\delta = \delta_0$.

Another extension is related to the limit of Bayes rules. A rule δ is said to be a *limit of Bayes rules* δ_c as $c \to \infty$ if $\delta_c(x) \to \delta(x)$ for almost all x.

As an example, consider the problem of finding the best estimator of the mean $\theta \in \Theta = \mathbb{R}^1$ of the normal random variable X with unit variance based on the sample size 1, assuming the Lebesgue improper prior $\pi(\mathrm{d}\theta) = \mathrm{d}\theta$ and quadratic loss $(\delta(X) - \theta)^2$. Then formally the posterior density of θ after observing X is normal with mean X and variance one, $p(\theta|X) = (2\pi)^{-1/2} e^{-(\theta-X)^2/2}$, so that the optimal generalized Bayes estimator is $\delta_0(X) = X$ and the Bayes risk in this scenario equals 1. Also, if the prior distribution of θ is taken to be the normal distribution with mean 0 and variance c^2, then the optimal Bayes estimator is $\delta_c(X) = Xc^2/(1+c^2)$ and it converges to X as $c \to \infty$. Clearly, the normal prior becomes more and more uniform when $c \to \infty$, so that this is expected.

These two notions are extremely useful in decision theory in general and for finding minimax or almost minimax rules in particular.

A typical scenario characteristic for game theory is that nature is dedicated to ruining the player. For this reason, the fundamental issue in game theory is establishing most general conditions under which it is true that the upper value coincides with the lower value, i.e., minimax is the same as maximin (equilibrium),

$$\inf_{\delta} \sup_{\pi} \rho(\pi, \delta) = \sup_{\pi} \inf_{\delta} \rho(\pi, \delta). \tag{2.232}$$

Since $\sup_{\pi} \rho(\pi, \delta) = \sup_{\theta \in \Theta} r_\theta(\delta)$ for every δ, the right-hand side of (2.232) is the upper value of the game, as defined above. Thus, the player can always choose a strategy that his average loss will not be greater than the upper value regardless of what prior distribution nature is using. On the other hand, nature would of course use a least favorable distribution to maximize a loss of the player and making it at least the lower value regardless of what strategy the player decides to use. Hence, in game theory it is of fundamental importance to know when equilibrium (2.232) holds, in which case this common quality is called the value of the game. This is the subject of the (fundamental) minimax theorem of game theory.

At the same time, in decision theory "nature" is not dedicated to ruining a decision-maker (statistician), so establishing general conditions under which fundamental equality (2.232) holds is of little interest. However, the minimax theorem allows one to find practical methods of discovering minimax rules as well as establishing the fact that minimax rules are also Bayes rules (more generally generalized Bayes rules). In fact, if (2.232) holds and the least favorable (proper) distribution π^* exists, then a minimax rule is π^*-Bayes, or generalized Bayes if π^* is an improper distribution.

We now discuss these issues in more detail. Note first that since $\sup_{\pi} \rho(\pi, \delta) = \sup_{\theta \in \Theta} r_\theta(\delta)$ for every δ, the equivalent definition of a minimax rule is: the decision rule is said to be minimax if $\sup_{\pi} \rho(\pi, \delta^*) = \overline{V}$, where $\overline{V} = \inf_{\delta} \sup_{\pi} \rho(\pi, \delta)$ is the upper value. Recall that prior distribution π^* is least favorable if $\inf_{\delta} \rho(\pi^*, \delta) = \underline{V}$, where $\underline{V} = \sup_{\pi} \inf_{\delta} \rho(\pi, \delta)$ is the lower value. The inequality $\underline{V} \le \overline{V}$ always holds because

$$\inf_{\tilde{\delta}} \rho(\pi, \tilde{\delta}) \le \sup_{\tilde{\pi}} \rho(\tilde{\pi}, \delta) \quad \text{for all } \pi, \delta.$$

The minimax theorem states that under certain general conditions the equality $\underline{V} = \overline{V} = V$ holds, where V is the value of the game, and that a least favorable distribution π^* as well as a minimax decision rule δ^* exist. A particular version of this theorem may be found in Ferguson's book [143, p. 82, Theorem 2.9.1].

A more important practical question is how to find or guess a minimax rule. The following theorem answers this important question giving practical prescriptions on how a minimax rule may be discovered. The first assertion of the theorem makes a connection of minimax and Bayes rules with respect to the least favorable distributions when these distributions are proper. Since this is not always so, especially in cases where the set Θ is infinite, the second assertion connects the minimax rules with generalized Bayes rules which are the limit of Bayes rules. Finally, the third assertion states that the minimax rules are equalizers and Bayes or generalized Bayes. A decision rule δ is said to be *an equalizer rule* if its conditional risk $r_\theta(\delta) = C$ is constant for all $\theta \in \Theta$.

Theorem 2.7.8. **(i)** *If δ^* is π^*-Bayes and*

$$r_\theta(\delta^*) \le \rho(\pi^*, \delta^*) \quad \textit{for all } \theta \in \Theta, \tag{2.233}$$

then the equality (2.232) *holds, δ^* is a minimax decision rule, and π^* is the least favorable distribution.*

(ii) *Let $\{\delta_c\}$ be a sequence of decision rules indexed by c and let $\{\pi_c\}$ be a sequence of prior distributions. Let δ_c be a π_c-Bayes rule. If $\rho(\pi_c, \delta_c) \to C$ as $c \to \infty$, if $\delta_c \to \delta^*$, and if $r_\theta(\delta^*) \le C$ for all $\theta \in \Theta$, then the equality* (2.232) *holds and δ^* is a minimax decision rule.*

(iii) *If δ^* is an equalizer and either Bayes or generalized Bayes, then it is a minimax decision rule.*

Proof. (i) Since δ^* is π^*-Bayes,

$$\rho(\pi^*, \delta^*) \le \inf_\delta \rho(\pi^*, \delta) \le \underline{V}.$$

Now, $\sup_{\theta \in \Theta} r_\theta(\delta^*) \ge \overline{V}$, so that by condition (2.233)

$$\overline{V} \le \sup_{\theta \in \Theta} r_\theta(\delta^*) \le \rho(\pi^*, \delta^*),$$

which yields $\overline{V} \le \underline{V}$. Since the reverse inequality is always true, this implies that $\underline{V} = \overline{V} = V$, proving the assertion (i).

(ii) The same argument as in the proof of (i) yields

$$\overline{V} \le \sup_{\theta \in \Theta} r_\theta(\delta^*) \le C = \lim_{c \to \infty} \rho(\pi_c, \delta_c) \le \underline{V}.$$

(iii) By the equalizer property, $r_\theta(\delta^*) = C$ for all $\theta \in \Theta$. If δ^* is a Bayesian rule, then $\rho(\delta^*, \pi^*) = C$ and by (i) it is minimax. If δ^* is a generalized Bayesian rule, then the assertion follows from (ii). This completes the proof. □

Theorem 2.7.8 has very important practical implications. In particular, it follows that in order to find a minimax rule we may use the following method: find an equalizer decision rule and check if it is either Bayes or generalized Bayes. Another method is to guess the least favorable distribution and then find a rule δ^* which is Bayes in cases where the least favorable distribution is proper or generalized Bayes if it is improper. After that, one is also encouraged to check that it is indeed minimax, e.g., to verify that it is an equalizer rule.

For illustration, consider again the normal example of estimating the mean θ of a normal population with unit variance (based on a single observation) with quadratic loss. Therefore, we have $L(\theta, \delta) = (\theta - \delta)^2$, $\Theta = \mathbb{R}^1$. The estimate $\delta(X) = X$ is equalizer since $r_\theta(X) = \mathsf{E}_\theta(\theta - X)^2 = 1$ for all $\theta \in \Theta$. Above, when discussing the generalized Bayes concept, we showed that this estimate is a generalized Bayesian rule (with respect to the Lebesgue prior measure), so that from Theorem 2.7.8(iii) we conclude that X is the minimax estimate. On the other hand, it is intuitively appealing that the least favorable prior distribution in this problem is improper uniform on real line. Also, as we established above, if the prior distribution $\pi_c(\theta)$ is normal with mean 0 and variance c^2, then the optimal Bayes estimator $\delta_c(X) = Xc^2/(1+c^2)$ converges to X as $c \to \infty$, while $\pi_c(\theta)$ converges to the improper uniform. Furthermore, the Bayes risk $\rho(\pi_c, \delta_c)$ converges to 1. Hence, X is the minimax estimate by Theorem 2.7.8(ii).

We close this subsection noting that it is not always possible to find strictly minimax decision rules. In this case, asymptotic approaches are in order, e.g., when the sample size or the expected sample size goes to infinity. It then may be possible to show that the conditional risk of a rule δ^* is asymptotically equal to

$$r_\theta(\delta^*) = C + \varepsilon(\theta),$$

where the first (leading) term C does not depend on θ (typically goes to infinity) and the second term $\varepsilon(\theta)$, which depends on θ, is either bounded (i.e., $O(1)$) or vanishes (i.e., $o(1)$). In this case the rule δ^* is nearly (asymptotically) minimax if it is generalized Bayes (or nearly generalized Bayes). We will deal with such problems in Sections 5.5, 8.3, 8.4, 8.5, and 9.2.

2.8 Information

In this section, we introduce two definitions of information, the Kullback–Leibler information and the Fisher information. The first information concept plays an important role in establishing asymptotic optimality of sequential hypothesis tests and changepoint detection procedures, as well as in the performance analysis of the algorithms. The second information concept plays an important role in asymptotic expansions of the likelihood ratio processes which are useful for designing locally optimal hypothesis testing and change detection algorithms.

2.8.1 *Kullback–Leibler Information*

We begin with defining the Kullback–Leibler information that plays an important role in statistical decision theory and especially in hypothesis testing [240].

Let P_0 and P_1 be two probability distributions of a random variable X with densities $p_0(x)$ and $p_1(x)$ with respect to some non-degenerate sigma-finite measure $\mu(\mathrm{d}x)$, and let E_i denote expectation with respect to P_i.

Definition 2.8.1 (Kullback–Leibler information). The *Kullback–Leibler (K–L) information* between the distributions P_0 and P_1 is defined as

$$I(\mathsf{P}_0,\mathsf{P}_1) = \int_{\mathscr{X}} \log\left[\frac{p_0(x)}{p_1(x)}\right] p_0(x)\mu(\mathrm{d}x) = \mathsf{E}_0\left[\log\frac{p_0(X)}{p_1(X)}\right]. \tag{2.234}$$

Often $I(\mathsf{P}_0,\mathsf{P}_1)$ is called the *K–L information number.*

The K–L information number has the following properties: $I(\mathsf{P}_0,\mathsf{P}_1) \ge 0$ and $I(\mathsf{P}_0,\mathsf{P}_1) = 0$ if, and only if, $p_0(x) = p_1(x)$ μ-almost everywhere. The K–L number is finite whenever P_0 is absolutely continuous with respect to P_1, i.e., the set where $p_1(x) = 0$ and $p_0(x) \neq 0$ has the μ-measure zero. The larger $I(\mathsf{P}_0,\mathsf{P}_1)$ is, the more distant the distributions P_0 and P_1 are. For example, if P_θ is the normal distribution with mean θ and unit variance, then $I(\mathsf{P}_0,\mathsf{P}_\theta) = \theta^2/2$. In this particular case the K–L number is symmetric, $I(\mathsf{P}_0,\mathsf{P}_\theta) = I(\mathsf{P}_\theta,\mathsf{P}_0) = \theta^2/2$. An example of when it is asymmetric is the exponential distribution with the pdf $p_\theta(x) = 1/(1+\theta)\ \exp\{-x/(1+\theta)\}$, $\theta, x > 0$. Then

$$I(\mathsf{P}_0,\mathsf{P}_\theta) = \log(1+\theta) - \theta/(1+\theta) \quad \text{but} \quad I(\mathsf{P}_\theta,\mathsf{P}_0) = \theta - \log(1+\theta).$$

Remark 2.8.1. Sometimes the K–L information number $I(\mathsf{P}_0,\mathsf{P}_1)$ is referred to as the Kullback–Leibler "distance" to stress its fundamental role in defining the separability of the distributions P_0 and P_1. Nevertheless, the K–L information number is not a distance in the topological sense because in general it is not symmetric; see, e.g., the above exponential example.

Similarly, in the case of a *random process* $\{X_n\}_{n\in\mathbb{Z}_+}$, we define the *local K–L information* contained in a sample $\mathbf{X}_1^n = (X_1, \dots, X_n)^\top$ of size n by

$$I_n(\mathsf{P}_0, \mathsf{P}_1) = \frac{1}{n} \int \log \left[\frac{p_0(\mathbf{X}_1^n)}{p_1(\mathbf{X}_1^n)} \right] p_0(\mathbf{X}_1^n) \mu_n(\mathrm{d}\mathbf{X}_1^n)$$
$$= \frac{1}{n} \sum_{i=1}^n \int Z_i \, p_0(\mathbf{X}_1^n) \mu_n(\mathrm{d}\mathbf{X}_1^n) = \frac{1}{n} \sum_{i=1}^n \mathsf{E}_0[Z_i], \tag{2.235}$$

where $Z_i = \log[p_0(X_i|\mathbf{X}_1^{i-1})/p_1(X_i|\mathbf{X}_1^{i-1})]$. The K–L information is defined to be the limit

$$I(\mathsf{P}_0, \mathsf{P}_1) = \lim_{n\to\infty} I_n(\mathsf{P}_0, \mathsf{P}_1) \tag{2.236}$$

if it exists, positive and finite. Note that if the variables $X_1, X_2, \dots$ are iid, then $I(\mathsf{P}_0, \mathsf{P}_1) = I_n(\mathsf{P}_0, \mathsf{P}_1)$ for every $n \ge 1$.

The following asymptotic approximation is of interest. As $n \to \infty$, by the law of large numbers under P_0 we have

$$\frac{1}{n} \sum_{i=1}^n Z_i \xrightarrow[n\to\infty]{\mathsf{P}_0} \lim_{n\to\infty} \frac{1}{n} \sum_{i=1}^n \int \log \frac{p_0(X_i|\mathbf{X}_1^{i-1})}{p_1(X_i|\mathbf{X}_1^{i-1})} p_0(\mathbf{X}_1^n) d\mathbf{X}_1^n = \frac{1}{n} \sum_{i=1}^n \mathsf{E}_0[Z_i],$$

assuming that the limit on the right-hand side exists, positive and finite. Therefore, for a sufficiently large n, the K–L information (2.235) can be approximated by

$$I(\mathsf{P}_0, \mathsf{P}_1) \approx \frac{1}{n} \sum_{i=1}^n Z_i. \tag{2.237}$$

In a variety of applications normalization by n in (2.235) and (2.237) is the right factor, which means that the mean of the cumulative LLR $\sum_{i=1}^n Z_i$ is approximately a linear function $I(\mathsf{P}_0, \mathsf{P}_1)n$ for large n. However, this is not always the case. For example, consider the following nonstationary model

$$X_n = \theta n^r + \xi_n, \quad n \ge 1, \; r \ne 0,$$

where $\xi_n, n = 1, 2, \dots$ are iid zero-mean normal random variables with unit variance. Then

$$\frac{1}{n} \sum_{i=1}^n \int Z_i \, p_0(\mathbf{X}_1^n) \mu_n(\mathrm{d}\mathbf{X}_1^n) = \frac{\theta^2}{2n} \sum_{i=1}^n i^{2r},$$

which, as $n \to \infty$, approaches ∞ for any $r > 0$ and 0 for any $r < 0$. Therefore, with the normalizing factor n^{-1} the definition of the K–L information does not make any sense. But if we normalize by $\psi(n) = n^{2r+1}$ for $r > -1/2$ and by $\log(1+n)$ for $r = -1/2$, then a positive and finite limit, $\theta^2/2$, will exist.

Therefore, more generally, we define the K–L information in the non-iid case as the limit (2.236) (if it exists) with

$$I_n(\mathsf{P}_0, \mathsf{P}_1) = \frac{1}{\psi(n)} \sum_{i=1}^n \int Z_i \, p_0(\mathbf{X}_1^n) \mu_n(\mathrm{d}\mathbf{X}_1^n) = \frac{1}{\psi(n)} \sum_{i=1}^n \mathsf{E}_0[Z_i], \tag{2.238}$$

where $\psi(n)$ is a positive increasing function, $\psi(\infty) = \infty$.

As we mentioned above, the K–L information is not a distance, since it is not symmetric. A symmetrized version, which is a distance, $\mathbf{J}(\mathsf{P}_0, \mathsf{P}_1) = I(\mathsf{P}_0, \mathsf{P}_1) + I(\mathsf{P}_1, \mathsf{P}_0)$ is called the *Kullback–Leibler divergence*. However, often the K–L information is also called the K–L divergence.

2.8.2 *Fisher Information*

Consider a parametric family of distributions $\{\mathsf{P}_\theta, \theta \in \Theta\}$ with density $p_\theta(x)$. For simplicity, we start with a scalar parameter θ, and then we extend to the vector case.

2.8.2.1 Scalar Parameter

Let

$$\ell_\theta(x) = \log p_\theta(x) \tag{2.239}$$

denote the log-likelihood function, where $p_\theta(x)$ is the parametrized probability density or the mass probability function of the random variable X. The log-likelihood ratio (LLR) is

$$Z(x) = \log \frac{p_{\theta_1}(x)}{p_{\theta_0}(x)} = \ell_{\theta_1}(x) - \ell_{\theta_0}(x). \tag{2.240}$$

Definition 2.8.2 (Efficient score). When θ is a scalar parameter, we define the *efficient score* for the *random variable* X as the quantity

$$s(x) = \frac{\partial \ell_\theta(x)}{\partial \theta}. \tag{2.241}$$

Similarly, the *efficient score for a sample of size n,* $\mathbf{X}_1^n = (X_1, \dots, X_n)$*, of a random process* $(X_n)_{n\geq 1}$ is defined by

$$S_n = \frac{\partial \ell_\theta(\mathbf{X}_1^n)}{\partial \theta}. \tag{2.242}$$

If we denote

$$s_i = \frac{\partial \ell_\theta(x_i|\mathbf{X}_1^{i-1})}{\partial \theta}, \tag{2.243}$$

we get

$$S_n = \sum_{i=1}^{n} s_i. \tag{2.244}$$

This concept was introduced by Fisher [148]. When the dependence on θ is of interest, we very often use the following notation

$$Z^*(x) = \log \frac{p_\theta(x)}{p_{\theta^*}(x)}, \quad s^*(x) = \left. \frac{\partial \ell_\theta(x)}{\partial \theta} \right|_{\theta=\theta^*}.$$

Now it is obvious that the efficient score is zero mean when $\theta = \theta^*$:

$$\mathsf{E}_{\theta^*}[s^*(X)] = 0.$$

A simple approximation to $\mathsf{E}_\theta[s^*(X)]$ for small values of $\theta - \theta^*$ will be given after the definition of the Fisher information. Note that in the particular case of the mean of a Gaussian random variable $X \sim \mathcal{N}(\theta, \sigma^2)$, the efficient score is $s(X) = (X - \theta)/\sigma^2$.

We now introduce the notion of Fisher information.

Definition 2.8.3 (Fisher information). Let us assume that $X \sim \mathsf{P}_\theta$, where $\mathcal{P} = \{\mathsf{P}_\theta\}_{\theta \in \Theta}$ is a parametric family of distributions. Suppose that distribution P_θ has the pdf $p_\theta(x)$. The *Fisher information* about θ contained in the *random variable* X is

$$\mathcal{F}(\theta) = \mathsf{E}_\theta \left(\frac{\partial \log p_\theta(X)}{\partial \theta} \right)^2 (> 0) \tag{2.245}$$

$$= \mathsf{var}_\theta \left(\frac{\partial \log p_\theta(X)}{\partial \theta} \right) = \mathsf{var}_\theta(s(X)). \tag{2.246}$$

Similarly, the Fisher information about the parameter θ contained in a sample of size n, $\mathbf{X}_1^n = (X_1, \dots, X_n)$, of a *random process* $(X_n)_{n\geq 1}$ is

$$\mathcal{F}_n(\theta) = \frac{1}{n} \mathsf{var}_\theta \left(\frac{\partial \log p_\theta(\mathbf{X}_1^n)}{\partial \theta} \right) = \frac{1}{n} \mathsf{var}_\theta (S_n), \tag{2.247}$$

and in this case the Fisher information is defined to be the limit (if it exists)

$$\mathcal{F}(\theta) = \lim_{n\to\infty} \mathcal{F}_n(\theta). \tag{2.248}$$

In several chapters of this book, we make use of the following properties of the LLR $Z(X)$ and the efficient score $s(X)$ for probability densities that do not need to belong to an exponential family of distributions, and especially do not need to be Gaussian.

Lemma 2.8.1 (Approximations for the expectation of LLR). *Let p_θ be any continuous probability density twice differentiable in θ. For small values of $(\theta_1 - \theta_0)$, we have*

$$\mathsf{E}_{\theta_0}(Z) \approx -\frac{1}{2}\,\mathcal{F}(\theta_0)\,(\theta_1-\theta_0)^2, \tag{2.249}$$

$$\mathsf{E}_{\theta_1}(Z) \approx \frac{1}{2}\,\mathcal{F}(\theta_1)\,(\theta_1-\theta_0)^2 \approx \frac{1}{2}\,\mathcal{F}(\theta_0)\,(\theta_1-\theta_0)^2 \approx -\mathsf{E}_{\theta_0}(Z), \tag{2.250}$$

$$\mathsf{E}_{\theta_0}(Z^2) \approx \mathcal{F}(\theta_0)\,(\theta_1-\theta_0)^2 \approx \mathsf{E}_{\theta_1}(Z^2). \tag{2.251}$$

The proof of this lemma relies upon the following second-order Taylor expansion of ℓ_θ:

$$Z(x) = \ell_{\theta_1}(x) - \ell_{\theta_0}(x) \approx (\theta_1-\theta_0)\,\left.\frac{\partial \ell_\theta(x)}{\partial\theta}\right|_{\theta=\theta_0} + \frac{1}{2}(\theta_1-\theta_0)^2\,\left.\frac{\partial^2 \ell_\theta(x)}{\partial\theta^2}\right|_{\theta=\theta_0}. \tag{2.252}$$

Taking the expectation E_{θ_0} of both sides of (2.252) leads to (2.249) [79] because

$$\mathsf{E}_{\theta_0}\left(\left.\frac{\partial \ell_\theta(X)}{\partial\theta}\right|_{\theta=\theta_0}\right) = 0. \tag{2.253}$$

The first approximation in (2.250) is deduced by symmetry, and the second one follows from the approximation $\mathcal{F}(\theta_1) \approx \mathcal{F}(\theta_0)$. Finally, raising (2.252) to the power 2 and keeping only second-order terms results in (2.251).

Lemma 2.8.2 (Approximation for the expectation of the efficient score). *Let p_θ be as before. For small values of $(\theta - \theta^*)$, we have*

$$\mathsf{E}_\theta(S^*) \approx \mathcal{F}(\theta^*)\,(\theta-\theta^*). \tag{2.254}$$

The proof of this lemma relies upon the first term of the Taylor expansion (2.252), which we rewrite as

$$Z^* = \ell_\theta - \ell_{\theta^*} \approx (\theta-\theta^*)\,s^*. \tag{2.255}$$

Using (2.250), we obtain (2.254).

The Fisher and K–L information do have strong connections in several particular cases of interest in this book. In the parametric case, we write $I(\theta_0,\theta_1)$ for the K–L information $I(\mathsf{P}_{\theta_0},\mathsf{P}_{\theta_1})$. A basic general connection is the following. From the approximation (2.249) and the definition (2.234), we find that for small values of $(\theta_1-\theta_0)$ the K–L information can be approximated as

$$I(\theta_0,\theta_1) \approx \frac{1}{2}(\theta_1-\theta_0)^2\,\mathcal{F}(\theta_0). \tag{2.256}$$

Note again that this approximation is fairly general and does not require that the distribution belong to an exponential family.

2.8.2.2 Vector Case

The previous results can be extended to the case of a vector observation in a trivial manner, and to the case of a vector parameter $\theta = (\theta_1,\ldots,\theta_r)$ of dimension r. In the latter case, the efficient score is defined as an r-dimensional vector-gradient of $\theta \mapsto \ell_\theta(x)$,

$$s = \nabla[\ell_\theta] = \left[\frac{\partial \ell_\theta(x)}{\partial\theta_1},\ldots,\frac{\partial \ell_\theta(x)}{\partial\theta_r}\right]^\top, \tag{2.257}$$

and the Fisher information is defined as an $r \times r$ matrix with elements

$$\mathcal{F}_{ij}(\theta) = \mathsf{E}_\theta \left[\frac{\partial \ell_\theta(X)}{\partial \theta_i} \frac{\partial \ell_\theta(X)}{\partial \theta_j} \right] = \int_{-\infty}^{+\infty} \frac{\partial \ell_\theta(x)}{\partial \theta_i} \frac{\partial \ell_\theta(x)}{\partial \theta_j} p_\theta(x) \mathrm{d}x, \tag{2.258}$$

which is an obvious extension of (2.245). If the observation X is a vector, the elements of the Fisher information matrix are expressed as multiple integrals.

Similarly, the efficient score for a sample of size n of a *random process* $(X_n)_{n\geq 1}$ is defined by

$$S_n = \frac{\partial \ell_\theta(\mathbf{X}_1^n)}{\partial \theta}. \tag{2.259}$$

Denoting

$$s_i = \frac{\partial \ell_\theta(X_i | \mathbf{X}_1^{i-1})}{\partial \theta}, \tag{2.260}$$

we get

$$S_n = \sum_{i=1}^n s_i. \tag{2.261}$$

The Fisher information matrix is then defined as an $r \times r$ matrix

$$\mathcal{F}_n(\theta) = \frac{1}{n} \mathsf{E}_\theta (S_n S_n^\top). \tag{2.262}$$

The following lemma provides useful approximations [71, 240].

Lemma 2.8.3 (Approximations). *For any continuous probability density twice differentiable in θ and for small values of $\|\theta_1 - \theta_0\|_2$, we have the following approximations:*

$$\mathsf{E}_{\theta_1}(Z) \approx \frac{1}{2} (\theta_1 - \theta_0)^\top \mathcal{F}(\theta_0)(\theta_1 - \theta_0), \tag{2.263}$$

$$\mathsf{E}_{\theta_0}(Z^2) \approx (\theta_1 - \theta_0)^\top \mathcal{F}(\theta_0)(\theta_1 - \theta_0), \tag{2.264}$$

$$\mathsf{E}_\theta(S^*) \approx \mathcal{F}(\theta^*)(\theta - \theta^*). \tag{2.265}$$

Furthermore, the K–L information (2.234) *can be approximated as*

$$I(\theta_0, \theta_1) \approx \frac{1}{2} (\theta_0 - \theta_1)^\top \, \mathcal{F}(\theta_0) \, (\theta_0 - \theta_1), \tag{2.266}$$

which is the extension of (2.256).

It is also of interest that the maximum likelihood estimate $\hat{\theta}$ of θ minimizes the K–L information $I(\theta, \hat{\theta})$ [4, 240]. This can be seen as a consequence of (2.266).

2.9 Hypothesis Testing: Performance Evaluation and Optimality Criteria

In Section 2.7, we considered general aspects of sequential decision-making and optimal stopping. In particular, we established that a sequential strategy $\delta = (T, d)$ consists of an optimal stopping rule T and a terminal decision $d = d_T$ that is made after stopping observations at time T. The sample sizes of sequential rules are random. A special case of this general decision rule is a *fixed sample size* (FSS) or nonsequential rule when the sample size is fixed in advance, i.e., $T = N$ w.p. 1, where N is a fixed positive number.

The focus of this book is not on general sequential decision strategies but on hypothesis testing and changepoint detection.[6] In this section, we outline the main ideas of fixed sample size hypothesis testing.

[6]Change detection is also a specific hypothesis testing problem, as we will see in the following chapters.

2.9.1 Notation and Main Criteria

Let us introduce the main definitions and criteria of the hypothesis testing framework.

Definition 2.9.1 (Simple hypothesis). We call any assumption concerning the distribution P, which can be reduced to a single value in the space of probability distributions, a *simple hypothesis* H.

Assume we are given M distinct distributions $\mathsf{P}_0, \dots, \mathsf{P}_{M-1}$, and let $\mathbf{X}_1^n \in \mathcal{X}^n$ be a n-size sample generated by one of these distributions, where $\mathcal{X}^n$ is the sample space. The problem of hypothesis testing is to decide which distribution is the true one, i.e., to test M simple hypotheses "$H_i : \mathsf{P} = \mathsf{P}_i$", $i = 0, 1, \dots, M-1$. The parametric version of this testing problem is the following. Let $\mathsf{P}_\theta \in \mathcal{P} = \{\mathsf{P}_\theta\}_{\theta \in \Theta}$ and consider the simple hypotheses $H_i : \theta = \theta_i$, where $\theta_0, \dots, \theta_{M-1}$ are fixed points in the parameter space.

There are two types of statistical tests: nonrandomized and randomized.

Definition 2.9.2 (Nonrandomized test). We call any measurable mapping $d_n : \mathcal{X}^n \to \{H_0, \dots, H_{M-1}\}$ a *nonrandomized statistical test* for testing between hypotheses $H_0, \dots, H_{M-1}$.

In other words, $d_n = d(\mathbf{X}_1^n)$ is a random variable which takes values in the set of hypotheses. If $d_n = k$, then the hypothesis H_k is accepted. In the parametric case, we simply say that $\theta = \theta_k$. We also call the function $d_n = d(\mathbf{X}_1^n)$ a *decision function*. If we forget for the time being about the random character of the observation vector $\mathbf{X}_1^n$, we recognize that the mapping $d_n : \mathcal{X}^n \to \{H_0, \dots, H_{M-1}\}$ is completely deterministic. Giving the nonrandomized decision function d_n is equivalent to giving a partition of $\mathcal{X}^n$ into M non-intersecting Borel sets $\Omega_0, \dots, \Omega_{M-1}$ inside which exactly one of the hypotheses is accepted. When $M = 2$, the set $\Omega_1 \subset \mathcal{X}^n$ is said to be the *critical region* of the test d_n.

Sometimes, for instance in the case of discrete distributions, the absence of randomization represents a serious obstacle to get an optimal rule in a particular class with certain constraints since these constraints cannot be satisfied due to discreteness. To overcome such a problem a more general type of test can be introduced.

Definition 2.9.3 (Randomized test). We call any probability distribution $d_n^*(\mathbf{X}_1^n, H_i)$ defined on $\mathcal{H} = \{H_0, \dots, H_{M-1}\}$, where $d_n^*(\mathbf{X}_1^n, H_i)$ is interpreted for a given vector $\mathbf{X}_1^n$ as the probability that the hypothesis H_i will be chosen, a *randomized statistical test* for testing between hypotheses $H_0, \dots, H_{M-1}$.

The quality of a hypothesis test d_n can be measured by the following sets of error probabilities :

$$
\begin{aligned}
\alpha_{ij}(d_n) &= \mathsf{P}(\mathbf{X}_1^n \in \Omega_j | H_i) = \mathsf{P}_i(d_n(\mathbf{X}_1^n) = j), \quad i \neq j; \\
\alpha_i(d_n) &= \mathsf{P}(\mathbf{X}_1^n \notin \Omega_i | H_i) = \mathsf{P}_i(d_n(\mathbf{X}_1^n) \neq i|), \quad i = 0, 1, \dots. M-1,
\end{aligned}
\tag{2.267}
$$

where $\alpha_{ij}(d_n)$ is the probability of accepting the hypothesis H_j when in reality H_i is true and $\alpha_i(d_n)$ is the probability of rejecting the hypothesis H_i when it is true. Clearly, $\alpha_i = \sum_{j \neq i} \alpha_{ij}$. Here we used the notation $\mathsf{P}_i(\cdot) = \mathsf{P}(\cdot | H_i)$ for the probability measure when the hypothesis H_i is true.

Let $W = [w_{ij}]$ be a $(M \times M)$ matrix of positive weights, except for w_{ii} which may be zero. The following weighted error probabilities

$$
\beta_j(d_n) = \sum_{\substack{i=0 \\ i \neq j}}^{M-1} w_{ij} \alpha_{ij}(d_n)
$$

are also of interest.

Obviously, the probabilities of errors should be relatively small. However, because the sample is of finite size, all the probabilities α cannot be made arbitrary small. The question then arises of how to compare two different statistical tests. We now consider three possible optimality criteria. For the sake of simplicity, in the following we omit the subscript n in d_n.

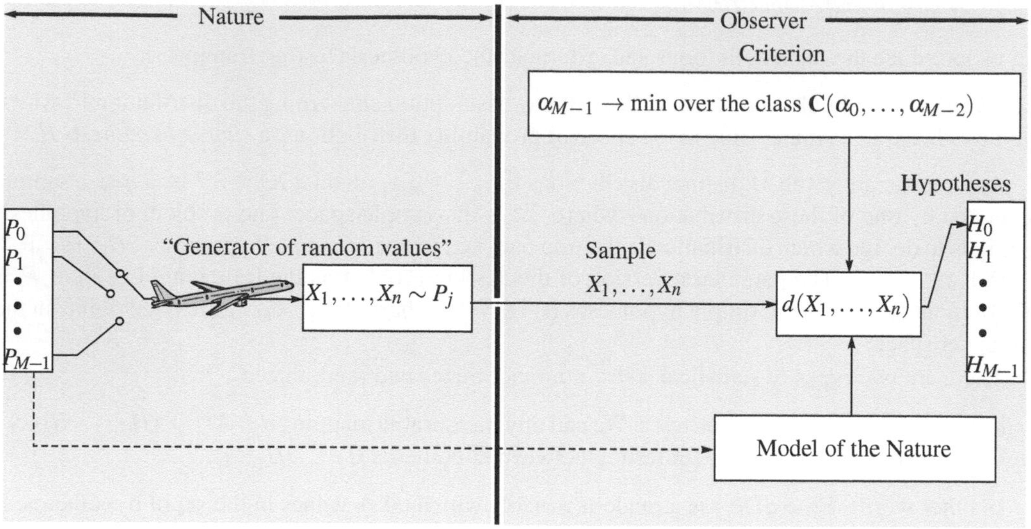

Figure 2.1: *The philosophical background of the most powerful approach.*

2.9.1.1 *Most Powerful Approach*

Define a class of tests with $M-1$ given upper bounds for probabilities of errors of rejecting the true hypotheses:

$$\mathbf{C}(\alpha_0,\ldots,\alpha_{M-2}) = \{d : \alpha_i(d) \le \alpha_i, \quad i = 0,\ldots,M-2\} \tag{2.268}$$

Definition 2.9.4 (Most powerful test). We say that the test $d^* \in \mathbf{C}(\alpha_0,\ldots,\alpha_{M-2})$ is the *most powerful* (MP) in this class if, for all $d \in \mathbf{C}(\alpha_0,\ldots,\alpha_{M-2})$, the following inequality holds for the M-th error probability:

$$\alpha_{M-1}(d^*) \le \alpha_{M-1}(d). \tag{2.269}$$

The philosophical background of the most powerful approach is "the nature is neutral." This situation is shown in Figure 2.1.

2.9.1.2 *Bayesian Approach*

Assume that the hypotheses H_i have known *a priori* probabilities $\pi_i = \mathsf{P}(H_i)$, $i = 0,\ldots,M-1$, $\sum_{i=0}^{M-1} \pi_i = 1$. Also, let $L_{ij} = L(d = i, H_j)$ be the loss associated with accepting the hypothesis H_i when H_j is true. As in the previous section, we define the average (integrated) risk

$$\rho(d) = \sum_{i=0}^{M-1}\sum_{j=0}^{M-1} L(d(\mathbf{X}_1^n) = i, H_j)\mathsf{P}(H_i)\mathsf{P}(d(\mathbf{X}_1^n) = i|H_j) = \sum_{i=0}^{M-1}\sum_{j=0}^{M-1} L_{ij}\pi_i\alpha_{ij}(d).$$

Definition 2.9.5 (Bayes test). The test $\bar{d}$ is said to be a *Bayes test* if it minimizes the average risk $\rho(d)$ for the given *a priori* probabilities $(\pi_i)_{i=0,\ldots,M-1}$ and losses $(L_{ij})_{i,j=0,\ldots,M-1}$, i.e.,

$$\bar{d} = \arg\inf_d \rho(d),$$

where the infimum is taken over all FSS tests.

In the particular case of the $0-1$ loss function when $L_{ij} = 1$ for $i \ne j$ and 0 for $i = j$, the average

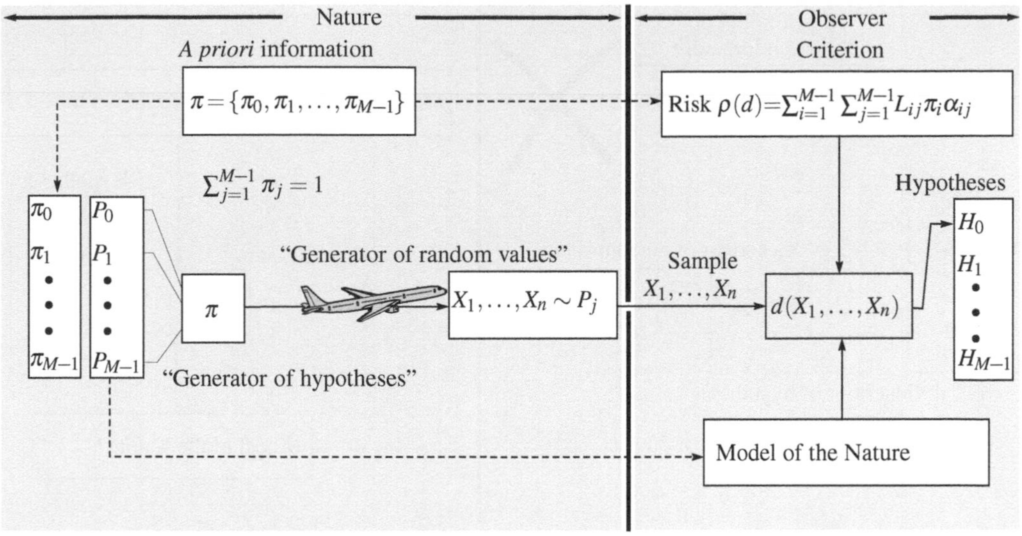

Figure 2.2: *The philosophical background of the Bayesian approach.*

risk is nothing but the average (weighted) probability of error $\bar{\alpha}(d)$ with the weights equal to the prior probabilities π_i:

$$\rho(d)=\bar{\alpha}(d)=\sum_{i=0}^{M-1}\sum_{\substack{j=0\\ j\neq i}}^{M-1}\pi_i\alpha_{ij}(d)=\sum_{i=0}^{M-1}\pi_i\alpha_i(d). \tag{2.270}$$

Note that if $w_{ij}=\pi_i$ for all $j\neq i$, then

$$\bar{\alpha}(d)=\sum_{j=0}^{M-1}\beta_j(d).$$

The philosophical background of the Bayesian approach is "the nature is gentle." This situation is shown in Figure 2.2.

2.9.1.3 Minimax Approach

For the given losses (L_{ij}) the conditional risk (loss) conditioned on hypothesis H_j being true is defined as

$$\rho_i(d)=\sum_{j=0}^{M-1}L_{ij}\alpha_{ij}(d).$$

In particular, for the $0-1$ losses, the conditional risk is equal to the probability of rejecting the hypotheses H_i erroneously, $\rho_i(d)=\alpha_i(d)$.

Define the *maximum error probability* of a test d

$$\alpha_{\max}(d)=\max_{i=0,\dots,M-1}\alpha_i(d). \tag{2.271}$$

Definition 2.9.6 (Minimax test). We say that the test $\tilde{d}$ is *minimax* if it minimizes $\alpha_{\max}(d)$, i.e.,

$$\alpha_{\max}(\tilde{d})=\inf_d \alpha_{\max}(d), \tag{2.272}$$

where the infimum is taken over all fixed sample size tests.

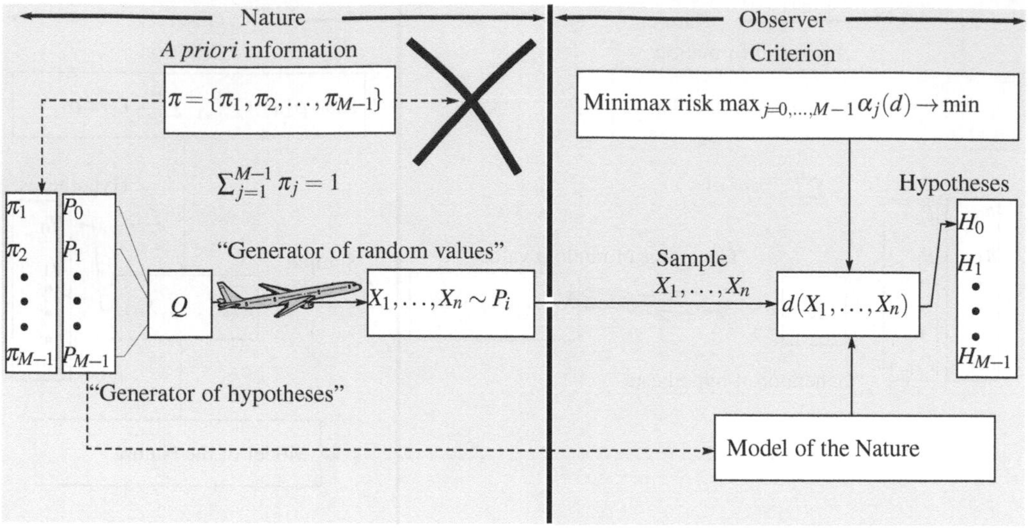

Figure 2.3: *The philosophical background of the minimax approach.*

The philosophical background of the minimax approach is "the nature is cruel." This situation is shown in Figure 2.3.

Minimax and Bayes tests do have strong connections. Specifically, sometimes it is possible to find the *a priori* distribution $\{\pi_i\}$ which maximizes the average error probability of all Bayesian tests. Such a set of *a priori* probabilities is called a *least favorable distribution*. Then the Bayes test which corresponds to this least favorable distribution is the minimax test. Similarly MP and Bayes tests also have strong connections. For an appropriate choice of the *a priori* probabilities $\{\pi_i\}$, a Bayes test $\bar{d}$ is an MP test in a certain class $\mathbf{C}(\alpha_0, \dots, \alpha_{M-2})$. See [79, 261] for further details.

2.9.2 *Testing between Two Simple Hypotheses*

Testing between two simple hypotheses H_0 and H_1 is an important special case of the problem of testing between M simple hypotheses. In this case, the error probability of type I (or false alarm probability), $\alpha_0(d)$, is called the *size* of the test or the *level of significance* of the test d. The value $\beta(d) = 1 - \alpha_1(d)$ is called the *power* of the test, where $\alpha_1(d)$ is the error probability of type II (or missed detection probability). Let us define the *critical function* $0 \le d(\mathbf{X}_1^n) \le 1$, for which we use the same notation as for the statistical test, because this function completely characterizes the test d. The test assigns a real number $d(\mathbf{X}_1^n)$ $(0 \le d(\mathbf{X}_1^n) \le 1)$, to the conditional probability $\mathsf{P}_\theta(\text{test } d(\mathbf{X}_1^n) \text{ accepts } H_1 | \mathbf{X}_1^n)$ for each point $\mathbf{X}_1^n \in \Omega_1$. This function defines the probability of acceptance of the hypothesis H_1. In the nonrandomized case, $d(\mathbf{X}_1^n)$ takes only two possible values, $d(\mathbf{X}_1^n) \in \{0, 1\}$; in the randomized case, it takes any value between 0 and 1, $d(\mathbf{X}_1^n) \in [0, 1]$.

The size and the power of the test d can obviously be computed as follows:

$$\alpha_0(d) = \mathsf{E}_0[d(\mathbf{X}_1^n)], \tag{2.273}$$

$$\beta(d) = 1 - \alpha_1(d) = \mathsf{E}_1[d(\mathbf{X}_1^n)], \tag{2.274}$$

where E_0 and E_1 are the expectations under the hypotheses H_0 and H_1, respectively.

2.9.2.1 MP Test

The following theorem follows from the fundamental Neyman–Pearson lemma whose proof can be found in [79, 261].

Theorem 2.9.1. *Consider the problem of testing two hypotheses* $H_0 : \mathsf{P} = \mathsf{P}_0$ *and* $H_1 : \mathsf{P} = \mathsf{P}_1$, *where* P_0 *and* P_1 *are two probability distributions with densities (or with pms)* p_0 *and* p_1 *with respect to some probability measure* μ, *for example* $\mathsf{P}_0 + \mathsf{P}_1$. *Let*

$$\Lambda(\mathbf{X}_1^n) = \frac{p_1(\mathbf{X}_1^n)}{p_0(\mathbf{X}_1^n)} \tag{2.275}$$

be the likelihood ratio between these hypotheses. The MP test is given by

$$d^*(\mathbf{X}_1^n) = \begin{cases} 1 & \textit{if } \Lambda(\mathbf{X}_1^n) > h \\ p & \textit{if } \Lambda(\mathbf{X}_1^n) = h \\ 0 & \textit{if } \Lambda(\mathbf{X}_1^n) < h, \end{cases} \tag{2.276}$$

where the threshold $h = h(\alpha)$ *and the randomizing probability* p *are selected so that*

$$\alpha_0(d^*) = \mathsf{E}_0[d^*(\mathbf{X}_1^n)] = \alpha. \tag{2.277}$$

Therefore, Theorem 2.9.1 implies that the MP test is necessarily based on the *likelihood ratio* (LR) between the hypotheses H_1 and H_0. The LR test $d^*(\mathbf{X}_1^n)$ given by (2.276) is usually referred to as the *Neyman–Pearson (NP) test.*

Remark 2.9.1. Note that equation (2.277) can be written as

$$\mathsf{P}_0\{\Lambda(\mathbf{X}_1^n) > h(\alpha)\} + p\,\mathsf{P}_0\{\Lambda(\mathbf{X}_1^n) = h(\alpha)\} = \alpha.$$

Randomization on the boundary h is needed only in certain discrete cases when the equation

$$\mathsf{P}_0\{\Lambda(\mathbf{X}_1^n) \ge h(\alpha)\} = \alpha \tag{2.278}$$

does not have any solution. If the P_0-distribution of the LR is continuous, this equation always has a solution and randomization is not required, i.e., in this case the MP test has the form

$$d^*(\mathbf{X}_1^n) = \begin{cases} 1 & \text{if } \Lambda(\mathbf{X}_1^n) \ge h \\ 0 & \text{if } \Lambda(\mathbf{X}_1^n) < h \end{cases} \tag{2.279}$$

with the threshold h found from the equation (2.278).

The NP test is also optimal with respect to the two other above-mentioned criteria, namely Bayes and minimax [79, 261].

2.9.2.2 Bayesian Test

Theorem 2.9.2. *Consider the two-hypothesis Bayesian testing problem with the* $0-1$ *loss function and the prior distribution* $\pi_0 = \pi$, $\pi_1 = 1-\pi$, *where* $0 < \pi < 1$. *The test* $\bar{d}_\pi(\mathbf{X}_1^n)$ *minimizing the average error probability* $\bar{\alpha}_\pi(d) = \pi\alpha_0(d) + (1-\pi)\alpha_1(d)$ *is given by*

$$\bar{d}_\pi(\mathbf{X}_1^n) = \begin{cases} 1 & \textit{if} \quad \Lambda(\mathbf{X}_1^n) \ge \pi/(1-\pi) \\ 0 & \textit{otherwise} \end{cases}.$$

The proof can be found, e.g., in [79, 261].

Unlike the MP test the optimal Bayes test is always nonrandomized.

2.9.2.3 Minimax Test

Theorem 2.9.3. *In the two-hypothesis testing problem the minimax test* $\tilde{d}(\mathbf{X}_1^n)$ *minimizing the maximal error probability* $\max\{\alpha_0(d), \alpha_1(d)\}$ *is the NP test* (2.276) *with the threshold h chosen so that* $\alpha_0(d^*) = \alpha_1(d^*)$.

The proof can be found, e.g., in [79, 261].

Note that for the same reason as in Remark 2.9.1 randomization is required only in certain discrete cases when the equality $\alpha_0(d^*) = \alpha_1(d^*)$ cannot be attained for $p = 1$.

2.9.2.4 Examples

Example 2.9.1 (Mean in a Gaussian sequence). Consider testing the mean value θ in an independent Gaussian sequence $\mathbf{X}_1^n$ with variance σ^2. The two hypotheses are $H_i : \theta = \theta_i, i = 0, 1$. It follows from Theorem 2.9.1 that the MP test with significance level α can be written as

$$d^* : \Lambda(\mathbf{X}_1^n) = \prod_{i=1}^{n} \frac{p_{\theta_1}(X_i)}{p_{\theta_0}(X_i)} = \prod_{i=1}^{n} \frac{\varphi\left(\frac{X_i - \theta_1}{\sigma}\right)}{\varphi\left(\frac{X_i - \theta_0}{\sigma}\right)} \underset{H_0}{\overset{H_1}{\gtrless}} A(\alpha), \tag{2.280}$$

where $\varphi(x) = 1/\sqrt{2\pi}\, \exp\{-x^2/2\}$ is the Gaussian density, or equivalently as

$$d^* : \log \Lambda(\mathbf{X}_1^n) = \frac{\theta_1 - \theta_0}{\sigma^2} \left[\sum_{i=1}^{n} X_i - n\, \frac{\theta_1 + \theta_0}{2}\right] \underset{H_0}{\overset{H_1}{\gtrless}} h = \log A(\alpha). \tag{2.281}$$

Therefore, $d^*(S_n) = \mathbb{1}_{\{S_n \geq h^*(\alpha)\}}$, where $S_n = \sum_{i=1}^n X_i$ is the sufficient statistic, and the threshold $h^*(\alpha)$ is found from the equation $\mathsf{P}_{\theta_0}(S_n \geq h^*) = \alpha$.

The error probabilities $\alpha_0(d^*)$ and $\alpha_1(d^*)$ are

$$\alpha_0(d^*) = \mathsf{P}_{\theta_0}\{\log \Lambda(S_n) \geq h\} = 1 - \Phi\left(\frac{h + n\, I(\theta_0, \theta_1)}{\sqrt{2n\, I(\theta_0, \theta_1)}}\right),$$
$$\alpha_1(d^*) = \mathsf{P}_{\theta_1}\{\log \Lambda(S_n) < h\} = \Phi\left(\frac{h - n\, I(\theta_0, \theta_1)}{\sqrt{2n\, I(\theta_0, \theta_1)}}\right),$$

where $I(\theta_0, \theta_1) = (\theta_0 - \theta_1)^2/2\sigma^2$ and $\Phi(x) = \int_{-\infty}^{x} 1/\sqrt{2\pi}\, \exp\{-y^2/2\}\, dy$. Often, especially in engineering applications, an operating characteristic of a test d is represented in the form of the power $\beta(d) = 1 - \alpha_1(d)$ as a function of the Type I error probability α_0. Such a function is usually called the *Receiver Operating Characteristic* (ROC). The ROC of the NP test is given by

$$\beta(\alpha_0, d^*) = 1 - \Phi\left(-\sqrt{2n\, I(\theta_0, \theta_1)} + \Phi^{-1}(1 - \alpha_0)\right).$$

A typical behavior of the ROC curves is shown in Figure 2.4 for $n = 10$. Note that if $I(\theta_0, \theta_1) = 0$, then $\beta(d^*) = 1 - \alpha_1(d^*) = \alpha_0(d^*)$.

Example 2.9.2 (AR Gaussian sequence). Often the observations $X_1, X_2, \ldots, X_n$ are dependent. Let $\mathbf{X}_1^n$ be a stationary autoregressive (AR) Gaussian sequence

$$(X_1, \ldots, X_p) \sim \mathcal{N}(0, \Gamma_p), \;\; X_k = \sum_{i=1}^{p} a_i X_{k-i} + \varepsilon_k, \;\; \varepsilon_k \sim \mathcal{N}(0, \sigma^2), \;\; k \geq p+1, \tag{2.282}$$

where $\theta^\top = (a_1, \ldots, a_p, \sigma)$ is a parameter vector, $\Gamma_p = \mathsf{Toepl}\{R_0, R_1, \ldots, R_{p-1}\}$ is a Toeplitz matrix

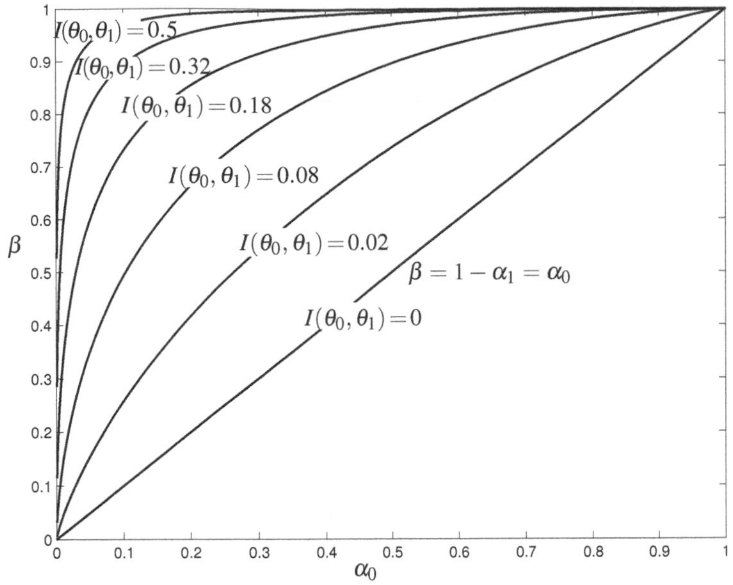

Figure 2.4: *ROC curves for different values of the K–L distance* $I(\theta_0,\theta_1)$ *that varies from* 0 *to* 0.5.

of size p, $R_j = \mathsf{E}(X_k X_{k+j})$ and $(\varepsilon_k)_{k\geq 1}$ is the sequence of iid Gaussian variables. In the case of two simple hypotheses $H_i : \theta = \theta_i$, $(i = 0,1)$, the LLR is given by

$$\log \Lambda(\mathbf{X}_1^n) = \log \frac{\varphi_{\theta_1}(\mathbf{X}_1^n)}{\varphi_{\theta_0}(\mathbf{X}_1^n)},$$

where

$$\log \varphi_\theta(\mathbf{X}_1^n) = -\frac{n}{2}\log 2\pi\sigma^2 + \frac{1}{2}\log\det \mathbf{T}_p^{-1}(\theta) - \frac{1}{2\sigma^2} S_n(\mathbf{X}_1^n,\theta), \tag{2.283}$$

$$S_n(\mathbf{X}_1^n,\theta) = (\mathbf{X}_1^p)^\top \mathbf{T}_p^{-1}(\theta)\mathbf{X}_1^p + \sum_{k=p+1}^{n}\left(X_k - \sum_{i=1}^{p} a_i X_{k-i}\right)^2, \mathbf{T}_p^{-1}(\theta) = \sigma^2\Gamma_p^{-1},$$

$\mathbf{X}_1^p = (X_1,\dots,X_p)^\top$, and the inverse of the Toeplitz matrix is computed with the aid of the Göhberg–Semmencul formula:

$$\mathbf{T}_p^{-1}(\theta) = \mathbf{T}_p^{-1}(a_1,\dots,a_p) = T_1 T_1^\top - T_2 T_2^\top \tag{2.284}$$

by using the AR parameters (2.282) and

$$T_1 = \begin{pmatrix} 1 & 0 & \dots & 0 & 0 \\ -a_1 & 1 & \dots & 0 & 0 \\ \vdots & \vdots & \vdots & \vdots & \vdots \\ -a_{p-1} & -a_{p-2} & \dots & -a_1 & 1 \end{pmatrix} \tag{2.285}$$

$$T_2 = \begin{pmatrix} -a_p & 0 & \dots & 0 & 0 \\ -a_{p-1} & -a_p & \dots & 0 & 0 \\ \vdots & \vdots & \vdots & \vdots & \vdots \\ -a_1 & -a_2 & \dots & -a_{p-1} & -a_p \end{pmatrix}. \tag{2.286}$$

In the particular case of AR(1), $\mathbf{T}_p^{-1}(\theta) = 1 - a_1^2$ and hence

$$S_n(\mathbf{X}_1^n, \theta) = (1 - a_1^2)X_1^2 + \sum_{k=2}^{n} (X_k - a_1 X_{k-1})^2. \tag{2.287}$$

Example 2.9.3 (Exponential family). Let $X_1, \ldots, X_n$ be the iid random variables from a Pitman–Koopman–Darmois (exponential) family of distributions with density

$$p_\theta(x) = h(x)\exp\{c(\theta)T(x) - b(\theta)\}. \tag{2.288}$$

We would like to test the null hypothesis $H_0 : \theta = \theta_0$ against the alternative $H_1 : \theta = \theta_1$. It is easy to see that the LR is a monotone function of the statistic $S_n(\mathbf{X}_1^n) = \sum_{i=1}^n T(X_i)$, which therefore is the sufficient statistic. The MP test (NP test) is based on thresholding this statistic, $d^*(S_n) = \mathbb{1}_{\{S_n \geq h\}}$, where the threshold $h = h(\alpha)$ is found from the equation $\mathsf{P}_{\theta_0}(S_n \geq h) = \alpha$, assuming that the solution exists, which is always the case if the distribution is continuous. In discrete cases a randomization may be needed; see Remark 2.9.1.

Example 2.9.4 (Testing between two χ^2 distributions). Assume that the observation X follows a non-central χ_k^2 distribution with k degrees of freedom.

Testing the noncentrality parameter λ of this χ_k^2 distribution against zero, i.e., $H_0 : \lambda = 0$ against $H_1 : \lambda = \lambda_1$, can be achieved again by using the LR of the two densities [237, Sec. 2.6],

$$p_0(x) = \frac{x^{\frac{k}{2}-1} e^{-\frac{x}{2}}}{2^{\frac{k}{2}} \Gamma\left(\frac{k}{2}\right)}, \quad x \in \mathbb{R}_+, \tag{2.289}$$

$$p_\lambda(x) = p_0(x) e^{-\frac{\lambda}{2}} {}_0F_1\left(\frac{k}{2}, \frac{\lambda x}{4}\right), \quad x \in \mathbb{R}_+, \tag{2.290}$$

where the generalized hypergeometric function ${}_0F_1(\kappa, x)$ [283, Ch. 5] is given by

$${}_0F_1(\kappa, x) = \sum_{p=0}^{\infty} \frac{x^p}{(\kappa)_p p!} = \sum_{p=0}^{\infty} \frac{\Gamma(\kappa) x^p}{\Gamma(\kappa + p) p!}, \tag{2.291}$$

and where $(\kappa)_p = \kappa(\kappa+1)\cdots(\kappa+p-1) = \Gamma(\kappa+p)/\Gamma(\kappa)$, $(\kappa)_0 = 1$, denotes the Pochhammer's symbol. Finally,

$$\begin{aligned}\Lambda(X) = \frac{p_\lambda(X)}{p_0(X)} &= e^{-\frac{\lambda}{2}} \sum_{p=0}^{\infty} \frac{\Gamma(\frac{k}{2})}{\Gamma(\frac{k}{2}+p)p!} \left(\frac{\lambda X}{4}\right)^p \\ &= e^{-\frac{\lambda}{2}} \left[1 + \sum_{p=1}^{\infty} \frac{1}{\frac{k}{2}(\frac{k}{2}+1)\ldots(\frac{k}{2}+p-1)p!} \left(\frac{\lambda X}{4}\right)^p\right].\end{aligned} \tag{2.292}$$

2.9.3 *Composite Hypothesis Testing Problems*

Definition 2.9.7 (Composite hypothesis). Any non-simple hypothesis is called a composite hypothesis.

Let $\{\mathsf{P}_\theta, \theta \in \Theta\}$ be a family of distributions and let Θ_0 and Θ_1 be two non-intersecting subsets in Θ. Let us define a composite hypothesis testing problem in the following manner:

$$H_i = \{\mathcal{L}(\mathbf{X}_1^n) = \mathsf{P}_\theta; \theta \in \Theta_i\}, \; \Theta_i \subset \Theta, \; i = 0, 1, \; \Theta_0 \cap \Theta_1 = \varnothing. \tag{2.293}$$

In the sequel, for brevity we write $H_i : \theta \in \Theta_i$. The quality of a composite hypothesis test can be defined by generalizing the criteria used for the simple hypothesis case. The *size* $\alpha_0(d)$ of a test is defined as the maximal probability of rejecting the hypothesis H_0 when it is true,

$$\alpha_0(d) = \sup_{\theta \in \Theta_0} \mathsf{E}_\theta[d(\mathbf{X}_1^n)] = \sup_{\theta \in \Theta_0} \mathsf{P}_\theta(d = 1). \tag{2.294}$$

Let $\mathbf{C}(\alpha)$ denote the class of tests with *level of significance* α, $0 < \alpha < 1$, i.e.,

$$\mathbf{C}(\alpha) = \left\{ d : \sup_{\theta \in \Theta_0} \mathsf{E}_\theta[d(\mathbf{X}_1^n)] \le \alpha \right\}. \tag{2.295}$$

The *power* of a test $d(\mathbf{X}_1^n)$ is now a function of θ and is defined as

$$\beta(d;\theta) = \mathsf{E}_\theta[d(\mathbf{X}_1^n)] = \mathsf{P}_\theta(d = 1), \quad \theta \in \Theta_1. \tag{2.296}$$

This function is often called the *power function* of the test. Obviously, $\beta(d;\theta)$ is the probability of correct acceptance of the hypothesis H_1 when the true parameter value is θ.

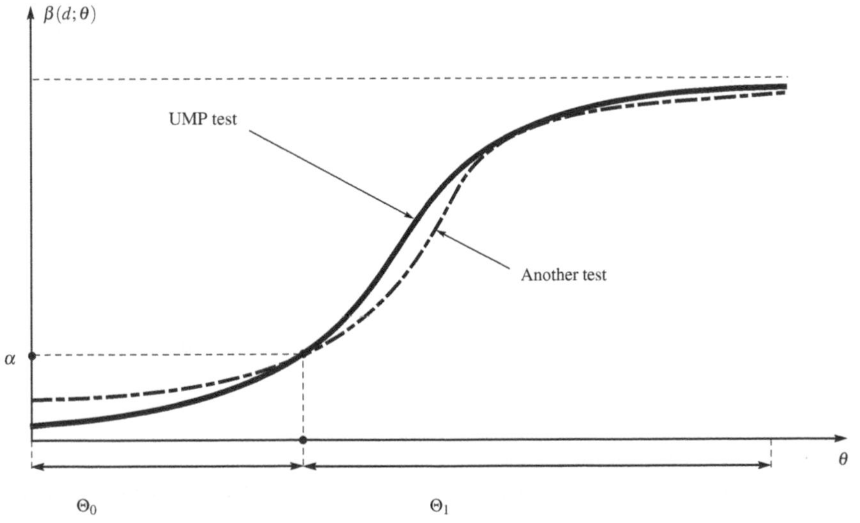

Figure 2.5: *The power function of a UMP test (solid line) and of another test (dash-dotted line).*

Definition 2.9.8 (UMP test). A test $d^*(\mathbf{X}_1^n)$ is said to be *uniformly most powerful* (UMP) in the class of tests $\mathbf{C}(\alpha) = \{d : \alpha_0(d) \le \alpha\}$ if, for all the other tests $d \in \mathbf{C}(\alpha)$,

$$\beta(d^*;\theta) \ge \beta(d;\theta) \quad \text{for all } \theta \in \Theta_1. \tag{2.297}$$

This definition is illustrated in Figure 2.5 where the power function $\beta(d^*;\theta)$ of the UMP test corresponds to the solid line, and the dotted line corresponds to the power function of any other test in the class $\mathbf{C}(\alpha)$.

2.9.3.1 *Monotone Likelihood Ratio and UMP Test*

Definition 2.9.9 (Monotone LR). Let $\mathbf{X}_1^n = (X_1, \ldots, X_n)^\top$ be a random sample belonging to a parametric family of densities (or mass probability functions) $\mathcal{P} = \{p_\theta, \theta \in \Theta\}$ with *scalar* parameter θ.

A family $\mathcal{P}$ is said to be with *monotone likelihood ratio* (LR) if there exists a function $T_n(\mathbf{X}_1^n)$ such that, for all θ_1 and θ_0, $\theta_1 > \theta_0$, the LR

$$\Lambda_n(\mathbf{X}_1^n) = \frac{p_{\theta_1}(\mathbf{X}_1^n)}{p_{\theta_0}(\mathbf{X}_1^n)} = \Lambda_n(T_n(\mathbf{X}_1^n)) \tag{2.298}$$

is a nondecreasing or nonincreasing function of $T_n(\mathbf{X}_1^n)$.

Note that $T_n(\mathbf{X}_1^n)$ is a sufficient statistic.

In the following we omit the index n and write Λ and T for brevity.

Theorem 2.9.4. *Assume that the random sample $\mathbf{X}_1^n = (X_1, \dots, X_n)^\top$ has a pdf (or a pmf) $p_\theta(\mathbf{X}_1^n)$, where θ is scalar, and the family $\{p_\theta\}_{\theta \in \Theta}$ has a monotone LR $\Lambda(\mathbf{X}_1^n) = \Lambda(T(\mathbf{X}_1^n))$. Assume that the function $T \mapsto \Lambda(T)$ is nondecreasing. Then the following assertions hold:*

(i) *For testing between the hypothesis $H_0 : \theta \le \theta_0$ and the* one-sided *alternative hypothesis $H_1 : \theta > \theta_0$, a UMP test exists in the class $\mathbf{C}(\alpha)$ and is given by*

$$d^*(T) = \begin{cases} 1 & \text{if } \; T(\mathbf{X}_1^n) > h \\ p & \text{if } \; T(\mathbf{X}_1^n) = h \\ 0 & \text{if } \; T(\mathbf{X}_1^n) < h \end{cases}, \tag{2.299}$$

where the constants p and h satisfy the equation

$$\mathsf{E}_{\theta_0}[d^*(T)] = \mathsf{P}_{\theta_0}(T > h) + p\, \mathsf{P}_{\theta_0}(T = h) = \alpha. \tag{2.300}$$

(ii) *The power function $\beta(d^*;\theta)$ is a strictly increasing function for all points θ for which $\beta(d^*;\theta) < 1$.*

(iii) *For all $\tilde{\theta}$, the test $d^*(T)$ is UMP in the class $\mathbf{C}(\beta(d^*;\tilde{\theta}))$ for testing between $H_0 : \theta \le \tilde{\theta}$ and $H_1 : \theta > \tilde{\theta}$.*

(iv) *For any $\theta < \theta_0$, the test $d^*(T)$ minimizes the power function $\beta(d;\theta) = \mathsf{E}_\theta(d(\mathbf{X}_1^n))$ in the class $\mathbf{C}(\alpha)$.*

See Lehmann [261, pp. 135–137] for the proof.

Let us discuss an important consequence of Theorem 2.9.4. Consider a single parameter exponential family with the pdf

$$p_\theta(x) = h(x)\, \exp\{c(\theta)\check{T}(x) - b(\theta)\}. \tag{2.301}$$

In the case of an iid sample $\mathbf{X}_1^n$, the LR is

$$\Lambda(\mathbf{X}_1^n) = \frac{p_{\theta_1}(\mathbf{X}_1^n)}{p_{\theta_0}(\mathbf{X}_1^n)} = \exp\left\{(c(\theta_1) - c(\theta_0)) \sum_{i=1}^n \check{T}(X_i) - n(b(\theta_1) - b(\theta_0))\right\}. \tag{2.302}$$

Clearly, it is a monotone function of $T(\mathbf{X}_1^n) = \sum_{i=1}^n \check{T}(X_i)$ provided that $(c(\theta_1) - c(\theta_0))$ has a constant sign for all θ_1, θ_0 such that $\theta_1 > \theta_0$. From Theorem 2.9.4 we deduce that there exists a UMP test given by (2.299) and (2.300) for testing between the hypotheses $H_0 : \theta \le \theta_0$ and $H_1 : \theta > \theta_0$ when $c(\theta)$ is an increasing function. When this function is decreasing, the three inequalities in (2.299) and (2.300) should be replaced by their converse.

So far we discussed only one-sided alternative hypotheses. Another important case is testing the two-sided alternatives, for example the hypothesis $H_0 : \theta = \theta_0$ against $H_1 : \theta \ne \theta_0$. However, no UMP test exists in this case [79, 261] except for particular examples.

2.9.3.2 Unbiased Tests

It follows from the previous subsection that the UMP test exists only in special cases. Let us introduce the subclass $\overline{\mathbf{C}}(\alpha)$ of the so-called unbiased tests in the class of UMP tests (2.295).

Definition 2.9.10 (Unbiased test). A test $d \in \mathbf{C}(\alpha)$ is said to be *unbiased* if the following condition holds:

$$\inf_{\theta\in\Theta_1} \mathsf{E}_\theta[d(\mathbf{X}_1^n)] \geq \sup_{\theta\in\Theta_0} \mathsf{E}_\theta[d(\mathbf{X}_1^n)], \tag{2.303}$$

or equivalently

$$\inf_{\theta\in\Theta_1} \beta(d;\theta)) \geq \sup_{\theta\in\Theta_0} \mathsf{P}_\theta(d=1).$$

Note that this condition is very natural, because the probability of rejection of the hypothesis H_0 when H_0 is false must be not less than the probability of rejection of H_0 when it is true.

It turns out that for the exponential family (2.301) there exists a UMP unbiased test $d^\star(\mathbf{X}_1^n)$ in the subclass $\overline{\mathbf{C}}(\alpha)$ with the level of significance α for two-sided alternative hypotheses. Assume that the pdf (or pmf) $p_\theta(x)$ belongs to the family (2.301) and that we want to test between the hypotheses $H_0 : \theta \in [\theta_0, \theta_1]$ and $H_1 : \theta \notin [\theta_0, \theta_1]$, where $\theta_0 \leq \theta_1$. Then the UMP unbiased test is given by

$$d^\star(T) = \begin{cases} 1 & \text{if} \quad T(\mathbf{X}_1^n) \notin (h_0, h_1) \\ p_i & \text{if} \quad T(\mathbf{X}_1^n) = h_i \ (i=0,1) \\ 0 & \text{if} \quad h_0 < T(\mathbf{X}_1^n) < h_1 \end{cases}, \tag{2.304}$$

where the constants h_i and p_i are determined from the equations

$$\begin{array}{rclll} \mathsf{E}_{\theta_0}[d^\star(T)] = \mathsf{E}_{\theta_1}[d^\star(T)] & = & \alpha & \text{if} & \theta_0 < \theta_1 \\ \left\{ \begin{array}{r} \mathsf{E}_{\theta_0}[d^\star(T)] \\ \mathsf{E}_{\theta_0}[T(\mathbf{X}_1^n)d^\star(T)] \end{array} \right. & \begin{array}{c} = \\ = \end{array} & \begin{array}{l} \alpha \\ \mathsf{E}_{\theta_0}[T(\mathbf{X}_1^n)]\,\alpha \end{array} & \text{if} & \theta_0 = \theta_1. \end{array} \tag{2.305}$$

It follows from Subsections 2.9.3.1 and 2.9.3.2 that the UMP test exists only in special cases and the UMP unbiased tests is less restrictive. Let us now compare the two tests for a simple Gaussian case which nevertheless reflects a typical practical situation.

Example 2.9.5 (Mean in a Gaussian sequence – Contd). Let us again consider testing the mean value θ in an iid Gaussian sequence $\{X_n\}$ with variance σ^2. Now, the two hypotheses are defined as follows: $\widetilde{H}_0 : \theta = 0$ against $\widetilde{H}_1 : \theta \neq 0$. It follows from Subsection 2.9.3.2 that the UMP unbiased test $\widetilde{d}$ with level of significance α can be written as

$$\widetilde{d}(S_n) = \begin{cases} 1 & \text{if} \quad |S_n| \geq \widetilde{h}(\alpha,n) \\ 0 & \text{if} \quad |S_n| < \widetilde{h}(\alpha,n) \end{cases}, \tag{2.306}$$

where $S_n = \sum_{k=1}^n X_k$ and $\widetilde{h} = \widetilde{h}(\alpha,n)$ is chosen so that $\mathsf{P}_0(|S_n| \geq \widetilde{h}) = \alpha$. Let us also consider the UMP test for testing the one-sided hypotheses $H_0 : \theta \leq 0$ and $H_1 : \theta > 0$ (see Subsection 2.9.3.1):

$$d(S_n) = \begin{cases} 1 & \text{if} \quad S_n \geq h(\alpha,n) \\ 0 & \text{if} \quad S_n < h(\alpha,n) \end{cases}. \tag{2.307}$$

The thresholds h and $\widetilde{h}$ of these tests are chosen to get the same level of significance α,

$$\mathsf{P}_0\left(|S_n| \geq \widetilde{h}\right) = \mathsf{P}_0(S_n \geq h) = \alpha,$$

which results in

$$\frac{h}{\sqrt{n}} = \Phi^{-1}(1-\alpha), \quad \frac{\widetilde{h}}{\sqrt{n}} = \Phi^{-1}\left(1-\frac{\alpha}{2}\right).$$

The power functions of the tests (2.306)–(2.307) are given by

$$\beta(d;\theta) = \mathsf{P}_\theta\left(\frac{1}{\sqrt{n}}\sum_{k=1}^{n}(X_k-\theta) \geq \frac{h-n\theta}{\sqrt{n}}\right) = 1-\Phi\left[\Phi^{-1}(1-\alpha)-\sqrt{n}\theta\right], \tag{2.308}$$

$$\begin{aligned}\beta(\widetilde{d};\theta) &= \mathsf{P}_\theta\left(\left|\frac{1}{\sqrt{n}}\sum_{k=1}^{n}X_k\right| \geq \frac{\widetilde{h}}{\sqrt{n}}\right)\\ &= 1+\Phi\left[-\Phi^{-1}\left(1-\frac{\alpha}{2}\right)+\sqrt{n}\theta\right]-\Phi\left[\Phi^{-1}\left(1-\frac{\alpha}{2}\right)+\sqrt{n}\theta\right].\end{aligned} \tag{2.309}$$

These functions are illustrated in Figure 2.6 for $\alpha = 0.1$ and $n = 5$. The power function $\beta_{\widetilde{d}}(\theta)$ of the

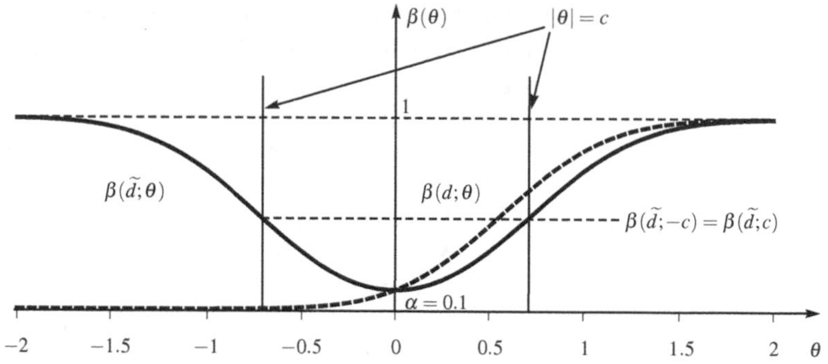

Figure 2.6: *Power functions of the UMP and unbiased UMP tests (Reprinted from : Décision et reconnaissance des formes en signal (Traité IC2, série Signal et image) Lengellé Régis, coord. ©Lavoisier, 2002, with kind permission from Hermes Sciences Publications, Lavoisier SAS).*

test $\widetilde{d}$ given by equation (2.309) is drawn by a solid line and the power function $\beta_d(\theta)$ of the test d given by (2.308) is drawn by a dashed line. The UMP test is slightly better than the unbiased UMP test for positive alternatives $\theta > 0$, but does not work for negative alternatives $\theta < 0$, as expected.

2.9.4 *Bayesian and Minimax Approaches for Two Composite Hypotheses*

Let $H_0 : \theta \in \Theta_0$ and $H_1 : \theta \in \Theta_1$ be two composite hypotheses to be tested, where Θ_0 and Θ_1 are two non-intersecting subsets of a set Θ. The Bayesian approach consists in introducing *a priori* probabilities for these hypotheses, $\pi_i = \mathsf{P}(\theta \in \Theta_i)$, $i = 0,1$ and the *a priori* distributions $\mathsf{P}_i(\theta)$, $i = 0,1$, for the parameter θ on the sets Θ_0 and Θ_1.

Let us consider, as previously, the two-decision Bayesian problem with the $0-1$ loss function, setting $\pi_0 = \pi, \pi_1 = 1-\pi$, where $0 < \pi < 1$. Introduce densities

$$p_i(\mathbf{X}_1^n) = \int_{\Theta_i} p_\theta(\mathbf{X}_1^n)\mathrm{d}\mathsf{P}_i(\theta), \quad i = 0,1. \tag{2.310}$$

The *Bayesian test* $\bar{d}_{\pi,\mathsf{P}(\theta)}(\mathbf{X}_1^n)$ minimizing the average error probability $\bar{\alpha}_\pi(d) = \pi\bar{\alpha}_0(d)+(1-\pi)\bar{\alpha}_1(d)$ for testing composite hypotheses can be shown to be the LR test of the form [79, 261]

$$\bar{d}_{\pi,\mathsf{P}(\theta)}(\mathbf{X}_1^n) = \begin{cases} 1 & \text{if} \quad \Lambda(\mathbf{X}_1^n) \geq \pi/(1-\pi) \\ 0 & \text{otherwise} \end{cases}, \tag{2.311}$$

where $\Lambda(\mathbf{X}_1^n) = p_1(\mathbf{X}_1^n)/p_0(\mathbf{X}_1^n)$ is the LR and

$$\bar{\alpha}_0(d) = \int_{\Theta_0} \mathsf{P}_\theta(d=1)\mathrm{d}\mathsf{P}_0(\theta), \quad \bar{\alpha}_1(d) = \int_{\Theta_1} \mathsf{P}_\theta(d=0)\mathrm{d}\mathsf{P}_1(\theta).$$

We now consider a minimax problem for composite hypotheses. Recall that the class $\mathbf{C}(\alpha)$ defined in (2.295) includes the tests d for which the maximal Type I error probability $\sup_{\theta\in\Theta_0}\alpha_d(\theta)$ does not exceed a given number $0<\alpha<1$, where $\alpha_d(\theta)=\mathsf{P}_\theta(d=1)$, $\theta\in\Theta_0$.

Definition 2.9.11 (Minimax test). The test $\tilde{d}$ is said to be *minimax* in the class $\mathbf{C}(\alpha)$ if it maximizes the minimum power

$$\sup_{d\in\mathbf{C}(\alpha)}\inf_{\theta\in\Theta_1}\mathsf{E}_\theta[d(\mathbf{X}_1^n)] = \sup_{d\in\mathbf{C}(\alpha)}\inf_{\theta\in\Theta_1}\beta(d;\theta), \tag{2.312}$$

or alternatively minimizes the maximal Type II error probability,

$$\inf_{d\in\mathbf{C}(\alpha)}\sup_{\theta\in\Theta_1}\mathsf{P}_\theta(d=0).$$

If the sets Θ_0 and Θ_1 touch each other, and if the power function is continuous, then the inequality $\sup_{d\in\mathbf{C}(\alpha)}\inf_{\theta\in\Theta_1}\beta(d;\theta)>\alpha$ cannot hold. In this situation, it is of interest to introduce an indifference (dead) zone as shown in Figure 2.9, or in other words, to separate the sets Θ_0 and Θ_1. The loss in the indifference zone is always 0, and this is appealing from a practical point of view, since discriminating between too close points does not make much sense (they have the same or almost the same likelihood).

If the sets Θ_0 and Θ_1 contact, any unbiased test d is minimax. The converse statement is true in general. Furthermore, another important property is that the UMP unbiased test d, i.e., the test optimal in the class $\bar{\mathbf{C}}(\alpha)$ is a minimax test in the class $\mathbf{C}(\alpha)$.

Now let us recall that by the minimax Theorem 2.7.8, the Bayes test given by (2.311) is minimax if there exists a pair of distributions $\mathsf{P}_0(\theta)$ and $\mathsf{P}_1(\theta)$ which is least favorable.[7] The main difficulty in using this theorem is to find or guess these least favorable distributions $\mathsf{P}_0(\theta)$ and $\mathsf{P}_1(\theta)$.

Define two simple hypotheses

$$\widetilde{H}_i=\left\{p_i(\mathbf{X}_1^n)=\int_{\Theta_i}p_\theta(\mathbf{X}_1^n)d\mathsf{P}_i(\theta)\right\},\quad i=0,1$$

to design an auxiliary NP test $d_{\mathsf{P}_0,\mathsf{P}_1}$ which is most powerful in the class $\mathbf{C}(\alpha)$. Note that this MP test coincides with the Bayesian test (2.311) which minimizes the average error probability $\bar{\alpha}_\pi(d)=\pi\bar{\alpha}_0(d)+(1-\pi)\bar{\alpha}_1(d)$ with specially selected $0<\pi<1$. The power of this test $d_{\mathsf{P}_0,\mathsf{P}_1}$ is $\beta_{\mathsf{P}_0,\mathsf{P}_1}=\mathsf{P}_{\mathsf{P}_1}\left\{d_{\mathsf{P}_0,\mathsf{P}_1}(\mathbf{X}_1^n)=1\right\}$. The main result is given by the following theorem whose proof can be found in [261].

Theorem 2.9.5. *Assume that there exists a prior distribution* $\mathsf{P}_0(\theta)$ *(resp.* $\mathsf{P}_1(\theta)$*) defined on the set* Θ_0 *(resp.* Θ_1*) and the auxiliary NP test* $d_{\mathsf{P}_0,\mathsf{P}_1}$ *such that*

$$\sup_{\theta\in\Theta_0}\mathsf{P}_\theta\left\{d_{\mathsf{P}_0,\mathsf{P}_1}(\mathbf{X}_1^n)=1\right\}\le\alpha \quad \text{and} \quad \inf_{\theta\in\Theta_1}\mathsf{P}_\theta\left\{d_{\mathsf{P}_0,\mathsf{P}_1}(\mathbf{X}_1^n)=1\right\}=\beta_{\mathsf{P}_0,\mathsf{P}_1}.$$

Then the test $d_{\mathsf{P}_0,\mathsf{P}_1}$ *is minimax in the class* $\mathbf{C}(\alpha)$ *for testing the hypothesis* $H_0:\theta\in\Theta_0$ *against the alternative* $H_1:\theta\in\Theta_1$.

As shown in [261], the prior distributions $\mathsf{P}_0(\theta)$ and $\mathsf{P}_1(\theta)$ can be defined on subsets $\Theta_0^0\subseteq\Theta_0$ and $\Theta_1^0\subseteq\Theta_1$, respectively. Often the prior distributions $\mathsf{P}_i(\theta)$ are concentrated on the boundaries of Θ_i.

2.9.5 Invariant Tests

For some families of distributions, it is very useful to use invariance properties with respect to some transformations in order to design an optimal test or to guess least favorable distributions [79, 143, 261].

[7]These distributions need not be proper (i.e., probability distributions) if the sets Θ_i are not compact.

2.9.5.1 Basic Notation

In this section, we introduce the notion of a family of distributions invariant under a certain group of transformations. Suppose that $X \in \mathcal{P}$, where $\mathcal{P} = \{P_\theta\}_{\theta \in \Theta}$ is a parametric family of distributions. It is supposed that this family satisfies the following condition: $\theta_1 \neq \theta_2$ implies $P_{\theta_1} \neq P_{\theta_2}$. Let us also consider a group G of measurable one-to-one transformations from $\mathcal{X}$ into itself.

Definition 2.9.12 (Invariance). A parametric family of distributions $\mathcal{P} = \{\mathsf{P}_\theta\}_{\theta \in \Theta}$ is invariant under a group of transformation G if

$$\text{for all } g \in G \text{ and for all } \theta \in \Theta \text{ there exists } \theta_g \in \Theta \text{ such that } \mathsf{P}_\theta(X \in A) = \mathsf{P}_{\theta_g}(X \in gA). \quad (2.313)$$

We write $\theta_g = \bar{g}\theta$, where the "generated" transformation $\bar{g}$ forms a group $\bar{G}$.

Example 2.9.6 (Multidimensional Gaussian distribution). Consider the case of n-dimensional Gaussian distribution $\mathcal{N}(\theta, \Sigma)$. It is well known that the covariance matrix Σ (positive definite) can be decomposed as follows:

$$\Sigma = PDP^{-1} = PD^{\frac{1}{2}}D^{\frac{1}{2}}P^{-1} = PD^{\frac{1}{2}}P^{-1}PD^{\frac{1}{2}}P^{-1} = RR \text{ and } \Sigma^{-1} = (R^{-1})R^{-1}, \quad (2.314)$$

where the matrix P is orthogonal ($\det P \neq 0$ and $P^{-1} = P^\top$), R is symmetric, $D = \mathrm{diag}(d_1, \ldots, d_n)$ and $D^{\frac{1}{2}} = \mathrm{diag}(\sqrt{d_1}, \ldots, \sqrt{d_n})$. Assume the following group G of linear non-degenerate transformations $g(X) = RX$. By using the change of variables $Y = g(X) = RX$ in a multiple integral with the Jacobian determinant $J = \det R$ and the fact that $(\det \Sigma)^{\frac{1}{2}} = |\det R|$, we get

$$\begin{aligned}
\mathsf{P}_{\bar{\theta},\Sigma}(X \in A) &= \int \cdots \int_{Y \in A} \frac{1}{(2\pi)^{\frac{n}{2}} (\det \Sigma)^{\frac{1}{2}}} \exp\left\{-\frac{1}{2}(Y - \bar{\theta})^\top \Sigma^{-1} (Y - \bar{\theta})\right\} \mathrm{d}y_1 \cdots \mathrm{d}y_n \\
&= \int \cdots \int_{X \in g^{-1}(A)} \frac{1}{(2\pi)^{\frac{n}{2}}} \exp\left\{-\frac{1}{2}(X - \theta)^\top (X - \theta)\right\} \mathrm{d}x_1 \cdots \mathrm{d}x_n \\
&= \mathsf{P}_{\theta,I}(g(X) \in A),
\end{aligned}$$

where $\bar{\theta} = R\theta$. It can be concluded that the family $\mathcal{N}(\theta, I)$ remains invariant under the group G of linear transformations $g(X) = RX$ and the generated transformations $\bar{g}(\theta, I) = (R\theta, RR) = (\bar{\theta}, \Sigma)$ form the group $\bar{G}$.

2.9.5.2 Design of Invariant Tests

Definition 2.9.13. The hypothesis testing problem $H_0 = \{\Xi_n \sim P_{\theta_0} | \theta_0 \in \Theta_0\}$ against $\mathcal{H}_1 = \{\Xi_n \sim P_{\theta_1} | \theta_1 \in \Theta_1\}$ is called *invariant* under a group of transformations G if

(i) The family $\mathcal{P} = \{P_\theta\}_{\theta \in \Theta}$ remains invariant under the group G;

(ii) The sets Θ_0 and Θ_1 remain invariant under $\bar{g}$, i.e., $\bar{g}(\Theta_i) = \Theta_i$, $i = 0, 1$.

If the hypothesis testing problem is invariant, then it is natural to solve such a problem by using an invariant test.

Definition 2.9.14. A statistic $T(X)$ is called *invariant* under a group of transformations G if

$$T(g(X)) = T(X) \text{ for all } X \in \mathbb{R}^n \text{ and for all } g \in G.$$

Definition 2.9.15. A statistical test d is called *invariant* if the critical function $d(X)$ is an invariant statistic.

Definition 2.9.16. A statistic $T(X)$ is called *maximal invariant* under a group of transformations G if

(i) T is an invariant statistic under G;

(ii) For all X_1 and X_2, $T(X_1) = T(X_2) \Rightarrow \exists g \in G : X_2 = g(X_1)$.

If it turns out that a hypothesis testing problem remains invariant under a group of transformations G, then the standard approach to design an optimal test is to find a maximal invariant under the group of transformations G and, next, to find an optimal (in some sense) invariant test by using the maximal invariant statistic. This results from the following proposition [79].

Proposition 2.9.1. *Let $T(X)$ be a maximal invariant. A statistic $S(X)$ is invariant if, and only if, it depends on the observations via $T(X)$, i.e., there exists a function φ such that $S(X) = \varphi(T(X))$.*

2.9.6 *Testing between Multiple Simple Hypotheses*

2.9.6.1 *Bayesian Test for Multiple Simple Hypotheses*

Assume we are given M distinct distributions $\mathsf{P}_0, \dots, \mathsf{P}_{M-1}$, and let $\mathbf{X}_1^n = (X_1, \dots, X_n)^\top$ be a n-size sample of random variables or vectors generated by one of these distributions. For the sake of simplicity, assume that the distributions $\mathsf{P}_0, \dots, \mathsf{P}_{M-1}$ are absolutely continuous with densities $f_0(x_1^n), \dots, f_{M-1}(x_1^n)$. In this subsection, we consider the problem of testing between M alternative hypotheses $H_0, \dots, H_{M-1}$ such that $H_j : \mathsf{P} = \mathsf{P}_j$, $j = 0, 1, \dots, M-1$. An optimal solution to the general FSS multiple hypothesis testing problem is unknown. Even for simple hypotheses optimal solutions are available only in some special cases [54, 143, 145, 147, 194, 252, 388]. First, the optimal solution is available in the Bayesian problem. In particular, consider the case of the "$0-1$" loss function when the loss is zero in the case of a correct decision and one in the case of a wrong decision; see Subsection 2.9.1.2. The *a priori* probabilities $\pi_i = \mathsf{P}(H_i) > 0$, $i = 0, \dots, M-1$, $\sum_{i=0}^{M-1} \pi_i = 1$ of the hypotheses $H_0, \dots, H_{M-1}$ are known. We refer to [194, Theorem 1] and [143, Sec. 6.2] for a complete proof of the following result.

Theorem 2.9.6. *Consider the M-hypothesis Bayesian testing problem with the $0-1$ loss function and the prior distribution $(\pi_i)_{i=0,\dots,M-1}$. The Bayesian test minimizing the average error probability*

$$\bar{d}_\pi(\mathbf{X}_1^n) = \arg\inf_d \rho(d) = \arg\inf_d \left\{ \sum_{i=0}^{M-1} \pi_i \alpha_i(d) \right\}.$$

is given by

$$\bar{d}_\pi(\mathbf{X}_1^n) = \arg \max_{0 \le \ell \le M-1} \pi_\ell f_\ell(\mathbf{X}_1^n) \tag{2.315}$$

Under the above assumption that the distributions $\mathsf{P}_0, \dots, \mathsf{P}_{M-1}$ are absolutely continuous, the event $f_\ell(x_1^n) = f_j(x_1^n)$ has the μ-measure zero for $\ell \neq j$; hence the maximum in (2.315) is unique with probability 1.

2.9.6.2 *A Slippage Problem for Multiple Simple Hypotheses*

Sometimes the existence of natural symmetry or invariance of the hypothesis testing problem under permutations of $H_1, \dots, H_n$ among themselves leads to a relatively simple design of an MP test in the class $\mathbf{C}(\alpha)$. One such problem is known under the name of a "slippage problem" [143, 304, 454]. In the subsequent chapters, we shall discuss the sequential version of the slippage problem, but here its classical version is presented.

Let us consider the observations $\mathbf{X}_1^n = (X_1, \dots, X_n)^\top$ and the hypotheses $H_0, \dots, H_n$. It is assumed that under hypothesis H_0 the observations $X_1, \dots, X_n$ are iid with common density $f_0(x)$ but under hypothesis H_j, the $X_1, \dots, X_n$ are independent, $X_1, \dots, X_{j-1}, X_{j+1}, \dots, X_n$ are identically distributed with common density $f_0(x)$ and X_j has density $f_1(x)$. An optimal solution to the slippage problem for multiple hypothesis testing problem is based on the Bayesian decision rule of level $\alpha = \mathsf{P}_0(d(X_1, \dots, X_n) \neq 0)$ subject to the assumption that the alternative hypotheses H_i, $i = 1, \dots, n$, are equally probable [143]. Indeed, the prior distribution $(\pi_i)_{i=1,\dots,n}$ invariant under the group of permutations of the set $\{H_1, \dots, H_n\}$ necessarily leads to equally probable alternative hypotheses.

Hence, the invariant prior distributions are of the form $\pi_0 = 1 - n\pi$ and $\pi_j = \pi$ for $j = 1,\ldots,n$. Due to this invariant property, the Bayesian test is equivalent to the MP test which maximizes the common power (i.e., the probability to accept the hypothesis H_ℓ when this hypothesis is true)

$$\beta = \mathsf{P}_\ell(d(\mathbf{X}_1^n) = \ell) = \mathsf{P}_j(d(\mathbf{X}_1^n) = j), \ \forall \ell, j \neq 0$$

in the class $\mathbf{C}(\alpha) = \{d : \mathsf{P}_0(d(\mathbf{X}_1^n) \neq 0) \leq \alpha\}$. We refer to [143] for a proof of the following result.

Theorem 2.9.7. *The MP test which maximizes the common power β in the class $\mathbf{C}(\alpha)$ is given by*

$$d(\mathbf{X}_1^n) = \begin{cases} \ell & \text{if} \ \ \max_{1\leq j\leq n}\left(\frac{f_1(X_j)}{f_0(X_j)}\right) \geq h \\ 0 & \text{if} \ \ \max_{1\leq j\leq n}\left(\frac{f_1(X_j)}{f_0(X_j)}\right) < h \end{cases}, \quad \ell = \arg\max_{1\leq j\leq n}\left(\frac{f_1(X_j)}{f_0(X_j)}\right), \tag{2.316}$$

where $h = h(\alpha)$ is selected from the equation

$$\mathsf{P}_0\left\{\max_{1\leq j\leq n}\left(\frac{f_1(X_j)}{f_0(X_j)}\right) \geq h\right\} = \alpha. \tag{2.317}$$

For the same reason as in Remark 2.9.1 equation (2.317) may not have a solution in discrete cases where the P_0-distribution of the LR $f_1(X)/f_0(X)$ is not continuous. In these cases, a randomization on h is needed, i.e., deciding with probability p whether to accept H_ℓ when

$$\frac{f_1(X_\ell)}{f_0(X_\ell)} = \max_{1\leq j\leq n}\left(\frac{f_1(X_j)}{f_0(X_j)}\right) = h$$

or to accept H_0 with probability $1-p$.

Example 2.9.7 (Slippage problem with a nuisance parameter). Let us consider one particular but more realistic case, a Gaussian model with an additive nuisance parameter: $Y_i = x + \xi_i$, $i = 1,\ldots,n$, where x is the (unknown) mean and ξ_i is a zero-mean Gaussian noise with variance σ^2. Let $\mathbf{Y}_1^n = (Y_1,\ldots,Y_n)^\top$ be an n-size sample of independent random variables Y_i. The goal is to test between $(n+1)$ alternative hypotheses

$$H_0 = \{\mathbf{Y}_1^n \sim \mathcal{N}(\mathbf{1}_n x, \sigma^2 \mathbb{I}_n)\}, \ H_j = \{\mathbf{Y}_1^n \sim \mathcal{N}(\mathbf{1}_n x + \theta_j, \sigma^2 \mathbb{I}_n)\}, \ j = 1,\ldots,n,$$

where $\mathbf{1}_n$ is an n-dimensional vector composed of ones, $\theta_j = (0,\ldots,0,\vartheta_j,0,\ldots,0)^\top$, $\vartheta_1 = \cdots = \vartheta_n = \vartheta > 0$. First, this problem is invariant under the group of permutations of the set $\{H_1,\ldots,H_n\}$. Second, to eliminate the negative impact of the unknown mean x, the test should be invariant under the group G of translations $\{\mathbf{Y}_1^n \mapsto g(\mathbf{Y}_1^n) = \mathbf{Y}_1^n + \mathbf{1}_n x\}$ generated by x. The statistic

$$\mathbf{Z}_1^{n-1} = W\mathbf{Y}_1^n = \begin{pmatrix} 1 & 0 & \cdots & 0 & -1 \\ 0 & 1 & \cdots & 0 & -1 \\ \vdots & \vdots & \vdots & \vdots & \vdots \\ 0 & 0 & \cdots & 1 & -1 \end{pmatrix} \begin{pmatrix} Y_1 \\ Y_2 \\ \vdots \\ Y_n \end{pmatrix} \sim \mathcal{N}(W\mathsf{E}(\mathbf{Y}_1^n), \sigma^2 W W^\top) \tag{2.318}$$

is maximal invariant under the group of translations $G = \{\mathbf{Y}_1^n \mapsto g(\mathbf{Y}_1^n) = \mathbf{Y}_1^n + \mathbf{1}_n x\}$. The previously introduced *invariant Bayesian test* (2.316) has a more complicated structure:

$$d\left(\mathbf{Z}_1^{n-1}\right) = \begin{cases} \ell & \text{when} \ \ \max_{1\leq j\leq n}\left(\frac{f_j(\mathbf{Z}_1^{n-1})}{f_0(\mathbf{Z}_1^{n-1})}\right) \geq h \\ 0 & \text{when} \ \ \max_{1\leq j\leq n}\left(\frac{f_j(\mathbf{Z}_1^{n-1})}{f_0(\mathbf{Z}_1^{n-1})}\right) < h \end{cases}, \ \ell = \arg\max_{1\leq j\leq n}\left(\frac{f_j(\mathbf{Z}_1^{n-1})}{f_0(\mathbf{Z}_1^{n-1})}\right), \tag{2.319}$$

where f_0 (resp. f_j) is the density of the normal distribution $\mathcal{N}(0, \sigma^2 W W^\top)$ (resp.

$\mathcal{N}(W\theta_j, \sigma^2 WW^\top)$). Taking into account that $(WW^\top)^{-1} = (\mathbb{I}_{n-1} + \mathbf{1}_{n-1}\mathbf{1}_{n-1}^\top)^{-1} = \mathbb{I}_{n-1} - \frac{1}{n}\mathbf{1}_{n-1}\mathbf{1}_{n-1}^\top$, after simple algebra, we obtain from equation (2.319)

$$d(\mathbf{Y}_1^n) = \begin{cases} \ell & \text{when } \max_{1\le j\le n}(Y_j - \overline{Y}) \ge h \\ 0 & \text{when } \max_{1\le j\le n}(Y_j - \overline{Y}) < h \end{cases}, \quad \ell = \arg\max_{1\le j\le n}(Y_j - \overline{Y}), \tag{2.320}$$

where $h = h(\alpha)$, $\alpha = \mathsf{P}_0(d(\mathbf{Y}_1^n) \neq 0)$ and $\overline{Y} = \frac{1}{n}\sum_{i=1}^n Y_i$. This test maximizes the common power

$$\beta = \mathsf{P}_\ell(d(\mathbf{Y}_1^n) = \ell) = \mathsf{P}_j(d(\mathbf{Y}_1^n) = j), \ \forall \ell, j \neq 0$$

in the class $\mathbf{C}(\alpha)$. It is worth noting that the MP (or Bayesian) test is independent of the value ϑ, hence it is also UMP.

2.9.6.3 *Constrained Minimax Test for Multiple Simple Hypotheses*

We now continue a discussion of the M-hypothesis testing problem started in Subsection 2.9.6.1. Let $\mathbf{X}_1^n = (X_1, \dots, X_n)^\top$ be an n-size sample of random variables or vectors generated by one of M distinct distributions $\mathsf{P}_0, \dots, \mathsf{P}_{M-1}$ with densities $f_0(x_1^n), \dots, f_{M-1}(x_1^n)$. Consider the class $\mathbf{C}(\alpha) = \{d : \mathsf{P}_0(d(\mathbf{X}_1^n) \neq 0) \le \alpha\}$. The constrained minimax test has been introduced in [54] as a solution to the M-hypothesis testing problem in the case where this problem does not possess any natural symmetry or invariance under the group of permutations of the set $\{H_1, \dots, H_{M-1}\}$.

Definition 2.9.17. A test $\overline{d}(\mathbf{X}_1^n)$ is constrained minimax of level α between hypotheses $H_0, \dots, H_{M-1}$ if $\overline{d}(\mathbf{X}_1^n) \in \mathbf{C}(\alpha)$ and if for any other test $d(\mathbf{X}_1^n) \in \mathbf{C}(\alpha)$ the following inequality is satisfied

$$\max_{1\le\ell\le M-1} \alpha_i(\overline{d}) \le \max_{1\le\ell\le M-1} \alpha_i(d).$$

The following result can be easily proved using Theorem 2.7.8; see also [54].

Theorem 2.9.8. *The weighted GLR test between the hypotheses $H_0, \dots, H_{M-1}$*

$$\overline{d}(\mathbf{X}_1^n) = \begin{cases} \ell \ \textit{if} \ \max_{1\le j\le M-1}\left(q_j \frac{f_j(\mathbf{X}_1^n)}{f_0(\mathbf{X}_1^n)}\right) \ge h \\ 0 \ \textit{if} \ \max_{1\le j\le M-1}\left(q_j \frac{f_j(\mathbf{X}_1^n)}{f_0(\mathbf{X}_1^n)}\right) < h \end{cases}, \quad \ell = \arg\max_{1\le j\le M-1}\left(q_j \frac{f_j(\mathbf{X}_1^n)}{f_0(\mathbf{X}_1^n)}\right), \tag{2.321}$$

where $q_1 \ge 0, \dots, q_{M-1} \ge 0$ are the weight coefficients, $\sum_{j=0}^{M-1} q_j = 1$, is constrained minimax if the threshold $h = h(\alpha)$ is selected so that

$$\mathsf{P}_0\left\{\max_{1\le j\le M-1}\left(q_j \frac{f_j(\mathbf{X}_1^n)}{f_0(\mathbf{X}_1^n)}\right) \ge h\right\} = \alpha \tag{2.322}$$

and the weight coefficients $q_1, \dots, q_{M-1}$ are selected so that the probability of false classification

$$\alpha_\ell(\overline{d}) = \alpha_j(\overline{d}), \ \forall \ell, j \neq 0$$

is constant over the set of alternative hypotheses $H_1, \dots, H_{M-1}$.

In the other words, this "equalizer test" maximizes the common power

$$\beta = \mathsf{P}_\ell(\overline{d}(\mathbf{X}_1^n) = \ell) = \mathsf{P}_j(\overline{d}(\mathbf{X}_1^n) = j), \ \forall \ell, j \neq 0$$

in the class $\mathbf{C}(\alpha)$.

Remark 2.9.2. Note that equation (2.322) always has a solution in the continuous case where the P_0-distributions of the LRs $f_j(\mathbf{X}_1^n)/f_0(\mathbf{X}_1^n)$, $j = 1, \dots, M-1$ are continuous. However, it may not have a solution in certain discrete cases. Then a randomization on the boundary h is required, that is, the hypothesis H_ℓ is accepted if

$$q_\ell \frac{f_\ell(\mathbf{X}_1^n)}{f_0(\mathbf{X}_1^n)} = \max_{1\le j\le M-1} \left(q_j \frac{f_j(\mathbf{X}_1^n)}{f_0(\mathbf{X}_1^n)} \right) > h$$

and the hypothesis H_0 is accepted with probability $1-p$ when

$$q_\ell \frac{f_\ell(\mathbf{X}_1^n)}{f_0(\mathbf{X}_1^n)} = \max_{1\le j\le M-1} \left(q_j \frac{f_j(\mathbf{X}_1^n)}{f_0(\mathbf{X}_1^n)} \right) = h.$$

In this more general randomized case, the threshold is the solution of the equation

$$\mathsf{P}_0\left\{ \max_{1\le j\le M-1} \left(q_j \frac{f_j(\mathbf{X}_1^n)}{f_0(\mathbf{X}_1^n)} \right) > h \right\} + p\,\mathsf{P}_0\left\{ \max_{1\le j\le M-1} \left(q_j \frac{f_j(\mathbf{X}_1^n)}{f_0(\mathbf{X}_1^n)} \right) = h \right\} = \alpha.$$

2.10 Hypothesis Testing: Gaussian Linear Model

Now we consider a Gaussian linear model — an important particular model which is widely used in the following chapters for illustrating general results.

2.10.1 Uniformly Best Constant Power Test

Let us now continue our discussion of hypothesis testing problems in the case of n-dimensional Gaussian distribution $X \sim \mathcal{N}(\theta, \Sigma)$ and a composite alternative hypothesis H_1. As follows from Example 2.9.5, even in the scalar parameter case this problem is not simple because the composite alternative hypothesis H_1 is too "rich." Hence, to solve such a testing problem some additional constraints (more or less natural) on the class of tests should be imposed. Let us now discuss the solution proposed by Wald in his fundamental paper [490].

2.10.1.1 Unit Covariance Matrix

For the sake of simplicity, assume first that $X \sim \mathcal{N}(\theta, \mathbb{I}_n)$. The hypothesis testing problem consists in deciding between $H_0 : \theta = 0$ and $H_1 : \theta \neq 0$. Note that a UMP test does not exist in the vector parameter case. To find an optimal test, Wald [490] proposes to impose an additional constraint on the class of considered tests, namely, a *constant power function* over a family of surfaces $\mathcal{S}$ defined on the parameter space $\Theta = \{\theta \in \mathbb{R}^n\}$.

Definition 2.10.1. A test $d^* \in \mathbf{C}(\alpha) = \{d : \mathsf{P}_0(d \neq 0) \le \alpha\}$, where α is the prescribed probability of false alarm, is said to have uniformly best constant power (UBCP) on the family of surfaces $\mathcal{S}$, if the following conditions are fulfilled [490]:

(i) For any pair of points θ_1 and θ_2 which lies on the same surface $S_c \in \mathcal{S}$, $\beta(d^*;\theta_1) = \beta(d^*;\theta_2)$, where $\beta(d;\theta) = \mathsf{P}_\theta(d=1)$ is the power function of the test d.

(ii) For another test $d \in \mathbf{C}(\alpha)$, which satisfies the previous condition, we have $\beta(d^*;\theta) \ge \beta(d;\theta)$.

This situation is illustrated in Figure 2.7 in the simplest case where $X \sim \mathcal{N}(\theta, \mathbb{I}_n)$. As it will be shown, in this case it is natural to define a family of concentric spherical surfaces $S_c = \{\theta : \|\theta\|_2^2 = c^2\}$ around $\theta = 0$. If we define a certain direction $\theta = c\Upsilon$, $c > 0$, by using a unit directional vector $\|\Upsilon\|_2 = 1$, a UMP test can be designed. As it has been shown in Example 2.9.5, this test is very efficient to test the alternatives $\theta = c\Upsilon$ against $H_0 : \theta = 0$ but it is not efficient for all other directions. Hence, the constraint $\beta(d;\theta \in S_c) = \mathtt{constant}$ excludes from the consideration such UMP tests which are very efficient over a certain subspace $\overline{\Theta}$ of Θ and very inefficient over $\Theta \setminus \overline{\Theta}$. This looks natural enough for practical applications.

The main result of this subsection is the following theorem by Wald [490].

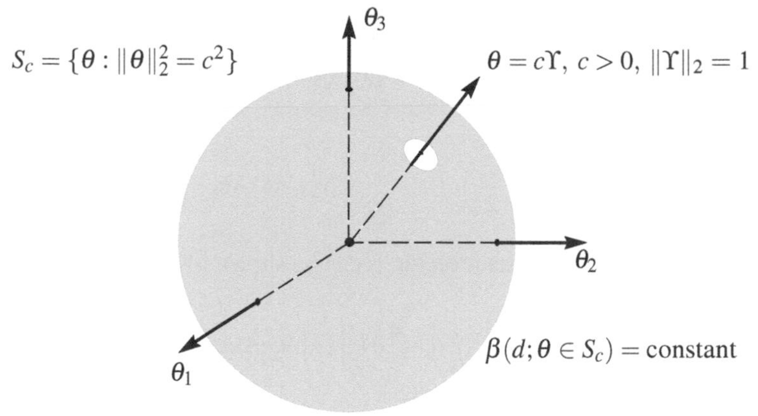

Figure 2.7: *Constant power $\beta(d;\theta)$ on a spherical surface (Reprinted from : Décision et reconnaissance des formes en signal (Traité IC2, série Signal et image) Lengellé Régis, coord. ©Lavoisier, 2002, with kind permission from Hermes Sciences Publications, Lavoisier SAS).*

Theorem 2.10.1. *Let $X \sim \mathcal{N}(0,\mathbb{I}_n)$. The test $d^*(X) \in \mathbf{C}(\alpha)$, given by*

$$d^*(Y) = \begin{cases} 1 & \text{if} \quad \Lambda(X) = \|X\|_2 \ \geq \ h(\alpha) \\ 0 & \text{if} \quad \Lambda(X) = \|X\|_2 \ < \ h(\alpha) \end{cases}, \tag{2.323}$$

is UBCP for testing the hypothesis $H_0 : \theta = 0$ versus the alternative $H_1 : \theta \neq 0$ over the family of concentric spherical surfaces

$$\mathcal{S} = \left\{S_c : \|\theta\|_2^2 = c^2,\ c > 0\right\}. \tag{2.324}$$

Proof. It is necessary to show that the proposed test coincides with a specially designed NP test which realizes a constant power function over the family of surfaces $\mathcal{S}$ independently of c. The alternative hypothesis $H_{1,\,c}$ corresponds to the sphere S_c given by equation $\|\theta\|_2^2 = c^2$ from the family $\mathcal{S}$. Define the following *a priori* density $p(\theta) = b = 1/A(S_c)$ of the parameter θ over the sphere S_c, where $A(S_c)$ is the area of the surface S_c. Hence, the LR between $H_0 = \{f_0(X)\}$ and $H_{1,\,c} = \left\{\int \cdots \int_{S_c} f_\theta(X) p(\theta) dS(\theta)\right\}$ is given by the following surface integral

$$\Lambda_{\mathrm{NP}}(X) = \int \cdots \int_{S_c} \frac{f_\theta(X)}{f_0(X)} p(\theta) dS(\theta) \tag{2.325}$$

where $f_\theta(X) = (2\pi)^{-\frac{n}{2}} \exp\{-\frac{1}{2}\|X-\theta\|_2^2\}$ is density of the observation vector X. The NP test to chose between the hypotheses H_0 and $H_{1,\,c}$ is given by

$$d_{\mathrm{NP}}(X) = \begin{cases} 1 & \text{when} \quad \Lambda_{\mathrm{NP}}(X) \geq h(\alpha) \\ 0 & \text{when} \quad \Lambda_{\mathrm{NP}}(X) \leq h(\alpha) \end{cases}, \tag{2.326}$$

where $h(\alpha)$ is the threshold chosen so that the probability of false alarm is equal to a prescribed significance level, $\mathsf{P}_0(d_{\mathrm{NP}}(X) \neq 0) = \alpha$. The LR $\Lambda_{\mathrm{NP}}(X)$ can be written as follows:

$$\begin{aligned} \Lambda_{\mathrm{NP}}(X) &= b \int \cdots \int_{S_c} \exp\left\{\frac{1}{2}\left[2\theta^\top X - \|\theta\|_2^2\right]\right\} dS(\theta) \\ &= b \exp\left(-\frac{c^2}{2}\right) \int \cdots \int_{S_c} \exp\{\theta^\top X\} dS(\theta) \\ &= b \exp\left(-\frac{c^2}{2}\right) \int \cdots \int_{S_c} \exp\left\{c\Lambda(X)\cos\left[\widehat{\theta X}\right]\right\} dS(\theta), \end{aligned} \tag{2.327}$$

where $\Lambda(X) = \|X\|_2$ and $\widehat{\theta X}$ denotes an angle between the vectors θ and X. To prove the theorem, it is necessary to show that the LR $\Lambda_{\rm NP}(X)$ of the NP test is a nondecreasing function of the statistic $\Lambda(X) = \|X\|_2$ of the test $d^*(X)$ in (2.323) for any value of $c > 0$. Due to the symmetry of the problem, we have

$$\Lambda_{\rm NP}(X) = b\exp\left(-\frac{c^2}{2}\right)\int\cdots\int_{S_c}\exp\left\{c\Lambda(X)\cos\left[\widehat{\theta U}\right]\right\}dS(\theta), \tag{2.328}$$

where U is an arbitrarily chosen constant vector. It can be shown [490] that

$$\frac{d\Lambda_{\rm NP}(\Lambda)}{d\Lambda} = cb\exp\left(-\frac{c^2}{2}\right)\int\cdots\int_{S_c}\cos\left[\widehat{\theta U}\right]\exp\left\{c\Lambda\cos\left[\widehat{\theta U}\right]\right\}dS(\theta) > 0 \tag{2.329}$$

and, therefore, the function $\|\Lambda\|_2 \mapsto \Lambda_{\rm NP}(\|\Lambda\|_2)$ is increasing for any $c > 0$. This proves that the test $d^*(X)$ given by (2.323) has uniformly best constant power with respect to S_c and, hence, the test $d^*(X)$ is UBCP over the family of spheres $\mathcal{S}$ given by equation (2.324). □

Remark 2.10.1. It follows from Subsection 2.9.5 that the family of normal distributions $\mathcal{N}(\theta,\mathbb{I})$ remains invariant under every orthogonal transformation $g : X' = CX$, where C is the matrix of the orthogonal transformation. In this case, the corresponding transformation $\bar{g}$ of the parameter set Θ can be defined by $\bar{g} : \theta' = C\theta$, and the hypotheses H_0 and H_1 remain invariant under the transformation $\bar{g}$. Therefore, Wald's choice of the *a priori* density $p(\theta)$ (constant on a sphere) seems to be natural since the hypothesis testing problem should be invariant under every orthogonal transformation $g : X' = CX$ and, hence, the *a priori* density $p(\theta)$ of the parameter θ must also remain invariant under the transformation $\bar{g}$.

2.10.1.2 General Covariance Matrix

Consider now the case of a general covariance matrix. The observation $X \in \mathbb{R}^n$ is generated by one of the two Gaussian distributions: $\mathcal{N}(0,\Sigma)$ or $\mathcal{N}(\theta \neq 0,\Sigma)$, where θ is the (unknown) mean vector and Σ is a known positive definite covariance matrix. Define the statistic $\Lambda(X) = X^\top\Sigma^{-1}X$.

Theorem 2.10.2. *Let $X \sim \mathcal{N}(0,\Sigma)$. The test $d^*(X) \in \mathbf{C}(\alpha)$ given by*

$$d^*(X)=\begin{cases}1 & \text{if } \Lambda(X) \geq h(\alpha)\\ 0 & \text{if } \Lambda(X) < h(\alpha)\end{cases} \tag{2.330}$$

is UBCP for testing the hypotheses $H_0 : \theta = 0$ and $H_1 : \theta \neq 0$ over the family of ellipsoids

$$\mathcal{S} = \left\{S_c : \theta^\top\Sigma^{-1}\theta = c^2,\ c > 0\right\}. \tag{2.331}$$

Proof. The alternative hypothesis $H_{1,c}$ corresponds to the ellipsoid S_c given by equation (2.331) from the family of ellipsoids $\mathcal{S}$, and *a priori* density of the parameter θ over the ellipsoid S_c is $p(\theta) = b = 1/A(S_c)$, where $A(S_c)$ is the area of the surface S_c. The covariance matrix Σ is positive definite, hence it can be decomposed as

$$\Sigma = RR^\top \quad \text{and} \quad \Sigma^{-1} = (R^{-1})^\top R^{-1} = R^{-\top}R^{-1}. \tag{2.332}$$

Define the vector $Z = R^{-1}X$. Using the change of variables $\theta' = R\theta$ ($\det R \neq 0$) and putting $X = RZ$, we get the same expression as in (2.327) with Z instead of X and with θ' instead of θ. The rest of the proof is identical to the proof of Theorem 2.10.1. □

2.10.1.3 Linear Model

Consider the following Gaussian linear model:

$$Y = M\theta + \xi, \tag{2.333}$$

where $Y \in \mathbb{R}^n$ is the observation vector, $\theta \in \mathbb{R}^r$ is an unknown parameter, M is a full column rank matrix of size $(n \times r)$ with $r < n$ and ξ is a zero mean Gaussian noise $\xi \sim \mathcal{N}(0, \sigma^2 \mathbb{I}_n)$, $\sigma^2 > 0$. As in the previous case, the hypothesis testing problem consists in deciding between $H_0 : \theta = 0$ and $H_1 : \theta \neq 0$. Let us apply the general Wald idea to design a test that can potentially be UBCP [490]. The Wald statistic is given by $\Lambda(Y) = \widehat{\theta}^\top \mathcal{F}_{\widehat{\theta}} \widehat{\theta}$, where $\widehat{\theta} = (M^\top M)^{-1} M^\top Y$ is the least square (LS) estimator of θ and $\mathcal{F}_{\widehat{\theta}} = \frac{1}{\sigma^2} M^\top M$ is the Fisher matrix. Hence, the test is

$$d^*(Y) = \begin{cases} 1 & \text{if} \quad \Lambda(Y) = \frac{1}{\sigma^2} Y^\top M (M^\top M)^{-1} M^\top Y \; \geq \; h(\alpha) \\ 0 & \text{if} \quad \Lambda(Y) = \frac{1}{\sigma^2} Y^\top M (M^\top M)^{-1} M^\top Y \; < \; h(\alpha) \end{cases}, \tag{2.334}$$

where the threshold $h(\alpha)$ is selected from the equation $\mathsf{P}_0(\Lambda(Y) \geq h) = \alpha$.

This test is UBCP over the following family of ellipsoids:

$$\mathcal{S}_M = \left\{ S_c : \theta^\top \mathcal{F}_{\widehat{\theta}} \, \theta = \frac{1}{\sigma^2} \|M\theta\|_2^2 = c^2, \; c > 0 \right\}. \tag{2.335}$$

Optimality of the test (2.334) is established in the following theorem [150].

Theorem 2.10.3. *Consider the regression model* (2.333). *The test* $d^*(Y) \in \mathbf{C}(\alpha)$ *given by* (2.334) *is UBCP for testing the hypotheses* $H_0 : \theta = 0$ *and* $H_1 : \{\theta \neq 0\}$ *over the family of ellipsoids* (2.335).

Proof. The alternative hypothesis $H_{1,c}$ of the auxiliary NP test corresponds to the ellipsoid S_c from the family $\mathcal{S}_M$ given by (2.335). Let us again define the uniform *a priori* density $p(\theta) = b = 1/A(S_c)$ over the ellipsoid S_c, where $A(S_c)$ is the area of the surface S_c. Hence, density of Y is given by the surface integral $\int \cdots \int_{S_c} f_\theta(Y) p(\theta) dS(\theta)$ when the hypothesis $H_{1,c}$ is true. The NP test for testing the hypothesis $H_0 = \{f_0(Y)\}$ against $H_{1,c} = \left\{ \int \cdots \int_{S_c} f_\theta(Y) p(\theta) dS(\theta) \right\}$ is given by

$$\Lambda_{\text{NP}}(Y) = b \int \cdots \int_{S_c} \exp\left\{ \frac{1}{2\sigma^2} \left[2\theta^\top M^\top Y - \theta^\top (M^\top M) \theta \right] \right\} dS(\theta). \tag{2.336}$$

The matrix $\Sigma = M^\top M$ is positive definite. It follows from (2.332) that the matrix Σ can be rewritten as $\Sigma = RR^\top$ and $\Sigma^{-1} = R^{-\top} R^{-1}$. Define the vector $Z = R^{-1} M^\top Y$. The statistic $\Lambda(Y)$ can be represented as a function of Z as $\Lambda(Z) = \frac{1}{\sigma^2} \parallel Z \parallel_2^2$. Since the matrix R is non-singular, the family of surfaces $\mathcal{S}_H$ can be rewritten as $\mathcal{S}' = \{ S'_c : \parallel \theta' \parallel_2^2 = c^2 \sigma^2, \; c > 0 \}$, where $\theta' = R^\top \theta$ and $\theta = R^{-\top} \theta'$. Using the change of variables $\theta' = R\theta$ ($\det R \neq 0$) and putting $M^\top Y = RZ$, we get

$$\Lambda_{\text{NP}}(Z) = b' \exp\left\{ -\frac{c^2}{2\sigma^2} \right\} \int \cdots \int_{S'_c} \exp\left\{ \frac{1}{\sigma^2} \theta'^\top Z \right\} dS'(\theta'),$$

where b' is a constant and the second term under the sign of integral $\exp\{-\frac{1}{2\sigma^2} \parallel \theta' \parallel_2^2\}$ is constant over the surface S'_c. The rest of the proof is identical to the proof of Theorem 2.10.1. □

Remark 2.10.2. The more general case where noise ξ in (2.333) follows the Gaussian distribution $\mathcal{N}(0, \Sigma)$ with a known (positive definite) covariance matrix Σ can be easily reduced to the model (2.333) with independent components treated in Theorem 2.10.3. To this end it suffices to use the change of variables $g(X) = R^{-1} X$, where the symmetric matrix R is defined by $\Sigma = RR$, along with the invariance properties of the Gaussian family $\mathcal{N}(\theta, \Sigma)$ [79].

2.10.1.4 Power Function

Suppose that $Y \in \mathbb{R}^n$. As it follows from [237, Ch. 2.7] :

(i) If $Y \sim \mathcal{N}(\theta,\Sigma)$ then the statistic $\Lambda(Y) = Y^\top\Sigma^{-1}Y$ is distributed according to the χ^2_n law with n degrees of freedom. This χ^2_n law is central under H_0 and noncentral under H_1 with noncentrality parameter $\lambda = c^2 = \theta^\top\Sigma^{-1}\theta$.

(ii) If $Y \sim \mathcal{N}(M\theta,\sigma^2\mathbb{I}_n)$ then the statistic $\Lambda(Y) = 1/\sigma^2\ Y^\top M(M^\top M)^{-1}M^\top Y$ is distributed according to the χ^2_r law with r degrees of freedom. This χ^2_r law is central under H_0 and noncentral under H_1 with noncentrality parameter $\lambda = c^2 = \|M\theta\|_2^2/\sigma^2$.

Hence, the power function is constant on each surface S_c : $\|M\theta\|_2^2/\sigma^2 = c^2$; see (2.335). For this reason, it is reasonable to present the power as a function of $c^2 = \lambda$, i.e., $\beta(d^*;\lambda)$. We obtain,

$$\beta(d^*;\lambda) = \mathsf{P}_\lambda(\Lambda(Y) \geq h(\alpha))$$
$$= \int_h^\infty p_\lambda(y)dy = \int_h^\infty p_0(y)e^{-\frac{\lambda}{2}}\,{}_0F_1\left(\frac{n}{2},\frac{\lambda y}{4}\right)dy, \tag{2.337}$$

where $p_0(y) = 2^{-n/2}\Gamma(n/2)^{-1}y^{n/2-1}e^{-y/2}$. The power function $\beta(d^*;\lambda)$ can be also represented as

$$\beta(d^*;\lambda) = \exp\left(-\frac{\lambda}{2}\right)\sum_{p=0}^{\infty}\frac{\lambda^p}{2^p p!}\,\mathsf{P}\left(\xi^2_{n+2p} \geq h\right), \tag{2.338}$$

Therefore, it is easy to see that it is monotonically increasing in λ, since

$$\frac{d\beta(d^*;\lambda)}{d\lambda} = \frac{1}{2}\exp\left(-\frac{\lambda}{2}\right)\left[\sum_{p=0}^{\infty}\frac{\lambda^p}{2^p p!}\left(\mathsf{P}\left(\xi^2_{n+2p+2} \geq h\right) - \mathsf{P}\left(\xi^2_{n+2p} \geq h\right)\right)\right] > 0. \tag{2.339}$$

A typical family of concentric ellipses and a power function $\beta(d^*,\lambda)$ are depicted in Figure 2.8.

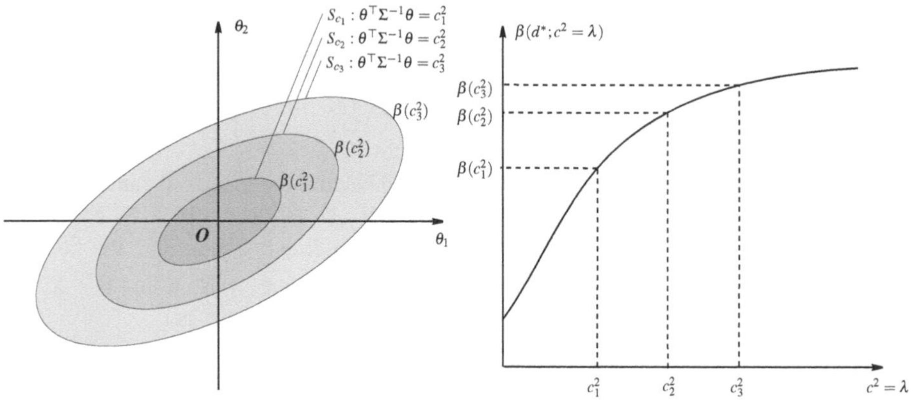

Figure 2.8: *Family of concentric ellipses (left) and the power function $\lambda \mapsto \beta(d^*;c^2=\lambda)$ (right) (Reprinted from : Décision et reconnaissance des formes en signal (Traité IC2, série Signal et image) Lengellé Régis, coord. ©Lavoisier, 2002, with kind permission from Hermes Sciences Publications, Lavoisier SAS).*

2.10.2 Minimax Test

Let us apply Theorem 2.9.5 for designing a minimax test for testing the mean of the Gaussian model.

2.10.2.1 Unit Covariance Matrix

Consider first an n-dimensional Gaussian random vector $X \sim \mathcal{N}(\theta, \mathbb{I}_n)$ and the problem of testing between the two following hypotheses (see [79] for further details) :

$$H_0 = \{\theta : \|\theta\|_2 \leq a\} \text{ and } H_1 = \{\theta : \|\theta\|_2 \geq b\} \text{ where } b > a. \tag{2.340}$$

In other words, we have an indifference zone between H_0 and H_1. These hypotheses are depicted

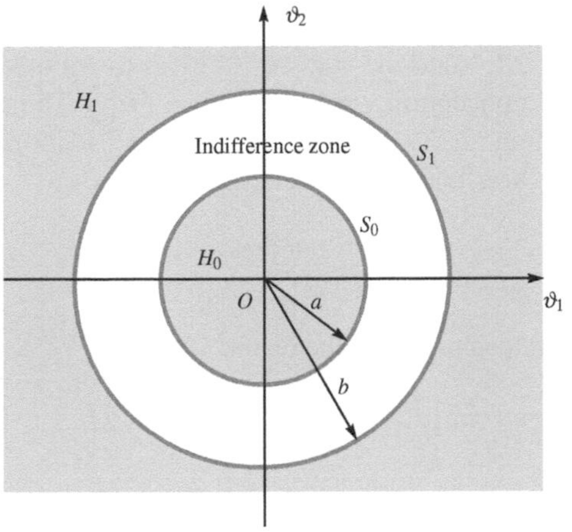

Figure 2.9: *Two hypotheses and an indifference zone.*

in Figure 2.9. As in Subsection 2.10.1, it turns out that, first, the least favorable distribution $\mathsf{P}_i(\theta)$ must remain invariant under the transformation $\bar{g}$, and second, this distribution must be concentrated on the boundary of Θ_i. Therefore, it follows that the least favorable distributions $\mathsf{P}_0(\theta)$ and $\mathsf{P}_1(\theta)$ are uniform distributions on the spheres $S_0 = \{\theta : \|\theta\|_2 = a\}$ and $S_1 = \{\theta : \|\theta\|_2 = b\}$; see Figure 2.9. In this case, *the minimax test is Bayesian*. Let us write the weighted LR between $H_{0,a} = \left\{\int \cdots \int_{S_0} f_\theta(X)p(\theta)dS(\theta)\right\}$ and $H_{1,b} = \left\{\int \cdots \int_{S_1} f_\theta(X)p(\theta)dS(\theta)\right\}$ by analogy with equation (2.325). In contrast to the case of the UBCP test, now the weighted LR is given by a ratio of two surface integrals

$$\Lambda_{\mathrm{NP}}(X) = \frac{\int \cdots \int_{S_1} f_\theta(X)p_1(\theta)dS(\theta)}{\int \cdots \int_{S_0} f_\theta(X)p_0(\theta)dS(\theta)}, \tag{2.341}$$

where $dS(\theta)$ is the surface element of the spheres S_i, $p_i(\theta) = 1/A(S_i)$ and $A(S_i)$ is the area of the surface S_i. After straightforward computations, we get

$$\Lambda_{\mathrm{NP}}(X) = \frac{A(S_0)}{A(S_1)} \exp\left(-\frac{b^2 - a^2}{2}\right) \frac{\int \cdots \int_{S_1} \exp\{\theta^\top X\} dS(\theta)}{\int \cdots \int_{S_0} \exp\{\theta^\top X\} dS(\theta)}. \tag{2.342}$$

Each integral in (2.342) can be re-written by analogy with Subsection 2.10.1.1 as

$$\Omega(c\Lambda) = \frac{1}{A(S_u)} \int \cdots \int_{S_u} \exp\left\{c\Lambda(X)\cos\left[\widehat{\theta U}\right]\right\} dS(\theta),$$

where $c = a$ or $c = b$ and S_u is the unit sphere. Hence, the LR can be re-written as

$$\Lambda_{\mathrm{NP}}(\Lambda) = \frac{\Omega(\Lambda b)}{\Omega(\Lambda a)}. \tag{2.343}$$

It turns out that $\Lambda_{\mathrm{NP}}(\Lambda)$ is a monotonically increasing function of $\Lambda = \|Y\|_2$ [79, 490]. Hence, the minimax test can be written finally as

$$\tilde{d}(X) = \begin{cases} 1 & \text{if} \quad \|X\|_2^2 \ \geq \ h(\alpha) \\ 0 & \text{if} \quad \|X\|_2^2 \ < \ h(\alpha) \end{cases} \tag{2.344}$$

It follows from Subsection 2.10.1.4 that the statistic $\|X\|_2^2$ is distributed according to the χ_n^2 law with n degrees of freedom and noncentrality parameter $\lambda = \|\theta\|_2^2$. The power function $\beta(\tilde{d};\lambda)$ of the test $\tilde{d}(X)$ is given by equation (2.337). The function $\beta(\tilde{d};\lambda)$ is monotonically increasing in λ for all $h(\alpha)$. It follows that the threshold $h(\alpha)$ is determined by

$$\sup_{\theta\in\Theta_0} \mathsf{E}_\theta(\tilde{d}(X)) = \sup_{\lambda\leq a^2} \mathsf{P}_\lambda(\|X\|_2^2 \geq h(\alpha)) = \mathsf{P}_{a^2}(\|X\|_2^2 \geq h(\alpha)) = \alpha. \tag{2.345}$$

Under this condition, the guaranteed power of the test is

$$\inf_{\theta\in\Theta_1} \mathsf{E}_\theta(\tilde{d}(X)) = \mathsf{P}_{b^2}(\|X\|_2^2 \geq h(\alpha)). \tag{2.346}$$

Therefore, all the conditions of Theorem 2.9.5 hold.

2.10.2.2 General Covariance Matrix

We now tackle the general covariance matrix case (see [79] for further details). Assume we have an r-dimensional random vector $X \sim \mathcal{N}(\theta,\Sigma)$. Consider the problem of testing the two hypotheses

$$\widetilde{H}_0 = \{\theta : \theta^\top\Sigma^{-1}\theta \leq a^2\} \text{ and } \widetilde{H}_1 = \{\theta : \theta^\top\Sigma^{-1}\theta \geq b^2\} \text{ where } b > a. \tag{2.347}$$

Let us show that this hypothesis testing problem can be transformed into the previous one. It follows from (2.332) that $\Sigma = RR^\top$ and also $\Sigma^{-1} = R^{-\top}R^{-1}$. We also know that the family $\mathcal{N}(\theta,\mathbb{I})$ remains invariant under the transformation $gX = RX$. Therefore, equation (2.313) can be written as

$$\Phi_{\theta,\mathbb{I}}(A) = \Phi_{\bar{g}(\theta,\mathbb{I})}(gA), \tag{2.348}$$

where $\bar{g}(\theta,\mathbb{I}) = (R\,\theta,\Sigma)$, $A = \{X : X^\top X < c^2\}$, $g(A) = \{Y = RX : X^\top X < c^2\} = \{Y : Y^\top\Sigma^{-1}Y < c^2\}$.

Thus, using $\bar{g}$ the problem of testing the hypotheses

$$\widetilde{H}_0 = \{\widetilde{\theta} = R\,\theta : \theta^\top\theta \leq a^2\} \text{ and } \widetilde{H}_1 = \{\widetilde{\theta} = R\,\theta : \theta^\top\theta \geq b^2\},$$

which are obviously equivalent to

$$\widetilde{H}_0 = \{\widetilde{\theta} : \widetilde{\theta}^\top\Sigma^{-1}\widetilde{\theta} \leq a^2\} \text{ and } \widetilde{H}_1 = \{\widetilde{\theta} : \widetilde{\theta}^\top\Sigma^{-1}\widetilde{\theta} \geq b^2\},$$

is equivalent to testing the hypotheses H_0 and H_1 in (2.340).

Therefore, the minimax test (2.344) with the critical region $X^\top X \geq h(\alpha)$, which was derived under the assumption that $X \sim \mathcal{N}(\theta,\mathbb{I})$ for testing the hypotheses (2.340), is also minimax under $Y \sim \mathcal{N}(\theta,\Sigma)$, where $Y = RX$, for testing the hypotheses (2.347). In other words, the minimax test for testing the hypotheses (2.347) can be written as

$$\widetilde{d}(X) = \begin{cases} 1 & \text{if} \quad X^\top\Sigma^{-1}X \geq h(\alpha) \\ 0 & \text{if} \quad X^\top\Sigma^{-1}X < h(\alpha) \end{cases}. \tag{2.349}$$

2.10.3 The Generalized Likelihood Ratio Test

The generalized likelihood ratio (GLR) test is one of the most popular and important methods for solving composite hypothesis testing problems [79, 47, 490]. Again consider the parametric family $\{\mathsf{P}_\theta\}_{\theta\in\Theta}$ and two composite hypotheses $H_i : \theta \in \Theta_i, i = 0, 1$. Let $\mathbf{X}_1^n = (X_1, \dots, X_n)^\top$ be the observed sample of the fixed size n and let $p_\theta(\mathbf{X}_1^n)$ be a pdf (or pmf) of $\mathbf{X}_1^n$.

Define the statistic

$$\hat{\Lambda}(\mathbf{X}_1^n) = \frac{\sup_{\theta\in\Theta_1} p_\theta(\mathbf{X}_1^n)}{\sup_{\theta\in\Theta_0} p_\theta(\mathbf{X}_1^n)} \tag{2.350}$$

and the decision rule

$$\hat{d}(\mathbf{X}_1^n) = \begin{cases} 1 & \text{if} \quad \hat{\Lambda}(\mathbf{X}_1^n) \geq h(\alpha) \\ 0 & \text{if} \quad \hat{\Lambda}(\mathbf{X}_1^n) < h(\alpha), \end{cases} \tag{2.351}$$

where the threshold $h(\alpha)$ is found from the equation

$$\sup_{\theta\in\Theta_0} \mathsf{E}_\theta(\hat{d}(\mathbf{X}_1^n)) = \sup_{\theta\in\Theta_0} \mathsf{P}_\theta(\hat{\Lambda}(\mathbf{X}_1^n) \geq h(\alpha)) = \alpha. \tag{2.352}$$

Definition 2.10.2 (GLR test). We shall call a test $\hat{d}$ defined in (2.350)–(2.352) the *generalized likelihood ratio test* of level α for testing between the hypotheses $H_0 = \theta : \theta \in \Theta_0$ and $H_1 : \theta \in \Theta_1$.

The statistic $\hat{\Lambda}(\mathbf{X}_1^n)$ is usually referred to as the *generalized likelihood ratio.*

Therefore, by definition the test $\hat{d}$ belongs to the class $\mathbf{C}(\alpha)$. The precise optimal properties of the GLR test in the general case are unknown, but for many special cases, the GLR test is optimal in a certain sense. In particular, the next example shows that the GLR test is the minimax test for testing two hypotheses on the mean of the normal population.

Example 2.10.1 (Gaussian vector sequence – Contd). Let us show that the minimax test given by (2.344) and (2.349) is a GLR test, in the case where $X \sim \mathcal{N}(\theta, \mathbb{I})$. Specifically, consider the problem of testing the hypotheses $H_0 : \|\theta\| \leq a$ and $H_1 : \|\theta\| \geq b$, where $b > a$. In this case, the critical region of the GLR test can be written as

$$\begin{aligned} S(\mathbf{X}_1^n) = \log \hat{\Lambda}(\mathbf{X}_1^n) &= \log \sup_{\|\theta\|\geq b} \exp\left\{-\frac{1}{2}\sum_{i=1}^n \|X_i - \theta\|_2^2\right\} \\ &- \log \sup_{\|\theta\|\leq a} \exp\left\{-\frac{1}{2}\sum_{i=1}^n \|X_i - \theta\|_2^2\right\} \geq \log h(\alpha). \end{aligned} \tag{2.353}$$

After simple transformations, we get

$$S(\mathbf{X}_1^n) = \sup_{\|\theta\|\geq b} \left\{-\frac{n}{2}\left\|\theta - \frac{1}{n}\sum_{i=1}^n X_i\right\|_2^2\right\} - \sup_{\|\theta\|\leq a} \left\{-\frac{n}{2}\left\|\theta - \frac{1}{n}\sum_{i=1}^n X_i\right\|_2^2\right\} \geq \log h(\alpha) \tag{2.354}$$

Clearly, the statistic $S(\mathbf{X}_1^n)$ can be re-written as

$$S(\mathbf{X}_1^n) = \begin{cases} -\frac{n}{2}(\|\overline{X}_n\|_2 - b)^2 & \text{if} \quad \|\overline{X}_n\|_2 \leq a \\ -\frac{n}{2}(\|\overline{X}_n\|_2 - b)^2 + \frac{n}{2}(\|\overline{X}_n\| - a)^2 & \text{if} \quad a \leq \|\overline{X}_n\|_2 \leq b \\ +\frac{n}{2}(\|\overline{X}_n\|_2 - a)^2 & \text{if} \quad \|\overline{X}_n\|_2 \geq b \end{cases}, \tag{2.355}$$

where $\overline{X}_n = \frac{1}{n}\sum_{i=1}^n X_i$. Therefore, $S(\mathbf{X}_1^N)$ is a continuous increasing function of $\|\overline{X}_n\|$, so that the GLR test $\hat{d}(\mathbf{X}_1^n)$ of (2.351) coincides with the minimax test $\widetilde{d}(\mathbf{X}_1^n)$ of (2.344) for a suitable constant $h(\alpha)$.

2.10.3.1 Linear Model with Nuisance Parameters: UBCP Test

Consider the Gaussian linear model

$$Y = HX + M\theta + \xi, \tag{2.356}$$

where $Y \in \mathbb{R}^n$ is the observation vector, $\theta \in \mathbb{R}^r$ is the informative parameter, $X \in \mathbb{R}^m$ is an unknown and non-random state vector (nuisance parameter), M is a full column rank matrix of size $n \times r$ with $r < n$ and H is a matrix of size $n \times m$ with $\mathsf{rank}(H) = q$. It is assumed that $n \geq q + r$ and ξ is a zero mean Gaussian noise $\xi \sim \mathcal{N}(0, \sigma^2 \mathbb{I}_n)$, $\sigma^2 > 0$. In contrast to the model given by (2.333), the new model (2.356) includes a nuisance parameter X. Hence, the hypothesis testing problem consists in deciding between

$$H_0 : \theta = 0, \ X \in \mathbb{R}^m \text{ and } H_1 : \theta \neq 0, \ X \in \mathbb{R}^m, \tag{2.357}$$

considering X as an unknown nuisance vector (see [150] for further details). The hypotheses (2.357) are defined by the following parametric sets

$$\Omega_0 = \{\theta = 0, \ X \in \mathbb{R}^m\} \text{ and } \Omega_1 = \{\theta \neq 0, \ X \in \mathbb{R}^m\}. \tag{2.358}$$

Since the nuisance parameter X is non-random and its values are not bounded ($X \in \mathbb{R}^m$), it is desir-

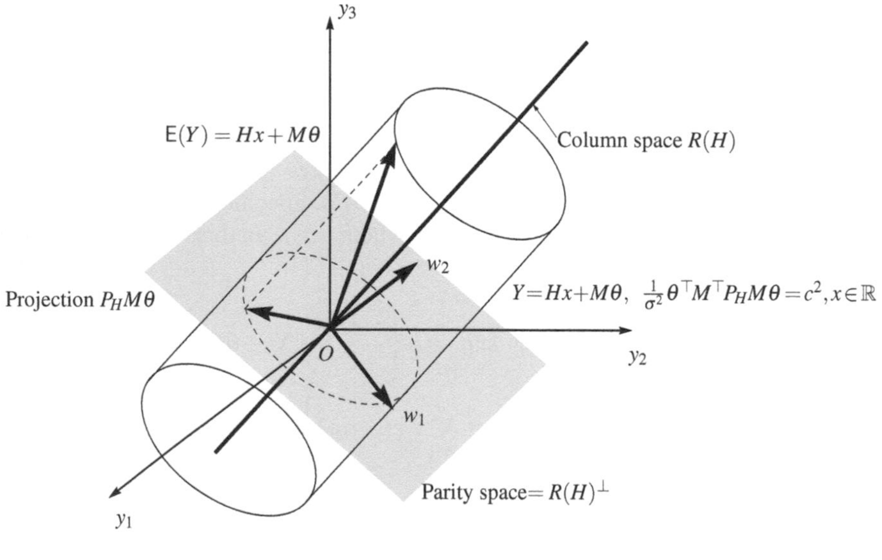

Figure 2.10: *The column space and its orthogonal complement (parity space) (Reprinted from Automatica, Vol. 41(7), M. Fouladirad, I. Nikiforov, "Optimal statistical fault detection with nuisance parameters", 1157–1171, Copyright (2005), with kind permission from Elsevier).*

able to eliminate the impact of X on the decision statistic $\Lambda(Y)$ using the invariance principle. Note that the family $\mathcal{N}(HX + M\theta, \sigma^2 \mathbb{I}_n)$ and the hypothesis testing problem (2.357) remain invariant under the group of translations $G = \{g : g(Y) = Y + HC\}$, $C \in \mathbb{R}^m$. The generated group $\bar{G}$ is also the group of translations $\bar{G} = \{\bar{g} : \bar{g}(Y) = Y + HC\}$, $C \in \mathbb{R}^m$. The sets Ω_i are preserved by this group, namely $\bar{g}\Omega_i = \Omega_i$, $i = 0, 1$.

To apply the invariance principle, define the column space $R(H)$ of the matrix H; see Figure 2.10. The solution is projecting Y on the orthogonal complement $R(H)^\perp$ of the column space $R(H)$. The space $R(H)^\perp$ is usually referred to as the "parity space" in the analytical redundancy literature [151]. Consider the group of translations $G = \{g : g(Y) = Y + HX\}$ and the statistic $M(Y) = Z = WY$. The parity vector $Z = WY$ is the transformation of the measured output Y into a set of $n-q$ linearly independent variables by projection onto the left null space of the matrix H. The matrix $W^\top = (w_1,\ldots,w_{n-q})$ of size $n\times(n-q)$ is composed of the eigenvectors $w_1,\ldots,w_{n-q}$ of the projection matrix $P_H = \mathbb{I}_n - H(H^\top H)^- H^\top$, where A^- is a generalized inverse of A [237, Sec. 1.5.3], corresponding to the eigenvalue 1. If the matrix H is full column rank, i.e., $q=m$, then $P_H = \mathbb{I}_n - H(H^\top H)^{-1}H^\top$. The matrix W satisfies the following conditions:

$$WH = 0,\ \ W^\top W = P_H,\ \ WW^\top = \mathbb{I}_{n-q}. \tag{2.359}$$

First, it immediately follows from the properties of the matrix W that $W(g(Y)) = W(Y+HX) = WY + WHX = WY$. Second, $WY_1 = WY_2$ implies $W^\top WY_1 = W^\top WY_2$ or $(\mathbb{I}_n - H(H^\top H)^{-1}H^\top)Y_1 = (\mathbb{I}_n - H(H^\top H)^{-1}H^\top)Y_2$. This leads to $Y_1 = Y_2 + H(H^\top H)^{-1}H^\top(Y_1-Y_2)$ or $Y_1 = Y_2 + HX$, where $X = (H^\top H)^{-1}H^\top(Y_1 - Y_2)$. Hence the statistic $Z = WY$ is a maximal invariant statistic. As follows from Proposition 2.9.1, any invariant test should be designed using maximal invariant, i.e., in our case the statistic

$$Z = WY = WM\theta + W\xi = WM\theta + \zeta. \tag{2.360}$$

Next, we apply Theorem 2.10.3 to design the UBCP invariant test and additionally assume that the $(n-q)\times r$ matrix WM is full column rank. Using (2.360), we obtain that the LS estimator of θ is given by $\widehat{\theta} = \left((WM)^\top(WM)\right)^{-1}(WM)^\top Z$ and the Fisher matrix is $\mathcal{F}_{\widehat{\theta}} = \frac{1}{\sigma^2}(WM)^\top WM$. It directly follows from (2.334) that the test given by

$$d^*(Y) = \begin{cases} 1 & \text{when} \quad \Lambda(Y) \geq h(\alpha) \\ 0 & \text{when} \quad \Lambda(Y) < h(\alpha) \end{cases}, \tag{2.361}$$

where $\Lambda(Y) = \frac{1}{\sigma^2}Y^\top P_H M\left(M^\top P_H M\right)^{-1}M^\top P_H Y$, is UBCP over the following family of surfaces (see Figure 2.10)

$$\mathcal{S}_{WM} = \left\{ \mathcal{S}_c : \frac{1}{\sigma^2}\|WM\theta\|_2^2 = c^2,\ c>0 \right\}. \tag{2.362}$$

2.10.3.2 Linear Model with Nuisance Parameters: Constrained Minimax Test

We continue a discussion of the constrained minimax test for multiple hypotheses presented in Subsection 2.9.6.3 in more detail for the Gaussian linear model with the nuisance parameter X,

$$Y = HX + \mu + \xi, \tag{2.363}$$

and M hypotheses $H_0,\ldots,H_{M-1}$ defined in the following manner [147]:

$$H_0 : \{\mu = 0,\ \ X\in\mathbb{R}^m\} \text{ and } H_j : \{\mu = \rho\,\theta_j,\ \rho\geq\rho_j,\ \ X\in\mathbb{R}^m\},\ j = 1,\ldots,M-1, \tag{2.364}$$

where $\rho_j > 0$ are known constants and $\theta_j \in \mathbb{R}^n$ are known vectors. In the context of simultaneous detection and classification problems, the goal is to detect the presence of a vector $\rho\,\theta_j$ with an amplitude greater than $\rho_j\,\|\theta_j\|_2$ and to identify the hypothesis of type j, $H_j : \mu = \rho\theta_j$ given that $\rho\geq\rho_j$.

As in the previous subsection, the hypothesis testing problem (2.364) includes a nuisance parameter X, and it remains invariant under the group of translations $G = \{g : g(Y) = Y + HC\}$, $C\in\mathbb{R}^m$. The generated group is also $\bar{G} = \{\bar{g} : \bar{g}(Y) = Y + HC\}$, $C\in\mathbb{R}^m$ and the sets $\Omega_j : \{\mu = \rho\,\theta_j,\ \rho\geq\rho_j,\ \ X\in\mathbb{R}^m\}$ are preserved by this group of translations, $\bar{g}\Omega_j = \Omega_j$, $j = 0,\ldots,M-1$.

Using the maximal invariant $Z = WY = W[HX + \mu + \xi]$ introduced in (2.360), the initial hypothesis testing problem (2.364) can be re-written for the maximal invariant Z. Let us additionally assume that the parameters $\rho_1,\ldots,\rho_{M-1}$ satisfy $\rho_{\min} = \rho_j\left\|W^\top\theta_j\right\|_2$, $j = 1,\ldots,M-1$, where $\rho_{\min}$ is a given positive constant. This means that all projections $W^\top[HX + \rho\theta_j] = \rho W^\top\theta_j$ onto the subspace of the maximal invariant statistic have the same minimum norm $\rho_{\min}$.

Hence, the initial multiple decision problem of testing the hypotheses $H_0,\dots,H_{M-1}$ is reduced to testing the following hypotheses associated with the maximal invariant statistic:

$$\underline{H_0}:\{Z\sim\mathcal{N}(0,\sigma^2,\mathbb{I}_n)\}\text{ and }\underline{H_j}:\{Z\sim\mathcal{N}(\rho\varphi_j,\sigma^2,\mathbb{I}_n),\rho\geq\rho_{\min}\},\ j=1,\dots,M-1,\qquad(2.365)$$

where $\varphi_j=\rho_j W^\top\theta_j/\rho_{\min}$, $\|\varphi_j\|_2=1$, $j=1,\dots,M-1$.

It is worth noting that the number of alternative hypotheses, $(M-1)$, can be greater than the dimension of the observation vector, $M-1>n$. In contrast to the slippage problem discussed in Subsection 2.9.6.2 and Example 2.9.7, the hypothesis testing problem (2.364) does not possess any natural symmetry or invariance under the group of permutations of the set $\{H_1,\dots,H_{M-1}\}$. For this reason, we cannot find an optimal minimax test and propose to design an asymptotically constrained minimax test.

Remark 2.10.3. To find a solution to the multiple hypothesis testing problem two additional assumptions on the mutual geometry between the anomalies φ_j are necessary. First, let $I_{\ell,j}=\|\varphi_\ell-\varphi_j\|_2$, $1\leq\ell,j\leq M-1$, be the distance between two unit vectors φ_ℓ and φ_j. The minimum classification distance I_j for the vector φ_j is given by $I_j=\min_{1\leq\ell\neq j\leq M-1}I_{j,\ell}$ and the minimum classification distance over all vectors is $I^*=\min_{1\leq j\leq M-1}I_j$. Let us define for a vector φ_j the number η_j of vectors φ_ℓ such that $I_{j,\ell}=I_j$. Hence η_j is the number of equidistant closest vectors from φ_j. It is assumed that $\eta_i=1$ for all $1\leq i\leq M-1$. Second, it is assumed that $0<I^*<2$, which means that the case of two alternative hypotheses $(M-1=2)$ such that $\varphi_1=-\varphi_2$ is excluded.

Definition 2.10.3. As $\rho_{\min}\to\infty$ a test $\overline{d}(Z)$ is *constrained asymptotically uniformly minimax* of level α between the hypotheses $H_0,\dots,H_{M-1}$ if $\overline{d}(Z)\in\mathbf{C}(\alpha)$ and if for any other test $d(Z)\in\mathbf{C}(\alpha)$ the following inequality for the probability of false classification is satisfied

$$\alpha_{\max}(\rho;\overline{d})\leq(1+\varepsilon(\rho_{\min}))\,\alpha_{\max}(\rho;d),\ \ \forall\rho\geq\rho_{\min},$$

where $\alpha_{\max}(\rho;d)=\max_{1\leq j\leq M-1}\alpha_j(\rho;d)$ and $\varepsilon(\rho_{\min})\to0$ as $\rho_{\min}\to\infty$.

For many practical situations, the vector Y defined by regression model (2.363) is observed N times, $Y_1,\dots,Y_N$. In such a case the previous definition is valid for $N\geq N_{\min}$ instead of $\rho\geq\rho_{\min}$ as $N_{\min}\to\infty$.

Hence, the criterion of optimality consists in finding a test with the smallest maximum probability of false classification in the class $\mathbf{C}(\alpha)$ of G-invariant tests, whatever the value $\rho\geq\rho_{\min}$. We refer to [147] for a complete proof of the following result. Let $Q^{-1}(x)$ denote the inverse function of the tail probability of the standard normal distribution $Q(x)=1-\Phi(x)$.

Theorem 2.10.4. *Consider the multiple hypothesis testing problem specified in* (2.365)*. Suppose that the two assumptions defined in Remark 2.10.3 hold: i)* $\eta_i=1$ *for all* $1\leq i\leq M-1$ *and ii)* $0<I^*<2$*. Let* $0<\alpha<1$*. The constrained asymptotically uniformly minimax test* $\overline{d}(Z)\in\mathbf{C}(\alpha)$ *is given by*

$$\overline{d}(Z)=\begin{cases}\ell & \text{if}\ \ \max_{1\leq j\leq M-1}\left(\varphi_j^\top Z\right)\geq h\\ 0 & \text{if}\ \ \max_{1\leq j\leq M-1}\left(\varphi_j^\top Z\right)<h\end{cases},\ \ell=\arg\max_{1\leq j\leq M-1}\left(\varphi_j^\top Z\right),\qquad(2.366)$$

where the threshold h satisfies the following equality

$$h=\sigma Q^{-1}(\alpha/m).\qquad(2.367)$$

The maximal false classification probability of the test $\overline{d}(Z)$ *is asymptotically equal to*

$$\alpha_{\max}(\rho;\overline{d})\sim Q\left(\frac{\rho I^*}{2\sigma}\right)\quad\text{for all}\ \rho\geq\rho_{\min}\ \ \text{as}\ \ \rho_{\min}\to\infty.$$

2.11 Hypothesis Testing: Asymptotic Approaches

2.11.1 Motivation

As it follows from the previous sections, the optimal tests can be easily obtained under essential restrictions: a scalar parameter, an exponential model, *etc.* However, there exists one case in which the design and analysis of optimal tests is simpler than in the general case. This case is encountered in the so-called asymptotic approach to hypothesis testing. Let us briefly discuss the main ideas of this approach.

We have shown in the previous sections that the design of optimal tests are especially simple in the case of a Gaussian linear model. Without a "Gaussian" assumption the design of optimal tests may be difficult. Moreover, the investigation of the properties of a test consists in computing a threshold and the probabilities of errors. Obviously, these computations require knowledge of the cdf of the LLR $\log \Lambda(\mathbf{X}_1^n)$ or a certain function of $\log \Lambda(\mathbf{X}_1^n)$. However, in general the distribution of the LLR is difficult to compute, and thus tuning of the threshold and the computation of the error probabilities are tricky.

Thus, it is of interest to discuss some approximate solutions to hypothesis testing for large samples of iid random variables. The asymptotic point of view for getting approximations for these probabilities consists in replacing the test $d(\mathbf{X}_1^n)$ by a sequence of tests for each n. There are two possible asymptotic approaches [79]. The first approach is called the *large deviation approach* and consists in assuming that the distributions P_0 and P_1 are fixed. In other words, the "distance" between them does not depend upon the sample size n. The second approach is called the *local (hypotheses) approach* and consists in assuming that the "distance" between P_0 and P_1 depends upon n in such a way that the two hypotheses get closer to each other when n grows to infinity [47, 79, 109, 202, 260, 400]. We now concentrate on this second approach because we use it throughout the book.

2.11.2 Local Asymptotic Expansion of the Likelihood Ratio

The LR has been shown to be a central tool in hypothesis testing problems. Thus the investigation of its properties under local asymptotic conditions, namely close hypotheses, is of interest. On the other hand, a theory of contiguity or closeness of probability measures was developed in [109, 202, 260, 400]. Let us introduce its main features.

We consider a parametric family of distributions $\mathcal{P} = \{\mathsf{P}_\theta\}_{\theta \in \Theta}$, $\Theta \subset \mathbb{R}^\ell$, satisfying some regularity assumptions [109, 202, 400] and a sample of size n. Let $(\nu_n \Upsilon)_{n \geq 1}$ be a convergent sequence of points in the space $\mathbb{R}^\ell$ such that $\nu_n \to \nu \in \mathbb{R}$ and where $\|\Upsilon\|_2 = 1$ defines the direction. Let $\theta_n = \theta + \nu_n/\sqrt{n}\,\Upsilon$. Therefore, the distance between the hypotheses :

$$H_0 = \{\mathcal{L}(X) = \mathsf{P}_\theta\} \text{ and } H_1 = \left\{\mathcal{L}(X) = \mathsf{P}_{\theta + \nu_n/\sqrt{n}\,\Upsilon}\right\} \tag{2.368}$$

depends upon n in such a way that the two probability measures get closer to each other when n grows to infinity. The LLR between these hypotheses for the sample $\mathbf{X}_1^n$ is

$$\log \Lambda_n(\theta, \theta_n) = \log \frac{p_{\theta_n}(\mathbf{X}_1^n)}{p_\theta(\mathbf{X}_1^n)}.$$

Definition 2.11.1 (LAN family of distributions). The parametric family of distributions $\mathcal{P} = \{\mathsf{P}_\theta\}_{\theta \in \Theta}$ is called *locally asymptotic normal* (LAN) if the LLR between the hypotheses H_0 and H_1 can be written as

$$\log \Lambda_n(\theta, \theta_n) = \nu\ \Upsilon^\top \Delta_n(\theta) - \nu^2/2\ \Upsilon^\top\ \mathcal{F}_n(\theta)\ \Upsilon + \varepsilon_n(\mathbf{X}_1^n, \theta, \nu\Upsilon), \tag{2.369}$$

where $\Delta_n(\theta) = 1/\sqrt{n}\,\nabla\,[\log p_\theta(\mathbf{X}_1^n)] = S_n/\sqrt{n}$, $\mathcal{F}_n(\theta)$ is the Fisher information matrix for the sample $\mathbf{X}_1^n$, $\varepsilon_n(\mathbf{X}_1^n, \theta, \nu\Upsilon) \xrightarrow[n\to\infty]{\mathsf{P}_\theta-\text{a.s.}} 0$, and where the following asymptotic normality holds

$$\Delta_n(\theta) \xrightarrow[n\to\infty]{\text{law}} \mathcal{N}(0, \mathcal{F}(\theta)).$$

Let $\mathcal{P}$ be a LAN family of distributions. Let $\Upsilon_\nu = \nu\Upsilon$, and denote $\hat{\Upsilon}_\nu = \sqrt{N}(\hat{\theta} - \theta)$ the value of the parameter Υ_ν for which $\log \Lambda_n(\theta, \theta_n)$ is maximum. This value is

$$\hat{\Upsilon}_\nu = \mathcal{F}_n^{-1}(\theta)\, \Delta_n(\theta) + \varepsilon_n(\mathbf{X}_1^n, \theta, \Upsilon_\nu) \tag{2.370}$$

and the LLR $\log \Lambda_n(\theta, \theta + \hat{\Upsilon}_\nu/\sqrt{n})$ can be re-written as

$$2\, \log \Lambda_n\left(\theta, \theta + \hat{\Upsilon}_\nu/\sqrt{n}\right) = \Delta_n^\top(\theta)\, \mathcal{F}_n^{-1}(\theta)\, \Delta_n(\theta) + \varepsilon_n(\mathbf{X}_1^n, \theta, \Upsilon_\nu).$$

As proven in [109, 202, 400], LAN properties exist for some important special cases. More precisely, the asymptotic local expansion (2.369) can be derived if

(i) $(X_n)_{n\geq 1}$ is a sequence of iid random variables;

(ii) $(X_n)_{n\geq 1}$ is a stationary Markov process of order p;

(iii) $(X_n)_{n\geq 1}$ is a stationary Gaussian random process, in particular $(X_n)_{n\geq 1}$ is an *autoregressive moving average* (ARMA) process.

Furthermore, we have the following asymptotic normality of $\log \Lambda_n(\theta, \theta_n)$, $\Delta_n(\theta)$ and $\hat{\Upsilon}_\nu$:

$$\log \Lambda_n(\theta, \theta_n) \xrightarrow[n\to\infty]{\text{law}} \begin{cases} \mathcal{N}\left(-\nu^2/2\ \Upsilon^\top\, \mathcal{F}(\theta)\, \Upsilon, \Upsilon^\top\, \mathcal{F}(\theta)\, \Upsilon\right) & \text{under} \quad \mathsf{P}_\theta \\ \mathcal{N}\left(+\nu^2/2\ \Upsilon^\top\, \mathcal{F}(\theta)\, \Upsilon, \Upsilon^\top\, \mathcal{F}(\theta)\, \Upsilon\right) & \text{under} \quad \mathsf{P}_{\theta+\nu/\sqrt{N}\,\Upsilon} \end{cases}, \tag{2.371}$$

$$\Delta_n(\theta) \xrightarrow[n\to\infty]{\text{law}} \mathcal{N}(\mathcal{F}(\theta)\,\Upsilon, \mathcal{F}(\theta)) \text{ under } \mathsf{P}_{\theta+\nu/\sqrt{n}\,\Upsilon},\ \hat{\Upsilon}_\nu \xrightarrow[n\to\infty]{\text{law}} \mathcal{N}(0, \mathcal{F}^{-1}(\theta)) \text{ under } \mathsf{P}_\theta.$$

We also have the following convergence

$$2 \log \Lambda_n\left(\theta, \theta + \hat{\Upsilon}_\nu/\sqrt{n}\right) \xrightarrow[n\to\infty]{\text{law}} \chi_\ell^2 \text{ under } \mathsf{P}_\theta, \tag{2.372}$$

where χ_ℓ^2 is a standard chi-squared random variable with ℓ degrees of freedom. Note that the random variable $\varepsilon_n(\mathbf{X}_1^n, \theta, \Upsilon_\nu)$ converges to zero almost surely as $n \to \infty$ under $\mathsf{P}_{\theta+\nu/\sqrt{n}\,\Upsilon}$.

The important corollary of the LAN properties for a parametric family $\mathcal{P}$ satisfying the regularity conditions, is that the LR $\Lambda_n(\theta, \theta_n)$ behaves approximately as if the family were exponential. It results that the *efficient score* $\Delta_n(\theta)$ is an *asymptotic sufficient statistic*. Moreover, it follows from the above asymptotic normality that it is possible to transform the asymptotic local hypothesis testing problem $H_0 : \{\mathcal{L}(X) = \mathsf{P}_\theta\}$ against $H_1 : \left\{\mathcal{L}(X) = \mathsf{P}_{\theta+\nu/\sqrt{n}\,\Upsilon}\right\}$ into a much simpler hypothesis testing problem for the mean of a Gaussian law. We continue to investigate this result in the next subsections.

Finally, we comment on the reason why the speed of convergence of the hypotheses H_0 and H_1 was chosen of order $1/\sqrt{n}$. In hypothesis testing problems, the contiguity of the probability measures must be compensated by the growth of the sample size n. If $\theta_n = \theta_0 + \nu/\sqrt{n}\,\Upsilon$, then the quantity of information for distinguishing between the hypotheses H_0 and H_1 remains constant when n goes to infinity. For this reason, the probabilities of errors of first and second types tend to fixed values.

2.11.3 The Main Idea of the Asymptotically Optimal Tests

Let us consider the following two hypothesis testing problems.

First problem. Assume that $\mathcal{P}$ is a LAN family, and that we want to test the local hypotheses (2.368) by using $\mathbf{X}_1^n$ when $n \to \infty$.

Second problem. Assume that the family $\mathcal{P}$ is such that $\mathcal{L}(X) = \mathcal{N}(\Upsilon_\nu, \Sigma)$, where $\Sigma = \mathcal{F}^{-1}(\theta^*)$, and that we want to test between the hypotheses $H_0 = \{\Upsilon_\nu \in \Gamma_0\}$ and $H_1 = \{\Upsilon_\nu \in \Gamma_1\}$ about the mean of a Gaussian law by using one sample point X_1.

Now, assume that the second problem can be solved by an optimal (UMP, Bayes, minimax, *etc.*) test $d(X_1)$. Let us denote the maximum likelihood estimate as $\hat{\theta}$, and let $\hat{\Upsilon}_\nu = \sqrt{n}(\hat{\theta} - \theta^*)$. Then the test $g(\hat{\Upsilon}_\nu)$ for the first problem has *asymptotically* the same properties as the optimal test $d(X_1)$ for the second problem.

This idea can be explained as follows. Let $\theta = \theta^* + \Upsilon_\nu/\sqrt{n}$. It results from (2.370) and (2.372) that

$$\sqrt{n}(\hat{\theta} - \theta^*) \xrightarrow[n\to\infty]{\text{law}} \mathcal{N}(\Upsilon_\nu, \mathcal{F}^{-1}(\theta^*)).$$

This normal distribution with mean vector Υ_ν and covariance matrix $\mathcal{F}^{-1}(\theta^*)$ is precisely the distribution in the second problem.

2.11.4 *Asymptotically Optimal Tests for Two Simple Hypotheses*

It follows from Theorem 2.9.1 that in the case of two simple hypotheses we have to compute the two following probabilities:

$$\alpha_0(d^*) = \mathsf{P}_0\left(\sum_{i=1}^{n} \log\frac{p_1(X_i)}{p_0(X_i)} \geq \log h\right), \quad \alpha_1(d^*) = \mathsf{P}_1\left(\sum_{i=1}^{n} \log\frac{p_1(X_i)}{p_0(X_i)} < \log h\right) \tag{2.373}$$

Let $n \to \infty$. As follows from Subsection 2.11.2, the distribution of the random variable $S_n/\sqrt{n}$ weakly converges to the normal random variable as $n \to \infty$. This normal approximation along with the expansion (2.369) lead to the following approximations for the error probabilities $\alpha_0(d^*)$ and $\alpha_1(d^*)$ defined in (2.373):

$$\alpha_0(d^*) \approx 1 - \Phi\left(\frac{\nu^2\mathcal{F}(\theta_0) + 2\log h}{2|\nu|\sqrt{\mathcal{F}(\theta_0)}}\right), \quad \alpha_1(d^*) \approx \Phi\left(\frac{-\nu^2\mathcal{F}(\theta_0) + 2\log h}{2|\nu|\sqrt{\mathcal{F}(\theta_0)}}\right).$$

Definition 2.11.2 (Asymptotic equivalence of tests). The two tests $d_0(\mathbf{X}_1^n)$ and $d_1(\mathbf{X}_1^n)$ are said to be *asymptotically equivalent* if

$$\limsup_{n\to\infty} |\alpha_j(d_0) - \alpha_j(d_1)| = 0, \quad j = 0, 1.$$

A test which is asymptotically equivalent to an MP test is called an *asymptotically MP test* (AMP).

Therefore, using the Gaussian approximation (2.371), we deduce that the test defined by

$$\operatorname{sign}(\nu)\,\frac{1}{\sqrt{n}}\,S_n \underset{H_0}{\overset{H_1}{\gtrless}} \operatorname{sign}(\nu)\,\frac{\nu^2\,\mathcal{F}(\theta) + 2\log h}{2|\nu|}$$

is asymptotically equivalent to the MP test $d^*(\mathbf{X}_1^n)$ from Theorem 2.9.1, and thus it is an AMP test.

Part I

Sequential Hypothesis Testing

Chapter 3

Sequential Hypothesis Testing: Two Simple Hypotheses

3.1 Sequential Probability Ratio Test

In this chapter we use the notation of Subsection 2.7.1, assuming that $\theta_n = \theta$, where θ is an unknown parameter or more generally a situation (not necessarily random) taking on the two values 0 or 1, so that $\Theta_n = \Theta = \{0,1\}$ and $p_{\theta=i}(\mathbf{X}_1^n) = p_i(\mathbf{X}_1^n)$, $i = 0,1$ are two distinct densities. The goal is to test two hypotheses $H_0 : \theta = 0$ and $H_1 : \theta = 1$, i.e., to identify which one of the two densities p_0 or p_1 is the true density. In applications, often $\{p_\theta(\mathbf{X}_1^n)\}$ is a parametric family, and the hypotheses are of the form $H_0 : \theta = \theta_0$ and $H_1 : \theta = \theta_1$. We would like to test the hypotheses sequentially, taking advantage of the fact that as an observation arrives, our knowledge of the true state of the process becomes more refined, so that we may decide whether or not more data are needed to make a final decision. Accordingly if we decide to stop at the stage n the terminal decision takes on two values $d_n = 0$ or 1, where $d_n = i$ means that the hypothesis H_i is accepted and therefore the alternative hypothesis is rejected.

In the mid 1940s, Wald [492, 494] introduced the *Sequential Probability Ratio Test* (SPRT) on the basis of the sequence of iid observations $X_1, X_2, \dots$, which have a common pdf $f_\theta(X)$ with respect to a sigma-finite measure $\mu(x)$, in which case

$$p_\theta(\mathbf{X}_1^n) = \prod_{k=1}^n f_\theta(X_k), \qquad \theta = 0,1.$$

Let

$$\Lambda_n(\mathbf{X}_1^n) = \prod_{k=1}^n \frac{f_1(X_k)}{f_0(X_k)}$$

be the likelihood ratio (LR) between the hypotheses H_1 and H_0 for the sample $\mathbf{X}_1^n = (X_1, \dots, X_n)^\top$ and let $A_0 \le 1 \le A_1$ be two constants (thresholds). After n observations have been made for each $n \ge 1$:

$$\begin{aligned} &\text{Stop and accept } H_1 \quad \text{if} \quad \Lambda_n \ge A_1. \\ &\text{Stop and accept } H_0 \quad \text{if} \quad \Lambda_n \le A_0. \\ &\text{Continue sampling} \quad \text{if} \quad A_0 < \Lambda_n < A_1. \end{aligned}$$

We denote by P_i and E_i the probability and the expectation under the hypothesis H_i. Also, let $\delta = (d, T)$ denote a generic decision rule (hypothesis test). More specifically, T is a Markov stopping time with respect to the filtration $\{\mathscr{F}_t\}_{t \ge 0}$, i.e., $\{T \le t\} \in \mathscr{F}_t$, and $d = d(X_0^T)$ is an $\mathscr{F}_T$-measurable terminal decision function with values in the set $\{0,1\}$. Therefore, $d = i$ is identified with accepting the hypothesis H_i, that is, $\{d = i\} = \{T < \infty,\ \delta \text{ accepts } H_i\}$. Further, let $\alpha_0(\delta) = \mathsf{P}_0(d = 1)$ and $\alpha_1(\delta) = \mathsf{P}_1(d = 0)$ be the error probabilities of the test δ.

Let $Z_k = \log[f_1(X_k)/f_0(X_k)]$ be the log-likelihood ratio (LLR) for the observation X_k and let

$$\lambda_n = \log \Lambda_n = \sum_{k=1}^n Z_k, \qquad n = 1, 2, \dots$$

be the LLR process. Let $a_0 = -\log A_0 (\geq 0)$ and $a_1 = \log A_1 (\geq 0)$. Clearly, the SPRT can be represented in the form

$$\delta^* = (d^*, T^*), \quad T^* = \inf\{n \geq 0 : \lambda_n \notin (-a_0, a_1)\}, \quad d^* = \begin{cases} 1 & \text{if } \lambda_{T^*} \geq a_1 \\ 0 & \text{if } \lambda_{T^*} \leq -a_0. \end{cases} \tag{3.1}$$

We take the convention that before the observations become available the value of the LLR is $\lambda_0 = 0$. Therefore, if $-a_0 < 0 < a_1$, then the SPRT never stops at $n = 0$. On the other hand, if $a_0 = 0 < a_1$, then the hypothesis H_0 is accepted w.p. 1 before the sampling begins, and if $-a_0 < 0 = a_1$, then in contrast the hypothesis H_1 is accepted immediately. Clearly, this is a sort of degenerate case from the practitioners' point of view.

As we shall see, in the iid case Wald's SPRT δ^* is optimal not only in the Bayesian context but it also has the following surprisingly strong optimality property: *For given probabilities of errors* $\alpha_0(\delta)$ *and* $\alpha_1(\delta)$, *this test minimizes the expected sample size under both hypotheses* H_0 *and* H_1, *i.e., it simultaneously minimizes* $\mathsf{E}_0 T$ *and* $\mathsf{E}_1 T$.

The following result implies that the SPRT terminates w.p. 1 and that all the moments of the stopping time T^* are finite (under both hypotheses), provided the hypotheses H_0 and H_1 are distinguishable, i.e., as long as $f_0(X)$ is not equal to $f_1(X)$ almost everywhere.

Lemma 3.1.1 (Stein's lemma). *Let* $\{Y_n\}_{n \geq 1}$ *be a sequence of iid random variables under* P, *let* $S_n = \sum_{k=1}^{n} Y_k$ *and let*

$$\tau = \inf\{n \geq 1 : S_n \notin (-a_0, a_1)\}.$$

If $\mathsf{P}(Y_1 = 0) \neq 1$, *then the stopping time* τ *is exponentially bounded, i.e., there exist constants* $C > 0$ *and* $0 < \rho < 1$ *such that* $\mathsf{P}(\tau > n) \leq C\rho^n$ *for all* $n \geq 1$. *Therefore,* $\mathsf{P}(\tau < \infty) = 1$ *and* $\mathsf{E}[\tau^m] < \infty$ *for all* $m > 0$.

Proof. If $\mathsf{P}(Y_1 = 0) \neq 1$, then there is $y > 0$ such that either $\mathsf{P}(Y_1 \geq y) > 0$ or $\mathsf{P}(Y_1 \leq -y) > 0$, and there is no loss of generality to assume that $\mathsf{P}(Y_1 \geq y) = \varepsilon > 0$. Let m be an integer for which $my > a_1 + a_0$. Then

$$\mathsf{P}(S_m \geq a_0 + a_1) \geq \mathsf{P}(S_m \geq my) \geq \mathsf{P}(Y_1 \geq y, \dots, Y_m \geq y) = \varepsilon^m,$$

and hence for all $k \geq 1$

$$\mathsf{P}(\tau > mk) = \mathsf{P}(-a_0 < S_n < a_1, n = 1, \dots, mk) \leq (1 - \varepsilon^m)^k.$$

For any n, let k be such that $mk < n \leq (k+1)m$. Then

$$\mathsf{P}(\tau > n) \leq \mathsf{P}(\tau > km) \leq (1 - \varepsilon^m)^k \leq (1 - \varepsilon^m)^{\frac{n}{m} - 1} = \frac{1}{1 - \varepsilon^m}(1 - \varepsilon^m)^{\frac{n}{m}} = C\rho^n,$$

where $C = 1/(1 - \varepsilon^m)$ and $\rho = (1 - \varepsilon^m)^{1/m}$, and the lemma follows. □

Returning to the SPRT, we note that if $\mathsf{P}_0 \not\equiv \mathsf{P}_1$, then $\mathsf{P}_i(Z_k \neq 0) = 1$, $i = 0, 1$. Thus, by Lemma 3.1.1, $\mathsf{P}_i(T^* < \infty) = 1$ and $\mathsf{E}_i T^* < \infty$, $i = 0, 1$ for all $a_0, a_1 < \infty$.

3.1.1 *Wald's Approximations for the Operating Characteristic and the Expected Sample Size*

Although this chapter deals with the case of testing two simple hypotheses, in many applications there is a parametric family $\{p_\theta(\mathbf{X}_1^n)\}$, the hypotheses are $H_0 : \theta = \theta_0$ and $H_1 : \theta = \theta_1$, and it is of interest to consider parameter values θ different from the putative values θ_0 and θ_1. Two functions of the parameter θ characterize the SPRT in the general case when the true value θ can be different from θ_0 or θ_1. These are the operating characteristic (OC) and the expected sample size (ESS) often also called the average sample number (ASN) or expected duration of the experiment.

Definition 3.1.1 (OC). The probability $\beta(\delta,\theta) = \mathsf{P}_\theta(d=0)$ of accepting the hypothesis H_0 treated as a function of $\theta \in \Theta$ is called the *operating characteristic* (OC) of the test δ.

Definition 3.1.2 (ESS). The *expected sample size* (ESS) is the mean number of sample points $\mathsf{E}_\theta T$ which is necessary for testing the hypotheses when the true parameter value is θ. The ESS is often referred to as the *average sample number* (ASN).

We are now ready to derive useful inequalities and approximations for the OC and the ESS of the SPRT, as suggested by Wald [491, 492, 493, 494]. See also [47, 160, 162, 428, 452]. In the case of two simple hypotheses these characteristics are defined by four quantities: two probabilities of errors $\alpha_0(\delta^*) = 1-\beta(\theta_0) = \mathsf{P}_0(d^*=1)$ and $\alpha_1(\delta^*) = \beta(\theta_1) = \mathsf{P}_1(d^*=0)$ and two conditional ESSs E_0T^* and E_1T^*.

3.1.1.1 *The Case of Putative Values $\theta = \theta_0$ and $\theta = \theta_1$*

We begin with analyzing the probabilities of errors $\alpha_0(\delta^*) = \mathsf{P}_0(\lambda_{T^*} \ge a_1)$ and $\alpha_1(\delta^*) = \mathsf{P}_1(\lambda_{T^*} \le -a_0)$. We write $\alpha_i(\delta^*) = \alpha_i^*$, $i=0,1$. Since $\mathsf{P}_i(T^*<\infty)=1$, by Wald's likelihood ratio identity (2.56), we obtain

$$
\begin{aligned}
\alpha_1^* &= \mathsf{E}_0\left[\Lambda_{T^*};\Lambda_{T^*} \le A_0\right] = \mathsf{E}_0\left[\exp\{\lambda_{T^*}\}\mathbb{1}_{\{\lambda_{T^*}\le -a_0\}}\right] \\
&\le e^{-a_0}\mathsf{E}_0\left[\mathbb{1}_{\{\lambda_{T^*}\le -a_0\}}\right] = e^{-a_0}\mathsf{P}_0(\lambda_{T^*}\le -a_0);
\end{aligned}
$$

$$
\begin{aligned}
\alpha_0^* &= \mathsf{E}_1\left[\Lambda_{T^*}^{-1};\Lambda_{T^*} \ge A_1\right] = \mathsf{E}_1\left[\exp\{-\lambda_{T^*}\}\mathbb{1}_{\{\lambda_{T^*}\ge a_1\}}\right] \\
&\le e^{-a_1}\mathsf{E}_1\left[\mathbb{1}_{\{\lambda_{T^*}\ge a_1\}}\right] = e^{-a_1}\mathsf{P}_1(\lambda_{T^*}\ge a_1),
\end{aligned}
$$

which imply that

$$
\alpha_1^* \le e^{-a_0}(1-\alpha_0^*), \quad \alpha_0^* \le e^{-a_1}(1-\alpha_1^*). \tag{3.2}
$$

These simple inequalities do not account for the threshold overshoots and they hold not only in the iid case but also in the general non-iid case. Wald suggested to ignore the overshoots and replace the inequalities with approximate equalities, that is, to use the approximate formulas

$$
\alpha_1^* \approx e^{-a_0}(1-\alpha_0^*), \quad \alpha_0^* \approx e^{-a_1}(1-\alpha_1^*), \tag{3.3}
$$

which yield the following approximations for error probabilities

$$
\alpha_0^* \approx \frac{e^{a_0}-1}{e^{a_1+a_0}-1}, \quad \alpha_1^* \approx \frac{e^{a_1}-1}{e^{a_1+a_0}-1}. \tag{3.4}
$$

Note that asymptotically when a_0 and a_1 are large, the approximations (3.4) yield

$$
\alpha_0^* \sim e^{-a_1}, \quad \alpha_1^* \sim e^{-a_0}.
$$

As shown in Subsection 3.1.3, these asymptotic approximations are asymptotically incorrect by a constant factor which is close to 1 only when the hypotheses are close, in which case ignoring the overshoots is reasonable.

Inverting (3.3), we obtain Wald's formulas for the thresholds

$$
-a_0 = \log\left(\frac{\alpha_1}{1-\alpha_0}\right), \quad a_1 = \log\left(\frac{1-\alpha_1}{\alpha_0}\right). \tag{3.5}
$$

The next important step is to obtain approximations for the expected sample sizes E_0T^* and E_1T^*. Let $I_1 = \mathsf{E}_1Z_1$ and $I_0 = \mathsf{E}_0(-Z_1)$ be Kullback–Leibler (K–L) information numbers. Recall that we established the finiteness of moments $\mathsf{E}_i(T^*)^m$ under the very weak condition that

$\mathsf{P}_i(Z_1 = 0) < 1$, which guarantees that $I_1 > 0$ and $I_0 > 0$. Assume now in addition that the K–L numbers are finite, $I_i < \infty$. Then, by Wald's identity (2.53),

$$\mathsf{E}_1 T^* = \mathsf{E}_1[\lambda_{T^*}]/I_1, \quad \mathsf{E}_0 T^* = \mathsf{E}_0[\lambda_{T^*}]/(-I_0). \tag{3.6}$$

Now, under the hypothesis H_1, the LLR λ_{T^*} takes on the values $a_1 + \kappa_1$ and $-a_0 - \kappa_0$ with probabilities $1-\alpha_1^*$ and α_1^*, respectively, where κ_1 and κ_0 are the corresponding overshoots. Again, ignoring the overshoots, we obtain that $\lambda_{T^*} \approx a_1$ w.p. $1-\alpha_1^*$ and $\lambda_{T^*} \approx -a_0$ w.p. α_1^*, so that

$$\mathsf{E}_1[\lambda_{T^*}] \approx a_1(1-\alpha_1^*) - a_0\alpha_1^*.$$

A similar argument shows that

$$\mathsf{E}_0[\lambda_{T^*}] \approx a_1\alpha_0^* - a_0(1-\alpha_0^*).$$

These last two approximate formulas along with the equalities (3.6) yield

$$\begin{aligned} \mathsf{E}_1 T^* &\approx [a_1(1-\alpha_1^*) - a_0\alpha_1^*]/I_1, \\ \mathsf{E}_0 T^* &\approx [a_0(1-\alpha_0^*) - a_1\alpha_0^*]/I_0. \end{aligned} \tag{3.7}$$

Finally, combining (3.3) with (3.7) yields

$$\begin{aligned} \mathsf{E}_1 T^* &\approx \frac{1}{I_1}\left[(1-\alpha_1^*)\log\left(\frac{1-\alpha_1^*}{\alpha_0^*}\right) - \alpha_1^*\log\left(\frac{1-\alpha_0^*}{\alpha_1^*}\right)\right], \\ \mathsf{E}_0 T^* &\approx \frac{1}{I_0}\left[(1-\alpha_0^*)\log\left(\frac{1-\alpha_0^*}{\alpha_1^*}\right) - \alpha_0^*\log\left(\frac{1-\alpha_1^*}{\alpha_0^*}\right)\right]. \end{aligned} \tag{3.8}$$

Introducing the function

$$\beta(x,y) = (1-x)\log\left(\frac{1-x}{y}\right) + x\log\left(\frac{x}{1-y}\right), \tag{3.9}$$

the approximations (3.8) can be written in the following compact form

$$\mathsf{E}_1 T^* \approx \frac{1}{I_1}\beta(\alpha_1^*, \alpha_0^*), \quad \mathsf{E}_0 T^* \approx \frac{1}{I_0}\beta(\alpha_0^*, \alpha_1^*). \tag{3.10}$$

Remark 3.1.1. The approximate formulas (3.3) for the error probabilities hold not only in the iid case but also in the general non-iid case as long as the SPRT terminates w.p. 1. However, the approximations for the expected sample sizes (3.7) and (3.8) are valid only in the iid case or slightly more generally when the LLR process $\{\lambda_n\}$ is a random walk (see Subsection 3.2.4).

3.1.1.2 The Case of Arbitrary Parameter Values

The goal of this subsection is to derive Wald's approximations for the OC and ESS in the case of an arbitrary value θ. We assume that the LLR $Z_1 = \log[f_{\theta_1}(X_1)/f_{\theta_0}(X_1)]$ satisfies the two following conditions:

(i) The moment generating function (mgf)

$$M_\theta(t) = \mathsf{E}_\theta\left[e^{tx}\right] = \int_{-\infty}^{\infty} e^{tx}\, \mathrm{d}F_\theta(x), \tag{3.11}$$

where F_θ is the cdf of Z_1 under P_θ, exists for all real t.

(ii) There exists $\delta > 0$ such that

$$\mathsf{P}_\theta(e^{Z_1} > 1+\delta) > 0 \text{ and } \mathsf{P}_\theta(e^{Z_1} < 1-\delta) > 0. \tag{3.12}$$

Then the equation

$$\mathsf{E}_\theta\left[e^{-\omega_0 Z_1}\right] = 1 \tag{3.13}$$

has only one positive real root $\omega_0 > 0$ if $\mathsf{E}_\theta(Z_1) > 0$, one negative root $\omega_0 < 0$ if $\mathsf{E}_\theta(Z_1) < 0$, and no non-zero real root if $\mathsf{E}_\theta(Z_1) = 0$.

Let us begin with a very useful and important result which is called the *fundamental identity of sequential analysis* [491].

Theorem 3.1.1 (Fundamental identity). *Let $\delta^* = (d^*, T^*)$ be an SPRT with boundaries $-a_0$ and a_1 for testing between the hypotheses $H_0 : \theta = \theta_0$ and $H_1 : \theta = \theta_1$ defined in* (3.1). *Assume that the increments $Z_k = \log[f_{\theta_1}(X_k)/f_{\theta_0}(X_k)]$ of the cumulative LLR $\lambda_n = \sum_{k=1}^n Z_k$ are iid and that the conditions* (3.11)–(3.12) *hold. Then, for every real ω,*

$$\mathsf{E}_\theta\left\{e^{-\omega\lambda_n}[M_\theta(-\omega)]^{-n}\right\} = 1. \tag{3.14}$$

If we replace ω in (3.14) with the solution $\omega_0(\theta) \neq 0$ of the equation $\mathsf{E}_\theta(e^{-\omega_0 s}) = 1$, we get the OC function

$$\beta(\theta) = \frac{\mathsf{E}_\theta(e^{-\omega_0\lambda_{T^*}} \mid \lambda_{T^*} \geq a_1) - 1}{\mathsf{E}_\theta(e^{-\omega_0\lambda_{T^*}} \mid \lambda_{T^*} \geq a_1) - \mathsf{E}_\theta(e^{-\omega_0\lambda_{T^*}} \mid \lambda_{T^*} \leq -a_0)}. \tag{3.15}$$

Then, under the assumption that *both excesses* of λ_n over the boundaries $-a_0$ and a_1 are *small*, we have

$$\mathsf{E}_\theta(e^{-\omega_0\lambda_T} \mid \lambda_T \leq -a_0) \approx e^{\omega_0 a_0} \quad \text{and} \quad \mathsf{E}_\theta(e^{-\omega_0\lambda_T} \mid \lambda_T \geq a_1) \approx e^{-\omega_0 a_1}, \tag{3.16}$$

which yield Wald's approximation for the OC

$$\beta(\theta) \approx \frac{e^{-\omega_0(\theta)a_1} - 1}{e^{-\omega_0(\theta)a_1} - e^{\omega_0(\theta)a_0}} \quad \text{when} \quad \mathsf{E}_\theta(Z_1) \neq 0. \tag{3.17}$$

The approximation of $\beta(\theta^*)$ when $\mathsf{E}_{\theta^*} Z_1 = 0$ can be obtained by taking the limit $\theta \to \theta^*$. Note that, when $\omega_0 \to 0$, we have $e^{-\omega_0 h} \approx 1 - \omega_0 h$, and thus

$$\beta(\theta^*) \approx \frac{a_1}{a_0 + a_1}. \tag{3.18}$$

Note that for $\theta = \theta_0$ or $\theta = \theta_1$, we have $\omega_0(\theta_0) = -1$ and $\omega_0(\theta_1) = 1$, respectively, and thus

$$\beta(\theta_0) \approx \frac{e^{a_1} - 1}{e^{a_1} - e^{-a_0}} \quad \text{and} \quad \beta(\theta_1) \approx \frac{e^{-a_1} - 1}{e^{-a_1} - e^{a_0}}. \tag{3.19}$$

When the SPRT terminates w.p. 1, we have by definition

$$\alpha_0^* = 1 - \beta(\theta_0) \quad \text{and} \quad \alpha_1^* = \beta(\theta_1). \tag{3.20}$$

Therefore, we get that (3.19) is nothing but *Wald's approximations* (3.4) for putative values discussed above.

Under the assumptions that both excesses of λ_n over the boundaries $-a_0$ and a_1 are small, using Wald's identity (2.53) we obtain as before

$$\begin{aligned} \mathsf{E}_\theta T^* &\approx \frac{1}{\mathsf{E}_\theta(Z_1)}\left[-a_0\beta(\theta) + a_1(1-\beta(\theta))\right] \quad \text{when} \quad \mathsf{E}_\theta(Z_1) \neq 0, \\ \mathsf{E}_\theta T^* &\approx \frac{1}{\mathsf{E}_\theta(Z_1^2)}\left[a_0^2\beta(\theta) + a_1^2(1-\beta(\theta))\right] \quad \text{when} \quad \mathsf{E}_\theta(Z_1) = 0. \end{aligned} \tag{3.21}$$

Putting together (3.19) and (3.21) and taking into account that $\mathsf{E}_{\theta_0}(-Z_1) = I_0$ and $\mathsf{E}_{\theta_1}(Z_1) = I_1$, we deduce that the approximation for the ESS at θ_0 or θ_1 can be re-written as in (3.10) using the function $\beta(x,y)$ defined in (3.9):

$$\mathsf{E}_1 T^* \approx \frac{1}{I_1}\beta(\alpha_1^*, \alpha_0^*), \quad \mathsf{E}_0 T^* \approx \frac{1}{I_0}\beta(\alpha_0^*, \alpha_1^*).$$

From these results, we deduce the following fact: for given putative values θ_0 and θ_1 and error probabilities α_0 and α_1, the ESS of the SPRT is a function of the K–L information. Thus, the K–L information can be used as a weak performance index for hypothesis testing algorithms.

Example 3.1.1 (Mean of a Gaussian sequence). Let the iid observations $X_1, X_2, \dots$ be Gaussian, $X_k \sim \mathcal{N}(\theta, \sigma^2)$. For testing the hypotheses $H_0 : \theta = \theta_0$ and $H_1 : \theta = \theta_1$, the LLR is

$$Z_k = \frac{\theta_1 - \theta_0}{\sigma^2}\left(X_k - \frac{\theta_0 + \theta_1}{2}\right), \quad \lambda_n = \frac{\theta_1 - \theta_0}{\sigma^2}\sum_{k=1}^{n} X_k - \frac{\theta_1^2 - \theta_0^2}{2\sigma^2} n, \tag{3.22}$$

and thus Z_k is distributed as a Gaussian random variable with the following mean and variance:

$$\mu_z = \frac{\theta_1 - \theta_0}{\sigma^2}\left(\theta - \frac{\theta_0 + \theta_1}{2}\right), \quad \sigma_z^2 = \frac{(\theta_1 - \theta_0)^2}{\sigma^2}. \tag{3.23}$$

Therefore, the solution of the equation

$$\mathsf{E}_\theta(e^{-\omega_0 Z_1}) = \int_{-\infty}^{\infty} e^{-\omega_0 s} \frac{1}{\sigma_z\sqrt{2\pi}} e^{-(s-\mu_z)^2/2\sigma_z^2}\, ds = 1$$

is given by

$$\omega_0 = \frac{2\mu_z}{\sigma_z^2} = \frac{2}{\theta_1 - \theta_0}\left(\theta - \frac{\theta_0 + \theta_1}{2}\right). \tag{3.24}$$

It follows from (3.17) and (3.21) that the approximations for the OC and the ESS functions are given by

$$\beta(\theta) \approx \frac{e^{-2\mu_z a_1/\sigma_z^2} - 1}{e^{-2\mu_z a_1/\sigma_z^2} - e^{2\mu_z a_0/\sigma_z^2}} \quad \text{if} \quad \mu_z \neq 0,$$

$$\beta(\theta) \approx \frac{a_1}{a_1 + a_0} \qquad \text{if} \quad \mu_z = 0.$$

and

$$\mathsf{E}_\theta T = \frac{1}{\mu_z}\left[\frac{1 - e^{2\mu_z a_0/\sigma_z^2}}{e^{-2\mu_z a_1/\sigma_z^2} - e^{2\mu_z a_0/\sigma_z^2}}\, a_1 - \frac{e^{-2\mu_z a_1/\sigma_z^2} - 1}{e^{-2\mu_z a_1/\sigma_z^2} - e^{2\mu_z a_0/\sigma_z^2}}\, a_0\right] \text{ if } \mu_z \neq 0$$
$$\mathsf{E}_\theta T = a_0 a_1/\sigma_z^2 \text{ if } \mu_z = 0. \tag{3.25}$$

3.1.2 *Bounds for the Operating Characteristic and the Expected Sample Size*

Let us now discuss the bounds for the OC and ESS functions which have been developed by Wald [491, 494]. It results from the fundamental identity (3.14) and the formula (3.15) that bounds for the conditional expectations $\mathsf{E}_\theta\left[e^{-\lambda_{T^*}\omega(\theta)} | \lambda_{T^*} \le -a_0\right]$ and $\mathsf{E}_\theta\left[e^{-\lambda_{T^*}\omega(\theta)} | \lambda_{T^*} \ge a_1\right]$ are necessary to calculate bounds for the OC $\beta(\theta)$. Let us start with bounds for $\mathsf{E}_\theta\left[e^{\lambda_{T^*}\omega(\theta)} | \lambda_{T^*} \le -a_0\right]$ when $\mathsf{E}_\theta(Z_1) < 0$ and thus $\omega(\theta) < 0$. Consider the random variable $e^{-\lambda_{T^*-1}\omega(\theta)}$, where λ_{T^*-1} is the LLR at the instant $T^* - 1$, and note that for $x \in [1, \infty)$

$$F_0(x) := \mathsf{P}_0(Z_1 \le x) = \mathsf{P}_\theta\left[e^{-\lambda_{T^*-1}\omega(\theta)} \le x e^{a_0\omega(\theta)}\right].$$

Recall that $\lambda_{T^*} = \lambda_{T^*-1} + Z_{T^*}$. Next, we get

$$\mathsf{E}_\theta\left[e^{-\lambda_{T^*}\omega(\theta)} | \lambda_{T^*} \le -a_0\right] = \int_1^\infty x e^{a_0\omega(\theta)} f_0(x) \mathsf{E}_\theta\left[e^{-Z_{T^*}\omega(\theta)} | e^{-Z_{T^*}\omega(\theta)} \le 1/x\right] \mathrm{d}x \tag{3.26}$$

where $f_0(x)$ is density of $F_0(x)$. Finally, a lower bound for $\mathsf{E}_\theta\left[e^{-\lambda_{T^*}\omega(\theta)}|\lambda_{T^*}\le -a_0\right]$ is given by

$$\mathsf{E}_\theta\left[e^{-\lambda_{T^*}\omega(\theta)}|\lambda_{T^*}\le -a_0\right]\ge e^{a_0\omega(\theta)}\eta(\theta), \tag{3.27}$$

where $\eta(\theta)=\inf_{x>1}\left\{x\,\mathsf{E}_\theta\left[e^{-Z_1\omega(\theta)}|e^{-Z_1\omega(\theta)}\le 1/x\right]\right\}$. Hence,

$$\eta(\theta)e^{a_0\omega(\theta)}\le \mathsf{E}_\theta\left[e^{-\lambda_{T^*}\omega(\theta)}|\lambda_{T^*}\le -a_0\right]\le e^{a_0\omega(\theta)}. \tag{3.28}$$

To calculate the bounds for $\mathsf{E}_\theta\left[e^{-\lambda_{T^*}\omega(\theta)}|\lambda_{T^*}\ge a_1\right]$, define the distribution $F_1(x):=\mathsf{P}_\theta\left[e^{-\lambda_{T^*-1}\omega(\theta)}\le xe^{-a_1\omega(\theta)}\right]$ with density $f_1(x)$, where $x\in(0,1)$. Analogously to equation (3.26), we get

$$\mathsf{E}_\theta\left[e^{-\lambda_{T^*}\omega(\theta)}|\lambda_{T^*}\ge a_1\right]=\int_0^1 x\,e^{-a_1\omega(\theta)}f_1(x)\mathsf{E}_\theta\left[e^{-Z_1\omega(\theta)}|e^{-Z_1\omega(\theta)}\ge 1/x\right]\mathrm{d}x. \tag{3.29}$$

Finally,

$$e^{-a_1\omega(\theta)}\le \mathsf{E}_\theta\left[e^{-\lambda_{T^*}\omega(\theta)}|\lambda_{T^*}\ge a_1\right]\le \delta(\theta)e^{-a_1\omega(\theta)}, \tag{3.30}$$

where $\delta(\theta)=\sup_{0<x<1}\left\{x\,\mathsf{E}_\theta\left[e^{-\omega_0 Z_k}|e^{-\omega_0 Z_k}\ge 1/x\right]\right\}$. By (3.15), (3.28), and (3.30), the bounds for the OC function are

$$\frac{e^{-\omega_0 a_1}-1}{e^{-\omega_0 a_1}-\eta(\theta)e^{\omega_0 a_0}}\le\beta(\theta)\le\frac{\delta(\theta)e^{-\omega_0 a_1}-1}{\delta(\theta)e^{-\omega_0 a_1}-e^{\omega_0 a_0}}\quad\text{when } \omega_0<0. \tag{3.31}$$

To calculate the bounds for the OC when $\omega_0>0$, Wald [494] proposes to define another SPRT $\delta^{*\prime}$ with $a_0'=a_1$, $a_1'=a_0$ and $Z_k'=-Z_k$. Then in a similar way the bounds for $1-\beta(\theta)$ can be obtained when $\omega_0>0$, which yields

$$\frac{e^{-\omega_0 a_1}-1}{e^{-\omega_0 a_1}-\delta(\theta)e^{\omega_0 a_0}}\le\beta(\theta)\le\frac{\eta(\theta)e^{-\omega_0 a_1}-1}{\eta(\theta)e^{-\omega_0 a_1}-e^{\omega_0 a_0}}\quad\text{when } \omega_0>0. \tag{3.32}$$

For $\theta=\theta^*$ such that $\mathsf{E}_{\theta^*}(Z_1)=0$, the bounds for $\beta(\theta^*)$ can be obtained by taking the limit $\theta\to\theta^*$ in (3.31) or (3.32).

Consider now the ESS function. Since the expectation $\mathsf{E}_\theta(\lambda_{T^*})$ can be written as

$$\mathsf{E}_\theta(\lambda_{T^*})=(1-\beta(\theta))\mathsf{E}_\theta(\lambda_{T^*}|\lambda_{T^*}\ge a_1)+\beta(\theta)\mathsf{E}_\theta(\lambda_{T^*}|\lambda_{T^*}\le -a_0), \tag{3.33}$$

by Wald's identity, when $\mathsf{E}_\theta(Z_1)\ne 0$ we may write

$$\mathsf{E}_\theta(T^*)=\frac{\mathsf{E}_\theta(\lambda_{T^*}|\lambda_{T^*}\le -a_0)\beta(\theta)+\mathsf{E}_\theta(\lambda_{T^*}|\lambda_{T^*}\ge a_1)(1-\beta(\theta))}{\mathsf{E}_\theta(Z_1)}. \tag{3.34}$$

To calculate an upper bound for the first expectation on the right-hand side of (3.33), we define the random variable $\xi=a_1-\lambda_{T^*-1}$ (undershoot). Let $Q_1(x)$ denote density of $\xi\in(0,\infty)$. Then

$$\begin{aligned}\mathsf{E}_\theta(\lambda_{T^*}|\lambda_{T^*}\ge a_1)&=\int_0^\infty Q_1(x)\mathsf{E}_\theta\left[a_1+Z_{T^*}-x|Z_{T^*}-x\ge 0\right]\mathrm{d}x\\&\le[a_1+\psi_1(\theta)]\int_0^\infty Q_1(x)\mathrm{d}x=a_1+\psi_1(\theta),\end{aligned} \tag{3.35}$$

where $\psi_1(\theta)=\sup_{x>0}\mathsf{E}_\theta(Z_1-x|Z_1-x\ge 0)$. Therefore,

$$a_1\le\mathsf{E}_\theta(\lambda_{T^*}|\lambda_{T^*}\ge a_1)\le a_1+\psi_1(\theta). \tag{3.36}$$

To find a lower bound for the second expectation in the right-hand side of (3.33), we define $\xi_0 = a_0 + \lambda_{T^*-1}$, and let $Q_0(x)$ denote density of $\xi_0 \in (0,\infty)$. Then

$$\begin{aligned}\mathsf{E}_\theta(\lambda_{T^*}|\lambda_{T^*} \le -a_0) &= \int_0^{+\infty} Q_0(x)\mathsf{E}_\theta\left[-a_0 + Z_{T^*} + x|Z_{T^*} + x \le 0\right]\mathrm{d}x \\ &\ge [-a_0 + \psi_0(\theta)], \end{aligned} \tag{3.37}$$

where $\psi_0(\theta) = \inf_{x>0} \mathsf{E}_\theta(Z_1 + x|Z_1 + x \ge 0)$. Therefore,

$$-a_0 + \psi_0(\theta) \le \mathsf{E}_\theta(\lambda_{T^*}|\lambda_{T^*} \le -a_0) \le -a_0. \tag{3.38}$$

Putting together (3.34), (3.36) and (3.38), when $\mathsf{E}_\theta(Z_1) > 0$ we obtain the following bounds for the ESS

$$\frac{(-a_0 + \psi_0(\theta))\beta(\theta) + a_1(1-\beta(\theta))}{\mathsf{E}_\theta(Z_1)} \le \mathsf{E}_\theta(T^*) \le \frac{-a_0\beta(\theta) + (a_1 + \psi_1(\theta))(1-\beta(\theta))}{\mathsf{E}_\theta(Z_1)}. \tag{3.39}$$

For $\mathsf{E}_\theta(Z_1) < 0$, these inequalities must be written the other way around.

For $\theta = \theta^*$ such that $\mathsf{E}_{\theta^*}(Z_1) = 0$, Wald's identity yields

$$\mathsf{E}_{\theta^*}(T^*) = \frac{\mathsf{E}_\theta(\lambda_{T^*}^2|\lambda_{T^*} \le -a_0)\beta(\theta) + \mathsf{E}_\theta(\lambda_{T^*}|\lambda_{T^*}^2 \ge a_1)(1-\beta(\theta))}{\mathsf{E}_{\theta^*}(Z_1^2)}. \tag{3.40}$$

By analogy with the equation (3.35), using the same notation we get

$$\begin{aligned}\mathsf{E}_\theta(\lambda_{T^*}^2|\lambda_{T^*} \ge a_1) &= \int_0^{\infty} Q_1(x)\mathsf{E}_\theta\left[a_1^2 + 2a_1(Z_{T^*} - x) + (Z_{T^*} - x)^2|Z_{T^*} - x \ge 0\right]\mathrm{d}x \\ &\le a_1^2 + 2a_1\psi_1^* + \varsigma_1^*, \end{aligned} \tag{3.41}$$

where $\psi_1^* = \psi(\theta^*)$ and $\varsigma_1^* = \sup_{x>0} \mathsf{E}_{\theta^*}\left[(Z_1 - x)^2|Z_1 - x \ge 0\right]$, and

$$\mathsf{E}_\theta(\lambda_{T^*}^2|\lambda_{T^*} \le -a_0) \le a_0^2 - 2a_0\psi_0^* + \varsigma_0^*, \tag{3.42}$$

where $\psi_0^* = \psi_0(\theta^*)$ and $\varsigma_0^* = \sup_{x>0} \mathsf{E}_{\theta^*}\left[(Z_1 + x)^2|Z_1 + x \le 0\right]$. Finally, we obtain the following bounds for the ESS at the point θ^*

$$\begin{aligned}\frac{a_0^2\beta(\theta^*) + a_1^2(1-\beta(\theta^*)}{\mathsf{E}_{\theta^*}(Z_1^2)} &\le \mathsf{E}_{\theta^*}(T^*) \\ &\le \frac{(a_0^2 - 2a_0\psi_0(\theta^*) + \varsigma_0^*)\beta(\theta^*) + (a_1^2 + 2a_1\psi_1^* + \varsigma_1^*)(1-\beta(\theta^*))}{\mathsf{E}_{\theta^*}(Z_1^2)}. \end{aligned} \tag{3.43}$$

3.1.3 *Asymptotically Accurate Approximations for the OC and the ESS*

In this subsection, we use a renewal-theoretic argument as well as a corrected Brownian motion method to obtain asymptotically accurate approximations for the operating characteristics of the SPRT when the values of a_1 and a_0 are large.

3.1.3.1 *Asymptotic Approximations for the Error Probabilities — A Renewal-Theoretic Approach*

Let $\kappa_1 = \kappa_1(a_0, a_1) = \lambda_{T^*} - a_1$ on $\{\lambda_{T^*} \ge a_1\}$ and $\kappa_0 = \kappa_0(a_0, a_1) = -(a_0 + \lambda_{T^*})$ on $\{\lambda_{T^*} \le -a_0\}$ denote the overshoots of a_1 and a_0 at stopping. Let $e_1 = \mathsf{E}_1[e^{-\kappa_1}|\lambda_{T^*} \ge a_1]$ and $e_0 = \mathsf{E}[e^{-\kappa_0}| - \lambda_{T^*} \ge a_0]$. It is easily shown (see, e.g., [452, p. 71]) that when accounting for the overshoots the probabilities of errors are equal to

$$\alpha_0^* = \frac{e_1 e^{a_0} - e_1 e_0}{e^{a_1 + a_0} - e_1 e_0}, \quad \alpha_1^* = \frac{e_0 e^{a_1} - e_1 e_0}{e^{a_1 + a_0} - e_1 e_0}. \tag{3.44}$$

In general, the expectations e_0 and e_1 cannot be evaluated for arbitrary thresholds $-a_0$ and a_1. However, this is possible asymptotically when $a_i \to \infty$, $i = 0, 1$.

Indeed, introduce the one-sided stopping times

$$T_0(a_0) = \inf\{n \geq 1 : -\lambda_n \geq a_0\}, \quad T_1(a_1) = \inf\{n \geq 1 : \lambda_n \geq a_1\} \tag{3.45}$$

and the associated overshoots $\tilde{\kappa}_0 = -\lambda_{T_0} - a_0$ on $\{T_0 < \infty\}$ and $\tilde{\kappa}_1 = \lambda_{T_1} - a_1$ on $\{T_1 < \infty\}$. If the K–L numbers I_0 and I_1 are positive and finite, then by Theorem 2.5.1 the limiting distributions $\mathsf{H}_0(y) = \lim_{a_0 \to \infty} \mathsf{P}_0(\tilde{\kappa}_0 \leq y)$ and $\mathsf{H}_1(y) = \lim_{a_1 \to \infty} \mathsf{P}_1(\tilde{\kappa}_1 \leq y)$ exist and are given by (2.88) in the nonarithmetic case and by (2.90) in the d-arithmetic case. Thus,

$$\lim_{a_i \to \infty} \mathsf{E}_i\left[e^{-\tilde{\kappa}_i}\right] = \zeta_i = \int_0^\infty e^{-y}\, d\mathsf{H}_i(y), \quad i = 0, 1. \tag{3.46}$$

Now, the stopping time of the SPRT can be represented as $T^* = \min(T_0, T_1)$. Since $\alpha_0^* = \mathsf{P}_0(T^* = T_1) \to 0$ and $\alpha_1^* = \mathsf{P}_1(T^* = T_0) \to 0$ as $a_i \to \infty$, it is reasonable to assume that $e_1 \to \zeta_1$ and $e_0 \to \zeta_0$ as $a_0, a_1 \to \infty$, and hence for sufficiently large thresholds the approximation (3.44) yields

$$\alpha_0^* \approx \frac{\zeta_1 e^{a_0} - \zeta_1\zeta_0}{e^{a_1 + a_0} - \zeta_1\zeta_0}, \quad \alpha_1^* \approx \frac{\zeta_0 e^{a_1} - \zeta_1\zeta_0}{e^{a_1 + a_0} - \zeta_1\zeta_0}. \tag{3.47}$$

Therefore, we may conjecture that asymptotically as $a_i \to \infty$

$$\alpha_0^* \sim \zeta_1 e^{-a_1}, \quad \alpha_1^* \sim \zeta_0 e^{-a_0}.$$

The following theorem makes this conjecture precise.

Theorem 3.1.2. *Assume that* $0 < I_0 < \infty$ *and* $0 < I_1 < \infty$. *Let* ζ_0 *and* ζ_1 *be given as in* (3.46).

(i) *If the LLR* $\{\lambda_n\}$ *is nonarithmetic, then*

$$\alpha_0^* = \zeta_1 e^{-a_1}(1 + o(1)) \quad \text{and} \quad \alpha_1^* = \zeta_0 e^{-a_0}(1 + o(1)) \quad \text{as } a_0, a_1 \to \infty. \tag{3.48}$$

(ii) *If the LLR* $\{\lambda_n\}$ *is d-arithmetic, then*

$$\alpha_0^* = \zeta_1 e^{-di}(1 + o(1)) \quad \text{and} \quad \alpha_1^* = \zeta_0 e^{-dj}(1 + o(1)) \quad \text{as } i, j \to \infty. \tag{3.49}$$

Proof. (i) Obviously,

$$\alpha_0^* = \mathsf{P}_0(\lambda_{T^*} \geq a_1) = \mathsf{P}_0(T^* = T_1),$$

so that changing the measure and applying Wald's likelihood ratio identity, we obtain

$$\alpha_0^* = \mathsf{E}_1\left[\Lambda_{T^*}^{-1}; T^* = T_1\right] = \mathsf{E}_1\left[e^{-\lambda_{T^*}} \mathbb{1}_{\{T^* = T_1\}}\right] = e^{-a_1}\mathsf{E}_1\left[e^{-\tilde{\kappa}_1} \mathbb{1}_{\{T^* = T_1\}}\right].$$

Since $\mathsf{P}_1(T^* = T_1) = 1 - \alpha_1^* \to 1$ as $a_0, a_1 \to \infty$, the expectation $\mathsf{E}_1\left[e^{-\tilde{\kappa}_1} \mathbb{1}_{\{T^* = T_1\}}\right]$ converges to ζ_1 defined by (3.46) whenever $0 < I_1 < \infty$ by Theorem 2.5.1. This establishes the first formula in (3.48). The second assertion is established analogously.

(ii) The assertions (3.49) are proved similarly as in (i). The only difference is that in the d-arithmetic case the values of a_0 and a_1 tend to infinity through multiples of the span d. □

Remark 3.1.2. Siegmund [425] combined a renewal-theoretic consideration with Wald's approximations to obtain the approximations (3.47) for the exponential family, but did not provide estimates of the residual terms. Tartakovsky [452] arrived at the same conclusion based on the general formula (3.44) for arbitrary distributions. Lotov [279] obtained very detailed approximations in a form of series, from which residual terms and a rate of convergence can be derived. In particular, under certain general conditions

$$\begin{aligned} \alpha_0^* &= \frac{\zeta_1 e^{a_0} - \zeta_1\zeta_0}{e^{a_1+a_0} - \zeta_1\zeta_0} + o\left(e^{-a_0(1+c_0)}\right) + o\left(e^{-a_1(1+c_1)}\right), \\ \alpha_1^* &= \frac{\zeta_0 e^{a_1} - \zeta_1\zeta_0}{e^{a_1+a_0} - \zeta_1\zeta_0} + o\left(e^{-a_0(1+c_0)}\right) + o\left(e^{-a_1(1+c_1)}\right), \end{aligned} \tag{3.50}$$

where c_0, c_1 are positive constants. These approximations show that the approximate formulas (3.47) are expected to have a high precision. The values of c_i grow when the distance between the hypotheses decreases, so that the rate of convergence becomes higher and higher, which of course is expected.

The most important issue in designing the SPRT is to choose the thresholds in order to guarantee given probabilities of errors α_0 and α_1 exactly or at least almost exactly. Inverting the formulas (3.48) or, better, the more precise formulas (3.50), we obtain

$$-a_0 = \log\frac{\alpha_1}{\zeta_1(1-\alpha_0)}, \quad a_1 = \log\frac{\zeta_0(1-\alpha_1)}{\alpha_0}. \tag{3.51}$$

It is expected that with this choice of thresholds the accuracy of the approximations $\alpha_i^* \approx \alpha_i$, $i=0,1$ will be high since the residual terms are of order $o(\alpha_i^{c_i})$, so that the accuracy increases exponentially fast when α_i decreases.

We now derive a useful relationship between the constants ζ_0 and ζ_1 that have to be computed in order to implement the approximations (3.47) for the error probabilities. Let

$$\mathscr{H}_i(\beta) = \int_0^\infty e^{-\beta y}\, \mathrm{dH}_i(y), \quad \beta > 0, \quad i = 0,1$$

be Laplace's transforms for the limiting distribution functions of the overshoots $\tilde{\kappa}_i$, $i=0,1$. Clearly, $\zeta_i = \mathscr{H}_i(\beta = 1)$. By Theorem 2.5.3(i), as long as $0 < I_i < \infty$

$$\begin{aligned} \mathscr{H}_1(1) &= \frac{1}{I_1}\exp\left\{-\sum_{n=1}^\infty \frac{1}{n}\mathsf{E}_1\left[e^{-\lambda_n^+}\right]\right\} \\ &= I_1^{-1}\exp\left\{-\sum_{n=1}^\infty \frac{1}{n}\mathsf{E}_1\left[e^{-\lambda_n}1\!\!1_{\{\lambda_n>0\}}\right] - \sum_{n=1}^\infty \frac{1}{n}\mathsf{P}_1(\lambda_n \le 0)\right\}, \\ \mathscr{H}_0(1) &= \frac{1}{I_0}\exp\left\{-\sum_{n=1}^\infty \frac{1}{n}\mathsf{E}_0\left[e^{-(-\lambda_n)^+}\right]\right\} \\ &= I_0^{-1}\exp\left\{-\sum_{n=1}^\infty \frac{1}{n}\mathsf{E}_0\left[e^{\lambda_n}1\!\!1_{\{\lambda_n<0\}}\right] - \sum_{n=1}^\infty \frac{1}{n}\mathsf{P}_0(\lambda_n \ge 0)\right\}. \end{aligned}$$

Since, by Wald's likelihood ratio identity,

$$\mathsf{E}_1\left[e^{-\lambda_n}1\!\!1_{\{\lambda_n>0\}}\right] = \mathsf{P}_0(\lambda_n > 0) \quad \text{and} \quad \mathsf{E}_0\left[e^{\lambda_n}1\!\!1_{\{\lambda_n<0\}}\right] = \mathsf{P}_1(\lambda_n < 0),$$

we obtain

$$\begin{aligned} \mathscr{H}_1(1) &= \frac{1}{I_1}\exp\left\{-\sum_{n=1}^\infty \frac{1}{n}[\mathsf{P}_0(\lambda_n > 0) + \mathsf{P}_1(\lambda_n \le 0)]\right\}, \\ \mathscr{H}_0(1) &= \frac{1}{I_0}\exp\left\{-\sum_{n=1}^\infty \frac{1}{n}[\mathsf{P}_0(\lambda_n \ge 0) + \mathsf{P}_1(\lambda_n < 0)]\right\}. \end{aligned}$$

Finally, note that for every $n \geq 1$

$$\mathsf{P}_1(\lambda_n = 0) = \mathsf{E}_1\left[e^{\lambda_n} 1\!\!1_{\{\lambda_n=0\}}\right] = \mathsf{P}_0(\lambda_n = 0),$$

so that

$$\sum_{n=1}^{\infty} \frac{1}{n}[\mathsf{P}_0(\lambda_n > 0) + \mathsf{P}_1(\lambda_n \leq 0)] = \sum_{n=1}^{\infty} \frac{1}{n}[\mathsf{P}_0(\lambda_n \geq 0) + \mathsf{P}_1(\lambda_n < 0)].$$

Thus, we proved the following theorem.

Theorem 3.1.3. *Define*

$$\Upsilon = \exp\left\{-\sum_{n=1}^{\infty} \frac{1}{n}[\mathsf{P}_0(\lambda_n > 0) + \mathsf{P}_1(\lambda_n \leq 0)]\right\}. \tag{3.52}$$

If $0 < I_0 < \infty$ *and* $0 < I_1 < \infty$, *then*

$$\zeta_0 = \Upsilon/I_0 \quad \text{and} \quad \zeta_1 = \Upsilon/I_1. \tag{3.53}$$

Therefore,

$$\zeta_0 I_0 = \zeta_1 I_1. \tag{3.54}$$

It is worth emphasizing that this result holds only for LLR-based random walks but not for arbitrary random walks.

3.1.3.2 Asymptotic Approximations for the ESS – A Renewal-Theoretic Approach

We now turn to the asymptotic approximations for the expected sample sizes. An elementary argument [452, p. 71] shows that when accounting for the overshoots the expectations of the stopping time can be written as

$$\begin{aligned}
\mathsf{E}_1 T^* &= I_1^{-1}\{[a_1 + \mathsf{E}_1(\kappa_1 | T^* = T_1)](1 - \alpha_1^*) - [a_0 + \mathsf{E}_1(\kappa_0 | T^* = T_0)]\alpha_1^*\}, \\
\mathsf{E}_0 T^* &= I_0^{-1}\{[a_0 + \mathsf{E}_0(\kappa_0 | T^* = T_0)](1 - \alpha_0^*) - [a_1 + \mathsf{E}_0(\kappa_1 | T^* = T_1)]\alpha_0^*\}.
\end{aligned} \tag{3.55}$$

In particular, ignoring the overshoots, we obtain the approximate formulas (3.7).

Since $\alpha_0^*, \alpha_1^* \to 0$ as $a_0, a_1 \to \infty$ with an exponential rate, the values of $[a_1 + \mathsf{E}_1(\kappa_0 | T^* = T_0)]\alpha_1^*$ and $[a_0 + \mathsf{E}_0(\kappa_1 | T^* = T_1)]\alpha_0^*$ are small. Denote by

$$\varkappa_i = \lim_{a_i \to \infty} \mathsf{E}_i[\tilde{\kappa}_i] = \int_0^{\infty} y \, \mathsf{dH}_i(y), \quad i = 0, 1$$

the limiting average overshoots in the one-sided tests (3.45). It may be expected that $\mathsf{E}_i(\kappa_i | T^* = T_i) \to \varkappa_i$. Therefore, we may conjecture that, as $a_0, a_1 \to \infty$,

$$\mathsf{E}_1 T^* = I_1^{-1}(a_1 + \varkappa_1) + o(1) \quad \text{and} \quad \mathsf{E}_0 T^* = I_0^{-1}(a_0 + \varkappa_0) + o(1). \tag{3.56}$$

The following theorem justifies this conjecture.

Theorem 3.1.4. *Assume that* $\mathsf{E}_0|Z_1|^2 < \infty$ *and* $\mathsf{E}_1|Z_1|^2 < \infty$.

(i) *If the random walk* $\{\lambda_n\}$ *is nonarithmetic, then the asymptotic approximations* (3.56) *hold as long as* a_0 *and* a_1 *tend to infinity in such a way that* $a_0 e^{-a_1} \to 0$ *and* $a_1 e^{-a_0} \to 0$.

(ii) *If the random walk* $\{\lambda_n\}$ *is d-arithmetic, then the approximations* (3.56) *hold with* $a_0 = id$ *and* $a_1 = jd$ *as* $i, j \to \infty$ *assuming that* $ie^{-j} \to 0$ *and* $je^{-i} \to 0$.

Proof. Consider the hypothesis H_1. Since $\mathsf{E}_1 T^* < \infty$, by Wald's identity

$$\mathsf{E}_1 T^* = \frac{1}{I_1}\mathsf{E}_1[\lambda_{T^*}] = \frac{1}{I_1}\left\{\mathsf{E}_1\left[\lambda_{T^*}\mathbb{1}_{\{\lambda_{T^*}\geq a_1\}}\right] + \mathsf{E}_1\left[\lambda_{T^*}\mathbb{1}_{\{\lambda_{T^*}\leq -a_0\}}\right]\right\}. \tag{3.57}$$

For $a_0 > 1$, the absolute value of the second term is upper-bounded as

$$\left|\mathsf{E}_1\left[\lambda_{T^*}\mathbb{1}_{\{\lambda_{T^*}\leq -a_0\}}\right]\right| = \left|\mathsf{E}_0\left[\lambda_{T^*}e^{\lambda_{T^*}}\mathbb{1}_{\{\lambda_{T^*}\leq -a_0\}}\right]\right| \leq a_0 e^{-a_0} \xrightarrow[a_0\to\infty]{} 0,$$

so that

$$\mathsf{E}_1\left[\lambda_{T^*}\mathbb{1}_{\{\lambda_{T^*}\leq -a_0\}}\right] = o(1) \quad \text{as } a_0 \to \infty.$$

Now, the first term writes

$$\begin{aligned}\mathsf{E}_1\left[\lambda_{T^*}\mathbb{1}_{\{\lambda_{T^*}\geq a_1\}}\right] &= \mathsf{E}_1\left[\lambda_{T_1}\mathbb{1}_{\{T^*=T_1\}}\right] = \mathsf{E}_1\left[(a_1+\kappa_1)\mathbb{1}_{\{\lambda_{T^*}\geq a_1\}}\right] \\ &= a_1(1-\alpha_1^*) + \mathsf{E}_1\left[\kappa_1\mathbb{1}_{\{\lambda_{T^*}\geq a_1\}}\right].\end{aligned}$$

By (3.2), we have

$$\alpha_1^* \leq e^{-a_0}(1-\alpha_0^*) \leq e^{-a_0},$$

so that $a_1\alpha_1^* \leq a_1 e^{-a_0} = o(1)$. Also $\mathsf{P}_1(\lambda_{T^*} \geq a_1) = 1-\alpha_1^* \to 1$, which implies that

$$\mathsf{E}_1\left[\kappa_1\mathbb{1}_{\{\lambda_{T^*}\geq a_1\}}\right] \xrightarrow[a_1\to\infty]{} \mathsf{E}_1[\kappa_1] = \varkappa_1.$$

Therefore,

$$\mathsf{E}_1\left[\lambda_{T^*}\mathbb{1}_{\{\lambda_{T^*}\geq a_1\}}\right] = a_1 + \varkappa_1 + o(1) \quad \text{as } a_0, a_1 \to \infty$$

and the proof of the first asymptotic expansion in (3.56) is complete.

The proof for the hypothesis H_0 is essentially the same. □

Note that in this theorem we require that $a_0 e^{-a_1} \to 0$ and $a_1 e^{-a_0} \to 0$ as $a_0, a_1 \to \infty$. This natural requirement is equivalent to the assumption that a_0 and a_1 go to infinity in such a way that the ratio a_0/a_1 is bounded away from 0 and from infinity, i.e., there is a constant $0 < c < \infty$ such that $a_0/a_1 \sim c$ as $a_0, a_1 \to \infty$. If $c = 1$, then the rate of increase of both thresholds is the same, and we will refer to this case as the *asymptotically symmetric case*. If $c = 0$ or ∞, then the problem becomes degenerate.

By (2.89) and (2.91), the average overshoots are found as

$$\varkappa_i = \frac{\mathsf{E}_i[\lambda_{T_+}^2]}{2\mathsf{E}_i[\lambda_{T_+}]}, \quad i = 0,1 \tag{3.58}$$

in the nonarithmetic case and

$$\varkappa_i = \frac{\mathsf{E}_i[\lambda_{T_+}^2]}{2\mathsf{E}_i[\lambda_{T_+}]} + \frac{d}{2}, \quad i = 0,1 \tag{3.59}$$

in the d-arithmetic case. Also, by Theorem 2.5.3(ii), if $\mathsf{E}_i|Z_1|^2 < \infty$, then

$$\varkappa_i = \frac{\mathsf{E}_i Z_1^2}{2\mathsf{E}_i Z_1} - \sum_{n=1}^{\infty}\frac{1}{n}\mathsf{E}[\lambda_n^-]. \tag{3.60}$$

Note also that while Wald's approximations for the error probabilities do not depend on the model and Wald's approximations for the average sample sizes depend on the model only through the K–L information numbers, the asymptotic approximations that take into account the overshoots are of course strongly model-dependent.

Remark 3.1.3. With an additional effort the asymptotic approximations (3.56) for the expected sample sizes may be slightly improved, especially in cases where one threshold is high but the other is not. Lotov [278] showed that if the LLR is nonarithmetic, then

$$\mathsf{E}_0 T^* = \frac{1}{I_0}\left[a_0\left(1-\gamma_1^{(0)}(a_0)\right)+\varkappa_0-\gamma_2^{(0)}(a_0)\right]+O\left((a_0+a_1)e^{-a_1}\right) \quad \text{as } a_1\to\infty,$$
$$\mathsf{E}_1 T^* = \frac{1}{I_1}\left[a_1\left(1-\gamma_1^{(1)}(a_1)\right)+\varkappa_1-\gamma_2^{(1)}(a_1)\right]+O\left((a_0+a_1)e^{-a_0}\right) \quad \text{as } a_0\to\infty,$$

where $\gamma_1^{(i)}(a_i)$ and $\gamma_2^{(i)}(a_i)$ tend to zero as $a_i\to\infty$ $(i=0,1)$.

3.1.3.3 Corrected Brownian Motion Approximations

An alternative way of approximating the operating characteristics is to correct the exact formulas for Brownian motion – the so-called *corrected Brownian motion* approximations. This method was proposed by Siegmund [426, 428]. The general idea is that, when the hypotheses H_0 and H_1 become close, the LLR process $\{\lambda_n\}$ can be approximated by the Brownian motion. For the Brownian motion the exact formulas for the operating characteristics can be obtained based on Wald's methods since there are no overshoots. However, in the cases where the hypotheses are not too close these approximations are not expected to be very accurate. To make them more accurate some corrections are needed.

Let $\{W_t\}_{t\ge 0}$ be a standard Brownian motion, and let the observed process be of the form

$$X_t = \theta t + \sigma W_t, \quad t\ge 0, \tag{3.61}$$

where the drift θ can take on the two values θ_0 and θ_1 and $\sigma>0$. The LLR for testing $H_0:\theta=\theta_0$ against $H_1:\theta=\theta_1$ is given by

$$\lambda_t = \frac{\theta_1-\theta_0}{\sigma}X_t - \frac{\theta_1^2-\theta_0^2}{2\sigma^2}t, \quad t\ge 0.$$

Since there are no overshoots, Wald's approximations become exact. Specifically, we set $e_0=e_1=1$ in (3.44) and $\kappa_1=\kappa_0=0$ in (3.55) to obtain

$$\alpha_0^* = \frac{e^{a_0}-1}{e^{a_1+a_0}-1}, \quad \alpha_1^* = \frac{e^{a_1}-1}{e^{a_1+a_0}-1} \tag{3.62}$$

and

$$\mathsf{E}_1 T^* = \frac{1}{I}[(1-\alpha_1^*)a_1-\alpha_1^* a_0], \quad \mathsf{E}_0 T^* = \frac{1}{I}[(1-\alpha_0^*)a_0-\alpha_0^* a_1], \tag{3.63}$$

where $I=(\theta_1-\theta_0)^2/(2\sigma^2)$. Comparing these two approximations for the error probabilities and the ESS with the formulas (3.4)–(3.7), we conclude that the approximation of the cumulative sum by a Brownian motion and Wald's approximations are equivalent in the case of putative values θ_0 and θ_1. Moreover, taking into account the formulas of Example 3.1.1 we recognize that this equivalence holds for any θ in the Gaussian case.

If the hypotheses are close, then the overshoots are small. We thus expect that

$$\zeta_i = \lim_{a_i\to\infty}\mathsf{E}_i\left[e^{-\tilde{\kappa}_i}\right] \approx \exp\left\{-\lim_{a_i\to\infty}\mathsf{E}_i\tilde{\kappa}_i\right\} = e^{-\varkappa_i}, \quad i=0,1.$$

Replacing ζ_i by $e^{-\varkappa_i}$ in (3.47), we obtain

$$\alpha_0^* \approx \frac{e^{a_0+\varkappa_0}-1}{e^{(a_1+\varkappa_1)+(a_0+\varkappa_0)}-1}, \quad \alpha_1^* \approx \frac{e^{a_1+\varkappa_1}-1}{e^{(a_1+\varkappa_1)+(a_0+\varkappa_0)}-1}. \tag{3.64}$$

Comparing these formulas with (3.62), we see that they are identical except for the thresholds that are corrected by adding the estimates of the average overshoots. The same idea applies to the expected sample sizes: adding the values of the average overshoots $\varkappa_0$ and $\varkappa_1$ to a_0 and a_1, respectively, we obtain

$$\begin{aligned} \mathsf{E}_1 T^* &\approx \frac{1}{I_1}[(1-\alpha_1^*)(a_1+\varkappa_1)-\alpha_1^*(a_0+\varkappa_0)], \\ \mathsf{E}_0 T^* &\approx \frac{1}{I_0}[(1-\alpha_0^*)(a_0+\varkappa_0)-\alpha_0^*(a_1+\varkappa_1)], \end{aligned} \tag{3.65}$$

where α_0^* and α_1^* are as in (3.64).

Siegmund [428, Theorem 10.13] justifies these approximations for the one-parameter exponential family when the difference $\delta = \theta_1 - \theta_0$ between the parameters corresponding to the hypotheses H_1 and H_0 is small. Specifically, Siegmund assumes that $\delta \to 0$ and $a_0, a_1 \to \infty$ in such a way that $\delta a_1 = \xi \in (0,\infty)$ and $a_1/a_0 = c \in (0,\infty)$. Also, Siegmund [428, Theorem 10.55] gives an expression for a suitable computation of the overshoots through the universal constant. This method turns out to have a reasonable accuracy as long as the hypotheses are relatively close.

3.1.4 *Integral Equations and Numerical Techniques for Performance Evaluation*

Consider testing the parametric hypotheses $H_i : \theta = \theta_i$, $i = 0,1$, where the true value of the parameter $\theta \in \Theta$ may differ from the putative values θ_0 and θ_1; that is, the true pdf f_θ is now associated with the measure P_θ, and E_θ is the corresponding expectation. In this case, the LLR is $Z_n = \log[f_{\theta_1}(X_n)/f_{\theta_0}(X_n)]$.

As mentioned above, the performance of the SPRT is evaluated via the operating characteristic (OC) $\beta_\theta = \mathsf{P}_\theta(d^* = 0)$, the probability of accepting the hypothesis H_0, and via the expected sample size $\mathsf{ESS}_\theta = \mathsf{E}_\theta[T^*]$, when the value of the parameter is θ. Note that $\beta_{\theta_0} = 1-\alpha_0^*$ and $\beta_{\theta_1} = \alpha_1^*$. The main question tackled in this subsection is the computation of β_θ and ESS_θ. For a given pair of thresholds $-a_0$ and a_1, β_θ and ESS_θ can be computed by solving Fredholm integral equations, following the suggestion in [223] and further developed in [104, 219, 345] and many other works.

Let $x \in \mathbb{R}$ be an arbitrary fixed number and consider the SPRT that is based on the LLR $\lambda_n^x = x + \lambda_n$, which starts at x ($\lambda_n = \sum_{k=1}^n Z_k$). Clearly, this new SPRT is equivalent to the ordinary SPRT whose decision region is the interval $(-a_0 - x, a_1 - x)$. Let $\beta_\theta(x)$ and $\mathsf{ESS}_\theta(x)$ be the OC and ESS functions of this SPRT, respectively.

Using the Markov property of the LLR $\{\lambda_n\}$, it can be shown that

$$\beta_\theta(x) = F_\theta(-a_0 - x) + \int_{-a_0}^{a_1} \mathcal{K}_\theta(x,y)\, \beta_\theta(y)\, \mathrm{d}y, \tag{3.66}$$

$$\mathsf{ESS}_\theta(x) = 1 + \int_{-a_0}^{a_1} \mathcal{K}_\theta(x,y)\, \mathsf{ESS}_\theta(y)\, \mathrm{d}y, \tag{3.67}$$

where $F_\theta(z) = \mathsf{P}_\theta(Z_1 \le z)$ and $\mathcal{K}_\theta(x,y) = \frac{\partial}{\partial y} F_\theta(y-x) = f_\theta(y-x)$. A detailed proof of (3.66) and (3.67) is given in [104, Ch. 2]. First, observe that (3.66) and (3.67) are special cases of the more general equation

$$u(x) = v(x) + \int_{-a_0}^{a_1} \mathcal{K}_\theta(x,y)\, u(y)\, \mathrm{d}y. \tag{3.68}$$

Indeed assuming $v(x) = F_\theta(-a_0 - x)$ for all x in (3.68) results in (3.66). Similarly, equation (3.67) can be derived from (3.68) merely by taking $v(x) \equiv 1$ for all x. Let

$$\mathcal{K}_\theta\, \varphi = \int_{-a_0}^{a_1} \mathcal{K}_\theta(x,y)\, \varphi(y)\, \mathrm{d}y$$

be the linear operator induced by the kernel $\mathcal{K}_\theta(x,y)$, and rewrite (3.68) equivalently in the operator form as $u = v + \mathcal{K}_\theta\, u$.

It is also worth noting that equations (3.66) and (3.67) are renewal equations written in the

Fredholm integral form of the second kind. Such equations rarely allow for an exact analytical solution, so that in general one has to resort to a numerical scheme. To this end, many numerical techniques are available; see, e.g., [17, 217]. The two most popular ones are the following:

- Solve the integral equations iteratively [345], namely

$$\beta_{n;\theta}(x) = F_\theta(-a_0 - x) + \int_{-a_0}^{a_1} f_\theta(y-x)\beta_{n-1;\theta}(y)\,\mathrm{d}y, \quad n \ge 1, \tag{3.69}$$

  with some initial condition $\beta_{0;\theta}(x)$, where the second integral on the right-hand side is replaced with a finite sum. The convergence of this recursive method is addressed in [221].

- Replace the integral equations with a system of linear algebraic equations, considering the equation only at a number of nodes of the quadrature, and solve this system with respect to the unknown variables [29, 47, 167, 234, 236, 282, 301, 310, 384, 482], i.e.,

$$\begin{cases} \widetilde{\beta}_\theta(y_1) & = & F_\theta(-a_0 - y_1) & + & \sum_{k=1}^{m} \rho_k f_\theta(y_k - y_1)\widetilde{\beta}_\theta(y_k) \\ \vdots & \vdots & \vdots & \vdots & \vdots \\ \widetilde{\beta}_\theta(y_m) & = & F_\theta(-a_0 - y_m) & + & \sum_{k=1}^{m} \rho_k f_\theta(y_k - y_m)\widetilde{\beta}_\theta(y_k) \end{cases}, \tag{3.70}$$

  where $\widetilde{\beta}_\theta(y)$ is an approximation of $\beta_\theta(y)$, $-a_0 \le y_1 < \ldots < y_m \le a_1$ $(m > 0)$ is a partition of the interval $[-a_0, a_1]$ and the integral in (3.66) and (3.67) is replaced with an appropriate quadrature

$$\int_{-a_0}^{a_1} f_\theta(y-x)\beta_\theta(y)\,\mathrm{d}y \approx \sum_{k=1}^{m} \rho_k f_\theta(y_k - x)\beta_\theta(y_k).$$

When the kernel $f_\theta(z)$ of this integral equation is a continuous smooth function in $[-a_0, a_1]$, an effective solution can be found by using the method of Gaussian quadrature [29, 167, 217, 482]. Specifically, assume that the LLR Z_n follows the Gaussian distribution $\mathcal{N}(\theta, \sigma^2)$, so

$$\mathcal{K}_\theta(x,y) = \frac{\partial}{\partial y} F_\theta(y-x) = f_\theta(y-x) = \frac{1}{\sigma\sqrt{2\pi}}\, e^{-(y-x-\theta)^2/2\sigma^2}. \tag{3.71}$$

Then the numerical solution of the Fredholm integral equation can be effectively obtained by approximating the integral on the right-hand side of the equation (3.66) as

$$\int_{-a_0}^{a_1} \frac{1}{\sigma\sqrt{2\pi}} e^{-(y-x-\theta)^2/2\sigma^2} \beta_\theta(y)dx \approx \sum_{i=1}^{m} \rho_i \frac{1}{\sigma\sqrt{2\pi}} e^{(y_i-y-\theta)^2/2\sigma^2} \beta_\theta(y_i),$$

where $y_i \in [-a_0, a_1]$ are the "Gaussian" points (namely the roots of the Legendre polynomial) and ρ_i are the weights of the Gaussian quadrature for the interval $[-a_0, a_1]$. Finally, let us replace (3.66) with the following system of linear equations

$$(\mathbb{I}_m - A) \cdot \widetilde{\beta} = B. \tag{3.72}$$

The matrix A $(m \times m)$ and column vectors $\tilde{\beta}$ $(m \times 1)$ and B $(m \times 1)$ are defined by

$$\begin{aligned} A &= [a_{ij}],\ a_{ij} = \frac{1}{\sigma\sqrt{2\pi}} e^{-(y_j-y_i-\theta)^2/2\sigma^2},\ i,j = 1,\ldots,m \\ \widetilde{\beta} &= (\widetilde{\beta}_\theta(y_1),\ldots,\widetilde{\beta}_\theta(y_m))^\top \\ B &= \left(\Phi\left(\frac{-a_0 - y_1 - \theta}{\sigma}\right),\ldots,\Phi\left(\frac{-a_0 - y_m - \theta}{\sigma}\right)\right)^\top. \end{aligned}$$

When the kernel $\mathcal{K}_\theta(x,y)$ of the integral equation has discontinuity points, the Gaussian quadrature method cannot be recommended. In such a case a piecewise-constant (zero-order polynomial) or piecewise linear (first-order polynomial) collocation method performs better. For instance, assume that the LLR Z_n follows a noncentral $\chi^2(1)$ distribution with one degree of freedom,

$$\mathcal{K}_\theta(x,y) = f_\theta(y-x) = \begin{cases} \frac{1}{\sqrt{2}\Gamma(1/2)}(y-x+\theta)^{-1/2}e^{-(y-x+\theta)/2} & \text{if } y-x+\theta \geq 0 \\ 0 & \text{if } y-x+\theta < 0. \end{cases} \tag{3.73}$$

This particular case has been discussed in [47, 234, 236, 301, 384]. By introducing a uniform partition $-a_0 = y_0 < y_1 < \ldots < y_m = a_1$ of the interval $[-a_0, a_1]$ with a small step $y_{i+1} - y_i$, the integral equation (3.66) can be approximated in the following manner

$$\begin{aligned} \beta_\theta(y) &= F_\theta(-a_0 - y) + \sum_{k=1}^{m} \int_{y_{k-1}}^{y_k} f_\theta(y_k - y)\beta_\theta(y)\,\mathrm{d}y \\ &\approx F_\theta(-a_0 - y) + \sum_{k=1}^{m} \beta_\theta(z_k) \int_{y_{k-1}}^{y_k} f_\theta(y_k - y)\,\mathrm{d}y, \end{aligned}$$

where $z_k = (y_k + y_{k-1})/2$, $k = 1, \ldots, m$. It is assumed that the function $z \mapsto \beta_\theta(z)$ is almost constant on each elementary subinterval $[y_k, y_{k+1}]$, i.e., $\beta_\theta(z) \approx \beta_\theta(z_k)$ for $z \in [y_k, y_{k+1}]$. To improve the precision of the solution with a moderate number m of partitions, a piecewise linear approximation around the point of discontinuity is suggested in [47, 234, 301].

A simple but efficient and sufficiently accurate numerical scheme has been proposed in [310] for a slightly different problem, but all the techniques can be applied to the SPRT as well. The numerical scheme of Moustakides *et al.* [310] is a piecewise-constant (zero-order polynomial) collocation method. This is a special case of the piecewise collocation method (an interpolation-projection technique) [18, 116, 239]. The method starts with introducing $-a_0 = y_0 < y_1 < \ldots < y_N = a_1$, $N > 0$, a partition of the interval $[-a_0, a_1]$; in general, the nodes $\{y_i\}$ need not be equidistant. Next, the sought function $u(x)$ is approximated as $u_N(x) = \sum_{j=1}^{N} u_{j,N}\chi_j(x)$, where $\{u_{j,N}\}_{1 \leq j \leq N}$ are constant coefficients to be determined and $\{\chi_j(x)\}_{1 \leq j \leq N}$ are suitably chosen basis functions.

The idea of the method is to seek a prescription for choosing the coefficients $\{u_{j,N}\}_{1 \leq j \leq N}$. For any such choice, substituting $u_N(x)$ into the equation (3.68) will give a residual $u_N - \mathcal{K}_\theta u_N - v$. Unless the true solution $u(x)$ itself is a linear combination of the basis functions $\{\chi_j(x)\}_{1 \leq j \leq N}$, no choice of the coefficients $\{u_{j,N}\}_{1 \leq j \leq N}$ makes the residual identically zero. However, by requiring the residual $u_N - \mathcal{K}_\theta u_N - v$ to be zero at $\{x_j\}_{1 \leq j \leq N}$, where $x_j \in [-a_0, a_1]$ for all $j = 1, 2, \ldots, N$, one can achieve a certain level of closeness of the residual to zero. As a result, we arrive at the following system of N algebraic equations for the coefficients $u_{j,N}$:

$$\mathbf{u}_N = \mathbf{v} + \mathbf{K}_\theta \mathbf{u}_N, \tag{3.74}$$

where $\mathbf{u}_N = [u_{1,N}, u_{2,N}, \ldots, u_{N,N}]^\top$, $\mathbf{v} = [v(x_1), v(x_2), \ldots, v(x_N)]^\top$, and $\mathbf{K}_\theta = (K_\theta^{i,j})$ is an N-by-N matrix whose (i,j)-th element is

$$K_\theta^{i,j} = \int_{-a_0}^{a_1} \mathcal{K}_\theta(x_i, y)\chi_j(y)\,\mathrm{d}y, \quad i, j = 1, 2, \ldots, N.$$

For the system (3.74) to be consistent, the functions $\{\chi_j(x)\}_{1 \leq j \leq N}$ have to be chosen so as to form a basis in the appropriate functional space, i.e., the $\{\chi_j(x)\}_{1 \leq j \leq N}$ need to be linearly independent. Set $\chi_j(x) = \mathbb{1}_{\{y_{j-1} \leq x < y_j\}}$ with $y_j = -a_0 + j(a_0 + a_1)/N$ (i.e., equidistant nodes) for $j = 1, \ldots, N$. For this basis, the (i,j)-th element of the matrix $\mathbf{K}_\theta$ is $F_\theta(y_j - x_i) - F_\theta(y_{j-1} - x_i)$. As for the collocation points $\{x_i\}_{1 \leq i \leq N}$, we suggest to take the middle points $x_j = (y_{j-1} + y_j)/2$, $j = 1, \ldots, N$.

It is well known that the rate of convergence of the collocation method is at worst that of the underlying (piecewise) interpolation scheme [18, 116]. Therefore, since in our case the interpolation is of order zero, the rate is linear or better. In fact, it can be shown that the choice of the collocation points being the middle points improves the convergence rate to quadratic. This superconvergence effect is discussed, e.g., in [18].

3.1.5 *Evaluation of the Operating Characteristics and Comparison with the Neyman–Pearson Test for the Gaussian Model*

Consider testing two hypotheses $H_0 : \theta = \theta_0$ and $H_1 : \theta = \theta_1$ for the iid Gaussian model

$$X_n = \theta + \xi_n, \quad n \geq 1, \quad \theta \in \mathbb{R}^1,$$

where $\xi_n \sim \mathcal{N}(0, \sigma^2)$, $n = 1, 2, \dots$ are iid zero-mean normal random variables (noise) with variance $\sigma^2 > 0$. Then

$$Z_k = \frac{\theta_1 - \theta_0}{\sigma^2} X_k - \frac{\theta_1^2 - \theta_0^2}{2\sigma^2}, \quad \lambda_n = \frac{\theta_1 - \theta_0}{\sigma^2} \sum_{k=1}^{n} X_k - \frac{\theta_1^2 - \theta_0^2}{2\sigma^2} n.$$

Write $q = (\theta_1 - \theta_0)^2/\sigma^2$ for the signal-to-noise ratio (SNR). It is easily seen that $I_0 = I_1 = q/2$ and $\mathsf{var}_0[Z_1] = \mathsf{var}_1[Z_1] = q$, so that the problem is symmetric and therefore $\zeta_0 = \zeta_1 = \zeta(q)$ and $\varkappa_0 = \varkappa_1 = \varkappa(q)$. By (3.50), we have

$$\alpha_0^*(q) \approx \frac{\zeta(q)[e^{a_0} - \zeta(q)]}{e^{a_1 + a_0} - \zeta^2(q)}, \quad \alpha_1^*(q) \approx \frac{\zeta(q)[e^{a_1} - \zeta(q)]}{e^{a_1 + a_0} - \zeta^2(q)} \tag{3.75}$$

and by (3.65),

$$\begin{aligned} \mathsf{ESS}_1(a_0, a_1, q) &\approx \frac{2}{q}[(1 - \alpha_1^*(q))(a_1 + \varkappa(q)) - \alpha_1^*(q)(a_0 + \varkappa(q))], \\ \mathsf{ESS}_0(a_0, a_1, q) &\approx \frac{2}{q}[(1 - \alpha_0^*(q))(a_0 + \varkappa(q)) - \alpha_0^*(q)(a_1 + \varkappa(q))], \end{aligned} \tag{3.76}$$

where $\mathsf{ESS}_i = \mathsf{E}_i T^*$.

Applying Theorem 2.5.3, we obtain

$$\zeta(q) = \frac{2}{q} \exp\left\{ -2 \sum_{n=1}^{\infty} \frac{1}{n} \Phi\left(-\frac{1}{2}\sqrt{qn} \right) \right\} \tag{3.77}$$

and

$$\varkappa(q) = 1 + \frac{q}{4} + \sqrt{q} \sum_{n=1}^{\infty} \frac{1}{\sqrt{n}} \left[\frac{1}{2}\sqrt{qn}\, \Phi\left(-\frac{1}{2}\sqrt{qn} \right) - \varphi\left(\frac{1}{2}\sqrt{qn} \right) \right]. \tag{3.78}$$

Also, Siegmund's corrected Brownian motion approximations yield

$$\begin{aligned} \zeta(q) &\approx \zeta_{\mathrm{app}}(q) = \exp\{-\rho\sqrt{q}\}, \\ \varkappa(q) &\approx \varkappa_{\mathrm{app}}(q) = q/8 + \rho q^{1/2}, \end{aligned} \tag{3.79}$$

where

$$\rho = -\frac{1}{\pi} \int_0^\infty \frac{1}{t^2} \log\left\{ \frac{2\left(1 - e^{-t^2/2}\right)}{t^2} \right\} dt \approx 0.582597.$$

The remaining terms in the approximations (3.79) are of order $o(q)$ for small q, and therefore their accuracy grows when the SNR decreases. The values of ζ, $\varkappa$, ζ_{app} and $\varkappa_{\mathrm{app}}$ for $q = 0.1(0.1)2.0$ are

Table 3.1: *The values of ζ, $\zeta_{\rm app}$, $\varkappa$, and $\varkappa_{\rm app}$ for different values of the SNR q.*

q	ζ	$\zeta_{\rm app}$	$\varkappa$	$\varkappa_{\rm app}$	q	ζ	$\zeta_{\rm app}$	$\varkappa$	$\varkappa_{\rm app}$
0.1	0.83183	0.83174	0.19706	0.19673	1.1	0.54495	0.54279	0.76046	0.74853
0.2	0.77087	0.77063	0.28647	0.28555	1.2	0.53063	0.52824	0.80179	0.78820
0.3	0.72721	0.72680	0.35830	0.35660	1.3	0.51728	0.51465	0.84208	0.82676
0.4	0.69240	0.69179	0.42109	0.41847	1.4	0.50477	0.50191	0.88145	0.86434
0.5	0.66316	0.66235	0.47812	0.47446	1.5	0.49301	0.48991	0.92000	0.90103
0.6	0.63784	0.63681	0.53109	0.52628	1.6	0.48191	0.47858	0.95783	0.93693
0.7	0.61544	0.61420	0.58100	0.57494	1.7	0.47142	0.46785	0.99499	0.97211
0.8	0.59534	0.59387	0.62849	0.62109	1.8	0.46146	0.45766	1.03156	1.00664
0.9	0.57709	0.57539	0.67403	0.66520	1.9	0.45200	0.44796	1.06757	1.04055
1.0	0.56037	0.55845	0.71794	0.70760	2.0	0.44298	0.43871	1.10309	1.07392

given in Table 3.1. It is seen that the approximations (3.79) are quite accurate and can be used even for $q > 1$ as long as it is not very large.

Note that using the expansions of Lotov [279], it can be shown that the residual terms in (3.75) are of order $O\left(e^{-a_0(1+c)}\right) + O\left(e^{-a_1(1+c)}\right)$ as in (3.50), where

$$c = c(q) = \frac{\sqrt{2}}{4}\left[1 + \left(1 + \frac{256\pi^2}{q^2}\right)^{1/2}\right]^{1/2} - \frac{1}{2}.$$

When q decreases the value of $c(q)$ increases, so the accuracy of the approximations increases, as expected. For example, $c(1) \approx 2.03$, $c(0.5) \approx 4.53$, and $c(0.1) \approx 24.57$.

It is interesting to compare the SPRT with the best nonsequential fixed sample size test, which is nothing but the classical Neyman–Pearson (NP) test having the form

$$d_n = \begin{cases} 1 & \text{if } \lambda_n \ge h, \\ 0 & \text{otherwise.} \end{cases} \tag{3.80}$$

See Theorem 2.9.1 and Remark 2.9.1. Here the sample size $n = n_q(\alpha_0, \alpha_1)$ and the threshold $h = h(\alpha_0, \alpha_1)$ are selected to satisfy the conditions $\mathsf{P}_{\theta_0}\left(\lambda_{n_q} \ge h\right) = \alpha_0$ and $\mathsf{P}_{\theta_1}\left(\lambda_{n_q} < h\right) = \alpha_1$, which lead to the following equations

$$1 - \Phi\left(\frac{h + qn_q/2}{\sqrt{qn_q}}\right) = \alpha_0, \quad \Phi\left(\frac{h - qn_q/2}{\sqrt{qn_q}}\right) = \alpha_1.$$

Denoting by $Q_\alpha = \Phi^{-1}(\alpha)$ the α-quantile of the standard normal distribution and solving these equations, we obtain that

$$n_q(\alpha_0, \alpha_1) = (Q_{1-\alpha_0} + Q_{1-\alpha_1})^2/q \tag{3.81}$$

and $h(\alpha_0, \alpha_1) = (Q_{1-\alpha_0}^2 - Q_{1-\alpha_1}^2)/2$.

Thus, if we define the efficiency of the SPRT with respect to the NP test as the ratio $\mathcal{E}_i = \mathsf{ESS}_i/n_q$, then from (3.10) and (3.81) we obtain

$$\mathcal{E}_0(\alpha_0, \alpha_1) \approx \frac{2\beta(\alpha_0, \alpha_1)}{(Q_{1-\alpha_0} + Q_{1-\alpha_1})^2}, \quad \mathcal{E}_1(\alpha_0, \alpha_1) \approx \frac{2\beta(\alpha_1, \alpha_0)}{(Q_{1-\alpha_0} + Q_{1-\alpha_1})^2}. \tag{3.82}$$

These approximations are asymptotically accurate as $\alpha_0, \alpha_1 \to 0$.

The computations show that $\mathcal{E}_0(\alpha_0,\alpha_1)=\mathcal{E}_1(\alpha_0,\alpha_1)\geq 17/30$ for $\alpha_0=\alpha_1\leq 0.03$. In the asymptotic case as $\alpha=\max(\alpha_0,\alpha_1)\to 0$,

$$n_q(\alpha_0,\alpha_1)=\frac{2}{q}\left(\sqrt{|\log\alpha_0|}+\sqrt{|\log\alpha_1|}\right)^2(1+o(1))$$

and

$$\beta(\alpha_0,\alpha_1)=|\log\alpha_1|(1+o(1)),\quad \beta(\alpha_1,\alpha_0)=|\log\alpha_0|(1+o(1)),$$

and we obtain

$$\mathcal{E}_0(\alpha_0,\alpha_1)\sim\frac{|\log\alpha_1|}{\left(\sqrt{|\log\alpha_0|}+\sqrt{|\log\alpha_1|}\right)^2},\quad \mathcal{E}_1(\alpha_0,\alpha_1)\sim\frac{|\log\alpha_0|}{\left(\sqrt{|\log\alpha_0|}+\sqrt{|\log\alpha_1|}\right)^2}.$$

Here we used the asymptotic approximation for the quantile

$$Q_{1-p}=-Q_p\sim\sqrt{2|\log p|},\quad p\to 0, \tag{3.83}$$

which follows from the approximation

$$\Phi(-x)=\frac{1}{x\sqrt{2\pi}}e^{-x^2/2}(1+O(x^{-2})),\quad x\to\infty.$$

Let $(\log\alpha_1)/(\log\alpha_0)\to c$, $c\in(0,\infty)$. Then

$$\lim_{\alpha\to 0}\mathcal{E}_0(\alpha_0,\alpha_1)=\frac{c}{(1+\sqrt{c})^2},\quad \lim_{\alpha\to 0}\mathcal{E}_1(\alpha_0,\alpha_1)=\frac{1}{(1+\sqrt{c})^2}.$$

Hence, in the asymptotically symmetric case where $c=1$ (i.e., α_0 and α_1 decrease with the same rate, $\alpha_0=c_1\alpha_1$, $c_1\in(0,\infty)$), we have

$$\lim_{\alpha\to 0}\mathcal{E}_i(\alpha_0,\alpha_1)=1/4,\quad i=0,1,$$

that is, the SPRT's expected sample sizes are four times smaller than the sample size of the NP test. However, the rate of convergence is slow. For example, $\mathcal{E}_0=\mathcal{E}_1=0.36$ if $\alpha_0=\alpha_1=10^{-3}$ and $\mathcal{E}_0=\mathcal{E}_1=0.33$ if $\alpha_0=\alpha_1=10^{-4}$, so that in the symmetric case one may expect that the SPRT beats the NP test about three times for the probabilities of errors of interest in the most practical applications.

In the asymmetric case where α_0 and α_1 decrease with different rates, the situation changes dramatically. For example, if $c\ll 1$, then $\mathcal{E}_0\approx c\ll 1$ and $\mathcal{E}_1\approx(1+2\sqrt{c})^{-1}\approx 1-c\approx 1$. This means that in the asymmetric case where $\alpha_0\ll\alpha_1$, the efficiency of the SPRT is high for the hypothesis H_0 but low for the hypothesis H_1. If, for instance, $\alpha_0=10^{-4}$ and $\alpha_1=10^{-1}$, then $\mathcal{E}_0=0.18$ (82%) and $\mathcal{E}_1=0.647$ (35%).

Similar results are also valid for close hypotheses in a general non-Gaussian case under some regularity conditions [452].

We now perform a more accurate performance evaluation using the integral equations (3.66) and (3.67) and the numerical techniques described in Subsection 3.1.4. It is easily seen that under the hypothesis that the true value of the parameter is θ, the LLR Z_1 has the normal distribution $\mathcal{N}(qr_\theta,q)$, where $r_\theta=[\theta-(\theta_1+\theta_0)/2]/(\theta_1-\theta_0)$. In particular, $r_{\theta_1}=-r_{\theta_0}=1/2$. Therefore,

$$F_\theta(z)=\Phi((z-qr_\theta)/q),\quad \mathcal{K}_\theta(x,y)=q^{-1}\,\varphi((y-x-qr_\theta)/q).$$

Write $\beta_i(x)=\beta_{\theta_i}(x)$ and $\mathsf{ESS}_i(x)=\mathsf{ESS}_{\theta_i}(x)$. For solving numerically the integral equations (3.66) and (3.67) for $\beta_i(x)$ and $\mathsf{ESS}_i(x)$, we used the method described in Subsection 3.1.4 with equidistant breakdown of the interval $[-a_0,a_1]$ into 10^4 points, which is sufficient to obtain a very small error

Table 3.2: *Results for* $\alpha_0 = 10^{-6}$ *and* $\alpha_1 = 10^{-3}$*; the thresholds are selected according to* (3.84).

q	Approximations (3.76)		Integral Equations			
	ESS_0	ESS_1	ESS_0	ESS_1	α_0	α_1
0.1	138.41	276.13	138.41	276.13	1.0000×10^{-6}	1.0000×10^{-3}
0.3	46.32	92.22	46.32	92.22	1.0000×10^{-6}	1.0000×10^{-3}
0.5	27.9	55.44	27.9	55.44	1.0000×10^{-6}	1.0000×10^{-3}
1.0	14.09	27.86	14.09	27.87	1.0000×10^{-6}	1.0000×10^{-3}
2.0	6.53	13.42	7.2	14.08	1.0000×10^{-6}	1.0000×10^{-3}
5.0	3.07	5.83	3.07	5.83	1.0000×10^{-6}	1.0040×10^{-3}
10.0	1.72	3.09	1.69	3.09	1.0000×10^{-6}	1.0200×10^{-3}
20.0	1.06	1.74	1.12	1.72	1.1000×10^{-6}	5.9410×10^{-4}

Table 3.3: *Results for* $\alpha_0 = 10^{-3}$ *and* $\alpha_1 = 10^{-1}$*; the thresholds are selected according to* (3.84).

q	Approximations (3.76)		Integral Equations			
	ESS_0	ESS_1	ESS_0	ESS_1	α_0	α_1
0.1	46.11	118.05	46.10	118.1	1.0000×10^{-3}	1.0000×10^{-1}
0.3	15.55	39.49	15.55	39.54	1.0000×10^{-3}	1.0000×10^{-1}
0.5	9.44	23.78	9.44	23.83	1.0000×10^{-3}	1.0000×10^{-1}
1.0	4.86	12.01	4.85	12.05	1.0000×10^{-3}	1.0029×10^{-1}
2.0	1.92	5.59	2.59	6.19	1.0000×10^{-3}	9.8443×10^{-2}
5.0	1.23	2.61	1.38	2.74	1.0313×10^{-3}	7.6954×10^{-2}
10.0	0.79	1.45	1.09	1.58	1.0775×10^{-3}	4.3733×10^{-2}
20.0	0.59	0.88	1.02	1.13	4.0090×10^{-4}	9.3300×10^{-3}

(less than a fraction of a percent). Recall that $\beta_0(0) = 1 - \alpha_0^*$, $\beta_1(0) = \alpha_1^*$, and $\mathsf{ESS}_i(0) = \mathsf{ESS}_i = \mathsf{E}_i T^*$, $i = 0, 1$.

Tables 3.2–3.6 summarize the results for three sets of error probabilities ($\alpha_0 = 10^{-3}, \alpha_1 = 10^{-6}$), ($\alpha_0 = 10^{-1}, \alpha_1 = 10^{-3}$), and ($\alpha_0 = 10^{-2}, \alpha_1 = 10^{-2}$) and for the SNR q varying from as small as 0.1 to as large as 20. In all tables, except for Table 3.4, the thresholds $-a_0$ and a_1 are selected according to the formulas

$$a_1(\alpha_0, \alpha_1, q) = \log\left[\frac{\zeta(q)(1-\alpha_1)}{\alpha_0}\right], \quad -a_0(\alpha_0, \alpha_1, q) = -\log\left[\frac{\zeta(q)(1-\alpha_0)}{\alpha_1}\right] \tag{3.84}$$

that are obtained by inverting the approximations (3.75) for the error probabilities; see also (3.51). In Table 3.4, the thresholds are chosen according to Wald's approximations (3.5) that ignore the overshoots.

As we can see, the approximations (3.75) for the error probabilities and (3.76) for the expected sample sizes accounting for the overshoots are very accurate as long as the SNR is not too high. Their accuracy certainly depends on the given probabilities of errors — it is higher for small probabilities of errors. For example, in the case where ($\alpha_0 = 10^{-3}, \alpha_1 = 10^{-6}$), the accuracy is perfect for $q \le 10$ and reasonable even for $q = 20$. For the other cases the approximations are accurate for $q \le 2$ and become less accurate for larger values of q. As far as Wald's approximations are concerned, comparing the numbers in Table 3.4 with those in Table 3.3 shows that Wald's approximations are not accurate as long as the SNR q is not very small. Even for $q = 0.3$ the precision is questionable.

Table 3.6 shows the efficiency $\mathcal{E}_i(\alpha_0, \alpha_1, q) = \mathsf{ESS}_i(\alpha_0, \alpha_1, q)/n_q(\alpha_0, \alpha_1)$ of the SPRT with re-

Table 3.4: *Accuracy of Wald's approximations for* $\alpha_0 = 10^{-3}$ *and* $\alpha_1 = 10^{-1}$*; the thresholds are selected according to Wald's approximations* (3.5).

	Wald's Approximations		Integral Equations			
q	ESS_0	ESS_1	ESS_0	ESS_1	α_0	α_1
0.1	45.85	127.05	49.81	124.23	8.4740×10^{-4}	8.3196×10^{-2}
0.3	15.28	42.35	17.68	43.01	7.4920×10^{-4}	7.2740×10^{-2}
0.5	9.17	25.41	11.09	26.49	6.8800×10^{-4}	6.6332×10^{-2}
1.0	4.58	12.7	6.03	13.89	5.8760×10^{-4}	5.6102×10^{-2}
2.0	2.29	6.35	3.39	7.42	4.7030×10^{-4}	4.4578×10^{-3}
5.0	0.92	2.54	1.75	3.41	3.0450×10^{-4}	2.6718×10^{-3}
10.0	0.46	1.27	1.24	1.99	2.0110×10^{-4}	1.2587×10^{-3}
20.0	0.23	0.64	1.04	1.25	9.5100×10^{-5}	3.0931×10^{-3}

Table 3.5: *Results for* $\alpha_0 = \alpha_1 = 10^{-2}$*; the thresholds are selected according to* (3.84).

	Approximations (3.76)		Integral Equations			
q	ESS_0	ESS_1	ESS_0	ESS_1	α_0	α_1
0.1	90.32	90.32	90.32	90.32	1.0000×10^{-2}	1.0000×10^{-2}
0.3	30.11	30.11	30.29	30.29	1.0000×10^{-2}	1.0000×10^{-2}
0.5	18.17	18.17	18.28	18.28	1.0000×10^{-2}	1.0000×10^{-2}
1.0	9.14	9.14	9.28	9.28	1.0000×10^{-2}	1.0000×10^{-2}
2.0	4.64	4.64	4.79	4.79	1.0000×10^{-2}	9.9994×10^{-3}
5.0	1.65	1.65	2.1	2.1	1.0078×10^{-2}	1.0078×10^{-3}
10.0	1.05	1.05	1.29	1.29	7.9826×10^{-3}	7.9826×10^{-3}
20.0	0.81	0.81	1.04	1.04	3.1443×10^{-3}	3.1443×10^{-3}

Table 3.6: *Efficiency of the SPRT with respect to the NP test.*

	$\alpha_0 = 10^{-6}$, $\alpha_1 = 10^{-3}$		$\alpha_0 = 10^{-3}$, $\alpha_1 = 10^{-1}$		$\alpha_0 = \alpha_1 = 10^{-2}$
q	$\mathcal{E}_0(\alpha_0, \alpha_1)$	$\mathcal{E}_1(\alpha_0, \alpha_1)$	$\mathcal{E}_0(\alpha_0, \alpha_1)$	$\mathcal{E}_1(\alpha_0, \alpha_1)$	$\mathcal{E}_0(\alpha_0, \alpha_1) = \mathcal{E}_1(\alpha_0, \alpha_1)$
0.1	0.22 (78%)	0.45 (55%)	0.24 (76%)	0.61 (39%)	0.41 (59%)
10.0	0.24 (76%)	0.44 (56%)	0.55 (45%)	0.79 (21%)	0.43 (67%)

spect to the NP test. The values of $\mathsf{ESS}_i(\alpha_0, \alpha_1, q)$ were computed by solving the integral equations numerically as described above and the value of $n_q(\alpha_0, \alpha_1, q)$ according to (3.81). It is seen that the numbers are different from what is expected from the above asymptotic theory that ignores the overshoots. However, the trend is as expected.

We remark that if the parameter value differs from the putative values θ_0 and θ_1, then the efficiency of the SPRT decreases. It becomes especially low in the vicinity of the least favorable point $\theta^* = (\theta_0 + \theta_1)/2$ where the ESS of the SPRT can be even larger than the sample size of the NP test. This issue is further discussed in Section 5.2.

3.1.6 *Evaluation of the Operating Characteristics and Comparison with the Neyman–Pearson Test for the Exponential Model*

Consider an exponential model where the pdf of the observations is $f_\theta(x) = \theta e^{-\theta x} \mathbb{1}_{\{x \ge 0\}}$, $\theta > 0$. This model has a wide spectrum of practical applications. In particular, it arises while testing the hypotheses $H_0 : \theta = \theta_0$ and $H_1 : \theta = \theta_1$ on the intensity θ of the Poisson process when observing time intervals between events. Also, this model is important in radar applications when detecting a target (in white Gaussian noise) that fluctuates slowly within radar pulses based on some optimal preliminary processing that includes matched filtering and square detection within pulses [22, 452], followed by non-coherent accumulation of the results of this preliminary processing. In this case, the hypothesis H_0 corresponds to the situation where there is no target (signal) and we set $\theta_0 = 1$, while the hypothesis H_1 means that there is a target with the signal-to-noise ratio (SNR) $q > 0$, and we set $\theta_1 = 1/(1+q)$. Namely this latter application motivates the detailed discussion presented in this subsection; see also Tartakovsky and Ivanova [450].

It is easy to see that with these assumptions

$$Z_k = \frac{q}{1+q} X_k - \log(1+q), \quad X_k \ge 0 \tag{3.85}$$

and

$$\mathsf{P}_1(Z_1 > z) = (1+q)^{-1/q} e^{-z/q}, \quad \mathsf{P}_0(Z_1 > z) = (1+q)^{-(1+q)/q} e^{-z(1+q)/q}, \tag{3.86}$$

for $z \ge -\log(1+q)$ and 1 otherwise, that is, the distributions of the LLR Z_1 have exponential right tails with parameters $C_0 = (1+q)^{-1/q}$ and $C_1 = 1/q$ under the hypothesis H_1 and with parameters $\widetilde{C}_0 = (1+q)^{-(1+q)/q}$ and $\widetilde{C}_1 = (1+q)/q$ under the hypothesis H_0, respectively (cf. Example 2.5.2). Using the argument similar to the one that has led to (2.100) and (2.101), we obtain that for all $y \ge 0$ and $a_1, a_0 \ge 0$,

$$\begin{aligned} \mathsf{P}_1(\kappa_1(a_0,a_1) > y \mid T^* = T_1) &= e^{-y/q}, \\ \mathsf{P}_0(\kappa_1(a_0,a_1) > y \mid T^* = T_1) &= e^{-y(1+q)/q}. \end{aligned} \tag{3.87}$$

Thus, for all $a_0, a_1 \ge 0$,

$$\begin{aligned} \varkappa_1(q) &= \mathsf{E}_1[\kappa_1(a_0,a_1) \mid T^* = T_1] = q, \\ \zeta_1(q) &= e_1(q) = \mathsf{E}_1\left[e^{-\kappa_1(a_0,a_1)} \mid T^* = T_1\right] = 1/(1+q). \end{aligned} \tag{3.88}$$

Furthermore, in this particular example we can also find the value of the average overshoot of the upper boundary a_1 under H_0 using the second distribution in (3.87):

$$\varkappa_{10}(q) = \mathsf{E}_0[\kappa_1(a_0,a_1) \mid T^* = T_1] = q/(1+q) \quad \text{for all } a_0, a_1 \ge 0. \tag{3.89}$$

We stress that the equalities (3.88) and (3.89) are exact, not just asymptotic for large a_0, a_1. Note also that these results can be extended to the case of composite hypotheses when the parameter q is not equal to the putative value [450].

Unfortunately, finding the exact distributions of the overshoot $\kappa_0(a_0,a_1)$ and hence the values of e_0 and $\mathsf{E}_0[\kappa_0 \mid T^* = T_0]$ is not possible and we have to resort to the asymptotic estimates $e_0 \approx \zeta_0$ and $\mathsf{E}_0[\kappa_0 \mid T^* = T_0] \approx \varkappa_0$. First, the K–L information numbers are

$$I_0 = \log(1+q) - q/(1+q), \quad I_1 = q - \log(1+q), \tag{3.90}$$

so that, from (3.54) in Theorem 3.1.3,

$$\zeta_0 = \zeta_1 I_1 / I_0 = \frac{q - \log(1+q)}{(1+q)\log(1+q) - q}. \tag{3.91}$$

Next, we use the formula (2.96) for the limiting density of the overshoot $\tilde{\kappa}_0(a_0)$ in the one-sided test:

$$h(y) = I_0^{-1}\mathsf{P}_0\left(\min_{n\geq 1}(-\lambda_n) > y\right) = I_0^{-1}\mathsf{P}_0\left(\tilde{\lambda}_n > y, n \geq 1\right),$$

where $\tilde{\lambda}_n = -\lambda_n$. Let

$$\tau = \tau(y,b) = \inf\left\{n \geq 1 : \tilde{\lambda}_n \notin (y,b)\right\}$$

Observe that

$$\mathsf{P}_0\left(\tilde{\lambda}_n > y, n \geq 1\right) = \lim_{b\to\infty} \mathsf{P}_0(\tilde{\lambda}_\tau \geq b) = 1 - \lim_{b\to\infty} \mathsf{P}_0(\tilde{\lambda}_\tau \leq y),$$

where

$$\lim_{b\to\infty} \mathsf{P}_0(\tilde{\lambda}_\tau \leq y) = \lim_{b\to\infty} \frac{e_1 e^b - e_0 e_0}{e^{b-y} - e_1 e_0} = e_1 e^y = (1+q)^{-1} e^y$$

for $0 \leq y \leq \log(1+q)$ and 1 otherwise (since $\tilde{\lambda}_1 \leq \log(1+q)$). Hence

$$h(y) = I_0^{-1}\left[1 - e^y/(1+q)\right] \mathbb{1}_{\{0\leq y\leq \log(1+q)\}}. \tag{3.92}$$

Using (3.88), (3.90) and (3.92), we obtain

$$\varkappa_0 = \frac{[\log(1+q)]^2}{2I_0} - 1 = \frac{(1+q)[\log(1+q)]^2}{2[(1+q)\log(1+q) - q]} - 1. \tag{3.93}$$

Also, the expression for ζ_0 in (3.91) can be obtained by direct computation using (3.92).

The only expectation that remains to evaluate is $\mathsf{E}_1[\kappa_0(a_0,a_1)|T^* = T_0]$. The lower bound is obviously 0, while the upper bound can be obtained by noticing that

$$\mathsf{E}_1[\kappa_0(a_0,a_1)|T^* = T_0] \leq \lim_{b\to -a_0} \mathsf{E}_1[\kappa_0 \mid \lambda_{T^*} \leq -a_0, \lambda_{T^*-1} = b] = \overline{\varkappa}_{01}.$$

It can be easily verified that, for all $n \geq 1$,

$$\begin{aligned}
&\mathsf{P}_1(-\lambda_n - a_0 \leq y \mid -\lambda_n > a_0, -\lambda_{n-1} = a_0) \\
&= \mathsf{P}_1\left(-\tfrac{q}{1+q}X_n + \log(1+q) \leq y \mid -\tfrac{q}{1+q}X_n + \log(1+q) > 0\right) \\
&= \frac{\mathsf{P}_1\left\{\frac{q}{1+q}[\log(1+q) - y] \leq X_n < \frac{q}{1+q}\log(1+q)\right\}}{\mathsf{P}_1\left\{X_n < \frac{q}{1+q}\log(1+q)\right\}} \\
&= \frac{e^{y/q} - 1}{(1+q)^{1/q} - 1} \quad \text{for } 0 \leq y \leq \log(1+q).
\end{aligned}$$

It follows that

$$\overline{\varkappa}_{01} = \frac{(1+q)^{1/q}\log(1+q)}{(1+q)^{1/q} - 1} - q. \tag{3.94}$$

Note that $\overline{\varkappa}_{01} \sim \log(1+q)$ as $q \to \infty$ and $\overline{\varkappa}_{01} \sim q/(e-1)$ as $q \to 0$.

Therefore, inverting the approximations (3.50), we obtain that in order to guarantee the given probabilities of errors α_i, $i = 0,1$ the thresholds in the SPRT are selected as

$$\begin{aligned}
a_1(\alpha_0,\alpha_1,q) &= \log\left[\frac{1-\alpha_1}{(1+q)\alpha_0}\right], \\
-a_0(\alpha_0,\alpha_1,q) &= -\log\left[\frac{1-\alpha_0}{\alpha_1}\,\frac{q - \log(1+q)}{(1+q)\log(1+q) - q}\right].
\end{aligned} \tag{3.95}$$

Recall that Wald's approximations lead to the formulas (3.5).

The exact formulas for the expected sample sizes $\mathsf{E}_i T^* = \mathsf{ESS}_i(\alpha_0, \alpha_1, q)$ in (3.55) along with the estimates for the average overshoots given in (3.88), (3.89), (3.93), and (3.94) suggest the following approximations for expected sample sizes $\mathsf{ESS}_i(\alpha_0, \alpha_1, q)$:

$$\begin{aligned} \mathsf{ESS}_1 &\approx \frac{(1-\alpha_1)[a_1(\alpha_0,\alpha_1,q)+q] - \alpha_1[a_0(\alpha_0,\alpha_1,q)+\overline{\varkappa}_{01}(q)]}{q-\log(1+q)}, \\ \mathsf{ESS}_0 &\approx \frac{(1-\alpha_0)[a_0(\alpha_0,\alpha_1,q)+\varkappa_0(q)] - \alpha_0[a_1(\alpha_0,\alpha_1,q)+q/(1+q)]}{\log(1+q)-q/(1+q)}. \end{aligned} \tag{3.96}$$

We now turn to the evaluation of the SPRT performance using the integral equations (3.66) and (3.67). In the notation of Subsection 3.1.4, the true density is $f_\theta(x) = \theta e^{-\theta x}\mathbb{1}_{\{x\geq 0\}}$, $\theta \in (0,\infty)$ and $\theta_0 = 1$, $\theta_1 = 1/(1+q)$. Note that $\theta_0 > \theta_1$. Using (3.85), we obtain

$$P_\theta^Z(z) = 1 - \exp\left\{-\theta\frac{1+q}{q}[z+\log(1+q)]\right\} \quad \text{for } z \geq -\log(1+q),$$

and $P_\theta^Z(z) = 0$ otherwise. Next, given $P_\theta^Z(z)$, it can be seen that

$$\mathcal{K}_\theta(x,y) = \theta\frac{1+q}{q}\exp\left\{-\theta\frac{1+q}{q}[y-x+\log(1+q)]\right\}\mathbb{1}_{\{y-x\geq -\log(1+q)\}},$$

and therefore, one can employ the numerical technique outlined in Subsection 3.1.4 to solve the equations (3.66) and (3.67); see also Moustakides *et al.* [310].

However, for this exponential example, the equations (3.66) and (3.67) can be solved analytically producing both $\beta_\theta(x)$ and $\mathsf{ESS}_\theta(x)$ in a closed form. We now present the exact formulae, which we obtained by adapting those derived by Kohlruss [238, Example 1 & Corollary 3], who found $\beta_\theta(x)$ and $\mathsf{ESS}_\theta(x)$ for the exponential scenario with $\theta_0 < \theta_1$; see also Stadje [441]. The idea is to observe that the two cases $\theta_0 < \theta_1$ and $\theta_0 > \theta_1$ can be mapped into one another by merely swapping the hypotheses H_0 and H_1. This is equivalent to employing the change-of-measure technique.

Specifically, suppose that we are interested in computing the functions $\beta_\theta(x)$ and $\mathsf{ESS}_\theta(x)$ of the SPRT with the decision region being the interval $(-a_0 - x, a_1 - x)$, $x \in \mathbb{R}$ and $\theta_0 > \theta_1$. Obviously, the ESS function of this SPRT is the same as the ESS of the SPRT with $\theta_0 < \theta_1$ and with the decision region being the interval $(-a_1 + x, a_0 + x)$, so that the formulas of Kohlruss [238] can be easily modified for our case. Let

$$\phi(t,s) = \sum_{i=0}^{\lfloor t\rfloor}\frac{(-1)^i}{i!}s^i(t-i)^i, \tag{3.97}$$

$$\psi(t,s) = e^{s(t-1)}\sum_{i=0}^{\lfloor t-1\rfloor}e^{-si}\sum_{j=0}^{i}\frac{(-1)^j}{j!}s^i(t-1-i)^i - \lfloor t\rfloor, \tag{3.98}$$

where hereafter $\lfloor x\rfloor$ is the integer part of $x \geq 0$. Using [238], we obtain

$$\beta_\theta(x) = 1 - \frac{e^{\rho_\theta(x-a_1)}\phi(\lfloor(a_0+x)/\gamma\rfloor, \delta_\theta e^{-\delta_\theta})}{e^{\delta_\theta}\phi(\lfloor(a_0+a_1)/\gamma\rfloor+1, \delta_\theta e^{-\delta_\theta})}, \tag{3.99}$$

$$\mathsf{ESS}_\theta(x) = 1 - \delta_\theta\Big\{\psi(\lfloor(a_0+x)/\gamma\rfloor, \delta_\theta) - \psi(\lfloor(a_0+a_1)/\gamma\rfloor+1, \delta_\theta)[1-\beta_\theta(x)]\Big\}, \tag{3.100}$$

where $\gamma = \log(\theta_0/\theta_1) = \log(1+q)$, $\rho_\theta = \theta/(\theta_0-\theta_1) = \theta(1+q)/q$ and $\delta_\theta = \gamma\rho_\theta = [\theta(1+q)/q]\log(1+q)$. We stress that we are interested in $\beta_\theta = \beta_\theta(0)$ and $\mathsf{ESS}_\theta = \mathsf{ESS}_\theta(0)$.

However, it turns out that if the ratio $(a_0+a_1)/q$ is even moderately large (say $a_0, a_1 > 2$ with $q = 0.1$), these exact formulas raise formidable computational problems. Thus, Albert [8] admits that the formulas for β_θ and ESS_θ are "so nearly indeterminate that the writer obtained absurd results from them using modest computing facilities." In a nutshell, the issue is entirely numerical. As one can see, the formulas involve sums of alternating series whose elements are of considerably diverse magnitudes. Such sums are known to be highly unstable numerically; this circumstance nearly nullifies the practical value of these exact solutions. A remedy was proposed by DeLucia and Poor [115] who suggested to rewrite and evaluate the sums in a recursive way thereby making them far less sensitive to perturbations, and therefore, more practical. Specifically, rewrite $\phi(t,s)$ and $\psi(t,s)$ defined in (3.97) and (3.98) as

$$\begin{aligned} \phi(t,s) &= (-s)^{\lfloor t \rfloor} e^{-s\lfloor t \rfloor} \sum_{i=0}^{\lfloor t \rfloor} V_{\lfloor t \rfloor - i}(s) \frac{(t-\lfloor t \rfloor)^i}{i!}, \\ \psi(t,s) &= (-s)^{\lfloor t \rfloor} e^{s(t-\lfloor t \rfloor)} \sum_{i=0}^{\lfloor t \rfloor} U_{\lfloor t \rfloor - i}(s) \frac{(t-\lfloor t \rfloor)^i}{i!} - \frac{\lfloor t \rfloor}{s}, \end{aligned} \tag{3.101}$$

where for $k = 1, 2, \dots$

$$V_k(s) = -se^{-s} \sum_{i=0}^{k-1} \frac{V_{k-1-i}(s)}{i!} \quad \text{and} \quad U_k(s) = -se^{-s} \sum_{i=0}^{k-1} \frac{U_{k-1-i}(s)}{i!} - (-s)^{-k-1}$$

with $V_0(s) = 1$ and $U_0(s) = 0$. The corresponding recursions can be effectively used for computations.

We now present performance evaluation results obtained using Wald's approximations (3.5) that ignore the overshoots, the approximations (3.95) and (3.96) that account for the overshoots, and the exact formulas (3.99) and (3.100) where the functions ϕ and ψ are computed recursively according to (3.101).

First, we evaluate the accuracy of Wald's approximations (3.5) for the thresholds $-a_0$ and a_1, given the desired error probabilities α_0 and α_1, as well as the accuracy of the approximations (3.95) proposed by Tartakovsky and Ivanova [450]. Tables 3.7 and 3.8 report the exact error probabilities $\alpha_0^* = 1 - \beta_{\theta_0}(0)$ and $\alpha_1^* = \beta_{\theta_1}(0)$ of the SPRT computed using the exact formulas (3.99) and (3.100) with the thresholds chosen from Wald's approximations (3.5) (two left columns) and from the approximations (3.95) (two right columns). Specifically, Table 3.7 assumes $\alpha_0 = 10^{-6}$ and $\alpha_1 = 10^{-3}$, while Table 3.8 assumes $\alpha_0 = 10^{-3}$ and $\alpha_1 = 10^{-1}$. We see that Wald's approximations are not accurate, while the approximations (3.95) are very accurate even for large values of q.

Table 3.7: *Exact error probabilities α_0^* and α_1^* obtained with the thresholds chosen from Wald's approximations (3.5) and the approximations (3.95) assuming that $\alpha_0 = 10^{-6}$ and $\alpha_1 = 10^{-3}$.*

	Wald's Approximations (3.5)		Approximations (3.95)	
q	α_0	α_1	α_0	α_1
0.1	9.0912×10^{-7}	9.6873×10^{-4}	1.0000×10^{-6}	1.0000×10^{-3}
0.3	7.6930×10^{-7}	9.1630×10^{-4}	1.0000×10^{-6}	1.0000×10^{-3}
0.5	6.6675×10^{-7}	8.7372×10^{-4}	1.0000×10^{-6}	1.0000×10^{-3}
1.0	5.0010×10^{-7}	7.9435×10^{-4}	1.0000×10^{-6}	1.0000×10^{-3}
5.0	1.6674×10^{-7}	5.5787×10^{-4}	1.0000×10^{-6}	9.9846×10^{-4}
10.0	9.0959×10^{-8}	4.5692×10^{-4}	9.9999×10^{-7}	1.0079×10^{-3}
20.0	4.7648×10^{-8}	3.9821×10^{-4}	1.0001×10^{-6}	8.9322×10^{-4}

We now turn to the evaluation of the accuracy of the approximations for the expected sample sizes $\mathsf{ESS}_0 = \mathsf{ESS}_{\theta_0}(0)$ and $\mathsf{ESS}_1 = \mathsf{ESS}_{\theta_1}(0)$ given in (3.96). Specifically, the thresholds $-a_0$ and a_1 are found from the approximations (3.95), which we already know as quite accurate. Tables 3.9 and 3.10 report the results obtained for $\alpha_0 = 10^{-6}, \alpha_1 = 10^{-3}$ and $\alpha_0 = 10^{-3}, \alpha_1 = 10^{-1}$, respectively. It can be seen that the approximations are nearly perfect. Tables 3.11 and 3.12 report the

Table 3.8: *Exact error probabilities α_0^* and α_1^* obtained with the thresholds chosen from Wald's approximations (3.5) and the approximations (3.95) assuming that $\alpha_0 = 10^{-3}$ and $\alpha_1 = 10^{-1}$.*

	Wald's Approximations (3.5)		Approximations (3.95)	
q	α_0	α_1	α_0	α_1
0.1	9.1224×10^{-4}	9.6882×10^{-2}	1.0000×10^{-3}	1.0000×10^{-1}
0.3	7.7637×10^{-4}	9.1651×10^{-2}	1.0000×10^{-3}	1.0000×10^{-1}
0.5	6.7500×10^{-4}	8.7401×10^{-2}	1.0000×10^{-3}	1.0000×10^{-1}
1.0	5.1140×10^{-4}	7.9486×10^{-2}	1.0000×10^{-4}	1.0000×10^{-1}
5.0	1.7487×10^{-4}	5.5699×10^{-2}	1.0054×10^{-3}	9.5164×10^{-2}
10.0	9.6662×10^{-5}	4.3043×10^{-2}	9.8484×10^{-4}	1.1364×10^{-1}
20.0	5.0327×10^{-5}	4.8828×10^{-2}	1.0078×10^{-3}	9.2950×10^{-2}

Table 3.9: *Exact and approximate (approximations (3.96)) expected sample sizes with the thresholds chosen from the approximations (3.95) assuming that $\alpha_0 = 10^{-6}$ and $\alpha_1 = 10^{-3}$.*

	Approximations (3.96)		Exact		Efficiency SPRT vs. NP		
q	ESS_0	ESS_1	ESS_0	ESS_1	N_{NP}	$\mathcal{E}_0$	$\mathcal{E}_1$
0.1	1569.609	2942.213	1569.609	2942.218	6729	0.23	0.44
0.3	218.694	367.505	218.694	367.507	879	0.25	0.42
0.5	95.829	146.910	95.829	146.9112	365	0.26	0.40
1.0	35.834	45.951	35.834	45.952	123	0.29	0.37
5.0	7.3026	5.298	7.305	5.299	18	0.41	0.29
10.0	4.750	2.813	4.748	2.814	10	0.48	0.28
20.0	3.428	1.813	3.484	1.813	7	0.50	0.26

Table 3.10: *Exact and approximate (approximations (3.96)) expected sample sizes with the thresholds chosen from the approximations (3.95) assuming that $\alpha_0 = 10^{-3}$ and $\alpha_1 = 10^{-1}$.*

	Approximations (3.96)		Exact		Efficiency SPRT vs. NP		
q	ESS_0	ESS_1	ESS_0	ESS_1	N_{NP}	$\mathcal{E}_0$	$\mathcal{E}_1$
0.1	520.948	1256.740	520.948	1257.243	2077	0.25	0.61
0.3	72.620	157.287	72.620	157.458	269	0.27	0.59
0.5	31.85	63.13	31.847	63.231	111	0.29	0.57
1.0	11.934	20.051	11.940	20.106	37	0.32	0.54
5.0	2.488	2.725	2.563	2.757	6	0.43	0.46
10.0	1.651	1.669	1.536	1.649	3	0.51	0.55
20.0	1.222	1.244	1.203	1.259	2	0.61	0.63

same results when using Wald's approximations. We note that these approximations that are somewhat reasonable for small values of q when the overshoots are relatively small, produce nonsense results for large values of q; recall that the expected sample size cannot be less than 1. More importantly, choosing the thresholds from Wald's formulas leads to an increase in the expected sample size compared with the case of the more accurate approximations (3.95). For example, in the case where $\alpha_0 = 10^{-3}$, $\alpha_1 = 10^{-1}$ and $q = 0.5$, the expected sample size $\mathsf{ESS}_1 \approx 63$ *versus* 68.

Finally, we provide a comparison of the SPRT with the best fixed sample size test which is the NP test having the form (3.80) where the threshold $h = h(\alpha_0, \alpha_1)$ and the sample size $n = n(\alpha_0, \alpha_1)$

Table 3.11: *Exact and approximate (Wald's approximations (3.5)) expected sample sizes with the thresholds chosen from Wald's approximations assuming that* $\alpha_0 = 10^{-6}$ *and* $\alpha_1 = 10^{-3}$.

	Wald's Approximations		Exact	
q	ESS_0	ESS_1	ESS_0	ESS_1
0.1	1569.551	2941.219	1576.827	2962.653
0.3	218.634	366.508	221.461	374.516
0.5	95.766	145.912	97.701	151.223
1.0	35.764	44.953	37.026	48.222
5.0	7.207	4.300	7.911	5.860
10.0	4.640	1.815	5.275	3.131
20.0	3.302	0.814	3.857	1.993

Table 3.12: *Exact and approximate (Wald's approximations (3.5)) expected sample sizes with the thresholds chosen from Wald's approximations assuming that* $\alpha_0 = 10^{-3}$ *and* $\alpha_1 = 10^{-1}$.

	Wald's Approximations		Exact	
q	ESS_0	ESS_1	ESS_0	ESS_1
0.1	520.889	1256.338	528.321	1280.997
0.3	72.558	156.553	75.442	165.605
0.5	31.782	62.326	33.754	68.246
1.0	11.869	19.201	13.151	22.754
5.0	2.392	1.837	3.095	3.425
10.0	1.540	0.775	2.257	2.098
20.0	1.096	0.348	1.560	1.495

are selected in such a way that $\mathsf{P}_0(d_n = 1) = \alpha_0$ and $\mathsf{P}_1(d_n = 0) = \alpha_1$. It is more convenient to rewrite the NP test in terms of the statistic $S_n = X_1 + \cdots + X_n$, i.e., $d_n = 1$ if $S_n \ge h$ and 0 otherwise. Clearly, under both hypotheses the statistic S_n has χ^2-distributions with $2n$ degrees of freedom with densities:

$$p_0(S_n) = \frac{S_n^{n-1}}{\Gamma(n)} \exp\{-S_n\} \, \mathbb{1}_{\{S_n \ge 0\}},$$

$$p_1(S_n) = \frac{S_n^{n-1}}{\Gamma(n)(1+q)^n} \exp\left\{-\frac{S_n}{1+q}\right\} \mathbb{1}_{\{S_n \ge 0\}},$$

where $\Gamma(n)$ is a Gamma function. Therefore, the probabilities of errors are given by

$$\mathsf{P}_0(d_n = 1) = \mathsf{P}_0(S_n \ge h) = 1 - G_{2n}(2h), \quad \mathsf{P}_1(d_n = 0) = \mathsf{P}_1(S_n < h) = G_{2n}(2h/(1+q)),$$

where

$$G_m(y) = \frac{1}{2^{m/2}\Gamma(m/2)} \int_0^y x^{m/2-1} e^{-x/2} \, \mathrm{d}x$$

is the standard χ^2 cdf with m degrees of freedom. Let $\chi^2_m(p)$ stand for the p-quantile of the χ^2-distribution with m degrees of freedom. Then the number of observations $n = n_q(\alpha_0, \alpha_1)$ is selected from the equation

$$\chi^2_{2n}(1-\alpha_0) = (1+q)\, \chi^2_{2n}(\alpha_1) \tag{3.102}$$

and the threshold according to the formula $h = \chi^2_{2n}(1-\alpha_0)/2$.

For computing the efficiency of the SPRT with respect to the NP test $\mathcal{E}_i(\alpha_0, \alpha_1, q) = \mathsf{ESS}_i(\alpha_0, \alpha_1, q)/n_q(\alpha_0, \alpha_1)$, the fixed sample size $n_q(\alpha_0, \alpha_1)$ in the NP test has been calculated by solving the equation (3.102) numerically, and the expected sample sizes $\mathsf{ESS}_i(\alpha_0, \alpha_1, q)$ in the SPRT have been computed using the exact formulas. The comparison against the NP test was performed only for the approximations of the thresholds given by (3.95) which are very accurate. For the results, see the last three columns in Tables 3.9 and 3.10. Observe that in all cases the SPRT reaches the decision substantially (often 4 times) more quickly.

All in all, although for the exponential scenario the operating characteristics of the SPRT can be evaluated analytically in closed form, the exact formulae are of little assistance in practice since they are quite complex mathematically, rather unstable numerically and offer only an insignificant improvement over the approximations given in (3.95) and (3.96), which are easily computable and fairly accurate unless the probabilities of errors are too high. Therefore, we can conclude that from the practical viewpoint the approximations (3.95) and (3.96) are preferable over the exact formulae.

3.2 SPRT Optimality in the iid Case

In Section 3.1, we mentioned that Wald's SPRT has an extraordinary optimality property: it minimizes both expected sample sizes $\mathsf{E}_0 T$ and $\mathsf{E}_1 T$ in the class of sequential (and nonsequential) tests with given error probabilities as long as the observations are iid under both hypotheses. This strong result is due to Wald and Wolfowitz [496]. In this section, we provide precise statements.

Let

$$\mathbf{C}(\alpha_0, \alpha_1) = \{\delta : \mathsf{P}_0(d=1) \le \alpha_0, \mathsf{P}_1(d=0) \le \alpha_1, \mathsf{E}_0 T < \infty, \mathsf{E}_1 T < \infty\}$$

be the class of (sequential or nonsequential) tests with error probabilities that do not exceed the given values α_0 and α_1, $0 < \alpha_0, \alpha_1 < 1$ and with finite expected sample sizes $\mathsf{E}_0 T$ and $\mathsf{E}_1 T$. Recall that for the SPRT we use the notation with asterisks: $\delta^* = (d^*, T^*)$, $\alpha_0^* = \mathsf{P}_0(d^* = 1)$, $\alpha_1^* = \mathsf{P}_1(d^* = 0)$, *etc.*

Theorem 3.2.1 (SPRT optimality). *Let the observations X_n, $n = 1, 2, \dots$ be iid with (relatively to some sigma-finite measure) density f_0 under H_0 and with density f_1 under H_1, where $f_0 \not\equiv f_1$. Assume that $\alpha_0 + \alpha_1 < 1$. If the thresholds $-a_0$ and a_1 can be selected in such a way that $\alpha_0^*(a_0, a_1) = \alpha_0$ and $\alpha_1^*(a_0, a_1) = \alpha_1$, then the SPRT $\delta^* = (d^*, T^*)$ is optimal in the class $\mathbf{C}(\alpha_0, \alpha_1)$ in the sense of minimizing both expected values $\mathsf{E}_0 T$ and $\mathsf{E}_1 T$,*

$$\inf_{\delta \in \mathbf{C}(\alpha_0, \alpha_1)} \mathsf{E}_0 T = \mathsf{E}_0 T^* \quad \textit{and} \quad \inf_{\delta \in \mathbf{C}(\alpha_0, \alpha_1)} \mathsf{E}_1 T = \mathsf{E}_1 T^*.$$

A rigorous proof of this theorem is tedious and involves a number of delicate technical details that are spelled out below. The proof is split into two parts presented in Subsections 3.2.2 and 3.2.3. We begin with obtaining useful lower bounds for the expected sample sizes in the class $\mathbf{C}(\alpha_0, \alpha_1)$ that allow us to conclude that the SPRT is at least approximately optimal when ignoring the overshoots.

3.2.1 Lower Bounds for the Expected Sample Sizes and Approximate Optimality

We start with a useful lemma.

Lemma 3.2.1. *Let $(\mathcal{X}, \mathscr{F})$ be a measurable space, and let P and Q be two mutually absolutely continuous probability measures defined on this space. Define $\Lambda(x) = \mathrm{d}\mathsf{P}(x)/\mathrm{d}\mathsf{Q}(x)$. For arbitrary nonintersecting sets $\mathcal{Y}_i \in \mathscr{F}$, $i \ge 0$ such that $\bigcup_{i \ge 0} \mathcal{Y}_i = \mathcal{X}$, the following inequality holds*

$$\int_{\mathcal{X}} \log[\Lambda(x)] \, \mathrm{d}\mathsf{P}(x) \ge \sum_{i \ge 0} \mathsf{P}(\mathcal{Y}_i) \log \frac{\mathrm{d}\mathsf{P}(\mathcal{Y}_i)}{\mathrm{d}\mathsf{Q}(\mathcal{Y}_i)}, \tag{3.103}$$

where the equality holds only if $\Lambda(x) = \mathsf{P}(\mathcal{Y}_i)/\mathsf{Q}(\mathcal{Y}_i)$ almost everywhere on $\mathcal{Y}_i$ for all $i \ge 0$.

Proof. Clearly,

$$\int_{\mathcal{X}} \log[\Lambda(x)]\, \mathrm{dP}(x) = \sum_{i\geq 0} \int_{\mathcal{Y}_i} \log[\Lambda(x)]\, \mathrm{dP}(x).$$

For each term on the right-hand side we have

$$\begin{aligned}\int_{\mathcal{Y}_i} \log[\Lambda(x)]\, \mathrm{dP}(x) &= \int_{\mathcal{Y}_i} \log \frac{\mathrm{dP}(x)}{\mathrm{dQ}(x)}\, \mathrm{dP}(x) \\ &\geq \left(\int_{\mathcal{Y}_i} \mathrm{dP}(x)\right) \log \frac{\int_{\mathcal{Y}_i} \mathrm{dP}(x)}{\int_{\mathcal{Y}_i} \mathrm{dQ}(x)} = \mathrm{P}(\mathcal{Y}_i) \log \frac{\mathrm{P}(\mathcal{Y}_i)}{\mathrm{Q}(\mathcal{Y}_i)},\end{aligned}$$

where the inequality follows immediately from Jensen's inequality since $\log y$ is strictly concave. The equality is achieved only if $\Lambda(x)$ is constant almost everywhere on $\mathcal{Y}_i$, i.e., when $\Lambda(x) = \mathrm{P}(\mathcal{Y}_i)/\mathrm{Q}(\mathcal{Y}_i)$. Thus, to achieve the equality in (3.103) the equalities $\Lambda(x) = \mathrm{P}(\mathcal{Y}_i)/\mathrm{Q}(\mathcal{Y}_i)$ should hold for almost all $x \in \mathcal{Y}_i$ and all $i \geq 0$. □

The following theorem establishes the lower bounds for the expected sample sizes.

Theorem 3.2.2 (Lower bounds). *Let the function $\beta(x,y)$ be as in* (3.9) *and let the K–L information numbers be positive and finite,* $0 < I_0, I_1 < \infty$. *If $\alpha_0 + \alpha_1 \leq 1$, then*

$$\inf_{\delta\in\mathbf{C}(\alpha_0,\alpha_1)} \mathrm{E}_1 T \geq \frac{\beta(\alpha_1,\alpha_0)}{I_1} \quad \textit{and} \quad \inf_{\delta\in\mathbf{C}(\alpha_0,\alpha_1)} \mathrm{E}_0 T \geq \frac{\beta(\alpha_0,\alpha_1)}{I_0}. \tag{3.104}$$

Proof. Write $\alpha_0(\delta) = \mathrm{P}_0(d=1)$ and $\alpha_1(\delta) = \mathrm{P}_1(d=0)$ for the probabilities of errors of an arbitrary test $\delta = (d,T)$ that has finite expectations $\mathrm{E}_i T$, $i = 0,1$. Setting $\mathrm{P} = \mathrm{P}_1$, $\mathrm{Q} = \mathrm{P}_0$ and $\mathcal{Y}_i = \{d = i\}$, $i = 0,1$ in Lemma 3.2.1, we obtain

$$\begin{aligned}\mathrm{E}_1[\lambda_T] &\geq \sum_{i=0}^{1} \mathrm{P}_1(d=i) \log \frac{\mathrm{P}_1(d=i)}{\mathrm{P}_0(d=i)} \\ &= \alpha_1(\delta) \log \frac{\alpha_1(\delta)}{1-\alpha_0(\delta)} + (1-\alpha_1(\delta)) \log \frac{1-\alpha_1(\delta)}{\alpha_0(\delta)} \quad = \beta(\alpha_1(\delta), \alpha_0(\delta)),\end{aligned}$$

and similarly, inverting the roles of P_1 and P_0,

$$\mathrm{E}_0[\lambda_T] \leq -\beta(\alpha_0(\delta), \alpha_1(\delta)).$$

By Wald's identity, $\mathrm{E}_1[\lambda_T] = I_1 \mathrm{E}_1 T$ and $\mathrm{E}_0[\lambda_T] = -I_0 \mathrm{E}_0[T]$, which along with the previous inequalities yield

$$\mathrm{E}_1 T \geq \frac{\beta(\alpha_1(\delta), \alpha_0(\delta))}{I_1}, \quad \mathrm{E}_0 T \geq \frac{\beta(\alpha_0(\delta), \alpha_1(\delta))}{I_0}.$$

Now the assertions (3.104) follow from the fact that the function $\beta(x,y)$ is decreasing in the domain $x + y \leq 1$. □

Comparing the lower bounds (3.104) with the approximate equalities (3.10) for the SPRT that ignore the overshoots, we conclude that the SPRT is at least approximately optimal when the overshoots are small. Also, the SPRT is asymptotically optimal as $\alpha_0, \alpha_1 \to 0$ since it follows from (3.104) that

$$\inf_{\delta\in\mathbf{C}(\alpha_0,\alpha_1)} \mathrm{E}_1 T \geq \frac{\log(1/\alpha_0) + o(1)}{I_1} \quad \text{and} \quad \inf_{\delta\in\mathbf{C}(\alpha_0,\alpha_1)} \mathrm{E}_0 T \geq \frac{\log(1/\alpha_1) + o(1)}{I_0} \tag{3.105}$$

and from Theorems 3.1.2 and 3.1.4 that

$$\mathrm{E}_1 T^* = \frac{\log(1/\alpha_0)}{I_1}(1 + o(1)) \quad \text{and} \quad \mathrm{E}_0 T^* = \frac{\log(1/\alpha_1)}{I_0}(1 + o(1)) \tag{3.106}$$

when $a_0 = \log(\zeta_1/\alpha_1)$ and $a_1 = \log(\zeta_0/\alpha_0)$ in which case $\alpha_i^* = \alpha_i(1+o(1))$, $i = 0, 1$.

Finally, completing the discussion of the issues related to the lower bounds for the expected sample sizes, we remark that the lower bounds in (3.104) are attained for the Brownian motion model (3.61). Indeed, using (3.62) and (3.63), it is easily shown that if $a_0 = \log[(1-\alpha_0)/\alpha_1]$ and $a_1 = \log[(1-\alpha_1)/\alpha_0]$, then $\alpha_0^* = \alpha_0$, $\alpha_1^* = \alpha_1$ and

$$\mathsf{E}_1 T^* = 2\beta(\alpha_1, \alpha_0)/q, \quad \mathsf{E}_0 T^* = 2\beta(\alpha_0, \alpha_1)/q.$$

These equalities coincide with the lower bounds since $I_0 = I_1 = (\theta_1 - \theta_0)^2/(2\sigma^2) = q$. Therefore, in the Brownian motion case the SPRT is exactly optimal. This is additional evidence that it may be also optimal in the general discrete-time case which is not limited by the Gaussian model.

The first main step in the formal proof of SPRT's optimality in the strong sense of Theorem 3.2.1 is to consider a Bayesian version of the problem assuming that the parameter (situation) θ is random and takes on two values 0 and 1 with probabilities $1-\pi$ and π, respectively. This is natural since the variational problem of minimizing both expectations can be reduced to the Bayesian one by introducing the Lagrangian. We therefore continue with the Bayesian setup and prove that a specially designed SPRT is Bayes-optimal.

3.2.2 *SPRT Optimality in a Bayesian Problem*

The main purpose of considering the Bayesian problem is to prove optimality of the SPRT in the strong sense. However it is of independent interest as well. For this reason, we provide a more detailed characterization of the Bayes solution, including the truncated case.

Note that the hypothesis testing problem of interest is a particular case of the general sequential decision-making problem of Section 2.7. More specifically, consider the following Bayesian sequential hypothesis testing problem. Suppose that $\theta \in \{0, 1\}$ is a random variable with $\mathsf{P}(\theta = 1) = \pi$, $0 \le \pi \le 1$, and the loss function associated with stopping at time $T = n$ is

$$L_n(\theta, u_n, \mathbf{X}_1^n) = L(\theta, u_n) + cn,$$

where $c > 0$ is the cost of making one observation (or the cost of the delay in making a decision) and $L(\theta = i, u_n = j)$ $(i, j = 0, 1)$ are the losses associated with making a terminal decision $u_n = 0$ (accept H_0) or $u_n = 1$ (accept H_1) that have the form

$$L(\theta = i, u_n = j) = \begin{cases} 0 & \text{if } i = j = 0, 1, \\ L_0 & \text{if } i = 0, j = 1, \\ L_1 & \text{if } i = 1, j = 0, \end{cases}$$

where $L_0, L_1 > 0$. In other words, we pay nothing for accepting a correct hypothesis, we are being charged with L_i for rejecting the hypothesis H_i when it is true $(i = 0, 1)$, and we pay c for sampling. The corresponding sequential Bayes problem will be denoted by $\mathcal{B}(\pi, L_0, L_1, c)$.

By the Bayes formula, the posterior probability $\pi_n(\mathbf{X}_1^n) = \mathsf{P}(\theta = 1 | \mathbf{X}_1^n)$ is

$$\pi_n = \frac{\pi \prod_{k=1}^n f_1(X_k)}{\pi \prod_{k=1}^n f_1(X_k) + (1-\pi) \prod_{k=1}^n f_0(X_k)}, \quad n \ge 0. \tag{3.107}$$

Certainly, $\mathsf{P}(\theta = 0 | \mathbf{X}_1^n) = 1 - \pi_n$. If $n = 0$, the products in (3.107) are treated as 1, so that $\pi_0 = \pi$. The APR $R_n(\mathbf{X}_1^n, u_n)$ is given by

$$R_n(\mathbf{X}_1^n, u_n = 0) = L_1 \pi_n + cn, \quad R_n(\mathbf{X}_1^n, u_n = 1) = L_0(1 - \pi_n) + cn, \quad n \ge 0,$$

and the minimal APR (2.181) when stopping at the n-th stage and making the best terminal decision is

$$\begin{aligned} R_n^0(\mathbf{X}_1^n) = R_n^0(\pi_n) &= \min\{R_n(\mathbf{X}_1^n, u_n = 0), R_n(\mathbf{X}_1^n, u_n = 1)\} \\ &= \min\{L_1\pi_n, L_0(1-\pi_n)\} + cn. \end{aligned} \tag{3.108}$$

In Subsection 2.7.1, we established that the optimal terminal decision rule d^0 does not depend on the stopping rule T and when $T = n$ the value of d^0 is $d_n^0 = u_n^0$, where u_n^0 is the element in the decision space that minimizes the APR. Hence, if follows from (3.108) that $d^0 = 1$ if $L_0(1-\pi_T) \le L_1\pi_T$ and 0 otherwise, which is equivalent to

$$d^0 = \begin{cases} 1 & \text{if } \pi_T \ge L_0/(L_0+L_1), \\ 0 & \text{if } \pi_T < L_0/(L_0+L_1). \end{cases} \tag{3.109}$$

First, consider optimization in the class of N-truncated sequential tests. Write

$$R_n^{\mathrm{st}}(\pi_n) = R_n^0(\pi_n) = \min\{L_1\pi_n, L_0(1-\pi_n)\} + cn$$

for the minimal APR associated with the best terminal decision when stopping at n. Let $\mathsf{P}^\pi = \pi\mathsf{P}_1 + (1-\pi)\mathsf{P}_0$ and let E^π stand for the corresponding expectation. By the results of Section 2.7 and (3.109), the problem is reduced to the optimal stopping problem

$$\inf_{T\in\mathcal{M}^N} \rho(d^0, T) = \inf_{T\in\mathcal{M}^N} \mathsf{E}^\pi\left[\min\left(L_1\pi_T, L_0(1-\pi_T)\right) + cT\right].$$

Now, using (3.107), it is straightforward to show that

$$\pi_{n+1} = \frac{\pi_n/(1-\pi_n)\, e^{Z_{n+1}}}{1+\pi_n/(1-\pi_n)\, e^{Z_{n+1}}}, \quad n \ge 0 \tag{3.110}$$

As before, $Z_n = \log[f_1(X_n)/f_0(X_n)]$ is the LLR, and we assume that $Z_0 = 0$. Thus, the statistic π_n is transitive (and P-Markov). Furthermore, the conditions (2.219) and (2.220) hold, so that π_n is a sufficient statistic and, by (2.207), the MAPR $R_n^N(\mathbf{X}_1^n) = R_n^N(\pi_n)$ satisfies the dynamic programming recursion

$$R_N^N(\pi_N) = R_N^{\mathrm{st}}(\pi_N), \quad R_n^N(\pi_n) = \min\left\{R_n^{\mathrm{st}}(\pi_n), \mathsf{E}^\pi[R_{n+1}^N(\pi_{n+1}) \mid \pi_n]\right\}, \quad 1 \le n < N,$$

where $\mathsf{E}^\pi[R_{n+1}^N(\pi_{n+1}) \mid \pi_n] = \tilde{R}_n^N(\pi_n)$ is the APR associated with continuation of observations.

Since $\mathsf{E}^\pi[\max_{n\le N} R_n^{\mathrm{st}}] \le \min(L_0, L_1) + cN$, Theorem 2.7.4 applies to show that in the optimal Bayesian test $\delta_0^N = (d^0, T_0^N)$ the terminal decision rule d^0 is defined by (3.109) with $T = T_0^N$ and the optimal stopping rule is

$$T_0^N = \min\left\{1 \le n \le N : R_n^{\mathrm{st}}(\pi_n) \le \tilde{R}_n^N(\pi_n)\right\}.$$

Obviously, the cost cn of n observations is common to both posterior risks R_n^{st} and $\tilde{R}_n^N$, so that when defining the optimal stopping time we can subtract cn. Specifically, let $G_n^N(\pi_n) = R_n^N(\pi_n) - cn$ and

$$G^{\mathrm{st}}(\pi_n) = R_n^{\mathrm{st}}(\pi_n) - cn = \min\{L_1\pi_n, L_0(1-\pi_n)\}.$$

Clearly, the function $G_n^N(\pi_n)$ satisfies the recursive equation

$$\begin{aligned} &G_N^N(\pi_N) = G^{\mathrm{st}}(\pi_N), \\ &G_n^N(\pi_n) = \min\left\{G^{\mathrm{st}}(\pi_n), \mathsf{E}^\pi[G_{n+1}^N(\pi_{n+1}) \mid \pi_n] + c\right\}, \quad 1 \le n < N \end{aligned} \tag{3.111}$$

and the optimal stopping time can be expressed as

$$T_0^N = \min\left\{1 \le n \le N : G^{\mathrm{st}}(\pi_n) \le \widetilde{G}_n^N(\pi_n)\right\}, \tag{3.112}$$

where $\widetilde{G}_n^N(\pi_n) = \mathsf{E}^\pi[G_n^N(\pi_{n+1})|\pi_n] + c$.

The following lemma is important for finding the structure of the optimal Bayes test.

Lemma 3.2.2. *$G_n^N(\pi_n)$ and $\widetilde{G}_n^N(\pi_n)$ are concave functions of π_n in $[0,1]$ for every $N \ge n$, including the nontruncated case $N = \infty$.*

Proof. To prove the first assertion it suffices to show that $G_0^N(\pi)$ is concave for each $1 \le N \le \infty$. By Theorem 2.7.4(ii), $G_0^N(\pi)$ is nothing but the minimum average Bayes risk:

$$G_0^N(\pi) = \rho_N^0(\pi) = \inf_{T \in \mathcal{M}^N} \rho_N(\pi, T).$$

Furthermore, $\rho_N(\pi, T)$ is linear in π. Indeed,

$$\begin{aligned}\rho_N(\pi,T) &= \mathsf{E}^\pi\left[L(\theta,d^0)+cT\right] \\ &= \pi\mathsf{E}_1[L(\theta=1,d^0=0)+cT]+(1-\pi)\mathsf{E}_0[L(\theta=0,d^0=1)+cT] \\ &= \pi L_0\mathsf{P}_1(d^0=0)+(1-\pi)L_1\mathsf{P}_0(d^0=1)+\pi c\mathsf{E}_1T+(1-\pi)c\mathsf{E}_0T.\end{aligned}$$

Hence, for every $0 < \alpha < 1$, π_1 and π_2,

$$\begin{aligned}G_0^N(\alpha\pi_1+(1-\alpha)\pi_2) &= \inf_{T\in\mathcal{M}^N} \rho_N(\alpha\pi_1+(1-\alpha)\pi_2,T) \\ &= \inf_{T\in\mathcal{M}^N}[\alpha\rho_N(\pi_1,T)+(1-\alpha)\rho_N(\pi_2,T)] \\ &\ge \alpha \inf_{T\in\mathcal{M}^N}\rho_N(\pi_1,T)+(1-\alpha)\inf_{T\in\mathcal{M}^N}\rho_N(\pi_2,T) \\ &= \alpha G_0^N(\pi_1)+(1-\alpha)G_0^N(\pi_2).\end{aligned}$$

For $N = \infty$ the proof is identical. Therefore, we proved that $G_n^N(\pi_n)$ is a concave function of π_n for each $1 \le N \le \infty$.

Now,

$$\widetilde{G}_n^N(\pi_n) = c + \mathsf{E}^\pi[G_{n+1}^N(\pi_{n+1}) \mid \pi_n],$$

where $G_{n+1}^N(\pi_{n+1})$ is concave by the first assertion. It is easy to see that the conditional expectation $\mathsf{E}^\pi[G_{n+1}^N(\pi_{n+1}) \mid \pi_n]$ is also a concave function, so that both assertions of the lemma follow. □

Now everything is prepared to obtain the detailed structures of the optimal N-truncated and nontruncated tests. Let $\Pi_n^N = \left\{\pi_n : G_n^{\mathrm{st}}(\pi_n) > \widetilde{G}_n^N(\pi_n)\right\}$ denote the region of continuation of observations at the n-th stage, so the optimal stopping rule (3.112) can be written as

$$T_0^N = \min\left\{1 \le n \le N : \pi_n \notin \Pi_n^N\right\}.$$

By Lemma 3.2.2, $\widetilde{G}_n^N(\pi_n)$ is concave in $\pi_n \in [0,1]$ and hence continuous in the open interval $\pi_n \in (0,1)$. Also, $\widetilde{G}_n^N(0) = \widetilde{G}_n^N(1) = c$ and $\lim_{\pi_n\to 0}\widetilde{G}_n^N(\pi_n) = \lim_{\pi_n \to 1}\widetilde{G}_n^N(\pi_n) = c$. To see this note that according to (3.111) $G_n^N(\pi_n) \le \min\{L_1\pi_n, L_0(1-\pi_n)\}$, so that

$$c \le \widetilde{G}_n^N(\pi_n) = c + \mathsf{E}^\pi[G_{n+1}^N(\pi_{n+1}) \mid \pi_n] \le c + L_1\mathsf{E}^\pi[\pi_{n+1} \mid \pi_n] = c + L_1\pi_n,$$

where we used the fact that $\{\pi_n\}$ is a P^π-martingale, i.e., $\mathsf{E}^\pi[\pi_{n+1} \mid \pi_n] = \pi_n$, which can be easily established. Hence $\widetilde{G}_n^N(0) = c$ and $\lim_{\pi_n\to 0}\widetilde{G}_n^N(\pi_n) = c$. Similarly,

$$c \le \widetilde{G}_n^N(\pi_n) \le c + L_0\mathsf{E}^\pi[(1-\pi_{n+1}) \mid \pi_n] = c + L_0(1-\pi_n),$$

implying that $\widetilde{G}_n^N(1) = c$ and $\lim_{\pi_n \to 1}\widetilde{G}_n^N(\pi_n) = c$.

Thus, the function $\widetilde{G}_n^N(\pi_n)$ characterizing the posterior loss associated with the continuation of observations is concave and continuous in the closed interval $[0,1]$, with $\widetilde{G}_n^N(0) = \widetilde{G}_n^N(1) = c$. The function $G^{\mathrm{st}}(\pi_n) = \min\{L_1\pi_n, L_0(1-\pi_n)\}$ characterizing the posterior loss associated with stopping is piecewise linear: $G^{\mathrm{st}}(\pi_n) = L_1\pi_n$ on $0 \le \pi_n \le h$ and $G^{\mathrm{st}}(\pi_n) = L_0(1-\pi_n)$ on $h \le \pi_n \le 1$, where $h = L_0/(L_0+L_1)$. The typical plots of the functions $G^{\mathrm{st}}(\pi_n)$ and $\widetilde{G}_n^N(\pi_n)$ are shown in Figure 3.1. Since $\widetilde{G}_n^N(\pi_n)$ is concave and continuous there may be only one root of the equation $\widetilde{G}_n^N(\pi_n) = L_1\pi_n$ in the interval $[0,h]$, which we denote as A_n^N. Similarly, there is at most one root of the equation $\widetilde{G}_n^N(\pi_n) = L_0(1-\pi_n)$ in the interval $[h,1]$, which we denote as B_n^N. If there are no roots, we set $A_n^N = B_n^N = h$. This may happen only if the cost of experimentation is too high. On the N-th stage we always have $A_N^N = B_N^N = h$ and the observations are stopped w.p. 1. Therefore, the continuation region is the interval (A_n^N, B_n^N) where A_n^N and B_n^N are two thresholds that depend on the maximal number of stages N and the current stage n. These two thresholds converge to a single one h at the N-th stage.

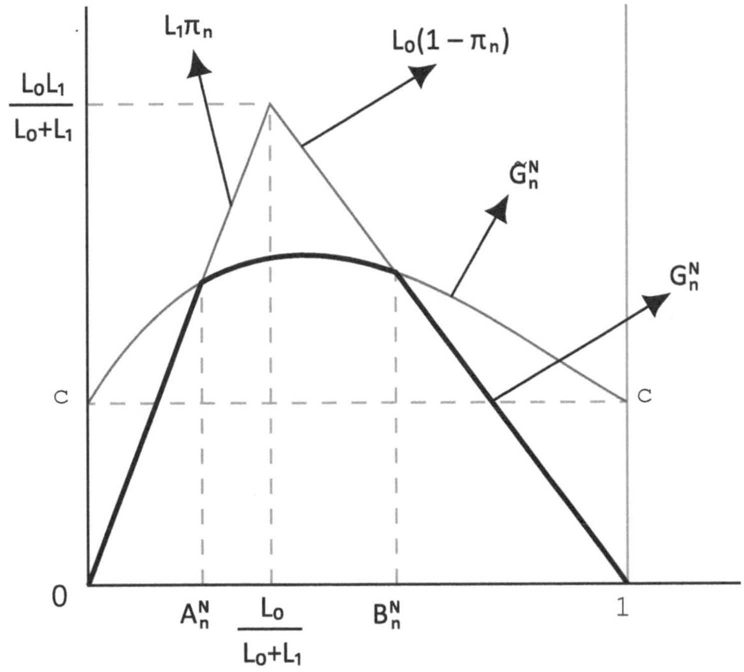

Figure 3.1: *Typical plots of the posterior losses. The MAPR function* $G_n^N(\pi_n) = R_n^N(\pi_n) - cn$ *is shown in bold.*

Now, note that the posterior probability π_n is related to the LLR λ_n via the formula

$$\pi_n = \frac{\chi e^{\lambda_n}}{1+\chi e^{\lambda_n}},$$

where $\chi = \pi/(1-\pi)$. Therefore,

$$\lambda_n = \log\frac{\pi_n}{(1-\pi_n)\chi} \tag{3.113}$$

and the continuation region can be equivalently represented in the space of the LLR as $(a_0(n,N), a_1(n,N))$, where

$$a_0(n,N) = \log\frac{A_n^N}{(1-A_n^N)\chi}, \quad a_1(n,N) = \log\frac{B_n^N}{(1-B_n^N)\chi}, \quad 1 \le n < N,$$

and $a_0(N,N) = a_1(N,N) = \log[L_0/(L_1\chi)]$.

A typical behavior of the truncated sequential probability ratio test is illustrated in Figure 3.2.

We therefore proved the following theorem that defines the structure of the optimal truncated sequential test.

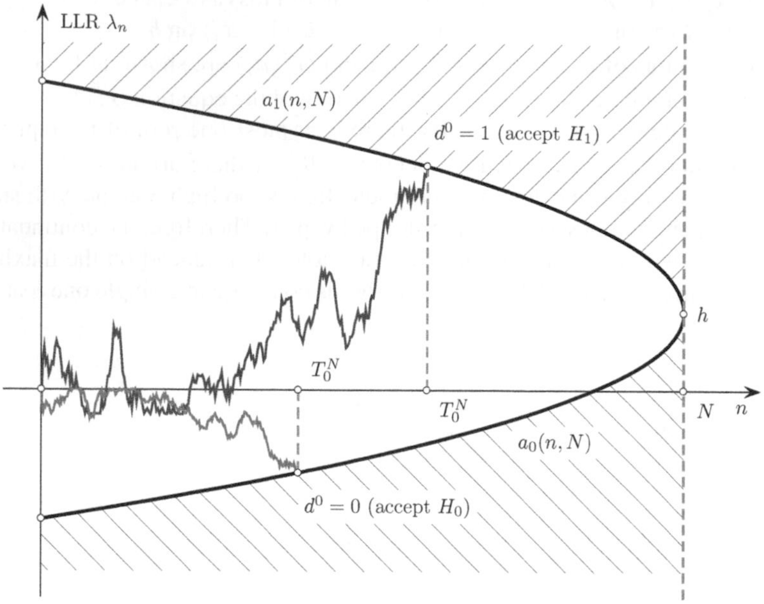

Figure 3.2: *Typical behavior of the truncated SPRT.*

Theorem 3.2.3 (Optimal truncated test). *Let $a = \log[L_0/(L_1\chi)]$. The optimal truncated sequential test in the iid $\mathcal{B}(\pi, L_0, L_1, c)$ problem is the truncated sequential probability ratio test with the curved boundaries $a_0(n,N) \le a$ and $a_1(n,N) \ge a$, that is,*

$$T_0^N = \min\left\{1 \le n \le N : \lambda_n \notin (a_0(n,N), a_1(n,N))\right\},$$
$$d^0 = \begin{cases} 1 & \text{if } \lambda_{T_0^N} \ge a_1(T_0^N, N), \\ 0 & \text{if } \lambda_{T_0^N} \le a_0(T_0^N, N), \end{cases} \tag{3.114}$$

where $a_0(N,N) = a_1(N,N) = a$, and for $n < N$ the boundaries $a_0(n,N)$ and $a_1(n,N)$ are found from the equations

$$\widetilde{G}_n^N(\lambda) = \frac{L_1 \chi e^{\lambda}}{1 + \chi e^{\lambda}} \quad \text{and} \quad \widetilde{G}_n^N(\lambda) = \frac{L_0}{1 + \chi e^{\lambda}}. \tag{3.115}$$

For $a_0(n,N) \le \lambda \le a_1(n,N)$, the function $\widetilde{G}_n^N(\lambda) = G_n^N(\lambda)$, where $G_n^N(\lambda)$ satisfies the recursive equation

$$\begin{aligned} G_N^N(\lambda) &= G^{\text{st}}(\lambda), \\ G_n^N(\lambda) &= \min\left\{G^{\text{st}}(\lambda), \mathsf{E}^{\pi}[G_{n+1}^N(\lambda_{n+1}) \mid \lambda_n = \lambda] + c\right\}, \quad 1 \le n < N, \end{aligned} \tag{3.116}$$

where

$$G^{\text{st}}(\lambda) = \frac{\min\left\{L_1 \chi e^{\lambda}, L_0\right\}}{1 + \chi e^{\lambda}}.$$

It is worth noting that solving the recursion (3.116) analytically and hence finding the thresholds is usually impossible. However, this can be done numerically starting from the end ($n = N$) and moving backward step by step.

Note also that the N-truncated SPRT (3.114) is optimal only if $a_0(n,N) < h < a_1(n,N)$ for all $1 \le n < N$. If this inequality holds for $n = 1, \dots, N_1$, $N_1 < N$, but is violated for $N > N_1$, then for finding the thresholds one has to repeat calculations replacing N with N_1, and then check this condition again. If the inequality holds, then the N_1-truncated procedure is optimal.

We now proceed with the nontruncated case. Letting $N \to \infty$ in (3.111) and taking into consideration the fact that $\{\pi_n\}_{n\geq 1}$ is a *homogeneous* Markov sequence, we obtain the following equation for the limiting function $G_n^\infty(\pi_n) = \lim_{N\to\infty} G_n^N(\pi_n) = G(\pi_n)$:

$$G(\pi_n) = \min\left\{G^{\rm st}(\pi_n), \mathsf{E}^\pi[G(\pi_{n+1}) \mid \pi_n] + c\right\}, \quad n \geq 0. \tag{3.117}$$

We stress that this function does not depend on n because of the homogeneity of π_n. Let $A = A(\pi, L_0, L_1, c)$ and $B = B(\pi, L_0, L_1, c)$ be found from the equations

$$\widetilde{G}(y) = L_1 y \quad \text{and} \quad \widetilde{G}(y) = L_0(1-y), \tag{3.118}$$

respectively, where $\widetilde{G}(y) = \mathsf{E}^\pi[G(\pi_{n+1}) \mid \pi_n = y] + c$. Using the previous consideration it is straightforward to show that the nontruncated sequential test with these thresholds is optimal in the Bayesian problem and that it coincides with the SPRT. The exact statement is given in the next theorem.

Theorem 3.2.4 (SPRT Bayesian optimality). *Let A and B satisfy the equations* (3.118). *If $A < \pi < B$, then the optimal nontruncated sequential test in the iid $\mathcal{B}(\pi, L_0, L_1, c)$ problem is the SPRT δ^* with the boundaries*

$$-a_0 = \log\left[\frac{A}{(1-A)\chi}\right] \quad \text{and} \quad a_1 = \log\left[\frac{B}{(1-B)\chi}\right]. \tag{3.119}$$

The minimal average risk $\rho_0 = \rho(\delta^*) = \inf_\delta \rho(\delta)$ is equal to $G(\pi)$.

Proof. In our case, the condition (2.213) of Theorem 2.7.5 has the form

$$\mathsf{E}\left[\sup_{n\geq 1} G^{\rm st}(\pi_n)\right] < \infty.$$

Since $G^{\rm st}(\pi_n) = \min\{L_1\pi_n, L_0(1-\pi_n)\} \leq L_1$, it is satisfied. By this theorem, the optimal stopping rule is

$$T_0 = \inf\left\{n \geq 1 : G^{\rm st}(\pi_n) \leq \widetilde{G}(\pi_n)\right\}.$$

Using Lemma 3.2.2, in just the same way as above we conclude that the function $\widetilde{G}(\pi_n) = c + \mathsf{E}^\pi[G(\pi_{n+1} \mid \pi_n]$ is concave and continuous in $[0,1]$, with $\widetilde{G}(0) = \widetilde{G}(1) = c$. Since the function $G^{\rm st}(\pi_n)$ is piecewise linear, there is at most one root A of the equation $\widetilde{G}(y) = L_1 y$ in the interval $[0, h]$, where $h = L_0/(L_0 + L_1)$, and there is at most one root B of the equation $\widetilde{G}(y) = L_0(1-y)$ in the interval $[h, 1]$– see Figure 3.1. Therefore, the continuation region is the interval (A, B), where A and B satisfy (3.118), and the optimal sequential test is of the form

$$T_0 = \inf\{n \geq 1 : \pi_n \notin (A, B)\},$$

$$d^0 = \begin{cases} 1 & \text{if } \pi_{T_0} \geq B, \\ 0 & \text{if } \pi_{T_0} \leq A. \end{cases}$$

By (3.113), this rule can be equivalently written as

$$T_0 = \inf\{n \geq 1 : \lambda_n \notin (-a_0, a_1)\},$$

$$d^0 = \begin{cases} 1 & \text{if } \lambda_{T_0} \geq a_1, \\ 0 & \text{if } \pi_{T_0} \leq -a_0, \end{cases}$$

with the thresholds defined in (3.119). This is nothing but the SPRT. Finally, recall that by definition the thresholds in the SPRT satisfy $-a_0 \leq 0 \leq a_1$. Clearly, the condition $A < \pi < B$ guarantees these inequalities.

In order to prove that $\rho^0 = G(\pi)$ we apply Theorem 2.7.5, which states that $\rho^0 = \mathsf{E}^\pi[R_0]$. However, $\mathsf{E}^\pi[R_0] = \mathsf{E}^\pi[G(\pi_1)] = G(\pi)$. □

The Bayesian optimality properties of the SPRT have been studied by many authors. In particular, Arrow, Blackwell, and Girshick [15] presented a proof of the existence of an optimal Bayesian sequential test using backward induction for the N-truncated case and then letting $N \to \infty$. See also Ferguson's book [143, Ch. 7] for an excellent presentation of such a proof. In fact, our proof is similar while the details are different. The idea of using a Bayesian approach for proving optimality of the SPRT in the strong sense appeared for the first time in the original paper of Wald and Wolfowitz [496]. The final step in the proof is given in the next subsection.

3.2.3 Strong Optimality of the SPRT

Now everything is prepared to prove Theorem 3.2.1. The proof is based on the following converse of Theorem 3.2.4. Fix $-\infty < -a_0 \le 0$, $0 \le a_1 < \infty$, and $0 < \pi < 1$. Then there exists a sequential Bayesian problem $\mathcal{B}(\pi, L_0, L_1, c)$ such that the SPRT with thresholds $-a_0, a_1$ is the optimal Bayes test with respect to π. In other words, there are $c, L_0, L_1 > 0$ such that the equalities (3.119) hold for given π, a_0, a_1. Several versions of the proof may be found in Wald and Wolfowitz [496], Burkholder and Wijsman [90], Matthes [289], Ferguson [143], Lehmann [261], Shiryaev [420], and Tartakovsky [452], among others. We present a proof sketch following [143] and omitting technicalities.

Note that, using the formulas (3.44), one can easily establish that $-a_0(\alpha_0^*, \alpha_1^*) \le 0 \le a_1(\alpha_0^*, \alpha_1^*)$ whenever $\alpha_0^* + \alpha_1^* < 1$. Therefore, the condition $\alpha_0 + \alpha_1 < 1$, postulated in Theorem 3.2.1, guarantees the required inequalities for the thresholds.

The first important step is to establish that the thresholds $A(\pi, L_0, L_1, c)$ and $B(\pi, L_0, L_1, c)$ in the Bayes test are continuous functions of c for fixed π, L_0, L_1 and furthermore that $A(\pi, L_0, L_1, c) \to 0$ and $B(\pi, L_0, L_1, c) \to 1$ as $c \to 0$. The proof can be found in [143, Lemma 7.6.3]. Now, using the continuity of the thresholds $A(c)$ and $B(c)$ in c, we can establish that for any given values of $-a_0 \le 0 \le a_1$ there are c, L_0, L_1 and $0 < \pi < 1$ satisfying (3.119). We stress that this is true for $0 < \pi < \varepsilon$ and $1 - \varepsilon < \pi < 1$, where ε is an arbitrary small positive number [143, Lemma 7.6.4]. This means that the SPRT $\delta^*(a_0, a_1)$ is the optimal Bayes test for suitably chosen c, L_0, L_1 and arbitrary π.

Finally, for any test $\delta \in \mathbf{C}(\alpha_0, \alpha_1)$, write $L_{\text{err}}^{\pi}(\delta) = \pi L_0 \alpha_0(\delta) + (1-\pi) L_1 \alpha_1(\delta)$ for the average loss associated with making wrong decisions. Note that the average risk is

$$\rho(\pi, \delta) = L_{\text{err}}^{\pi}(\delta) + c[\pi \mathsf{E}_0 T + (1-\pi)\mathsf{E}_1 T].$$

Let the stopping boundaries $-a_0$ and a_1 in the SPRT $\delta^*(a_0, a_1)$ be selected so that $\alpha_0(\delta^*) = \alpha_0$ and $\alpha_1(\delta^*) = \alpha_1$. Then, for any $\pi \in (0,1)$, the SPRT $\delta^*(a_0, a_1)$ is Bayes-optimal, so that

$$L_{\text{err}}^{\pi}(\delta^*) + c[\pi \mathsf{E}_0 T^* + (1-\pi)\mathsf{E}_1 T^*] \le L_{\text{err}}^{\pi}(\delta) + c[\pi \mathsf{E}_0 T + (1-\pi)\mathsf{E}_1 T] \quad \text{for any } \delta.$$

Obviously, $L_{\text{err}}^{\pi}(\delta^*) \ge L_{\text{err}}^{\pi}(\delta)$ for any $\delta \in \mathbf{C}(\alpha_0, \alpha_1)$, so for the second terms we have the inequality

$$\pi \mathsf{E}_0 T^* + (1-\pi)\mathsf{E}_1 T^* \le \pi \mathsf{E}_0 T + (1-\pi)\mathsf{E}_1 T.$$

Since π is arbitrary (including π arbitrarily close to 0 and 1), this inequality holds for all $0 < \pi < 1$. This implies that necessarily $\mathsf{E}_0 T^* \le \mathsf{E}_0 T$ and $\mathsf{E}_1 T^* \le \mathsf{E}_1 T$ for all $\delta \in \mathbf{C}(\alpha_0, \alpha_1)$, i.e., both expected values are minimized in the class $\mathbf{C}(\alpha_0, \alpha_1)$. This completes the proof of Theorem 3.2.1.

Remark 3.2.1. In Theorem 3.2.1, when defining the class $\mathbf{C}(\alpha_0, \alpha_1)$, we require the finiteness of the expected values of the stopping time, which is a natural requirement. The SPRT remains optimal among all tests with given probabilities of errors α_0 and α_1, including the ones with infinite expectations $\mathsf{E}_i T$, $i = 0, 1$.

Remark 3.2.2. For practical purposes, the condition $-a_0 \le 0 \le a_1$ has to be replaced with the strict inequalities $-a_0 < 0 < a_1$, since otherwise one stops with no observations made since $\lambda_0 = 0$. For example, if $a_0 = 0$ and $a_1 > 0$, then the hypothesis H_0 is accepted without sampling and rejected with no sampling if $a_1 = 0$ and $-a_0 < 0$. If $a_0 = a_1 = 0$, then any of the hypotheses can be accepted, say H_1.

Remark 3.2.3. If the stopping thresholds in the SPRT cannot be chosen so that the equalities $\alpha_i(\delta^*) = \alpha_i$, $i = 0,1$ are satisfied (i.e., only the inequalities $\alpha_i(\delta^*) < \alpha_i$, $i = 0,1$ hold), then the conventional SPRT is not generally optimal. In this case, a randomization on the thresholds (i.e., when $\lambda_n = -a_0$ or a_1) is needed, and the corresponding extended SPRT may be optimal. This situation rules out if the distributions of the LLR λ_1 are continuous.

3.2.4 *Generalization to a Special Non-iid Case*

Note that for constructing optimal truncated and nontruncated sequential Bayes tests in Theorems 3.2.3 and 3.2.4, the key fact is the Markovian property of the posterior probability π_n. This property is preserved whenever the LLR process $\{\lambda_n\}_{n\geq 1}$ is a process with iid increments Z_k, $k = 1,2,\dots$ or, in other words, λ_n is a random walk.

Assume that the observations X_n, $n \geq 1$ are dependent, but there exists a "whitening" transform $\mathbf{W}(\mathbf{X}) = \{W_n(\mathbf{X}_1^n), n \geq 1\}$, not depending on the hypotheses, such that the transformed random variables $\widetilde{X}_n = W_n(\mathbf{X}_1^n)$, $n \geq 1$ are iid for both hypotheses:

$$p(\widetilde{\mathbf{X}}_1^n \mid H_i) = \prod_{k=1}^n f_i(\widetilde{X}_k), \quad i = 0,1, \quad n \geq 1. \tag{3.120}$$

Some additional conditions may be needed to make sure that this transformation does not lead to any loss of information. For example, some simple conditions for the Jacobian of the inverse transformation are sufficient [452]. The SPRT test in the case of a Markov sequence has been studied in [91] and in the case of an AR(p) model in [316].

The LLR of the transformed data is equal to $\widetilde{Z}_k = \log[f_1(\widetilde{X}_k)/f_0(\widetilde{X}_k)]$ and the cumulative LLR $\widetilde{\lambda}_n = \widetilde{Z}_1 + \cdots + \widetilde{Z}_n$ is a random walk since the $\widetilde{Z}_k$, $k \geq 1$ are iid under both hypotheses. The $\widetilde{\lambda}$-based SPRT is optimal in this more general case.

To be more precise, let

$$Z_n = \log \frac{f_1(X_n \mid \mathbf{X}_1^{n-1})}{f_0(X_n \mid \mathbf{X}_1^{n-1})},$$

where $f_i(X_n \mid \mathbf{X}_1^{n-1})$ $(i = 0,1)$ are the conditional densities of X_n conditioned on the past $\mathbf{X}_1^{n-1} = (X_1, \dots, X_{n-1})$. The (cumulative) LLR process is $\lambda_n = \sum_{k=1}^n Z_k$. Suppose that the following condition holds:

$$\mathsf{P}_0(Z_n \leq z \mid \mathbf{X}_1^{n-1}) = F_0(z) \quad \text{for every} \quad -\infty < z < \infty \quad \text{and} \quad n \geq 1. \tag{3.121}$$

In other words, the conditional distribution of the LLR Z_n does not depend on the past $\mathbf{X}_1^{n-1}$. Then it is easily seen that

$$\mathsf{P}_0(Z_1 \leq z_1, \dots, Z_n \leq z_n \mid \mathbf{X}_1^{n-1}) = \prod_{k=1}^n F_0(z_k),$$

and changing the measure we obtain that

$$\mathsf{P}_1(Z_1 \leq z_1, \dots, Z_n \leq z_n \mid \mathbf{X}_1^{n-1}) = \prod_{k=1}^n F_1(z_k)$$

with $F_1(z) = \widetilde{F}_1(\log z)$, where

$$\widetilde{F}_1(z) = \int_0^z y \, \mathrm{d}\widetilde{F}_0(y), \quad \widetilde{F}_0(y) = F_0(e^y).$$

For illustration we consider three examples of dependent observations for which the SPRT is strictly optimal.

The first example is testing two hypotheses regarding the mean of the Markov sequence (AR(1) model) assuming that under H_i the data follow the recursion

$$X_n = \rho X_{n-1} + \mu_i + \xi_n, \quad X_0 = 0, \quad i = 0, 1, \quad n \geq 1, \tag{3.122}$$

where μ_0 and μ_1 are two distinct constants. Here $|\rho| < 1$ and $\{\xi_n\}_{n\geq 1}$ is a sequence of zero-mean iid random variables with density $f(\xi)$. The whitening transform has the form $W_n = X_n - \rho X_{n-1}$, $n \geq 1$, so that $W_n = \mu_i + \xi_n$ under H_i, $i = 0, 1$ for all $n \geq 1$ and the LLR process

$$\lambda_n = \sum_{k=1}^{n} \log \frac{f(W_k - \mu_1)}{f(W_k - \mu_0)}, \quad n \geq 1$$

is a random walk under both hypotheses. If the $\xi_n \sim \mathcal{N}(0, \sigma^2)$ are normal random variables, then the LLR λ_n is a Gaussian random walk with drift I under H_1 and $-I$ under H_0, where $I = (\mu_1 - \mu_0)^2/(2\sigma^2)$ is the K–L information number.

The second example consists in testing two hypotheses $H_i : \sigma = \sigma_i$, $i = 0, 1$ regarding the variance of the stationary AR(1) model

$$X_1 \sim \mathcal{N}(\mu, \sigma^2/(1-\rho^2)), \; X_n = \rho X_{n-1} + \xi_n, \; \xi_n \sim \mathcal{N}(0, \sigma^2), \; n \geq 2. \tag{3.123}$$

It follows from (2.283)– (2.287) that

$$Z_1 = \frac{1}{2} \log \frac{\sigma_0^2}{\sigma_1^2} + \frac{1}{2} \left[\frac{1}{\sigma_0^2} - \frac{1}{\sigma_1^2} \right] X_1^2 (1 - \rho^2) \tag{3.124}$$

and

$$Z_k = \frac{1}{2} \log \frac{\sigma_0^2}{\sigma_1^2} + \frac{1}{2} \left[\frac{1}{\sigma_0^2} - \frac{1}{\sigma_1^2} \right] (X_k - \rho X_{k-1})^2, \quad k \geq 2. \tag{3.125}$$

It is easy to see that

$$X_1 \sqrt{1 - \rho^2} \sim \mathcal{N}(0, \sigma_i^2) \text{ and } X_k - \rho X_{k-1} \sim \mathcal{N}(0, \sigma_i^2). \tag{3.126}$$

Hence, the LLR process $\{\lambda_n = \sum_{k=1}^{n} Z_k\}_{n \geq 1}$ is a random walk for all $n \geq 1$. Moreover, because the distribution of Z_k, $k \geq 1$ is independent of ρ, there is no impact of the autocorrelation on the ESS and the OC. Assume that the discrete-time AR(1) process is obtained from a continuous-time exponentially correlated Markov Gaussian process $\{V_t\}_{t \geq 0}$, $\mathsf{E}_i[V_t] = 0$, $\mathsf{E}_i[V_t V_{t+u}] = \sigma_{\mathrm{v},i}^2 e^{-\alpha|u|}$ ($\alpha > 0$), $i = 0, 1$ and that we can vary the sampling rate, i.e., the time interval Δ between the observations. Then the reduction of the sampling period Δ increases the efficiency of the SPRT when the AR(1) model is perfectly known because the expected durations $\Delta \cdot \mathsf{E}_i T^*$, $i = 0, 1$ measured in units of time are linear functions of Δ [320]. Hence, theoretically the expected durations of the SPRT can be unlimitedly reduced.

Yet another example is a two-state Markov chain $X_n \in \{0, 1\}$ with the transition matrices

$$\mathbf{P}_i = \begin{bmatrix} p_i & 1 - p_i \\ 1 - p_i & p_i \end{bmatrix}$$

under H_i, $i = 0, 1$. In other words, $\mathsf{P}_i(X_n = 1 | X_{n-1} = 1) = p_i$, $\mathsf{P}_i(X_n = 1 | X_{n-1} = 0) = 1 - p_i$, $\mathsf{P}_i(X_n = 0 | X_{n-1} = 0) = p_i$, $\mathsf{P}_i(X_n = 0 | X_{n-1} = 1) = 1 - p_i$. If the initial distribution is $\mathsf{P}_i(X_n = 1) = 1 - \mathsf{P}_i(X_k = 0) = p_i$, then the LLR is a random walk for all $n \geq 1$. Indeed, it is easily seen that Z_k takes on two values $\log(p_1/p_0)$ and $\log[(1 - p_1)/(1 - p_0)]$ and that

$$\mathsf{P}_i\left(Z_k = \log \frac{p_1}{p_0} \mid \mathbf{X}_1^{n-1} \right) = p_i, \quad \mathsf{P}_i\left(Z_k = \log \frac{1 - p_1}{1 - p_0} \mid \mathbf{X}_1^{n-1} \right) = 1 - p_i, \quad k \geq 1$$

regardless of the past $\mathbf{X}_1^{n-1} = (X_1, \dots, X_{n-1})$. Clearly, the condition (3.121) holds. The LLR process can be written as

$$\lambda_n = \sum_{k=1}^{n} \left[\widetilde{X}_k \log \frac{p_1}{p_0} + (1-\widetilde{X}_k) \log \frac{1-p_1}{1-p_0} \right],$$

where $\{\widetilde{X}_k\}_{k\geq 1}$ is an iid Bernoulli sequence, $\mathsf{P}_i(\widetilde{X}_k = 1) = 1 - \mathsf{P}_i(\widetilde{X}_k = 0) = p_i$, $i = 0, 1$.

3.3 Extended Optimality of the SPRT in the General Non-iid Case

In the previous section, we established optimality properties of the SPRT in the discrete-time iid case. In Subsection 3.2.1, making use of the lower bounds (3.104), we also justified that the SPRT is optimal in the continuous-time Brownian motion problem. The latter optimality result certainly carries over to continuous-time processes with iid increments [204]. In this section, we provide a generalization to general stochastic models with no (or almost no) assumptions on the distributions of the observations, and we treat this problem simultaneously for the discrete- and continuous-time cases.

Let $(\Omega, \mathscr{F}, \mathscr{F}_t, \mathsf{P})$, $t \in \mathbb{Z}_+ = \{0, 1, \dots\}$ or $t \in \mathbb{R}_+ = [0, \infty)$, be a stochastic basis with standard assumptions about monotonicity and in the continuous-time case $t \in \mathbb{R}_+$ also right-continuity of the σ-algebras $\mathscr{F}_t$. The sub-σ-algebra $\mathscr{F}_t = \mathscr{F}_t^X = \sigma(\mathbf{X}_0^t)$ of $\mathscr{F}$ is assumed to be generated by the process $\mathbf{X}_0^t = \{X(u), 0 \leq u \leq t\}$ observed up to time t, which is defined on the space $(\Omega, \mathscr{F})$. Consider the problem of sequential testing of two simple hypotheses $H_i:\ \mathsf{P} = \mathsf{P}_i$, $i = 0, 1$, where P_0 and P_1 are known probability measures. For simplicity, we assume that P_0 and P_1 are locally mutually absolutely continuous, i.e., for all $0 \leq t < \infty$ the restrictions P_0^t and P_1^t of these measures to the sub-σ-algebras $\mathscr{F}_t^X$ are equivalent. For $t \geq 0$, define the LR and LLR processes

$$\Lambda_t = \frac{d\mathsf{P}_1^t}{d\mathsf{P}_0^t}(\mathbf{X}_0^t), \quad \lambda_t = \log \Lambda_t \quad (\Lambda_0 = 1,\ \lambda_0 = 0\ \mathsf{P}_i-\text{a.s.}).$$

Let

$$\mathcal{I}_1(t) = \mathsf{E}_1 \left[\log \frac{d\mathsf{P}_1^t}{d\mathsf{P}_0^t}(\mathbf{X}_0^t) \right] = \mathsf{E}_1 \lambda_t, \quad \mathcal{I}_0(t) = \mathsf{E}_0 \left[\log \frac{d\mathsf{P}_0^t}{d\mathsf{P}_1^t}(\mathbf{X}_0^t) \right] = \mathsf{E}_0(-\lambda_t). \tag{3.127}$$

The value of $\mathcal{I}_1(t)$ can be interpreted as the total accumulated K–L information in the trajectory $\mathbf{X}_0^t$ for the hypothesis H_1 *vs.* H_0, and $\mathcal{I}_0(t)$ as the total K–L information in the trajectory $\mathbf{X}_0^t$ for the hypothesis H_0 *vs.* H_1. Note that in the iid case (including the continuous-time processes with iid increments) $\mathcal{I}_1(t) = I_1 t$ and $\mathcal{I}_0(t) = I_0 t$, where $I_1 = \mathsf{E}_1 \lambda_1$, $I_0 = \mathsf{E}_0(-\lambda_1)$ are the K–L information numbers. Thus, for a stopping time T, the objects $\mathsf{E}_i[\mathcal{I}_i(T)]$, $i = 0, 1$, can be regarded as the average accumulated K–L information when observing the stopped trajectory $\mathbf{X}_0^T$.

In this section, we argue that for general non-iid models the SPRT is optimal with respect to the average K–L information as long as there are no overshoots over the boundaries $-a_0$ and a_1. This latter condition is very restrictive for discrete-time models but holds for a large class of continuous-time processes with continuous paths.

The following lemma establishes the lower bounds for the expected K–L information in the class $\mathbf{C}(\alpha_0, \alpha_1) = \{\delta : \alpha_0(\delta) \leq \alpha_0, \alpha_1(\delta) \leq \alpha_1\}$.

Lemma 3.3.1. *Let the function $\beta(x,y)$ be as in (3.9). If $\alpha_0 + \alpha_1 \leq 1$, then*

$$\inf_{\delta \in \mathbf{C}(\alpha_0, \alpha_1)} \mathsf{E}_1[\mathcal{I}_1(T)] \geq \beta(\alpha_1, \alpha_0) \quad \text{and} \quad \inf_{\delta \in \mathbf{C}(\alpha_0, \alpha_1)} \mathsf{E}_0[\mathcal{I}_0(T)] \geq \beta(\alpha_0, \alpha_1). \tag{3.128}$$

Proof. Using Lemma 3.2.1 with $\mathsf{P} = \mathsf{P}_1$, $\mathsf{Q} = \mathsf{P}_0$ and $\mathcal{Y}_i = \{d = i\}$, $i = 0, 1$, we obtain

$$\begin{aligned} \mathsf{E}_1[\mathcal{I}_1(T)] \equiv \mathsf{E}_1[\lambda_T] &\geq \sum_{i=0}^{1} \mathsf{P}_1(d = i) \log \frac{\mathsf{P}_1(d = i)}{\mathsf{P}_0(d = i)} \\ &= \alpha_1(\delta) \log \frac{\alpha_1(\delta)}{1 - \alpha_0(\delta)} + (1 - \alpha_1(\delta)) \log \frac{1 - \alpha_1(\delta)}{\alpha_0(\delta)} = \beta(\alpha_1(\delta), \alpha_0(\delta)). \end{aligned}$$

Now, inverting the roles of P_1 and P_0, we similarly obtain

$$\mathsf{E}_0[\mathcal{I}_0(T)] \equiv \mathsf{E}_0[-\lambda_T] \ge \beta(\alpha_0(\delta), \alpha_1(\delta)).$$

The assertions (3.128) follow since the function $\beta(x,y)$ is decreasing in the domain $x+y \le 1$. □

Assume now that the thresholds $-a_0$ and a_1 in the SPRT $\delta^* = (T^*, d^*)$ are selected so that $\alpha_i^* = \alpha_i$ and there are no overshoots at stopping, i.e., $\lambda_{T^*} = -a_0$ or a_1 w.p. 1. Then

$$\mathsf{E}_1[\mathcal{I}_1(T^*)] = \mathsf{E}_1[\lambda_{T^*}] = \alpha_1^* \log \frac{\alpha_1^*}{1-\alpha_0^*} + (1-\alpha_1^*) \log \frac{1-\alpha_1^*}{\alpha_0^*} = \beta(\alpha_1, \alpha_0),$$
$$\mathsf{E}_0[\mathcal{I}_0(T^*)] = \mathsf{E}_0[-\lambda_{T^*}] = \alpha_0^* \log \frac{\alpha_0^*}{1-\alpha_1^*} + (1-\alpha_0^*) \log \frac{1-\alpha_0^*}{\alpha_1^*} = \beta(\alpha_0, \alpha_1).$$

Therefore, we proved the following result regarding optimality of the SPRT in the general non-iid case.

Theorem 3.3.1. *Assume that the SPRT terminates w.p.* 1, $\mathsf{P}_i(T^* < \infty) = 1$, $i = 0, 1$. *Let the boundaries* $-a_0$ *and* a_1 *in the SPRT be chosen in such a way that* $\alpha_i^* = \alpha_i$, $i = 0, 1$. *If* $\lambda_{T^*} \in \{-a_0, a_1\}$ *w.p.* 1 *under both measures* P_0 *and* P_1 *and* $\alpha_0 + \alpha_1 < 1$, *then the SPRT is optimal in the class* $\mathbf{C}(\alpha_0, \alpha_1)$ *with respect to the average K–L information,*

$$\inf_{\delta \in \mathbf{C}(\alpha_0,\alpha_1)} \mathsf{E}_1[\mathcal{I}_1(T)] = \mathsf{E}_1[\mathcal{I}_1(T^*)] = \beta(\alpha_1, \alpha_0),$$
$$\inf_{\delta \in \mathbf{C}(\alpha_0,\alpha_1)} \mathsf{E}_0[\mathcal{I}_0(T)] = \mathsf{E}_0[\mathcal{I}_0(T^*)] = \beta(\alpha_0, \alpha_1). \tag{3.129}$$

In the following discussion we consider the continuous-time case. Note first that Theorem 3.3.1 implies optimality of the SPRT in the Brownian motion case (3.61) with respect to the expected sample sizes, since $\mathcal{I}_i(t) = qt/2$, $i = 0, 1$ and, by the continuity in time and space, there are no overshoots, so that if

$$a_0 = \log[(1-\alpha_0)/\alpha_1] \quad \text{and} \quad a_1 = \log[(1-\alpha_1)/\alpha_0], \tag{3.130}$$

then $\alpha_0^* = \alpha_0$, $\alpha_1^* = \alpha_1$. This fact has been already discussed in Subsection 3.2.1.

Next, assume that the LLR process λ_t has continuous trajectories and that P_1^∞ and P_0^∞ are orthogonal, so that $\mathsf{P}_1(\lambda_\infty = \infty) = \mathsf{P}_0(\lambda_\infty = -\infty) = 1$. Recall that we assumed that $\lambda_0 = 0$ w.p. 1, which means that $\mathsf{P}_1^0 = \mathsf{P}_0^0$. Then Theorem 3.3.1 holds, since there are no overshoots and the SPRT terminates w.p. 1. Moreover, by Jacod [210, V.1] there exists a local continuous P_0-martingale M_t, $M_0 = 0$ such that $\lambda_t = M_t - \frac{1}{2}\langle M, M\rangle_t$, where $\{\langle M, M\rangle_t\}_{t\in\mathbb{R}_+}$ is the previsible increasing process under P_0. It can be shown (see, e.g., Irle [203]) that $\mathsf{E}_1\lambda_t \equiv \mathcal{I}_1(t) = \frac{1}{2}\mathsf{E}_1\langle M, M\rangle_t$ and $\mathsf{E}_0(-\lambda_t) \equiv \mathcal{I}_0(t) = \frac{1}{2}\mathsf{E}_0\langle M, M\rangle_t$. By the orthogonality of P_1^∞ and P_0^∞, $\mathsf{P}_i(\langle M, M\rangle_\infty = \infty) = 1$, $i = 0, 1$ and moreover $\mathsf{E}_i\langle M, M\rangle_{T^*} < \infty$. Therefore, Theorem 3.3.1 implies the following general result for continuous-time processes.

Theorem 3.3.2. *Suppose that the LLR process* $\{\lambda_t\}_{t\in\mathbb{R}_+}$ *has continuous trajectories and that* P_1^∞ *and* P_0^∞ *are orthogonal. Let* $\alpha_0 + \alpha_1 < 1$. *If the boundaries* $-a_0$ *and* a_1 *in the SPRT are chosen as in* (3.130), *then* $\alpha_i^* = \alpha_i$, $i = 0, 1$ *and the SPRT is optimal in the class* $\mathbf{C}(\alpha_0, \alpha_1)$ *with respect to the average K–L information, that is, the equalities* (3.129) *hold with* $\mathsf{E}_i[\mathcal{I}_i(T)] = \frac{1}{2}\mathsf{E}_i[\langle M, M\rangle_T]$, $i = 0, 1$.

Example 3.3.1 (Testing for Itô processes). Suppose that $S_0(t) = S_0(t, \omega)$ and $S_1(t) = S_1(t, \omega)$ are two distinct stochastic processes (signals) and that, under the hypothesis H_i ($i = 0, 1$), the observed process X_t has the Itô stochastic differential

$$\mathrm{d}X_t = S_i(t)\mathrm{d}t + \sigma(t)\,\mathrm{d}W_t, \quad t \ge 0, \tag{3.131}$$

where $\{W_t\}_{t\geq 0}$ is a standard Brownian motion and $\sigma(t)$ is a positive deterministic function (intensity of noise). Assume that

$$\int_0^t \mathsf{E} S_i^2(u)\,du < \infty,\ i=0,1, \quad \int_0^t \sigma^2(u)\,du < \infty \quad \text{for every } t<\infty. \tag{3.132}$$

Define the functionals $\hat{S}_i(t) = \mathsf{E}_i[S_i(t) \mid \mathscr{F}_t^X]$ $(i=0,1)$ which are the optimal Bayes filtering estimates (in the mean-square-error sense) of $S_i(t)$ observed in white Gaussian noise $\dot{W}_t$. Indeed, by virtue of the conditions (3.132), $\inf_{\tilde{S}_i(t)} \mathsf{E}_i[S_i(t)-\tilde{S}_i(t)]^2$ is attained by the posterior mean $\hat{S}_i(t)$ and its expected square loss is finite. By Lemma 2.1.1, there exist standard Brownian motions $\{\widetilde{W}_t^{(i)}\}_{t\geq 0}$, $i=0,1$ such that under H_i the process X_t allows for the following minimal representation in the form of a diffusion-type process

$$\mathrm{d}X_t = \hat{S}_i(t)\,\mathrm{d}t + \sigma(t)\,\mathrm{d}\widetilde{W}_t^{(i)}, \quad t\geq 0,\ i=0,1. \tag{3.133}$$

The innovation processes $\widetilde{W}_t^{(i)}$ are statistically equivalent to the original Brownian motion W_t.

Suppose that

$$\mathsf{P}_i\left(\int_0^t [\hat{S}_1^2(u)+\hat{S}_0^2(u)]\,du < \infty\right) = 1 \quad \text{for } t<\infty \text{ and } i=0,1.$$

Note that under H_i the estimates $\hat{S}_1(t)$ and $\hat{S}_0(t)$ of the processes $S_1(t)$ and $S_0(t)$ can be written as $\hat{S}_1^{(i)}(t) = \mathsf{E}_1[S_1(t) \mid \mathscr{F}_t^{X^{(i)}}]$ and $\hat{S}_0^{(i)}(t) = \mathsf{E}_0[S_0(t) \mid \mathscr{F}_t^{X^{(i)}}]$, where $X^{(i)}$ stands for the process X when the hypothesis H_i is true.[1] Then we may use the representation (3.133) along with Theorem 2.1.4 on the absolute continuity of probability measures for diffusion-type processes to obtain

$$\lambda_t = \int_0^t \frac{\hat{S}_1(u)-\hat{S}_0(u)}{\sigma^2(u)}\,\mathrm{d}X_u - \frac{1}{2}\int_0^t \frac{\hat{S}_1^2(u)-\hat{S}_0^2(u)}{\sigma^2(u)}\,du. \tag{3.134}$$

Now, using (3.133) and (3.134), we get that under H_i the LLR $\lambda_t^{(i)}$ has the form

$$\lambda_t^{(i)} = \int_0^t \frac{\hat{S}_1^{(i)}(u)-\hat{S}_0^{(i)}(u)}{\sigma(u)}\,\mathrm{d}\widetilde{W}_u^{(i)} + (-1)^{i+1}\frac{1}{2}\int_0^t \frac{[\hat{S}_1^{(i)}(u)-\hat{S}_0^{(i)}(u)]^2}{\sigma^2(u)}\,du, \tag{3.135}$$

so that

$$\mathcal{I}_i(t) = \frac{1}{2}\int_0^t \mathsf{E}_i\left\{\frac{[\hat{S}_1(u)-\hat{S}_0(u)]^2}{\sigma^2(u)}\right\}du, \quad i=0,1.$$

Therefore, in the case of Itô processes,

$$M_t = \int_0^t \frac{\hat{S}_1(u)-\hat{S}_0(u)}{\sigma(u)}\,\mathrm{d}\widetilde{W}_u^{(i)} \quad \text{and} \quad \langle M,M\rangle_t = \int_0^t \frac{[\hat{S}_1(u)-\hat{S}_0(u)]^2}{\sigma^2(u)}\,du.$$

If

$$\mathsf{P}_i\left(\int_0^\infty [\hat{S}_1^2(u)+\hat{S}_0^2(u)]\,du = \infty\right) = 1 \quad \text{for } i=0,1,$$

then $\mathsf{P}_i(\langle M,M\rangle_\infty = \infty) = 1$ and $\mathsf{P}_1(\lambda_\infty = \infty) = \mathsf{P}_0(\lambda_\infty = -\infty) = 1$. In this case, $\mathsf{E}_i[\mathcal{I}_i(T^*)] < \infty$ and the SPRT is optimal with respect to the total K–L information metric.

It is worth pointing out that the representation of the LLR via the signal estimates $\hat{S}_i(t)$ given in (3.134), but not via the original random signals $S_i(t)$ is extremely important since $\hat{S}_i(t)$, $i=0,1$ depend on the observed trajectory $\mathbf{X}_0^t$. Thus, the LLR can be computed in all cases where the signal

[1] The estimates $\hat{S}_1^{(1)}(t)$ and $\hat{S}_0^{(0)}(t)$ are true estimates while $\hat{S}_1^{(0)}(t)$ and $\hat{S}_0^{(1)}(t)$ are pseudo-estimates.

estimates $\hat{S}_i(t)$ can be computed. However, in general the $\hat{S}_i(t)$, $i=0,1$ are nonlinear functionals solving a nonlinear filtering problem. This problem has no exact closed form solution, except for a handful of particular cases. The exceptions are the Gaussian processes for which the $\hat{S}_i(t)$ are linear functionals and the conditionally Gaussian processes for which the $\hat{S}_i(t)$ are nonlinear but computable [267]. In the other cases, approximations are in order, e.g., using on-line Markov Chain Monte Carlo (MCMC) techniques such as particle filtering methods [122].

Consider a particular scenario when $\sigma(t)=\sigma>0$, $S_0(t)=V_t$ and $S_1(t)=S_t+V_t$, where V_t is a L_2-continuous Gaussian process that can be interpreted as clutter in target detection applications and S_t is a deterministic and square integrable function (signal). Let $\hat{V}_t^{(i)}=\mathsf{E}_i[V_t|\mathscr{F}_t]$ be an optimal (in the mean-square-error sense) filtering estimate of the process V_t under H_i observed in white Gaussian noise. Since V_t is a Gaussian process, the $\hat{V}_t^{(i)}$ are linear functionals

$$\hat{V}_t^{(i)}=\int_0^t C(t,u)\,[\mathrm{d}X_u-iS_u\,\mathrm{d}u],\quad i=0,1,$$

where $C(t,u)$ is the impulse response of an optimal filter that satisfies the well-known Wiener–Hopf equation. Define

$$\widetilde{S}_t=S_t-\int_0^t S_uC(t,u)\,\mathrm{d}u,\quad \mathrm{d}\widetilde{X}_t=\mathrm{d}X_t-\left[\int_0^t C(t,u)\,\mathrm{d}X_u\right]\mathrm{d}t.$$

Using (3.134), it can be shown that the LLR has the form

$$\lambda_t=\frac{1}{\sigma^2}\int_0^t \widetilde{S}_u\,\mathrm{d}\widetilde{X}_u-\frac{1}{2\sigma^2}\int_0^t \widetilde{S}_u^2\,\mathrm{d}u, \tag{3.136}$$

where the process $\widetilde{X}_t$ can be represented via an innovation (standard Brownian motion) process $\widetilde{W}_t$ as

$$\widetilde{X}_t=i\int_0^t \widetilde{S}_u\,\mathrm{d}u+\sigma\widetilde{W}_t,\quad i=0,1$$

(see Lemma 2.1.1). Therefore, in the latter case,

$$\mathcal{I}_i(t)=\frac{1}{2\sigma^2}\int_0^t \widetilde{S}_u^2\,\mathrm{d}u,\quad i=0,1,$$

and in order to guarantee optimality of the SPRT we have to assume that

$$\int_0^t \widetilde{S}_u^2\,\mathrm{d}u<\infty\quad\text{for } t<\infty\quad\text{and}\quad\int_0^\infty \widetilde{S}_u^2\,\mathrm{d}u=\infty.$$

Note that $\widetilde{S}_u$ is nothing but a signal at the output of the whitening filter that transforms correlated clutter V_t into white Gaussian noise, and that $\int_0^t \widetilde{S}_u^2\,du$ is the energy of this signal. Thus, the value of

$$Q(t)=\frac{1}{2\sigma^2}\int_0^t \widetilde{S}_u^2\,\mathrm{d}u$$

is interpreted as the signal-to-noise ratio (SNR) , so that in this example the K–L information is equal to the cumulative SNR at the output of the whitening filter. Note also that often in the applications

$$\lim_{t\to\infty}\frac{1}{t}Q(t)=q,\quad 0<q<\infty$$

(cf. Baron and Tartakovsky [28] and Subsection 3.4.3).

It is worth stressing that the problem of whether the SPRT is optimal in terms of minimizing the average K–L information when there are overshoots (e.g., in the discrete-time non-iid case) is still open.

3.4 Asymptotic Optimality of the SPRT in the General Non-iid Case

We continue to consider the general non-iid case, formulated in the previous section, assuming that the time index may be either discrete or continuous, $t \in \mathbb{Z}_+$ or $t \in \mathbb{R}_+$. While Section 3.3 suggests that the SPRT with no overshoots is strictly optimal with respect to the average K–L information, it is of interest to investigate whether the SPRT has some optimality properties with respect to the expected sample sizes. In this section, we show that, as the error probabilities tend to zero, the SPRT minimizes asymptotically the ESS as well as higher moments of the stopping time distribution under very general conditions. In Section 3.6 we shall show that these results can be applied to testing composite hypotheses in the problems that allow for an invariant structure.

Recall first that the inequalities for the error probabilities of the SPRT that do not account for threshold overshoots

$$\alpha_1^* \le e^{-a_0}(1-\alpha_0^*), \quad \alpha_0^* \le e^{-a_1}(1-\alpha_1^*) \tag{3.137}$$

hold in the most general non-iid case; see (3.2).

As before, let $\mathbf{C}(\alpha_0,\alpha_1) = \{\delta : \mathsf{P}_0(d=1) \le \alpha_0, \mathsf{P}_1(d=0) \le \alpha_1)\}$. Recall that in the iid case, by (3.105), the following asymptotic lower bounds hold

$$\inf_{\delta \in \mathbf{C}(\alpha_0,\alpha_1)} \mathsf{E}_1 T \ge \frac{|\log \alpha_0|}{I_1} + o(1), \qquad \inf_{\delta \in \mathbf{C}(\alpha_0,\alpha_1)} \mathsf{E}_0 T \ge \frac{|\log \alpha_1|}{I_0} + o(1) \quad \text{as } \alpha_{\max} \to 0,$$

where $\alpha_{\max} = \max(\alpha_0,\alpha_1)$. Also, by (3.106), if

$$a_0 \sim \log(1/\alpha_1) \quad \text{and} \quad a_1 \sim \log(1/\alpha_0), \tag{3.138}$$

then

$$\mathsf{E}_1 T^* \sim \frac{|\log \alpha_0|}{I_1}, \quad \mathsf{E}_0 T^* \sim \frac{|\log \alpha_1|}{I_0} \quad \text{as } \alpha_{\max} \to 0. \tag{3.139}$$

Hereafter we always assume that $\alpha_1|\log \alpha_0| \to 0$ and $\alpha_0|\log \alpha_1| \to 0$ as $\alpha_{\max} \to 0$, i.e., that there exists a constant $0 < c < \infty$ such that the ratio satisfies $|\log \alpha_0|/|\log \alpha_1| \sim c$, namely it is bounded away from 0 and infinity.

In the iid case, by the SLLN, the LLR process $\{\lambda_t\}$ has the following stability property

$$t^{-1}\lambda_t \xrightarrow[t\to\infty]{\mathsf{P}_1-\text{a.s.}} I_1, \quad t^{-1}\lambda_t \xrightarrow[t\to\infty]{\mathsf{P}_0-\text{a.s.}} -I_0. \tag{3.140}$$

This allows us to conjecture that if in the general non-iid case the LLR is also stable in the sense that the almost sure convergence conditions (3.140) are satisfied with some positive and finite constants I_1 and I_0, then the asymptotic formulas (3.139) still hold. Note that, with some abuse of notation, hereafter in this section we use for these constants the same notation I_i as for the K–L numbers. In the iid case, I_1 and I_0 are indeed the (true) K–L numbers. In the general case, these numbers represent the local K–L information in the sense that usually $I_1 = \lim_{t\to\infty} t^{-1}\mathsf{E}_1[\lambda_t]$ and $I_0 = \lim_{t\to\infty} t^{-1}\mathsf{E}_0[-\lambda_t]$.

Indeed, making use of (3.128) along with the almost sure convergence (3.140), we expect that, as $\alpha_{\max} \to 0$,

$$\inf_{\delta \in \mathbf{C}(\alpha_0,\alpha_1)} \mathsf{E}_1[\mathcal{I}_1(T)] \sim I_1 \inf_{\delta \in \mathbf{C}(\alpha_0,\alpha_1)} \mathsf{E}_1 T \ge |\log \alpha_0|(1+o(1)),$$
$$\inf_{\delta \in \mathbf{C}(\alpha_0,\alpha_1)} \mathsf{E}_0[\mathcal{I}_0(T)] \sim I_0 \inf_{\delta \in \mathbf{C}(\alpha_0,\alpha_1)} \mathsf{E}_0 T \ge |\log \alpha_1|(1+o(1)).$$

These bounds are asymptotically attained by the SPRT with the thresholds (3.138) assuming that $\mathsf{E}_i T^* < \infty$ and ignoring the overshoots.

Note, however, that in the general non-iid case the strong law of large numbers does not even guarantee the finiteness of the expected sample sizes $\mathsf{E}_i T^*$ of the SPRT, so some additional conditions are needed, such as a certain rate of convergence in the strong law, say complete convergence.

More generally, let $\psi(t)$ be a nonnegative increasing function, $\psi(\infty)=\infty$, and assume that

$$\lim_{t\to\infty}[\mathcal{I}_i(t)/\psi(t)] = I_i, \quad \frac{\lambda_t - \mathcal{I}_1(t)}{\psi(t)} \xrightarrow[t\to\infty]{\mathsf{P}_1-\text{a.s.}} 0, \quad \frac{\lambda_t + \mathcal{I}_0(t)}{\psi(t)} \xrightarrow[t\to\infty]{\mathsf{P}_0-\text{a.s.}} 0, \tag{3.141}$$

where $\mathcal{I}_i(t)$ is the total accumulated K–L information defined in (3.127). That is, we assume that

$$\frac{\lambda_t}{\psi(t)} \xrightarrow[t\to\infty]{\mathsf{P}_1-\text{a.s.}} I_1, \quad \frac{\lambda_t}{\psi(t)} \xrightarrow[t\to\infty]{\mathsf{P}_0-\text{a.s.}} -I_0 \tag{3.142}$$

with some constants $0 < I_i < \infty$, $i = 0,1$. Let $\Psi(t)$ stand for the function inverse to $\psi(t)$.

Below we show that under some quite general conditions the SPRT has asymptotic optimality properties with respect to the moments of the stopping time $\mathsf{E}_i[T^m]$, $m>0$, $i=0,1$. The method of proof is based on the lower-upper bound technique. Specifically, we first derive asymptotic lower bounds for $\mathsf{E}_i[T^m]$ in the class $\mathbf{C}(\alpha_0,\alpha_1)$ and then show that these bounds are attained for the SPRT.

3.4.1 *Lower Bounds for Moments of the Stopping Time and Weak Asymptotic Optimality*

We begin with deriving lower bounds. Recall that $\{d=i\} = \{T<\infty, \text{accept } H_i\}$.

Lemma 3.4.1. *Assume there exists a nonnegative increasing function $\psi(t)$ and positive finite constants I_0 and I_1 such that, for all $\varepsilon > 0$,*

$$\begin{aligned} &\lim_{L\to\infty} \mathsf{P}_1\left\{\frac{1}{\psi(L)}\sup_{0\le t\le L}\lambda_t \ge (1+\varepsilon)I_1\right\} = 1, \\ &\lim_{L\to\infty} \mathsf{P}_0\left\{\frac{1}{\psi(L)}\sup_{0\le t\le L}(-\lambda_t) \ge (1+\varepsilon)I_0\right\} = 1. \end{aligned} \tag{3.143}$$

Then, for all $0<\varepsilon<1$,

$$\begin{aligned} &\lim_{\alpha_{\max}\to 0}\inf_{\delta\in\mathbf{C}(\alpha_0,\alpha_1)} \mathsf{P}_1\left\{T > (1-\varepsilon)\Psi\left(I_1^{-1}|\log\alpha_0|\right)\right\} = 1, \\ &\lim_{\alpha_{\max}\to 0}\inf_{\delta\in\mathbf{C}(\alpha_0,\alpha_1)} \mathsf{P}_0\left\{T > (1-\varepsilon)\Psi\left(I_0^{-1}|\log\alpha_1|\right)\right\} = 1, \end{aligned} \tag{3.144}$$

and for all $m>0$, as $\alpha_{\max}\to 0$,

$$\begin{aligned} &\inf_{\delta\in\mathbf{C}(\alpha_0,\alpha_1)} \mathsf{E}_1[T^m] \ge \left[\Psi\left(I_1^{-1}|\log\alpha_0|\right)\right]^m(1+o(1)), \\ &\inf_{\delta\in\mathbf{C}(\alpha_0,\alpha_1)} \mathsf{E}_0[T^m] \ge \left[\Psi\left(I_0^{-1}|\log\alpha_1|\right)\right]^m(1+o(1)). \end{aligned} \tag{3.145}$$

Proof. Write $\Omega_{i,L} = \{d=i\}\cap\{T\le L\}$. By Wald's likelihood ratio identity

$$\mathsf{P}_0(d=1) = \mathsf{E}_1\left\{\mathbb{1}_{\{d=1\}}e^{-\lambda_T}\right\},$$

and hence, for any $L>0$ and $B>0$,

$$\begin{aligned} \mathsf{P}_0(d=1) &\ge \mathsf{E}_1\left[\mathbb{1}_{\{\Omega_{1,L},\lambda_T<B\}}e^{-\lambda_T}\right] \\ &\ge e^{-B}\mathsf{P}_1\left(\Omega_{1,L}, \sup_{t\le L}\lambda_t < B\right) \\ &\ge e^{-B}\left[\mathsf{P}_1(\Omega_{1,L}) - \mathsf{P}_1\left(\sup_{t\le L}\lambda_t \ge B\right)\right]. \end{aligned}$$

Since $\mathsf{P}_1(\Omega_{1,L}) \ge \mathsf{P}_1(d=1) - \mathsf{P}_1(T > L)$, it follows that

$$\mathsf{P}_1(T > L) \ge \mathsf{P}_1(d=1) - \mathsf{P}_0(d=1)\, e^B - \mathsf{P}_1\left(\sup_{t \le L} \lambda_t \ge B\right),$$

which for any stopping time in the class $\mathbf{C}(\alpha_0, \alpha_1)$ yields

$$\mathsf{P}_1(T > L) \ge 1 - \alpha_1 - \alpha_0 e^B - \mathsf{P}_1\left(\sup_{t \le L} \lambda_t \ge B\right). \tag{3.146}$$

Let $B = (1+\varepsilon) I_1 \psi(L)$ and $L = \Psi((1-\varepsilon) I_1^{-1} |\log \alpha_0|)$. Then, using (3.146), one obtains

$$\mathsf{P}_1\left\{T > (1-\varepsilon)\Psi\left(I_1^{-1}|\log \alpha_0|\right)\right\} \ge 1 - \alpha_1 - \alpha_0^{\varepsilon^2} - \mathsf{P}_1\left(\sup_{t \le L} \lambda_t \ge (1+\varepsilon) I_1 \psi(L)\right). \tag{3.147}$$

Since (3.147) holds regardless of a specific choice of the stopping time T in the class $\mathbf{C}(\alpha_0, \alpha_1)$, it follows that

$$\inf_{\delta \in \mathbf{C}(\alpha_0, \alpha_1)} \mathsf{P}_1\left\{T > (1-\varepsilon)\Psi\left(I_1^{-1}|\log \alpha_0|\right)\right\} \ge 1 - \alpha_1 - \alpha_0^{\varepsilon^2} - \mathsf{P}_1\left(\sup_{t \le L} \lambda_t \ge (1+\varepsilon) I_1 \psi(L)\right). \tag{3.148}$$

Note that $L \to \infty$ as $\alpha_{\max} \to 0$, so that by (3.143) the right-hand side approaches 1, implying the first assertion in (3.144). The second one is proved similarly.

Denoting $N_1 = T/\Psi(I_1^{-1}\, |\log \alpha_0|)$ and applying Chebyshev's inequality, we obtain

$$\mathsf{E}_1[N_1^m] \ge (1-\varepsilon)^m \mathsf{P}_1\left\{N_1 > (1-\varepsilon)\right\} \quad \text{for any } m > 0 \text{ and } 0 < \varepsilon < 1,$$

where by (3.144)

$$\lim_{\alpha_{\max} \to 0} \inf_{\delta \in \mathbf{C}(\alpha_0, \alpha_1)} \mathsf{P}_1\left\{N_1 > (1-\varepsilon)\right\} = 1 \quad \text{for every } 0 < \varepsilon < 1.$$

This shows that

$$\liminf_{\alpha_{\max} \to 0} \inf_{\delta \in \mathbf{C}(\alpha_0, \alpha_1)} \mathsf{E}_1[N_1^m] \ge 1 \quad \text{for any } m > 0,$$

and the first asymptotic inequality in (3.145) follows. The second inequality is proved in an analogous manner. □

Remark 3.4.1. It is worth noting that if the almost sure convergence conditions (3.142) hold and if in addition for all finite L,

$$\mathsf{P}_1\left(\sup_{t \le L} \lambda_t^+ < \infty\right) = 1, \quad \mathsf{P}_0\left(\sup_{t \le L} (-\lambda_t)^+ < \infty\right) = 1, \tag{3.149}$$

then the conditions (3.143) are satisfied; see the proof of Lemma 2.1 in [455].

It is natural to ask whether the SPRT with certain thresholds is asymptotically optimal in terms of minimizing the expected sample sizes or, more generally, in terms of minimizing moments of the stopping time, i.e., whether the asymptotic lower bounds (3.145) are attained for the SPRT under the conditions (3.142) and (3.149). The following counterexample shows that the answer is negative in general.

Let $X_t = W_t$ under H_0 and $X_t = (1+t)^{-1/2} + W_t$ under H_1, where W_t is a standard Brownian motion. Then

$$\lambda_t = \int_0^t (1+u)^{-1/2} \mathrm{d}X_u - \frac{1}{2}\log(1+t),$$

so that λ_t is the normal random variable with mean values $\frac{1}{2}\log(1+t)$ and $-\frac{1}{2}\log(1+t)$ under H_1 and H_0, respectively, and with variance $\log(1+t)$ under both hypotheses. The almost sure convergence conditions (3.142) are satisfied with $\psi(t) = \log(1+t)$, but $\mathsf{E}_i T^* = \infty$ if $\alpha_0 = \alpha_1 < 0.301$, as calculations in Subsection 3.4.3 show.

The following theorem establishes asymptotic optimality of the SPRT in a weak sense, which is perhaps of little interest for most applications.

Theorem 3.4.1 (Weak asymptotic optimality). *Assume that there exist finite positive numbers I_0 and I_1 such that the a.s. convergence conditions* (3.142) *hold with $\psi(t) = t^k$ for some $k > 0$ and, in addition, the conditions* (3.149) *hold. Suppose the thresholds $-a_0$ and a_1 are selected in such a way that the inequalities* (3.137) *are satisfied and the asymptotic approximations* (3.138) *hold. Then*

$$\frac{T^*}{|\log \alpha_0|^{1/k}} \xrightarrow[\alpha_{\max}\to 0]{\mathsf{P}_1-a.s.} \left(\frac{1}{I_1}\right)^{1/k}, \quad \frac{T^*}{|\log \alpha_1|^{1/k}} \xrightarrow[\alpha_{\max}\to 0]{\mathsf{P}_0-a.s.} \left(\frac{1}{I_0}\right)^{1/k}, \tag{3.150}$$

and, for every $0 < \varepsilon < 1$,

$$\inf_{\delta\in\mathbf{C}(\alpha_0,\alpha_1)} \mathsf{P}_i(T > \varepsilon T^*) \to 1 \quad \text{as } \alpha_{\max} \to 0, \quad i = 0, 1. \tag{3.151}$$

Proof. Assuming the condition (3.138), namely $a_1 \sim |\log \alpha_0|$ and $a_0 \sim |\log \alpha_1|$, by Lemma 3.4.1, $T^* \to \infty$ in probability as $\alpha_{\max} \to 0$, so by the a.s. convergence (3.142) with $\psi(t) = t^k$,

$$T^* \lambda_{T^*}^{-1/k} \xrightarrow[\alpha_{\max}\to 0]{\mathsf{P}_1-\text{a.s.}} I_1^{-1/k}, \quad T^*(-\lambda_{T^*})^{-1/k} \xrightarrow[\alpha_{\max}\to 0]{\mathsf{P}_0-\text{a.s.}} I_0^{-1/k}.$$

Since on $\{d^* = 1\}$ (whose probability is $1 - \alpha_1^* \sim 1 - \alpha_1 \to 1$ under P_1)

$$\frac{T^*}{\lambda_{T^*}^{1/k}} \le \frac{T^*}{a_1^{1/k}} \le \frac{T^*}{\lambda_{T^*-1}^{1/k}}$$

and on $\{d^* = 0\}$ (whose probability is $1 - \alpha_0^* \sim 1 - \alpha_0 \to 1$ under P_0)

$$\frac{T^*}{(-\lambda_{T^*})^{1/k}} \le \frac{T^*}{a_0^{1/k}} \le \frac{T^*}{(-\lambda_{T^*-1})^{1/k}},$$

and since $a_1 \sim |\log \alpha_0|$, $a_0 \sim |\log \alpha_1|$, the two asymptotic convergences (3.150) follow.

Finally, the weak asymptotic optimality (3.151) follows immediately from (3.144) and (3.150). □

3.4.2 Asymptotic Optimality of the SPRT with Respect to Moments of the Stopping Time

From the previous discussion it is clear that in order to establish an asymptotic optimality property of the SPRT with respect to the expected sample size, or more generally, with respect to higher moments of the sample size, the almost sure convergence (3.142) of the LLR has to be strengthened. It is natural to strengthen this convergence into the r-quick convergence

$$\frac{\lambda_t}{\psi(t)} \xrightarrow[t\to\infty]{\mathsf{P}_1-r-\text{quickly}} I_1 \quad \text{and} \quad \frac{\lambda_t}{\psi(t)} \xrightarrow[t\to\infty]{\mathsf{P}_0-r-\text{quickly}} -I_0. \tag{3.152}$$

In other words, according to the definition given in Subsection 2.4.3, we require that for some $r > 0$ and all $\varepsilon > 0$ the expectations $\mathsf{E}_1[L_1^r(\varepsilon)]$ and $\mathsf{E}_0[L_0^r(\varepsilon)]$ are finite, where

$$L_1(\varepsilon) = \sup\{t : |\lambda_t/\psi(t) - I_1| > \varepsilon\}, \ L_0(\varepsilon) = \sup\{t : |\lambda_t/\psi(t) + I_0| > \varepsilon\} \ (\sup\{\varnothing\} = 0).$$

The following theorem establishes asymptotic optimality of the SPRT in the general non-iid case with respect to moments of the sample size.

Theorem 3.4.2 (Asymptotic optimality). *Assume that there exist finite positive numbers I_0 and I_1 and an increasing nonnegative function $\psi(t)$ such that the r-quick convergence conditions* (3.152) *hold. Then the following three assertions are true.*

(i) $\mathsf{E}_i[(T^*)^r] < \infty$, $i = 0, 1$, *for any finite a_0 and a_1.*

(ii) *Asymptotically as* $\max(a_0, a_1) \to \infty$ *for all* $0 < m \le r$

$$\mathsf{E}_i[(T^*)^m] \sim \left[\Psi\left(\frac{a_i}{I_i}\right)\right]^m. \tag{3.153}$$

(iii) *If the thresholds $-a_0$ and a_1 are selected so that the inequalities* (3.137) *are satisfied and the asymptotic approximations* (3.138) *hold, then for all $0 < m \le r$, as $\alpha_{\max} \to 0$,*

$$\begin{aligned} \inf_{\delta \in \mathbf{C}(\alpha_0,\alpha_1)} \mathsf{E}_1[T^m] \sim \mathsf{E}_1[(T^*)^m] \sim \left[\Psi\left(\frac{|\log\alpha_0|}{I_1}\right)\right]^m, \\ \inf_{\delta \in \mathbf{C}(\alpha_0,\alpha_1)} \mathsf{E}_0[T^m] \sim \mathsf{E}_0[(T^*)^m] \sim \left[\Psi\left(\frac{|\log\alpha_1|}{I_0}\right)\right]^m. \end{aligned} \tag{3.154}$$

Consequently, the SPRT asymptotically minimizes the moments of the stopping time distribution up to the order r.

Proof. We provide detailed proofs only for the hypothesis H_1 since for H_0 the proofs are essentially the same.

Proof of (i). Let $T_1 = \inf\{t : \lambda_t \ge a_1\}$ and $T_0 = \inf\{t : \lambda_t \le -a_0\}$. Since $T^* = \min(T_0, T_1)$ it suffices to prove that $\mathsf{E}_i[T_i^r] < \infty$.

On one hand $\lambda_{T_1 - 1} < a_1$ and on the other hand, by the condition (3.152),

$$\lambda_{T_1-1} > \psi(T_1 - 1)(I_1 - \varepsilon) \quad \text{on } \{T_1 > L_1(\varepsilon) + 1\},$$

so that

$$T_1 < 1 + \Psi\left(\frac{a_1}{I_1 - \varepsilon}\right) \quad \text{on } \{L_1(\varepsilon) + 1 < T_1 < \infty\}.$$

Hence,

$$\begin{aligned} T_1 &\le 1 + \mathbb{1}_{\{T_1 > L_1(\varepsilon)+1\}} \Psi\left(\frac{a_1}{I_1 - \varepsilon}\right) + \mathbb{1}_{\{T_1 \le L_1(\varepsilon)+1\}} L_1(\varepsilon) \\ &\le 1 + \Psi\left(\frac{a_1}{I_1 - \varepsilon}\right) + L_1(\varepsilon). \end{aligned} \tag{3.155}$$

Similarly,

$$T_0 \le 1 + \Psi\left(\frac{a_0}{I_0 - \varepsilon}\right) + L_0(\varepsilon).$$

Since, by (3.152), $\mathsf{E}_i[L_i^r(\varepsilon)] < \infty$, it follows that $\mathsf{E}_i[T^*]^r < \infty$.

Proof of (ii). It follows from (3.155) and the r-quick convergence condition (3.152) that

$$\mathsf{E}_1[T^*]^r \le \left[\Psi\left(\frac{a_1}{I_1 - \varepsilon}\right)\right]^r (1 + o(1)) \quad \text{for all } 0 < \varepsilon < I_1,$$

so that when $\varepsilon \to 0$, we obtain the upper estimate

$$\mathsf{E}_1[T^*]^r \le \left[\Psi\left(\frac{a_1}{I_1}\right)\right]^r (1 + o(1)) \quad \text{as } \max(a_0, a_1) \to \infty, \tag{3.156}$$

and similarly

$$\mathsf{E}_0[T^*]^r \le \left[\Psi\left(\frac{a_0}{I_0}\right)\right]^r (1 + o(1)) \quad \text{as } \max(a_0, a_1) \to \infty. \tag{3.157}$$

Now, by the same argument as in the proof of Lemma 3.4.1 that has led to the lower bounds (3.145), replacing α_0 with e^{-a_1} and α_1 with e^{-a_0}, as $\max(a_0, a_1) \to \infty$ we obtain the following asymptotic lower estimates

$$\mathsf{E}_1[T^*]^r \geq \left[\Psi\left(\frac{a_1}{I_1}\right)\right]^r (1+o(1)), \quad \mathsf{E}_0[T^*]^r \geq \left[\Psi\left(\frac{a_0}{I_0}\right)\right]^r (1+o(1)). \tag{3.158}$$

Comparing (3.156) and (3.157) with (3.158) implies (3.153).

Proof of (iii). The asymptotic equalities (3.154) follow immediately from the asymptotic lower bounds (3.145) and upper bounds (3.156) and (3.157) with $a_1 \sim |\log \alpha_0|$ and $a_0 \sim |\log \alpha_1|$ in (3.138). This completes the proof of the theorem. □

We emphasize that while the asymptotic lower bounds for the moments of the stopping time (3.145) in Lemma 3.4.1 hold under the almost sure convergence (3.142), the r-quick convergence condition (3.152) cannot be weakened into the a.s. convergence in Theorem 3.4.2. (Note that $\lambda_t/\psi(t) \to I_1$ P_1-a.s. and $\lambda_t/\psi(t) \to -I_0$ P_0-a.s. iff $\mathsf{P}_i\{L_i(\varepsilon) < \infty\} = 1$ for all $\varepsilon > 0$.) In fact, the a.s. convergence does not even guarantee the finiteness of the moments of the stopping time T^*.

The case where $\psi(t) = t^k$, $k > 0$ (asymptotically power nonhomogeneity) is especially important for many practical applications; see the examples below. In this case, Theorem 3.4.2 implies the following corollary.

Corollary 3.4.1. *Assume that, for some $r > 0$ and $k > 0$,*

$$t^{-k}\lambda_t \xrightarrow[t\to\infty]{\mathsf{P}_1-r-quickly} I_1 \quad \text{and} \quad t^{-k}\lambda_t \xrightarrow[t\to\infty]{\mathsf{P}_0-r-quickly} -I_0. \tag{3.159}$$

Then, for all $0 < m \leq r$,

$$\mathsf{E}_i[(T^*)^m] \sim \left(\frac{a_i}{I_i}\right)^{m/k} \quad \text{as } \max(a_0, a_1) \to \infty, \; i = 0, 1. \tag{3.160}$$

Moreover, if the thresholds $-a_0$ and a_1 are selected so that the inequalities (3.137) *are satisfied and the asymptotics* (3.138) *hold, then for all $0 < m \leq r$, as $\alpha_{\max} \to 0$,*

$$\begin{aligned} \inf_{\delta\in\mathbf{C}(\alpha_0,\alpha_1)} \mathsf{E}_1[T^m] \sim \mathsf{E}_1[(T^*)^m] \sim \left(\frac{|\log \alpha_0|}{I_1}\right)^{m/k}, \\ \inf_{\delta\in\mathbf{C}(\alpha_0,\alpha_1)} \mathsf{E}_0[T^m] \sim \mathsf{E}_0[(T^*)^m] \sim \left(\frac{|\log \alpha_1|}{I_0}\right)^{m/k}. \end{aligned} \tag{3.161}$$

Furthermore, in a wide variety of statistical applications $k = 1$, i.e., $\psi(t) = t$. This means that the total accumulated K–L information in the trajectory $\{X_u\}_{0\leq u\leq t}$ is a linear function of t at least asymptotically: $\mathcal{I}_i(t) \sim I_i t$, $t \to \infty$.

While in the iid case the SPRT is optimal with respect to the expected sample size and there are accurate approximations to $\mathsf{E}_i[T^*]$ presented in the previous sections, it is interesting to ask whether it has also some optimality property with respect to higher moments of the sample size. Theorem 3.4.2 and Corollary 3.4.1 allow us to deduce that higher moments of the sample size are also asymptotically minimized.

Indeed, let $t = n \in \mathbb{Z}_+$ and $r \geq 1$. By Theorem 2.4.4, the $(r+1)$-th moment conditions $\mathsf{E}_i|\lambda_1|^{r+1} < \infty$, $i = 0, 1$ are both necessary and sufficient for the r-quick convergence (3.159) with $k = 1$, where in this case $I_1 = \mathsf{E}_1\lambda_1$ and $I_0 = \mathsf{E}_0[-\lambda_1]$ are the K–L information numbers. Therefore, the following result holds true.

Corollary 3.4.2. *Consider the discrete-time iid case. Assume that the K–L information numbers I_1, I_0 are positive and $\mathsf{E}_i|\lambda_1|^{r+1} < \infty$, $i = 0, 1$ for some $r \geq 1$. If the thresholds $-a_0$ and a_1 are* *selected so that the inequalities* (3.137) *are satisfied and the asymptotics* (3.138) *hold, then for all* $1 \le m \le r$, *as* $\alpha_{\max} \to 0$,

$$\inf_{\delta \in \mathbf{C}(\alpha_0,\alpha_1)} \mathsf{E}_1[T^m] \sim \mathsf{E}_1[(T^*)^m] \sim \left(\frac{|\log \alpha_0|}{I_1}\right)^m, \\ \inf_{\delta \in \mathbf{C}(\alpha_0,\alpha_1)} \mathsf{E}_0[T^m] \sim \mathsf{E}_0[(T^*)^m] \sim \left(\frac{|\log \alpha_1|}{I_0}\right)^m. \tag{3.162}$$

Note that in the case $m = r = 1$ we have already obtained better higher-order approximations, but for $r, m > 1$ the result is new.

Remark 3.4.2. It can be shown that the following conditions

$$\operatorname{ess\,sup} \mathsf{P}_0\left(n^{-1}\sum_{i=k}^{n} Z_i < -I_0(1-\varepsilon) | \mathscr{F}_k\right) \xrightarrow[n\to\infty]{} 0, \\ \operatorname{ess\,sup} \mathsf{P}_1\left(n^{-1}\sum_{i=k}^{n} Z_i < I_1(1-\varepsilon) | \mathscr{F}_k\right) \xrightarrow[n\to\infty]{} 0 \quad \text{for all } 0<\varepsilon<1 \text{ and } k \ge 1 \tag{3.163}$$

are sufficient for the upper bounds

$$\mathsf{E}_1[(T^*)^m] \le \left(\frac{|\log \alpha_0|}{I_1}\right)^m (1+o(1)), \quad \mathsf{E}_0[(T^*)^m] \le \left(\frac{|\log \alpha_1|}{I_0}\right)^m (1+o(1)), \quad \alpha_{\max} \to 0$$

to hold for all $m > 0$ (compare with the condition (8.201) and see the proof of Theorem 8.2.6). In the iid case, this condition holds whenever the K–L numbers I_0, I_1 are finite. Indeed,

$$\mathsf{P}_1\left(n^{-1}\sum_{i=k}^{n} Z_i < I_1(1-\varepsilon) | \mathscr{F}_k\right) = \mathsf{P}_1\left(n^{-1}\sum_{i=1}^{n} Z_i < I_1(1-\varepsilon)\right) \xrightarrow[n\to\infty]{} 0$$

since Z_i, $i = 1, 2, \dots$ are iid with mean I_1 under P_1 (and similarly under P_0). Therefore, the assertions of Corollary 3.4.2 (asymptotic relations (3.162)) hold for all $m \ge 1$ under the sole first moment condition $0 < I_i < \infty$ ($i = 0, 1$), so that under this latter condition the SPRT asymptotically minimizes all positive moments of the sample size. See also [129] for a different proof.

We now provide several examples that illustrate the usefulness of the general theory.

3.4.3 Detection of a Deterministic Signal in Gaussian Noise

Let $t \in \mathbb{R}_+$ and assume that the process $\{X_t\}$ admits a stochastic differential equation

$$\mathrm{d}X_t = \theta S_t \mathrm{d}t + V_t \mathrm{d}t + \sigma \mathrm{d}W_t, \quad t \ge 0, \quad \theta = 0, 1,$$

where S_t is a deterministic signal, W_t is a standard Brownian motion, and V_t is an L_2-continuous Gaussian process ($\sigma > 0$). In target detection applications, S_t characterizes a signal from a target, $\dot{W}_t$ models white Gaussian noise of a sensor and V_t is clutter. The hypotheses are $H_0 : \theta = 0$ (no signal) and $H_1 : \theta = 1$ (the signal is present).

This model has been already considered in Example 3.3.1 at the end of the previous section, where we showed in (3.136) that the LLR has the form

$$\lambda_t = \frac{1}{\sigma^2}\int_0^t \widetilde{S}_u \mathrm{d}\widetilde{X}_u - \frac{1}{2\sigma^2}\int_0^t \widetilde{S}_u^2 \mathrm{d}u \tag{3.164}$$

with

$$\widetilde{S}_t = S_t - \int_0^t S_u C(t,u)\mathrm{d}u, \quad \mathrm{d}\widetilde{X}_t = \mathrm{d}X_t - \left[\int_0^t C(t,u)\,\mathrm{d}X_u\right]\mathrm{d}t.$$

Here $C(t,u)$ is the characteristic of an optimal "whitening" filter that satisfies the Wiener–Hopf equation. Also, the process $\widetilde{X}_t$ may be represented via an innovation (standard Brownian motion) process $\widetilde{W}_t$ as

$$\widetilde{X}_t = \int_0^t \theta \widetilde{S}_u \, \mathrm{d}u + \sigma \widetilde{W}_t, \quad \theta = 0,1. \tag{3.165}$$

Using (3.164) and (3.165), we obtain that under H_θ

$$\lambda_t^\theta = \frac{1}{\sigma} \int_0^t \widetilde{S}_u \, \mathrm{d}\widetilde{W}_u + (-1)^{\theta+1} \frac{1}{2} \mu(t), \tag{3.166}$$

where $\mu(t) = \sigma^{-2} \int_0^t \widetilde{S}_u^2 \, du$ is the cumulative SNR at the output of the whitening filter over the time interval $[0,t]$.

In Section 3.3 we argued that the SPRT is strictly optimal in the sense of minimizing the total expected K–L information $\mathsf{E}_i[\mu(T^*)/2]$. Here we show that the SPRT asymptotically minimizes all the moments of the stopping time under quite general assumptions on the behavior of the signal $\widetilde{S}_t$.

Write

$$W_t^* = \frac{1}{\sigma} \int_0^t \widetilde{S}_u \, \mathrm{d}\widetilde{W}_u,$$

and assume that $\mu(t) = q\psi(t)$ with $q > 0$ and $\psi(t)$ being an increasing function ($\psi(\infty) = \infty$). Then the r-quick convergence conditions (3.152) hold (with $I_1 = I_0 = I = q/2$) if, and only if,

$$\frac{1}{\psi(t)} W_t^* \xrightarrow[t\to\infty]{\mathsf{P}_i-r-\text{quickly}} 0, \quad i = 0,1. \tag{3.167}$$

By Lemma 2.4.1, the following implication holds:

$$\left\{ \int_0^\infty t^{r-1} \mathsf{P}_i \left(\sup_{u \le t} |W_u^*| > \varepsilon \psi(t) \right) \mathrm{d}t < \infty \; \forall \varepsilon > 0 \right\} \Longrightarrow \frac{1}{\psi(t)} W_t^* \xrightarrow[t\to\infty]{\mathsf{P}_i-r\text{-quickly}} 0. \tag{3.168}$$

Assume that $\psi(t) = t^k$, $k > 0$, i.e., $\widetilde{S}_t^2/\sigma^2 = kqt^{k-1}$ or, more generally,

$$\lim_{t\to\infty} \frac{1}{\sigma^2 t^k} \int_0^t \widetilde{S}_u^2 \, du = q. \tag{3.169}$$

For instance, this condition holds if $\widetilde{S}_t^2 = \sum_{j=1}^{k-1} c_j t^j$ is a polynomial of the $(k-1)$-th order. Then, denoting by $\Phi(y)$ the standard normal distribution function, we obtain

$$\begin{aligned} \int_0^\infty t^{r-1} \mathsf{P}_i \left\{ |W_t^*| > \varepsilon t^k \right\} \mathrm{d}t &= 2 \int_0^\infty t^{r-1} \Phi\left(-\varepsilon t^{k/2} / \sqrt{q} \right) \mathrm{d}t \\ &\le \frac{4q^{(r-k/2+1)/2}}{k\varepsilon^{r-k/2+1}} \int_{-\infty}^\infty t^{r-k/2} \, \mathrm{d}\Phi(t) < \infty \quad \text{for all } \varepsilon, k, r > 0. \end{aligned} \tag{3.170}$$

Therefore, under condition (3.169), the r-quick convergence conditions (3.159) are satisfied for all $r > 0$ with $I_1 = I_0 = I = q/2$. By Corollary 3.4.1, as $\alpha_{\max} \to 0$, for all $m > 0$

$$\begin{aligned} \inf_{\delta \in \mathbf{C}(\alpha_0,\alpha_1)} \mathsf{E}_1 T^m &\sim \mathsf{E}_1[T^*]^m \sim \left(\frac{2|\log \alpha_0|}{q} \right)^{m/k}, \\ \inf_{\delta \in \mathbf{C}(\alpha_0,\alpha_1)} \mathsf{E}_0 T^m &\sim \mathsf{E}_0[T^*]^m \sim \left(\frac{2|\log \alpha_1|}{q} \right)^{m/k}, \end{aligned} \tag{3.171}$$

so that the SPRT asymptotically minimizes all positive moments of the stopping time. The condition (3.169), which turns out to be sufficient for asymptotic optimality of the SPRT, is satisfied for most applications. Furthermore, often $k = 1$, i.e., $\mu(t) \sim qt$ as $t \to \infty$.

3.4.3.1 The Case of Markov Gaussian Noise

Consider the particular case where $\{V_t\}$ is a zero-mean exponentially correlated Markov Gaussian process, $\mathsf{E}[V_t] = 0$, $\mathsf{E}[V_t V_{t+u}] = \sigma_{\rm v}^2 e^{-\rho|u|}$ $(\rho > 0)$. Let $S_t = S_0$ for all $t \ge 0$. Then it can be shown that the condition (3.169) is satisfied with $k = 1$ and

$$q = \frac{S_0^2}{\sigma^2[1+2\sigma_{\rm v}^2/(\sigma^2\rho)]}.$$

Hence, $I_1 = I_0 = S_0^2/2\sigma^2[1+2\sigma_{\rm v}^2/(\sigma^2\rho)]$. Recall that in many applications the process V_t represents clutter, so that the parameter $Q_{\rm v} = 2\sigma_{\rm v}^2/(\sigma^2\rho)$ can be interpreted as the clutter-to-noise ratio, and $S_0^2/\sigma^2 = Q_s$ is the SNR. Thus, $I_1 = I_0 = Q_s/2(1+Q_{\rm v})$.

Now, suppose that $S_t = A\sin\omega t$ is a harmonic signal with the amplitude $A > 0$ and frequency ω. Then the condition (3.169) is again satisfied with $k = 1$ and

$$q = Q_s \frac{\omega^2+\rho^2}{\omega^2+\rho^2(1+Q_{\rm v})},$$

where $Q_{\rm v}$ is the clutter-to-noise ratio as above and $Q_s = A^2/2\sigma^2$ is the SNR, so that $I_1 = I_0 = q/2$. Note that $q \approx Q_s$ when $Q_{\rm v} \ll 1$ as well as when $\omega \gg \rho(1+Q_{\rm v})^{1/2}$. In these cases clutter does not affect the performance. This can be expected since in the first case clutter is weak compared to noise, and in the second case the frequency of the signal is much bigger than the effective bandwidth of the clutter spectrum which is defined by ρ. A similar result can be obtained for the sequence of harmonic pulses — the signal that is typically used in radar applications.

3.4.3.2 Exact Formulas for the Operating Characteristics

It is interesting that under quite general assumptions on the signal $\widetilde{S}_t$ one may find exact expressions for the distribution functions $F_i(t) = \mathsf{P}_i(T^* \le t)$ and the moments $\mathsf{E}_i[T^*]^m$ of the stopping time T^* in the form of infinite series, assuming that $\mu(t) < \infty$ for $t < \infty$ and $\mu(\infty) = \infty$. To obtain exact expressions for the moments we also have to assume that $\mu(t)$ increases sufficiently fast so that the moments are finite. We now provide a sketch of the argument that leads to the corresponding exact expressions. Further details can be found in Tartakovsky [449, 452].

Note that, by (3.166), the LLR process λ_t^i under H_i is a nonhomogeneous Gaussian process with independent increments and with drift and diffusion coefficients $d_i(t) = (-1)^{i+1}\widetilde{S}_t^2/2\sigma^2$ and $D_i(t) = \widetilde{S}_t^2/\sigma^2$ $(i = 0, 1)$. Let $\{\tilde{\lambda}_t = \lambda_{\gamma_t}, t \ge 0\}$ be the process obtained by a change of time

$$t \mapsto \gamma_t = \inf\left\{v : \sigma^{-2}\int_0^v \widetilde{S}_u^2 \mathrm{d}u \ge t\right\} = \inf\{v : \mu(v) \ge t\}.$$

Denoting

$$\widehat{W}_t = \sigma^{-1}\int_0^{\gamma_t} \widetilde{S}_u \mathrm{d}\widetilde{W}_u$$

and using (3.166), we obtain

$$\tilde{\lambda}_t^i = \widehat{W}_t + (-1)^{i+1} t/2, \quad t \ge 0,$$

where $\{\widehat{W}_t\}_{t\ge 0}$ is the standard Brownian motion; see Gikhman and Skorokhod [164, Sec. 4, Theorem 3]. Thus, the processes $\{\tilde{\lambda}_t^i\}$, $i = 0, 1$ are homogeneous Brownian motions with drift and diffusion coefficients $d_1 = -d_0 = \frac{1}{2}$, $D_1 = D_0 = 1$. From Lemma 1 in [449], it is easy to show that the trajectories $\{\lambda_t, 0 \le t \le T\}$ and $\{\widetilde{\lambda}_t, 0 \le t \le \mu(T)\}$ coincide, so that $\widetilde{T}^* = \inf\{t : \widetilde{\lambda}_t \notin (-a_0, a_1)\} = \mu(T^*)$, and for studying the SPRT we can use the homogeneous Markov process $\widetilde{\lambda}_t$ in place of the nonhomogeneous process λ_t. In fact, $\mathsf{P}_i\{T^* \le t\} = \mathsf{P}_i\{\tilde{T}^* \le \mu(t)\}$, i.e., if we denote by $\widetilde{F}_i(t) = \mathsf{P}_i(\widetilde{T}^* \le t)$ the distribution function of the first exit time of the homogeneous Brownian motion $\widetilde{\lambda}_t$ from the interval $(-a_0, a_1)$, then

$$F_i(t) = \widetilde{F}_i(\mu(t)) \quad \text{for } t \ge 0 \text{ and } i = 0, 1. \tag{3.172}$$

Next, introduce the function $G_i(t,z) = \mathsf{P}_i\{\widetilde{\lambda}_s \in (-a_0, a_1), s \in (u,t] | \widetilde{\lambda}_u = z\}$, which is the conditional probability that the homogeneous Brownian motion $\widetilde{\lambda}_s^i$ does not leave the region $(-a_0, a_1)$ in the interval $u < s \le t$ conditioned on $\widetilde{\lambda}_u = z$, $z \in (-a_0, a_1)$ $(u \ge 0)$. It can be shown [449] that the function $g_i(t,z) = G_i(t,z) \exp\{d_i z + t/8\}$ satisfies the following linear parabolic boundary problem

$$\frac{\partial g_i(t,z)}{\partial t} = \frac{1}{2} \frac{\partial^2 g_i(t,z)}{\partial z^2}, \quad z \in (-a_0, a_1),\ t \ge 0;$$
$$g_i(0,z) = e^{d_i z}, \quad g_i(t, -a_0) = g_i(t, a_1) = 0, \quad t > 0.$$

Solving this problem by the Fourier method and taking into account that by (3.172),

$$F_i(t) = 1 - G_i(\mu(t), 0) = 1 - g_i(\mu(t), 0) e^{-\mu(t)/8},$$

we obtain [449]

$$\begin{aligned} F_i(t) &= 1 - \sum_{n=1}^{\infty} C_{n,i} \exp\left\{-\frac{1}{2}\left[d_i^2 + \frac{\pi^2 n^2}{(a_1 + a_0)^2}\right]\mu(t)\right\} \sin\left[\frac{\pi n a_0}{a_1 + a_0}\right], \\ C_{n,i} &= \frac{2\pi n}{(a_1 + a_0)^2} \frac{e^{-d_i a_0} + (-1)^{n+1} e^{d_i a_1}}{d_i^2 + \pi^2 n^2/(a_1 + a_0)^2}, \end{aligned} \tag{3.173}$$

where $d_1 = -d_0 = 1/2$. Now, it is not difficult to show that the series (3.173) converges uniformly on $(0, \infty)$, and hence it can be integrated on a term-by-term basis. Taking into account that $\mathsf{E}_i T^* = \int_0^\infty [1 - F_i(t)]\, \mathrm{d}t$ and performing integration, we obtain the expression for the expected sample size

$$\mathsf{E}_i T^* = \sum_{n=1}^{\infty} C_{n,i} \sin\left[\frac{\pi n a_0}{a_1 + a_0}\right] \int_0^\infty \exp\left\{-\frac{1}{2}\left[d_i^2 + \frac{\pi^2 n^2}{(a_1 + a_0)^2}\right]\mu(t)\right\} \mathrm{d}t. \tag{3.174}$$

Note that this equality holds when the series on the right hand-side converges, which is the case unless $\mu(t)$ is very slowly increasing as in the case $\mu(t) \sim \log t$ considered below.

In a similar way, using (3.173), we can obtain expressions for higher moments of the stopping time. The details are omitted.

It is interesting to ask whether the simple asymptotic approximations (3.153) and (3.154) are sufficiently accurate or not. To answer this question, consider a particular case of power non-homogeneity assuming that $\mu(t) = q t^k$ for some $k > 0$. By (3.161), for sufficiently small $\alpha_{\max}$,

$$\mathsf{E}_1 T^* \approx \left(\frac{2|\log \alpha_0|}{q}\right)^{1/k}, \quad \mathsf{E}_0 T^* \approx \left(\frac{2|\log \alpha_1|}{q}\right)^{1/k}. \tag{3.175}$$

Also, if the thresholds are selected as in (3.130), then by (3.129),

$$\mathsf{E}_1[\mu(T^*)] = 2\beta(\alpha_1, \alpha_0), \quad \mathsf{E}_0[\mu(T^*)] = 2\beta(\alpha_0, \alpha_1).$$

Applying Jensen's inequality, we obtain that if $\mu(t)$ is a convex function, then

$$\mathsf{E}_1 T^* \le \mu^{-1}(2\beta(\alpha_1, \alpha_0)), \quad \mathsf{E}_1 T^* \le \mu^{-1}(2\beta(\alpha_0, \alpha_1)),$$

where μ^{-1} is the inverse function for μ. As $\alpha_{\max} \to 0$ these upper bounds are asymptotically attained, which is particularly true for the considered "power" example. Clearly, another approximation worth trying is

$$\mathsf{E}_1 T^* \approx \left(\frac{2\beta(\alpha_1, \alpha_0)}{q}\right)^{1/k}, \quad \mathsf{E}_0 T^* \approx \left(\frac{2\beta(\alpha_0, \alpha_1)}{q}\right)^{1/k}. \tag{3.176}$$

By virtue of (3.174) and the fact that $d_1 = -d_0 = 1/2$, we obtain the exact expressions for the expected observation time

$$\mathsf{E}_i T^* = k^{-1}\Gamma(k^{-1}) \left(\frac{2}{q}\right)^{1/k} \sum_{n=1}^{\infty} C_{n,i} \left[\frac{1}{4} + \frac{\pi^2 n^2}{(a_1+a_0)^2}\right]^{-1/k} \sin\left[\frac{\pi n a_0}{a_1+a_0}\right]. \tag{3.177}$$

Assume that the signal $\widetilde{S}_t = At$ is a linear function, in which case $k = 3$ and $q = A^2/3\sigma^2$. The results of computations in the symmetric case $\alpha_0 = \alpha_1 = \alpha$ are shown in Table 3.13 for $q = 1/\sqrt{2}$, i.e., for the values normalized by $(2/q)^{1/3}$. It is seen that the accuracy of the asymptotic approximation (3.175) is fairly high already for $\alpha = 0.01$, while it is not especially accurate for $\alpha = 0.1$.

Table 3.13: *Accuracy of the asymptotic approximation and the upper bound for the expected observation time for* $k = 3$.

Expected Observation Time	Probabilities of Errors, $\alpha_0 = \alpha_1 = \alpha$			
$\mathsf{E}_0 T^* = \mathsf{E}_1 T^*$	10^{-1}	10^{-2}	10^{-3}	10^{-6}
Approximation (3.175)	1.48221	1.86747	2.13772	2.69336
Approximation (3.176)	1.35465	1.85359	2.13619	2.69336
Exact Formula (3.177)	1.28186	1.78286	2.07656	2.65306

It is also of interest to compare the efficiency of the SPRT with respect to the NP test. By (3.166), the LLR λ_t is the normal random variable with parameters

$$\mathsf{E}_1 \lambda_t = -\mathsf{E}_0 \lambda_t = \mu(t)/2, \quad \mathsf{var}_1[\lambda_t] = \mathsf{var}_0[\lambda_t] = \mu(t),$$

so that we can use the argument of Subsection 3.1.5 with simply replacing N with $\mu(t)$ to obtain that the fixed observation time t_{NP} in the NP test required to guarantee the given error probabilities α_0 and α_1 is

$$t_{\mathrm{NP}}(\alpha_0, \alpha_1) = \mu^{-1}\left((Q_{1-\alpha_0} + Q_{1-\alpha_1})^2\right), \tag{3.178}$$

where μ^{-1} is the inverse for μ and Q_p is the p-quantile of the normal distribution ($\Phi(Q_p) = p$). Therefore, in the case where $\mu(t) = qt^k$, $k > 0$, we have from (3.178)

$$t_{\mathrm{NP}}(\alpha_0, \alpha_1, q) = \left[\frac{(Q_{1-\alpha_0} + Q_{1-\alpha_1})^2}{q}\right]^{1/k}. \tag{3.179}$$

Taking into account that $Q_{1-\alpha} \sim (2|\log\alpha|)^{1/2}$ as $\alpha \to 0$ (see (3.83)), we obtain that in the symmetric case $\log\alpha_0 \sim \log\alpha_1 \sim \log\alpha$,

$$t_{\mathrm{NP}}(\alpha, q) \sim \left[\frac{8|\log\alpha|}{q}\right]^{1/k}, \quad \alpha \to 0.$$

Thus, in the symmetric case the asymptotic efficiency of the SPRT with respect to the NP test is

$$\mathcal{E}_i(\alpha) = \frac{\mathsf{E}_i T^*(\alpha, q)}{t_{\mathrm{NP}}(\alpha, q)} = \mathcal{E}(\alpha) \sim \left(\frac{1}{4}\right)^{1/k} \quad \text{as } \alpha \to 0, \quad i = 0, 1.$$

In particular, for $k = 3$, we obtain $\mathcal{E}(\alpha) \sim 0.63$. Table 3.14 shows the exact values of $\mathcal{E}(\alpha)$ as a function of the error probabilities in the symmetric case $\alpha_0 = \alpha_1 = \alpha$ that were calculated using the exact formulas (3.177) and (3.179).

Table 3.14: *Efficiency $\mathcal{E}(\alpha)$ of the SPRT vs. the NP test in the symmetric case $\alpha_0 = \alpha_1 = \alpha$.*

Efficiency $\mathcal{E}(\alpha) = \mathsf{E}_i T^*(\alpha,q)/t_{\mathrm{NP}}(\alpha,q)$	Probabilities of Errors, $\alpha_0 = \alpha_1 = \alpha$			
	10^{-1}	10^{-2}	10^{-3}	10^{-6}
Exact Formulas (3.177) and (3.179)	0.76825	0.71805	0.69209	0.66358

3.4.3.3 The Case of Infinite Expected Sample Size

If the function $\mu(t)$ grows too slowly, then the expected observation time may be infinite. Indeed, let $V_t \equiv 0$ and $S_t = [A^2(1+t)]^{-1/2}$, so that $\mu(t) = q\psi(t) = q\log(1+t)$, where $q = A^2/\sigma^2$. Then

$$\frac{1}{\log(1+t)}\lambda_t \xrightarrow[t\to\infty]{\mathsf{P}_1-\text{a.s.}} q/2, \qquad \frac{1}{\log(1+t)}\lambda_t \xrightarrow[t\to\infty]{\mathsf{P}_0-\text{a.s.}} -q/2,$$

but not r-quickly, since the condition (3.168) is not satisfied and, as a result, $\mathsf{E}_i[L_i(\varepsilon)] = \infty$. Indeed,

$$\int_0^\infty \mathsf{P}_i(|W_t| > \varepsilon\log(1+t))\,\mathrm{d}t = 2\int_0^\infty \Phi\left(-\varepsilon\sqrt{\frac{2\log(1+t)}{q}}\right)\mathrm{d}t$$
$$= 4\int_0^\infty u\left(e^{u^2}-1\right)\Phi\left(-\varepsilon u/\sqrt{q/2}\right)\mathrm{d}u = \infty \quad \text{for sufficiently small } \varepsilon.$$

Thus, Theorem 3.4.2 cannot be applied. In fact, setting $A^2/\sigma^2 = 1$, $-a_0 = a_1 = a$ and using [168, Lemma 6.2], we obtain

$$\begin{aligned} &\mathsf{E}_i[T^*] = \frac{\cosh(a/2)}{\cos(\sqrt{7}a/2) - 1} \quad \text{if } 0 < a < 7^{-1/2}\pi, \\ &\mathsf{E}_i[T^*] = \infty \quad \text{if } a \ge 7^{-1/2}\pi. \end{aligned} \tag{3.180}$$

Since $\alpha^* = e^{-a}(1-\alpha^*)$, it follows that $\mathsf{E}_i[T^*] = \infty$ for $\alpha \le (1+e^{\pi/\sqrt{7}})^{-1} \approx 0.301$. At the same time, the NP test has a finite sample size for all $0 < \alpha < 1$: $t_{\mathrm{NP}} = \exp\{4Q_{1-\alpha}^2\}$. In particular, if $\alpha \to 0$, then $t_{\mathrm{NP}} \sim \alpha^{-8}$. This value is extremely large for small α but still finite.

3.4.4 Detection of a Gaussian Markov Signal in White Noise

Consider the general model defined by the stochastic differential equation (3.131) with $\sigma(t) = \sigma > 0$, $S_0(t) = 0$ and $S_1(t) = S_t$, i.e., we suppose that under the hypothesis H_i the observed process has the stochastic differential

$$\mathrm{d}X_t = iS_t\,\mathrm{d}t + \sigma\,\mathrm{d}W_t, \quad t \ge 0, \quad i = 0,1,$$

where $\{W_t\}_{t\ge 0}$ is a standard Brownian motion and $\{S_t\}_{t\ge 0}$ is a stationary Gaussian Markov process, $\mathsf{E}[S_t] = 0$, $\mathsf{E}[S_tS_{t+u}] = \sigma_s^2 e^{-\rho|u|}$ ($\rho > 0$). This latter process in turn satisfies the Itô stochastic equation

$$\mathrm{d}S_t = -\rho S_t\,\mathrm{d}t + \varkappa\,\mathrm{d}w_t, \quad t \ge 0, \quad S_0 \sim \mathcal{N}(0,\sigma_s^2),$$

where $\sigma_s^2 = \varkappa^2/2\rho$ and $\{w_t\}$ is a standard Brownian motion that may depend on W_t. (All asymptotic results are valid for an arbitrary initialization, including $S_0 = 0$, in which case S_t is not stationary.) In other words, we are interested in detecting the Markov Gaussian signal in white noise, i.e., the hypotheses are

$$H_i : X_t = i\int_0^t S_u\,\mathrm{d}u + \sigma W_t, \quad i = 0,1. \tag{3.181}$$

Let $\hat{S}_t(\mathbf{X}_0^t) = \mathsf{E}_1[S_t \mid \mathscr{F}_t^X]$. Despite the fact that it is straightforward to obtain the representation of the LLR process in the form

$$\lambda_t = \frac{1}{\sigma^2}\int_0^t \hat{S}_u \,\mathrm{d}X_u - \frac{1}{2\sigma^2}\int_0^t \hat{S}_u^2 \,\mathrm{d}u \tag{3.182}$$

as a particular case of (3.134), it is instructive to use the minimal representation of the Itô process in Lemma 2.1.1 and Theorem 2.1.3 on the absolute continuity of the measures corresponding to Itô processes with respect to the Wiener measure directly. Since

$$\int_0^t \mathsf{E}|S_u|^2 \,\mathrm{d}u = \sigma_s^2 t < \infty,$$

the condition (2.26) holds and, by Lemma 2.1.1, under the hypothesis H_1 the observed process X_t can be represented as

$$X_t = \int_0^t \hat{S}_u(\mathbf{X}_0^u)\,\mathrm{d}u + \sigma\widetilde{W}_t, \tag{3.183}$$

where $\widetilde{W}_t$ is some standard Brownian motion statistically indistinguishable from W_t. Note that under H_0 the process $X_t = \sigma W_t$ is a Brownian motion.

Next, since $S_t \sim \mathcal{N}(0, \sigma^2 t)$, it is easily seen that

$$\mathsf{P}\left(\int_0^t S_u^2 \,\mathrm{d}u < \infty\right) = 1 \quad \text{and} \quad \mathsf{E}\left[\exp\left(\frac{1}{2}\int_0^t S_u^2 \,\mathrm{d}u\right)\right] < \infty,$$

so that the conditions (2.30) and (2.31) hold. Hence, applying Theorem 2.1.3, we obtain (3.182) as well as the following representation for the LLR $\lambda_t^{(i)}$ under H_i:

$$\begin{aligned}
\lambda_t^{(1)} &= \frac{1}{\sigma}\int_0^t \hat{S}_u(\mathbf{X}_0^{u(1)})\,\mathrm{d}\widetilde{W}_u + \frac{1}{2\sigma^2}\int_0^t \hat{S}_u^2(\mathbf{X}_0^{u(1)})\,\mathrm{d}u,\\
\lambda_t^{(0)} &= \frac{1}{\sigma}\int_0^t \hat{S}_u(\mathbf{X}_0^{u(0)})\,\mathrm{d}W_u - \frac{1}{2\sigma^2}\int_0^t \hat{S}_u^2(\mathbf{X}_0^{u(0)})\,\mathrm{d}u,
\end{aligned} \tag{3.184}$$

where $\mathbf{X}_0^{u(i)} = \{X_v^{(i)}, 0 \le v \le u\}$ and $X_t^{(i)}$ denotes the process X_t under the hypothesis H_i, i.e., $X_t^{(0)} = \sigma W_t$ and $X_t^{(1)}$ is given by (3.181) with $i = 1$ or alternatively by (3.183).

Note first that since the process X_t is Gaussian, the functional $\hat{S}_t$, the estimator of the signal S_t, is linear and can be computed either using the Wiener–Hopf integral formula or the Kalman differential approach. Specifically, write $D_t = \mathsf{E}_1[S_t - \hat{S}_t]^2$ for the mean-square error of the estimator $\hat{S}_t$. As shown in, e.g., [266, Theorem 10.1], the estimator $\hat{S}_t$ satisfies the system of Kalman–Bucy equations

$$\mathrm{d}\hat{S}_t = -(\rho + D_t/\sigma^2)\hat{S}_t\,\mathrm{d}t + (D_t/\sigma^2)\,\mathrm{d}X_t, \quad t \ge 0, \quad \widehat{S}_0 = 0, \tag{3.185}$$

$$\dot{D}_t = -2\rho D_t - D_t^2/\sigma^2 + 2\rho\sigma_s^2, \quad t \ge 0, \quad D_0 = \sigma_s^2, \tag{3.186}$$

where $2\rho\sigma_s^2 = \varkappa^2$. These equations provide an efficient scheme for computing $\hat{S}_t$ and therefore for implementing the LLR λ_t.

The second observation is that the equalities (3.184) imply

$$\mathcal{I}_i(t) = \frac{1}{2\sigma^2}\int_0^t \mathsf{E}_i\left[\hat{S}_u^2\right]\mathrm{d}u, \quad i = 0, 1.$$

We now show that

$$\lim_{t\to\infty} \frac{1}{t}\mathcal{I}_i(t) = I_i, \quad i = 0, 1 \tag{3.187}$$

with some finite and positive constants I_i.

It is easy to see that

$$\mathsf{E}_1[\hat{S}_t^2] = \mathsf{E}_1[S_t^2] - D_t = \sigma_s^2 - D_t. \tag{3.188}$$

Solving the Ricatti equation (3.186), after simple algebra we obtain

$$\frac{1}{t}\int_0^t D_u \, du = \frac{2\sigma_s^2}{1+\sqrt{1+2\sigma_s^2/(\sigma^2\rho)}} + o(1) \quad \text{as } t \to \infty. \tag{3.189}$$

It follows from (3.188) and (3.189) that

$$\lim_{t\to\infty} \frac{1}{\sigma^2 t}\int_0^t \mathsf{E}_1[\hat{S}_u^2]\, du = \frac{\sigma_s^2}{\sigma^2}\frac{Q}{\left(1+\sqrt{1+Q}\right)^2} = q_1, \tag{3.190}$$

where $Q = 2\sigma_s^2/(\rho\sigma^2)$ is a parameter which characterizes a SNR. Analogously, one can show that for $i=0$

$$\lim_{t\to\infty} \frac{1}{\sigma^2 t}\int_0^t \mathsf{E}_0[\hat{S}_u^2]\, du = \frac{q_1}{\sqrt{1+Q}} = q_0. \tag{3.191}$$

Thus, the equalities (3.187) hold with $I_i = q_i/2$, $i=0,1$, where q_1 and q_0 are given by (3.190) and (3.191).

Finally, using the fact that $\{\hat{S}_t\}_{t\ge 0}$ is a Gaussian process it can be shown [455, Sec. 3.2] that for all $r>0$

$$\frac{1}{t}\int_0^t \hat{S}_u \, d\widetilde{W}_u \xrightarrow[t\to\infty]{\mathsf{P}_1-r\text{-quickly}} 0, \quad \frac{1}{t}\int_0^t \hat{S}_u \, dW_u \xrightarrow[t\to\infty]{\mathsf{P}_0-r\text{-quickly}} 0$$

and

$$\frac{1}{t}\int_0^t \hat{S}_u^2 \, du \xrightarrow[t\to\infty]{\mathsf{P}_i-r\text{-quickly}} \lim_{t\to\infty}\frac{1}{t}\int_0^t \mathsf{E}_i[\hat{S}_u^2]\, du$$

which along with (3.184), (3.190) and (3.191) prove that for all $r>0$

$$\frac{1}{t}\lambda_t \xrightarrow[t\to\infty]{\mathsf{P}_1-r\text{-quickly}} I_1, \quad \frac{1}{t}\lambda_t \xrightarrow[t\to\infty]{\mathsf{P}_0-r\text{-quickly}} -I_0,$$

where

$$I_1 = \frac{\sigma_s^2}{2\sigma^2}\frac{Q}{\left(1+\sqrt{1+Q}\right)^2}, \quad I_0 = \frac{\sigma_s^2}{2\sigma^2}\frac{Q}{\left(1+\sqrt{1+Q}\right)^2\sqrt{1+Q}}.$$

Applying Corollary 3.4.1, we may conclude that the asymptotic equalities (3.161) hold for all $m>0$ with $k=1$ and that the SPRT asymptotically minimizes all positive moments of the stopping time distribution.

Note also that, as shown by Sosulin *et al.* [437], the SPRT is asymptotically optimal as $Q \to 0$ for arbitrary and not necessarily small error probabilities α_i.

3.4.5 Testing for a Nonhomogeneous Poisson Process

Let, under the hypotheses H_i, the observed process $\{X_t\}_{t\ge 0}$ be a nonstationary Poisson random process with intensity $\gamma_i(t)$ $(i=0,1)$, where $\gamma_0(t)$ and $\gamma_1(t)$ are continuous functions. Since the compensators are $A_i(t) = \int_0^t \gamma_i(u)\,du$ $(i=0,1)$, by (2.39), the LLR process can be represented as

$$\lambda_t = \int_0^t \log\left(\frac{\gamma_1(u)}{\gamma_0(u)}\right) dX_u - \Upsilon_t, \quad t \ge 0,$$

where $\Upsilon_t = \int_0^t [\gamma_1(u) - \gamma_0(u)]\,du$. Therefore, we have

$$\mathsf{E}_1\lambda_t = \int_0^t \gamma_1(u)\log\left(\frac{\gamma_1(u)}{\gamma_0(u)}\right) du - \Upsilon_t, \quad \mathsf{E}_0\lambda_t = \int_0^t \gamma_0(u)\log\left(\frac{\gamma_1(u)}{\gamma_0(u)}\right) du - \Upsilon_t,$$

$$\mathsf{var}_1[\lambda_t] = \int_0^t \gamma_1(u)\left(\log\frac{\gamma_1(u)}{\gamma_0(u)}\right)^2 du, \quad \mathsf{var}_0[\lambda_t] = \int_0^t \gamma_0(u)\left(\log\frac{\gamma_1(u)}{\gamma_0(u)}\right)^2 du.$$

Note that λ_t is a process with independent but nonstationary increments $\Delta\lambda_t = \lambda_t - \lim_{s\uparrow t} \lambda_t$.

Assume that $\gamma_i(t)$ is a power function, $\gamma_i(t) = Q_i t^{k-1}$, where k and Q_i are positive numbers. Then

$$\mathsf{E}_1\lambda_t = \frac{1}{k}\left[Q_1 \log \frac{Q_1}{Q_0} - (Q_1 - Q_0)\right] t^k, \quad \mathsf{E}_0\lambda_t = \frac{1}{k}\left[Q_0 \log \frac{Q_1}{Q_0} - (Q_1 - Q_0)\right] t^k,$$

$$\mathsf{var}_1[\lambda_t] = \frac{1}{k}\left[Q_1 \left(\log \frac{Q_1}{Q_0}\right)^2\right] t^k, \quad \mathsf{var}_0[\lambda_t] = \frac{1}{k}\left[Q_0 \left(\log \frac{Q_1}{Q_0}\right)^2\right] t^k,$$

and the increments $\Delta\lambda_t$ are bounded as $|\Delta\lambda_t| \le |\log(Q_1/Q_0)|$. Thus,

$$\frac{1}{t^k}\lambda_t \xrightarrow[t\to\infty]{\mathsf{P}_1-r\text{-quickly}} I_1, \quad \frac{1}{t^k}\lambda_t \xrightarrow[t\to\infty]{\mathsf{P}_0-r\text{-quickly}} -I_0,$$

where

$$I_1 = \frac{1}{k}\left(Q_1 \log \frac{Q_1}{Q_0} - Q_1 + Q_0\right), \quad I_0 = \frac{1}{k}\left(Q_1 - Q_0 - Q_0 \log \frac{Q_1}{Q_0}\right). \tag{3.192}$$

By Corollary 3.4.1, the SPRT is asymptotically optimal relative to any positive moment of the stopping time distribution.

3.4.6 Testing the Mean of AR Processes

First, assume that the observed sequence follows the AR(1) model

$$X_n = \mu + \eta_n, \quad n \ge 1, \tag{3.193}$$

where the mean is $\mu = \mu_i$ under the hypothesis H_i, $i = 0, 1$, and $\{\eta_n\}_{n\ge 0}$ is a zero-mean Markov sequence satisfying the recursion

$$\eta_{n+1} = \rho \eta_n + \xi_{n+1}, \quad n \ge 0, \quad \eta_0 = 0.$$

Here $|\rho| < 1$ and $\{\xi_n\}_{n\ge 1}$ is a sequence of zero-mean iid random variables with density $f(\xi)$. Since $X_n - \rho X_{n-1} = \mu_i(1-\rho) + \xi_n$ for $n \ge 2$ and $X_1 = \mu_i + \xi_1$ under H_i, $i = 0, 1$, the LLR is $\lambda_n = \sum_{k=1}^n Z_k$, where

$$Z_1 = \log \frac{f(X_1 - \mu_1)}{f(X_1 - \mu_0)}, \quad Z_k = \log \frac{f(X_k - \rho X_{k-1} - \mu_1(1-\rho))}{f(X_k - \rho X_{k-1} - \mu_0(1-\rho))}, \quad k \ge 2.$$

Obviously, the increments of the LLR Z_k are iid for $k = 2, 3, \dots$ and independent of Z_1. Therefore, if $\mathsf{E}_i|Z_k|^{r+1} < \infty$, then $n^{-1}\lambda_n \to I_i$ r-quickly under P_i, where $I_i = \mathsf{E}_i Z_k$, and by Corollary 3.4.1 the SPRT asymptotically minimizes the moments of the stopping time up to order r. Note that one of the interesting practical applications of the model (3.193) is the detection of a signal $\mu = \mu_1$ in the presence of a correlated Markov noise η_n ($\mu_0 = 0$), in which case it is often assumed that $\{\xi_n\}$ are iid Gaussian, $\xi_n \sim \mathcal{N}(0, \sigma^2)$. Then

$$Z_1 = \mu_1 X_1 - \mu_1^2/2\sigma^2, \quad Z_k = \mu_1(1-\rho)(X_k - \rho X_{k-1}) - \mu_1^2(1-\rho)^2/2\sigma^2, \quad k \ge 2$$

and $I_1 = I_0 = \mu_1^2(1-\rho^2)/2\sigma^2$, and the SPRT minimizes all positive moments of the stopping time distribution.

Consider now a different AR(1) scheme given by the recursion

$$X_n = \mu(1-\rho) + \rho X_{n-1} + \xi_n, \quad n \ge 1, \quad X_0 = 0. \tag{3.194}$$

This scheme can be useful in control theory to describe a linear dynamic system. To explain the first term on the right-hand side of (3.194), let us recall that the expectation of X_n is given by

$\mathsf{E}(X_n) = \mu(1-\rho)[1+\rho+\cdots+\rho^{n-1}]$. Because $|\rho| < 1$, we get the following asymptotic expectation $\lim_{n\to\infty} \mathsf{E}(X_n) = \lim_{n\mapsto\infty} \mu(1-\rho)[1+\rho+\cdots+\rho^{n-1}] = \mu$. The goal is to test two hypotheses concerning the asymptotic expectation $H_0 : \mu = \mu_0$ and $H_1 : \mu = \mu_1$.

In this case, the LLR λ_n is a random walk for all $n \geq 1$ since the increments

$$Z_k = \log \frac{f(X_k - \rho X_{k-1} - \mu_1(1-\rho))}{f(X_k - \rho X_{k-1} - \mu_0(1-\rho))}$$

are iid for all $k = 1, 2, \ldots$, so the SPRT is strictly optimal for any error probabilities with respect to the ESS and also minimizes all positive moments of the sample size asymptotically as long as $\mathsf{E}_i|Z_1| < \infty$.

Let us now consider the Gaussian case $\xi_n \sim \mathcal{N}(0, \sigma^2)$ and analyze the impact of the autoregressive coefficient ρ on the ESS. Expressing as functions of ρ the K–L information

$$I_i = \frac{(\mu_1 - \mu_0)^2}{2\sigma_{\mathrm{x}}^2} \frac{1-\rho}{1+\rho}, \quad i = 0, 1 \tag{3.195}$$

and the asymptotic variance $\sigma_{\mathrm{x}}^2 = \lim_{n\to\infty} \mathsf{var}(X_n) = \sigma^2/(1-\rho^2)$, and using (3.10), we obtain that for sufficiently small error probabilities an approximation for the ESS is given by

$$\mathsf{E}_1 T^* \approx \frac{2\sigma_{\mathrm{x}}^2}{(\mu_1 - \mu_0)^2} \frac{(1+\rho)}{(1-\rho)} \beta(\alpha_1^*, \alpha_0^*), \quad \mathsf{E}_0 T^* \approx \frac{2\sigma_{\mathrm{x}}^2}{(\mu_1 - \mu_0)^2} \frac{(1+\rho)}{(1-\rho)} \beta(\alpha_0^*, \alpha_1^*). \tag{3.196}$$

It follows from (3.196) that for fixed error probabilities α_0^* and α_1^* the ESS of the SPRT is a function of the autoregressive coefficient ρ. For a positive correlation $\rho > 0$ the ESS $\mathsf{E}_i T^*$ is approximately $(1+\rho)/(1-\rho)$ times greater than the ESS in the iid case $\rho = 0$. For a negative correlation $\rho < 0$, the ESS $\mathsf{E}_i T^*$ is approximately $(1+\rho)/(1-\rho)$ times less than the ESS in the iid case.

Remark 3.4.3. In some practical applications the sampling period Δ may be selected arbitrarily. Let us consider an exponentially correlated Markov Gaussian process $\{V_t\}_{t\geq 0}$, $\mathsf{E}[V_t] = \mu_i$, $i = 0, 1$, $\mathsf{E}[(V_t - \mu_i)(V_{t+u} - \mu_i)] = \sigma_{\mathrm{v}}^2 e^{-\alpha|u|}$ ($\alpha > 0$). We continue the above-mentioned example of testing two hypotheses regarding the mean of the Markov sequence obtained by sampling the process $\{V_t\}_{t\geq 0}$ with the period Δ, i.e., $X_k = V_{(k)\cdot\Delta}$, $k = 0, 1, \ldots$. The autoregressive coefficient of such an AR(1) model is $\rho(\Delta) = e^{-\alpha\Delta}$ and $\sigma_{\mathrm{x}} = \sigma_{\mathrm{v}}$. The impact of the sampling period Δ on the expected durations $\Delta \cdot \mathsf{E}_i T^*$, $i = 0, 1$, measured in units of time, is of interest. It results from (3.196) that, for the fixed values of α_0^*, α_1^*, μ_0 , μ_1, and σ_{x}^2, the expected durations

$$\Delta \cdot \mathsf{E}_1 T^* \approx \frac{2\sigma_{\mathrm{x}}^2}{(\mu_1 - \mu_0)^2} \beta(\alpha_1^*, \alpha_0^*)\, R(\Delta), \quad \Delta \cdot \mathsf{E}_0 T^* \approx \frac{2\sigma_{\mathrm{x}}^2}{(\mu_1 - \mu_0)^2} \beta(\alpha_0^*, \alpha_1^*)\, R(\Delta) \tag{3.197}$$

are functions of Δ via the following expression:

$$R(\Delta) = \Delta \frac{1+\rho(\Delta)}{1-\rho(\Delta)} = \Delta \frac{1+e^{-\alpha\Delta}}{1-e^{-\alpha\Delta}}. \tag{3.198}$$

This function is increasing. For large Δ ($\Delta \to \infty$), $R(\Delta) \approx \Delta$. Since $\lim_{\Delta\to 0} R(\Delta) = 2/\alpha$ and $\lim_{\Delta\to 0} \frac{dR(\Delta)}{d\Delta} = 0$, for small Δ we have

$$R(\Delta) \approx \frac{2}{\alpha}\left(1 + \frac{\alpha^2\Delta^2}{6}\right).$$

It is worth noting that the expected durations $\Delta \cdot \mathsf{E}_i T^*$ are almost linear functions of Δ for large Δ but, starting from a certain value of Δ, the further reduction of the sampling period has practically no effect on the expected durations of the SPRT $\Delta \cdot \mathsf{E}_i T^*$.

In some applications it may be reasonable to start the process X_n off at a random stationary point $X_0 \sim \mathcal{N}(\mu, \sigma_x^2)$ instead of $X_0 = 0$, assuming that the system worked for a long time before the measurements became available. In this case the observed AR(1) process becomes stationary:

$$X_1 \sim \mathcal{N}(\mu, \sigma_x^2), \; X_n = \rho X_{n-1} + (1-\rho)\mu + \xi_n, \; \xi_n \sim \mathcal{N}(0, \sigma^2), \; n \geq 2, \tag{3.199}$$

where $\sigma^2 = (1-\rho^2)\sigma_x^2$ and $\mathsf{E}(X_n) = \mu$. Using (2.283) and (2.287) we obtain that the log-likelihood function for the first n observations $\mathbf{X}_1^n$ is given by

$$\log \varphi(\mathbf{X}_1^n) = -\frac{n}{2}\log 2\pi\sigma^2 + \frac{1}{2}\log(1-\rho^2) - \frac{1}{2\sigma^2} S_n(\mathbf{X}_1^n; \rho, \mu), \tag{3.200}$$

$$S_n(\mathbf{X}_1^n; \rho, \mu) = (1-\rho^2)(X_1 - \mu)^2 + \sum_{k=2}^{n} (X_k - \rho X_{k-1} - (1-\rho)\mu)^2$$

and, hence, the LLR at the first point

$$Z_1 = \frac{(\mu_1 - \mu_0)}{\sigma_x^2}\left[X_1 - \frac{\mu_1 + \mu_0}{2}\right], \; Z_1 \sim \mathcal{N}\left(\frac{(\mu_1 - \mu_0)(2\mu - \mu_1 - \mu_0)}{2\sigma_x^2}, \frac{(\mu_1 - \mu_0)^2}{\sigma_x^2}\right) \tag{3.201}$$

is different from

$$Z_k = \frac{(1-\rho)(\mu_1 - \mu_0)}{\sigma^2}\left[X_k - \rho X_{k-1} - \frac{(1-\rho)(\mu_1 + \mu_0)}{2}\right], \quad k \geq 2, \tag{3.202}$$

$$Z_k \sim \mathcal{N}\left(\frac{(\mu_1 - \mu_0)(2\mu - \mu_1 - \mu_0)}{2\sigma_x^2} \cdot \frac{1-\rho}{1+\rho}, \frac{(\mu_1 - \mu_0)^2}{\sigma_x^2} \cdot \frac{1-\rho}{1+\rho}\right),$$

so that the LLR process $\{\lambda_n = \sum_{k=1}^n Z_k\}_{n \geq 1}$ is not a random walk. However, it is almost a random walk and the SPRT is nearly optimal as long as the expected sample size is not too small. The corrective terms can be easily calculated by using the standard equations for the OC and the ESS of the SPRT. Indeed for the OC $\beta(\mu) = \mathsf{P}_\mu(d^* = 0)$ and the ESS, $\mathsf{ESS}(\mu) = \mathsf{E}_\mu[T^*]$, we have

$$\beta_{\text{AR}}(\mu) = F_{\mu,1}(-a_0) + \int_{-a_0}^{a_1} \beta_{\mu,k}(x) f_{\mu,1}(x) \mathrm{d}x, \tag{3.203}$$

$$\mathsf{ESS}_{\text{AR}}(\mu) = 1 - [F_{\mu,1}(a_1) - F_{\mu,1}(-a_0)] + \int_{-a_0}^{a_1} [\mathsf{ESS}_{\mu,k}(x) + 1] f_{\mu,1}(x) \mathrm{d}x, \tag{3.204}$$

where $\beta_{\text{AR}}(\mu)$ (resp. $\mathsf{ESS}_{\text{AR}}(\mu)$) denotes the OC (resp. the ESS) calculated by taking into account that the first term Z_1 in (3.201) is different from Z_k in (3.202), $F_{\mu,1}(x)$ is the cdf of Z_1, $\beta_{\mu,k}(x)$ (resp. $\mathsf{ESS}_{\mu,k}(x)$) is the OC (resp. the ESS) of the SPRT started from $Z_0 = x$ with the LLR increments corresponding to Z_k, $k \geq 2$ without taking into account that the first term Z_1 is different.

As an illustration consider the example of Gaussian observations $X_1, X_2, \ldots$ generated by a stationary autoregressive model with $\rho = 0.9$ or $\rho = -0.9$. The two hypotheses are $H_0 : \mu = \mu_0 = 0$ and $H_1 : \mu = \mu_1 = 2$, $\sigma_x^2 = 1$. Figure 3.3 displays the OC and ESS functions as functions of the expectation μ. The plots of $\beta_{\text{AR}}(\mu)$ and $\mathsf{ESS}_{\text{AR}}(\mu)$ are compared with the curves $\beta_{\mu,k}(0)$ and $\mathsf{ESS}_{\mu,k}(0)$ defined in (3.203) and (3.204). The thresholds are chosen relatively small: $a_0 = a_1 = 2$. The plots in Figure 3.3 show that even for autoregressive coefficients close to the stationarity bounds, i.e., to 1 or -1, the impact of the first term Z_1 on the OC is small. The impact of the first term on the ESS is more visible but for larger values of the thresholds a_0, a_1 it is also small. This is illustrated in Figure 3.4. Now the two hypotheses are $H_0 : \mu = \mu_0 = 0$ and $H_1 : \mu = \mu_1 = 1$, $\sigma_x^2 = 1$. The thresholds are chosen larger than previously: $a_0 = 6$ and $a_1 = 10$. It follows from Figure 3.4 that the impact of the first term Z_1 on the OC and ESS is negligible. The curve $\beta_{\mu,k}(0)$ is invisible because it lies almost exactly on the curve $\beta_{\text{AR}}(\mu)$. This result can be easily generalized to the case of autoregressive models of order p [47, 316].

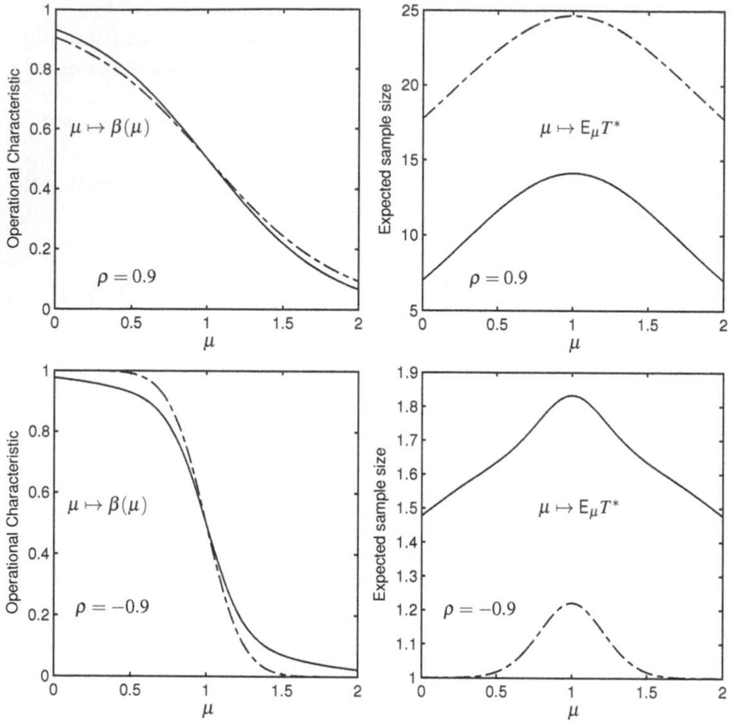

Figure 3.3: *Comparison of the operating characteristic* $\beta_{AR}(\mu)$ *and the expected sample size* $\mathsf{ESS}_{AR}(\mu)$ *when taking into account that the first term* Z_1 *is different from* Z_k, $k \geq 2$, *with* $\beta_{\mu,k}(0)$ *and* $\mathsf{ESS}_{\mu,k}(0)$ *without taking into account this fact for the thresholds* $a_0 = a_1 = 2$. $\beta_{AR}(\mu)$ *and* $\mathsf{ESS}_{AR}(\mu)$ *(solid),* $\beta_{\mu,k}(0)$ *and* $\mathsf{ESS}_{\mu,k}(0)$ *(dash-dotted),* $\rho = 0.9$ *(top),* $\rho = -0.9$ *(bottom).*

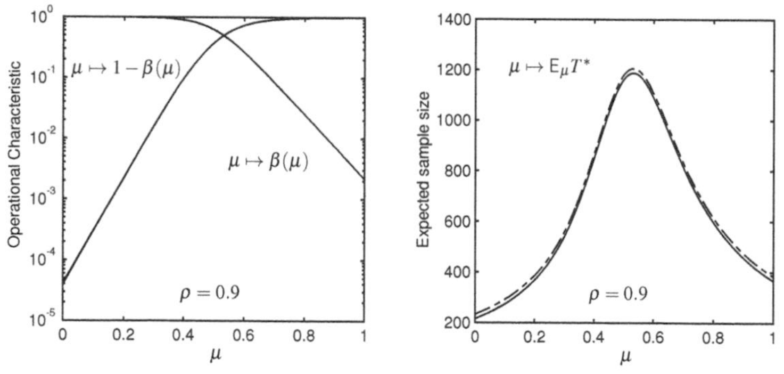

Figure 3.4: *Comparison of the operating characteristic* $\beta_{AR}(\mu)$ *and the expected sample size* $\mathsf{ESS}_{AR}(\mu)$ *when taking into account that the first term* Z_1 *is different from* Z_k, $k \geq 2$, *with* $\beta_{\mu,k}(0)$ *and* $\mathsf{ESS}_{\mu,k}(0)$ *without taking into account this fact for the thresholds* $a_0 = 6$ *and* $a_1 = 10$. $\beta_{AR}(\mu)$ *and* $\mathsf{ESS}_{AR}(\mu)$ *(solid),* $\beta_{\mu,k}(0)$ *and* $\mathsf{ESS}_{\mu,k}(0)$ *(dash-dotted).*

3.4.7 Testing the Mean in Linear State-Space Models

Consider the linear state-space model where, under the hypotheses H_i, $i = 0,1$, for $n = 1,2,\ldots$, the unobserved m-dimensional Markov vector θ_n and the observed r-dimensional vector X_n are given by

$$\theta_n = \mathbf{F}\theta_{n-1} + W_{n-1} + i\,\mathbf{b}_\theta, \quad \theta_0 = 0,$$
$$X_n = \mathbf{H}\theta_n + V_n + i\,\mathbf{b}_x.$$

Here W_n and V_n are zero-mean Gaussian iid vectors having covariance matrices $\mathbf{K}_W$ and $\mathbf{K}_V$, respectively; $\mathbf{b}_\theta = (b_\theta^1, \dots, b_\theta^m)^\top$ and $\mathbf{b}_x = (b_x^1, \dots, b_x^r)^\top$ are the mean values; $\mathbf{F}$ is the state transition matrix with size $m \times m$; and $\mathbf{H}$ is the observation matrix with size $r \times m$.

It can be shown that under the null hypothesis H_0 the observed sequence $\{X_n\}_{n \ge 1}$ has an equivalent representation with respect to the innovation process $\xi_n = X_n - \mathbf{H}\hat{\theta}_n$:

$$X_n = \mathbf{H}\hat{\theta}_n + \xi_n, \quad n \ge 1,$$

where $\xi_n \sim \mathcal{N}(0, \boldsymbol{\Sigma}_n)$, $n = 1, 2, \dots$ are independent Gaussian vectors and $\hat{\theta}_n = \mathsf{E}(\theta_n \mid \mathscr{F}_{n-1})$. Note that $\hat{\theta}_n$ is the optimal one-step ahead predictor in the mean-square sense, i.e., the estimate of θ_n based on $\mathbf{X}_1^{n-1} = (X_1, \dots, X_{n-1})^\top$, which can be obtained by the Kalman filter. Under the hypothesis H_1 we have

$$X_n = \Upsilon_n + \mathbf{H}\hat{\theta}_n + \xi_n, \quad n \ge 1,$$

where Υ_n depends on n and can be computed using relations given in [47, pp. 282-283], [507]. It follows that the LLR λ_n is represented in the form

$$\lambda_n = \sum_{k=1}^{n} \Upsilon_k^\top \boldsymbol{\Sigma}_k^{-1} \xi_k - \frac{1}{2} \sum_{k=1}^{n} \Upsilon_k^\top \boldsymbol{\Sigma}_k^{-1} \Upsilon_k,$$

where $\boldsymbol{\Sigma}_k$, $k = 1, 2, \dots$ are given by Kalman's equations; see, e.g., [47, Eq. (3.2.20)]. Therefore, the original testing problem for the constant mean values in the sequence of dependent observations is equivalent to testing the hypotheses $H_0 : \Upsilon_n = 0$ *versus* $H_1 : \Upsilon_n > 0$ in the sequence of independent Gaussian innovations ξ_n with covariance matrices $\boldsymbol{\Sigma}_n$, $n \ge 1$.

As $n \to \infty$, the normalized LLR $n^{-1}\lambda_n$ converges P_i-a.s. to the positive constant

$$I = \frac{1}{2} \lim_{n \to \infty} \frac{1}{n} \sum_{k=1}^{n} \Upsilon_k^\top \boldsymbol{\Sigma}_k^{-1} \Upsilon_k.$$

Using [47], we obtain that this constant is equal to

$$I = \frac{1}{2} \left\{ \mathbf{b}_\theta^\top (\mathbb{I}_m - \mathbf{F}^*)^{-\top} \mathbf{H}^\top \boldsymbol{\Sigma}^{-1} \mathbf{H} (\mathbb{I}_m - \mathbf{F}^*)^{-1} \mathbf{b}_\theta \right\}$$

when there is only one non-zero mean $\mathbf{b}_\theta$ and

$$I = \frac{1}{2} \left\{ \mathbf{b}_x^\top \left[\mathbb{I}_r - \mathbf{H}(\mathbb{I}_m - \mathbf{F}^*)^{-1} \mathbf{F}\mathbf{K} \right]^\top \boldsymbol{\Sigma}^{-1} \left[\mathbb{I}_r - \mathbf{H}(\mathbb{I}_m - \mathbf{F}^*)^{-1} \mathbf{F}\mathbf{K} \right] \mathbf{b}_x \right\}$$

when there is only one non-zero mean $\mathbf{b}_x$. Here $\mathbf{K}$ is the Kalman filter gain in the stationary regime, $\mathbb{I}_m$ is the $(m \times m)$ identity matrix, and $\mathbf{F}^* = \mathbf{F}(\mathbb{I}_m - \mathbf{K}\mathbf{H})$.

Moreover, since the process λ_n, $n \ge 1$, is Gaussian with independent increments, $n^{-1}\lambda_n$ converges r-quickly to I for all $r > 0$. Therefore, Corollary 3.4.1 shows that the SPRT is asymptotically optimal as $\alpha_{\max} \to 0$ with respect to all positive moments of the sample size.

3.5 SPRT: Local Approach

As in Section 2.11, let us now consider the case of iid observations and *local hypotheses* where the distance between the parameter values $\|\theta_1 - \theta_0\|_2$ is small. In this case, we first continue to use the SPRT as defined in (3.1) and we use the local approach for the investigation of its statistical properties, OC and ESS. Second, we use the local approach for the design of the algorithm through the efficient score. In this case, the criteria are slightly different — now we want to have maximum *local* slope of the SPRT power function at θ^* to test $H_0 : \theta \leq \theta^*$ against $H_1 : \theta > \theta^*$ [64].

It turns out that, in the case of sequential analysis, the local approach has some specific features. From Section 2.11 we know that for investigating the power function $\beta(\theta)$ of FSS tests and for computing the thresholds it is necessary to know the distribution of the LLR. In general, finding the distributions of the LLR is a complex problem. But, in some sense, the situation in sequential analysis is a little bit better than in the FSS case. Actually, it follows from (3.7)–(3.8) that if the parameter is exactly equal to θ_0 or θ_1, then the approximated OC and ESS do *not* depend upon the distribution of the increment Z_k of the cumulative sum $\lambda_n = \sum_{k=1}^{n} Z_k$. Another problem of interest from the robustness point of view is the computation of the OC and ESS for some θ in the neighborhood of θ_0 or θ_1. In this case, the OC and the ESS can be computed using the local approach as we explain now.

3.5.1 ESS Function

Let $\mathcal{P} = \{\mathsf{P}_\theta\}_{\theta \in \Theta}$ $(\Theta \subset \mathbb{R})$ be a family of distributions with scalar parameter θ. We consider the following simple *local* hypotheses

$$H_0 : \theta = \theta_0 \text{ and } H_1 : \theta = \theta_1 \text{ where } \theta_1 = \theta_0 + \nu, \ \ \nu \to 0.$$

As pointed out in Section 2.8, under some general regularity assumptions about the pdf of the family $\mathcal{P}$, we have the following approximation for the expectations of the LLR $Z_k = \log f_{\theta_1}(X_k)/f_{\theta_0}(X_k)$:

$$\begin{array}{rcrcr} I(\theta_0,\theta_1) & = & -\mathsf{E}_{\theta_0}(Z_k) & = & 1/2\, \nu^2\, \mathcal{F}(\theta_0) + o(\nu^2) \\ I(\theta_1,\theta_0) & = & \mathsf{E}_{\theta_1}(Z_k) & = & 1/2\, \nu^2\, \mathcal{F}(\theta_1) + o(\nu^2) \\ & & \mathsf{E}_{\theta}(Z_k) & = & \nu\,(\widetilde{\nu} - 1/2\, \nu)\, \mathcal{F}(\theta) + o(\nu^2) \end{array} \tag{3.205}$$

where $\widetilde{\nu} = \theta - \theta_0$. Then it results from the approximations (3.10) that the ESS for the putative values θ_0 or θ_1 can be approximated as

$$\mathsf{E}_1 T^* \approx \frac{2\beta(\alpha_1^*,\alpha_0^*)}{\nu^2\, \mathcal{F}(\theta_1)}, \quad \mathsf{E}_0 T^* \approx \frac{2\beta(\alpha_0^*,\alpha_1^*)}{\nu^2\, \mathcal{F}(\theta_0)}.$$

In Section 2.11, we explained that the speed of convergence of the hypotheses H_0 and H_1 in the case of local FSS tests should be of the order of magnitude of $\mathsf{constant}/\sqrt{n}$. For the SPRT, the sample size is a random variable and the above approximations for the ESS provide us with the relevant sequential counterpart for the order of magnitude for the deviation ν between the two hypotheses:

$$\nu \approx \mathsf{constant}/\sqrt{\mathsf{E}_i T^*}, \quad i = 0, 1. \tag{3.206}$$

As it follows from (3.21), for an arbitrary value of θ, the approximation of the ESS is

$$\mathsf{E}_\theta T^* \approx \frac{-a_0\, \beta(\theta) + a_1(1-\beta(\theta))}{\nu(\widetilde{\nu} - 1/2\, \nu)\, \mathcal{F}(\theta)}. \tag{3.207}$$

3.5.2 OC Function

To compute the OC in the local case we assume again as in Section 2.8 that the first three derivatives of the pdf $f_\theta(X)$ with respect to θ exist in the neighborhood of θ_0 or θ_1 and that the following integrals for the LLR Z_k and for the efficient score $\partial \log f_\theta(X_k)/\partial\theta$ are finite: $\mathsf{E}_\theta(|Z_k|^3) < \infty$ and $\mathsf{E}_\theta(|\partial \log f_\theta(X_k)/\partial\theta|^3) < \infty$. Further details can be found in [34, 494]. Under these assumptions, and using Taylor's expansion exactly as we did in Section 2.8, we have

$$
\begin{aligned}
e^{-\omega_0 Z_k} \approx\; & 1-\omega_0 \nu \left.\frac{\partial \log f_\theta(x)}{\partial \theta}\right|_{\theta=\theta_0} - \frac{\omega_0 \nu^2}{2}\left[\left.\frac{\partial^2 \log f_\theta(x)}{\partial \theta^2}\right|_{\theta=\theta_0} + \left(\left.\frac{\partial \log f_\theta(x)}{\partial \theta}\right|_{\theta=\theta_0}\right)^2\right] \\
+\; & \frac{\omega_0(1+\omega_0)\nu^2}{2}\left(\left.\frac{\partial \log f_\theta(x)}{\partial \theta}\right|_{\theta=\theta_0}\right)^2 .
\end{aligned} \tag{3.208}
$$

Taking the expectation E_θ on both sides in (3.208), we get

$$
\mathsf{E}_\theta(e^{-\omega_0 Z_k}) \approx 1-\omega_0\left(\widetilde{\nu}-\nu\,\frac{\omega_0+1}{2}\right)\nu\,\mathcal{F}(\theta_0).
$$

Finally, we get the non-zero root ω_0 of the equation $\mathsf{E}_\theta(e^{-\omega_0 Z_k})=1$ for some θ in the neighborhood of θ_0 or θ_1:

$$
\omega_0(\theta) \approx \frac{2}{\theta_1-\theta_0}\left(\theta-\frac{\theta_1+\theta_0}{2}\right). \tag{3.209}
$$

The OC can then be estimated using (3.17).

Comparing (3.205) with (3.23) on one hand and (3.209) with (3.24) on the other hand, we conclude that in the case of local hypotheses the ESS and OC can be computed exactly as in the Gaussian independent case, replacing μ_z with the values given in (3.205) and ω_0 with (3.209). In other words, in the case of sequential testing the local approach follows exactly the same lines as the non-sequential FSS hypothesis test.

3.5.3 *Locally Most Powerful Sequential Test*

It is shown in Section 2.11 that optimal FSS tests for local one-sided hypotheses are based upon the *efficient score*

$$
s_k = \frac{\partial \log f_\theta(X_k)}{\partial \theta}. \tag{3.210}
$$

Berk [64] has established one particular optimal property for a local sequential test. Consider the following composite hypotheses:

$$
H_0 : \theta \le \theta^* \text{ against } H_1 : \theta > \theta^* \tag{3.211}
$$

where θ is the scalar parameter of a family $\mathcal{P}=\{\mathsf{P}_\theta\}_{\theta\in\Theta}$. Let (d,T) be a sequential test, and denote by $\dot{\beta}(\theta^*) = d\beta(\theta)/d\theta|_{\theta=\theta^*}$ the derivative of the power function. The *local characteristic* of the sequential test (d,T) is the triplet $(\alpha_0(d), \mathsf{E}_{\theta^*}T, \dot{\beta}(\theta^*))$, with very natural and obvious interpretation. In the class $\mathbf{C}(\alpha_0)$ of sequential tests with fixed size α_0, we want to find the test (d,T) with maximum *power* $\beta(\theta)$ for the local alternative $\theta > \theta^*$. In other words, we want to have maximum *local* slope $\dot{\beta}(\theta^*)$ of the power function $\beta(\theta)$ at θ^*.

Under some regularity conditions, the SPRT test (3.1) based on the cumulative sum

$$
\lambda_n = \sum_{k=1}^{n} \left.\frac{\partial \log f_\theta(X_k)}{\partial \theta}\right|_{\theta=\theta^*} \tag{3.212}
$$

is *locally most powerful* (LMP) [64]. This means that, for any other sequential test $(\widetilde{d},\widetilde{T})$ of H_0 against H_1 with local triplet $(\alpha_0(\widetilde{d}), \mathsf{E}_{\theta^*}\widetilde{T}, \dot{\widetilde{\beta}}(\theta^*))$ and $\mathsf{E}_{\theta^*}\widetilde{T} \le \mathsf{E}_{\theta^*}T$, the following inequality holds

$$
\dot{\widetilde{\beta}}(\theta^*) \le \dot{\beta}(\theta^*). \tag{3.213}
$$

Thus the sequential test (3.212) is LMP among all sequential tests with size $\widetilde{\alpha}_0 \le \alpha_0$ and satisfying $\mathsf{E}_{\theta^*}\widetilde{T} \le \mathsf{E}_{\theta^*}T$.

Under some conditions, this local sequential test coincides with the usual SPRT. Assume that the family $\mathcal{P}$ is exponential with density

$$f_\theta(y) = h(y)\, e^{\theta^\top y - b(\theta)} \tag{3.214}$$

where $b(\theta)$ is infinitely differentiable and strictly convex. The LLR in this case is

$$Z_k = \log \frac{f_{\theta_1}(X_k)}{f_{\theta_0}(X_k)} = (\theta_1 - \theta_0)^\top X_k + b(\theta_0) - b(\theta_1), \tag{3.215}$$

and the efficient score is

$$s_k^* = \left.\frac{\partial \log f_\theta(X_k)}{\partial \theta}\right|_{\theta=\theta^*} = X_k - \dot{b}(\theta^*). \tag{3.216}$$

Denote by (d', T') the SPRT (3.1) for testing the simple hypotheses $H_0' : \theta = \theta_0$ and $H_1' : \theta = \theta_1$, where $\theta_0 < \theta^* < \theta_1$. Berk [64] has shown the existence of a continuum of pairs (θ_0, θ_1) of points in the neighborhood of θ^* for which the following equation holds:

$$\dot{b}(\theta^*) = \frac{b(\theta_1) - b(\theta_0)}{\theta_1 - \theta_0}. \tag{3.217}$$

Then, the SPRT for the simple hypotheses H_0' and H_1' is *simultaneously LMP sequential test* for the local hypotheses H_0 and H_1 in (3.211). Note here that the point θ^* is a particular point for the SPRT for the hypotheses H_0' and H_1', because it satisfies $\mathsf{E}_{\theta^*}(Z_k) = 0$.

3.6 Nuisance Parameters and an Invariant SPRT

So far we considered the case where the measures P_0 and P_1 corresponding to the hypotheses H_0 and H_1 are completely known. However, a more practical situation is when these measures are unknown, partially (parametric uncertainty) or completely (nonparametric uncertainty). In many cases, prior uncertainty can be overcome with the principle of invariance with respect to nuisance parameters. In this section, we first explain how the results of Section 3.4 can be used for sequential testing of composite hypotheses in the presence of nuisance parameters, and then we present several particular problems with parametric and nonparametric uncertainty that illustrate the procedure.

Let $\{X_t\}$, $t \in \mathbb{R}_+$ or $t = n \in \mathbb{Z}_+$, be a stochastic process defined on the probability space $(\Omega, \mathscr{F}, \mathsf{P})$, where the measure P is assumed to belong to a parametric or nonparametric family $\mathscr{P}$, and the hypotheses to be tested are $H_i : \mathsf{P} \in \mathscr{P}_i$, where $\mathscr{P}_i \subset \mathscr{P}$, $i = 0, 1$. Next, suppose that the family $\mathscr{P}$ is invariant under a group of measurable transformations $\mathcal{G}$ on the sample space $\mathcal{X}$. The invariance property implies that the distribution of any invariant statistic depends on P only through its orbits. Therefore, if there exists a group $\mathcal{G}$ that leaves the problem invariant, and we are interested only in invariant sequential tests $\delta \in \mathscr{I}$, for which $\delta(G(X)) = \delta(X)$ for all $X \in \mathcal{X}$, $G \in \mathcal{G}$, then the hypotheses become simple. For more details, see Ferguson [143], Ghosh [160], Lehmann [261].

Let $\mathbb{M}(X)$ be a maximal invariant statistic with respect to the group $\mathcal{G}$, i.e., $\mathbb{M}$ is a function on $\mathscr{X}$ such that

(a) $\mathbb{M}(G(X)) = \mathbb{M}(X)$ for all $X \in \mathcal{X}$ and $G \in \mathcal{G}$;

(b) $\mathbb{M}(X) = \mathbb{M}(Y)$ implies $X = G(Y)$ for some $G \in \mathcal{G}$.

Instead of restricting attention to the class of invariant tests $\mathscr{I}$, by Theorem 5.6.1 in [143], we can restrict attention to the class of all tests that are functions of the maximal invariant $\mathbb{M}$: these two classes are equivalent. Thus, we can use all the results obtained in Section 3.4 by simply replacing the observed data X with a maximal invariant statistic $\mathbb{M}(X)$. In other words, we can again consider the SPRT with the LLR λ_t now defined by $\lambda_t = \log[d\mathsf{Q}_1^t/d\mathsf{Q}_0^t](\mathbb{M}_t)$, where Q_i is the measure corresponding to the maximal invariant under H_i and Q_i^t is the restriction of this measure to the sigma-algebra $\mathscr{F}_t^M = \sigma(\mathbb{M}_u, u \le t)$ generated by the maximal invariant. This SPRT is referred to as the *invariant SPRT* (ISPRT).

It is worth mentioning that the invariant LR $\Lambda_t = dQ_1^t/dQ_0^t](\mathbb{M}_t)$ can be also obtained by averaging the LR Λ_t^η, where η is a nuisance parameter, over the Haar measure [522]. For example, when $\eta \in \mathbb{R}^1$ is an unknown shift, the integration is performed over the Lebesgue measure. Therefore, the invariant LR can be viewed as a weighted LR with a specifically chosen weight function. In this sense the invariant approach has some similarity with the method of mixtures of LRs or weighted LRs proposed by Wald [494] in the context of testing composite hypotheses, not necessarily for models with nuisance parameters. See Section 5.4.3.

Theorem 3.4.2 and Corollary 3.4.1 remain true for the ISPRT, i.e., under the corresponding r-quick convergence conditions for the invariant LLR, as $\alpha_{\max} \to 0$ the ISPRT is asymptotically optimal in the class $\mathbf{C}(\alpha_0, \alpha_1)$ of sequential and non-sequential invariant tests for which $\delta \in \mathscr{I}$ and $\Pr(\text{rejecting } H_i \mid H_i \text{ true}) \le \alpha_i$, $i = 0, 1$.

We now consider several challenging special cases and show that the ISPRT is asymptotically optimal for these problems. Other classical ISPRTs, such as the sequential F-test and the sequential T^2-test, may be treated in a similar way [248]. Further theoretical study and interesting examples can be found in [160, 177, 209]. Several interesting multihypotheses generalizations will be considered in Section 4.5.

3.6.1 Testing the Variance of a Normal Population with Unknown Mean

Let $X_n \sim \mathcal{N}(\mu, \sigma^2)$, $n = 1, 2, \dots$ be iid normal variables with unknown mean μ and variance σ^2. The hypotheses to be tested are $H_0 : \sigma^2 = \sigma_0^2$ and $H_1 : \sigma^2 = \sigma_1^2$, where σ_i, $i = 0, 1$ are pre-specified distinct numbers, and the mean μ is thus a nuisance parameter. The problem is invariant under the group of shifts $\mathcal{G} : G_b(x) = x + b$, $-\infty < b < \infty$, and the maximal invariant is $\mathbb{M}_n = (Y_1, \dots, Y_n)$, where $Y_k = X_k - X_1$, $Y_1 = 0$. It can be easily established that under H_i the pdf of $\mathbb{M}_n$ is

$$\begin{aligned} p(\mathbb{M}_n|H_i) &= \frac{1}{(2\pi\sigma_i^2)^{n/2}} \int_{-\infty}^{\infty} \exp\left\{ -\frac{1}{2\sigma_i^2} \sum_{k=1}^n (Y_k + u)^2 \right\} du \\ &= \frac{n^{-1/2}}{(2\pi\sigma_i^2)^{(n-1)/2}} \exp\left\{ -\frac{1}{2\sigma_i^2} \sum_{k=1}^n (Y_k - \overline{Y}_n)^2 \right\}, \end{aligned} \tag{3.218}$$

where $\overline{Y}_n = n^{-1} \sum_{k=1}^n Y_k$.

Write $q = \sigma_1/\sigma_0$, $\overline{X}_n = n^{-1} \sum_{k=1}^n X_k$, and $s_n^2 = (n-1)^{-1} \sum_{k=1}^n (X_k - \overline{X}_n)^2$. Using (3.218) and noting that $\sum_{k=1}^n (Y_k - \overline{Y}_n)^2 = (n-1)s_n^2$, we obtain that the invariant LLR is

$$\lambda_n = \log \frac{p(\mathbb{M}_n|H_1)}{p(\mathbb{M}_n|H_0)} = (n-1)\left(\frac{q^2 - 1}{2\sigma_1^2} s_n^2 - \log q \right), \quad n \ge 2 \ (\lambda_1 = 0).$$

Since $\mathsf{E}_i |X_1|^r < \infty$ for all $r > 0$, it follows from Theorem 2.4.4 that $n^{-1} \sum_{k=1}^n X_k^2 \to \mu^2 + \sigma_i^2$, $\overline{X}_n \to \mu$ and $\overline{X}_n^2 \to \mu^2$ r-quickly for all $r > 0$ under P_i, so that $s_n^2 \to \sigma_i^2$ r-quickly under P_i and

$$n^{-1}\lambda_n \xrightarrow[n\to\infty]{\mathsf{P}_1-r-\text{quickly}} \frac{q^2}{2} - \frac{1}{2} - \log q, \quad n^{-1}\lambda_n \xrightarrow[n\to\infty]{\mathsf{P}_0-r-\text{quickly}} \frac{1}{2} - \frac{1}{2q^2} - \log q.$$

Thus, the r-quick convergence conditions (3.152) hold for all $r > 0$ with $\psi(n) = n$ and

$$I_1(q) = \frac{q^2}{2} - \frac{1}{2} - \log q, \quad -I_0(q) = \frac{1}{2} - \frac{1}{2q^2} - \log q.$$

It is easy to check that $I_1(q) > 0$ and $-I_0(q) < 0$ for all $q \ne 1$ $(q > 0)$. By Theorem 3.4.2(iii), the ISPRT asymptotically minimizes all positive moments of the stopping time distribution.

Note that the exponential boundedness of the distribution of the stopping time in this problem has been established by Wijsman [504].

3.6.2 *Testing a Normal Population with Unknown Mean and Variance (t-SPRT)*

Suppose $X_n \sim \mathcal{N}(\mu, \sigma^2)$, $n = 1, 2, \dots$ are iid normal random variables with unknown mean μ and variance σ^2, and the hypotheses are $H_i : \mu/\sigma = q_i$, $i = 0, 1$, where q_0 and q_1 are given numbers. This problem is invariant under the group of scale changes $\mathcal{G} : G_b(x) = bx$, $b > 0$, and the maximal invariant is $\mathbb{M}_n = (Y_1, \dots, Y_n)$, where $Y_k = X_k/X_1$. It can be shown that under H_i the pdf of $\mathbb{M}_n$ is

$$\begin{aligned} p(\mathbb{M}_n|H_i) &= \frac{1}{\sqrt{2\pi(n-1)ns_n^{2(n-1)}}} \int_0^\infty u^{n-1} \exp\left\{-\frac{n}{2}u^2 + q_i \frac{\sum_{k=1}^n Y_k}{s_n} u\right\} du \\ &= \frac{1}{\sqrt{2\pi(n-1)ns_n^{2(n-1)}}} \int_0^\infty u^{-1} \exp\left\{n\left(-\frac{1}{2}u^2 + q_i t_n u + \log u\right)\right\} du \end{aligned} \tag{3.219}$$

where $s_n^2 = n^{-1} \sum_{k=1}^n Y_k^2$ and

$$t_n = \frac{\overline{Y}_n}{s_n} = \frac{n^{-1}\sum_{k=1}^n X_k}{\{n^{-1}\sum_{k=1}^n X_k^2\}^{1/2}}. \tag{3.220}$$

Define

$$f(u,z) = -\tfrac{1}{2}u^2 + zu + \log u, \quad J_n(z) = \int_0^\infty u^{-1} \exp\{nf(u,z)\}\, du. \tag{3.221}$$

By (3.219), the LLR for the maximal invariant is of the form $\lambda_n = \log[J_n(q_1 t_n)/J_n(q_0 t_n)]$. Note that the statistic (3.220) is the famous Student t-statistic which is the basis for Student's t-test in the fixed sample size setting. For this reason, the ISPRT based on $\lambda_n(t_n)$ is referred to as the t-SPRT or the sequential t-test.

The LLR λ_n is too complicated for direct use. In fact, it is rather difficult, if not impossible, to calculate λ_n explicitly, although it can be done locally for small differences $|q_1 - q_0|$, i.e., for close hypotheses. However, for such general questions as convergence, it is possible to replace λ_n with a suitable approximation, say λ_n^*. If $\mathsf{P}_i(|\lambda_n - \lambda_n^*| < C) = 1$ for some $n \ge n_0$ ($n_0 \ge 1$) with C a constant, then the r-quick convergence of $n^{-1}\lambda_n^*$ to I_1 under P_1 and $n^{-1}\lambda_n^*$ to $-I_0$ under P_0 imply the corresponding convergence $n^{-1}\lambda_n \to I_1$ under P_1 and $n^{-1}\lambda_n \to -I_0$ under P_0.

We now show that

$$\left|n^{-1}\lambda_n - g(t_n)\right| \le C/n, \quad n \ge 1, \tag{3.222}$$

where C is a finite positive constant and

$$\begin{aligned} g(t_n) &= \phi(q_1 t_n) - \phi(q_0 t_n) - \tfrac{1}{2}\left(q_1^2 - q_0^2\right), \\ \phi(t_n) &= \tfrac{1}{4}\, t_n \left(t_n + \sqrt{4 + t_n^2}\right) + \log\left(t_n + \sqrt{4 + t_n^2}\right). \end{aligned} \tag{3.223}$$

The proof is based, in essence, on the application of the Laplace asymptotic integration method; see, e.g., Copson [103], and its uniform version due to Wijsman [503]. Specifically, using the conventional Laplace asymptotic method, we obtain

$$J_n(q_i t_n) \sim \sqrt{\frac{2\pi u_m^{-1}}{-nf''(u_m, q_i t_n)}} \exp\{n[f(u_m, q_i t_n) - q_i/2]\}, \quad n \to \infty, \tag{3.224}$$

where $u_m = u_m(q_i t_n)$ maximizes $f(u, q_i t_n)$: $f(u_m, q_i t_n) = \max_{u \ge 0} f(u, q_i t_n)$. This formula is valid for every fixed t_n. However, we are interested in the behavior of $J_n(q_i t_n)$ for a random t_n and hence, we need the convergence in (3.224) to be uniform for $|t_n| \le 1$. Thus, we need the extension of the conventional Laplace method to the uniform version that gives [503, Theorem 4.1]. It is easily checked that all the conditions of this theorem hold in our case, so that (3.224) holds uniformly in $|t_n| \le 1$. This means that uniformly in $|q_i t_n| \le |q_i|$

$$\log J_n(q_i t_n) = n f(u_m(q_i t_n), q_i t_n) - \tfrac{1}{2}\log n + C(q_i t_n) + o(1) \quad \text{as } n \to \infty, \tag{3.225}$$

where

$$C(q_i t_n) = \frac{1}{2}\log\left(\frac{2\pi}{-u_m(q_i t_n) f''(u_m, q_i t_n)}\right).$$

It is easily verified that the equation $\dot{f}(u, q_i t_n) = -u + q_i t_n + 1/u = 0$ has one positive root $u_m = u_m(q_i t_n) = 1/2\ [q_i t_n + (4 + q_i^2 t_n^2)^{1/2}]$ that maximizes the corresponding function. Also, $-\ddot{f}(u_m, q_i t_n) = 1 + u_m^{-2}$, so that

$$C(q_i t_n) = \frac{1}{2}\log\left(\frac{2\pi}{u_m(q_i t_n)[1 + 1/u_m^2(q_i t_n)]}\right).$$

Using (3.225) and noting that $f(u_m(q_i t_n), q_i t_n) = \max_{u \ge 0} f(u, q_i t_n) = \phi(q_i t_n) - 1/2 - \log 2$, where $\phi(z)$ is defined in (3.223), we obtain that

$$\log\frac{J_n(q_1 t_n)}{J_n(q_0 t_n)} = n\left[\phi(q_1 t_n) - \phi(q_0 t_n) - (q_1^2 - q_0^2)/2\right] + \Delta(t_n) + o(1), \tag{3.226}$$

where $o(1) \to 0$ as $n \to \infty$ uniformly in $|t_n| \le 1$ and

$$\Delta(t_n) = \frac{1}{2}\log\left[\frac{\left(q_1 t_n + \sqrt{4 + (q_1 t_n)^2}\right)\left[1 + 1/4\left(q_0 t_n + \sqrt{4 + (q_0 t_n)^2}\right)\right]}{\left(q_0 t_n + \sqrt{4 + (q_0 t_n)^2}\right)\left[1 + 1/4\left(q_1 t_n + \sqrt{4 + (q_1 t_n)^2}\right)\right]}\right]. \tag{3.227}$$

By (3.226), $|\lambda_n - n g(t_n) - \Delta(t_n)| = o(1)$ as $n \to \infty$ uniformly in t_n. Since $|t_n| \le 1$, the term $\Delta(t_n)$ is bounded by a constant. Hence, there is a finite positive constant C for which the approximation (3.222) holds with the function g defined by (3.223).

Thus, if under P_i the statistic t_n converges r-quickly to some constant Q_i, the normalized LLR $n^{-1}\lambda_n$ converge to $g(Q_i)$, and to obtain the final result it suffices to investigate the limiting behavior of t_n. Since $\mathsf{E}_i|X_1|^r < \infty$ for all $r > 0$, it follows that

$$t_n \xrightarrow[n\to\infty]{\mathsf{P}_i - r-\text{quickly}} \frac{\mathsf{E}_i X_1}{\sqrt{\mathsf{E}_i X_1^2}} = \frac{q_i}{\sqrt{1 + q_i^2}} \quad \text{for } i = 0, 1 \text{ and for all } r > 0.$$

Therefore, the r-quick convergence conditions (3.152) hold for all $r > 0$ with $\psi(n) = n$ and

$$I_1 = g(Q_1), \quad -I_0 = g(Q_0), \quad Q_i = q_i/(1 + q_i^2)^{1/2}. \tag{3.228}$$

It remains to verify that $I_1 > 0$ and $-I_0 < 0$. To this end, note that for any fixed $|t| \le 1$, the maximum of the function $\tilde{\phi}(q, t) = \phi(qt) - q^2/2$ over q is attained at $q^* = t/(1 - t^2)^{1/2}$, so that $q^* = q_1$ if $t = Q_1 = q_1/(1 + q_1)^{1/2}$ and $q^* = q_0$ if $t = Q_0 = q_0/(1 + q_0)^{1/2}$. Hence,

$$g(Q_1) = \tilde{\phi}(q_1, Q_1) - \tilde{\phi}(q_0, Q_1) > 0, \quad g(Q_0) = \tilde{\phi}(q_1, Q_0) - \tilde{\phi}(q_0, Q_0) < 0.$$

By Theorem 3.4.2(iii), the ISPRT asymptotically minimizes all positive moments of the stopping time and, as $\alpha_{\max} \to 0$,

$$\begin{aligned} \inf_{\delta \in \mathbf{C}(\alpha_0, \alpha_1)} \mathsf{E}_1 T^m &\sim \mathsf{E}_1[T^*]^m \sim \left(\frac{|\log \alpha_0|}{g(Q_1)}\right)^m, \\ \inf_{\delta \in \mathbf{C}(\alpha_0, \alpha_1)} \mathsf{E}_0 T^m &\sim \mathsf{E}_0[T^*]^m \sim \left(\frac{|\log \alpha_1|}{-g(Q_0)}\right)^m, \end{aligned} \tag{3.229}$$

where the Q_i, $i = 0, 1$ are defined in (3.228).

3.6.3 Rank-Order Nonparametric ISPRT for Lehmann's Alternatives

In this subsection, we apply Corollary 3.4.1 to a two-sample nonparametric problem with Lehmann hypotheses (or proportional hazards, if we replace F with $1-F$ below) and show that Savage's sequential test is asymptotically optimal with respect to any positive moment of the stopping time distribution.

Let $\mathbf{X}_n = (X_{1,n}, X_{2,n})$, where $\{X_{1,n}\}_{n\geq 1}$ are iid with a continuous cdf F_0, and independent of $\{X_{2,n}\}_{n\geq 1}$, which are iid with a cdf F_1. The hypotheses are $H_0 : F_0 = F_1$ and $H_1 : F_1 = (F_0)^A$, where A is a specified positive constant, $A \neq 1$, and F_0 is completely unknown.

An interesting application of this problem is in target detection in the presence of clutter with unknown continuous distribution F_0, based on the unclassified data $X_{n,2}$ when a pre-classified training sequence $X_{n,1}$ is also available. That is, it is known in advance that clutter only generates the data $X_{n,1}$, but the data $X_{n,2}$ may or may not contain a target along with noise and clutter.

Let g be an arbitrary continuous increasing function. A maximal invariant with respect to the group $\mathbf{X}_n \to (g(X_{1,n}), g(X_{2,n}))$ is the vector of ranks of $\mathbf{X}_{1,n}$ among $(\mathbf{X}_{1,n}, \mathbf{X}_{2,n})$, where $\mathbf{X}_{i,n} = (X_{i,1}, \dots, X_{i,n})$. Let $\hat{F}_{n,i}(x) = n^{-1}\sum_{k=1}^n \mathbb{1}_{\{X_{i,k}\leq x\}}$, $i = 1,2$, be the empirical distribution functions. As shown in Savage and Sethuraman [405], the LLR of the maximal invariant is

$$\lambda_n = \log\left[\frac{A^n(2n)!}{n^{2n}}\right] - \sum_{k=1}^{n} \log\left([\hat{F}_{n,1}(X_{1,k}) + A\hat{F}_{n,2}(X_{1,k})][\hat{F}_{n,1}(X_{2,k}) + A\hat{F}_{n,2}(X_{2,k})]\right).$$

Define

$$q(A, F_0, F_1) = \log(4A) - 2 - \int \log[F_0(x) + AF_1(x)]\,(\mathrm{d}F_0(x) + \mathrm{d}F_1(x)).$$

It follows from [405] that $q_0(A) = q(A, F_0, F_0) < 0$ and $q_1(A) = q(A, F_0, F_0^A) > 0$. Also, by [405, Lemma 2], for any $\varepsilon > 0$ there exists a number $\rho < 1$ such that

$$\mathsf{P}_i\left(\left|n^{-1}\lambda_n - q_i(A)\right| > \varepsilon\right) \leq O(\rho^n),$$

which obviously implies the r-quick convergence of $n^{-1}\lambda_n$ to $q_i(A)$ as $n \to \infty$ under P_i for all $r > 0$.

Thus, based on Corollary 3.4.1, one may conclude that Savage's rank-order ISPRT asymptotically minimizes all moments of the stopping time distribution in the sense that the asymptotic equalities (3.161) hold for all $m \geq 1$ with $k = 1$, $I_0 = -q_0(A)$, and $I_1 = q_1(A)$.

3.6.4 Linear Model with Nuisance Parameters

Suppose that the observations X_n, $n = 1,2,\dots$, are described by the Gaussian linear model with nuisance parameters (2.356)

$$X_n = HY_n + M\theta + \xi_n, \quad \xi_n \sim \mathcal{N}(0, \sigma^2\mathbb{I}_\ell), \quad n = 1,2,\dots,$$

discussed in Subsection 2.10.3.1 in the framework of FSS hypothesis testing. Here $X_n \in \mathbb{R}^\ell$ is the observation vector, $\theta \in \mathbb{R}^r$ is the informative parameter, $Y_n \in \mathbb{R}^m$ is an unknown and non-random state vector (nuisance parameter), M is a full column rank matrix with size $\ell \times r$ where $r < \ell$ and H is a matrix with size $\ell \times m$ where $m < \ell$ and $\mathsf{rank}(H) = q$. The hypothesis testing problem consists in deciding between

$$H_0 = \{\theta = \theta_0,\ Y_n \in \mathbb{R}^m\} \text{ and } H_1 = \{\theta = \theta_1,\ Y_n \in \mathbb{R}^m\} \tag{3.230}$$

while considering θ_0 and θ_1 as given vectors and Y_n as an unknown vector. As shown in Subsection 2.10.3.1, this hypothesis testing problem is invariant under the group of translations $G = \{g : g(X) = X + HC\}$, $C \in \mathbb{R}^m$, and the maximal invariant is

$$\mathbb{M}_n = WX_n = WM\theta + W\xi_n = WM\theta + \zeta_n, \ \zeta_n \sim \mathcal{N}(0, \sigma^2\mathbb{I}_{\ell-q}), \ n = 1,2,\dots,$$

where the matrix W satisfies the conditions (2.359). As previously, it is additionally assumed that the matrix WM with size $(\ell - q) \times r$ is full column rank and that $WM\theta_0 \neq WM\theta_1$. The hypotheses (3.230) reduce to the hypotheses

$$H_0 = \{\mathbb{M}_n \sim \mathcal{N}(WM\theta_0, \sigma^2 \mathbb{I}_{\ell-q})\} \text{ and } H_1 = \{\mathbb{M}_n \sim \mathcal{N}(WM\theta_1, \sigma^2 \mathbb{I}_{\ell-q})\} \tag{3.231}$$

concerning the maximal invariant statistics $\mathbb{M}_n$, $n = 1, 2, \ldots$. The invariant LLR is

$$\lambda_n = \sum_{k=1}^{n} \mathbb{M}_k = \frac{1}{\sigma^2} [WM(\theta_1 - \theta_0)]^\top W \sum_{k=1}^{n} X_k - \frac{n}{2\sigma^2} \left[\|WM\theta_1\|_2^2 - \|WM\theta_0\|_2^2 \right]. \tag{3.232}$$

Hence, the LLR $\lambda_n = \lambda_{n-1} + \mathbb{M}_n$ is a random walk since $\mathbb{M}_n$, $n = 1, 2, \ldots$, are iid normal random variables under both hypotheses (and even for an arbitrarily chosen vector θ, different from putative values θ_0 and θ_1) with mean and variance

$$\mu_Z(\theta) = \frac{1}{\sigma^2} [WM(\theta_1 - \theta_0)]^\top WM\theta - \frac{1}{2\sigma^2} \left[\|WM\theta_1\|_2^2 - \|WM\theta_0\|_2^2 \right], \tag{3.233}$$

$$\sigma^2 = \frac{1}{\sigma^2} \|WM(\theta_1 - \theta_0)\|_2^2. \tag{3.234}$$

If $\theta = \theta_i$, $i = 0, 1$, then $\mu_i = \mu_Z(\theta_i) = (-1)^{i+1} \frac{1}{2\sigma^2} \|WM(\theta_1 - \theta_0)\|_2^2$. It is worth noting that the negative impact of the nuisance parameters Y_n leads to a reduction of the K–L information numbers and, hence, to an increase of the ESS of the SPRT. Indeed, let I_i be the K–L information numbers for the linear model $X_n = HY_n + M\theta + \xi_n$ with nuisance parameters and $\widetilde{I}_i$ be the K–L information numbers for the linear model $X_n = M\theta + \xi_n$ without nuisance parameters. It can be shown that

$$I_i = \frac{1}{2\sigma^2} \|WM(\theta_1 - \theta_0)\|_2^2 \leq \widetilde{I}_i = \frac{1}{2\sigma^2} \|M(\theta_1 - \theta_0)\|_2^2.$$

Chapter 4

Sequential Hypothesis Testing: Multiple Simple Hypotheses

In this chapter we extend the results of Chapter 3 to the case of multiple hypotheses. Note, however, that we consider only a classical "classification-type" problem setting. This problem is to be distinguished from simultaneous/parallel testing multiple hypotheses, which is important in certain modern applications, e.g., in sequential clinical trials with multiple treatment groups and multiple endpoints (see [30, 111, 448]).

4.1 The Matrix Sequential Probability Ratio Test

For $N > 1$, we now generalize the problem considered in Section 3.1 to the case of testing $N+1$ hypotheses concerning the values of a parameter θ, i.e., $\theta \in \Theta = \{0, 1, \dots, N\}$ and assuming that the $N+1$ densities $p_{\theta=i}(\mathbf{X}_1^n) = p_i(\mathbf{X}_1^n)$, $i = 0, 1, \dots, N$ are distinct. The goal is to test $N+1$ hypotheses $H_i : \theta = i$, $i = 0, 1, \dots, N$, namely to identify which one of the $N+1$ densities $p_0, p_1, \dots, p_N$ is the true one.

Even more generally, as in Section 3.3, we consider the following set-up without distinguishing the discrete- and continuous-time cases. Let $(\Omega, \mathscr{F}, \mathscr{F}_t, \mathsf{P})$, $t \in \mathbb{Z}_+ = \{0, 1, \dots\}$ or $t \in \mathbb{R}_+ = [0, \infty)$, be a filtered probability space with standard assumptions about the monotonicity, and also, in the continuous-time case, the right-continuity of the σ-algebras $\mathscr{F}_t$. The sub-σ-algebra $\mathscr{F}_t = \mathscr{F}_t^X = \sigma(\mathbf{X}_0^t)$ of $\mathscr{F}$ is assumed to be generated by the process $\mathbf{X}_0^t = \{X(u), 0 \le u \le t\}$ observed up to time t, which is defined on the space $(\Omega, \mathscr{F})$. The hypotheses are $H_i :\ \mathsf{P} = \mathsf{P}_i$, $i = 0, 1, \dots, N$, where $\mathsf{P}_0, \mathsf{P}_1, \dots, \mathsf{P}_N$ are given probability measures assumed to be locally mutually absolutely continuous, i.e., their restrictions P_i^t and P_j^t to $\mathscr{F}_t^X$ are equivalent, for all $0 \le t < \infty$ and all $i, j = 0, 1, \dots, N$, $i \neq j$. For $t \ge 0$, we define the LR and LLR processes

$$\Lambda_i(t) = \frac{\mathrm{d}\mathsf{P}_i^t}{\mathrm{d}\mathsf{Q}^t}(\mathbf{X}_0^t), \quad \lambda_i(t) = \log \Lambda_i(t) \quad (\Lambda_i(0) = 1,\ \lambda_i(0) = 0\ \mathsf{P}_i - \text{a.s.}),$$

where Q^t is some dominating measure which may not be equal to one of the measures P_j^t, $j = 0, 1, \dots, N$, and we denote by $\Lambda_{ij}(t)$ the LR between the hypotheses H_i and H_j for the sample trajectory $\mathbf{X}_0^t$, i.e., $\mathsf{Q}^t = \mathsf{P}_j^t$ in this case.

As before, E_i denotes the expectation under the hypothesis H_i. A multihypothesis sequential test is a pair $\delta = (d, T)$, where T is a stopping time with respect to the filtration $\{\mathscr{F}_t\}_{t \ge 0}$ and $d = d(\mathbf{X}_0^T)$ is an $\mathscr{F}_T$-measurable terminal decision function with values in the set $\{0, 1, \dots, N\}$. Hence, $d = i$ means that the hypothesis H_i is accepted, i.e., $\{d = i\} = \{T < \infty,\ \delta \text{ accepts } H_i\}$.

Let $\alpha_{ij}(\delta) = \mathsf{P}_i(d = j)$, $i \neq j$, $i, j = 0, 1, \dots, N$, be the error probabilities of the test δ, i.e., the probabilities of accepting a specific H_j when H_i is true. Note that the probabilities of rejecting the hypothesis H_i when it is true, $\alpha_i(\delta) = \mathsf{P}_i(d \neq i)$, $i = 0, 1, \dots, N$, are also of interest. Obviously, $\alpha_i(\delta) = \sum_{j \neq i} \alpha_{ij}(\delta)$. In addition, we are interested in the weighted error probabilities defined as $\beta_j(\delta) = \sum_{i=0}^N w_{ij} \mathsf{P}_i(d = j)$, where $[w_{ij}]$ is a given matrix of weights, all positive except for the diagonal elements w_{ii} which are zero. Correspondingly, we introduce three classes of tests

$$\begin{aligned}\mathbf{C}([\alpha_{ij}]) &= \{\delta : \alpha_{ij}(\delta) \le \alpha_{ij},\, i,j = 0,1,\dots,N,\, i \ne j\},\\ \mathbf{C}(\boldsymbol{\alpha}) &= \{\delta : \alpha_i(\delta) \le \alpha_i,\, i = 0,1,\dots,N\},\\ \mathbf{C}(\boldsymbol{\beta}) &= \{\delta : \beta_j(\delta) \le \beta_j,\, j = 0,1,\dots,N\}.\end{aligned} \tag{4.1}$$

Here $[\alpha_{ij}]$ is a matrix of given error probabilities that are positive numbers less than 1; $\boldsymbol{\alpha} = (\alpha_0, \alpha_1, \dots, \alpha_N)$ is a vector of positive numbers α_i less than 1 and $\boldsymbol{\beta} = (\beta_0, \beta_1, \dots, \beta_N)$ is a vector of positive numbers β_i not necessarily less than 1. In other words, the classes $\mathbf{C}([\alpha_{ij}])$ and $\mathbf{C}(\boldsymbol{\alpha})$ include the tests for which the error probabilities $\Pr(\text{accept } H_j | H_i \text{ true})$ and $\Pr(\text{reject } H_i | H_i \text{ true})$ are not greater than the predefined numbers $0 < \alpha_{ij} < 1$ and $0 < \alpha_i < 1$, respectively, and the class $\mathbf{C}(\boldsymbol{\beta})$ includes the tests for which the weighted error probabilities do not exceed the given values $\beta_i > 0$. In particular, $\beta_j(\delta)$ may represent the average loss in making the decision $d = j$ if $w_{ij} = \pi_i L_{ij}$, where L_{ij} is the loss when the hypothesis H_i is true and the decision $d = j$ is made and π_i is the prior probability of H_i. To avoid trivialities, without special emphasis in the sequel we consider only tests with finite ESSs, $\mathsf{E}_i T < \infty$ for $i = 0,1,\dots,N$.

For a *threshold* or *boundary matrix* $\mathbf{A} = [A_{ij}]$, with $A_{ij} > 0$ and the A_{ii} are immaterial say 0, define the *matrix* SPRT (MSPRT) $\delta_N^* = (T_N^*, d_N^*)$, built on $(N+1)N/2$ one-sided SPRTs between the hypotheses H_i and H_j, as follows:

$$\text{Stop at the first } t \ge 0 \text{ such that, for some } i, \quad \Lambda_{ij}(t) \ge A_{ji} \quad \text{for all } j \ne i, \tag{4.2}$$

and accept the unique i that satisfies these inequalities. Note that for $N = 1$ this test coincides with Wald's SPRT. Note also that if $A_{ji} = A_i$ do not depend on j, then δ_N^* stops at the first time such that for some i

$$\frac{\Lambda_i(t)}{\max\limits_{j \ne i} \Lambda_j(t)} \ge A_i,$$

which is known in statistics as the generalized likelihood ratio test (GLRT).

In the following we omit the subscript N in $\delta_N^* = (T_N^*, d_N^*)$ for brevity. Obviously, with $a_{ji} = \log A_{ji}$, the MSPRT in (4.2) can be written as

$$T^* = \inf\{t \ge 0 : \lambda_{ij}(t) \ge a_{ji} \text{ for all } j \ne i \text{ and some } i\}, \tag{4.3}$$

$$d^* = i \text{ for which (4.3) holds.} \tag{4.4}$$

For further study, it is convenient to introduce the Markov accepting time for the hypothesis H_i defined as

$$T_i = \inf\left\{ t \ge 0 : \lambda_i(t) \ge \max_{\substack{1 \le j \le N \\ j \ne i}} [\lambda_j(t) + a_{ji}] \right\}, \tag{4.5}$$

and to write the test in (4.3)–(4.4) in the following form:

$$T^* = \min(T_0, T_1, \dots, T_N), \qquad d^* = i \quad \text{if} \quad T^* = T_i. \tag{4.6}$$

Thus, in the MSPRT, each component SPRT is extended until, for some $i = 0,1,\dots,N$, all N SPRTs involving H_i accept H_i.

The following lemma shows how the thresholds can be selected in order to embed the MSPRT in the classes of tests defined in (4.1).

Lemma 4.1.1. *The following assertions hold:*

(i) $\alpha_{ij}^* \le e^{-a_{ij}}$ *for* $i,j = 0,1,\dots,N,\ i \ne j$;

(ii) $\alpha_i^* \le \sum_{j \ne i} e^{-a_{ij}}$ *for* $i = 0,1,\dots,N$;

(iii) $\beta_j^* \le \sum_{i\neq j} w_{ij} e^{-a_{ij}}$.

Proof. (i) By the definition of T_j,

$$\Lambda_{ij}(T_j) \le e^{-a_{ij}} \quad \text{for all } i \neq j \text{ on } T_j < \infty,$$

and hence, using Wald's likelihood ratio identity, we obtain

$$\alpha_{ij}^* = \mathsf{P}_i(T^* = T_j, T_j < \infty) \le \mathsf{P}_i(T_j < \infty) = \mathsf{E}_j\left[\Lambda_{ij}(T_j)\mathbb{1}_{\{T_j<\infty\}}\right] \le e^{-a_{ij}}. \tag{4.7}$$

(ii) Using the inequality (4.7), we obtain

$$\alpha_i^* = \sum_{j\neq i} \alpha_{ij}^* \le \sum_{j\neq i} e^{-a_{ij}}.$$

(iii) By the inequality (4.7), we have

$$\beta_j^* = \sum_{i\neq j} w_{ij}\alpha_{ij}^* \le \sum_{i\neq j} w_{ij} e^{-a_{ij}}.$$

□

Therefore, we have the following implications:

$$a_{ji} = \log(1/\alpha_{ji}) \quad \text{implies } \delta^* \in \mathbf{C}([\alpha_{ij}]); \tag{4.8}$$

$$a_{ji} = a_j = \log(N/\alpha_j) \quad \text{implies } \delta^* \in \mathbf{C}(\boldsymbol{\alpha}); \tag{4.9}$$

$$a_{ji} = a_i = \log\left(\frac{\sum_{k\neq i} w_{ki}}{\beta_i}\right) \quad \text{implies } \delta^* \in \mathbf{C}(\boldsymbol{\beta}). \tag{4.10}$$

In the following sections we show that the MSPRT with these thresholds is first-order asymptotically optimal both in the iid and non-iid cases.

We emphasize that the upper bounds in Lemma 4.1.1 and the implications (4.8)–(4.10) hold in the most general non-iid case, and that these bounds are distribution-free.

4.2 The Structure of the Optimal Multihypothesis Sequential Test in the iid Case

Consider the discrete-time iid case where the observations $X_1, X_2, \dots$ are independent and have common density $f_\theta(X)$ with respect to a sigma-finite measure $\mu(x)$, so that

$$p_\theta(\mathbf{X}_1^n) = \prod_{k=1}^n f_\theta(X_k), \quad \theta = 0, 1, \dots, N$$

and

$$\Lambda_i(n) = \prod_{k=1}^n \frac{f_i(X_k)}{q(X_k)},$$

where q is some density which may not be equal to one of the densities f_j, $j = 0, 1, \dots, N$. As before, we denote by $\Lambda_{ij}(n)$ and $\lambda_{ij}(n)$ the LR and the LLR between the hypotheses H_i and H_j for the sample $\mathbf{X}_1^n = (X_1, \dots, X_n)$, i.e., when $q = f_j$.

Consider first the Bayesian multihypothesis problem that generalizes the two-hypothesis problem of Subsection 3.2.2. Suppose that $\theta \in \{0, 1, \dots, N\}$ is a random variable with prior distribution $\mathsf{P}(\theta = i) = \pi_i$, $i = 0, 1, \dots, N$, and that the loss incurred when stopping at time $T = n$ and making the decision $u_n = j$ while $\theta = i$ is true is $L_n(\theta = i, u_n = j, \mathbf{X}_1^n) = L_{ij} + cn$, where $c > 0$ is the cost of making one observation or sampling cost and where $0 < L_{ij} < \infty$ for $i \neq j$ and 0 if $i = j$.

By the Bayes formula, the posterior distribution given n observations, $\pi_{in} = \mathsf{P}(\theta = i|\mathbf{X}_1^n)$, is

$$\pi_{in} = \frac{\pi_i \prod_{k=1}^{n} f_i(X_k)}{\sum_{s=0}^{N} \pi_s \prod_{k=1}^{n} f_s(X_k)}, \quad n \geq 0. \tag{4.11}$$

As usual, for $n = 0$ the products in (4.11) are treated as 1. The APR is given by

$$R_n(\mathbf{X}_1^n, u_n = j) = \sum_{i=0}^{N} L_{ij}\pi_{in} + cn, \quad n \geq 0,$$

and the minimal APR when stopping at the n-th stage and making the best terminal decision is

$$R_n^{\mathrm{st}}(\boldsymbol{\pi}_n) = \min_{0\leq j\leq N} R_n(\boldsymbol{\pi}_n, u_n = j) = \min_{0\leq j\leq N} \sum_{i=0}^{N} L_{ij}\pi_{in} + cn, \tag{4.12}$$

where $\boldsymbol{\pi}_n = (\pi_{0n}, \dots, \pi_{Nn})$. As shown in Subsection 2.7.1, the optimal terminal decision rule d^0 does not depend on the stopping rule T and minimizes the APR, so that it follows from (4.12) that

$$d^0 = k \quad \text{if} \quad \sum_{i=0}^{N} L_{ik}\pi_{iT} \leq \min_{\substack{0\leq j\leq N \\ j\neq k}} \sum_{i=0}^{N} L_{ij}\pi_{iT}. \tag{4.13}$$

Let $\mathsf{P}^\pi = \sum_{i=0}^{N} \pi_i \mathsf{P}_i$ and let E^π stand for the corresponding expectation. By the results of Section 2.7 and (4.13), the problem is reduced to the optimal stopping problem

$$\inf_{T\in\mathcal{M}} \rho(d^0, T) = \inf_{T\in\mathcal{M}} \mathsf{E}^\pi \left[\min_{0\leq j\leq N} \sum_{i=0}^{N} L_{ij}\pi_T(i) + cT \right].$$

In just the same way as in Subsection 3.2.2, the cost cn is common to the two posterior risks R_n^{st} (stopping) and $\widetilde{R}_n$ (continuation), so that we can subtract cn and define the minimal *a posteriori* risk associated with stopping without sampling cost

$$G^{\mathrm{st}}(\boldsymbol{\pi}_n) = \min_{0\leq j\leq N} \sum_{i=0}^{N} L_{ij}\pi_{in}.$$

Also, by (4.11) the vector statistic $\boldsymbol{\pi}_n = \boldsymbol{\pi}_n(\boldsymbol{\pi}_{n-1}, X_{n+1})$ is transitive and P^π-Markov. The conditions (2.219) and (2.220) hold, so that $\boldsymbol{\pi}_n$ is a sufficient statistic and the MAPR $G_n(\boldsymbol{\pi}_n) = R_n(\boldsymbol{\pi}_n) - cn$ satisfies the dynamic programming recursion

$$G_n(\boldsymbol{\pi}_n) = \min\left\{G^{\mathrm{st}}(\boldsymbol{\pi}_n), \mathsf{E}^\pi[G_{n+1}(\boldsymbol{\pi}_{n+1}) \mid \boldsymbol{\pi}_n] + c\right\}.$$

Thus the optimal stopping time can be expressed as

$$T_0 = \inf\left\{n \geq 1 : G^{\mathrm{st}}(\boldsymbol{\pi}_n) = G_n(\boldsymbol{\pi}_n)\right\}, \tag{4.14}$$

and the optimal stopping region is $\{\boldsymbol{\pi}_n : G^{\mathrm{st}}(\boldsymbol{\pi}_n) = G_n(\boldsymbol{\pi}_n)\}$. This region can be also represented as $\pi_n(k) \geq B_k(\boldsymbol{\pi}_n^{(k)})$, where $k = k_n$ is the number for which $\min_{0\leq j\leq N} \sum_{i=0}^{N} L_{ij}\pi_{in}$ is attained, $\boldsymbol{\pi}_n^{(k)} = (\pi_n(0), \dots, \pi_n(k-1), \pi_n(k+1), \dots, \pi_n(N))$ and $B_k(\cdot)$ is a nonlinear function. Finding this function is an extremely difficult problem, so that approximations are needed.

4.3 Asymptotic Optimality of the MSPRT in the iid Case

From the above discussion we may conclude that the MSPRT, while optimal for $N=1$, is not strictly optimal for $N>1$. However, the MSPRT is a good approximation for an optimal multihypothesis test. Under certain conditions and with some choice of the threshold matrix **A**, it minimizes the expected sample sizes $\mathsf{E}_i T$ for all $i=0,1,\dots,N$ to within a vanishing term $o(1)$ for a small cost c and for small error probabilities.

We begin with establishing a much weaker first-order optimality property of the MSPRT and deriving first-order approximations to its operating characteristics. To this end, we first generalize to the case of multiple hypotheses the lower bounds for the ESS given in Theorem 3.2.2 for two hypotheses. Recall that $\alpha_i = \sum_{k\neq i}\alpha_{ik}$.

Lemma 4.3.1 (Lower bounds)**.** *Let the K–L information numbers $I_{ij}=\mathsf{E}_i[\lambda_{ij}(1)]$ be positive and finite, $0<I_{ij}<\infty$, $i,j=0,1,\dots,N$. If $\alpha_0+\dots+\alpha_N\le 1$, then*

$$\inf_{\delta\in\mathbf{C}([\alpha_{ij}])}\mathsf{E}_i T \ge \max_{j\neq i}\left[\frac{1}{I_{ij}}\sum_{k=0}^{N}\alpha_{ik}\log\frac{\alpha_{ik}}{\alpha_{jk}}\right] \quad \textit{for all}\ \ i=0,1,\dots,N. \tag{4.15}$$

Proof. Setting $\mathsf{P}=\mathsf{P}_i$, $\mathsf{Q}=\mathsf{P}_j$ and $\mathcal{Y}_k=\{d=k\}$, $k=0,1,\dots,N$, in Lemma 3.2.1, we obtain

$$\mathsf{E}_i[\lambda_{ij}(T)] \ge \sum_{k=0}^{N}\mathsf{P}_i(d=k)\log\frac{\mathsf{P}_i(d=k)}{\mathsf{P}_j(d=k)} = \sum_{k=0}^{N}\alpha_{ik}(\delta)\log\frac{\alpha_{ik}(\delta)}{\alpha_{jk}(\delta)}.$$

By Wald's identity, $\mathsf{E}_i[\lambda_{ij}(T)] = I_{ij}\mathsf{E}_i T$, so that

$$\mathsf{E}_i T \ge \frac{1}{I_{ij}}\sum_{k=0}^{N}\alpha_{ik}(\delta)\log\frac{\alpha_{ik}(\delta)}{\alpha_{jk}(\delta)} \quad \text{for all}\ \ j\neq i. \tag{4.16}$$

The inequality in (4.15) follows from the fact that the sum on the right-hand side of the inequality (4.16) is a decreasing function in the domain $\alpha_0+\dots+\alpha_N\le 1$, so that it attains its maximum when $\alpha_{ij}(\delta)=\alpha_{ij}$. □

Unfortunately the lower bound (4.15) is unattainable at least in general. To see this recall that in the case of two hypotheses it is attainable only if there are no overshoots over the boundaries in the SPRT. However, this bound can be used to establish the first-order asymptotic optimality property.

4.3.1 *First-Order Asymptotic Optimality of the MSPRT in the iid Case*

Let $\alpha_{ij}\to 0$ in such a way that $\alpha_{ik}\log\alpha_{jk}\to 0$ and define $\alpha_{\max}=\max_{i,j}\alpha_{ij}$. Then it follows from (4.15) that

$$\begin{aligned}\inf_{\delta\in\mathbf{C}([\alpha_{ij}])}\mathsf{E}_i T &\ge \max_{j\neq i}\left[\frac{1}{I_{ij}}\left(\sum_{k\neq i}\alpha_{ik}\log\frac{\alpha_{ik}}{\alpha_{jk}}+(1-\alpha_i)\log\frac{1-\alpha_i}{\alpha_{ji}}\right)\right]\\ &= \max_{j\neq i}\left[\frac{|\log\alpha_{ji}|}{I_{ij}}\right]+o(1) \quad \text{as}\ \alpha_{\max}\to 0.\end{aligned} \tag{4.17}$$

Now, it follows from [128, Theorem 4.1] that for any $m>0$ and all $i=0,1,\dots,N$,

$$\mathsf{E}_i[T^*]^m \sim \max_{\substack{1\le j\le N\\ j\neq i}}(a_{ji}/I_{ij})^m \quad \text{as}\ \min_{i,j} a_{ji}\to\infty \tag{4.18}$$

whenever $0<I_{ij}<\infty$.

Therefore, we conclude that the MSPRT with the thresholds

$$a_{ji} = |\log \alpha_{ji}| \quad \text{for } i,j = 0,1,\dots,N, i \neq j \tag{4.19}$$

(see (4.8)) is asymptotically first-order optimal in the sense of minimizing the expected sample sizes for all hypotheses:

$$\inf_{\delta \in \mathbf{C}([\alpha_{ij}])} \mathsf{E}_i T \sim \mathsf{E}_i T^* \sim \max_{\substack{1 \le j \le N \\ j \neq i}} \left[\frac{|\log \alpha_{ji}|}{I_{ij}} \right] \quad \text{as } \alpha_{\max} \to 0 \quad \text{for all } i = 0,1,\dots,N.$$

Furthermore, this result obviously holds if the thresholds a_{ji} are selected in such a way that $\alpha_{ij}^* \le \alpha_{ij}$ and $a_{ji} \sim |\log \alpha_{ji}|$ as $\alpha_{\max} \to 0$. This last observation is important when accounting for the overshoots which are ignored in Lemma 4.1.1.

From now on we consider a general case for the error probabilities or risk constraints where it is assumed that the numbers α_{ij} and α_{ks} approach zero in such a way that for all i,j,k,s,

$$\lim_{\alpha_{\max} \to 0} \frac{\log \alpha_{ij}}{\log \alpha_{ks}} = c_{ijks}, \quad 0 < c_{ijks} < \infty. \tag{4.20}$$

That is, we assume that the ratios $\log \alpha_{ij} / \log \alpha_{ks}$, or more generally the ratios a_{ij}/a_{ks}, are bounded away from 0 and ∞. This guarantees that any α_{ij} does not go to zero at an exponentially faster (slower) rate than any other α_{ks}. Note that we do not require that the α_{ij}'s go to zero at the same rate—if this were the case all the c_{ijks}'s would equal 1, which is the asymptotically symmetric case. The reason for allowing c_{ijks}'s different from 1 is that there are many interesting problems for which the risks for the various hypotheses may be orders of magnitude different.

If we are interested in the class $\mathbf{C}(\boldsymbol{\alpha})$ that confines the rejection error probabilities $\alpha_i(\delta) = \mathsf{P}_i(d \neq i)$, the following argument allows us to obtain the asymptotic optimality property of the MSPRT. Since $\alpha_i(\delta) = \sum_{j \neq i} \alpha_{ij}(\delta) \ge \alpha_{ik}(\delta)$ for all $k \neq i$, by (4.16),

$$\begin{aligned}
\mathsf{E}_i T &\ge \max_{j \neq i} \left[\frac{1}{I_{ij}} \left(\sum_{k \neq i} \alpha_{ik}(\delta) \log \frac{\alpha_{ik}(\delta)}{\alpha_{jk}(\delta)} + (1 - \alpha_i(\delta)) \log \frac{1 - \alpha_i(\delta)}{\alpha_{ji}(\delta)} \right) \right] \\
&\ge \max_{j \neq i} \left[\frac{1}{I_{ij}} \left(\sum_{k \neq i} \alpha_{ik}(\delta) \log \frac{\alpha_{ik}(\delta)}{\alpha_{jk}(\delta)} + (1 - \alpha_i(\delta)) \log \frac{1 - \alpha_i(\delta)}{\alpha_{j}(\delta)} \right) \right] \\
&= \max_{j \neq i} \frac{|\log \alpha_j(\delta)|}{I_{ij}} + o(1) \quad \text{as } \max_j \alpha_j(\delta) \to 0.
\end{aligned}$$

Hence,

$$\inf_{\delta \in \mathbf{C}(\boldsymbol{\alpha})} \mathsf{E}_i T \ge \max_{j \neq i} \frac{|\log \alpha_j|}{I_{ij}} + o(1) \quad \text{as } \max_j \alpha_j \to 0.$$

Now, taking $a_{ji} = \log(N/\alpha_j)$ (see (4.9)) and using (4.18), we obtain

$$\inf_{\delta \in \mathbf{C}(\boldsymbol{\alpha})} \mathsf{E}_i T \sim \mathsf{E}_i T^* \sim \max_{j \neq i} (|\log \alpha_j| / I_{ij}) \quad \text{as } \max_j \alpha_j \to 0 \quad \text{for all } i = 0,1,\dots,N,$$

i.e., the MSPRT is first-order asymptotically optimal in the class $\mathbf{C}(\boldsymbol{\alpha})$ for all hypotheses.

It is interesting to ask whether the MSPRT also asymptotically minimizes to first order the higher moments of the stopping time distribution. The answer is affirmative, as the following theorem shows.

Theorem 4.3.1 (First-order AO wrt moments). *Let* $0 < I_{ij} < \infty$.

(i) *If the thresholds are selected as* $a_{ji} = \log(1/\alpha_{ji})$ *for* $i, j = 0, 1, \dots, N$, $i \neq j$, *then* $\delta^* \in \mathbf{C}([\alpha_{ij}])$ *and for all* $m > 0$ *and* $i = 0, 1, \dots, N$,

$$\inf_{\delta \in \mathbf{C}([\alpha_{ij}])} \mathsf{E}_i[T]^m \sim \mathsf{E}_i[T^*]^m \sim \max_{\substack{1 \le j \le N \\ j \neq i}} \left[\frac{|\log \alpha_{ji}|}{I_{ij}}\right]^m \quad \text{as } \alpha_{\max} \to 0. \tag{4.21}$$

The asymptotic formulas (4.21) *remain true if the thresholds are chosen so that* $\alpha_{ij}(\delta^*) \le \alpha_{ij}$ *and* $a_{ji} \sim \log(1/\alpha_{ji})$.

(ii) *If the thresholds are selected as* $a_{ji} = a_j = \log(N/\alpha_j)$ *for* $i, j = 0, 1, \dots, N$, *then* $\delta^* \in \mathbf{C}(\boldsymbol{\alpha})$ *and for all* $m > 0$ *and* $i = 0, 1, \dots, N$,

$$\inf_{\delta \in \mathbf{C}(\boldsymbol{\alpha})} \mathsf{E}_i[T]^m \sim \mathsf{E}_i[T^*]^m \sim \max_{\substack{1 \le j \le N \\ j \neq i}} \left[\frac{|\log \alpha_j|}{I_{ij}}\right]^m \quad \text{as } \max_i \alpha_i \to 0. \tag{4.22}$$

The asymptotic formulas (4.22) *remain true if the thresholds are chosen so that* $\alpha_i(\delta^*) \le \alpha_i$ *and* $a_j \sim \log(1/\alpha_j)$.

(iii) *If the thresholds are selected as* $a_{ji} = a_i = \log(\beta_i^{-1} \sum_{k \neq i} w_{ki})$ *for* $i, j = 0, 1, \dots, N$, *then* $\delta^* \in \mathbf{C}(\boldsymbol{\beta})$ *and for all* $m > 0$ *and* $i = 0, 1, \dots, N$,

$$\inf_{\delta \in \mathbf{C}(\boldsymbol{\beta})} \mathsf{E}_i[T]^m \sim \mathsf{E}_i[T^*]^m \sim \left[\frac{|\log \beta_i|}{\min\limits_{\substack{1 \le j \le N \\ j \neq i}} I_{ij}}\right]^m \quad \text{as } \max_i \beta_i \to 0. \tag{4.23}$$

The asymptotic formulas (4.23) *remain true if the thresholds are chosen so that* $\beta_i(\delta^*) \le \beta_i$ *and* $a_i \sim \log(1/\beta_i)$.

Proof. Proof of (i) and (ii). By the SLLN, $n^{-1}\lambda_{ij}(n) \to I_{ij}$ a.s. under P_i. Therefore, we can apply Theorem 2.2 of Tartakovsky [455] to obtain the lower bounds

$$\inf_{\delta \in \mathbf{C}([\alpha_{ij}])} \mathsf{E}_i[T]^m \ge \max_{j \neq i} \left[\frac{|\log \alpha_{ji}|}{I_{ij}}\right]^m (1 + o(1)) \quad \text{as } \max_{i,j} \alpha_{ij} \to 0,$$

$$\inf_{\delta \in \mathbf{C}(\boldsymbol{\alpha})} \mathsf{E}_i[T]^m \ge \max_{j \neq i} \left[\frac{|\log \alpha_j|}{I_{ij}}\right]^m (1 + o(1)) \quad \text{as } \max_i \alpha_i \to 0.$$

Now, substituting $a_{ji} = \log(1/\alpha_{ji})$ and $a_{ji} = \log(N/\alpha_j)$ in (4.18), we obtain

$$\mathsf{E}_i[T^*]^m \sim \max_{j \neq i}(|\log \alpha_{ji}|/I_{ij})^m, \quad \mathsf{E}_i[T^*]^m \sim \max_{j \neq i}(|\log \alpha_j|/I_{ij})^m,$$

which along with the lower bounds yield (4.21) and (4.22).

Proof of (iii). Substituting $a_{ji} = \log(\beta_i^{-1} \sum_{k \neq i} w_{ki})$ in (4.18) yields

$$\mathsf{E}_i[T^*]^m \sim (|\log \beta_i| / \min_{j \neq i} I_{ij})^m,$$

while, by [128, Theorem 3.1],

$$\inf_{\delta \in \mathbf{C}(\boldsymbol{\beta})} \mathsf{E}_i[T]^m \ge (|\log \beta_i| / \min_{j \neq i} I_{ij})^m (1 + o(1)).$$

Combining both asymptotic relations, we obtain (4.23). □

We stress that the MSPRT asymptotically minimizes all positive moments of the stopping time under the sole first moment condition, $0 < I_{ij} < \infty$.

4.3.2 Near Optimality of the MSPRT in the iid Case

It turns out that the results of the previous section can be substantially improved if we assume the second moment condition

$$\mathsf{E}_i[\lambda_{ij}(1)]^2 < \infty, \quad i,j = 0,1,\dots,N. \tag{4.24}$$

In this case the MSPRT can be designed in such a way that it is third-order asymptotically optimal with respect to the expected sample sizes in the class $\mathbf{C}([\alpha_{ij}])$, i.e.,

$$\inf_{T \in \mathbf{C}([\alpha_{ij}])} \mathsf{E}_i T = \mathsf{E}_i T^* + o(1) \quad \text{as } \alpha_{\max} \to 0.$$

By analogy with (3.52), define the numbers

$$\Upsilon_{ij} = \exp\left\{ -\sum_{n=1}^{\infty} \frac{1}{n} [\mathsf{P}_j(\lambda_{ij}(n) > 0) + \mathsf{P}_i(\lambda_{ij}(n) \le 0)] \right\}, \quad i,j = 0,1,\dots,N, \; i \neq j. \tag{4.25}$$

Clearly, these numbers are symmetric, $\Upsilon_{ij} = \Upsilon_{ji}$, and $0 < \Upsilon_{ij} \le 1$. Furthermore, $\Upsilon_{ij} = 1$ only if the measures P_i^n and P_j^n are singular, so that the absolute continuity assumption is violated. Also, by Theorem 3.1.3, these numbers are tightly related to the overshoots in the one-sided tests

$$\tau_{ij}(a) = \inf\{n \ge 0 : \lambda_{ij}(n) \ge a\}. \tag{4.26}$$

Specifically, $\zeta_{ij} = \Upsilon_{ij}/I_{ij}$, where $\zeta_{ij} = \lim_{a\to\infty} \mathsf{E}_i\{\exp[-(\lambda_{ij}(\tau_{ij}(a)) - a)]\}$ – see (3.53).

Since the MSPRT represents a matrix combination of one-sided tests (4.26), it is reasonable first to try to solve a simpler optimization problem assuming that there are only two hypotheses H_i and H_j with prior probabilities $\pi = \mathsf{P}(H_i)$, $1-\pi = \mathsf{P}(H_j)$, and that under the hypothesis H_j the observations cost nothing, but the loss is equal to L if $T < \infty$, while under H_i each observation costs c but there is no loss upon stopping. In other words, we want to stop as soon as possible if H_i is true but to continue indefinitely if H_j is true. In this case, the average risk of a stopping rule T is $\rho(T) = (1-\pi)L\mathsf{P}_j(T<\infty) + \pi c \mathsf{E}_i T$. It appears that the optimal stopping rule in this problem is nothing but the one-sided test (4.26) with the threshold $a = \log[(1-\pi)L\Upsilon_{ij}/\pi c]$, as the following lemma due to Lorden [275] shows.

Lemma 4.3.2. *For $L > 0$, $c > 0$, and $\pi > 0$, let $\rho(T) = (1-\pi)L\mathsf{P}_j(T<\infty) + \pi c \mathsf{E}_i T$. If $I_{ij} < \infty$ and $a = \log[(1-\pi)L\Upsilon_{ij}/\pi c]$ in (4.26), then*

$$\inf_T \rho(T) = \rho(\tau_{ij}(a)),$$

where the infimum is taken over all stopping times.

Proof. Define the first ascending and descending ladder epochs for the random walk $\lambda_{ij}(n)$,

$$\tau^+ = \inf\{n \ge 1 : \lambda_{ij}(n) > 0\}, \quad \tau^- = \inf\{n \ge 1 : \lambda_{ij}(n) \le 0\}.$$

Setting $L_0 = L$, $L_1 = 0$ in Theorem 3.2.4 on SPRT optimality, we obtain that the optimal procedure in the considered problem is to stop at the first time $n \ge 0$ when the posterior probability $\pi_n = \mathsf{P}(H_i|\mathbf{X}_1^n)$ of the hypothesis H_i exceeds a threshold $B(c,L,\pi)$. Obviously, this stopping time is nothing but $\tau_{ij}(a)$ if we take

$$a = \log[B/(1-B)\chi], \quad \chi = \pi/(1-\pi).$$

Now, if we let $B = \pi$ (i.e., $a = 0$), then on one hand the average risk associated with the stopping time $\tau_{ij}(a=0)$ is

$$\rho(\tau_{ij}(0)) = (1-B)L\mathsf{P}_j(\tau^+ < \infty) + Bc\,\mathsf{E}_i\tau^+,$$

and on the other hand immediate stopping (i.e., $T_0 = 0$) is also optimal, the risk of which is $\rho(T_0) = L(1-\pi) = L(1-B)$, so that we have $\rho(T_0) = \rho(\tau_{ij}(0))$, i.e.,

$$(1-B)L = (1-B)L\,\mathsf{P}_j(\tau^+ < \infty) + Bc\,\mathsf{E}_i\tau^+$$

or, dividing by $(1-B)L$,

$$\frac{B}{1-B}\frac{c}{L}\,\mathsf{E}_i\tau^+ = \mathsf{P}_j(\tau^+ = \infty).$$

Solving for B yields

$$\frac{B}{1-B} = \frac{L}{c}\frac{\mathsf{P}_j(\tau^+ = \infty)}{\mathsf{E}_i\tau^+}.$$

By Lemma 2.5.1,

$$\mathsf{E}_i\tau^+ = \frac{1}{\mathsf{P}_i(\tau^- = \infty)} = \exp\left\{\sum_{n=1}^{\infty}\frac{1}{n}\mathsf{P}_i(\lambda_{ij}(n) \le 0)\right\},$$

$$\mathsf{E}_j\tau^- = \frac{1}{\mathsf{P}_j(\tau^+ = \infty)} = \exp\left\{\sum_{n=1}^{\infty}\frac{1}{n}\mathsf{P}_j(\lambda_{ij}(n) > 0)\right\},$$

so that $\mathsf{P}_j(\tau^+ = \infty)/\mathsf{E}_i\tau^+ = \Upsilon_{ij}$ and $B/(1-B) = L\Upsilon_{ij}/c$. Therefore,

$$a = \log\left[\frac{B}{(1-B)\chi}\right] = \log\left[\frac{L\Upsilon_{ij}(1-\pi)}{c\pi}\right]$$

and the proof is complete. □

Note that if instead of the Bayesian setup with the loss L, cost c, and prior π, one is interested in the frequentist problem of minimizing the expected sample size E_iT in the class of power-one test procedures $\mathbf{C}(\alpha) = \{T : \mathsf{P}_j(T < \infty) \le \alpha\}$ with the error probability constraint $\alpha < 1$, then the one-sided test $\tau_{ij}(a_\alpha)$ is optimal as long as a_α is selected so that $\mathsf{P}_j(\tau_{ij}(a_\alpha) < \infty) = \alpha$. This is not easy to achieve exactly if there is an overshoot of a when crossing. But with the $\zeta_{ij} = \Upsilon_{ij}/I_{ij}$ being correction factors to the error probability bound $\mathsf{P}_j(\tau_{ij}(a) < \infty) \le e^{-a}$, the asymptotic approximation

$$\mathsf{P}_j(\tau_{ij}(a) < \infty) \sim \zeta_{ij}e^{-a}, \quad a \to \infty,$$

works well even for moderate values of a. So that taking $a = \log(I_{ij}/\Upsilon_{ij}\alpha)$ allows one to attain a nearly optimal solution in this latter frequentist problem. Hence, the Υ-numbers play a significant role both in the Bayes and the frequentist settings, allowing to induce corrections to the boundaries needed to attain optimum.

Lemma 4.3.2 is a key for proving almost optimality of the MSPRT with specially designed thresholds a_{ji} that depend on the numbers Υ_{ij}. As for establishing optimality of the SPRT, the proof relies upon a Bayesian method. Let $\pi_i = \mathsf{P}(H_i)$, $i = 0, 1, \dots, N$, be prior probabilities of the hypotheses and let L_{ij} be the loss when the hypothesis H_i is true and the decision $d = j$ is made, i.e., the hypothesis H_j is accepted. Assume $L_{ii} = 0$. Assuming the observation cost c, the average (integrated) risk of the test $\delta = (d, T)$ is

$$\rho_c^\pi(\delta) = \sum_{i=0}^{N}\pi_i\left[\sum_{j=0}^{N}L_{ij}\mathsf{P}_i(d = j) + c\,\mathsf{E}_iT\right].$$

Lorden [275, Theorem 1] shows that, as $c \to 0$, the MSPRT δ^* defined in (4.2) with the thresholds $A_{ji}(c) = (\pi_j/\pi_i)L_{ji}\Upsilon_{ij}/c$ is asymptotically nearly optimal to within $o(c)$ under the second moment condition (4.24):

$$\rho_c^\pi(\delta^*) = \inf_\delta \rho_c^\pi(\delta) + o(c) \quad \text{as } c \to 0.$$

Using this Bayes asymptotic optimality result, it can be proven that the MSPRT is also nearly optimal to within $o(1)$ with respect to the expected sample sizes $\mathsf{E}_i T$ for all hypotheses in the classes of tests with constraints imposed on the error probabilities. In other words, the MSPRT has an asymptotic property similar to the exact optimality of the SPRT for two hypotheses. This result is more practical than the above Bayes optimality.

The following theorem spells out the details. Recall that $\alpha_{ij}(\delta) = \mathsf{P}_i(d = j)$ is the probability of erroneous acceptance of the hypothesis H_j when H_i is true, $\alpha_i(\delta) = \mathsf{P}_i(d \neq i)$ is the probability of erroneous rejection of H_i when it is true, and $\beta_j(\delta) = \sum_{i=0}^N w_{ij} \mathsf{P}_i(d = j)$ is the weighted probability of or risk in accepting H_i, where $[w_{ij}]$ is a given matrix of positive weights. Recall also the definition of the classes of tests (4.1). If $A_{ij} = A_{ij}(c)$ is a function of the small parameter c, then the error probabilities $\alpha_{ij}^*(c)$, $\alpha_i^*(c)$ and $\beta_j^*(c)$ of the MSPRT are also functions of this parameter, and if $A_{ji}(c) \to \infty$, then $\alpha_{ij}^*(c), \beta_j^*(c) \to 0$ as $c \to 0$. We write $\boldsymbol{\beta}^*(c)$ for the vector $(\beta_0^*(c), \dots, \beta_N^*(c))$ and $\boldsymbol{\alpha}^*(c)$ for the vector $(\alpha_0^*(c), \dots, \alpha_N^*(c))$.

Theorem 4.3.2 (MSPRT near optimality). *Assume that the second moment condition* (4.24) *holds.*

(i) *If the thresholds in the MSPRT are selected as $A_{ji}(c) = w_{ji}\Upsilon_{ij}/c$, $i, j = 0, 1, \dots, N$, then*

$$\mathsf{E}_i T^*(c) = \inf_{T \in \mathbf{C}(\boldsymbol{\beta}^*(c))} \mathsf{E}_i T + o(1) \quad \text{as } c \to 0 \quad \text{for all } i = 0, 1, \dots, N, \tag{4.27}$$

i.e., the MSPRT minimizes to within $o(1)$ the expected sample sizes among all tests whose weighted error probabilities are less than or equal to those of δ^*.

(ii) *For any $[B_{ij}]$ ($B_{ij} > 0$, $i \neq j$), let $A_{ji} = B_{ji}/c$. The MSPRT δ^* asymptotically minimizes the expected sample sizes for all hypotheses to within $o(1)$ as $c \to 0$ among all tests whose error probabilities $\alpha_{ij}(\delta)$ are less than or equal to those of δ^* as well as those whose error probabilities $\alpha_i(\delta)$ are less than or equal to those of δ^*, i.e.,*

$$\mathsf{E}_i T^*(c) = \inf_{T \in \mathbf{C}([\alpha_{ij}^*(c)])} \mathsf{E}_i T + o(1) \quad \text{as } c \to 0 \quad \text{for all } i = 0, 1, \dots, N, \tag{4.28}$$

and

$$\mathsf{E}_i T^*(c) = \inf_{T \in \mathbf{C}(\boldsymbol{\alpha}^*(c))} \mathsf{E}_i T + o(1) \quad \text{as } c \to 0 \quad \text{for all } i = 0, 1, \dots, N. \tag{4.29}$$

Proof. Part (i) follows from [275, Theorem 4]. Part (ii) is an immediate consequence of (i). Indeed, for a given B_{ij}, one can always select the weights w_{ij} to satisfy the equalities $A_{ji}(c) = w_{ji}\Upsilon_{ij}/c$. □

Remark 4.3.1. Theorem 4.3.2 covers only the asymptotically symmetric case where

$$\lim_{c \to 0} \frac{\log \beta_j^*(c)}{\log \beta_k^*(c)} = 1, \quad \lim_{c \to 0} \frac{\log \alpha_i^*(c)}{\log \alpha_k^*(c)} = 1 \quad \text{and} \quad \lim_{c \to 0} \frac{\log \alpha_{ij}^*(c)}{\log \alpha_{ks}^*(c)} = 1. \tag{4.30}$$

Introducing for the hypotheses H_i different observation costs c_i that may go to 0 at different rates, i.e., setting $A_{ji} = B_{ji}/c_i$, the results of Theorem 4.3.2 can be generalized to the more general asymmetric case where the ratios in (4.30) are bounded away from zero and infinity; see (4.20). This generalization is important for certain applications such as the detection of objects when the hypothesis H_0 is associated with the object absence and H_i with its presence in a specific location. Then $\alpha_{0j} = \alpha_0$ is the false alarm probability, $\alpha_{i0} = \alpha_1$ is the misdetection probability, and $\alpha_{ij} = \alpha_2$ ($i, j \neq 0$) is the misidentification probability. Usually, the required false alarm probability is much smaller than α_1 and α_2, say $\alpha_0 = 10^{-6}$ and $\alpha_1 = \alpha_2 = 10^{-2}$, so that the ratio is 3 but not 1.

Clearly, the MSPRT is not the unique asymptotically optimal test. There are many others. For example, define the Markov times

$$\widetilde{T}_i = \inf\left\{ n \geq 1 : \Lambda_i(n) \geq \sum_{j=0}^{N} \widetilde{A}_{ji}\, \Lambda_j(n) \right\}, \quad i = 0, 1, \dots, N, \tag{4.31}$$

and let $\widetilde{\delta} = (\widetilde{T}, \widetilde{d})$, where the stopping time $\widetilde{T}$ is the minimum of $\widetilde{T}_0, \widetilde{T}_1, \dots, \widetilde{T}_N$ and the terminal decision is $\widetilde{d} = i$ if $\widetilde{T} = \widetilde{T}_i$. Note that if $\widetilde{A}_{ji} = (\pi_j/\pi_i)\widetilde{A}/c$, where $\pi_k = \Pr(H_k)$ is the prior probability of the hypothesis H_k, then

$$\widetilde{T} = \inf\left\{n \ge 1 : \max_{0 \le i \le N} \pi_{in} \ge \widetilde{A}/c\right\}, \tag{4.32}$$

where $\pi_{in} = \pi_i \Lambda_i(n) / \sum_{j=0}^N \pi_j \Lambda_j(n)$ is the posterior probability of the hypothesis H_i given the sample $\mathbf{X}_1^n$. For this reason we will refer to this test as the quasi-Bayes test.

The test $\widetilde{\delta}$ is also third-order asymptotically optimal if the "thresholds" $\widetilde{A}_{ij}$ are selected appropriately. This fact is established in the next subsection.

4.3.3 *Higher-Order Asymptotic Approximations for the Expected Sample Sizes*

By Theorem 4.3.1, if $I_{ij} < \infty$ then the MSPRT is first-order asymptotically optimal and the following first-order approximations hold for the expected sample sizes

$$\mathsf{E}_i[T^*] = \max_{\substack{1 \le j \le N \\ j \ne i}} (a_{ji}/I_{ij})\,(1 + o(1)) \quad \text{as } \min_{i,j} a_{ji} \to \infty \tag{4.33}$$

(see also (4.18)). On the other hand, adding the second moment condition allows us to conclude that the MSPRT is third-order asymptotically optimal to within the negligible term $o(1)$ if the thresholds $a_{ji} = \log A_{ji}$ are selected appropriately.

It is interesting to obtain higher-order asymptotic approximations to the ESSs assuming higher-order moment conditions. This is the topic of this subsection. We also give some insight on how the thresholds are related to the error probabilities. Instead of giving rigorous proofs which are quite involved and lengthy, we provide a heuristic argument. We refer to Dragalin *et al.* [129] for formal proofs (see also Baum and Veeravalli [50]).

Recall that, by (4.6), $T^* = \min_{0 \le i \le N} T_i$, where T_i is given by (4.5). Setting $a_{ji} = \log(B_{ji}/c_i)$ (see Theorem 4.3.2) and extracting $b_i = \log(1/c_i)$ from the maximization, we obtain that the Markov times T_i can be written as

$$T_i = \inf\left\{n \ge 1 : \lambda_i(n) \ge b_i + \max_{\substack{1 \le j \le N \\ j \ne i}} [\lambda_j(n) + \log B_{ji}]\right\}, \quad i = 0, 1, \dots, N. \tag{4.34}$$

Here we assume that the cost of observation, c_i, depends on the true hypothesis and may go to zero with different rates for different hypotheses to cover the asymptotically asymmetric case $b_i/b_j \sim C_{ij}$, $0 < C_{ij} < \infty$; see Remark 4.3.1. The matrix $[B_{ij}]$ is arbitrary with positive elements, except for the B_{ii}'s which are immaterial. When we are interested in the weighted error probabilities $\beta_j(\delta) = \sum_{i=0}^N w_{ij} \mathsf{P}_i(d = j)$, as in Theorem 4.3.2(i), we set $B_{ji} = w_{ji}\Upsilon_{ij}$. Not that in all cases the B_{ji}'s are pre-fixed positive numbers and only the values b_i are large parameters ($c_i \to 0$).

The asymptotic approximations turn out to be very different for the asymmetric and symmetric or partially symmetric cases with respect to the hypotheses. More importantly, the techniques for obtaining these asymptotics differ substantially. For this reason, we consider the two cases separately.

4.3.3.1 *The Asymmetric Case*

First, consider the asymmetric case where the hypothesis H_{j^*} ($j^* = j^*(i)$) for which the K–L "distance," I_{ij}, attains its minimum over $j \ne i$ is unique. Specifically, assume that

$$j^*(i) = \arg\min_{j \ne i} I_{ij} \quad \text{is unique for any} \quad i \in \{0, 1, \dots, N\}. \tag{4.35}$$

Therefore, if we write $I_i = \min_{j \neq i} I_{ij}$, then in this asymmetric case the set $\{0 \le j \le N : I_{ij} = I_i\}$ consists of the single point that is denoted by $j^*(i)$.

We now show that in this case we may apply relevant results from the classical nonlinear renewal theory for perturbed random walks provided in Subsection 2.6.2. To this end, we rewrite the stopping times of (4.34) in the form of a random walk crossing a constant boundary plus a nonlinear term that is slowly changing in the sense of Definition 2.6.1. By subtracting $\lambda_{j^*}(n)$ on both sides of the inequalities in (4.34), we see that

$$T_i = \inf\{n \ge 1 : \lambda_{ij^*}(n) - Y_i(n) \ge b_i\}, \tag{4.36}$$

where

$$Y_i(n) = \log\Big(\max\Big[B_{j^*i}, \max_{j \neq i, j^*} B_{ji} \exp\{-\lambda_{j^*j}(n)\}\Big]\Big). \tag{4.37}$$

First note that $\mathsf{P}_i(T^* \neq T_i) \to 0$ as $b_{\min} = \min_i b_i \to \infty$, so that we expect that

$$\mathsf{E}_i T^* = \mathsf{E}_i T_i + o(1) \quad \text{as } b_{\min} \to \infty. \tag{4.38}$$

The details of proving the validity of (4.38) can be found in the proof of Theorem 3.1 in [129]. Therefore, it suffices to find an asymptotic approximation for the expected value of the stopping time T_i.

Since $\mathsf{E}_i \lambda_{j^*j}(1) = \mathsf{E}_i[\lambda_{ij}(1) - \lambda_{ij^*}(1)] = I_{ij} - I_{ij^*} > 0$, it is easily seen that $Y_i(n)$ converges almost surely under P_i to $\log B_{j^*i}$ as $n \to \infty$ [129, Lemma 3.1], which implies that $\{Y_i(n)\}$ is a slowly changing sequence. As discussed in Subsection 2.6.1, an important consequence of the slowly changing property is that the limiting distributions of the overshoot of a random walk with positive mean over a fixed threshold are unchanged by the addition of a slowly changing nonlinear term. See Theorem 2.6.1, the First Nonlinear Renewal Theorem. Now note that from (4.36),

$$\lambda_{ij^*}(T_i) = b_i + Y_i(T_i) + \chi_{ij^*} \quad \text{on} \quad \{T_i < \infty\}, \tag{4.39}$$

where χ_{ij^*} is the overshoot of the process $\lambda_{ij^*}(n) - Y_i(n)$ over the level b_i at time T_i. Taking the expectations in both sides of this last equality and applying Wald's identity, we obtain

$$I_{ij^*}\, \mathsf{E}_i T_i = b_i + \mathsf{E}_i Y_i(T_i) + \mathsf{E}_i \chi_{ij^*}.$$

The key point is that since the sequence $\{Y_i(n)\}$ is slowly changing, the overshoot χ_{ij^*} has the same limiting distribution as the overshoot $\kappa_{ij^*}(a) = \lambda_{ij^*}(\tau_{ij^*}(a)) - a$ (as $a \to \infty$) in the one-sided test (4.26). Furthermore, since $Y_i(n)$ converges to the constant $\log B_{j^*i}$ as $n \to \infty$, for large b_i, the ESS $\mathsf{E}_i T^*$ is approximately equal to $I_{ij^*}^{-1}(b_i + \log B_{j^*i} + \varkappa_{ij^*})$, where

$$\varkappa_{ij^*} = \int_0^\infty y \, \mathrm{d}G_{ij^*}(y) = \lim_{a \to \infty} \mathsf{E}_i[\kappa_{ij^*}(a)]$$

is the expected limiting overshoot ($G_{ij^*}(y) = \lim_{a \to \infty} \mathsf{P}_i\{\kappa_{ij^*}(a) \le y\}$). The following theorem is a formal statement of this result, whose proof is based on Theorem 2.6.4(i), the Second Nonlinear Renewal Theorem. All the required conditions of this theorem hold, as shown in [129, Theorem 3.1].

Theorem 4.3.3. *Suppose that* (4.35) *holds and that the b_i/b_j are bounded away from zero and infinity, i.e., for some* $0 < C_{ij} < \infty$,

$$b_i/b_j \sim C_{ij} \quad \text{as } b_{\min} \to \infty. \tag{4.40}$$

In addition, assume that $\mathsf{E}_i|\lambda_{ij}(1)|^2 < \infty$ *and that the LLRs* $\lambda_{ij}(1)$ *are* P_i*–nonarithmetic. Then*

$$\mathsf{E}_i T^* = \frac{1}{\min_{j \neq i} I_{ij}} (b_i + \log B_{j^*i} + \varkappa_{ij^*}) + o(1) \quad \text{as } b_{\min} \to \infty. \tag{4.41}$$

Remark 4.3.2. The condition that $\lambda_{ij}(1)$ are P_i–nonarithmetic is imposed due to the necessity of considering certain discrete cases separately in the nonlinear renewal theorem. If $\lambda_{ij^*}(1)$ are P_i–arithmetic with span $d_i > 0$, the result of Theorem 4.3.3 holds true as $b_i \to \infty$ through multiples of d_i (i.e., $b_i = kd_i$, $k \to \infty$) and with the corresponding modification in the definition of $\varkappa_{ij^*}$ as in Theorem 2.6.4(ii). Specifically, applying Theorem 2.6.4(ii), we may show that in the d_i-arithmetic case

$$\mathsf{E}_i T^* = \frac{1}{\min\limits_{j \neq i} I_{ij}} \left(kd_i + \log B_{j^* i} + \varkappa_{ij^*}\right) + o(1) \quad \text{as } k \to \infty.$$

4.3.3.2 The General Case

We now relax the condition (4.35) that the minimum K–L distance $I_i = \min_{j \neq i} I_{ij}$ is achieved uniquely. For fixed i, let $I_{i[1]} \le I_{i[2]} \le \cdots \le I_{i[N]}$ be the ordered values of I_{ij}, $j \neq i$ ($I_{i[N+1]} = +\infty$). Assume that

$$I_{i[1]} = \cdots = I_{i[r]} < I_{i[r+1]} \quad \text{for some } r \in \{1, \dots, N\}. \tag{4.42}$$

Note that condition (4.42) includes the fully symmetric case

$$I_{ij} = I_i \quad \text{for all } j \in \{0, 1, \dots, N\} \setminus i \tag{4.43}$$

when $r = N$. Note also that the previous asymmetric case is covered by setting $r = 1$.

The derivation of the asymptotics for the general case is much more complicated than in the asymmetric case. The reason is that if we write the Markov times T_i in the form given in (4.36), the sequences $\{Y_i(n)\}$ are not necessarily slowly changing in the general case. As a result, a different approach and stronger conditions are required.

Let $\mu_{ij} = \mathsf{E}_i \lambda_j(1)$, $j \in \{0, 1, \dots, N\} \setminus i$. (In particular, one may set $\lambda_i(n) = \sum_{k=1}^n \log f_i(X_k)$ to be the log-likelihood function of the observations up to time n.) Let $\mu_{i[1]} \le \mu_{i[2]} \le \cdots \le \mu_{i[M-1]}$ be the ordered values of μ_{ij}, $j \neq i$, with $\langle j \rangle$ denoting the index of the ordered values, i.e.,

$$\langle j \rangle = k \quad \text{if} \quad \mu_{i[j]} = \mu_{ik}, \tag{4.44}$$

with an arbitrary index assignment in case of ties.

Now, under the assumption (4.42), there are r values that achieve the minimum K–L distance I_i, and since $I_{ij} = \mathsf{E}_i \lambda_i(1) - \mu_{ij}$ there are r values that achieve the maximum of $\{\mu_{ij}, j \neq i\}$, i.e.,

$$\mu_{i[N-r]} < \mu_{i[N-r+1]} = \cdots = \mu_{i[N]} \quad (\mu_{i[0]} = -\infty).$$

Next, define the r-dimensional vector $\mathbf{Y}_i = (Y_{1,i}, Y_{2,i}, \dots, Y_{r,i})^\top$ with components

$$Y_{k,i} = \lambda_{\langle N-r+k \rangle}(1) - \mu_{i[N-r+k]} = \lambda_{\langle N-r+k \rangle}(1) - \mu_{i[N]}, \quad k = 1, \dots, r.$$

Obviously, $\mathbf{Y}_i$ is zero-mean. Let $\mathbf{V}_i = \mathsf{cov}_i(\mathbf{Y}_i)$ denote its covariance matrix with respect to P_i. Now let

$$\phi_{0,V_i}(\mathbf{x}) = \left[(2\pi)^r (\det \mathbf{V}_i)\right]^{-1/2} \exp\left\{-\frac{1}{2}\,\mathbf{x}^\top \mathbf{V}_i^{-1} \mathbf{x}\right\}$$

be density of the multivariate normal distribution with covariance matrix $\mathbf{V}_i$. The asymptotic expansions in the general case are derived using normal approximations [66]. In this context, we introduce the variables $h_{r,i}$ and $C_{r,i}$. The variable $h_{r,i}$ is the expected value of the maximum of r zero-mean normal random variables with the pdf $\phi_{0,V_i}(\mathbf{x})$, i.e.,

$$h_{r,i} = \int_{\mathbb{R}^r} \left(\max_{1 \le k \le r} x_k\right) \phi_{0,V_i}(\mathbf{x})\, d\mathbf{x}. \tag{4.45}$$

Write

$$\boldsymbol{\ell}_i = (\ell_{1,i}, \dots, \ell_{r,i})^\top; \quad \ell_{k,i} = \log B_{\langle M-1-r+k \rangle i}. \tag{4.46}$$

The variable $C_{r,i}$ is given by

$$C_{r,i} = \int_{\mathbb{R}^r} \left(\max_{1 \le k \le r} x_k \right) \left(\mathcal{P}_i(\mathbf{x}) + \boldsymbol{\ell}_i^\top \mathbf{V}_i^{-1} \mathbf{x} \right) \phi_{0,V_i}(\mathbf{x}) \, \mathrm{d}\mathbf{x}, \tag{4.47}$$

where $\mathcal{P}_i(\mathbf{x})$ is a polynomial in $\mathbf{x} \in \mathbb{R}^r$ with degree 3 whose coefficients involve $\mathbf{V}_i$ and the P_i-cumulants of $\mathbf{Y}_i$ up to order 3 and is given explicitly in [66, Eq. (7.19)].

Computing the constant $h_{r,i}$ for a given application is relatively straightforward, since the integral in (4.45) involves only the covariance matrix of the vector $\mathbf{Y}_i$; a further simplification occurs when $\mathbf{V}_i$ is diagonal. Computing the constant $C_{r,i}$ is, in general, quite difficult due to the fact that the polynomial $\mathcal{P}_i(\mathbf{x})$ is a complicated function of the cumulants of $\mathbf{Y}_i$. However, in the symmetric situation of frequent interest, where $\mathbf{V}_i$ is of the form $v_i^2 \mathbb{I} + \varepsilon$, and $\ell_{k,i} = \ell_i$, for $k = 1, \dots, r$, a considerable simplification is possible; see below. Note that the application of the normal approximations requires a Cramér-type condition on the joint characteristic function of the vector $\mathbf{Y}_i$ given in (4.51) below.

We now present an outline of our approach to finding asymptotics in the general case.

Note that T_i in (4.34) may be written in the form

$$T_i = \inf\left\{ n : S_i(n) \ge b_i + \xi_{r,i}(n) + h_{r,i}\sqrt{n} \right\}, \tag{4.48}$$

where $S_i(n) = \lambda_i(n) - n\mu_{i[N]}$ is a random walk with increments having positive means $\mathsf{E}_i \lambda_i(1) - \mu_{i[N]} = I_i$, and

$$\xi_{r,i}(n) = \max_{k \ne i} \left[\log B_{ki} + \lambda_k(n) - n\mu_{i[N]} \right] - h_{r,i}\sqrt{n}. \tag{4.49}$$

From (4.48) we have

$$S_i(T_i) = b_i + \xi_{r,i}(T_i) + h_{r,i}\sqrt{T_i} + \chi_i(b_i) \quad \text{on } \{T_i < \infty\}, \tag{4.50}$$

where $\chi_i(b_i)$ is the overshoot of the process $S_i(n) - \xi_{r,i}(n) - h_{r,i}\sqrt{n}$ over the level b_i at time T_i. It can be established that $\{\xi_{r,i}(n)\}$ is a slowly changing sequence that converges in distribution to a random variable ξ and that $\lim_{n\to\infty} \mathsf{E}_i[\xi_{r,i}(n)] = \mathsf{E}_i \xi = C_{r,i}$, where $C_{r,i}$ is defined in (4.47); for a formal proof, see the proof of Theorem 3.3 in [129]. Comparing with the corresponding relation for the asymmetric case given in (4.34), we see that in addition to the slowly changing term there is a nonlinear deterministic term that is added to the threshold in (4.48).

Thus, in the limit as $b_i \to \infty$, we expect T_i to approximately satisfy the equation

$$T_i I_i = b_i + \xi + h_{r,i}\sqrt{T_i} + \chi,$$

where χ is the limiting average overshoot. Solving this equation for T_i gives

$$T_i \approx \frac{1}{I_i} \left[b_i + \frac{h_{r,i}^2}{2I_i} + h_{r,i}\sqrt{\frac{b_i}{I_i} + \frac{h_{r,i}^2}{4I_i^2} + \chi + \xi} \right].$$

Finally, using uniform integrability arguments and the fact that the asymptotic relation (4.38) holds in the general case too, we expect that

$$\mathsf{E}_i T^* \approx \mathsf{E}_i T_i \approx \frac{1}{I_i} \left[b_i + \frac{h_{r,i}^2}{2I_i} + h_{r,i}\sqrt{\frac{b_i}{I_i} + \frac{h_{r,i}^2}{4I_i^2} + \varkappa_i + C_{r,i}} \right],$$

where $\varkappa_i$ is the expected limiting overshoot, and $C_{r,i}$ is the expectation of the limit of the slowly changing sequence $\{\xi_{r,i}(n)\}$.

The following theorem formalizes this result, and the detailed proof based on the general nonlinear renewal theory of Subsection 2.6.3 is given in [129].

Theorem 4.3.4. *Suppose that* (4.40) *holds, $\lambda_i(1)$ is P_i–nonarithmetic, the covariance matrix $\mathbf{V}_i$ of the vector $\mathbf{Y}_i$ is positive definite, $\mathsf{E}_i\|\mathbf{Y}_i\|^3 < \infty$, and the Cramér condition*

$$\limsup_{\|\mathbf{t}\|\to\infty} \mathsf{E}_i \exp\{\imath \mathbf{t}^\top \mathbf{Y}_i)\} < 1 \tag{4.51}$$

on the joint characteristic function of $\mathbf{Y}_i$ is satisfied. Then

$$\mathsf{E}_i T^* = \frac{1}{I_i}\left[b_i + h_{r,i}\sqrt{\frac{b_i}{I_i} + \frac{h_{r,i}{}^2}{4I_i^2} + \frac{h_{r,i}{}^2}{2I_i}} + \varkappa_i + C_{r,i}\right] + o(1) \quad \text{as } b_{\min}\to\infty, \tag{4.52}$$

where $\varkappa_i$ is the expectation of the limiting overshoot in the one-sided SPRT based on the LLR $\lambda_{i\langle N\rangle}(n) = \lambda_i(n) - \lambda_{\langle N\rangle}(n)$.

Proof. We include only a very short sketch of the proof presenting the main steps. This is instrumental for illustrating the importance of the general nonlinear renewal theory and especially of Theorem 2.6.7, the Second General Nonlinear Renewal Theorem.

Note first that it is not difficult to prove that $\mathsf{E}_i T^* = \mathsf{E}_i T_i + o(1)$ as $b_{\min}\to\infty$, so that it is sufficient to establish (4.52) for $\mathsf{E}_i T_i$.

Clearly, the stopping time T_i of (4.48) is equivalent to the stopping time τ_λ defined in (2.141) with $Z_n = S_i(n) - \xi_{r,i}(n)$ and $a_n(\lambda) = \lambda + h_{r,i}\sqrt{n}$, where $S_i(n) = \lambda_i(n) - n\mu_{i[N]}$ is a random walk with positive drift I_i and $\xi_{r,i}(n)$ is defined in (4.49). In the following we set $\lambda = b_i$ in $a_n(\lambda)$ to be consistent with the notation used in the present section.

By Theorem 2.6.7, under certain conditions

$$\mathsf{E}_i T_i = N_{b_i} + \frac{1}{I_i}\left(\varkappa_i + \lim_{n\to\infty} \mathsf{E}_i[\xi_{r,i}(n)]\right) + o(1) \quad \text{as } b_i\to\infty, \tag{4.53}$$

where

$$N_{b_i} = \sup\{t\ge 1 : a_t(b_i) \ge I_i t\} = \frac{b_i}{I_i} + \frac{h_{r,i}}{I_i}\sqrt{\frac{b_i}{I_i} + \frac{h_{r,i}^2}{4I_i^2} + \frac{h_{r,i}^2}{2I_i^2}}. \tag{4.54}$$

The verification of all the required conditions of this theorem is a tedious task [129]. Combining (4.53) with (4.54), we obtain

$$\mathsf{E}_i T_i = \frac{1}{I_i}\left\{b_i + h_{r,i}\sqrt{\frac{b_i}{I_i} + \frac{h_{r,i}{}^2}{4I_i^2} + \frac{h_{r,i}{}^2}{2I_i}} + \varkappa_i + \lim_{n\to\infty}\mathsf{E}_i[\xi_{r,i}(n)]\right\} + o(1) \quad \text{as } b_i\to\infty.$$

In order to verify that $\lim_{n\to\infty}\mathsf{E}_i[\xi_{r,i}(n)] = C_{r,i}$, it may be first shown that the maximization over $k\ne i$ in (4.49) can be replaced with a maximization over only those k corresponding to the r nearest hypotheses. Specifically, for a sufficiently large (but finite) n,

$$\xi_{r,i}(n) = \sqrt{n}\left[\frac{1}{\sqrt{n}}\max_{1\le k\le r}\left[\ell_{k,i} + \lambda_{\langle N-r+k\rangle}(n) - n\mu_{i[N]}\right] - h_{r,i}\right].$$

Now, using [66, Theorem 20.1] with $f(\mathbf{x}) = \max_{1\le k\le r}\{x_k\}$ and $s=3$, after some manipulations the desired convergence can be established, which completes the proof. □

Note that when $r=1$ we are in the asymmetric case (4.35), so that it is expected that the result of Theorem 4.3.4 will coincide with the asymptotic approximation (4.41) obtained in Theorem 4.3.3 under a weaker second moment condition. Indeed, it is easily shown that $h_{1,i}=0$ and $\mathcal{P}_i(x) = \mathsf{E}_i Y_{1,i}^3\,(x^3-3x)/6$ in this case; see [66, Eq. (7.21)]. Finally, since for the standard Gaussian random variable $\mathsf{E}X^4 = 3\mathsf{E}X^2$, we see from (4.47) that $C_{1,i} = \log B_{j^*i}$. Thus, the resulting expression for the ESS is consistent with the result of Theorem 4.3.3, as expected.

As mentioned above, computing the constants $C_{i,r}$ is a difficult task in general. We now consider a special symmetric case where it is possible.

Assume that $\{Y_{k,i}, k = 1,\dots,r\}$ are identically distributed but not necessarily independent, that $\mathbf{V}_i$ is of the form $\mathrm{v}_i^2\mathbb{I} + \varepsilon$ (with $\mathbb{I}$ being the identity matrix), and that $\ell_{k,i} = \ell_i$ for $k = 1,\dots,r$. This case often arises in practice; see the examples in [129, Section IV] as well as Example 4.3.3.4 below.

First suppose that $\varepsilon = \lambda_i = 0$. In this special case, it is easy to see from (4.45) that

$$h_{r,i} = \mathrm{v}_i h_r^*, \tag{4.55}$$

where h_r^* is the expected value of the standard normal order statistic, whose computed values are displayed in Table 4.1.

Table 4.1: *Expected values of standard normal order statistics. The first four values coincide with the known "exact" values.*

r	h_r^*	r	h_r^*	r	h_r^*
2	0.56418 95835	14	1.70338 1555	80	2.42677 4421
3	0.84628 43753	15	1.73591 3445	90	2.46970 0479
4	1.02937 5373	16	1.76599 1393	100	2.50759 3639
5	1.16296 4473	17	1.79394 1981	200	2.74604 2451
6	1.26720 6361	18	1.82003 1880	300	2.87776 6853
7	1.35217 8376	19	1.84448 1512	400	2.96817 8187
8	1.42360 0306	20	1.86747 5060	500	3.03669 9351
9	1.48501 3162	30	2.04276 0846	600	3.09170 2266
10	1.53875 2731	40	2.16077 7180	700	3.13754 7901
11	1.58643 6352	50	2.24907 3631	800	3.17679 1412
12	1.62922 7640	60	2.31927 8210	900	3.21105 5997
13	1.66799 0177	70	2.37735 9241	1000	3.24143 5777

Note that the second term inside the integral in (4.47) is zero since $\boldsymbol{\ell} = \mathbf{0}$ which leads to the following relatively simple expression for $C_{r,i}$:

$$C_{r,i} = \left(\mathsf{E}_i Y_{1,i}^3\right) \frac{C_r^*}{6\mathrm{v}_i^2}, \tag{4.56}$$

where

$$C_r^* = r\int_{-\infty}^{\infty} x\varphi(x)\Phi(x)^{r-2}\Big[(r-1)\varphi(x)(1-x^2) + (x^3-3x)\Phi(x)\Big]\,\mathrm{d}x. \tag{4.57}$$

The computed values of C_r^* are included in Table 4.2.

If $\ell_i \neq 0$, then we simply subtract it from $\xi_{r,i}(n)$ and add it to b_i in (4.48). Clearly, $\lim_{n\to\infty} \mathsf{E}_i[\xi_{r,i}(n) - \ell_i] = C_{r,i}$ where $C_{r,i}$ is given by (4.56).

Finally, the restriction that $\varepsilon = 0$ can also be removed, as explained in Section III.C of [129].

In summary, for $\varepsilon > -\mathrm{v}_i^2/r$ and $\ell_i \in \mathbb{R}$,

$$\mathsf{E}_i T^* = \frac{1}{I_i}\left[F_r(b_i + \ell_i, I_i, \mathrm{v}_i, \mathsf{E}_i\widetilde{Y}_{1,i}^3) + \varkappa_i\right] + o(1), \tag{4.58}$$

where

$$F_r(x,q,u,g) = x + uh_r^*\sqrt{\frac{x}{q} + \frac{u^2(h_r^*)^2}{4q^2}} + \frac{u^2(h_r^*)^2}{2q} + \frac{gC_r^*}{6u^2} \tag{4.59}$$

and

$$\widetilde{Y}_{k,i} = Y_{k,i} - \frac{(1+\varepsilon r/\mathrm{v}_i^2)^{1/2} - 1}{(1+\varepsilon r/\mathrm{v}_i^2)^{1/2}}\left(\frac{1}{r}\sum_{k=1}^{r} Y_{k,i}\right). \tag{4.60}$$

Table 4.2: *Values of the absolute constant C_r^* for the case $\boldsymbol{\ell} = \mathbf{0}$, $\mathbf{V} = \mathbb{I}$.*

r	C_r^*	r	C_r^*	r	C_r^*
2	0.0	14	2.20924	80	5.08274
3	0.27566	15	2.31444	90	5.28802
4	0.55133	16	2.41374	100	5.47243
5	0.80002	17	2.50776	200	6.70147
6	1.02174	18	2.59705	300	7.43096
7	1.22030	19	2.68205	400	7.95237
8	1.39953	20	2.76316	500	8.35874
9	1.56262	30	3.41871	600	8.69193
10	1.71210	40	3.89695	700	8.97438
11	1.85003	50	4.27404	800	9.21958
12	1.97802	60	4.58561	900	9.43625
13	2.09740	70	4.85120	1000	9.63036

4.3.3.3 Error Probabilities

An important practical question is how to choose the thresholds b_i for the given weights B_{ji} in order to guarantee the prespecified constraints on the error probabilities. Unfortunately, this problem is open in general since there are no analogs to the approximations (3.48) for the MSPRT when $N \geq 2$. The distribution-free upper bounds given in Lemma 4.1.1 are usually not too accurate due to neglecting the overshoots and, more importantly, also because of the dramatic difference between the structure of the one-sided stopping times τ_{ij} and the stopping times T_i, due to the presence of the nonlinear term that depends on the number of hypotheses. However, in the asymmetric case (4.35) the stopping times τ_{ij^*} and T_i behave similarly, so there is some hope to obtain some improvements in this case.

We now show that if $B_{ji} = w_{ji}$ then in the asymmetric case (4.35) the following asymptotic approximations hold for the weighted error probabilities $\beta_i^* = \sum_{j \neq i} w_{ji}\, \alpha_{ji}(\delta^*)$:

$$\beta_i^* = M_N \zeta_{ij^*} e^{-b_i}(1+o(1)) \quad \text{as } b_{\min} \to \infty,\ i = 0,1,\dots,N, \tag{4.61}$$

where $1 \leq M_N \leq N$ and $\zeta_{ij^*} = \Upsilon_{ij^*}/I_{ij^*}$.

Indeed, we have

$$\begin{aligned}\beta_i^* &= \sum_{j \neq i} w_{ji} \mathsf{P}_j(T^* = T_i) = \sum_{j \neq i} w_{ji} \int_{T^*=T_i} \Lambda_{ji}(T_i)\, \mathrm{d}\mathsf{P}_i \\ &= \mathsf{E}_i \left\{ \mathbb{1}_{\{T^*=T_i\}} \sum_{j \neq i} w_{ji} e^{\lambda_{ji}(T_i)} \right\} = \mathsf{E}_i \left\{ \mathbb{1}_{\{T^*=T_i\}} e^{-\lambda_{ij^*}(T_i)} \sum_{j \neq i} w_{ji} e^{-\lambda_{j^*j}(T_i)} \right\} \\ &= \mathsf{E}_i \left\{ \mathbb{1}_{\{T^*=T_i\}} e^{-\lambda_{ij^*}(T_i) + \xi_i(T_i)} \right\},\end{aligned} \tag{4.62}$$

where

$$\xi_i(n) = \log \left[w_{j^*i} + \sum_{j \neq i, j^*} w_{ji} e^{-\lambda_{j^*j}(n)} \right].$$

It follows that

$$\beta_i^* = \mathsf{E}_i \left\{ \Delta_i(T_i) e^{-(\lambda_{ij^*}(T_i) - Y_i(T_i))} \mathbb{1}_{\{T^*=T_i\}} \right\},$$

where $Y_i(n)$ is defined in (4.37) with $B_{ji} = w_{ji}$ and $\Delta_i(T_i) = \exp\{\xi_i(T_i) - Y_i(T_i)\}$. Now, by (4.39),

$$\lambda_{ij^*}(T_i) - Y_i(T_i) = b_i + \chi_{ij^*} \quad \text{on} \quad T_i < \infty,$$

where χ_{ij^*} is the overshoot of $\lambda_{ij^*}(n) - Y_i(n)$ over the level b_i at time T_i, and we obtain

$$\beta_i^* = e^{-b_i} \mathsf{E}_i \left\{ \Delta_i(T_i) e^{-\chi_{ij^*}} \mathbb{1}_{\{T^*=T_i\}} \right\}.$$

Recall that the $Y_i(n)$, $n \geq 1$, are slowly changing and that $\mathsf{P}_i(T^* = T_i) \to 1$ as $b_{\min} \to \infty$, so by Theorem 2.6.1, $\mathsf{E}_i[\mathbb{1}_{\{T^*=T_i\}} e^{-\chi_{ij^*}}] \to \zeta_{ij^*}$. Furthermore, it is easily seen that $1 \leq \Delta_i(T_i) \leq N$. Hence we have

$$\zeta_{ij^*} \leq \lim_{b_i \to \infty} \beta_i^* e^{b_i} \leq N \zeta_{ij^*},$$

which implies (4.61).

Note that $\Delta_i(T_i) \to 1$ as $b_i \to \infty$ P_i-a.s. Therefore, it is expected that for sufficiently high thresholds and not too large N the quantity M_N is close to 1, especially in substantially asymmetric situations. This is confirmed by Monte Carlo simulations [129]. Note also that the upper bound that ignores the overshoots is

$$\beta_i^* \leq e^{-b_i} \sum_{j \neq i} (w_{ji}/B_{ji}) = N e^{-b_i}.$$

This bound holds in the most general case too, and it is usually fairly inaccurate.

Now, recall the definition of the quasi-Bayes test $\widetilde{\delta}$ in (4.31). If we set $\widetilde{A}_{ji} = w_{ji}/c_i$ for $j \neq i$ and $\widetilde{A}_{ii} = 0$ and write $\widetilde{b}_i = \log(1/c_i)$, then $\widetilde{T}_i$ in (4.31) can be written as

$$\widetilde{T}_i = \inf \left\{ n \geq 1 : \lambda_i(n) \geq \widetilde{b}_i + \log \left(\sum_{j \neq i} w_{ji} e^{\lambda_j(n)} \right) \right\}. \tag{4.63}$$

Replacing T^* with $\widetilde{T}$ and T_i with $\widetilde{T}_i$ in (4.62), we obtain

$$\widetilde{\beta}_i := \beta_i(\widetilde{\delta}) = \mathsf{E}_i \left\{ e^{-(\lambda_{ij^*}(\widetilde{T}_i) - \xi_i(\widetilde{T}_i))} \mathbb{1}_{\{\widetilde{T}=\widetilde{T}_i\}} \right\}.$$

Denoting the corresponding overshoot by $\widetilde{\chi}_{ij^*} = (\lambda_{ij^*}(\widetilde{T}_i) - \xi_i(\widetilde{T}_i)) - \widetilde{b}_i$, we obtain

$$\widetilde{\beta}_i = e^{-\widetilde{b}_i} \mathsf{E}_i \left\{ e^{-\widetilde{\chi}_{ij^*}} \mathbb{1}_{\{\widetilde{T}=\widetilde{T}_i\}} \right\}.$$

Since $\mathsf{E}_i \lambda_{j^*j}(1) = \mathsf{E}_i[\lambda_{ij}(1) - \lambda_{ij^*}(1)] = I_{ij} - I_{ij^*} > 0$, it follows that $\xi_i(n) \to \log w_{j^*i}$ P_i-a.s. Hence the $\xi_i(n)$, $n \geq 1$, are slowly changing and by the same argument as above

$$\widetilde{\beta}_i = \zeta_{ij^*} e^{-\widetilde{b}_i} (1 + o(1)) \quad \text{as } \widetilde{b}_{\min} \to \infty, \; i = 0, 1, \ldots, N. \tag{4.64}$$

Comparing (4.61) with (4.64), we see that the quasi-Bayes (mixture-based) test has certain advantages over the MSPRT. Indeed, the exact value of the number M_N is not known. It is only definite that it belongs to the interval $[1, N]$, and we conjectured that often it is close to 1. On the other hand, based on (4.64) we may set $\widetilde{b}_i = \log(\zeta_{ij^*}/\widetilde{\beta}_i)$ to guarantee the given error probability $\widetilde{\beta}_i$ almost precisely for sufficiently large thresholds.

Also, the same argument as above shows that the asymptotic approximations (4.41) and (4.52) are valid for the ESS of the test $\widetilde{\delta}$. It is therefore clear that both tests are asymptotically third-order optimal at least in the asymmetric case. Simulations show that both tests have almost identical performance in the symmetric case too [129].

Finally, set $\widetilde{A}_{ji} = (\pi_j/\pi_i) A$ in (4.31), in which case the stopping time $\widetilde{T}$ is given by (4.32) with $A = \widetilde{A}/c$, where $\pi_k = \Pr(H_k)$ is the prior probability of the hypothesis H_k. It may be shown that if there is no overshoot over A, say for the continuous-time BM model, then the Bayes average error probability is

$$\sum_{k=0}^{N} \pi_k \mathsf{P}_k(\widetilde{d} \neq k) = 1 - A.$$

If there is an overshoot, then the inequality holds instead of equality. This is only due to the overshoot but not to the presence of a nonlinear term as in the case of the MSPRT. Therefore, in the completely symmetric case where $I_{ij} = I$ for all $i, j = 0, 1, \dots, N$ ($i \neq j$) and the error probabilities $\alpha_{ij}(\delta) = \widetilde{\alpha}$ are the same, the equality $\widetilde{\alpha} = (1-A)/N$ holds, assuming that there is no overshoot. Note that $\widetilde{T}_i$ can be written in the form (4.63) by setting $\widetilde{b}_i = \widetilde{b} = \log[A/(1-A)]$ and $w_{ji} = \pi_j/\pi_i$. Hence, in order to guarantee the given error probability $\widetilde{\alpha}_{ij} = \alpha$ the threshold $\widetilde{b}$ should be set as

$$\widetilde{b} = \log \frac{1 - N\alpha}{N\alpha}.$$

Also, $\widetilde{\alpha}_i = N\widetilde{\alpha}$ and if the prior distribution is uniform, then $\widetilde{\beta}_i = \sum_{k \neq i} \pi_k \widetilde{\alpha}_{ki} = (N-1)\widetilde{\alpha}/N$.

We stress that there are no exact equalities for the MSPRT even when there is no overshoot. Thus, we can conclude that the quasi-Bayes test suits better for matching the given error probabilities than the MSPRT.

4.3.3.4 An Example: Multipopulation Model

Consider a multipopulation hypothesis testing problem which is usually referred to as the *Multisample Slippage Problem*. This problem is considered in great detail in Section 4.6. Suppose there are N mutually independent populations. Let $\mathbf{X}_n = (X_{1,n}, \dots, X_{N,n})$, $n \geq 1$, be the associated vector of observations, where $X_{i,n}$ is the observation from the i-th population at the time n. Under the hypothesis H_0, the observations $X_{1,n}, \dots, X_{N,n}$ are mutually independent and distributed with common density $g_0(x)$. Under H_i, all the $X_{k,n}$ are mutually independent, $X_{1,n}, \dots, X_{i-1,n}, X_{i+1,n}, \dots, X_{N,n}$ are distributed with common density $g_0(x)$ and $X_{i,n}$ has density $g_i(x)$.

For the sake of simplicity we focus on the symmetric case where $g_i(x) = g_1(x)$ for $i = 1, \dots, N$. Then

$$f_0(\mathbf{X}_n) = \prod_{k-1}^{N} g_0(X_{k,n}); \quad f_i(\mathbf{X}_n) = g_1(x_{i,n}) \prod_{\substack{k=1 \\ k \neq i}}^{N} g_0(X_{k,n}), \quad i = 1, \dots, N. \tag{4.65}$$

Therefore, the log-likelihood ratios are given by

$$\lambda_{i0}(n) = -\lambda_{0i}(n) = \sum_{k=1}^{n} \log \frac{g_1(X_{i,k})}{g_0(X_{i,k})}, \quad i = 1, \dots, N;$$

$$\lambda_{ij}(n) = \lambda_{i0}(n) + \lambda_{0j}(n), \quad i \neq j;\ i, j \neq 0.$$

By the symmetry of the problem, the K–L distances $I_{i0} = D_1$ are the same for $i = 1, \dots, N$, and so are $I_{0i} = D_0$, where

$$D_1 = \int \log \left[\frac{g_1(x)}{g_0(x)}\right] g_1(x)\, \mathrm{d}\mu(x) \quad \text{and} \quad D_0 = \int \log \left[\frac{g_0(x)}{g_1(x)}\right] g_0(x)\, \mathrm{d}\mu(x). \tag{4.66}$$

Also the K–L distances between the non-null hypotheses are given by $I_{ij} = D_1 + D_0$ for all $i, j \neq 0$. Recall that I_i stands for the minimum of I_{ij} over $j \neq i$. Hence $I_0 = D_0$ and $I_i = D_1$, $i = 1, \dots, N$. This means that for the hypothesis H_0 we have the fully symmetric case with $I_0 = \min_{j \neq 0} I_{0j} = D_0$, $i = 1, \dots, N$ (i.e., $r = N$); while for any other hypothesis H_i, $i \neq 0$, the asymmetric condition (4.35) holds with $j^*(i) = 0$.

In the following we are interested in the weighted error probabilities $\beta_j(\delta) = \sum_{i \neq j} \pi_i \alpha_{ij}(\delta)$ with weights $w_{ij} = \pi_i$, where $\pi_i = \Pr(H_i)$, $i = 0, 1, \dots, N$ are the prior probabilities of the hypotheses. Further, we assume that the conditional prior distribution of the populations to be affected conditioned on the event that one of them is affected is uniform, i.e., $\Pr(H_i \mid H_0 \text{ is incorrect}) = 1/N$. In other words, if $\pi_0 = \Pr(H_0)$ is the prior probability of the event that none of the populations is affected, then $\pi_i = \Pr(H_i) = (1 - \pi_0)/N$ for $i = 1, \dots, N$.

To compute the ESSs for $i \neq 0$ we apply Theorem 4.3.3 to get

$$\mathsf{E}_i T^* \approx \frac{1}{D_1}(b_1 + \ell_1 + \varkappa_1), \quad i = 1, \ldots, N, \tag{4.67}$$

where $\ell_1 = \log w_{0i} = \log \pi_0$.

To compute the ESS under H_0 we need to use Theorem 4.3.4. By the symmetry of the problem, $r = N$ in this case. In order to compute the constants $h_{r,i}$ and $C_{r,i}$, it is convenient to use the measure $\mathsf{Q}(\mathrm{d}\mathbf{x}_n) = \prod_{k=1}^{n} g_1(x_{k,n})\mathrm{d}\mathbf{x}_n$ as the dominating measure for defining the densities. Then the likelihood functions of (4.65) get modified to

$$f_0(\mathbf{X}_n) = \prod_{k=1}^{N} \frac{g_0(X_{k,n})}{g_1(X_{k,n})}; \quad f_i(\mathbf{X}_n) = g_1(X_{i,n}) \prod_{\substack{k=1 \\ k \neq i}}^{N} \frac{g_0(X_{k,n})}{g_1(X_{k,n})}, \quad i = 1, \ldots, N.$$

With these likelihood functions, the vector $\mathbf{Y}_0$ has components given by

$$Y_{k,0} = \sum_{\substack{m=1 \\ m \neq k}}^{N} \frac{g_0(X_{m,n})}{g_1(X_{m,n})} - \mathsf{E}_0 \left[\sum_{\substack{m=1 \\ m \neq k}}^{N} \frac{g_0(X_{m,n})}{g_1(X_{m,n})} \right].$$

It is easy to show that the covariance matrix $\mathbf{V}_0$ has the form $\mathrm{v}_0^2 \mathbb{I} + \varepsilon$ considered as a special case above, where $\mathrm{v}_0^2 = \mathsf{var}_0\{\log[g_1(X)/g_0(X)]\}$ with $\mathsf{var}_0[\cdot]$ being the variance relative to $g_0(x)$, and where $\varepsilon = (N-2)\mathrm{v}_0^2$. Furthermore, $\ell_{k,0} = \ell_0 = \log w_{k,0} = \log[(1-\pi_0)/N]$. Thus, (4.58) may be applied to get

$$\mathsf{E}_0 T^* \approx \frac{1}{D_0} \left[F_N(b_0 + \ell_0, D_0, \mathrm{v}_0, \mathsf{E}_0 \widetilde{Y}_{1,0}^3) + \varkappa_0 \right], \tag{4.68}$$

where $F_N(\cdot,\cdot,\cdot,\cdot)$ is as defined in (4.59).

We performed a detailed analysis using Monte Carlo experiments for the slippage problem with normal populations, i.e.,

$$g_0(x) = \frac{1}{\sigma\sqrt{2\pi}} \exp\left\{-\frac{x^2}{2\sigma^2}\right\}, \quad g_1(x) = \frac{1}{\sigma\sqrt{2\pi}} \exp\left\{-\frac{(x-\theta)^2}{2\sigma^2}\right\}.$$

Write $q = \theta^2/\sigma^2$. Then it is easy to show that $D_0 = D_1 = q/2$. The constants $\zeta_1 = \zeta_0 = \zeta$ and $\varkappa_1 = \varkappa_0 = \varkappa$ are found numerically from the formulas

$$\zeta = \frac{2}{q} \exp\left\{ -2 \sum_{k=1}^{\infty} \frac{1}{k} \Phi\left(-\frac{1}{2}\sqrt{qk}\right) \right\};$$

$$\varkappa = 1 + \frac{q}{4} - \sqrt{q} \sum_{k=1}^{\infty} \left[\frac{1}{\sqrt{k}} \varphi\left(\frac{1}{2}\sqrt{qk}\right) - \frac{1}{2}\sqrt{q}\, \Phi\left(-\frac{1}{2}\sqrt{qk}\right) \right],$$

which can be derived from Theorem 2.5.3. The vectors $\mathbf{Y}_0$ and $\widetilde{\mathbf{Y}}_0$ are zero mean Gaussian. Hence $\mathsf{E}_0 Y_{1,0}^3 = 0$. The constant v_0^2 is equal to q. Using these constants, we can compute the ESS of the MSPRT. Sample results are given in Table 4.3, where $\hat{\beta}_i$ and $\hat{\mathsf{E}}_i T^*$ are the estimates of the error probabilities and the ESS obtained by Monte Carlo techniques. The number of Monte Carlo trials used in the simulations was chosen so that less than a 1% error was guaranteed in estimating the error probabilities and the ESS. Note that the higher-order asymptotics are considerably more accurate than the first-order asymptotics, particularly for the hypothesis H_0. The quantities $\varepsilon_{\mathrm{fo}}$ and $\varepsilon_{\mathrm{ho}}$ are the relative errors of first-order (fo) and higher-order (ho) approximations. It is seen that for the higher-order approximations the error was always smaller than 2%.

Table 4.3: *Comparison of Monte Carlo simulations with the asymptotic approximations (the number of trials used in the simulations was 10^6).*

Results for $N = 10$, $\pi_0 = 0.5$, and $q = 0.25$									
	Error Probabilities & Thresholds			ESS & Accuracy of Approximations					
	β_i	b_i	$\hat{\beta}_i$	$\hat{\mathsf{E}}_i T^*$	$(\mathsf{E}_i T^*)_{\text{fo}}$	$\varepsilon_{\text{fo}}\%$	$(\mathsf{E}_i T^*)_{\text{ho}}$	$\varepsilon_{\text{ho}}\ \%$
H_0	0.01	4.31	0.011	58.65	28.97	50.60	59.62	1.65
H_1	0.001	4.31	0.001	49.98	28.97	42.04	49.98	0.00
H_0	0.001	6.61	0.0011	87.10	47.39	45.59	88.66	1.79
H_1	0.0001	6.61	0.0001	69.30	47.39	31.61	68.40	1.29

4.4 Asymptotic Optimality of the MSPRT in the General Non-iid Case

We now generalize the results of Section 3.4 to the case of multiple hypotheses and very general non-iid models. Since the techniques are similar to the case of two hypotheses, most of the proofs are omitted. The details may be found in Tartakovsky [455].

Let $\mathcal{I}_i(t) = \mathsf{E}_i[\lambda_i(t)]$ be the accumulated K–L information in the trajectory $\mathbf{X}_0^t$. As in (3.141) and (3.142), assume that there are a nonnegative increasing function $\psi(t)$ ($\psi(\infty) = \infty$) and positive and finite numbers I_i, $i = 0, 1, \dots N$ such that the following conditions hold

$$\lim_{t\to\infty} [\mathcal{I}_i(t)/\psi(t)] = I_i, \quad \frac{\lambda_i(t)}{\psi(t)} \xrightarrow[t\to\infty]{\mathsf{P}_i-\text{a.s.}} I_i \quad \text{for } i = 0, 1, \dots, N. \tag{4.69}$$

Write $\lambda_{ij}(t) = \lambda_i(t) - \lambda_j(t) (= \log[d\mathsf{P}_i^t/d\mathsf{P}_j^t])$ and $I_{ij} = I_i - I_j$, and assume that $I_{ij} > 0$. Then, by assumption (4.69),

$$\frac{\lambda_{ij}(t)}{\psi(t)} \xrightarrow[t\to\infty]{\mathsf{P}_i-\text{a.s.}} I_{ij} \quad \text{for } i, j = 0, 1, \dots, N,\ i \neq j, \tag{4.70}$$

where the numbers I_{ij} are positive and finite.

Let $\Psi(t)$ be the inverse function for $\psi(t)$.

We begin with obtaining asymptotic lower bounds for arbitrary moments of the stopping time in the corresponding classes of tests.

Lemma 4.4.1. *Assume there exist a nonnegative increasing function $\psi(t)$, $\psi(\infty) = \infty$, and positive finite constants I_{ij}, $i, j = 0, 1, \dots N$, $i \neq j$ such that*

$$\lim_{L\to\infty} \mathsf{P}_i\left\{\frac{1}{\psi(L)} \sup_{0\le t\le L} \lambda_{ij}(t) \ge (1+\varepsilon) I_{ij}\right\} = 1 \quad \textit{for all } \varepsilon > 0 \textit{ and } i, j = 0, 1, \dots, N. \tag{4.71}$$

Then for all $m > 0$ and all $i = 0, 1, \dots, N$

$$\begin{aligned}
\inf_{\delta\in\mathbf{C}([\alpha_{ij}])} \mathsf{E}_i[T]^m &\ge \left[\Psi\left(\max_{\substack{0\le j\le N\\ j\neq i}} \frac{|\log\alpha_{ji}|}{I_{ij}}\right)\right]^m (1+o(1)) \quad \textit{as } \max_{i,j}\alpha_{ij}\to 0,\\
\inf_{\delta\in\mathbf{C}(\boldsymbol{\alpha})} \mathsf{E}_i[T]^m &\ge \left[\Psi\left(\max_{\substack{0\le j\le N\\ j\neq i}} \frac{|\log\alpha_j|}{I_{ij}}\right)\right]^m (1+o(1)) \quad \textit{as } \max_{i}\alpha_i\to 0,\\
\inf_{\delta\in\mathbf{C}(\boldsymbol{\beta})} \mathsf{E}_i[T]^m &\ge \left[\Psi\left(\frac{|\log\beta_i|}{\min_{\substack{0\le j\le N\\ j\neq i}} I_{ij}}\right)\right]^m (1+o(1)) \quad \textit{as } \max_{i}\beta_i\to 0.
\end{aligned} \tag{4.72}$$

Proof. The proof runs along the lines of the proof of Lemma 3.4.1. See [455] for details. □

Note that if the almost sure convergence conditions (4.70) hold and, for all finite L, $\mathsf{P}_i\{\sup_{0\le t\le L}[\lambda_{ij}(t)]^+ < \infty\} = 1$, then the conditions (4.71) are satisfied; see Remark 3.4.1.

Next, strengthening the almost sure convergence (4.70) into the r-quick convergence, i.e., requiring $\mathsf{E}_i[L_{ij}(\varepsilon)]^r < \infty$ for all $\varepsilon > 0$, where $L_{ij}(\varepsilon) = \sup\{t : |\lambda_{ij}(t)/\psi(t) - I_{ij}| > \varepsilon\}$, in a similar way as in the proof of Theorem 3.4.2, we may obtain the upper bounds for moments of the stopping time and prove asymptotic optimality of the MSPRT. The exact result is given in the next theorem.

Theorem 4.4.1 (MSPRT asymptotic optimality). *Assume that there exist finite positive numbers I_{ij}, $i, j = 0, 1, \dots, N$, $i \ne j$ and an increasing nonnegative function $\psi(t)$ such that for some $r > 0$*

$$\frac{\lambda_{ij}(t)}{\psi(t)} \xrightarrow[t\to\infty]{\mathsf{P}_i-r-quickly} I_{ij} \quad \text{for all } i, j = 0, 1, \dots, N, i \ne j. \tag{4.73}$$

Then the following assertions are true.

(i) *For all $0 < m \le r$ and $i = 0, 1, \dots, N$,*

$$\mathsf{E}_i[T^*]^m \sim \left[\Psi\left(\max_{\substack{0\le j\le N\\ j\ne i}} \frac{a_{ji}}{I_{ij}}\right)\right]^m \quad \text{as } \min_{j,i} a_{ji} \to \infty. \tag{4.74}$$

(ii) *If the thresholds are selected so that $\alpha_{ij}(\delta^*) \le \alpha_{ij}$ and $a_{ji} \sim \log(1/\alpha_{ji})$, in particular $a_{ji} = \log(1/\alpha_{ji})$, then for all $0 < m \le r$ and $i = 0, 1, \dots, N$,*

$$\inf_{\delta\in\mathbf{C}([\alpha_{ij}])} \mathsf{E}_i[T]^m \sim \mathsf{E}_i[T^*]^m \sim \left[\Psi\left(\max_{\substack{0\le j\le N\\ j\ne i}} \frac{|\log \alpha_{ji}|}{I_{ij}}\right)\right]^m \quad \text{as } \max_{i,j} \alpha_{ij} \to 0. \tag{4.75}$$

(iii) *If the thresholds are selected so that $\alpha_i(\delta^*) \le \alpha_i$ and $a_{ji} \sim \log(1/\alpha_j)$, in particular $a_{ji} = \log(N/\alpha_j)$, then for all $0 < m \le r$ and $i = 0, 1, \dots, N$,*

$$\inf_{\delta\in\mathbf{C}(\boldsymbol{\alpha})} \mathsf{E}_i[T]^m \sim \mathsf{E}_i[T^*]^m \sim \left[\Psi\left(\max_{\substack{0\le j\le N\\ j\ne i}} \frac{|\log \alpha_j|}{I_{ij}}\right)\right]^m \quad \text{as } \max_i \alpha_i \to 0. \tag{4.76}$$

(iv) *If the thresholds are selected so that $\beta_i(\delta^*) \le \beta_i$ and $a_{ji} \sim \log(1/\beta_i)$, in particular $a_{ji} = \log(\beta_i^{-1}\sum_{k\ne i} w_{ki})$, then for all $0 < m \le r$ and $i = 0, 1, \dots, N$,*

$$\inf_{\delta\in\mathbf{C}(\boldsymbol{\beta})} \mathsf{E}_i[T]^m \sim \mathsf{E}_i[T^*]^m \sim \left[\Psi\left(\frac{|\log \beta_i|}{\min_{\substack{0\le j\le N\\ j\ne i}} I_{ij}}\right)\right]^m \quad \text{as } \max_i \beta_i \to 0. \tag{4.77}$$

Consequently, the MSPRT minimizes asymptotically the moments of the stopping time distribution up to order r for all hypotheses $H_0, \dots, H_N$ in the corresponding classes of tests.

Proof. Proof of (i). Replacing α_{ji} in (4.72) with $e^{-a_{ji}}$, we obtain the lower bound

$$\mathsf{E}_i[T^*]^m \ge \left[\Psi\left(\max_{j\ne i} \frac{a_{ji}}{I_{ij}}\right)\right]^m (1 + o(1)) \quad \text{as } \min_{j,i} a_{ji} \to \infty.$$

Thus, to prove (4.74) it suffices to show that this lower bound is also an upper bound.

Define $\widetilde{\lambda}_{ij}(t) = \lambda_{ij}(t)/I_{ij}$, $\widetilde{a}_{ji} = a_{ji}/I_{ij}$,

$$\widetilde{T}_i = \inf\left\{t : \min_{j\neq i} \widetilde{\lambda}_{ij}(t) \geq \max_{j\neq i} \widetilde{a}_{ji}\right\},$$

$\widetilde{L}_{ij}(\varepsilon) = \sup\left\{t : |\widetilde{\lambda}_{ij}(t)/\psi(t) - 1| > \varepsilon\right\}$, and $\widetilde{L}_i(\varepsilon) = \max_{j\neq i} \widetilde{L}_{ij}(\varepsilon)$.

Note that the Markov time T_i given in (4.5) can be written as

$$T_i = \inf\left\{t : \min_{j\neq i}[\widetilde{\lambda}_{ij}(t) - \widetilde{a}_{ji}] \geq 0\right\}.$$

Clearly, $T^* = \min_k T_k \leq T_i \leq \widetilde{T}_i$, so it suffices to show that for all $0 < m \leq r$

$$\mathsf{E}_i[\widetilde{T}_i]^m \leq \left[\Psi\left(\max_{j\neq i} \widetilde{a}_{ji}\right)\right]^m (1 + o(1)) \quad \text{as } \min_{j,i} a_{ji} \to \infty. \tag{4.78}$$

To this end, observe that on one hand

$$\min_{j\neq i} \widetilde{\lambda}_{ij}(\widetilde{T}_i - 1) < \max_{j\neq i} \widetilde{a}_{ji} \quad \text{on } \{\widetilde{T}_i < \infty\},$$

and on the other hand

$$\widetilde{\lambda}_{ij}(\widetilde{T}_i - 1) > \psi(\widetilde{T}_i - 1)(1 - \varepsilon) \quad \text{on } \{\widetilde{T}_i - 1 > \widetilde{L}_i(\varepsilon)\}.$$

(In the continuous-time case, $\widetilde{T}_i - 1$ can be replaced with $\widetilde{T}_i - \Delta$, where Δ is a small positive number.) Hence,

$$\widetilde{T}_i < 1 + \Psi\left(\frac{\max_{j\neq i} \widetilde{a}_{ji}}{1 - \varepsilon}\right) \quad \text{on } \{\widetilde{L}_i(\varepsilon) + 1 < \widetilde{T}_i < \infty\},$$

and we obtain that for every $0 < \varepsilon < 1$,

$$\begin{aligned}\widetilde{T}_i &\leq 1 + \mathbb{1}_{\{\widetilde{T}_i > \widetilde{L}_i(\varepsilon)+1\}} \Psi\left(\frac{\max_{j\neq i} \widetilde{a}_{ji}}{1 - \varepsilon}\right) + \mathbb{1}_{\{\widetilde{T}_i \leq \widetilde{L}_i(\varepsilon)+1\}} \widetilde{L}_i(\varepsilon) \\ &\leq 1 + \Psi\left(\frac{\max_{j\neq i} \widetilde{a}_{ji}}{1 - \varepsilon}\right) + \widetilde{L}_i(\varepsilon),\end{aligned} \tag{4.79}$$

which implies that for all $0 < \varepsilon < 1$

$$\mathsf{E}_i[\widetilde{T}_i]^m \leq \left\{1 + \Psi\left(\frac{\max_{j\neq i} \widetilde{a}_{ji}}{1 - \varepsilon}\right) + \mathsf{E}_i[\widetilde{L}_i(\varepsilon)]\right\}^m.$$

Since by the r-quick convergence condition (4.73)

$$\mathsf{E}_i[\widetilde{L}_i(\varepsilon)]^r \leq N \max_{j\neq i} \mathsf{E}_i[\widetilde{L}_{ij}(\varepsilon)] < \infty,$$

the upper bound (4.78) follows and the proof of (i) is complete.

Proof of (ii)–(iv). The asymptotic equalities (4.75), (4.76), and (4.77) follow from the asymptotic lower bounds (4.72) and the asymptotic equalities (4.74) with $a_{ji} \sim \log(1/\alpha_{ji})$, $a_{ji} \sim \log(1/\alpha_j)$, and $a_{ji} \sim \log(1/\beta_i)$, respectively. □

Remark 4.4.1. As the example given in Subsection 3.4.3 shows, the r-quick convergence conditions in Theorem 4.4.1 cannot be generally relaxed into the almost sure convergence. However, the weak asymptotic optimality result established in Theorem 3.4.1 for the SPRT also holds for the MSPRT. Specifically, in the conditions of Lemma 4.4.1 with $\psi(t) = t^k$, $k > 0$, for every $0 < \varepsilon < 1$,

$$\inf_{\delta \in \mathbf{C}([\alpha_{ij}])} \mathsf{P}_i(T > \varepsilon T^*) \to 1 \quad \text{as } \max_{j,i} \alpha_{ji} \to 0 \text{ for all } \quad i = 0, 1, \ldots, N \tag{4.80}$$

whenever the thresholds a_{ji} are selected as in Theorem 4.4.1(ii). These asymptotic relations also hold for the classes $\mathbf{C}(\boldsymbol{\alpha})$ and $\mathbf{C}(\boldsymbol{\beta})$ if the thresholds are selected as prescribed in Theorem 4.4.1(iii)–(iv).

4.5 Invariant Multihypothesis Sequential Probability Ratio Test

So far we considered the case of simple multiple hypotheses. However, as we already mentioned in Section 3.6, in many applications the models are only known partially up to a set of unknown nuisance parameters or the distributions may even be unknown. In this section, we assume that the probability measure P belongs to a family $\mathscr{P}$ that is invariant under a group of measurable transformations $\mathcal{G}$ on the sample space $\mathcal{X}$, and that the hypotheses are described by $H_i : \mathsf{P} \in \mathscr{P}_i$, $i = 0,1,\dots,N$, where $\mathscr{P}_i \in \mathscr{P}$. In just the same way as in Section 3.6, we restrict attention to the invariant sequential tests $\delta \in \mathscr{I}$ with respect to the group $\mathcal{G}$ that leaves the problem invariant, i.e., we consider the classes $\mathbf{C}([\alpha_{ij}])$, $\mathbf{C}(\boldsymbol{\alpha})$, and $\mathbf{C}(\boldsymbol{\beta})$ of invariant tests with the constraints (4.1). With some abuse of notation, we use the same notation for the classes of invariant tests with the corresponding constraints.

Let $\{\mathbb{M}_t\}$ be a maximal invariant statistic with respect to the group $\mathcal{G}$, and let $\lambda_i(t) = \log[\mathrm{d}\mathsf{P}_i^t(\mathbb{M}_t)/\mathrm{d}\mathsf{Q}^t(\mathbb{M}_t)]$ and $\lambda_{ij}(t) = \lambda_i(t) - \lambda_j(t) = \log[\mathrm{d}\mathsf{P}_i^t(\mathbb{M}_t)/\mathrm{d}\mathsf{P}_j^t(\mathbb{M}_t)]$, $i,j = 0,\dots,N$, be the LLRs constructed based on the maximal invariant. If now we use these invariant versions of the LLR processes in the MSPRT, then the resulting invariant MSPRT (IMSPRT) $\delta^* = (d^*, T^*)$ is asymptotically optimal in the corresponding classes of invariant tests as long as the r-quick convergence conditions (4.73) hold for the invariant LLRs. More specifically, the following result is true.

Theorem 4.5.1 (Asymptotic optimality of the IMSPRT). *Consider the problem of testing the hypotheses $H_i : \mathsf{P} \in \mathscr{P}_i$, $i = 0,1,\dots,N$, $\mathscr{P}_i \in \mathscr{P}$. Assume that the family $\mathscr{P}$ is invariant under a group of measurable transformations $\mathcal{G}$. Let $\lambda_{ij}(t) = \log[\mathrm{d}\mathsf{P}_i^t(\mathbb{M}_t)/\mathrm{d}\mathsf{P}_j^t(\mathbb{M}_t)]$, $i,j = 0,1,\dots,N$ be the "invariant" LLRs constructed based on the maximal invariant $\{\mathbb{M}_t\}$ with respect to the group $\mathcal{G}$. If the r-quick convergence conditions* (4.73) *are satisfied for the invariant LLRs $\lambda_{ij}(t)$, then the statements (i)–(iv) of Theorem 4.4.1 are true in the classes of invariant multihypothesis tests, so that the IMSPRT asymptotically minimizes the moments of the stopping time distribution up to the order r for all hypotheses $H_0,\dots,H_N$ in the corresponding classes of invariant tests.*

The examples considered in Subsections 3.6.1–3.6.3 can be relatively easily extended to the case of multiple hypotheses. In particular, consider a multihypothesis version of the binary problem studied in Subsection 3.6.2, assuming that $X_n \sim \mathcal{N}(\mu, \sigma^2)$, $n = 1,2,\dots$ are iid normal with unknown mean μ and unknown variance σ^2, and the hypotheses to be tested are $H_i : \mu/\sigma = q_i$, $i = 0,1,\dots,N$, where $q_0, q_1, \dots, q_N$ are given numbers. The problem is invariant under the group of scale changes, and the maximal invariant is $\mathbb{M}_n = (1, X_2/X_1, \dots, X_n/X_1)$. For $n \geq 1$, write

$$Y_n = X_n/X_1, \quad s_n^2 = n^{-1}\sum_{k=1}^n Y_k^2, \quad \text{and } t_n = \overline{Y}_n/s_n = \left[n^{-1}\sum_{k=1}^n X_k\right] \Big/ \left[n^{-1}\sum_{k=1}^n X_k^2\right]^{1/2}.$$

By (3.219),

$$p(\mathbb{M}_n|H_i) = \frac{1}{\sqrt{2\pi(n-1)ns_n^{2(n-1)}}} \int_0^\infty u^{-1} \exp\{nf(u, q_i t_n)\}\, \mathrm{d}u, \quad i = 0,1,\dots,N,$$

where $f(u,z) = -u^2/2 + zu + \log u$. Therefore, the invariant LLRs are given by

$$\lambda_{ij}(n) = \log\left[\frac{\int_0^\infty u^{-1} \exp\{nf(u, q_i t_n)\}\, \mathrm{d}u}{\int_0^\infty u^{-1} \exp\{nf(u, q_j t_n)\}\, \mathrm{d}u}\right], \quad i,j = 0,1,\dots,N.$$

Using the same argument as in Subsection 3.6.2, we obtain (see (3.226))

$$\lambda_{ij}(n) = n\left[\phi(q_i t_n) - \phi(q_j t_n) - (q_i^2 - q_j^2)/2\right] + \Delta_{ij}(t_n) + o(1),$$

where $\phi(t_n)$ is defined in (3.223), $o(1) \to 0$ as $n \to \infty$ uniformly in $|t_n| \leq 1$ and $\Delta_{ij}(t_n)$ is given in (3.227) with q_1 replaced with q_i and q_0 with q_j. Clearly, $|\Delta_{ij}(t_n)|$ is bounded by a finite constant. Hence, for some positive constants C_{ij},

$$\left|n^{-1}\lambda_{ij}(n) - g_{ij}(t_n)\right| \le C_{ij}/n, \quad n \ge 1, \;\; i,j = 0,1,\dots,N, \tag{4.81}$$

where

$$g_{ij}(t_n) = \phi(q_i t_n) - \phi(q_j t_n) - \tfrac{1}{2}(q_i^2 - q_j^2).$$

It follows from (4.81) that $n^{-1}\lambda_{ij}(n)$ converges r-quickly to $g_{ij}(Q_i)$ as long as t_n converges r-quickly to Q_i. Since $\mathsf{E}_i|X_1|^r < \infty$ for all $r > 0$, we have

$$t_n \xrightarrow[n\to\infty]{\mathsf{P}_i-r-\text{quickly}} \frac{\mathsf{E}_i X_1}{\sqrt{\mathsf{E}_i X_1^2}} = \frac{q_i}{\sqrt{1+q_i^2}} = Q_i \quad \text{for all } i = 0,1,\dots,N \text{ and } r > 0.$$

Thus, the conditions (4.73) hold for all $r > 0$ with $\psi(n) = n$ and

$$I_{ij} = g_{ij}(Q_i), \quad i,j = 0,1,\dots,N, \tag{4.82}$$

where $g_{ij}(Q_i) > 0$ by the same argument as that provided in the end of Subsection 3.6.2. By Theorem 4.5.1, the IMSPRT is asymptotically optimal with respect to all positive moments of the sample size in the corresponding classes. In particular, for all $m > 0$ and $i = 0,1,\dots,N$,

$$\inf_{\delta\in\mathbf{C}([\alpha_{ij}])} \mathsf{E}_i T^m \sim \mathsf{E}_i[T^*]^m \sim \left(\max_{j\neq i} \frac{|\log \alpha_{ji}|}{g_{ij}(Q_i)}\right)^m \quad \text{as } \max_{i,j} \alpha_{ij} \to 0.$$

Further examples related to the IMSPRT optimality are considered in the next subsection (Examples 4.6.3 and 4.6.4).

4.6 Multisample Slippage Problems

In this section we apply the general results obtained above to a specific multiple decision problem called the *slippage problem*; see Ferguson [143], Mosteller [304], Tartakovsky [454]. Suppose there are N populations whose distribution functions $F(x-\theta_1),\dots,F(x-\theta_N)$ are identical except possibly for different shifts $\theta_1,\dots,\theta_N$. On the basis of samples from the N populations we wish to decide whether or not the populations are equal or one of these populations has slipped to the right of the rest and, if so, which one is the odd one. In other words, we may ask whether or not for some i, we have $\theta_i > \theta_k = \theta$, for all $k \neq i$. In the language of hypothesis testing we want to test the null hypothesis $H_0 : \theta_1 = \theta_2 = \cdots = \theta_N = \theta$ against N alternatives $H_i : \theta_i = \theta + \Delta$, $i = 1,\dots,N$, where $\Delta \neq 0$. The slippage problem is of considerable practical importance and closely related to the so-called ranking and selection problem in which the goal is to select the best population [56].

A problem of this kind was first discussed by Mosteller [304] in a non-sequential setting for the nonparametric case when both the form of distribution function $F(\cdot)$ and the values of θ and Δ are unknown. Ferguson [143] generalized this result for the case of arbitrary but known distributions $F_0(x)$ and $F_1(x)$ not necessarily with different means, i.e., when, under hypothesis H_0, all N populations have the same distribution $F_0(\cdot)$ and, under H_i, the i-th representative has a different distribution $F_1(\cdot)$. Tartakovsky [454] considered the case of possibly different and unknown distribution functions $F_i(\cdot)$, $i = 1,\dots,N$, in a minimax non-sequential setting. As a result, the minimax–invariant solution to this problem has been obtained.

In addition to ranking/selection problems, another interesting application of this model is object or signal detection in a multichannel system. There may be no object (signal) at all (hypothesis H_0) or an object may be present in one of the N channels, in the i-th, say (hypothesis H_i). It is necessary to detect an object as soon as possible and to identify the channel where it is located. This important practical problem will be emphasized in the subsequent examples. Moreover, we will consider a more general case of possibly correlated and non-identically distributed observations in each population while the populations will be assumed statistically mutually independent.

To be specific, let $\mathbf{X}_t = (X_{1,t}, \dots, X_{N,t})$, $N \geq 2$, be an N-component process observed either in discrete or continuous time. The component $X_{j,t}$ corresponds to the observation in the j-th channel or corresponds to the j-th population, and it is assumed that all components may be observed simultaneously and may have a fairly general structure. We also suppose that they are mutually independent. Write $\mathbf{X}^t = \{\mathbf{X}_u, 0 \leq u \leq t\}$ and $\mathbf{X}_k^t = \{X_{k,u}, 0 \leq u \leq t\}$. Therefore, if we select $\mathsf{Q} = \mathsf{P}_0$ to be the measure associated with the null hypothesis H_0, then the LLR writes

$$\lambda_k(t) := \log \frac{d\mathsf{P}_k^t}{d\mathsf{P}_0^t}(\mathbf{X}^t) = \log \frac{d\mathsf{P}_k^t}{d\mathsf{P}_0^t}(\mathbf{X}_k^t),$$

so that the LLR $\lambda_{ij}(t) = \lambda_i(t) - \lambda_j(t)$ depends on the observation process $\mathbf{X}_t$ through only the components $X_{i,t}$ and $X_{j,t}$.

We now consider several particular examples.

Example 4.6.1 (Detection–identification of a signal in Gaussian noise in a multichannel system—continuous time). We begin with generalizing the examples considered in Subsections 3.4.3 and 3.4.4 to the multichannel case.

Suppose that for $t \in \mathbb{R}_+$ the i-th component admits a stochastic differential equation

$$dX_{i,t} = \begin{cases} S_i(t)dt + V_i(t)dt + \sigma_i dW_i(t) & \text{if } H_i \text{ is true} \\ V_i(t)dt + \sigma_i dW_i(t) & \text{if } H_i \text{ is wrong,} \end{cases}$$

where $S_i(t)$ is a deterministic signal, $W_i(t)$ is a standard Brownian motion, $V_i(t)$ is an L_2-continuous Gaussian process, and $\sigma_i > 0$. The processes $W_i(t)$ and $W_j(t)$ are independent, so are $V_i(t)$ and $V_j(t)$.

By (3.164),

$$\lambda_k(t) = \frac{1}{\sigma_k^2} \int_0^t \widetilde{S}_k(u)\, d\widetilde{X}_{k,u} - \frac{1}{2\sigma_i^2} \int_0^t \widetilde{S}_k^2(u)\, du, \quad k = 1, \dots, N, \tag{4.83}$$

where

$$\widetilde{S}_k(t) = S_k(t) - \int_0^t S_k(u) C_k(t,u)\, du, \quad d\widetilde{X}_{k,t} = dX_{k,t} - \left[\int_0^t C_k(t,u)\, dX_{k,u}\right] dt.$$

Using the same argument that led to (3.166), we obtain that the LLR $\lambda_k^j(t)$ under the hypothesis H_j can be written as

$$\lambda_k^j(t) = \begin{cases} \frac{1}{\sigma_k} \int_0^t \widetilde{S}_k(u)\, d\widetilde{W}_k(u) + \frac{1}{2}\mu_k(t) & \text{if } k = j \\ \frac{1}{\sigma_k} \int_0^t \widetilde{S}_k(u)\, d\widetilde{W}_k(u) - \frac{1}{2}\mu_k(t) & \text{if } k \neq j, \end{cases} \tag{4.84}$$

where $\mu_k(t) = \sigma_k^{-2} \int_0^t \widetilde{S}_k^2(u)\, du$ is the cumulative SNR in the k-th channel at the output of the whitening filter over the time interval $[0,t]$ and $\widetilde{W}_k(t)$ is the Brownian motion statistically indistinguishable from $W_k(t)$.

Using (4.84) and assuming that for some $\ell > 0$

$$\lim_{t \to \infty} \frac{1}{\sigma_k^2 t^\ell} \int_0^t \widetilde{S}_k^2(u)\, du = q_k, \tag{4.85}$$

where $0 < q_k < \infty$, it can be shown in just the same way as in Subsection 3.4.3 that for all $r > 0$

$$t^{-\ell} \lambda_{ij}(t) \xrightarrow[t \to \infty]{\mathsf{P}_i - r - \text{quickly}} \frac{1}{2}(q_i + q_j), \quad i,j = 0,1,\dots,N, i \neq j,$$

where $q_0 = 0$. Thus, by Theorem 4.4.1, the MSPRT asymptotically minimizes all positive moments of the stopping time. In particular, for all $m > 0$

$$\begin{aligned}
&\inf_{\delta \in \mathbf{C}([\alpha_{ij}])} \mathsf{E}_i T^m \sim \mathsf{E}_i [T^*]^m \sim \max\left(\max_{1\le j\le N} \frac{2|\log \alpha_{0j}|}{q_j}, \max_{\substack{1\le j\le N\\ j\neq i}} \frac{2|\log \alpha_{ji}|}{q_i + q_j} \right)^{m/\ell}, \quad i \neq 0, \\
&\inf_{\delta \in \mathbf{C}([\alpha_{ij}])} \mathsf{E}_0 T^m \sim \mathsf{E}_0 [T^*]^m \sim \left(\max_{1\le j\le N} \frac{2|\log \alpha_{j0}|}{q_j} \right)^{m/\ell}.
\end{aligned} \tag{4.86}$$

In this problem, α_{0i} can be interpreted as the probability of a false alarm raising in the i-th channel, α_{j0} as the probability of missing the object when it is located in the j-th channel and α_{ij} $(i, j = 1, \dots, N,\ i \neq j)$ as the probabilities of misidentification of channels where the object is located. If we set $\alpha_{0i} = \alpha_0$, $\alpha_{j0} = \alpha_1$, and $\alpha_{ij} = \alpha_2$ (which is reasonable) and take into account that usually $\alpha_0 \ll \min(\alpha_1, \alpha_2)$, then we get from (4.86)

$$\begin{aligned}
&\inf_{\delta \in \mathbf{C}([\alpha_{ij}])} \mathsf{E}_i T^m \sim \mathsf{E}_i [T^*]^m \sim \left(\frac{2|\log \alpha_0|}{\min_{1\le j\le N} q_j} \right)^{m/\ell}, \quad i = 1, \dots, N \\
&\inf_{\delta \in \mathbf{C}([\alpha_{ij}])} \mathsf{E}_0 T^m \sim \mathsf{E}_0 [T^*]^m \sim \left(\frac{2|\log \alpha_1|}{\min_{1\le j\le N} q_j} \right)^{m/\ell}.
\end{aligned} \tag{4.87}$$

Consider now a multichannel generalization of the problem treated in Subsection 3.4.4. In other words, we assume that the i-th component has the stochastic differential

$$dX_{i,t} = \begin{cases} S_i(t)dt + \sigma_i dW_i(t) & \text{if } H_i \text{ is true} \\ \sigma_i dW_i(t) & \text{if } H_i \text{ is wrong,} \end{cases}$$

where $W_i(t)$, $i = 1, \dots, N$ are mutually independent standard Brownian motions, and the signals $S_1(t), \dots, S_N(t)$ are statistically independent stationary Markov Gaussian processes,

$$\mathsf{E}[S_i(t)] = 0, \quad \mathsf{E}[S_i(t)S_i(t+u)] = d_i^2 \exp(-\rho_i |u|), \quad \mathsf{E}[S_i(t)S_j(t)] = 0, \quad i \neq j,$$

where $\rho_i > 0$, $d_i > 0$.

Let $\hat{S}_i(t) = \mathsf{E}_i[S_i(t) | \mathscr{F}_t^{X_i}]$. Using (3.182) along with the fact that $X_{i,t}$ and $X_{j,t}$ are mutually independent, we obtain

$$\begin{aligned}
\lambda_{ij}(t) = {} & \frac{1}{\sigma_i^2} \int_0^t \hat{S}_i(u)\, \mathrm{d}X_i(u) - \frac{1}{\sigma_j^2} \int_0^t \hat{S}_j(u)\, \mathrm{d}X_j(u) \\
& - \frac{1}{2} \int_0^t \{[\hat{S}_i^2(u)/\sigma_i^2] - [\hat{S}_j^2(u)/\sigma_j^2]\}\, \mathrm{d}u,
\end{aligned} \tag{4.88}$$

where $\hat{S}_0(t) = 0$. Again since the processes $X_i(t)$ are Gaussian, the functionals $\hat{S}_i(t)$ (estimators of the signals $S_i(t)$) are linear and can be computed using the systems of Kalman–Bucy equations (3.185)–(3.186). For the sake of simplicity, we provide the final results for the symmetric case assuming that $\rho_i = \rho, \sigma_i = \sigma, d_i = d$. Using (4.88) and a similar quite cumbersome argument as in Subsection 3.4.4, it may be shown that for all $r > 0$

$$\frac{1}{t}\lambda_{ij}(t) \xrightarrow[t\to\infty]{\mathsf{P}_i - r-\text{quickly}} I_{ij} = \begin{cases} \frac{Q}{\sqrt{1+Q}(1+\sqrt{1+Q})} \frac{d^2}{2\sigma^2} & \text{for } i \neq 0 \\ \frac{Q}{\sqrt{1+Q}(1+\sqrt{1+Q})^2} \frac{d^2}{2\sigma^2} & \text{for } i = 0, \end{cases} \tag{4.89}$$

where $Q = 2d^2/(\rho\sigma^2)$ is the SNR. All the details can be found in Tartakovsky [455].

Therefore, the MSPRT is asymptotically optimal with respect to all positive moments of the stopping time distribution.

Example 4.6.2 (Detection–identification of a deterministic signal in Gaussian correlated noise in a multichannel system—discrete time). Let $t = n \in \mathbb{Z}_+$ and

$$X_{i,n} = \begin{cases} S_i(n) + \xi_i(n) & \text{if } H_i \text{ is true} \\ \xi_i(n) & \text{if } H_i \text{ is wrong,} \end{cases}$$

where the functions $S_1(n), \dots, S_N(n)$ are deterministic and the noise processes $\xi_i(n)$, $n = 0, 1, \dots,$ are stable first-order autoregressive Gaussian processes, i.e.,

$$\xi_i(n) = \gamma_i \xi_i(n-1) + \zeta_i(n), \quad n \geq 1, \; \xi_i(0) = 0,$$

where $\zeta_i(1), \zeta_i(2), \dots$ are iid Gaussian variables with zero mean and variance σ_i^2, $\zeta_i(n)$ and $\zeta_j(n)$ are independent, and $|\gamma_i| < 1$. It is easy to show that the LLRs are of the form

$$\lambda_{ij}(n) = \sum_{k=1}^{n} \widetilde{S}_i(k)\widetilde{X}_{i,k} - \sum_{k=1}^{n} \widetilde{S}_j(k)\widetilde{X}_{j,k} - \frac{1}{2}\sum_{k=1}^{n} \left[\widetilde{S}_i^2(k) - \widetilde{S}_j^2(k)\right], \quad i,j = 0,1,\dots,N, \; i \neq j,$$

where $\widetilde{X}_{i,k} = \sigma_i^{-1}[X_{i,k} - \gamma_i X_{i,k-1}]$, $\widetilde{S}_i(k) = \sigma_i^{-1}[S_i(k) - \gamma_i S_i(k-1)]$ for $k \geq 2$ and $\widetilde{X}_{i,1} = X_{i,1}$, $\widetilde{S}_i(1) = \sigma_i^{-1} S_i(1)$ ($\widetilde{S}_0(k) = 0$). Hence, under H_i,

$$\lambda_{ij}^{(i)}(n) = \frac{1}{2}\sum_{k=1}^{n} \left[\widetilde{S}_i^2(k) + \widetilde{S}_j^2(k)\right] + \frac{1}{\sigma_i}\sum_{k=1}^{n} \widetilde{S}_i(k)\zeta_i(k) - \frac{1}{\sigma_j}\sum_{k=1}^{n} \widetilde{S}_j(k)\zeta_j(k).$$

Let the accumulated SNR for the channel i up to time n be denoted by $\mu_i(n) = \sum_{k=1}^{n} \widetilde{S}_i^2(k)$. Assume that

$$\lim_{n\to\infty} n^{-\ell}\mu_i(n) = q_i \quad \text{for some } \ell > 0, \tag{4.90}$$

where q_i, $i = 1, \dots, N$, are finite positive numbers. Now we establish that

$$n^{-\ell}\lambda_{ij}(n) \to (q_i + q_j)/2 \quad \mathsf{P}_i - r - \text{quickly for all } r > 0.$$

It is sufficient to show that

$$\sum_{n=1}^{\infty} n^{r-1}\mathsf{P}_i\left(|W_i(n)| > \varepsilon n^{\ell}\right) < \infty \quad \text{for some } \varepsilon > 0 \text{ and all } r > 0, \tag{4.91}$$

where $W_i(n) = \sigma_i^{-1}\sum_{k=1}^{n} \widetilde{S}_i(k)\zeta_i(k)$. Since $W_i(n)$ is a weighted Gaussian sample sum with mean zero and variance $\sim q_i n^{\ell}$ for large n, it is easy to show that there is a number $\gamma < 1$ such that

$$\mathsf{P}_i\left(|W_i(n)| > \varepsilon n^{\ell}\right) \leq O(\gamma^n),$$

and hence (4.91) is fulfilled. Thus, by Theorem 4.4.1, the asymptotic equalities (4.75)–(4.77) hold with $I_{ij} = (q_i + q_j)/2$, $i, j \neq 0$; $I_{i0} = q_i/2$, $I_{0j} = q_j/2$, $i, j = 1, \dots, N$, and the MSPRT is asymptotically optimal.

Considering the symmetry assumptions with respect to the false alarm, misdetection, and misclassification probabilities, as in the previous example, the asymptotic formulas (4.86) become

$$\begin{aligned} \inf_{\delta \in \mathbf{C}([\alpha_{ij}])} \mathsf{E}_i T^m &\sim \mathsf{E}_i[T^*]^m \sim \left(\frac{2|\log \alpha_0|}{q_{\min}}\right)^{m/\ell}, \quad i = 1, \dots, N, \\ \inf_{\delta \in \mathbf{C}([\alpha_{ij}])} \mathsf{E}_0 T^m &\sim \mathsf{E}_0[T^*]^m \sim \left(\frac{2|\log \alpha_1|}{q_{\min}}\right)^{m/\ell}, \end{aligned} \tag{4.92}$$

where $q_{\min} = \min_{1 \leq i \leq N} q_i$ is the minimal SNR in channels.

In particular, if $S_i(n) = \theta_i$, then it is easy to see that $\ell = 1$ works in (4.92). It is also easy to see that the corresponding SNR values are $q_i = \theta_i^2(1-\gamma_i)^2/\sigma_i^2$. Hence, in this case the expected values of stopping times are proportional to the constraints imposed on the error probabilities and inversely proportional to the minimal SNR $q_{\min}$.

Example 4.6.3 (Generalized multisample t-test). Consider the following generalization of the model of Subsection 3.6.2. Under H_0, the components $X_{k,n}$, $n=1,2,\dots$ of the vector $\mathbf{X}_n=(X_{1,n},\dots,X_{N,n})$ are of the form

$$X_{k,n}=\theta+\xi_k(n), \quad k=1,\dots,N,$$

while under H_i, $i=1,\dots,N$,

$$X_{i,n}=\theta+\mu_i+\xi_i(n) \quad \text{and} \quad X_{k,n}=\theta+\xi_k(n) \quad \text{for } k\neq i,$$

where $\xi_i(n)$, $n=1,2,\dots$ are iid Gaussian random variables with zero mean and variance σ^2. Both parameters, the mean θ and the variance σ^2, are unknown nuisance parameters, and we wish to test $N+1$ hypotheses

$$H_0: \mu_k/\sigma=0 \quad \text{for all } k=1,\dots N;$$

$$H_i: \mu_k/\sigma=0 \text{ for } k\neq i \quad \text{and} \quad \mu_i/\sigma=q_i, \; i=1,\dots,N,$$

where $q_1,\dots,q_N$ are given positive numbers. (The extension to arbitrary, possibly negative numbers is straightforward.)

This problem is invariant under changes in the location and the scale, and the maximal invariant is $\mathbb{M}_n=(Y_1^n,\dots,Y_N^n)$, where

$$Y_k^n=(Y_{k,1},\dots,Y_{k,n}), \quad Y_{k,n}=[X_{k,n}-X_{1,1}]/[X_{2,1}-X_{1,1}], \quad Y_{1,1}=0, \quad Y_{2,1}=1.$$

It may be shown that under H_i the distribution of $\mathbb{M}_n$ has density

$$\begin{aligned} p(\mathbb{M}_n|H_i) &= \frac{1}{\sqrt{(2\pi)^{Nn-1}Nns_{nN}^{2(Nn-1)}}}\exp\left\{-\frac{q_i^2(N-1)n}{2N}\right\} \\ &\times\int_0^\infty u^{Nn-2}\exp\left\{-\frac{Nn}{2}u^2+q_i\frac{\sum_{k=1}^n[Y_{i,k}-\overline{Y}_{nN}]}{s_{nN}}u\right\}\,du, \end{aligned} \tag{4.93}$$

where

$$\overline{Y}_{nN}=(Nn)^{-1}\sum_{k=1}^n\sum_{j=1}^N Y_{j,k}, \quad s_{nN}^2=(Nn)^{-1}\sum_{k=1}^n\sum_{j=1}^N[Y_{j,k}-\overline{Y}_{nN}]^2.$$

Define

$$\overline{X}_{nN}=\frac{1}{nN}\sum_{k=1}^n\sum_{j=1}^N X_{j,k}, \quad t_{i,n}^N=\frac{(Nn)^{-1}\sum_{k=1}^n[X_{i,k}-\overline{X}_{nN}]}{\left\{(Nn)^{-1}\sum_{k=1}^n\sum_{j=1}^N[X_{j,k}-\overline{X}_{nN}]^2\right\}^{1/2}},$$

$$J_n^N(z)=\int_0^\infty u^{-2}\exp[nNf(u,z)]\,du,$$

where $f(u,z)$ is defined in (3.221). By (4.93), the LLRs for the maximal invariant statistic are of the form

$$\lambda_{ij}(n)=\log\frac{J_n^N(q_it_{i,n}^N)}{J_n^N(q_jt_{j,n}^N)}-\frac{N-1}{2N}(q_i^2-q_j^2)n, \quad i,j=0,1,\dots,N,\; i\neq j.$$

It is difficult to compute the LLRs exactly, even more difficult than for the t-test in Subsection 3.6.2. However, again using the uniform version of the Laplace integration method and computations similar to but slightly more complicated than those in Subsection 3.6.2, it may be shown that there are positive constants C_{ij} such that

$$\left|\frac{1}{n}\lambda_{ij}(n)-g_{ij}^N(t_{i,n}^N,t_{j,n}^N)\right|\leq\frac{1}{n}C_{ij}, \quad i\neq j, \quad n=1,2,\dots, \tag{4.94}$$

where

$$g_{ij}^N(x,y) = N\left[\phi(x) - \phi(y)\right] - \frac{N-1}{2N}(q_i^2 - q_j^2), \quad i \neq j \tag{4.95}$$

and where $\phi(z)$ is defined in (3.223). The details are omitted and can be found in [455, Sec. 4.2].

It is easily seen that under H_i for any positive r

$$t_{k,n}^N \xrightarrow[n\to\infty]{r-\text{quickly}} \begin{cases} 0, & i=0,\ k=1,\dots,N, \\ -q_i/[N^2\{1+(N-1)q_i^2/N^2\}^{1/2}], & k\neq i,\ i\neq 0, \\ q_i(N-1)/[N^2\{1+(N-1)q_i^2/N^2\}^{1/2}], & k=i,\ i\neq 0, \end{cases} \tag{4.96}$$

which along with (4.94) implies that $n^{-1}\lambda_{ij}(n)$ converge r-quickly for all $r>0$ (under $\mathsf{P}_{i,\theta,\mu,\sigma}$) to positive numbers

$$I_{ij} = g_{ij}^N(Q_{ii}, Q_{ij}) = N\left[\phi(Q_{ii}) - \phi(Q_{ij})\right] - \frac{N-1}{2N}(q_i^2 - q_j^2), \quad i,j \neq 0; \tag{4.97}$$

$$I_{0j} = \frac{N-1}{2N} q_j^2, \quad I_{i0} = N\left[\phi(Q_{ii}) - \log 2\right] - \frac{N-1}{2N} q_i^2, \tag{4.98}$$

where

$$Q_{ii} = \frac{(N-1)q_i^2}{N^2\sqrt{1+(N-1)q_i^2/N^2}}, \quad Q_{ij} = -\frac{q_i q_j}{N^2\sqrt{1+(N-1)q_i^2/N^2}}. \tag{4.99}$$

By Theorem 4.5.1, the invariant MSPRT is asymptotically optimal with respect to all positive moments of the stopping time distribution among all invariant tests within the corresponding classes. Furthermore, it is easily seen that $\min_{j\neq i} q_{ij} = q_{i0}$ for $i=1,\dots,N$, and hence, in the case where $\alpha_i = \alpha$ for $i=1,\dots,N$ and $\alpha_0 \le \alpha$, we obtain from (4.76)

$$\begin{aligned} \inf_{\delta\in\mathbf{C}(\boldsymbol{\alpha})} \mathsf{E}_i T^m \sim \mathsf{E}_i[T^*]^m \sim \left|\frac{2N\log\alpha_0}{(N-1)q_i^2(\rho_{i,N}-1)}\right|^m, \quad i\neq 0, \\ \inf_{\delta\in\mathbf{C}(\boldsymbol{\alpha})} \mathsf{E}_0 T^m \sim \mathsf{E}_0[T^*]^m \sim \left|\frac{2N\log\alpha}{(N-1)(\min_j q_j)^2}\right|^m, \end{aligned} \tag{4.100}$$

where $\rho_{i,N} = 2N^2[\phi(Q_{ii}) - \log 2]/[(N-1)q_i^2]$.

Let us compare the characteristics (4.100) of the IMSPRT with the characteristics of the MSPRT when both parameters θ and σ^2 are known. Using Theorem 4.3.1 and taking into account that $I_{ij} = (q_i^2 + q_j^2)/2$ in (4.22), we see that as $\max_i \alpha_i \to 0$

$$\frac{\mathsf{E}_i[T^*]^m(\mu, \sigma^2 \text{ known})}{\mathsf{E}_i[T^*]^m(\mu, \sigma^2 \text{ unknown})} \sim \begin{cases} [(N-1)/N]^m, & i=0, \\ [(N-1)(\rho_{i,N}-1)/N]^m, & i\neq 0. \end{cases} \tag{4.101}$$

It follows from (4.101) that the performance in the case of known parameters is always superior to the case of unknown parameters, as expected. However, if $N\to\infty$, then $I_{i0} \sim q_i^2/2$, $I_{0j} \sim q_j^2/2$, and hence,

$$\frac{\mathsf{E}_i[T^*]^m(\theta, \sigma^2 \text{ known})}{\mathsf{E}_i[T^*]^m(\theta, \sigma^2 \text{ unknown})} \sim 1, \quad i=0,1,\dots,N.$$

Note that N may approach infinity only in such a way that $\log(N)/|\log\alpha| \to 0$, since $\mathsf{E}_i[T^*]^m = O([\log(N/\alpha)]^m)$. In the case of close hypotheses when $q_i \to 0$ and N is fixed, we obtain

$$\frac{\mathsf{E}_i[T^*]^m(\theta, \sigma^2 \text{ known})}{\mathsf{E}_i[T^*]^m(\theta, \sigma^2 \text{ unknown})} \sim \left(\frac{N-1}{N}\right)^m, \quad i=0,1,\dots,N.$$

If $q_i \to \infty$ and N is fixed, then $\mathsf{E}_0[T^*]^m(\theta,\sigma^2 \text{ known})$ and $\mathsf{E}_0[T^*]^m(\theta,\sigma^2 \text{ unknown})$ differ by the factor of $[(N-1)/N]^m$, but

$$\frac{\mathsf{E}_i[T^*]^m(\theta,\sigma^2 \text{ known})}{\mathsf{E}_i[T^*]^m(\theta,\sigma^2 \text{ unknown})} \sim (N \log q_i)^{-m} \to 0, \quad i=1,\dots,N.$$

Note that for $N=1$ (one population/channel) the problem considered in this example does not have an invariant solution. Thus, the condition $N \geq 2$ appears to be substantial.

We also point out that the problem of testing the hypotheses $H_i : \mu = \mu_i$ does not have an invariant solution when there is the additional nuisance parameter, namely the variance. In fact, following the lines of the proof of [264, Theorem 3.3.1], one can establish that the hypothesis $H : (\mu/\sigma^\varepsilon) = d$ does not admit an invariant testing for $\varepsilon \in [0,1)$. To avoid this complication, adaptive sequential tests may be constructed based on estimates of the nuisance parameters.

Example 4.6.4 (The slippage problem for the exponential distribution with a scale nuisance parameter). Let $\mathcal{E}(\theta)$ be the exponential distribution with density

$$p_\theta(x) = \theta \exp(-\theta x) \mathbb{1}_{\{[0,\infty)\}}(x), \tag{4.102}$$

where θ is an unknown positive scale parameter. The hypotheses H_i, $i=0,1,\dots,N$, are of the form

$$H_0 : X_{k,n} \sim \mathcal{E}(\theta),\ k=1,\dots,N;$$

$$H_i : X_{i,n} \sim \mathcal{E}(\theta_i) \quad \text{and} \quad X_{k,n} \sim \mathcal{E}(\theta),\ k \neq i, \quad i=1,\dots,N.$$

Write $\theta_i/\theta = c_i$, and assume that c_i, $i=1,\dots,N$, are given numbers in the interval $(0,1)$. The case $c_i > 1$ is treated similarly. Thus, under H_0, the observations $X_{1,n},\dots,X_{N,n}$, $n=1,2,\dots$ are iid $\mathcal{E}(\theta)$. Under H_i, the $X_{j,n}$, $n=1,2,\dots$, $j \neq i$, are iid again according to $\mathcal{E}(\theta)$ and $X_{i,n}$, $n=1,2,\dots$ are iid according to $\mathcal{E}(c_i\theta)$, where θ is unknown, $c_i \in (0,1)$, $i=1,\dots,N$, are given, and $X_{i,n}$ and $X_{j,n}$, $i \neq j$, are mutually independent.

Write

$$Y_{k,n} = X_{k,n}/X_{1,1}, \quad Y_k^n = (Y_{k,1},\dots,Y_{k,n}), \quad \mathbb{M}_n = (Y_1^n,\dots,Y_N^n).$$

The problem is invariant under the group of scale changes $\mathcal{G} : G_b(x) = bx$, $b>0$ and $\mathbb{M}_n$ is the maximal invariant. Under H_i, the statistic $\mathbb{M}_n$ has the distribution with density

$$\begin{aligned} p(\mathbb{M}_n|H_i) &= c_i^n \int_0^\infty u^{Nn-1} \exp\left(-uc_i \sum_{j=1}^n Y_{i,j} - u \sum_{j=1}^n \sum_{\substack{k=1 \\ k\neq i}}^N Y_{k,j}\right) \mathrm{d}u \\ &= \frac{c_i^n \Gamma(Nn)}{\left[\sum_{k=1}^N \sum_{j=1}^n Y_{k,j} - (1-c_i)\sum_{j=1}^n Y_{i,j}\right]^{Nn}}, \quad i=0,1,\dots,N, \end{aligned}$$

where $\Gamma(\cdot)$ is the gamma function and where $c_0 = 1$ $(Y_{1,1}=1)$. Using these relations, we obtain that the LLRs are equal to

$$\lambda_{ij}(n) = n\left\{\log(c_i/c_j) + N \log\left[\frac{1-(1-c_j)M_j(n)}{1-(1-c_i)M_i(n)}\right]\right\}, \tag{4.103}$$

where

$$M_k(n) = \frac{n^{-1}\sum_{j=1}^n X_{k,j}}{\sum_{m=1}^N n^{-1}\sum_{j=1}^n X_{m,j}}.$$

Since $\mathsf{E}_i\left|X_{k,1}\right|^s < \infty\ \forall\, s>0$, we have

$$M_k(n) \xrightarrow[n\to\infty]{\mathsf{P}_{i,\theta}-r-\text{quickly}} \frac{\mathsf{E}_{i,\theta} X_{k,1}}{\mathsf{E}_{i,\theta} X_{i,1} + \sum_{m\neq i} \mathsf{E}_{i,\theta} X_{m,1}} \tag{4.104}$$

for every $r > 0$ and $i = 0, 1, \dots, N$. Next, since $\mathsf{E}_{i,\theta} X_{i,1} = (c_i\theta)^{-1}$ and $\mathsf{E}_{i,\theta} X_{m,1} = \theta^{-1}$ for $m \neq i$, we obtain from (4.104) that for all $r > 0$

$$M_k(n) \xrightarrow[t\to\infty]{\mathsf{P}_{i,\theta}-r-\text{quickly}} \begin{cases} [1+(N-1)c_i]^{-1}, & k=i, \\ c_i[1+(N-1)c_i]^{-1}, & k\neq i,\ i\neq 0, \\ N^{-1}, & i=0. \end{cases} \tag{4.105}$$

Now, using (4.103) and (4.105), it is easy to show that the conditions (4.73) are fulfilled for every $r > 0$ with $\ell = 1$ and

$$I_{ij} = \begin{cases} \log(c_i/c_j) + N\log(1+((1-c_i)^2 + c_i(c_j-c_i))/c_iN), & i\neq 0,\ j\neq 0, \\ N\log(1+(1-c_i)/c_iN) - |\log c_i|, & i\neq 0,\ j=0, \\ N\log(1-(1-c_j)/N) + |\log c_j|, & i=0,\ j\neq 0. \end{cases} \tag{4.106}$$

It is also easy to check that I_{ij} are positive and finite for all $i \neq j$ and for arbitrary $0 < c_i < 1$, $i = 1, \dots, N$. Thus, applying Theorem 4.5.1, we conclude that the invariant MSPRT asymptotically minimizes all moments of the sample size within the corresponding classes of invariant tests. The asymptotic relations (4.75)–(4.77) are true with $\Psi(y) = y$ and I_{ij} defined in (4.106). If, furthermore, N tends to infinity in such a way that $\log(N)/|\log\alpha_{ij}| = o(1)$, then

$$I_{ij} = \begin{cases} \log(c_i/c_j) + (1-c_i)^2 c_i^{-1} + c_j - c_i + O(1/N), & i,j\neq 0, \\ (1-c_i)c_i^{-1} - |\log c_i| + O(1/N), & i\neq 0,\ j=0, \\ -(1-c_j) + |\log c_j| + O(1/N), & i=0,\ j\neq 0. \end{cases} \tag{4.107}$$

It is interesting to compare these asymptotic operating characteristics with those in the case when one knows the value of θ in advance. Consider the class $\mathbf{C}(\boldsymbol{\alpha})$ and suppose, for simplicity, that $\alpha_1 = \dots = \alpha_N = \alpha$ and $\alpha_0 \le \alpha$; this case is typical for many applications. Also set $c_i = 1/(1+Q_i)$ ($Q_i > 0$) and consider the symmetric scenario where $Q_i = Q$ for all i. Then the operating characteristics of the IMSPRT are

$$\mathsf{E}_i[T^*]^m \sim \left[\frac{\log(N/\alpha_0)}{N\log(1+Q/N) - \log(1+Q)}\right]^m, \quad i = 1, \dots, N;$$

$$\mathsf{E}_0[T^*]^m \sim \left|\frac{\log(N/\alpha)}{\log(1+Q) + N\log[1-Q/(1+Q)N]}\right|^m.$$

Using Theorem 4.3.1 and taking into account that $I_{0j} = Q - \log(1+Q)$ and $I_{j0} = \log(1+Q) - Q/(1+Q)$ in (4.22), we obtain that the operating characteristics of the MSPRT in the latter case with known θ are

$$\mathsf{E}_i[T^*]^m \sim \left[\frac{\log(N/\alpha_0)}{Q - \log(1+Q)}\right]^m, \quad i = 1, \dots, N;$$

$$\mathsf{E}_0[T^*]^m \sim \left[\frac{\log(N/\alpha)}{\log(1+Q) - Q/(1+Q)}\right]^m.$$

One can see that the asymptotic performance of the MSPRT in the case of complete prior knowledge is always better than that of the IMSPRT, as expected. However, if $N \to \infty$ such that $\log(N)/|\log\alpha| = o(1)$, then by (4.107) the IMSPRT has asymptotically the same characteristics as the asymptotically optimal sequential test in the case of known θ. At the same time, if N is fixed and $Q \to 0$, then the IMSPRT provides $[(N-1)/N]^m$ times worth characteristics compared with the case of completely known θ (for all H_i, $i = 0, 1, \dots, N$). By contrast, as $Q \to \infty$ and N is fixed, then

$$\mathsf{E}_0[T^*]^m(\theta \text{ unknown}) \sim \mathsf{E}_0[T^*]^m(\theta \text{ known}),$$

$$\frac{\mathsf{E}_i[T^*]^m(\theta \text{ unknown})}{\mathsf{E}_i[T^*]^m(\theta \text{ known})} \sim (Q/[(N-1)\log Q])^m \to \infty, \quad i = 1, \dots, N.$$

Finally, we note that for $N = 1$ the problem considered does not have an invariant solution.

Chapter 5

Sequential Hypothesis Testing: Composite Hypotheses

5.1 Introduction

In Chapters 3 and 4 we addressed the problems of testing binary and multiple simple hypotheses, with the exception of the problems with nuisance parameters where the hypotheses are composite but can be reduced to simple ones using the invariance principle. In this chapter, we deal with more practical scenarios that involve composite hypotheses. We restrict ourselves to parametric cases associated with parametric families of distributions.

In his 1947 book [494, Sec. 6], Wald suggested two approaches for modifying the SPRT for testing a simple null hypothesis $H_0 : \theta = \theta_0$ against a composite alternative $H_1 : \theta \in \Theta_1$. One method is to replace the LR $\Lambda_n^\theta = \prod_{k=1}^n [p_\theta(X_k)/p_{\theta_0}(X_k)]$ with a weighted LR

$$\bar{\Lambda}_n = \int_{\Theta_1} w(\theta)\Lambda_n^\theta \, \mathrm{d}\theta,$$

using a suitably selected weight function $w(\theta)$ on the hypothesis H_1. This leads to the *Weighted SPRT* (WSPRT) $\bar{\delta} = (\bar{T}, \bar{d})$ with the stopping time

$$\bar{T}(A_0,A_1) = \inf\{n \ge 1 : \bar{\Lambda}_n \notin (A_0,A_1)\}, \quad 0 < A_0 < 1, \, A_1 > 1. \tag{5.1}$$

The other method is to apply the generalized likelihood ratio (GLR) approach of the classical fixed sample size theory, employing the GLR statistic

$$\hat{\Lambda}_n = \sup_{\theta \in \Theta_1} \Lambda_n^\theta$$

in place of the LR Λ_n^θ with *a priori* fixed parameters. This leads to the *Generalized Sequential Likelihood Ratio Test* (GSLRT) $\hat{\delta} = (\widehat{T}, \hat{d})$ with the stopping time

$$\widehat{T}(A_0,A_1) = \inf\{n \ge 1 : \hat{\Lambda}_n \notin (A_0,A_1)\}.$$

The WSPRT has the advantage over the GSLRT that the upper bound on the error probability of Type I can be obtained in just the same way as for the SPRT, and that the average (weighted) Type II error probability

$$\bar{\alpha}_1(\bar{\delta}) = \int_{\Theta_1} \mathsf{P}_\theta(\bar{d} = 0)\, w(\theta)\, \mathrm{d}\theta$$

can also be controlled. Indeed, using the change-of-measure $\mathsf{P}_{\theta_0} \to \mathsf{P}^w = \int_{\Theta_1} \mathsf{P}_\theta w(\theta) \mathrm{d}\theta$ and Wald's likelihood ratio identity, we obtain

$$\alpha_0(\bar{\delta}) = \mathsf{E}_{\theta_0}\left[\mathbb{1}_{\{\bar{d}=1\}}\right] = \mathsf{E}^w\left[\bar{\Lambda}_{\bar{T}}^{-1}\mathbb{1}_{\{\bar{d}=1\}}\right] \le A_1^{-1}[1 - \bar{\alpha}_1(\bar{\delta})] \le 1/A_1, \tag{5.2}$$

and the change-of-measure $\mathsf{P}^w \to \mathsf{P}_{\theta_0}$ yields

$$\bar{\alpha}_1(\bar{\delta}) = \mathsf{E}^w\left[\mathbb{1}_{\{\bar{d}=0\}}\right] = \mathsf{E}_{\theta_0}\left[\bar{\Lambda}_{\bar{T}}\mathbb{1}_{\{\bar{d}=0\}}\right] \le A_0[1-\bar{\alpha}_0(\bar{\delta})] \le A_0.$$

Here without loss of generality we assume that the weight function $w(\theta)$ is normalized to 1, $\int_{\Theta_1} w(\theta)\mathrm{d}\theta = 1$. Unfortunately, the GLR approach has no such flexibility.

If one is interested in the one-sided SPRT-type tests (the so-called *open-ended tests*), which continue sampling indefinitely with a given probability $\mathsf{P}_{\theta_0}(T<\infty)$ under H_0 and accept the alternative H_1 when stopping, then an upper bound on and an accurate approximation for the probability $\alpha_0(\bar{T}) = \mathsf{P}_{\theta_0}(\bar{T}<\infty)$ can be readily obtained for the one-sided WSPRT $\bar{T}_A = \inf\{n : \bar{\Lambda}_n \ge A\}$. This has been exploited by Robbins [390] for obtaining approximations for the stopping threshold A. Indeed, similarly to (5.2), $\alpha_0(\bar{\delta}) \le 1/A$ and a nonlinear renewal theoretic argument yields

$$\alpha_0(\bar{\delta}) \sim A^{-1}\int_{\Theta_1} \zeta_\theta \mathrm{d}\theta \quad \text{as } A\to\infty,$$

where $0<\zeta_\theta<1$ is a constant accounting for the overshoot that is a subject of renewal theory. See Subsection 5.5.1 for further details. Again, there is no such upper bound on the error probability of a one-sided GSLRT, and obtaining asymptotic approximations is a much more delicate and challenging task [94, 272, 274].

In the more general case where the null hypothesis is also composite, $H_0 : \theta \in \Theta_0$, Wald [494] proposed to use the WSPRT (5.1) with the following weighted LR

$$\bar{\Lambda}_n = \frac{\int_{\Theta_1} w_1(\theta)\prod_{k=1}^n p_\theta(X_k)\,\mathrm{d}\theta}{\int_{\Theta_0} w_0(\theta)\prod_{k=1}^n p_\theta(X_k)\,\mathrm{d}\theta}.$$

Applying Wald's likelihood ratio identity, we again obtain the upper bounds on the average error probabilities:

$$\bar{\alpha}_0(\bar{\delta}) = \int_{\Theta_0} \mathsf{P}_\theta(\bar{d}=1)w_0(\theta)\mathrm{d}\theta \le A_0, \quad \bar{\alpha}_1(\bar{\delta}) = \int_{\Theta_1} \mathsf{P}_\theta(\bar{d}=0)w_1(\theta)\mathrm{d}\theta \le 1/A_1.$$

Clearly, for practical purposes, one would strongly prefer to upper-bound not the average error probabilities, which depend on the particular choice of weights, but rather the maximal error probabilities of Type I and Type II, i.e., to consider the class of tests

$$\mathbf{C}(\alpha_0,\alpha_1) = \left\{\delta : \sup_{\theta\in\Theta_0} \mathsf{P}_\theta(d=1) \le \alpha_0, \quad \sup_{\theta\in\Theta_1} \mathsf{P}_\theta(d=0) \le \alpha_1\right\}, \quad \alpha_0+\alpha_1<1.$$

However, in general it is not clear how to obtain the upper bounds on maximal error probabilities of the WSPRT, and of the GSLRT of course too. Wald, recognizing this fact, attempted to find particular examples when this is possible. He showed that, when testing the simple null hypothesis $H_0 : \theta = 0$ *versus* the composite alternative $H_1 : |\theta/\sigma| = q$ for the mean of a normal population $\mathcal{N}(\theta,\sigma^2)$ with unknown variance σ^2, it is possible to select the weight functions w_0, w_1 such that

$$\sup_{\theta\in\Theta_0} \mathsf{P}_\theta(\bar{d}=1) = \int_{\Theta_0} \mathsf{P}_\theta(\bar{d}=1)w_0(\theta)\mathrm{d}\theta, \quad \sup_{\theta\in\Theta_1} \mathsf{P}_\theta(\bar{d}=0) = \int_{\Theta_1} \mathsf{P}_\theta(\bar{d}=0)w_1(\theta)\mathrm{d}\theta.$$

Notice a similarity between this problem and the one considered in Subsection 3.6.2 where we built and studied the sequential t-test (t-SPRT). The t-SPRT exploits the weight function $w_1(u) = u^{-1}\mathbb{1}_{\{u>0\}}$ which is not integrable. In fact, the invariant approach may be considered as a particular case of Wald's weighted-based integral approach with specially chosen weights that are Haar measures; see, e.g., Zacks [522].

The weighted-based tests are also often called *mixture-based* tests or simply *mixtures*.

More generally, one may be interested in multihypothesis models with composite hypotheses. Specifically, consider the following fairly general continuous- or discrete-time multihypothesis scenario. Let $(\Omega, \mathscr{F}, \mathscr{F}_t, \mathsf{P}_\theta)$, $t \in \mathbb{Z}_+ = \{0,1,\dots\}$ or $t \in \mathbb{R}_+ = [0,\infty)$, be a filtered probability space with standard assumptions about monotonicity and in the continuous-time case $t \in \mathbb{R}_+$ also right-continuity of the σ-algebras $\mathscr{F}_t$. The parameter $\theta = (\theta_1, \dots, \theta_l)$ belongs to a subset $\widetilde{\Theta}$ of the l-dimensional Euclidean space $\mathbb{R}^l$. The sub-σ-algebra $\mathscr{F}_t = \mathscr{F}_t^X = \sigma(\mathbf{X}_0^t)$ of $\mathscr{F}$ is generated by the stochastic process $\mathbf{X}_0^t = \{X(u), 0 \le u \le t\}$ defined on $(\Omega, \mathscr{F})$ and observed up to time t. The hypotheses to be tested are $H_i: \ \theta \in \Theta_i$, $i = 0,1,\dots,N$ $(N \ge 1)$, where Θ_i are disjoint subsets of $\widetilde{\Theta}$. We also suppose that there is an *indifference* zone $\mathrm{I}_{\mathrm{in}} \in \widetilde{\Theta}$ in which there are no constraints on the error probabilities imposed. The indifference zone, where any decision is acceptable, is usually introduced keeping in mind that the correct action is not critical and often not even possible when the hypotheses are too close, which is perhaps the case in most, if not all, practical applications. However, in principle I_{in} may be an empty set. The probability measures P_θ and $\mathsf{P}_{\widetilde{\theta}}$ are assumed to be locally mutually absolutely continuous, i.e., the restrictions P_θ^t and $\mathsf{P}_{\widetilde{\theta}}^t$ of these measures to the sub-σ-algebras $\mathscr{F}_t$ are equivalent for all $0 \le t < \infty$ and all $\theta, \widetilde{\theta} \in \widetilde{\Theta}$.

As in Chapter 4, a multihypothesis sequential test δ consists of the pair (T, d), where T is a stopping time with respect to the filtration $\{\mathscr{F}_t\}_{t \ge 0}$, and $d = d_T(\mathbf{X}_0^T) \in \{0,1,\dots,N\}$ is an $\mathscr{F}_T$-measurable terminal decision rule specifying which hypothesis is to be accepted once observations have stopped: the hypothesis H_i is accepted if $d = i$ and rejected if $d \ne i$, i.e., $\{d = i\} = \{T < \infty, \ \delta \text{ accepts } H_i\}$.

The quality of a sequential test is judged on the basis of its error probabilities and expected sample sizes or more generally on the moments of the sample size. For $\theta \in \Theta_i$, let $\alpha_{ij}(\delta, \theta) = \mathsf{P}_\theta(d = j)$, $i \ne j$, $i, j = 0,1,\dots,N$, be the probability that the test δ accepts the hypothesis H_j when the true value of the parameter θ is fixed within the subset Θ_i. Also, let $\alpha_i(\delta, \theta) = \mathsf{P}_\theta(d \ne i)$, $\theta \in \Theta_i$, $i = 0,1,\dots,N$, denote the probability that the procedure δ terminates with an incorrect decision when the hypothesis H_i is true, i.e., rejects the hypothesis H_i when it is true. Clearly, if the test is finite w.p. 1, then $\alpha_i(\delta, \theta) = \sum_{j \ne i} \alpha_{ij}(\delta, \theta)$. Introduce the two following classes of tests

$$
\begin{aligned}
\mathbf{C}([\alpha_{ij}]) &= \left\{ \delta : \sup_{\theta \in \Theta_i} \alpha_{ij}(\delta, \theta) \le \alpha_{ij} \text{ for all } i, j = 0,1,\dots,N, i \ne j \right\}, \\
\mathbf{C}(\boldsymbol{\alpha}) &= \left\{ \delta : \sup_{\theta \in \Theta_i} \alpha_i(\delta, \theta) \le \alpha_i \text{ for all } i = 0,1,\dots,N \right\},
\end{aligned}
\tag{5.3}
$$

where α_{ij} and α_i, the predefined maximal error probabilities, are positive numbers less than 1 and $\boldsymbol{\alpha} = (\alpha_0, \alpha_1, \dots, \alpha_N)$. To avoid trivialities we always assume that $\sum_{i,j,i \ne j} \alpha_{ij} < 1$ and $\sum_i \alpha_i < 1$.

Note that the classes $\mathbf{C}([\alpha_{ij}])$ and $\mathbf{C}(\boldsymbol{\alpha})$ confine the error probabilities in the regions Θ_i but not in the indifference zone I_{in} where the hypotheses are too close to be distinguished with the given and relatively low error probabilities. In other words, there is no loss associated with wrong decisions in the indifference zone. However, it may be reasonable to take into account that the multihypothesis tests considered in the previous chapter, in particular the matrix SPRT, tuned to putative parameter values being nearly optimal for these putative parameter values perform poorly for $\theta \in \mathrm{I}_{\mathrm{in}}$. The ESS $\mathsf{E}_\theta T$ of these tests increases dramatically in the vicinity of the worst point θ^* where the ESS attains its maximum. This is particularly true for the SPRT, as we will see from the further discussion. An alternative is to find a minimax test that would minimize the ESS in the worst situation. The desirability of reducing the ESS for parameter values between the hypotheses and, in particular, of minimizing the maximal ESS over all parameter values, which is usually attained at the least desirable point in the indifference zone (i.e., between the hypotheses), was recognized in the late 1950s and the beginning of the 1960s (see, e.g., Kiefer and Weiss [232], Anderson [9], Bechhoffer [55], Weiss [498]). This problem, known as the *Kiefer–Weiss problem*, is considered in Section 5.3 in the case of multiple and two composite hypotheses for both non-iid and iid data models.

Ideally we would like to minimize the ESS for all possible parameter values in order to guarantee optimality at the true parameter value that we do not know. Unfortunately, there is no such test, since the structure of the test that minimizes the ESS $\mathsf{E}_\theta T$ at a specific parameter value $\theta = \widetilde{\theta}$ depends on $\widetilde{\theta}$. However, this problem may be solved asymptotically when the error probabilities are small. More specifically, in the following sections we construct the tests for which the ESS is close to the best possible $\inf_{\delta \in \mathbf{C}([\alpha_{ij}])} \mathsf{E}_\theta T$ for all $\theta \in \Theta$ as $\alpha_{ij} \to 0$, where $\Theta = \sum_{i=0}^N \Theta_i + \mathrm{I}_{\text{in}}$.

5.2 Critique of the SPRT

Even though Wald's SPRT has the remarkable optimality property of minimizing the ESS at the putative values θ_0 and θ_1 of the parameter θ to which it is tuned, it loses the optimality property for the values of θ other than θ_0 or θ_1. The performance of the SPRT degrades in the indifference zone for values of θ between θ_0 and θ_1, and especially in the vicinity of the least favorable point. In some cases, the SPRT's ESS $\mathsf{E}_\theta T^*$ is larger than the number of observations required by a fixed sample size test with the same error probabilities. To see this consider the Gaussian example of Subsection 3.1.5 where $X_n = \theta + \xi_n$ and $\xi_n \sim \mathcal{N}(0, \sigma^2)$, $n = 1, 2, \dots$, are iid zero-mean normal random variables. Assume that we are interested in testing the two one-sided composite hypotheses $H_0 : \theta \le \theta_0$ and $H_1 : \theta \ge \theta_1$ with the indifference interval $\mathrm{I}_{\text{in}} = (\theta_0, \theta_1)$. On one hand, the SPRT still has some useful optimality properties since $\sup_{\theta \le \theta_0} \mathsf{E}_\theta T^* = \mathsf{E}_{\theta_0} T^*$ and $\sup_{\theta \ge \theta_1} \mathsf{E}_\theta T^* = \mathsf{E}_{\theta_1} T^*$ and the error probabilities $\alpha_0^*(\theta)$ and $\alpha_1^*(\theta)$ decrease within $(-\infty, \theta_0]$ and $[\theta_1, \infty)$, respectively. Hence, the SPRT has a minimax property — it minimizes the expected sample sizes in the worst case scenarios with respect to the hypotheses in the class $\mathbf{C}(\alpha_0, \alpha_1)$ of tests with error probabilities at most α_0 and α_1. However, if we account for the indifference zone, then the situation changes dramatically.

We now present the results of comparison of the SPRT's $\mathsf{ESS}_\theta(\alpha_0, \alpha_1) = \mathsf{E}_\theta T^*$ with the sample size of the FSS NP test. These results show that the NP test may have a smaller sample size when $\theta \in (\theta_0, \theta_1)$. Note that the latter test is based on thresholding the sum $S_n = \sum_{k=1}^n X_k$, and that it is a uniformly most powerful test, i.e., the best FSS test one can get. By (3.81), in order to guarantee for the NP test the same error probabilities $\alpha_{0n}(\theta_0) = \alpha_0$ and $\alpha_{1n}(\theta_1) = \alpha_1$ at the points θ_0 and θ_1 as for the SPRT, the sample size $n = n_q(\alpha_0, \alpha_1)$ should be taken as $(Q_{1-\alpha_0} + Q_{1-\alpha_1})^2/q$, where $q = (\theta_1 - \theta_0)^2/\sigma^2$ and Q_p is the p-quantile of the standard normal distribution.

Define the relative efficiency of the SPRT and the NP test as the ratio $\mathcal{E}_\theta = \mathsf{ESS}_\theta(\alpha_0, \alpha_1)/n_q(\alpha_0, \alpha_1)$. Consider the symmetric case that $\alpha_0 = \alpha_1 = \alpha$. Then $n_q(\alpha) = 4Q_{1-\alpha}^2/q$ and, according to the computations presented in Subsection 3.1.5, $\mathcal{E}_{\theta_i}(\alpha) \sim 1/4$, $i = 0, 1$ asymptotically as $\alpha \to 0$. In other words, the asymptotic efficiency of the SPRT is four times higher than that of the NP test at the putative points. However, using (3.25) and (3.83), at the middle point $\theta^* = (\theta_0 + \theta_1)/2$, namely the worst point where the ESS of the SPRT attains its maximum, we obtain

$$\mathcal{E}_{\theta^*}(\alpha) \sim \frac{\{\log[(1-\alpha)/\alpha]\}^2}{4Q_{1-\alpha}^2} \sim |\log \alpha|/8, \quad \alpha \to 0.$$

This implies that the NP test performs much better than the SPRT at the worst point of the indifference interval for sufficiently small error probabilities.

Figure 5.1 displays the SPRT's ESS, $\mathsf{ESS}_\theta(T^*) = \mathsf{E}_\theta T^*$, as a function of θ obtained by solving the integral equations (3.66) and (3.67) using the numerical techniques described in Subsection 3.1.4 for $\alpha_0 = \alpha_1 = \alpha = 10^{-3}$ and $\theta_0 = 0, \theta_1 = 0.5$. The horizontal line also shows the sample size $n_q(\alpha_0, \alpha_1)$ of the NP test. Table 5.1 contains numerical data for different values of the parameter θ and the error probability α. It is seen that, in the vicinity of the worst point $\theta^* = \theta_1/2 = 0.25$, the NP test performs significantly better for $\alpha \le 0.001$. In particular, for $\alpha = 0.001$ the SPRT requires

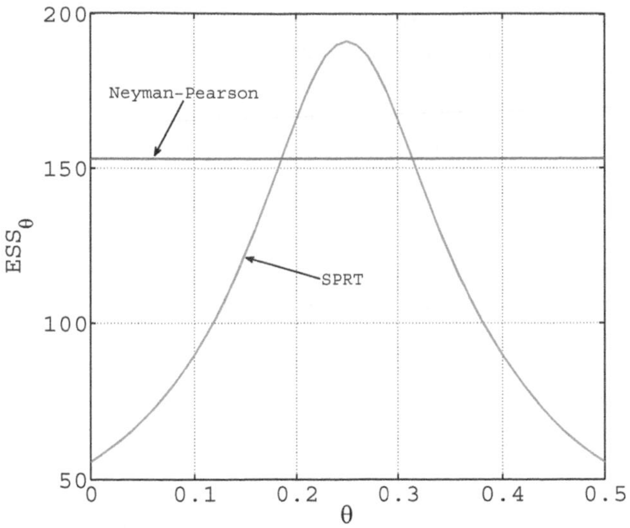

Figure 5.1: *The SPRT's ESS as a function of θ for $\alpha = 0.001$, $\theta_0 = 0, \theta_1 = 0.5$. The horizontal line corresponds to the fixed sample size of the NP test achieving the same error probabilities as the SPRT.*

on average 25% more observations than the NP test and the efficiency of the NP test increases when α decreases.

Table 5.1: *The SPRT's ESS for $\theta_0 = 0$, $\theta_1 = 0.5$, and different values of the parameter θ and the error probability α.*

θ	$\alpha_0 = \alpha_1 = \alpha$					
	10^{-1}	10^{-2}	10^{-3}	10^{-4}	10^{-5}	10^{-6}
0	14.32	36.29	55.41	73.93	92.37	110.8
0.1	17.19	54.22	89.48	122.1	153.5	184.4
0.2	19.26	79.24	165.6	267.9	377.1	487.3
0.25	19.57	84.73	191.1	339.6	530.5	763.8
n_{NP}	27	87	153	222	292	362
$\mathcal{E}_{\theta^*}(\alpha)$	0.71	0.97	1.25	1.53	1.82	2.11

Note that in this example

$$F_\theta(z) = \Phi\left(\frac{z - qr_\theta}{q}\right), \quad \mathcal{K}_\theta(x,y) = \frac{1}{q}\varphi\left(\frac{y - x - qr_\theta}{q}\right),$$

where $r_\theta = [\theta - (\theta_1 + \theta_0)/2]/(\theta_1 - \theta_0)$ and $q = (\theta_1 - \theta_0)^2$; see (3.66), (3.67).

5.3 The Kiefer–Weiss Problem

As follows from the discussion in the previous section, when testing two composite hypotheses the SPRT performs poorly in the vicinity of the least favorable parameter value θ^*. Therefore, it stands to reason to find a test that minimizes the expected sample size in the least favorable situation, i.e., $\sup_{\theta \in \Theta} \mathsf{E}_\theta T$. The problem of finding such a test in the class of tests with upper bounds on the error probabilities at two other fixed points, θ_0 and θ_1, is called the *Kiefer–Weiss problem*. Usually the maximum of the ESS is attained at some point θ^* belonging to the indifference zone $\mathrm{I}_{\mathrm{in}} = (\theta_0, \theta_1)$, and the maximal values of the error probabilities are attained at the boundary points θ_0 and θ_1. Therefore, it is instructive to consider first the related problem of minimizing the ESS at a fixed point θ for the given bounds on the error probabilities at θ_0, θ_1. This problem is called the *modified Kiefer–Weiss problem*. This problem has been first considered by Kiefer and Weiss [232], who showed that it is equivalent to the Bayes problem of minimizing an average risk that represents a weighted average of $\mathsf{E}_\theta T$ and the two error probabilities $\mathsf{P}_{\theta_0}(d=1)$ and $\mathsf{P}_{\theta_1}(d=0)$. Furthermore, in the discrete-time case and for an exponential family of distributions, the corresponding Bayes problem is truncated and, hence, can be solved exactly using Bellman's backward iterative induction; see Theorem 2.7.4. In particular, Weiss [498] showed that the Kiefer–Weiss problem reduces to the modified problem in symmetric situations involving Gaussian and binomial distributions that can be solved numerically using the backward induction; Lai [243] studied the continuous-time Brownian motion case and found asymptotic approximations for the optimal stopping boundaries; and Lorden [276] characterized the basic structure of the optimal test for the modified problem, with particularly useful results for exponential families. However, finding the exact solution involves quite heavy computations. Therefore, approximations are in order which we discuss next.

In the discrete-time iid case, Lorden [273] proposed a procedure that he called the *2-SPRT*, which is based on a parallel running of two one-sided SPRTs. One of these SPRTs exploits the LR between the points θ and θ_0, the other — between θ and θ_1. Lorden showed that the 2-SPRT is nearly optimal for small error probabilities. His results along with some extensions are discussed in Subsection 5.3.2. Below we extend the Kiefer–Weiss problem to multiple hypotheses. Moreover, we consider a general non-iid case similar to that discussed in Chapter 4, and we show that the r-quick convergence techniques developed in Chapters 3 and 4 can be effectively used for proving first-order asymptotic optimality of some procedures in the modified Kiefer–Weiss problem.

Consider the general multihypothesis model described in Section 5.1 and the classes of procedures (5.3). Then the Kiefer–Weiss problem can be formulated as finding a multihypothesis test that minimizes $\sup_{\theta\in\Theta} \mathsf{E}_\theta T$ under the constraints $\sup_{\theta\in\Theta_i} \mathsf{P}_\theta(d=j) \le \alpha_{ij}$ and $\sup_{\theta\in\Theta_i} \mathsf{P}_\theta(d\neq i) \le \alpha_i$:

$$\inf_{\delta\in\mathbf{C}([\alpha_{ij}])(\mathbf{C}(\boldsymbol{\alpha}))} \sup_{\theta\in\Theta} \mathsf{E}_\theta T \longrightarrow T_{\mathrm{opt}},$$

where $\Theta = \sum_{i=0}^N \Theta_i + \mathrm{I}_{\mathrm{in}}$. An extension of the argument used in Section 4.2 shows that finding the minimax test is usually not possible or at least very difficult even in the iid case. For this reason, we focus on the asymptotic setting assuming that the error probabilities vanish.

More generally, the modified Kiefer–Weiss problem in the context of testing multiple hypotheses can be formulated as follows. Let $\mathsf{P}_0, \mathsf{P}_1, \dots, \mathsf{P}_N$ be mutually locally absolutely continuous probability measures and let P be another probability measure locally absolutely continuous with respect to all the P_i's. By E_i and E, we denote the expectations with respect to P_i and P, respectively. In other words, we again consider the multihypothesis problem of Section 4.1, but now we are interested in constructing a test that would at least approximately minimize the expected sample size $\mathsf{E}T$ at an intermediate point P in the classes of tests $\mathbf{C}([\alpha_{ij}]) = \{\delta : \mathsf{P}_i(d=j) \le \alpha_{ij}, i, j = 0, 1, \dots, N, i \neq j\}$ and $\mathbf{C}(\boldsymbol{\alpha}) = \{\delta : \mathsf{P}_i(d\neq i) \le \alpha_i, i = 0, 1, \dots, N\}$; see (4.1). In the next subsection we present two constructions that minimize not only the ESS but also the higher moments of the stopping time to first order for general non-iid models.

5.3.1 *Asymptotically Optimal Tests at an Intermediate Point in the General Non-iid Case*

Let Q^t be a dominating measure. For $t \ge 0$, define

$$\Lambda_i(t) = \frac{\mathrm{d}\mathsf{P}_i^t}{\mathrm{d}\mathsf{Q}^t}(\mathbf{X}_0^t), \quad \Lambda(t) = \frac{\mathrm{d}\mathsf{P}^t}{\mathrm{d}\mathsf{Q}^t}(\mathbf{X}_0^t), \quad \lambda_i(t) = \log\Lambda_i(t), \quad \lambda(t) = \log\Lambda(t),$$

with $\Lambda_i(0) = \Lambda(0) = 1$, $\lambda_i(0) = \lambda(0) = 0$ almost surely.

Define the Markov times

$$T_i = \inf\left\{t : \lambda(t) \geq \max_{\substack{0\leq j\leq N \\ j\neq i}} [\lambda_j(t) + a_{ji}]\right\}, \quad i = 0,1,\dots,N, \tag{5.4}$$

where a_{ji} are positive numbers. The first test $\delta^\star = (T^\star, d^\star)$ is defined as follows:

$$T^\star = \min_{0\leq i\leq N} T_i, \qquad d^\star = i \quad \text{if} \quad T^\star = T_i. \tag{5.5}$$

Note that T_i is the time of accepting the hypothesis H_i, so that it is natural to call this test the *accepting* test. Note also that it is a straightforward modification of the MSPRT (4.5), (4.6). Indeed, the procedure (5.5) can be equivalently represented as:

$$\text{Stop at the first } t \geq 0 \text{ such that for some } i \quad \frac{\mathrm{dP}^t}{\mathrm{dP}_j^t}(\mathbf{X}_0^t) \geq e^{a_{ji}} \quad \text{for all } j \neq i, \tag{5.6}$$

and accept the unique i that satisfies these inequalities. Comparing with the MSPRT (4.2) we see that the LRs between the hypotheses H_j and H_i are now replaced with the LRs between the measures P and P_i. Hence, we will call it the *modified accepting MSPRT*.

We now construct another test, the *modified rejecting MSPRT* $\delta_\star = (T_\star, d_\star)$, in which the observations are continued up to rejection of all except one hypothesis, and at this instant, the remaining hypothesis is accepted. For $i, j = 0,1,\dots,N$ $(i \neq j)$, define the Markov times

$$\tau_{ij} = \inf\{t : \lambda(t) \geq \lambda_i(t) + b_{ij}\}, \quad i,j = 0,1,\dots,N, i \neq j, \tag{5.7}$$

where b_{ij} are positive thresholds. The Markov rejecting time for the hypothesis H_i is

$$\tau_i = \max_{\substack{0\leq j\leq N \\ j\neq i}} \tau_{ij},$$

and the test $\delta_\star$ is defined as

$$T_\star = \min_{0\leq j\leq N} \max_{\substack{0\leq i\leq N \\ i\neq j}} \tau_i, \qquad d_\star = i \quad \text{if} \quad \max_{0\leq j\leq N} \tau_j = \tau_i. \tag{5.8}$$

In other words, this test stops at the time $T_\star = \tau_{(N-1)}$, where $\tau_{(0)} \leq \tau_{(1)} \leq \cdots \leq \tau_{(N-1)} \leq \tau_{(N)}$ is the time-ordered set of rejecting times $\tau_0, \tau_1, \dots, \tau_N$.

In the case of two hypotheses ($N = 1$), both tests coincide and can be written as $T^\star = \min(T_0, T_1)$, $d^\star = \arg\min_i T_i$, where

$$T_0 = \inf\left\{t : \log\frac{\mathrm{dP}^t}{\mathrm{dP}_1^t}(\mathbf{X}_0^t) \geq a_1\right\}, \quad T_1 = \inf\left\{t : \log\frac{\mathrm{dP}^t}{\mathrm{dP}_0^t}(\mathbf{X}_0^t) \geq a_0\right\}.$$

This test, called the 2-SPRT, has been first introduced and studied by Lorden [273] in the discrete-time iid case.

In the sequel, it is convenient to introduce the following notation

$$\mathcal{L}_i(t) = \frac{\mathrm{dP}^t}{\mathrm{dP}_i^t}(\mathbf{X}_0^t), \quad \ell_i(t) = \log\mathcal{L}_i(t), \quad \mathcal{L}_i(0) = 1, \quad \ell_i(0) = 0 \ \mathsf{P}_i, \mathsf{P}-\text{a.s.}.$$

With this notation, the Markov times T_i and τ_{ij} in (5.4) and (5.7) can be rewritten as

$$T_i = \inf\left\{t : \min_{\substack{0\leq j\leq N \\ j\neq i}} [\ell_j(t) - a_{ji}] \geq 0\right\}, \tag{5.9}$$

and

$$\tau_{ij} = \inf\{t : \ell_i(t) \ge b_{ij}\}. \tag{5.10}$$

First, we obtain the upper bounds for the error probabilities $\alpha_{ij}(\delta) = \mathsf{P}_i(d = j)$ and $\alpha_i(\delta) = \mathsf{P}_i(d \neq i)$ of the procedures $\delta^\star$ and $\delta_\star$.

Lemma 5.3.1. *The following inequalities hold:*

(i) $\alpha_{ij}(\delta^\star) \le e^{-a_{ij}}\mathsf{P}(d^\star = j)$ *for* $i, j = 0, 1, \dots, N$, $i \neq j$;

(ii) $\alpha_i(\delta^\star) \le \sum_{j\neq i} e^{-a_{ij}}\mathsf{P}(d^\star = j)$ *for* $i = 0, 1, \dots, N$;

(iii) $\alpha_{ij}(\delta_\star) \le e^{-b_{ji}}$ *for* $i, j = 0, 1, \dots, N$, $i \neq j$;

(iv) *If* $b_{ij} = b_i$, *then* $\alpha_i(\delta_\star) \le e^{-b_i}$ *for* $i = 0, 1, \dots, N$.

Proof. Proof of (i) and (ii). By the definition of T_j in (5.9), $\ell_i(T_j) \ge a_{ij}$ for all $i \neq j$ on $T_j < \infty$, so that

$$1/\mathcal{L}_i(T_j) \le e^{-a_{ij}} \quad \text{for all } \; i \neq j \; \text{ on } \; T_j < \infty.$$

Hence, using Wald's likelihood ratio identity, we obtain

$$\alpha_{ij}(\delta^\star) = \mathsf{E}_i \mathbb{1}_{\{T^\star = T_j\}} = \mathsf{E}\left[\mathcal{L}_i^{-1}(T^\star)\mathbb{1}_{\{T^\star = T_j\}}\right] \le e^{-a_{ij}}\mathsf{P}(T^\star = T_j) = e^{-a_{ij}}\mathsf{P}(d^\star = j), \tag{5.11}$$

which proves (i).

Using the inequality in (5.11), we obtain

$$\alpha_i(\delta^\star) = \sum_{j\neq i} \alpha_{ij}(\delta^\star) \le \sum_{j \neq i} e^{-a_{ij}}\mathsf{P}(d^\star = j),$$

namely (ii).

Proof of (iii) and (iv). Note first that

$$\alpha_{ij}(\delta_\star) = \mathsf{P}_i\left(\max_k \tau_k = \tau_j, \tau_j < \infty\right) \le \mathsf{P}_i(\tau_j < \infty) \le \mathsf{P}(\tau_{ji} < \infty).$$

Now, using Wald's likelihood ratio identity and the fact that $1/\mathcal{L}_j(\tau_{ji}) \le e^{-b_{ji}}$ by (5.10), we obtain

$$\mathsf{P}(\tau_{ji} < \infty) = \mathsf{E}\left[\mathcal{L}_j^{-1}(\tau_{ji})\mathbb{1}_{\{\tau_{ji} < \infty\}}\right] \le e^{-b_{ji}},$$

which proves (iii).

To establish (iv) we note that if the thresholds b_{ij} do not depend on j, that is $b_{ij} = b_i$, then $\tau_{ij} = \tau_i$ is the time of rejection of the hypothesis H_i, so that

$$\mathsf{P}_i(d_\star \neq i) = \mathsf{P}_i\left(\max_{0\le j \le N} \tau_j \neq \tau_i\right) \le \mathsf{P}_i(\tau_i < \infty) = \mathsf{E}\left[\mathcal{L}_i^{-1}(\tau_i)\mathbb{1}_{\{\tau_i < \infty\}}\right] \le e^{-b_i}.$$

□

Therefore, we have the following implications:

$$a_{ji} = \log(1/\alpha_{ji}) \quad \text{implies} \quad \delta^\star \in \mathbf{C}([\alpha_{ij}]); \tag{5.12}$$

$$a_{ji} = a_j = \log(N/\alpha_j) \quad \text{implies} \quad \delta^\star \in \mathbf{C}(\boldsymbol{\alpha}); \tag{5.13}$$

$$b_{ij} = \log(1/\alpha_{ji}) \quad \text{implies} \quad \delta_\star \in \mathbf{C}([\alpha_{ij}]); \tag{5.14}$$

$$b_{ij} = b_i = \log(1/\alpha_i) \quad \text{implies} \quad \delta_\star \in \mathbf{C}(\boldsymbol{\alpha}). \tag{5.15}$$

Note that the formulas (5.12) and (5.13) for the thresholds can be improved when the probabilities $\mathsf{P}(d^\star = j)$ can be estimated. For example, setting $a_{ij} = \log(\mathsf{P}(d^\star = j)/\alpha_{ji})$ implies $\alpha_{ij}(\delta^\star) \le \alpha_{ij}$. In particular, in certain symmetric cases, $\mathsf{P}(d^\star = j) = 1/(N+1)$.

The two proposed tests perform very similarly. Below we establish asymptotic optimality properties of these tests under very general conditions. Since the results and proofs are essentially analogous for both tests, for brevity we formulate the results only for the first test $\delta^\star$. We also stress once again that these tests coincide in the case of testing two hypotheses, i.e., when $N = 1$, in which case they are nothing but the 2-SPRT.

Remark 5.3.1. It is seen from (5.12)–(5.15) that the accepting test $\delta^\star$ is better suited to control the probabilities of accepting wrong hypotheses, i.e., to the class $\mathbf{C}([\alpha_{ij}])$, than the rejecting test $\delta_\star$, and *vice versa* — the rejecting test is better suited to control the probabilities of rejecting the correct hypotheses, i.e., to the class $\mathbf{C}(\boldsymbol{\alpha})$. Indeed, the accepting test requires the thresholds $a_j = \log(N/\alpha_j)$ that are higher than the thresholds $b_i = \log(1/\alpha_i)$ required by the rejecting test. Also, the thresholds $b_{ij} = \log(1/\alpha_{ji})$ typically provide an extremely conservative estimate. To see this, consider the symmetric case that $\alpha_{ji} = \alpha$ and $\alpha_i = N\alpha$. Clearly, in this case the choice $b_{ij} = \log(1/\alpha)$ according to (5.14) leads to the error probability $\alpha_{ij}(\delta_\star) \le \alpha/N$.

Define $\alpha_{\max} = \max_{i,j,i\neq j} \alpha_{ij}$ and $\tilde{\alpha}_{\max} = \max_i \alpha_i$. As in Chapter 4, we assume that for the classes $\mathbf{C}([\alpha_{ij}])$ and $\mathbf{C}(\boldsymbol{\alpha})$ the following conditions hold:

$$\lim_{\alpha_{\max}\to 0} \frac{|\log \alpha_{ij}|}{|\log \alpha_{\max}|} = c_{ij}, \qquad \lim_{\tilde{\alpha}_{\max}\to 0} \frac{|\log \alpha_i|}{|\log \tilde{\alpha}_{\max}|} = c_i, \tag{5.16}$$

where $1 \le c_{ij} < \infty$ and $1 \le c_i < \infty$. This allows us to consider asymptotically asymmetric situations with respect to the error probabilities when some of them may decay faster than the others.

Let $\mathcal{I}_{\mathsf{P},\mathsf{P}_i}(t) = \mathsf{E}[\ell_i(t)]$ be the total K–L information between the probability measures P and P_i accumulated in the trajectory $\mathbf{X}_0^t$. Assume that there are a nonnegative increasing function $\psi(t)$ ($\psi(\infty) = \infty$) and finite nonnegative numbers $I(\mathsf{P},\mathsf{P}_i) = I_i(\mathsf{P})$, $i = 0, 1, \dots N$, such that $\lim_{t\to\infty}[\mathcal{I}_{\mathsf{P},\mathsf{P}_i}(t)/\psi(t)] = I_i(\mathsf{P})$ and the following almost sure convergence conditions hold

$$\frac{\ell_i(t)}{\psi(t)} \xrightarrow[t\to\infty]{\mathsf{P}-\text{a.s.}} I_i(\mathsf{P}) \quad \text{for } i = 0, 1, \dots, N \tag{5.17}$$

(cf. (4.69)). Let $\Psi(t)$ be the inverse function for $\psi(t)$.

It can be shown that if the a.s. convergence condition (5.17) holds and the numbers $I_i(\mathsf{P})$ are not simultaneously equal to 0, i.e., $I_i(\mathsf{P}) \ge 0$ and $\max_{0\le i\le N} I_i(\mathsf{P}) > 0$, then

$$\inf_{\delta\in\mathbf{C}([\alpha_{ij}])} \mathsf{P}\left\{T > \varepsilon\Psi\left(\min_{0\le i\le N}\max_{\substack{0\le j\le N\\ j\neq i}} \frac{|\log \alpha_{ji}|}{I_j(\mathsf{P})}\right)\right\} \to 1 \quad \text{as } \alpha_{\max}\to 0,$$

$$\inf_{\delta\in\mathbf{C}(\boldsymbol{\alpha})} \mathsf{P}\left\{T > \varepsilon\Psi\left(\min_{0\le i\le N}\max_{\substack{0\le j\le N\\ j\neq i}} \frac{|\log \alpha_{j}|}{I_j(\mathsf{P})}\right)\right\} \to 1 \quad \text{as } \tilde{\alpha}_{\max}\to 0$$

for any $0 < \varepsilon < 1$. Thus, introducing the notation

$$J = \max_{0\le i\le N}\min_{\substack{0\le j\le N\\ j\neq i}} [I_j(\mathsf{P})/c_{ji}], \quad \tilde{J} = \max_{0\le i\le N}\min_{\substack{0\le j\le N\\ j\neq i}} [I_j(\mathsf{P})/c_{j}]$$

and using the Chebyshev inequality, we obtain that for all $m > 0$

$$\begin{aligned} &\inf_{\delta\in\mathbf{C}([\alpha_{ij}])} \mathsf{E}[T]^m \ge \left[\Psi\left(\frac{|\log \alpha_{\max}|}{J}\right)\right]^m (1+o(1)) \quad \text{as } \alpha_{\max}\to 0, \\ &\inf_{\delta\in\mathbf{C}(\boldsymbol{\alpha})} \mathsf{E}[T]^m \ge \left[\Psi\left(\frac{|\log \tilde{\alpha}_{\max}|}{\tilde{J}}\right)\right]^m (1+o(1)) \quad \text{as } \tilde{\alpha}_{\max}\to 0 \end{aligned} \tag{5.18}$$

(cf. Lemma 4.4.1).

Let $L_i(\varepsilon) = \sup\{t : |\ell_i(t) - I_i(\mathsf{P})\psi(t)| > \psi(t)\varepsilon\}$. If we now strengthen the almost sure convergence (5.17) into the r-quick convergence, i.e., if we assume that $\mathsf{E}[L_i(\varepsilon)]^r < \infty$ for all $\varepsilon > 0$ and some $r > 0$, we may obtain the upper bounds for the moments of the stopping time up to order r and prove asymptotic optimality of the modified accepting and rejecting MSPRTs. For the sake of compactness, we present the results only for the accepting MSPRT $\delta^\star$, keeping in mind that all the results hold for the rejecting MSPRT $\delta_\star$ as well.

Theorem 5.3.1 (Asymptotic optimality of the modified MSPRT). *Assume that there exist an increasing nonnegative function $\psi(t)$ and finite nonnegative numbers $I_0(\mathsf{P}), I_1(\mathsf{P}), \dots, I_N(\mathsf{P})$ such that $\max\{I_0(\mathsf{P}), \dots, I_N(\mathsf{P})\} > 0$ and for some $r > 0$*

$$\frac{\ell_i(t)}{\psi(t)} \xrightarrow[t\to\infty]{\mathsf{P}-r-quickly} I_i(\mathsf{P}) \quad \text{for all } i = 0, 1, \dots, N. \tag{5.19}$$

(i) *For all $0 < m \le r$,*

$$\mathsf{E}[T^\star]^m \sim \left[\Psi\left(\min_{0\le i\le N} \max_{\substack{0\le j\le N \\ j\ne i}} \frac{a_{ji}}{I_j(\mathsf{P})}\right)\right]^m \quad \text{as } \min_{j,i} a_{ji} \to \infty. \tag{5.20}$$

(ii) *If the thresholds are selected so that $\alpha_{ij}(\delta^\star) \le \alpha_{ij}$ and $a_{ji} \sim \log(1/\alpha_{ji})$, in particular $a_{ji} = \log(1/\alpha_{ji})$, then for all $0 < m \le r$,*

$$\inf_{\delta\in\mathbf{C}([\alpha_{ij}])} \mathsf{E}[T]^m \sim \mathsf{E}[T^\star]^m \sim \left[\Psi\left(\frac{|\log \alpha_{\max}|}{J}\right)\right]^m \quad \text{as } \alpha_{\max} \to 0. \tag{5.21}$$

(iii) *If the thresholds are selected so that $\alpha_i(\delta^\star) \le \alpha_i$ and $a_{ji} \sim \log(1/\alpha_j)$, in particular $a_{ji} = \log(N/\alpha_j)$, then for all $0 < m \le r$,*

$$\inf_{\delta\in\mathbf{C}(\boldsymbol{\alpha})} \mathsf{E}[T]^m \sim \mathsf{E}[T^\star]^m \sim \left[\Psi\left(\frac{|\log \tilde{\alpha}_{\max}|}{\tilde{J}}\right)\right]^m \quad \text{as } \tilde{\alpha}_{\max} \to 0. \tag{5.22}$$

Consequently, the modified accepting MSPRT minimizes asymptotically the moments of the stopping time distribution up to order r under the intermediate measure P in the classes of tests $\mathbf{C}([\alpha_{ij}])$ and $\mathbf{C}(\boldsymbol{\alpha})$.

Proof. Define $\tilde{\ell}_j(t) = \ell_j(t)/I_j(\mathsf{P})$, $\tilde{a}_{ji} = a_{ji}/I_j(\mathsf{P})$,

$$\widetilde{T}_i = \inf\left\{t : \min_{j\ne i} \tilde{\ell}_j(t) \ge \max_{j\ne i} \tilde{a}_{ji}\right\}, \quad \widetilde{T} = \min_{0\le i\le N} \widetilde{T}_i,$$
$$\tilde{L}(\varepsilon) = \sup\left\{t : \max_{0\le i\le N} \left|\tilde{\ell}_i(t) - \psi(t)\right| > \psi(t)\varepsilon\right\}.$$

Observe that

$$(1-\varepsilon)\psi(\widetilde{T} - \Delta) < \min_{j\ne i} \tilde{\ell}_j(\widetilde{T} - \Delta) < \max_{j\ne i} \tilde{a}_{ji} \quad \text{on } \{\tilde{L}(\varepsilon) + \Delta < \widetilde{T} < \infty\},$$

where in the continuous-time case Δ is a small positive number and in the discrete-time case $\Delta = 1$. Hence, for all $i = 0, 1, \dots, N$,

$$\widetilde{T} < \Delta + \Psi\left(\frac{\min_i \max_{j\ne i} \tilde{a}_{ji}}{1-\varepsilon}\right) \quad \text{on } \{\tilde{L}(\varepsilon) + \Delta < \tilde{T} < \infty\},$$

so that for every $0 < \varepsilon < 1$,

$$\begin{aligned}\widetilde{T} &\le \Delta + 1\!\!1_{\{\widetilde{T} > \tilde{L}(\varepsilon)+\Delta\}} \Psi\left(\frac{\min_i \max_{j\neq i} \tilde{a}_{ji}}{1-\varepsilon}\right) + 1\!\!1_{\{\widetilde{T} \le \tilde{L}(\varepsilon)+\Delta\}} \tilde{L}(\varepsilon) \\ &\le \Delta + \Psi\left(\frac{\min_i \max_{j\neq i} \tilde{a}_{ji}}{1-\varepsilon}\right) + \tilde{L}(\varepsilon).\end{aligned} \tag{5.23}$$

By the r-quick convergence condition (5.19), $\mathsf{E}[\tilde{L}(\varepsilon)]^r < \infty$. Hence (5.23) yields for $m \le r$

$$\mathsf{E}[\widetilde{T}]^m \le \left[\Psi\left(\min_i \max_{j\neq i} \tilde{a}_{ji}\right)\right]^m (1+o(1)) \quad \text{as} \quad \min_{j,i} a_{ji} \to \infty. \tag{5.24}$$

Obviously, the Markov time T_i given in (5.9) can be written as

$$T_i = \inf\left\{t : \min_{j\neq i}[\tilde{\ell}_{ij}(t) - \tilde{a}_{ji}] \ge 0\right\},$$

so $T_i \le \widetilde{T}_i$ and $T^\star = \min_i T_i \le \widetilde{T} = \min_i \widetilde{T}_i$. Therefore, (5.24) implies that $\mathsf{E}[T^\star]^r < \infty$ and that for all $0 < m \le r$,

$$\mathsf{E}[T^\star]^m \le \left[\Psi\left(\min_{0\le i\le N} \max_{\substack{0\le j\le N \\ j\neq i}} \frac{a_{ji}}{I_i(\mathsf{P})}\right)\right]^m (1+o(1)) \quad \text{as} \quad \min_{j,i} a_{ji} \to \infty. \tag{5.25}$$

The lower bound

$$\mathsf{E}[T^\star]^m \ge \left[\Psi\left(\min_{0\le i\le N} \max_{\substack{0\le j\le N \\ j\neq i}} \frac{a_{ji}}{I_i(\mathsf{P})}\right)\right]^m (1+o(1)) \quad \text{as} \quad \min_{j,i} a_{ji} \to \infty$$

results from (5.18) if we replace α_{ji} by $e^{-a_{ji}}$. Therefore, the asymptotic approximation (5.20) follows.

To prove (ii) and (iii), it suffices to set $a_{ji} = |\log \alpha_{ji}|$ and $a_{ji} = |\log(N/\alpha_j)$ in (5.20) and then use the asymptotic lower bounds (5.18). □

Remark 5.3.2. As the example given in Subsection 3.4.3 shows, the r-quick convergence conditions in Theorem 5.3.1 cannot be relaxed into the almost sure convergence ones. However, the following weak asymptotic optimality result holds for the proposed accepting and rejecting MSPRTs. Assume that $\psi(t) = t^k$ for some $k > 0$ and the almost sure convergence condition (5.17) is satisfied with some nonnegative finite constants $I_0(P), I_1(P), \dots, I_N(P)$ not simultaneously equal to zero. Then

$$\frac{T^\star}{\min_{0\le i\le N} \max_{\substack{0\le j\le N \\ j\neq i}} (a_{ji}/I_j(\mathsf{P}))} \xrightarrow[a_{ji}\to 0]{\mathsf{P}-\text{a.s.}} 1$$

and, for every $0 < \varepsilon < 1$,

$$\begin{aligned}&\inf_{\delta\in \mathbf{C}([\alpha_{ij}])} \mathsf{P}(T > \varepsilon T^\star) \to 1 \quad \text{as} \quad \alpha_{\max} \to 0; \\ &\inf_{\delta\in \mathbf{C}(\boldsymbol{\alpha})} \mathsf{P}(T > \varepsilon T^\star) \to 1 \quad \text{as} \quad \tilde{\alpha}_{\max} \to 0\end{aligned} \tag{5.26}$$

when the thresholds a_{ji} are selected as in Theorem 5.3.1(ii) and (iii).

As discussed above, the main motivation for considering the modified Kiefer–Weiss problem is that it is tightly related to the problem of finding a minimax solution, namely the Kiefer–Weiss problem. Returning back to the general parametric model we would like to find a test that minimizes the maximal ESS, $\sup_{\theta\in\Theta} \mathsf{E}_\theta T$ with $\Theta = \sum_{i=0}^N \Theta_i + \mathrm{I}_{\mathrm{in}}$. Therefore, if we select $\mathsf{P} = \mathsf{P}_\theta$ with $\theta = \theta^*$, where θ^* is the almost least favorable point in the sense that $\mathsf{E}_\theta[T^\star(\theta^*)]^m$ is nearly maximized at $\theta = \theta^*$, then the proposed asymptotic solutions to the modified Kiefer–Weiss problem will allow us to obtain asymptotic solutions to the Kiefer–Weiss problem, i.e.,

$$\inf_{\delta\in\mathbf{C}([\alpha_{ij}])} \sup_{\theta\in\Theta} \mathsf{E}_\theta[T]^m \sim \sup_{\theta\in\Theta} \mathsf{E}_\theta[T^\star(\theta^*)]^m \sim \mathsf{E}_{\theta^*}[T^\star(\theta^*)]^m$$

$$\sim \left[\Psi\left(\frac{|\log \alpha_{\max}|}{\inf_{\theta\in\Theta} J(\theta)}\right)\right]^m \quad \text{as } \alpha_{\max} \to 0; \tag{5.27}$$

$$\inf_{\delta\in\mathbf{C}(\boldsymbol{\alpha})} \sup_{\theta\in\Theta} \mathsf{E}_\theta[T]^m \sim \sup_{\theta\in\Theta} \mathsf{E}_\theta[T^\star(\theta^*)]^m \sim \mathsf{E}_{\theta^*}[T^\star(\theta^*)]^m$$

$$\sim \left[\Psi\left(\frac{|\log \tilde{\alpha}_{\max}|}{\inf_{\theta\in\Theta} \tilde{J}(\theta)}\right)\right]^m \quad \text{as } \tilde{\alpha}_{\max} \to 0. \tag{5.28}$$

Here $T^\star(\theta^*)$ corresponds to the accepting MSPRT with $\mathsf{P} = \mathsf{P}_{\theta^*}$.

In the general multiparameter case this plan is difficult to carry out using the solution to the modified Kiefer–Weiss problem since it is not clear what putative values of the parameter $\theta = \theta_i$ have to be selected in the regions Θ_i. Several approaches solving the Kiefer–Weiss problem in the general multiparameter case are considered in Section 5.4. However, this can be easily done in scalar one-parameter cases as well as for multisample slippage problems that can be reduced to testing scalar one-sided composite hypotheses. For the time being assume that the parameters $\theta_0, \theta_1, \dots, \theta_N$ have been somehow selected. Theorem 5.3.2 below provides sufficient conditions under which the asymptotic optimality results (5.27), (5.28) hold.

We write $\ell_i(t) = \ell_i^\theta(t)$, $I_i(\mathsf{P}) = I_i(\theta)$ and $\delta^\star(\theta) = (T^\star(\theta), d^\star(\theta))$ when $\mathsf{P} = \mathsf{P}_\theta$, $\theta \in \Theta$.

Theorem 5.3.2 (Asymptotic minimaxity of the modified MSPRT). *Let $\mathsf{P} = \mathsf{P}_\theta$ and let the r-quick convergence conditions* (5.19) *be satisfied with $I_i(\mathsf{P}) = I_i(\theta)$.*

(i) *Let $\theta^* = \arg\inf_{\theta\in\Theta} J(\theta)$. Assume that there exist a function $\psi(t)$ and finite numbers $I_i(\theta^*, \theta)$ such that*

$$\frac{\ell_i^{\theta^*}(t)}{\psi(t)} \xrightarrow[t\to\infty]{\mathsf{P}_\theta-a.s.} I_i(\theta^*, \theta) \quad \text{for all } i = 0, 1, \dots, N. \tag{5.29}$$

For $\varepsilon > 0$, let

$$L_{\theta^*,\theta}(\varepsilon) = \sup\left\{t : \max_{0\le i\le N} \left|\ell_i^{\theta^*}(t) - I_i(\theta^*, \theta)\psi(t)\right| > \psi(t)\varepsilon\right\} \quad (\sup\{\varnothing\} = 0).$$

Let the thresholds be selected so that $\alpha_{ij}(\delta^\star) \le \alpha_{ij}$ and $a_{ji} \sim \log(1/\alpha_{ji})$, in particular $a_{ji} = \log(1/\alpha_{ji})$. If $\inf_{\theta\in\Theta} J(\theta) > 0$ and $\sup_{\theta\in\Theta} \mathsf{E}_\theta[L_{\theta^*,\theta}(\varepsilon)]^r < \infty$ for all $\varepsilon > 0$ and some $r > 0$, then for all $0 < m \le r$ the asymptotic relations (5.27) *hold, and therefore, the modified accepting MSPRT $\delta^\star(\theta^*)$ is asymptotically minimax in the sense of minimizing the supremum of the moments of the stopping time distribution up to order r in the class of tests $\mathbf{C}([\alpha_{ij}])$ as $\alpha_{\max} \to 0$.*

(ii) *Let $\theta^* = \arg\inf_{\theta\in\Theta} \tilde{J}(\theta)$. Assume that the conditions* (5.29) *hold. Let the thresholds be selected so that $\alpha_i(\delta^\star) \le \alpha_i$ and $a_{ji} \sim \log(1/\alpha_j)$, in particular $a_{ji} = \log(N/\alpha_j)$. If $\inf_{\theta\in\Theta} \tilde{J}(\theta) > 0$ and if $\sup_{\theta\in\Theta} \mathsf{E}_\theta[L_{\theta^*,\theta}(\varepsilon)]^r < \infty$ for all $\varepsilon > 0$ and some $r > 0$, then for all $0 < m \le r$ the asymptotic relations* (5.28) *hold, and therefore, the modified accepting MSPRT $\delta^\star(\theta^*)$ is asymptotically minimax in the sense of minimizing the supremum of the moments of the stopping time distribution up to order r in the class of tests $\mathbf{C}(\boldsymbol{\alpha})$ as $\tilde{\alpha}_{\max} \to 0$.*

Proof. We give a sketch of the proof omitting the details. By Theorem 5.3.1(ii),

$$\mathsf{E}_{\theta^*}[T^\star(\theta^*)]^m \sim \left[\Psi\left(\frac{|\log \alpha_{\max}|}{J(\theta^*)}\right)\right]^m \quad \text{as } \alpha_{\max} \to 0.$$

Therefore, to prove that $\sup_{\theta\in\Theta} \mathsf{E}_\theta[T^\star(\theta^*)]^m \sim \mathsf{E}_{\theta^*}[T^\star(\theta^*)]^m$, it suffices to show that

$$\sup_{\theta\in\Theta} \mathsf{E}_\theta[T^\star(\theta^*)]^m \sim \left[\Psi\left(\frac{|\log \alpha_{\max}|}{J(\theta^*)}\right)\right]^m \quad \text{as } \alpha_{\max} \to 0.$$

For the sake of simplicity take $a_{ji} \sim a$. The general case is handled analogously by scaling. Using the same argument that has led to (5.23), we obtain that when the true parameter value is θ,

$$T^\star(\theta^*) \le \Delta + \Psi\left(\frac{a}{\max_i I_i(\theta^*, \theta) - \varepsilon}\right) + L_{\theta^*,\theta}(\varepsilon). \tag{5.30}$$

Since $J(\theta^*) \le \max_i \inf_\theta I_i(\theta^*, \theta)$, it follows that for every $\varepsilon < J(\theta^*)$

$$\begin{aligned}\sup_\theta \mathsf{E}_\theta[T^\star(\theta^*)]^r &\le \sup_\theta \mathsf{E}_\theta\left[L_{\theta^*,\theta}(\varepsilon) + \Delta + \Psi\left(\frac{a}{J(\theta^*) - \varepsilon}\right)\right]^r \\ &= \left[\Psi\left(\frac{|\log \alpha_{\max}|}{J(\theta^*)}\right)\right]^r (1 + o(1)),\end{aligned}$$

where the latter equality is true thanks to the condition $\sup_{\theta\in\Theta} \mathsf{E}_\theta[L_{\theta^*,\theta}(\varepsilon)]^r < \infty$. The lower bound

$$\inf_{\delta\in\mathbf{C}([\alpha_{ij}])} \sup_{\theta\in\Theta} \mathsf{E}_\theta[T]^r \ge \left[\Psi\left(\frac{|\log \alpha_{\max}|}{\inf_{\theta\in\Theta} J(\theta)}\right)\right]^r (1 + o(1)) \quad \text{as } \alpha_{\max} \to 0$$

follows from (5.18). □

For illustration purposes we now apply this theorem to two examples.

Example 5.3.1 (Detection–identification of a signal with unknown amplitude in Gaussian noise). As in Subsection 3.4.3, suppose that for $t \in \mathbb{R}_+$ the observed process $\{X_t\}$ admits a stochastic differential

$$\mathrm{d}X_t = \theta S_t \mathrm{d}t + V_t \mathrm{d}t + \sigma \mathrm{d}W_t, \quad X_0 = 0,$$

where now $\theta \ge 0$ is the unknown amplitude (intensity) of the deterministic signal S_t, W_t is a standard Brownian motion (noise), $\sigma > 0$, and V_t is an L_2-continuous Gaussian process (clutter). The hypotheses are $H_i : \theta = \theta_i$ for $i = 0, 1, \dots, N$, where $\theta_i = \theta_{i-1} + \Delta$, $\Delta > 0$, $\theta_0 = 0$. In other words, under the null hypothesis H_0 there is no signal at all ($\theta = 0$), while under H_i, $i = 1, \dots, N$, the signal is present and has the strength $\theta = i\Delta$. The goal is to detect the signal and to identify its strength. The true intensity θ may take on any nonnegative value.

Assuming $\theta = \vartheta \in (0, \infty)$, similarly to (4.83) we obtain

$$\ell_i^\vartheta(t) = \frac{\vartheta - \theta_i}{\sigma^2} \int_0^t \widetilde{S}_u \mathrm{d}\widetilde{X}_u - \frac{\vartheta^2 - \theta_i^2}{2\sigma^2} \int_0^t \widetilde{S}_u^2 \mathrm{d}u, \quad i = 0, 1, \dots, N, \tag{5.31}$$

where $\theta_i = i\Delta$,

$$\widetilde{S}_t = S_t - \int_0^t S_u C(t,u) \mathrm{d}u, \quad \mathrm{d}\widetilde{X}_t = \mathrm{d}X_t - \left[\int_0^t C(t,u) \mathrm{d}X_u\right] \mathrm{d}t,$$

and where $C(u,t)$ is the impulse response of the whitening filter. Using (5.31), it is easily shown that under the measure P_θ, i.e., when the true parameter value is θ, the LLR $\ell_i^{\tilde\theta}(t)$ tuned to $\theta = \tilde\theta$ can be written as

$$\ell_i^{\tilde\theta}(t) = \frac{\tilde\theta - \theta_i}{\sigma} \int_0^t \widetilde{S}_u \, d\widetilde{W}_u + \frac{(\tilde\theta - \theta_i)[2\theta - (\tilde\theta + \theta_i)]}{2\sigma^2} \int_0^t \widetilde{S}_u^2 \, du, \tag{5.32}$$

where $\widetilde{W}_t$ is a Brownian motion statistically indistinguishable from W_t. In particular, if $\tilde\theta = \theta$, then

$$\ell_i^{\theta}(t) = \frac{\theta - \theta_i}{\sigma} \int_0^t \widetilde{S}_u \, d\widetilde{W}_u + \mu_i^\theta(t), \tag{5.33}$$

where $\mu_i^\theta(t) = (\theta - \theta_i)^2 (2\sigma^2)^{-1} \int_0^t \widetilde{S}_u^2 \, du$ is the cumulative SNR at the output of the whitening filter over the time interval $[0,t]$.

Assume that for some $k > 0$

$$\lim_{t\to\infty} \frac{1}{t^k} \int_0^t \widetilde{S}_u^2 \, du = S^2, \tag{5.34}$$

where $0 < S^2 < \infty$. Then, using (5.33), it can be shown in just the same way as in Subsection 3.4.3 that for all $r > 0$,

$$t^{-k} \ell_i^\theta(t) \xrightarrow[t\to\infty]{\mathsf{P}_\theta - r - \text{quickly}} I_i(\theta) = \frac{S^2(\theta - \theta_i)^2}{2\sigma^2}, \quad i = 0, 1, \dots, N. \tag{5.35}$$

Thus, by Theorem 5.3.1, the modified accepting and rejecting MSPRTs asymptotically minimize all positive moments of the stopping time distribution at any point $\theta \ge 0$. Note that the probability $\alpha_0(\delta^\star) = \mathsf{P}_0(d^\star \neq 0)$ is the probability of a false alarm (i.e., raising an alarm when there is no signal) and for $i = 1, \dots, N$, the probability $\alpha_i(\delta^\star) = \mathsf{P}_{\theta_i}(d^\star \neq i) = \mathsf{P}_{\theta_i}(d^\star = 0) + \sum_{j \neq i,0} \mathsf{P}_{\theta_i}(d^\star = j)$ is the sum of the probability of signal missing and the cumulative probability of misidentification. Set $\alpha_i = \alpha_1$ for all $i = 1, \dots, N$. Usually $\alpha_0 \ll \alpha_1$, so that we are dealing with the asymmetric case (5.16) with $\tilde\alpha_{\max} = \alpha_1$, $c_0 > 1$, and $c_1 = \cdots = c_N = 1$. Using (5.22) and (5.35), we obtain that, for $\theta \in [(i-1)\Delta, i\Delta]$,

$$\inf_{\delta \in \mathbf{C}(\boldsymbol{\alpha})} \mathsf{E}_\theta T^m \sim \mathsf{E}_\theta[T^\star(\theta)]^m \sim \left(\frac{|\log \alpha_1|}{\widetilde{J}(\theta)} \right)^{m/k} \quad \text{as } \alpha_1 \to 0,$$

where

$$\widetilde{J}(\theta) = \frac{S^2}{2\sigma^2} \max\left\{ \theta^2/c_0, (\theta - (i-1)\Delta)^2, (\theta - i\Delta)^2 \right\},$$

or alternatively, for $\theta \in [(i-1)\Delta, i\Delta]$,

$$\mathsf{E}_\theta[T^\star(\theta)]^m \sim \left[\frac{2\sigma^2}{S^2} \min\left\{ \frac{|\log \alpha_0|}{\theta^2}, \frac{|\log \alpha_1|}{\max\left[(\theta - (i-1)\Delta)^2, (\theta - i\Delta)^2\right]} \right\} \right]^{m/k}. \tag{5.36}$$

The maximum of the right-hand side in (5.36) occurs at the point

$$\theta^* = \Delta \frac{\sqrt{c_0}}{1 + \sqrt{c_0}}$$

and

$$\mathsf{E}_{\theta^*}[T^\star(\theta^*)]^m \sim \left[\frac{|\log \alpha_1| (1 + \sqrt{c_0})^2}{\widetilde{\Delta}^2} \right]^{m/k}, \tag{5.37}$$

where $\widetilde{\Delta}^2 = S^2 \Delta^2 / 2\sigma^2$.

Next, using (5.31), similarly to (5.35) one can show that

$$t^{-k} \ell_i^{\theta^*}(t) \xrightarrow[t\to\infty]{\mathsf{P}_\theta - r - \text{quickly}} I_i(\theta^*, \theta) = \frac{S^2(\theta^* - \theta_i)[2\theta - (\theta^* + \theta_i)]}{2\sigma^2}, \quad i = 0, 1, \dots, N, \tag{5.38}$$

i.e., $\mathsf{E}_\theta[L_i^{\theta^*,\theta}(\varepsilon)]^r < \infty$ for all $i = 0,1,\dots,N$ and all $\varepsilon > 0, r > 0$, where

$$L_i^{\theta^*,\theta}(\varepsilon) = \sup\left\{t : \left|t^{-k}\ell_i^{\theta^*}(t) - I_i(\theta^*,\theta)\right| > \varepsilon\right\}.$$

Write

$$Q_{i,t}(\theta^*,\theta) = I_i(\theta^*,\theta)\left(\frac{1}{t^k}\int_0^t \widetilde{S}_u^2\,\mathrm{d}u - 1\right), \quad V_{i,t} = \frac{(\theta^* - \theta_i)}{t^k}\int_0^t \widetilde{S}_u\,\mathrm{d}\widetilde{W}_u.$$

Then the last entry time $L_i^{\theta^*,\theta}(\varepsilon)$ can be written as

$$L_i^{\theta^*,\theta}(\varepsilon) = \sup\{t : |V_{i,t} - Q_{i,t}(\theta^*,\theta)| > \varepsilon\}.$$

Since $\sup_\theta Q_{i,t}(\theta^*,\theta) \to 0$ as $t \to \infty$ and $V_{i,t}$ converges r-quickly to 0 for all $r > 0$ regardless of the value of θ ($V_{i,t}$ does not depend on θ), it follows that $\sup_\theta \mathsf{E}_\theta[L_i^{\theta^*,\theta}(\varepsilon)]^r < \infty$. Finally, since $L_{\theta^*,\theta}(\varepsilon) \le \min_i L_i^{\theta^*,\theta}(\varepsilon)$, the condition $\sup_\theta \mathsf{E}_\theta[L_{\theta^*,\theta}(\varepsilon)]^r < \infty$ is satisfied for all $r > 0$.

Applying Theorem 5.3.2, we can conclude that the test $\delta^\star(\theta^*)$ tuned to the parameter $\theta^* = \Delta\sqrt{c_0}/(1+\sqrt{c_0})$ is asymptotically minimax in the sense of minimizing all moments of the stopping time in the worst-case scenario:

$$\begin{aligned}\inf_{\delta\in\mathbf{C}(\boldsymbol{\alpha})}\sup_{\theta\in\Theta}\mathsf{E}_\theta[T]^m &\sim \sup_{\theta\in\Theta}\mathsf{E}_\theta[T^\star(\theta^*)]^m \sim \mathsf{E}_{\theta^*}[T^\star(\theta^*)]^m \\ &\sim \left[\frac{|\log\alpha_1|(1+\sqrt{c_0})^2}{\widetilde{\Delta}^2}\right]^{m/k} \quad \text{as } \alpha_1 \to 0 \quad \text{for all } m > 0.\end{aligned} \tag{5.39}$$

It is worth noting that the key condition (5.34) is satisfied in a variety of applications. For example, in radar applications, namely for the detection of objects in noise/clutter, typically $S_t = \sin\omega t$ is a harmonic signal with frequency ω or more generally a sequence of harmonic pulses. Assume that clutter V_t is a stationary Markov Gaussian process described by the Itô stochastic equation

$$\mathrm{d}V_t = -\gamma V_t\,\mathrm{d}t + \sigma_w\,\mathrm{d}w_t \quad (\gamma > 0,\ \sigma_w > 0),$$

where $\{w_t\}_{t\ge 0}$ is a standard Wiener process. In this case, $\mathsf{E}[V_t] = 0$, $\mathsf{E}[V_tV_{t+u}] = \sigma_\mathrm{v}^2 e^{-\gamma|u|}$, $\sigma_\mathrm{v}^2 = \sigma_w^2/2\gamma$ and it can be verified that the condition (5.34) holds with $k = 1$ and

$$S^2 = \frac{\gamma^2 + \omega^2}{\omega^2 + \gamma^2(1 + 2\sigma_\mathrm{v}^2/\sigma^2\gamma)}.$$

If $S_t = 1$ for $t \ge 0$, then the condition (5.34) holds with $k = 1$ and $S^2 = 1/(1 + 2\sigma_\mathrm{v}^2/\sigma^2\gamma)$.

Suppose now that $V_t \equiv 0$ (no clutter), so that a deterministic signal is being detected in white Gaussian noise $\dot{W}_t$ with intensity σ, in which case $\widetilde{S}_t \equiv S_t$. If $S_t = \sin\omega t$, then the condition (5.34) holds with $k = 1$ and $S^2 = 1/2$. Finally, if $S_t = d_0 + d_1 t + \dots + d_l t^l$ is a polynomial, then the condition (5.34) holds with $k = 2l+1$ and $S^2 = d_l^2/(2l+1)$.

Example 5.3.2 (2-t-SPRT). Let us obtain the minimax solution to the problem considered in Subsection 3.6.2. Specifically, now we consider the problem of testing two composite hypotheses $H_0 : \theta \le \theta_0$ and $H_1 : \theta \ge \theta_1$ ($\theta = \mu/\sigma$) with the indifference interval (θ_0, θ_1) for the iid normal sequence $X_n \sim \mathcal{N}(\mu, \sigma^2)$, $n = 1, 2, \dots$ with unknown mean μ and variance σ^2. The argument similar to that used in Subsection 3.6.2 yields $\ell_i^\theta(n) = \log[J_n(\theta t_n)/J_n(\theta_i t_n)]$, where $J_n(z)$ is defined in (3.221), and that the r-quick convergence conditions (5.19) are satisfied under $\mathsf{P} = \mathsf{P}_\theta$ with $\psi(n) = n$ and

$$I_i(\theta) = \phi(\theta^2/(1+\theta^2)^{1/2}) - \phi(\theta_i\theta/(1+\theta^2)^{1/2}) - \frac{1}{2}(\theta^2 - \theta_i^2), \tag{5.40}$$

where $\phi(z)$ is defined in (3.223). By the same argument as in Subsection 3.6.2, it can be verified that $I_i(\theta) > 0$ as long as $\theta \neq \theta_i$. Therefore, by Theorem 5.3.1, the 2-t-SPRT $\delta^\star(\theta)$ tuned to θ with the thresholds $a_i = |\log \alpha_i|$, $i = 0, 1$ asymptotically minimizes all positive moments of the stopping time at the point θ, namely for all $m > 0$,

$$\inf_{\delta \in \mathbf{C}(\alpha_0,\alpha_1)} \mathsf{E}_\theta[T]^m \sim \mathsf{E}_\theta[T^\star(\theta)]^m \sim \min\left\{\frac{|\log \alpha_0|}{I_0(\theta)}, \frac{|\log \alpha_1|}{I_1(\theta)}\right\}^m \quad \text{as } \alpha_{\max} \to 0.$$

The right-hand side of this asymptotic relation is maximized at the point θ^* which is the root of the equation $|\log \alpha_0|/I_0(\theta) = |\log \alpha_1|/I_1(\theta)$ or $cI_1(\theta) = I_0(\theta)$, where $c = \lim_{\alpha_{\max}\to 0} |\log \alpha_0/|\log \alpha_1|$. In particular, in the asymptotically symmetric case where $c = 1$, θ^* is the root of the equation

$$\phi(\theta_1\theta/(1+\theta^2)^{1/2}) - \phi(\theta_0\theta/(1+\theta^2)^{1/2}) + \theta_1^2/2 - \theta_0^2/2 = 0$$

where $\phi(z)$ is defined in (3.223). Thus, for all $m > 0$,

$$\inf_{\delta \in \mathbf{C}(\alpha_0,\alpha_1)} \mathsf{E}_{\theta^*}[T]^m \sim \mathsf{E}_{\theta^*}[T^\star(\theta^*)]^m \sim \left(\frac{|\log \alpha|}{I_0(\theta^*)}\right)^m \quad \text{as } \alpha \to 0 \tag{5.41}$$

and it is reasonable to conjecture that the 2-t-SPRT $\delta^\star(\theta^*)$ tuned to θ^* is asymptotically minimax.

To show this, define

$$g_i^*(z) = \phi(\theta^* z) - \phi(\theta_i z) - \frac{1}{2}[(\theta^*)^2 - \theta_i^2], \quad i = 0, 1.$$

A slight extension of the argument in Subsection 3.6.2 shows that, for all $r > 0$,

$$n^{-1}\ell_i^{\theta^*}(n) \xrightarrow[t\to\infty]{\mathsf{P}_\theta - r - \text{quickly}} I_i(\theta^*, \theta) = g_i^*(\theta/(1+\theta)^{1/2}), \quad i = 0, 1,$$

i.e., $\mathsf{E}_\theta[L_i^{\theta^*,\theta}(\varepsilon)]^r < \infty$ for all $\varepsilon > 0, r > 0$, where

$$L_i^{\theta^*,\theta}(\varepsilon) = \sup\left\{n : \left|n^{-1}\ell_i^{\theta^*}(n) - g_i^*(\theta/(1+\theta)^{1/2})\right| > \varepsilon\right\}.$$

Furthermore, it can be established that for every $\varepsilon > 0$ there is $0 < \widetilde{\varepsilon}(\varepsilon) < 1/2$ such that

$$\sup_\theta \max_i L_i^{\theta^*,\theta}(\varepsilon) \le L(\widetilde{\varepsilon}) = \sup\left\{n : |\bar{Y}_n| > \widetilde{\varepsilon} \text{ or } |n^{-1}\sum_{k=1}^n (Y_k - \bar{Y}_n)^2 - 1| > \widetilde{\varepsilon}\right\},$$

where $Y_n \sim \mathcal{N}(0,1)$. Clearly,

$$L_{\theta^*,\theta}(\varepsilon) = \sup\left\{n : \max_{i=0,1}\left|n^{-1}\ell_i^{\theta^*}(n) - I_i(\theta^*,\theta)\right| > \varepsilon\right\} \le \max_i L_i^{\theta^*,\theta}(\varepsilon),$$

so that $\sup_\theta L_{\theta^*,\theta}(\varepsilon) \le \sup_\theta \max_i L_i^{\theta^*,\theta}(\varepsilon) \le L(\widetilde{\varepsilon})$. Since $\mathsf{E}[L(\widetilde{\varepsilon})]^r < \infty$ for all $r > 0$, the required condition $\sup_\theta \mathsf{E}_\theta[L_{\theta^*,\theta}(\varepsilon)]^r < \infty$ holds for all $r > 0$. Applying Theorem 5.3.2 and (5.41), we obtain that the 2-t-SPRT $\delta^\star(\theta^*)$ is asymptotically minimax in the sense of minimizing all moments of the stopping time:

$$\inf_{\delta \in \mathbf{C}(\alpha_0,\alpha_1)} \sup_{\theta\in\Theta} \mathsf{E}_\theta[T]^m \sim \sup_{\theta\in\Theta} \mathsf{E}_\theta[T^\star(\theta^*)]^m \sim \mathsf{E}_{\theta^*}[T^\star(\theta^*)]^m$$
$$\sim \left(\frac{|\log \alpha|}{I_0(\theta^*)}\right)^m \quad \text{as } \alpha \to 0 \quad \text{for all } m > 0,$$

where $I_0(\theta^*) = g_0^*(\theta^*/(1+\theta^*)^{1/2})$ is given by (5.40).

5.3.2 *Asymptotically Optimal Tests at an Intermediate Point in the iid Case*

Consider now the iid case that the observed stochastic process $\{X_t\}$ ($t \in \mathbb{Z}_+$ or $t \in \mathbb{R}_+$) has iid increments under all the measures $\mathsf{P}, \mathsf{P}_0, \dots, \mathsf{P}_N$. By Theorem 2.4.4, the $(r+1)$-th moment condition $\mathsf{E}|\ell_i(1)|^{r+1} < \infty$ is both necessary and sufficient for the r-quick convergence (3.159) with $\psi(t) = t$ and

$$I_i(\mathsf{P}) = I(\mathsf{P}, \mathsf{P}_i) = \int \log\left(\frac{d\mathsf{P}}{d\mathsf{P}_i}(\mathbf{X}_0^1)\right) d\mathsf{P}, \quad i = 0, 1, \dots, N$$

being the corresponding K–L information numbers. Therefore, Theorem 5.3.1 implies the following result in the iid case.

Corollary 5.3.1. *Let the observed stochastic process $\{X_t\}$ ($t \in \mathbb{Z}_+$ or $t \in \mathbb{R}_+$) be a process with iid increments under $\mathsf{P}, \mathsf{P}_0, \dots, \mathsf{P}_N$. Assume that the K–L information numbers $I(\mathsf{P}, \mathsf{P}_i)$ are nonnegative and finite and that $\max_i I(\mathsf{P}, \mathsf{P}_i) > 0$. Further, for some $r \geq 1$, assume the $(r+1)$-th moment conditions $\mathsf{E}|\ell_i(1)|^{r+1} < \infty$, $i = 0, 1, \dots, N$.*

(i) *If the thresholds are selected so that $\alpha_{ij}(\delta^\star) \leq \alpha_{ij}$ and $a_{ji} \sim \log(1/\alpha_{ji})$, in particular $a_{ji} = \log(1/\alpha_{ji})$, then for all $0 < m \leq r$,*

$$\inf_{\delta \in \mathbf{C}([\alpha_{ij}])} \mathsf{E}[T]^m \sim \mathsf{E}[T^\star]^m \sim \left(\frac{|\log \alpha_{\max}|}{J}\right)^m \quad \text{as } \alpha_{\max} \to 0, \tag{5.42}$$

where $J = \max_{0 \leq i \leq N} \min_{j \neq i}[I(\mathsf{P}, \mathsf{P}_j)/c_{ji}]$.

(ii) *If the thresholds are selected so that $\alpha_i(\delta^\star) \leq \alpha_i$ and $a_{ji} \sim \log(1/\alpha_j)$, in particular $a_{ji} = \log(N/\alpha_j)$, then for all $0 < m \leq r$,*

$$\inf_{\delta \in \mathbf{C}(\boldsymbol{\alpha})} \mathsf{E}[T]^m \sim \mathsf{E}[T^\star]^m \sim \left(\frac{|\log \tilde{\alpha}_{\max}|}{\widetilde{J}}\right)^m \quad \text{as } \tilde{\alpha}_{\max} \to 0, \tag{5.43}$$

where $\widetilde{J} = \max_{0 \leq i \leq N} \min_{j \neq i}[I(\mathsf{P}, \mathsf{P}_j)/c_j]$.

Remark 5.3.3. Using an argument similar to that used by Dragalin *et al.* [129] for proving asymptotic optimality of the MSPRT under P_i, it can be shown that the asymptotic relations (5.42) and (5.43) hold for all $m \geq 1$ under the sole first moment condition $0 \leq I(\mathsf{P}, \mathsf{P}_i) < \infty$ ($i = 0, 1, \dots, N$), $\max_i I(\mathsf{P}, \mathsf{P}_i) > 0$. Therefore, the proposed accepting and rejecting MSPRTs asymptotically minimize all positive moments of the sample size to first order when $0 \leq I(\mathsf{P}, \mathsf{P}_i) < \infty$, and in Corollary 5.3.1 the conditions $\mathsf{E}|\ell_i(1)|^{r+1} < \infty$, $i = 0, 1, \dots, N$, can be removed. However, to obtain higher-order asymptotics and optimality these conditions are needed, as shown next.

5.3.2.1 *Near Optimality of the Modified MSPRT with Respect to the Expected Sample Size*

Recall that by Theorem 4.3.2 the MSPRT minimizes the expected sample sizes $\mathsf{E}_i T$ in the class $\mathbf{C}([\alpha_{ij}])$ under all hypotheses as long as the second moments $\mathsf{E}_i|\lambda_{ij}(1)|^2$ of the LLRs are finite and the thresholds A_{ij} are selected so that the error probabilities $\alpha_{ij}(\delta^*)$ are exactly equal to α_{ij}, and the same is true for the class $\mathbf{C}(\boldsymbol{\alpha})$. Now note that the MSPRT δ^* is a particular case of the modified MSPRT $\delta^\star$ when in (5.6) we set $\mathsf{P} = \mathsf{P}_i$. Therefore, it stands to reason that the assertions of Theorem 4.3.2 also hold for the modified MSPRT under the measure P. In other words, one could expect that in the iid case the modified MSPRT is also nearly optimal in the sense of minimizing the ESS $\mathsf{E}T$ in the corresponding classes of tests within a negligible term $o(1)$ under the second moment condition $\mathsf{E}|\ell_i(1)|^2 < \infty$. The proof of this fact can be built based on a slight extension of Lorden's Bayesian argument in [275] used for the proof of Theorem 4.3.2 in the discrete-time iid case and, for the continuous-time processes with iid increments, by extending the method presented by Dragalin and Novikov in [126] in the case of two hypotheses. The following theorem contains the details.

Theorem 5.3.3 (Near optimality of the modified MSPRT). *Let the observed stochastic process $\{X_t\}$ ($t \in \mathbb{Z}_+$ or $t \in \mathbb{R}_+$) be a process with iid increments under $\mathsf{P}, \mathsf{P}_0, \dots, \mathsf{P}_N$. Assume that $\max_i \mathsf{E}|\ell_i(1)|^2 < \infty$ and $\max_i I(\mathsf{P}, \mathsf{P}_i) > 0$. If the thresholds a_{ij} are selected so that the error probabilities of the modified MSPRT, $\alpha_{ij}(\delta^\star)$, are exactly equal to α_{ij}, then*

$$\inf_{\delta \in \mathbf{C}([\alpha_{ij}])} \mathsf{E}[T] = \mathsf{E}[T^\star] + o(1) \quad \text{as } \alpha_{\max} \to 0. \tag{5.44}$$

The same assertion is true for the class $\mathbf{C}(\boldsymbol{\alpha})$.

5.3.2.2 *Asymptotic Minimaxity of the 2-SPRT for the Exponential Family*

The case of two hypotheses ($N = 1$) is of special interest. Then the modified MSPRT reduces to two parallel one-sided SPRTs,

$$T_0 = \inf\{t : \ell_1(t) \geq a_1\}, \quad T_1 = \inf\{t : \ell_0(t) \geq a_0\}, \tag{5.45}$$

and is called the 2-SPRT [273]. Its stopping time is $T^\star = \min(T_0, T_1)$ and the terminal decision is $d^\star = \arg\min_{i=0,1} T_i$. By Lemma 5.3.1, if $a_i = \log(1/\alpha_i)$, $i = 0, 1$, then $\alpha_i(\delta^\star) \leq \alpha_i$, i.e., this test belongs to the class $\mathbf{C}(\alpha_0, \alpha_1) = \{\delta : \alpha_0(\delta) \leq \alpha_0, \alpha_1(\delta) \leq \alpha_1\}$. These upper bounds may be rather conservative. For example, in the symmetric case $\mathsf{P}(d^\star = 1) = \mathsf{P}(d^\star = 0) = 1/2$, we have $\alpha_i(\delta^\star) \leq \alpha_i/2$.

Consider the parametric case $\mathsf{P} = \mathsf{P}_\theta$, $\mathsf{P}_i = \mathsf{P}_{\theta_i}$ where the hypotheses are $H_0 : \theta = \theta_0$ and $H_1 : \theta = \theta_1$, $\theta_0 < \theta_1$. Let θ be an arbitrary point belonging to the interval (θ_0, θ_1) and let $\delta^\star(\theta) = (d^\star(\theta), T^\star(\theta))$ denote the 2-SPRT tuned to θ. In other words, $T^\star(\theta) = \min(T_0^\theta, T_1^\theta)$, where the T_i^θ's are defined by (5.45) with the LLRs $\ell_i^\theta(t) = \log[\mathrm{d}\mathsf{P}_\theta^t / \mathrm{d}\mathsf{P}_{\theta_i}^t](X_t)$, $i = 0, 1$, tuned to θ. Here and in the rest of this section we use the superscript θ to emphasize that the 2-SPRT and the respective quantities correspond to a particular choice of the parameter θ.

Theorem 5.3.3 implies that the 2-SPRT $\delta^\star(\theta)$ is third-order asymptotically optimal in terms of minimizing the ESS $\mathsf{E}_\theta T$ at the intermediate point $\theta \in (\theta_0, \theta_1)$ when the second moments $\mathsf{E}_\theta|\ell_0^\theta(1)|^2$ and $\mathsf{E}_\theta|\ell_1^\theta(1)|^2$ are finite and the thresholds a_i are selected in such a way that the error probabilities are exactly equal to the given numbers α_i. The latter requirement is a difficult task. However, setting $a_i = |\log \alpha_i|$ embeds the 2-SPRT into the class $\mathbf{C}(\alpha_0, \alpha_1)$ and Theorem 5.3.3 suggests that if one can find a nearly least favorable point θ^*, i.e., θ^* can be selected so that $\sup_\theta \mathsf{E}_\theta[T^\star(\theta^*)] \approx \mathsf{E}_{\theta^*}[T^\star(\theta^*)]$, then $\delta^\star(\theta^*)$ is an approximate solution to the Kiefer–Weiss problem of minimizing $\sup_\theta \mathsf{E}_\theta[T]$.

We now proceed with considering this problem for the single-parameter exponential family assuming that in the continuous-time case the process $\{X_t\}_{t \in \mathbb{R}_+}$ has iid increments, for all $\theta, \widetilde{\theta} \in \Theta \subset \mathbb{R}$ the measures P_θ^t and $\mathsf{P}_{\widetilde{\theta}}^t$ are equivalent and

$$\frac{\mathrm{d}\mathsf{P}_\theta^t}{\mathrm{d}\mathsf{P}_{\widetilde{\theta}}^t}(X_t) = \exp\left\{(\theta - \widetilde{\theta})X_t - (b(\theta) - b(\widetilde{\theta}))t\right\}, \quad t \in \mathbb{R}_+, \tag{5.46}$$

where $b(\theta)$ is a convex and infinitely differentiable function on $\widetilde{\Theta} \subset \Theta$. A simple calculation shows that $\mathsf{E}_\theta X_1 = \dot{b}(\theta)$, $\sigma_\theta^2 = \mathsf{var}_\theta(X_1) = \ddot{b}(\theta)$ and the K–L numbers are

$$I(\theta, \theta_i) = (\theta - \theta_i)\dot{b}(\theta) - (b(\theta) - b(\theta_i)).$$

Without loss of generality we assume that $\inf_{\theta \in \widetilde{\Theta}} \sigma_\theta^2 > 0$, $\sup_{\theta \in \widetilde{\Theta}} \sigma_\theta^2 < \infty$.

In the discrete-time case $t = n \in \mathbb{Z}_+$, we assume that the observations $X_1, X_2, \dots$ are iid with common density $p_\theta(x)$ such that

$$\frac{p_\theta(X_n)}{p_{\widetilde{\theta}}(X_n)} = \exp\left\{(\theta - \widetilde{\theta})X_n - (b(\theta) - b(\widetilde{\theta}))\right\}, \quad n = 1, 2, \dots \tag{5.47}$$

Note that both discrete- and continuous-time models can be merged if (with a certain abuse of notation) we replace X_t in (5.46) with the sum $S_n = X_1 + \cdots + X_n$, where X_k, $k = 1, 2, \dots$ are iid random variables.

We now describe a method for determining the nearly least favorable point $\theta_\star(\alpha_0, \alpha_1, \theta_0, \theta_1)$ such that the 2-SPRT with thresholds $a_i = \log(1/\alpha_i)$ is second-order asymptotically minimax for the exponential family (5.46).

For the sake of concreteness, for the time being consider the discrete-time case (5.47), while all final asymptotic results hold for the continuous-time case too. For an arbitrary $\theta \in (\theta_0, \theta_1)$, let $\ell_i^\theta(n) = \log[\mathrm{dP}_\theta^n/\mathrm{dP}_{\theta_i}^n](\mathbf{X}_1^n)$, $i = 0, 1$, denote the LLRs tuned to θ. Writing $S_n^\theta = S_n - (\mathsf{E}_\theta X_1)n \equiv S_n - \dot{b}(\theta)n$ and noting that $\ell_i^\theta(n) = (\theta - \theta_i)S_n^\theta + I(\theta, \theta_i)n$, it is easy to see that the stopping times T_i^θ can be written as

$$T_0^\theta = \inf\left\{n : \ell_1^\theta(n) \ge |\log \alpha_1|\right\} = \inf\left\{n : S_n^\theta \le (\theta_1 - \theta)^{-1}[-|\log \alpha_1| + I(\theta, \theta_1)n]\right\},$$
$$T_1^\theta = \inf\left\{n : \ell_0^\theta(n) \ge |\log \alpha_0|\right\} = \inf\left\{n : S_n^\theta \ge (\theta - \theta_0)^{-1}[|\log \alpha_0| - I(\theta, \theta_0)n]\right\}.$$

Therefore, the stopping time $T^\star(\theta) = \min(T_0^\theta, T_1^\theta)$ of the 2-SPRT tuned to θ can be represented as

$$T^\star(\theta) = \min\left\{n \ge 1 : S_n^\theta \notin (h_0^\theta(n), h_1^\theta(n))\right\},$$

where the boundaries $h_1^\theta(n)$ and $h_0^\theta(n)$ are linear functions of n:

$$h_1^\theta(n) = \frac{|\log \alpha_0| - I(\theta, \theta_0)\, n}{\theta - \theta_0}, \quad h_0^\theta(n) = \frac{-|\log \alpha_1| + I(\theta, \theta_1)\, n}{\theta_1 - \theta}.$$

Define $\gamma(\theta, \theta_i) = (\theta - \theta_i)/I(\theta, \theta_i)$. Since $\gamma(\theta, \theta_1) < 0 < \gamma(\theta, \theta_0)$, it is easily verified that these boundaries intersect at the point

$$n^\star(\theta) = \frac{1}{\gamma(\theta, \theta_0) + |\gamma(\theta, \theta_1)|}\left[\frac{\gamma(\theta, \theta_0)}{I(\theta, \theta_1)}|\log \alpha_1| + \frac{|\gamma(\theta, \theta_1)|}{I(\theta, \theta_0)}|\log \alpha_0|\right]. \tag{5.48}$$

Therefore, the region of continuation of observations is a triangle, as shown in Figure 5.2. This means that the 2-SPRT is a truncated test with the maximal number of steps $n^\star(\theta)$. Figure 5.2 also shows the typical curved boundaries of the optimal truncated test that can be computed using Bellman's backward induction. We also note that in the continuous-time case the optimal test is not truncated, which makes it practically impossible to compute the optimal stopping regions.

By Corollary 5.3.1, the 2-SPRT $\delta^\star(\theta)$ minimizes all positive moments of the stopping time at the point θ to first order (as $\alpha_{\max} \to 0$) in the class $\mathbf{C}(\alpha_0, \alpha_1)$ as long as the ratio $|\log \alpha_0|/|\log \alpha_1|$ is bounded away from zero and infinity,

$$\lim_{\alpha_{\max} \to 0} \frac{|\log \alpha_0|}{|\log \alpha_1|} = c, \quad 0 < c < \infty, \tag{5.49}$$

and for all $m > 0$

$$\inf_{\delta \in \mathbf{C}(\alpha_0, \alpha_1)} \mathsf{E}_\theta[T]^m \sim \mathsf{E}_\theta[T^\star(\theta)]^m \sim \min\left[\frac{|\log \alpha_0|}{I(\theta, \theta_0)}, \frac{|\log \alpha_1|}{I(\theta, \theta_1)}\right]^m \quad \text{as } \alpha_{\max} \to 0. \tag{5.50}$$

The right-hand side of the asymptotic equality (5.50) is maximized at $\theta = \theta^*$ that satisfies the equation

$$\frac{|\log \alpha_1|}{I(\theta^*, \theta_0)} = \frac{|\log \alpha_0|}{I(\theta^*, \theta_1)} (\equiv n^\star(\theta^*)) \tag{5.51}$$

in which case

$$\inf_{\delta \in \mathbf{C}(\alpha_0, \alpha_1)} \mathsf{E}_{\theta^*}[T]^m \sim \mathsf{E}_{\theta^*}[T^\star(\theta^*)]^m \sim [n^\star(\theta^*)]^m \quad \text{as } \alpha_{\max} \to 0$$

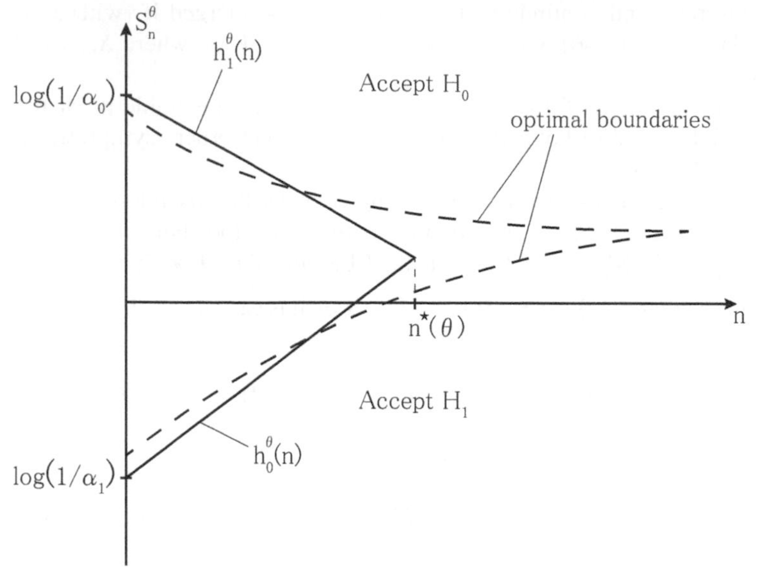

Figure 5.2: *The boundaries $h_1^\theta(n)$ and $h_0^\theta(n)$ of the* 2-*SPRT (solid) and optimal boundaries (dashed) as functions of n.*

and $\sup_\theta \mathsf{E}_\theta[T^\star(\theta^*)]^m \approx [n^\star(\theta^*)]^m$. Next, we observe that the function $n^\star(\theta)$ attains its minimum at the point $\theta = \theta^*$. Thus, the point θ^* has two properties simultaneously — it minimizes the time of truncation $\min_{\theta\in(\theta_0,\theta_1)} n^\star(\theta) = n^\star(\theta^*)$ on one hand, and approximately maximizes the ESS of the 2-SPRT on the other hand. Hence, we may expect that $\delta^\star(\theta^*)$ is asymptotically minimax. This is indeed the case, which can be established using Theorem 5.3.2:

$$\inf_{\delta\in\mathbf{C}(\alpha_0,\alpha_1)} \sup_{\theta\in\Theta} \mathsf{E}_\theta[T]^m \sim \sup_{\theta\in\Theta} \mathsf{E}_\theta[T^\star(\theta^*)]^m \sim \mathsf{E}_{\theta^*}[T^\star(\theta^*)]^m \sim [n^\star(\theta^*)]^m \quad \text{for all } m>0.$$

This proves that the 2-SPRT tuned to the point θ^*, which is found from (5.51), is first-order minimax. Moreover, this is true not only for the ESS but also for the higher moments of the stopping time distribution. However, with this choice the supremum of the ESS $\mathsf{E}_\theta[T^\star(\theta^*)]$ over θ is attained at the point θ^* only to within the term $o(n^\star(\theta^*))$. This result can be improved to within $O(1)$ by choosing the tuning point in a slightly different manner as $\theta_\star = \theta^* + \mathsf{constant}\cdot n^\star(\theta^*)^{-1/2}$.

Specifically, Huffman [198] considered the discrete-time exponential family and obtained the first two terms of the asymptotic expansion for the ESS of the 2-SPRT with the tuning point

$$\theta_\star = \theta^* + \frac{r^*}{\sigma_{\theta^*}\sqrt{n^\star(\theta^*)}}, \tag{5.52}$$

where r^* is a solution of the equation

$$\Phi(r^*) = \frac{|\gamma(\theta^*,\theta_1)|}{|\gamma(\theta^*,\theta_1)| + \gamma(\theta^*,\theta_0)}.$$

The residual term in Huffman's approximation is of order $o(n^\star(\theta^\star)^{1/2})$. Dragalin and Novikov [126] considered this problem in both the continuous- and discrete-time cases and found a more accurate estimate of the residual term in the expansion for the ESS showing that it is of order $O(1)$.

The following theorem states the final result without separating the continuous- and discrete-time scenarios. Note again that the discrete-time model (5.47) can be merged with the continuous-time model (5.46) by replacing the observed process X_t with the sum S_n of n iid random variables

$X_1,\dots,X_n$. We write $t^\star(\theta)$ for $n^\star(\theta)$ defined in (5.48) and, as usual, we write $\varphi(x) = \mathrm{d}\Phi(x)/\mathrm{d}x$ for density of the standard normal distribution $\Phi(x)$.

Theorem 5.3.4. *Let the observed process X_t ($t \in \mathbb{R}_+$ or $\mathbb{Z}_+$) be a process with iid increments from the exponential family* (5.46)*. Let the thresholds in the 2-SPRT be selected as $a_1 = |\log \alpha_1|$ and $a_0 = |\log \alpha_0|$, and let the tuning point $\theta_\star$ be as in* (5.52)*. If the condition* (5.49) *holds, then as $\alpha_{\max} \to 0$*

$$\mathsf{E}_{\theta_\star}[T^\star(\theta_\star)] = t^\star(\theta^*) - \sigma_{\theta^*}\left[(\gamma(\theta^*,\theta_0) + |\gamma(\theta^*,\theta_1)|)\varphi(r^*)\sqrt{t^\star(\theta^*)}\right] + O(1),$$
$$\sup_{\theta\in\Theta} \mathsf{E}_\theta[T^\star(\theta_\star)] = \mathsf{E}_{\theta_\star}[T^\star(\theta_\star)] + O(1),$$
$$\inf_{\delta\in\mathbf{C}(\alpha_0,\alpha_1)} \sup_{\theta\in\Theta} \mathsf{E}_\theta[T] = \mathsf{E}_{\theta_\star}[T^\star(\theta_\star)] + O(1),$$

where

$$t^\star(\theta^*) = \frac{|\log\alpha_1|}{I(\theta^*,\theta_0)} = \frac{|\log\alpha_0|}{I(\theta^*,\theta_1)}.$$

Therefore, the 2-SPRT $\delta^\star(\theta_\star)$ is second-order asymptotically minimax and

$$\frac{\inf_{\delta\in\mathbf{C}(\alpha_0,\alpha_1)} \sup_{\theta\in\Theta} \mathsf{E}_\theta[T]}{\sup_{\theta\in\Theta} \mathsf{E}_\theta[T^\star(\theta_\star)]} = 1 - O(|\log\alpha_{\max}|^{-1}).$$

As we have already pointed out, the formulas $a_i = |\log\alpha_i|$, which guarantee the inequalities $\alpha_i(\delta^\star(\theta)) \le \alpha_i$, are rather conservative. An improvement can be obtained by observing that

$$\alpha_1(\delta^\star(\theta)) = \mathsf{P}_\theta(T^\star = T_0^\theta)e^{-a_1}\mathsf{E}_\theta\left\{e^{-(a_1-\ell_1^\theta(T_0^\theta))} | T^\star = T_0^\theta\right\},$$
$$\alpha_0(\delta^\star(\theta)) = \mathsf{P}_\theta(T^\star = T_1^\theta)e^{-a_0}\mathsf{E}_\theta\left\{e^{-(a_0-\ell_0^\theta(T_1^\theta))} | T^\star = T_1^\theta\right\}$$

and by noticing that asymptotically as $a_i \to \infty$

$$\mathsf{P}_\theta(T^\star = T_1^\theta) \to \frac{|\gamma(\theta,\theta_1)|}{|\gamma(\theta,\theta_1)| + \gamma(\theta,\theta_0)}, \quad \mathsf{P}_\theta(T^\star = T_0^\theta) \to \frac{\gamma(\theta,\theta_0)}{|\gamma(\theta,\theta_1)| + \gamma(\theta,\theta_0)},$$
$$\mathsf{E}_\theta\left\{e^{-(a_0-\ell_0^\theta(T_1^\theta))} | T^\star = T_1^\theta\right\} \to \zeta_0^\theta, \quad \mathsf{E}_\theta\left\{e^{-(a_1-\ell_1^\theta(T_0^\theta))} | T^\star = T_0^\theta\right\} \to \zeta_1^\theta,$$

where for the nonarithmetic case ζ_i^θ can be computed as in Theorem 2.5.3(i) letting $\lambda = 1$. This yields

$$\alpha_0(\delta^\star(\theta)) \sim \frac{|\gamma(\theta,\theta_1)|}{|\gamma(\theta,\theta_1)| + \gamma(\theta,\theta_0)}\zeta_0^\theta e^{-a_0}, \quad \alpha_1(\delta^\star(\theta)) \sim \frac{\gamma(\theta,\theta_0)}{|\gamma(\theta,\theta_1)| + \gamma(\theta,\theta_0)}\zeta_1^\theta e^{-a_1}.$$

Lorden [273] performed an extensive performance analysis while testing the mean θ in the normal symmetric case where $\alpha_0 = \alpha_1$ and $\theta_\star = (\theta_0+\theta_1)/2$. The conclusion is that the 2-SPRT performs almost as the optimal minimax test with the curved boundaries obtained using the backward induction. The efficiency depends on the error probabilities, but it is always more than 99% in all the performed experiments. Similar results were obtained by Huffman [198] for the exponential example $f_\theta(x) = \theta e^{-\theta x}$, $x \ge 0$, $\theta_0 = 1$, $\theta_1 = 2$. The 2-SPRT has an efficiency of more than 98% and almost always over 99% for a broad range of error probabilities studied in this work.

5.4 Uniformly First-Order Asymptotically Optimal Sequential Tests

As already mentioned above, in order to obtain sequential tests that are uniformly optimal (at least approximately) for all possible parameter values, ideally the point $\theta_\star$ in the 2-SPRT should be taken to be equal to the true parameter value θ that is not known. One way of resolving this issue is to replace it with an estimate $\hat{\theta}_n$. This idea leads to a family of adaptive tests that are considered in Subsections 5.4.1 and 5.4.2. Another way is to use a quasi-Bayesian approach based on mixtures, which is considered in Subsection 5.4.3. These methods lead to efficient tests that have certain optimality properties for all parameter values and can be further optimized for specific parameter values, for example in a minimax sense.

5.4.1 The Generalized Sequential Likelihood Ratio Test

Recall that our ultimate goal, formulated in Section 5.1, is to design multihypothesis sequential tests that are at least approximately optimal for testing multiple composite hypotheses of a general structure. Specifically, assume that the sequence of iid observations $X_1, X_2, \dots$ comes from a common distribution with density $p_\theta(x)$ with respect to some non-degenerate sigma-finite measure $\mu(\mathrm{d}x)$, where the l-dimensional parameter $\theta = (\theta_1, \dots, \theta_l)$ belongs to a subset Θ of the Euclidean space $\mathbb{R}^l$. The parameter space Θ is split into $N+2$ disjoint sets $\Theta_0, \Theta_1, \dots \Theta_N$ and $\mathrm{I_{in}}$, i.e., $\Theta = \Theta_0 \cup \cdots \Theta_N \cup \mathrm{I_{in}}$, and we write $\Theta = \sum_{i=0}^N \Theta_i + \mathrm{I_{in}}$, where $N \geq 1$. The composite hypotheses to be tested are

$$H_i : \theta \in \Theta_i, \quad i = 0, 1, \dots, N. \tag{5.53}$$

The subset $\mathrm{I_{in}}$ of Θ represents an indifference zone where the loss $L(\theta, d)$ associated with correct or incorrect decisions d is zero, i.e., no constraints on the probabilities $\mathsf{P}_\theta(d = i)$ are imposed if $\theta \in \mathrm{I_{in}}$. As already mentioned in the introduction, the indifference zone is usually introduced since in most applications the correct action is not critical and often not even possible when the hypotheses are very close. However, in principle $\mathrm{I_{in}}$ may be an empty set, in which case the loss takes the value zero only at some boundary points separating the hypotheses. For example, if the hypotheses are $H_0 : \theta < 0$ and $H_1 : \theta > 0$, then an appropriate loss is $L(\theta) = |\theta|$. If the hypotheses are $H_0 : \theta \leq -\Delta$ and $H_1 : \theta > +\Delta$, then one can take $L(\theta) = \mathbb{1}_{\{\theta \in (-\infty, -\Delta] \cup [\Delta, \infty)\}}$ or zero in $(-\Delta, \Delta)$ and an increasing function otherwise.

We are interested in finding multihypothesis tests $\delta = (T, d)$ that minimize the expected sample size $\mathsf{E}_\theta T$ uniformly for all $\theta \in \Theta$ in the classes of tests $\mathbf{C}([\alpha_{ij}])$ and $\mathbf{C}(\boldsymbol{\alpha})$ defined in (5.3), i.e., in which the maximal error probabilities $\sup_{\theta \in \Theta_i} \alpha_{ij}(\delta, \theta)$ and $\sup_{\theta \in \Theta_i} \alpha_i(\delta, \theta)$ are upper-bounded by the given numbers. In other words, we are interested in the following frequentist problems:

$$\inf_{\delta \in \mathbf{C}([\alpha_{ij}])} \mathsf{E}_\theta T \quad \text{and} \quad \inf_{\delta \in \mathbf{C}(\boldsymbol{\alpha})} \mathsf{E}_\theta T \quad \text{uniformly in } \theta \in \Theta. \tag{5.54}$$

Unfortunately, such a uniformly optimal solution does not exist, and one has to resort to finding asymptotic approximations for small error probabilities.

This setup can be further generalized to cover general non-iid models as in Section 5.1. The methods developed in Subsection 5.3.1 can still be used to prove asymptotic optimality of the adaptive and mixture-based tests considered below. A general theory is addressed in Subsection 5.4.4.

In the particular case of two hypotheses ($N = 1$), the problem is to find tests that approximately solve the problem (5.54) in the class

$$\mathbf{C}(\alpha_0, \alpha_1) = \left\{ \delta : \sup_{\theta \in \Theta_0} \mathsf{P}_\theta(d = 1) \leq \alpha_0, \ \sup_{\theta \in \Theta_1} \mathsf{P}_\theta(d = 0) \leq \alpha_1 \right\}, \tag{5.55}$$

where α_i are given numbers such that $\alpha_0 + \alpha_1 < 1$.

5.4.1.1 A One-Dimensional Case

There are two classical sequential approaches in testing composite hypotheses — with and without an indifference zone, and the solutions are generally different. To begin with, consider the problem of testing the null hypothesis $H_0 : \theta \le \theta_0$ *versus* the alternative hypothesis $H_1 : \theta \ge \theta_1$ ($\theta_0 \le \theta_1$) for a scalar parameter. Note that when $\theta_0 = \theta_1$ (no indifference zone) the hypotheses should be reformulated as $H_0 : \theta < \theta_0$ and $H_1 : \theta > \theta_0$.

If instead of the frequentist problem (5.54) we adopt a Bayesian approach putting a prior distribution $W(\theta)$ on Θ with a cost c per observation and a loss function $L(\theta)$ at the point θ associated with accepting the incorrect hypothesis, then the Bayes integrated risk of a sequential test $\delta = (T, d)$ is

$$\rho_c^W(\delta) = \int_{\theta \le \theta_0} L(\theta) \mathsf{P}_\theta(d = 1) W(\mathrm{d}\theta) + \int_{\theta \ge \theta_1} L(\theta) \mathsf{P}_\theta(d = 0) W(\mathrm{d}\theta) + c \int_\Theta \mathsf{E}_\theta[T] W(\mathrm{d}\theta).$$

In the case of the discrete-time one-parameter exponential family (5.47), using optimal stopping theory (see Theorem 2.7.5), it can be shown that the optimal decision-making strategy is

$$T^0 = \inf\{n \ge 1 : (S_n, n) \in \mathcal{B}_c\}, \quad d_0 = 0 \text{ if } (S_n, n) \in \mathcal{B}_c^0, \quad d_0 = 1 \text{ if } (S_n, n) \in \mathcal{B}_c^1,$$

where $S_n = X_1 + \cdots + X_n$ and $\mathcal{B}_c = \mathcal{B}_c^0 \cup \mathcal{B}_c^1$ is a set that can be found numerically.

In the problem with an indifference zone where $\theta_0 < \theta_1$, $L(\theta) = 0$ for $\theta \in (\theta_0, \theta_1)$ and again for the family (5.47), Schwarz [407] derived the test $\delta^\star(\hat{\theta})$ with $\hat{\theta} = \{\hat{\theta}_n\}$ being the maximum likelihood estimator (MLE) of θ, as an asymptotic solution as $c \to 0$ to the Bayesian problem with the $0-1$ loss function. Specifically, the *a posteriori* risk of stopping is

$$R_n^{\mathrm{st}}(S_n) = \min_{i=0,1} \left\{ \frac{\int_{\Theta_i} \exp\{\theta S_n - nb(\theta)\} W(\mathrm{d}\theta)}{\int_\Theta \exp\{\theta S_n - nb(\theta)\} W(\mathrm{d}\theta)} \right\}, \tag{5.56}$$

where $\Theta_0 = \{\theta \le \theta_0\}$, $\Theta_1 = \{\theta \ge \theta_1\}$. Schwarz showed that $\mathcal{B}_c / |\log c| \to \mathcal{B}_0$ as $c \to 0$ and proposed a simple procedure: continue sampling until $R_n^{\mathrm{st}}(S_n)$ is less than c and upon stopping accept the hypothesis for which the minimum is attained in (5.56). Denote this procedure by $\widetilde{\delta}(c) = (\widetilde{T}(c), \widetilde{d}(c))$. Applying Laplace's asymptotic integration method to evaluate the integrals in (5.56) leads to an approximation that prescribes to stop sampling at the time $\widehat{T}(\hat{\theta}) = \min(\widehat{T}_0(\hat{\theta}), \widehat{T}_1(\hat{\theta}))$, where

$$\begin{aligned} \widehat{T}_i(\hat{\theta}) &= \inf\left\{ n : \sum_{k=1}^n \log \frac{p_{\hat{\theta}_n}(X_k)}{p_{\theta_i}(X_k)} \ge |\log c| \right\} \\ &= \inf\left\{ n : \sup_{\theta \in \Theta} [\theta S_n - nb(\theta)] - [\theta_i S_n - nb(\theta_i)] \ge |\log c| \right\}, \end{aligned} \tag{5.57}$$

i.e., to the likelihood ratio test where the true parameter is replaced by the MLE $\hat{\theta}_n$. The terminal decision rule $\hat{d}(\hat{\theta})$ of the test $\hat{\delta}(\hat{\theta}) = (\widehat{T}(\hat{\theta}), \hat{d}(\hat{\theta}))$ accepts H_0 if $\hat{\theta}_{\widehat{T}} < \theta^*$, where θ^* is such that $I(\theta^*, \theta_0) = I(\theta^*, \theta_1)$. Note also that

$$\widehat{T} = \inf\{n \ge 1 : n \max[I(\hat{\theta}_n, \theta_0), I(\hat{\theta}_n, \theta_1)] \ge |\log c|\}. \tag{5.58}$$

The tests that use the MLEs are usually referred to as the *Generalized Sequential Likelihood Ratio Tests* (GSLRT). Wong [508] showed that the GSLRT $\hat{\delta}$ is first-order asymptotically Bayes as $c \to 0$:

$$\rho_c^W(\hat{\delta}) \sim \inf_\delta \rho_c^W(\delta) \sim c|\log c| \int_\Theta \frac{W(\mathrm{d}\theta)}{I_{\max}(\theta)},$$

and also

$$\mathsf{E}_\theta[\widehat{T}] \sim \frac{|\log c|}{I_{\max}(\theta)} \quad \text{for every } \theta \in \Theta,$$

where $I_{\max}(\theta) = \max\{I(\theta,\theta_0), I(\theta,\theta_1)\}$.

Kiefer and Sacks [231] showed that the quasi-Bayesian procedure $\widetilde{\delta}(c) = (\widetilde{T}(c), \widetilde{d}(c))$ with the stopping time

$$\widetilde{T}(c) = \inf\{n \ge 1 : R_n^{\mathrm{st}}(S_n) \le c\}$$

initially proposed by Schwarz [407], is also first-order asymptotically Bayes, i.e., for any prior distribution W, $\rho_c^W(\widetilde{\delta}(c)) \sim \inf_\delta \rho_c^W(\delta)$ as $c \to 0$. Lorden [269] refined this result by introducing the stopping region as the first n such that $R_n^{\mathrm{st}}(S_n) \le Qc$, where Q is a positive constant, and showed that it can be made second-order asymptotically optimal, i.e., $\inf_\delta \rho_c^W(\delta) = \rho_c^W(\widetilde{\delta}(Qc)) + O(c)$ as $c \to 0$, while $\inf_\delta \rho_c^W(\delta) = O(c|\log c|)$. It should be noted that the problem addressed by Lorden is much more general than that we are discussing here. It may involve general iid models, not necessarily exponential families, as well as multiple-decision cases.

A breakthrough in the Bayesian theory for testing separated hypotheses about the parameter of the one-dimensional exponential family (5.47) was achieved by Lorden [274] in 1977 in an unpublished manuscript, where he showed that the family of GSLRTs can be designed so as to attain the Bayes risk to within $o(c)$ as $c \to 0$, i.e., the third order asymptotic optimality. Lorden gave sufficient conditions for families of tests to be third-order asymptotically Bayes and examples of such procedures based not only on the GLR approach but also on mixtures of likelihood ratios. In addition, the error probabilities of the GSLRTs have been evaluated asymptotically as a consequence of a general theorem on boundary-crossing probabilities. Due to the importance of this work we give a more detailed overview of Lorden's theory.

The hypotheses to be tested are $H_0 : \underline{\theta} \le \theta \le \theta_0$ and $H_1 : \overline{\theta} \ge \theta \ge \theta_1$, where $\underline{\theta}$ and $\overline{\theta}$ are interior points of the natural parameter space Θ. Let $\hat{\theta}_n \in [\underline{\theta}, \overline{\theta}]$ be the MLE that maximizes the likelihood over θ in $[\underline{\theta}, \overline{\theta}]$. Lorden's GSLRT stops at $\widehat{T}$ which is the minimum of the stopping times $\widehat{T}_0, \widehat{T}_1$ defined as

$$\begin{aligned} \widehat{T}_0(\hat{\theta}) &= \inf\left\{ n \ge 1 : \sum_{k=1}^n \log\left[\frac{p_{\hat{\theta}_n}(X_k)}{p_{\theta_0}(X_k)} h_0(\hat{\theta}_n)\right] \ge a \text{ and } \hat{\theta}_n \ge \theta^* \right\}, \\ \widehat{T}_1(\hat{\theta}) &= \inf\left\{ n \ge 1 : \sum_{k=1}^n \log\left[\frac{p_{\hat{\theta}_n}(X_k)}{p_{\theta_1}(X_k)} h_1(\hat{\theta}_n)\right] \ge a \text{ and } \hat{\theta}_n \le \theta^* \right\}, \end{aligned} \tag{5.59}$$

where a is a threshold, θ^* satisfies $I(\theta^*,\theta_0) = I(\theta^*,\theta_1)$, and h_0, h_1 are positive continuous functions on $[\theta^*, \overline{\theta}], [\underline{\theta}, \theta^*]$, respectively. The hypothesis H_i is rejected when $\widehat{T} = \widehat{T}_i$. To summarize, Lorden's family of GSPRTs is defined as

$$\widehat{T}(\hat{\theta}) = \min\left\{\widehat{T}_0(\hat{\theta}), \widehat{T}_1(\hat{\theta})\right\}, \quad \hat{d} = \begin{cases} 0 & \text{if } \widehat{T}(\hat{\theta}) = \widehat{T}_1(\hat{\theta}) \\ 1 & \text{if } \widehat{T} = \widehat{T}_0(\hat{\theta}) \end{cases}, \tag{5.60}$$

with the $\widehat{T}_i(\hat{\theta})$'s in (5.59).

Lorden assumes that the prior distribution has a continuous density $w(\theta)$ positive on $[\underline{\theta}, \overline{\theta}]$, and that the loss $L(\theta)$ equals zero in the indifference zone (θ_0, θ_1) and is continuous and positive elsewhere and bounded away from 0 on $[\underline{\theta}, \theta_0] \cup [\theta_1, \overline{\theta}]$. The main results in [274, Theorem 1] can be briefly outlined as follows.

(i) Under these assumptions the family of GSLRTs defined by (5.59)–(5.60) with $a = |\log c| - \frac{1}{2}\log|\log c|$ is second-order asymptotically optimal, i.e.,

$$\rho_c^w(\hat{\delta}) = \inf_\delta \rho_c^w(\delta) + O(c) \quad \text{as } c \to 0,$$

where $\rho_c^w(\delta)$ is the average risk of the test $\delta = (T,d)$:

$$\rho_c^w(\delta) = \int_{\underline{\theta}}^{\theta_0} L(\theta)\mathsf{P}_\theta(d=1)w(\theta)\,\mathrm{d}\theta + \int_{\theta_1}^{\overline{\theta}} L(\theta)\mathsf{P}_\theta(d=0)w(\theta)\,\mathrm{d}\theta + c\int_{\underline{\theta}}^{\overline{\theta}} \mathsf{E}_\theta[T]w(\theta)\,\mathrm{d}\theta.$$

(ii) This result can be improved from $O(c)$ to $o(c)$, i.e., to the third order

$$\rho_c^w(\hat{\delta}) = \inf_{\delta} \rho_c^w(\delta) + o(c) \quad \text{as } c \to 0,$$

making the right choice of the functions h_0 and h_1 by setting

$$h_i(\theta) = \sqrt{\frac{2\pi I(\theta,\theta_i)}{\ddot{b}(\theta)}} \, \frac{w(\theta)|\dot{b}(\theta) - \dot{b}(\theta_i)|}{w(\theta_i)L(\theta_i)\zeta(\theta,\theta_i)}, \quad i = 0,1,$$

where $\zeta(\theta,\theta_i) = \Upsilon(\theta,\theta_i)/I(\theta,\theta_i)$ is a correction for the overshoot over the boundary, the factor which is the subject of renewal theory. Specifically,

$$\zeta(\theta,\theta_i) = \lim_{a\to\infty} \mathsf{E}_\theta \exp\{-[\lambda_{\tau_a}(\theta,\theta_i) - a]\}, \quad \tau_a = \inf\{n : \lambda_n(\theta,\theta_i) \geq a\}, \tag{5.61}$$

where

$$\lambda_n(\theta,\theta_i) = \sum_{k=1}^{n} \log\left[\frac{p_\theta(X_k)}{p_{\theta_i}(X_k)}\right] = (\theta - \theta_i)S_n - [b(\theta) - b(\theta_i)]n.$$

In particular, $\Upsilon(\theta,\theta_i)$ can be computed as

$$\Upsilon(\theta,\theta_i) = \exp\left\{-\sum_{n=1}^{\infty} \frac{1}{n}\left[\mathsf{P}_\theta(\lambda_n(\theta,\theta_i) \leq 0) + \mathsf{P}_{\theta_i}(\lambda_n(\theta,\theta_i) > 0)\right]\right\}, \tag{5.62}$$

which can be easily obtained using Lemma 2.5.1 and Theorem 2.5.3. Since the Bayes average risk $\inf_\delta \rho_c^w(\delta)$ is of order $c|\log c|$, this implies that the asymptotic relative efficiency $\mathcal{E}_c = [\rho_c^w(\hat{\delta}) - \inf_\delta \rho_c^w(\delta)]/\rho_c^w(\hat{\delta})$ of Lorden's test is of order $1 - o(1/|\log c|)$ as $c \to 0$.

Note the crucial difference between Schwarz's GSLRT (5.57) and Lorden's GSLRT (5.60). While in the Schwarz test $h_i \equiv 1$ and the threshold is $a = |\log c|$, in the Lorden test there are two innovations. First, the threshold is smaller by $\frac{1}{2}\log|\log c|$, and second, there are adaptive weights $h_i(\hat{\theta}_n)$ in the GLR statistic. Since the stopping times $\widehat{T}_i$ can be obviously written as

$$\begin{aligned} \widehat{T}_0(\hat{\theta}) &= \inf\{n \geq 1 : \lambda_n(\hat{\theta}_n,\theta_0) \geq a - \log h_0(\hat{\theta}_n) \text{ and } \hat{\theta}_n \geq \theta^*\}, \\ \widehat{T}_1(\hat{\theta}) &= \inf\{n \geq 1 : \lambda_n(\hat{\theta}_n,\theta_1) \geq a - \log h_1(\hat{\theta}_n) \text{ and } \hat{\theta}_n \leq \theta^*\}, \end{aligned} \tag{5.63}$$

alternatively Lorden's GSLRT can be viewed as the GSLRT with the curved adaptive boundaries

$$a_i(\hat{\theta}_n) = |\log c| - \tfrac{1}{2}\log|\log c| - \log h_i(\hat{\theta}_n), \quad i = 0,1,$$

that depend on the behavior of the MLE $\hat{\theta}_n$. These two innovations make this modification of the GLR test nearly optimal.

The formal mathematical proof by Lorden is very involved, so we give only a heuristic sketch that fixes the main ideas of the approach. The Bayesian consideration naturally leads to the mixture LR statistics

$$\bar{\Lambda}_n^i = \frac{\int_{\underline{\theta}}^{\overline{\theta}} e^{\theta S_n - nb(\theta)} w(\theta)\,\mathrm{d}\theta}{\int_{\Theta_i} L(\theta) e^{\theta S_n - nb(\theta)} w(\theta)\,\mathrm{d}\theta}, \quad i = 0,1,$$

where $\Theta_0 = [\underline{\theta},\theta_0]$, $\Theta_1 = [\theta_1,\overline{\theta}]$ and $L(\theta) = 1$ for the simple $0-1$ loss function. Indeed, the *a posteriori* stopping risk is given by

$$R_n^{\mathrm{st}}(S_n) = \min_{i=0,1}\left\{\frac{\int_{\Theta_i} L(\theta) e^{\theta S_n - nb(\theta)} w(\theta)\,\mathrm{d}\theta}{\int_{\underline{\theta}}^{\overline{\theta}} e^{\theta S_n - nb(\theta)} w(\theta)\,\mathrm{d}\theta}\right\}. \tag{5.64}$$

A candidate for the approximate optimum is the procedure that stops as soon as $R_n^{\rm st}(S_n) \le A_c$ with some $A_c \approx c$. This is equivalent to stop as soon as $\max_{i=0,1} \hat{\Lambda}_n^i \ge 1/A_c$. The GLR statistics are given by

$$\hat{\Lambda}_n^i = \frac{\max\limits_{\theta\in[\underline{\theta},\overline{\theta}]} e^{\theta S_n - nb(\theta)}}{\max\limits_{\theta\in\Theta_i} e^{\theta S_n - nb(\theta)}} \approx \frac{\max\limits_{\theta\in[\underline{\theta},\overline{\theta}]} e^{\theta S_n - nb(\theta)}}{e^{\theta_i S_n - nb(\theta_i)}}, \quad i=0,1,$$

which follows from the approximations

$$\begin{aligned} &\int_{\underline{\theta}}^{\overline{\theta}} e^{\theta S_n - nb(\theta)} w(\theta)\, \mathrm{d}\theta \sim \sqrt{\frac{2\pi}{n\ddot{b}(\hat{\theta}_n)}}\, e^{\hat{\theta}_n S_n - nb(\hat{\theta}_n)}\, w(\hat{\theta}_n), \\ &\int_{\Theta_i} L(\theta) e^{\theta S_n - nb(\theta)} w(\theta)\, \mathrm{d}\theta \sim \frac{e^{\theta_i S_n - nb(\theta_i)}}{n\dot{b}(\theta_i) - S_n}\, L(\theta_i)\, w(\theta_i). \end{aligned} \tag{5.65}$$

The approximations (5.65) follow from the well-known asymptotic expansions for integrals (Laplace's integration method and its variations). Using the asymptotics (5.65), the stopping posterior risk (5.64) can be approximated as

$$R_n^{\rm st}(S_n) \approx \min_{i=0,1} \frac{w(\theta_i) L(\theta_i) [\ddot{b}(\hat{\theta}_n)/2\pi n]^{1/2}}{w(\hat{\theta}_n)|\dot{b}(\theta_i) - \dot{b}(\hat{\theta}_n)|}\, e^{-\lambda_n(\hat{\theta}_n, \theta_i)}, \tag{5.66}$$

where $i=0$ if $\hat{\theta}_n \le \theta^*$ and $i=1$ otherwise.

Next, Lorden showed that there exists $Q>1$ such that if the stopping risk exceeds Qc, then the continuation risk is smaller than the stopping risk, and hence, it is approximately optimal to stop at the first time such that $R_n^{\rm st}$ falls below Qc. This result along with the approximation (5.66) yields $T^0 \approx \min(\tau_0, \tau_1)$, where

$$\tau_i = \inf\left\{n : e^{-\lambda_n(\hat{\theta}_n,\theta_i)}/\widetilde{h}_i(\hat{\theta}_n) n^{1/2} \le Qc\right\} = \inf\left\{n : \lambda_n(\hat{\theta}_n, \theta_i) \ge -\log[n^{1/2}\widetilde{h}_i(\hat{\theta}_n) Qc]\right\}$$

with $\widetilde{h}_i(\hat{\theta}_n)$ given by

$$\widetilde{h}_i(\hat{\theta}_n) = \sqrt{\frac{2\pi}{\ddot{b}(\hat{\theta}_n)}}\frac{w(\theta)|\dot{b}(\hat{\theta}_n) - \dot{b}(\theta_i)|}{w(\theta_i) L(\theta_i)}.$$

For small c, the expected value $\mathsf{E}_\theta \tau_i$ is of order $|\log c|$, so $n^{1/2}$ can be replaced by $|\log c|^{1/2}$, which yields

$$\tau_i \approx \widehat{T}_i = \inf\left\{n : \lambda_n(\hat{\theta}_n, \theta_i) \ge -\log[c|\log c|^{1/2} Q\widetilde{h}_i(\hat{\theta}_n)]\right\}.$$

Note that these stopping times look exactly like the ones defined in (5.63) with the stopping boundaries

$$a_i(\hat{\theta}_n) = |\log c| - \tfrac{1}{2}\log|\log c| - \log[Q\widetilde{h}_i(\hat{\theta}_n)], \quad i=0,1.$$

The test based on these stopping times is already optimal to second order. However, to make it third-order optimal the constant Q should be selected in a special way in order to account for the overshoots. Specifically, using this result, Lorden proves that the risks of an optimal rule and of the GSLRT are both connected to the risks of the family of one-sided tests $\tau_a(\theta) = \inf\{n : \lambda_n(\theta, \theta_i) \ge a\}$, which according to Lemma 4.3.2 are strictly optimal in the problem

$$\rho(\theta, v) = \inf_T \{\mathsf{E}_\theta T + v\mathsf{P}_{\theta_i}(T<\infty)\} = \inf_T \mathsf{E}_\theta\left\{T + v\prod_{n=1}^T \frac{p_{\theta_i}(X_n)}{p_\theta(X_n)}\right\}$$

if we set $a = \log[v\Upsilon(\theta, \theta_i)]$. Therefore, if we now take $Q = 1/\Upsilon(\theta, \theta_i)$, then the resulting test will be nearly optimal up to $o(c)$. Since θ is not known, we need to replace it with the estimate $\hat{\theta}_n$ to obtain

$$a_i(\hat{\theta}_n) = |\log c| - \tfrac{1}{2}\log|\log c| - \log[\widetilde{h}_i(\hat{\theta}_n)/\Upsilon(\hat{\theta}_n, \theta_i)] = |\log c| - \tfrac{1}{2}\log|\log c| - \log[h_i(\hat{\theta}_n)].$$

The mixtures of LRs arise most naturally in the Bayesian context. Therefore, it is clear that mixture-based rules should also be asymptotically third-order optimal. This is indeed so if, for example, we replace the generalized LLR statistics in (5.63) with the mixtures over the priors $\widetilde{w}_i(\theta) = w(\theta)H_i(\theta)$. Note that the normalization is not necessary, i.e.,

$$T_i = \inf\left\{n : \widetilde{\Lambda}_n^i \geq L(\theta_i)w(\theta_i)/c|\log c|\right\}, \quad i = 0, 1,$$

where we first use the second approximation in (5.65) to obtain

$$\bar{\Lambda}_n^i \approx \frac{n}{L(\theta_i)w(\theta_i)} \int_{\underline{\theta}}^{\overline{\theta}} e^{\lambda_n(\theta,\theta_i)} w(\theta)|\dot{b}(\theta) - \dot{b}(\theta_i)|\, \mathrm{d}\theta$$

and then replace n with $|\log c|$ and add the factor $\zeta(\theta, \theta_i)$ to account for the overshoots, to finally obtain

$$\widetilde{\Lambda}_n^i = \int_{\underline{\theta}}^{\overline{\theta}} e^{\lambda_n(\theta,\theta_i)} w(\theta)H_i(\theta)\, \mathrm{d}\theta, \quad H_i(\theta) = |\dot{b}(\theta) - \dot{b}(\theta_i)|/\zeta(\theta, \theta_i).$$

In addition, it is interesting to compare Lorden's approach with the Kiefer–Sacks test that stops at the first time when R_n^{st} becomes smaller than c. Lorden's approach allows us to show that the test with the stopping time

$$\widehat{T} = \inf\left\{n : R_n^{\mathrm{st}}(S_n) \leq c/\Upsilon(\hat{\theta}_n)\right\},$$

where $\Upsilon(\hat{\theta}_n) = \Upsilon(\hat{\theta}_n, \theta_1)$ if $\hat{\theta}_n < \theta^*$ and $\Upsilon(\hat{\theta}_n) = \Upsilon(\hat{\theta}_n, \theta_0)$ otherwise, is nearly optimal to within $o(c)$. Recall that the factor $\zeta(\theta, \theta_i) = I(\theta, \theta_i)\Upsilon(\theta, \theta_i)$ provides a necessary correction for the excess over the thresholds at stopping. This gives a considerable improvement over the Kiefer–Sacks test that ignores the overshoots, and not necessarily in the case of testing close hypotheses when $\Upsilon(\theta, \theta_i) \ll 1$, but it may be important even if the parameter values are well separated. For example, in the binomial case with the success probabilities $\theta_1 = 0.6$ and $\theta_0 = 0.4$, $\Upsilon(\theta_1, \theta_0) \approx 1/15$, so Lorden's test will stop much earlier.

Observe, however, that the implementation of Lorden's fully optimized GSLRT is problematic since usually computing the numbers $\zeta(\theta, \theta_i)$ is not possible analytically except for some particular models such as the exponential one. For example, when testing the mean in the Gaussian case, these numbers can be computed only numerically. Siegmund's corrected Brownian motion approximations can be used but these approximations are of sufficient accuracy only when the difference between θ and θ_i is relatively small. Therefore, for practical purposes only partially optimized solutions, which provide $O(c)$-optimality, are feasible. A way around is a discretization of the parameter space that is discussed in Subsection 5.5.1.2.

By $\hat{\alpha}_0(\theta) = \mathsf{P}_\theta(\hat{d} = 1)$, $\theta \in \Theta_0 = [\underline{\theta}, \theta_0]$ and $\hat{\alpha}_1(\theta) = \mathsf{P}_\theta(\hat{d} = 0)$, $\theta \in \Theta_1 = [\theta_1, \overline{\theta}]$ denote the error probabilities of the GSLRT $\hat{\delta}_a$. Note that due to the monotonicity of $\hat{\alpha}_i(\theta)$, $\sup_{\theta \in \Theta_i} \hat{\alpha}_i(\theta) = \hat{\alpha}_i(\theta_i)$. In addition to the Bayesian third-order optimality property, Lorden established asymptotic approximations to the error probabilities of the GSLRT. Specifically, by [274, Theorem 2],

$$\hat{\alpha}_i(\theta_i) \sim \sqrt{a}e^{-a}C_i(\theta_i), \quad i = 0, 1 \quad \text{as } a \to \infty, \tag{5.67}$$

where

$$C_0(\theta_0) = \int_{\theta^*}^{\overline{\theta}} \zeta(\theta, \theta_0)h_0(\theta)\sqrt{\frac{\ddot{b}(\theta)}{2\pi I(\theta, \theta_0)}}\, \mathrm{d}\theta,$$

$$C_1(\theta_1) = \int_{\underline{\theta}}^{\theta^*} \zeta(\theta, \theta_1)h_1(\theta)\sqrt{\frac{\ddot{b}(\theta)}{2\pi I(\theta, \theta_1)}}\, \mathrm{d}\theta$$

and where $\zeta(\theta, \theta_i)$, $i = 0, 1$, are defined in (5.61)–(5.62). These approximations are important for frequentist problems, of main interest for most applications. Despite the fact that there are no upper bounds on the error probabilities, so in general there is no prescription on how to embed the GSLRT into the class $\mathbf{C}(\alpha_0, \alpha_1)$, the asymptotic approximations (5.67) allow us to select the thresholds a_i in the stopping times $\widehat{T}_i$ so that $\hat{\alpha}_i(\theta_i) \approx \alpha_i$, $i = 0, 1$ at least for sufficiently small α_i. Note that in this latter case a in (5.63) should be replaced with a_i, where a_i are the roots of the transcendental equations

$$a_i - \frac{1}{2} \log a_i = \log[C_i(\theta_i)/\alpha_i], \quad i = 0, 1.$$

With this choice the GSLRT is asymptotically uniformly first-order optimal with respect to the expected sample size:

$$\inf_{\delta \in \mathbf{C}(\alpha_0, \alpha_1)} \mathsf{E}_\theta T = \mathsf{E}_\theta \widehat{T}(1 + o(1)) \quad \text{as } \alpha_{\max} \to 0 \quad \text{for all } \theta \in [\underline{\theta}, \overline{\theta}],$$

where $o(1)$ may be shown to be of order $O(\log|\log \alpha_{\max}|/|\log \alpha_{\max}|)$. We should note that this result is correct not necessarily in the asymptotically symmetric case where $\log \alpha_0 \sim \log \alpha_1$ and $a_0 \sim a_1$ as $\alpha_{\max} \to 0$, but also in the asymmetric case where a_0 and a_1 go to infinity with different rates as long as $a_1 e^{-a_0} \to 0$.

Note that the Schwarz–Lorden asymptotic theory assumes a fixed indifference zone that does not allow for local alternatives, i.e., θ_1 cannot approach θ_0 as $c \to 0$. In other words, this theory is limited to the case where the width of the indifference zone $\theta_1 - \theta_0$ is considerably larger than $c^{1/2}$. This assumption may be problematic in applications where the length of the indifference interval $\theta_1 - \theta_0$ is relatively small. Indeed, let X_n be normal with mean θ and unit variance and $H_0 : \theta \le -\Delta$, $H_1 : \theta \ge \Delta$ for some $\Delta > 0$. Then $I_{\max}(\theta) = (|\theta| + \Delta)^2/2$ and the approximation $\mathsf{E}_\theta[\widehat{T}] \approx 2|\log c|/(|\theta| + \Delta)^2$ is valid only if $|\log c| \gg 2/(|\theta| + \Delta)^2$. For example, if $\theta = 0$ then we should require $c \ll \exp\{-2/\Delta^2\}$, which for $\Delta = \sqrt{2} \cdot 0.1$ yields $c \ll \exp\{-100\} \approx 3.7 \cdot 10^{-44}$. Obviously, for practical purposes this is not a reasonable number.

On the other hand, Chernoff [98] considered testing the mean of a normal distribution with no indifference zone ($\theta_0 = \theta_1$) and with a loss $L(\theta) = |\theta|$ associated with the wrong decision, and he derived a different and more complicated approximation to the Bayes test, which exploits a time-varying boundary, as opposed to the Schwarz test. As a result, setting $\theta_0 = \theta_1$ in Schwartz's test (and in Lorden's adaptive modification too) does not yield Chernoff's test, while intuitively the Bayesian test with an indifference zone should approach the Bayesian test without an indifference zone when $\theta_0 \to \theta_1$. This issue along with another problem concerning the adequacy of Schwarz's theory related to the lack of uniformity in the indifference zone in the convergence of the normalized quasi-Bayes continuation region has been resolved by Lai [249], who suggested a unified solution to both problems — with and without the indifference zone — that uses the stopping rule where the threshold $|\log c|$ in the definition of $\widehat{T}_i(\hat{\theta})$ is replaced with the time-varying boundary $g(cn)$, which satisfies $g(t) \sim |\log t|$ as $t \to 0$. The stopping time $\widehat{T}_c(\hat{\theta})$ of Lai's GSLRT $\hat{\delta}_c(\hat{\theta}) = (\widehat{T}_c(\hat{\theta}), \hat{d}_c(\hat{\theta}))$ can be written as

$$\widehat{T}_c(\hat{\theta}) = \inf\left\{n \ge 1 : n \max[I(\hat{\theta}_n, \theta_0), I(\hat{\theta}_n, \theta_1)] \ge g(cn)\right\} \tag{5.68}$$

(cf. (5.58)). The boundary $g(t) = g_\gamma(t)$ has the form $g_\gamma(t) = [h(t) + \gamma t]^2/2t$, where $\gamma = (\theta_1 - \theta_0)[\ddot{b}(\theta_0)/c]^{1/2}/2$ and

$$h(t) = \sqrt{2t\left(|\log t| + \frac{1}{2}\log|\log t| - \frac{1}{2}\log 4\pi + o(1)\right)} \quad \text{as } t \to 0.$$

Lai proved that the GSLRT $\hat{\delta}_c(\hat{\theta})$ with this time-varying boundary is first-order asymptotically Bayes and also uniformly first-order optimal with respect to the expected sample size in the class $\mathbf{C}(\alpha_0, \alpha_1)$. Specifically, the following statements hold [249, Theorems 1 and 2]:

(i) For fixed θ_0, θ_1, as $c \to 0$

$$\rho_c^W(\hat{\delta}_c) \sim \inf_{\delta} \rho_c^W(\delta) \sim c|\log c| \int_{\Theta} \frac{W(\mathrm{d}\theta)}{I_{\max}(\theta)},$$

$$\log \alpha_0(\hat{\delta}_c) \sim \log \alpha_1(\hat{\delta}_c) \sim \log c,$$

$$\inf_{\delta \in \mathbf{C}(\alpha_0,\alpha_1)} \mathsf{E}_\theta[T] \sim \mathsf{E}_\theta[\widehat{T}(\hat{\theta})] \sim \frac{|\log c|}{I_{\max}(\theta)} \quad \text{uniformly for all } \theta \text{ in a bounded subset of } \Theta;$$

(ii) If $c \to 0$ and $\theta_1 \to \theta_0$ such that $\Delta_c = c/(\theta_1 - \theta_0)^2 \to 0$ and $W(\theta)$ has positive continuous density $w(\theta)$ in a neighborhood of θ_0, then

$$\rho_c^w(\hat{\delta}_c) \sim \inf_{\delta} \rho_c^w(\delta) \sim \frac{8w(\theta_0)}{\ddot{b}(\theta_0)} (\Delta_c)^{1/2} |\log \Delta_c|,$$

$$\log \alpha_i(\hat{\delta}_c) \sim \log \Delta_c,$$

$$\inf_{\delta \in \mathbf{C}(\alpha_0,\alpha_1)} \sup_{\theta \in \Theta} \mathsf{E}_\theta[T] \sim \sup_{\theta \in \Theta} \mathsf{E}_\theta[\widehat{T}(\hat{\theta})] \sim \frac{8|\log \Delta_c|}{\ddot{b}(\theta_0)(\theta_1 - \theta_0)^2}.$$

Hence for fixed alternatives $\sup_{\theta \in \Theta} \mathsf{E}_\theta[\widehat{T}(\hat{\theta})]/|\log c| = O(1)$, while for local alternatives $\sup_{\theta \in \Theta} \mathsf{E}_\theta[\widehat{T}(\hat{\theta})]/|\log c| \to \infty$ as $c \to 0$. Therefore, Lai's GSLRT with a time-varying boundary guarantees the asymptotic optimality properties over a broad range of parameter values and covers both fixed alternatives and local alternatives approaching θ_0. Besides, setting $\theta_0 = \theta_1$ yields the stopping time $\widehat{T}_c = \inf\{n : nI(\theta_n, \theta_0) \geq g_0(cn)\}$, so that the two cases with and without the indifference zone are unified, as opposed to the Schwartz and Chernoff theories.

5.4.1.2 *The Multidimensional Case*

The above results can be generalized to the multiparameter exponential family with density

$$p_\theta(x) = h(x) \exp\{\theta^\top x - b(\theta)\}, \quad \left\{\theta \in \Theta \subset \mathbb{R}^l : b(\theta) = \log \int_{\Theta} e^{\theta^\top x} \mu(\mathrm{d}x) < \infty\right\} \tag{5.69}$$

with respect to some non-degenerate measure $\mu(\mathrm{d}x)$. The hypotheses are $H_i : \theta \in \Theta_i$, $i = 0, 1$, where Θ_0 and Θ_1 are disjoint subsets of Θ such that $\inf_{\theta \in \Theta_0, \widetilde{\theta} \in \Theta_1} ||\theta - \widetilde{\theta}||_2 = \Delta > 0$. For $i = 0, 1$, define the GLR statistics

$$\hat{\Lambda}_n^i = \frac{\sup_{\theta \in \Theta} \prod_{k=1}^n p_\theta(X_k)}{\sup_{\theta \in \Theta_i} \prod_{k=1}^n p_\theta(X_k)} \tag{5.70}$$

and, with the MLE $\hat{\theta}_n = \sup_{\theta \in \Theta} \prod_{k=1}^n p_\theta(X_k)$, the corresponding generalized LLRs

$$\hat{\lambda}_n^i = \sum_{k=1}^n \log p_{\hat{\theta}_n}(X_k) - \sup_{\theta \in \Theta_i} \sum_{k=1}^n \log p_\theta(X_k) = n \inf_{\theta \in \Theta_i} I(\hat{\theta}_n, \theta), \tag{5.71}$$

where

$$I(\theta, \widetilde{\theta}) = \mathsf{E}_\theta\left[\log \frac{p_\theta(X_1)}{p_{\widetilde{\theta}}(X_1)}\right] = (\theta - \widetilde{\theta})^\top \nabla b(\theta) - (b(\theta) - b(\widetilde{\theta})) \tag{5.72}$$

is the K–L information number, and ∇ stands for the gradient. Let $\hat{\delta} = (\hat{T}, \hat{d})$ be the multivariate GSLRT with the stopping time

$$\widehat{T} = \inf\left\{n \geq 1 : \max_{i=0,1} \hat{\lambda}_n^i \geq g(cn)\right\} \tag{5.73}$$

and the terminal decision $\hat{d} = i$ if $\hat{\lambda}^i_{\widehat{T}} \geq g(c\widehat{T})$, where

$$g(t) \sim |\log t| \quad \text{and} \quad g(t) \geq |\log t| + \xi \log|\log t| \quad \text{as } t \to 0$$

and ξ is a positive number. If $\xi > l/2$, then it follows from Lai and Zhang [257] that the GSLRT is asymptotically as $c \to 0$ first-order Bayes optimal for fixed Δ as well as for $\Delta \to 0$ such that $c/\Delta^2 \to 0$.

The advantage of the GSLRT is that the GLR statistics are adaptive and self-tuned to the unknown parameter values using the maximal available information from the data. However, the basic issue with the GSLRT is how to choose the threshold (or c) in order to upper-bound the error probabilities $\hat{\alpha}_i$ if one is interested in the frequentist approach of minimizing the ESS in the class $\mathbf{C}(\alpha_0, \alpha_1)$. This problem comes from the fact that the GLR statistics $\hat{\Lambda}^i_n = \prod_{k=1}^n [p_{\hat{\theta}_n}(X_k)/p_{\theta_i}(X_k)]$ are not martingales with unit expectation under P_{θ_i} as in the case of the fixed parameter θ. As a result, one cannot apply Wald's likelihood ratio identity for this purpose. Despite the fact that certain asymptotic estimates for the error probabilities can be obtained — see (5.67) for the one-dimensional case and Chan and Lai [94] for multidimensional generalizations, the GSLRT is not flexible in this sense. Besides, the error probabilities $\hat{\alpha}_i(c)$ of the GSLRT (5.73) are asymptotically symmetric in the sense that $\log \hat{\alpha}_0(c)/\log \hat{\alpha}_1(c) \to 1$ as $c \to 0$, which may be too restrictive for some applications, as previously discussed. For this reason in the following sections we discuss alternative approaches that allow for overcoming this drawback.

We now return to the general frequentist multihypothesis problem (5.53), (5.54) and propose two multihypothesis GLR tests (MGSLRT) which are asymptotically first-order optimal in the classes (5.3) as the error probabilities vanish. Recall the construction of the modified accepting MSPRT, which tests a fixed parameter θ *versus* other values $\theta_0, \dots, \theta_N$, given in (5.6). Since now the parameter θ is unknown, it is reasonable to replace it with an estimate $\hat{\theta}_n$, e.g., with the MLE. Also, since *a priori* it is not clear what θ_i should be selected in the set Θ_i corresponding to the hypothesis H_i, we replace $p_{\theta_i}(\mathbf{X}_1^n)$ with $\sup_{\theta \in \Theta_i} p_\theta(\mathbf{X}_1^n)$. Hence, for $i = 0, 1, \dots, N$, we define the GLR statistics as in (5.70), i.e.,

$$\hat{\Lambda}^i_n = \frac{\prod_{k=1}^n p_{\hat{\theta}_n}(X_k)}{\sup\limits_{\theta \in \Theta_i} \prod_{k=1}^n p_\theta(X_k)}, \tag{5.74}$$

where $\hat{\theta}_n = \arg\sup_{\theta \in \Theta} \prod_{k=1}^n p_\theta(X_k)$ is the MLE. With these replacements the test of (5.6) takes the form

$$\text{Stop at the first } n \geq 1 \text{ such that for some } i \quad \hat{\Lambda}^j_n \geq e^{a_{ji}} \quad \text{for all } j \neq i \tag{5.75}$$

and accept the unique H_i that satisfies these inequalities. Here the a_{ji}'s are positive numbers (thresholds). This test is referred to as the *accepting MGSLRT* and denoted by $\hat{\delta} = (\widehat{T}, \hat{d})$, as before for Lai's GSLRT of two hypotheses. Taking logarithms and using the notation

$$\hat{\ell}_n = \sum_{k=1}^n \log p_{\hat{\theta}_n}(X_k), \quad \ell^i_n = \sup_{\theta \in \Theta_i} \sum_{k=1}^n \log p_\theta(X_k),$$

the accepting MGSLRT can be also expressed via the Markov times

$$\widehat{T}_i = \inf\left\{ n \geq 1 : \hat{\ell}_n \geq \max_{\substack{0 \leq j \leq N \\ j \neq i}} [\ell^j_n + a_{ji}] \right\}, \quad i = 0, 1, \dots, N \tag{5.76}$$

as

$$\widehat{T} = \min_{0 \leq i \leq N} \widehat{T}_i, \qquad \hat{d} = i \quad \text{if} \quad \widehat{T} = \widehat{T}_i. \tag{5.77}$$

Note that $\widehat{T}_i$ is the time of accepting the hypothesis H_i.

An alternative construction referred to as the *rejecting MGSLRT* and denoted by $\hat{\delta}_\star = (\widehat{T}_\star, \hat{d}_\star)$

is also of interest. In this test, the hypotheses are rejected one-by-one and the observations are continued up to the rejection of all except one hypothesis. This remaining hypothesis is accepted. Specifically, define the Markov times

$$\hat{\tau}_{ij} = \inf\left\{t : \hat{\ell}_n \geq \ell_n^i + b_{ij}\right\}, \quad i,j = 0,1,\ldots,N, i \neq j, \tag{5.78}$$

where b_{ij} are positive thresholds. The Markov rejecting time for the hypothesis H_i is

$$\hat{\tau}_i = \max_{\substack{0 \leq j \leq N \\ j \neq i}} \hat{\tau}_{ij},$$

and the test $\hat{\delta}_\star$ is defined as

$$\widehat{T}_\star = \min_{0 \leq j \leq N} \max_{\substack{0 \leq i \leq N \\ i \neq j}} \hat{\tau}_i, \qquad \hat{d}_\star = i \quad \text{if} \quad \max_{0 \leq j \leq N} \hat{\tau}_j = \hat{\tau}_i. \tag{5.79}$$

That is, the rejecting MGSLRT stops at the time $\widehat{T}_\star = \hat{\tau}_{(N-1)}$, where $\hat{\tau}_{(0)} \leq \hat{\tau}_{(1)} \leq \cdots \leq \hat{\tau}_{(N-1)}$ is the time-ordered set of rejecting times $\hat{\tau}_0, \hat{\tau}_1, \ldots, \hat{\tau}_N$.

Note that if $N = 1$ (two hypotheses) these tests coincide. Moreover, when $a_{10} = a_{01} = a$ the test has the same structure as Lai's GSLRT if the time-varying boundary $g(cn)$ is replaced with a constant threshold a. In fact, in the latter case this test yields Schwarz's test if $a = \log(1/c)$. In the following we assume that the indifference zone is fixed and its size is not too small while the error probabilities approach zero, in which case the time-varying boundary $g(cn)$ can be replaced with a constant threshold to obtain an asymptotically optimal GSLRT.

We consider a general asymptotically asymmetric case with respect to the error probabilities when some of them may decay faster than the others assuming that for the classes $\mathbf{C}([\alpha_{ij}])$ and $\mathbf{C}(\boldsymbol{\alpha})$ the conditions (5.16) hold, i.e., $|\log \alpha_{ij}| \sim c_{ij} |\log \alpha_{\max}|$ as $\alpha_{\max} = \max_{k,l} \alpha_{kl} \to 0$ and $|\log \alpha_i| \sim c_i |\log \tilde{\alpha}_{\max}|$ as $\tilde{\alpha}_{\max} = \max_k \alpha_k \to 0$, where $1 \leq c_{ij} < \infty$ and $1 \leq c_i < \infty$ are some constants. Unfortunately, as we already mentioned, we cannot obtain an analogue of Lemma 5.3.1, so that the implications (5.12)–(5.15) are not valid in general for the MGSLRT, due to the fact that the GLR statistics are not martingales anymore and Wald's likelihood ratio identity cannot be applied.

From now on we consider the l-parameter exponential family (5.69) in the canonical form with the natural parametrization and the natural parameter space $\widetilde{\Theta} = \{\theta : \int e^{\theta^\top x} \mu(\mathrm{d}x) < \infty\}$, and we always assume without special emphasis that the parameter space Θ on which the hypotheses are specified is a compact subset of $\widetilde{\Theta}$. Recall that the K–L information numbers $I(\theta, \widetilde{\theta})$ are given by (5.72). Let $I(\theta, \Theta_i) = \inf_{\widetilde{\theta} \in \Theta_i} I(\theta, \widetilde{\theta})$ be the minimal K–L distance between the point $\theta \notin \Theta_i$ and the subset Θ_i. For brevity we write $I(\theta, \Theta_i) = I_i(\theta)$. Note that $I_i(\theta) = 0$ for $\theta \in \Theta_i$. Let $k(\theta)$ denote the index of the nearest subset Θ_i from the point θ, i.e., $\min_{0 \leq i \leq N} I_i(\theta) = I_{k(\theta)}(\theta)$. In what follows we suppose that the sets $\Theta_0, \Theta_1, \ldots, \Theta_N$ are separated enough in the sense that the K–L distance between any point θ and the nearest subset is positive: $\min_{i \neq k(\theta)} I_i(\theta) > 0$. In other words, we require that

$$\min_{\substack{0 \leq j \leq N \\ j \neq i}} I_j(\theta) > 0 \quad \text{for all} \quad \theta \in \Theta_i, \quad i = 0,1,\ldots,N; \quad \min_{0 \leq i \leq N} I_i(\theta) > 0 \quad \text{for all} \quad \theta \in \mathrm{I}_{\mathrm{in}}. \tag{5.80}$$

To deal with the Kiefer–Weiss problem of minimizing the maximal ESS we need in addition to assume that

$$\widetilde{I}_{\inf}(\theta) = \inf_{\theta \in \Theta} \widetilde{I}(\theta) > 0, \quad \text{where } \widetilde{I}(\theta) = \begin{cases} \min\limits_{j \neq i} I_j(\theta) & \text{if } \theta \in \Theta_i, \\ \max\limits_{0 \leq i \leq N} \min\limits_{j \neq i} I_j(\theta) & \text{if } \theta \in \mathrm{I}_{\mathrm{in}}. \end{cases} \tag{5.81}$$

Note that for any $\theta \in \Theta$,

$$\widetilde{I}(\theta) = \min_{i \neq k(\theta)} I_i(\theta) = \inf_{\widetilde{\theta} \in \cup_{i \neq k(\theta)} \Theta_i} I(\theta, \widetilde{\theta}),$$

i.e., $\widetilde{I}(\theta)$ is equal to the second shortest distance between θ and the sets Θ_i. If, for example, $\theta \in \mathrm{I}_{\mathrm{in}}$ and $I_0(\theta) < I_1(\theta) < \cdots < I_N(\theta)$, then $\widetilde{I}(\theta) = I_1(\theta)$.

It is also convenient to introduce the scaled K–L distances $I_i(\theta)/c_{ij}$ and $I_i(\theta)/c_i$ that account for the asymptotic relative differences between the error probabilities in the classes $\mathbf{C}([\alpha_{ij}])$ and $\mathbf{C}(\boldsymbol{\alpha})$. That said, let

$$
\begin{aligned}
J_i(\theta) &= \min_{\substack{0\le j\le N\\ j\neq i}} [I_j(\theta)/c_{ji}] \quad \text{for} \quad \theta \in \Theta_i,\\
J(\theta) &= \max_{0\le i\le N} \min_{\substack{0\le j\le N\\ j\neq i}} [I_j(\theta)/c_{ji}] = \max_{0\le i\le N} J_i(\theta) \quad \text{for} \quad \theta \in \mathrm{I}_{\mathrm{in}},
\end{aligned}
\tag{5.82}
$$

for the class $\mathbf{C}([\alpha_{ij}])$ and

$$
\begin{aligned}
J_i^*(\theta) &= \min_{\substack{0\le j\le N\\ j\neq i}} [I_j(\theta)/c_j] \quad \text{for} \quad \theta \in \Theta_i,\\
J^*(\theta) &= \max_{0\le i\le N} \min_{\substack{0\le j\le N\\ j\neq i}} [I_j(\theta)/c_j] = \max_{0\le i\le N} J_i^*(\theta) \quad \text{for} \quad \theta \in \mathrm{I}_{\mathrm{in}}
\end{aligned}
\tag{5.83}
$$

for the class $\mathbf{C}(\boldsymbol{\alpha})$. Finally, let the functions $\widetilde{J}(\theta)$ and $\widetilde{J}^*(\theta)$ be defined as

$$
\widetilde{J}(\theta) = \begin{cases} J_i(\theta) & \text{if } \theta \in \Theta_i \\ J(\theta) & \text{if } \theta \in \mathrm{I}_{\mathrm{in}} \end{cases}, \qquad \widetilde{J}^*(\theta) = \begin{cases} J_i^*(\theta) & \text{if } \theta \in \Theta_i \\ J^*(\theta) & \text{if } \theta \in \mathrm{I}_{\mathrm{in}}. \end{cases}
$$

Note that the conditions (5.80) guarantee that $J(\theta)$ and $J^*(\theta)$ are strictly positive for all $\theta \in \Theta = \sum_{i=0}^N \Theta_i + \mathrm{I}_{\mathrm{in}}$ and the condition (5.81) implies that $\widetilde{J}_{\inf}(\theta) = \inf_{\theta\in\Theta} \widetilde{J}(\theta) > 0$ and $\widetilde{J}_{\inf}^*(\theta) = \inf_{\theta\in\Theta} \widetilde{J}^*(\theta) > 0$.

In order to obtain meaningful results we of course need some conditions on the rate of convergence of the MLE $\hat{\theta}_n$ to the true parameter value θ as $n \to \infty$. Intuitively, by Theorem 5.3.1, one may expect that the modified MSPRT tuned to the true parameter and the MGSLRT will perform similarly only if $\hat{\theta}_n$ is close to θ for $n \ll a/\min_j I_j(\theta)$ (assuming $a_{ji} = a$) in the sense that $I(\theta, \hat{\theta}_n)$ converges to 0 fast enough so that $I(\theta, \hat{\theta}_n) \approx 0$ for sufficiently large but not too large n, in which case $\mathsf{E}_\theta[\widehat{T}] \approx \mathsf{E}_\theta[T^\star] \approx a/\min_j I_j(\theta)$ (see (5.20)). The following condition

$$
\sum_{n=1}^{\infty} \mathsf{E}_\theta[I(\theta, \hat{\theta}_n)]^r < \infty \quad \text{for some } r \ge 1 \tag{5.84}
$$

is sufficient for this purpose. Note that for many typical examples (Gaussian, Poisson, Bernoulli, etc.) $I(\theta, \widetilde{\theta}) \le C|\theta - \widetilde{\theta}|^2$, so that this condition holds for every $r > 1$.

We now show that the accepting and rejecting MGSLRTs are asymptotically optimal to first order in the frequentist sense. We formulate and prove this result only for the accepting MGSLRT since, for approximately the same thresholds, $\widehat{T}_\star \le \widehat{T}$, so that asymptotic optimality of $\hat{\delta}$ implies optimality of $\hat{\delta}_\star$. The proof is partially based on the comparison of the MGSLRT with another adaptive test, the MASLRT, defined in the next section in (5.93)–(5.96). We define $a_{\min} = \min_{j,i,j\neq i} a_{ji}$.

Theorem 5.4.1 (MGSLRT asymptotic optimality). *Let the observations $X_1, X_2, \dots$ be iid from the multiparameter exponential family* (5.69). *Assume that the conditions* (5.80) *and* (5.84) *hold.*

(i) *Then, as $a_{\min} \to \infty$,*

$$
\mathsf{E}_\theta \widehat{T} \sim \begin{cases} \max\limits_{\substack{0\le j\le N\\ j\neq i}} [a_{ji}/I_j(\theta)] & \text{for all } \theta \in \Theta_i \text{ and all } i = 0, 1, \dots, N, \\ \min\limits_{0\le i\le N} \max\limits_{\substack{0\le j\le N\\ j\neq i}} [a_{ji}/I_j(\theta)] & \text{for all } \theta \in \mathrm{I}_{\mathrm{in}} \end{cases} \tag{5.85}
$$

and, if in addition the condition (5.81) *is satisfied, then*

$$\sup_{\theta\in\Theta} \mathsf{E}_\theta \widehat{T} \sim \sup_{\theta\in\Theta} \min_{0\le i\le N} \max_{\substack{0\le j\le N\\ j\neq i}} [a_{ji}/I_j(\theta)], \quad \text{as } a_{\min}\to\infty. \tag{5.86}$$

(ii) *If the thresholds are selected so that* $\sup_{\theta\in\Theta_i} \alpha_{ij}(\hat{\delta},\theta) \le \alpha_{ij}$ *and* $a_{ji} \sim \log(1/\alpha_{ji})$, *then as* $\alpha_{\max}\to 0$

$$\inf_{\delta\in\mathbf{C}([\alpha_{ij}])} \mathsf{E}_\theta T \sim \mathsf{E}_\theta \widehat{T} \sim \begin{cases} |\log\alpha_{\max}|/J_i(\theta) & \text{for all } \theta\in\Theta_i \text{ and } i=0,1,\dots,N,\\ |\log\alpha_{\max}|/J(\theta) & \text{for all } \theta\in \mathrm{I}_{\rm in}.\end{cases} \tag{5.87}$$

(iii) *If the thresholds are selected so that* $\sup_{\theta\in\Theta_i} \alpha_i(\hat{\delta},\theta)\le\alpha_i$ *and* $a_{ji}\sim\log(1/\alpha_j)$, *then as* $\tilde{\alpha}_{\max}\to 0$

$$\inf_{\delta\in\mathbf{C}(\boldsymbol{\alpha})} \mathsf{E}_\theta T \sim \mathsf{E}_\theta \widehat{T} \sim \begin{cases} |\log\tilde{\alpha}_{\max}|/J_i^*(\theta) & \text{for all } \theta\in\Theta_i \text{ and all } i=0,1,\dots,N,\\ |\log\tilde{\alpha}_{\max}|/J^*(\theta) & \text{for all } \theta\in \mathrm{I}_{\rm in}.\end{cases} \tag{5.88}$$

(iv) *If instead of* (5.80) *we assume a stronger separability condition* (5.81), *then the statements (ii) and (iii) hold in the worst-case scenario, i.e.,*

$$\begin{aligned} &\inf_{\delta\in\mathbf{C}([\alpha_{ij}])} \sup_{\theta\in\Theta} \mathsf{E}_\theta T \sim \sup_{\theta\in\Theta} \mathsf{E}_\theta\widehat{T} \sim |\log\alpha_{\max}|/J_{\inf} \quad \text{as } \alpha_{\max}\to 0;\\ &\inf_{\delta\in\mathbf{C}(\boldsymbol{\alpha})} \sup_{\theta\in\Theta} \mathsf{E}_\theta T \sim \sup_{\theta\in\Theta} \mathsf{E}_\theta\widehat{T} \sim |\log\tilde{\alpha}_{\max}|/J^*_{\inf} \quad \text{as } \tilde{\alpha}_{\max}\to 0.\end{aligned} \tag{5.89}$$

Consequently, the accepting MGSLRT minimizes asymptotically the expected sample size uniformly in $\theta\in\Theta$ *in the classes of tests* $\mathbf{C}([\alpha_{ij}])$ *and* $\mathbf{C}(\boldsymbol{\alpha})$ *and also solves the Kiefer–Weiss problem to first order.*

Proof. (i) Obviously, for the same set of thresholds $(a_{ij})_{i,j}$ the stopping time of the MGSLRT $\widehat{T}$ does not exceed the stopping time T^* of the MASLRT defined in (5.95) below in Subsection 5.4.2. Therefore, by Theorem 5.4.3(i) in Subsection 5.4.2,

$$\mathsf{E}_\theta\widehat{T} \le \begin{cases} \max\limits_{\substack{0\le j\le N\\ j\neq i}} [a_{ji}/I_j(\theta)](1+o(1)) & \text{for } \theta\in\Theta_i,\ i=0,1,\dots,N,\\ \min\limits_{0\le i\le N}\max\limits_{\substack{0\le j\le N\\ j\neq i}} [a_{ji}/I_j(\theta)](1+o(1)) & \text{for } \theta\in\mathrm{I}_{\rm in}.\end{cases}$$

Since for the exponential family all moments $\mathsf{E}_\theta|X_1|^m$ are finite, the normalized LLR $n^{-1}\sum_{k=1}^n \log[p_\theta(X_k)/p_{\widetilde{\theta}}(X_k)]$ converges almost surely under P_θ to $I(\theta,\widetilde{\theta})$. Using large deviation approximations, it can be shown [94, Theorem 1] that

$$\sup_{\theta\in\Theta_i} \mathsf{P}_\theta(\hat{d}=j) = a_{ji}^{l/2} e^{-a_{ji}+O(1)} \quad \text{as } \min_{ij} a_{ij}\to\infty. \tag{5.90}$$

Hence, $a_{ij}\sim\log(1/\alpha_{ij})$ and setting $a_{ij}\sim\log(1/\alpha_{ij})$ in (5.147) and (5.148), we obtain that the right-hand side in the previous inequalities are also the lower bounds, which yields (5.85).

The proof of (5.86) is similar to that in Theorem 5.4.3 and is omitted.

(ii) While the asymptotic approximation (5.90) does not yield an exact prescription of how the thresholds can be selected in order to guarantee the inequalities $\sup_{\theta\in\Theta_i}\alpha_{ij}(\hat{\delta},\theta)\le\alpha_{ij}$, it is clear that if these inequalities hold, then $a_{ij}\sim|\log\alpha_{ij}|$. Hence, (5.85) yields

$$\mathsf{E}_\theta\widehat{T} \sim \begin{cases} \max\limits_{\substack{0\le j\le N\\ j\neq i}} [|\log\alpha_{ji}|/I_j(\theta)] & \text{for } \theta\in\Theta_i,\ i=0,1,\dots,N,\\ \min\limits_{0\le i\le N}\max\limits_{\substack{0\le j\le N\\ j\neq i}} [|\log\alpha_{ji}|/I_j(\theta)] & \text{for } \theta\in\mathrm{I}_{\rm in},\end{cases}$$

as $\alpha_{\max} \to 0$. Since by Theorem 5.4.3(ii),

$$\inf_{\delta \in \mathbf{C}([\alpha_{ij}])} \mathsf{E}_\theta T \sim \begin{cases} \max\limits_{\substack{0 \le j \le N \\ j \ne i}} [|\log \alpha_{ji}|/I_j(\theta)] & \text{for } \theta \in \Theta_i,\ i = 0, 1, \dots, N, \\ \min\limits_{0 \le i \le N} \max\limits_{\substack{0 \le j \le N \\ j \ne i}} [|\log \alpha_{ji}|/I_j(\theta)] & \text{for } \theta \in \mathrm{I}_{\text{in}}, \end{cases}$$

the asymptotic approximations (5.87) follow for all $\theta \in \sum_{i=0}^N \Theta_i + \mathrm{I}_{\text{in}}$.

(iii) The proof of (iii) is essentially the same as that of (ii) and is omitted.

(iv) Using (5.86) and the argument similar to that used in the proof of (ii), we obtain that if the a_{ij}'s are selected so that $\sup_{\theta \in \Theta_i} \alpha_{ij}(\hat{\delta}, \theta) \le \alpha_{ij}$ and $a_{ji} \sim \log(1/\alpha_{ji})$, then as $\alpha_{\max} \to 0$

$$\sup_{\theta \in \Theta} \mathsf{E}_\theta \widehat{T} \sim \sup_{\theta \in \Theta} \min_{0 \le i \le N} \max_{\substack{0 \le j \le N \\ j \ne i}} [|\log \alpha_{ji}|/I_j(\theta)] = |\log \alpha_{\max}|/J_{\text{inf}}.$$

Since (5.112) implies that $\inf_{\delta \in \mathbf{C}([\alpha_{ij}])} \sup_{\theta \in \Theta} \mathsf{E}_\theta T \sim |\log \alpha_{\max}|/J_{\text{inf}}$, the first asymptotic equalities in (5.89) follow. The second ones are proved analogously. □

It is worth noting that for Theorem 5.4.1 to be useful in the applications the separability condition for the hypotheses $\widetilde{I}_{\text{inf}} > 0$ should be actually strengthened into $\widetilde{I}_{\text{inf}} > \Delta$, where Δ is not too small, so that $|\log \tilde{\alpha}_{\max}|$ (say) is smaller than Δ. Otherwise, a local approach is in order.

Finally, we stress that while the GSLRT is perhaps the most efficient test, definitely more efficient than the adaptive test considered in the next subsection, it has a substantial drawback — it is difficult, if at all possible, to design its thresholds to guarantee the given upper bounds for the error probabilities. In this sense, the adaptive approach considered in the next subsection is much more convenient.

5.4.2 *Adaptive Likelihood Ratio Tests with One-Stage Delayed Estimators*

Let $\hat{\theta}_n = \hat{\theta}_n(X_1, \dots, X_n)$ be an estimator of θ, not necessarily the MLE. If in the pdf $p_\theta(X_k)$ for the k-th observation we replace the parameter with the estimate $\hat{\theta}_{k-1}$ built upon the sample $(X_1, \dots, X_{k-1})$ that includes not k, but $k-1$ observations, then $p_{\hat{\theta}_{k-1}}(X_k)$ is still a viable probability density, in contrast to the case of the GLR approach where $p_{\hat{\theta}_n}(X_k)$ is not a probability density anymore for $k \le n$. Therefore, the statistic

$$\Lambda_n^*(\theta_i) = \prod_{k=1}^n \frac{p_{\hat{\theta}_{k-1}}(X_k)}{p_{\theta_i}(X_k)} = \Lambda_{n-1}^*(\theta_i) \times \frac{p_{\hat{\theta}_{n-1}}(X_n)}{p_{\theta_i}(X_n)} \tag{5.91}$$

is a viable likelihood ratio, which we call the *adaptive LR*, and it is the nonnegative P_{θ_i}-martingale with unit expectation, since $\mathsf{E}_{\theta_i}[\Lambda_n^*(\theta_i) | \mathbf{X}_1^{n-1}] = \Lambda_n^*(\theta_i)$. Therefore, one can use Wald's likelihood ratio identity for finding bounds for the error probabilities if $\Lambda_n^*(\theta_i)$ is used instead of the LR with the true parameter value θ as in the MSPRT. Specifically, the martingale property of the adaptive LR $\Lambda_n^*(\theta)$ with respect to P_θ implies the Wald–Doob identity:

$$\mathsf{E}_\theta \left[\Lambda_T^*(\theta) \mathbb{1}_{\{T < \infty\}}\right] = 1 \quad \text{for an arbitrary stopping time } T \text{ and all } \theta \in \Theta. \tag{5.92}$$

Because of exactly this very convenient property as well as the simple recursive structure (5.91), the hypothesis tests based on the adaptive LRs represent a very attractive alternative to the GLR tests as well to the mixture-based tests that are considered in Subsection 5.4.3. Robbins and Siegmund [391, 392] were the first who suggested the idea of using such tests in the context of power one tests.

Hence, instead of the GLR statistics (5.70) we now exploit the statistics

$$\Lambda_n^*(\Theta_i) = \frac{\prod_{k=1}^n p_{\hat{\theta}_{k-1}}(X_k)}{\sup\limits_{\theta \in \Theta_i} \prod_{k=1}^n p_\theta(X_k)}, \quad i = 0, 1, \dots, N \tag{5.93}$$

and the new multihypothesis test, which we refer to as the *accepting Multihypothesis Adaptive Sequential Likelihood Ratio Test* (MASLRT), similarly to (5.75) has the form

$$\text{Stop at the first } n \geq 1 \text{ such that for some } i \quad \Lambda_n^*(\Theta_j) \geq e^{a_{ji}} \quad \text{for all } j \neq i \tag{5.94}$$

and accept the unique H_i that satisfies these inequalities.

Next, introducing the following adaptive statistic

$$\ell_n^* = \sum_{k=1}^{n} \log p_{\hat{\theta}_{k-1}}(X_k),$$

and recalling that $\ell_n^i = \sup_{\theta \in \Theta_i} \sum_{k=1}^n \log p_\theta(X_k)$, similarly to (5.76), (5.77) the accepting MASLRT can be also written as

$$T^* = \min_{0 \leq i \leq N} T_i^*, \qquad d^* = i \quad \text{if} \quad T^* = T_i^*, \tag{5.95}$$

where

$$T_i^* = \inf\left\{ n \geq 1 : \ell_n^* \geq \max_{\substack{0 \leq j \leq N \\ j \neq i}} [\ell_n^j + a_{ji}] \right\}, \qquad i = 0, 1, \dots, N. \tag{5.96}$$

Also, by analogy with (5.79), we can construct the rejecting MASLRT $\delta^\star = (T^\star, d^\star)$ as

$$T^\star = \min_{0 \leq j \leq N} \max_{\substack{0 \leq i \leq N \\ i \neq j}} T_i^\star, \qquad d^\star = i \quad \text{if} \quad \max_{0 \leq j \leq N} T_j^\star = T_i^\star, \tag{5.97}$$

where

$$T_{ij}^\star = \inf\left\{ t : \ell_n^* \geq \ell_n^i + b_{ij} \right\}, \quad i, j = 0, 1, \dots, N, i \neq j$$

and

$$T_i^\star = \max_{\substack{0 \leq j \leq N \\ j \neq i}} T_{ij}^\star$$

is the rejecting time of the hypothesis H_i.

We begin with obtaining the upper bounds for the error probabilities of the adaptive tests δ^* and $\delta^\star$ in the general case without assuming the exponential family model. We denote with asterisks $\alpha_{ij}^*(\theta) = \mathsf{P}_\theta(d^* = j)$ and $\alpha_i^*(\theta) = \mathsf{P}_\theta(d^* \neq i)$, $\theta \in \Theta$ the error probabilities of the accepting MASLRT δ^* and similarly with stars $\alpha_{ij}^\star(\theta)$ and $\alpha_i^\star(\theta)$ the error probabilities of the rejecting MASLRT $\delta^\star$.

Lemma 5.4.1. *The following inequalities hold:*

(i) $\sup_{\theta \in \Theta_i} \alpha_{ij}^*(\theta) \leq e^{-a_{ij}}$ *for* $i, j = 0, 1, \dots, N$, $i \neq j$;

(ii) $\sup_{\theta \in \Theta_i} \alpha_i^*(\theta) \leq \sum_{j \neq i} e^{-a_{ij}}$ *for* $i = 0, 1, \dots, N$;

(iii) $\sup_{\theta \in \Theta_i} \alpha_{ij}^\star(\theta) \leq e^{-b_{ji}}$ *for* $i, j = 0, 1, \dots, N$, $i \neq j$;

(iv) *If* $b_{ji} = b_i$, *then* $\sup_{\theta \in \Theta_i} \alpha_i^\star(\theta) \leq e^{-b_i}$ *for* $i = 0, 1, \dots, N$.

Proof. Proof of (i) and (ii). Since $\{d^* = j\} = \{T^* = T_j^*\}$ implies $\{T_j^* < \infty\}$, we have for all $\theta \in \Theta_i$

$$\alpha_{ij}^*(\theta) = \mathsf{E}_\theta \mathbb{1}_{\{d^* = j\}} \leq \mathsf{E}_\theta \mathbb{1}_{\{T_j^* < \infty\}} = \mathsf{E}_\theta \left[\mathbb{1}_{\{T_j^* < \infty\}} \Lambda_{T_j^*}^*(\theta) / \Lambda_{T_j^*}^*(\theta) \right].$$

By the definition of T_j^*, $\Lambda_{T_j^*}^*(\Theta_i) \geq e^{a_{ij}}$ and clearly $\Lambda_{T_j^*}^*(\theta) \geq \Lambda_{T_j^*}^*(\Theta_i)$ for all $\theta \in \Theta_i$. Therefore, for all $\theta \in \Theta_i$,

$$\alpha_{ij}^*(\theta) \leq \mathsf{E}_\theta \left[\mathbb{1}_{\{T_j^* < \infty\}} \Lambda_{T_j^*}^*(\theta) / \Lambda_{T_j^*}^*(\theta) \right] \leq e^{-a_{ij}} \mathsf{E}_\theta \left[\mathbb{1}_{\{T_j^* < \infty\}} \Lambda_{T_j^*}^*(\theta) \right] = e^{-a_{ij}},$$

where the last equality follows from the identity (5.92). This proves (i). Part (ii) follows immediately from the fact that

$$\alpha_i^*(\theta) \le \sum_{j \ne i} \mathsf{P}_\theta(T_j^* < \infty) \le \sum_{j \ne i} e^{-a_{ij}}.$$

Proof of (iii) and (iv). Note first that for all $\theta \in \Theta_i$

$$\alpha_{ij}^\star(\theta) \le \mathsf{P}_\theta(T_j^\star < \infty) \le \mathsf{P}_\theta(T_{ji}^\star < \infty) = \mathsf{E}_\theta\left[\mathbb{1}_{\{T_{ji}^\star < \infty\}} \Lambda_{T_{ji}^\star}^*(\theta) / \Lambda_{T_{ji}^\star}^*(\theta)\right].$$

Next, using the fact that

$$\Lambda_{T_{ji}^\star}^*(\theta) \ge \Lambda_{T_{ji}^\star}^*(\Theta_i) \ge e^{b_{ji}} \quad \text{on } \{T_{ji}^\star < \infty\}$$

and that, by (5.92), $\mathsf{E}_\theta\left[\Lambda_{T_{ji}^\star}^*(\theta) \mathbb{1}_{\{T_{ji}^\star < \infty\}}\right] = 1$, we obtain

$$\alpha_{ij}^\star(\theta) \le \mathsf{E}_\theta\left[\mathbb{1}_{\{T_{ji}^\star < \infty\}} \Lambda_{T_{ji}^\star}^*(\theta) / \Lambda_{T_{ji}^\star}^*(\theta)\right] \le e^{-b_{ji}} \mathsf{E}_\theta\left[\mathbb{1}_{\{T_{ji}^\star < \infty\}} \Lambda_{T_{ji}^\star}^*(\theta)\right] = e^{-b_{ji}}.$$

Finally, (iv) follows from the analogous argument:

$$\alpha_i^\star(\theta) \le \mathsf{P}_\theta(T_i^\star < \infty) \le \mathsf{E}_\theta\left[\mathbb{1}_{\{T_i^\star < \infty\}} \Lambda_{T_i^\star}^*(\theta) / \Lambda_{T_i^\star}^*(\theta)\right] \le e^{-b_i} \mathsf{E}_\theta\left[\mathbb{1}_{\{T_i^\star < \infty\}} \Lambda_{T_i^\star}^*(\theta)\right] = e^{-b_i}.$$

□

Therefore, we have the following important implications:

$$a_{ij} = \log(1/\alpha_{ij}) \Longrightarrow \delta^* \in \mathbf{C}([\alpha_{ij}]); \tag{5.98}$$
$$a_{ij} = a_i = \log(N/\alpha_i) \Longrightarrow \delta^* \in \mathbf{C}(\boldsymbol{\alpha}); \tag{5.99}$$
$$b_{ij} = \log(1/\alpha_{ji}) \Longrightarrow \delta^\star \in \mathbf{C}([\alpha_{ij}]); \tag{5.100}$$
$$b_{ij} = b_i = \log(1/\alpha_i) \Longrightarrow \delta^\star \in \mathbf{C}(\boldsymbol{\alpha}). \tag{5.101}$$

Note that again, as with the accepting and rejecting MSPRTs, the accepting MASLRT is better suited for the class $\mathbf{C}([\alpha_{ij}])$ and the rejecting MASLRT is better suited for the class $\mathbf{C}(\boldsymbol{\alpha})$; see Remark 5.3.1. Note also that Lemma 5.4.1 and the implications (5.98)–(5.101) hold in the most general non-iid case too. In fact, the proof of Lemma 5.4.1 does not involve any assumptions on the iid structure of the data.

The following two theorems establish the asymptotic performance and asymptotic optimality of the adaptive accepting and rejecting tests. We formulate them only for the accepting test. In fact, the rejecting MASLRT outperforms the accepting test for the same thresholds since it can be seen that $T^\star \le T^*$.

Theorem 5.4.2 (MASLRT asymptotic performance). *Let the observations $X_1, X_2, \dots$ be iid from the multiparameter exponential family* (5.69), *and let the parameter space Θ be a compact subset of the natural space $\widetilde{\Theta} \subset \mathbb{R}^l$. Assume that the conditions* (5.80) *are satisfied and $\{\hat{\theta}_n\}_{n \ge 1}$ is a sequence of estimates of θ such that the condition* (5.84) *holds. Then, as $a_{\min} = \min_{i,j} a_{ij} \to \infty$,*

$$\mathsf{E}_\theta T^* \sim \begin{cases} \max\limits_{\substack{0 \le j \le N \\ j \ne i}} [a_{ji}/I_j(\theta)] & \text{for all } \theta \in \Theta_i \text{ and } i = 0, 1, \dots, N, \\ \min\limits_{0 \le i \le N} \max\limits_{\substack{0 \le j \le N \\ j \ne i}} [a_{ji}/I_j(\theta)] & \text{for all } \theta \in \mathrm{I}_{\mathrm{in}}, \end{cases} \tag{5.102}$$

and if in addition the condition (5.81) *is satisfied, then*

$$\sup_{\theta \in \Theta} \mathsf{E}_\theta T^* \sim \sup_{\theta \in \Theta} \min_{0 \le i \le N} \max_{\substack{0 \le j \le N \\ j \ne i}} [a_{ji}/I_j(\theta)] \quad \text{as } a_{\min} \to \infty. \tag{5.103}$$

Proof. The proof is based on the conventional upper–lower bounding technique. We first show that asymptotically as $a_{\min} \to \infty$

$$\mathsf{E}_\theta T^* \ge A(\theta)(1+o(1)) \quad \text{for all } \theta \in \Theta, \tag{5.104}$$

where we set

$$A(\theta) = \begin{cases} \max\limits_{\substack{0\le j\le N\\ j\ne i}} [a_{ji}/I_j(\theta)] & \text{for } \theta \in \Theta_i,\ i=0,1,\dots,N, \\ \min\limits_{0\le i\le N} \max\limits_{\substack{0\le j\le N\\ j\ne i}} [a_{ji}/I_j(\theta)] & \text{for } \theta \in \mathrm{I}_{\rm in}. \end{cases}$$

Let θ and $\widetilde{\theta}$ be two arbitrary distinct points such that $I(\theta,\widetilde{\theta}) > 0$. Then, by the Hoeffding–Simons inequality [192, 432], for any multihypothesis test $\delta = (T,d)$

$$I(\theta,\widetilde{\theta})\mathsf{E}_\theta T \ge \sum_{j=0}^N \mathsf{P}_\theta(d=j)\log\left[\frac{\mathsf{P}_\theta(d=j)}{\mathsf{P}_{\widetilde{\theta}}(d=j)}\right]. \tag{5.105}$$

Taking $\theta \in \Theta_i$ and $\widetilde{\theta} \in \Theta_j$, after elementary calculations we obtain that for every $\theta \in \Theta_i$

$$\mathsf{E}_\theta T \ge \max_{\substack{0\le j\le N\\ j\ne i}} \frac{|\log \sup_{\widetilde{\theta}\in\Theta_j} \alpha_{ji}(\delta,\widetilde{\theta})|}{I_j(\theta)} + o(1) \quad \text{as } \max_{j\ne i} \sup_{\widetilde{\theta}\in\Theta_j} \alpha_{ji}(\delta,\widetilde{\theta}) \to 0. \tag{5.106}$$

Now, by Lemma 5.4.1(i), $\sup_{\widetilde{\theta}\in\Theta_j} \alpha_{ji}(\delta^*,\widetilde{\theta}) \le e^{-a_{ji}}$, which along with (5.106) yields

$$\mathsf{E}_\theta T^* \ge \max_{\substack{0\le j\le N\\ j\ne i}} \frac{a_{ji}}{I_j(\theta)} + o(1) \quad \text{for all } \theta \in \Theta_i,\ i=0,1,\dots,N \quad \text{as } a_{\min} \to \infty.$$

Hence, the lower bound (5.104) follows for $\theta \in \Theta \setminus \mathrm{I}_{\rm in} = \sum_{i=1}^N \Theta_i$.

For θ in the indifference zone $\mathrm{I}_{\rm in}$, the Hoeffding inequality does not yield the desired asymptotic inequality since the probabilities $\mathsf{P}_\theta(d=j)$ may not go to 0, and some of them do not, so that the situation is more delicate. However, the asymptotic lower bound

$$\mathsf{E}_\theta T^* \ge \min_{0\le i\le N} \max_{\substack{0\le j\le N\\ j\ne i}} \frac{a_{ji}}{I_j(\theta)}(1+o(1)) \quad \text{for all } \theta \in \mathrm{I}_{\rm in} \quad \text{as } a_{\min} \to \infty, \tag{5.107}$$

which is equivalent to the lower bound (5.104) for $\theta \in \mathrm{I}_{\rm in}$, can be easily proved as a particular case of the lower bounds (5.148) for the moments of the stopping time derived in Subsection 5.4.4 in a general, not necessarily iid case; see Lemma 5.4.2 for the proof when $N=1$. Indeed, let $\ell_n(\theta) = \sum_{k=1}^n \log p_\theta(X_k)$ and

$$\lambda_n(\theta,\widetilde{\theta}) = \ell_n(\theta) - \ell_n(\widetilde{\theta}) = \sum_{k=1}^n \log\frac{p_\theta(X_k)}{p_{\widetilde{\theta}}(X_k)}.$$

For the exponential family the function $b(\theta)$ is bounded, so

$$n^{-1}\lambda_n(\theta,\widetilde{\theta}) \xrightarrow[n\to\infty]{\mathsf{P}_\theta\text{-a.s.}} I(\theta,\widetilde{\theta}),$$

i.e., in our case the almost sure convergence conditions (5.153) hold with $I(\theta,\widetilde{\theta}) > 0$ being the K–L numbers. Thus, the inequalities (5.148) with $\alpha_{ij} = e^{-a_{ij}}$ can be applied to obtain that for any $m>0$ and $\theta \in \mathrm{I}_{\rm in}$

$$\mathsf{E}_\theta[T^*]^m \ge \left(\min_{0\le i\le N} \max_{\substack{0\le j\le N\\ j\ne i}} \frac{a_{ji}}{I_j(\theta)}\right)^m (1+o(1)) \quad \text{as } a_{\min} \to \infty,$$

which implies (5.107).

Once the lower bound (5.104) is established for all $\theta \in \Theta$, it remains to show that it is also an upper bound. Write $\tilde{\ell}_n^j = \ell_n^j / I_j(\theta)$, $\tilde{\ell}_n^* = \ell_n^* / I_j(\theta)$, $\tilde{a}_{ji} = a_{ji}/I_j(\theta)$. Since

$$\min_{j \neq i} \{\ell_n^* - \ell_n^j - a_{ji}\} = \min_{j \neq i} \{I_j(\theta)[\tilde{\ell}_n^* - \tilde{\ell}_n^j - \tilde{a}_{ji}]\} \geq \min_{j \neq i} I_j(\theta) \times \left[\min_{j \neq i}(\tilde{\ell}_n^* - \tilde{\ell}_n^j) - \max_{j \neq i} \tilde{a}_{ji}\right],$$

we have

$$T_i^* = \inf\left\{n : \min_{j \neq i}[\ell_n^* - \ell_n^j - a_{ji}] \geq 0\right\} \leq \inf\left\{n : \min_{j \neq i}[\tilde{\ell}_n^* - \tilde{\ell}_n^j] \geq \max_{j \neq i} \tilde{a}_{ji}\right\} := \tau_i.$$

Obviously, $T^* \leq \min_i \tau_i$, and it suffices to show that

$$\mathsf{E}_\theta \tau_i \leq \max_{\substack{0 \leq j \leq N \\ j \neq i}} [a_{ji}/I_j(\theta)](1 + o(1)) \quad \text{for all } \theta \in \Theta_i + \mathrm{I}_{\text{in}}, \quad i = 0, 1, \ldots, N. \tag{5.108}$$

Define

$$Z_n^i = \tilde{\ell}_n^* - \min_{j \neq i} \tilde{\ell}_n^j, \quad m_n(\theta, \widetilde{\theta}) = \lambda_n(\theta, \widetilde{\theta}) - nI(\theta, \widetilde{\theta}),$$
$$R_n(\theta) = \sum_{k=1}^{n} I(\theta, \hat{\theta}_{k-1}), \quad M_n(\theta) = \ell_n^* - \ell_n(\theta) + R_n(\theta).$$

We have

$$\ell_n^* - \ell_n^j = M_n(\theta) - R_n(\theta) - [\ell_n^j - \ell_n(\theta)] = M_n(\theta) - R_n(\theta) - \sup_{\widetilde{\theta} \in \Theta_j} [-\lambda_n(\theta, \widetilde{\theta})]$$
$$= M_n(\theta) - R_n(\theta) + nI_j(\theta) - \sup_{\widetilde{\theta} \in \Theta_j} [-m_n(\theta, \widetilde{\theta})].$$

The processes $\{m_n(\theta, \widetilde{\theta})\}$ and $\{M_n(\theta)\}$ are zero-mean P_θ-martingales, $\mathsf{E}_\theta m_n(\theta, \widetilde{\theta}) = \mathsf{E}_\theta M_n(\theta) = 0$, and since for the exponential family the function $b(\theta)$ is bounded, by the inequalities for martingales [265],

$$\mathsf{E}_\theta |M_n(\theta)| \leq \mathsf{constant} \cdot n^{1/2}, \quad \mathsf{E}_\theta \left[\sup_{\widetilde{\theta} \in \Theta} |m_n(\theta, \widetilde{\theta})|\right] \leq \mathsf{constant} \cdot n^{1/2}.$$

Using these last inequalities and the condition (5.84) it can be established that $n^{-1}R_n(\theta)$, $n^{-1}M_n(\theta)$ and $n^{-1}\sup_{\widetilde{\theta} \in \Theta_j} |m_n(\theta, \widetilde{\theta})|$ converge almost surely as well as P_θ-r-quickly to 0 for $r = 1$. Therefore, $n^{-1}Z_n^i$ converge 1-quickly to $\widetilde{I}_i(\theta) = \min_{j \neq i} I_j(\theta)$, $i = 0, 1, \ldots, N$. In other words, if for $\varepsilon > 0$ we define the last entry times

$$L_i(\varepsilon) = \sup\left\{n \geq 1 : \left|Z_n^i - \widetilde{I}_i(\theta)n\right| > \varepsilon n\right\}, \quad i = 0, 1, \ldots, N,$$

then $\mathsf{E}_\theta[L_i(\varepsilon)] < \infty$ for all $\varepsilon > 0$.

Observe that

$$(1 - \varepsilon)(\tau_i - 1) < Z_{\tau_i - 1}^i < \max_{j \neq i} \tilde{a}_{ji} \quad \text{on } \{L_i(\varepsilon) + 1 < \tau_i < \infty\},$$

and hence, for all $i = 0, 1, \ldots, N$,

$$\tau_i < 1 + \frac{\max_{j \neq i} \tilde{a}_{ji}}{1 - \varepsilon} \quad \text{on } \{L_i(\varepsilon) + 1 < \tau_i < \infty\},$$

which yields, for every $0 < \varepsilon < 1$,

$$\tau_i \le 1 + \mathbb{1}_{\{\tau_i > L_i(\varepsilon)+1\}} \frac{\max_{j \ne i} \tilde{a}_{ji}}{1-\varepsilon} + \mathbb{1}_{\{\tau_i \le L_i(\varepsilon)+1\}} L_i(\varepsilon) \le 1 + \frac{\max_{j\ne i} \tilde{a}_{ji}}{1-\varepsilon} + L_i(\varepsilon).$$

Since $\mathsf{E}_\theta[L_i(\varepsilon)] < \infty$, the upper bound (5.108) follows. Since $T^* \le \tau_i$ for all $i = 0, 1, \dots, N$, this yields

$$\mathsf{E}_\theta T^* \le \min_i \mathsf{E}_\theta \tau_i \le A(\theta)(1+o(1)) \quad \text{for all } \theta \in \Theta \text{ as } a_{\min} \to \infty, \tag{5.109}$$

and since this upper bound is asymptotically the same as the lower bound (5.104) the proof of (5.102) is complete.

In order to prove (5.103), set $\tilde{a}_{ji} = a_{\min} c_{ji}/I_j(\theta) = a_{\min}/J(\theta)$, where $J(\theta)$ is defined in (5.83). Using Lemma 5.7 in [352], it can be shown that if $J_{\inf}(\theta) > 0$, then for any $\gamma > 0$ there is a value of a such that, for $a_{\min} < a$,

$$\mathsf{E}_\theta[T^*]/a_{\min} \le 1/J(\theta) + \gamma \quad \text{for all } \theta \in \Theta,$$

which implies that

$$\limsup_{a_{\min}\to\infty} \left(\sup_{\theta\in\Theta} \mathsf{E}_\theta[T^*]/a_{\min} \right) \le 1/J_{\inf}.$$

On the other hand, the lower bound (5.104) obviously yields

$$\liminf_{a_{\min}\to\infty} \left(\sup_{\theta\in\Theta} \mathsf{E}_\theta[T^*]/a_{\min} \right) \ge 1/J_{\inf},$$

which along with the previous inequality completes the proof. □

We now use Theorem 5.4.2 to establish first-order asymptotic optimality of the accepting and rejecting adaptive tests in the classes $\mathbf{C}([\alpha_{ij}])$ and $\mathbf{C}(\boldsymbol{\alpha})$ for small error probabilities.

Theorem 5.4.3 (MASLRT asymptotic optimality). *Let the observations $X_1, X_2, \dots$ be iid from the multiparameter exponential family* (5.69). *Assume that the conditions* (5.80) *are satisfied and $\{\hat{\theta}_n\}_{n\ge1}$ is a sequence of estimates of θ such that the condition* (5.84) *holds.*

(i) *If the thresholds a_{ij} are selected so that $\sup_{\theta\in\Theta_i} \alpha_{ij}(\delta^*, \theta) \le \alpha_{ij}$ and $a_{ij} \sim \log(1/\alpha_{ij})$, in particular $a_{ij} = \log(1/\alpha_{ij})$, then as $\alpha_{\max} \to 0$*

$$\inf_{\delta\in\mathbf{C}([\alpha_{ij}])} \mathsf{E}_\theta T \sim \mathsf{E}_\theta T^* \sim \begin{cases} |\log\alpha_{\max}|/J_i(\theta) & \text{for all } \theta \in \Theta_i \text{ and } i = 0, 1, \dots, N, \\ |\log\alpha_{\max}|/J(\theta) & \text{for all } \theta \in \mathrm{I}_{\mathrm{in}}. \end{cases} \tag{5.110}$$

(ii) *If the thresholds $a_{ij} = a_i$ are selected so that $\sup_{\theta\in\Theta_i} \alpha_i(\delta^*, \theta) \le \alpha_i$ and $a_i \sim \log(1/\alpha_i)$, in particular $a_i = \log(N/\alpha_i)$, then as $\tilde{\alpha}_{\max} \to 0$*

$$\inf_{\delta\in\mathbf{C}(\boldsymbol{\alpha})} \mathsf{E}_\theta T \sim \mathsf{E}_\theta T^* \sim \begin{cases} |\log\tilde{\alpha}_{\max}|/J_i^*(\theta) & \text{for all } \theta \in \Theta_i \text{ and } i = 0, 1, \dots, N, \\ |\log\tilde{\alpha}_{\max}|/J^*(\theta) & \text{for all } \theta \in \mathrm{I}_{\mathrm{in}}. \end{cases} \tag{5.111}$$

(iii) *If the condition* (5.81) *is satisfied, then the assertions (ii) and (iii) hold in the worst-case scenario, i.e.,*

$$\begin{aligned} \inf_{\delta\in\mathbf{C}([\alpha_{ij}])} \sup_{\theta\in\Theta} \mathsf{E}_\theta T &\sim \sup_{\theta\in\Theta} \mathsf{E}_\theta T^* \sim |\log\alpha_{\max}|/J_{\inf} \quad \text{as } \alpha_{\max} \to 0; \\ \inf_{\delta\in\mathbf{C}(\boldsymbol{\alpha})} \sup_{\theta\in\Theta} \mathsf{E}_\theta T &\sim \sup_{\theta\in\Theta} \mathsf{E}_\theta T^* \sim |\log\tilde{\alpha}_{\max}|/J^*_{\inf} \quad \text{as } \tilde{\alpha}_{\max} \to 0. \end{aligned} \tag{5.112}$$

Consequently, the accepting MASLRT minimizes asymptotically the expected sample size uniformly in $\theta \in \Theta$ in the classes $\mathbf{C}([\alpha_{ij}])$ and $\mathbf{C}(\boldsymbol{\alpha})$ and also solves the Kiefer–Weiss problem to first order.

Proof. (i) Setting $a_{ij} = \log(1/\alpha_{ij})$ and using Theorem 5.4.2, we immediately obtain asymptotic approximation (5.110) for $\mathsf{E}_\theta T^*$. Denote the right-hand side of (5.110) by $A_\theta(\alpha_{\max})$. In order to show that

$$\inf_{\delta \in \mathbf{C}([\alpha_{ij}])} \mathsf{E}_\theta T \sim A_\theta(\alpha_{\max}) \quad \text{as} \quad \alpha_{\max} \to 0, \tag{5.113}$$

it suffices to prove that

$$\inf_{\delta \in \mathbf{C}([\alpha_{ij}])} \mathsf{E}_\theta T \geq A_\theta(\alpha_{\max})(1+o(1)) \quad \text{as} \quad \alpha_{\max} \to 0. \tag{5.114}$$

Replacing $\sup_{\widetilde{\theta} \in \Theta_j} \alpha_{ji}(\delta, \widetilde{\theta})$ with α_{ji} in the inequality (5.106), we obtain that for every $\theta \in \Theta_i$ and any test in the class $\mathbf{C}([\alpha_{ij}])$

$$\mathsf{E}_\theta T \geq \max_{\substack{0 \le j \le N \\ j \neq i}} \frac{|\log \alpha_{ji}|}{I_j(\theta)} + o(1) \quad \text{as} \quad \alpha_{\max} \to 0,$$

which implies (5.114) for $\theta \in \Theta_i$, $i = 0, 1, \dots, N$.

Next, the same sort of argument that has led to (5.107) along with the inequalities (5.18) yields

$$\inf_{\delta \in \mathbf{C}([\alpha_{ij}])} \mathsf{E}_\theta T \geq \min_{0 \le i \le N} \max_{\substack{0 \le j \le N \\ j \neq i}} \frac{|\log \alpha_{ji}|}{I_j(\theta)} (1+o(1)) \quad \text{for all} \quad \theta \in \mathrm{I}_{\mathrm{in}} \quad \text{as} \quad \alpha_{\max} \to 0,$$

which is clearly the same as (5.114).

(ii) The proof of (ii) is essentially the same as that of (i) and is omitted.

(iii) The inequality (5.114) implies that

$$\liminf_{\alpha_{\max} \to 0} \left(\inf_{\delta \in \mathbf{C}([\alpha_{ij}])} \sup_{\theta \in \Theta} \mathsf{E}_\theta [T] / |\log \alpha_{\max}| \right) \geq 1/J_{\inf},$$

and (5.103) that

$$\lim_{\alpha_{\max} \to 0} \left(\sup_{\theta \in \Theta} \mathsf{E}_\theta [T^*] / |\log \alpha_{\max}| \right) = 1/J_{\inf}$$

whenever the thresholds a_{ij} are selected so that $\sup_{\theta \in \Theta_i} \alpha_{ij}(\delta^*, \theta) \leq \alpha_{ij}$ and $a_{ji} \sim \log(1/\alpha_{ji})$, in particular when $a_{ij} = \log(1/\alpha_{ij})$. Hence, the first assertions in (5.112) follow. The second ones are proved analogously. The proof is complete. □

Pavlov [352] considered the symmetric rejecting MASLRT with $b_i = b$ and proved its asymptotic optimality in the class $\mathbf{C}(\boldsymbol{\alpha})$ with $\alpha_i = t_i \alpha$, where t_i are positive constants (i.e., in the asymptotically symmetric case $c_i = 1$), using different conditions on the behavior of the estimates and more general conditions on the distributions. These more general conditions are not easy to check. Restricting the consideration to the exponential family seems to be a reasonable compromise. The condition (5.84) on the behavior of the estimates $\{\hat{\theta}_n\}_{n \geq 1}$ was proposed by Dragalin and Novikov [127], who considered both the rejecting and the accepting tests in a general asymmetric situation. While our proof is different from that used in [127] that is based on bounding the overshoots, the main ideas are similar. All the results are also valid for continuous-time processes with iid increments from the multidimensional exponential family [127].

While Theorem 5.4.2 was primarily used for proving first-order asymptotic optimality of the MASLRT and the MGSLRT in the frequentist setup, it is of importance in its own right. For example, it may be used for establishing the asymptotic optimality properties of the tests in the decision-theoretic formulation as well. Indeed, assume we are now interested in the Bayesian problem with the cost c per observation and the loss function $L(\theta, d = j) = L_j(\theta)$ if $\theta \in \Theta_i$, $i \neq j$ when making a wrong decision and $L(\theta, d = j) = 0$ if $\theta \in \Theta_j + \mathrm{I}_{\text{in}}$ when making a correct decision and in the indifference zone. Then the average risk assuming the prior distribution $W(\theta)$ is

$$\rho_c^W(\delta) = \sum_{\substack{0 \le i,j \le N \\ i \ne j}} \int_{\Theta_i} L_j(\theta) \mathsf{P}_\theta(d = j) W(\mathrm{d}\theta) + c \int_\Theta \mathsf{E}_\theta[T] W(\mathrm{d}\theta).$$

In particular, if we assume the zero–one loss function setting $L_j(\theta) = 1$, then

$$\rho_c^W(\delta) = \sum_{i=0}^N \int_{\Theta_i} \alpha_i(\theta, \delta) W(\mathrm{d}\theta) + c \int_\Theta \mathsf{E}_\theta[T] W(\mathrm{d}\theta).$$

In the latter case, Theorem 5.4.2 and Lemma 5.4.1 can be used to show that the MASLRT with $a_{ij} = \log(1/c)$ is asymptotically Bayes as $c \to 0$:

$$\inf_\delta \rho_c^W(\delta) \sim \rho_c^W(\delta^*) \sim \rho_c^W(\delta^\star) \sim c|\log c| \int_\Theta \frac{W(\mathrm{d}\theta)}{\widetilde{I}(\theta)}.$$

Finally, note that in the case of two hypotheses the MASLRT is nothing but an adaptive version of the 2-SPRT. Indeed, for $N = 1$ both the accepting and the rejecting MASLRTs coincide and can be written as

$$\begin{aligned} T^* &= \min(T_0^*, T_1^*), \quad d^* = \arg\min_{i=0,1} T_i^*, \\ T_0^* &= \inf\{n : \lambda_n^*(\Theta_1) \ge a_1\}, \quad T_1^* = \inf\{n : \lambda_n^*(\Theta_0) \ge a_0\}, \end{aligned} \tag{5.115}$$

where $\lambda_n^*(\Theta_i) = \log \Lambda_n(\Theta_i)$. We naturally refer to this test as the *Adaptive 2-SPRT* and write 2-ASPRT for brevity.

Asymptotic relations (5.110) and (5.112) of Theorem 5.4.3 yield the following corollary.

Corollary 5.4.1 (2-ASPRT asymptotic optimality). *Let $N = 1$ and let α_i go to zero in such a way that $\lim_{\alpha_{\max} \to 0}(\log \alpha_0 / \log \alpha_1) = c$, where c is bounded away from zero and infinity. Suppose that $I_0(\theta) > 0$ for $\theta \in \Theta_1$, $I_1(\theta) > 0$ for $\theta \in \Theta_0$, and $\min[I_0(\theta), I_1(\theta)] > 0$ for $\theta \in \mathrm{I}_{\text{in}}$ and that the conditions (5.84) hold. If $a_i = \log(1/\alpha_i)$ or the thresholds a_i are selected so that $\sup_{\theta \in \Theta_i} \alpha_i(\delta^*, \theta) \le \alpha_i$ and $a_i \sim \log(1/\alpha_i)$, then as $\alpha_{\max} = \max(\alpha_0, \alpha_1) \to 0$,*

$$\inf_{\delta \in \mathbf{C}(\alpha_0, \alpha_1)} \mathsf{E}_\theta T \sim \mathsf{E}_\theta T^* \sim \begin{cases} |\log \alpha_0| / I_0(\theta) & \text{for all } \theta \in \Theta_1 \\ |\log \alpha_1| / I_1(\theta) & \text{for all } \theta \in \Theta_0, \\ \min_{i=0,1} |\log \alpha_i| / I_i(\theta) & \text{for all } \theta \in \mathrm{I}_{\text{in}}. \end{cases} \tag{5.116}$$

If, in addition, $\inf_{\theta \in \Theta} \max[I_0(\theta), cI_1(\theta)] > 0$, then as $\alpha_{\max} \to 0$,

$$\inf_{\delta \in \mathbf{C}(\alpha_0, \alpha_1)} \sup_{\theta \in \Theta} \mathsf{E}_\theta T \sim \sup_{\theta \in \Theta} \mathsf{E}_\theta T^* \sim \sup_{\theta \in \Theta} \min\{|\log \alpha_0| / I_0(\theta), |\log \alpha_1| / I_1(\theta)\}. \tag{5.117}$$

Therefore, the 2-ASPRT asymptotically minimizes the expected sample size uniformly in $\theta \in \Theta$ in the class of tests $\mathbf{C}(\alpha_0, \alpha_1)$ and also solves the Kiefer–Weiss problem to first order.

Example 5.4.1 (Testing for the Gaussian mean with unknown variance). Consider the Gaussian example assuming that $X_n \sim \mathcal{N}(\mu, \sigma^2)$, $n = 1, 2, \dots$ are iid normal random variables with unknown mean μ and unknown variance σ^2 and the hypotheses are $H_0 : \mu \le \mu_0, \sigma^2 > 0$ and $H_1 : \mu \ge \mu_1, \sigma^2 > 0$, where μ_1, μ_0 are given numbers, $\mu_1 > \mu_0$. The variance σ^2 is a nuisance parameter. In Subsection 3.6.2, this model was treated in the context of invariant tests when the hypotheses are $H_i : \mu / \sigma = q_i$, $i = 0, 1$, where q_0 and q_1 are given numbers. In particular, it was shown that the ISPRT, which is the sequential t-test (t-SPRT), is asymptotically optimal to first order among the tests invariant with respect to the unknown σ^2. However, for different values of $q = \mu / \sigma$ this test is not optimal. An analysis given below shows that it behaves especially poorly in the indifference zone (q_0, q_1). To overcome this drawback in Example 5.3.2 a 2-ISPRT (t-2-SPRT) has been constructed, which minimizes the ESS at the worst point $q^* \in (q_0, q_1)$. But this test is also not optimal for any other point and performs not great at the points located far from q^*, as we will see. In general, the invariant versions of the SPRT are not adaptive and flexible. On the other hand, the GSLRT and the adaptive versions of the SPRT are flexible and asymptotically efficient at any point q.

In the following we consider a specific case $H_0 : \mu = 0$ and $H_1 : \mu \geq \mu_1$ ($\mu_1 > 0$). This problem is of special interest in certain applications. For example, when detecting targets in noise the observations have the form $X_n = \mu + V_n$ if there is a target and $X_n = V_n$ if there is no target. The value of μ, $\mu > 0$ characterizes the intensity of the signal from the target; V_n is sensor noise or clutter plus noise. Assuming that $\{V_n\}_{n \geq 1}$ is a zero-mean white Gaussian process with unknown variance σ^2, we arrive at this problem. In radar applications, μ usually represents the result of the preprocessing by attenuation and matched filtering of the modulated pulses and, also, is not known. The value of $\mu_1 > 0$ is a prespecified limit or cut-off intensity of the target. In this interpretation the value of $q = \mu/\sigma$ represents an unknown signal-to-noise ratio and $q_1 = \mu_1/\sigma$ is a given cut-off SNR level . Thus, we are dealing with the two-hypotheses problem ($N = 1$) for the two-dimensional exponential model with parameter $\theta = (\mu, \sigma^2)$ and parameter space $\Theta = [0, \infty) \times (0, \infty)$. Also, $\Theta_0 = \{0\} \times (0, \infty)$, $\Theta_1 = [\mu_1, \infty) \times (0, \infty)$, and $\mathrm{I}_{\mathrm{in}} = (0, \mu_1) \times (0, \infty)$.

We now show that all conditions of Corollary 5.4.1 are satisfied when $\{\hat{\theta}_n\}$ is a sequence of MLEs, which implies uniform asymptotic optimality of the 2-ASPRT with $a_i = \log(1/\alpha_i)$, $i = 0, 1$.

Let $\widetilde{\theta} = (\widetilde{\mu}, \widetilde{\sigma}^2)$, where $\widetilde{\mu}$ and $\widetilde{\sigma}$ are arbitrary numbers, $\widetilde{\mu} \geq 0$, $\widetilde{\sigma} > 0$. Then the LLR $\lambda_n(\theta, \widetilde{\theta}) = \ell_n(\theta) - \ell_n(\widetilde{\theta}) = \sum_{k=1}^n \log[p_\theta(X_k)/p_{\widetilde{\theta}}(X_k)]$ is given by

$$\begin{aligned}\lambda_n(\theta, \widetilde{\theta}) &= \frac{n}{2} \log\left(\frac{\widetilde{\sigma}^2}{\sigma^2}\right) + \frac{\sigma^2 - \widetilde{\sigma}^2}{2\widetilde{\sigma}^2\sigma^2} \sum_{k=1}^n X_k^2 \\ &\quad + \frac{\mu\widetilde{\sigma}^2 - \widetilde{\mu}\sigma^2}{\widetilde{\sigma}^2\sigma^2} \sum_{k=1}^n X_k - \frac{\mu^2\widetilde{\sigma}^2 - \widetilde{\mu}^2\sigma^2}{2\widetilde{\sigma}^2\sigma^2} n.\end{aligned} \tag{5.118}$$

Using (5.118), it is not difficult to show that

$$I(\theta, \widetilde{\theta}) = \frac{1}{2}\left\{[(\mu - \widetilde{\mu})^2 + \sigma^2]/\widetilde{\sigma}^2 + \log(\widetilde{\sigma}^2/\sigma^2) - 1\right\}. \tag{5.119}$$

The minimum

$$\min_{\widetilde{\sigma}>0} I(\theta, \widetilde{\theta}) = \frac{1}{2} \log\left[1 + (\mu - \widetilde{\mu})^2/\sigma^2\right]$$

is achieved at the point $\widetilde{\sigma}^* = [\sigma^2 + (\mu - \widetilde{\mu})^2]^{1/2}$ and $I_1(\theta) = \min_{\widetilde{\mu} \geq \mu_1} \min_{\widetilde{\sigma}>0} I(\theta, \widetilde{\theta})$ and $I_0(\theta) = \min_{\widetilde{\mu} \in \{0\}} \min_{\widetilde{\sigma}>0} I(\theta, \widetilde{\theta})$ equal

$$\begin{aligned} I_1(q) &= \begin{cases} \frac{1}{2}\log[1 + (q_1 - q)^2] & \text{for } 0 \leq q < q_1 \\ 0 & \text{for } q \geq q_1 \end{cases}, \\ I_0(q) &= \frac{1}{2}\log(1 + q^2) \quad \text{for } q \geq 0, \end{aligned} \tag{5.120}$$

where $q = \mu/\sigma$ and $q_1 = \mu_1/\sigma$. Thus, as expected, the initial two-dimensional hypothesis testing problem is reduced to the equivalent single-parameter testing problem $H_0 : q = 0$ against $H_1 : q \geq q_1$ with the parameter space $Q = [0, \infty)$ and subsets $Q_0 = \{0\}$, $Q_1 = [q_1, \infty)$, $\mathrm{I}_{\mathrm{in}} = (0, q_1)$.

Clearly, $I_0(q) > 0$ for $q \in Q_1 + \mathrm{I}_{\mathrm{in}} = (0, \infty)$ and $I_1(q) > 0$ for $q \in Q_0 + \mathrm{I}_{\mathrm{in}} = [0, q_1)$, and hence, $\min[I_0(q), I_1(q)] > 0$ for $q \in \mathrm{I}_{\mathrm{in}} = (0, q_1)$. Also, it is easily verified that $\inf_{q \in Q} \max[I_0(q), I_1(q)c] > 0$ for any $0 < c < \infty$. In fact, the maximum is attained at the point $q^* \in (0, q_1)$ for which $I_0(q^*) = I_1(q^*)c$, and it is a solution of the equation

$$(1 + q^2)^{1/c} = 1 + (q_1 - q)^2. \tag{5.121}$$

In particular, $q^* = q_1/2$ and $\inf_{q\in Q}\max[I_0(q), I_1(q)c] = \log(1+q_1^2/4)$ for $c=1$. Therefore, the conditions related to the minimal K–L distances for the corresponding sets hold, and it remains to deal with the condition (5.84) related to the rate of convergence of an estimate $\hat{\theta}_n$ to the true parameter value.

Not surprisingly we choose $\hat{\theta}_n = (\hat{\mu}_n, \hat{\sigma}_n^2)$ as the maximum likelihood estimator,

$$(\hat{\mu}_n, \hat{\sigma}_n^2) = \arg \sup_{\substack{\mu\geq 0,\\ \sigma^2>0}} \lambda_n(\mu, \sigma^2, \widetilde{\mu}, \widetilde{\sigma}^2),$$

which is of course a combination of the positive part of the sample mean and sample variance, i.e.,

$$\hat{\mu}_n = \max\{0, \overline{X}_n\}, \quad \hat{\sigma}_n^2 = n^{-1}\sum_{k=1}^{n}(X_k - \hat{\mu}_n)^2,$$

where $\overline{X}_n = n^{-1}\sum_{k=1}^n X_k$ is the sample mean. After certain manipulations it can be shown that $\mathsf{E}_\theta[I(\theta, \hat{\theta}_n)]^2 = O(n^{-2})$, so that the required condition holds with $r=2$.

Therefore, all conditions of Corollary 5.4.1 are satisfied. According to this corollary, the 2-ASPRT is uniformly (point-wise) asymptotically optimal to first order, and the relations (5.116), (5.117), and (5.120) can be used to compute its asymptotic performance. Specifically,

$$\inf_{\delta\in\mathbf{C}(\alpha_0,\alpha_1)} \mathsf{E}_\theta T \sim \mathsf{E}_\theta T^* \sim \begin{cases} |\log\alpha_0|/I_0(q) & \text{if } q\geq q_1 \\ |\log\alpha_1|/I_1(q) & \text{if } q=0 \\ \min_{i=0,1}|\log\alpha_i|/I_i(q) & \text{if } 0<q<q_1, \end{cases} \tag{5.122}$$

where the $I_i(q)$'s are given by (5.120). Denoting by $\mathsf{ESS}_q(T^*) = \mathsf{E}_{\mu,\sigma^2}T^*$ the ESS of the 2-ASPRT, (5.122) yields the approximate relations

$$\begin{aligned} \mathsf{ESS}_q(T^*) &\approx \frac{2|\log\alpha_1|}{\log[1+(q_1-q)^2]} \quad \text{for } 0\leq q\leq q^*, \\ \mathsf{ESS}_q(T^*) &\approx \frac{2|\log\alpha_0|}{\log(1+q^2)} \quad \text{for } q\geq q^*, \end{aligned} \tag{5.123}$$

where $q^*\in(0,q_1)$ is the point of the (asymptotic) maximum of $\mathsf{ESS}_q(T^*)$, which is the solution of the equation

$$\frac{|\log\alpha_1|}{\log[1+(q_1-q)^2]} = \frac{|\log\alpha_0|}{\log(1+q^2)} \tag{5.124}$$

in the interval $(0,q_1)$. Of course equation (5.124) is equivalent to (5.121) with $c = |\log\alpha_0|/|\log\alpha_1|$. Thus, we also have

$$\inf_{\delta\in\mathbf{C}(\alpha_0,\alpha_1)} \sup_{\theta\in\Theta} \mathsf{E}_\theta T \sim \mathsf{ESS}_{q^*}(T^*) \sim \frac{2|\log\alpha_0|}{\log(1+(q^*)^2)} \quad \text{as } \alpha_{\max}\to 0.$$

Let $\mu_{n,1} = \max\{\mu_1, \overline{X}_n\}$, $\sigma_{n,1}^2 = n^{-1}\sum_{k=1}^n (X_k-\mu_{n,1})^2$, $\sigma_{n,0}^2 = n^{-1}\sum_{k=1}^n X_k^2$. Obviously, $(\mu_{n,1}, \sigma_{n,1}^2)$ and $\sigma_{n,0}^2$ are the conditional maximum likelihood estimators, conditioned on the hypotheses H_1 and H_0, respectively. That is,

$$(\mu_{n,1}, \sigma_{n,1}^2) = \arg\sup_{\substack{\mu\geq\mu_1,\\ \sigma^2>0}} \lambda_n(\mu, \sigma^2, \widetilde{\mu}, \widetilde{\sigma}^2), \quad \sigma_{n,0}^2 = \arg\sup_{\sigma^2>0} \lambda_n(\mu=0, \sigma^2, \widetilde{\mu}, \widetilde{\sigma}^2).$$

Without loss of generality we may take $\widetilde{\mu} = 0, \widetilde{\sigma}^2 = 1$ to obtain

$$\lambda_n^*(\widetilde{\theta}) = \sum_{k=1}^{n} \left[\frac{1}{2} \log\left(\frac{1}{\hat{\sigma}_{k-1}^2}\right) + \frac{\hat{\sigma}_{k-1}^2 - 1}{2\hat{\sigma}_{k-1}^2} X_k^2 + \frac{\hat{\mu}_{k-1} X_k}{\hat{\sigma}_{k-1}^2} - \frac{\hat{\mu}_{k-1}^2}{2\hat{\sigma}_{k-1}^2} \right],$$
$$\lambda_n^1(\widetilde{\theta}) = n \left[\frac{1}{2} \log\left(\frac{1}{\sigma_{n,1}^2}\right) + \frac{(\sigma_{n,1}^2 - 1)\sigma_{n,0}^2}{2\sigma_{n,1}^2} + \frac{\mu_{n,1} \overline{X}_n}{\sigma_{n,1}^2} - \frac{\mu_{n,1}^2}{2\sigma_{n,1}^2} \right], \tag{5.125}$$
$$\lambda_n^0(\widetilde{\theta}) = \frac{n}{2} \left[\log\left(\frac{1}{\sigma_{n,0}^2}\right) + \sigma_{n,0}^2 - 1 \right],$$

where we used the notation $\lambda_n^*(\widetilde{\theta}) = \ell_n - \ell_n(\widetilde{\theta})$, $\lambda_n^i(\widetilde{\theta}) = \ell_n^i - \ell_n(\widetilde{\theta})$. These statistics allow for an efficient recursive computation. Using (5.125), we can now easily compute the statistics $\lambda_n^*(\Theta_i) = \lambda_n^*(\widetilde{\theta}) - \lambda_n^i(\widetilde{\theta})$, and therefore, implement the 2-ASPRT (5.115). Note that λ_1^* requires initial conditions for the estimates $\hat{\theta}_0 = (\hat{\mu}_0, \hat{\sigma}_0^2)$. In general, these conditions are the design parameters that can be deterministic or even random. Setting $\lambda_1^* = 0$ is a particular and typical choice.

We now compare the 2-ASPRT with the t-SPRT and the t-2-SPRT described in Subsections 3.6.2 and 5.3.1 in detail. For the t-SPRT $\delta_t = (T_t, d_t)$, according to (3.229)

$$\mathsf{E}_{q_1} T_t \sim 2|\log \alpha_0| / \log(1 + q_1^2), \quad \mathsf{E}_0 T_t \sim 2|\log \alpha_1| / q_1^2. \tag{5.126}$$

To obtain the expression for the ESS at any point q, we note that for an arbitrary value of $q \geq 0$ the normalized invariant LLR $n^{-1}\lambda_n$ converges P_q-r-quickly to $g(q_1 q/(1+q^2)^{1/2})$ as $n \to \infty$, where $g(t)$ is defined in (3.223). As a result, we have

$$\mathsf{E}_q T_t \sim \begin{cases} |\log \alpha_1| / |g(q_1 q/(1+q^2)^{1/2})| & \text{for } 0 \leq q < q_t^* \\ |\log \alpha_0| / g(q_1 q/(1+q^2)^{1/2}) & \text{for } q > q_t^* \end{cases} \quad \text{as } \alpha_{\max} \to 0, \tag{5.127}$$

where $q_t^* \in (0, q_1)$ is the unique root of the equation

$$g\left(q q_1 / \sqrt{1+q^2}\right) = 0. \tag{5.128}$$

To determine the ESS at the point q_t^*, we apply Theorem 3 of Lai [245] and obtain

$$\mathsf{E}_{q_t^*} T_t \sim C |\log \alpha_0|^2 (1+c)^2 / c^2, \tag{5.129}$$

where C is a constant that depends on c and q_1 and can be computed numerically.

The asymptotic relative efficiency of the t-SPRT δ_t with respect to the 2-ASPRT δ^* is defined as the limit of the ratio of the ESSs:

$$\mathsf{ARE}_q(\delta_t : \delta^*) = \lim_{\alpha_{\max} \to 0} [\mathsf{E}_q T_t / \mathsf{E}_q T^*].$$

Recall that q^* is the root of the equation (5.124) and q_t^* is the root of the equation (5.128). Assume that $q_t^* < q^*$. Then, using (5.123)–(5.129), we get

$$\mathsf{ARE}_q(\delta_t : \delta^*) = \begin{cases} \log[(1 + (q_1 - q)^2] / 2|g(q_1 q/(1+q^2)^{1/2})| & \text{if } 0 \leq q < q_t^* \\ \log[(1 + (q_1 - q)^2] c / 2g(q_1 q/(1+q^2)^{1/2}) & \text{if } q_t^* < q \leq q^* \\ \log(1+q^2) / 2g(q_1 q/(1+q^2)^{1/2}) & \text{if } q > q^* \\ \infty & \text{if } q = q_t^* . \end{cases}$$

Practically speaking, this formula shows that in the vicinity of the point q_t^* the ESS of the t-SPRT is much bigger than that of the 2-ASPRT. The ARE as a function of q for $q_1 = 1$ is shown in Figure 5.3.

It is seen that the 2-ASPRT is substantially more efficient than the t-SPRT for all values of q except for $q \in [0, 0.3)$ and in the vicinity of $q = q_1 = 1$. In the interval $q \in [0, 0.25)$, the t-test outperforms the 2-ASPRT. This is not surprising, since the former test is specifically designed to be optimal at the points $q = 0$ and $q = q_1$. However, if the true value of q differs from q_1, the t-SPRT loses its optimality property very quickly. Let us compare the tests at the points of optimality of the t-SPRT $q = 0$ and $q = q_1$ in more detail. Using (5.126) yields $\mathsf{ARE}_0 = q_1^{-2} \log(1 + q_1^2)$ and $\mathsf{ARE}_{q_1} = 1$. Therefore, at the point $q = q_1$ both tests have exactly the same asymptotic performance, while at the point $q = 0$ the performance of the t-SPRT is better. The efficiency of this test increases when q_1 increases and tends to 1 when $q_1 \to 0$.

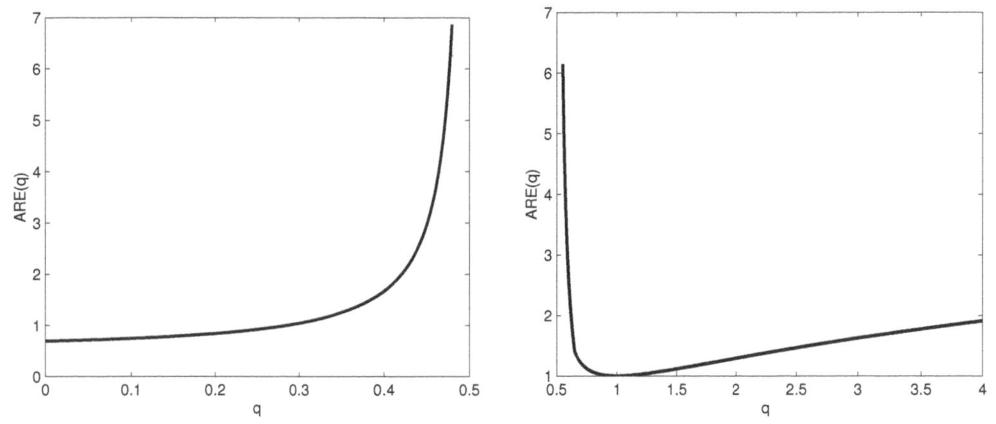

Figure 5.3: *Asymptotic relative efficiency* $\mathsf{ARE}_q(\delta_t : \delta^*)$ *as a function of* q *(*$q_0 = 0, q_1 = 1$*). The plot is split into two parts since in the vicinity of the point* q_t^* *the ARE goes to infinity.*

We now turn to comparing with the invariant 2-SPRT test (t-2-SPRT) considered in Example 5.3.2. The normalized LLRs $n^{-1}\ell_i^{q^\star}(n) = n^{-1} \log[J_n(q^\star t_n)/J_n(\theta_i t_n)]$, where $J_n(z)$ is defined in (3.221), converge r-quickly under P_q as $n \to \infty$ to the numbers

$$I_i^\star(q^\star, q) = \phi(q^\star q/(1+q^2)^{1/2}) - \phi(q_i q/(1+q^2)^{1/2}) - \frac{1}{2}[(q^\star)^2 - q_i^2], \quad i = 0, 1, \qquad (5.130)$$

which can be interpreted as the analogs of the K–L numbers ($\phi(z)$ is defined in (3.223)). As a result, the t-2-SPRT $\delta^\star(q^\star) = (T^\star, d^\star)$ tuned to the worst point $q^\star$ is asymptotically optimal at $q^\star$, i.e., asymptotically minimax among all invariant tests belonging to the class $\mathbf{C}(\alpha_0, \alpha_1)$. This worst point $q^\star$ is the unique root of the equation $cI_1^\star(q) = I_0^\star(q)$, $q \in (q_0, q_1)$, where we write $I_i^\star(q) = I_i^\star(q, q)$, and as $\alpha_{\max} \to 0$

$$\mathsf{E}_q[T^\star(q^\star)] \sim \begin{cases} |\log \alpha_1|/I_1^\star(q^\star, q) & \text{if } q < q^\star \\ |\log \alpha_0|/I_0^\star(q^\star, q) & \text{if } q > q^\star \\ |\log \alpha_1|/I_1^\star(q^\star) = |\log \alpha_0|/I_0^\star(q^\star) & \text{if } q = q^\star. \end{cases} \qquad (5.131)$$

Recall that $c = \lim_{\alpha_{\max} \to 0} |\log \alpha_0/|\log \alpha_1|$. Also, in the special scenario that we consider where $q_0 = 0$ and $H_0 : q = 0$, using (5.130) and the fact that $\phi(0) = \log 2$, we obtain

$$I_0^\star(q^\star, q) = \phi(q^\star q/(1+q^2)^{1/2}) - \log 2 - (q^\star)^2/2, \quad I_1^\star(q^\star, 0) = [q_1^2 - (q^\star)^2]/2.$$

These expressions can be used to compute the ESS at the points $q = 0$ and $q = q_1$, which along with the point $q^\star$ are of special interest.

The first-order approximations to the expected sample sizes of 2-ASPRT and t-2-SPRT given by formulas (5.123) and (5.130)–(5.131) are depicted in Figure 5.4 for the error probabilities $\alpha_0 = 10^{-3}$ and $\alpha_1 = 10^{-1}$, and $q_1 = 1$. The maximum value of the 2-ASPRT's ESS $\mathsf{E}_{q^*} T^* \approx 39.4$ is attained at the point $q^* \approx 0.6479$, which is the solution of the equation (5.121), and the maximum value of the t-2-SPRT's ESS, $\mathsf{E}_{q^\star} T^\star \approx 41.08$, is attained at the point $q^\star \approx 0.6323$, which is the solution of the equation (5.128). Note that $c = |\log \alpha_0|/|\log \alpha_1| = 3$ in the case considered. Table 5.2 contains more exact numbers for the ESSs as well as the asymptotic relative efficiency $\mathsf{ARE}_q(\delta^\star; \delta^*) = \lim_{\alpha_{\max} \to 0} [\mathsf{E}_q T^\star / \mathsf{E}_q T^*]$. It is seen that the 2-ASPRT is uniformly better. The difference for $0 \le q \le 1$ is relatively small, while it becomes visible for $q > q_1 = 1$, as expected. Note that we choose the probability α_0 to be much smaller than α_1, keeping in mind the target detection applications where α_0 is the false alarm probability and α_1 is the misdetection probability.

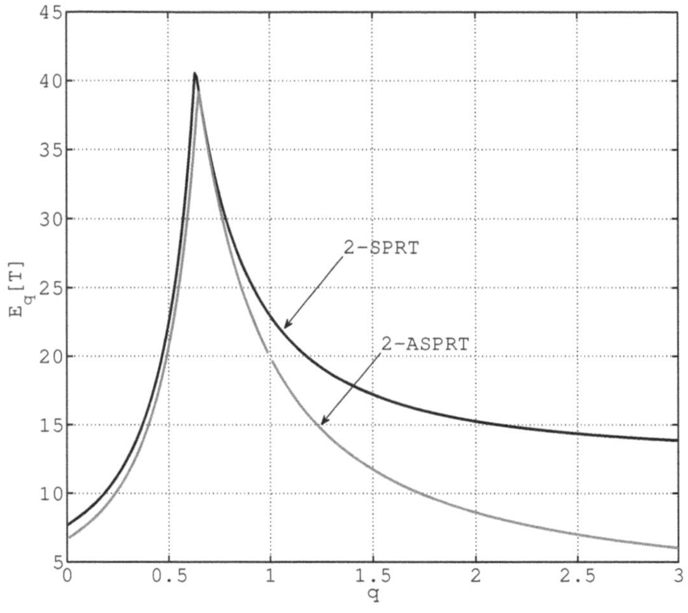

Figure 5.4: *Asymptotic approximations for the expected sample sizes of the t-2-SPRT and the 2-ASPRT as functions of q* ($\alpha_0 = 10^{-3}$, $\alpha_1 = 10^{-1}$, $q_1 = 1$).

Table 5.2: *The ESSs and the ARE of the t-2-SPRT w.r.t. the 2-ASPRT as functions of q* ($\alpha_0 = 10^{-3}$, $\alpha_1 = 10^{-1}$, $q_1 = 1$; $q = 0.65/0.63$ *denotes the least favorable points for the 2-ASPRT / t-2-SPRT*).

q	0.0	0.25	0.5	0.65 / 0.63	0.75	1.0	1.5	2.0	2.5	3.0
$\mathsf{E}_q T^*$	6.64	10.32	20.64	39.40	30.96	19.93	11.72	8.58	6.97	6.00
$\mathsf{E}_q T^\star$	7.67	11.42	22.38	41.08	31.73	22.96	17.21	15.25	14.34	13.85
ARE_q	1.16	1.11	1.08	1.04	1.03	1.15	1.46	1.78	2.06	2.31

The fact that the 2-ASPRT performs slightly better at the worst point than the t-2-SPRT, while both tests are asymptotically minimax solving the Kiefer–Weiss problem, is not surprising since the 2-ASPRT is asymptotically optimal in the class of all tests with upper bounds on the error probabilities α_0, α_1 but the t-2-SPRT is asymptotically optimal only in the class of invariant tests with the same error probability constraints.

5.4.3 Mixture-Based Sequential Likelihood Ratio Tests

Yet another approach is to use mixtures of LRs (weighted LRs) in test constructions, as was briefly discussed in the introduction. This approach was proposed by Wald [494] in his seminal work on the SPRT and its extensions to two composite hypotheses.

Define the weighted LRs

$$\bar{\Lambda}_n^i = \frac{\int_\Theta \prod_{k=1}^n p_\theta(X_k)\, W(\mathrm{d}\theta)}{\int_{\Theta_i} \prod_{k=1}^n p_\theta(X_k) W_i(\mathrm{d}\theta)}, \quad i = 0,1,\dots,N, \tag{5.132}$$

where the weight functions $W(\theta)$, $W_i(\theta)$, $i = 0,1,\dots,N$, are not necessarily normalized to 1. If the weights are normalized to 1, then they can be regarded as probability distributions. Let $\{a_{ij}\}$ $(i \neq j)$ be positive numbers. The multihypothesis weighted SPRT (MWSPRT) $\bar{\delta} = (\bar{T},\bar{d})$ is of the form

$$\text{Stop at the first } n \geq 1 \text{ such that for some } i \quad \bar{\Lambda}_n^j \geq e^{a_{ji}} \quad \text{for all } j \neq i \tag{5.133}$$

and accept the H_i that satisfies these inequalities. Taking logarithms and writing

$$\bar{\ell}_n = \log \int_\Theta \prod_{k=1}^n p_\theta(X_k)\, W(\mathrm{d}\theta), \quad \bar{\ell}_n^i = \log \int_{\Theta_i} \prod_{k=1}^n p_\theta(X_k) W_i(\mathrm{d}\theta),$$

the MWSPRT can be also expressed as

$$\bar{T} = \min_{0 \leq i \leq N} \bar{T}_i, \qquad \bar{d} = i \quad \text{if} \quad \bar{T} = \bar{T}_i \tag{5.134}$$

where

$$\bar{T}_i = \inf\left\{ n \geq 1 : \bar{\ell}_n \geq \max_{\substack{0 \leq j \leq N \\ j \neq i}} [\bar{\ell}_n^j + a_{ji}] \right\}, \quad i = 0,1,\dots,N. \tag{5.135}$$

Using an argument analogous to that in Subsection 5.4.1.2, it may be shown that Theorem 5.4.1 holds for the MWSPRT, i.e., that the MWSPRT is uniformly asymptotically first-order optimal for the exponential family under quite general assumptions on the weight functions. This is true, for example, when $W(\theta)$ and $W_i(\theta)$ have continuous densities $w(\theta)$ and $w_i(\theta)$ positive on Θ and Θ_i, respectively. The main bottleneck is again finding the upper bounds on the maximal error probabilities $\alpha_{ij}(\bar{\delta})$ of the mixture-based tests. So far there is no prescription how this can be done in general.

Example 5.4.2 (χ^2-SPRT). Let X_n, $n \geq 1$, be a sequence of ℓ-dimensional random vectors with distribution $\mathcal{L}(X) = \mathcal{N}(\theta,\mathbb{I})$. Consider the problem of sequential testing between the two hypotheses

$$H_0 = \{\theta : \|\theta\|_2 = a\} \text{ and } H_1 = \{\theta : \|\theta\|_2 = b\}, \text{ where } b > a. \tag{5.136}$$

In Subsection 2.10.1 it is shown how Wald's idea of weighted LR test leads to an optimal solution, namely, to the UBCP test in the fixed sample size case. By analogy with Subsection 2.10.1, we start with the choice of the weight functions $w_i(\theta)$: these are constant functions concentrated on the spheres $S_0 = \{\theta : \|\theta\|_2 = a\}$ and $S_1 = \{\theta : \|\theta\|_2 = b\}$. Note that $w_i(\theta)$ can be regarded as densities of the least favorable prior distributions $W_i(\theta)$, which are uniform distributions on the spheres S_i. In Subsection 2.10.2 it is shown that $\|\overline{X}_n\|_2^2$, with $\overline{X}_n = 1/n\sum_{k=1}^n X_k$, is a sufficient statistic for testing between the hypotheses H_0 and H_1. Since $\mathcal{L}(\sqrt{n}\,\overline{X}_n) = \mathcal{N}(\sqrt{n}\theta,\mathbb{I})$, it follows that

$$\mathcal{L}(n\|\overline{X}_n\|_2^2) = \chi^2(\ell, n\|\theta\|_2^2) \tag{5.137}$$

where $\chi^2(\ell,c)$ is the χ^2-distribution with ℓ degrees of freedom and noncentrality parameter c. Therefore, after transformation the initial problem (5.136) can be reduced to the following hypothesis testing problem concerning the noncentrality parameter c of the χ^2-distribution:

$$H_0 : c = a^2 \text{ and } H_1 : c = b^2, \text{ where } b > a. \tag{5.138}$$

In other words, for solving the initial problem we first have to transform the initial data $(X_1,\dots,X_n)$ into the sequence of sufficient statistics $(\|\overline{X}_1\|_2^2, 2\|\overline{X}_2\|_2^2,\dots,n\|\overline{X}_n\|_2^2)$. Second, we have to compute the LLR

$$\widetilde{\lambda}_n = \log \frac{f_{b^2}(\|\overline{X}_1\|_2^2, 2\|\overline{X}_2\|_2^2,\dots,n\|\overline{X}_n\|_2^2)}{f_{a^2}(\|\overline{X}_1\|_2^2, 2\|\overline{X}_2\|_2^2,\dots,n\|\overline{X}_n\|_2^2)}, \tag{5.139}$$

where f_c is the joint density of the statistics $(\|\overline{X}_1\|_2^2, 2\|\overline{X}_2\|_2^2,\dots,n\|\overline{X}_n\|_2^2)$. It turns out that direct computation of this ratio using the joint densities is very difficult, and that it is of key importance to use the following factorization:

$$f_c(\|\overline{X}_1\|_2^2, 2\|\overline{X}_2\|_2^2,\dots,n\|\overline{X}_n\|_2^2) = f_c(n\|\overline{X}_n\|_2^2)\, \beta(\|\overline{X}_1\|_2^2, 2\|\overline{X}_2\|_2^2,\dots,n\|\overline{X}_n\|_2^2), \tag{5.140}$$

which follows from Cox's theorem [209]. Using this factorization, we obtain from Example 2.9.4 that the LLR of the hypotheses (5.138) can be written as

$$\widetilde{\lambda}_n = -n\,\frac{b^2-a^2}{2} + \log {}_0F_1\left(\frac{\ell}{2}, \frac{b^2 n^2 \chi_n^2}{4}\right) - \log {}_0F_1\left(\frac{\ell}{2}, \frac{a^2 n^2 \chi_n^2}{4}\right), \quad \chi_n^2 = \|\overline{X}_n\|_2^2, \tag{5.141}$$

where ${}_0F_1(\kappa, x)$ is the generalized hypergeometric function (2.291). Let us define the following SPRT

$$\widetilde{\delta} = (\widetilde{d}, \widetilde{T}), \quad \widetilde{T} = \inf\left\{n \ge 0 : \widetilde{\lambda}_n \notin (-a_0, a_1)\right\}, \quad \widetilde{d} = \begin{cases} 1 & \text{if } \widetilde{T} \ge a_1 \\ 0 & \text{if } \widetilde{T} \le -a_0. \end{cases} \tag{5.142}$$

Note that the LLR of the χ^2-distribution is a monotone function of the sufficient statistic $\|\overline{X}_n\|_2^2$. For this reason the sequential test given by (5.142) is called the *sequential* χ^2*-test* or χ^2-SPRT.

We now extend the previous results to the case of a general covariance matrix. Assume we have a sequence of ℓ-dimensional random vectors X_n, $n \ge 1$, with distribution $\mathcal{L}(X) = \mathcal{N}(\theta, \Sigma)$. Consider the problem of testing between the hypotheses $H_0 : \theta^\top \Sigma^{-1}\theta = a^2$ and $H_1 : \theta^\top \Sigma^{-1}\theta = b^2$, where $b > a$. It is shown in Subsection 2.10.2 that it is possible to transform the hypothesis testing problem with a general covariance matrix into the previous one. The formula for the LLR (5.141) holds true, but the χ_n^2 statistic (5.141) must be replaced with $\chi_n^2 = \overline{X}_n^\top \Sigma^{-1} \overline{X}_n$.

Extending the argument in [321, 329] it can be verified that for any $r > 0$ and $\varepsilon > 0$

$$\sum_{n=1}^{\infty} n^{r-1} \mathsf{P}_1\left\{|n^{-1}\widetilde{\lambda}_n - (b-a)^2/2| > \varepsilon\right\} < \infty, \quad \sum_{n=1}^{\infty} n^{r-1} \mathsf{P}_0\left\{|n^{-1}\widetilde{\lambda}_n + (b-a)^2/2| > \varepsilon\right\} < \infty,$$

so that the r-complete convergence conditions hold for all $r > 0$ with $\psi(n) = n$ and $I_1 = I_0 = (b-a)^2/2$. By Theorem 3.4.2(iii), the invariant χ^2-SPRT asymptotically minimizes all positive moments of the stopping time and, as $\alpha_{\max} \to 0$,

$$\begin{aligned} \inf_{\delta \in \mathbf{C}(\alpha_0,\alpha_1)} \mathsf{E}_1 T^m &\sim \mathsf{E}_1[T^*]^m \sim \left(\frac{2|\log \alpha_0|}{(b-a)^2}\right)^m, \\ \inf_{\delta \in \mathbf{C}(\alpha_0,\alpha_1)} \mathsf{E}_0 T^m &\sim \mathsf{E}_0[T^*]^m \sim \left(\frac{2|\log \alpha_1|}{(b-a)^2}\right)^m. \end{aligned} \tag{5.143}$$

We now present the results of the comparison of the ESS of the χ^2-SPRT, $\mathsf{E}_i\widetilde{T}$, $i = 0,1$, with the sample size of the best possible χ^2 FSS test, the so-called UBCP test for testing between the hypotheses $H_0 : \theta = 0$ and $H_1 : \theta^\top \Sigma^{-1}\theta = b^2$, $b > 0$ (see Subsection 2.10.1). Consider the symmetric case $\alpha_0 = \alpha_1 = \alpha$. Note that the χ^2 FSS test is based on thresholding the statistic $\chi_n^2 = \overline{X}_n^\top \Sigma^{-1} \overline{X}_n$, and it is a uniformly best constant power test on the surface $S_b = \left\{\theta^\top \Sigma^{-1}\theta = b^2\right\}$, i.e., the best FSS test one can get. By (2.337), in order to guarantee the same error probabilities in the FSS test

$\alpha_{0n}(\theta=0)=\alpha_0$ and $\alpha_{1n}(\theta:\theta^\top\Sigma^{-1}\theta=b^2)=\alpha_1$ at the points $\theta=0$ and $\theta:\theta^\top\Sigma^{-1}\theta=b^2$ as in the SPRT, the sample size $N=N_{\text{FSS}}(\alpha)$ should be chosen as

$$N_{\text{FSS}}(\alpha)=\min\{n\ge 1:\alpha_{0n}(h_n)=\alpha_{1n}(h_n)\le\alpha\} \tag{5.144}$$

where the threshold h_n is the unique solution of the equation

$$\int_{h_n}^{\infty}p_0(x)\mathrm{d}x=\int_0^{h_n}p_0(x)e^{-\frac{nb^2}{2}}{}_0F_1\left(\frac{\ell}{2},\frac{nb^2x}{4}\right)\mathrm{d}x,\;\; p_0(x)=\frac{x^{\frac{\ell}{2}-1}e^{-\frac{x}{2}}}{2^{\frac{\ell}{2}}\Gamma\left(\frac{n}{2}\right)}. \tag{5.145}$$

Define the relative efficiency of the SPRT and the FSS test as the ratio $\mathcal{E}(\alpha)=\max_{i=0,1}\{\mathsf{ESS}_i(\alpha)/N_{\text{FSS}}(\alpha)\}$. Table 5.3 contains numerical data for $\ell=1,5,10$ and $\alpha=10^{-1},10^{-2},10^{-3}$, obtained by using the equations (5.143)–(5.145) and from a 10^5-repetition Monte Carlo (MC) simulation. It is seen that for a relatively small error probability the efficiency of the sequential test is about twice better than that of the FSS test.

Table 5.3: *The relative efficiency of the SPRT and the FSS test for $a=0$, $b=0.5$.*

	$\ell=1,\ \alpha_0=\alpha_1=\alpha$			$\ell=5,\ \alpha_0=\alpha_1=\alpha$			$\ell=10,\ \alpha_0=\alpha_1=\alpha$		
$\mathsf{ESS}_i(\alpha)$	10^{-1}	10^{-2}	10^{-3}	10^{-1}	10^{-2}	10^{-3}	10^{-1}	10^{-2}	10^{-3}
$\mathsf{E}_i\widetilde{T}$ (5.143)	18.42	36.84	55.26	18.42	36.84	55.26	18.42	36.84	55.26
$\mathsf{E}_0\widetilde{T}$ MC	27.43	53.45	76.60	42.36	77.06	106.4	52.76	93.63	127.4
$\mathsf{E}_1\widetilde{T}$ MC	23.10	44.28	63.20	38.13	66.56	89.92	48.21	82.01	108.8
$N_{\text{FSS}}(\alpha)$	34	96	163	55	134	211	69	160	245
$\mathcal{E}(\alpha)$	0.81	0.56	0.47	0.77	0.57	0.51	0.77	0.59	0.52

5.4.4 Generalization to the Non-iid Case

We now generalize previous results obtained for iid models to the general non-iid case along the lines of Section 4.3 and Subsection 5.3.1. To be specific, consider the general continuous- or discrete-time multihypothesis problem assuming that the observed process X_t and the parametric probability measure P_θ are associated with the filtered probability space $(\Omega,\mathscr{F},\mathscr{F}_t,\mathsf{P}_\theta)$ with the usual monotonicity and right-continuity properties, where $t=n\in\mathbb{Z}_+=\{0,1,\dots\}$ or $t\in\mathbb{R}_+=[0,\infty)$. As before, the l-dimensional parameter $\theta=(\theta_1,\dots,\theta_l)$ belongs to a subset Θ of the Euclidean space $\mathbb{R}^l$ which is split into $N+2$ ($N\ge 1$) disjoint subsets $\Theta_0,\Theta_1,\dots,\Theta_N$ and I_{in}, so $\Theta=\sum_{i=0}^N\Theta_i+\mathrm{I}_{\text{in}}$. The hypotheses are $H_i:\ \theta\in\Theta_i$, $i=0,1,\dots,N$ and I_{in} is an indifference zone which may be an empty set. The probability measures P_θ and $\mathsf{P}_{\widetilde{\theta}}$ are assumed to be locally mutually absolutely continuous, i.e., the restrictions P_θ^t and $\mathsf{P}_{\widetilde{\theta}}^t$ of these measures to the sub-σ-algebras $\mathscr{F}_t$ are equivalent for all $0\le t<\infty$ and all $\theta,\widetilde{\theta}\in\Theta$.

Let $\lambda_t(\theta,\widetilde{\theta})=\log[\mathrm{d}\mathsf{P}_\theta^t/\mathrm{d}\mathsf{P}_{\widetilde{\theta}}^t]$ denote the corresponding LLR, and let $\mathcal{I}_{\theta,\widetilde{\theta}}(t)=\mathsf{E}_\theta[\lambda_t(\theta,\widetilde{\theta})]$ be the total K–L information in the trajectory $\mathbf{X}_0^t$ between the probability measures P_θ and $\mathsf{P}_{\widetilde{\theta}}$. The following condition is sufficient for obtaining the asymptotic lower bounds for the moments of the stopping time in the classes $\mathbf{C}([\alpha_{ij}])$ and $\mathbf{C}(\boldsymbol{\alpha})$: there are finite positive numbers $I(\theta,\widetilde{\theta})$ such that $\lim_{t\to\infty}[\mathcal{I}_{\theta,\widetilde{\theta}}(t)/t]=I(\theta,\widetilde{\theta})$ and

$$\frac{1}{t}\lambda_t(\theta,\widetilde{\theta})\xrightarrow[t\to\infty]{\mathsf{P}_\theta-\text{a.s.}}I(\theta,\widetilde{\theta})\quad\text{for all }\theta,\widetilde{\theta}\in\Theta,\theta\ne\widetilde{\theta}. \tag{5.146}$$

Here with a certain abuse of notation we denote the corresponding limits by $I(\theta,\widetilde{\theta})$, i.e., as the K–L numbers in the iid case. The reason is that in the iid case these conditions hold with $I(\theta,\widetilde{\theta})$ being the true K–L numbers. In the general case, the limiting values $I(\theta,\widetilde{\theta})$ can also be regarded as the local K–L information.

Let $I_i(\theta) = \inf_{\widetilde{\theta}\in\Theta_i} I(\theta,\widetilde{\theta})$ and assume that $I_i(\theta) > 0$ for all $i = 0,1,\dots,N$. Defining the functions $J_i(\theta), J(\theta), J_i^*(\theta), J^*(\theta)$ as in (5.82) and (5.83), it may be proved that if the a.s. convergence condition (5.146) holds, then for every $0 < \varepsilon < 1$ and $\theta \in \Theta$

$$
\begin{aligned}
&\inf_{\delta\in\mathbf{C}([\alpha_{ij}])} \mathsf{P}_\theta\{T > \varepsilon A_\theta(\alpha_{\max})\} \to 1 \quad \text{as } \alpha_{\max} \to 0, \\
&\inf_{\delta\in\mathbf{C}(\boldsymbol{\alpha})} \mathsf{P}_\theta\{T > \varepsilon A_\theta(\tilde{\alpha}_{\max})\} \to 1 \quad \text{as } \tilde{\alpha}_{\max} \to 0,
\end{aligned} \tag{5.147}
$$

where

$$
A_\theta(\alpha_{\max}) = \begin{cases} |\log\alpha_{\max}|/J_i(\theta) & \text{for } \theta\in\Theta_i,\ i=0,1,\dots,N \\ |\log\alpha_{\max}|/J(\theta) & \text{for } \theta\in\mathrm{I}_{\mathrm{in}} \end{cases}
$$

and

$$
A_\theta(\tilde{\alpha}_{\max}) = \begin{cases} |\log\tilde{\alpha}_{\max}|/J_i^*(\theta) & \text{for } \theta\in\Theta_i,\ i=0,1,\dots,N \\ |\log\tilde{\alpha}_{\max}|/J^*(\theta) & \text{for } \theta\in\mathrm{I}_{\mathrm{in}}. \end{cases}
$$

Applying the Chebyshev inequality we readily obtain that for all $m > 0$

$$
\begin{aligned}
&\inf_{\delta\in\mathbf{C}([\alpha_{ij}])} \mathsf{E}[T]^m \ge [A_\theta(\alpha_{\max})]^m(1+o(1)) \quad \text{as } \alpha_{\max}\to 0, \\
&\inf_{\delta\in\mathbf{C}(\boldsymbol{\alpha})} \mathsf{E}[T]^m \ge [A_\theta(\tilde{\alpha}_{\max})]^m(1+o(1)) \quad \text{as } \tilde{\alpha}_{\max}\to 0.
\end{aligned} \tag{5.148}
$$

A detailed proof of (5.147) and hence the asymptotic lower bounds (5.148) in the case of two hypotheses ($N = 1$) is given below; see Lemma 5.4.2, which is a refined version of Lemma 1 in Tartakovsky [458]. The case $N > 1$ can be handled in a similar way.

While all the results hold for both continuous and discrete time, from now on we focus on the discrete-time case. Let $p_\theta(X_n|\mathbf{X}_1^{n-1})$ be the conditional density of the n-th observation conditioned on the previous observations $\mathbf{X}_1^{n-1} = (X_1,\dots,X_{n-1})^\top$. Since the joint density of $\mathbf{X}_1^n = (X_1,\dots,X_n)^\top$ can be written as

$$
p_\theta(\mathbf{X}_1^n) = \prod_{k=1}^n p_\theta(X_k|\mathbf{X}_1^{k-1}),
$$

the MLE is $\hat{\theta}_n = \arg\sup_{\theta\in\Theta}\prod_{k=1}^n p_\theta(X_k|\mathbf{X}_1^{k-1})$. Let $\ell_n(\theta) = \sum_{k=1}^n \log p_\theta(X_k|\mathbf{X}_1^{k-1})$. The accepting MGSLRT is again defined by (5.76)–(5.77) with $\hat{\ell}_n$ and ℓ_n^i now defined as

$$
\hat{\ell}_n = \sum_{k=1}^n \log p_{\hat{\theta}_n}(X_k|\mathbf{X}_1^{k-1}), \quad \ell_n^i = \sup_{\theta\in\Theta_i}\ell_n(\theta).
$$

Also, the accepting MASLRT is defined as in (5.95)–(5.96) with

$$
\ell_n^* = \sum_{k=1}^n \log p_{\hat{\theta}_{k-1}}(X_k|\mathbf{X}_1^{k-1})
$$

and ℓ_n^i as above. Recall that here $\hat{\theta}_n(\mathbf{X}_1^n)$ is some estimate, not necessarily the MLE.

We now focus on the MASLRT. Moreover, we begin with the case of two hypotheses and provide proofs only for the 2-ASPRT defined in (5.115), where

$$
\lambda_n^*(\Theta_i) = \ell_n^* - \ell_n^i, \quad i = 0,1.
$$

We also write $\lambda_n(\theta,\widetilde{\theta}) = \ell_n(\theta) - \ell_n(\widetilde{\theta})$ and $\lambda_n^*(\widetilde{\theta}) = \ell_n^* - \ell_n(\widetilde{\theta})$.

As usual, to derive upper bounds, we have to strengthen the almost sure convergence condition (5.146). Indeed, this condition generally does not even guarantee the finiteness of the ESS $\mathsf{E}_\theta T^*$. We now strengthen the strong law (5.146) into the r-quick version assuming that

$$\frac{1}{n}\lambda_n(\theta,\widetilde{\theta}) \xrightarrow[n\to\infty]{\mathsf{P}_\theta-r-\text{quickly}} I(\theta,\widetilde{\theta}) \quad \text{for all } \theta,\widetilde{\theta}\in\Theta, \theta\neq\widetilde{\theta}. \tag{5.149}$$

In addition, we certainly need some conditions on the behavior of the estimate $\hat{\theta}_n$ for large n, which should converge to the true value θ in a proper way. To this end, we require the following natural condition on the adaptive LLR process:

$$\frac{1}{n}\lambda_n^*(\widetilde{\theta}) \xrightarrow[n\to\infty]{\mathsf{P}_\theta-r-\text{quickly}} I(\theta,\widetilde{\theta}) \quad \text{for all } \theta,\widetilde{\theta}\in\Theta, \theta\neq\widetilde{\theta}, \tag{5.150}$$

so that the normalized LLR tuned to the true parameter value and its adaptive version converge to the same constants. In certain cases, but not always, the conditions (5.149) and (5.150) imply the following conditions

$$\frac{1}{n}\lambda_n^*(\Theta_i) \xrightarrow[n\to\infty]{\mathsf{P}_\theta-r-\text{quickly}} I_i(\theta) \quad \text{for all } \theta\in\Theta\setminus\Theta_i,\ i=0,1. \tag{5.151}$$

These latter conditions are sufficient for asymptotic optimality of the 2-ASPRT in the class $\mathbf{C}(\alpha_0,\alpha_1)$ as $\alpha_{\max}=\max(\alpha_0,\alpha_1)\to 0$. In other words, for $\varepsilon>0$, define the random variables

$$L_i(\varepsilon,\theta)=\sup\{n\geq 1: |\lambda_n^*(\Theta_i)/n - I_i(\theta)|>\varepsilon\}, \quad i=0,1 \ (\sup\{\varnothing\}=0). \tag{5.152}$$

In terms of the random variables $L_i(\varepsilon,\theta)$, the r-quick convergence conditions (5.151) are rewritten as $\mathsf{E}_\theta[L_i(\varepsilon,\theta)]^r<\infty$ for all $\varepsilon>0$, $\theta\in\Theta\setminus\Theta_i$ and some $r\geq 1$.

In Theorem 5.4.4 below we show that under conditions (5.152) the right-hand sides in (5.148) are the upper bounds for the expectation $\mathsf{E}_\theta[T^*]^r$ of the 2-ASPRT with thresholds $a_i=\log(1/\alpha_i)$, therefore proving first-order optimality with respect to the moments of the stopping time up to order r.

We now proceed with establishing the asymptotic lower bounds (5.148) in the case of two composite hypotheses.

Lemma 5.4.2. *Assume that there are positive and finite numbers $I(\theta,\widetilde{\theta})$ such that*

$$\frac{1}{n}\lambda_n(\theta,\widetilde{\theta}) \xrightarrow[t\to\infty]{\mathsf{P}_\theta-a.s.} I(\theta,\widetilde{\theta}) \quad \textit{for all } \theta,\widetilde{\theta}\in\Theta, \theta\neq\widetilde{\theta}. \tag{5.153}$$

Let $I_i(\theta)=\inf_{\widetilde{\theta}\in\Theta_i} I(\theta,\widetilde{\theta})$ and suppose $\min_{i=0,1} I_i(\theta)>0$. Then for all $\theta\in\Theta$ and $0<\varepsilon<1$

$$\inf_{\delta\in\mathbf{C}(\alpha_0,\alpha_1)} \mathsf{P}_\theta\{T>\varepsilon A_\theta(\alpha_0,\alpha_1)\}\to 1 \quad \textit{as } \alpha_{\max}\to 0, \tag{5.154}$$

and therefore, for all $m>0$ and $\theta\in\Theta$,

$$\inf_{\delta\in\mathbf{C}(\alpha_0,\alpha_1)} \mathsf{E}_\theta T^m \geq [A_\theta(\alpha_0,\alpha_1)]^m(1+o(1)) \quad \textit{as } \alpha_{\max}\to 0, \tag{5.155}$$

where

$$A_\theta(\alpha_0,\alpha_1)=\begin{cases} |\log\alpha_0|/I_0(\theta) & \textit{for } \theta\in\Theta_1\\ |\log\alpha_1|/I_1(\theta) & \textit{for } \theta\in\Theta_0\\ \min\{|\log\alpha_0|/I_0(\theta), |\log\alpha_1|/I_1(\theta)\} & \textit{for } \theta\in \mathrm{I}_{\text{in}}. \end{cases}$$

Proof. Let $\delta = (T, d)$ be an arbitrary sequential test of two hypotheses from the class $\mathbf{C}(\alpha_0, \alpha_1)$. We may restrict our attention only to tests that terminate w.p. 1, $\mathsf{P}_\theta(d=0) + \mathsf{P}_\theta(d=1) = \mathsf{P}_\theta(T < \infty) = 1$, since otherwise $\mathsf{E}_\theta[T^m] = \infty$ and the statement follows trivially. By changing the measure, we obtain that for any $s \ge 1$, $C > 0$, and any two distinct points θ and $\widetilde{\theta}$

$$
\begin{aligned}
\mathsf{P}_{\widetilde{\theta}}(d=1) &= \mathsf{E}_\theta\{\mathbb{1}_{\{d=1\}} e^{-\lambda_T(\theta,\widetilde{\theta})}\} \\
&\ge \mathsf{E}_\theta\{\mathbb{1}_{\{d=1, T \le s, e^{\lambda_T(\theta,\widetilde{\theta})} < e^C\}} e^{-\lambda_T(\theta,\widetilde{\theta})}\} \\
&\ge e^{-C} \mathsf{P}_\theta\left(d=1, T \le s, \max_{n \le s} e^{\lambda_n(\theta,\widetilde{\theta})} < e^C\right) \\
&\ge e^{-C}\left\{\mathsf{P}_\theta(d=1, T \le s) - \mathsf{P}_\theta\left(\max_{n \le s} \lambda_n(\theta,\widetilde{\theta}) \ge C\right)\right\} \\
&\ge e^{-C}\left\{\mathsf{P}_\theta(T \le s) - \mathsf{P}_\theta(d=0) - \mathsf{P}_\theta\left(\max_{n \le s} \lambda_n(\theta,\widetilde{\theta}) \ge C\right)\right\}.
\end{aligned}
\tag{5.156}
$$

Note that the last two inequalities follow from the inequality $\mathsf{P}(A \cap B) \ge \mathsf{P}(A) - \mathsf{P}(B^c)$, where B^c is a complement of the event B, if we first set $A = \{d = 1, T \le s\}$ and $B = \{\max_{n \le s} \lambda_n(\theta,\widetilde{\theta}) < C\}$, and then, $A = \{T \le s\}$ and $B = \{d = 1\}$. It follows that

$$
\mathsf{P}_\theta(T \le s) \le \mathsf{P}_\theta(d=0) + e^C \mathsf{P}_{\widetilde{\theta}}(d=1) + \mathsf{P}_\theta\left\{\max_{n \le s} \lambda_n(\theta,\widetilde{\theta}) \ge C\right\}. \tag{5.157}
$$

A similar argument applies to $\mathsf{P}_{\widetilde{\theta}}(d=0)$ in order to show that

$$
\mathsf{P}_\theta(T \le s) \le \mathsf{P}_\theta(d=1) + e^C \mathsf{P}_{\widetilde{\theta}}(d=0) + \mathsf{P}_\theta\left\{\max_{n \le s} \lambda_n(\theta,\widetilde{\theta}) \ge C\right\}. \tag{5.158}
$$

Using the inequality (5.157), we obtain that for $\theta \in \Theta_1$

$$
\begin{aligned}
\mathsf{P}_\theta(T \le s) &\le \sup_{\theta \in \Theta_1} \mathsf{P}_\theta(d=0) + e^C \sup_{\theta \in \Theta_0} \mathsf{P}_\theta(d=1) + \mathsf{P}_\theta\left\{\max_{n \le s} \lambda_n(\theta,\widetilde{\theta}) \ge C\right\} \\
&\le \alpha_1 + \alpha_0 e^C + \mathsf{P}_\theta\left\{\max_{n \le s} \lambda_n(\theta,\widetilde{\theta}) \ge C\right\},
\end{aligned}
\tag{5.159}
$$

and using (5.158), we obtain that for $\theta \in \Theta_0$

$$
\mathsf{P}_\theta(T \le s) \le \alpha_0 + \alpha_1 e^C + \mathsf{P}_\theta\left\{\max_{n \le s} \lambda_n(\theta,\widetilde{\theta}) \ge C\right\}. \tag{5.160}
$$

Let $\theta \in \Theta_1$ and $\widetilde{\theta} \in \Theta_0$. To prove (5.154) in this case, we put $s = s(\alpha_0) = (1-\varepsilon)|\log \alpha_0|/I(\theta,\widetilde{\theta})$ and $C = (1+\varepsilon) I(\theta,\widetilde{\theta}) s(\alpha_0) = (1-\varepsilon^2)|\log \alpha_0|$ and use (5.159) to obtain

$$
\mathsf{P}_\theta\left\{T \le (1-\varepsilon)|\log \alpha_0|/I(\theta,\widetilde{\theta})\right\} \le \alpha_1 + \alpha_0^{\varepsilon^2} + \mathsf{P}_\theta\left\{\max_{n \le s(\alpha_0)} \lambda_n(\theta,\widetilde{\theta}) \ge (1-\varepsilon^2)|\log \alpha_0|\right\}.
$$

Since the right-hand side does not depend on δ, we have

$$
\begin{aligned}
\sup_{\delta \in \mathbf{C}(\alpha_0,\alpha_1)} \mathsf{P}_\theta\left\{T \le (1-\varepsilon)|\log \alpha_0|/I(\theta,\widetilde{\theta})\right\} &\le \alpha_1 + \alpha_0^{\varepsilon^2} \\
&+ \mathsf{P}_\theta\left\{\max_{n \le s(\alpha_0)} \lambda_n(\theta,\widetilde{\theta}) \ge (1-\varepsilon^2)|\log \alpha_0|\right\}.
\end{aligned}
\tag{5.161}
$$

To complete the proof, we only need to show that the third term on the right-hand side in the above inequality goes to 0 as $\alpha_0 \to \infty$ for all $0 < \varepsilon < 1$. By the almost sure convergence condition (5.153),

$$
s^{-1} \max_{n \le s} \lambda_n(\theta,\widetilde{\theta}) \xrightarrow[s \to \infty]{\text{in } \mathsf{P}_\theta-\text{probability}} I(\theta,\widetilde{\theta}). \tag{5.162}
$$

Therefore, for every $0 < \varepsilon < 1$ as $\alpha_0 \to 0$

$$\mathsf{P}_\theta\left\{\max_{n \le s(\alpha_0)} \lambda_n(\theta,\widetilde{\theta}) \ge (1-\varepsilon^2)|\log\alpha_0|\right\} = \mathsf{P}_\theta\left\{\max_{n \le s(\alpha_0)} \lambda_n(\theta,\widetilde{\theta}) \ge (1+\varepsilon)I(\theta,\widetilde{\theta})s(\alpha_0)\right\} \to 0,$$

which along with (5.161) proves that for all $\theta \in \Theta_1$ and all $\widetilde{\theta} \in \Theta_0$

$$\sup_{\delta \in \mathbf{C}(\alpha_0,\alpha_1)} \mathsf{P}_\theta\left\{T \le (1-\varepsilon)|\log\alpha_0|/I(\theta,\widetilde{\theta})\right\} \xrightarrow[\alpha_{\max}\to 0]{} 0,$$

and therefore,

$$\sup_{\delta \in \mathbf{C}(\alpha_0,\alpha_1)} \mathsf{P}_\theta\left\{T \le (1-\varepsilon)|\log\alpha_0|/I_0(\theta)\right\} \xrightarrow[\alpha_{\max}\to 0]{} 0,$$

where $I_0(\theta) > 0$ for $\theta \in \Theta_1$ by the conditions of the lemma. This proves (5.154) for $\theta \in \Theta_1$.

To prove (5.154) for $\theta \in \Theta_0$ we take $C = (1+\varepsilon)I(\theta,\widetilde{\theta})s$ with $s = s(\alpha_1) = (1-\varepsilon)|\log\alpha_1|/I(\theta,\widetilde{\theta})$ and arbitrary $\widetilde{\theta} \in \Theta_1$. Then using (5.160) yields for all $\theta \in \Theta_0$, $\widetilde{\theta} \in \Theta_1$ and every $0 < \varepsilon < 1$,

$$\sup_{\delta \in \mathbf{C}(\alpha_0,\alpha_1)} \mathsf{P}_\theta\left\{T \le (1-\varepsilon)|\log\alpha_1|/I(\theta,\widetilde{\theta})\right\} \le \alpha_0 + \alpha_1^{\varepsilon^2}$$
$$+\mathsf{P}_\theta\left\{\max_{n \le s(\alpha_1)} \lambda_n(\theta,\widetilde{\theta}) \ge (1-\varepsilon^2)|\log\alpha_1|\right\},$$

where by the same argument as above all three terms on the right-hand side go to 0, implying that for all $\theta \in \Theta_0$

$$\sup_{\delta \in \mathbf{C}(\alpha_0,\alpha_1)} \mathsf{P}_\theta\{T \le (1-\varepsilon)|\log\alpha_1|/I_1(\theta)\} \to 0 \quad \text{as } \alpha_{\max} \to 0.$$

It remains to prove (5.154) for the indifference zone, $\theta \in \mathrm{I}_{\text{in}}$. For any $\theta_0 \in \Theta_0$ and $\theta_1 \in \Theta_1$, let $K_\theta(\alpha_0,\alpha_1) = \min\{|\log\alpha_0|/I(\theta,\theta_0), |\log\alpha_1|/I(\theta,\theta_1)\}$. By (5.156), for every $\theta \in \mathrm{I}_{\text{in}}$ and $\theta_0 \in \Theta_0$,

$$\mathsf{P}_\theta(T \le s, d=1) \le \mathsf{P}_{\theta_0}(d=1)e^C + \mathsf{P}_\theta\left\{\max_{n \le s} \lambda_n(\theta,\theta_0) \ge C\right\},$$

so taking

$$s = s(\alpha_0,\alpha_1) = (1-\varepsilon)K_\theta(\alpha_0,\alpha_1), \quad C = (1+\varepsilon)I(\theta,\theta_0)s = (1-\varepsilon^2)I(\theta,\theta_0)K_\theta(\alpha_0,\alpha_1),$$

we obtain

$$\mathsf{P}_\theta\left\{d=1, T \le (1-\varepsilon)K_\theta\right\} \le \alpha_0^{\varepsilon^2} + \mathsf{P}_\theta\left\{\max_{n \le (1-\varepsilon)K_\theta} \lambda_n(\theta,\theta_0) \ge (1-\varepsilon^2)I(\theta,\theta_0)K_\theta\right\}.$$

Analogously, for every $\theta \in \mathrm{I}_{\text{in}}$ and $\theta_1 \in \Theta_1$

$$\mathsf{P}_\theta\left\{d=0, T \le (1-\varepsilon)K_\theta\right\} \le \alpha_1^{\varepsilon^2} + \mathsf{P}_\theta\left\{\max_{n \le (1-\varepsilon)K_\theta} \lambda_n(\theta,\theta_1) \ge (1-\varepsilon^2)I(\theta,\theta_1)K_\theta\right\}.$$

Therefore, for all $\theta_0 \in \Theta_0$ and $\theta_1 \in \Theta_1$,

$$\sup_{\delta \in \mathbf{C}(\alpha_0,\alpha_1)} \mathsf{P}_\theta\{T \le (1-\varepsilon)K_\theta\} \le \alpha_0^{\varepsilon^2} + \alpha_1^{\varepsilon^2} + \mathsf{P}_\theta\left\{\max_{n \le (1-\varepsilon)K_\theta} \lambda_n(\theta,\theta_0) \ge (1-\varepsilon^2)I(\theta,\theta_0)K_\theta\right\}$$
$$+\mathsf{P}_\theta\left\{\max_{n \le (1-\varepsilon)K_\theta} \lambda_n(\theta,\theta_1) \ge (1-\varepsilon^2)I(\theta,\theta_1)K_\theta\right\},$$

where all terms on the right-hand side go to 0, so that for every $0 < \varepsilon < 1$ and $\theta \in \mathrm{I}_{\mathrm{in}}$,

$$\sup_{\delta \in \mathbf{C}(\alpha_0,\alpha_1)} \mathsf{P}_\theta\{T \le (1-\varepsilon)A_\theta(\alpha_0,\alpha_1)\} \to 0 \quad \text{as } \alpha_{\max} \to 0,$$

where $A_\theta(\alpha_0,\alpha_1) = \min\{|\log \alpha_0|/I_0(\theta), |\log \alpha_1|/I_1(\theta)\}$. The proof of (5.154) for all $\theta \in \Theta$ is complete.

To prove the asymptotic lower bounds (5.155) for the moments of the stopping time, it suffices to use Chebyshev's inequality: for any $0 < \varepsilon < 1$, any $m > 0$, and all $\theta \in \Theta$,

$$\inf_{\delta \in \mathbf{C}(\alpha_0,\alpha_1)} \mathsf{E}_\theta[T/A_\theta(\alpha_0,\alpha_1)]^m \ge \varepsilon^m \inf_{\delta \in \mathbf{C}(\alpha_0,\alpha_1)} \mathsf{P}_\theta\{T/A_\theta(\alpha_0,\alpha_1) > \varepsilon\} \xrightarrow[\alpha_{\max}\to 0]{} \varepsilon^m,$$

where the latter limit follows from (5.154). Since ε is arbitrary, it follows that

$$\liminf_{\alpha_{\max}\to 0} \inf_{\delta \in \mathbf{C}(\alpha_0,\alpha_1)} \mathsf{E}_\theta[T/A_\theta(\alpha_0,\alpha_1)]^m \ge 1,$$

which completes the proof. □

We are now prepared to prove asymptotic optimality of the 2-ASPRT.

Theorem 5.4.4 (2-ASPRT asymptotic optimality). *Let α_i go to zero in such a way that $\lim_{\alpha_{\max}\to 0}(\log \alpha_0/\log \alpha_1) = c$, where c is bounded away from zero and infinity. Suppose that $I_0(\theta) > 0$ for $\theta \in \Theta_1$, $I_1(\theta) > 0$ for $\theta \in \Theta_0$, and $\min[I_0(\theta), I_1(\theta)] > 0$ for $\theta \in \mathrm{I}_{\mathrm{in}}$ and that r-quick convergence conditions (5.149) and (5.151) hold. If the thresholds a_0 and a_1 are selected so that $\sup_{\theta\in\Theta_i} \alpha_i(\delta^*,\theta) \le \alpha_i$ and $a_i \sim \log(1/\alpha_i)$, $i = 0,1$, in particular $a_i = \log(1/\alpha_i)$, then for $m \le r$ as $\alpha_{\max} \to 0$*

$$\inf_{\delta \in \mathbf{C}(\alpha_0,\alpha_1)} \mathsf{E}_\theta T^m \sim \mathsf{E}_\theta[T^*]^m \sim \begin{cases} [|\log \alpha_0|/I_0(\theta)]^m & \text{for all } \theta \in \Theta_1 \\ [|\log \alpha_1|/I_1(\theta)]^m & \text{for all } \theta \in \Theta_0 \\ [\min_{i=0,1} |\log \alpha_i|/I_i(\theta)]^m & \text{for all } \theta \in \mathrm{I}_{\mathrm{in}}. \end{cases} \tag{5.163}$$

Therefore, the 2-ASPRT minimizes asymptotically the moments of the stopping time distribution up to order r uniformly in $\theta \in \Theta$ in the class of tests $\mathbf{C}(\alpha_0,\alpha_1)$.

Proof. To prove the theorem, we have to show that the lower bounds (5.155) are also the upper bounds (asymptotically) for $\mathsf{E}_\theta[T^*]^r$. Since $T^* \le T_i^*$, $i = 0,1$, in order to derive the upper bounds for $\mathsf{E}_\theta[T^*]^r$ one can operate with the stopping times T_0^* and T_1^*.

Consider the region $\Theta_0 + \mathrm{I}_{\mathrm{in}}$ and the stopping time T_0^*. Recall the definition of the random variables $L_i(\varepsilon,\theta)$ given in (5.152). By the condition (5.151), for every $\theta \in \Theta_0 + \mathrm{I}_{\mathrm{in}}$

$$\lambda^*_{T_0^*-1}(\Theta_1) \ge (T_0^* - 1)[I_1(\theta) - \varepsilon] \quad \text{on } \{T_0^* > L_1(\varepsilon,\theta) + 1\}$$

and, by the definition of the stopping time T_0^*, $\lambda^*_{T_0^*-1}(\Theta_1) < a_1$ on the set $\{T_0^* < \infty\}$. Therefore, for any $0 < \varepsilon < I_1(\theta)$,

$$\begin{aligned} T_0^* &< \left(1 + \frac{a_1}{I_1(\theta) - \varepsilon}\right) \mathbb{1}_{\{T_0^* > 1 + L_1(\varepsilon,\theta)\}} + [1 + L_1(\varepsilon,\theta)]\, \mathbb{1}_{\{T_0^* \le 1 + L_1(\varepsilon,\theta)\}} \\ &\le 1 + L_1(\varepsilon,\theta) + \frac{a_1}{I_1(\theta) - \varepsilon}. \end{aligned} \tag{5.164}$$

Since $T^* \le T_0^*$ and by (5.151) $\mathsf{E}_\theta[L_1(\varepsilon,\theta)]^r < \infty$, by letting $\varepsilon \to 0$ it follows from (5.164) that for all $\theta \in \Theta_0 + \mathrm{I}_{\mathrm{in}}$

$$\mathsf{E}_\theta[T^*]^r \le \left(\frac{|\log \alpha_1|}{I_1(\theta)}\right)^r (1 + o(1)) \quad \text{as } \alpha_{\max} \to 0 \tag{5.165}$$

when $a_i = \log(1/\alpha_i)$ or $a_i \sim \log(1/\alpha_i)$ and $\sup_{\theta \in \Theta_i} \alpha_i(\delta^*, \theta) \le \alpha_i$. Since for $\theta \in \Theta_0$ this upper bound is asymptotically the same as the lower bound (5.155), the asymptotic equality (5.163) follows for $\theta \in \Theta_0$.

Now, consider the region $\theta \in \Theta_1 + \mathrm{I}_{\text{in}}$ and the stopping time T_1^*. Exactly the same argument applies in order to show that for all $\theta \in \Theta_1 + \mathrm{I}_{\text{in}}$

$$\mathsf{E}_\theta[T^*]^r \le \left(\frac{|\log \alpha_0|}{I_0(\theta)}\right)^r (1+o(1)) \quad \text{as } \alpha_{\max} \to 0. \tag{5.166}$$

One sees that, for $\theta \in \Theta_1$, the upper bound (5.165) coincides with the lower bound (5.163) yielding (5.163) for $\theta \in \Theta_1$.

It remains to prove (5.163) for the indifference zone I_{in}. Toward this end, we use the upper bounds (5.165) and (5.166) jointly, which yields

$$\mathsf{E}_\theta[T^*]^r \le \left\{\min\left[\frac{|\log \alpha_1|}{I_1(\theta)}, \frac{|\log \alpha_0|}{I_0(\theta)}\right]\right\}^r (1+o(1)) \quad \text{for all } \theta \in \mathrm{I}_{\text{in}} \quad \text{as } \alpha_{\max} \to 0.$$

Combining this upper bound with the lower bound (5.155) yields (5.163) for $\theta \in \mathrm{I}_{\text{in}}$. Thus, the asymptotic assertions (5.163) are proved for all $\theta \in \Theta$. □

In the iid case, the I-numbers are nothing but the K–L numbers $I(\theta, \widetilde{\theta}) = \mathsf{E}_\theta[\lambda_1(\theta, \widetilde{\theta})]$, and the lower bounds (5.155) hold under the first moment condition $\mathsf{E}_\theta|X_1| < \infty$, i.e., when $0 < I(\theta, \widetilde{\theta}) < \infty$. Theorem 5.4.4 holds if we assume the r-quick convergence conditions (5.151) with K–L numbers. In particular, this gives us an alternative version of Corollary 5.4.1 for the exponential family where the condition (5.84) is replaced with (5.151). Note that this version does not necessarily assume an exponential family.

The proofs presented for two hypotheses can be carried over to multiple hypotheses, in which case the notation and argument become more difficult. We present the final result omitting the proofs. The following theorem is a complete generalization of Theorem 5.4.3 to the non-iid case and a generalization of Theorem 5.4.4 to the multiple hypothesis case.

Theorem 5.4.5 (MASLRT asymptotic optimality). *Assume that the r-quick convergence conditions* (5.149) *are satisfied and*

$$\frac{1}{n}\lambda_n^*(\Theta_i) \xrightarrow[n\to\infty]{\mathsf{P}_\theta - r - quickly} I_i(\theta) \quad \textit{for all } \theta \in \Theta \setminus \Theta_i,\ i = 0, 1, \dots, N. \tag{5.167}$$

(i) *If the thresholds a_{ij} are selected so that $\sup_{\theta \in \Theta_i} \alpha_{ij}(\delta^*, \theta) \le \alpha_{ij}$ and $a_{ij} \sim \log(1/\alpha_{ij})$, in particular $a_{ij} = \log(1/\alpha_{ij})$, then for $m \le r$ as $\alpha_{\max} \to 0$*

$$\inf_{\delta \in \mathbf{C}([\alpha_{ij}])} \mathsf{E}_\theta T^m \sim \mathsf{E}_\theta[T^*]^m \sim \begin{cases} [|\log \alpha_{\max}|/J_i(\theta)]^m & \textit{for all } \theta \in \Theta_i \textit{ and } i = 0, 1, \dots, N \\ [|\log \alpha_{\max}|/J(\theta)]^m & \textit{for all } \theta \in \mathrm{I}_{\text{in}}, \end{cases} \tag{5.168}$$

where the functions $J_i(\theta)$ and $J(\theta)$ are defined as in (5.82).

(ii) *If the thresholds $a_{ij} = a_i$ are selected so that $\sup_{\theta \in \Theta_i} \alpha_i(\delta^*, \theta) \le \alpha_i$ and $a_i \sim \log(1/\alpha_i)$, in particular $a_i = \log(N/\alpha_i)$, then for $m \le r$ as $\tilde{\alpha}_{\max} \to 0$*

$$\inf_{\delta \in \mathbf{C}(\boldsymbol{\alpha})} \mathsf{E}_\theta T^m \sim \mathsf{E}_\theta[T^*]^m \sim \begin{cases} [|\log \tilde{\alpha}_{\max}|/J_i^*(\theta)]^m & \textit{for all } \theta \in \Theta_i \textit{ and } i = 0, 1, \dots, N \\ [|\log \tilde{\alpha}_{\max}|/J^*(\theta)]^m & \textit{for all } \theta \in \mathrm{I}_{\text{in}}, \end{cases} \tag{5.169}$$

where the functions $J_i^*(\theta)$ and $J^*(\theta)$ are defined as in (5.83).

Consequently, the MASLRT minimizes asymptotically the moments of the stopping time up to order r uniformly in $\theta \in \Theta$ in the classes of tests $\mathbf{C}([\alpha_{ij}])$ and $\mathbf{C}(\boldsymbol{\alpha})$.

The proof is based again on showing that the lower bounds (5.148) are attained for the MASLRT whenever the conditions (5.167) hold.

Remark 5.4.1. The assertions of Theorem 5.4.5 of course also hold for the MGSLRT and MWSPRT when the r-quick convergence conditions (5.167) are satisfied for the GLR statistics $\hat{\lambda}_n(\Theta_i) = \hat{\ell}_n - \ell_n^i$ and the mixtures $\bar{\lambda}_n^i$. However, we recall that there are no simple upper bounds for the error probabilities of the MGSLRT and the MWSPRT. Furthermore, while for iid exponential families certain asymptotic approximations for the error probabilities can be obtained based on the boundary-crossing framework and large deviations [94, 274], for general non-iid models no such results exist.

Remark 5.4.2. The assertions of Theorem 5.4.5 remain true if the normalization by n in (5.167) is replaced with the normalization by $\psi(n)$, where $\psi(t)$ is an increasing function, $\psi(\infty) = \infty$, in which case $[|\log \alpha_{\max}|/J_i(\theta)]^m$ in (5.168) should be replaced with $\Psi([|\log \alpha_{\max}|/J_i(\theta)]^m)$, where Ψ is inverse to ψ, and similarly in (5.169). The argument is analogous to that used, e.g., in Theorem 4.4.1.

We now consider two examples that illustrate Theorem 5.4.5.

Example 5.4.3 (Testing for the Gaussian mean with unknown variance, continued). In Example 5.4.1 of Subsection 5.4.2 we established uniform asymptotic optimality of the 2-ASPRT with the MLE using Corollary 5.4.1 by verifying the condition (5.84). Alternatively, the validity of (5.122) can be verified using Theorem 5.4.4. Moreover, this theorem allows us to extend (5.122) to higher moments of the stopping time. Toward this end, we have to check the conditions (5.149) and (5.151). We follow the notation introduced in Example 5.4.1.

First, since the LLR $\{\lambda_n(\theta, \widetilde{\theta})\}_{n \ge 1}$ defined in (5.118) is a random walk with drift $I(\theta, \widetilde{\theta})$ and $\mathsf{E}_\theta|\lambda_1(\theta, \widetilde{\theta})|^r < \infty$ for all positive r, by Theorem 2.4.4,

$$n^{-1}\lambda_n(\theta, \widetilde{\theta}) \xrightarrow[n\to\infty]{\mathsf{P}_\theta\text{-}r\text{-quickly}} I(\theta, \widetilde{\theta}) \quad \text{for all } r > 0.$$

Hence, the conditions (5.149) and of course (5.153) are satisfied with $I(\theta, \widetilde{\theta})$ equal the K–L numbers given by (5.119) and $\theta = (\mu, \sigma^2)$.

Recall that $\lambda_n^*(\Theta_i) = \lambda_n^* - \lambda_n^i$, $i = 0, 1$, where λ_n^* and λ_n^i are defined in (5.125). Since $X_1, X_2, \dots$ are iid and $\mathsf{E}_\theta|X_1|^r < \infty$ for all $r > 0$, the following r-quick convergence conditions hold as $n \to \infty$ under P_θ:

$$\hat{\mu}_n \to \mu, \quad \hat{\mu}_n^2 \to \mu^2, \quad \hat{\sigma}_n^2 \to \sigma^2, \quad \sigma_{n,0}^2 \to \sigma^2 + \mu^2, \quad \forall \mu \ge 0, \sigma^2 > 0,$$

$$\mu_{n,1} \to \begin{cases} \mu & \text{if } \mu \ge \mu_1 \\ \mu_1 & \text{if } 0 \le \mu < \mu_1 \end{cases}, \qquad \sigma_{n,1}^2 \to \begin{cases} \sigma^2 & \text{if } \mu \ge \mu_1 \\ \sigma^2 + (\mu - \mu_1)^2 & \text{if } 0 \le \mu < \mu_1. \end{cases}$$

Using these relations along with (5.125), it can be verified that P_θ-r-quickly as $n \to \infty$

$$\begin{aligned} n^{-1}\lambda_n^* &\to \left(\mu^2 + \sigma^2 - \log \sigma^2 - 1\right)/2, \quad \forall \mu \ge 0, \sigma^2 > 0; \\ n^{-1}\lambda_n^0 &\to \left[\mu^2 + \sigma^2 - \log(\mu^2 + \sigma^2) - 1\right]/2, \quad \forall \mu \ge 0, \sigma^2 > 0; \\ n^{-1}\lambda_n^1 &\to \begin{cases} \left(\mu^2 + \sigma^2 - \log \sigma^2 - 1\right)/2 & \text{if } \mu \ge \mu_1, \sigma^2 > 0 \\ \left\{\mu^2 + \sigma^2 + \log[\sigma^2 + (\mu - \mu_1)^2] - 1\right\}/2 & \text{if } 0 \le \mu < \mu_1, \sigma^2 > 0. \end{cases} \end{aligned}$$

Combining these formulas yields

$$n^{-1}\lambda_n^*(\Theta_i) \xrightarrow[n\to\infty]{\mathsf{P}_\theta\text{-}r\text{-quickly}} I_i(\theta), \quad \Theta \in \Theta \setminus \Theta_i, \; i = 0, 1 \; \text{for all } r > 0, \tag{5.170}$$

where $I_1(\theta) = \min_{\widetilde{\mu} \ge \mu_1} \min_{\widetilde{\sigma} > 0} I(\theta, \widetilde{\theta}) \equiv I_1(q)$ and $I_0(\theta) = \min_{\widetilde{\mu} \in \{0\}} \min_{\widetilde{\sigma} > 0} I(\theta, \widetilde{\theta}) \equiv I_0(q)$ are given by (5.120) ($q = \mu/\sigma$, $q_1 = \mu_1/\sigma$).

Therefore, the conditions (5.151) are satisfied with $I_i(\theta) = I_i(q)$. By Theorem 5.4.4, the

2-ASPRT is asymptotically uniformly optimal in the sense of minimizing all positive moments of the stopping time distribution: for all $r \geq 1$ as $\alpha_{\max} \to 0$

$$\inf_{\delta \in \mathbf{C}(\alpha_0,\alpha_1)} \mathsf{E}_\theta T^r \sim \mathsf{E}_\theta[T^*]^r \sim \begin{cases} \{2|\log \alpha_1|/\log[1+(q_1-q)^2]\}^r & \text{if } 0 \leq q \leq q^* \\ \{2|\log \alpha_0|/\log[1+q^2]\}^r & \text{if } q \geq q^*. \end{cases}$$

In addition,

$$\inf_{\delta \in \mathbf{C}(\alpha_0,\alpha_1)} \sup_{\theta \in \Theta} \mathsf{E}_\theta T^r \sim \mathsf{E}_{\theta^*}[T^*]^r \sim \{2|\log \alpha_0|/\log[1+(q^*)^2]\}^r,$$

where $q^* \in (0, q_1)$ is the solution of the equation (5.124).

These asymptotics are also true for the GSLRT and the WSPRT if the thresholds are selected so that the logarithms of the error probabilities of these tests are asymptotic to $\log \alpha_i$.

Example 5.4.4 (Testing for a nonhomogeneous AR sequence). Let

$$X_n = \theta \cdot S_n + \xi_n, \quad n = 1, 2, \dots,$$

where S_n is a deterministic function and $\{\xi_n\}_{n \geq 1}$ is a stable first-order AR Gaussian sequence given by the recursion

$$\xi_n = \gamma\, \xi_{n-1} + \zeta_n, \quad n \geq 1,$$

where $\zeta_1, \zeta_2, \dots$ are iid Gaussian variables, $\zeta_k \sim \mathcal{N}(0, \sigma^2)$, and $|\gamma| < 1$. For the sake of concreteness we set $\xi_0 = 0$ and $S_0 = 0$, while all the results are true for arbitrary deterministic or random initial conditions. The hypotheses are $H_0 : \theta \leq \theta_0$ and $H_1 : \theta \geq \theta_1$, where $\theta_0 < \theta_1$ are given numbers. That is, $\Theta = (-\infty, \infty)$, $\Theta_0 = (-\infty, \theta_0]$, $\Theta_1 = [\theta_1, \infty)$, $\mathrm{I}_{\text{in}} = (\theta_0, \theta_1)$. In this case, the LLR can be written in the form

$$\lambda_n(\theta, \widetilde{\theta}) = \frac{\theta - \widetilde{\theta}}{\sigma^2} \sum_{k=1}^n \widetilde{S}_k \widetilde{X}_k - \frac{\theta^2 - \widetilde{\theta}^2}{2\sigma^2} \sum_{k=1}^n \widetilde{S}_k^2,$$

where $\widetilde{X}_k = X_k - \gamma X_{k-1}$, $\widetilde{S}_k = S_k - \gamma S_{k-1}$ $(X_0 = S_0 = 0)$. Direct computation shows that

$$\mathsf{E}_\theta[\lambda_n(\theta, \widetilde{\theta})] = \frac{(\theta - \widetilde{\theta})^2}{2\sigma^2} \sum_{k=1}^n \widetilde{S}_k^2.$$

Suppose that

$$\lim_{n \to \infty} n^{-1} \sum_{k=1}^n \widetilde{S}_k^2 = \widetilde{S}^2, \tag{5.171}$$

where $\widetilde{S}^2$ is a positive and finite number. Then, for all $\theta, \widetilde{\theta} \in (-\infty, \infty)$ $(\theta \neq \widetilde{\theta})$,

$$n^{-1} \lambda_n(\theta, \widetilde{\theta}) \to \frac{(\theta - \widetilde{\theta})^2 \widetilde{S}^2}{2\sigma^2} \quad \mathsf{P}_\theta - r - \text{quickly for all } r > 0. \tag{5.172}$$

Indeed, under P_θ, the whitened observations $\widetilde{X}_n$ can be written as $\widetilde{X}_n = \theta \widetilde{S}_n + \zeta_n$ and the LLR as

$$\lambda_n(\theta, \widetilde{\theta}) = \frac{\theta - \widetilde{\theta}}{\sigma^2} V_n + \frac{(\theta - \widetilde{\theta})^2}{2\sigma^2} \sum_{k=1}^n \widetilde{S}_k^2,$$

where $V_n = \sum_{k=1}^n \widetilde{S}_k \zeta_k$ is a weighted sum of iid normal random variables. Since $\mathsf{E} V_n = 0$ and, by (5.171), for large n $\mathsf{E} V_n^2 \sim \widetilde{S}^2 n$, it is obvious that there exists a number $\delta < 1$ such that $\mathsf{P}(|V_n| > \varepsilon n) \leq O(\delta^n)$, which yields

$$\sum_{n=1}^\infty n^{r-1} \mathsf{P}(|V_n| > \varepsilon n) < \infty \quad \text{for some } \varepsilon > 0 \text{ and all } r > 0. \tag{5.173}$$

In other words, V_n/n converges r-quickly to 0, which implies (5.172). Hence, the conditions (5.149) and therefore (5.153) hold with $I(\theta,\widetilde{\theta}) = (\theta-\widetilde{\theta})^2\widetilde{S}^2/2\sigma^2$, and it remains to check the conditions (5.151), where

$$I_1(\theta) = \inf_{\widetilde{\theta}\ge\theta_1} I(\theta,\widetilde{\theta}) = \frac{(\theta_1-\theta)^2\widetilde{S}^2}{2\sigma^2} \quad \text{for } \theta<\theta_1,$$
$$I_0(\theta) = \inf_{\widetilde{\theta}\le\theta_0} I(\theta,\widetilde{\theta}) = \frac{(\theta-\theta_0)^2\widetilde{S}^2}{2\sigma^2} \quad \text{for } \theta>\theta_0. \tag{5.174}$$

Let $\hat{\theta}_n$ be the MLE $\sum_{k=1}^n \widetilde{S}_k\widetilde{X}_k/\sum_{k=1}^n \widetilde{S}_k^2$. Moreover let $\hat{\theta}_{n,1} = \max(\theta_1,\hat{\theta}_n)$ and $\hat{\theta}_{n,0} = \min(\theta_0,\hat{\theta}_n)$. Then the statistics $\lambda_n^*(\Theta_i)$ write as

$$\lambda_n^*(\Theta_i) = \frac{1}{\sigma^2}\sum_{k=1}^n (\hat{\theta}_{k-1}-\hat{\theta}_{n,i})\widetilde{S}_k\widetilde{X}_k - \frac{1}{2\sigma^2}\sum_{k=1}^n(\hat{\theta}_{k-1}^2-\hat{\theta}_{n,i}^2)\widetilde{S}_k^2, \quad i=0,1.$$

Using an argument similar to that has led to (5.173) with a minor generalization, we conclude that, r-quickly under P_θ,

$$\hat{\theta}_n\to\theta,\quad \hat{\theta}_{n,1}\to\max(\theta_1,\theta),\quad \hat{\theta}_{n,0}\to\max(\theta_0,\theta),\quad \hat{\theta}_n^2\to\theta^2,$$
$$\hat{\theta}_{n,1}^2\to\max(\theta_1^2,\theta^2),\quad \hat{\theta}_{n,0}^2\to\max(\theta_0^2,\theta^2),\quad n^{-1}\sum_{k=1}^n\widetilde{S}_k\widetilde{X}_k\to\theta^2\widetilde{S}^2,$$

which after some manipulations yield

$$n^{-1}\lambda_n^*(\Theta_i) \xrightarrow[n\to\infty]{\mathsf{P}_\theta\text{-}r\text{-quickly}} I_i(\theta), \quad \theta\in\Theta\setminus\Theta_i,\ i=0,1 \text{ for all } r>0,$$

where the $I_i(\theta)$'s are given by (5.174).

Thus, by Theorem 5.4.4, the 2-ASPRT is asymptotically optimal, minimizing all positive moments of the sample size: for all $r\ge 1$ as $\alpha_{\max}\to 0$,

$$\inf_{\delta\in\mathbf{C}(\alpha_0,\alpha_1)} \mathsf{E}_\theta T^r \sim \mathsf{E}_\theta[T^*]^r \sim \begin{cases} \{2|\log\alpha_1|/[(q_1-q)^2\widetilde{S}^2]\}^r & \text{if } q\le q^* \\ \{2|\log\alpha_0|/[(q-q_0)^2\widetilde{S}^2]\}^r & \text{if } q\ge q^*, \end{cases}$$

where $q=\theta/\sigma, q_i=\theta_i/\sigma$, and q^* is a solution of the equation

$$|\log\alpha_0|/[(q-q_0)^2 = |\log\alpha_1|/(q_1-q)^2.$$

In particular, if $S_n=1$, then $\widetilde{S}^2=(1-\gamma)^2$.

5.5 Nearly Minimax Sequential Tests of Composite Hypotheses with Information Cost

Let $X_1,X_2,\dots$ be iid observations with a common pdf $p_\theta(x)$. Consider testing the simple null hypothesis $H_0:\theta=\theta_0$ *versus* the composite alternative $H_1:\theta\in\Theta_1$. It follows from the discussion in Section 5.4 that, at least for the exponential family, the GSLRT, ASPRT, and WSPRT asymptotically minimize the ESS $\mathsf{E}_\theta T$ to first order uniformly for all $\theta\in\Theta_1$ in the class $\mathbf{C}(\alpha_0,\alpha_1)=\{\delta:\mathsf{P}_0(d=1)\le\alpha_0,\sup_{\theta\in\Theta_1}\mathsf{P}_\theta(d=0)\le\alpha_1\}$ as $\alpha_{\max}\to 0$. In particular, this is true for the WSPRT with an arbitrary prior weight function $W(\theta)$. While this first-order uniform asymptotic optimality property is undoubtedly important, the natural question arises of how to choose the distribution $W(\theta)$ in order to further optimize the performance in some sense. Since this is not possible for all parameter values, one solution is to consider a minimax approach and minimize $\sup_{\theta\in\Theta_1}\mathsf{E}_\theta T$ in the class $\mathbf{C}(\alpha_0,\alpha_1)$. However, this minimax approach for selecting the weights may be very conservative and may lead to a very inefficient design. Indeed, for $X_n \sim \mathcal{N}(\theta, 1)$ and $\Theta_1 = [\theta_1, \infty)$, where $\theta_1 > \theta_0 = 0$, the minimax solution is obviously the SPRT tuned to θ_1, i.e., the WSPRT with the degenerate prior concentrated at θ_1, and for this test

$$\mathsf{E}_\theta T \sim 2|\log \alpha_0|/[\theta_1(2\theta - \theta_1)], \quad \theta \geq \theta_1.$$

The minimax approach thus leads to an efficiency loss for the parameter values $\theta > \theta_1$ and especially for $\theta \gg \theta_1$ as compared with the WSPRT and other uniformly asymptotically optimal tests, since for the latter tests the ESS asymptotically equals $2|\log \alpha_0|/\theta^2$. Therefore, a different criterion is needed for preserving the first-order asymptotic optimality property with respect to the ESS for all parameter values.

Another, more fruitful idea is to follow a minimax approach with respect to the expected K–L distance, i.e.,

$$\inf_{\delta \in \mathbf{C}(\alpha_0, \alpha_1)} \sup_{\theta \in \Theta_1} (I_\theta \mathsf{E}_\theta T), \tag{5.175}$$

where $I_\theta \mathsf{E}_\theta T = \mathsf{E}_\theta[\lambda_T(\theta, \theta_0)]$ is the K–L distance between p_θ and p_{θ_0} accumulated in the trajectory X_1^T. Therefore, the criterion (5.175) corresponds to minimizing the expected accumulated K–L information in the least favorable situation. This criterion, being natural in its own right, turns out to be consistent since, as shown below, for the WSPRT $\tau(\alpha_0, \alpha_1)$,

$$I_\theta \; \mathsf{E}_\theta[\tau(\alpha_0, \alpha_1)] = \log \alpha_0^{-1} + \frac{1}{2} \log\log \alpha_0^{-1} + C_\theta(W) + o(1) \quad \text{as } \alpha_{\max} \to 0,$$

where $C_\theta(W)$ is a constant depending on the parameter θ and the prior distribution W. Hence, $I_\theta \mathsf{E}_\theta[\tau(\alpha_0, \alpha_1)]$ does not depend on θ to within $O(1)$, i.e., it is an approximate equalizer if we neglect the constant $C_\theta(W)$, so it is a good candidate for the minimax test for any prior, which turns out to be the case. Moreover, under certain assumptions and depending on the model at hand, the prior distribution W can be selected in such a way that the constant C_θ does not depend on θ, which makes the WSPRT an equalizer to within the negligible term $o(1)$. The details are presented in the next subsections.

5.5.1 Nearly Minimax Open Ended Sequential Tests

We begin with considering the simplest problem of optimizing power-one (open-ended) tests, which continue sampling indefinitely with prescribed probability $\mathsf{P}_{\theta_0}(T < \infty) = \alpha$, $0 < \alpha < 1$, when the hypothesis H_0 is true and accept the alternative hypothesis H_1 at stopping, which has to be accepted as soon as possible if H_1 is true. In other words, the problem is to improve the performance of uniformly asymptotically optimal one-sided tests that minimize the ESS $\mathsf{E}_\theta T$ for all $\theta \in \Theta_1$ to first order in the class $\mathbf{C}(\alpha) = \{T : \mathsf{P}_0(T < \infty) \leq \alpha\}$, say the GLR or mixture-based tests. This improvement is achieved by using the minimax criterion

$$\inf_{\delta \in \mathbf{C}(\alpha)} \sup_{\theta \in \Theta_1} (I_\theta \mathsf{E}_\theta T). \tag{5.176}$$

In Subsection 5.5.1.1, we show that the weight function $W(\theta)$ of the one-sided WSPRT $T_a(W)$ can be selected in such a way that it is third-order asymptotically minimax for the exponential family:

$$\inf_{\delta \in \mathbf{C}(\alpha)} \sup_{\theta \in \Theta_1} (I_\theta \mathsf{E}_\theta T) = \sup_{\theta \in \Theta_1} (I_\theta \mathsf{E}_\theta T_a) + o(1) \quad \text{as } \alpha \to 0. \tag{5.177}$$

In Subsections 5.5.1.2 and 5.5.1.3 we tackle the case where the alternative hypothesis $H_1 : \mathsf{P} \in \mathcal{P}$ is a finite set, $\mathcal{P} = \{\mathsf{P}_1, \dots, \mathsf{P}_N\}$. This is a more general framework than the exponential family in that the distributions P_i and P_0 are not required to belong to the same parametric family. Moreover, it can be seen as a discrete approximation to the continuous setup. Such an approximation is often necessary in practice, since the continuously weighted likelihood ratio may not be implementable without discretization. Another motivation for the discrete setup is that it arises naturally in many applications, for example in the multisample slippage problem, where there are N sources of observations (channels or populations) and there are two possibilities, in and out of control, for the distribution of each source as discussed in Subsection 5.5.1.3. This problem has a variety of important applications, in particular in target detection and cybersecurity; see, e.g., [465, 472, 475].

5.5.1.1 *Near Minimaxity for the Exponential Families*

Suppose that $\{p_\theta\}_{\theta\in\Theta}$ is the one-parameter exponential family (5.47). Without loss of generality, up to a linear transformation of the coordinates, we may assume that $\theta_0 = 0$ and $b(0) = \dot{b}(0) = 0$. Therefore, we are testing the simple hypothesis $H_0 : \theta = 0$ against the composite alternative $H_1 : \theta \in \Theta_1 \subset \Theta$ for the family

$$\frac{p_\theta(x)}{p_{\theta_0}(x)} = \exp\{\theta x - b(\theta)\}, \quad \theta \in \Theta, \tag{5.178}$$

where $\Theta = \{\theta \in \mathbb{R} : \mathsf{E}_0[e^{\theta X_1}] < \infty\}$ is the natural parameter space. While for establishing minimaxity we require Θ_1 to be a finite interval bounded away from 0, for the analysis of the mixture and GLR tests this assumption is not required. In particular, Θ_1 may be equal to $\Theta\setminus\{0\}$, i.e., $H_1 : \theta \neq 0$, which we assume first.

Let $\Lambda_n^\theta = \Lambda_n(\theta,\theta_0)$ and $\lambda_n^\theta = \lambda_n(\theta,\theta_0)$ be the LR and the LLR tuned to θ,

$$\Lambda_n^\theta = \prod_{k=1}^n \frac{p_\theta(X_k)}{p_0(X_k)}, \quad \lambda_n^\theta = \sum_{k=1}^n \log\frac{p_\theta(X_k)}{p_0(X_k)} = \theta S_n - b(\theta) n,$$

where $S_n = \sum_{k=1}^n X_k$. Since $\theta_0 = 0$, the K–L numbers are

$$I(\theta,\theta_0) = I_\theta = \theta\dot{b}(\theta) - b(\theta), \quad \theta \in \Theta\setminus\{0\}.$$

For every $\theta \in \Theta_1 = \Theta\setminus\{0\}$ and $a \geq 0$, define the one-sided SPRT and the overshoot

$$T_a^\theta = \inf\left\{n \geq 1 : \lambda_n^\theta \geq a\right\}, \quad \kappa_a^\theta = \lambda_{T_a^\theta}^\theta - a \quad \text{on } \{T_a^\theta < \infty\},$$

as well as

$$\zeta_\theta = \int_0^\infty e^{-x}\,\mathsf{H}_\theta(\mathrm{d}x), \quad \varkappa_\theta = \int_0^\infty x\,\mathsf{H}_\theta(\mathrm{d}x), \tag{5.179}$$

where $\mathsf{H}_\theta(x) = \lim_{a\to\infty} \mathsf{P}_\theta(\kappa_a^\theta \leq x)$ is the limiting distribution of the overshoot κ_a^θ under P_θ.

Using Lemma 4.3.2 it is readily shown that, if $\mathsf{P}_0(T_a^\theta < \infty) = \alpha$, then the stopping time T_a^θ minimizes the expected sample size $\mathsf{E}_\theta T$ at the point θ in the class $\mathbf{C}(\alpha)$. Now, by Theorem 2.5.6,

$$\mathsf{E}_\theta T_a^\theta = I_\theta^{-1}(a + \varkappa_\theta) + o(1) \quad \text{as } a \to \infty,$$

and by the same argument as in Theorem 3.1.2,

$$\mathsf{P}_0(T_a^\theta < \infty) \sim \zeta_\theta e^{-a}, \quad a \to \infty.$$

Thus, the optimal asymptotic performance under P_θ is

$$\inf_{T\in\mathbf{C}(\alpha)} \mathsf{E}_\theta T = \mathsf{E}_\theta T_a^\theta = I_\theta^{-1}\left(|\log\alpha| + \varkappa_\theta + \log\zeta_\theta\right) + o(1) \quad \text{as } \alpha \to 0. \tag{5.180}$$

We now introduce a class of mixture-based sequential tests that have power 1 against all the alternatives $\theta \neq 0$. Specifically, let $W(\theta)$ be a mixing distribution over Θ and let $a > 0$ be a threshold or a critical level. For $n \geq 1$, define the mixture LR

$$\bar{\Lambda}_n = \int_\Theta \exp\{\lambda_n^\theta\} W(\mathrm{d}\theta) = \int_\Theta \exp\{\theta S_n - b(\theta)n\} W(\mathrm{d}\theta).$$

Let $\bar{\lambda}_n = \log \bar{\Lambda}_n$ be the corresponding LLR. The one-sided mixture test T_a is defined as

$$T_a = \inf\{n \geq 1 : \bar{\lambda}_n \geq a\}, \quad a > 0.$$

In other words, the hypothesis H_0 is rejected at the first time that the mixture LR $\bar{\Lambda}_n$ exceeds the critical level $A = e^a > 1$, and if $\bar{\Lambda}_n < A$ for all n (i.e., $T_a = \infty$) then the sampling continues indefinitely and H_0 is never rejected.

Given a mixing distribution W, define the probability measure $\mathsf{P}^W(\cdot) = \int_\Theta \mathsf{P}_\theta(\cdot) W(\mathrm{d}\theta)$. Obviously, $\bar{\Lambda}_n = (\mathrm{d}\mathsf{P}^W/\mathrm{d}\mathsf{P}_0)|_{\mathscr{F}_n}$, so that the standard change-of-measure trick yields the upper bound

$$\mathsf{P}_0(T_a < \infty) = \mathsf{E}^W[\bar{\Lambda}_{T_a}^{-1} \mathbb{1}_{\{T_a < \infty\}}] \leq e^{-a} \mathsf{P}^W(T_a < \infty) \leq e^{-a}, \tag{5.181}$$

which holds for an arbitrary mixing distribution W and all $a > 0$. Also, $\mathsf{P}_\theta(T_a < \infty) = 1$ for all $\theta \neq 0$ if the support of W is Θ, which trivially follows from the fact that $\bar{\lambda}_n \to \infty$ P_θ-a.s. as $n \to \infty$ for all $\theta \neq 0$.

For large a, the inequality (5.181) can be refined using nonlinear renewal theory. Let $\chi_a = \bar{\lambda}_{T_a} - a$, on $\{T_a < \infty\}$, be the overshoot of the weighted LLR over the level a at stopping. We have

$$\begin{aligned} \mathsf{P}_0(T_a < \infty) &= \mathsf{E}^W\left[e^{-\bar{\lambda}_{T_a}} \mathbb{1}_{\{T_a < \infty\}}\right] = e^{-a} \mathsf{E}^W\left[e^{-\chi_a} \mathbb{1}_{\{T_a < \infty\}}\right] \\ &= e^{-a} \int_\Theta \mathsf{E}_\theta\left[e^{-\chi_a} \mathbb{1}_{\{T_a < \infty\}}\right] W(\mathrm{d}\theta). \end{aligned} \tag{5.182}$$

Theorem 2.6.1, the first nonlinear renewal theorem, can be exploited to prove that in the nonarithmetic case

$$\mathsf{E}_\theta\left[e^{-\chi_a} \mathbb{1}_{\{T_a < \infty\}}\right] \to \zeta_\theta \quad \text{as } a \to \infty,$$

and hence, it is expected that the error probability may be approximated as

$$\mathsf{P}_0(T_a < \infty) \sim e^{-a} \int_{\Theta_1} \zeta_\theta W(\mathrm{d}\theta) \quad \text{as } a \to \infty.$$

To apply this theorem, we decompose the LLR $\bar{\lambda}_n$ for any fixed $\theta \neq 0$ as

$$\bar{\lambda}_n = \lambda_n^\theta + \xi_n^\theta, \quad n \geq 1, \tag{5.183}$$

where $\lambda_n^\theta = \theta S_n - b(\theta)n$ is the random walk under P_θ with drift I_θ and

$$\xi_n^\theta = \log\left(\int_\Theta \exp\{(t-\theta)S_n - [b(t) - b(\theta)]n\} W(\mathrm{d}t)\right). \tag{5.184}$$

From now on we assume that the distribution $W(\theta)$ has a positive continuous density $w(\theta)$ with respect to the Lebesgue measure on Θ. It can be verified that in this case $\{\xi_n^\theta\}_{n\geq 1}$ is a slowly changing sequence under P_θ. Also, $\lambda_1^\theta = \theta X_1 - b(\theta)$ is nonarithmetic for almost all θ with respect to the Lebesgue measure. Hence, by Theorem 2.6.1, $\lim_{a\to\infty} \mathsf{P}_\theta(\chi_a \leq x) = \mathsf{H}_\theta(x)$, which along with (5.182) yields

$$e^a \mathsf{P}_0(T_a < \infty) \to \int_\Theta \zeta_\theta w(\theta) \mathrm{d}\theta \quad \text{as } a \to \infty. \tag{5.185}$$

Next, using Theorem 2.6.4(i), the second nonlinear renewal theorem, we establish that the following asymptotic approximation holds for every $\theta \in \Theta \setminus \{0\}$:

$$I_\theta \, \mathsf{E}_\theta T_a = a + \log\sqrt{a} - \frac{1+\log(2\pi)}{2} + \log\left(e^{\varkappa_\theta} \frac{\sqrt{\ddot{b}(\theta)/I_\theta}}{w(\theta)} \right) + o(1) \quad \text{as } a \to \infty. \tag{5.186}$$

In order to apply this theorem the conditions (2.126)–(2.132) have to be checked. We omit the detailed proof giving only several main steps that explain why this is expected to be true. The detailed proofs can be found in [367, 513]. The key is the representation (5.183) for the mixed LLR where the sequence ξ_n^θ, $n \ge 1$ defined in (5.184) is slowly changing. More specifically, let $\hat{\theta}_n = \arg\sup_{\theta\in\Theta}[\theta S_n - b(\theta)n]$ be the MLE of θ, which satisfies the equation $\dot{b}(\hat{\theta}_n) = S_n/n$. Then $\xi_n = \widetilde{\xi}_n + \xi_n^*$, where

$$\widetilde{\xi}_n = (\hat{\theta}_n - \theta)S_n - [b(\hat{\theta}_n) - b(\theta)]n = nI(\hat{\theta}_n, \theta),$$
$$\xi_n^* = \log\left(\int_\Theta e^{(t-\hat{\theta}_n)S_n - [b(t)-b(\hat{\theta}_n)]n} w(t)\, \mathrm{d}t \right) = -\log\sqrt{n} + \log V_n(S_n),$$
$$V_n(S_n) = \sqrt{n} \int_\Theta e^{(t-\hat{\theta}_n)S_n - [b(t)-b(\hat{\theta}_n)]n} w(t)\, \mathrm{d}t.$$

Therefore, $\xi_n = \ell_n + \eta_n$ can be written as in (2.127) with $\ell_n = -\log\sqrt{n}$ and

$$\eta_n = nI(\hat{\theta}_n, \theta) + \log V_n(S_n),$$

where $V_n(S_n)$ can be shown to converge to $[2\pi/\ddot{b}(\theta)]^{1/2} w(\theta)$ under P_θ, and $nI(\hat{\theta}_n, \theta)$ converges weakly to $\chi_1^2/2$ as $n \to \infty$, where χ_1^2 is chi-squared distributed with one degree of freedom, so that

$$\eta_n \xrightarrow[n\to\infty]{\text{law}} \chi_1^2/2 + \log\left[w(\theta)\sqrt{2\pi/\ddot{b}(\theta)} \right].$$

Hence, by Theorem 2.6.4(i), assuming that all other conditions are satisfied,

$$\mathsf{E}_\theta T_a = I_\theta^{-1}(a - \ell_{N_a} + \varkappa_\theta - \bar{\eta}) + o(1) \quad \text{as } a \to \infty,$$

where $N_a = a/I_\theta$, $\ell_{N_a} = -\log(a/I_\theta)^{1/2}$, and $\bar{\eta} = 1/2 + \log[w(\theta)(2\pi/\ddot{b}(\theta))^{1/2}]$. This yields (5.186).

Consequently, it follows from (5.185) that

$$a = |\log\alpha| + \log\left(\int_{\Theta_1} \zeta_\theta \, w(\theta)\, \mathrm{d}\theta \right) + o(1), \tag{5.187}$$

and substituting (5.187) into (5.186) yields

$$I_\theta \mathsf{E}_\theta T_a = |\log\alpha| + \log\sqrt{|\log\alpha|} - \frac{1+\log(2\pi)}{2} + C(\theta, w) + o(1) \quad \text{as } \alpha \to 0, \tag{5.188}$$

where

$$C(\theta, w) = \log\left(\frac{e^{\varkappa_\theta}\sqrt{\ddot{b}(\theta)/I_\theta}}{w(\theta)} \int_{\Theta_1} \zeta_t w(t)\, \mathrm{d}t \right) \tag{5.189}$$

is the constant depending on the parameter θ and prior density w. Recall that at the moment we assume that $\Theta_1 = \Theta \setminus \{0\}$.

The asymptotic approximations (5.180) and (5.188) imply that any mixture rule with positive and continuous density on Θ_1 minimizes the expected sample size to first order for every $\theta \in \Theta_1$, i.e.,

$$\mathsf{E}_\theta T_a = \inf_{T\in\mathbf{C}(\alpha)} \mathsf{E}_\theta T\,(1+o(1)) \quad \text{as } \alpha \to 0 \text{ for all } \theta \in \Theta_1,$$

and that the distance between $\mathsf{E}_\theta T_a$ and the optimal asymptotic performance (5.180) under P_θ is

$$\mathsf{E}_\theta T_a - \inf_{T \in \mathbf{C}(\alpha)} \mathsf{E}_\theta T = O\left(\frac{1}{2I_\theta} \log\log \alpha^{-1}\right) \quad \text{for all } \theta \in \Theta_1.$$

The latter order of magnitude cannot be improved for all $\theta \in \Theta_1$ by any test, mixture, GLR, or any other.

More importantly, the approximation (5.188) shows that if density w is selected so that the constant term $C(\theta, w)$ does not depend on θ, then the mixture test $T_a(w)$ becomes an almost equalizer in the sense that the risk $\mathcal{R}_\theta(T_a) = I_\theta \mathsf{E}_\theta T_a$ does not depend on θ to within a negligible term $o(1)$. It is easy to see that this happens when the mixing density is

$$w_0(\theta) = \frac{e^{\varkappa_\theta} \sqrt{\ddot{b}(\theta)/I_\theta}}{\int_{\Theta_1} e^{\varkappa_t} \sqrt{\ddot{b}(t)/I_t}\, \mathrm{d}t}, \quad \theta \in \Theta_1, \tag{5.190}$$

in which case, for every $\theta \in \Theta_1$,

$$\mathcal{R}_\theta(T_a(w_0)) = I_\theta \mathsf{E}_\theta[T_a(w_0)] = |\log \alpha| + \log(\sqrt{|\log \alpha|}) + M + o(1) \quad \text{as } \alpha \to 0, \tag{5.191}$$

where

$$M = \log\left(\int_{\Theta_1} \zeta_t e^{\varkappa_t} \sqrt{\ddot{b}(t)/I_t}\, \mathrm{d}t\right) - \frac{1+\log(2\pi)}{2} \tag{5.192}$$

is a constant that depends on Θ_1 and the model but not on θ. In order to avoid complications, from now on we assume that Θ_1 is a finite interval bounded away from zero. In this case, (5.191) yields

$$\sup_{\theta \in \Theta_1} \mathcal{R}_\theta(T_a(w_0)) = |\log \alpha| + \log(\sqrt{|\log \alpha|}) + M + o(1) \quad \text{as } \alpha \to 0. \tag{5.193}$$

The importance of the equalizer property can be understood if we recall that by Theorem 2.7.8(iii) a minimax rule should have two properties:

(i) It is an equalizer rule, i.e., $\mathcal{R}_\theta(T)$ is constant in $\theta \in \Theta_1$, and

(ii) It is either the Bayes rule or the generalized Bayes rule (i.e., the limit of a sequence of Bayes rules and a Bayes rule with respect to the improper prior if the parameter space is not finite).

Hence, the mixture one-sided SPRT $T_a(w_0)$ is a natural candidate for the almost minimax test, and in order to establish its near minimaxity we need to show that it is Bayes or almost Bayes. Lemma 5.5.1 below shows that T_a is a Bayes rule for a sufficiently small cost c in the following Bayes problem denoted by $\mathcal{B}(\gamma, w, c)$. Let $\gamma \in (0,1)$ be the prior probability of the null hypothesis $H_0 : \theta = 0$, and assume that the loss associated with stopping at time T is 1 if $T < \infty$ and the hypothesis H_0 is true and the sampling cost is $c \cdot I_\theta$ per observation if P_θ, $\theta \in \Theta_1$, is the true probability measure, where $c > 0$ is a fixed constant. Therefore, the cost of every observation under P_θ is proportional to the difficulty in discriminating between P_θ and P_0 measured by the K–L divergence I_θ. Since the prior probability of the alternative hypothesis $H_1 : \theta \in \Theta_1$ is $(1-\gamma)\int_{\Theta_1} w(\theta) \mathrm{d}\theta = 1-\gamma$, the Bayes risk associated with an arbitrary stopping time T is

$$\rho_c^w(T) = \gamma\, \mathsf{P}_0(T < \infty) + c(1-\gamma) \int_{\Theta_1} \mathcal{R}_\theta(T) w(\theta)\, \mathrm{d}\theta. \tag{5.194}$$

For any positive constant Q such that $Qc < \gamma$, let $T_{A_{Qc}} = \inf\{n : \bar{\Lambda}_n \ge A_{Qc}\}$ be the mixture one-sided SPRT with the critical level

$$A_{Qc} = \left(\frac{1-Qc}{Qc}\right) \Big/ \left(\frac{1-\gamma}{\gamma}\right). \tag{5.195}$$

This test has a natural Bayesian interpretation. Indeed, let $\mathsf{P}^w = \int_{\Theta_1} w(\theta)\mathsf{P}_\theta \mathrm{d}\theta$ and $\mathbf{P}^{\gamma,w} = \gamma\mathsf{P}_0 + (1-\gamma)\mathsf{P}^w$. Then $\mathbf{P}^{\gamma,w}(\cdot|H_0) = \gamma\mathsf{P}_0(\cdot)$, $\mathbf{P}^{\gamma,w}(\cdot|H_1) = (1-\gamma)\mathsf{P}^w(\cdot)$, and the posterior probability of the hypothesis H_0 can be written as

$$\mathbf{P}^{\gamma,w}(H_0|\mathbf{X}_1^n) = \frac{1}{1+\frac{1-\gamma}{\gamma}\bar{\Lambda}_n}, \quad n \geq 1.$$

Thus, $T_{A_{Qc}}$ is the first time that the posterior probability of the null hypothesis becomes smaller than Qc.

Solving the problem $\mathcal{B}(\gamma,w,c)$ requires minimizing the expected loss (5.194), and the mixture test $T_{A_{Qc}}$ turns out to be optimal for sufficiently small c. Specifically, the following result holds true, whose proof can be found in Fellouris and Tartakovsky [141].

Lemma 5.5.1. *Consider the Bayesian problem $\mathcal{B}(\gamma,w,c)$. Let $T_{A_{Qc}}$ be the mixture one-sided SPRT with critical level* (5.195)*. For any given $\gamma \in (0,1)$ and $Q > 1/e$, there exists c^* such that*

$$\rho_c^w(T_{A_{Qc}}) = \inf_T \rho_c^w(T) \quad \textit{for every } c < \gamma c^*, \tag{5.196}$$

where the infimum is taken over all stopping times.

Using the equalizer property of $T_a(w_0)$ and this lemma, it is straightforward to establish the third-order asymptotic minimaxity of $T_a(w_0)$ under the criterion (5.176), i.e., in the sense of minimizing the maximal K–L information $\sup_\theta(I_\theta \mathsf{E}_\theta T)$ in the class $\mathbf{C}(\alpha)$ as $\alpha \to 0$; see (5.177).

Theorem 5.5.1. *Assume that $\Theta_1 \subset \Theta$ is a finite interval bounded away from zero. Let $T_a(w_0)$ be the mixture one-sided SPRT with mixing density* (5.190) *and $\mathsf{P}_0(T_a(w_0) < \infty) = \alpha$. Let M be defined as in* (5.192)*. If the limiting average overshoot $\varkappa_\theta$ is a continuous function on Θ_1, then*

$$\inf_{T\in\mathbf{C}(\alpha)} \sup_{\theta\in\Theta_1} (I_\theta \mathsf{E}_\theta T) = \sup_{\theta\in\Theta_1} \{I_\theta \mathsf{E}_\theta[T_a(w_0)]\} + o(1) \quad \textit{as } \alpha \to 0, \tag{5.197}$$

and

$$\sup_{\theta\in\Theta_1} \{I_\theta \mathsf{E}_\theta[T_a(w_0)]\} = \log\alpha^{-1} + \frac{1}{2}\log\log\alpha^{-1} + M + o(1) \quad \textit{as } \alpha \to 0. \tag{5.198}$$

Therefore, the one-sided mixture SPRT $T_a(w_0)$ is third-order asymptotically minimax.

Proof. Let w be an arbitrary mixing density with support on Θ_1. Let $\gamma = 1/2$ and choose $c = c_\alpha < 1/2Q$ so that $\mathsf{P}_0(T_{A_{Qc}} < \infty) = \alpha$. Then it can be verified that $\alpha \leq 2Qc$ and from the definition of $\rho_c^w(T)$ we obtain the following inequality:

$$\frac{\alpha}{2} + \frac{c}{2} \inf_{T\in\mathbf{C}(\alpha)} \sup_{\theta\in\Theta_1} \mathcal{R}_\theta(T) \geq \inf_{T\in\mathbf{C}(\alpha)} \rho_c^w(T). \tag{5.199}$$

By Lemma 5.5.1, there exists $c^* < 1/Q$ such that for every $c < c^*/2$, and consequently for every $\alpha < Qc^*$,

$$\inf_{T\in\mathbf{C}(\alpha)} \rho_c^w(T) = \rho_c^w(T_{A_{Qc}}) = \frac{\alpha}{2} + \frac{c}{2}\int_{\Theta_1} \mathcal{R}_\theta(T_{A_{Qc}}) w(\theta)\,\mathrm{d}\theta. \tag{5.200}$$

Consequently, from (5.199) and (5.200) it follows that

$$\inf_{T\in\mathbf{C}(\alpha)} \sup_{\theta\in\Theta_1} \mathcal{R}_\theta(T) \geq \int_{\Theta_1} \mathcal{R}_\theta(T_{A_{Qc}}) w(\theta)\,\mathrm{d}\theta.$$

However, if $w = w_0$ is chosen according to (5.190), then by (5.191) the test $T_{A_{Qc}}(w_0)$ is almost equalizer and

$$\int_{\Theta_1} \mathcal{R}_\theta(T_{A_{Qc}}(w_0)) w(\theta)\,\mathrm{d}\theta = |\log\alpha| + \log\sqrt{|\log\alpha|} + M + o(1) \quad \text{as } \alpha \to 0,$$

so that

$$\inf_{T\in\mathbf{C}(\alpha)}\sup_{\theta\in\Theta_1}\mathcal{R}_\theta(T)\geq|\log\alpha|+\log\sqrt{|\log\alpha|}+M+o(1)\quad\text{as }\alpha\to 0. \tag{5.201}$$

Finally, it follows from (5.193) that

$$\sup_{\theta\in\Theta_1}\mathcal{R}_\theta(T_A)=|\log\alpha|+\log\sqrt{|\log\alpha|}+M+o(1)\quad\text{as }\alpha\to 0$$

whenever $A=A_\alpha$ is chosen so that $\mathsf{P}_0(T_A<\infty)=\alpha$. The proof is complete. □

Note that for (5.191) to hold, the mixing density (5.190) must be continuous, which requires $\varkappa_\theta$ to be a continuous function, since $b(\theta)$ and $I_\theta=\theta\dot{b}(\theta)-b(\theta)$ are continuous. This is true at least when the distribution of λ_1^θ is continuous.

Remark 5.5.1. It is usually not possible to choose the threshold $a=a_\alpha$ so that the equality $\mathsf{P}_0(T_{a_\alpha}(w_0)<\infty)=\alpha$ is guaranteed exactly. However, due to (5.185), setting

$$a_\alpha=\log\left(\int_{\Theta_1}\zeta_\theta w_0(\theta)\,\mathrm{d}\theta/\alpha\right)$$

implies $\mathsf{P}_0(T_{a_\alpha}(w_0)<\infty)\sim\alpha$ as $\alpha\to 0$.

Remark 5.5.2. The asymptotic lower bound (5.201) and the asymptotic approximation (5.188) imply that the one-sided mixture SPRT $T_a(w)$ is asymptotically second-order minimax for an arbitrary prior density w which is positive and finite on the finite interval Θ_1 bounded away from zero:

$$\inf_{T\in\mathbf{C}(\alpha)}\sup_{\theta\in\Theta_1}\mathcal{R}_\theta(T)=\sup_{\theta\in\Theta_1}\mathcal{R}_\theta(T_a(w))+O(1)\quad\text{as }\alpha\to 0. \tag{5.202}$$

This result was first obtained by Pollak [364].

Theorem 5.5.1 can be extended to the multidimensional exponential family (5.69). Specifically, if Θ_1 is a bounded l-dimensional subset of Θ, $\{\mathbf{0}\}\notin\Theta_1$, then for any positive continuous density on Θ_1 (5.202) holds and uniformly on $\theta\in\Theta_1$, as $\alpha\to 0$,

$$\mathcal{R}_\theta(T_a(w))=\log\alpha^{-1}+\frac{l}{2}\log\log\alpha^{-1}-\frac{l}{2}[1+\log(2\pi)]+C(\theta,w)+o(1), \tag{5.203}$$

where

$$C(\theta,w)=\log\left(\frac{e^{\varkappa_\theta}\sqrt{\det[\nabla^2 b(\theta)]/I_\theta^l}}{w(\theta)}\int_{\Theta_1}\zeta_t w(t)\,\mathrm{d}t\right) \tag{5.204}$$

and where $\nabla^2 b(\theta)$ is the Hessian matrix of the second partial derivatives. If

$$w(\theta)=w_0(\theta)=\frac{e^{\varkappa_\theta}\sqrt{\det[\nabla^2 b(\theta)]/I_\theta^l}}{\int_{\Theta_1}e^{\varkappa_t}\sqrt{\det[\nabla^2 b(\theta)]/I_t^l}\,\mathrm{d}t},\qquad\theta\in\Theta_1,$$

then

$$\inf_{T\in\mathbf{C}(\alpha)}\sup_{\theta\in\Theta_1}\mathcal{R}_\theta(T)=\sup_{\theta\in\Theta_1}\mathcal{R}_\theta(T_a(w_0))+o(1)\quad\text{as }\alpha\to 0,$$

and

$$\sup_{\theta\in\Theta_1}\mathcal{R}_\theta(T_a(w_0))=\log\alpha^{-1}+\frac{l}{2}\log\log\alpha^{-1}+M+o(1)\quad\text{as }\alpha\to 0,$$

where

$$M=\log\left(\int_{\Theta_1}\zeta_t e^{\varkappa_t}\sqrt{\det[\nabla^2 b(t)]/I_t^l}\,\mathrm{d}t\right)-\frac{l}{2}[1+\log(2\pi)].$$

Remark 5.5.3. Analogous third-order asymptotic optimality results can be also obtained for the weighted one-sided GSLRT $\widehat{T}_a(w)$, which is based on the weighted GLR statistic $\hat{\Lambda}_n = \sup_{\theta\in\Theta_1} w(\theta)\Lambda_n^\theta$ with an appropriate choice of the prior distribution $w(\theta)$ whose shape depends on the model. However, in contrast to the mixture test, upper-bounding the error probability $\mathsf{P}_0(\widehat{T}_a(w) < \infty)$ is a difficult task, while some asymptotic approximations such as $\mathsf{P}_0(\widehat{T}_a(w) < \infty) \sim \mathsf{constant}\, a^{l/2} e^{-a}$ can be obtained. For example, consider a one-parameter exponential family and let $\theta \in (\theta_1, \theta_2) \subseteq \Theta$, $\theta_1 < 0 < \theta_2$. Define $I_{\min}(\varepsilon) = \min\{I_{-\varepsilon}, I_\varepsilon\}$, $\bar{I}_{\min} = \min\{I_{\theta_1}, I_{\theta_2}\}$, and $\widetilde{\Theta}_\varepsilon = \{\theta : I_{\min}(\varepsilon) < I_\theta < \bar{I}_{\min}\}$. Note that the interval (θ_1, θ_2) need not be the entire natural parameter space Θ. A quite tedious argument based on calculating large and moderate deviations shows that [513, Theorem 7.1]

$$\mathsf{P}_\infty(\hat{T}_a < \infty) = K_\varepsilon \sqrt{a} e^{-a}(1 + o(1)) \quad \text{as } a \to \infty, \tag{5.205}$$

where

$$K_\varepsilon = \frac{1}{\sqrt{2\pi}} \int_{\widetilde{\Theta}_\varepsilon} \sqrt{\frac{\ddot{b}(\theta)}{I_\theta}}\, \zeta_\theta \, d\theta \tag{5.206}$$

Note that the GLR test with an arbitrary prior weight, in particular the conventional GLR test with uniform $w(\theta)$, is only second-order asymptotically minimax.

Typically, the computation of optimal mixing density (5.190) requires a discretization. An example where such a discretization is not necessary is the exponential distribution. More specifically, suppose that $p_0(x) = e^{-x}$ and $p_\theta(x) = e^{-(1-\theta)x}$, $x \ge 0$, $0 < \theta < 1$. Then $b(\theta) = -\log(1-\theta)$, $I_\theta = \theta/(1-\theta) + \log(1-\theta)$ and the distribution of the overshoot κ_a^θ is exponential with rate $(1-\theta)/\theta$ for every $a > 0$. Therefore, $\varkappa_\theta = \theta/(1-\theta)$ and $\zeta_\theta = \theta$. As a result, mixing density (5.190) is completely specified up to the normalizing constant

$$\int_{\Theta_1} e^{\varkappa_\theta} \sqrt{\ddot{b}(\theta)/I_\theta}\, d\theta = \int_{\Theta_1} \frac{\exp\{\theta/(1-\theta)\}}{\sqrt{(1-\theta)[\theta + (1-\theta)\log(1-\theta)]}}\, d\theta, \tag{5.207}$$

which can be computed numerically.

Unfortunately, $\varkappa_\theta$ and ζ_θ do not have analogous closed-form expressions in terms of θ in general. Therefore, it is typically difficult to compute optimal mixing density w_0. Thus, in practice it may be more convenient to choose a mixing density w from the class of probability density functions on the whole parameter space Θ that are *conjugate* to p_θ, so that the resulting mixture rule is easily computable. However, we recall that such a mixture rule is only second-order asymptotically minimax over Θ_1.

In the following subsection, we consider another alternative to the nearly minimax continuous mixture rule.

5.5.1.2 Near Minimaxity for Discrete Spaces

A practical alternative to the optimal continuous mixture rule is to approximate the interval Θ_1 by a discrete set $\{\theta_1, \dots, \theta_N\} \subset \Theta_1$. In this case, the discrete mixture LR statistic

$$\bar{\Lambda}_n = \sum_{i=1}^N W_i e^{\lambda_n^i}, \quad n \ge 1 \tag{5.208}$$

is easily computable ($W_i = W(\theta_i)$, $\lambda_n^i = \theta_i S_n - b(\theta_i)$). However, this is not the only motivation for considering a discrete case. It naturally occurs in a number of classification applications associated with multisensor/multichannel systems that give rise to multiple isolated populations, and the goal is to test whether one of the N populations is anomalous or all populations are identical. This problem, referred to as the multisample slippage problem, was considered in Section 4.6 in the multiple decision context and will be further considered in Subsection 5.5.1.3 in more detail.

Therefore, assume that the hypotheses are $H_0 : p = p_0$ and $H_1 : p \in \mathcal{P}$, where $\mathcal{P} = \{p_1, \dots, p_N\}$ and $p_0, p_1, \dots p_N$ are arbitrary densities. Let $\{W_i\}$ be an arbitrary prior distribution, $W_i \ge 0$ for every $i = 1, \dots, N$ and $\sum_{i=1}^N W_i = 1$. Let

$$\lambda_n^i = \sum_{k=1}^n \log \frac{p_i(X_k)}{p_0(X_k)}, \quad n \ge 1, \; i = 1, \dots, N$$

denote the LLRs between the hypotheses H_i and H_0 and let $I_i = \mathsf{E}_i \lambda_1^i$, $i = 1, \dots, N$ denote the corresponding K–L numbers.

To avoid trivialities, we always assume that the I_i's are positive and finite. Most of the notation used in the previous section remains the same by replacing θ with i. In particular, $T_a^i = \inf\{n : \lambda_n^i \ge a\}$ stands for the one-sided SPRT tuned to p_i, $\kappa_a^i = \lambda_{T_a^i}^i - a$ on $\{T_a^i < \infty\}$ is the excess over a and $\mathsf{H}_i(x) = \lim_{a\to\infty} \mathsf{P}_i(\kappa_a^i \le x)$ is the limiting distribution of κ_a^i under P_i, so that

$$\zeta_i = \int_0^\infty e^{-x} \mathsf{H}_i(\mathrm{d}x) = \lim_{a\to\infty} \mathsf{E}_i \left[e^{-\kappa_a^i}\right], \quad \varkappa_i = \int_0^\infty x \mathsf{H}_i(\mathrm{d}x) = \lim_{a\to\infty} \mathsf{E}_i \kappa_a^i$$

are the corresponding limit values which are the subject of renewal theory. For the sake of simplicity, we suppose that the LLRs λ_1^i are P_j-nonarithmetic, $j = 0, i$.

Assuming in addition the second moment condition $\mathsf{E}_i |\lambda_1^i|^2 < \infty$ and using Theorem 2.5.6, in just the same way as in the previous section we obtain that (5.180) holds with $\theta = i$, i.e., for all $i = 1, \dots, N$

$$\inf_{T \in \mathbf{C}(\alpha)} \mathsf{E}_i T = \mathsf{E}_i T_a^i = I_i^{-1} (|\log \alpha| + \varkappa_i + \log \zeta_i) + o(1) \quad \text{as } \alpha \to 0. \tag{5.209}$$

Hereafter we suppose that $W_i > 0$ for all $i = 1, \dots, N$. Clearly, the asymptotic approximation (5.209) provides a benchmark for the performance of any stopping time under P_i.

Next, for every $i = 1, \dots, N$, we have the decomposition $\bar{\lambda}_n = \lambda_n^i + Y_n^i$, where λ_n^i is a random walk with positive drift I_i under P_i and the variables

$$Y_n^i = \log W_i + \log \left(1 + \sum_{j \ne i} \frac{W_j}{W_i} e^{\lambda_n^j - \lambda_n^i}\right), \quad n \ge 1 \tag{5.210}$$

are slowly changing, so that we are able to use nonlinear renewal theory to understand the asymptotic behavior of the mixture rule T_a. The argument is even simpler than in the continuous case. Indeed, it is easily seen that $e^{\lambda_n^j - \lambda_n^i} \to 0$ as $n \to \infty$ P_i-a.s., so $\mathsf{P}_i(Y_n^i \downarrow \log W_i) = 1$, and hence the sequence $\{Y_n^i\}$ is slowly changing under P_i.

Using the first nonlinear renewal theorem, we obtain that

$$e^a \mathsf{P}_0(T_a < \infty) \to \sum_{i=1}^N W_i \zeta_i \quad \text{as } a \to \infty. \tag{5.211}$$

Note that the inequality $\mathsf{P}_0(T_a < \infty) \le e^{-a}$ always holds, even in the degenerate case. Now, the second nonlinear renewal theorem yields

$$I_i \, \mathsf{E}_i T_a = a + \varkappa_i - \log W_i + o(1) \quad \text{as } a \to \infty. \tag{5.212}$$

Indeed all conditions of this theorem are satisfied; see the proof of Lemma 2.4 in Fellouris and Tartakovsky [141]. Combining (5.211) with (5.212) yields

$$I_i \, \mathsf{E}_i T_a = |\log \alpha| + \log \left(\sum_{j=1}^N W_j \zeta_j\right) + \varkappa_i - \log W_i + o(1) \quad \text{as } a \to \infty. \tag{5.213}$$

Comparing (5.212) with (5.209) shows that if $a = a_\alpha$ is selected so that $\mathsf{P}_0(T_{a_\alpha} < \infty) = \alpha$, then the one-sided WSPRT T_{a_α} is second-order asymptotically optimal with respect to the ESS:

$$\mathsf{E}_i T_{a_\alpha} = \inf_{T \in \mathbf{C}(\alpha)} \mathsf{E}_i T + O(1) \quad \text{as } \alpha \to 0 \text{ for all } i = 1, \dots, N.$$

Note the difference with the continuous-parameter case where

$$\mathsf{E}_\theta T_{a_\alpha} - \inf_{T \in \mathbf{C}(\alpha)} \mathsf{E}_\theta T = O\big((\log\log \alpha^{-1})/2I_\theta\big) \to \infty.$$

The final step is to select the best prior in order to make the test T_a nearly minimax. To this end, note first that if in (5.212) one takes

$$W_i = W_i^0 = \frac{e^{\varkappa_i}}{\sum_{j=1}^N e^{\varkappa_j}}, \quad i = 1, \dots, N, \tag{5.214}$$

then $I_i \mathsf{E}_i T_a$ becomes independent of i to within a small term $o(1)$ which vanishes as $\alpha \to 0$:

$$I_i \mathsf{E}_i[T_a(W^0))] = |\log \alpha| + \log\left(\sum_{j=1}^N \zeta_j e^{\varkappa_j}\right) + o(1).$$

Thus, it is tempting to conjecture that this test is third-order asymptotically minimax. To complete the proof, we note that Lemma 5.5.1 holds in the discrete case too, which allows one to prove the main result that establishes near minimaxity in the discrete-space case given in the next theorem.

Theorem 5.5.2. *Suppose that $\mathsf{E}_i|\lambda_1^i|^2 < \infty$ and that λ_1^i are P_i-nonarithmetic for $i = 1, \dots, N$. If the mixing distribution in the one-sided WSPRT $T_a = T_a(W^0)$ is chosen as in (5.214) and the threshold $a = a_\alpha$ so as to guarantee the exact equality $\mathsf{P}_0(T_a < \infty) = \alpha$, then as $\alpha \to 0$*

$$\begin{aligned} \inf_{T \in \mathbf{C}(\alpha)} \max_{i=1,\dots,N} (I_i \mathsf{E}_i T) &= \max_{i=1,\dots,N} \{I_i \mathsf{E}_i[T_a(W^0))]\} + o(1), \\ \max_{i=1,\dots,N} \{I_i \mathsf{E}_i[T_a(W^0))]\} &= |\log \alpha| + \log\left(\sum_{j=1}^N \zeta_j e^{\varkappa_j}\right) + o(1). \end{aligned} \tag{5.215}$$

We refer the reader to [141] for the complete detailed proofs of all the results.

We remark that similar results hold for the weighted one-sided GSPRT based on thresholding the weighted GLR statistic $\max_{1 \le i \le N} W_i e^{\lambda_n^i}$.

5.5.1.3 *A Special Multipopulation Discrete Case*

We emphasize that the problem of testing the null hypothesis $H_0 : p = p_0$ against the discrete alternative hypothesis $H_1 : p \in \{p_1, \dots, p_N\}$ in addition to being useful as an approximation to a continuous-parameter testing problem, also arises naturally in the context of multisample slippage or multichannel/multisensor problems, which have a wide range of applications. See [465, 472, 475] and Section 4.6. In this context, there are N populations some of which may be anomalous. For example, N sensors monitor different areas and a signal may be present in one or more of these areas. The problem is to decide whether there is actually a signal without identifying its location. More specifically, the sensor i takes a sequence of iid observations $\{X_n^i\}_{n \ge 1}$, whose common density is f_0^i when there is no signal and f_1^i when a signal is present. Suppose for simplicity that the observations are statistically independent across the sensors conditionally on the correct hypothesis and that the signal may appear in only one sensor which is the hardest case to detect. Then, this setup turns out to be a special case of the problem with $X_n = (X_n^1, \dots, X_n^N)$ and

$$p_0(X_n) = \prod_{j=1}^N f_0^j(X_n^j), \quad p_i(X_n) = f_1^i(X_n^i) \prod_{\substack{j=1 \\ j \ne i}}^N f_0^j(X_n^j), \quad i = 1, \dots, N,$$

so the LLRs λ_n^i take the form

$$\lambda_n^i = \sum_{k=1}^{n} \log \frac{f_1^i(X_k^i)}{f_0^i(X_k^i)}, \quad i = 1, \dots, N.$$

We close this subsection by explaining once more why we chose to work with a modified, K–L information-based minimax criterion, instead of the straightforward minimax approach $\inf_{T \in \mathbf{C}(\alpha)} \max_i \mathsf{E}_i T$, which at first glance looks seemingly natural. The reason is that such a minimax rule can be very inefficient — it is not even uniformly first-order asymptotically optimal unless we are dealing with the symmetric case $I_1 = \cdots = I_N$. Indeed, suppose $I_1 \ll I_2 = \cdots = I_N$. Then, to attain $\inf_{T \in \mathbf{C}(\alpha)} \max_i \mathsf{E}_i T$, even asymptotically, one should use the one-sided SPRT T_a^1 tuned to f_1^1, which is optimal under P_1, but ignores all the other states of the alternative hypothesis. This is clearly not a meaningful answer, so the straightforward minimax criterion is not appropriate.

5.5.1.4 *Monte Carlo Simulations*

We now study the accuracy of the asymptotic approximations with the help of simulation experiments in the Gaussian case where $p_0(x) = \varphi(x)$ and $p_i(x) = \varphi(x-i)$ for $i = 1,2,3$, where $\varphi(x) = (2\pi)^{-1/2} e^{-x^2/2}$ is the standard normal pdf. Thus, the observations are normally distributed with unit variance and zero mean under H_0 and with mean either 1 or 2 or 3 under H_1 ($N = 3$). In this example, $I_i = i^2/2$ and the quantities $\varkappa_i$ and ζ_i can be computed with any precision using the following expressions:

$$\varkappa_i = 1 + \frac{i^2}{4} - i \sum_{n=1}^{\infty} \left[\frac{1}{\sqrt{n}} \varphi\left(\frac{i}{2}\sqrt{n}\right) - \frac{i}{2} \Phi\left(-\frac{i}{2}\sqrt{n}\right) \right],$$
$$\zeta_i = \frac{2}{i^2} \exp\left\{ -2 \sum_{n=1}^{\infty} \frac{1}{n} \Phi\left(-\frac{i}{2}\sqrt{n}\right) \right\}. \tag{5.216}$$

We also study the efficiency loss when several other priors are used instead of the optimal prior $W_i^0 \propto e^{\varkappa_i}$, $i = 1,2,3$.

The performance loss of the mixture one-sided SPRT $T_a = T_a(W)$ with an arbitrary mixing prior $W = \{W_i\}$, assuming that the error probability $\mathsf{P}_0(T_a < \infty) = \alpha$, compared with the optimal minimax test can be defined as

$$\mathcal{L}_\alpha(W) = \max_{1 \le i \le N} [I_i \mathsf{E}_i T_a(W)] - \inf_{T \in \mathbf{C}(\alpha)} \max_{1 \le i \le N} (I_i \mathsf{E}_i T). \tag{5.217}$$

By (5.213), as $\alpha \to 0$

$$\max_{1 \le i \le N} [I_i \mathsf{E}_i T_a(W)] = |\log \alpha| + \log \left[\left(\sum_{j=1}^{N} W_j \zeta_j \right) \left(\max_{1 \le i \le N} (e^{\varkappa_i}/W_i) \right) \right] + o(1), \tag{5.218}$$

which along with (5.215) yields

$$\mathcal{L}(W) = \log \frac{(\sum_{j=1}^{N} W_j \zeta_j)\ (\max_{1 \le i \le N} (e^{\varkappa_i}/W_i))}{\sum_{j=1}^{N} e^{\varkappa_j} \zeta_j}, \tag{5.219}$$

where $\mathcal{L}(W) = \lim_{\alpha \to 0} \mathcal{L}_\alpha(W)$ is the asymptotic loss.

Clearly, $\mathcal{L}(W) > \mathcal{L}(W^0) = 0$ for any W, where $W^0 = \{W_i^0\}$ is the best mixing distribution defined in (5.214). Along with the uniform mixing distribution $W^u = \{W_i^u\}$, $W_i^u = 1/N$, which would be perhaps the first choice for practical implementation, consider the following mixing distributions:

$$W_i^{KL} = \frac{I_i}{\sum_{j=1}^{N} I_j}, \quad W_i^{1/\zeta} = \frac{1/\zeta_i}{\sum_{j=1}^{N} (1/\zeta_j)}, \quad W_i^{e^{\varkappa}/\zeta} = \frac{e^{\varkappa_i}/\zeta_i}{\sum_{j=1}^{N} (e^{\varkappa_j}/\zeta_j)}, \quad 1 \le i \le N, \tag{5.220}$$

which resemble W^0 in that they all give more weight to those members of $\mathcal{P}$ that are further apart from P_0. Notice also that in the completely symmetric case where the P_i-distribution of λ_1^i does not depend on i, these mixing distributions reduce to the uniform mixing W^u. Using (5.219), we obtain

$$\mathcal{L}(W^{KL}) = \log \frac{(\sum_{j=1}^N \zeta_j I_j)\ (\max_{1\le i\le N}(e^{\varkappa_i}/I_i))}{\sum_{j=1}^N \zeta_j e^{\varkappa_j}}, \quad \mathcal{L}(W^{1/\zeta}) = \log \frac{N(\max_{1\le i\le N}(\zeta_i e^{\varkappa_i}))}{\sum_{j=1}^K \zeta_j e^{\varkappa_j}},$$

$$\mathcal{L}(W^u) = \log \frac{(\sum_{j=1}^N \zeta_j)\ (\max_{1\le i\le N} e^{\varkappa_i})}{\sum_{j=1}^N \zeta_j e^{\varkappa_j}}, \quad \mathcal{L}(W^{e^{\varkappa}/\zeta}) = \log \frac{(\sum_{j=1}^N e^{\varkappa_j})\ (\max_{1\le i\le N} \zeta_i)}{\sum_{j=1}^N \zeta_j e^{\varkappa_j}}.$$

In Table 5.4, we present the quantities $\varkappa_i$ and ζ_i given by (5.216), the optimal mixing distribution (5.214), as well as the mixing distributions defined in (5.220). Using Table 5.4, we can compute the asymptotic performance loss (5.219) for each of the corresponding mixture rules:

$$\mathcal{L}(W^{KL}) = 0.21, \quad \mathcal{L}(W^{1/\zeta}) = 0.58, \quad \mathcal{L}(W^{e^{\varkappa}/\zeta}) = 0.85, \quad \mathcal{L}(W^u) = 1.21.$$

It is seen that the uniform distribution gives the highest loss and the KL-distribution the smallest.

Table 5.4: *Different mixing distributions.*

i	I_i	$\varkappa_i$	ζ_i	$W^{e^{\varkappa}/\zeta}$	W^0	W^{KL}	$W^{1/\zeta}$	W^u
1	0.5	0.718	0.560	0.25	0.066	0.071	0.176	0.33
2	2	1.747	0.320	0.125	0.185	0.286	0.307	0.33
3	4.5	3.146	0.190	0.85	0.749	0.643	0.517	0.33

The asymptotic approximation (5.211) suggests that if we set the threshold a as

$$a = \log\left(\frac{\sum_{i=1}^N W_i \zeta_i}{\alpha}\right), \tag{5.221}$$

then the error probability $\mathsf{P}_0(T_a(W) < \infty)$ is expected to be approximately equal to α for sufficiently small values of α. In Table 5.5, we present the actual probabilities computed using Monte Carlo simulations. An importance sampling technique was used in these experiments, taking advantage of the representation $\mathsf{P}_0(T_a(W) < \infty) = \sum_i W_i \mathsf{E}_i[e^{-\bar{\lambda}_{T_a}}]$. This allowed us to evaluate a very low error probability with a reasonable number of Monte Carlo runs. It is seen that the formula (5.221) ensures an extremely high accuracy of the approximation of the desired error probability for all mixing distributions as long as $\alpha \le 10^{-2}$.

Table 5.5: *Error probability $\mathsf{P}_0(T_a(W) < \infty)$ for different mixing distributions. The first column represents the desired error probabilities; the other columns represent the actual error probabilities obtained by Monte Carlo simulations when the threshold is chosen according to* (5.221).

α	$W^{e^{\varkappa}/\zeta}$	W^0	W^{KL}	$W^{1/\zeta}$	W^u
10^{-1}	$5.9979\,10^{-2}$	$6.7037\,10^{-2}$	$8.0337\,10^{-2}$	$8.0029\,10^{-2}$	$8.9314\,10^{-2}$
10^{-2}	$9.1127\,10^{-3}$	$9.4317\,10^{-3}$	$9.8754\,10^{-3}$	$9.8885\,10^{-3}$	$1.0049\,10^{-2}$
10^{-4}	$1.0104\,10^{-4}$	$1.0107\,10^{-4}$	$1.0027\,10^{-4}$	$1.0038\,10^{-4}$	$1.0011\,10^{-4}$
10^{-6}	$1.0017\,10^{-6}$	$1.0006\,10^{-6}$	$1.0009\,10^{-6}$	$1.0004\,10^{-6}$	$1.0008\,10^{-6}$
10^{-8}	$1.0008\,10^{-8}$	$1.0033\,10^{-8}$	$1.0002\,10^{-8}$	$1.0017\,10^{-8}$	$1.0006\,10^{-8}$

Finally, Table 5.6 allows us to verify the accuracy of the asymptotic approximation (5.218) for the maximal expected accumulated K–L information $\max_i(I_i\mathsf{E}_i[T_a(W)])$ for the optimal and uniform mixing distributions $W = W^0$ and $W = W^u$. For the optimal mixing distribution W^0, the asymptotic approximation (5.218) is reasonably accurate for all studied error probabilities $\alpha \le 0.01$. For the uniform mixing distribution, this approximation is considerably less accurate, but improves significantly as the error probability goes to 0.

Table 5.6: *The maximal expected accumulated K–L information* $\max_i(I_i\mathsf{E}_i[T_a(W)])$ *for the optimal and uniform mixing distributions* W^0 *and* W^u*. The threshold a is selected according to* (5.221).

(a) Optimal mixing distribution

α	Monte Carlo	Approx. (5.218)
10^{-1}	4.99	4.31
10^{-2}	6.36	6.61
10^{-4}	10.99	11.21
10^{-6}	15.65	15.82
10^{-8}	20.33	20.42

(b) Uniform mixing distribution

α	Monte Carlo	Approx. (5.218)
10^{-1}	5.04	5.52
10^{-2}	6.88	7.82
10^{-4}	11.87	12.42
10^{-6}	16.59	17.03
10^{-8}	21.29	21.63

5.5.2 Nearly Minimax Double-Sided Sequential Tests

We now generalize the results of Subsection 5.5.1.2 for the class $\mathbf{C}(\alpha_0,\alpha_1) = \{\delta : \mathsf{P}_0(d=1) \le \alpha_0, \max_{1\le i\le N}\mathsf{P}_i(d=0) \le \alpha_1\}$ in the problem of testing the simple null hypothesis $H_0 : p = p_0$ *versus* the discrete composite hypothesis $H_1 : p \in \{p_1,\dots,p_N\}$, showing that the specially designed two-sided WSPRT and weighted GSLRT (WGSLRT) are almost minimax to third-order in the sense of the criterion

$$\inf_{\delta\in\mathbf{C}(\alpha_0,\alpha_1)} \max_{1\le i\le N}(I_i\mathsf{E}_iT) \quad \text{and} \quad \inf_{\delta\in\mathbf{C}(\alpha_0,\alpha_1)} \mathsf{E}_0T. \tag{5.222}$$

Let $\mathbf{w}_0 = (w_1^0,\dots,w_N^0)^\top$ and $\mathbf{w}_1 = (w_1^1,\dots,w_N^1)^\top$ be the vectors of positive weights (normalization to 1 is optional) and define the mixed LRs and weighted GLRs

$$\bar{\Lambda}_n^i(\mathbf{w}_i) = \sum_{k=1}^N w_k^i\Lambda_n^k, \quad \hat{\Lambda}_n^i(\mathbf{w}_i) = \max_{1\le k\le N} w_k^i\Lambda_n^k, \quad i=0,1.$$

Let $\bar{\lambda}_n^i(\mathbf{w}_i) = \log\bar{\Lambda}_n^i(\mathbf{w}_i)$, $\hat{\lambda}_n^i(\mathbf{w}_i) = \log\hat{\Lambda}_n^i(\mathbf{w}_i)$. The two-sided WSPRT $\bar{\delta} = (\bar{T},\bar{d})$ is defined as

$$\bar{T}(a_1,a_0) = \min\{\bar{T}_0(a_0,\mathbf{w}_0),\bar{T}_1(a_1,\mathbf{w}_1)\}, \quad \bar{d} = i \ \text{ if } \bar{T} = \bar{T}_i, \quad i=0,1, \tag{5.223}$$

where

$$\bar{T}_1(a_1,\mathbf{w}_1) = \inf\{n\ge 1 : \bar{\lambda}_n^1 \ge a_1\}, \quad \bar{T}_0(a_0,\mathbf{w}_0) = \inf\{n\ge 1 : \bar{\lambda}_n^0 \le a_0\} \tag{5.224}$$

and $a_0 < 0$, $a_1 > 1$ are fixed thresholds. The two-sided WGSLRT $\hat{\delta} = (\hat{T},\hat{d})$ is defined as

$$\hat{T}(a_1,a_0) = \min\{\hat{T}_0(a_0,\mathbf{w}_0),\hat{T}_1(a_1,\mathbf{w}_1)\}, \quad \hat{d} = i \ \text{ if } \hat{T} = \hat{T}_i, \quad i=0,1, \tag{5.225}$$

where

$$\hat{T}_1(a_1,\mathbf{w}_1) = \inf\left\{n\ge 1 : \hat{\lambda}_n^1 \ge a_1\right\}, \quad \hat{T}_0(A_0,\mathbf{w}_0) = \inf\left\{n\ge 1 : \hat{\lambda}_n^0 \le a_0\right\}. \tag{5.226}$$

The main issue is to show how to select the weights $\mathbf{w}_0$, $\mathbf{w}_1$ so that both sequential tests attain the infimums in (5.222) up to $o(1)$ terms as $\alpha_{\max} = \max(\alpha_0, \alpha_1) \to 0$. The proof of this strong asymptotic optimality property relies on the asymptotic solution of an auxiliary Bayesian problem and higher-order asymptotic expansions for the expected sample sizes.

For every $i = 1, \dots, N$, define the one-sided stopping times

$$\tau_c^i = \inf\{n \ge 1 : \lambda_n^i \ge c\}, \quad \sigma_c^i = \inf\{n \ge 1 : \lambda_n^i \le -c\}, \quad c > 0, \tag{5.227}$$

where $\lambda_n^i = \sum_{k=1}^n \log[p_i(X_k)/p_0(X_k)]$. Throughout the rest of this section we assume that the LLRs λ_1^i are P_i-nonarithmetic, $i = 0, 1, \dots, N$, and that $\mathsf{E}_i|\lambda_1^i|^2 < \infty$ and $\mathsf{E}_0|\lambda_1^i|^2 < \infty$. Thus, the K–L numbers $I_i = \mathsf{E}_i\lambda_1^i$ and $I_0^i = \mathsf{E}_0(-\lambda_1^i)$ are finite. Let

$$\mathsf{H}_i^0(x) = \lim_{c\to\infty} \mathsf{P}_0(-\lambda_{\sigma_c^i}^i - c \le x), \quad \mathsf{H}_i^1(x) = \lim_{c\to\infty} \mathsf{P}_i(\lambda_{\tau_c^i}^i - c \le x), \quad x > 0 \tag{5.228}$$

denote the limiting distributions of the overshoots $\lambda_{\tau_c^i}^i - c$ and $\lambda_{\sigma_c^i}^i + c$ and define the Laplace transforms

$$\zeta_i^0 = \int_0^\infty e^{-x}\,\mathsf{H}_i^0(\mathrm{d}x), \quad \zeta_i^1 = \int_0^\infty e^{-x}\,\mathsf{H}_i^1(\mathrm{d}x) \tag{5.229}$$

and the limiting average overshoots

$$\varkappa_i^0 = \int_0^\infty x\,\mathsf{H}_i^0(\mathrm{d}x)\,, \quad \varkappa_i^1 = \int_0^\infty x\,\mathsf{H}_i^1(\mathrm{d}x). \tag{5.230}$$

Obviously, the mixed LLR and the GLLR can be written as

$$\bar{\lambda}_n(\mathbf{w}) = \lambda_n^i + \log w_i + Y_n^i(\mathbf{w}), \quad \hat{\lambda}_n(\mathbf{w}) = \lambda_n^i + \log w_i + \hat{Y}_n^i(\mathbf{w}), \quad n \ge 1, \tag{5.231}$$

where

$$Y_n^i(\mathbf{w}) = \log\left(1 + \sum_{1\le j\ne i\le N} \frac{w_j}{w_i}\exp\left(\lambda_n^j - \lambda_n^i\right)\right), \quad n \ge 1, \tag{5.232}$$

$$\hat{Y}_n^i(\mathbf{w}) = \log\left(\max\left\{1, \max_{1\le j\ne i\le N} \frac{w_j}{w_i}\exp\left(\lambda_n^j - \lambda_n^i\right)\right\}\right), \quad n \ge 1. \tag{5.233}$$

We need several auxiliary results. The following lemma is elementary.

Lemma 5.5.2. *The following assertions hold for any weight* $\mathbf{w}$ *and any* $i = 1, \dots, N$*:*

(i) *For every* $n \ge 1$,

$$0 \le \hat{Y}_n^i(\mathbf{w}) \le Y_n^i(\mathbf{w}) = \hat{Y}_n^i(\mathbf{w}) + C_N, \tag{5.234}$$

and consequently,

$$\lambda_n^i + \log w_i \le \hat{\lambda}_n(\mathbf{w}) \le \bar{\lambda}_n(\mathbf{w}) = \hat{\lambda}_n(\mathbf{w}) + C_N, \tag{5.235}$$

where $0 < C_N < 2(N-1)$.

(ii) *As* $n \to \infty$, $\hat{Y}_n^i(\mathbf{w})$ *and* $Y_n^i(\mathbf{w})$ *converge to* 0 *almost surely and in mean under* P_i. *Therefore, the sequences* $\{\hat{Y}_n^i(\mathbf{w})\}_{n\ge1}$ *and* $\{Y_n^i(\mathbf{w})\}_{n\ge1}$ *are slowly changing.*

Hereafter without special emphasis we always assume that a_1 and $|a_0|$ approach infinity in such a way that $a_1 e^{-|a_0|} \to 0$ and $|a_0| e^{-a_1} \to 0$.

Lemma 5.5.3. *As* $|a_0| \to \infty$ *for all* $m > 0$

$$\left(\bar{T}_0(a_0, \mathbf{w}_0) e^{-|a_0|}\right)^m \to (1/I_0)^m \quad \text{and} \quad \left(\hat{T}_0(a_0, \mathbf{w}_0) e^{-|a_0|}\right)^m \to (1/I_0)^m$$

P_0*-almost surely and in mean and as* $a_1 \to \infty$ *for all* $m > 0$

$$\left(\bar{T}_1(a_1, \mathbf{w}_1) e^{-a_1}\right)^m \to (1/I_i)^m \quad \text{and} \quad \left(\hat{T}_1(a_1, \mathbf{w}_1) e^{-a_1}\right)^m \to (1/I_i)^m$$

P_i*-almost surely and in mean for every* $i = 1, \dots, N$.

Proof. The proof follows the same steps as in Appendix A of [465]. □

Define the overshoots

$$\chi(a_0,a_1) = \bar{\lambda}_{\bar{T}}(\mathbf{w}_1) - a_1]\mathbb{1}_{\{\bar{d}=1\}} - [\bar{\lambda}_{\bar{T}}(\mathbf{w}_0) + |a_0|]\mathbb{1}_{\{\bar{d}=0\}},$$
$$\hat{\chi}(a_0,a_1) = [\hat{\lambda}_{\hat{T}}(\mathbf{w}_1) - a_1]\mathbb{1}_{\{\hat{d}=1\}} - [\hat{\lambda}_{\hat{T}}(\mathbf{w}_0) + |a_0|]\mathbb{1}_{\{\hat{d}=0\}}.$$

The following lemma provides bounds for the error probabilities of the tests $\bar{\delta}$ and $\hat{\delta}$. Write $\alpha_0(\delta) = \mathsf{P}_0(d=1)$ and $\alpha_1(\delta) = \max_{1\le i\le N} \mathsf{P}_i(d=0)$.

Lemma 5.5.4. (i) *For any* a_1, a_0,

$$\alpha_0(\bar{\delta}) \le e^{-a_1}\sum_{j=1}^{N} w_j^1, \quad \alpha_0(\hat{\delta}) \le e^{-a_1}\sum_{j=1}^{N} w_j^1, \tag{5.236}$$

$$\alpha_1(\bar{\delta}) \le \frac{1}{\min\limits_{1\le i\le N} w_i^0} e^{-|a_0|}, \quad \alpha_1(\hat{\delta}) \le \frac{1}{\min\limits_{1\le i\le N} w_i^0} e^{-|a_0|}. \tag{5.237}$$

(ii) *As* $a_1, |a_0| \to \infty$,

$$\alpha_0(\bar{\delta}) = e^{-a_1}\left(\sum_{j=1}^{N} w_j^1 \zeta_j^1\right)(1+o(1)), \tag{5.238}$$

$$\alpha_0(\hat{\delta}) \le e^{-a_1}\left(\sum_{j=1}^{N} w_j^1 \zeta_j^1\right)(1+o(1)). \tag{5.239}$$

Proof. Define the probability measure $\mathsf{P}^w = [\sum_{j=1}^N w_j^1]^{-1}\sum_{i=1}^N w_i^1 \mathsf{P}_i$ and let E^w denote the corresponding expectation. Changing the measure $\mathsf{P}_0 \mapsto \mathsf{P}^w$, we obtain

$$\alpha_0(\bar{\delta}) = \mathsf{E}^w\left[e^{-\{\bar{\lambda}_{\bar{T}}(\mathbf{w}_1) - \log\sum_{j=1}^N w_j^1\}}\mathbb{1}_{\{\bar{d}=1\}}\right] = \sum_{i=1}^{N} w_i^1 \mathsf{E}_i\left[e^{-\bar{\lambda}_{\bar{T}}(\mathbf{w}_1)}\mathbb{1}_{\{\bar{d}=1\}}\right]$$
$$= e^{-a_1}\sum_{i=1}^{N} w_i^1 \mathsf{E}_i\left[e^{-\chi}\mathbb{1}_{\{\bar{d}=1\}}\right].$$

The first inequality in (5.236) follows immediately since $\mathsf{E}_i[e^{-\chi}\mathbb{1}_{\{\bar{d}=1\}}] \le 1$. In order to establish (5.238) recall that $\bar{\lambda}_n(\mathbf{w}_1)$ can be written as in (5.231) where by Lemma 5.5.2(ii) $Y_n^i(\mathbf{w}_1)$, $n \ge 1$ are slowly changing. Since $\mathsf{P}_i(\bar{d}=1) \to 1$ as $a_{\min} \to \infty$, by the first nonlinear renewal theorem,

$$\mathsf{E}_i[e^{-\chi}\mathbb{1}_{\{\bar{d}=1\}}] \to \zeta_i^1 \quad \text{as } a_{\min} \to \infty \tag{5.240}$$

and (5.238) follows.

For the rest of the proof we refer to Fellouris and Tartakovsky [142]. □

It follows from Lemma 5.5.4(i) that if the thresholds a_0, a_1 are chosen as

$$a_0 = \log\left(\alpha_1 \min_{1\le i\le N} w_i^0\right), \quad a_1 = \log\left(\frac{\sum_{j=1}^N w_j^1}{\alpha_0}\right), \tag{5.241}$$

then $\bar{\delta}, \hat{\delta} \in \mathbf{C}(\alpha_0, \alpha_1)$. From Lemma 5.5.4(ii) it follows that if a_1 is chosen as

$$a_1(\alpha_0, \mathbf{w}_1) = \log\left(\frac{\sum_{j=1}^N w_j^1 \zeta_i^1}{\alpha_0}\right), \tag{5.242}$$

then $\alpha_0(\bar{d}) = \alpha_0(1+o(1))$ and $\alpha_0(\hat{d}) \le \alpha_0(1+o(1))$ as $\alpha_{\max} \to 0$. Hence, the WSPRT has the advantage over the WGSLRT.

The next step is to obtain approximations for the ESS of the tests. Recall that we suppose that the LLRs λ_1^i are nonarithmetic under P_0 and P_i, $i = 1,\dots,N$, and that $\mathsf{E}_i|\lambda_1^i|^2 < \infty$.

First, consider the hypothesis H_1. Clearly,

$$T_1(a_1,\mathbf{w}_1) - \bar{T}(a_0,a_1) = [T_1(a_1,\mathbf{w}_1) - T_0(a_0,\mathbf{w}_0)]\,\mathbb{1}_{\{\bar{d}=0\}} \le T_1(a_1,\mathbf{w}_1)\,\mathbb{1}_{\{\bar{d}=0\}}.$$

Applying the Cauchy–Schwartz inequality and using Lemmas 5.5.3 and 5.5.4, we obtain

$$\mathsf{E}_i[T_1(a_1,\mathbf{w}_1)\mathbb{1}_{\{\bar{d}=0\}}] \le \sqrt{\mathsf{E}_i[(T_1(a_1,\mathbf{w}_1)^2]\mathsf{P}_i(\bar{d}=0)} = O(a_1^2 e^{-|a_0|}) \to 0,$$

so $\mathsf{E}_i T_1(a_1,\mathbf{w}_1) - \mathsf{E}_i\bar{T}(a_0,a_1) = o(1)$, which along with (5.212) yields[1]

$$I_i\,\mathsf{E}_i\bar{T} = a_1 + \varkappa_i^1 - \log w_i^1 + o(1), \quad I_i\,\mathsf{E}_i\hat{T} = a_1 + \varkappa_i^1 - \log w_i^1 + o(1) \tag{5.243}$$

for all $i = 1,\dots,N$.

Two observations are in order. First, if the thresholds are chosen according to (5.241), then both sequential tests $\bar{\delta}$ and $\hat{\delta}$ minimize the expected sample size to within a constant under every P_i, $i = 1,\dots,N$, i.e., as $\alpha_{\max} \to 0$ so that $\alpha_1|\log\alpha_0| = o(1)$,

$$\mathsf{E}_i\bar{T} = \inf_{\delta\in\mathbf{C}(\alpha_0,\alpha_1)} \mathsf{E}_i T + O(1), \quad \mathsf{E}_i\hat{T} = \inf_{\delta\in\mathbf{C}(\alpha_0,\alpha_1)} \mathsf{E}_i T + O(1). \tag{5.244}$$

Second, if a_1 is chosen as in (5.242), then

$$\begin{aligned} I_i\,\mathsf{E}_i\bar{T} &= |\log\alpha_0| + \log\left(\sum_{j=1}^N w_j^1\zeta_j^1\right) + \varkappa_i^1 - \log w_i^1 + o(1),\\ I_i\,\mathsf{E}_i[\hat{T}] &= |\log\alpha_0| + \log\left(\sum_{j=1}^N w_j^1\zeta_j^1\right) + \varkappa_i^1 - \log w_i^1 + o(1). \end{aligned} \tag{5.245}$$

Consider now the hypothesis H_0. Write $I_0^i = \mathsf{E}_0(-\lambda_1^i)$ for the K–L distance between p_0 and p_i and $I_0 = \min_{1\le i\le N} I_0^i$ for the minimal K–L distance. Without loss of generality assume that the densities $p_1,\dots,p_N$ are ordered with respect to the K–L distances I_0^i,

$$I_0 = I_0^1 \le I_0^2 \le \cdots \le I_0^N.$$

Notice that $I_0^i = \mathsf{E}_0[\log p_0(X_1)] - \mu_i$ and $I_0 = \mathsf{E}_0[\log p_0(X_1)] - \mu$, where

$$\mu = \max_{1\le i\le N}\mu_i, \quad \mu_i = \mathsf{E}_0[\log p_i(X_1)], \quad i = 1,\dots,N. \tag{5.246}$$

By r denote the number of densities whose K–L distance from p_0 attains its minimum, i.e.,

$$I_0 = \min_{1\le i\le N} I_0^i = I_0^1 = \cdots = I_0^r < I_0^{r+1} \le \cdots \le I_0^N. \tag{5.247}$$

The asymptotic analysis of the operating characteristics depends on whether $r = 1$ or $r > 1$. The first case is the completely asymmetric case where there is a unique density for which the K–L distance is the smallest. The case $r = N$ corresponds to the fully symmetric situation where all the densities have the same K–L distance to p_0.

We begin with the asymmetric case $r = 1$. Obviously, this case is no different from the previous

[1]For the WGSLRT the argument is essentialy the same.

consideration for the expectation E_i under P_i, so that we can just borrow the previous results. Indeed, the stopping time $\bar{T}_0(a_0, \mathbf{w}_0)$ can be written as

$$\bar{T}_0(a_0, \mathbf{w}_0) = \inf\{n \geq 1 : -\lambda_n^1 - Y_n^1(\mathbf{w}_0) - \log w_1^0 \geq |a_0|\},$$

where $\{-\lambda_n^1\}$ is a random walk with mean I_0 under P_0, whereas the SLLN and (5.247) imply that $Y_n^1(\mathbf{w}_0)$ converges to 0 almost surely under P_0 as $n \to \infty$. This implies that $\{Y_n^1(\mathbf{w}_0)\}$ is a slowly changing sequence under P_0. Changing the measure $\mathsf{P}_i \mapsto \mathsf{P}_0$ and using the first nonlinear renewal theorem, we obtain that as $a_{\min} \to \infty$

$$\mathsf{P}_i(\bar{d} = 0) \leq (\zeta_1^0 / w_i^0) e^{-|a_0|}(1 + o(1)), \quad \mathsf{P}_i(\hat{d} = 0) \leq (\zeta_1^0 / w_i^0) e^{-|a_0|}(1 + o(1)). \tag{5.248}$$

Hence, it follows from (5.248) that if a_0 is chosen as

$$a_0(\alpha_1, \mathbf{w}_0) = \log\left(\alpha_1 \min_{1 \leq i \leq N} w_i^0 / \zeta_1^0\right), \tag{5.249}$$

then as $a_{\min} \to \infty$

$$\max_{1 \leq i \leq N} \mathsf{P}_i(\bar{d} = 0) \leq \alpha_1(1 + o(1)), \quad \max_{1 \leq i \leq N} \mathsf{P}_i(\hat{d} = 0) \leq \alpha_1(1 + o(1)).$$

Also, using the second nonlinear renewal theorem, analogously to (5.243) as $a_{\min} \to \infty$ we obtain

$$I_0 \, \mathsf{E}_0 \bar{T} = |a_0| + \varkappa_1^0 - \log w_1^0 + o(1), \quad I_0 \, \mathsf{E}_0 \hat{T} = |a_0| + \varkappa_1^0 - \log w_1^0 + o(1). \tag{5.250}$$

Consider now the general case $1 < r \leq N$. In order to obtain higher-order asymptotic expansions for the ESSs in this case, we need stronger integrability conditions and additional notation. To apply nonlinear renewal theory, it is convenient to write the stopping time $\hat{T}_0(a_0, \mathbf{w}_0)$ in the following form:

$$\hat{T}_0(a_0, \mathbf{w}_0) = \inf\left\{n \geq 1 : S_n^0 \geq |a_0| + \max_{1 \leq i \leq N} (\log w_i^0 + S_n^i)\right\}, \tag{5.251}$$

where

$$S_n^i = \sum_{k=1}^n \log p_i(X_k), \quad n \geq 1, \; i = 0, 1, \dots, N.$$

Recall that $\mu = \max_i \mu_i = \mu_1 = \cdots = \mu_r$ and define the r-dimensional random vector $\mathbf{V} = (S_1^1 - \mu, \dots, S_1^r - \mu)^\top$. Let Σ be the covariance matrix of $\mathbf{V}$ under P_0, φ_Σ the joint density of r zero-mean Gaussian random variables with covariance matrix Σ and set

$$d_r = h_r / 2\sqrt{I_0}, \quad h_r = \int_{\mathbb{R}^r} \left(\max_{1 \leq i \leq r} x_i\right) \varphi_\Sigma(\mathbf{x}) \, \mathrm{d}\mathbf{x},$$

where $\mathbf{x} = (x_1, \dots, x_r)^\top$, i.e., h_r is the expected value of the maximum of r zero-mean Gaussian random variables with covariance matrix Σ. Also define the following quantities:

$$D_r(\mathbf{w}_0) = \frac{h_r^2}{2I_0} + \varkappa_1^0 + \int_{\mathbb{R}^r} \left[\max_{1 \leq i \leq r} (x_i) \left(\mathcal{P}(\mathbf{x}) + \boldsymbol{\ell}^\top(\mathbf{w}_0) \Sigma^{-1} \mathbf{x}\right)\right] \varphi_\Sigma(\mathbf{x}) \, \mathrm{d}\mathbf{x},$$

where $\boldsymbol{\ell}(\mathbf{w}_0) = (\log w_1^0, \dots, \log w_r^0)^\top$ and $\mathcal{P}$ a third-degree polynomial whose coefficients depend on the P_0-cumulants of $\mathbf{V}$ [66]. Applying the general nonlinear renewal theorem (Theorem 2.6.6) to the stopping time (5.251), it can be proved that as $a_{\min} \to \infty$

$$I_0 \, \mathsf{E}_0 \bar{T} = |a_0| + 2d_r \sqrt{|a_0|} + O(1), \quad I_0 \, \mathsf{E}_0 \hat{T} = |a_0| + 2d_r \sqrt{|a_0|} + O(1). \tag{5.252}$$

If in addition the following conditions are satisfied:

(A1) The distribution of $\log p_0(X_1)$ is nonarithmetic under P_0;

(A2) The covariance matrix Σ is positive-definite;

(A3) $\mathsf{E}_0[||\mathbf{V}||_2^3] < \infty$ and $\limsup_{||\mathbf{t}||_2 \to \infty} \mathsf{E}_0[e^{j\mathbf{t}^\top \mathbf{V}}] < 1$, where j is the imaginary unit and $\mathbf{t} = (t_1, \dots, t_r)$,

then using the second general nonlinear renewal theorem (Theorem 2.6.7) it can be established that as $a_{\min} \to \infty$

$$\begin{aligned} I_0\,\mathsf{E}_0\bar{T} &= |a_0| + 2d_r\sqrt{|a_0| + C_N + d_r^2} + D_r(\mathbf{w}_0) + C_N + o(1), \\ I_0\,\mathsf{E}_0\hat{T} &= |a_0| + 2d_r\sqrt{|a_0| + d_r^2} + D_r(\mathbf{w}_0) + o(1), \end{aligned} \tag{5.253}$$

where $0 < C_N < 2(N-1)$ is a constant as in Lemma 5.5.2. The proof of the higher-order approximations (5.253) is quite tedious and runs along the lines of the argument presented in Subsection 4.3.3.2 for the multiple hypothesis case; see the proof sketch of Theorem 4.3.4. The details can be found in Tartakovsky *et al.* [465] for the GSLRT with uniform weights.

The final step is to show that the tests $\bar{\delta}$ and $\hat{\delta}$ are almost Bayes and equalizers with a special choice of the weights. We provide only a heuristic argument and the final result without proof, referring the reader to the paper of Fellouris and Tartakovsky [142] for all mathematical details. Note that it follows from the asymptotic expansions (5.245) that both tests are almost equalizers to within $o(1)$ with respect to the accumulated expected K–L information $I_i\mathsf{E}_iT$, $i \in \{1,\dots,N\}$ if we select $w_i^1 = e^{\varkappa_i^1}$, so that with these weights the tests are candidates for nearly minimax tests. This of course can be expected since for small error probabilities the one-sided WSPRT is very close to the two-sided WSPRT (to within $o(1)$). In addition, we have to select the weights $\mathbf{w}_0$ to make it almost optimal with respect to E_0T. To do this recall that by Lemma 4.3.2 the local one-sided SPRT

$$T_0^i = \inf\left\{n \ge 1 : W_i\Upsilon_i\Lambda_n^i \le B_c\right\},$$

where $\Upsilon_i = I_i^0\zeta_i^0 = I_i\zeta_i^1$ and W_i is the prior probability of the i-th local hypothesis $H_i : p = p_i$, is exactly Bayes. Hence, it is reasonable to conjecture that the one-sided mixture SPRT,

$$T_0 = \inf\left\{n \ge 1 : \sum_{i=1}^N W_i\Upsilon_i\Lambda_n^i \le B_c\right\},$$

is asymptotically optimal as $c \to 0$, and the weighted GSLRT too. Thus, $w_i^0 = W_i\Upsilon_i$, where according to the previous discussion $W_i \propto w_i^1\zeta_i^1$, yielding $w_i^0 = e^{\varkappa_i^1}\zeta_i^1\Upsilon_i$.

The following theorem formalizes all the details. We formulate the theorem only for the two-sided WSPRT. The same result holds for the WGSLRT.

Theorem 5.5.3. *Suppose that $\mathsf{E}_i|\lambda_1^i|^2 < \infty$ and $\mathsf{E}_0|\lambda_1^i|^2 < \infty$ for $i = 1,\dots,N$ and that λ_1^i are P_i-nonarithmetic. Let $\bar{\delta}^* = (\bar{T}^*, \bar{d}^*)$ denote the two-sided WSPRT when the weights are selected as*

$$w_i^1 = e^{\varkappa_i^1}, \quad w_i^0 = e^{\varkappa_i^1} I_i(\zeta_i^1)^2, \quad i = 1,\dots,N.$$

If the thresholds a_0 and a_1 are so selected that

$$\mathsf{P}_0(\bar{d}^* = 1) = \alpha_0, \quad \max_{1\le i\le N} \mathsf{P}_i(\bar{d}^* = 0) = \alpha_1, \tag{5.254}$$

then as $\alpha_{\max} \to 0$

$$\mathsf{E}_0\bar{T}^* = \inf_{\delta\in\mathbf{C}(\alpha_0,\alpha_1)} \mathsf{E}_0T + o(1), \tag{5.255}$$

$$\max_{1\le i\le N}(I_i\,\mathsf{E}_i\bar{T}^*) = \inf_{\delta\in\mathbf{C}(\alpha_0,\alpha_1)} \max_{1\le i\le N}(I_i\,\mathsf{E}_iT) + o(1). \tag{5.256}$$

Part II

Changepoint Detection

Chapter 6

Statistical Models with Changes: Problem Formulations and Optimality Criteria

In this chapter, we discuss optimality criteria and several Bayesian and non-Bayesian approaches to quickest changepoint detection. We also describe different types of changes and address several modeling issues.

6.1 Introduction

Sequential changepoint detection, or quickest change detection, or quickest disorder detection, is concerned with the design and analysis of techniques for on-line *quickest* detection of a change in the state of a process, subject to a tolerable limit on the risk of false detection. Specifically, a time process of interest may unexpectedly undergo an abrupt or gradual change-of-state from normal to abnormal, each defined as deemed appropriate given the physical context. The inference about the current state of the process is drawn by virtue of quantitative observations (measurements). In the *sequential setting*, the observations are obtained one at a time and, as long as their behavior is consistent with the initial normal or target state, one is content to let the process continue. If the state changes and becomes abnormal, then one is interested in detecting that a change is in effect, usually *as soon as possible* after its occurrence, so that an appropriate response can be provided in a timely manner. Thus, with the arrival of every new observation one is faced with the question of whether to let the process continue, or to stop and raise an alarm and, e.g., investigate. The decision has to be made in real time based on the available data. The time instance at which the process's state changes is referred to as the *changepoint*, and the challenge is that it is not known in advance.

More specifically, the changepoint problem posits that one obtains a series of observations $X_1, X_2, \dots$ such that, for some value ν, $\nu \geq 0$ (the changepoint), $X_1, X_2, \dots, X_\nu$ have one distribution and $X_{\nu+1}, X_{\nu+2}, \dots$ have another distribution; $\nu = \infty$ means that all the observations are under the nominal regime and $\nu = 0$ means that all the observations are under the abnormal regime. The changepoint ν is unknown, and the sequence $X = \{X_i\}_{i \geq 1}$ is being monitored for detecting a change. A sequential detection policy is defined by a stopping time T with respect to the Xs, so that after observing $X_1, X_2, \dots, X_T$ it is declared that apparently a change is in effect.

Sometimes the changepoint problem has a multiple hypothesis nature, i.e., the observations $X_{\nu+1}, X_{\nu+2}, \dots$ obey one of several possible post-change distributions, in which case both the changepoint ν and the post-change distribution are unknown, and the sequence $X = \{X_i\}_{i \geq 1}$ is being monitored for detecting and isolating, or classifying, or identifying a change. In this case, a sequential decision policy $\delta = (T, d)$ is a pair defined by a stopping time T and a final decision d on the post-change hypothesis, so that after observing $X_1, X_2, \dots, X_T$ it is declared that apparently a change is in effect and the post-change hypothesis is d.

Historically, the subject of changepoint detection first began to emerge in the 1920–1930s motivated by quality control considerations. Shewhart's charts were popular in the past [411]. Optimal and quasi-optimal sequential detection procedures were developed much later in the 1950–1960s, after the emergence of Sequential Analysis [494]. The ideas set in motion by Shewhart and Wald have formed a platform for a vast research on both the theory and practice of sequential changepoint detection. See, e.g., [47, 86, 165, 346, 394, 414, 415, 420, 428, 452].

The desire to detect the change quickly causes one to be trigger-happy, which, on one hand, leads to an unacceptably high level of *false alarm risk* — terminating the process prematurely before a real change has occured. On the other hand, attempting to avoid false alarms too strenuously causes a long delay between the actual onset time of the change (the true changepoint) and the time it is detected. Hence, the essence of the problem is to attain a tradeoff between two contradicting performance measures — the loss associated with the delay in detecting a true change and that associated with raising a false alarm. A good sequential detection policy is expected to minimize the average loss associated with the detection delay, subject to a constraint on the loss associated with false alarms, or *vice versa*.

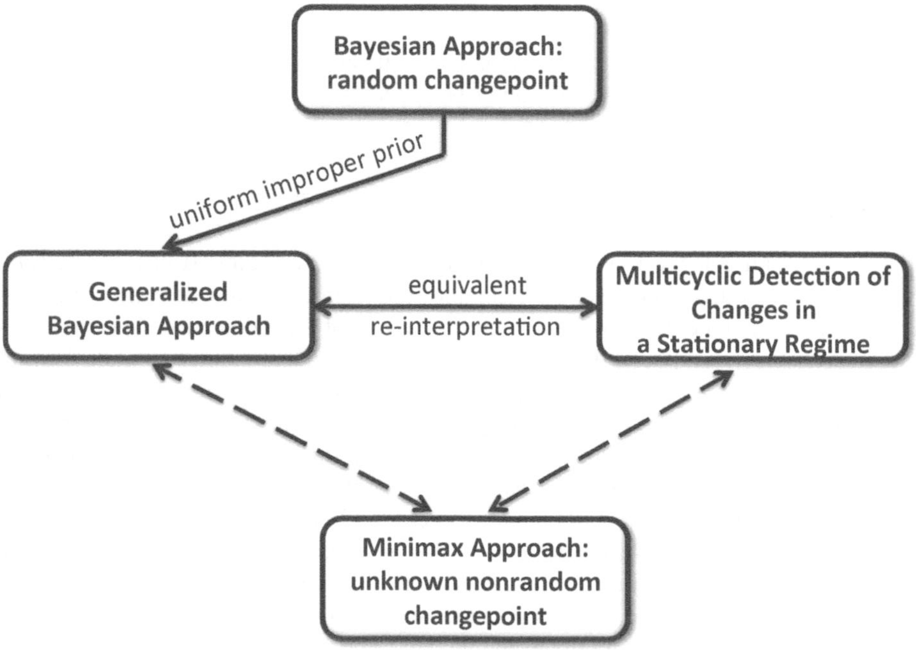

Figure 6.1: *Four approaches to sequential quickest changepoint detection.*

Putting this idea on a rigorous mathematical basis requires a formal definition of both the detection delay and the false alarm risk. To this end, we distinguish four different approaches: the Bayesian approach, the generalized Bayesian approach, the minimax approach, and the approach related to multicyclic detection of a distant change in a stationary regime. While each of those approaches has its own history and areas of application, nevertheless they are connected and fit together into one big picture shown in Figure 6.1.

Further details on these four approaches to quickest change detection are presented in Section 6.3 and Chapters 7 and 8.

The problem of change detection with multiple hypotheses is considered in Section 6.3 and Chapter 9.

6.2 Changepoint Models

To formally state the general quickest changepoint detection problem, we first have to introduce a changepoint model. A changepoint model is characterized by the probabilistic structure of the monitored process — independent, identically or nonidentically distributed observations, dependent observations, *etc.* — as well as by the model adopted for the changepoint — unknown deterministic, random completely or partially dependent on the observed data, random fully independent from the observations.

6.2.1 Models for Observations

Let $X_1, X_2, \dots$ denote the series of observations, and let ν be the serial number of the *last pre-change* observation, i.e., $X_{\nu+1}$ is the first post-change observation. Let P_k and E_k denote the probability measure and expectation when $\nu = k$, and let P_∞ and E_∞ denote the same when $\nu = \infty$, i.e., there is no change. Another practice popular in the literature is to define ν as the serial number of the first post-change observation. Although the two definitions map into one another, throughout the remainder of the book we follow the former convention.

Let $p_k(\mathbf{X}_1^n) = p(X_1, \dots, X_n | \nu = k)$ be the joint probability density of $X_1, \dots, X_n$ when the changepoint ν equals k. Let $\{f_{0,n}(X_n|\mathbf{X}_1^{n-1})\}_{n\geq 1}$ and $\{f_{1,n}(X_n|\mathbf{X}_1^{n-1})\}_{n\geq 1}$ be two sequences of conditional densities of X_n given $\mathbf{X}_1^{n-1}$, with the convention $f_{j,1}(X_1|X_0) \equiv f_{j,1}(X_1)$. In the most general non-iid case, we can write

$$
\begin{aligned}
p_k(\mathbf{X}_1^n) &= p_\infty(\mathbf{X}_1^n) = \prod_{i=1}^{n} f_{0,i}(X_i|\mathbf{X}_1^{i-1}) \quad \text{for } k \geq n, \\
p_k(\mathbf{X}_1^n) &= p_\infty(\mathbf{X}_1^k) \times p_0(\mathbf{X}_{k+1}^n|\mathbf{X}_1^k) \\
&= \prod_{i=1}^{k} f_{0,i}(X_i|\mathbf{X}_1^{i-1}) \times \prod_{i=k+1}^{n} f_{1,i}(X_i|\mathbf{X}_1^{i-1}) \quad \text{for } k < n.
\end{aligned}
\tag{6.1}
$$

In other words, under the measure P_∞ the conditional density of X_n given $\mathbf{X}^{n-1}$ is $f_{0,n}(X_n|\mathbf{X}^{n-1})$ for every $n \geq 1$ and under the measure P_k for any $0 \leq k < \infty$ the conditional density of X_n is $f_{0,n}(X_n|\mathbf{X}^{n-1})$ if $n \leq k$ and is $f_{1,n}(X_n|\mathbf{X}^{n-1})$ if $n > k$. Therefore, if the change occurs at time $\nu = k$, then the conditional density of the $(k+1)$-th observation changes from $f_{0,k+1}(X_{k+1}|\mathbf{X}^k)$ to $f_{1,k+1}(X_{k+1}|\mathbf{X}^k)$.

The densities $f_{0,n}(X_n|\mathbf{X}^{n-1})$ and $f_{1,n}(X_n|\mathbf{X}^{n-1})$, $n \geq 1$, are referred to as the *pre-change* and *post-change* conditional densities, respectively.

Note that in general the post-change densities may and often do depend on the changepoint ν, namely $f_{1,n}(X_n|\mathbf{X}^{n-1}) = f_{1,n}^{(\nu)}(X_n|\mathbf{X}^{n-1})$, $n > \nu$. This is typically the case for state–space hidden models due to the propagation of the change and the fact that the pre-change model affects the post-change model [466]. A classical example is the case of iid observations whose pdf is controlled by a hidden Markov model (HMM). Even if the transition matrix of the HMM remains the same before and after the change, the pdf of the observations $p_\nu(X_1, \dots, X_n)$ satisfies (6.1) only if post-change densities depend on the changepoint, as can be easily verified. See the linear state–space model example in Subsection 7.4.2.

Therefore, using the convention that $\prod_{i=m}^n Z_i = 1$ for $m > n$, for any $0 \leq \nu \leq \infty$, the most general changepoint model that allows the observations to be arbitrary dependent and nonidentically distributed can be written as

$$
p_\nu(\mathbf{X}_1^n) = \prod_{i=1}^{\nu} f_{0,i}(X_i|\mathbf{X}_1^{i-1}) \times \prod_{i=\nu+1}^{n} f_{1,i}^{(\nu)}(X_i|\mathbf{X}_1^{i-1}). \tag{6.2}
$$

This model is indeed very general. In addition to assuming neither independence nor homogeneity of the observations, it allows the post-change distribution to depend on the changepoint. The model (6.2) includes practically all possible scenarios. If, for example, there is a switch from one non-iid model to another non-iid model, which are mutually independent, then the two segments, pre- and post-change, of the observed process are independent, and in (6.2) the post-change conditional densities $f_{1,i}^{(\nu)}(X_i|\mathbf{X}_1^{i-1})$, $i \ge \nu+1$ are replaced with $f_{1,i}^{(\nu)}(X_i|\mathbf{X}_\nu^{i-1})$,

$$p_\nu(\mathbf{X}_1^n) = p_\infty(\mathbf{X}_1^\nu) \times p_0^{(\nu)}(\mathbf{X}_{\nu+1}^n) = \prod_{i=1}^{\nu} f_{0,i}(X_i|\mathbf{X}_1^{i-1}) \times \prod_{i=\nu+1}^{n} f_{1,i}^{(\nu)}(X_i|\mathbf{X}_{\nu+1}^{i-1}). \tag{6.3}$$

Such a model can be interpreted in the following way: it is assumed that the "Nature" is equipped with two generators producing two mutually independent pre- and post-change random sequences. When the "Nature" switches one generator to the other one, the memory of the observations is not kept after the change.

Suppose now that the observations $\{X_n\}_{n\ge1}$ are *independent* and such that $X_1,\dots,X_\nu$ are each distributed according to a common density $f_0(x)$, while $X_{\nu+1},X_{\nu+2},\dots$ each follows a common density $f_1(x) \not\equiv f_0(x)$. This is the simplest and most prevalent case. For convenience and with some abuse of terminology, from now on it is referred to as the *iid case*. It can be seen that in this case, the model (6.2) reduces to

$$p_\nu(\mathbf{X}_1^n) = \prod_{i=1}^{\nu} f_0(X_i) \times \prod_{i=\nu+1}^{n} f_1(X_i). \tag{6.4}$$

So far we have considered the case of simple hypotheses assuming that the pre- and post-change distributions are known, i.e., that conditional densities $f_{0,i}(X_i|\mathbf{X}_1^{i-1})$ and $f_{1,i}^{(\nu)}(X_i|\mathbf{X}_1^{i-1})$ are completely specified and only the value of ν, the changepoint, is unknown. In practice, pre-change densities $f_{0,i}(X_i|\mathbf{X}_1^{i-1})$ are often known. Realistically, post-change densities $f_{1,i}^{(\nu)}(X_i|\mathbf{X}_1^{i-1})$ are not completely specified. While the simple setting yields a benchmark for the best one can hope, it is of interest to consider a more realistic scenario where the post-change parameters are not known. Specifically, consider a parametric family $f_\theta(X_i|\mathbf{X}_1^{i-1})$ and assume that the change occurs in the parameter θ that changes from a known value θ_0 to some unknown value $\vartheta \neq \theta_0$. Note that in general the post-change parameter value $\vartheta = \vartheta(i,\nu)$ may depend on time and the changepoint. This problem is considered in Chapter 8.

In certain applications, it is of importance to generalize the single-hypothesis changepoint model to the case of multiple post-change hypotheses, which was first done by Nikiforov [322]. Unlike the classical changepoint model, where there is only one post-change model or a sequence of conditional densities $\{f_{1,n}^{(\nu)}(X_n|\mathbf{X}_1^{n-1})\}_{n\ge1}$, in the case of multiple hypotheses there are several possible post-change hypotheses H_j, that is, several sequences of conditional densities $\{f_{j,n}^{(\nu)}(X_n|\mathbf{X}_1^{n-1})\}_{n\ge1}$, $j=1,2,\dots,M$. The goal is then not only to detect a change as quickly as possible but also to identify the correct post-change hypothesis H_j. This multihypothesis change detection problem is also called sequential *change detection and isolation* problem. Obviously, there are three contradicting performance measures in the detection–isolation problem — the loss associated with the delay in detecting a true change, the loss associated with raising a false alarm, and that producing false identifications or isolations. The false isolation event means that after the detection of a true change the true post-change hypothesis H_j is rejected and another wrong post-change hypothesis H_k ($k \neq j$) is accepted.

6.2.2 *Models for the Changepoint*

The changepoint ν may be either considered as an unknown deterministic number or as a random variable.

If the changepoint is treated as a random variable, which is the ground assumption of the Bayesian approach (see Figure 6.1), then the model has to be supplied with the *prior distribution* of the changepoint. There may be several changepoint mechanisms and, as a result, a random variable ν may be partially or completely dependent on the observations or independent of the observations. To account for these possibilities at once, let $q=\Pr(\nu<0)$ and $\pi_k=\Pr(\nu=k|\mathbf{X}_1^k)$, $k\ge0$, and observe that π_k, $k = 1, 2, \ldots$ are $\mathscr{F}_k$-adapted. That is, the probability of a change occurring at the time instant $\nu = k$ depends on $\mathbf{X}_1^k$, the observations' history accumulated up to and including the time $k \geq 1$. The probability $q + \pi_0 = \Pr(\nu \leq 0)$ represents the probability of the "atom" associated with the event that the change already took place before the observations became available. With the so-defined prior distribution one can describe very general changepoint models, including those that assume ν to be a $\{\mathscr{F}_n\}$-adapted stopping time; see Moustakides [307].

Finally, we note that when the sequence of probabilities $\{\pi_k\}_{k \geq 1}$ depends on the observed data $\{X_k\}_{k \geq 1}$, it is arguable whether $\{\pi_k\}_{k \geq 1}$ can be referred to as the prior distribution: it can just as well be viewed as the *a posteriori* distribution. However, a deeper discussion of this subject is out of scope to this book, and from now on we assume that $\{\pi_k\}_{k \geq 0}$ do not depend on $\{X_k\}_{k \geq 1}$, in which case it represents the true prior distribution.

We also note that while we allow the changepoint ν to take on negative values, the detailed distribution $\Pr(\nu = k)$ for $k = -1, -2, \ldots$ is not important. The only value we need is the cumulative probability $q = \sum_{k=-\infty}^{-1} \Pr(\nu = k)$.

6.2.3 *Different Types of Changes*

Let us now describe the two different problems associated with additive and nonadditive changes. Most change detection problems can be classified into one of these two categories. Of course, these two types of changes can occur simultaneously, but it is of interest to investigate their main features separately. We describe these two types of changes using both intuitive and theoretical points of view.

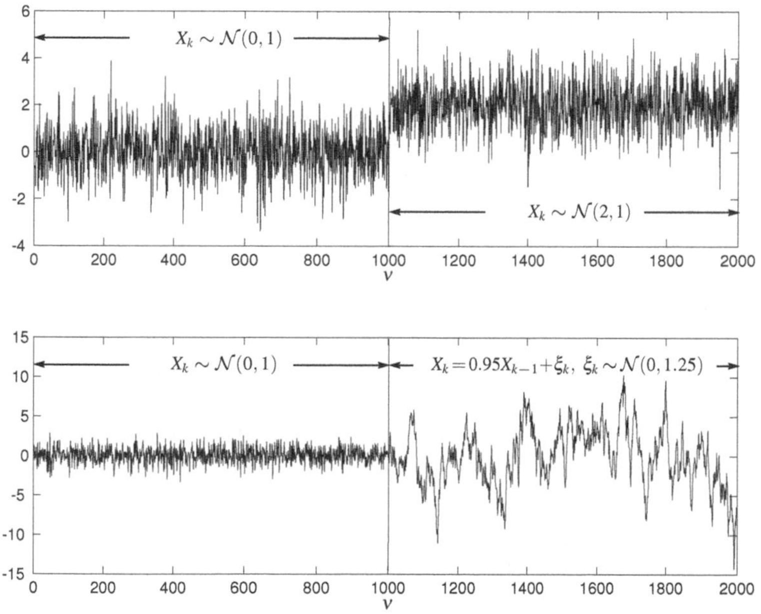

Figure 6.2: *Examples with changepoint* $\nu = 1000$. *An additive change (top), a nonadditive change (bottom).*

Intuitively, anyone who has ever looked at a real signal knows the qualitative difference between a jump, namely a change in the mean value of the signal and a change in the behavior around a mean level, as depicted in Figure 6.2. It is thus easy to guess that the two corresponding change detection problems do not have the same characteristics or the same degree of difficulty.

When we speak about additive changes, we refer to any changes in a signal or a linear system that result in *changes only in the mean value* of the sequence of observations. An important case is a linear system excited by a Gaussian white noise and perturbed by additive changes. An additive change can occur in the mean value θ of the system input, or in the mean value of the system output. Thus these changes do not affect the system dynamics. For such systems, an extension of the methods and theoretical results available in the iid case is possible and relatively easy. The main reason for this is that the problem of detecting additive changes remains unchanged under the transformation from the observations to the innovations, which corresponds to a whitening filter. Often the system dynamics generates a transient process in the mean value, which typically stabilizes quite rapidly. That is, after whitening the observations are independent but nonidentically distributed for some limited time, becoming iid with a relatively small delay.

When we speak about nonadditive or spectral changes, we refer to more general and difficult cases where the changes occur in the variance, correlations, spectral characteristics, dynamics of the signal or system. Two sub-cases must be distinguished. In the first case, the change is assumed to occur in the energy of the input excitation. Thus it does not affect the dynamics of the system and the change detection problem remains unchanged under the transformation from the observations to the innovations, as before. In the second case where the change affects the dynamics of the system itself, the problem is more complex because the log-likelihood ratio process cannot be transformed into a random walk with different parameters before and after the change. A more precise analysis of this case is presented later on.

6.3 Optimality Criteria

As discussed in Section 6.1, we consider four changepoint problem settings — the Bayesian approach, the generalized Bayesian approach, the minimax approach, and the approach related to multicyclic detection of a disorder in a stationary regime; see Figure 6.1. The multihypothesis changepoint problem setting is also considered. The objective of this section is to briefly discuss each problem setting.

A *sequential change detection procedure* is a stopping time T on $X_1,X_2,\dots$, i.e., $\{T \le n\} \in \mathscr{F}_n$, where $\mathscr{F}_n = \sigma(X_1,\dots,X_n)$ is the sigma-algebra induced by the first n observations, with the convention $\mathscr{F}_0 = \{\varnothing,\Omega\}$. We assume that all random objects are defined on a complete probability space $(\Omega,\mathscr{F},\mathsf{P})$, $\mathscr{F} = \vee_{n\ge 0}\mathscr{F}_n$, and that $\{\mathscr{F}_n\}_{n\ge 0}$ satisfies the usual conditions. After observing $X_1,\dots,X_T$ it is declared that the change is in effect. That may or may not be the case. If it is not, then $T \le \nu$, and it is said that a *false alarm* has been raised. Note also that since $\mathscr{F}_0$ is the trivial sigma-algebra, any $\{\mathscr{F}_n\}$-adapted stopping time T is either strictly positive w.p. 1, or $T = 0$ w.p. 1. The latter case is clearly degenerate, and to preclude it from now on we assume $T > 0$ w.p. 1.

A *sequential change detection and isolation procedure* is a pair (T,d), where T is again a stopping time and d is a $\{\mathscr{F}_T\}$- measurable final decision with values in the set $\{1,2,\dots,M\}$. Hence, $d = j$ means that the post-change hypothesis H_j is accepted. Again that may or may not be the case. Assume that H_k is the true post-change hypothesis. If the final decision is $d \ne k$, then the true post-change hypothesis H_k is rejected and it is said that a *false isolation* or *false classification* is produced.

We distinguish between two different scenarios — single-run detection procedures and multicyclic procedures. In the first scenario, which is common in most theoretical studies related to the Bayesian, generalized Bayesian, and minimax approaches, it is assumed that the detection procedure is applied *only once*; the result is either a false alarm or a correct possibly delayed detection. What takes place beyond the stopping point T is of no concern. We will refer to this as the *single-run paradigm*, which is shown in Figure 6.3. Figure 6.3(a) shows an example of the behavior of the observations $\{X_n\}_{n\ge 1}$. It can be seen that the process undergoes a shift in the mean at some time instant ν, the changepoint. The grey line in Figure 6.3(b) gives an example of the corresponding detection statistic trajectory that exceeds the detection threshold prematurely, before the change occurs. This is a false alarm situation, and T can be regarded as the random run length to false alarm. Another possibility is depicted by the black line in Figure 6.3(b). This is an example where the detection statistic exceeds the detection threshold after the changepoint. Note that the detection delay $T - \nu$ is random.

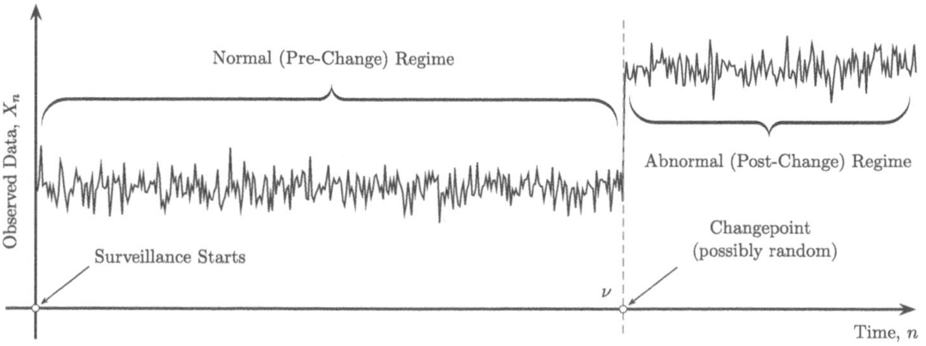

(a) An example of the behavior of the observed process $\{X_n\}_{n\geq 1}$.

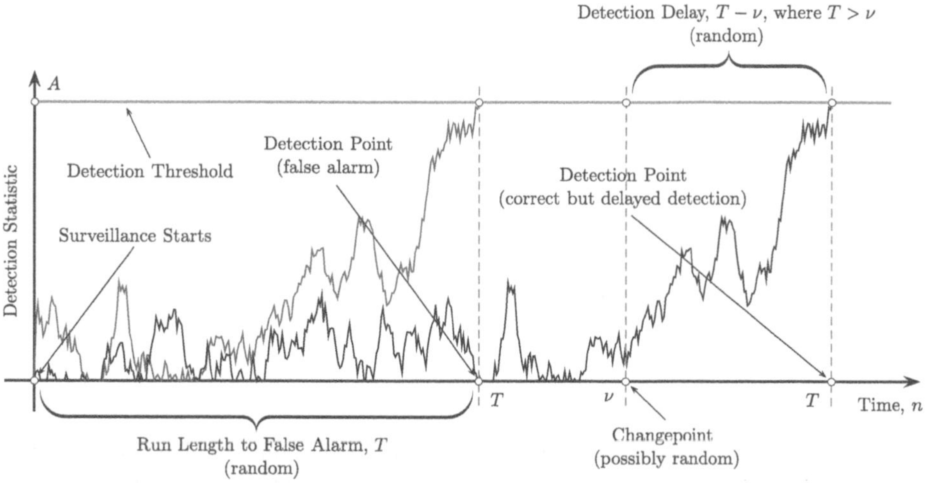

(b) Two possibilities in the detection process: false alarm (left) and correct detection (right).

Figure 6.3: *Single-run sequential changepoint detection.*

In a variety of surveillance applications the detection procedure should be applied *repeatedly*. This requires the specification of a renewal mechanism after each false or true alarm. The simplest renewal strategy is to restart from scratch, in which case the procedure becomes multicyclic with similar cycles in a statistical sense if the process is homogeneous. In the following sections, we consider such an approach related to the detection of a distant change in a stationary regime, assuming that the detection procedure is applied repeatedly starting anew after each time the detection statistic exceeds the threshold. This scenario is illustrated in Figure 6.4 where we can see a repeated detection process with multiple false alarms prior to the true detection. After each alarm the detection procedure is renewed from scratch.

Figure 6.3 suggests that it is reasonable to measure the risk associated with a false alarm by the mean time to false alarm $\mathsf{E}_\infty T$ and the risk associated with a true change detection by the conditional average delay to detection $\mathsf{CADD}_\nu(T) = \mathsf{E}_\nu(T-\nu|T>\nu)$, $\nu = 0, 1, \dots$. A good detection procedure should guarantee small values of the expected detection delay $\mathsf{CADD}_\nu(T)$ for all $\nu \geq 0$ when $\mathsf{E}_\infty T$ is fixed at a certain level. However, if the false alarm risk is measured in terms of the mean time to false alarm, i.e., it is required that $\mathsf{E}_\infty T \geq \gamma$ for some $\gamma > 1$, then a procedure that minimizes the conditional average delay to detection $\mathsf{CADD}_\nu(T)$ uniformly over all ν does not exist.

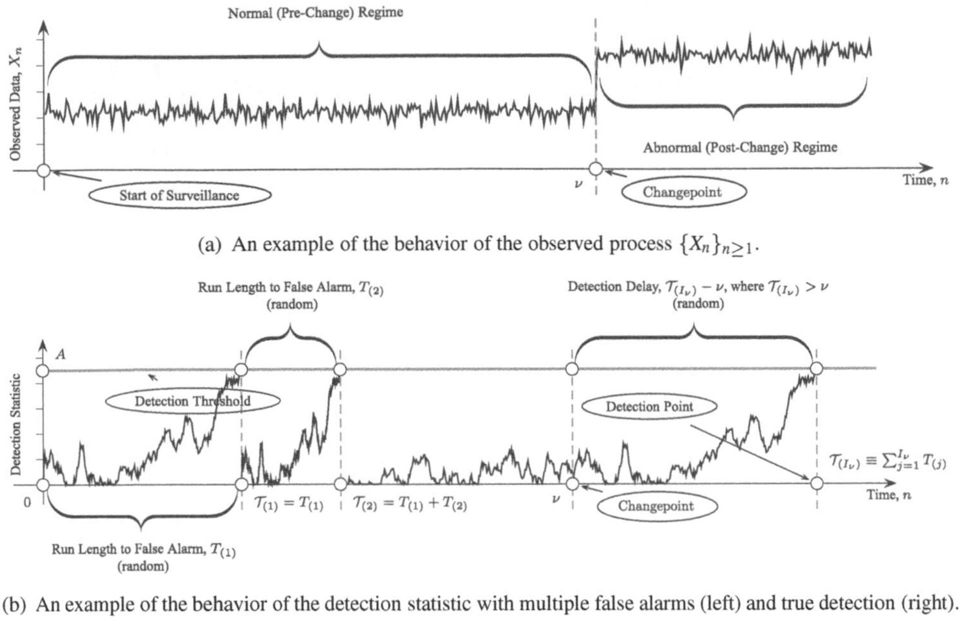

Figure 6.4: *Multicyclic changepoint detection in a stationary regime.*

For this reason, we have to resort to different optimality criteria, e.g., to Bayes or minimax criteria, which are considered in the next subsections.

6.3.1 *Bayesian Formulation*

The characteristic feature of the Bayes criterion is the assumption that the changepoint is a random variable possessing a prior distribution. This is instrumental in certain applications [422, 423, 476], but mostly of interest since the limiting versions of Bayesian solutions lead to useful procedures, which are optimal or asymptotically optimal in more practical problems.

Let $\{\pi_k\}_{-\infty<k<+\infty}$ be the prior distribution of the changepoint ν, $\pi_k = \Pr(\nu = k)$, which we write without specification of the probabilities for $k < 0$ as

$$\pi_k = (1-q)\widetilde{\pi}_k \quad \text{for } k = 0, 1, 2 \ldots,$$

where $q = \Pr(\nu < 0)$ and $\widetilde{\pi}_k = \Pr(\nu = k | \nu \geq 0)$, $k \geq 0$. The probability $\Pr(\nu \leq 0) = q + (1-q)\widetilde{\pi}_0$ is interpreted as the probability that the change has already occurred before the observations became available. Clearly, there is no need to specify the detailed distribution of the changepoint for negative integers since $T > 0$ w.p. 1.

For any measurable event $\mathcal{A}$, define the probability measure

$$\mathsf{P}^\pi(\mathcal{A}) = \sum_{k=0}^{\infty} \pi_k \mathsf{P}_k(\mathcal{A}),$$

where the π in the superscript emphasizes the dependence on the prior distribution.

From the Bayesian point of view, it is reasonable to measure the false alarm risk with the weighted *Probability of False Alarm* (PFA), which is defined as

$$\mathsf{PFA}^\pi(T) := \mathsf{P}^\pi(T \leq \nu) = \sum_{k=1}^{\infty} \pi_k \mathsf{P}_k(T \leq k). \tag{6.5}$$

Note that the summation in (6.5) is over $k \geq 1$ since by convention $\mathsf{P}_k(T \geq 1) = 1$, so that

$\mathsf{P}_k(T \le 0) = 0$. A reasonable way to benchmark the detection delay is through the *Average Delay to Detection* (ADD), which is defined as

$$\mathsf{ADD}^\pi(T) = \mathsf{E}^\pi(T - \nu | T > \nu) = \frac{\mathsf{E}^\pi(T-\nu)^+}{\mathsf{P}^\pi(T > \nu)}, \tag{6.6}$$

where hereafter $x^+ = \max\{0, x\}$ and E^π denotes the expectation with respect to P^π.

We are now in a position to formally introduce the notion of Bayesian optimality. Let $\mathbf{C}_\alpha = \{T\colon \mathsf{PFA}^\pi(T) \le \alpha\}$ be the class of detection procedures or stopping times for which the PFA does not exceed a preset level $\alpha \in (0,1)$. Then under the Bayesian approach the goal is to

$$\text{Find } T_{\text{opt}} \in \mathbf{C}_\alpha \text{ such that } \mathsf{ADD}^\pi(T_{\text{opt}}) = \inf_{T \in \mathbf{C}_\alpha} \mathsf{ADD}^\pi(T) \text{ for every } \alpha \in (0,1). \tag{6.7}$$

For the iid model (6.4) and under the assumption that the changepoint ν has a *geometric* prior distribution, this problem was solved by Shiryaev [414, 415, 420]. Specifically, Shiryaev assumed that ν is distributed according to the zero-modified geometric distribution

$$\Pr(\nu < 0) = q \text{ and } \Pr(\nu = k) = (1-q)p(1-p)^k \quad \text{for } k = 0, 1, 2, \dots, \tag{6.8}$$

where $q \in [0,1)$ and $p \in (0,1)$. Note that in our previous notation $\Pr(\nu \le 0) = q + (1-q)p$, $\pi_k = (1-q)p(1-p)^k$, $k \ge 0$, and $\widetilde{\pi}_k = p(1-p)^k$, $k \ge 0$.

Observe now that if $\alpha \ge 1 - q$, then there is a trivial solution to the optimization problem (6.7) since we can simply stop right away without any observation. Indeed, this strategy produces $\mathsf{ADD}^\pi = 0$ and $\mathsf{PFA}^\pi = \Pr(\nu \ge 0) = 1 - q$, which satisfies the constraint $\mathsf{PFA}^\pi(T) \le \alpha$. Therefore, to avoid trivialities we have to assume that $\alpha < 1 - q$. In this case, Shiryaev [414, 415, 420] proved that the optimal detection procedure is based on comparing the posterior probability of a change currently being in effect with a certain detection threshold: the procedure stops as soon as $\Pr(\nu < n | \mathscr{F}_n)$ exceeds the threshold. We refer to this strategy as the *Shiryaev procedure*. To guarantee its strict optimality the detection threshold should be set so as to guarantee that the PFA is exactly equal to the selected level α, which is rarely possible.

We should note that there seems to be a slight difference in our test statistic as compared with the statistic $\Pr(\nu \le n | \mathscr{F}_n)$ originally proposed by Shiryaev. However, this difference is only notational and it is due to the fact that in our case ν is the last time instant under the nominal regime while in Shiryaev's approach it is the first instant under the alternative regime. Actually the two procedures are exactly the same as we realize next by expressing them in terms of an alternative test statistic.

The Shiryaev procedure plays an important role in the sequel when considering non-Bayesian criteria as well. It is more convenient to express Shiryaev's procedure through the statistic

$$R_{n,p} = \frac{q}{(1-q)p} \prod_{j=1}^{n} \left(\frac{\mathcal{L}_j}{1-p}\right) + \sum_{k=1}^{n} \prod_{j=k}^{n} \left(\frac{\mathcal{L}_j}{1-p}\right), \tag{6.9}$$

where $\mathcal{L}_j = f_1(X_j)/f_0(X_j)$ is the LR for the j-th data point X_j. Indeed, by using the Bayes rule, one can show that

$$\Pr(\nu < n | \mathscr{F}_n) = \frac{R_{n,p}}{R_{n,p} + 1/p}, \tag{6.10}$$

whence it is readily seen that thresholding the posterior probability $\Pr(\nu < n | \mathscr{F}_n)$ is the same as thresholding the process $\{R_{n,p}\}_{n \ge 1}$. Therefore, the Shiryaev detection procedure has the form

$$T_{\mathrm{S}}(A) = \inf\{n \ge 1\colon R_{n,p} \ge A\}, \tag{6.11}$$

and if $A = A_\alpha$ can be selected in such a way that the PFA is exactly equal to α, i.e., $\mathsf{PFA}^\pi(T_{\mathrm{S}}(A_\alpha)) = \alpha$, then it is strictly optimal in the class $\mathbf{C}_\alpha$,

$$\inf_{T \in \mathbf{C}_\alpha} \mathsf{ADD}^\pi(T) = \mathsf{ADD}^\pi(T_{\mathrm{S}}(A_\alpha)) \quad \text{for any } 0 < \alpha < 1 - q.$$

Note that Shiryaev's statistic $R_{n,p}$ can be rewritten in the recursive form

$$R_{n,p} = (1+R_{n-1,p})\frac{\mathcal{L}_n}{1-p}, \quad n \geq 1, \quad R_{0,p} = \frac{q}{(1-q)p}. \tag{6.12}$$

We also note that (6.9) and (6.10) remain true under the geometric prior distribution (6.8) even in the general non-iid case (6.2) with $\mathcal{L}_n = f_{1,n}(X_n|\mathbf{X}_1^{n-1})/f_{0,n}(X_n|\mathbf{X}_1^{n-1})$. However, in order for the recursion (6.12) to hold in this case, $\{\mathcal{L}_n\}_{n\geq 1}$ should be independent of the changepoint ν, which is not typically the case, as discussed in Subsection 6.2.1.

As $p \to 0$, where p is the parameter of the geometric prior distribution (6.8), the Shiryaev detection statistic (6.12) converges to the so-called *Shiryaev–Roberts (SR) detection statistic*. The latter is the basis for the *SR procedure*. As we will see, the SR procedure is a bridge between the four different approaches to changepoint detection mentioned above.

Further details on Bayes and asymptotically Bayes procedures for iid and general non-iid models will be presented in Chapter 7. In particular, for a general asymptotic Bayesian changepoint detection theory that covers practically arbitrary non-iid models and prior distributions, see Section 7.2. Specifically, in a general non-iid case, we assume only that the prior distribution is independent of the observations and show that the Shiryaev procedure is asymptotically (as $\alpha \to 0$) optimal in a very broad class of changepoint models and prior distributions.

6.3.2 Generalized Bayesian Formulation

The generalized Bayesian approach is the limiting case of the Bayesian formulation where the changepoint ν is assumed to be a generalized random variable with a uniform improper prior distribution.

First, return to the Bayesian constrained minimization problem (6.7) for the iid model (6.4), assuming that the changepoint ν is distributed according to the zero-modified geometric distribution (6.8). Suppose now that $q = 0$ and $p \to 0$, in which case the scaled by p^{-1} geometric prior distribution (6.8) turns into an improper uniform distribution. It can be seen that in this case $\{R_{n,p}\}_{n\geq 0}$ converges to the statistic $\{R_{n,0} = R_n\}_{n\geq 0}$, where $R_0 = 0$ and $R_n = (1+R_{n-1})\mathcal{L}_n$, $n \geq 1$, with $\mathcal{L}_n = f_1(X_n)/f_0(X_n)$. The limit $\{R_n\}_{n\geq 0}$ is known as the SR statistic.

Next, when $q = 0$ and $p \to 0$ it can also be shown that, for an arbitrary stopping time T,

$$\frac{\mathsf{P}^p(T > \nu)}{p} \to \mathsf{E}_\infty T \quad \text{and} \quad \frac{\mathsf{E}^p(T-\nu)^+}{p} \to \sum_{k=0}^{\infty} \mathsf{E}_k(T-k)^+. \tag{6.13}$$

Here and in the following we use the notation P^p and E^p indexed by p in place of π when considering the geometric prior. As a result, one may conjecture that the SR procedure minimizes the relative *Integral Average Detection Delay* (IADD)

$$\mathsf{IADD}(T) = \frac{\sum_{k=0}^{\infty} \mathsf{E}_k(T-k)^+}{\mathsf{E}_\infty T} \tag{6.14}$$

over all detection procedures for which the *Average Run Length (ARL) to false alarm* (ARL2FA), $\mathsf{ARL2FA}(T) = \mathsf{E}_\infty T$, is no less than $\gamma > 1$, an *a priori* set level.

Let

$$\mathbf{C}_\gamma = \{T : \mathsf{E}_\infty T \geq \gamma\} \tag{6.15}$$

be the class of detection procedures or stopping times for which the ARL to false alarm $\mathsf{ARL2FA}(T)$ is at least $\gamma > 1$. Then under the generalized Bayesian formulation the goal is to

$$\text{Find } T_{\text{opt}} \in \mathbf{C}_\gamma \text{ such that } \mathsf{IADD}(T_{\text{opt}}) = \inf_{T\in\mathbf{C}_\gamma} \mathsf{IADD}(T) \text{ for every } \gamma > 1. \tag{6.16}$$

We have already hinted that this problem is solved by the SR procedure. This was formally demonstrated in [370] in the discrete-time iid case and in [139, 415] in continuous time for detecting a shift in the mean of a Brownian motion. The details are given in Section 7.6.

We conclude this subsection with two remarks. First, observe that if the assumption $q = 0$ is replaced with $q = rp$, where $r \geq 0$ is a fixed number, then, as $p \to 0$, the Shiryaev statistic $\{R_{n,p}\}_{n\geq 0}$ converges to $\{R_n^r\}_{n\geq 0}$, where $R_n^r = (1+R_{n-1}^r)\mathcal{L}_n$, $n \geq 1$ with $R_0^r = r \geq 0$. This is the so-called *Shiryaev–Roberts–r (SR-r) detection statistic*, and it is the basis for the SR-r changepoint detection procedure that starts from an arbitrary deterministic point r. This procedure is due to Moustakides *et al.* [310]. The SR-r procedure possesses certain minimax properties [373, 470]. We discuss this procedure at greater length in Section 8.4.3. Secondly, though the generalized Bayesian formulation is the limit case of the Bayesian approach as $p \to 0$, it may also be equivalently reinterpreted as a completely different approach — *multicyclic disorder detection in a stationary regime* [370, 374]. We address this approach in Section 6.3.4.

6.3.3 *Minimax Formulation*

Contrary to the Bayes formulation the minimax approach posits that the changepoint is an unknown not necessarily random number. Even if it is random, its distribution is unknown. The minimax approach has multiple optimality criteria.

The first minimax theory is due to Lorden [271] who proposed to measure the false alarm risk by the ARL to false alarm $\mathsf{ARL2FA}(T) = \mathsf{E}_\infty T$; recall the false alarm scenario shown in Figure 6.3(b). As far as the risk associated with detection delay is concerned, Lorden suggested to use the worst-worst-case ADD defined as

$$\mathsf{ESADD}(T) = \sup_{0\leq \nu<\infty} \left\{ \operatorname{ess\,sup} \mathsf{E}_\nu[(T-\nu)^+|\mathscr{F}_\nu] \right\}. \tag{6.17}$$

In other words, the conditional ADD is first maximized over all possible trajectories of observations up to the changepoint and then over the changepoint. The Lorden's minimax optimization problem seeks to

$$\text{Find } T_{\text{opt}} \in \mathbf{C}_\gamma \text{ such that } \mathsf{ESADD}(T_{\text{opt}}) = \inf_{T\in\mathbf{C}_\gamma} \mathsf{ESADD}(T) \text{ for every } \gamma > 1, \tag{6.18}$$

where $\mathbf{C}_\gamma$ is the class of detection procedures with lower bound γ on the ARL to false alarm defined in (6.15).

For the iid scenario (6.4), Lorden [271] showed that Page's Cumulative Sum (CUSUM) procedure [346] is first-order asymptotically minimax as $\gamma \to \infty$. For any $\gamma > 1$, this problem was solved by Moustakides [305], who showed that CUSUM is exactly optimal. See also Ritov [389] for a different decision-theoretic argument.

Though strict $\mathsf{ESADD}(T)$-optimality of the CUSUM procedure is a strong result, it is more natural to construct a procedure that minimizes the conditional average delay to detection $\mathsf{E}_\nu(T-\nu|T > \nu)$ for all $\nu \geq 0$ simultaneously. As no such uniformly optimal procedure is possible, Pollak [365] suggested to revise Lorden's version of minimax optimality by replacing $\mathsf{ESADD}(T)$ with the maximal conditional average delay to detection, namely the supremum ADD or SADD:

$$\mathsf{SADD}(T) = \sup_{0\leq\nu<\infty} \mathsf{E}_\nu(T-\nu|T>\nu). \tag{6.19}$$

Thus, Pollak's version of the minimax optimization problem seeks to

$$\text{Find } T_{\text{opt}} \in \mathbf{C}_\gamma \text{ such that } \mathsf{SADD}(T_{\text{opt}}) = \inf_{T\in\mathbf{C}_\gamma} \mathsf{SADD}(T) \text{ for every } \gamma > 1. \tag{6.20}$$

It is our opinion that $\mathsf{SADD}(T)$ is better suited for practical purposes for two reasons. First, Lorden's criterion is effectively a double-minimax approach, and therefore, is overly pessimistic in the sense that $\mathsf{SADD}(T) \leq \mathsf{ESADD}(T)$. Second, Pollak's minimax criterion is directly connected to the conventional decision theoretic approach: the optimization problem (6.20) can be solved by finding the least favorable prior distribution. More specifically, since by the general decision theory the minimax solution corresponds to either the Bayes or the generalized Bayes solution with the least favorable prior distribution — see Theorem 2.7.8 in Section 2.7.4 — it can be shown that $\sup_\pi \mathsf{ADD}^\pi(T) = \mathsf{SADD}(T)$, where $\mathsf{ADD}^\pi(T)$ is defined in (6.6).

Unlike Lorden's minimax problem (6.18), Pollak's minimax problem (6.20) is still not solved in general. Some light as to the possible solution in the iid case is shed in [310, 373, 470] and discussed in Sections 8.4.2 and 8.4.3.

6.3.4 Multicyclic Detection of a Disorder in a Stationary Regime

The common feature of all of the above approaches is that the detection procedure is applied only once. This is the single-run paradigm. The result is either a correct, though usually delayed, detection or a false alert; recall Figure 6.3. In a variety of surveillance applications, however, such as target detection and tracking, rapid detection of intrusions in computer networks, and environmental monitoring, it is of utmost importance to detect very rapidly changes that may occur in a distant future, in which case the true detection of a real change may be preceded by a long interval with frequent false alarms that are being *filtered by a separate mechanism or algorithm.* For example, falsely initiated target tracks are usually filtered by a track confirmation/deletion algorithm; false detections of attacks in computer networks in anomaly-based Intrusion Detection Systems (IDS) may be filtered by Signature-based IDS algorithms, *etc.* See [471, 472, 475] and Section 11.3.

Specifically, consider a context in which it is of utmost importance to detect the change as quickly as possible, even at the expense of raising many false alarms before the change occurs, using a repeated application of the same stopping rule. This means that the changepoint ν is substantially larger than the mean time between false alarms given by the tolerable level γ. That is, the change occurs at a distant time horizon and is preceded by a *stationary flow of false alarms*. This scenario is schematically shown in Figure 6.4. As one can see, the ARL to false alarm in this case is the mean time between consecutive false alarms, and therefore, may be regarded as the false alarm rate or average frequency.

Formally, let $T_{(1)}, T_{(2)}, \dots$ denote sequential independent repetitions of the same stopping time T, and let $\mathcal{T}_{(j)} = T_{(1)} + T_{(2)} + \cdots + T_{(j)}$ be the time of the j-th alarm. Define $N_\nu = \min\{j \geq 1 \colon \mathcal{T}_{(j)} > \nu\}$. Put otherwise, $\mathcal{T}_{(N_\nu)}$ is the time of detection of the true change that occurs at the time instant ν after $N_\nu - 1$ false alarms have been raised. Write

$$\mathsf{STADD}(\mathcal{T}) = \lim_{\nu\to\infty} \mathsf{E}_\nu[\mathcal{T}_{(N_\nu)} - \nu] \tag{6.21}$$

for the limiting value of the ADD of the multicyclic detection procedure $\mathcal{T} = \{\mathcal{T}_{N_\nu}\}$ that we refer to as the *stationary ADD* (STADD).

We are now in a position to formalize the notion of optimality in the multicyclic setup:

$$\text{Find } \mathcal{T}_{\text{opt}} \in \mathbf{C}_\gamma \text{ such that } \mathsf{STADD}(\mathcal{T}_{\text{opt}}) = \inf_{\mathcal{T}\in\mathbf{C}_\gamma} \mathsf{STADD}(\mathcal{T}) \text{ for every } \gamma > 1 \tag{6.22}$$

among all multicyclic procedures.

For the iid model (6.4), this problem was solved by Pollak and Tartakovsky [370], who showed that the solution is the multicyclic SR procedure by arguing that $\mathsf{STADD}(\mathcal{T})$ is the same as $\mathsf{IADD}(T)$ defined in (6.14). This suggests that the optimal solution of the problem of multicyclic changepoint detection in a stationary regime is completely equivalent to the solution of the generalized Bayesian problem. The exact result is stated in Theorem 8.4.1.

As we argued above, this multicyclic approach is instrumental in many surveillance applications, in particular in the areas concerned with intrusion/anomaly detection, e.g., cybersecurity and particularly detection of attacks in computer networks. See Section 11.3 for a more detailed discussion.

6.3.5 Uniform Optimality Criterion

While the Bayesian and minimax formulations are reasonable and can be justified in many applications, it would be most desirable to guarantee small values of the conditional average detection delay $\mathsf{CADD}_\nu(T) = \mathsf{E}_\nu(T-\nu|T \ge \nu)$ uniformly for all $\nu \ge 0$ when the false alarm risk is fixed at a certain level. However, if the false alarm risk is measured in terms of the ARL to false alarm, i.e., if it is required that $\mathsf{ARL2FA}(T) \ge \gamma$ for some $\gamma > 1$, then a procedure that minimizes $\mathsf{CADD}_\nu(T)$ for all ν does not exist, as we previously discussed. More importantly, the requirement of having large values of the $\mathsf{ARL2FA}(T) = \mathsf{E}_\infty T$ generally does not guarantee small values of the local probability of false alarm (PFA) $\mathsf{P}_\infty(k < T \le k+m)$ in a time interval (window) with fixed length $m \ge 1$ for all $k \ge 0$ or small values of the corresponding conditional local PFA $\mathsf{P}_\infty(k < T \le k+m|T > k) = \mathsf{P}_\infty(T \le k+m|T > k)$, $k \ge 0$. Indeed, the following lemma shows that the condition $\mathsf{E}_\infty T \ge \gamma$ only guarantees, for any $m \ge 1$, the existence of some k, that possibly depends on γ, for which

$$\mathsf{P}_\infty(T \le k+m|T > k) < m/\gamma. \tag{6.23}$$

Lemma 6.3.1. *Let $T \in \mathbf{C}_\gamma$. Then for any $m \ge 1$ there exists a k, possibly depending on γ, for which the inequality* (6.23) *holds.*

Proof. We prove the inequality (6.23) by contradiction. Without loss of generality we may assume that $\mathsf{P}_\infty\{T > k\} > 0$, since otherwise $\mathsf{P}_\infty(T > k) = 0$ for all k and $\mathsf{E}_\infty T = 0$, which contradicts the inequality $\mathsf{E}_\infty T \ge \gamma$. Assume that

$$\mathsf{P}_\infty(T > k+m|T > k) < 1 - m/\gamma \quad \text{for all } k \ge 0. \tag{6.24}$$

It suffices to consider only integer m, $m \le \gamma$. Due to (6.24) and the fact that

$$\mathsf{P}_\infty(T > i+km) = \mathsf{P}_\infty(T > i)\mathsf{P}_\infty(T > i+km|T > i),$$

we have

$$\begin{aligned}\mathsf{E}_\infty T &= \sum_{i=0}^{\infty} \mathsf{P}_\infty\{T > i\} = \sum_{i=0}^{m-1}\sum_{k=0}^{\infty} \mathsf{P}_\infty\{T > i+km\} \\ &< \sum_{i=0}^{m-1}\sum_{k=0}^{\infty} \mathsf{P}_\infty\{T > i\}(1-m/\gamma)^k = (\gamma/m)\sum_{i=0}^{m-1} \mathsf{P}_\infty\{T > i\} \le \gamma,\end{aligned}$$

which contradicts the assumption $\mathsf{E}_\infty T \ge \gamma$. □

This means that, for a given $0 < \beta < 1$, the PFA constraint

$$\sup_{k \ge 0} \mathsf{P}_\infty(T \le k+m|T > k) \le \beta \quad \text{for a certain } m \ge 1 \tag{6.25}$$

is stronger than the ARL constraint $\mathsf{E}_\infty T \ge \gamma$. In particular, it is easily verified that (6.25) implies $\mathsf{E}_\infty T \ge \gamma$ for $\gamma = \gamma(\beta, m) = \beta^{-1}\sum_{i=0}^{m-1} \mathsf{P}_\infty(T > i)$.

At the same time, for many practical applications, it is desirable to control the maximal local PFA, which is given by $\mathsf{MPFA}_m(T) = \sup_{k \ge 0} \mathsf{P}_\infty(T \le k+m|T > k)$, at a certain (usually low) level β. For this reason, we introduce the class of detection procedures that satisfy (6.25),

$$\mathbf{C}^m(\beta) = \{T : \mathsf{MPFA}_m(T) \le \beta\}, \tag{6.26}$$

i.e., for which $\mathsf{MPFA}_m(T)$, the maximal conditional probability of raising a false alarm inside a sliding window of $m \ge 1$ observations, does not exceed a certain *a priori* chosen level $0 < \beta < 1$ for some $m \ge 1$.

The goal is to

$$\text{Find } T_{\text{opt}} \in \mathbf{C}^m(\beta) \text{ such that } \mathsf{CADD}_\nu(T_{\text{opt}}) = \inf_{T\in\mathbf{C}^m(\beta)} \mathsf{CADD}_\nu(T) \quad \text{for every } 0<\beta<1 \text{ and some } m\geq 1. \tag{6.27}$$

Such an optimal procedure may exist, while its structure is not known unless β is small. We discuss asymptotically uniformly optimal solutions as $\beta \to 0$ in Sections 8.2.8 and 8.3. It will also become apparent that the size of the window $m = m(\beta)$ should go to infinity when $\beta \to 0$.

Another reason for considering the PFA constraint (6.25) is that the appropriateness of the ARL to false alarm $\mathsf{E}_\infty T$ as an exhaustive measure of the risk of raising a false alarm is questionable, unless the P_∞-distribution of T is geometric (exponential), at least approximately. The geometric distribution is characterized entirely by a single parameter, which uniquely determines $\mathsf{E}_\infty T$ and is uniquely determined by $\mathsf{E}_\infty T$. As a result, if T is geometric, one can evaluate $\mathsf{P}_\infty(k<T\leq k+m|T>k)$ for any $k\geq 0$ (in fact, for all $k\geq 0$ at once) and both constraints are quite similar and can be recalculated in each other.

For the iid model (6.4), Pollak and Tartakovsky [369] showed that under mild assumptions the P_∞-distribution of the stopping times associated with detection schemes from a certain class is asymptotically exponential with parameter $1/\mathsf{E}_\infty T$ as $\mathsf{E}_\infty T \to \infty$. The class includes all most popular procedures, including CUSUM and SR. Hence, for the iid model (6.4), the ARL to false alarm is an acceptable measure of the false alarm rate. However, for a general non-iid model this is not necessarily true, which suggests that alternative measures of the false alarm rate are in order.

We stress once again that in general $\sup_k \mathsf{P}_\infty(k<T\leq k+m|T>k)\leq\beta$ is a *stronger* condition than $\mathsf{E}_\infty T\geq\gamma$. Hence, in general, $\mathbf{C}^m(\beta)\subset\mathbf{C}_\gamma$.

In Section 8.4.3, we present a procedure that solves the optimization problem (6.20) in the class (6.26) in a specific case; see Example 8.4.1.

6.3.6 *Sequential Change Detection and Isolation*

We now discuss the formulation of the quickest multiple hypothesis change detection and isolation problem, which was first suggested by Nikiforov [322]. As before, let $X_1, X_2, \dots$ denote the series of observations, and let ν be the serial number of the *last pre-change* observation. In the case of multiple hypothesis, there are several possible post-change hypotheses H_j, $j=1,2,\dots,M$. Let P_k^j and E_k^j denote the probability measure and the expectation when $\nu=k$ and H_j is the true post-change hypothesis, and let P_∞ and E_∞ denote the same when $\nu=\infty$, i.e., there is no change. The Bayesian and minimax approaches to quickest change detection are considered in the previous subsections. For this reason, we mainly study now the multiple hypothesis nature of this new problem and new performance measures related to sequential change detection and isolation.

6.3.6.1 *Sequential Change Detection and Isolation: Bayesian Formulation*

As mentioned in Subsection 6.3.1, a characteristic feature of the Bayesian formulation is the assumption that the changepoint ν is a random variable possessing a prior distribution $\{\pi_k\}_{0\leq k<\infty}$. In the case of the sequential change detection and isolation problem, it is also necessary to assume a prior distribution $p(j)$, $j=1,2,\dots,M$, of the post-change hypotheses $H_1,\dots,H_M$. From the Bayesian point of view, and by analogy with the case of pure change detection, the probability constraint imposed on the pair $\delta=(T,d)$ can be formulated using the weighted *Probability of False Alarm and Isolation* (PFAI), which is defined by [252]

$$\mathsf{PFAI}^\pi(\delta) = \sum_{j=1}^M\sum_{k=0}^\infty p(j)\pi_k\mathsf{P}_k^j(T>k\,\&\,d\neq j) + \sum_{k=0}^\infty \pi_k\mathsf{P}_\infty(T\leq k)\leq\alpha. \tag{6.28}$$

As in the case of pure change detection, a practically reasonable way to benchmark the detection delay is through the ADD for every post-change hypothesis H_j, $j = 1,2,\dots,M$,

$$\mathsf{ADD}_j^\pi(\delta) = \sum_{k=0}^{\infty} \pi_k \mathsf{E}_k^j (T-k)^+. \tag{6.29}$$

We are now in a position to formally introduce the notion of Bayesian optimality; see, e.g., Lai [252]. Let

$$\mathbf{C}_\alpha = \{\delta \colon \mathsf{PFAI}^\pi(\delta) \le \alpha\}$$

be the class of detection and isolation procedures, namely stopping times and final decisions, for which the PFAI does not exceed a preset level $\alpha \in (0,1)$. Then under the Bayesian approach the goal is to

$$\begin{aligned} &\text{Find } \delta_{\text{opt}} \in \mathbf{C}_\alpha \text{ such that } \mathsf{ADD}_j^\pi(\delta_{\text{opt}}) = \inf_{\delta \in \mathbf{C}_\alpha} \mathsf{ADDI}_j^\pi(\delta) \\ &\text{for every } \alpha \in (0,1) \text{ and for every } j = 1,2,\dots,M. \end{aligned} \tag{6.30}$$

An asymptotic (as $\alpha \to 0$) solution to this problem has been established by Lai [252, Theorem 5]. The existence of a nonasymptotic solution still seems to be an open problem.

6.3.6.2 *Sequential Change Detection and Isolation: Minimax Formulation*

Contrary to the Bayes formulation the minimax change detection and isolation approach posits that the changepoint ν and the post-change hypotheses H_j, $j = 1,2,\dots,M$, are unknown and not necessarily random. Even if they are random their distributions are unknown. It is intuitively obvious that the optimality criterion should favor fast detection with few false alarms and false isolations.

The first minimax theory of change detection and isolation is due to Nikiforov [322], where it is proposed to measure the false alarm and false isolation risks by the ARL to false alarm or to false isolation. It is assumed that when $\nu = 0$ and the hypothesis H_j is true the observations $X_1, X_2, \dots$ obey the distribution P_0^j, $j = 0,1,2,\dots,M$. Assume also that all observations are either under the nominal regime, in which case they obey the measure $\mathsf{P}_0^0 = \mathsf{P}_\infty$, or under the abnormal regime of type j, in which case they follow the measure P_0^j. Consider the following sequences of alarm times and final decisions (T_r, d_r):

$$T_0 = 0 < T_1 < T_2 < \cdots < T_r < \cdots \quad \text{and} \quad d_1,\ d_2,\ \cdots,\ d_r,\ \cdots,$$

where T_r is the alarm time of the detection–isolation algorithm which is applied to $X_{T_{r-1}+1}, X_{T_{r-1}+2}, \dots$. It is assumed that after a false alarm the observation process immediately restarts from scratch.

Let the observations $X_1, X_2, \dots$ obey the distribution P_∞. The ARL to first false alarm of type j is defined by $\mathsf{E}_0^0(\inf_{r\ge1}\{T_r : d_r = j\}) = \mathsf{E}_\infty(\inf_{r\ge1}\{T_r : d_r = j\})$. Analogously, let the observations $X_1, X_2, \dots$ obey the distribution P_0^ℓ, $\ell = 1,2,\dots,M$. Then the ARL to the first false isolation of type $j \ne \ell$ is $\mathsf{E}_0^\ell(\inf_{r\ge1}\{T_r : d_r = j\})$. Let

$$\mathbf{C}_\gamma = \left\{\delta = (T,d) \colon \min_{0\le\ell\le M} \min_{1\le j\ne\ell\le M} \mathsf{E}_0^\ell\left(\inf_{r\ge1}\{T_r : d_r = j\}\right) \ge \gamma\right\} \tag{6.31}$$

be the class of detection and isolation procedures for which the ARL to false alarm and false isolation is at least $\gamma > 1$.

The risk associated with the detection delay is defined analogously to Lorden's worst-worst-case ADD introduced in (6.17). In the case of detection–isolation procedures this risk is

$$\mathsf{ESADD}(\delta) = \max_{1\le j\le M} \sup_{0\le\nu<\infty} \left\{\operatorname{ess\,sup} \mathsf{E}_\nu^j[(T-\nu)^+ | \mathscr{F}_\nu]\right\}. \tag{6.32}$$

Hence, the minimax optimization problem seeks to

$$\text{Find } \delta_{\text{opt}} \in \mathbf{C}_\gamma \text{ such that } \mathsf{ESADD}(\delta_{\text{opt}}) = \inf_{\delta \in \mathbf{C}_\gamma} \mathsf{ESADD}(\delta) \text{ for every } \gamma > 1, \tag{6.33}$$

where $\mathbf{C}_\gamma$ is the class of detection and isolation procedures with the lower bound γ on the ARL to false alarm and false isolation defined in (6.31).

Another minimax approach to change detection and isolation is as follows; see Nikiforov [328, 331]. First, unlike the definition of the class $\mathbf{C}_\gamma$ in (6.31), where we fixed *a priori* the changepoint $\nu = 0$ in the definition of false isolation to simplify theoretical analysis, the false isolation rate is now expressed by the maximal probability of false isolation $\sup_{\nu \ge 0} \mathsf{P}_\nu^\ell(d = j \neq \ell | T > \nu)$. As usual, we measure the level of false alarms by the ARL2FA $\mathsf{E}_\infty T$. Hence, define the class

$$\mathbf{C}_{\gamma,\beta} = \left\{ \delta : \mathsf{E}_\infty T \ge \gamma, \max_{1 \le \ell \le M} \max_{1 \le j \neq \ell \le M} \sup_{\nu \ge 0} \mathsf{P}_\nu^\ell(d = j | T > \nu) \le \beta \right\}. \tag{6.34}$$

As mentioned in Subsection 6.3.3, sometimes Lorden's worst-worst-case ADD is too conservative, especially for recursive change detection and isolation procedures, and another measure of the detection speed, namely the maximal conditional average delay to detection $\mathsf{SADD}(T) = \sup_\nu \mathsf{E}_\nu(T - \nu | T > \nu)$, is better suited for practical purposes. In the case of change detection and isolation, the SADD is given by

$$\mathsf{SADD}(\delta) = \max_{1 \le j \le M} \sup_{0 \le \nu < \infty} \mathsf{E}_\nu^j(T - \nu | T > \nu). \tag{6.35}$$

We require that the $\mathsf{SADD}(\delta)$ should be *as small as possible* subject to the constraints on the ARL to false alarm and the maximum probability of false isolation. Therefore, this version of the minimax optimization problem seeks to

$$\begin{gathered} \text{Find } \delta_{\text{opt}} \in \mathbf{C}_{\gamma,\beta} \text{ such that } \mathsf{SADD}(\delta_{\text{opt}}) = \inf_{\delta \in \mathbf{C}_{\gamma,\beta}} \mathsf{SADD}(\delta) \\ \text{for every } \gamma > 1 \text{ and } \beta \in (0,1). \end{gathered} \tag{6.36}$$

Note that this problem formulation is fully minimax: it seeks to find a procedure that minimizes the risk maximized over both the changepoint ν and the hypothesis H_j. It is also reasonable to consider the following relaxed version with respect to the hypotheses suggested by Tartakovsky [456, 461]. Let

$$\mathbf{C}_{\gamma,\{\alpha_j\}} = \left\{ \delta : \mathsf{E}_\infty T \ge \gamma, \sup_{\nu \ge 0} \mathsf{P}_\nu^j(d \neq j | T > \nu) \le \alpha_j \text{ for all } j = 1, \dots, N \right\} \tag{6.37}$$

and let $\mathsf{SADD}_j(T) = \sup_{\nu \ge 0} \mathsf{E}_\nu^j(T - \nu | T > \nu)$ be the maximal average delay to detection over the unknown changepoint associated with the hypothesis H_j. We seek to

$$\begin{gathered} \text{Find } \delta_{\text{opt}} \in \mathbf{C}_{\gamma,\{\alpha_j\}} \text{ such that } \mathsf{SADD}_j(\delta_{\text{opt}}) = \inf_{\delta \in \mathbf{C}_{\gamma,\{\alpha_j\}}} \mathsf{SADD}_j(\delta) \\ \text{for all } j = 1, \dots, M \text{ and for every } \gamma > 1 \text{ and } \alpha_j \in (0,1). \end{gathered} \tag{6.38}$$

Usually it is almost impossible to find an optimal procedure that solves the optimization problems (6.36) and (6.38) for arbitrary values of γ, β and α_j. For this reason, in Chapter 9 we address an asymptotic setting when $\gamma \to \infty$, $\beta \to 0$ and $\alpha_j \to 0$ and find asymptotically optimal change detection–isolation procedures.

Chapter 7

Sequential Changepoint Detection: Bayesian Approach

In this chapter, we discuss in detail the Bayesian approach to changepoint detection for iid and general non-iid models, which was outlined in Chapter 6.

7.1 Optimality and Operating Characteristics of the Shiryaev Procedure in the iid Case

7.1.1 The Shiryaev Procedure and Its Optimality

We begin with considering the iid model (6.3) and the Bayesian changepoint detection problem (6.7), assuming the zero-modified geometric prior distribution for the changepoint given in (6.8). In Subsection 6.3.1, we stated that the solution of this problem is given by the Shiryaev procedure defined in (6.11) and (6.9). Specifically, let $T_s(A)$ be the stopping time of the Shiryaev change detection procedure given by

$$T_s(A) = \inf\{n \ge 1 : R_{n,p} \ge A\}, \tag{7.1}$$

where A is a positive threshold and $R_{n,p}$ is the Shiryaev detection statistic defined in (6.9) that can be written in the recursive form (6.12).

It is worth noting that the statistic $R_{n,p}$ can be expressed via the average likelihood ratio of the hypotheses $H_k : \nu = k$ and $H_\infty : \nu = \infty$

$$\bar{\Lambda}_n := \sum_{k=-\infty}^{\infty} \pi_k \prod_{i=k+1}^{n} \mathcal{L}_i = q\prod_{i=1}^{n}\mathcal{L}_i + \sum_{k=0}^{n-1} \pi_k \prod_{i=k+1}^{n} \mathcal{L}_i + \Pr(\nu \ge n),$$

where $\mathcal{L}_i = f_1(X_i)/f_0(X_i)$. For the zero-modified geometric prior distribution (6.8),

$$\bar{\Lambda}_n = q\prod_{i=1}^{n}\mathcal{L}_i + (1-q)p\sum_{k=0}^{n-1}(1-p)^k \prod_{i=k+1}^{n} \mathcal{L}_i + (1-q)(1-p)^n,$$

so that

$$R_{n,p} = \frac{\bar{\Lambda}_n}{(1-q)p(1-p)^n} - 1.$$

Note also that for any prior distribution the posterior probability $\Pi_n = \mathsf{P}^\pi(\nu < n|\mathbf{X}_1^n)$ is equal to $[\bar{\Lambda}_n - \Pr(\nu \ge n)]/\bar{\Lambda}_n$.

The following theorem establishes optimality of the Shiryaev procedure in the class of detection procedures $\mathbf{C}_\alpha = \{T : \mathsf{PFA}^\pi(T) \le \alpha\}$ for which the weighted false alarm probability $\mathsf{PFA}^\pi(T) = \mathsf{P}^\pi(T \le \nu) = \sum_{k=1}^{\infty} \pi_k \mathsf{P}_\infty(T \le k)$ is upper-bounded by the *a priori* specified level α.

Theorem 7.1.1. *Consider the iid case* (6.3) *and assume the geometric prior distribution* (6.8). *If* $0 < \alpha < 1 - q$ *and the threshold* $A = A_\alpha$ *is selected so that* $\mathsf{PFA}^\pi(T_s(A_\alpha)) = \alpha$, *then*

$$\inf_{T \in \mathbf{C}_\alpha} \mathsf{ADD}^\pi(T) = \mathsf{ADD}^\pi(T_s(A_\alpha)) \text{ for any } 0 < \alpha < 1 - q. \tag{7.2}$$

If $\alpha \geq 1-q$, the optimal procedure is degenerate, $T_{\rm opt} = 0$ w.p. 1, i.e., it prescribes making the decision on the change with no sampling.

The proof of this theorem is quite tedious and can be found in the book of Shiryaev [420, Ch. 4, Sec. 3]. It is based on the "purely" Bayes solution since, as usual, using the standard Lagrangian technique, the constrained optimization problem (6.7) reduces to the unconstrained Bayes problem with some Lagrange multiplier $c = c_\alpha$ that depends on the given constraint α:

$$\rho_c^\pi(T) = \mathsf{PFA}^\pi(T) + c\,\mathsf{ADD}^\pi(T)\mathsf{P}^\pi(T > \nu) = \mathsf{PFA}^\pi(T) + c\,\mathsf{E}^\pi(T-\nu)^+. \tag{7.3}$$

It is therefore instrumental to consider the "purely" Bayes, unconstrained problem with the loss $1\!\!1_{\{T\leq\nu\}}$ associated with a false alarm and the cost c ($c > 0$) per one-step delay of detecting the true change, in which case the expected loss or average risk of the procedure T is given by (7.3). This Bayes problem is denoted as $\mathcal{B}(q,p,c)$. The goal is to find an optimal Bayes procedure that minimizes the average risk over all stopping times,

$$\rho_c^\pi(T_{\rm opt}) = \inf_{T\geq 0} \rho_c^\pi(T). \tag{7.4}$$

The following result holds true whose proof may be found in [420].

Theorem 7.1.2. *Consider the iid case* (6.4). *Let $c > 0$, $0 < p < 1$, and $0 \leq q < 1$. The Shiryaev detection procedure $T_{\rm s}(A_0)$ given in* (7.1) *is optimal in the Bayes problem $\mathcal{B}(q,p,c)$, where $A_0(c,p)$ is a threshold that depends on c,p, in such a way that $A_0(c,p) > q/[(1-q)p]$. If $A_0(c,p) \leq q/[(1-q)p]$, then the optimal procedure prescribes stopping with no sampling, i.e., $T_{\rm opt} = 0$ w.p. 1.*

Note that strict optimality of the Shiryaev procedure in the class $\mathbf{C}_\alpha$ requires the exact match of the false alarm probability which is related to the estimation of the overshoot in the stopping rule (7.1). This is problematic in general. However, a simple upper bound, which ignores the overshoot, can be easily obtained. Indeed, using the Bayes formula, the posterior probability $\Pi_n = \mathsf{P}^\pi(\nu < n|\mathbf{X}_1^n)$ can be written as

$$\Pi_n = \frac{R_{n,p}}{R_{n,p}+1/p}, \tag{7.5}$$

so the stopping time (7.1) can be equivalently represented as

$$T_{\rm s}(B) = \inf\{n \geq 1 \colon \Pi_n \geq B\}, \quad B = A/(A+1/p). \tag{7.6}$$

Now, since $\mathsf{P}^\pi\{T_{\rm s}(B) \leq \nu\} = \mathsf{E}^\pi\{1-\Pi_{T_{\rm s}(B)}\}$ and $1-\Pi_{T_{\rm s}(B)} \leq 1-B$ on $\{T_{\rm s}(B) < \infty\}$, we obtain the inequalities

$$\mathsf{PFA}^\pi(T_{\rm s}(B)) \leq 1-B \quad \text{and} \quad \mathsf{PFA}^\pi(T_{\rm s}(A)) \leq 1/(1+Ap). \tag{7.7}$$

Therefore,

$$A = A_\alpha = (1-\alpha)/\alpha p \quad \text{implies} \quad \mathsf{PFA}^\pi(T_{\rm s}(A_\alpha)) \leq \alpha. \tag{7.8}$$

Note also that the first inequality in (7.7) holds true for an arbitrary proper and not necessarily geometric prior distribution as well as for general non-iid models.

In Subsection 7.1.2, we discuss how to numerically compute the PFA almost precisely and we present asymptotic approximations for large A or equivalently small α.

Remark 7.1.1. In Theorem 7.1.1 we assume that the stopping boundary $A = A_\alpha$ can be selected so that the strict equality $\mathsf{PFA}^\pi(T_{\rm s}(A_\alpha)) := \mathsf{PFA}^\pi(A_\alpha) = \alpha$ holds. This is always possible when $\mathsf{PFA}^\pi(A)$ is continuous in A, which is in turn true if the P^π-distribution of the LR $\mathcal{L}_1$ is continuous. The case where the LR is not continuous may require randomization on the stopping boundary A, i.e., deciding whether to stop or continue when $R_{n,p} = A$, since the nonrandomized Shiryaev procedure may not guarantee the strict equality $\mathsf{PFA}^\pi(A_\alpha) = \alpha$.

In the next subsection, we also show that, when the K–L information number

$$I = \int \left[\log \frac{f_1(x)}{f_0(x)} \right] f_1(x) \, \mathrm{d}\mu(x)$$

is positive and finite, then the Shiryaev procedure asymptotically (as $\alpha \to 0$) minimizes all positive moments of the delay to detection $\mathsf{E}^{\pi}[(T - \nu)^m | T > \nu]$, $m > 0$.

7.1.2 *Operating Characteristics*

Clearly, the main operating characteristics of detection procedures in the Bayes setup are the average delay to detection $\mathsf{ADD}^{\pi}(T)$ and the false alarm probability $\mathsf{PFA}^{\pi}(T)$ defined in (6.6) and (6.5), respectively. Thus, we have to be able to compute both quantities for various procedures and for the Shiryaev procedure in the first place as functions of the threshold A. Below we suggest numerical and asymptotic techniques that allow for such analysis.

7.1.2.1 *Integral Equations and Numerical Techniques*

In this subsection, we provide integral equations for the operating characteristics of a generic changepoint detection procedure given by the stopping time

$$T_A^r = \inf\{n \geq 1 : \mathcal{S}_n^r \geq A\}, \quad A > 0, \tag{7.9}$$

where $\mathcal{S}_n^r$ is a Markov statistic satisfying the recursion

$$\mathcal{S}_n^r = \Phi(\mathcal{S}_{n-1}^r)\mathcal{L}_n, \quad n \geq 1, \quad \mathcal{S}_0^r = r \geq 0, \tag{7.10}$$

and $\Phi(\mathcal{S})$ is a positive-valued function. Using numerical techniques one can then obtain efficient numerical approximations.

Note that many procedures considered in this book satisfy particular cases of the recursion (7.10). Indeed, because of (6.12) for the Shiryaev procedure we have $\Phi(\mathcal{S}) = (1+\mathcal{S})/(1-p)$, and for the SR procedure, briefly discussed in Subsection 6.3.2 and further discussed in Section 7.6, because of (7.149) we have $\Phi(\mathcal{S}) = 1 + \mathcal{S}$. Also, for the CUSUM procedure defined in (7.155) and further considered in detail in the next chapter, $\Phi(\mathcal{S}) = \max\{1, \mathcal{S}\}$.

We now derive a set of equations for the performance metrics of the generic detection procedure given by (7.9) and (7.10) in the Bayesian context and for the geometric prior model (6.8). These equations can be easily adapted to special cases by substituting the proper function $\Phi(\mathcal{S})$. For simplicity we assume that the P_i-distributions $F_i(x) = \mathsf{P}_i(\mathcal{L}_1 \leq x)$ of the likelihood ratio $\mathcal{L}_1$ are continuous for $i = \{\infty, 0\}$.

For $k \geq 0$, let $\delta_k(r) = \mathsf{E}_k(T_A^r - k)^+$ and $\rho_k(r) = \mathsf{P}_\infty(T_A^r > k)$ ($\rho_0^r = 1$).

Observe that the PFA can be written as

$$\begin{aligned}
\mathsf{PFA}^{\pi}(T_A^r) &= (1-q)p \sum_{k=1}^{\infty} (1-p)^k \mathsf{P}_\infty(T_A^r \leq k) \\
&= (1-q) \left\{ 1 - p \sum_{k=0}^{\infty} (1-p)^k \rho_k(r) \right\},
\end{aligned} \tag{7.11}$$

and the numerator $\mathsf{N}(T_A^r) = \sum_{k=0}^{\infty} \pi_k \mathsf{E}_k(T_A^r - k)^+$ and the denominator $\mathsf{D}(T_A^r) = 1 - \mathsf{PFA}^{\pi}(T_A^r)$ of the

$\mathsf{ADD}^\pi(T_A^r)$ as

$$\begin{aligned}
\mathsf{N}(T_A^r) &= [q+(1-q)p]\mathsf{E}_0[T_A^r] + (1-q)p\sum_{k=1}^{\infty}(1-p)^k \mathsf{E}_k(T_A^r-k)^+ \\
&= q\delta_0(r) + (1-q)p\sum_{k=0}^{\infty}(1-p)^k\delta_k(r),
\end{aligned} \tag{7.12}$$

$$\mathsf{D}(T_A^r) = q+(1-q)p\sum_{k=0}^{\infty}(1-p)^k\rho_k(r). \tag{7.13}$$

The equations (7.11)–(7.13) suggest that we need to find suitable equations for the evaluation of the two series

$$\psi_p(r) = \sum_{k=0}^{\infty}(1-p)^k\delta_k(r) \;\text{ and }\; \chi_p(r) = \sum_{k=0}^{\infty}(1-p)^k\rho_k(r),$$

as well as for the average run length to detection $\delta_0(r)$. Using the Markov property of the statistic $\mathcal{S}_n^r$ it is readily seen that $\delta_0(r)$, $\psi_p(r)$, and $\chi_p(r)$ satisfy the following Fredholm integral equations

$$\delta_0(r) = 1 + \int_0^A \delta_0(x)\mathcal{K}_0(x,r)\,\mathrm{d}x, \tag{7.14}$$

$$\psi_p(r) = \delta_0(r) + (1-p)\int_0^A \psi_p(x)\mathcal{K}_\infty(x,r)\,\mathrm{d}x, \tag{7.15}$$

$$\chi_p(r) = 1 + (1-p)\int_0^A \chi_p(x)\mathcal{K}_\infty(x,r)\,\mathrm{d}x \tag{7.16}$$

with the kernels

$$\mathcal{K}_i(x,r) = \frac{\partial}{\partial x}F_i\left(\frac{x}{\Phi(r)}\right), \quad i=0,\infty. \tag{7.17}$$

Using the solutions of the equations (7.14)–(7.16), we can compute the two quantities of interest as follows

$$\mathsf{PFA}^\pi(T_A^r) = (1-q)\{1-p\chi_p(r)\} \;\text{ and }\; \mathsf{ADD}^\pi(T_A^r) = \frac{q\delta_0(r)+(1-q)p\psi_p(r)}{q+(1-q)p\chi_p(r)}.$$

In addition, it is of interest to compute the conditional average detection delays $\mathsf{E}_k(T_A^r - k | T_A^r > k) = \delta_k(r)/\rho_k(r)$ for any $k \ge 1$. The functions $\delta_k(r)$ and $\rho_k(r)$ can be computed recursively; see [309, 310] and Subsection 8.2.6.1.

Let $\{f_\theta(x), \theta \in \Theta\}$ be a parametric family and $f_1(x) = f_{\theta_1}(x)$. In practice, typically the parameter of the post-change distribution θ is not known, so we are also interested in finding the average delay to detection ADD_θ^π at a parameter value different from the putative value θ_1. In this case, the integral equations (7.14)–(7.16) hold with the kernel $\mathcal{K}_0(x,r)$ replaced with $\mathcal{K}_\theta(x,r) = \frac{\partial}{\partial x}F_\theta\left(x/\Phi(r)\right)$, where $F_\theta(y) = \mathsf{P}_\theta(\mathcal{L}_1 \le y)$.

The equations (7.11)–(7.13) are Fredholm equations of the second kind. Since generally no analytical solutions are possible we have to resort to numerical techniques. A number of numerical techniques for solving this kind of integral equation are described in detail in Subsection 3.1.4. In particular, a simple numerical scheme using a quadrature rule with $N \gg 1$ breakpoints to approximate the integrals can be used to provide an approximate solution. The resulting systems of linear equations can be solved either directly or iteratively. The accuracy of such numerical schemes is directly related to the number N of breakpoints, and one expects the accuracy to improve while this number increases. Clearly, when the false alarm constraint becomes very stringent, i.e., the threshold A takes large values, this requires an exceedingly large number of points resulting in unrealistic processing times. Thus, numerical techniques can be used for moderate to small values of the false alarm constraint. For very small values of the PFA numerical methods cannot be used and asymptotic approximations are needed. This issue is discussed in the next subsection.

7.1.2.2 *Asymptotic Approximations*

By Theorem 7.1.1, the Shiryaev procedure $T_s(A)$ is optimal when the threshold A can be chosen in such a way that $\mathsf{PFA}(T_s(A)) = \alpha$. Since it is difficult to meet this exact requirement, we first study the asymptotic properties of the detection procedure $T_s(A)$ for a large A and establish first-order asymptotic optimality of this procedure with $A = (1-\alpha)/(\alpha p)$, which guarantees the inequality $\mathsf{PFA}(T_s(A)) \le \alpha$ outlined in (7.8). Since this choice neglects the overshoot, it is expected that the actual false alarm probability may be substantially smaller than α; see Subsection 7.1.2.4. Theorem 7.1.5 below determines an asymptotic approximation for the PFA that accounts for the overshoot. In Subsections 7.1.2.3 and 7.1.2.4 making use of numerical techniques of the previous subsection this approximation is shown to be very accurate.

Before getting to asymptotic approximations in the iid case, which is the main topic of this subsection, we prove the following fundamental result in the general non-iid case and for a general prior distribution. Specifically, consider the most general non-iid model (6.2) where the data follow the conditional densities $f_{0,n}(X_n|\mathbf{X}_1^{n-1})$, $1 \le n \le \nu$, in the pre-change mode and the conditional densities $f_{1,n}^{(\nu)}(X_n|\mathbf{X}_1^{n-1})$, $n \ge \nu+1$, in the post-change mode. Let $Z_i = \log[f_{1,i}(X_i|\mathbf{X}_1^{i-1})/f_{0,i}(X_i|\mathbf{X}_1^{i-1})]$ and let

$$\lambda_n^k = \log \frac{\mathsf{dP}_k}{\mathsf{dP}_\infty}(\mathbf{X}_1^n) = \sum_{i=k+1}^{n} Z_i$$

denote the LLR between the hypotheses $H_k : \nu = k$ and $H_\infty : \nu = \infty$ given the sample $\mathbf{X}_1^n$. Note that generally the post-change densities $f_{1,i}(X_i|\mathbf{X}_1^{i-1}) = f_{1,i}^{(\nu)}(X_i|\mathbf{X}_1^{i-1})$, $i > \nu$, depend on the changepoint ν, but we omit the index ν for brevity. We consider general prior distributions, $\pi_k = \Pr(\nu = k)$,

$$\Pr(\nu < 0) = q, \quad \pi_k = (1-q)\widetilde{\pi}_k, \; k = 0,1,2,\ldots, \tag{7.18}$$

where $q \in [0,1)$ and $\widetilde{\pi}_k = \Pr(\nu = k|\nu \ge 0)$, $k = 0,1,2,\ldots$, assuming that

$$\lim_{k\to\infty} \frac{\log \Pr(\nu > k)}{k} = -d \quad \text{for some } d \ge 0. \tag{7.19}$$

Note that if $d > 0$, then this is the prior model with bounded hazard rate, e.g., the geometric distribution, in which case $d = |\log(1-p)|$ for all k. The case $d = 0$ relates to the prior models with vanishing hazard rate, e.g., the Zipf distribution. The case $d = 0$ can be further generalized by replacing 0 with a negative number $-d_\alpha$ that vanishes when $\alpha \to 0$. Practitioners further narrow the set of the considered prior distributions based on the specificities of the particular problem at hand. If the occurrence of a changepoint becomes approximately memoryless in the remote future, although its hazard rate may still vary, the prior distribution is selected from the class with $d > 0$, which includes the distributions with an exponential tail. However, if the process has a memory, and longer periods of no change make a practitioner decide that a changepoint during the next minute is *a priori* less likely, then $d = 0$, i.e., heavy-tailed distributions. The set of models excluded from our consideration consists of priors with unboundedly increasing hazard rates. In such models, time intervals containing a changepoint with the same probability, given that it has not occurred earlier, decrease to 0. Hence the change is likely to be detected at early stages, and the asymptotic study is impractical.

Theorem 7.1.3. *Consider the general non-iid model given in* (6.2). *Let the prior distribution satisfy the condition* (7.19). *Let $K_\alpha = |\log \alpha|/(I+d)$. Assume that there exists a finite number $I > 0$ such that*

$$\mathsf{P}_k\left\{\frac{1}{M}\max_{0\le n<M} \lambda_{k+n}^k \ge (1+\varepsilon)I\right\} \xrightarrow[M\to\infty]{} 0 \quad \text{for all } \varepsilon > 0 \text{ and } k \ge 0. \tag{7.20}$$

Then, for all $0 < \varepsilon < 1$,

$$\lim_{\alpha\to 0} \sup_{T\in \mathbf{C}_\alpha} \mathsf{P}^\pi\{\nu < T < \nu + (1-\varepsilon)K_\alpha\} = 0, \tag{7.21}$$

and for every $m > 0$

$$\liminf_{\alpha \to 0} \frac{\inf_{T \in \mathbf{C}_\alpha} \mathsf{E}^\pi[(T-\nu)^m | T > \nu]}{|\log \alpha|^m} \ge \left(\frac{1}{I+d}\right)^m. \tag{7.22}$$

Proof. Write $\gamma^{(k)}_{\varepsilon,\alpha}(T) = \mathsf{P}_k\{k < T < k + (1-\varepsilon)K_\alpha\}$. By the change of measure $\mathsf{P}_\infty \to \mathsf{P}_k$, we obtain that for any $C > 0$ and $\varepsilon \in (0,1)$,

$$\begin{aligned}
\mathsf{P}_\infty\{k < T < k + (1-\varepsilon)K_\alpha\} &= \mathsf{E}_k\left\{\mathbb{1}_{\{k<T<k+(1-\varepsilon)K_\alpha\}} e^{-\lambda_T^k}\right\} \\
&\ge \mathsf{E}_k\left\{\mathbb{1}_{\{k<T<k+(1-\varepsilon)K_\alpha, \lambda_T^k < C\}} e^{-\lambda_T^k}\right\} \\
&\ge e^{-C}\mathsf{P}_k\left\{k < T < k+(1-\varepsilon)K_\alpha, \max_{k<n<k+(1-\varepsilon)K_\alpha} \lambda_n^k < C\right\} \\
&\ge e^{-C}\left[\mathsf{P}_k\{k < T < k+(1-\varepsilon)K_\alpha\} - \mathsf{P}_k\left\{\max_{1\le n<(1-\varepsilon)K_\alpha} \lambda_{k+n}^k \ge C\right\}\right],
\end{aligned} \tag{7.23}$$

where the last inequality follows trivially from the fact that for any events $\mathcal{A}$ and $\mathcal{B}$, $\Pr(\mathcal{A} \cap \mathcal{B}) \ge \Pr(\mathcal{A}) - \mathsf{P}(\mathcal{B}^c)$. Setting $C = (1-\varepsilon^2)IK_\alpha$ yields

$$\begin{aligned}
\gamma^{(k)}_{\varepsilon,\alpha}(T) \le\ & e^{(1-\varepsilon^2)IK_\alpha}\mathsf{P}_\infty\{k < T < k+(1-\varepsilon)K_\alpha\} \\
&+ \mathsf{P}_k\left\{\max_{1\le n<(1-\varepsilon)K_\alpha} \lambda_{k+n}^k \ge (1-\varepsilon^2)IK_\alpha\right\}.
\end{aligned} \tag{7.24}$$

Define

$$p_k(\alpha,\varepsilon) = e^{(1-\varepsilon^2)IK_\alpha}\mathsf{P}_\infty\{k < T < k+(1-\varepsilon)K_\alpha\}$$

and

$$\beta_k(\alpha,\varepsilon) = \mathsf{P}_k\left\{\max_{1\le n<(1-\varepsilon)K_\alpha} \lambda_{k+n}^k \ge (1-\varepsilon^2)IK_\alpha\right\}.$$

By the condition (7.20), for every $0 < \varepsilon < 1$ and all $k \ge 0$,

$$\beta_k(\alpha,\varepsilon) = \mathsf{P}_k\left\{\frac{1}{(1-\varepsilon)K_\alpha}\max_{1\le n<(1-\varepsilon)K_\alpha} \lambda_{k+n}^k \ge (1+\varepsilon)I\right\} \xrightarrow[\alpha\to 0]{} 0. \tag{7.25}$$

Next, for any $T \in \mathbf{C}_\alpha$ and $n \ge 1$,

$$\begin{aligned}
\alpha \ge \mathsf{PFA}(T) &\ge \Pr\{T \le n \cap \nu > n\} = \Pr\{T \le n | \nu > n\}\Pr\{\nu > n\} \\
&= \mathsf{P}_\infty\{T \le n\}\Pr(\nu > n),
\end{aligned}$$

and therefore,

$$\mathsf{P}_\infty\{T \le n\} \le \alpha / \Pr(\nu > n), \quad n \ge 1. \tag{7.26}$$

It follows that the first term in the inequality (7.24) is upper-bounded as

$$p_k(\alpha,\varepsilon) \le \alpha e^{(1-\varepsilon^2)IK_\alpha} / \Pr\{\nu > m_\alpha(k)\}, \tag{7.27}$$

where $m_\alpha(k) = \lfloor k + (1-\varepsilon)K_\alpha \rfloor$ is the greatest integer number $\le k + (1-\varepsilon)K_\alpha$. Since $\alpha = e^{-(I+d)K_\alpha}$ and $(m_\alpha(k) - k - 1)/(1-\varepsilon) \le K_\alpha \le (m_\alpha(k) - k)/(1-\varepsilon)$, we obtain

$$p_k(\alpha,\varepsilon) \le \exp\left\{-\frac{d+\varepsilon^2 I}{1-\varepsilon}(m_\alpha(k) - k - 1)\right\} / \Pr\{\nu > m_\alpha(k)\}$$

and

$$\frac{\log p_k(\alpha,\varepsilon)}{m_\alpha(k)} \le -\frac{\log \Pr\{\nu > m_\alpha(k)\}}{m_\alpha(k)} - \frac{d+\varepsilon^2 I}{1-\varepsilon}\frac{m_\alpha(k) - k - 1}{m_\alpha(k)}.$$

By the condition (7.19),

$$\lim_{\alpha\to 0} \frac{\log p_k(\alpha,\varepsilon)}{m_\alpha(k)} \le d - \frac{d+\varepsilon^2 I}{1-\varepsilon} = -\frac{\varepsilon}{1-\varepsilon}(d+\varepsilon I),$$

where $d \ge 0$. It follows that

$$p_k(\alpha,\varepsilon) \to 0 \quad \text{as } \alpha \to 0 \text{ for all } k \ge 0 \text{ and } 0 < \varepsilon < 1. \tag{7.28}$$

Therefore, we obtain that for every $T \in \mathbf{C}_\alpha$ and $\varepsilon > 0$,

$$\gamma^{(k)}_{\varepsilon,\alpha}(T) \le p_k(\alpha,\varepsilon) + \beta_k(\alpha,\varepsilon), \tag{7.29}$$

where by (7.25) and (7.28), $\beta_k(\alpha,\varepsilon)$ and $p_k(\alpha,\varepsilon)$ go to zero as $\alpha \to 0$.

Let $N_\alpha = \lfloor \varepsilon K_\alpha \rfloor$ be the greatest integer number less than or equal to εK_α. Evidently,

$$\mathsf{P}^\pi\{\nu < T < \nu + (1-\varepsilon)K_\alpha\} = \sum_{k=0}^{\infty} \pi_k \gamma^{(k)}_{\varepsilon,\alpha}(T) \le \sum_{k=1}^{N_\alpha} \pi_k \gamma^{(k)}_{\varepsilon,\alpha}(T) + \Pr\{\nu > N_\alpha\}. \tag{7.30}$$

Using (7.29) and (7.30), we obtain

$$\mathsf{P}^\pi\{\nu < T < \nu + (1-\varepsilon)K_\alpha\} \le \Pr\{\nu > N_\alpha\} + \sum_{k=1}^{N_\alpha} \pi_k \beta_k(\alpha,\varepsilon) + \sup_{k \le N_\alpha} p_k(\alpha,\varepsilon). \tag{7.31}$$

Clearly, $\Pr\{\nu > N_\alpha\} \to 0$ as $\alpha \to 0$. The second term goes to zero as $\alpha \to 0$ by the condition (7.20) (see (7.25)) and Lebesgue's dominated convergence theorem. It suffices to show that the third term vanishes as $\alpha \to 0$.

By (7.27),

$$\sup_{k \le N_\alpha} p_k(\alpha,\varepsilon) \le \alpha e^{(1-\varepsilon^2)IK_\alpha} / \Pr\{\nu > m_\alpha(N_\alpha)\} = \exp\{-(d+\varepsilon^2 I)K_\alpha\} / \Pr\{\nu > m_\alpha(N_\alpha)\},$$

where $m_\alpha(N_\alpha) = \lfloor N_\alpha + (1-\varepsilon)K_\alpha \rfloor$. Obviously, $K_\alpha \le m_\alpha(N_\alpha) \le K_\alpha + 2$, and hence,

$$\frac{\log \sup_{k \le N_\alpha} p_k(\alpha,\varepsilon)}{m_\alpha(N_\alpha)} \le -\frac{\log \Pr\{\nu > m_\alpha(N_\alpha)\}}{m_\alpha(N_\alpha)} - (d+\varepsilon^2 I)\frac{K_\alpha}{K_\alpha + 2}.$$

Since

$$-\frac{\log \Pr\{\nu > m_\alpha(N_\alpha)\}}{m_\alpha(N_\alpha)} \to d \quad \text{as } \alpha \to 0,$$

we obtain

$$\lim_{\alpha \to 0} \frac{\log \sup_{k \le N_\alpha} p_k(\alpha,\varepsilon)}{m_\alpha(N_\alpha)} \le -\varepsilon^2 I,$$

showing that $\sup_{k \le N_\alpha} p_k(\alpha,\varepsilon) \to 0$ as $\alpha \to 0$. Since the right-hand side in (7.31) does not depend on a particular stopping time T, (7.21) follows.

To prove the asymptotic lower bound (7.22) we use the Chebyshev inequality, according to which for any $0 < \varepsilon < 1$ and any $T \in \mathbf{C}_\alpha$,

$$\mathsf{E}^\pi[(T-\nu)^+]^m \ge [(1-\varepsilon)K_\alpha]^m \mathsf{P}^\pi\{T - \nu \ge (1-\varepsilon)K_\alpha\},$$

where

$$\mathsf{P}^\pi\{T - \nu \ge (1-\varepsilon)K_\alpha\} = \mathsf{P}^\pi\{T > \nu\} - \mathsf{P}^\pi\{\nu < T < \nu + (1-\varepsilon)K_\alpha\}.$$

Since $\mathsf{P}^\pi(T > \nu) \ge 1-\alpha$ for any $T \in \mathbf{C}_\alpha$, it follows that

$$\begin{aligned}\mathsf{E}^\pi[(T-\nu)^m|T>\nu] &= \frac{\mathsf{E}^\pi[(T-\nu)^+]^m}{\mathsf{P}^\pi\{T>\nu\}} \\ &\ge [(1-\varepsilon)K_\alpha]^m\left[1-\frac{\mathsf{P}^\pi\{\nu<T<\nu+(1-\varepsilon)K_\alpha\}}{\mathsf{P}^\pi\{T>\nu\}}\right] \\ &\ge [(1-\varepsilon)K_\alpha]^m\left[1-\frac{\mathsf{P}^\pi\{\nu<T<\nu+(1-\varepsilon)K_\alpha\}}{1-\alpha}\right].\end{aligned} \tag{7.32}$$

Since ε can be arbitrarily small and, by (7.21), $\sup_{T\in\mathbf{C}_\alpha}\mathsf{P}^\pi\{\nu<T<\nu+(1-\varepsilon)K_\alpha\}\to 0$ as $\alpha\to 0$, the asymptotic lower bound (7.22) follows, and the proof is complete. □

Theorem 7.1.3 is of great importance. It is repeatedly used in the sequel for obtaining asymptotic approximations and establishing asymptotic optimality properties of the Shiryaev procedure in the iid and non-iid cases.

We now return to the iid model. Then $Z_i = \log\mathcal{L}_i = \log[f_1(X_i)/f_0(X_i)]$ and $\lambda_n^k = \sum_{i=k+1}^n Z_i$, $n>k$ is a random walk with mean $\mathsf{E}_k\lambda_n^k = I(n-k)$. Recall that $I = \mathsf{E}_0 Z_1$ stands for the K–L information number.

For $A>1$, introduce the sequence of one-sided stopping times

$$\eta_A^{(k)} = \inf\left\{n\ge 1 : S_{k+n}^k(p) \ge \log A\right\}, \quad k=0,1,2,\dots, \tag{7.33}$$

where

$$S_{k+n}^k(p) = \lambda_{k+n}^k + n|\log(1-p)|. \tag{7.34}$$

The following theorem provides first-order approximations for all positive moments of the detection delay $\mathsf{E}^\pi[(T_s-\nu)^m|T_s>\nu]$, $m>0$, which for $m=1$ yield the approximation for the average delay to detection $\mathsf{ADD}^\pi(T_s)$.

Theorem 7.1.4 (First-order approximations). *Let, conditioned on $\nu=k$, the observations $X_1,X_2,\dots,X_k$ be iid with the pdf $f_0(x)$ and $X_{k+1},X_{k+2},\dots$ be iid with the pdf $f_1(x)$. Let the prior distribution of the changepoint ν be geometric* (6.8). *Suppose that the K–L number is positive and finite,*

$$0<I<\infty. \tag{7.35}$$

(i) *Then for any $m>0$,*

$$\mathsf{E}^\pi[(T_s(A)-\nu)^m|T_s(A)>\nu] \sim \left(\frac{\log A}{I+|\log(1-p)|}\right)^m \quad \text{as } A\to\infty. \tag{7.36}$$

(ii) *If $A=A_\alpha$ is selected so that $\mathsf{PFA}(T_s(A_\alpha))\le\alpha$ and $\log A_\alpha\sim\log(1/\alpha)$, in particular $A_\alpha = (1-\alpha)/\alpha p$, then for all $m>0$,*

$$\begin{aligned}\inf_{T\in\mathbf{C}_\alpha}\mathsf{E}^\pi[(T-\nu)^m|T>\nu] &\sim \mathsf{E}^\pi[(T_s(A_\alpha)-\nu)^m|T_s(A_\alpha)>\nu] \\ &\sim \left(\frac{|\log\alpha|}{I+|\log(1-p)|}\right)^m \quad \text{as } \alpha\to 0.\end{aligned} \tag{7.37}$$

Proof. The proof is based on the upper–lower bounding technique.

Proof of (i). Let $q=0$. A generalization for $q>0$ is straightforward.

Note that under P_k the statistic defined in (7.34),

$$S_{k+n}^k(p) = \sum_{i=k+1}^{n+k} Z_i + n|\log(1-p)|,$$

is a random walk with mean $\mathsf{E}_k S_{k+1}^k(p) = I + |\log(1-p)|$.

By the condition (7.35), I is positive and finite and hence $\mathsf{E}_k\{-\min(0,Z_k)\}^m < \infty$ for all $m > 0$. Indeed,

$$\mathsf{E}_k \exp\{-\min(0,Z_k)\} = \mathsf{E}_k e^{-Z_k} \mathbb{1}_{\{Z_k<0\}} + \mathsf{E}_k \mathbb{1}_{\{Z_k \ge 0\}} \le \mathsf{E}_k e^{-Z_k} + 1 = 2.$$

Therefore, it follows from (4.18) that for all $m > 0$

$$\mathsf{E}_k\left[\eta_A^{(k)}\right]^m = \mathsf{E}_0\left[\eta_A^{(0)}\right]^m = \left(\frac{\log A}{I + |\log(1-p)|}\right)^m (1+o(1)) \quad \text{as } A \to \infty;$$

see also [128, Theorem 4.1]. Since $\sup_k \mathsf{E}_k[(T_s(A)-k)^m | T_s(A) > k] = \mathsf{E}_0[T_s(A)]^m$, it follows

$$\sup_{0 \le k < \infty} \mathsf{E}_k[(T_s(A)-k)^m | T_s(A) > k] \le \left(\frac{\log A}{I + |\log(1-p)|}\right)^m (1+o(1)) \quad \text{as } A \to \infty,$$

where the term $o(1)$ in the right-hand side does not depend on k. This yields the asymptotic upper bound

$$\begin{aligned} \mathsf{E}^\pi[(T_s(A)-\nu)^m | T_s(A) > \nu] &= \sum_{k=0}^{\infty} p(1-p)^k \mathsf{E}_k[(T_s(A)-k)^m | T_s(A) > k] \\ &\le \left(\frac{\log A}{I + |\log(1-p)|}\right)^m (1+o(1)) \quad \text{as } A \to \infty. \end{aligned} \tag{7.38}$$

We now show that the right-hand side of this inequality is also the lower bound, which proves the desired result. To this end, Theorem 7.1.3 plays a fundamental role. Indeed, in the iid case the condition (7.20) holds trivially with I being the K–L information number, since λ_{k+n}^k/n converges to I (P_k–a.s.) by the strong law of large numbers and since

$$\mathsf{P}_k\left\{M^{-1} \max_{0 \le n < M} \lambda_{k+n}^k \ge (1+\varepsilon)I\right\}$$

does not depend on k. Thus, the lower bound follows from (7.22) if we replace α with $1/A$:

$$\mathsf{E}^\pi[(T_s(A)-\nu)^m | T_s(A) > \nu] \ge \left(\frac{\log A}{I + |\log(1-p)|}\right)^m (1+o(1)) \quad \text{as } A \to \infty.$$

which completes the proof of (7.36).

Proof of (ii). Since the condition (7.20) holds, on one hand the lower bound (7.22) of Theorem 7.1.3 applies to show that

$$\liminf_{\alpha \to 0} \frac{\inf_{T \in \mathbf{C}_\alpha} \mathsf{E}^\pi[(T-\nu)^m | T > \nu]}{|\log \alpha|^m} \ge \left(\frac{1}{I + |\log(1-p)|}\right)^m.$$

On the other hand, setting in (7.38) $A = (1-\alpha)/\alpha p$ or in any other way such that $\log A \sim \log(1/\alpha)$, we obtain the upper bound

$$\limsup_{\alpha \to 0} \frac{\mathsf{E}^\pi[(T_s(A)-\nu)^m | T_s(A) > \nu]}{|\log \alpha|^m} \le \left(\frac{1}{I + |\log(1-p)|}\right)^m.$$

The proof of (7.37) is complete. □

Theorem 7.1.4 implies that for large values of the threshold A or equivalently small α the average delay to detection of the Shiryaev procedure can be approximated to first order as

$$\mathsf{ADD}^\pi(T_s(A)) \sim \frac{\log A}{I + |\log(1-p)|} \quad \text{as } A \to \infty. \tag{7.39}$$

We now use nonlinear renewal theory to improve the first-order approximation for the ADD. Furthermore, we provide an accurate approximation for the PFA that takes into account the excess over the boundary.

With a minor loss of generality, suppose that $q = 0$, i.e., in the rest of this subsection the pure geometric prior distribution, $\pi_k = p(1-p)^k$, $k \ge 0$, is considered. We first observe that $\mathsf{CADD}_0(T_s) \ge \mathsf{CADD}_k(T_s)$ for all $k \ge 0$. To understand why, it is sufficient to consider the recursion (6.12) and to note that, for $\nu = k = 0$, the initial condition is $R_{0,p} = 0$ while, for $\nu = k \ge 1$, $0 \le R_{k,p} < A$ on $T_s > k$. Moreover, for large A, the difference between $\mathsf{E}_0 T_s$ and $\mathsf{E}_k(T_s - k | T_s > k)$ is a constant that is approximately equal to the mean of the initial condition, $\mathsf{E}_\infty \log R_{k,p}$; for $k = 0$ this value is equal to 0. This constant varies for different models and its calculation is usually problematic. For this reason, we concentrate on the evaluation of the worst-case delay $\mathsf{CADD}_0(T_s)$, which also provides an upper bound for the $\mathsf{ADD}^\pi(T_s)$.

In order to apply relevant results from nonlinear renewal theory, we rewrite the stopping time T_s in the form of a random walk crossing a constant threshold plus a nonlinear term that is slowly changing in the sense defined in Subsection 2.6.1, Definition 2.6.1. Indeed, the stopping time T_s can be written in the following form

$$T_s(A) = \inf\{n \ge 1 : S_n(p) + \ell_n \ge \log A\}, \tag{7.40}$$

where $S_n(p) = \lambda_n + n|\log(1-p)|$ is a random walk with drift $\mathsf{E}_0 S_1(p) = I + |\log(1-p)|$ and

$$\ell_n = \log\left\{1 + \sum_{i=1}^{n-1} (1-p)^i \prod_{s=1}^{i} \mathcal{L}_s^{-1}\right\} = \log\left\{1 + \sum_{i=1}^{n-1} (1-p)^i e^{-\lambda_i}\right\}. \tag{7.41}$$

Here and in the rest of this subsection, we write $\lambda_n = \sum_{i=1}^n Z_i$ in the place of λ_n^0.

For $A > 1$, define $\eta_A^{(0)} = \eta_A$ as in (7.33), i.e.,

$$\eta_A = \inf\{n \ge 1 : S_n(p) \ge \log A\}, \tag{7.42}$$

and let $\kappa_A = S_{\eta_A}(p) - \log A$ on $\{\eta_A < \infty\}$ denote the overshoot. Let

$$\mathsf{H}(y) = \mathsf{H}(y, p, I) = \lim_{A\to\infty} \mathsf{P}_0\{\kappa_A \le y\} \tag{7.43}$$

be the limiting distribution of the overshoot and let

$$\varkappa(p, I) = \lim_{A\to\infty} \mathsf{E}_0 \kappa_A = \int_0^\infty y \, d\mathsf{H}(y)$$

denote the related limiting average overshoot. Define also

$$\zeta(p, I) = \lim_{A\to\infty} \mathsf{E}_0[e^{-\kappa_A}] = \int_0^\infty e^{-y} \, d\mathsf{H}(y)$$

and

$$C(p, I) = \mathsf{E}_0\left\{\log\left[1 + \sum_{i=1}^\infty (1-p)^i e^{-\lambda_i}\right]\right\}. \tag{7.44}$$

Note that by (7.40),

$$S_{T_s}(p) = \log A - \ell_{T_s} + \chi_A \quad \text{on } \{T_s(A) < \infty\},$$

where $\chi_A = S_{T_s}(p) + \ell_{T_s} - \log A$ is the excess of the process $S_n(p) + \ell_n$ over the level $\log A$ at time T_s. Taking the expectations on both sides and applying Wald's identity, we obtain

$$(I + |\log(1-p)|)\mathsf{E}_0 T_s(A) = \log A - \mathsf{E}_0 \ell_{T_s} + \mathsf{E}_0 \chi_A. \tag{7.45}$$

The crucial observations are that the sequence $\{\ell_n, n \ge 1\}$ is slowly changing, and that ℓ_n converges P_0-a.s. as $n \to \infty$ to the random variable

$$\ell = \log\left\{1 + \sum_{i=1}^{\infty} (1-p)^i e^{-\lambda_i}\right\}$$

with finite expectation $\mathsf{E}_0 \ell = C(p,I)$. In fact, applying Jensen's inequality yields

$$C(p,I) = \mathsf{E}_0 \ell \le \log\left(1 + \sum_{k=1}^{\infty} (1-p)^k\right) = \log(1/p). \tag{7.46}$$

Moreover, $\lim_{n\to\infty} \mathsf{E}_0 \ell_n = C(p,I)$ due to the fact that $\ell_n \le \ell$.

An important consequence of the slowly changing property is that, under mild conditions, the limiting distribution of the excess of a random walk over a threshold does not change by the addition of a slowly changing nonlinear term; see Theorem 2.6.1, the First Nonlinear Renewal Theorem. Furthermore, since $\ell_n \to \ell$ and $\mathsf{E}_0 \ell_n \to C(p,I)$, using (7.45) we expect that for large A,

$$\mathsf{E}_0 T_s(A) \approx \frac{1}{I + |\log(1-p)|}\left[\log A - C(p,I) + \varkappa(p,I)\right].$$

The mathematical details are given in Theorem 7.1.5 below.

More importantly, nonlinear renewal theory allows us to obtain an asymptotically accurate approximation for $\mathsf{PFA}(T_s(A))$ that takes the overshoot into account.

Theorem 7.1.5 (Higher-order approximations). *Assume that Z_1 is nonarithmetic with respect to P_∞ and P_0.*

(i) *If the condition* (7.35) *holds, then*

$$\mathsf{PFA}(T_s(A)) = \frac{\zeta(p,I)}{Ap}(1+o(1)) \quad \text{as } A \to \infty. \tag{7.47}$$

(ii) *If, in addition, the second moment $\mathsf{E}_0|Z_1|^2$ is finite, then as $A \to \infty$*

$$\mathsf{E}_0 T_s(A) = \frac{1}{I + |\log(1-p)|}\left[\log A - C(p,I) + \varkappa(p,I)\right] + o(1). \tag{7.48}$$

Proof. (i) Obviously,

$$\begin{aligned}\mathsf{PFA}(T_s(A)) &= \mathsf{E}^\pi(1 - \Pi_{T_s}) = \mathsf{E}^\pi(1 + pR_{T_s,p})^{-1} = \mathsf{E}^\pi[1 + Ap(R_{T_s,p}/A)]^{-1} \\ &= \frac{1}{Ap}\mathsf{E}^\pi[e^{-\chi_A}](1+o(1)) \quad \text{as } A \to \infty,\end{aligned}$$

where $\chi_A = \log R_{T_s,p} - \log A = S_{T_s}(p) + \ell_{T_s} - \log A$ is the overshoot. Since $\chi_A \ge 0$ and $\mathsf{PFA}(T_s(A)) < 1/pA$, it follows that

$$\begin{aligned}\mathsf{E}^\pi[e^{-\chi_A}] &= \mathsf{E}^\pi\left[e^{-\chi_A} | T_s(A) \le \nu\right] \mathsf{PFA}(T_s(A)) + \mathsf{E}^\pi\left[e^{-\chi_A} | T_s(A) > \nu\right]\left[1 - \mathsf{PFA}(T_s(A))\right] \\ &= \mathsf{E}^\pi\left[e^{-\chi_A}, T_s(A) > \nu\right] + o(1) \quad \text{as } A \to \infty.\end{aligned}$$

Therefore, it suffices to evaluate the value of

$$\mathsf{E}^\pi\left[e^{-\chi_A}, T_s(A) > \nu\right] = \sum_{k=0}^{\infty} \pi_k \mathsf{E}_k\left[e^{-\chi_A}, T_s(A) > k\right].$$

Since $\mathsf{P}_k(T_s > k) = \mathsf{P}_\infty(T_s > k) \to 1$ as $A \to \infty$ and since $0 < I < \infty$, we can apply Theorem 2.6.1 to obtain

$$\lim_{A\to\infty} \mathsf{E}^\pi\left[e^{-\chi_A}, T_s(A) > \nu\right] = \lim_{A\to\infty} \mathsf{E}^\pi\left[e^{-\chi_A}\right] = \zeta(p,I),$$

which completes the proof of (7.47).

(ii) The proof of (7.48) rests on Theorem 2.6.4, the Second Nonlinear Renewal Theorem. Indeed, by (7.40), the stopping time $T_s(A)$ is based on thresholding the sum of the random walk $S_n(p)$ and the nonlinear term ℓ_n. Since

$$\ell_n \xrightarrow[n\to\infty]{\mathsf{P}_0\text{-a.s.}} \ell \text{ and } \mathsf{E}_0\ell_n \xrightarrow[n\to\infty]{} \mathsf{E}_0\ell = C(p,I),$$

ℓ_n, $n \ge 1$, are slowly changing under P_0. In order to apply Theorem 2.6.4 we have to check the validity of the following three conditions:

$$\sum_{n=1}^{\infty} \mathsf{P}_0\{\ell_n \le -\varepsilon n\} < \infty \text{ for some } 0 < \varepsilon < I; \tag{7.49}$$

$$\max_{0\le k\le n} |\ell_{n+k}|,\ n \ge 1, \text{ are } \mathsf{P}_0 - \text{uniformly integrable;} \tag{7.50}$$

$$\lim_{A\to\infty} (\log A)\, \mathsf{P}_0\{T_s(A) \le \varepsilon K_A\} = 0 \text{ for some } 0 < \varepsilon < 1, \tag{7.51}$$

where $K_A = I_p^{-1}\log A$, $I_p = I + |\log(1-p)|$.

The condition (7.49) holds trivially since $\ell_n \ge 0$. Since ℓ_n, $n = 1,2\dots$, are non-decreasing, $\max_{0\le k\le n}|\ell_{n+k}| = \ell_{2n}$ and to prove (7.50) it suffices to show that ℓ_n, $n \ge 1$, are P_0-uniformly integrable. Since $\ell_n \le \ell$ and, by (7.46), $\mathsf{E}_0\ell < \infty$, the desired uniform integrability follows. Therefore, the condition (7.50) is satisfied.

We now turn to checking the condition (7.51). Using the inequalities (7.24) and (7.27) with $\alpha = 1/A$ in the proof of Theorem 7.1.3, we obtain

$$\mathsf{P}_0\{T_s(A) < (1-\varepsilon)K_A\} \le A^{-y_\varepsilon} + \mathsf{P}_0\left\{\max_{1\le n<(1-\varepsilon)K_A} \lambda_n \ge (1-\varepsilon^2)IK_A\right\},$$

where $y_\varepsilon > 0$ for all $\varepsilon > 0$. The first term in the above inequality is of order $o(1/\log A)$ as $A \to \infty$. Since by the conditions of the theorem we have $\mathsf{E}_0|Z_1|^2 < \infty$, the second term is also of order $o(1/\log A)$ by Lemma 2.6.2. Hence, the condition (7.51) holds for all $0 < \varepsilon < 1$.

Thus, all conditions of Theorem 2.6.4 are satisfied. The use of this theorem yields (7.48) for large A. □

Remark 7.1.2. The constants $\varkappa(p,I)$ and $\zeta(p,I)$ are the subject of renewal theory; see, e.g., Theorem 2.5.3. The constant $C(p,I)$ is not easy to compute in general. For p close to 1, the upper bound (7.46) may be useful. Obviously, this bound is asymptotically accurate when $I \to 0$. Monte Carlo experiments may be used to estimate C with a reasonable accuracy; see, e.g., Table 7.1 in Subsection 7.1.2.4.

The usefulness of Theorem 7.1.5 is twofold. First, it provides accurate approximations for both the $\mathsf{PFA}^{\pi}(T_s)$ and the upper bound on $\mathsf{ADD}^{\pi}(T_s)$. Second, it allows us to study the important limiting case of $p \to 0$, i.e., the asymptotic properties of the Shiryaev–Roberts procedure. This is the subject of Section 8.4.

7.1.2.3 Example: Change Detection in the Parameter of the Exponential Distribution

In this subsection, we test the accuracy of the asymptotic approximations derived in the previous subsection by comparing with numerical computations suggested in Subsection 7.1.2.1 for a particular example.

Consider an exponential example where the mean before the change equals 1 and after the change equals $1+q$,

$$f_0(x) = e^{-x}\mathbb{1}_{\{x\ge 0\}}, \quad f_1(x) = (1+q)^{-1}e^{-x/(1+q)}\mathbb{1}_{\{x\ge 0\}}, \quad q > 0.$$

We always assume the purely geometric prior distribution.

In this example, the K–L information number is equal to $I = q - \log(1+q)$.

In Figure 7.1 we present the false alarm probability $\mathsf{PFA}^{\pi}(T_s)$ of the Shiryaev procedure as a function of the threshold A. The solid curves correspond to the almost exact values obtained solving the integral equation (7.16) numerically and the dashed ones using the asymptotic formula (7.47), i.e., $\mathsf{PFA}^{\pi}(T_s) \approx \zeta(p,q)/(Ap)$. In the exponential case, $\zeta(p,q)$ can be computed analytically as $\zeta(p,q) = 1/(1+q)$; consequently, $\mathsf{PFA}(T_s) \approx [Ap(1+q)]^{-1}$. As we can see, the asymptotic formula provides a very efficient approximation as long as $\mathsf{PFA}(T_s) \leq 0.1$. Since for most practical applications larger values of the false alarm probability are of no interest, we can conclude that the approximation (7.47) provides a perfect fit.

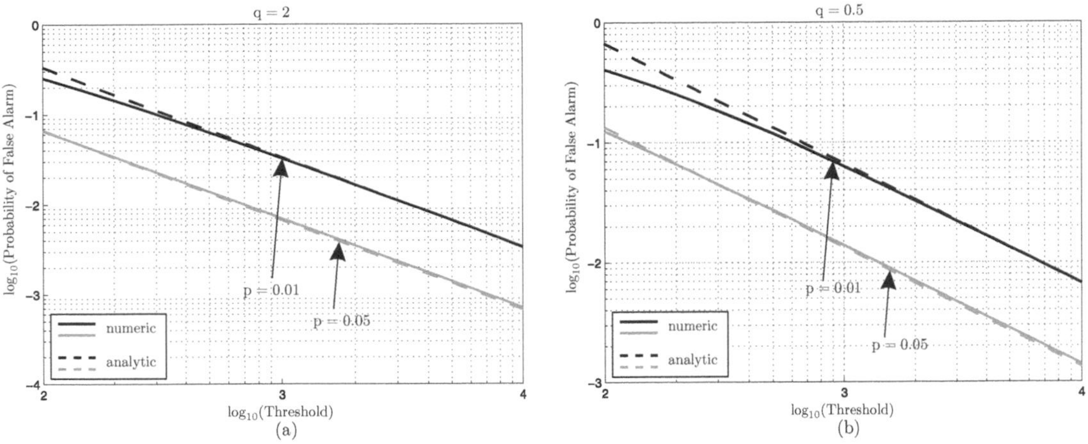

Figure 7.1: *The false alarm probability of Shiryaev's procedure as a function of the threshold A for* $p = 0.01, 0.05$ *and* $q = 2.0$ *(left),* $q = 0.5$ *(right).*

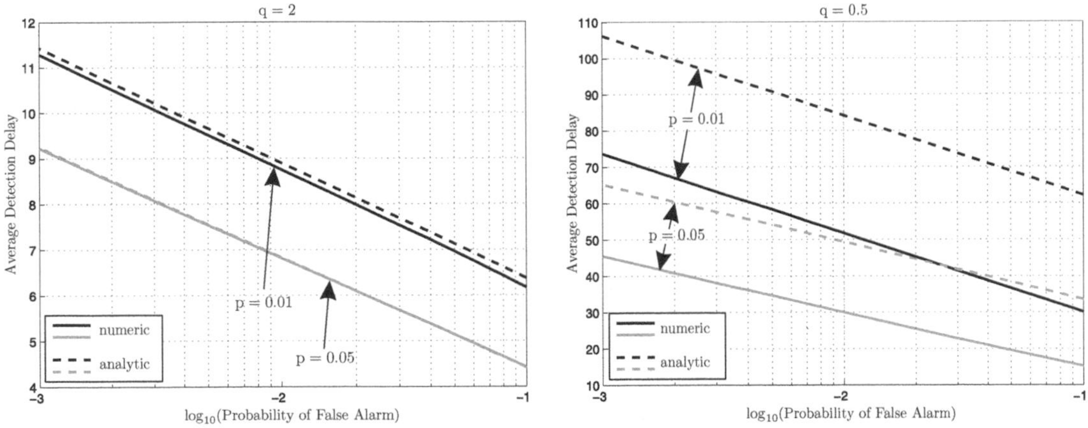

Figure 7.2: *The operating characteristics of Shiryaev's procedure computed numerically and using the asymptotic formula for* $p = 0.01, 0.05$ *and* $q = 2$ *(left),* $q = 0.5$ *(right).*

Figure 7.2 compares very accurate numerical approximations for the average delay to detection with the asymptotic formula (7.36) for $m = 1$, i.e.,

$$\mathsf{ADD}^{\pi}(T_s) \approx \frac{\log A_\alpha}{q - \log(1+q) - \log(1-p)},$$

where the threshold A_α relates to the false alarm probability α by the asymptotic formula (7.47), i.e., $A_\alpha = 1/[p(1+q)\alpha]$. Recall that according to Figure 7.1 the latter formula is fairly accurate. As we can see, the slopes exhibit a very good match between the numerical and the analytical asymptotic values for different values of p,q. There is, however, a constant shift especially for $q = 0.5$ that can be explained by the fact that first-order approximations neglect constants. The difference increases when the K–L number decreases. These constants are difficult to compute either analytically or numerically. Therefore, the integral equations and the numerical techniques proposed in Subsection 7.1.2.1 are valuable tools for achieving accurate performance evaluation.

In the next subsection, we present results of MC simulations for another, Gaussian example that allow us to compare the first-order approximations for the ADD and higher-order approximations for the upper bound on the ADD with actual ADD values.

7.1.2.4 Example: Change Detection for the Parameter of the Normal Distribution

Consider the signal plus noise model, in which we assume that $X_n = \mathbb{1}_{\{n>\nu\}}\theta + \xi_n$, $n \ge 1$, where θ is a constant signal that appears at an unknown point in time $\nu+1$, and $\{\xi_n, n \ge 1\}$ is a Gaussian iid zero-mean sequence, $\xi_n \sim \mathcal{N}(0,\sigma^2)$. As usual, $\varphi(x) = (2\pi)^{-1/2}e^{-x^2/2}$ denotes the pdf of the standard normal distribution $\Phi(x)$. Therefore,

$$f_0(X_n) = \frac{1}{\sigma}\varphi\left(\frac{X_n}{\sigma^2}\right), \quad f_1(X_n) = \frac{1}{\sigma}\varphi\left(\frac{X_n-\theta}{\sigma^2}\right), \quad Z_n = \frac{\theta}{\sigma^2}X_n - \frac{\theta^2}{2\sigma^2},$$

and $I = \theta^2/2\sigma^2$. Note that all moments of the LLR Z_1 are finite.

According to Theorem 7.1.4, the Shiryaev detection procedure with $A = (1-\alpha)/\alpha p$ asymptotically minimizes all positive moments of the detection delay in the class $\mathbf{C}_\alpha$, and the asymptotic formulas (7.36) hold with $I = \theta^2/2\sigma^2$, which can be interpreted as the signal-to-noise ratio (SNR) in a single observation. According to Theorem 7.1.5, the PFA can be approximated as

$$\mathsf{PFA}(T_s(A)) \approx \frac{\zeta(p,I)}{Ap} \tag{7.52}$$

and the upper bound on the ADD is given by

$$\begin{aligned}\mathsf{ADD}^\pi(T_s(A)) &\le \mathsf{CADD}_0(T_s(A)) \\ &\approx \max\left\{0, \frac{1}{I+|\log(1-p)|}\left[\log A - C(p,I) + \varkappa(p,I)\right]\right\}.\end{aligned} \tag{7.53}$$

According to Theorem 2.5.3, the constants $\zeta(p,I)$ and $\varkappa(p,I)$ are computed from the formulas

$$\zeta(p,I) = \frac{1}{I_p}\exp\left\{-\sum_{k=1}^\infty \frac{1}{k}F_k(p,I)\right\}, \tag{7.54}$$

$$\varkappa(p,I) = \frac{I_p^2+2I}{2I_p} - \sqrt{2I_p}\sum_{k=1}^\infty\left[k^{-1/2}\varphi\left(\frac{I_p\sqrt{k}}{\sqrt{2I}}\right) - \frac{I_p}{\sqrt{2I}}\Phi\left(-\frac{I_p\sqrt{k}}{\sqrt{2I}}\right)\right], \tag{7.55}$$

where we use the notation $I_p = I + |\log(1-p)|$ and

$$F_k(p,I) = \Phi\left(-\frac{I_p}{\sqrt{2I}}\sqrt{k}\right) + (1-p)^k\Phi\left(-\frac{I-|\log(1-p)|}{\sqrt{2I}}\sqrt{k}\right); \tag{7.56}$$

see also Example 2.5.1.

Note that computing $\mathsf{CADD}_0(T_s)$ from the higher-order (HO) approximation (7.53) requires the evaluation of the constant $C(p,I)$ using (7.44). As observed in Remark 7.1.2, we usually have to resort to MC methods to estimate $C(p,I)$. The values of C for various choices of I, p are given in

Table 7.1: *Values of the constant $C(p,I)$ for different I, p.*

I	0.5	0.5	0.5	0.25	0.25	0.25
p	0.3	0.1	0.01	0.3	0.1	0.01
$C(p,I)$	0.8366	1.2396	1.4647	0.9949	1.5859	2.0371
I	0.125	0.125	0.125	0.05	0.05	0.05
p	0.3	0.1	0.01	0.3	0.1	0.01
$C(p,I)$	1.0913	1.8694	2.5992	1.1630	2.1290	3.3528

Table 7.1. The number of trials was such that the estimate of the standard deviation of C was within 0.5% of the mean.

For the purpose of comparison, we also used the first-order (FO) approximation for the $\mathsf{ADD}^{\pi}(T_s)$ (see (7.36) with $m = 1$)

$$\mathsf{ADD}^{\pi}(T_s(A)) \approx \mathsf{CADD}_0(T_s(A)) \approx \max\left\{0, I_p^{-1} \log A\right\}, \tag{7.57}$$

which also gives the FO approximation for $\mathsf{CADD}_0(T_s)$. Note that $C(p,I)$ increases when I and p decrease, so we expect that the FO approximation performs especially poorly for small values of I and p. This is confirmed by MC simulations.

Extensive Monte Carlo simulations have been performed for different values of I, p, and α. The number of trials is given by $1000/\alpha$. Sample results are shown in Tables 7.2 and 7.3. In these tables, we present the MC estimates of the ADD along with the theoretical values computed according to (7.53) and (7.57). The abbreviations MC-ADD, MC-CADD, FO-ADD, and HO-CADD are used for $\mathsf{ADD}^{\pi}(T_s)$ obtained by the MC experiment, $\mathsf{CADD}_0(T_s)$ obtained by the MC experiment, the FO approximation (7.57), and the HO approximation (7.53) for $\mathsf{CADD}_0(T_s)$, respectively. We also display the MC estimates MC-PFA for the false alarm probability $\mathsf{PFA}(T_s)$.

Table 7.2: *Results for $A = (1-\alpha)/(p\alpha)$.*

$p = 0.1, I = 0.125$					
α	MC-PFA	MC-ADD	MC-CADD	FO-ADD	HO-CADD
0.1000	0.0768	9.2315	12.3424	18.5338	11.9508
0.0600	0.0464	11.1187	14.4889	20.9400	14.4484
0.0300	0.0215	14.0684	17.6605	24.0854	17.5509
0.0100	0.0070	18.7026	22.4509	28.9431	22.3195
0.0060	0.0043	20.8690	24.6155	31.1781	24.6034
0.0030	0.0024	23.7940	27.6285	34.2001	27.5586
0.0010	0.0007	28.5247	32.3746	38.9779	32.3523
$p = 0.1, I = 0.05$					
α	MC-PFA	MC-ADD	MC-CADD	FO-ADD	HO-CADD
0.1000	0.0858	11.9405	16.4280	27.9637	15.8823
0.0600	0.0460	14.7162	19.7314	31.5316	19.4116
0.0300	0.0240	18.9581	24.2873	36.1953	24.0955
0.0100	0.0083	25.6559	31.3594	43.3981	31.2017
0.0060	0.0048	28.8515	34.6259	46.7120	34.5589
0.0030	0.0025	33.2216	39.0867	51.1929	39.0017
0.0010	0.0008	40.2447	46.1591	58.2772	46.1774

Table 7.3: *Results for* $A = \zeta(p,I)/(p\alpha)$.

$p=0.1, I=0.5$					
α	MC-PFA	MC-ADD	MC-CADD	FO-ADD	HO-CADD
0.1000	0.0914	3.9388	4.9192	5.6139	4.8214
0.0600	0.0554	4.6407	5.7084	6.4577	5.6622
0.0300	0.0293	5.7191	6.8523	7.6027	6.8195
0.0100	0.0100	7.4474	8.6344	9.4175	8.6221
0.0060	0.0060	8.2627	9.4719	10.2614	9.4494
0.0030	0.0030	9.3973	10.6116	11.4064	10.6225
0.0010	0.0010	11.1895	12.4177	13.2212	12.4328
$p=0.01, I=0.5$					
α	MC-PFA	MC-ADD	MC-CADD	FO-ADD	HO-CADD
0.1000	0.0907	8.5173	9.9681	11.4037	9.9424
0.0600	0.0547	9.5119	11.0090	12.4052	10.9459
0.0300	0.0294	10.7933	12.2914	13.7642	12.3139
0.0100	0.0100	12.9459	14.4763	15.9181	14.4788
0.0060	0.0058	13.9602	15.4800	16.9196	15.4538
0.0030	0.0030	15.2986	16.8320	18.2786	16.8507
0.0010	0.0010	17.4523	18.9875	20.4325	18.9818
$p=0.1, I=0.125$					
α	MC-PFA	MC-ADD	MC-CADD	FO-ADD	HO-CADD
0.1000	0.0915	8.4385	11.4574	17.6206	11.0010
0.0600	0.0558	10.2942	13.5159	19.8381	13.2243
0.0300	0.0294	12.9266	16.3797	22.8471	16.2042
0.0100	0.0099	17.4060	21.0897	27.6162	21.0265
0.0060	0.0060	19.4797	23.2396	29.8337	23.2110
0.0030	0.0029	22.4640	26.2587	32.8426	26.2497
0.0010	0.0010	27.1694	31.0175	37.6117	31.0254
$p=0.1, I=0.05$					
α	MC-PFA	MC-ADD	MC-CADD	FO-ADD	HO-CADD
0.1000	0.0914	11.3236	15.7955	27.2882	15.1322
0.0600	0.0549	13.8997	18.8115	30.5762	18.3927
0.0300	0.0298	17.8510	23.1455	35.0377	22.8742
0.0100	0.0099	24.3888	30.0665	42.1091	29.9915
0.0060	0.0060	27.5188	33.2905	45.3971	33.1729
0.0030	0.0030	31.9391	37.7784	49.8586	37.7489
0.0010	0.0010	38.9244	44.8407	56.9300	44.8114

Table 7.2 contains the results for the threshold $A = (1-\alpha)/(p\alpha)$. This threshold value is based on the general upper bound that ignores the overshoot. It can be seen that the MC estimates for the PFA in this case are somewhat smaller than the design values α, but for the considered parameter values the accuracy is more or less satisfactory. Still this leads to an increase in the true values of the average detection delay, which is undesirable. It can also be seen that the FO approximations are inaccurate even for relatively small α, while the HO approximations for the CADD are very accurate.

The results in Table 7.3 correspond to the case where the threshold A is set by inverting the asymptotically accurate approximation (7.52) that accounts for the overshoot, i.e., $A =$

$\zeta(p,I)/(p\alpha)$. It is seen that the MC estimates for the PFA match α very closely, especially for values smaller than 0.03. Thus, (7.52) provides an accurate method for threshold design to meet the PFA constraint α. It is also seen that, as expected, MC-CADD exceeds MC-ADD in all cases. The FO-ADD values are not good approximations even when the PFA is small, especially for small values of I and p, as expected from Table 7.1. On the other hand, the higher-order approximation for CADD_0 given by HO-CADD is seen to be very accurate even for moderate values of the PFA.

7.2 Asymptotic Optimality of the Shiryaev Procedure in the Non-iid Case

We now turn to considering the general non-iid case (6.2). In other words, conditioned on $\nu = k$ the observations $\{X_n, 1 \le n \le k\}$ are distributed according to the conditional densities $f_{0,n}(X_n|\mathbf{X}_1^{n-1})$ and the observations $\{X_n, n \ge k+1\}$ are distributed according to the conditional densities $f_{1,n}(X_n|\mathbf{X}_1^{n-1})$, where the post-change densities $f_{1,n}(X_n|\mathbf{X}_1^{n-1}) = f_{1,n}^{(k)}(X_n|\mathbf{X}_1^{n-1})$ may depend on the changepoint k. Recall that we use

$$\lambda_n^k = \log \frac{\mathsf{d}\mathsf{P}_k}{\mathsf{d}\mathsf{P}_\infty}(\mathbf{X}_1^n) = \sum_{i=k+1}^{n} Z_i$$

to denote the LLR between the hypotheses $H_k : \nu = k$ and $H_\infty : \nu = \infty$, where $Z_i = \log[f_{1,i}(X_i|\mathbf{X}_1^{i-1})/f_{0,i}(X_i|\mathbf{X}_1^{i-1})]$.

In the following we consider the general prior distribution $\pi_k = \Pr(\nu = k)$ defined in (7.18), and we always assume without special emphasis that it satisfies the condition (7.19), i.e., $k^{-1} \log \Pr(\nu > k) \to -d$ as $k \to \infty$ for some $d \ge 0$. Note that the geometric prior satisfies this condition for all k with $d = |\log(1-p)|$. As discussed in Subsection 7.1.2.2, this condition excludes prior distributions with unboundedly increasing hazard rates for which the change is likely to be detected at early stages, and the asymptotic study is impractical.

In the general case, it is convenient to define the Shiryaev procedure $T_s(B) = \inf\{n : \Pi_n \ge B\}$ ($\Pi_n = \Pr(\nu < n|\mathbf{X}_1^n)$) as follows

$$T_s(A) = \inf\{n \ge 1 : \Lambda_n \ge A\}, \tag{7.58}$$

where $\Lambda_n = \Pi_n/(1-\Pi_n)$, $n \ge 0$ and $A = B/(1-B)$. It is easily verified that the statistic Λ_n can be written as

$$\Lambda_n = \frac{q}{1-q} \prod_{i=1}^{n} \mathcal{L}_i + \frac{1}{\Pr(\nu \ge n)} \sum_{k=0}^{n-1} \pi_k \prod_{i=k+1}^{n} \mathcal{L}_i, \quad n \ge 0, \tag{7.59}$$

where $\mathcal{L}_i = f_{1,i}(X_i|\mathbf{X}_1^{i-1})/f_{0,i}(X_i|\mathbf{X}_1^{i-1})$ and the product $\prod_{i=j}^{m} \mathcal{L}_i$ is treated as 1 for $m < j$. Note that for the geometric prior $\Lambda_n = pR_{p,n}$ where $R_{p,n}$ is given by (6.9).

Evidently, we have to require $A > q/(1-q) = \Lambda_0$ in order to avoid triviality, since otherwise this procedure terminates immediately with no observations, i.e., $T_s = 0$.

The following lemma provides an upper bound for the false alarm probability of the Shiryaev procedure in the general non-iid case.

Lemma 7.2.1. *Consider the general non-iid model* (6.2). *The false alarm probability of the Shiryaev procedure $T_s(A)$ defined in* (7.58) *satisfies the inequality*

$$\mathsf{PFA}^\pi(T_s(A)) \le 1/(1+A). \tag{7.60}$$

Therefore,

$$A_\alpha = (1-\alpha)/\alpha \quad \textit{implies} \quad T_s(A_\alpha) \in \mathbf{C}_\alpha. \tag{7.61}$$

Proof. Evidently,

$$\mathsf{PFA}^\pi(T_s(A)) = \mathsf{E}^\pi[1 - \Pi_{T_s(A)} \mathbb{1}_{\{T_s(A)<\infty\}}]$$

and since $1-\Pi_n = 1/(1+\Lambda_n)$ and $\Lambda_{T_s(A)} \geq A$ on $\{T_s(A) < \infty\}$, we obtain

$$\mathsf{PFA}^\pi(T_s(A)) = \mathsf{E}^\pi\left\{[1+\Lambda_{T_s(A)}]^{-1} 1\!\!1_{\{T_s(A)<\infty\}}\right\} \leq (1+A)^{-1},$$

so (7.60) follows. The implication (7.61) follows directly from the inequality (7.60). □

In general, we do not assume any particular model for the observations, and as a result, there is no particular structure of the LLR process. We hence have to impose some conditions on the behavior of the LLR process at least for large n. It is natural to assume that there exists a positive finite number I, which depends on the model, such that $\lambda_n^k/(n-k)$ converges almost surely to I,

$$\frac{1}{n}\lambda_{k+n}^k \xrightarrow[n\to\infty]{\mathsf{P}_k-\text{a.s.}} I \quad \text{for every } 0 \leq k < \infty. \tag{7.62}$$

This is always true for iid data models, in which case $I = \mathsf{E}_0 Z_1$ is the K–L information number. The a.s. convergence condition (7.62) is sufficient, but not necessary, for obtaining lower bounds for all positive moments of the detection delay. Indeed, the key condition (7.20) in Theorem 7.1.3 holds whenever $\lambda_n^k/(n-k)$ converges almost surely to the number I. Therefore, this number is of paramount importance in the general changepoint detection theory.

For every $k = 0,1,2,\dots$ and $\varepsilon > 0$, define the random variable

$$L_\varepsilon^{(k)} = \sup\left\{n \geq 1 : \left|\frac{1}{n}\lambda_{k+n}^k - I\right| > \varepsilon\right\}, \tag{7.63}$$

where $\sup\{\varnothing\} = 0$. In terms of $L_\varepsilon^{(k)}$, the almost sure convergence of (7.62) may be written as $\mathsf{P}_k\{L_\varepsilon^{(k)} < \infty\} = 1$ for all $\varepsilon > 0$ and $k \geq 0$, which implies the condition (7.20).

While according to Theorem 7.1.3 the almost sure condition (7.62) is sufficient for obtaining lower bounds for the moments of the detection delay, in particular for the average detection delay, they need to be strengthened in order to establish asymptotic optimality properties of the detection procedure $T_s(A)$, and to obtain asymptotic expansions for the moments of the detection delay. Indeed, in general these conditions do not even guarantee finiteness of $\mathsf{ADD}^\pi(T_s(A))$. However, see the Remark 7.2.2 regarding weak asymptotic optimality that holds under the conditions (7.62). In order to study the asymptotics for the average detection delay one may impose the following constraints on the rate of convergence in the strong law for λ_{k+n}^k/n:

$$\mathsf{E}_k\left[L_\varepsilon^{(k)}\right] < \infty \quad \text{for all } \varepsilon > 0 \text{ and } k \geq 0, \tag{7.64}$$

and

$$\sum_{k=0}^\infty \pi_k \mathsf{E}_k\left[L_\varepsilon^{(k)}\right] < \infty \quad \text{for all } \varepsilon > 0. \tag{7.65}$$

Note that (7.64) is closely related to the condition

$$\sum_{n=1}^\infty \mathsf{P}_k\left\{\left|\sum_{i=k+1}^{k+n} Z_i - In\right| > \varepsilon n\right\} < \infty \quad \text{for all } \varepsilon > 0 \text{ and } k \geq 0,$$

which is nothing but the complete convergence of λ_{k+n}^k/n to I under P_k, i.e.,

$$\frac{1}{n}\lambda_{k+n}^k \xrightarrow[n\to\infty]{\mathsf{P}_k-\text{completely}} I \quad \text{for every } k \geq 0. \tag{7.66}$$

The convergence condition (7.65) is a joint condition on the rates of convergence of λ_{k+n}^k/n for each $\nu = k$ and the prior distribution. We write this condition compactly as

$$\frac{1}{n}\lambda_{\nu+n}^\nu \xrightarrow[n\to\infty]{\mathsf{P}^\pi-\text{completely}} I. \tag{7.67}$$

To study the asymptotics for higher moments of the detection delay, the complete convergence conditions (7.66) and (7.67) should be further strengthened. A natural generalization is to require, for some $r \geq 1$, the following r-quick convergence conditions

$$\mathsf{E}_k\left[L_\varepsilon^{(k)}\right]^r < \infty \quad \text{for all } \varepsilon > 0 \text{ and } k \geq 0 \tag{7.68}$$

and

$$\sum_{k=0}^{\infty} \pi_k \mathsf{E}_k\left[L_\varepsilon^{(k)}\right]^r < \infty \quad \text{for all } \varepsilon > 0. \tag{7.69}$$

That is, in general we require the r-quick version of the strong law of large numbers:

$$\frac{1}{n}\lambda_{k+n}^k \xrightarrow[n\to\infty]{\mathsf{P}_k-r\text{-quickly}} I \quad \text{for every } k \geq 0 \tag{7.70}$$

and

$$\frac{1}{n}\lambda_{\nu+n}^{\nu} \xrightarrow[n\to\infty]{\mathsf{P}^\pi-r\text{-quickly}} I. \tag{7.71}$$

The complete and r-quick convergence conditions are used in Chapter 3 for proving asymptotic optimality of the SPRT and the MSPRT for general statistical models. Below we take advantage of these results and prove that the condition (7.71) is sufficient for asymptotic optimality of the Shiryaev changepoint detection procedure.

In the following theorem, we establish operating characteristics of the detection procedure $T_{\mathrm{s}}(A)$ for large values of the threshold A and its asymptotic optimality for small PFA when the r-quick convergence condition (7.71) holds.

Theorem 7.2.1 (FO asymptotic optimality). *Consider the general non-iid model* (6.2) *and assume that the prior distribution satisfies the condition* (7.19)*. Let the r-quick convergence condition* (7.71) *hold for some* $I > 0$ *and* $r \geq 1$.

(i) *Then for all* $m \leq r$

$$\mathsf{E}^\pi[(T_{\mathrm{s}}(A)-\nu)^m | T_{\mathrm{s}}(A) > \nu] \sim \left(\frac{\log A}{I+d}\right)^m \quad \text{as } A \to \infty. \tag{7.72}$$

(ii) *If* $A = A_\alpha$ *is selected so that* $\mathsf{PFA}^\pi(T_{\mathrm{s}}(A_\alpha)) \leq \alpha$ *and* $\log A_\alpha \sim \log(1/\alpha)$, *in particular* $A_\alpha = (1-\alpha)/\alpha$, *then for all* $m \leq r$

$$\begin{aligned} \inf_{T\in\mathbf{C}_\alpha} \mathsf{E}^\pi([(T-\nu)^m | T > \nu] &\sim \mathsf{E}^\pi[(T_{\mathrm{s}}(A_\alpha)-\nu)^m | T_{\mathrm{s}}(A_\alpha) > \nu] \\ &\sim \left(\frac{|\log \alpha|}{I+d}\right)^m \quad \text{as } \alpha \to 0. \end{aligned} \tag{7.73}$$

Proof. Define

$$S_{k+n}^k = \lambda_{k+n}^k + n w_{n,k}, \quad w_{n,k} = n^{-1}\log[\pi_k/\Pr(\nu \geq k+n)]$$

and

$$\eta_A^{(k)} = \inf\{n \geq 1 : S_{k+n}^k \geq \log A\}, \quad k = 0, 1, 2, \dots$$

It is easily verified that the statistic $\log \Lambda_n$ can be written in the form

$$\log \Lambda_n = \lambda_n^k + (n-k) w_{k+n,k} + \ell_{n,k} = S_n^k + \ell_{n,k}, \tag{7.74}$$

where the random variable $\ell_{n,k}$ is nonnegative. Thus,

$$T_{\mathrm{s}}(A) - k \leq \eta_A^{(k)} \quad \text{on } \{T_{\mathrm{s}}(A) > k\} \text{ for every } k \geq 0. \tag{7.75}$$

To obtain the upper bound, define

$$\widetilde{L}_{\varepsilon}^{(k)} = \sup\left\{n \ge 1 : \left|n^{-1} S_{k+n}^{k} - I_d\right| > \varepsilon\right\}.$$

It is easy to see that $S_{k+\eta_A^{(k)}-1}^{k} < \log A$ on $\{\eta_A^{(k)} < \infty\}$ and $S_{k+\eta_A^{(k)}-1}^{k} \ge (\eta_A^{(k)} - 1)(I_d - \varepsilon)$ on $\{\eta_A^{(k)} - 1 > \widetilde{L}_{\varepsilon}^{(k)}\}$. It follows that for every $0 < \varepsilon < I_d$

$$\begin{aligned} \eta_A^{(k)} &\le 1 + \frac{\log A}{I_d - \varepsilon} 1\!\!1_{\{\eta_A^{(k)} > \widetilde{L}_{\varepsilon}^{(k)} + 1\}} + (\widetilde{L}_{\varepsilon}^{(k)} + 1) 1\!\!1_{\{\eta_A^{(k)} \le \widetilde{L}_{\varepsilon}^{(k)} + 1\}} \\ &\le \widetilde{L}_{\varepsilon}^{(k)} + 2 + \frac{\log A}{I_d - \varepsilon}, \end{aligned} \tag{7.76}$$

so that for every $0 < \varepsilon < I_d$ and $k \ge 0$,

$$T_s(A) - k \le \widetilde{L}_{\varepsilon}^{(k)} + 2 + \frac{\log A}{I_d - \varepsilon} \quad \text{on } \{T_s(A) > k\}. \tag{7.77}$$

Applying (7.77) along with the fact that $\mathsf{P}^{\pi}(T_s(A) \ge \nu) \ge A/(1+A)$ yields

$$\mathsf{E}^{\pi}[(T_s(A) - \nu)^m | T_s(A) > \nu] \le \frac{A+1}{A} \sum_{k=1}^{\infty} \pi_k \mathsf{E}_k \left(\frac{\log A}{I_d - \varepsilon} + \widetilde{L}_{\varepsilon}^{(k)} + 2\right)^m.$$

Since $w_{n,k} \to d$ as $n \to \infty$ and by the assumption of the theorem $\sum_{k=1}^{\infty} \pi_k \mathsf{E}_k [L_{\varepsilon}^{(k)}]^m < \infty$, this implies $\sum_{k=1}^{\infty} \pi_k \mathsf{E}_k [\widetilde{L}_{\varepsilon}^{(k)}]^m < \infty$. Since ε can be arbitrarily small it follows that

$$\mathsf{E}^{\pi}[(T_s(A) - \nu)^m | T_s(A) > \nu] \le \left(\frac{\log A}{I_d}\right)^m (1 + o(1)) \quad \text{as } A \to \infty. \tag{7.78}$$

The lower bound

$$\mathsf{E}^{\pi}[(T_s(A) - \nu)^m | T_s(A) > \nu] \ge \left(\frac{\log A}{I_d}\right)^m (1 + o(1)) \quad \text{as } A \to \infty$$

follows from Theorem 7.1.3, which along with the upper bound (7.78) completes the proof of the asymptotic equality (7.72). The asymptotic equalities (7.73) follow immediately from (7.72) and (7.22) in Theorem 7.1.3. □

Remark 7.2.1. The log-likelihood ratio λ_n^k can be interpreted as a distance measure between the pre- and post-change distributions in the time interval $[k, n]$. Therefore, $(n-k)^{-1}\lambda_n^k$ can be regarded as an instantaneous (local) distance at time n, i.e., the amount of change. Substantial changes result in large values of I in (7.20) and (7.62), the quantity that always appears in the denominator of the first-order asymptotics of the ADD. Hence the delay is inversely proportional to the amount of change. According to this, larger changes are detected more promptly, which is intuitively obvious. Thus, I measures both the magnitude of the change and our ability in its fast detection. On the other hand, the value of d characterizes the amount of our prior knowledge about the change occurrence. As we can see from (7.72), it also impacts the average detection delay, which is intuitively appealing.

Remark 7.2.2. By Theorem 7.1.3, the almost sure convergence condition (7.62) is sufficient for obtaining the lower bound for the moments of the detection delay, but it is not sufficient for asymptotic optimality with respect to the average delay to detection. It is interesting to ask whether some asymptotic optimality result can still be obtained under this condition. The answer is affirmative. In fact, it is rather easy to prove the following weak asymptotic optimality property of the Shiryaev procedure: if the condition (7.62) holds and $A_{\alpha} = (1-\alpha)/\alpha$, then for every $0 < \varepsilon < 1$

$$\inf_{T \in \mathbf{C}_{\alpha}} \mathsf{P}^{\pi}\left\{(T - \nu)^+ > \varepsilon (T_s(A_{\alpha}) - \nu)^+\right\} \to 1 \quad \text{as } \alpha \to 0.$$

Assume now that instead of the constrained optimization problem we are interested in the unconstrained Bayes problem with the loss function

$$L_m(T,\nu) = 1\!\!1_{\{T\le\nu\}} + c\,(T-\nu)^m 1\!\!1_{\{T>\nu\}},$$

where $m, c > 0$. The expected loss or average risk associated with the detection procedure T is given by

$$\rho^{\pi}_{c,m}(T) = \mathsf{P}^{\pi}(T\le\nu) + c\mathsf{E}^{\pi}[(T-\nu)^+]^m = \mathsf{PFA}(T) + c[1-\mathsf{PFA}(T)]\mathsf{E}^{\pi}[(T-\nu)^m|T>\nu)],$$

and the goal is to find an asymptotically optimal procedure that minimizes $\rho^{\pi}_{c,m}(T)$ as $c\to 0$ over all stopping times. Clearly, the Shiryaev procedure $T_s(A)$ is a special candidate.

The first question is how to choose the threshold $A_{c,m}$ to optimize the performance for small values of the cost c. To answer this question, we observe that, ignoring the overshoot, $\mathsf{PFA}(T_s(A)) \approx 1/A$ and that by (7.72) $\mathsf{E}^{\pi}[(T_s(A)-\nu)^m|T_s(A)\ge\nu] \approx (I_d^{-1}\log A)^m$. Therefore, for large A the average risk is approximately equal to

$$\rho^{\pi}_{c,m}(T_s(A)) \approx 1/A + c(I_d^{-1}\log A)^m = g_{c,m}(A).$$

The optimal threshold value $A = A_{c,m}$ that minimizes $g_{c,m}(A)$, $A>0$, is solution of the equation

$$m(A/I_d)\left(\frac{\log A}{I_d}\right)^{m-1} = 1/c. \tag{7.79}$$

In particular, for $m=1$ we obtain $A_{c,1} = I_d/c$.

It is intuitively appealing that the procedure $T_s(A_{c,m})$ with the threshold $A_{c,m}$ that satisfies (7.79) is asymptotically optimal as $c\to 0$. In the next theorem, we establish sufficient conditions under which this is true.

Theorem 7.2.2. *Suppose that the condition* (7.71) *holds for some* $I>0$ *and* $r\ge 1$. *Let* $A = A_{c,m}$ *be the solution of the equation* (7.79). *Then, for all* $m\le r$,

$$\inf_{T\ge 0}\rho^{\pi}_{c,m}(T) \sim \rho^{\pi}_{c,m}(T_s(A_{c,m})) \sim c\left(\frac{-\log c}{I+d}\right)^m \quad \text{as } c\to 0. \tag{7.80}$$

Proof. By Theorem 7.2.1,

$$\mathsf{E}^{\pi}[(T_s(A)-\nu)^m|T_s(A)>\nu] = \left(\frac{\log A}{I_d}\right)^m (1+o(1)) \quad \text{as } A\to\infty.$$

Since $\mathsf{PFA}(T_s(A)) < 1/A$ and since for any stopping time T

$$\rho^{\pi}_{c,m}(T) = \mathsf{PFA}(T) + c[1-\mathsf{PFA}(T)]\mathsf{E}^{\pi}[(T-\nu)^m|T>\nu)],$$

we obtain the asymptotic relation

$$\rho^{\pi}_{c,m}(T_s(A)) \sim c\left(\frac{\log A}{I_d}\right)^m \quad \text{as } A\to\infty. \tag{7.81}$$

In the case of $m=1$, the threshold $A_{c,1}$ equals I_d/c and we immediately obtain

$$\rho^{\pi}_{c,1}(T_s(A_{c,1})) \sim c\frac{|\log c|}{I_d} \quad \text{as } c\to 0.$$

Next, it is easily seen that, for any $m>1$, the threshold $A_{c,m}$ goes to infinity as $c\to 0$ in such a way that $\log A_{c,m} \sim |\log c|$. Moreover, since $cA_{c,m}(\log A_{c,m})^{m-1} = I_d^m/m$, it follows that

$$cA_{c,m}(\log A_{c,m})^m \sim (I_d^m/m)|\log c| \quad \text{as } c\to 0.$$

Therefore,

$$\rho^{\pi}_{c,m}(T_s(A_{c,m})) \sim c\left(\frac{|\log c|}{I_d}\right)^m (1+1/[cA_{c,m}(\log A_{c,m})^m]) \sim c\left(\frac{|\log c|}{I_d}\right)^m \quad \text{as } c \to 0.$$

It remains to prove that

$$\inf_{T\ge 0} \rho^{\pi}_{c,m}(T) \ge c\left(\frac{|\log c|}{I_d}\right)^m (1+o(1)) \quad \text{as } c \to 0.$$

Moreover, since

$$\lim_{c\to o} \frac{g_{c,m}(A_{c,m})}{c(|\log c|/I_d)^m} = 1,$$

it suffices to prove that

$$\liminf_{c\to 0} \frac{\inf_{T\ge 0} \rho^{\pi}_{c,m}(T)}{g_{c,m}(A_{c,m})} \ge 1. \tag{7.82}$$

Suppose that (7.82) is not true, i.e., there exist stopping rules $T = T_c$ such that

$$\limsup_{c\to 0} \frac{\rho^{\pi}_{c,m}(T_c)}{g_{c,m}(A_{c,m})} < 1. \tag{7.83}$$

Let $\alpha_c = \mathsf{PFA}(T_c)$. Since

$$\alpha_c \le \rho^{\pi}_{c,m}(T_c) < g_{c,m}(A_{c,m})(1+o(1)) \to 0 \quad \text{as } c \to 0$$

and, by Theorem 7.1.3,

$$\mathsf{E}^{\pi}[(T_c-\nu)^m | T_c > \nu] \ge \left(\frac{\log(1/\alpha_c)}{I_d}\right)^m (1+o(1)),$$

it follows that

$$\rho^{\pi}_{c,m}(T_c) \ge \alpha_c + \left(\frac{\log(1/\alpha_c)}{I_d}\right)^m (1+o(1)) \quad \text{as } c \to 0.$$

Thus,

$$\frac{\rho^{\pi}_{c,m}(T_c)}{g_{c,m}(A_{c,m})} \ge \frac{g_{c,m}(1/\alpha_c)}{\min_{b>0} g_{c,m}(b)}(1+o(1)) \ge 1+o(1),$$

which contradicts (7.83). Hence, (7.82) follows and the proof is complete. □

Remark 7.2.3. The r-quick convergence condition (7.71) is sufficient but not necessary for the asymptotic optimality results in Theorems 7.2.1 and 7.2.2. In particular, the proofs of these theorems show that the last entry time $L^{(k)}_{\varepsilon}$ in the corresponding conditions can be replaced with the one-sided left-tail version:

$$\widetilde{L}^{(k)}_{\varepsilon} = \sup\left\{n \ge 1 : n^{-1}\lambda^k_{k+n} - I < -\varepsilon\right\},$$

i.e., it suffices to require $\mathsf{E}^{\pi}[\widetilde{L}^{(\nu)}_{\varepsilon}]^r < \infty$ for all $\varepsilon > 0$. It can be also shown that the following condition is sufficient for obtaining the upper bounds:

$$\sum_{k=0}^{\infty} \pi_k \sum_{n=0}^{\infty} n^{r-1} \mathsf{P}_k\left\{n^{-1}\lambda^k_{k+n} - I \le -\varepsilon\right\} < \infty \quad \forall\, \varepsilon > 0.$$

Therefore, if this one-sided condition is satisfied along with the condition (7.20), then the assertions of Theorems 7.2.1 and 7.2.2 hold. In particular examples it is often not that difficult to check the following slightly stronger left- and right-tail conditions:

$$\begin{aligned} &\sum_{n=0}^{\infty} n^{r-1} \sup_{0\le k<\infty} \mathsf{P}_k\left\{n^{-1}\lambda_{k+n}^k - I \le -\varepsilon\right\} < \infty \quad \text{for all } \varepsilon > 0, \\ &\sup_{0\le k<\infty} \mathsf{P}_k\left\{\frac{1}{M}\max_{0\le n<M} \lambda_{k+n}^k \ge (1+\varepsilon)I\right\} \xrightarrow[M\to\infty]{} 0 \quad \text{for all } \varepsilon > 0. \end{aligned} \tag{7.84}$$

Finally, the condition

$$\sum_{n=0}^{\infty} n^{r-1} \sup_{0\le k<\infty} \mathsf{P}_k\left\{n^{-1}|\lambda_{k+n}^k - I| \ge \varepsilon\right\} < \infty \quad \text{for all } \varepsilon > 0 \tag{7.85}$$

implies both conditions in (7.84), and therefore, it is sufficient for the assertions of Theorems 7.2.1 and 7.2.2 to hold.

So far we have considered the discrete-time case. The general continuous-time model can be handled in a similar manner, but the asymptotic optimality results do not directly follow from the corresponding results in discrete time. In fact continuous time is slightly more delicate and the proofs of certain results, in particular the upper bounds for the ADD, are more involved and tedious. We now formulate the main results with no proofs. For complete detailed proofs we refer the reader to the paper by Baron and Tartakovsky [28].

Specifically, let $(\Omega, \mathscr{F}, \mathscr{F}_t, \mathsf{P})$, $t \in \mathbb{R}_+ = [0,\infty)$, be a filtered probability space with standard assumptions about monotonicity and right-continuity of the σ-algebras $\mathscr{F}_t(\mathbf{X}_0^t)$. Let P_u and E_u denote the probability measure and the corresponding expectation when the change occurs at time $\nu = u$, for every measurable set $\mathcal{A}$ let $\mathsf{P}^\pi(\mathcal{A}) = \int_0^\infty \mathsf{P}_u(\mathcal{A})\,\mathrm{d}\pi_u$, and let E^π denote the expectation with respect to P^π. Here $\pi_u = \Pr(\nu \le u)$, $u \in \mathbb{R}_+$ is the prior distribution of the changepoint ν. We again restrict ourselves to prior distributions that satisfy the condition (7.19), i.e., $\lim_{t\to\infty} t^{-1}\log(1-\pi_t) = -d$ for some $d \ge 0$, and in the heavy-tailed priors case where $d = 0$ we assume in addition that $\limsup_{t\to\infty}(1-\pi_{kt})/(1-\pi_t) < 1$ for some $k > 1$. For $d = 0$, the study of the moments of the delay requires a finite r-th moment for the prior distribution, $\mathsf{E}[\nu^r] = \int_0^\infty u^r \mathrm{d}\pi_u < \infty$, which automatically holds for all priors with $d > 0$.

By

$$\mathcal{L}_t^u := \frac{\mathrm{d}\mathsf{P}_u^t}{\mathrm{d}\mathsf{P}_\infty^t}(\mathbf{X}_0^t) \quad \text{and} \quad \lambda_t^u = \log \mathcal{L}_t^u, \quad t \ge u, \tag{7.86}$$

we again denote the LR and the LLR for the hypotheses that the change occurred at the point $\nu = u$ and at $\nu = \infty$ (no change at all). As usual, P^t stands for a restriction of the probability measure P to the sigma-algebra $\mathscr{F}_t$. For $t \ge 0$, define the statistic

$$\Lambda_t = \frac{1}{1-\pi_t}\int_0^t \mathcal{L}_t^u \,\mathrm{d}\pi_u, \tag{7.87}$$

which is a continuous-time version of the statistic Λ_n defined in (7.59).

The Shiryaev procedure is the stopping time

$$T_{\mathrm{s}}(A) = \inf\{t \ge 0 : \Lambda_t \ge A\}, \quad A > 0. \tag{7.88}$$

Evidently, Lemma 7.2.1 holds true in the continuous-time case too, i.e., $\mathsf{PFA}(T_{\mathrm{s}}) := \mathsf{P}(T_{\mathrm{s}} \le \nu) \le 1/(1+A)$ for any $A > 0$, so that, by setting $A_\alpha = (1-\alpha)/\alpha$, we guarantee the constraint $\mathsf{PFA}(T_{\mathrm{s}}(A_\alpha)) \le \alpha$ for every $0 < \alpha < 1-q$, where $q = \Pr(\nu < 0)$.

We assume the strong law for the LLR process λ_{u+t}^u, $t \ge 0$:

$$\frac{1}{t}\lambda_{u+t}^u \xrightarrow[t\to\infty]{\mathsf{P}_u-\text{a.s.}} I \quad \text{for every } u < \infty, \tag{7.89}$$

where I is some positive and finite number. Moreover, we strengthen this condition by the r-quick version of the strong law assuming that, for some $r \geq 1$,

$$\mathsf{E}_u\left[L_\varepsilon^{(u)}\right]^r < \infty \quad \text{for all } \varepsilon > 0 \text{ and } u \geq 0, \tag{7.90}$$

and

$$\mathsf{E}^\pi\left[L_\varepsilon^{(\nu)}\right]^r = \int_0^\infty \mathsf{E}_u\left[L_\varepsilon^{(u)}\right]^r \mathrm{d}\pi_u < \infty \quad \text{for all } \varepsilon > 0. \tag{7.91}$$

We are now ready to formulate an analog of Theorem 7.2.1 that establishes the first-order asymptotic optimality of the Shiryaev procedure in the general continuous-time case.

Theorem 7.2.3. *Consider the general continuous-time stochastic model. Let the r-quick convergence condition* (7.91) *hold for some $I > 0$ and $r \geq 1$ and let $\mathsf{E}[\nu^r] < \infty$.*

(i) *For all $m \leq r$,*

$$\mathsf{E}^\pi[(T_{\mathrm{s}}(A) - \nu)^m | T_{\mathrm{s}}(A) > \nu] \sim \left(\frac{\log A}{I+d}\right)^m \quad \text{as } A \to \infty. \tag{7.92}$$

(ii) *If $A = A_\alpha$ is selected so that $\mathsf{PFA}^\pi(T_{\mathrm{s}}(A_\alpha)) \leq \alpha$ and $\log A_\alpha \sim \log(1/\alpha)$, in particular $A_\alpha = (1-\alpha)/\alpha$, then for all $m \leq r$*

$$\begin{aligned}\inf_{T \in \mathbf{C}_\alpha} \mathsf{E}^\pi([(T-\nu)^m | T > \nu] &\sim \mathsf{E}^\pi[(T_{\mathrm{s}}(A_\alpha) - \nu)^m | T_{\mathrm{s}}(A_\alpha) > \nu] \\ &\sim \left(\frac{|\log \alpha|}{I+d}\right)^m \quad \text{as } \alpha \to 0.\end{aligned} \tag{7.93}$$

An analog of Theorem 7.2.2 also holds; see Baron and Tartakovsky [28, Theorem 5].

Remark 7.2.4. We iterate that the r-quick convergence conditions (7.71) and (7.91) are very close to the following conditions, which can also be shown to be sufficient for Theorems 7.2.1 and 7.2.3:

$$\sum_{k=0}^\infty \sum_{n=1}^\infty n^{r-1} \mathsf{P}_k\left\{\left|\lambda_{k+n}^k - In\right| > \varepsilon n\right\} \pi_k < \infty \quad \text{for all } \varepsilon > 0 \tag{7.94}$$

and

$$\int_0^\infty \int_0^\infty t^{r-1} \mathsf{P}_u\left\{\left|\lambda_{u+t}^u - It\right| > \varepsilon t\right\} \mathrm{d}t\, \mathrm{d}\pi_u < \infty \quad \text{for all } \varepsilon > 0. \tag{7.95}$$

These conditions can be further relaxed into the left-tail conditions

$$\sum_{k=0}^\infty \sum_{n=1}^\infty n^{r-1} \mathsf{P}_k\left\{\lambda_{k+n}^k - In < -\varepsilon n\right\} \pi_k < \infty \quad \text{for all } \varepsilon > 0 \tag{7.96}$$

and

$$\int_0^\infty \int_0^\infty t^{r-1} \mathsf{P}_u\left\{\lambda_{u+t}^u - It < -\varepsilon t\right\} \mathrm{d}t\, \mathrm{d}\pi_u < \infty \quad \text{for all } \varepsilon > 0. \tag{7.97}$$

Recall that, according to Definition 2.4.3, the condition (7.94) means the r-complete convergence of $\lambda_{\nu+n}^\nu / n$ to I under P^π.

We complete this section with the simple but particularly interesting problem of detecting a change in the drift of the BM. Specifically, let the observed process obey the stochastic differential equation

$$\mathrm{d}X_t = \mathbb{1}_{\{\nu \leq t\}} S\, \mathrm{d}t + \sigma \mathrm{d}W_t, \quad t \geq 0,\ X_0 = 0,$$

where S and $\sigma > 0$ are given constants and $\{W_t\}_{t \geq 0}$ is the standard Wiener process. Assume that the prior distribution is zero-modified exponential with density $\pi_t' = (1-q)e^{-\beta t} \mathbb{1}_{\{t \geq 0\}}$, where $q = \Pr(\nu < 0)$. Since the trajectories of the statistic Λ_t are continuous, there is no overshoot over the boundary A, so $\mathsf{PFA}^\pi(T_{\mathrm{s}}) = 1/(1+A)$ and $A_\alpha = (1-\alpha)/\alpha$ guarantees the exact equality

$\mathsf{PFA}^{\pi}(T_s) = \alpha$. With this threshold the Shiryaev procedure $T_s(A_\alpha)$ is strictly optimal, minimizing the average delay to detection $\mathsf{ADD}^{\pi}(T)$ in the class $\mathbf{C}_\alpha$ for any $0 < \alpha < 1 - q$. The proof of this result is due to Shiryaev [420, Ch. 4, Sec. 4]. The LLR has the form

$$\lambda_{u+t}^{t} = \frac{S}{\sigma^2}(X_{u+t} - X_u) - \frac{S}{2\sigma^2}t, \quad t \geq 0,$$

and it is easily checked that, for all $r > 0$ and $\varepsilon > 0$ and with $I = S^2/2\sigma^2$,

$$\int_0^\infty \int_0^\infty t^{r-1} \mathsf{P}_u\left\{\left|\lambda_{u+t}^{u} - It\right| > \varepsilon t\right\} \mathrm{d}t \, e^{-\beta u} \mathrm{d}u < \infty,$$

so the condition (7.95) is satisfied and by Theorem 7.2.3 the procedure $T_s(A_\alpha)$ asymptotically minimizes all positive moments of the detection delay: for all $m > 0$, as $\alpha \to 0$,

$$\inf_{T \in \mathbf{C}_\alpha} \mathsf{E}^{\pi}([(T-\nu)^m | T > \nu] \sim \mathsf{E}^{\pi}[(T_s(A_\alpha) - \nu)^m | T_s(A_\alpha) > \nu] \sim \left(\frac{|\log \alpha|}{S^2/2\sigma^2 + \beta}\right)^m.$$

For $m = 1$ this approximation can be improved up to a negligible term:

$$\mathsf{ADD}^{\pi}(T_s(A_\alpha)) = \frac{1}{S^2/2\sigma^2 + \beta}(|\log \alpha| - C) + o(1), \quad \alpha \to 0,$$

where $C > 0$ is a constant that is not easily computable.

7.3 Asymptotically Optimal Detection Procedures under Global False Alarm Probability Constraint

Note once again that since the event $\{T \leq k\}$ is measurable with respect to the σ-algebra $\mathscr{F}_k = \sigma(\mathbf{X}_1^k)$ and, conditioned on the changepoint $\nu = k$, the measure P_∞ changes to P_k at $k+1$, this implies $\mathsf{P}_k(T \leq k) = \mathsf{P}_\infty(T \leq k)$. Therefore,

$$\mathsf{PFA}^{\pi}(T) = \sum_{k=0}^{\infty} \pi_k \mathsf{P}_k(T \leq k) = \sum_{k=0}^{\infty} \pi_k \mathsf{P}_\infty(T \leq k). \tag{7.98}$$

In the previous sections, we considered the class of detection procedures $\mathbf{C}_\alpha = \{T : \mathsf{PFA}^{\pi}(T) \leq \alpha\}$ with the constraint imposed on the weighted false alarm probability. Another possibility is to impose a stronger constraint on the maximal probability

$$\sup_{k \geq 0} \mathsf{P}_k(T \leq k) = \sup_{k \geq 0} \mathsf{P}_\infty(T \leq k) = \mathsf{P}_\infty(T < \infty) \leq \alpha,$$

i.e., to consider the class of stopping times $\mathbf{C}_\alpha^\infty = \{T : \mathsf{P}_\infty(T < \infty) \leq \alpha\}$ for which the *global* worst-case false alarm probability $\sup_{k \geq 0} \mathsf{P}_k(T \leq k)$ is restricted by a given number $\alpha < 1$. The goal is to find an optimal procedure from the following optimization problem:

$$\inf_{T \in \mathbf{C}_\alpha^\infty} \mathsf{E}^{\pi}(T - \nu \mid T > \nu) \to T_{\mathrm{opt}}. \tag{7.99}$$

As it will become apparent later, the minimax solution that minimizes $\sup_k \mathsf{E}_k(T - k | T > k)$ is not feasible under this strong constraint since the maximal expected delay to detection is infinitely large. In our opinion the only feasible solution is Bayesian.

It turns out that it is difficult to find an exact solution to the optimization problem (7.99) even in the iid case. For this reason, we focus on the asymptotic problem, letting α go to zero. Moreover, we address the problem of minimizing the higher moments of the detection delay

$$\inf_{T \in \mathbf{C}_\alpha^\infty} \mathsf{E}^{\pi}[(T - \nu)^m \mid T \geq \nu], \quad m > 1, \quad \text{as } \alpha \to 0.$$

The techniques developed in the previous sections can be effectively used for studying the asymptotic properties of changepoint detection procedures in the class $\mathbf{C}_\alpha^\infty$ when $\alpha \to 0$ for general stochastic models.

7.3.1 The Detection Method

Recall that by

$$\Lambda_n^k := \frac{p(\mathbf{X}_1^n \mid \nu = k)}{p(\mathbf{X}_1^n \mid \nu = \infty)} = \prod_{i=k+1}^{n} \frac{f_1(X_i \mid \mathbf{X}_1^{i-1})}{f_0(X_i \mid \mathbf{X}_1^{i-1})}, \quad n > k,$$

we denote the LR between the hypotheses $H_k : \nu = k$ and $H_\infty : \nu = \infty$ and we write $\mathcal{L}_i = f_1(X_i \mid \mathbf{X}_1^{i-1})/f_0(X_i \mid \mathbf{X}_1^{i-1})$, $\lambda_n^k = \log \Lambda_n^k$ and $Z_i = \log \mathcal{L}_i$. As before, we use the convention that for $n = 0$, i.e., before the observations become available, $\mathcal{L}_0 = f_1(X_0)/f_0(X_0) = 1$ almost everywhere.

Define the weighted LR statistic

$$\bar{\Lambda}_n = \sum_{k=-\infty}^{\infty} \pi_k \frac{p(\mathbf{X}_1^n \mid \nu = k)}{p(\mathbf{X}_1^n \mid \nu = \infty)} = \sum_{k=-\infty}^{\infty} \pi_k \prod_{i=k+1}^{n} \frac{f_1(X_i \mid \mathbf{X}_1^{i-1})}{f_0(X_i \mid \mathbf{X}_1^{i-1})},$$

which can be represented as

$$\bar{\Lambda}_n = q \prod_{i=1}^{n} \mathcal{L}_i + (1-q) \sum_{j=0}^{n-1} \widetilde{\pi}_j \prod_{i=j+1}^{n} \mathcal{L}_i + \Pr(\nu \geq n), \quad \bar{\Lambda}_0 = 1. \tag{7.100}$$

Introduce the stopping time

$$\tau_A = \inf\{n \geq 1 : \bar{\Lambda}_n \geq A\}, \quad A > 1. \tag{7.101}$$

It is useful to establish a relationship between the detection procedure τ_A and the Shiryaev procedure $T_{\text{SR}}(\widetilde{A}) = \inf\{n : \Lambda_n \geq \widetilde{A}\}$ with Λ_n in (7.59). Using (7.59) and (7.100), we obtain

$$\Lambda_n = \frac{\bar{\Lambda}_n - \Pr(\nu \geq n)}{\Pr(\nu \geq n)},$$

which shows that the stopping time τ_A can be written as

$$\tau_A = \inf\{n \geq 1 : \Lambda_n \geq A/\Pr(\nu \geq n) - 1\}, \quad A > 1.$$

Therefore, while in Shiryaev's procedure the statistic Λ_n is compared with a constant threshold, in the proposed detection procedure the stopping boundary is an increasing function in n. This is an unavoidable penalty for the very strong maximal PFA constraint.

7.3.2 Asymptotic Optimality and Asymptotic Performance

As usual, let $\mathsf{P}^{(n)}$ denote the restriction of the measure P to the σ-algebra $\mathscr{F}_n = \sigma(X_1, \dots, X_n)$. The following lemma gives a simple upper bound for the global PFA $\mathsf{P}_\infty(\tau_A < \infty)$ of the detection procedure (7.101) in the general case. This conservative bound can be improved in the iid case.

Lemma 7.3.1. *For any $A > 1$,*

$$\mathsf{P}_\infty(\tau_A < \infty) \leq A^{-1}. \tag{7.102}$$

Proof. Noting that $\bar{\Lambda}_n = d\mathsf{P}^{\pi(n)}/d\mathsf{P}_\infty^{(n)}$ and using the Wald likelihood ratio identity, we obtain

$$\mathsf{P}_\infty(\tau_A < \infty) = \mathsf{E}_\infty \mathbb{1}_{\{\tau_A < \infty\}} = \mathsf{E}^\pi[\bar{\Lambda}_{\tau_A}^{-1} \mathbb{1}_{\{\tau_A < \infty\}}].$$

By definition of the stopping time τ_A, the inequality $\bar{\Lambda}_{\tau_A} \geq A$ holds on the set $\{\tau_A < \infty\}$, which implies the inequality (7.102). □

Therefore, setting $A = A_\alpha = 1/\alpha$ guarantees $\mathsf{P}_\infty(\tau_A < \infty) \le \alpha$, i.e., $A_\alpha = \alpha^{-1}$ implies $\tau_{A_\alpha} \in \mathbf{C}_\alpha^\infty$.

We now tackle the question of asymptotic optimality of the procedure τ_A with $A = 1/\alpha$ in the class $\mathbf{C}_\alpha^\infty$ as $\alpha \to 0$. As before, the proof of asymptotic optimality is performed in two steps. The first step is to obtain an asymptotic lower bound for the moments of the detection delay $\mathsf{E}^\pi[(T-\nu)^m | T > \nu]$ for any procedure in the class $\mathbf{C}_\alpha^\infty$. The second step is to show that the procedure τ_{A_α} achieves this lower bound.

Our first observation is that, using an almost identical argument as in the proof of Theorem 7.1.3 with $K_\alpha = |\log \alpha|/I$, it can be shown that under the condition (7.20) the following asymptotic lower bound holds in the class $\mathbf{C}_\alpha^\infty$:

$$\liminf_{\alpha \to 0} \frac{\inf_{T \in \mathbf{C}_\alpha^\infty} \mathsf{E}^\pi[(T-\nu)^m | T > \nu]}{(|\log \alpha|/I)^m} \ge 1 \quad \text{for every } m > 0. \tag{7.103}$$

The detailed proof can be found in the paper by Tartakovsky [462].

Now, using a similar reasoning as in the proof of Theorem 7.2.1, we prove that the procedure τ_A asymptotically minimizes positive moments of the detection delay up to order r whenever the r-quick convergence conditions (7.71) hold. Define the random variable

$$L_{k,\varepsilon}^{\mathrm{ls}} = \sup\{n \ge 1 : n^{-1}\lambda_{k+n}^k - I < -\varepsilon\} \qquad \left(\sup\{\varnothing\} = 0\right) \tag{7.104}$$

which is the time after which $n^{-1}\lambda_{k+n}^k$ does not leave the region $[I-\varepsilon, \infty)$.

Theorem 7.3.1. *Consider the general discrete-time stochastic model. Let τ_A be defined by* (7.101)*. Let, for some positive and finite I, the condition* (7.20) *hold and let for some $r \ge 1$,*

$$\sum_{k=0}^\infty \pi_k \mathsf{E}_k[L_{k,\varepsilon}^{\mathrm{ls}}]^r < \infty \quad \text{for all } \varepsilon > 0. \tag{7.105}$$

Assume that

$$\sum_{k=0}^\infty |\ln \pi_k|^m \pi_k < \infty \quad \text{for } m \le r. \tag{7.106}$$

(i) *Then for all $m \le r$,*

$$\mathsf{E}^\pi[(\tau_A - \nu)^m | \tau_A > \nu] \sim \left(I^{-1} \log A\right)^m \quad \text{as } A \to \infty. \tag{7.107}$$

(ii) *If $A = A_\alpha$ is selected so that $\mathsf{P}_\infty(\tau_{A_\alpha} < \infty) \le \alpha$ and $\log A_\alpha \sim \log(1/\alpha)$, in particular $A_\alpha = 1/\alpha$, then for all $m \le r$,*

$$\begin{aligned}\inf_{T \in \mathbf{C}_\alpha^\infty} \mathsf{E}^\pi([(T-\nu)^m | T > \nu] &\sim \mathsf{E}^\pi[(\tau_{A_\alpha} - \nu)^m | \tau_{A_\alpha} > \nu] \\ &\sim \left(I^{-1} |\log \alpha|\right)^m \quad \text{as } \alpha \to 0.\end{aligned} \tag{7.108}$$

Proof. Proof of (i). Note first that setting $\alpha = 1/A$ in the inequality (7.103) implies the asymptotic lower bound

$$\mathsf{E}^\pi[(\tau_A - \nu)^m | \tau_A > \nu] \ge \left(I^{-1} \log A\right)^m (1 + o(1)) \quad \text{as } A \to \infty.$$

Thus, to prove (7.107) it suffices to show that this lower bound is asymptotically the same as the upper bound, i.e., we have to show that

$$\mathsf{E}^\pi[(\tau_A - \nu)^m | \tau_A > \nu] \le \left(I^{-1} \log A\right)^m (1 + o(1)) \quad \text{as } A \to \infty. \tag{7.109}$$

Extracting the term $\Lambda_n^k = e^{\lambda_n^k}$ in (7.100), the statistic $\bar{\Lambda}_n$ can be written as follows:

$$\bar{\Lambda}_n = (1-q)\tilde{\pi}_k e^{\lambda_n^k}(1 + Y_n^k), \tag{7.110}$$

where

$$Y_n^k = \widetilde{\pi}_k^{-1}\left[\frac{\Pr(\nu \geq n)}{1-q} e^{-\lambda_n^k} + \left(\widetilde{\pi}_0 + \frac{q}{1-q}\right) e^{\lambda_n^0} + \sum_{j=1}^{k-1} \widetilde{\pi}_j e^{\lambda_k^j} + \sum_{j=k+1}^{n-1} \widetilde{\pi}_j e^{-\lambda_k^j}\right].$$

We obtain that for every $k \geq 0$

$$\log \bar{\Lambda}_n = \lambda_n^k + \log(1+Y_n^k) + \log \pi_k. \tag{7.111}$$

Since $Y_n^k \geq 0$, it follows from the equality (7.111) that $\log \bar{\Lambda}_n \geq \lambda_n^k + \log \pi_k$, and therefore, for any $k \geq 0$

$$(\tau_A - k)^+ \leq \eta_A^{(k)} = \inf\{n \geq 1 : \lambda_{k+n}^k \geq \log(A\pi_k^{-1})\}. \tag{7.112}$$

Thus,

$$\mathsf{E}^\pi[(\tau_A - \nu)^m | \tau_A > \nu] \leq \frac{\sum_{k=0}^\infty \pi_k \mathsf{E}_k[\eta_A^{(k)}]^m}{\mathsf{P}^\pi(\tau_A > \nu)}.$$

Since, by Lemma 7.3.1, $\mathsf{P}^\pi(\tau_A > \nu) \geq 1 - \mathsf{P}_\infty(\tau_A < \infty) \geq 1 - A^{-1}$, it is sufficient to prove that

$$\sum_{k=0}^\infty \pi_k \mathsf{E}_k[\eta_A^{(k)}]^m \leq \left(\frac{\log A}{I}\right)^m (1+o(1)) \quad \text{as } A \to \infty. \tag{7.113}$$

By the definition of the stopping time $\eta_A^{(k)}$, on $\{\eta_A^{(k)} < \infty\}$ the inequality $\lambda^k_{k+\eta_A^{(k)}-1} < \log(A/\pi_k)$ is satisfied. On the other hand, by the definition of the last entry time (7.104), on $\{\eta_A^{(k)} > 1 + L_{k,\varepsilon}^{\mathrm{ls}}\}$ the inequality $\lambda^k_{k+\eta_A^{(k)}-1} \geq (I-\varepsilon)(\eta_A^{(k)} - 1)$ holds. Hence,

$$(I-\varepsilon)(\eta_A^{(k)} - 1) \leq \log(A/\pi_k) \quad \text{on } \{L_{k,\varepsilon}^{\mathrm{ls}} + 1 < \eta_A^{(k)} < \infty\},$$

and we obtain

$$\begin{aligned} \mathsf{E}_k\left[\eta_A^{(k)}\right]^m &= \mathsf{E}_k\left[\eta_A^{(k)} \mathbb{1}_{\{L_{k,\varepsilon}^{\mathrm{ls}}+1<\eta_A^{(k)}<\infty\}}\right]^m + \mathsf{E}_k\left[\eta_A^{(k)} \mathbb{1}_{\{\eta_A^{(k)} \leq L_{k,\varepsilon}^{\mathrm{ls}}+1\}}\right]^m \\ &\leq \left(\frac{1+\log(A/\pi_k)}{I-\varepsilon}\right)^m + \mathsf{E}_k(1+L_{k,\varepsilon}^{\mathrm{ls}})^m. \end{aligned}$$

Averaging over the prior distribution yields

$$\sum_{k=0}^\infty \pi_k \mathsf{E}_k\left[\eta_A^{(k)}\right]^m \leq \sum_{k=0}^\infty \pi_k \left(\frac{1+\log(A/\pi_k)}{I-\varepsilon}\right)^m + \sum_{k=1}^\infty \pi_k \mathsf{E}_k(1+L_{k,\varepsilon}^{\mathrm{ls}})^m.$$

By the conditions (7.106) and (7.105), $\sum_{k=0}^\infty |\log \pi_k|^m \pi_k < \infty$ and $\sum_{k=0}^\infty \pi_k \mathsf{E}_k(L_{k,\varepsilon}^{\mathrm{ls}})^m < \infty$ for $m \leq r$. Since ε can be arbitrarily small, the asymptotic upper bound (7.113) follows, which implies the upper bound (7.109), and the proof of (7.107) is complete.

Proof of (ii). The asymptotic relations (7.108) immediately follow from (7.107) and the asymptotic lower bound (7.103). This completes the proof of the theorem. □

If instead of the condition (7.105), we impose the r-quick convergence condition (7.69) as in Theorem 7.2.1 that limits the behavior of both tails of the distribution of the LLR $\lambda_{\nu+n}^\nu$, then both conditions (7.20) and (7.105) are satisfied and, therefore, the following statement holds.

Corollary 7.3.1. *Suppose that the condition* (7.69) *is satisfied for some* $0 < I < \infty$. *Then the asymptotic relations* (7.107) *and* (7.108) *hold.*

We also recall that in some particular cases, instead of checking the condition (7.69) one may check the r-complete convergence condition (7.94), which is sufficient for the asymptotic optimality property; see Remark 7.2.4.

In the iid case, the finiteness of the $(r+1)$-th absolute moment of the LLR, $\mathsf{E}_0|Z_1|^{r+1} < \infty$, is both a necessary and a sufficient condition for the r-quick convergence (7.71) since $\mathsf{E}_0[L_\varepsilon^{(0)}]^r = \mathsf{E}_k[L_\varepsilon^{(k)}]^r$ for every $k \ge 1$ and $\mathsf{E}_0|Z_1|^{r+1} < \infty$ implies $\mathsf{E}_0[L_\varepsilon^{(0)}]^r < \infty$ and *vice versa* by Theorem 2.4.4. Therefore, Theorem 7.3.1 implies the asymptotic relations (7.107) and (7.108) with $I = \mathsf{E}_0 Z_1$ for $m \le r$ under the finiteness of the $(r+1)$-th moment. However, we now show that in the iid case these relations hold for all $m > 0$ under the sole first moment condition $0 < I < \infty$.

Theorem 7.3.2. *Consider the iid discrete-time model. Let τ_A be defined by* (7.101)*. Let $0 < I < \infty$ and let the prior distribution be such that $\sum_{k=0}^\infty |\ln \pi_k|^m \pi_k < \infty$ for all $m > 0$.*

(i) *Then for all $m > 0$,*

$$\mathsf{E}^\pi[(\tau_A - \nu)^m | \tau_A > \nu] \sim \left(I^{-1} \log A\right)^m \quad \text{as } A \to \infty. \tag{7.114}$$

(ii) *If $A = A_\alpha$ is selected so that $\mathsf{P}_\infty(\tau_{A_\alpha} < \infty) \le \alpha$ and $\log A_\alpha \sim \log(1/\alpha)$, in particular $A_\alpha = 1/\alpha$, then for all $m > 0$,*

$$\begin{aligned} \inf_{T \in \mathbf{C}_\alpha^\infty} \mathsf{E}^\pi([(T-\nu)^m | T > \nu] &\sim \mathsf{E}^\pi[(\tau_{A_\alpha} - \nu)^m | \tau_{A_\alpha} > \nu] \\ &\sim \left(I^{-1} |\log \alpha|\right)^m \quad \text{as } \alpha \to 0. \end{aligned} \tag{7.115}$$

Proof. (i) By (7.112), $(\tau_A - k)^+ \le \eta_A^{(k)}$ for every $k \ge 0$. By the iid property of the data, the random variables Z_i, $i = 1, 2, \dots$, are also iid, and hence the P_k-distribution of $\eta_A^{(k)}$ is the same as the P_0-distribution of the stopping time

$$\widetilde{\eta}_A(\pi_k) = \inf\left\{n \ge 1 : \lambda_n^0 \ge \ln(A\pi_k^{-1})\right\}. \tag{7.116}$$

Therefore, for all $k \ge 0$,

$$\mathsf{E}_k\left[(\tau_A - k)^+\right]^m \le \mathsf{E}_0\left[\widetilde{\eta}_A(\pi_k)\right]^m. \tag{7.117}$$

Now, the same reasoning as in the proof of Theorem 7.1.4 yields

$$\mathsf{E}_0\left[\widetilde{\eta}_A(\pi_k)\right]^m = \left(I^{-1} \log(\pi_k^{-1} A)\right)^m (1 + o(1)) \quad \text{as } A \to \infty \quad \text{for all } m > 0, \tag{7.118}$$

which along with (7.117) implies

$$\mathsf{E}_k[(\tau_A - k)^+]^m \le \left(I^{-1} \ln(\pi_k^{-1} A)\right)^m \left[1 + \varepsilon(k, m, A)\right], \tag{7.119}$$

where $\varepsilon(k, m, A) \to 0$ as $A \to \infty$.

Let $a = \log A$. Averaging in (7.119) over the prior distribution, we obtain

$$\begin{aligned} \sum_{k=1}^\infty \pi_k \mathsf{E}_k[(\tau_A - k)^+]^m \le \left(\frac{a}{I}\right)^m \Bigg\{ &\sum_{k=0}^\infty \pi_k \left(1 + \frac{|\ln \pi_k|}{a}\right)^m \\ &+ \sum_{k=0}^\infty \pi_k \left(1 + \frac{|\ln \pi_k|}{a}\right)^m \varepsilon(k, m, A) \Bigg\}. \end{aligned} \tag{7.120}$$

Since, by the conditions of the theorem, $\sum_{k=0}^\infty |\ln \pi_k|^m \pi_k < \infty$, we have

$$\sum_{k=0}^\infty \pi_k \left(1 + \frac{|\log \pi_k|}{a}\right)^m = 1 + o(1) \quad \text{as } A \to \infty. \tag{7.121}$$

The important observation is that, since $|\log \pi_k| \to \infty$ as $k \to \infty$, the asymptotic equality (7.118) and hence the inequality (7.119) also hold for any $A > 1$ as $k \to \infty$. This means that $\varepsilon(k,m,A) \to 0$ as $k \to \infty$ for any fixed $A > 1$ and also as $A \to \infty$. Therefore, $\sum_{k=0}^{\infty} \pi_k |\log \pi_k|^m \varepsilon(k,m,A) < \infty$ for any $A > 1$, and hence,

$$\sum_{k=0}^{\infty} \pi_k \left(1 + \frac{|\log \pi_k|}{a}\right)^m \varepsilon(k,m,A) \to 0 \quad \text{as } A \to \infty. \tag{7.122}$$

Combining (7.120), (7.121), and (7.122) yields the asymptotic inequality

$$\mathsf{E}^{\pi}\left[(\tau_A - \nu)^+\right]^m \le \left(I^{-1}\log A\right)^m (1 + o(1)) \quad \text{as } A \to \infty.$$

Finally, noting that $\mathsf{P}^{\pi}(\tau_A > \nu) \ge \mathsf{P}_{\infty}(\tau_A = \infty) \ge 1 - A^{-1}$ (cf. Lemma 7.3.1) and

$$\mathsf{E}^{\pi}[(\tau_A - \nu)^+]^m = \mathsf{P}^{\pi}(\tau_A > \nu)\mathsf{E}^{\pi}[(\tau_A - \nu)^m | \tau_A > \nu],$$

we obtain the upper bound

$$\mathsf{E}^{\pi}[(\tau_A - \nu)^m | \tau_A > \nu] \le \left(I^{-1}\log A\right)^m (1 + o(1)).$$

Comparing this asymptotic upper bound with the lower bound

$$\mathsf{E}^{\pi}[(\tau_A - \nu)^m | \tau_A > \nu] \ge \left(I^{-1}\log A\right)^m (1 + o(1)), \quad A \to \infty,$$

which follows trivially from (7.103) by setting $\alpha = 1/A$, and noting that the condition (7.20) holds in the iid case by the strong law of large numbers, complete the proof of (7.114).

(ii) The fact that $\tau_{A_\alpha} \in \mathbf{C}_{\alpha}^{\infty}$ as $A_\alpha = 1/\alpha$ follows from Lemma 7.3.1. The asymptotic relation (7.115) follows from (7.114) and the lower bound (7.103). The theorem is proven. □

The upper bound (7.102) for the global PFA, $\mathsf{P}_{\infty}(\tau_A < \infty) \le A^{-1}$, which neglects the threshold overshoot, holds in the most general non-iid case. In the iid case, an accurate approximation for $\mathsf{P}_{\infty}(\tau_A < \infty)$ can be obtained by taking into account the overshoot using a nonlinear renewal theory argument as in Theorem 7.1.5. This is important in situations where the upper bound (7.102) that ignores the overshoot is conservative, which is always the case when the densities $f_1(x)$ and $f_0(x)$ are not close enough.

We now present an intuitive argument. A rigorous proof can be found in [462, Lemma 4.2]. In order to apply relevant results from nonlinear renewal theory, we have to rewrite the stopping time τ_A in the form of a random walk crossing a constant threshold plus a nonlinear term that is slowly changing. For the sake of simplicity, assume that $q = 0$. Using (7.110) and denoting

$$\ell_n^k = \log\left(\pi_k + \Pr(\nu \ge n) e^{-\sum_{i=k+1}^{n} Z_i} + \sum_{j=0}^{k-1} \pi_j e^{\sum_{i=j+1}^{k} Z_i} + \sum_{j=k+1}^{n-1} \pi_j e^{-\sum_{i=j+1}^{k} Z_i}\right), \tag{7.123}$$

we obtain that for every $k \ge 0$

$$\log \bar{\Lambda}_n = \lambda_n^k + \ell_n^k. \tag{7.124}$$

Therefore, on the set $\{\tau_A > k\}$ for any $k \ge 0$ the stopping time τ_A can be written in the following form:

$$\tau_A = \inf\{n \ge k : \lambda_n^k + \ell_n^k \ge a\}, \quad a = \log A, \tag{7.125}$$

where ℓ_n^k is given by (7.123) and $\{\lambda_n^k, n > k\}$ is a random walk with mean $\mathsf{E}_k \lambda_n^k = I(n-k)$.

For $b > 0$, define $\eta_b = \inf\{n \ge 1 : \lambda_n^0 \ge b\}$ and let $\kappa_b = \lambda_{\eta_b}^0 - b$ on $\{\eta_b < \infty\}$ denote the corresponding overshoot. Let $\zeta(I) = \lim_{b\to\infty} \mathsf{E}_0[e^{-\kappa_b}]$.

The important observation is that ℓ_n^k, $n \geq k$, are slowly changing. To see this it suffices to realize that, as $n \to \infty$, the values of ℓ_n^k converge to the random variable

$$\ell_\infty^k = \log\left(\pi_k + \sum_{j=0}^{k-1} \pi_j e^{\sum_{i=j+1}^k Z_i} + \sum_{j=k+1}^{\infty} \pi_j e^{-\sum_{i=k+1}^j Z_i}\right),$$

which has a finite negative expectation. Indeed, on one hand, $\ell_\infty^k \geq \log \pi_k$, and on the other hand, by Jensen's inequality,

$$\begin{aligned}\mathsf{E}_k \ell_\infty^k &\leq \log\left(\pi_k + \sum_{j=0}^{k-1} \pi_j \mathsf{E}_k e^{\sum_{i=j+1}^k Z_i} + \sum_{j=k+1}^{\infty} \pi_j \mathsf{E}_k e^{-\sum_{i=k+1}^j Z_i}\right) \\ &= \log\left(\pi_k + \sum_{j=0}^{k-1} \pi_j + \sum_{j=k+1}^{\infty} \pi_j\right) = \log\left(\sum_{j=0}^{\infty} \pi_j\right) = 0,\end{aligned}$$

where we used the equalities

$$\mathsf{E}_k e^{\sum_{i=j+1}^k Z_i} = \prod_{i=j+1}^{k} \mathsf{E}_k \frac{f_1(X_i)}{f_0(X_i)} = 1, \quad \mathsf{E}_k e^{-\sum_{i=k+1}^j Z_i} = \prod_{i=k+1}^{j} \mathsf{E}_k \frac{f_0(X_i)}{f_1(X_i)} = 1.$$

Since

$$\mathsf{P}_\infty(\tau_A < \infty) = \mathsf{E}^\pi\left[\bar{\Lambda}_{\tau_A}^{-1} \mathbb{1}_{\{\tau_A<\infty\}}\right] = \mathsf{E}^\pi\left[A(A\bar{\Lambda}_{\tau_A})^{-1} \mathbb{1}_{\{\tau_A<\infty\}}\right] = A^{-1}\mathsf{E}^\pi\left[e^{-\chi_a} \mathbb{1}_{\{\tau_A<\infty\}}\right],$$

where $\chi_a = \log \bar{\Lambda}_{\tau_A} - a$, a routine application of the first nonlinear renewal theorem along with computations similar to those of Theorem 7.1.5 yield

$$\mathsf{P}_\infty(\tau_A < \infty) \sim A^{-1}\zeta(I) \quad \text{as } A \to \infty.$$

Remark 7.3.1. The global false alarm probability constraint $\sup_k \mathsf{P}_k(\tau < k\} = \mathsf{P}_\infty(\tau < \infty) \leq \alpha$ leads to an unbounded maximal expected detection delay $\sup_k \mathsf{E}_k(T - k | T > k)$ whenever $\alpha < 1$ due to the high price that should be paid for such a strong constraint. Note that to overcome this difficulty in a minimax setting a dynamic sampling technique can be used when it is feasible [16]. To the expense of a large amount of data that should be sampled, the maximal average delay to detection may then be made bounded, yet keeping the global PFA below the given small level. However, dynamic sampling is rarely possible in applications. We, therefore, considered a Bayesian problem with the prior distribution. The proposed asymptotically Bayesian detection procedure can be regarded as the Shiryaev detection procedure with a threshold that increases in time. The need for the increasing threshold is due to the strong constraint imposed on the global PFA in place of the weighted PFA constraint used in the classical problem setting.

Remark 7.3.2. While the results of the previous subsection may be used to devise a reasonably simple detection procedure to handle the global false alarms probability bound, our personal opinion is that this constraint is too strong to be useful in applications. In fact, the conditional ADD of the proposed detection procedure, $\mathsf{E}_k(\tau_A - k | \tau_A > k)$, grows fairly fast with k, and the nice property that the weighted ADD, $\mathsf{E}^\pi(\tau_{A_\alpha} - \nu | \tau_{A_\alpha} > \nu)$, is as small as possible for small α perhaps will not convince practitioners of the usefulness of the procedure. In addition, the mean time to false alarm in this detection procedure is unbounded, which is an unavoidable payoff for the very strong global PFA constraint.

Remark 7.3.3. The sufficient conditions for asymptotic optimality in Theorem 7.3.1 are quite general and hold in most applications. In Section 7.4 below, these conditions are verified for several examples that include both additive and non-additive changes in non-iid models.

Remark 7.3.4. Similar results can be proved for general continuous-time stochastic models, as in Theorem 7.2.3. The proof requires a more difficult argument as in [28] for the classical setting.

7.4 Examples

7.4.1 *Detection of a Change in the Mean of a Gaussian Autoregressive Process*

Let $X_n = \theta \mathbb{1}_{\{\nu<n\}} + V_n$, $n \ge 1$, where $\theta \neq 0$ is a constant signal that appears at an unknown time point ν and V_n, $n \ge 1$, is a zero-mean stable Gaussian AR(p) sequence (noise) that obeys the recursive relation

$$V_n = \sum_{j=1}^{p} \rho_j V_{n-j} + \xi_n, \quad n \ge 1, \quad V_j = 0 \text{ for } j \le 0,$$

where ξ_n, $n \ge 1$, are iid $\mathcal{N}(0, \sigma^2)$ random variables and the equation $1 - \sum_{j=1}^{p} \rho_j y^j = 0$ has no roots inside the unit circle. For $i \ge 1$, define

$$\widetilde{X}_i = \begin{cases} X_1, & \text{if } i = 1, \\ X_i - \sum_{j=1}^{i-1} \rho_j X_{i-j}, & \text{if } 2 \le i \le p, \\ X_i - \sum_{j=1}^{p} \rho_j X_{i-j}, & \text{if } i \ge p+1, \end{cases}$$

and for $i > k$ and $k = 0, 1, 2, \dots$, define

$$\widetilde{\theta}_i = \begin{cases} \theta, & \text{if } i = k+1, \\ \theta \left(1 - \sum_{j=1}^{i-k-1} \rho_j\right), & \text{if } k+1 < i \le p+k, \\ \theta \left(1 - \sum_{j=1}^{p} \rho_j\right), & \text{if } i > p+k. \end{cases}$$

The conditional pre-change pdf $f_0(X_i | \mathbf{X}_1^{i-1})$ is of the form

$$f_0(X_i \mid \mathbf{X}_1^{i-1}) = \sigma^{-1} \varphi(\sigma^{-1} \widetilde{X}_i) \quad \text{for all } i \ge 1,$$

and the conditional post-change pdf $f_1(X_i | \mathbf{X}_1^{i-1})$, conditioned on $\nu = k$, is given by

$$f_1(X_i \mid \mathbf{X}_1^{i-1}) = \sigma^{-1} \varphi(\sigma^{-1}(\widetilde{X}_i - \widetilde{\theta}_i)) \quad \text{for } i > k,$$

where $\varphi(y)$ is the standard normal pdf.

Using these formulas, we easily obtain the expression for the LLR:

$$\lambda_n^k = \frac{1}{\sigma^2} \sum_{i=k+1}^{n} \widetilde{\theta}_i \widetilde{X}_i - \frac{1}{2\sigma^2} \sum_{i=k+1}^{n} \widetilde{\theta}_i^2, \quad 0 \le k \le n-1, \quad n = 1, 2, \dots.$$

Let

$$I = \frac{\theta^2}{2\sigma^2} \left(1 - \sum_{j=1}^{p} \rho_j\right)^2.$$

Note that, under P_k, the LLR process λ_{k+n}^k, $n \ge 1$ has independent Gaussian increments Z_n. Moreover, the increments are iid for $n \ge p+1$ with mean $\mathsf{E}_k Z_n = I$ and variance $I/2$. Thus, λ_{k+n}^k / n converges r-quickly to I for all positive r under P_k.

Applying Theorem 7.2.1 we conclude that the Shiryaev detection procedure $T_s(A_\alpha)$ with $A_\alpha = (1-\alpha)/\alpha$ asymptotically minimizes all positive moments of the detection delay in the class $\mathbf{C}_\alpha$. Also, by Theorem 7.3.1, the detection procedure τ_{A_α} with $A_\alpha = 1/\alpha$ is asymptotically optimal in the class $\mathbf{C}_\alpha^\infty$ with respect to all positive moments of the detection delay.

Note that in the stationary mode when $T_s(A), \tau_A \gg \nu = k$, the original problem of detecting a change of the intensity θ in a correlated Gaussian noise is equivalent to detecting a change of the intensity $\theta(1 - \sum_{j=1}^{p} \rho_j)$ in a white Gaussian noise. This is primarily because the original problem allows for whitening without loss of information through the innovations $\widetilde{X}_n$, $n \ge 1$ that contain the same information about the hypotheses H_k and H_∞ as the original sequence X_n, $n \ge 1$.

7.4.2 Detection of Additive Changes in Linear State–Space Models

In this subsection we address the sequential change detection counterpart of the problem investigated in Subsection 3.4.7. Consider the linear state–space model where the unobserved m-dimensional Markov component θ_n is given by the recursion

$$\theta_n = F\theta_{n-1} + W_{n-1} + \eta_\theta \mathbb{1}_{\{\nu<n\}}, \quad n \geq 0, \quad \theta_0 = 0,$$

and the observed r-dimensional component

$$X_n = H\theta_n + V_n + \eta_x \mathbb{1}_{\{\nu<n\}}, \quad n \geq 1.$$

Here W_n and V_n are zero-mean Gaussian iid vectors having covariance matrices K_W and K_V, respectively; $\eta_\theta = (\eta_\theta^1, \dots, \eta_\theta^m)^\top$ and $\eta_x = (\eta_x^1, \dots, \eta_x^r)^\top$ are vectors of the corresponding change intensities; and F and H are full rank matrices with sizes $m \times m$ and $r \times m$, respectively.

It can be shown that under the no-change hypothesis the observed sequence X_n, $n \geq 1$ has an equivalent representation with respect to the innovation process $\xi_n = X_n - \mathsf{E}(\theta_n|\mathscr{F}_{n-1})$, $n \geq 1$:

$$X_n = H\hat{\theta}_n + \xi_n, \quad n \geq 1,$$

where $\xi_n \sim \mathcal{N}(0, \Sigma_n)$, $n = 1, 2, \dots$ are independent Gaussian vectors and $\hat{\theta}_n = \mathsf{E}(\theta_n|\mathbf{X}_1^{n-1})$. Note that $\hat{\theta}_n$ is the optimal (in the mean-square sense) one-step ahead predictor, i.e., the estimate of θ_n based on $\mathbf{X}_1^{n-1}$, which can be obtained by the Kalman filter. Under the hypothesis $H_k : \nu = k$ we have

$$X_n = \Upsilon_n(k) + H\hat{\theta}_n + \xi_n, \quad n \geq 1,$$

where $\Upsilon_n(k)$, the signature of the change on the innovation, depends on n and the changepoint k. The value of $\Upsilon_n(k)$ can be computed using relations given, e.g., in [47, pp. 282-283] and [507].

It follows that the LLR λ_n^k is represented in the form

$$\lambda_n^k = \sum_{i=k+1}^{n} \Upsilon_i(k)^\top \Sigma_i^{-1} \xi_i - \frac{1}{2} \sum_{i=k+1}^{n} \Upsilon_i(k)^\top \Sigma_i^{-1} \Upsilon_i(k),$$

where the Σ_i's are given by the Kalman equations; see, e.g., [47, Eq. (3.2.20)]. Therefore, the original detection problem for an abrupt change that occurs at $\nu = k$ is equivalent to detecting a gradual change from zero to $\Upsilon_i(k)$, $i > k$, in the sequence of independent Gaussian innovations ξ_i with covariance matrices Σ_i. These innovations can be formed by the Kalman filter. Note also that since the post-change distribution depends on the changepoint k through the value of $\Upsilon_n(k)$, there is no efficient recursive formulas for the statistics Λ_n and $\tilde{\Lambda}_n$ in contrast to the iid case.

As $n \to \infty$, the normalized LLR $n^{-1}\lambda_{k+n}^k$ converges P_k-a.s. to the positive constant

$$I = \frac{1}{2} \lim_{n\to\infty} \frac{1}{n} \sum_{i=k+1}^{k+n} \Upsilon_i(k)^\top \Sigma_i^{-1} \Upsilon_i(k).$$

Using [47], we obtain that this constant is equal to

$$I = \frac{1}{2} \left\{ \eta_\theta^\top (\mathbb{I}_m - F^*)^{-\top} H^\top \Sigma^{-1} H (\mathbb{I}_m - F^*)^{-1} \eta_\theta \right\},$$

when the change occurs only in η_θ and

$$I = \frac{1}{2} \left\{ \eta_x^\top \left[\mathbb{I}_r - H(\mathbb{I}_m - F^*)^{-1} FK\right]^\top \Sigma^{-1} \left[\mathbb{I}_r - H(\mathbb{I}_m - F^*)^{-1} FK\right] \eta_x \right\}$$

when the change occurs only in η_x. Here K is the Kalman filter gain in the stationary regime, $\mathbb{I}_m$ is the unit $(m \times m)$-matrix, and $F^* = F(\mathbb{I}_m - KH)$.

Moreover, since the process $\{\lambda_{k+n}^k, n \geq 1\}$ is Gaussian with independent increments, $n^{-1}\lambda_{k+n}^k$ converges r-quickly to I for all $r > 0$. Therefore, Corollary 7.3.1 shows that the detection procedure τ_{A_α} with $A_\alpha = 1/\alpha$ is, as $\alpha \to 0$, asymptotically optimal in the class $\mathbf{C}_\alpha^\infty$ with respect to all positive moments of the detection delay, and Theorem 7.2.1 shows the same property for the procedure $T_{\mathrm{s}}(A_\alpha)$ with $A_\alpha = (1-\alpha)/\alpha$ in the class $\mathbf{C}_\alpha$.

7.4.3 Detection of Nonadditive Changes in Mixture-Type Models and Hidden Markov Models

In the previous two examples we considered additive changes. Consider now an example with non-additive changes where the observations are iid in the abnormal mode and mixture-type dependent in the normal mode. This example can be used as a counterexample to disprove that the CUSUM and Shiryaev–Roberts detection procedures are asymptotically optimal in the minimax setting with a lower bound on the ARL to false alarm. However, we show below that the Bayesian procedures $T_s(A)$ and τ_A are asymptotically optimal. This primarily happens because the strong law of large numbers still holds, while a stronger essential supremum condition (8.84), which is required for obtaining a lower bound for the maximal average detection delay, fails to hold; this is addressed in Subsection 8.2.5.

Let $g_1(X_n)$, $g_2(X_n)$ and $f_1(X_n)$ be three distinct densities. The problem is to detect a change from the mixture density

$$f_0(\mathbf{X}_1^n) = \beta \prod_{i=1}^{n} g_1(X_i) + (1-\beta) \prod_{i=1}^{n} g_2(X_i)$$

to the density f_1, where $0 < \beta < 1$ is a mixing probability. Therefore, the observations are dependent with the joint pdf $f_0(\mathbf{X}_1^n)$ before the change occurs and iid with density f_1 after the change.

Let $R_j(n) = \log[f_1(X_n)/g_j(X_n)]$ and $I_j = \mathsf{E}_0 R_j(1)$, $j = 1,2$.

It is easy to show that

$$\frac{f_1(X_i)}{f_0(X_i \mid \mathbf{X}_1^{i-1})} = \frac{e^{R_2(i)}(\beta\xi_{i-1} + 1 - \beta)}{\beta\xi_i + 1 - \beta},$$

where $\xi_i = \prod_{m=1}^{i} \Delta\xi_m$, $\Delta\xi_m = g_1(X_m)/g_2(X_m)$. Next, note that

$$\prod_{i=k}^{n} \frac{1-\beta+\beta\xi_{i-1}}{1-\beta+\beta\xi_i} = \frac{1+w\xi_{k-1}}{1+w\xi_n},$$

where $w = \beta/(1-\beta)$, so that the LLR has the form

$$\lambda_n^k := \sum_{i=k+1}^{n} \log \frac{f_1(X_i)}{f_0(X_i \mid \mathbf{X}_1^{i-1})} = \sum_{i=k+1}^{n} R_2(i) + \log \frac{1+w\xi_k}{1+w\xi_n}. \tag{7.126}$$

Assume that $I_1 > I_2$, in which case $\mathsf{E}_k \log \Delta\xi_m < 0$ for $k < m$, and hence,

$$\xi_n = \xi_k \prod_{m=k+1}^{n} \Delta\xi_m \xrightarrow[n\to\infty]{\mathsf{P}_k\text{-a.s.}} 0 \quad \text{for all } 0 \le k < \infty.$$

The condition (7.20), which is necessary to lower-bound the moments of the detection delay, holds with the constant $I = I_2$. Indeed, since $R_2(i)$, $i > k$, are iid random variables with mean I_2 under P_k, and since $\xi_n \to 0$, the LLR obeys the strong law of large numbers: $n^{-1}\lambda_{n+k}^k \to I_2$ P_k-a.s. as $n \to \infty$, which implies (7.20) with $I = I_2$, and hence, the lower bounds:

$$\inf_{T \in \mathbf{C}_\alpha^\infty} \mathsf{E}^\pi[(T-\nu)^m | T > \nu] \ge (I_2^{-1}|\log\alpha|)^m (1+o(1)) \quad \text{as } \alpha \to 0 \text{ for all } m \ge 1,$$

$$\inf_{T \in \mathbf{C}_\alpha} \mathsf{E}^\pi[(T-\nu)^m | T > \nu] \ge [(I_2+d)^{-1}|\log\alpha|]^m (1+o(1)) \quad \text{as } \alpha \to 0 \text{ for all } m \ge 1.$$

Next, using (7.111) and (7.126), we can write the statistic $\log\bar{\Lambda}_n$ in the following form:

$$\log\bar{\Lambda}_n = \sum_{i=k+1}^{n} R_2(i) + \psi(k,n) + \log\left[\pi_k(1+Y_n^k)\right],$$

where $\psi(k,n) = \log[1+w\xi_k)/(1+w\xi_n)]$. Clearly, the sequence Y_n^k, $n \ge k$, is slowly changing.

The sequence $\psi(k,n)$, $n \ge k$, is also slowly changing. Indeed, since $\xi_n \to 0$ with probability 1, it converges to the finite random variable $\log(1+w\xi_k)$. Therefore, by the nonlinear renewal theorem

$$\mathsf{E}^\pi(\tau_A - \nu \mid \tau_A > \nu) = (I_2^{-1}\log A)\big(1+o(1)\big) \quad \text{as } A \to \infty,$$

and the detection procedure τ_{A_α} with $A_\alpha = 1/\alpha$ is asymptotically optimal. The same is true for the Shiryaev detection procedure $T_s(A_\alpha)$ with $A_\alpha = (1-\alpha)/\alpha$ under the traditional constraint on the weighted false alarm probability, i.e., in the class $\mathbf{C}_\alpha$, since by the nonlinear renewal theorem

$$\mathsf{E}^\pi[T_s(A) - \nu \mid T_s(A) > \nu] = [(I_2+d)^{-1}\log A]\big(1+o(1)\big) \quad \text{as } A \to \infty.$$

Finally, we note that the above simple mixture model is obviously a degenerate case of a more general model governed by a two-state hidden Markov model where the state transition probabilities are equal to zero and the initial distribution is given by the probability β. The proposed procedure τ_A as well as the Shiryaev procedure $T_s(A)$ in the conventional setting remain asymptotically optimal for the model where the pre-change distribution is controlled by a finite-state non-degenerate hidden Markov model, while the post-change model is iid.

7.4.4 Continuous-Time Changepoint Detection in Additive Itô Processes

We now consider a linear continuous-time model driven by a Brownian motion where the unobservable additive component is described by a general Itô process. Specifically, let the observed process X_t have the Itô stochastic differential

$$\mathrm{d}X_t = \begin{cases} S_0(t)dt + \sqrt{N}\,\mathrm{d}W(t) & \text{for } \nu \ge t, \\ S_1(t)dt + \sqrt{N}\,\mathrm{d}W(t) & \text{for } \nu < t, \end{cases} \tag{7.127}$$

where $S_0(t)$ and $S_1(t)$ are Itô stochastic processes. In what follows, we always assume that

$$\int_0^t \mathsf{E}S_i^2(y)\,\mathrm{d}y < \infty, \; i = 0,1 \quad \text{for every } t < \infty.$$

Define the functionals $\hat{S}_1(u,t) = \mathsf{E}_u[S_1(t) \mid \mathscr{F}_t]$ and $\hat{S}_0(t) = \mathsf{E}_\infty[S_0(t) \mid \mathscr{F}_t]$. By Lemma 2.1.1, there exist standard Wiener innovation processes $\{\widetilde{W}_i(t)\}$, $i = 0,1$ such that under the hypothesis $H_u : \nu = u$ the process X_t has the following minimal representation in the form of diffusion-type processes:

$$\mathrm{d}X_t = \begin{cases} \hat{S}_0(t)\mathrm{d}t + \sqrt{N}\,d\widetilde{W}_0(t) & \text{for } u \ge t, \\ \hat{S}_1(u,t)\mathrm{d}t + \sqrt{N}\,d\widetilde{W}_1(t) & \text{for } u < t. \end{cases} \tag{7.128}$$

The processes $\widetilde{W}_i(t)$ are statistically equivalent to the original Wiener process $W(t)$. Using this representation and Theorem 2.1.3 on the absolute continuity of probability measures of diffusion-type processes with respect to the Wiener measure P_W, we obtain

$$\log\frac{\mathrm{d}\mathsf{P}_u^t}{\mathrm{d}\mathsf{P}_W^t}(X^t) = \frac{1}{N}\int_0^u \hat{S}_0(\mathrm{v})\,\mathrm{d}X_\mathrm{v} - \frac{1}{2N}\int_0^u \hat{S}_0^2(\mathrm{v})\,\mathrm{d}\mathrm{v} + \frac{1}{N}\int_u^t \hat{S}_1(u,\mathrm{v})\,\mathrm{d}X_\mathrm{v} - \frac{1}{2N}\int_u^t \hat{S}_1^2(u,\mathrm{v})\,\mathrm{d}\mathrm{v};$$

$$\log\frac{\mathrm{d}\mathsf{P}_\infty^t}{\mathrm{d}\mathsf{P}_W^t}(X^t) = \frac{1}{N}\int_0^t \hat{S}_0(\mathrm{v})\,\mathrm{d}X_\mathrm{v} - \frac{1}{2N}\int_0^t \hat{S}_0^2(\mathrm{v})\,\mathrm{d}\mathrm{v}.$$

Applying these formulas yields

$$\lambda_t^u = \frac{1}{N}\int_u^t \left[\hat{S}_1(u,\mathrm{v}) - \hat{S}_0(\mathrm{v})\right]\mathrm{d}X_\mathrm{v} - \frac{1}{2N}\int_u^t \left[\hat{S}_1^2(u,\mathrm{v}) - \hat{S}_0^2(\mathrm{v})\right]\mathrm{d}\mathrm{v}. \tag{7.129}$$

Now, using (7.128) and (7.129), we get that under H_u,

$$\lambda_t^u = \frac{1}{\sqrt{N}} \int_u^t \left[\hat{S}_1(u,\mathrm{v}) - \hat{S}_0(\mathrm{v})\right] \mathrm{d}\widetilde{W}_1(\mathrm{v}) + \frac{1}{2N} \int_u^t \left[\hat{S}_1(u,\mathrm{v}) - \hat{S}_0(\mathrm{v})\right]^2 \mathrm{d}\mathrm{v}. \tag{7.130}$$

Assume that

$$\frac{1}{t} \int_\nu^{\nu+t} \hat{S}_i(\nu,\mathrm{v}) \,\mathrm{d}\widetilde{W}_1(\mathrm{v}) \xrightarrow[t\to\infty]{\mathsf{P}^\pi-\text{completely}} 0, \quad i = 0, 1,$$

and

$$\frac{1}{t} \int_\nu^{\nu+t} \left[\hat{S}_1(\nu,\mathrm{v}) - \hat{S}_0(\mathrm{v})\right]^2 \mathrm{d}\mathrm{v} \xrightarrow[t\to\infty]{\mathsf{P}^\pi\text{-completely}} \mu,$$

where μ is positive and finite. Then, obviously,

$$\frac{1}{t} \lambda_{\nu+t}^\nu \xrightarrow[t\to\infty]{\mathsf{P}^\pi-\text{completely}} I = \mu/2N$$

and it follows from Theorem 7.2.3 that the Shiryaev detection procedure $T_{\mathrm{s}}(A)$ asymptotically minimizes the average delay to detection $\mathsf{ADD}^\pi(T)$ in the class $\mathbf{C}_\alpha$ as $\alpha \to 0$.

In most cases

$$\mu = \lim_{t\to\infty} \frac{1}{t} \int_0^t \mathsf{E}_0 \left[\hat{S}_1(0,\mathrm{v}) - \hat{S}_0(\mathrm{v})\right]^2 \mathrm{d}\mathrm{v}.$$

This fact can be illustrated for Gaussian Markov processes. Specifically, consider a continuous-time Gaussian hidden Markov model with $S_0(t) \equiv 0$ and $S_1(t)$ being the stationary Markov Gaussian process which for $\nu = u$, $u \geq 0$, obeys the following Itô stochastic differential equation

$$\mathrm{d}S_1(t) = -\beta_1 S_1(t)\mathrm{d}t + \sqrt{\sigma_1}\mathrm{d}w_1(t), \quad t > u, \quad S_1(u) \sim \mathcal{N}(0, \delta_1^2),$$

where $\{w_1(t)\}_{t\geq 0}$ is a standard Wiener process independent of the Wiener process $W(t)$ and $\beta_1, \sigma_1 > 0$ are given constants. Note that $S_1(t) = S_1(u,t)$ depends on u, the changepoint, but its statistical properties are invariant to u since it is the stationary Gaussian process with the parameters

$$\mathsf{E}_u S_1(u,t) = 0, \quad \mathsf{E}_u[S_1(u,t)S_1(u,t+\mathrm{v})] = \delta_1^2 e^{-\beta_1|\mathrm{v}|}$$

for every $u \geq 0$. Thus, we can write $S_1(u,t) = S_1(t-u)$, $t > u$.

The functional $\hat{S}_1(u,t) = \hat{S}_1(t-u)$ is nothing but the optimal mean-square-error estimate of the Markov process $S_1(t-u)$ based on the data $\mathbf{X}_u^t = \{X_s, u < s \leq t\}$. For $t > u$, it satisfies the system of Kalman–Bucy equations

$$\mathrm{d}\hat{S}_1(t-u) = -\left(\beta_1 - \frac{K_1(t-u)}{N}\right)\hat{S}_1(t-u)\mathrm{d}t + \frac{K_1(t-u)}{N}\mathrm{d}X_t, \quad \hat{S}_1(0) = 0, \tag{7.131}$$

$$\frac{\mathrm{d}K_1(t-u)}{\mathrm{d}t} = -2\beta_1 K_1(t-u) - K_1^2(t-u)/N + \sigma_1, \quad K_1(0) = \delta_1^2, \tag{7.132}$$

where $K_1(t-u) = \mathsf{E}_u[S_1(t-u) - \hat{S}_1(t-u)]^2$ is the mean-square filtering error; see, e.g., [266, Theorem 10.1]. Clearly, the statistical properties of the LLR process λ_{u+t}^u do not depend on u, so

$$\mathsf{P}_u\left(\left|t^{-1}\lambda_{u+t}^u - I\right| > \varepsilon\right) = \mathsf{P}_0\left(\left|t^{-1}\lambda_t^0 - I\right| > \varepsilon\right),$$

where I is the constant (to be determined) to which $t^{-1}\lambda_t^0$ converges almost surely under P_0. Hence, in order to prove that $t^{-1}\lambda_{\nu+t}^\nu$ converges r-completely to I under P^π, i.e.,

$$\int_0^\infty \mathrm{d}\pi_u \left[\int_0^\infty t^{r-1}\mathsf{P}_u\left(\left|t^{-1}\lambda_{u+t}^u - I\right| > \varepsilon\right)\mathrm{d}t\right] < \infty,$$

it remains to show that

$$\int_0^\infty t^{r-1}\mathsf{P}_0\left(\left|t^{-1}\lambda_t^0 - I\right| > \varepsilon\right)\,\mathrm{d}t < \infty. \tag{7.133}$$

Observe that

$$\mathsf{E}_0\left[\hat{S}_1(t)\right]^2 = \mathsf{E}_0\left[S_1(t)\right]^2 - K_1(t) = \delta_1^2 t - K_1(t) \tag{7.134}$$

and

$$\frac{1}{t}\int_0^t K_1(s)\,\mathrm{d}s = N\beta_1\left(\sqrt{1+Q}-1\right) + o(1) \quad \text{as } t\to\infty, \tag{7.135}$$

where $Q = 1 + 2\delta_1^2/N\beta_1$. Combining (7.134) and (7.135) yields

$$I = \frac{1}{2N}\lim_{t\to\infty}\int_0^t S_1^2(y)\,\mathrm{d}y = \frac{\delta_1^2}{2N}\,\frac{Q}{\left(1+\sqrt{1+Q}\right)^2}. \tag{7.136}$$

Establishing the almost sure convergence $t^{-1}\lambda_t^0 \to I$ under P_0 as well as the r-complete convergence (7.133) is a tedious task. The details can be found in [455, Sec. 3.2].

Note that, since the trajectories of the statistic Λ_t are continuous, there is no overshoot and $A = A_\alpha = (1-\alpha)/\alpha$ implies $\mathsf{PFA}(T_\mathrm{s}(A_\alpha)) = \alpha$. By Theorem 7.2.3, the Shiryaev detection procedure $T_\mathrm{s}(A_\alpha)$ is asymptotically optimal in the class $\mathbf{C}_\alpha$: for all $r > 0$,

$$\inf_{T\in\mathbf{C}_\alpha}\mathsf{E}^\pi[(T-\nu)^r|T>\nu] \sim \mathsf{E}^\pi[(T_\mathrm{s}-\nu)^r|T_\mathrm{s}>\nu] \sim \left(\frac{|\log\alpha|}{I+d}\right)^r \quad \text{as } \alpha\to 0,$$

where I is defined in (7.136). And the procedure τ_{A_α} with $A_\alpha = 1/\alpha$ minimizes asymptotically all positive moments of the detection delay in the class $\mathbf{C}_\alpha^\infty$.

7.4.5 *Changepoint Detection in the Intensity of a Nonhomogeneous Poisson Process*

In the previous Gaussian model the trajectories of the decision statistic are continuous. Consider now the example of a process with jumps. Suppose that, under the hypothesis $H_u : \nu = u$, the observed process $\{X_t, t\ge 0\}$ is a non-stationary Poisson random process with intensity $\gamma_0(t)$ before the changepoint u and with intensity $\gamma_1(t)$ after the change. Then, for $t > u$,

$$\lambda_t^u = \int_u^t \log\left(\frac{\gamma_1(s)}{\gamma_0(s)}\right)\mathrm{d}X_s - \int_u^t (\gamma_1(s)-\gamma_0(s))\,\mathrm{d}s;$$

$$\mu_{u,t} = \mathsf{E}_u\lambda_t^u = \int_u^t\left\{\gamma_0(s) + \gamma_1(s)\left[\log\left(\frac{\gamma_1(s)}{\gamma_0(s)}\right) - 1\right]\right\}\mathrm{d}s;$$

$$\widetilde{\lambda}_t^u = \lambda_t^u - \mu_{u,t} = \int_u^t \log\left(\frac{\gamma_1(s)}{\gamma_0(s)}\right)\mathrm{d}\left[X_s - \int_u^s \gamma_1(\mathrm{v})\,\mathrm{dv}\right];$$

$$D_{u,t} = \mathsf{E}_u(\widetilde{\lambda}_t^u)^2 = \int_u^t \gamma_1(s)\left(\log\frac{\gamma_1(s)}{\gamma_0(s)}\right)^2\mathrm{d}s.$$

Suppose that $\gamma_i(t) = q_i f(t)$, where the q_i's are positive numbers and $f(t)$ is a function such that

$$\int_0^t f(s)\,\mathrm{d}s = Ct(1+o(1)) \quad \text{as } t\to\infty, \quad C>0.$$

Then, as $t\to\infty$,

$$\int_0^t \gamma_i(s)\,\mathrm{d}s \sim q_i t, \quad \mu_{0,t} \sim C\left(q_1\log\frac{q_1}{q_0} + q_0 - q_1\right)t, \quad D_{0,t} \sim Cq_1\left(\log\frac{q_1}{q_0}\right)^2 t. \tag{7.137}$$

We now show that

$$t^{-1}\widetilde{\lambda}_{\nu+t}^{\nu} \xrightarrow[t\to\infty]{\mathsf{P}^{\pi}-r-\text{completely}} 0. \tag{7.138}$$

To this end, we have to show that

$$\int_0^\infty \mathrm{d}\pi_u \left(\int_0^\infty t^{r-1} \mathsf{P}_u \left\{ |\widetilde{\lambda}_{u+t}^u| > \varepsilon t \right\} \mathrm{d}t \right) < \infty \quad \text{for all } \varepsilon > 0 \text{ and } r \geq 1. \tag{7.139}$$

Let $m > 2$. By the Chebyshev inequality,

$$\mathsf{P}_u \left\{ |\widetilde{\lambda}_{u+t}^u| > \varepsilon t \right\} \leq \frac{\mathsf{E}_u |\widetilde{\lambda}_{u+t}^u|^m}{\varepsilon^m t^m}.$$

Further, let $\Delta\widetilde{\lambda}_{u+t}^u = \widetilde{\lambda}_{u+t}^u - \lim_{s\uparrow t} \widetilde{\lambda}_{u+s}^u$ denote the jump of the process $\{\widetilde{\lambda}_{u+s}^u\}$ at the time $u + t$. Under P_u, the process $\widetilde{\lambda}_{u+t}^u$, $t \geq 0$, is a square integrable martingale with independent increments. By the moment inequalities for martingales, there exist universal constants C_m and c_m such that

$$\mathsf{E}_u |\widetilde{\lambda}_{u+t}^u|^m \leq C_m D_{u,u+t}^{m/2} + c_m \mathsf{E}_u |\Delta\widetilde{\lambda}_{u+t}^u|^m \quad \text{for any } m > 2; \tag{7.140}$$

see Theorem 2.3.5(iii). Using (7.137) and (7.140) along with the fact that

$$\mathsf{E}_u |\Delta\widetilde{\lambda}_{u+t}^u|^m = \left| \log \frac{\gamma_1(u+t)}{\gamma_0(u+t)} \right|^m \mathsf{E}_u |\Delta X_{u+t}|^m \leq |\log(q_1/q_0)|^m,$$

we obtain that, as $t \to \infty$ and with $\widetilde{C}_m = C C_m$,

$$\sup_{u \geq 0} \mathsf{E}_u |\widetilde{\lambda}_{u+t}^u|^m \leq \widetilde{C}_m |\log(q_1/q_0)|^m t^{m/2} (1 + o(1)) + c_m |\log q_1/q_0)|^m.$$

It follows that, for large t and every $m > 2$,

$$\sup_{u \geq 0} \mathsf{P}_u \left\{ |\widetilde{\lambda}_{u+t}^u| > \varepsilon t \right\} \leq \frac{c_m |\log q_1/q_0)|^m}{\varepsilon^m t^m} + \frac{\widetilde{C}_m |\log(q_1/q_0)|^m}{\varepsilon^m t^{m/2}} (1 + o(1)), \tag{7.141}$$

which implies (7.139) and thus (7.138).

Hence, $t^{-1}\lambda_{\nu+t}^{\nu}$ converges r–completely to $I = C(q_1 \log q_1/q_0 + q_0 - q_1)$ under P^{π} for all $r > 0$, so Theorem 7.2.3 applies to show that the Shiryaev procedure asymptotically minimizes all positive moments of the detection delay.

If, for example, $f(t) = A^2 \sin^2 t$, then the results are valid with $C = A^2/2$.

The above consideration can also be applied to the detection of a gradual change in the intensity when $\gamma_0(t) = q_0$ and $\gamma_1(t) = q_1 f(t-u)$, $t \geq u$ (conditioned on $\nu = u$), where $f(t)$ is a monotone nondecreasing function such that $f(0) = q_0/q_1$ and $f(\infty) = 1$.

7.5 Asymptotically Optimal Changepoint Detection Procedures for Composite Hypotheses

Let the observed random variables X_n be l-dimensional vectors belonging to the multiparameter exponential family of distributions with density

$$f_\theta(X_n) = h(X_n) \exp\left\{\theta^\top X_n - b(\theta)\right\},$$

where $\theta = (\theta_1, \dots, \theta_l)^\top$ is an l-dimensional vector and $b(\cdot)$ is the log-moment generating function of the observation X_n. In the normal mode, the observations $X_1, \dots, X_\nu$ are iid with known parameter θ_0. As in (5.69), by changing coordinates, we may set $\theta_0 = 0$ and $b(0) = \nabla b(0) = 0$, so that

$$\frac{f_\theta(X_n)}{f_{\theta_0}(X_n)} = \exp\left\{\theta^\top x - b(\theta)\right\}, \quad \left\{\theta \in \Theta \subset \mathbb{R}^l : b(\theta) = \log \int_\Theta e^{\theta^\top x} \mu(\mathrm{d}x) < \infty \right\}. \tag{7.142}$$

From now on $\nabla b(\theta)$ denotes the gradient vector. After the change occurs, the observations $X_{\nu+1}, X_{\nu+2}, \dots$ are again iid but with a non-zero parameter $\theta \in \Theta \setminus \mathbf{0}$, which is unknown. We use $\mathsf{P}_{k,\theta}$ to denote the probability measure when the change happens at the time $\nu = k$ and the post-change parameter value is θ. Given n observations $\mathbf{X}_1^n$, the LR of the hypothesis $H_{k,\theta}$ that the change occurs at $\nu = k < n$ with the post-change parameter $\theta \in \Theta \setminus \mathbf{0}$ *versus* the no-change hypothesis H_∞ is

$$\Lambda_n^k(\theta) = \prod_{i=k+1}^{n} \mathcal{L}_i(\theta) = \exp\left\{ \sum_{i=k+1}^{n} \left[\theta^\top X_i - b(\theta)\right] \right\},$$

where $\mathcal{L}_i(\theta) = f_\theta(X_i)/f_0(X_i)$ is given by (7.142). Therefore, the LLR is

$$\lambda_n^k(\theta) = \theta^\top S_n^k - (n-k) b(\theta),$$

where $S_n^k = \sum_{i=k+1}^n X_i$, and the K–L information number is

$$I_\theta = \mathsf{E}_{0,\theta}\left[\lambda_1^0(\theta)\right] = \theta^\top \nabla b(\theta) - b(\theta).$$

Assuming the zero-modified geometric prior, the detection statistic of the Shiryaev procedure tuned to the parameter value θ is

$$R_{n,p}(\theta) = \frac{q}{(1-q)p} \prod_{i=1}^{n} \frac{\mathcal{L}_i(\theta)}{1-p} + \sum_{j=1}^{n} \prod_{i=j}^{n} \frac{\mathcal{L}_j(\theta)}{1-p}.$$

Let $W(\theta)$ be a prior distribution on Θ with positive and continuous density $w(\theta)$ (with respect to the Lebesgue measure) and define the mixture

$$\bar{R}_{n,p} = \int_\Theta R_{n,p}(\theta) w(\theta) \, \mathrm{d}\theta$$

and the corresponding stopping time

$$T_{\mathrm{mix}}(A) = \inf\{n \ge 1 : \bar{R}_{n,p} \ge A\}, \quad A > 0. \tag{7.143}$$

We assume that θ is independent of ν.

We now show that as $\alpha \to 0$ the mixture-based detection procedure (7.143) is first-order asymptotically optimal in the class $\mathbf{C}_\alpha$ in the sense that

$$\mathsf{ADD}_\theta^p(T_{\mathrm{mix}}(A)) \sim \inf_{T \in \mathbf{C}_\alpha} \mathsf{ADD}_\theta^p(T) \sim \frac{|\log \alpha|}{I_\theta + |\log(1-p)|} \quad \text{for all } \theta \in \Theta \setminus \mathbf{0}, \tag{7.144}$$

where $\mathsf{ADD}_\theta^p(T) = \mathsf{E}_\theta^p(T - \nu | T > \nu)$ is the average delay to detection for the fixed θ.

Note first that, by the same argument, the implication (7.8) holds for the mixture rule $T_{\mathrm{mix}}(A)$.

For $k = 0, 1, \dots$, consider the power-one mixture tests

$$T_A^{(k)}(w,p) = \inf\left\{ n > k : \int_\Theta \exp\left[\theta^\top S_n^k - (n-k)b(\theta) + (n-k)|\log(1-p)|\right] w(\theta)\, \mathrm{d}\theta \ge A \right\}.$$

Let $\widetilde{\lambda}_n^k(\theta,p) = \lambda_n^k(\theta) + (n-k)|\log(1-p)|$ and define

$$\tau_A(\theta,p) = \inf\left\{ n \ge 1 : \widetilde{\lambda}_n^0(\theta,p) \ge \log A \right\},$$
$$\kappa_A(\theta,p) = \widetilde{\lambda}^0_{\tau_A(\theta,p)}(\theta,p) - \log A \text{ on } \{\tau_A(\theta,p) < \infty\},$$
$$\varkappa(\theta,p) = \lim_{A\to\infty} \mathsf{E}_{0,\theta}[\kappa_A(\theta,p)], \quad \zeta(\theta,p) = \lim_{A\to\infty} \mathsf{E}_{0,\theta}\left[e^{-\kappa_A(\theta,p)}\right].$$

In the following we assume that Θ is a compact set. Applying the same argument as in Subsection 5.5.1.1 that have led to (5.203), as $A \to \infty$ we obtain the asymptotic expansion

$$\begin{aligned}\mathsf{E}_{0,\theta}[T_A^{(0)}(w,p)] = \frac{1}{I_\theta + |\log(1-p)|}\Big\{\log A + \frac{l}{2}\log\log A \\ - \frac{l}{2}[1+\log(2\pi)] + C(\theta,p,w)\Big\} + o(1)\end{aligned} \tag{7.145}$$

that holds uniformly for all $\theta \in \Theta \setminus \mathbf{0}$, where

$$C(\theta,p,w) = \log\left(\frac{e^{\varkappa(\theta,p)}\sqrt{\det[\nabla^2 b(\theta)]/[I_\theta + |\log(1-p)|]^l}}{w(\theta)}\int_{\Theta\setminus\mathbf{0}} \zeta(t,p)w(t)\,\mathrm{d}t\right)$$

and $\nabla^2 b(\theta)$ is the Hessian matrix.

It is easily seen that for all $n > k$,

$$\bar{R}_{n,p} > \int_\Theta \exp\left[\theta^\top S_n^k - (n-k)b(\theta) + (n-k)|\log(1-p)|\right] w(\theta)\,\mathrm{d}\theta,$$

so $T_{\mathrm{mix}}(A) \le T_A^{(k)}(w,p)$ on $\{T_{\mathrm{mix}}(A) > k\}$. Since $T_A^{(k)}(w,p)$ is distributed as $T_A^{(0)}(w,p)$ for any $k = 1,2,\dots$, it follows that $\mathsf{E}_{k,\theta}[T_{\mathrm{mix}}(A) - k]^+ \le \mathsf{E}_{0,\theta}[T_A^{(0)}(w,p)]$. Since the right-hand side does not depend on k, using (7.145) we obtain that, as $A \to \infty$,

$$\mathsf{ADD}_\theta^p(T_{\mathrm{mix}}(A))[1 - \mathsf{PFA}^p(T_{\mathrm{mix}}(A))] \le \frac{1}{I_\theta + |\log(1-p)|}\Big\{\log A + \frac{l}{2}\log\log A\Big\} + O(1).$$

Therefore, setting $A = A_\alpha$ so that $\log A_\alpha \sim |\log\alpha|$ and $\mathsf{PFA}^\pi(A_\alpha) \le \alpha$ yields

$$\mathsf{ADD}_\theta^p(T_{\mathrm{mix}}(A_\alpha)) \le \frac{|\log\alpha|}{I_\theta + |\log(1-p)|}(1+o(1)) \quad \text{as } \alpha \to 0.$$

Since by Theorem 7.1.3,

$$\inf_{T \in \mathbf{C}_\alpha} \mathsf{ADD}_\theta(T) \ge \frac{|\log\alpha|}{I_\theta + |\log(1-p)|}(1+o(1)) \quad \text{as } \alpha \to 0,$$

the asymptotic relation (7.144) follows.

With a certain additional effort it can be proved that, for $\overline{\mathsf{ADD}}^{p,w}(T) = \int_\Theta \mathsf{ADD}^p(T)w(\theta)\,\mathrm{d}\theta$,

$$\overline{\mathsf{ADD}}^{p,w}(T_{\mathrm{mix}}(A)) \sim \inf_{T \in \mathbf{C}_\alpha} \overline{\mathsf{ADD}}^{p,w}(T) \sim |\log\alpha| \int_\Theta \frac{w(\theta)}{I_\theta + |\log(1-p)|}\,\mathrm{d}\theta. \tag{7.146}$$

These results are summarized in the following theorem.

Theorem 7.5.1. *Let the observations belong to the multivariate exponential family and the prior distribution of the changepoint be geometric. Let Θ be a compact set. If the threshold $A = A_\alpha$ in the mixture detection procedure* (7.143) *is selected so that $\mathsf{PFA}^p(T_{\mathrm{mix}}) \le \alpha$ and $\log A_\alpha \sim |\log\alpha|$, in particular $A_\alpha = (1-\alpha)/\alpha p$, then the asymptotic relations* (7.144) *and* (7.146) *hold. That is, the mixture procedure is first-order asymptotically optimal in the class $\mathbf{C}_\alpha$.*

Remark 7.5.1. The statement of Theorem 7.5.1 holds for higher moments of the detection delay. Specifically, under the conditions of the theorem, for all $m \ge 1$ and all $\theta \in \Theta \setminus \mathbf{0}$,

$$\inf_{T \in \mathbf{C}_\alpha} \mathsf{E}_\theta^p[(T-\nu)^m | T > \nu] \sim \mathsf{E}_\theta^p[(T_{\mathrm{mix}}(A)-\nu)^m | T_{\mathrm{mix}} > \nu] \sim \left(\frac{|\log\alpha|}{I_\theta + |\log(1-p)|}\right)^m,$$

and

$$\inf_{T \in \mathbf{C}_\alpha} \int_\Theta \mathsf{E}_\theta^p[(T-\nu)^m | T > \nu] w(\theta)\, \mathrm{d}\theta \sim \int_\Theta \mathsf{E}_\theta^p[(T_{\mathrm{mix}}(A)-\nu)^m | T_{\mathrm{mix}} > \nu] w(\theta)\, \mathrm{d}\theta$$

$$\sim (|\log \alpha|)^m \int_\Theta \frac{w(\theta)}{[I_\theta + |\log(1-p)|]^m}\, \mathrm{d}\theta.$$

While Theorem 7.5.1 establishes nice asymptotic optimality properties of the mixture rule for any prior distribution $W(\theta)$, the implementation of this procedure is problematic since computing the statistic $\bar{R}_{n,p}$ is not feasible in general. The computations become feasible for the conjugate prior, which for the multiparameter exponential family (7.142) is of the form

$$w_0(\theta; n_0, x_0) = c(n_0, x_0) \exp\left\{n_0 x_0^\top \theta - n_0 b(\theta)\right\}, \quad \theta \in \Theta,$$

where

$$c(n_0, x_0) = \left(\int_\Theta \exp\left\{n_0 x_0^\top \theta - n_0 b(\theta)\right\} \mathrm{d}\theta \right)^{-1}.$$

In this case, the posterior density of θ given $\mathbf{X}_1^n$ is [118]

$$w_n(\theta | \mathbf{X}_1^n) = w_0\left(\theta; n_0 + n, (n_0 x_0 + S_n)/(n_0 + n)\right)$$

and

$$\int_\Theta f_\theta(\mathbf{X}) w_0(\theta; m, a)\, \mathrm{d}\theta = \frac{c(m,a)}{c(m+1, (ma + \mathbf{X})/(m+1))}.$$

As a result, assuming $q = 0$, the mixture statistic $\bar{R}_{n,p}$ can be computed as

$$\bar{R}_{n,p} = \sum_{i=1}^n \frac{c(n_0, x_0) w_{1,n}}{(1-p)^{n-i} w_{1,i} w_{i,n}}, \tag{7.147}$$

where

$$w_{i,j} = c\left(n_0 + j - i + 1, \frac{n_0 x_0 + S_j^{i-1}}{n_0 + j - i + 1}\right), \quad i, j = 1, 2, \ldots$$

The computations can be further simplified by using a window-limited version where the summation is performed not over all available observations, i.e., from 1 to n as in (7.147), but in a sliding window with size m, i.e., from $n - m + 1$ to n. In order to make the resulting window-restricted mixture procedure asymptotically optimal, the size of the window $m = m_\alpha$ should be selected so that $|\log \alpha| \ll m_\alpha \ll 1/\alpha$, i.e., asymptotically as $\alpha \to 0$ the ratio $m_\alpha / |\log \alpha|$ should approach infinity, but sufficiently slowly so $\log m_\alpha / |\log \alpha| \to 0$. See Lai and Xing [256] for further details.

As usual, an alternative to mixtures is the GLR method based on maximizing the statistic $R_{n,p}(\theta)$ over $\theta \in \Theta$. It is not clear, however, how to select the threshold in this procedure to guarantee the inequality $\mathsf{PFA}^p(A) \le \alpha$. But with an appropriate threshold this procedure is most certainly also asymptotically optimal, i.e., Theorem 7.5.1 holds. A window-limited version is even more important since an iterative maximization is not possible and the complexity grows tremendously with the sample size n.

We complete this section by noting that the mixture-based SR and GLR-based CUSUM procedures are asymptotically optimal if the parameter of the geometric prior p approaches 0. For practical purposes it suffices to require that the K–L information $I_\theta \gg |\log(1-p)|$, so that the prior knowledge brings very little information compared to the observations.

7.6 A Generalized Bayesian Approach and the Shiryaev–Roberts Procedure

We now turn to the special limiting case associated with the Bayesian setup, which we refer to as the *generalized Bayesian setting*. This problem formulation was briefly considered and motivated in Subsection 6.3.2. Throughout this subsection we consider only the iid case. The goal is to solve the optimization problem (6.16).

Specifically, assume that the changepoint ν is a generalized random variable with a uniform improper prior distribution. The simplest limiting transition from the Bayesian setup is to consider the geometric prior distribution (6.8) assuming $q = 0$ and letting p, the parameter of the geometric prior, go to zero, in which case the geometric prior scaled by p^{-1} turns into an improper uniform distribution. It can be easily seen that the Shiryaev statistic $R_{n,p} \to R_n$ as $p \to 0$, where

$$R_n = \sum_{j=1}^{n} \prod_{i=j}^{n} \mathcal{L}_i, \tag{7.148}$$

with $\mathcal{L}_i = f_1(X_i)/f_0(X_i)$. Note that R_n can also be written recursively as

$$R_n = (1 + R_{n-1})\mathcal{L}_n, \quad n \geq 1, \; R_0 = 0. \tag{7.149}$$

We refer to R_n as the *Shiryaev–Roberts statistic* and to the corresponding change detection procedure $T_{\mathrm{SR}}(A)$ with $A > 0$,

$$T_{\mathrm{SR}}(A) = \inf\{n \geq 1 : R_n \geq A\}, \tag{7.150}$$

as the *Shiryaev–Roberts procedure*, using in the following the abbreviation SR for short.

In Subsection 6.3.2, based on a heuristic argument we conjectured that the optimal generalized Bayes procedure that solves the problem (6.16) is the SR procedure. We now formulate and prove the exact result. For the geometric prior, the measure $\mathsf{P}^\pi = \sum_{k=0}^\infty p(1-p)^k \mathsf{P}_k$ is denoted as P^p. Recall that $\mathsf{IADD}(T)$ stands for the relative *Integral Average Detection Delay* defined as

$$\mathsf{IADD}(T) = \frac{\sum_{k=0}^\infty \mathsf{E}_k(T-k)^+}{\mathsf{ARL2FA}(T)}, \tag{7.151}$$

where $\mathsf{ARL2FA}(T) = \mathsf{E}_\infty T$ is the *Average Run Length to False Alarm.*

For ease of exposition we assume throughout this subsection that the likelihood ratio $\mathcal{L}_1 = f_1(X_1)/f_0(X_1)$ is P_∞-continuous. The case where $\mathcal{L}_1$ is not continuous may require randomization on the stopping boundary A, i.e., deciding whether to continue or stop when $R_n = A$, in order to guarantee the prescribed ARL to false alarm. All the results are valid for this case too, but presenting the fine points clutters the exposition with details that obscure the main ideas.

Theorem 7.6.1. *Let the threshold $A = A_\gamma$ in the SR procedure $T_{\mathrm{SR}}(A)$ be chosen so that $\mathsf{ARL2FA}(T_{\mathrm{SR}}(A_\gamma)) = \gamma$. Then, for any $\gamma > 1$, the SR procedure minimizes the integral average delay to detection $\mathsf{IADD}(T)$ over all procedures with $\mathsf{E}_\infty T \geq \gamma$, i.e.,*

$$\inf_{T \in \mathbf{C}_\gamma} \mathsf{IADD}(T) = \mathsf{IADD}(T_{\mathrm{SR}}(A_\gamma)), \tag{7.152}$$

where $\mathbf{C}_\gamma = \{T : \mathsf{ARL2FA}(T) \geq \gamma\}$.

Proof. Consider the Bayesian problem $\mathcal{B}(p,c) = \mathcal{B}(q=0,p,c)$ formulated in Subsection 7.1.1. Specifically, suppose that ν is a random variable, independent of the observations, with the geometric distribution $\Pr(\nu = k) = p(1-p)^k$, $k \geq 0$, and that the losses associated with stopping at time T are 1 if $T \leq \nu$ and $c(T-\nu)$ if $T > \nu$, where $0 < p < 1$ and $c > 0$ are fixed constants.

Solving $\mathcal{B}(p,c)$ requires the minimization of the average risk

$$\rho_c^p(T) = \mathsf{PFA}^p(T) + c\,\mathsf{E}^p(T-\nu)^+$$

or, equivalently, the maximization of the expected gain $p^{-1}[1 - \rho_c^\pi(T)]$. By Theorem 7.1.2, the Bayes rule for this problem is given by the Shiryaev procedure $T_{\mathrm{s}}(p,c) = \inf\{n : R_{n,p} \geq A(p,c)\}$, where $A(p,c) > 0$ is an appropriate threshold. Observe first that $R_{n,p} \to R_n$ as $p \to 0$.

Now, it follows from [365] that there are a constant $0 < c^* < \infty$ and a sequence $\{p_i, c_i\}_{i \ge 1}$ with $p_i \to 0$, $c_i \to c^*$ as $i \to \infty$, such that $T_{\text{SR}}(A_\gamma)$ is the limit of the Bayes stopping times $T_{\text{s}}(p_i, c_i)$ as $i \to \infty$ and

$$\limsup_{p \to 0, c \to c^*} \frac{1 - \rho_c^p(T_{\text{s}}(p,c))}{1 - \rho_c^p(T_{\text{SR}}(A_\gamma))} = 1. \tag{7.153}$$

Next, for any stopping time T,

$$\begin{aligned}\frac{\mathsf{E}^p(T-\nu)^+}{p} &= \frac{1}{p}\sum_{k=0}^{\infty} p(1-p)^k \mathsf{E}_k(T-k)^+ \\ &= \sum_{k=0}^{\infty}(1-p)^k \mathsf{E}_k(T-k)^+ \xrightarrow[p\to 0]{} \sum_{k=0}^{\infty} \mathsf{E}_k(T-k)^+\end{aligned}$$

and

$$\begin{aligned}\frac{\mathsf{P}^p(T>\nu)}{p} &= \frac{1}{p}\sum_{k=0}^{\infty} p(1-p)^k \mathsf{P}_k(T>k) \\ &= \sum_{k=0}^{\infty}(1-p)^k \mathsf{P}_\infty(T>k) \xrightarrow[p\to 0]{} \sum_{k=0}^{\infty} \mathsf{P}_\infty(T>k) = \mathsf{E}_\infty T,\end{aligned}$$

where we used the fact that $\mathsf{P}_k(T>k) = \mathsf{P}_\infty(T>k)$ since by the definition of the stopping time the event $\{T \le k\}$ belongs to the σ-algebra $\mathscr{F}_k$ and at the time instant k the distribution is still f_0.

Since

$$\frac{1}{p}[1-\rho_c^p(T)] = \frac{\mathsf{P}^p(T>\nu)}{p} - c\frac{\mathsf{E}^p(T-\nu)^+}{p},$$

it follows that, for any stopping time T with $\mathsf{ARL2FA}(T) < \infty$,

$$\frac{1}{p}[1-\rho_c^p(T)] \xrightarrow[p\to 0]{} \mathsf{ARL2FA}(T) - c\sum_{k=0}^{\infty} \mathsf{E}_k(T-k)^+,$$

which together with (7.153) establishes that the SR procedure minimizes $\mathsf{IADD}(T)$ over all stopping times that satisfy $\mathsf{ARL2FA}(T) = \gamma$.

It remains to prove that (7.152) holds in the class $\mathbf{C}_\gamma$, i.e., for stopping times that may have $\mathsf{ARL2FA}(T) > \gamma$.

Let T be such that $\mathsf{E}_\infty T = \gamma_1 > \gamma$. Define a randomized stopping time τ that is equal to T with probability π and 0 with probability $1-\pi$, where $\pi = \gamma/\gamma_1$. Note that $\mathsf{IADD}(\tau) = \mathsf{IADD}(T)$ for every $0 < \pi \le 1$ since $\mathsf{E}_k(\tau-k)^+ = \pi\,\mathsf{E}_k(T-k)^+$ and $\mathsf{E}_\infty\tau = \pi\,\mathsf{E}_\infty T$. Therefore, for any stopping time T such that $\mathsf{ARL2FA}(T) > \gamma$, we can find another stopping time τ with $\mathsf{ARL2FA}(\tau) = \gamma$ and such that $\mathsf{IADD}(\tau) = \mathsf{IADD}(T)$, which means that it is sufficient to optimize over stopping times that satisfy the exact equality $\mathsf{ARL2FA}(T) = \gamma$. Finally, since the optimum over stopping times with $\mathsf{ARL2FA}(T) = \gamma$ is the SR procedure that does not randomize 0, it follows that this procedure is optimal in the class $\mathbf{C}_\gamma$. □

Remark 7.6.1. Theorem 7.6.1 can be alternatively proved using the optimal stopping theory without referring to the Bayes problem and considering the Shiryaev procedure. Specifically, it can be shown that, for any stopping time, $\sum_{k=0}^{\infty} \mathsf{E}_k(T-\nu)^+ = \mathsf{E}_\infty[\sum_{n=0}^{T-1} R_n]$, so that the generalized Bayes problem is reduced to minimizing the risk $\mathsf{E}_\infty[\sum_{n=0}^{T-1} R_n] - c\mathsf{E}_\infty T$ over all Markov times, where R_n is a time-homogeneous Markov sequence. Applying the optimal stopping theory to this latter problem yields the desired result. However, the detailed proof is not trivial, even tedious.

Remark 7.6.2. A similar result is true in continuous time for detecting a shift in the mean of a Brownian motion. Namely, the SR procedure with the threshold $A = \gamma$ is strictly optimal for every $\gamma > 0$ in terms of minimizing $\int_0^\infty \mathsf{E}_t[(T-t)^+]dt/\mathsf{E}_\infty T$ in the class $\mathbf{C}_\gamma$. This can be established by reducing the problem to the optimal stopping of a Markov process. See Feinberg and Shiryaev [139].

Remark 7.6.3. While Theorem 7.6.1 is of interest in its own right, it is extremely useful for proving another even more interesting result related to optimality of the multicyclic SR procedure with respect to the stationary average delay to detection among all multicyclic procedures with given ARL to false alarm, as outlined in Section 8.4.1. See also Subsection 6.3.4 for a preliminary discussion.

The optimality property of the SR procedure requires computing the threshold A_γ to match the ARL to false alarm $\mathsf{E}_\infty[T_{\rm SR}(A_\gamma)] = \gamma$. Unfortunately, due to the overshoot it is usually not feasible to obtain an analytic closed form for A_γ to guarantee the exact equality, and, if the precision is absolutely necessary, the calculation may be done either by Monte Carlo or by solving integral equations numerically using the methodology developed in Subsection 7.1.2.1. Specifically, in the notation of Subsection 7.1.2.1 the ARL to false alarm $\chi_0(r) = \mathsf{E}_\infty[T_{\rm SR}^r(A)]$ of the SR procedure that starts from $R_0 = r$ satisfies the Fredholm integral equation

$$\chi_0(r) = 1 + \int_0^A \chi_0(x)\mathcal{K}_\infty(x,r)\,\mathrm{d}x,$$

which is a special case of (7.16) with $p = 0$ and $\mathcal{K}_\infty(x,r) = \partial/\partial x F_\infty(x/1+r)$, where $F_\infty(y) = \mathsf{P}_\infty(\Lambda_1 \le y)$; see (7.17).

Moreover, still using the notation of Subsection 7.1.2.1, the integral average detection delay is written as $\mathsf{IADD}(T_{\rm SR}^r(A)) = \psi_0(r)/\chi_0(r)$, where $\psi_0(r) = \sum_{k=0}^\infty \delta_k(r)$. The function $\psi_0(r)$ satisfies the integral equation

$$\psi_0(r) = \delta_0(r) + \int_0^A \psi_0(x)\mathcal{K}_\infty(x,r)\,\mathrm{d}x,$$

which is a particular case of (7.15) with $p = 0$. Solving these integral equations numerically allows us to obtain very accurate approximations for the ARL to false alarm and the IADD.

Concluding this subsection we remark that an approximation $\mathsf{E}_\infty[T_{\rm SR}(A)] \approx A/\zeta$ can be obtained by noticing that $R_n - n$ is a P_∞-martingale with zero expectation, so that by the optional sampling theorem $\mathsf{E}_\infty(R_{T_{\rm SR}} - T_{\rm SR}) = 0$. Hence $\mathsf{E}_\infty T_{\rm SR} = \mathsf{E}_\infty R_{T_{\rm SR}}$, and, since $R_{T_{\rm SR}}$ is the first excess over A, renewal theory can be applied to the overshoot $\log R_{T_{\rm SR}} - \log A$ [366]. The constant $0 < \zeta < 1$ depends on the model and can be computed numerically. Therefore, $A = \gamma\zeta$ guarantees $\mathsf{E}_\infty T_{\rm SR} \approx \gamma$. This approximation is asymptotically accurate when $\gamma \to \infty$ and is reasonable already for relatively small values of γ [369]. Using Theorem 7.1.5(ii), we can also obtain the asymptotic approximation for the maximal average delay to detection of the SR procedure, $\mathsf{SADD}(T_{\rm SR}) = \mathsf{E}_0 T_{\rm SR}$. Indeed, letting $p \to 0$ in (7.48) yields

$$\mathsf{E}_0 T_{\rm SR}(A) = I^{-1}\left(\log A - C_0 + \varkappa\right) + o(1), \quad A \to \infty,$$

where $\varkappa = \varkappa(0,I)$ and $C_0 = C(0,I) = \mathsf{E}_0\{\log[1 + \sum_{i=1}^\infty e^{-\lambda_i}]\}$; see (7.44). Clearly, $\mathsf{IADD}(T_{\rm SR}(A)) < \mathsf{E}_0 T_{\rm SR}(A)$. A more detailed study shows that

$$\mathsf{IADD}(T_{\rm SR}(A)) = I^{-1}\left(\log A - C + \varkappa\right) + o(1), \quad A \to \infty,$$

where $C > C_0$. See the approximation (8.368) in Subsection 8.4.3.2 for the details.

7.7 Comparison of the Shiryaev Procedure with Other Procedures in the Bayesian Context

7.7.1 Asymptotic Analysis

In this section, we consider the SR and CUSUM procedures in the Bayesian context for the purpose of comparison with the Shiryaev procedure, which is optimal for the iid models and asymptotically optimal for general non-iid models. We recall that these two popular procedures have optimality properties in non-Bayesian settings, namely minimax and generalized Bayesian, regardless of the knowledge of the prior distribution for the changepoint ν. It is clear that, under the Bayesian setup, both detection procedures exhibit a performance which is inferior to the asymptotically optimal

Shiryaev procedure. Our goal is to examine whether this loss in performance is in fact essential as the PFA $\mathsf{P}^\pi(T < \nu)$ is small.

As we discussed in Section 7.6, the SR statistic $R_n = \sum_{k=1}^n \prod_{i=k}^n \mathcal{L}_i$ is the limiting form of Shiryaev's statistic $R_{n,p}$ when we select $q = 0$ and let $p \to 0$. It can be computed recursively as in (7.149). The corresponding stopping time $T_{\mathrm{SR}}(A)$ is defined in (7.150).

In contrast to the SR statistic which is based on averaging, the CUSUM procedure is motivated by the maximum likelihood argument. Specifically, the CUSUM statistic is defined as [1]

$$V_n = \max_{0 \le k < n} \prod_{j=k+1}^n \mathcal{L}_j$$

and the CUSUM stopping time as

$$T_{\mathrm{CS}}(A) = \inf\{n \ge 1 : V_n \ge A\}, \quad A > 0. \tag{7.154}$$

The statistic V_n can be computed recursively as

$$V_n = \max\{1, V_{n-1}\}\mathcal{L}_n, \quad n \ge 1, \quad U_0 = 1. \tag{7.155}$$

Recall that the two recursions in (7.149) and (7.155) are also applicable in the general non-iid case with $\mathcal{L}_n = f_1(X_n|\mathbf{X}^{n-1})/f_0(X_n|\mathbf{X}^{n-1})$ as long as the likelihood ratio $\mathcal{L}_n$ does not depend on the changepoint ν.

There exists a strong evidence that the SR and CUSUM procedures are asymptotically inferior to the Shiryaev procedure under the Bayesian setup, unless the prior distribution is heavy-tailed, i.e., $d = 0$. Indeed, it follows from the proof of [476, Theorem 6] that, as long as the condition (7.71) is satisfied for some $r \ge 1$, the following asymptotic approximations hold for $m \le r$,

$$\begin{aligned} \mathsf{E}^\pi[(T_{\mathrm{SR}}(A) - \nu)^m | T_{\mathrm{SR}}(A) > \nu] &\sim \mathsf{E}^\pi[(T_{\mathrm{CS}}(A) - \nu)^m | T_{\mathrm{CS}}(A) > \nu] \\ &\sim \left(\frac{\log A}{I}\right)^m \quad \text{as } A \to \infty. \end{aligned} \tag{7.156}$$

In order to compare the asymptotic performance of the SR and CUSUM procedures with the optimal performance given in (7.73), we need good estimates of the $\mathsf{PFA}(A)$ of these procedures or at least reasonable upper bounds that allow for the asymptotics $\log A_\alpha \sim \log(1/\alpha)$ similar to $\mathsf{PFA}^\pi(T_{\mathrm{S}}(A)) \le (1+A)^{-1}$ of the Shiryaev procedure, which is inaccurate only due to neglecting the overshoot. Unfortunately, no such good estimates are available. In particular, the upper bound $\mathsf{PFA}(T_{\mathrm{SR}}) \le O(1)/A$ suggested in [476], which can be easily derived from Doob's submartingale inequality, is not accurate unless d/I is small. As conjectured in [466], the asymptotically accurate approximations are of the form

$$\mathsf{PFA}^\pi(T_{\mathrm{SR}}(A)) \sim \frac{O(1)}{A^{s(d)}}, \quad \mathsf{PFA}^\pi(T_{\mathrm{CS}}(A)) \sim \frac{O(1)}{A^{s(d)}}, \quad A \to \infty, \tag{7.157}$$

with different constants $O(1)$, where $s(d) > 1$ and $s(d) \to 1$ as $d \to 0$. Setting $\mathsf{PFA}^\pi = \alpha$ and using (7.157) and (7.156), we obtain

$$\mathsf{E}^\pi[(T_{\mathrm{SR}} - \nu)^m | T_{\mathrm{SR}} > \nu] \sim \mathsf{E}^\pi[(T_{\mathrm{CS}} - \nu)^m | T_{\mathrm{CS}} > \nu] \sim \left(\frac{|\log \alpha|}{I s(d)}\right)^m \quad \text{as } \alpha \to 0.$$

Comparing with (7.73) shows that the asymptotic relative efficiency of the asymptotically optimal Shiryaev procedure compared with the SR and CUSUM procedures is given by

$$\mathsf{ARE} = \left(\frac{I s(d)}{I + d}\right)^m.$$

[1] Usually the CUSUM is defined as the logarithm of V_n, which explains its name cumulative sum.

Since $I+d$ corresponds to the optimal performance, we certainly have $I+d \geq I s(d)$. The question is whether this inequality is in fact strict. Again we conjecture that this is indeed the case, provided that $d > 0$. When $d = 0$ or tends to 0, then $s(d) \to 1$ and the asymptotic relative efficiency is 1, i.e., the SR and CUSUM procedures are asymptotically optimal. This is supported by numerical computations presented in the next subsection.

7.7.2 *Change Detection in an Exponential Distribution*

Again consider the exponential example of Subsection 7.1.2.3,

$$f_0(x) = e^{-x}\mathbb{1}_{\{x\geq 0\}}, \quad f_1(x) = (1+q)^{-1}e^{-x/(1+q)}\mathbb{1}_{\{x\geq 0\}}, \quad q > 0,$$

and the geometric prior distribution with parameter p.

In Figure 7.3 we depict the operating characteristics of the SR, CUSUM, and Shiryaev procedures in terms of the ADD^π as functions of the PFA^π for $p = 0.1$ and $q = 2$; 0.5. These operating characteristics were computed by solving numerically the integral equations (7.14)–(7.16) presented in Subsection 7.1.2.1. In the exponential case, the K–L information number is equal to $I = q - \log(1+q)$. It is seen from Figure 7.3(a) that for $q = 2$ when $I \gg |\log(1-p)|$ the SR procedure performs almost as well as the Shiryaev procedure, as expected. The CUSUM procedure performs somewhat worse but the difference is not dramatic. On the other hand, Figure 7.3(b) shows that for $q = 0.5$ when the values of I and $|\log(1-p)|$ are comparable, Shiryaev's procedure performs much better, also as expected.

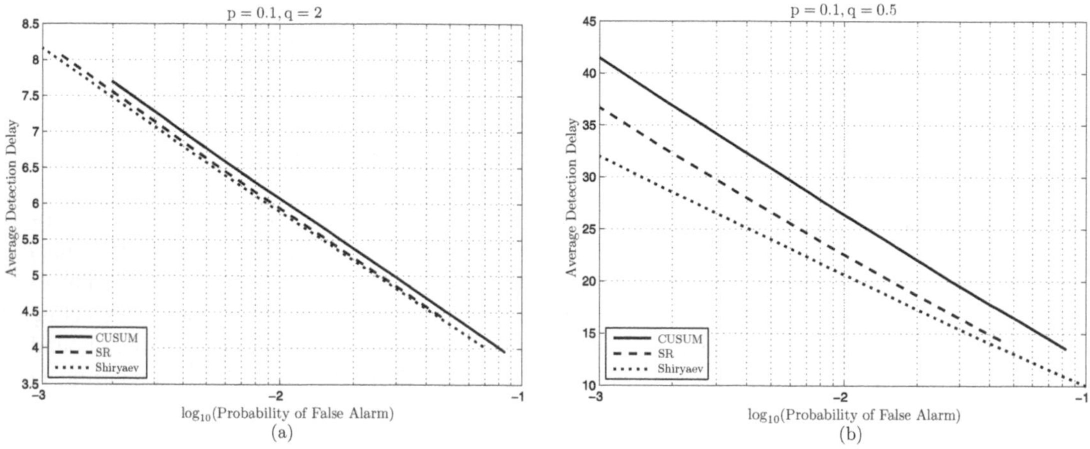

Figure 7.3: *The operating characteristics of the SR, CUSUM, and Shiryaev procedures for* $p = 0.1$ *and (a)* $q = 2$, *(b)* $q = 0.5$.

Figure 7.4 depicts the ADD as a function of the PFA for the SR procedure. We can see that the slope of the ADD depends on p which, according to our conjecture, comes from the exponent $s(p)$ in the approximations (7.157).

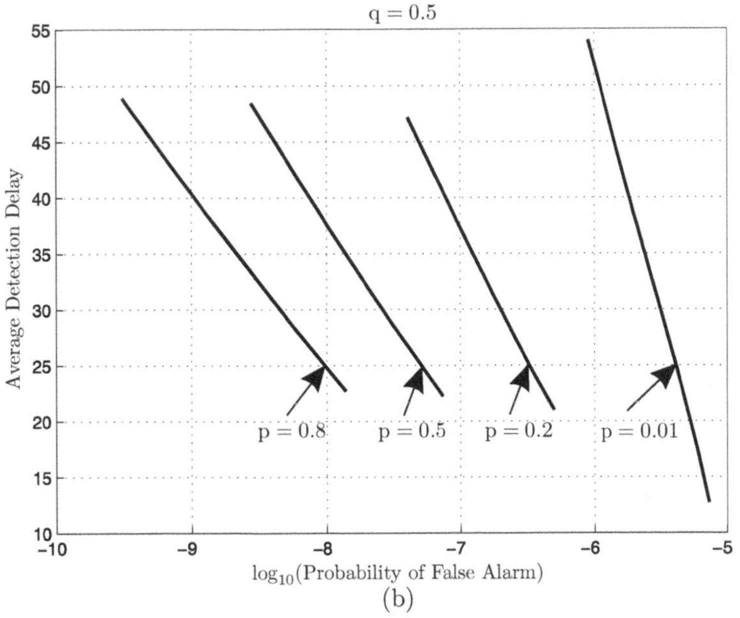

Figure 7.4: *The operating characteristics of the SR procedure for various p and* $q = 0.5$.

Chapter 8

Sequential Change Detection: Non-Bayesian Approaches

This chapter is devoted to sequential non-Bayesian changepoint detection procedures and study of their optimality and asymptotic optimality properties. We first discuss analytic formulas for investigating the properties of a number of on-line detectors, namely Shewhart control charts, geometric moving average charts, finite moving average charts, CUSUM-type algorithms, and the GLR detector. We then discuss optimality properties of the Shiryaev–Roberts-type detection procedures. We also discuss numerical methods for solving integral equations which are useful for estimating operating characteristics of change detection algorithms. Second, we present the results of comparison between different algorithms using analytical methods, numerical techniques and statistical simulations. We also discuss robustness issues of some algorithms with respect to *a priori* information.

Suppose that the observations $\{X_n\}_{n\geq 1}$ are *independent* and such that $X_1,\dots,X_\nu$ are each distributed according to a common distribution $F_0(x)$ (density $f_0(x)$), while $X_{\nu+1},X_{\nu+2},\dots$ each follow a common distribution $F_1(x)\not\equiv F_0(x)$ (density $f_1(x)\not\equiv f_0(x)$),

$$X_n \sim \begin{cases} F_0(x) & \text{if} \quad 1\leq n\leq \nu \\ F_1(x) & \text{if} \quad n\geq \nu+1, \end{cases} \tag{8.1}$$

where the changepoint ν is unknown deterministic (not random).

8.1 Elementary Algorithms

In this section, we describe several simple and well-known change detection algorithms, derive some analytical properties of these algorithms and discuss numerical approximations. Most of the algorithms presented here work on samples of data with *fixed* size window or moving fixed size window; only one uses a growing memory. On the contrary, in the next section, we shall deal basically with a random size sliding window algorithm. In quality control, these elementary algorithms are usually called *Shewhart control charts*, *Geometric Moving Average control charts* or *Exponentially Weighted Moving Average charts*, and *Finite Moving Average control charts*.

We begin with the following parametric case of (8.1):

$$X_n \sim \begin{cases} F_{\theta_0}(x) & \text{if} \quad 1\leq n\leq \nu \\ F_{\theta_1}(x) & \text{if} \quad n\geq \nu+1, \end{cases} \tag{8.2}$$

which allows us to investigate the properties of these elementary algorithms with the aid of the *Average Run Length* (ARL) function, which reasonably characterizes the performance of changepoint detection algorithms. As discussed in Subsection 2.9.3, the power function $\theta \mapsto \beta_d(\theta)$ for hypothesis testing contains the entire information about the statistical properties of the test d. In the sequential change detection, the analog of this function is the ARL function which was introduced in [14]. We have previously discussed the ARL to false alarm or to false isolation in Chapter 6, but the ARL can be also defined as a function of θ and provides us with the average delay to detection.

Definition 8.1.1 (ARL function). Let T be the *time of alarm or a stopping time* of a sequential change detection algorithm, i.e., the time at which the change is detected. We call the *ARL function* the following function of the parameter θ:

$$\mathsf{ARL}(\theta) = \mathsf{E}_\theta(T). \tag{8.3}$$

Sometimes it is useful to make explicit the dependence upon the starting value z of the decision statistic using the notation $\mathsf{ARL}(\theta; z)$.

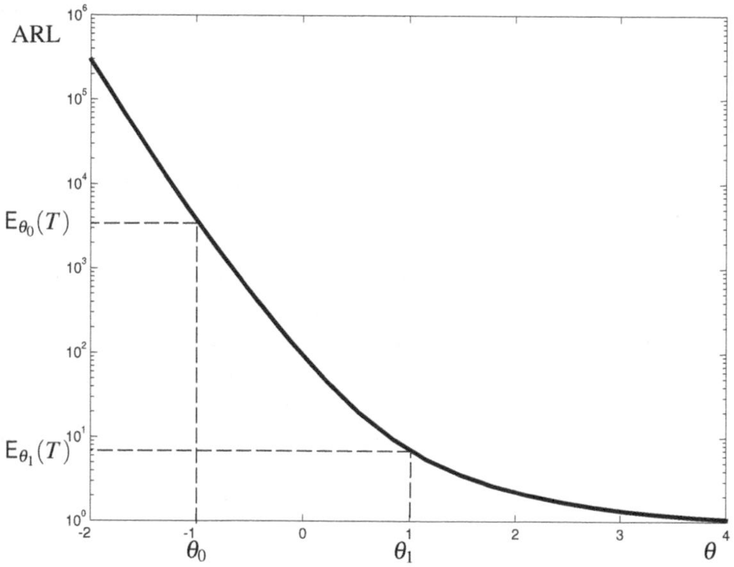

Figure 8.1: *The ARL function, as a function of the parameter θ, provides us with the mean time to false alarm* $\mathsf{ARL2FA}(T)$ *and with the average delay to detection* $\mathsf{E}_\theta T$, *and with other information regarding the robustness of the algorithm.*

The characteristic feature of the ARL function is the assumption that all observations $\{X_n\}_{n\geq 1}$ are distributed according to the law P_θ with the same constant parameter θ. It is obvious from Chapter 6 that the ARL function defines, at θ_0, the ARL2FA, and at θ_1 the conditional average delay to detection under the assumption that $\nu = 0$:

$$\mathsf{E}_\infty(T) = \mathsf{E}_{\theta_0}(T) \text{ and } \mathsf{E}_\nu(T-\nu|T>\nu)|_{\nu=0} = \mathsf{E}_{\theta_1}(T)$$

as depicted in Figure 8.1. Actually the ARL function (like the power function $\beta_d(\theta)$) contains much more information than the properties of the change detection algorithm for these two putative values, and this additional information is very useful in practice, because it is related to the behavior of the change detection algorithm for different parameter values θ before and after the change. The robustness of a particular algorithm can be studied by defining the ARL as a function of other tuning parameters, like the threshold, the standard deviation of noise, *etc.*

8.1.1 Fixed Sample Size Algorithms — Shewhart Control Charts

One of the simplest FSS change detection algorithms is a repeated Neyman–Pearson test (see Theorem 2.9.1) applied to samples with fixed size m. In quality control, this algorithm is also known as the *Shewhart control chart*. Specifically, let $\lambda(K)$ be the LLR corresponding to the K-th such sample. The stopping time of this algorithm is

$$T_{\text{NP}} = m \cdot K^* = m \cdot \inf\{K \geq 1 : d_K = 1\} \tag{8.4}$$

where K^* is the number of the first sample with $d_K = 1$ and the decision rule d_K is defined by

$$d_K = \begin{cases} 0 & \text{if} \quad \lambda(K) < h \\ 1 & \text{if} \quad \lambda(K) \geq h \end{cases}, \quad \lambda(K) = \sum_{i=(K-1)m+1}^{Km} \log \frac{f_{\theta_1}(X_i)}{f_{\theta_0}(X_i)}, \tag{8.5}$$

where h is a given threshold.

Let us compute the ARL function of this chart. Obviously, the number of samples K^* has a geometric distribution $\mathsf{P}_\theta(K^* = k) = (1 - \beta_d(\theta))^{k-1}\beta_d(\theta)$, where the power function $\beta_d(\theta)$ defines the properties of the Neyman–Pearson test not only for putative θ_0 and θ_1 but for any possible value θ. Therefore, the expectation of K^* is $\frac{1}{\beta_d(\theta)}$ and the ARL function of the Shewhart control chart at θ can be written as

$$\mathsf{ARL}(\theta) = \mathsf{E}_\theta(T_{\text{NP}}) = \frac{m}{\beta_d(\theta)}, \tag{8.6}$$

and, in particular,

$$\mathsf{ARL}(\theta_0) = \mathsf{E}_{\theta_0}(T_{\text{NP}}) = \frac{m}{\alpha_0}, \quad \mathsf{ARL}(\theta_1) = \mathsf{E}_{\theta_1}(T_{\text{NP}}) = \frac{m}{1-\alpha_1}. \tag{8.7}$$

Therefore, the ARL function of a Shewhart control chart can be computed directly from the properties of a Neyman–Pearson test which are described in detail in Subsection 2.9.2. Moreover, equation (8.6) shows that change detection algorithms should satisfy the same requirements as hypothesis testing algorithms. Actually, the power function must be close to zero before the change, leading to a large ARL2FA, and close to one after the change, leading to a small delay to detection.

When the hypotheses before and after the change are composite, the formula (8.6) for the ARL function is still valid, and in this case the use of the results of Sections 2.9 and 2.10 is relevant. For instance, let us assume that the density belongs to the one-parameter exponential family $f_\theta(x) = h(x)\exp\{c(\theta)T(x) - b(\theta)\}$ and add some comment about the corresponding two-sided change detection problem, namely the change from θ_0 to $\underline{\theta}_1$ or $\overline{\theta}_1$, $\underline{\theta}_1 < \theta_0 < \overline{\theta}_1$. As discussed in Subsection 2.9.3.2, for an exponential family of distributions, the optimal test in this situation is

$$d_K = \begin{cases} 0 & \text{if} \quad \lambda(K) \in (h_1, h_2) \\ 1 & \text{if} \quad \lambda(K) \notin (h_1, h_2), \end{cases} \tag{8.8}$$

where $\lambda(K) = \sum_{i=(K-1)m+1}^{Km} T(X_i)$. Other possible solutions to this problem are described in Sections 2.9 and 2.10. In all cases, the ARL function is computed as in (8.6) with the aid of the power function of the corresponding statistical test.

Example 8.1.1 (Change in the Gaussian mean). Let us continue the discussion of the FSS Neyman–Pearson test started in Example 2.9.1 of Section 2.9. Specifically, consider the case of a change in the mean of an independent Gaussian sequence $\mathcal{N}(\theta, \sigma^2)$ from θ_0 to $\theta_1 > \theta_0$ with known variance σ^2. By (2.281), the NP test can be rewritten as

$$\overline{X}(K) - \theta_0 \underset{H_0}{\overset{H_1}{\gtrless}} \kappa \frac{\sigma}{\sqrt{m}}, \quad \overline{X}(K) = \frac{1}{m} \sum_{i=(K-1)m+1}^{Km} X_i. \tag{8.9}$$

Hence, the alarm is raised at the first time at which $\overline{X}(K) - \theta_0 \geq \kappa \frac{\sigma}{\sqrt{m}}$, the power function is given by

$$\beta(\theta) = \mathsf{P}_\theta\left(\overline{X}(K) - \theta_0 \geq \kappa \frac{\sigma}{\sqrt{m}}\right) = 1 - \Phi\left(\kappa - \sqrt{m}\frac{\theta - \theta_0}{\sigma}\right),$$

and the ARL function is given by

$$\mathsf{ARL}(\theta) = \frac{m}{1 - \Phi\left(\kappa - \sqrt{m}\frac{\theta - \theta_0}{\sigma}\right)}. \tag{8.10}$$

In particular, the ARL function at θ_0 and θ_1 is

$$\mathsf{ARL}(\theta_0) = \frac{m}{1-\Phi(\kappa)}, \quad \mathsf{ARL}(\theta_1) = \frac{m}{1-\Phi\left(\kappa + \sqrt{m}\frac{\theta_1-\theta_0}{\sigma}\right)}. \tag{8.11}$$

In the two-sided case, namely for a change from θ_0 to either $\theta_1^+ = \theta_0 + \delta_\theta$ or $\theta_1^- = \theta_0 - \delta_\theta$, it follows from (8.8) that the alarm is raised when

$$|\overline{X}(K) - \theta_0| \geq \kappa \frac{\sigma}{\sqrt{m}}, \tag{8.12}$$

and thus the ARL function at θ is

$$\mathsf{ARL}(\theta) = \frac{m}{1-\Phi\left(\kappa - \sqrt{m}\frac{\theta-\theta_0}{\sigma}\right) + \Phi\left(-\kappa - \sqrt{m}\frac{\theta-\theta_0}{\sigma}\right)} \tag{8.13}$$

and, in particular,

$$\mathsf{ARL}(\theta_0) = \frac{m}{2(1-\Phi(\kappa))}. \tag{8.14}$$

Let us finally comment on the choice of the tuning parameters m and κ. The numerical optimization of Shewhart's algorithm with respect to the two criteria was performed in [347] for values of the SNR $\frac{|\theta_1-\theta_0|}{\sigma}$ between 0.2 and 1.8. The first criterion is to minimize the ARL to detection $\mathsf{ARL}(\theta_1)$ for a fixed $\mathsf{ARL2FA} = \mathsf{ARL}(\theta_0)$ with respect to the parameters m and κ. The second criterion is to maximize the $\mathsf{ARL2FA}(\theta_0)$ for a fixed ARL to detection $\mathsf{ARL}(\theta_1)$. In these optimizations, the change time ν is taken to be equal to a multiple of the sample size m.

The ARL function can be used for computing the average delay to detection provided that the change time ν is non-random and equal to a multiple of the sample size m. When this assumption fails to be true, namely when $(K-1)m+1 \leq \nu < Km$ (which is practically the most relevant situation), the computation of the mean delay with the aid of the ARL function induces an error. For computing the worst mean delay in this case, the conditional average delay to detection $\mathsf{CADD}_\nu(T) = \mathsf{E}_\nu(T-\nu|T>\nu)$ has to be computed and maximized with respect to ν.

Consider now two formulations of abrupt change detection introduced in Chapter 6: minimax and multicyclic detection of a disorder in a stationary regime. The optimization of the tuning parameters n and m in (8.4)–(8.5) in the case of simple alternative hypothesis has been discussed in [326, 356, 414, 415, 416, 417].

We first recall the "worst-worst-case" Lorden's criterion (6.17) defined as

$$\mathsf{ESADD}(T) = \sup_{0 \leq \nu < \infty} \left\{ \operatorname{ess\,sup} \mathsf{E}_\nu[(T-\nu)^+ | \mathcal{F}_\nu] \right\}. \tag{8.15}$$

The minimax optimization problem seeks to

$$\text{find } T_{\text{opt}} \in \mathbf{C}_\gamma \text{ such that } \mathsf{ESADD}(T_{\text{opt}}) = \inf_{T \in \mathbf{C}_\gamma} \mathsf{ESADD}(T) \text{ for every } \gamma > 1, \tag{8.16}$$

where $\mathbf{C}_\gamma$ is the class of detection procedures with lower bound γ on the ARL2FA defined in (6.15). Obviously, we now consider the restriction of this general optimization problem to a subclass of repeated Neyman–Pearson tests given by equations (8.4)–(8.5). It follows from [326] that the ESADD and ARL2FA for this FSS change detection algorithm are given by the following expressions

$$\begin{aligned} \mathsf{ESADD}(T_{\text{NP}}) &= \sup_{0 \leq \nu < \infty} \left\{ \operatorname{ess\,sup} \mathsf{E}_\nu[(T_{\text{NP}}-\nu)^+ | \mathcal{F}_\nu] \right\} = \max_{0 \leq \nu \leq m-1} \operatorname{ess\,sup} \mathsf{E}_\nu[(T_{\text{NP}}-\nu)^+ | \mathscr{F}_\nu] \\ &= \max_{1 \leq \nu \leq m-1} \left[(m-\nu) + \frac{m}{1-\alpha_1} \right] = m - 1 + \frac{m}{1-\alpha_1}, \end{aligned} \tag{8.17}$$

and $\mathsf{ARL2FA} = \mathsf{E}_{\theta_0}(T_{\mathrm{NP}}) = m/\alpha_0$.

We now optimize the FSS algorithm in the context of Example 8.1.1. It follows from (8.5) that the LLR is given by

$$\lambda(K) = \sum_{i=(K-1)m+1}^{Km} \log \frac{f_{\theta_1}(X_i)}{f_{\theta_0}(X_i)} = \frac{\theta_1 - \theta_0}{\sigma^2} \left(\sum_{i=(K-1)m+1}^{Km} X_i - m\frac{\theta_1 + \theta_0}{2} \right). \tag{8.18}$$

There are two tuning parameters of this FSS algorithm: the sample size m and the threshold h. To optimize the FSS algorithm it is necessary to find the probabilities α_0 and α_1 as functions of m and h and to minimize the ESADD subject to $\mathsf{ARL2FA} = \gamma$. To solve the optimization problem introduce the change of variables

$$x = \frac{h + m_\rho}{\sqrt{2m_\rho}}, \quad y = \frac{h - m_\rho}{\sqrt{2m_\rho}}, \quad m_\rho = mI_1,$$

where $I_1 = \mathsf{E}_{\theta_1}\left(\log \frac{f_{\theta_1}(X_k)}{f_{\theta_0}(X_k)}\right) = \frac{(\theta_1-\theta_0)^2}{2\sigma^2}$ is the K–L information. It can be shown that

$$\mathsf{ESADD}(T_{\mathrm{NP}}) = \frac{m}{1-\beta} + m - 1 = \frac{(x-y)^2}{2I_{1,0}(1-\Phi(y))} + \frac{(x-y)^2}{2I_1} - 1, \tag{8.19}$$

$$\mathsf{ARL2FA}(T_{\mathrm{NP}}) = \frac{m}{\alpha} = \frac{(x-y)^2}{2I_1(1-\Phi(x))}. \tag{8.20}$$

See [326] for a detailed proof.

The optimization problem seeks to minimize the ESADD of the repeated Neyman–Pearson test

$$\begin{cases} \mathsf{ESADD}(T_{\mathrm{NP}}; x^*, y^*) & = \min\limits_{x,y} \left\{ \dfrac{(x-y)^2}{2I_1(1-\Phi(y))} + \dfrac{(x-y)^2}{2I_1} - 1 \right\} \\ \text{subject to} & \dfrac{(x-y)^2}{2I_1(1-\Phi(x))} = \gamma \end{cases}. \tag{8.21}$$

The asymptotically optimal tuning of the repeated Neyman–Pearson test (8.4)–(8.5) is given by the following theorem.

Theorem 8.1.1. *Consider the case of a change, from θ_0 to $\theta_1 > \theta_0$, in the mean of an independent Gaussian sequence with known variance σ^2. Let T_{NP} be the stopping time of the FSS change detection algorithm* (8.4)–(8.5). *Then*

$$x^* \sim \sqrt{2\log\gamma}, \quad y^* \sim -\sqrt{\log\log\gamma} \quad \text{as} \quad \gamma \to \infty \tag{8.22}$$

and

$$\mathsf{ESADD}(T_{\mathrm{NP}}; x^*, y^*) \sim 2\frac{\log\gamma}{I_1} \quad \text{as} \quad \gamma \to \infty. \tag{8.23}$$

The detailed proof of Theorem 8.1.1 may be found in [326]. It follows from [326] that the optimal values of the probabilities α_0 and α_1 are given by the following asymptotic formulas:

$$\alpha_0^* \sim \frac{1}{\sqrt{2\pi}\ \gamma\sqrt{2\log\gamma}}, \quad \alpha_1^* \sim \frac{1}{\sqrt{2\pi\log\gamma \cdot \log\log\gamma}} \quad \text{as} \quad \gamma \to \infty \tag{8.24}$$

and the asymptotically optimal choice of the tuning parameters m and h is

$$m^* \sim \frac{\log\gamma}{I_1}, \quad h^* \sim \log\gamma \quad \text{as} \quad \gamma \to \infty. \tag{8.25}$$

Remark 8.1.1. Note that the $\mathsf{ESADD}(T_{\rm NP})$ is asymptotically twice as great as the $\mathsf{ARL}(\theta_1)$ for the FSS change detection algorithm (8.4)–(8.5). Indeed, it follows from equations (8.24) and (8.25) that

$$\mathsf{ARL}(\theta_1) = \mathsf{E}_{\theta_1}(T_{\rm NP}) = \frac{m}{1-\alpha_1} = \frac{\log\gamma}{I_1}(1+o(1)) \quad \text{as } \gamma\to\infty.$$

It is shown in Subsection 8.2.5 below that the asymptotic lower bound for ESADD in the class $\mathbf{C}_\gamma$ is given by $\frac{\log\gamma}{I_1}(1+o(1))$.

Another, slightly different measure of the "worst-worst-case" ADD is used in [356]. A characteristic feature of the statistical model proposed in [356] is that the observations are assumed to be L dependent. Hence, the autocorrelation function is defined only for the first L lags

$$R(\tau) = \frac{1}{\sigma^2}\mathsf{E}\left[(X_k-\theta)(X_{k+\tau}-\theta)\right], \quad -L\le\tau\le L. \tag{8.26}$$

The following asymptotic approach has been considered in [356]. Let the SNR be small, $\frac{|\theta_1-\theta_0|}{\sigma\sqrt{d}}\to 0$, where $d=\sum_{\tau=-L}^{L}R(\tau)$, and simultaneously the ARL to false alarm becomes large $\gamma\to\infty$ in such a way that the quantity $\frac{|\theta_1-\theta_0|}{\sigma\sqrt{d}}\sqrt{\gamma}$ is fixed. A numerical procedure to find the optimal tuning parameters $(m,\, h,\, \alpha_0,\, \alpha_1)$ of the repeated Neyman–Pearson test in order to

$$\text{minimize the ratio } \frac{\mathsf{SADD}(T_{\rm NP})}{\gamma} \text{ provided that } \frac{|\theta_1-\theta_0|}{\sigma\sqrt{d}}\sqrt{\gamma}\ge a>0, \tag{8.27}$$

where $\mathsf{SADD}(T)=\sup_{0\le\nu<\infty}\mathsf{E}_\nu(T-\nu|T>\nu)$, has been proposed in [356]. By using the above-mentioned asymptotic approach, it is shown that the optimal sample size satisfies $m\gg L$. For this reason, the contribution of L observations at the beginning and the end of the K-th sample to the total sum $\lambda(K)=\sum_{i=(K-1)m+1}^{Km}\log\frac{f_{\theta_1}(X_i)}{f_{\theta_0}(X_i)}$ is negligible. Hence, the sequence of sums $\{\lambda(K)\}_{K\ge1}$ can be assumed to be independent.

The case of multicyclic detection of a disorder in a stationary regime has been discussed in the pioneering papers [414, 415, 416, 417]. Recall Shiryaev's *stationary ADD* criterion (6.21) defined as the ADD corresponding to a certain stationary regime before the true change occurs. Let $\mathcal{T}_{(N_\nu)}$ be the time of detection of the change that occurs at the time instant ν after $N_\nu-1$ false alarms have been raised. The STADD is

$$\mathsf{STADD}(T) = \lim_{\nu\to\infty}\mathsf{E}_\nu[\mathcal{T}_{(N_\nu)}-\nu]. \tag{8.28}$$

The multicyclic optimization problem seeks to

$$\text{find } T_{\rm opt}\in\mathbf{C}_\gamma \text{ such that } \mathsf{STADD}(T_{\rm opt}) = \inf_{T\in\mathbf{C}_\gamma}\mathsf{STADD}(T) \text{ for every } \gamma>1 \tag{8.29}$$

among all multicyclic procedures. As previously, we consider now the restriction of this general optimization of the multicyclic procedures to a subclass of repeated Neyman–Pearson tests given by equation (8.4). It follows from [414, 415, 416, 417] that the optimum STADD in this subclass is given by

$$\mathsf{STADD}(T_{\rm NP}) \sim \frac{3}{2}\frac{\log\gamma}{I} \quad \text{as } \gamma\to\infty. \tag{8.30}$$

Again, in the case of L dependent observations, under the assumption that $\frac{|\theta_1-\theta_0|}{\sigma\sqrt{d}}\to 0$ and $\gamma\to\infty$ in such a way that the quantity $\frac{|\theta_1-\theta_0|}{\sigma\sqrt{d}}\sqrt{\gamma}$ is fixed, a numerical method which seeks to

$$\text{minimize } \mathsf{STADD}(T_{\rm NP}) \text{ provided that } \frac{|\theta_1-\theta_0|}{\sigma\sqrt{d}}\sqrt{\gamma}\ge a>0 \tag{8.31}$$

is proposed in [356].

Finally, we consider the FSS change detection test in the case of a composite post-change alternative for the Gaussian linear model. Let $\{\mathbf{X}_n\}_{n\geq 1}$ be the independent sequence of Gaussian vectors

$$\mathbf{X}_n \sim \begin{cases} \mathcal{N}(\theta_0, I_p) & \text{if} \quad 1 \leq n \leq \nu \\ \mathcal{N}(\theta_1, I_p) & \text{if} \quad n \geq \nu+1, \end{cases} \tag{8.32}$$

where $\mathbf{X}_n \in \mathbb{R}^p$, $\theta_0 \in \mathbb{R}^p$ is the known pre-change mean vector and the post-change mean vector θ_1 belongs to the spherical surface

$$\theta_1 \in S_c = \{\theta : \|\theta - \theta_0\|_2 = c\}. \tag{8.33}$$

As it follows from Section 2.10, more general Gaussian linear models with nuisance parameters can be reduced to the above simple case, due to the invariance properties of the Gaussian linear model. The UBCP Wald's test (2.323) maximizes the constant power over the family of concentric spherical surfaces. We call this algorithm χ^2-FSS test (or Shewhart χ^2-chart) because it is based upon a quadratic form of the observations. The stopping time of the χ^2-FSS change detection test is given by (8.4) where d_K is defined as

$$T_{\text{FSS}} = m \cdot \inf\{K \geq 1 : d_K = 1\}, \; d_K = \begin{cases} 0 & \text{if} \quad \left\|\overline{\mathbf{X}}_K\right\|_2 < mh \\ 1 & \text{if} \quad \left\|\overline{\mathbf{X}}_K\right\|_2 \geq mh \end{cases}, \; \overline{\mathbf{X}}_K = \sum_{i=(K-1)m+1}^{Km} (\mathbf{X}_i - \theta_0). \tag{8.34}$$

It follows from [327] that the ESADD and ARL2FA for this χ^2-FSS change detection algorithm are given by the following expressions

$$\mathsf{ESADD}(T_{\text{FSS}}) = \max\left\{\frac{m}{1-\beta}, \max_{1\leq\nu\leq m-1}\left[m - \nu + \frac{m}{1-\beta}\mathsf{P}\left(\chi^2_{p,0} < \frac{m^2h^2}{m-\nu}\right)\right]\right\}, \tag{8.35}$$

where $\beta = \mathsf{P}\left(\chi^2_{p,mc^2} < mh^2\right)$ and $\chi^2_{p,\lambda}$ is distributed according to a non-central χ^2 law with p degrees of freedom and noncentrality parameter λ, and

$$\mathsf{ARL2FA}(T_{\text{FSS}}) = \mathsf{E}_\infty(T_{\text{FSS}}) = m\left[1 - \mathsf{P}\left(\chi^2_{p,0} < mh^2\right)\right]^{-1}. \tag{8.36}$$

The optimization problem seeks to minimize the ESADD of the χ^2-FSS change detection algorithm

$$\begin{cases} \mathsf{ESADD}(T_{\text{FSS}}; m^*, h^*) & = \min\limits_{m,h}\max\left\{\dfrac{m}{1-\beta}, \max\limits_{1\leq\nu\leq m-1}\left[m-\nu+\dfrac{m\mathsf{P}\left(\chi^2_{p,0} < \frac{m^2h^2}{m-\nu}\right)}{1-\beta}\right]\right\} \\ \text{subject to} & m\left[1-\mathsf{P}\left(\chi^2_{p,0} < mh^2\right)\right]^{-1} = \gamma \end{cases}. \tag{8.37}$$

The asymptotically optimal tuning of the χ^2-FSS change detection algorithm is given by the following theorem [327].

Theorem 8.1.2. *Let $\{\mathbf{X}_n\}_{n\geq 1}$ be the independent sequence of Gaussian vectors* (8.32). *Consider the stopping time T_{FSS} of the FSS change detection algorithm* (8.4) *and* (8.34). *The optimal values of the parameters m^* and h^* are given by the following asymptotic formulas*

$$\begin{cases} m^* & \sim \; 2\log\gamma/c^2 \\ h^* & \sim \; c\left\{1 + 2\sqrt{\left[\log\left(\sqrt{2\log\gamma}/2\right) - \log 2\sqrt{2\pi}\right]/\log\gamma}\right\}^{-1/2} \end{cases} \quad \text{as } \gamma \to \infty \tag{8.38}$$

and

$$\mathsf{ESADD}(T_{\text{FSS}}) \leq \frac{4\log\gamma}{c^2}(1 + o(1)) \quad \text{as } \gamma \to \infty. \tag{8.39}$$

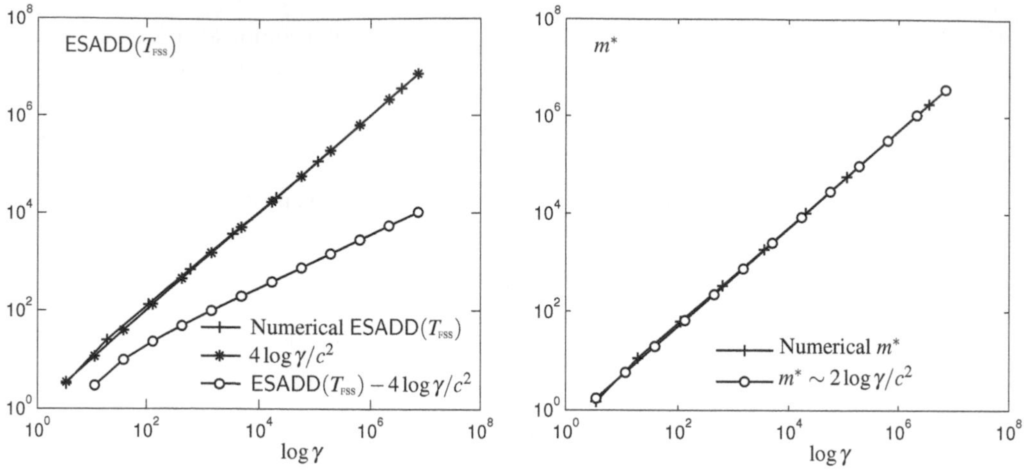

Figure 8.2: *The numerical* $\mathsf{ESADD}(T_{FSS})$*, the asymptotic upper bound* $4\log\gamma/c^2$ *for ESADD and the difference* $\mathsf{ESADD}(T_{FSS}) - 4\log\gamma/c^2$ *as functions of* $\log\gamma$ *(left). The numerical and asymptotic parameters* m^* *as functions of* $\log\gamma$ *(right).*

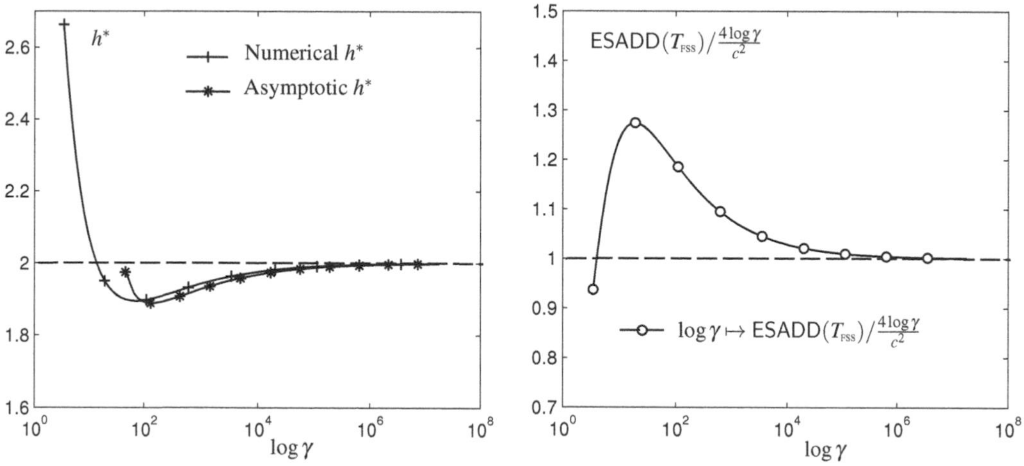

Figure 8.3: *The numerical and asymptotic parameters* h^* *as functions of* $\log\gamma$ *(left). The ratio* $\mathsf{ESADD}(T_{FSS})/(4\log\gamma/c^2)$ *as a function of* $\log\gamma$ *(right).*

The detailed proof of Theorem 8.1.2 may be found in [327].

Let us compare the *asymptotic* upper bound for $\mathsf{ESADD}(T_{\text{FSS}}) \leq (4\log\gamma/c^2)(1+o(1))$ with the ESADD computed by the *nonasymptotic* numerical optimization for the χ^2-FSS when $p=5$ and $c=2$. The nonasymptotic solution has been obtained by numerical constrained minimization of the objective function $\mathsf{ESADD}(T_{\text{FSS}};m^*,h^*)$ (8.37) for a given γ. The numerical and asymptotic parameters $\mathsf{ESADD}(T_{\text{FSS}})$ and m^* as functions of $\log\gamma$ are presented in Figure 8.2. The numerical and asymptotic parameters h^* and the ratio $\mathsf{ESADD}(T_{\text{FSS}})/(4\log\gamma/c^2)$ as functions of $\log\gamma$ are presented in Figure 8.3. It is worth noting a very slow convergence of the numerically computed functions ESADD, m^* and h^* to the asymptotic ones as $\gamma\to\infty$. The speed of the convergence is defined by the term $1/\sqrt{2\log\gamma}$. The explanation of this fact may be found in the proof of [327][Theorem 8.1.2]. It follows from Theorem 8.1.2 that $\mathsf{ESADD}(T_{\text{FSS}})/(4\log\gamma/c^2)-1 \leq o(1)$ as $\gamma\to\infty$. Figure 8.2

shows that the difference $\mathsf{ESADD}(T_{\text{FSS}}) - 4\log\gamma/c^2$ (dash-dot line) is an increasing function when $\gamma \to \infty$ but Figure 8.3 shows that the ratio $\mathsf{ESADD}(T_{\text{FSS}})/(4\log\gamma/c^2)$ tends to one as $\gamma \to \infty$ which is consistent with Theorem 8.1.2.

8.1.2 Exponentially Weighted Moving Average Control Charts

Traditionally the decision statistic of an *Exponentially Weighted Moving Average* (EWMA) control chart, also known as the *Geometric Moving Average* (GMA) control chart is defined with respect to observations, but we define the EWMA in slightly differently with respect to the LLR by the following recursion

$$g_n = (1-\alpha)g_{n-1} + \alpha Z_n, \quad \text{with } g_0 = 0, \tag{8.40}$$

where $0 < \alpha \leq 1$ is the tuning parameter and $Z_n = \log\frac{f_{\theta_1}(X_n)}{f_{\theta_0}(X_n)}$. Both definitions coincide in the case of changes in the Gaussian mean. The stopping rule is given by

$$T_{\text{EWMA}} = \inf\{n \geq 1 : g_n \geq h\}, \tag{8.41}$$

where h is a threshold. To understand the idea of the EWMA chart, let us rewrite (8.40) as

$$g_n = \alpha Z_n + (1-\alpha)\alpha Z_{n-1} + \cdots + (1-\alpha)^{n-1}\alpha Z_1.$$

It is easy to see that g_k represents the weighted LLR with exponentially decreasing weights over time. Hence, the EWMA approach consists in the adaptation of the LLR or another convenient statistic by exponential smoothing. As in the previous subsection consider the case of a two-sided change detection problem. It is again assumed that the density belongs to the exponential family $f_\theta(x) = h(x)\exp\{c(\theta)S(x) - b(\theta)\}$. As it follows from Subsection 2.9.3.2, for an exponential family of distributions, the optimal FSS unbiased test is given by (8.8). It is based on the statistic $Z_k = S(X_k)$. The two-sided EWMA stopping rule is

$$T_{\text{EWMA}} = \inf\{n \geq 1 \ : g_n \notin (h_1, h_2)\}. \tag{8.42}$$

There exist different methods for computing the average delay to detection and the ARL2FA, and more generally, for estimating the ARL function of the EWMA algorithm.

A simple and efficient numerical method for computing the ARL function is suggested in [105], where the derivation of the formula for the ARL function of EWMA (8.42) is very close to the computation of the expected sample size $\mathsf{ESS}_\theta(z)$ (3.67) of the SPRT with symmetric thresholds (absorbing boundaries) $(-h, h)$ and starting point $g_0 = z$, as discussed in Chapter 3. Let $\mathsf{ARL}_\theta(z)$ denote the ARL function when the parameter is θ and the initial value of the decision statistic g_0 is z, $g_0 = z$. Let $\mathsf{P}(\Omega_1)$ be the probability of the event $\Omega_1 = \{|g_1| \geq h\}$. The density of $g_1 = (1-\alpha)z + \alpha Z_1$ can be written as

$$\frac{d}{dx}\mathsf{P}(g_1 \leq x) = f_\theta\left(\frac{x-(1-\alpha)z}{\alpha}\right) \cdot \left|\frac{\partial}{\partial x}\left(\frac{x-(1-\alpha)z}{\alpha}\right)\right| = \frac{1}{\alpha} \cdot f_\theta\left(\frac{x-(1-\alpha)z}{\alpha}\right),$$

where $f_\theta(x)$ is the density of the iid increment Z_k of the decision function g_k. If the first observation Z_1 is such that the event Ω_1 occurs, the run length is equal to 1. If $|g_1| < h$, the run of the EWMA continues with new starting point $g_1 = (1-\alpha)z + \alpha Z_1$ and $\mathsf{ARL}_\theta(g_1)$. The $\mathsf{ARL}_\theta(z)$ function of the EWMA (8.42) is thus equal to the weighted sum of these two run lengths:

$$\begin{aligned} \mathsf{ARL}_\theta(z) &= \mathsf{P}(\Omega_1) + [1-\mathsf{P}(\Omega_1)]\left(1 + \frac{1}{[1-\mathsf{P}(\Omega_1)]\alpha}\int_{-h}^{h} \mathsf{ARL}_\theta(y)\, f_\theta\left(\frac{y-(1-\alpha)z)}{\alpha}\right) dy\right) \\ &= 1 + \frac{1}{\alpha}\int_{-h}^{h} \mathsf{ARL}_\theta(y)\, f_\theta\left(\frac{y-(1-\alpha)z}{\alpha}\right) dy. \end{aligned} \tag{8.43}$$

This integral equation for $\mathsf{ARL}_\theta(z)$ is a Fredholm integral equation. The numerical solution of this equation is discussed in Subsection 3.1.4.

Another numerical method for the approximation of the ARL function is described in [395]. The observations $(X_n)_{n\geq 1}$ are assumed to form an independent Gaussian sequence $\mathcal{N}(\mu,\sigma^2)$, and the EWMA statistic can be rewritten as

$$\widetilde{g}_n = (1-\alpha)\widetilde{g}_{n-1} + \alpha(X_n - \mu_0), \quad \text{with } \widetilde{g}_0 = z, \tag{8.44}$$

where μ_0 is the pre-change mean. If the post-change mean is $\mu_1 = \mu_0 + \delta_\mu$, the stopping rule is

$$T_{\text{EWMA}} = \inf\{n \geq 1 : \widetilde{g}_n \geq h\}. \tag{8.45}$$

If the post-change mean is either $\mu_1^+ = \mu_0 + \delta_\mu$ or $\mu_1^- = \mu_0 - \delta_\mu$, then the two-sided EWMA procedure

$$T_{\text{EWMA}} = \inf\{n \geq 1 : |\widetilde{g}_n| \geq h\} \tag{8.46}$$

should be applied. The computation of the ARL function is based upon the following idea. Define the probability $p_k = \mathsf{P}_\theta(T_{\text{EWMA}} > k | \widetilde{g}_0 = z)$, where $\theta = (\mu, \sigma^2, \alpha, h)$. By definition, the ARL function is given by $\mathsf{ARL}_\theta(z) = \sum_{k=0}^{\infty} p_k$. On the other hand, we have

$$p_k = p_{k-1}\, \mathsf{P}_\theta(\widetilde{g}_k \in (-h,h) | \widetilde{g}_{k-1} \in (-h,h)). \tag{8.47}$$

Let us define the limit probability $p = \lim_{k\to\infty} \mathsf{P}_\theta(\widetilde{g}_k \in (-h,h)|\widetilde{g}_{k-1} \in (-h,h))$ which can be approximated as follows

$$p \simeq \widetilde{p} = \mathsf{P}_\theta(\widetilde{g}_K \in (-h,h)|\widetilde{g}_{K-1} \in (-h,h)), \tag{8.48}$$

where K is the number of iterations empirically chosen as a function of α, h and $\mu_1 - \mu_0$. Then the following approximation holds

$$\mathsf{ARL}_\theta(z) \approx \sum_{k=0}^{K-1} p_k + p_{K-1} \cdot \frac{\widetilde{p}}{1-\widetilde{p}}. \tag{8.49}$$

The characteristic feature of this approach is the approximation of the probability p_k by an Edgeworth series [395]. In the problem of the multicyclic detection of a disorder in a stationary regime considered by Shiryaev in [415], it is assumed that the change occurs after an infinitely long waiting period. The stationary ADD (6.21) is used as a measure of the speed of detection in this case. An heuristic method to calculate the STADD is used in [395]. The $\mathsf{STADD}(T_{\text{EWMA}})$ is computed by weighting all the possible values of the random variable g_ν. To define the weights to calculate $\mathsf{STADD}(T_{\text{EWMA}})$ it is assumed that the random variable g_ν reaches its steady pre-change distribution. Tabulations of the ARL function for different values of the magnitude of change $\mu_1 - \mu_0$, the threshold h and the coefficient α for the one-sided and two-sided EWMA may be found in [395].

Analytical approaches to the investigation of the EWMA properties are suggested by Fishman [149] and Novikov [337]. It follows from [149, 337] that when detecting a change in the drift of the Brownian motion and in the mean of the iid Gaussian sequence the ARL to detection $\mathsf{E}_0 T$ of the EWMA procedure is asymptotically 23% worse than that of the CUSUM or SR procedures. Srivastava and Wu [440] also considered the problem of detecting a change in the drift of the Brownian motion, but analyzed a slightly different EWMA procedure. Their results show that asymptotically this modified procedure has competitive performance compared to the CUSUM and SR procedures with respect to the STADD.

Polunchenko *et al.* [372] recently showed numerically that the optimized EWMA with respect to both the headstart and the smoothing factor has almost indistinguishable STADD and SADD compared to the optimal or nearly optimal change detection procedures in the case of detecting a change in the parameter of the exponential distribution.

8.1.3 Finite Moving Average Charts

An idea similar to the previous control charts consists in replacing the exponential smoothing operation by a finite memory one, and thus in using a finite set of weights, which are no longer assumed to form a geometric sequence. This detector, called the finite moving average (FMA) algorithm, is based on the statistic

$$g_k = \sum_{i=0}^{n-1} c_i Z_{k-i}, \tag{8.50}$$

where c_i is chosen so that $c_0 > 0, \ldots, c_{n-1} > 0$ and $c_k = 0$ when $k \geq n$. The stopping rule is

$$T_{\text{FMA}} = \inf\{k \geq n \,:\, g_k \geq h\}, \tag{8.51}$$

where h is a threshold that controls the FAR. The estimation of the ARL function of the FMA algorithm is addressed in [75, 244]. Assume an iid sequence of LLRs $(Z_k)_{k\geq 1}$ with density $p_\mu(x)$ having expectation μ and finite second moment, $\mathsf{var}(Z) = \sigma^2 < \infty$. Let us slightly modify the formula (8.50) and construct the statistic

$$g_k = \sum_{i=1}^{k} c_{k-i}(Z_i - \mu_0), \quad k \geq n. \tag{8.52}$$

Under the assumption that the sequence $(Z_k)_{k\geq 1}$ is iid, the new random sequence $(g_k)_{k\geq n}$ is stationary also with mean

$$\mathsf{E}_\mu(g_k) = \sum_{i=0}^{n-1} c_i(\mu - \mu_0) \tag{8.53}$$

and covariance

$$R_\ell^g = \begin{cases} \sigma^2 \sum_{i=0}^{n-\ell-1} c_i c_{i+\ell} & \text{if} \quad \ell = 0, \ldots, n-1 \\ 0 & \text{if} \quad \ell \geq n \end{cases}. \tag{8.54}$$

Since Z_k, $k = 1, 2, \ldots$ are iid random variables and the function $g(Z_k)$ is nondecreasing (due to the fact that $c_0 > 0, \ldots, c_{n-1} > 0$), the random variables $(g_k)_{k\geq n}$ possess a property useful for proving a key inequality when computing bounds for the ARL function $\mathsf{ARL}(\mu) = \mathsf{E}_\mu(T_{\text{FMA}})$ of the FMA algorithm. The detailed analysis of such random variables may be found in [75, 134]. Let p_ℓ be the probability

$$\begin{aligned} p_0(h) &= 1 \\ p_\ell(h) &= \mathsf{P}_\mu(g_n < h, \ldots, g_{n+\ell} < h) = \mathsf{P}_\mu(T_{\text{FMA}} > \ell + n) \text{ when } \ell > 0. \end{aligned} \tag{8.55}$$

The ARL is obviously

$$\mathsf{ARL}(\mu) = n + \sum_{i=1}^{\infty} p_i(h). \tag{8.56}$$

Unfortunately, no analytical expression of the ARL function is available. For this reason, *upper and lower bounds* for the ARL are given in [75, 244].

Let q_ℓ be the probability that the decision function g_k exceeds the threshold h for the first time at instant $k = \ell + n$,

$$q_\ell = \mathsf{P}_\mu\left(g_n < h, \ldots, g_{n+\ell-1} < h, g_{n+\ell} \geq h\right) \text{ if } \ell > 0$$

Obviously,

$$q_\ell(h) = p_{\ell-1}(h) - p_\ell(h)$$

and $q_\ell(h)$ is a nonincreasing function of ℓ for any h. These probabilities have the following property, which is proved in [244] for the Gaussian case, and then generalized in [75] for any probability distribution with finite second moment.

Lemma 8.1.1. *Assume that the random sequence* $(Z_k)_{k\geq 1}$ *is iid and let* $(g_k)_{k\geq n}$ *be defined as in* (8.52). *Then the following inequality holds true for* $k \geq n$

$$q_k(h) \geq p_{k-n}(h) q_{k-1}(h). \tag{8.57}$$

This lemma and other results in [244] lead to the following lower and upper bounds for the ARL function

$$1 + \frac{q_n(h)}{p_n(h)} \leq \mathsf{ARL}(\mu) \leq n + \frac{q_n(h)}{p_n(h)}. \tag{8.58}$$

8.2 The CUSUM Algorithm

We now introduce the *Cumulative Sum* (CUSUM) algorithm, which was first proposed by Page [346]. First, we consider the iid model (6.4) and describe several different derivations in this case. The first derivation is more intuition-based and uses ideas related to a simple accumulation of samples with an *adaptive threshold*. The second derivation is based on a more formal on-line statistical approach similar to that used before for introducing control charts and based upon a *repeated implementation of the SPRT*. The third derivation is based upon the concept of open-ended tests, and the fourth method is based on the pure maximum likelihood (GLR) approach. We then extend the CUSUM algorithm for the general non-iid model given in (6.2).

8.2.1 Intuitive Derivation

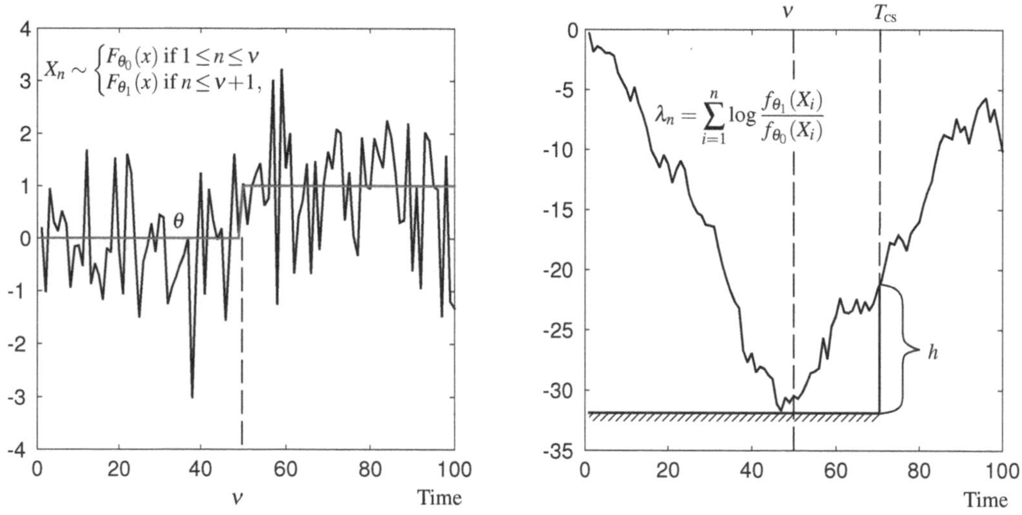

Figure 8.4: *A change in the mean (left) and the typical behavior of the LLR statistic* g_n *before and after the changepoint (right).*

As we already mentioned in the previous section, the typical behavior of the LLR $\lambda_n = \sum_{i=1}^{n} Z_i$ ($Z_i = \log[f_1(X_i)/f_0(X_i)]$) shows a negative drift before change and a positive drift after change, as depicted in Figure 8.4(right). Indeed,

$$\mathsf{E}_\nu Z_n = \begin{cases} -I_0 & < & 0 & \text{if} & 1 \leq n \leq \nu \\ I & > & 0 & \text{if} & n \geq \nu+1, \end{cases} \tag{8.59}$$

where I_0 and I are the K–L information numbers. Therefore, the relevant information, as far as the change is concerned, is in the difference between the value of the LLR $\lambda_n = \sum_{i=1}^n Z_i$ (the *cumulative sum*) and its current (running) minimum value; and the corresponding decision rule is then, at each time instant n, to compare this difference with a threshold:

$$g_n = \lambda_n - \min_{0\le j\le n} \lambda_j = \sum_{i=1}^n Z_i - \min_{0\le j\le n} \sum_{i=1}^j Z_i = \max_{1\le j\le n+1} \sum_{i=j}^n Z_i = \max\left\{0, \max_{1\le j\le n} \sum_{i=j}^n Z_i\right\} \gtrless h, \quad (8.60)$$

where $\sum_{i=n+1}^n Z_i = 0$. The CUSUM stopping time is

$$T_{\text{CS}} = \inf\{n \ge 1 \, : g_n \ge h\}, \quad (8.61)$$

which can be obviously rewritten as

$$T_{\text{CS}} = \inf\{n \ge 1 : \lambda_n \ge m_n + h\}, \quad m_n = \min_{0\le j\le n} \lambda_j. \quad (8.62)$$

Now it becomes clear that this detection rule is nothing but a comparison of the cumulative sum λ_n with an *adaptive threshold* $m_n + h$. This threshold is not only modified on-line, but also keeps *complete* memory of the entire useful information contained in the past observations. Sometimes it is more convenient to define the CUSUM procedure via the LR,

$$T_{\text{CS}} = \inf\{n \ge 1 \, : \exp(g_n) \ge \exp(h)\}, \quad \exp(g_n) = \max\left\{1, \max_{1\le j\le n} \prod_{i=j}^n \frac{f_1(X_i)}{f_0(X_i)}\right\}. \quad (8.63)$$

Clearly, this last definition is equivalent to the previous one (8.61).

Note also that the statistic g_n is the LLR with reflection from the zero barrier (and e^{g_n} is the LR with reflection from the unit barrier). In other words, when the cumulative sum of partial LLRs hits the zero barrier it reflects and starts all over again forgetting the past.

8.2.2 The CUSUM Algorithm as a Repeated SPRT

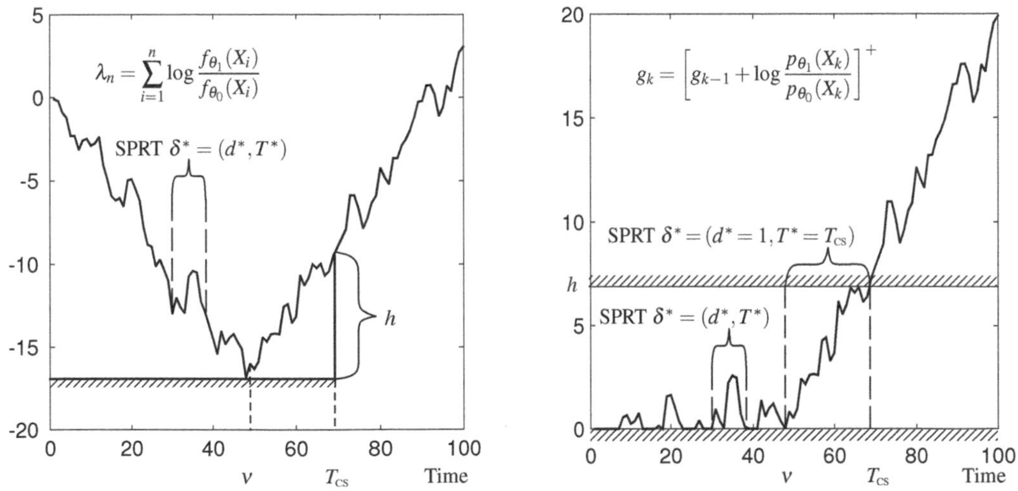

Figure 8.5: *CUSUM as a repeated SPRT: typical behavior of the cumulative sum λ_n (left) and the recursive detection statistic g_n of the repeated SPRT with thresholds 0 and h (right).*

In his pioneering paper [346], Page also suggested to consider the CUSUM procedure as a sequence of SPRTs (3.1) between two simple hypotheses $H_0 : \theta = \theta_0$ and $H_1 : \theta = \theta_1$ with thresholds 0 and h. The SPRT δ^* (3.1) is defined by the pair $\delta^* = (d^*, T^*)$, where d^* is the terminal decision rule and T^* is a stopping time of the form

$$T^* = \inf\{n \geq 1 : \lambda_n \notin (0, h)\}, \quad d^* = \begin{cases} 1 & \text{if } \lambda_{T^*} \geq h \\ 0 & \text{if } \lambda_{T^*} \leq 0. \end{cases} \tag{8.64}$$

The repeated SPRT is used until the decision $d^* = 1$ is made in a cyclic manner. Specifically, consider the sequence of SPRT cycles (d^*_K, T^*_K), $K = 1, 2, \ldots$. The typical behavior of the repeated SPRT is depicted in Figure 8.5. The CUSUM stopping time is defined as the first time at which $d_K = 1$, i.e., we stop observation and do not restart a new cycle of the SPRT at

$$T_{\text{CS}} = \inf\{T^*_K : d^*_K = 1 | K = 1, 2, \ldots\}. \tag{8.65}$$

Using an intuitive motivation, Page suggested that the optimal value of the lower threshold should be zero. It turns out that this conjecture is indeed correct under certain circumstances [271, 305, 389, 414, 416]. Clearly for the zero value of the lower threshold in the repeated SPRT the resulting detection procedure can be rewritten in a recursive manner as

$$g_n = (g_{n-1} + Z_n)^+, \quad n = 1, 2, \ldots, \quad T_{\text{CS}} = \inf\{n \geq 1 : g_n \geq h\}, \tag{8.66}$$

where $g_0 = 0$. Again we obtained that the CUSUM statistic is the cumulative LLR with reflection from the zero barrier! The typical behavior of this statistic is depicted in Figure 8.5(right). It is easy to see that this form of decision rule is equivalent to the other form which we presented in (8.60)–(8.62). The CUSUM stopping time given by (8.66) can be rewritten by using the recursive formula:

$$T_{\text{CS}} = \inf\{n \geq 1 : e^{g_n} \geq e^h\}, \quad e^{g_n} = \max\{e^{g_{n-1}} \mathcal{L}_n, 1\}, \quad e^{g_0} = 1, \tag{8.67}$$

where $\mathcal{L}_n = f_1(X_n)/f_0(X_n)$.

Consider now the two-sided problem where the post-change θ is either $\theta_1^u = \theta_0 + \delta_\theta$ or $\theta_1^\ell = \theta_0 - \delta_\theta$, with δ_θ known. Recall that in the parametric case we set $f_i = f_{\theta_i}$, $i = 0, 1$. In this case, it is reasonable to use the two parallel CUSUM algorithms; the first one for detecting an increase of θ, the second one for detecting a decrease of θ. The stopping time of such two-sided CUSUM is

$$T_{2\text{-CS}} = \inf\{n \geq 1 : (g_n^u \geq h) \cup (g_n^\ell \geq h)\}, \tag{8.68}$$

$$g_n^u = \left[g_{n-1}^u + \log \frac{f_{\theta_1^u}(X_n)}{f_{\theta_0}(X_n)}\right]^+, \quad g_n^\ell = \left[g_{n-1}^\ell - \log \frac{f_{\theta_1^\ell}(X_n)}{f_{\theta_0}(X_n}\right]^+.$$

In the Gaussian case of detecting a change in mean where the post-change mean value is either $\mu_1^u = \mu_0 + \delta_\mu$ or $\mu_1^\ell = \mu_0 - \delta_\mu$, the functions g_n^u and g_n^ℓ become

$$g_n^u = \left[g_{n-1}^u + X_n - \mu_0 - \frac{\delta_\mu}{2}\right]^+, \quad g_n^\ell = \left[g_{n-1}^\ell - X_n + \mu_0 - \frac{\delta_\mu}{2}\right]^+. \tag{8.69}$$

In these formulas, we canceled the multiplicative term $(\mu_1 - \mu_0)/\sigma^2$ which can be incorporated in the threshold in an obvious manner. The formula (8.69) corresponds to the well-known two-sided CUSUM chart widely used in continuous inspection for quality control. The geometric interpretation of the two-sided CUSUM chart in quality control, the so-called V-mask, is presented in Figure 8.6. Here two parallel open-ended SPRTs are applied at each step in reverse time, one to detect μ_1^u and another one to detect μ_1^ℓ.

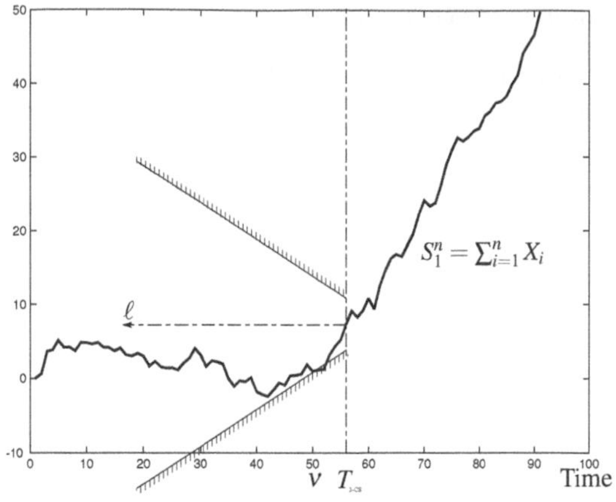

Figure 8.6: *The V-mask for the two-sided CUSUM algorithm.*

8.2.3 The CUSUM Algorithm as a GLR Test

Yet another recursion for the CUSUM statistic has been used in (7.155). In order to obtain this recursion, consider the changepoint detection problem as a problem of testing two hypotheses: H_ν that the change occurs at a fixed point $0 \le \nu < \infty$ against the alternative H_∞ that the change never occurs. The LR between these hypotheses is $\Lambda_n^\nu = \prod_{i=\nu+1}^n \mathcal{L}_i$ for $\nu < n$ and 1 for $\nu \ge n$. Since the hypothesis H_ν is composite, we may apply the GLR (maximum likelihood ratio) approach maximizing the LR Λ_n^ν over ν to obtain the GLR statistic

$$V_n = \max_{0 \le \nu < n} \prod_{i=\nu+1}^{n} \mathcal{L}_i, \quad n \ge 1.$$

It is easy to verify that this statistic follows the recursion (7.155),

$$V_n = \max\{1, V_{n-1}\}\mathcal{L}_n, \quad n \ge 1, \ V_0 = 1. \tag{8.70}$$

Note that this statistic is related to the statistic e^{g_n} defined in (8.67) as $e^{g_n} = \max(1, V_n)$. In fact, the statistic e^{g_n} can also be obtained via the GLR approach by maximizing the LR Λ_n^ν over $0 \le \nu < \infty$. However, since the hypotheses H_∞ and H_ν are indistinguishable for $\nu \ge n$ the maximization over $\nu \ge n$ does not make too much sense. Note also that the statistic V_n may take values smaller than 1, so the CUSUM procedure

$$T_{\rm CS} = \inf\{n \ge 1 : V_n \ge A\} \tag{8.71}$$

or its log-scale counterpart

$$T_{\rm CS} = \inf\{n \ge 1 : W_n \ge h\}, \quad W_n = \log V_n, \quad h = \log A \tag{8.72}$$

makes sense even for negative values of the threshold h. Thus, it is more general than Page's CUSUM. Note the recursion

$$W_n = W_{n-1}^+ + Z_n, \quad n \ge 1, \ W_0 = 0. \tag{8.73}$$

Since $W_n = Z_n$ on $(0, 1]$, for $h \le 0$ the CUSUM (8.72) coincides with the LLR-based Shewhart chart $\inf\{n \ge 1 : Z_n \ge h\}$.

Below in Subsection 8.2.5 we discuss exact optimality of CUSUM (8.72) in the class $\mathbf{C}_\gamma = \{T : \mathsf{E}_\infty T \ge \gamma\}$ for any $\gamma > 1$, the result established by Moustakides [305] in 1986.

Finally, we note that the stopping times (8.66) and (8.72) are equivalent if $h > 0$, since the trajectories of W_n and $g_n = \max(0, W_n)$ are the same on a positive half-plane.

8.2.4 *The CUSUM Algorithm in the General Non-iid Case*

Consider now the general non-iid model (6.2). That is, the observations $\{X_n\}_{n\geq 1}$ are distributed according to conditional densities $f_{0,n}(X_n|\mathbf{X}_1^{n-1})$ for $n=1,\dots,\nu$ and according to conditional densities $f_{1,n}(X_n|\mathbf{X}_1^{n-1})$ for $n>\nu$, where post-change densities $f_{1,n}(X_n|\mathbf{X}_1^{n-1}) = f_{1,n}^{(\nu)}(X_n|\mathbf{X}_1^{n-1})$ generally depend on the changepoint ν. Recall that in this general case the LR and LLR processes between the hypotheses H_ν and H_∞ are

$$\Lambda_n^\nu = \prod_{i=\nu+1}^{n} \mathcal{L}_i^{(\nu)}, \quad \lambda_n^\nu = \sum_{i=\nu+1}^{n} Z_i^{(\nu)},$$

where

$$\mathcal{L}_i^{(\nu)} = \frac{f_{1,n}^{(\nu)}(X_n|\mathbf{X}_1^{n-1})}{f_{0,n}(X_n|\mathbf{X}_1^{n-1})}, \quad Z_i^{(\nu)} = \log \frac{f_{1,n}^{(\nu)}(X_n|\mathbf{X}_1^{n-1})}{f_{0,n}(X_n|\mathbf{X}_1^{n-1})}. \tag{8.74}$$

In this general case, the CUSUM statistics V_n and W_n are defined as

$$V_n = \max_{0\leq \nu<n} \prod_{i=\nu+1}^{n} \mathcal{L}_i^{(\nu)}, \quad W_n = \max_{0\leq \nu<n} \sum_{i=\nu+1}^{n} Z_i^{(\nu)}, \quad n\geq 1. \tag{8.75}$$

Obviously, these statistics do not obey recursions similar to (8.70) and (8.73), which complicates their computation tremendously. However, if $\mathcal{L}_i^{(\nu)} = \mathcal{L}_i$ and $Z_i^{(\nu)} = Z_i$ do not depend on the changepoint ν, then these recursions still hold. Also, in this case the statistic g_n with reflection from the zero barrier can also be computed recursively as

$$g_n = \left\{ g_{n-1} + \log \frac{f_{1,n}(X_n|\mathbf{X}_1^{n-1})}{f_{0,n}(X_n|\mathbf{X}_1^{n-1})} \right\}^+, \quad g_0 = 0.$$

As before the CUSUM stopping time is defined as the first time of exceedance of a threshold that controls the false alarm rate.

Sometimes the LLR $\log[f_{\theta_1}(X_j^k)/f_{\theta_0}(X_j^k)]$ cannot be written in the recursive form even in the case of iid observations $X_1,\dots,X_k$. See for instance Example 5.4.2, where the χ^2-SPRT test is considered and where the LLR statistic is replaced with the weighted likelihood ratio (WLR) statistics. In such a case, the CUSUM statistic (as a repeated SPRT) can be written via the LLR (or WLR) as

$$g_n = \left[\lambda_{n-N_n+1}^n\right]^+, \quad \lambda_{n-N_n+1}^n = \log\left[\frac{f_1(\mathbf{X}_{n-N_n+1}^n)}{f_0(\mathbf{X}_{n-N_n+1}^n)}\right], \quad g_0 = 0, \tag{8.76}$$

where

$$N_n = N_{n-1}\cdot \mathbb{1}_{\{g_{n-1}>0\}} + 1. \tag{8.77}$$

In this formula, N_n is the number of observations after re-start of the SPRT. Formula (8.76) can be interpreted as an integration of the observations over a *sliding window with random size*. This size is chosen according to the behavior of the entire past observations.

8.2.5 *Optimal Properties of the CUSUM Algorithm*

In this subsection, we describe the statistical properties of CUSUM algorithm in the iid case (6.4). We begin with establishing strict optimality of CUSUM with respect to Lorden's minimax criterion and then generalize this result discussing asymptotic optimality properties with respect to higher moments of the detection delay, also in a minimax setting.

Consider the simple changepoint detection problem with iid observations having the pdf $f_0(x)$

in the pre-change mode and the pdf $f_1(x)$ in the post-change mode (with respect to some non-degenerate sigma-finite measure) and Lorden's essential supremum criterion (8.16). To avoid triviality we assume that the pre-specified $\mathsf{ARL2FA}(T) = \mathsf{E}_\infty T = \gamma$ is strictly bigger than 1, since otherwise the optimal procedure is degenerate. Indeed, for $0 < \gamma \le 1$, the optimal procedure $T_{\rm opt}$ prescribes stopping w.p. $1-\gamma$ immediately with no observations (at $n = 0$) or making one observation w.p. γ and stop at $n = 1$. For this procedure, $\mathsf{ARL2FA}(T_{\rm opt}) = \gamma$ and $\mathsf{ESADD}(T_{\rm opt}) = 1$, which is obviously the best one can do. It turns out that the CUSUM procedure defined in (8.70) and (8.71) with the threshold $A > 0$ selected so that $\mathsf{E}_\infty[T_{\rm CS}(A)] = \gamma$ is strictly minimax for every $\gamma > 1$. We give only a sketch of the proof referring for details to the paper of Moustakides [305].

We start with an intuitive argument on why it is expected that CUSUM is a minimax detection procedure for Lorden's criterion. Write $\mathsf{D}_\nu(T) = \operatorname{ess\,sup} \mathsf{E}_\nu[(T-\nu)^+ | \mathscr{F}_\nu]$. The reason is simple – the CUSUM procedure is an equalizer rule in the sense that $\mathsf{D}_\nu(T_{\rm CS})$ does not depend on ν, i.e., constant. Specifically,

$$\mathsf{D}_\nu(T_{\rm CS}(A)) = \mathsf{E}_0[T_{\rm CS}(A)] \quad \text{for all } \nu \ge 0 \text{ and for any } A > 0. \tag{8.78}$$

Since the minimax rule should be an equalizer rule, CUSUM is a natural candidate being also the LR-based procedure. For $A \le 1$ the equalizer property (8.78) follows almost trivially since in this case $T_{\rm CS} = \inf\{n : \mathcal{L}_n \ge A\}$ is memoryless, so $\mathsf{D}_\nu(T_{\rm CS}) = \mathsf{E}_\nu(T_{\rm CS} - \nu)^+$. Since Λ_n, $n > \nu$ are iid under P_ν and have the same distribution for any $\nu \ge 0$, (8.78) obviously holds. Let $A > 1$. Evidently, trajectories of the statistic V_n coincide with those of $\widetilde{V}_n = \max(1, V_n)$ on $[1, \infty)$, so $T_{\rm CS}$ can be alternatively written as $T_{\rm CS} = \inf\{n : \widetilde{V}_n \ge A\}$. However, for $n > \nu$, the statistic $\widetilde{V}_n = \max(1, \widetilde{V}_{n-1}\mathcal{L}_n)$ with the initial condition $\widetilde{V}_\nu$ is obviously monotone nondecreasing function of $\widetilde{V}_\nu$. This implies that on $\{T_{\rm CS} > \nu\}$ the stopping time $T_{\rm CS}$ is nonincreasing in $\widetilde{V}_\nu$, and therefore,

$$\operatorname{ess\,sup} \mathsf{E}_\nu[(T_{\rm CS} - \nu)^+ | \mathscr{F}_\nu] = \operatorname{ess\,sup} \mathsf{E}_\nu[(T_{\rm CS} - \nu)^+ | \widetilde{V}_\nu] = \mathsf{E}_\nu[(T_{\rm CS} - \nu)^+ | \widetilde{V}_\nu = 1],$$

where $\mathsf{E}_\nu[(T_{\rm CS} - \nu)^+ | \widetilde{V}_\nu = 1] = \mathsf{E}_0 T_{\rm CS}$ due to homogeneity of $\widetilde{V}_n$ and the fact that $\widetilde{V}_0 = 1$.

The first very important step in proving optimality of CUSUM is establishing the following lower bound for the risk: for any procedure with finite ARL2FA,

$$\mathsf{ESADD}(T) \ge \frac{\mathsf{E}_\infty[\sum_{n=0}^{T-1} \max(1, V_n)]}{\mathsf{E}_\infty[\sum_{n=0}^{T-1} (1 - V_n)^+]} = \mathcal{J}(T). \tag{8.79}$$

Furthermore, $\mathsf{ESADD}(T_{\rm CS}) = \mathcal{J}(T_{\rm CS})$. For the proof of (8.79), see [305, Lemma 3]. Therefore, we can replace the initial optimization problem with minimization of the lower bound $\inf_{T \in \mathbf{C}_\gamma} \mathcal{J}(T)$ and focus on showing that the optimal solution of this modified problem is CUSUM.

Second, using the argument with randomization at 0 similar to that in the proof of Theorem 7.6.1 it is easily shown that we can restrict ourselves to the class $\widetilde{\mathbf{C}}_\gamma = \{\mathsf{ARL2FA}(T) = \gamma\}$ with strict equality; see also [305, Lemma 2]. This is instructive and simplifies a proof of CUSUM minimaxity.

The third step is to show that $T_{\rm CS}(A)$ with $A = A_\gamma$ such that $\mathsf{ARL2FA}(T_{\rm CS}(A)) = \gamma$ simultaneously minimizes the numerator and maximizes the denominator of $\mathcal{J}(T)$. The proof can be reduced to optimal stopping of the homogeneous Markov process. The formal proof of this result is very involved [305]. Note that the fact that we can consider the class $\widetilde{\mathbf{C}}_\gamma$ instead of $\mathbf{C}_\gamma$ is critical, since in the class $\mathbf{C}_\gamma$ the "optimal" denominator is infinity, i.e., corresponds to the stopping time $T = \infty$.

The following theorem summarizes the above argument.

Theorem 8.2.1 (CUSUM minimaxity). *Let $T_{\rm CS}(A)$ be the CUSUM procedure defined in* (8.70) *and* (8.71). *If $A = A_\gamma$ is selected so that the ARL to false alarm is exactly equal to γ, $\mathsf{ARL2FA}(T_{\rm CS}(A_\gamma)) = \gamma$, then CUSUM is strictly optimal in the class $\mathbf{C}_\gamma$,*

$$\mathsf{ESADD}(T_{\rm CS}(A_\gamma)) = \inf_{T \in \mathbf{C}_\gamma} \mathsf{ESADD}(T) \quad \textit{for any } \gamma > 1.$$

Note that in general it is not always possible to select the threshold in such a way that the equality $\mathsf{ARL2FA}(T_{\text{CS}}(A_\gamma)) = \gamma$ is guaranteed. For example, if the P_∞-distribution of the LR $\mathcal{L}_1$ has atoms, then an additional randomization may be needed on the stopping boundary (when $V_n = A$). If we require the P_∞-distribution of the LR to be continuous this is potentially always possible. However, practically finding A_γ exactly is rarely possible, and approximations are needed. This problem is addressed in Subsection 8.2.6.

We now extend the above optimality result to higher moments of the detection delay defining, for $m \ge 1$,

$$\begin{aligned} \mathsf{ESM}_m(T) &= \sup_{0 \le \nu < \infty} \operatorname{ess\,sup} \mathsf{E}_\nu \left\{[(T-\nu)^+]^m | \mathscr{F}_\nu\right\}, \\ \mathsf{SM}_m(T) &= \sup_{0 \le \nu < \infty} \mathsf{E}_\nu \left\{(T-\nu)^m | T > \nu\right\} \end{aligned} \tag{8.80}$$

and considering the asymptotic setting of minimizing $\mathsf{ESM}_m(T)$ and $\mathsf{SM}_m(T)$ in the class $\mathbf{C}_\gamma$ as $\gamma \to \infty$.

The following lemma establishes a simple lower bound for the ARL to false alarm of CUSUM, which holds not only in the iid case but in a general non-iid case too. Recall that in the general case the CUSUM procedure is defined by the stopping time (8.71) where the CUSUM statistic is given by (8.74), (8.75). We iterate that in general there is no recursion for this statistic similar to (8.70) since $\mathcal{L}_i^{(\nu)}$ may depend on ν.

Lemma 8.2.1. *Consider the general non-iid model* (6.2). *For any $A > 0$, the ARL to false alarm of the CUSUM procedure $T_{\text{CS}}(A)$ defined in* (8.71) *satisfies the inequality*

$$\mathsf{ARL2FA}(T_{\text{CS}}(A)) \ge A. \tag{8.81}$$

Therefore,

$$A_\gamma = \gamma \quad \text{implies} \quad T_{\text{CS}}(A_\gamma) \in \mathbf{C}_\gamma. \tag{8.82}$$

Proof. Evidently, $T_{\text{CS}}(A) \ge T_{\text{SR}}(A)$, where $T_{\text{SR}}(A) = \inf\{n : R_n \ge A\}$ is the SR stopping time, $R_n = \sum_{k=0}^n \prod_{i=k+1}^n \mathcal{L}_i^{(k)}$, $R_0 = 0$. Since $\mathsf{E}_\infty[\mathcal{L}_n^{(k)} \mid \mathscr{F}_{n-1}] = 1$, it is easily verified that $\mathsf{E}_\infty(R_n \mid \mathscr{F}_{n-1}) = 1 + R_{n-1}, n \ge 1$. Thus, the statistic $\{R_n - n\}_{n \ge 1}$ is a $(\mathsf{P}_\infty, \mathscr{F}_n)-$martingale with zero mean. Suppose that $\mathsf{E}_\infty[T_{\text{SR}}(A)] < \infty$. (Otherwise the statement of the lemma is trivial.) Therefore $\mathsf{E}_\infty(R_{T_{\text{SR}}} - T_{\text{SR}})$ exists. Next, since $0 \le R_n < A$ on the event $\{T_{\text{SR}} > n\}$, it is seen that

$$\liminf_{n \to \infty} \int_{\{T_{\text{SR}} > n\}} |R_n - n| \, d\mathsf{P}_\infty = 0.$$

Hence, we can apply the optional sampling theorem, which yields $\mathsf{E}_\infty R_{T_{\text{SR}}} = \mathsf{E}_\infty T_{\text{SR}}$, while by definition of T_{SR}, $R_{T_{\text{SR}}} \ge A$. This shows that $\mathsf{E}_\infty[T_{\text{CS}}(A)] \ge \mathsf{E}_\infty[T_{\text{SR}}(A)] \ge A$ and the proof is complete. □

We now use the method of proof of Theorem 7.1.3 to derive asymptotic lower bounds for moments of the detection delay in the general non-iid case, which then will be used for establishing asymptotic optimality properties of CUSUM (and SR too) in both iid and non-iid cases.

Theorem 8.2.2 (Lower bounds)**.** *Consider the general non-iid model given in* (6.2). *Assume that the normalized LLR $n^{-1}\lambda_{\nu+n}^\nu$ converges in probability to a positive and finite number I under P_ν,*

$$\frac{1}{n}\lambda_{\nu+n}^\nu \xrightarrow[n \to \infty]{\mathsf{P}_\nu} I, \tag{8.83}$$

and in addition

$$\sup_{0 \le \nu < \infty} \operatorname{ess\,sup} \mathsf{P}_\nu \left\{ M^{-1} \max_{0 \le n < M} \lambda_{\nu+n}^\nu \ge (1+\varepsilon) I | \mathscr{F}_\nu \right\} \xrightarrow[M \to \infty]{} 0 \quad \text{for all } \varepsilon > 0. \tag{8.84}$$

Then, for all $0 < \varepsilon < 1$

$$\lim_{\gamma\to\infty} \sup_{T\in\mathbf{C}_\gamma} \sup_{0\le\nu<\infty} \mathsf{P}_\nu\left\{\nu < T < \nu + (1-\varepsilon)I^{-1}\log\gamma \,|\, T > \nu\right\} = 0, \tag{8.85}$$

and for all $m > 0$

$$\liminf_{\gamma\to\infty} \frac{\inf_{T\in\mathbf{C}_\gamma} \mathsf{SM}_m(T)}{(\log\gamma)^m} \ge \frac{1}{I^m} \quad \text{and} \quad \liminf_{\gamma\to\infty} \frac{\inf_{T\in\mathbf{C}_\gamma} \mathsf{ESM}_m(T)}{(\log\gamma)^m} \ge \frac{1}{I^m}. \tag{8.86}$$

Proof. Observe first that, for any $\nu \ge 0$ and any $T \in \mathbf{C}_\gamma$,

$$\mathsf{ESM}_m(T) \ge \mathsf{SM}_m(T) \ge \mathsf{E}_\nu\left\{[(T-\nu)^+]^m | T > \nu\right\}.$$

Thus, to prove the lower bounds (8.86) it is sufficient to prove that for any $T \in \mathbf{C}_\gamma$ and some $\nu \ge 0$

$$\mathsf{E}_\nu\left\{[(T-\nu)^+]^m | T > \nu\right\} \ge [I^{-1}\log\gamma]^m\left(1+o(1)\right) \quad \text{as } \gamma\to\infty, \tag{8.87}$$

where the right-hand side is uniform in ν, i.e., the residual term $o(1)$ does not depend on ν. This last asymptotic inequality follows from the Chebyshev inequality provided that (8.85) holds. Indeed, by the Chebyshev inequality, for any $0 < \varepsilon < 1$ and any $T \in \mathbf{C}_\gamma$

$$\mathsf{E}_\nu\left\{[(T-\nu)^+]^m | T > \nu\right\} \ge [(1-\varepsilon)K_\gamma]^m \mathsf{P}_\nu\left\{T - \nu \ge (1-\varepsilon)K_\gamma | T > \nu\right\},$$

where $K_\gamma = I^{-1}\log\gamma$. Since, by (8.85), $\inf_{T\in\mathbf{C}_\gamma}\inf_\nu \mathsf{P}_\nu\left\{T-\nu \ge (1-\varepsilon)K_\gamma | T > \nu\right\} \to 1$, it follows that for some $\nu \ge 0$

$$\mathsf{E}_\nu\left\{[(T-\nu)^+]^m | T > \nu\right\} \ge [(1-\varepsilon)I^{-1}\log\gamma]^m(1+o(1)) \quad \text{as } \gamma\to\infty,$$

where the right side does not depend on ν. Since this last inequality holds for an arbitrary small ε, inequality (8.87) follows, and all that remains to do is to show the validity of (8.85).

Changing the measure $\mathsf{P}_\infty \to \mathsf{P}_\nu$, in just the same way as in the proof of Theorem 7.1.3 (using the chain of inequalities (7.23) conditioned on the event $\{T > \nu\}$) we obtain that for any $\varepsilon \in (0,1)$

$$\mathsf{P}_\nu\left\{T < \nu + (1-\varepsilon)K_\gamma | T > \nu\right\} \le p_{\gamma,\varepsilon}^{(\nu)}(T) + \beta_{\gamma,\varepsilon}^{(\nu)}(T), \tag{8.88}$$

where

$$p_{\gamma,\varepsilon}^{(\nu)}(T) = e^{(1-\varepsilon^2)IK_\gamma}\mathsf{P}_\infty\left\{T < \nu + (1-\varepsilon)K_\gamma | T > \nu\right\},$$

$$\beta_{\gamma,\varepsilon}^{(\nu)}(T) = \mathsf{P}_\nu\left\{\max_{1\le n<(1-\varepsilon)K_\gamma} \lambda_{\nu+n}^\nu \ge (1-\varepsilon^2)IK_\gamma | T > \nu\right\}.$$

Note that

$$\beta_{\gamma,\varepsilon}^{(\nu)}(T) \le \operatorname{ess\,sup} \mathsf{P}_\nu\left\{\max_{1\le n<(1-\varepsilon)K_\gamma} \lambda_{\nu+n}^\nu \ge (1-\varepsilon^2)IK_\gamma | \mathscr{F}_\nu\right\} := \tilde{\beta}_{\gamma,\varepsilon}^{(\nu)}$$

and $\sup_{\nu\ge0}\tilde{\beta}_{\gamma,\varepsilon}^{(\nu)} \to 0$ as $\gamma\to\infty$ by condition (8.84). Next, by (6.23), for any $T \in \mathbf{C}_\gamma$ there is a $\nu \ge 0$, possibly depending on γ, such that

$$\mathsf{P}_\infty\left\{T < \nu + (1-\varepsilon)K_\gamma | T > \nu\right\} \le (1-\varepsilon)K_\gamma/\gamma,$$

and therefore,

$$p_{\gamma,\varepsilon}^{(\nu)}(T) \le (1-\varepsilon)K_\gamma e^{(1-\varepsilon^2)IK_\gamma}/\gamma = (1-\varepsilon)I^{-1}(\log\gamma)/\gamma^{\varepsilon^2}. \tag{8.89}$$

Hence, for every $\varepsilon < 1$,

$$\sup_{\nu\ge0}\mathsf{P}_\nu\left\{T < \nu + (1-\varepsilon)K_\gamma | T > \nu\right\} \le \sup_{\nu\ge0}\tilde{\beta}_{\gamma,\varepsilon}^{(\nu)} + (1-\varepsilon)I^{-1}(\log\gamma)/\gamma^{\varepsilon^2} \to 0.$$

An important observation is that the right-hand side in this inequality does not depend either on a particular stopping time T or on ν, which proves (8.85). □

Obviously, in the iid case condition (8.84) holds with $I = \mathsf{E}_0 Z_1$, the K–L information number, since for every $\nu \ge 0$ and all $\varepsilon > 0$

$$\mathsf{P}_\nu\left\{M^{-1}\max_{1\le n<M}\lambda_{\nu+n}^{\nu} \ge (1+\varepsilon)I|\mathscr{F}_\nu\right\} = \mathsf{P}_0\left\{\max_{1\le n<M} M^{-1}\lambda_n^0 \ge (1+\varepsilon)I\right\} \to 0 \quad \text{as } M\to\infty$$

by the SLLN. Therefore, we have the following corollary for the iid case.

Corollary 8.2.1. *Let the observations $X_1,\dots,X_\nu$ be iid with the pdf $f_0(x)$ and the observations $X_{\nu+1},X_{\nu+2},\dots$ be iid with the pdf $f_1(x)$. Assume that the K–L number is positive and finite, $0<I<\infty$. Then the asymptotic lower bounds* (8.86) *hold for all $m>0$.*

To establish asymptotic optimality of the CUSUM procedure we now have to show that the asymptotic lower bounds are also upper bounds for $\mathsf{ESM}_m(T_{\text{CS}})$ and $\mathsf{SM}_m(T_{\text{CS}})$, which is the subject of the next theorem.

Theorem 8.2.3 (Asymptotic optimality). *Let the observations $X_1,\dots,X_\nu$ be iid with the pdf $f_0(x)$ and $X_{\nu+1},X_{\nu+2},\dots$ be iid with the pdf $f_1(x)$. Assume that the K–L number is positive and finite, $0<I<\infty$.*

(i) *Then for any $m>0$,*

$$\mathsf{SM}_m(T_{\text{CS}}(A)) = \mathsf{ESM}_m(T_{\text{CS}}(A)) = \mathsf{E}_0[T_{\text{CS}}(A)]^m \sim \left(\frac{\log A}{I}\right)^m \quad \text{as } A\to\infty. \tag{8.90}$$

(ii) *If $A=A_\gamma$ is selected so that $\mathsf{ARL2FA}(T_{\text{CS}}(A_\gamma)) \ge \gamma$ and $\log A_\gamma \sim \log\gamma$, in particular $A_\gamma=\gamma$, then for all $m>0$,*

$$\inf_{T\in\mathbf{C}_\gamma}\mathsf{ESM}_m(T) \sim \inf_{T\in\mathbf{C}_\gamma}\mathsf{SM}_m(T) \sim \mathsf{E}_0[T_{\text{CS}}(A_\gamma)]^m \sim \left(\frac{\log\gamma}{I}\right)^m \quad \text{as } \gamma\to\infty. \tag{8.91}$$

Proof. (i) For $0<A\le 1$, the CUSUM statistic is memoryless, so in this case $\mathsf{E}_\nu\{[(T_{\text{CS}}(A)-\nu)^+]^m|\mathscr{F}_\nu\} = \mathsf{E}_\nu\{[T_{\text{CS}}(A)-\nu]^m|T_{\text{CS}}>\nu\} = \mathsf{E}_0[T_{\text{CS}}(A)]^m$ for all $\nu\ge 0$, hence $\mathsf{SM}_m(T_{\text{CS}}(A)) = \mathsf{ESM}_m(T_{\text{CS}}(A)) = \mathsf{E}_0[T_{\text{CS}}(A)]^m$. For $A>1$, it is equivalent to the statistic with reflection from the unit barrier, as we discussed above. For $\nu=0$, the latter statistic always starts from 1, while for $\nu\ge 1$ it starts from a random point $V_\nu\in[1,A)$ on $\{T_{\text{CS}}>\nu\}$. Hence, it is clear that $\mathsf{SM}_m(T_{\text{CS}}(A)) = \mathsf{ESM}_m(T_{\text{CS}}(A)) = \mathsf{E}_0[T_{\text{CS}}(A)]^m$ in this case too, so these equalities hold for all $A>0$.

For $A>1$, introduce the one-sided stopping time $\eta_A = \inf\{n\ge 1: \lambda_n \ge \log A\}$, where $\lambda_n = \lambda_n^0 = \sum_{i=1}^n Z_i$. Clearly, $\log V_n \ge \lambda_n$, so $T_{\text{CS}}(A)\le\eta_A$ and $\mathsf{E}_0[T_{\text{CS}}(A)]^m \le \mathsf{E}_0\eta_A^m$. By (4.18), for all $m>0$, $\mathsf{E}_0\eta_A^m \sim (I^{-1}\log A)^m$ as $A\to\infty$, and therefore,

$$\mathsf{E}_0[T_{\text{CS}}(A)]^m \le \left(I^{-1}\log A\right)^m(1+o(1)) \quad \text{as } A\to\infty. \tag{8.92}$$

Finally, replacing γ with A in Corollary 8.2.1, we obtain the lower bound

$$\mathsf{E}_0[T_{\text{CS}}(A)]^m \ge \left(I^{-1}\log A\right)^m(1+o(1)) \quad \text{as } A\to\infty,$$

which together with the upper bound (8.92) yields (8.90).

(ii) Asymptotic relations (8.91) follow from (8.90) and Corollary 8.2.1. □

The first-order asymptotic optimality property of CUSUM with respect to $\mathsf{ESM}_1(T) = \mathsf{ESADD}(T)$ and asymptotic approximation (8.91) for $\mathsf{ESADD}(T_{\text{CS}}(\gamma))$ was first established by Lorden [271] in 1971. Therefore, Theorem 8.2.3 can be considered as an extension of Lorden's first-order asymptotic theory to arbitrary essential supremum moments of the detection delay as well as arbitrary maximal moments of the detection delay $\mathsf{SM}_m(T)$. Our proof is different and more general

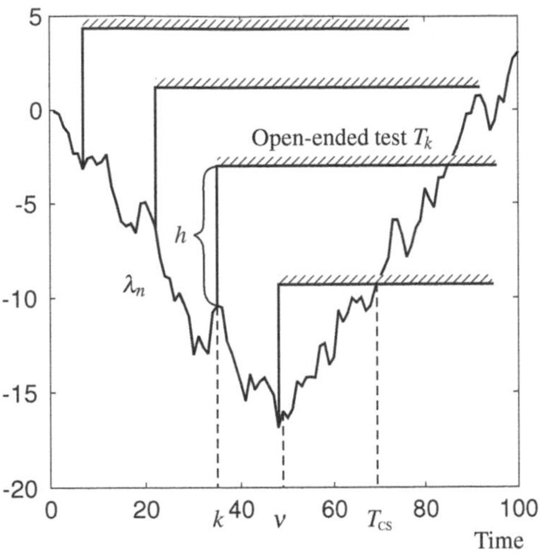

Figure 8.7: *The interpretation of the CUSUM test as a set of parallel open-ended SPRTs.*

than Lorden's. However, Lorden's approach has some interesting and useful features that we discuss next.

We first discuss useful results by Lorden [271] concerning a class of extended stopping times, which can be applied not only to the CUSUM algorithm, but also to other algorithms such as the GLR CUSUM, weighted CUSUM, and SR-type procedures.

Lorden noticed that the CUSUM stopping time can be represented via a sequence of one-sided (open-ended) SPRTs $\eta_h^{(k)} = \inf\{n \ge 1 : \lambda_{k+n}^k \ge h\}$, $k = 0,1,\dots$ (applied cyclically) as follows: $T_{\mathrm{CS}}(A) = \inf_{k \ge 0}[\eta_h^{(k)} + k]$ ($h = \log A$). See Figure 8.7. Therefore, operating characteristics of CUSUM can be approximated via those of the one-sided SPRT. More generally, let T be an arbitrary generic stopping time with respect to $\{X_n\}_{n \ge 1}$, and for $k = 0,1,2,\dots$ let $\overline{T}_k$ be the stopping time obtained by applying T to $X_{k+1}, X_{k+2}, \dots$. Let $T_k = \overline{T}_k + k$ and $T^* = \inf_{k \ge 0} T_k$. The following important result holds.

Theorem 8.2.4. **(i)** *Let T be a stopping time that is applied repeatedly. If $\mathsf{P}_\infty(T < \infty) \le \alpha$, then*

$$\mathsf{E}_\infty T^* \ge 1/\alpha \quad \text{and} \quad \mathsf{ESADD}(T^*) \le \mathsf{E}_0 T. \tag{8.93}$$

(ii) *Let the observations $X_1, \dots, X_\nu$ be iid with the pre-change pdf $f_0(x)$ and $X_{\nu+1}, X_{\nu+2}, \dots$ be iid with the post-change pdf $f_1(x)$. Assume that the K–L number is positive and finite, $0 < I < \infty$. If $A = A_\gamma$ is selected so that $\mathsf{ARL2FA}(T_{\mathrm{CS}}(A_\gamma)) \ge \gamma$ and $\log A_\gamma \sim \log \gamma$, in particular $A_\gamma = \gamma$, then*

$$\inf_{T \in \mathbf{C}_\gamma} \mathsf{ESADD}(T) \sim \mathsf{ESADD}(T_{\mathrm{CS}}(A_\gamma)) \sim (I^{-1} \log \gamma)\big(1 + o(1)\big), \quad \gamma \to \infty. \tag{8.94}$$

Proof. (i) See Lorden [271].

(ii) Set $T_k = \eta_h^{(k)} + k$. Then the CUSUM stopping time is $T_{\mathrm{CS}}(A) = T^* = \inf_{k \ge 0} T_k$. Applying Wald's identities yields (see Chapter 3),

$$\mathsf{P}_\infty\big(\eta_h^{(0)} < \infty\big) \le e^{-h} = 1/A \quad \text{and} \quad \mathsf{E}_0\big[\eta_h^{(0)}\big] \sim h/I, \quad h \to \infty.$$

Combining these relations with assertion (i) yields

$$\mathsf{E}_\infty T_{\mathrm{CS}}(A) \ge A, \quad \mathsf{ESADD}(T_{\mathrm{CS}}(A)) \le (I^{-1} \log A)\big(1 + o(1)\big), \quad A \to \infty.$$

The lower bound

$$\inf_{T \in \mathbf{C}_\gamma} \mathsf{ESADD}(T) \geq (I^{-1} \log \gamma)\big(1 + o(1)\big), \quad \gamma \to \infty \tag{8.95}$$

follows from (8.86) with $m = 1$, which completes the proof of (8.94). □

This theorem implies that the CUSUM test with $A = \gamma$ is first-order asymptotically minimax with respect to Lorden's essential supremum criterion. Note that this result is a particular case of Theorem 8.2.3(ii) with $m = 1$, which has been proven using different techniques.

Usefulness of Theorem 8.2.4 is far beyond establishing asymptotic optimality of the CUSUM algorithm with the threshold $A = \gamma$ for the simple post-change hypothesis. In some sense, *the lower bound* (8.95) *plays the same role in the change detection theory as the Cramer–Rao lower bound in estimation theory*. In fact, it can be applied to construct asymptotically optimal CUSUM-type procedures for a composite post-change hypothesis for parametric (especially exponential) families when the post-change parameter value $\theta \in \Theta$ is not known and may differ from the putative value θ_1. Indeed, if we select T to be a uniformly asymptotically optimal test in the hypothesis testing problem $H_0 : \theta = \theta_0$ *versus* $H_\theta : \theta \in \Theta$ with the level $\alpha = 1/\gamma$, then according to Theorem 8.2.4 the corresponding CUSUM will be uniformly (for all $\theta \in \Theta$) asymptotically minimax as $\gamma \to \infty$. For example, using the mixture-based one-sided test leads to the mixture (weighted) CUSUM and the GLR one-sided test to the GLR CUSUM that are both asymptotically optimal for all post-change parameter values $\theta \in \Theta$. See Section 8.3 for further details.

Previous results show that the CUSUM algorithm is either optimal (Theorem 8.2.1) or nearly optimal (Theorem 8.2.3) in the case of simple pre- and post-change hypotheses, i.e., when it is tuned to the true values of the parameters before and after change. When the algorithm is used in situations where the actual parameter values are different from the pre-assigned ones, this optimal property is *lost*. For this reason, it is of key interest to compute the ARL function for *other* parameter values, which we do in the next subsection.

8.2.6 *Operating Characteristics of the CUSUM Algorithm*

In this subsection, we study a number of operating characteristics of the CUSUM algorithm, in particular the ARL function $\mathsf{ARL}(\theta) = \mathsf{E}_\theta[T_{\text{CS}}]$, where E_θ stands for expectation when the true parameter value is θ. We iterate that $\mathsf{ARL}(\theta_0) = \mathsf{ARL2FA}(T_{\text{CS}})$ is the ARL to false alarm and $\mathsf{ARL}(\theta_1) = \mathsf{ESADD}(T_{\text{CS}})$ is the ALR to detection when the true parameter value coincides with the putative value. We discuss the numerical methods of calculating operating characteristics (exact operating characteristics), approximations, and bounds.

8.2.6.1 *Exact Operating Characteristics*

Assume that $\theta \in \Theta$, where Θ is some set of interest, and let $\Theta^* = \Theta \setminus \theta_0$ be a set of post-change parameter values. Let $\mathsf{P}_{\nu,\theta}$ ($\mathsf{E}_{\nu,\theta}$) denote probability (resp. expectation) when the change occurs at $0 \leq \nu < \infty$ and the true post-change parameter is θ. For the sake of brevity, we write P_θ (E_θ) for $\mathsf{P}_{0,\theta}$ ($\mathsf{E}_{0,\theta}$). Clearly, $\mathsf{P}_\infty = \mathsf{P}_{\theta_0}$ ($\mathsf{E}_\infty = \mathsf{E}_{\theta_0}$).

We first address an (almost) exact computation based upon the numerical solutions of Fredholm integral equations introduced in Subsection 3.1.4 as well as other Fredholm equations similar to those discussed in Subsection 7.1.2.1. Define the ARL function $\mathsf{ARL}_z(\theta) = \mathsf{E}_\theta[T_{\text{CS}}]$ when CUSUM is initialized from $g_0 = z \geq 0$. We begin with a particular case of a zero initial condition $\mathsf{ARL}_0(\theta)$. Recall that this case corresponds to the minimax solution at the point $\theta = \theta_1$ and is perhaps the most interesting as well as a customary way. However, starting off CUSUM at a non-zero point $g_0 = z > 0$ is also of interest, since it leads to a *fast initial response* (FIR) allowing to detect faster changes that occur soon after surveillance begins to a certain expense of detecting slower "far away" changes.

Recalling that the CUSUM algorithm (8.66) can be derived using a sequence of repeated SPRTs with zero lower bound, as explained in Subsection 8.2.2, we derive Page's formula which links the ARL and the statistical properties of the SPRT $\delta^* = (d^*, T^*)$ given by (8.64) with the lower threshold $-\varepsilon = 0$ and the upper threshold $h > 0$ [346]. It follows from Wald's identity that

$$\mathsf{ARL}_0(\theta) = \mathsf{E}_\theta(T^*_{0,h} | \lambda_{T^*} \le 0, \lambda_0 = 0) \cdot \mathsf{E}_\theta(c-1) + \mathsf{E}_\theta(T^*_{0,h} | \lambda_{T^*} \ge h, \lambda_0 = 0) \cdot 1, \tag{8.96}$$

where $\mathsf{E}_\theta(T^*_{-\varepsilon,h} | \lambda_{T^*} \le -\varepsilon, \lambda_0 = z) = \mathsf{E}_\theta(T^*_{-\varepsilon,h} | \lambda_{T^*} \le -\varepsilon, z)$ is the conditional ESS of one cycle of the SPRT when the cumulative sum starting from z reaches the lower threshold $-\varepsilon$. In (8.96), $\mathsf{E}_\theta(c-1)$ is the mean number of cycles before the last cycle when the final decision is made, as depicted in Figure 8.5. Obviously, $c-1$ has a geometric distribution $\mathsf{P}_\theta(c-1=k) = (1-p)p^k$, $k = 0, 1, 2, \ldots$, where $p = \mathsf{P}_\theta(\lambda_{T^*} \le 0 | \lambda_0 = 0)$ is the probability that the SPRT starting from $z = 0$ reaches the lower threshold $-\varepsilon = 0$. Moreover, the probability p is nothing but the operating characteristic (OC); see Subsection 3.1.1. Thus,

$$\mathsf{E}_\theta(c-1) = \frac{1}{1-p} - 1 \tag{8.97}$$

and it follows from (8.96) that

$$\begin{aligned} \mathsf{ARL}_0(\theta) &= \frac{\mathsf{E}_\theta(T^*_{0,h} | \lambda_{T^*} \le 0, 0) \cdot p + \mathsf{E}_\theta(T^*_{0,h} | \lambda_{T^*} \ge h, = 0) \cdot (1-p)}{1-p} \\ &= \frac{\mathsf{E}_\theta(T^*_{0,h} | 0)}{1 - \mathsf{P}_\theta(\lambda_{T^*} \le 0 | \lambda_0 = 0)}. \end{aligned} \tag{8.98}$$

It is obvious from formula (8.98) and the definitions of $\mathsf{E}_\theta(T^*_{-\varepsilon,h} | \lambda_0 = z)$ and $\mathsf{P}_\theta(\lambda_{T^*} \le -\varepsilon | \lambda_0 = z)$ that the ARL is

$$\mathsf{ARL}_0(\theta) = \frac{\mathsf{ESS}_\theta(0)}{1 - \beta_\theta(0)}, \tag{8.99}$$

where $\mathsf{ESS}_\theta(z) = \mathsf{E}_\theta(T^*_{0,h} | \lambda_0 = z)$ is the expected sample size of the SPRT (d^*, T^*) with thresholds at 0 and h starting from z and $\beta_\theta(z) = \mathsf{P}_\theta(\lambda_{T^*} \le 0 | \lambda_0 = z)$ is its OC.

It is also of interest to get a general formula for the ARL function $\mathsf{ARL}_z(\theta)$ when CUSUM is initialized from an arbitrary point $z \in [0, h)$. Therefore, let $g_0 = z \ge 0$ in (8.66). In this case, the formula for the ARL function can be written as

$$\begin{aligned} \mathsf{ARL}_z(\theta) &= \mathsf{E}_\theta(T^*_{0,h} | \lambda_0 = z) + \mathsf{P}_\theta(\lambda_{T^*} \le 0 | \lambda_0 = z)\, \mathsf{ARL}_0(\theta) \\ &= \mathsf{ESS}_\theta(z) + \beta_\theta(z) \mathsf{ARL}_0(\theta). \end{aligned} \tag{8.100}$$

Indeed, when the detection statistic g_k starts from z, there are two possible scenarios: either g_k goes down and reaches the lower boundary $-\varepsilon = 0$ of the SPRT first or g_k goes up and reaches the upper boundary h of the SPRT without reaching the lower one. The probabilities of these two events are $\mathsf{P}_\theta(\lambda_{T^*} \le 0 | \lambda_0 = z)$ and $1 - \mathsf{P}_\theta(\lambda_{T^*} \le 0 | \lambda_0 = z)$, respectively. For computing the ARL function $\mathsf{ARL}_z(\theta)$, it is necessary to weight the conditional means with these two probabilities, and thus

$$\begin{aligned} \mathsf{ARL}_z(\theta) = {} & [1 - \mathsf{P}_\theta(\lambda_{T^*} \le 0 | \lambda_0 = z)]\ \mathsf{E}_\theta(T^*_{0,h} | \lambda_{T^*} \ge h, \lambda_0 = z) \\ & + \mathsf{P}_\theta(\lambda_{T^*} \le 0 | \lambda_0 = z)\ \left[\mathsf{E}_\theta(T^*_{0,h} | \lambda_{T^*} \ge h, \lambda_0 = z) + \mathsf{ARL}_0(\theta)\right], \end{aligned}$$

which implies (8.100). Note that when $z = 0$, we recover (8.98).

For computing $\mathsf{ARL}_z(\theta)$ from (8.100) we need to compute the $\mathsf{ESS}_\theta(z)$ of the SPRT (d^*, T^*) with thresholds 0 and h and its OC $\beta_\theta(z)$, which are solutions of the Fredholm integral equations of the second kind obtained in Subsection 3.1.4. Specifically, using equations (3.66)–(3.67), we obtain

$$\mathsf{ESS}_\theta(z) = 1 + \int_0^h g_\theta(y-z) \mathsf{ESS}_\theta(y)\, \mathrm{d}y, \tag{8.101}$$

$$\beta_\theta(z) = \int_{-\infty}^{-z} g_\theta(x) \mathrm{d}x + \int_0^h g_\theta(y-z) \beta_\theta(y)\, \mathrm{d}y, \tag{8.102}$$

for $0 \le z \le h$, where the kernel $g_\theta(x)$ of the integral equations is the pdf of the LLR increment $Z_1 = \log[f_{\theta_1}(X_1)/f_{\theta_0}(X_1)]$ under the true parameter value θ. These integral equations and numerical techniques for performance evaluation are discussed in Subsection 3.1.4 in detail; see also [167, 221, 345, 482].

An alternative way of obtaining integral equations for operating characteristics, including not only the ARL function but also the conditional ADD $\mathsf{E}_\nu(T_{\rm cs} - \nu | T_{\rm cs} > \nu)$ for every $\nu \ge 0$, is to consider the CUSUM procedure at the exponential scale as in (8.70) and (8.71), i.e., for $A > 0$, we define

$$T_{\rm cs}(A) = \inf\{n \ge 1 : V_n \ge A\} \quad \text{with} \quad V_n = \max(1, V_{n-1})\mathcal{L}_n, \quad n \ge 1, \ V_0 = 1.$$

The approach is similar to that considered in Subsection 7.1.2.1 for the Bayesian problem. In order to cover not only the CUSUM procedure but also other procedures such as SR and EWMA, we again consider a generic changepoint detection procedure given by (7.9) and (7.10),

$$T_A^r = \inf\{n \ge 1 : \mathcal{S}_n^r \ge A\}, \quad \mathcal{S}_n^r = \Phi(\mathcal{S}_{n-1}^r)\mathcal{L}_n, \quad n \ge 1, \quad \mathcal{S}_0^r = r \ge 0, \tag{8.103}$$

where $\Phi(\mathcal{S})$ is a positive-valued function. Note that for the CUSUM procedure $\Phi(\mathcal{S}) = \max(1, \mathcal{S})$ and $r = 1$.

Assume that distribution $F_\theta(x) = \mathsf{P}_\theta(\mathcal{L}_1 \le x)$ of the LR is continuous for all $\theta \in \Theta$. (Recall that $\mathsf{P}_{\theta_0} = \mathsf{P}_{\nu=\infty}$ and $\mathsf{P}_\theta = \mathsf{P}_{0,\theta}$ for $\theta \in \Theta^*$.)

Write $\mathsf{E}_{\nu,\theta}$ for the expectation under $\mathsf{P}_{\nu,\theta}$ and introduce the following notation:

$$\delta_{\nu,r}(\theta) = \mathsf{E}_{\nu,\theta}(T_A^r - \nu)^+; \quad \rho_{\nu,r}(\theta_0) = \mathsf{P}_\infty(T_A^r > \nu), \quad \nu \ge 0;$$
$$\mathsf{ARL}_r(\theta_0) = \mathsf{E}_\infty[T_A^r]; \quad \psi_r(\theta) = \sum_{\nu=0}^{\infty} \mathsf{E}_{\nu,\theta}(T_A^r - \nu)^+,$$

where, obviously, $\rho_{0,r} = 1$, $\delta_{0,r}(\theta) = \mathsf{ARL}_r(\theta)$ is the ARL to detection and $\mathsf{ARL}_r(\theta_0) = \mathsf{E}_\infty[T_A^r]$ is the ARL to false alarm. Also, define the kernels

$$\mathcal{K}_\theta(x,r) = \frac{\partial}{\partial x} F_\theta\left(\frac{x}{\Phi(r)}\right), \quad x, r \in [0, A), \ \theta \in \Theta. \tag{8.104}$$

Using the Markov property of the statistic $\mathcal{S}_n^r$, the following Fredholm integral equations (of the second kind) for operating characteristics can be obtained [309, 310]:

$$\mathsf{ARL}_r(\theta_0) = 1 + \int_0^A \mathsf{ARL}_x(\theta_0)\mathcal{K}_{\theta_0}(x,r)\,\mathrm{d}x \tag{8.105}$$

$$\mathsf{ARL}_r(\theta) = 1 + \int_0^A \mathsf{ARL}_x(\theta)\mathcal{K}_\theta(x,r)\,\mathrm{d}x, \quad \theta \in \Theta^* \tag{8.106}$$

$$\delta_{\nu,r}(\theta) = \int_0^A \delta_{\nu-1,x}(\theta)\,\mathcal{K}_{\theta_0}(x,r)\,\mathrm{d}x, \quad \nu = 1, 2, \dots, \ \theta \in \Theta^* \tag{8.107}$$

$$\rho_{\nu,r}(\theta_0) = \int_0^A \rho_{\nu-1,x}(\theta_0)\,\mathcal{K}_{\theta_0}(x,r)\,\mathrm{d}x, \quad \nu = 1, 2, \dots \tag{8.108}$$

$$\psi_r(\theta) = \mathsf{ARL}_r(\theta) + \int_0^A \psi_x(\theta)\,\mathcal{K}_{\theta_0}(x,r)\,\mathrm{d}x, \quad \theta \in \Theta^*. \tag{8.109}$$

The conditional average delay to detection is computed as

$$\mathsf{E}_{\nu,\theta}(T_A^r - \nu | T_A^r > \nu) = \delta_{\nu,r}(\theta)/\rho_{\nu,r}(\theta_0), \quad \nu \ge 0$$

and the integral average delay to detection as

$$\mathsf{IADD}(T_A^r) = \frac{r\mathsf{ARL}_r(\theta) + \psi(r)}{r + \mathsf{ARL}_r(\theta_0)}.$$

The local conditional false alarm probabilities $\mathsf{P}_\infty(T_A^r \le k+m|T_A^r > k)$, $k=0,1,\dots$, inside a fixed "window" of size $m \ge 1$ can also be easily evaluated noticing that

$$\mathsf{P}_\infty(T_A^r \le k+m|T_A^r > k) = \frac{\mathsf{P}_\infty(k < T_A^r \le k+m)}{\mathsf{P}_\infty(T_A^r > k)} = 1 - \frac{\mathsf{P}_\infty(T_A^r > k+m)}{\mathsf{P}_\infty(T_A^r > k)} = 1 - \frac{\rho_{k+m,r}}{\rho_{k,r}}, \quad (8.110)$$

where $\rho_{0,r} = 1$ and $\rho_{j,r}$, $j = 1,2\dots$ are given in (8.108).

It is also of interest to compute the ADD at infinity

$$\mathsf{ADD}_{\infty,\theta}(T_A^r) = \lim_{\nu\to\infty} \mathsf{E}_{\nu,\theta}(T_A^r - \nu|T_A^r > \nu),$$

which allows us to evaluate the performance of detection procedures in the case where the change occurs at a far time horizon, long after surveillance begins. Note that the initialization point can be taken arbitrary, since $\mathsf{ADD}_{\infty,\theta}(T_A^r)$ does not depend on r (but the ARL2FA depends). Let

$$\mathsf{Q}_A(x) = \lim_{n\to\infty} \mathsf{P}_\infty\left(\mathcal{S}_n^r \le x|T_A^r > n\right) \quad (8.111)$$

denote the *quasi-stationary* distribution of the Markov statistic $\mathcal{S}_n^r$ and let $q_A(x) = \mathrm{d}\mathsf{Q}_A(x)/\mathrm{d}x$ be quasi-stationary density. As mentioned in Subsection 2.1.4, in the continuous case the quasi-stationary distribution always exists.

The ADD at infinity is computed as

$$\mathsf{ADD}_{\infty,\theta}(T_A^r) = \int_0^A \mathsf{ARL}_x(\theta) q_A(x)\mathrm{d}x, \quad (8.112)$$

and by (2.17) density $q_A(x)$ satisfies the following integral equation

$$\lambda_A\, q_A(x) = \int_0^A q_A(y)\mathcal{K}_{\theta_0}(x,y)\,\mathrm{d}y, \quad (8.113)$$

where λ_A is the leading eigenvalue of the linear operator associated with the kernel $\mathcal{K}_{\theta_0}(x,y)$. Thus, $q_A(x)$ is the corresponding (left) eigenfunction. It also satisfies the constraint

$$\int_0^A q_A(x)\,\mathrm{d}x = 1. \quad (8.114)$$

Equations (8.113) and (8.114) uniquely define λ_A and $q_A(x)$. The equations have unique solutions, since $\lambda_A < 1$, as follows from [310]. Once $q_A(x)$ is computed (numerically) from integral equation (8.113), the ADD at infinity is evaluated from (8.112).

Note also that if the procedure T_A^r is initialized not from the fixed point r but from a random point $\mathcal{S}_0 \sim \mathsf{Q}_A$ sampled from the quasi-stationary distribution, the ARL to false alarm of such randomized procedure is computed as

$$\mathsf{E}_\infty\left[T_A^{\mathcal{S}_0}\right] = \int_0^A \mathsf{ARL}_r(\theta_0) q_A(r)\,\mathrm{d}r. \quad (8.115)$$

Also, the conditional ADD of this randomized procedure does not depend on ν, $\mathsf{E}_{\nu,\theta}(T_A^{\mathcal{S}_0} - \nu|T_A^{\mathcal{S}_0} > \nu) = \mathsf{E}_{0,\theta}T_A^{\mathcal{S}_0}$ for all $\nu \ge 1$, and it is computed as

$$\mathsf{E}_{0,\theta}\left[T_A^{\mathcal{S}_0}\right] = \int_0^A \mathsf{ARL}_r(\theta) q_A(r)\,\mathrm{d}r. \quad (8.116)$$

Setting $\Phi(r) = \max(1,r)$ yields operating characteristics for the CUSUM procedure for any initialization point r, including $r=1$. Note also that for the SR procedure one has to set $\Phi(r) = 1+r$ and for the double sided EWMA procedure $\Phi(r) = r^{(1-\alpha)}$. The latter follows from the fact that the EWMA procedure given by (8.40) and (8.42) can be written as

$$G_n^r = \left(G_{n-1}^r\right)^{1-\alpha}\mathcal{L}_n, \;\; G_0^r = r; \quad T_{\text{EWMA}}^r(A) = \inf\{n \ge 1 : G_n^r \notin (1/A, A)\},$$

where $G_n^r = \exp\{g_n/\alpha\}$ and $A = \exp\{h/\alpha\}$. In the two-sided case, integration in integral equations (8.105)–(8.109) is performed not in $[0,A)$ but in $[1/A,A)$.

Since the accuracy of numerical schemes for solving these integral equations depends on the number of breakpoints, which increases when the ARL2FA and hence the threshold A may take large values, this will require exceedingly large number of points resulting in a very long computation time. Thus, numerical techniques can be used for moderate values of the ARL2FA. For very large values of the ARL2FA, numerical methods cannot be used unless the form of the solution is known or can be guessed, and other approximations are needed. This issue is discussed in the next subsections.

8.2.6.2 *Wald's Approximations*

We now derive approximations for the ARL function, using (8.99) and Wald's approximation for ESS and OC [219, 220, 229, 317, 318, 319, 387]. We assume that the LLR increment $Z_i = \log[f_{\theta_1}(X_i)/f_{\theta_0}(X_i)]$ of the cumulative sum $\lambda_n = \sum_{i=1}^n Z_i$ satisfies conditions (3.11) and (3.12). From (3.21) and (3.17), we deduce that Wald's approximations for the ESS $\mathsf{E}_\theta(T^*_{-\varepsilon,h}|0)$ and the OC $\mathsf{P}_\theta(-\varepsilon|\lambda_0=0)$ are (the symbol $\widetilde{\ }$ is used to stress the fact of approximation)

$$\widetilde{\mathsf{E}}_\theta(T^*_{-\varepsilon,h}|\lambda_0=0) = \frac{-\varepsilon\widetilde{\mathsf{P}}_\theta(\lambda_{T^*}\le-\varepsilon|\lambda_0=0)+h(1-\widetilde{\mathsf{P}}_\theta(\lambda_{T^*}\le-\varepsilon|\lambda_0=0))}{\mathsf{E}_\theta Z_1}, \tag{8.117}$$

$$\widetilde{\mathsf{P}}_\theta(\lambda_{T^*}\le-\varepsilon|\lambda_0=0) = \frac{e^{-\omega_0 h}-1}{e^{-\omega_0 h}-e^{\omega_0\varepsilon}}, \tag{8.118}$$

where $\omega_0=\omega_0(\theta)$ is the unique non-zero root of the equation

$$\mathsf{E}_\theta\left(e^{-\omega_0 Z_1}\right)=1. \tag{8.119}$$

However, in the case of $\varepsilon=0$ formulas (8.117)–(8.118) lead to a degenerate ratio in (8.98). To avoid this difficulty, the ARL function $\mathsf{ARL}_0(\theta)$ is defined by the limit

$$\mathsf{ARL}_0(\theta)\approx\lim_{\varepsilon\to0}\frac{\widetilde{\mathsf{E}}_\theta(T^*_{-\varepsilon,h}|\lambda_0=0)}{1-\widetilde{\mathsf{P}}_\theta(\lambda_{T^*}\le-\varepsilon|\lambda_0=0)} \tag{8.120}$$

Combining (8.117), (8.118), and (8.120) yields the following approximation

$$\mathsf{ARL}_0(\theta)\approx\frac{1}{J_\theta}\left(h+\frac{e^{-\omega_0 h}}{\omega_0}-\frac{1}{\omega_0}\right), \tag{8.121}$$

which holds for θ such that $J_\theta=\mathsf{E}_\theta(Z_1)\neq0$.

We now compute the ARL function $\mathsf{ARL}_0(\theta^*)$ at the point where $\mathsf{E}_{\theta^*}(Z_1)=0$. In this case, the only root of equation (8.119) is equal to zero, and Wald's approximations (3.17) and (3.21) for the quantities $\mathsf{E}_{\theta^*}(T^*_{-\varepsilon,h}|\lambda_0=0)$ and $\mathsf{P}_{\theta^*}(\lambda_{T^*}\le-\varepsilon|\lambda_0=0)$ are given by

$$\widetilde{\mathsf{E}}_{\theta^*}(T^*_{-\varepsilon,h}|\lambda_0=0) = \frac{\widetilde{\mathsf{P}}_{\theta^*}(\lambda_{T^*}\le-\varepsilon|\lambda_0=0)\varepsilon^2+(1-\widetilde{\mathsf{P}}_{\theta^*}(\lambda_{T^*}\le-\varepsilon|\lambda_0=0))h^2}{\mathsf{E}_{\theta^*}(Z_k^2)}, \tag{8.122}$$

$$\widetilde{\mathsf{P}}_{\theta^*}(-\varepsilon|\lambda_0=0) = \frac{h}{h+\varepsilon}. \tag{8.123}$$

After substitution of these approximations in the formula (8.120), we get

$$\mathsf{ARL}_0(\theta^*)\approx\frac{h^2}{\mathsf{E}_{\theta^*}(Z_1^2)}. \tag{8.124}$$

So far we assumed that $z=0$, but it is interesting to get an approximation of the ARL function $\mathsf{ARL}_z(\theta)$ when $z>0$. Obviously, in this case, Wald's approximations (8.117) and (8.118) can be rewritten as

$$\widetilde{\mathsf{E}}_\theta(T^*_{0,h}|\lambda_0=z) = \frac{-z\widetilde{\mathsf{P}}_\theta(\lambda_{T^*}\le 0|\lambda_0=z)+(h-z)(1-\widetilde{\mathsf{P}}_\theta(\lambda_{T^*}\le 0|\lambda_0=z))}{J_\theta}, \quad (8.125)$$

$$\widetilde{\mathsf{P}}_\theta(\lambda_{T^*}\le 0|\lambda_0=z) = \frac{e^{-\omega_0(h-z)}-1}{e^{-\omega_0(h-z)}-e^{\omega_0 z}} \quad (8.126)$$

After substitution of $\widetilde{\mathsf{E}}_\theta(T^*_{0,h}|\lambda_0=z)$ from this equation and of $\mathsf{ARL}_0(\theta)$ from (8.121) into (8.100) we obtain

$$\mathsf{ARL}_z(\theta)\approx \frac{-z\widetilde{\mathsf{P}}_\theta(\lambda_{T^*}\le 0|\lambda_0=z)+(h-z)(1-\widetilde{\mathsf{P}}_\theta(\lambda_{T^*}\le 0|\lambda_0=z))}{J_\theta} + \frac{\widetilde{\mathsf{P}}_\theta(\lambda_{T^*}\le 0|\lambda_0=z)}{J_\theta}\left(h+\frac{e^{-\omega_0 h}}{\omega_0}-\frac{1}{\omega_0}\right). \quad (8.127)$$

Finally the substitution of the previous estimate of $\mathsf{P}_\theta(\lambda_{T^*}\le 0|\lambda_0=z)$ results in

$$\mathsf{ARL}_z(\theta)\approx\frac{1}{\mathsf{E}_\theta(Z_1)}\left(h-z+\frac{e^{-\omega_0 h}}{\omega_0}-\frac{e^{-\omega_0 z}}{\omega_0}\right). \quad (8.128)$$

Let us now derive the approximation for the ARL function $\mathsf{ARL}_z(\theta)$ at $\theta=\theta^*$. Wald's approximations for $\mathsf{E}_{\theta^*}(T^*_{0,h}|\lambda_0=z)$ and $\mathsf{P}_{\theta^*}(0|\lambda_0=z)$ when $g_0=z$ are as follows

$$\widetilde{\mathsf{E}}_{\theta^*}(T^*_{0,h}|\lambda_0=z)=\frac{\widetilde{\mathsf{P}}_{\theta^*}(\lambda_{T^*}\le 0|\lambda_0=z)z^2+(1-\widetilde{\mathsf{P}}_{\theta^*}(\lambda_{T^*}\le 0|\lambda_0=z))(h-z)^2}{\mathsf{E}_{\theta^*}(Z_k^2)}, \quad (8.129)$$

$$\widetilde{\mathsf{P}}_{\theta^*}(\lambda_{T^*}\le 0|\lambda_0=z)=\frac{h-z}{h}. \quad (8.130)$$

After substitution of these approximations in (8.100) we get

$$\mathsf{ARL}_z(\theta^*)\approx\frac{h^2-z^2}{\mathsf{E}_{\theta^*}(Z_k^2)}. \quad (8.131)$$

Let us finally discuss the two approximations (8.121) and (8.128). One may expect that the ARL function $\mathsf{ARL}_z(\theta)$ is monotone in z, so in particular $\mathsf{ARL}_z(\theta)<\mathsf{ARL}_0(\theta)$ for $z>0$. The obtained approximations confirm this fact. Indeed, it follows from (8.121) and (8.128) that

$$\mathsf{ARL}_z(\theta)-\mathsf{ARL}_0(\theta)\approx\frac{\omega_0}{J_\theta}\left(-\omega_0 z-e^{-\omega_0 z}+1\right),$$

where the right-hand side is negative since the first term in the product is always positive because $\mathsf{E}_\theta(Z_1)$ and $\omega_0=\omega_0(\theta)$ have the same sign, and the second term of the product is always negative because $e^x\ge 1+x$ for all $x\in\mathbb{R}$. For θ^*, using (8.124) and (8.131), we obtain

$$\mathsf{ARL}_z(\theta^*)-\mathsf{ARL}_0(\theta^*)\approx\frac{-z^2}{\mathsf{E}_{\theta^*}(Z_1^2)}<0 \quad \text{for all } z>0.$$

8.2.6.3 Siegmund's Corrected Brownian Motion Approximation

In the previous subsection, we derived the approximations to the ARL function based upon Wald's formulas for ESS and OC that ignore overshoots over the boundaries, e.g., replacing $\mathsf{E}_\theta(\lambda_{T^*}|\lambda_{T^*}\le-\varepsilon)$ by $-\varepsilon$. Unfortunately these approximations are not very accurate especially when $\mathsf{E}_\theta(Z_1)<0$

and $\sqrt{\mathsf{var}(Z_1)}$ have the same order of magnitude as h. In this case, the error in $\mathsf{P}_\theta(-\varepsilon|\lambda_0=0)$ and $\mathsf{E}_\theta(T|\lambda_0=0)$ resulting from the omission of the overshoots is significant. Therefore the issue of accuracy and comparison between different approximations for ARL is of interest and discussed next.

In this subsection, we discuss another approximation that accounts for the overshoots introduced in Subsections 2.5.1 and 2.5.2 and applied to the SPRT in Subsection 3.1.3. This approximation, referred to as the *corrected Brownian motion approximation*, was suggested by Siegmund [425, 426, 427, 428]. The general theory of the corrected diffusion approximation is complex (see Subsection 3.1.3), and here we only present a short heuristic proof of the ARL formula and refer to Siegmund's works for exact mathematical results and details. The main idea of Siegmund's approximation is the following. The formula for the ARL is the same as before in (8.121), except that we replace the threshold h by $h+\varkappa$, where $\varkappa$ is some positive constant due to the overshoot.

Let us consider the SPRT with boundaries $-\varepsilon$ and h. The ESS $\mathsf{E}_\theta(T^*_{-\varepsilon,h})$ and the OC $\mathsf{P}_\theta(\lambda_{T^*}\leq -\varepsilon)=\mathsf{P}_\theta(-\varepsilon)$ can be written as

$$\mathsf{E}_\theta(T^*_{-\varepsilon,h})=\frac{\mathsf{E}_\theta(\lambda_{T^*}|\lambda_{T^*}\leq-\varepsilon)\mathsf{P}_\theta(\lambda_{T^*}\leq-\varepsilon)+\mathsf{E}_\theta(\lambda_{T^*}|\lambda_{T^*}\geq h)(1-\mathsf{P}_\theta(\lambda_{T^*}\leq-\varepsilon))}{J_\theta}, \tag{8.132}$$

$$\mathsf{P}_\theta(\lambda_{T^*}\leq-\varepsilon)=\frac{\mathsf{E}_\theta(e^{-\omega_0\lambda_{T^*}}|\lambda_{T^*}\geq h)-1}{\mathsf{E}_\theta(e^{-\omega_0\lambda_{T^*}}|\lambda_{T^*}\geq h)-\mathsf{E}_\theta(e^{-\omega_0\lambda_{T^*}}|\lambda_{T^*}\leq -\varepsilon)}. \tag{8.133}$$

It results from (8.98) that for $\varepsilon=0$

$$\begin{aligned}
\mathsf{ARL}_0(\theta) &= \frac{\mathsf{E}_\theta(T^*_{0,h})}{1-\mathsf{P}_\theta(\lambda_{T^*}\leq 0)}\\
&= \frac{\mathsf{E}_\theta(\lambda_{T^*}|\lambda_{T^*}\leq 0)\,\mathsf{P}_\theta(0)+\mathsf{E}_\theta(\lambda_{T^*}|\lambda_{T^*}\geq h)\,(1-\mathsf{P}_\theta(0))}{J_\theta(1-\mathsf{P}_\theta(0))}\\
&= \frac{1}{J_\theta}\left[\mathsf{E}_\theta(\lambda_{T^*}|\lambda_{T^*}\geq h)-\mathsf{E}_\theta(\lambda_{T^*}|\lambda_{T^*}\leq 0)+\frac{\mathsf{E}_\theta(\lambda_{T^*}|\lambda_{T^*}\leq 0)}{1-\mathsf{P}_\theta(0)}\right] \qquad (8.134)\\
&= \frac{1}{J_\theta}\left[h+\mathsf{E}_\theta(\lambda_{T^*}-h|\lambda_{T^*}-h\geq 0)-\mathsf{E}_\theta(\lambda_{T^*}|\lambda_{T^*}\leq 0)+\frac{\mathsf{E}_\theta(\lambda_{T^*}|\lambda_{T^*}\leq 0)}{1-\mathsf{P}_\theta(0)}\right].
\end{aligned}$$

Let us compute the fourth term on the right-hand side of the last equality. It is obvious from (8.132) that this term can be rewritten as

$$\begin{aligned}
\frac{\mathsf{E}_\theta(\lambda_{T^*}|\lambda_{T^*}\leq 0)}{1-\mathsf{P}_\theta(0)} &= \mathsf{E}_\theta(\lambda_{T^*}|\lambda_{T^*}\leq 0)\,\frac{\mathsf{E}_\theta(e^{-\omega_0\lambda_{T^*}}|\lambda_{T^*}\geq h)-\mathsf{E}_\theta(e^{-\omega_0\lambda_{T^*}}|\lambda_{T^*}\leq 0)}{1-\mathsf{E}_\theta(e^{-\omega_0\lambda_{T^*}}|\lambda_{T^*}\leq 0)}\\
&= \mathsf{E}_\theta(\lambda_{T^*}|\lambda_{T^*}\leq 0)\,\frac{e^{-\omega_0 h}\mathsf{E}_\theta(e^{-\omega_0(\lambda_{T^*}-h)}|\lambda_{T^*}-h\geq 0)-\mathsf{E}_\theta(e^{-\omega_0\lambda_{T^*}}|\lambda_{T^*}\leq 0)}{1-\mathsf{E}_\theta(e^{-\omega_0\lambda_{T^*}}|\lambda_{T^*}\leq 0)}.
\end{aligned}$$

Assume now that $h\to\infty$ and $\omega_0\to 0$ in such a way that $h\omega_0=\varsigma\in(0,\infty)$. In this case, the expansion of the expectation of $e^{-\omega_0\lambda_{T^*}}$ can be written as

$$\mathsf{E}_\theta(e^{-\omega_0\lambda_{T^*}})\approx 1-\omega_0\mathsf{E}_\theta(\lambda_{T^*})+\frac{1}{2}\omega_0^2\mathsf{E}_\theta(\lambda_{T^*}^2)+\ldots \tag{8.135}$$

This means that the above approximation is especially good in the case of θ close to θ^*. From (8.135) it results immediately that

$$\begin{aligned}
\frac{\mathsf{E}_\theta(\lambda_{T^*}|\lambda_{T^*}\leq 0)}{1-\mathsf{P}_\theta(0)} &\approx \frac{1}{\omega_0}\Big[e^{-\omega_0 h}(1-\omega_0\mathsf{E}_\theta(\lambda_{T^*}-h)|\lambda_{T^*}-h\geq 0)+\ldots) \qquad (8.136)\\
&- (1-\omega_0\mathsf{E}_\theta(\lambda_{T^*}|\lambda_{T^*}\leq 0)+\ldots)\Big].
\end{aligned}$$

Let us now recall the notion of the overshoot of the LLR λ_{T^*} over the boundaries introduced in Subsections 3.1.1 and 3.1.3. Let $\kappa_- = \kappa_0(0,h) = -\lambda_{T^*}$ and $\kappa_+ = \kappa_1(0,h) = \lambda_{T^*} - h$ be the overshoots over the boundaries h and 0 and

$$\varkappa_- = \mathsf{E}_\theta(-\lambda_{T^*}|\lambda_{T^*} \le 0), \quad \varkappa_+ = \mathsf{E}_\theta(\lambda_{T^*} - h|\lambda_{T^*} - h \ge 0) \tag{8.137}$$

be their expectations. By using the approximation $1 - x = e^x + o(x)$ as $x \to 0$, we obtain

$$\frac{\mathsf{E}_\theta(\lambda_{T^*}|\lambda_{T^*} \le 0)}{1 - \mathsf{P}_\theta(0)} \approx \frac{e^{-\omega_0(h+\varkappa_+ + \varkappa_-)} - 1}{\omega_0 e^{-\omega_0 \varkappa_-}} \approx \frac{e^{-\omega_0(h+\varkappa_+ + \varkappa_-)} - 1}{\omega_0}[1 + o(\omega_0)]. \tag{8.138}$$

Inserting (8.137) and (8.138) in (8.134), we get the following ARL approximation

$$\mathsf{ARL}_0(\theta) \approx \frac{1}{J_\theta}\left(h + \varkappa_+ + \varkappa_- + \frac{e^{-\omega_0(h+\varkappa_+ + \varkappa_-)} - 1}{\omega_0}\right). \tag{8.139}$$

The direct comparison of (8.121) and (8.139) shows that *Siegmund's approximation has the same form as Wald's approximation with h replaced with* $h + \varkappa_+ + \varkappa_-$.

Let us discuss how to incorporate average overshoots over the boundaries $\varkappa_+$ and $\varkappa_-$. Interested readers are referred to [426, Theorems 2 and 3] and [428, Theorems 10.13 and 10.16]. Let us assume that the pdf $p_\theta(x)$ of the LLR increment $Z_k = \log \frac{p_{\theta_1}(X_k)}{p_{\theta_0}(X_k)}$ belongs to the exponential family

$$p_\theta(x) = h(x)\exp\{\theta x - b(\theta)\}.$$

Also define $\tau_+ = \inf\{n \ge 1 : \lambda_n > 0\}$ and $\tau_- = \inf\{n \ge 1 : \lambda_n \le 0\}$, where $\lambda_n = \sum_{k=1}^n Z_k$. Let $\mu_0 = \mathsf{E}_{\theta_0}(Z_k) = b'(\theta_0)$ and $\mu_1 = \mathsf{E}_{\theta_1}(Z_k) = b'(\theta_1)$ be conjugate in the sense that $\mu_0 < 0 < \mu_1$ and $b(\theta_0) = b(\theta_1)$. Assuming that $\theta_1 - \theta_0 \to 0$ and $h \to +\infty$ in such a way that $(\theta_1 - \theta_0)h \in (0, +\infty)$ and using renewal theory, it can be proved that the limiting average overshoot is asymptotically

$$\varkappa_+ = \mathsf{E}_{\mu_1}(\lambda_{T^*} - h|\lambda_{T^*} - h \ge 0) \to \rho_+ = \frac{\mathsf{E}_0(\lambda_{\tau_+}^2)}{2\,\mathsf{E}_0(\lambda_{\tau_+})} \quad \text{as } h \to \infty. \tag{8.140}$$

For the Gaussian case $p(x) = \varphi(x) = \frac{1}{\sqrt{2\pi}}\exp\{-\frac{1}{2}x^2\}$,

$$\rho_+ = -\pi^{-1}\int_0^\infty x^{-2}\log\left(\frac{2}{x^2}\left[1 - e^{-\frac{1}{2}x^2}\right]\right)\mathrm{d}x \approx 0.583. \tag{8.141}$$

Analogously, $\varkappa_- \to -\rho_- = -\dfrac{\mathsf{E}_0(\lambda_{\tau_-}^2)}{2\,\mathsf{E}_0(\lambda_{\tau_-})} \approx 0.583.$

8.2.6.4 Renewal Theoretic Approximations

Write $T_h^* = T_{-\varepsilon=0,h}^*$ for the SPRT with the zero lower threshold and define

$$e_1(\theta) = \mathsf{E}_\theta\left[e^{-(\lambda_{T_h^*} - h)}|\lambda_{T_h^*} \ge h\right], \quad e_0 = \mathsf{E}_{\theta_0}\left[e^{\lambda_{T_h^*}}|\lambda_{T_h^*} \le 0\right], \quad \widetilde{e}_0(\theta_1) = \mathsf{E}_\theta\left[e^{-\lambda_{T_h^*}}|\lambda_{T_h^*} \le 0\right],$$
$$\delta_{00} = \mathsf{E}_{\theta_0}(\lambda_{T_h^*}|\lambda_{T_h^*} \le 0), \quad \delta_{10} = \mathsf{E}_{\theta_0}(\lambda_{T_h^*} - h|\lambda_{T_h^*} \ge h),$$
$$\delta_{11}(\theta) = \mathsf{E}_\theta(\lambda_{T_h^*} - h|\lambda_{T_h^*} \ge h), \quad \delta_{01}(\theta) = \mathsf{E}_\theta(\lambda_{T_h^*}|\lambda_{T_h^*} \le 0).$$

Setting $\theta = \theta_0$ and $\theta = \theta_1$ in equality (8.134) and noting that $\omega(\theta_0) = -1$ and $\omega_0(\theta_1) = 1$, we obtain the exact formulas for the ARL2FA and the ARL to detection (at the putative value θ_1) [451]

$$\mathsf{ARL2FA}(h) = \frac{1}{I_0}\left[\frac{|\delta_{00}|e^h}{e_1(\theta_1)(1 - e_0)} - h - \frac{|\delta_{00}|}{1 - e_0} - \delta_{10}\right], \tag{8.142}$$

$$\mathsf{ADD}_0(h) = \frac{1}{I}\left[h - \frac{e_1(\theta_1)|\delta_{01}(\theta_1)|}{1 - \widetilde{e}_0(\theta_1)}e^{-h} - \frac{|\delta_{01}(\theta_1)|}{1 - \widetilde{e}_0(\theta_1)} + \delta_{11}(\theta_1)\right]. \tag{8.143}$$

However, except for some rare cases such as the exponential example below, the values of e_i and δ_{ij} cannot be computed exactly. Asymptotic approximations can be obtained based on renewal theoretic considerations. We first provide an approximate formula for the ARL2FA that follows directly from (8.142). Note that the main interest for applications represent an accurate approximation for the ARL2FA that allows us to design the threshold to guarantee equality $\mathsf{ARL2FA}(h) = \gamma$ almost exactly. The approximations for the ADD are of secondary interest.

Let $T_+ = \inf\{n : \lambda_n > 0\}$ and $T_- = \inf\{n : \lambda_n \le 0\}$. It can be verified that, as $h \to \infty$,

$$e_1(\theta_1) \to \zeta, \quad e_0 \to \mathsf{E}_\infty\left[e^{\lambda_{T_-}}\right], \quad \delta_{00} \to \mathsf{E}_\infty \lambda_{T_-}.$$

Next, by Wald's identities,

$$\mathsf{E}_\infty \lambda_{T_-} = -I_0 \mathsf{E}_\infty T_-, \quad \mathsf{E}_\infty\left[e^{\lambda_{T_-}}\right] = \mathsf{P}_\infty(T_- < \infty),$$

and, by Lemma 2.5.1, $\mathsf{E}_\infty T_- = 1/\mathsf{P}_\infty(T_+ = \infty)$. Hence

$$\frac{|\delta_{00}|}{1-e_0} \to \frac{I_0}{\mathsf{P}_\infty(T_+ = \infty)\mathsf{P}_0(T_- = \infty)}.$$

Finally, it can be shown that

$$\zeta = \mathsf{P}_\infty(T_+ = \infty)\mathsf{P}_0(T_- = \infty)/I,$$

which along with the previous equality yields

$$\frac{|\delta_{00}|}{1-e_0} \to \frac{I_0}{I\zeta}.$$

Substitution into (8.142) yields

$$\mathsf{ARL2FA}(h) \approx I_0^{-1}\left[I_0/(I\zeta^2)e^h - h - I_0/(I\zeta) - \delta_{10}\right]. \tag{8.144}$$

Ignoring the linear term h as well as constants in this approximation, we obtain the first-order (FO) asymptotic approximation

$$\mathsf{ARL2FA}(h) = \frac{e^h}{I\zeta^2}(1+o(1)) \quad \text{as } h \to \infty. \tag{8.145}$$

We now proceed with evaluating the conditional ADD $\mathsf{E}_{\nu,\theta}[T_{\mathrm{CS}}(h) - \nu | T_{\mathrm{CS}}(h) > \nu]$ for large h and θ such that $J_\theta > 0$. However, evaluation of the CADD for arbitrary ν is not feasible. The only feasible cases seem to be $\nu = 0$ and large values of ν, i.e., the limiting value $\mathsf{CADD}_{\infty,\theta}(h) = \lim_{\nu\to\infty} \mathsf{E}_{\nu,\theta}[T_{\mathrm{CS}}(h) - \nu | T_{\mathrm{CS}}(h) > \nu]$.

Consider $\nu = 0$. As before, for the sake of brevity, we omit the index $\nu = 0$ and write P_θ and E_θ for $\mathsf{P}_{0,\theta}$ and $\mathsf{E}_{0,\theta}$. Let $T_{\mathrm{CS}}^z(h)$ be the CUSUM stopping time when $g_n = g_n^z$ is initialized from z, $g_0^z = z \ge 0$. Rewrite the stopping time $T_{\mathrm{CS}}^z(h)$ in the form of a random walk crossing a threshold plus a nonlinear slowly changing term. This allows us to apply nonlinear renewal theory. Specifically, observe that T_{CS}^z can be written in the form

$$T_{\mathrm{CS}}^z(h) = \inf\left\{n \ge 1 : \lambda_n - \min\left(-z, \min_{0\le k\le n} \lambda_k\right) \ge h\right\}$$

($\lambda_0 = 0$). Write $\xi_n^z = -\min(-z, \min_{0\le k\le n}\lambda_k)$. Then $\lambda_{T_{\mathrm{CS}}^z} = h - \xi_{T_{\mathrm{CS}}^z}^z + \chi_h$, where $\chi_h = \lambda_{T_{\mathrm{CS}}^z} + \xi_{T_{\mathrm{CS}}^z}^z - h$ is the excess over the level h. Observing that the sequence $\{\lambda_n\}_{n\ge1}$ is a random walk with mean $J_\theta n$, taking the expectation and applying Wald's identity, we obtain

$$J_\theta \mathsf{E}_\theta T_{\mathrm{CS}}^z = h - \mathsf{E}_\theta \xi_{T_{\mathrm{CS}}^z}^z + \mathsf{E}_\theta \chi_h.$$

The crucial observation is that the sequence $\{\xi_n^z\}_{n\ge 1}$ is slowly changing and, moreover, converges P_θ-a.s. as $n\to\infty$ to the random variable $\xi^z = -\min(-z, \min_{k\ge 0}\lambda_k)$ with finite expectation

$$\mathsf{E}_\theta \xi^z = \bar{\xi}(\theta,z) = -\int_{-\infty}^{-z} y \,\mathrm{d}\widetilde{\mathsf{Q}}_\theta(y),$$

where $\widetilde{\mathsf{Q}}_\theta(y) = \mathsf{P}_\theta(\min_{k\ge 0}\lambda_k \le y)$. By the Nonlinear Renewal Theorem, under certain conditions spelled out below in Theorem 8.2.5,

$$\mathsf{ARL}_z(\theta) = J_\theta^{-1}\left[h + \varkappa(\theta) - \bar{\xi}(\theta,z)\right] + o(1) \quad \text{as } h\to\infty,\ \theta\in\tilde{\Theta}, \tag{8.146}$$

where $\tilde{\Theta} = \{\theta : J_\theta > 0\}$ and $\varkappa(\theta) = \lim_{a\to\infty}\mathsf{E}_\theta(a-\lambda_{\tau_a})$ is the limiting average overshoot in the one-sided test $\tau_a = \inf\{n : \lambda_n \ge a\}$.

Consider now the case of large ν. Let $\mathsf{Q}_h(y) = \lim_{n\to\infty}\mathsf{P}_\infty(g_n^z \le y | T_{\mathrm{CS}}^z > n)$ be the quasi-stationary distribution of the CUSUM statistic and let $\mathsf{Q}_{\mathrm{st}}(y) = \lim_{n\to\infty}\mathsf{P}_\infty(g_n^z\le y)$ be the stationary distribution of g_n^z. Analogously to (8.112),

$$\mathsf{CADD}_{\infty,\theta}(h) = \int_0^h \mathsf{ARL}_z(\theta)\,\mathrm{d}\mathsf{Q}_h(z),$$

so taking into account (8.146) and the fact that $\mathsf{Q}_h\to\mathsf{Q}_{\mathrm{st}}$ as $h\to\infty$ (see Theorem 2.1.2) it stands to reason that

$$\mathsf{CADD}_{\infty,\theta}(h) = J_\theta^{-1}\left[h + \varkappa(\theta) - C(\theta)\right] + o(1) \quad \text{as } h\to\infty,\ \theta\in\tilde{\Theta}, \tag{8.147}$$

where

$$C(\theta) = -\int_0^\infty \int_{-\infty}^{-z} y\,\mathrm{d}\widetilde{\mathsf{Q}}_\theta(y)\,\mathrm{d}\mathsf{Q}_{\mathrm{st}}(z).$$

The exact result is given in the following theorem.

Theorem 8.2.5. *Assume that* $\mathsf{E}_\theta|Z_1|^2<\infty$ *and that* Z_1 *is* P_θ*-nonarithmetic. Let there exist a positive* $\omega_0 = \omega_0(\theta)$ *such that* $\mathsf{E}_\theta[e^{-\omega_0 Z_1}] = 1$. *Then the ARL function* $\mathsf{ARL}_z(\theta) = \mathsf{E}_\theta[T_{\mathrm{CS}}^z]$ *can be asymptotically approximated as in* (8.146) *and* $\mathsf{CADD}_{\infty,\theta}(h) = \lim_{\nu\to\infty}\mathsf{E}_{\nu,\theta}[T_{\mathrm{CS}}^z(h) - \nu | T_{\mathrm{CS}}^z(h) > \nu]$ *as in* (8.147).

Proof. The proof is based on the second Nonlinear Renewal Theorem 2.6.4. To prove (8.146) note that for CUSUM $\ell_n = 0$ and $\xi_n = \eta_n = -\min(-z, \min_{0\le k\le n}\lambda_k)$. Conditions (2.126)–(2.128) hold trivially. Condition (2.131) is satisfied with $\eta = \xi^z = -\min(-z,\min_{k\ge 0}\lambda_k)$. The second moment condition $\mathsf{E}_\theta|Z_1|^2<\infty$ implies uniform integrability in (2.129). It remains to check condition (2.132), i.e., in the notation of this section, that

$$\lim_{h\to\infty} h\,\mathsf{P}_{0,\theta}\{T_{\mathrm{CS}}^z(h) \le (1-\varepsilon)h/J_\theta\} = 0 \quad \text{for some } 0<\varepsilon<1. \tag{8.148}$$

Moreover, obviously, it suffices to check this condition for the standard CUSUM that starts from zero, $z=0$.

Since $\mathsf{E}_\theta[e^{-\omega_0 Z_1}] = 1$, it follows that $p_\theta(x) = e^{-\omega_0 Z_1(x)} f_\theta(x)$ is a bona fide probability density and $p_\theta(x)/f_\theta(x) = e^{-\omega_0 Z_1(x)}$ is the LR. Therefore, again, as in (7.23), we can use the change-of-measure argument to obtain that for any $C>0$ and $\varepsilon\in(0,1)$

$$\begin{aligned}
\mathsf{P}_\infty\{T_{\mathrm{CS}} \le (1-\varepsilon)K\} &= \mathsf{E}_\theta\left\{\mathbb{1}_{\{T_{\mathrm{CS}}\le(1-\varepsilon)K\}} e^{-\lambda_{T_{\mathrm{CS}}}\omega_0}\right\} \\
&\ge \mathsf{E}_\theta\left\{\mathbb{1}_{\{T_{\mathrm{CS}}\le(1-\varepsilon)K,\lambda_{T_{\mathrm{CS}}}\le C\}} e^{-\lambda_{T_{\mathrm{CS}}}\omega_0}\right\} \\
&\ge e^{-\omega_0 C}\mathsf{P}_\theta\left\{T_{\mathrm{CS}}\le(1-\varepsilon)K,\ \max_{n\le(1-\varepsilon)K}\lambda_n \le C\right\} \\
&\ge e^{-\omega_0 C}\left[\mathsf{P}_\theta\{T_{\mathrm{CS}}\le(1-\varepsilon)K\} - \mathsf{P}_\theta\left\{\max_{n\le(1-\varepsilon)K}\lambda_n > C\right\}\right].
\end{aligned} \tag{8.149}$$

Setting $C = (1-\varepsilon^2)h/\omega_0$ and $K = h/J_\theta$ yields

$$\begin{aligned}\mathsf{P}_\theta\{T_{\rm CS} \le (1-\varepsilon)h/J_\theta\} \le \mathsf{P}_\theta\left\{\max_{n\le(1-\varepsilon)h/J_\theta} \lambda_n > (1-\varepsilon^2)h/\omega_0\right\} \\ + e^{(1-\varepsilon^2)h}\mathsf{P}_\infty\{T_{\rm CS} \le (1-\varepsilon)h/J_\theta\}.\end{aligned} \tag{8.150}$$

Writing $N_h = (1-\varepsilon)h/(J_\theta\omega_0)$, for the first term we have

$$\begin{aligned}\mathsf{P}_\theta\left\{\max_{n\le(1-\varepsilon)h/J_\theta} \lambda_n > (1-\varepsilon^2)h/\omega_0\right\} &= \mathsf{P}_\theta\left\{\max_{n\le N_h}\lambda_n > (1+\varepsilon)J_\theta N_h\right\} \\ &= \mathsf{P}_\theta\left\{\max_{n\le N_h}(\lambda_n - J_\theta N_h) > \varepsilon J_\theta N_h\right\} \\ &\le \mathsf{P}_\theta\left\{\max_{n\le N_h}(\lambda_n - J_\theta n) > \varepsilon J_\theta N_h\right\}.\end{aligned}$$

By Lemma 2.6.2, for every $0<\varepsilon<1$

$$\mathsf{P}_\theta\left\{\max_{n\le N_h}(\lambda_n - J_\theta n) > \varepsilon J_\theta N_h\right\} = o(1/N_h) = o(1/h) \quad \text{as } h\to\infty$$

whenever $\mathsf{E}_\theta|Z_1|^2<\infty$. It remains to show that the second term in (8.150) is of order $o(1/h)$.

Since $e^{g_n} \le R_n = \sum_{k=1}^n \prod_{i=k}^n e^{Z_i}$ for all $n\ge 1$, we have

$$\mathsf{P}_\infty\{T_{\rm CS}(h) \le (1-\varepsilon)h/J_\theta\} = \mathsf{P}_\infty\left\{\max_{n\le(1-\varepsilon)h/J_\theta} g_n \ge h\right\} \le \mathsf{P}_\infty\left\{\max_{n\le(1-\varepsilon)h/J_\theta} R_n \ge e^h\right\}.$$

Noting that R_n is a nonnegative P_∞-submartingale with mean $\mathsf{E}_\infty R_n = n$ and applying Doob's submartingale inequality to the last probability, we obtain

$$\mathsf{P}_\infty\{T_{\rm CS}(h) \le (1-\varepsilon)h/J_\theta\} \le (1-\varepsilon)(h/J_\theta)e^{-h},$$

so the second term in (8.150) is bounded by

$$e^{(1-\varepsilon^2)h}(1-\varepsilon)(h/J_\theta)e^{-h} = e^{-\varepsilon^2 h}(1-\varepsilon)(h/J_\theta) = o(1/h) \quad \text{as } h\to\infty.$$

Therefore, condition (8.148) holds and the proof of (8.146) is complete.

A rigorous proof of approximation (8.147) is more difficult and lengthy, so we present only a proof sketch. Let $0<N_h<h$ be such that, for some $\delta\in(0,1)$, $N_h/(h^{1-\delta}\log h)\to\infty$ and $N_h = o(h/\log h)$ as $h\to\infty$. An extension of the above nonlinear renewal-theoretic argument (cf. Lemma 3.1 in Tartakovsky *et al.* [467] for the SR procedure) shows that, as $h\to\infty$,

$$\mathsf{ARL}_z(\theta) = J_\theta^{-1}[h + \varkappa(\theta) - \mathsf{E}_\theta\xi^z] + o(1), \tag{8.151}$$

where $o(1)\to 0$ as $h\to\infty$ uniformly on $\{z<h/N_h\}$. Finally, using (8.151), the fact that $\mathsf{CADD}_{\infty,\theta}(T_{\rm CS}) = \mathsf{E}\{\mathsf{E}_\theta[T_{\rm CS}^{g_0}|g_0]\}$, where $g_0\sim \mathsf{Q}_h$, and applying a rather tedious reasoning similar to that in the proof of Theorem 3.2 in [467] yields

$$\mathsf{CADD}_{\infty,\theta}(T_{\rm CS}) = J_\theta^{-1}\left\{h + \varkappa(\theta) + \mathsf{E}\left[\min(-g_\infty, \min_{k\ge 0}\lambda_k)\right]\right\} + o(1),$$

where g_∞ is the random variable that has stationary distribution $\mathsf{Q}_{\rm st}(y) = \lim_{n\to\infty}\mathsf{P}_\infty(g_n\le y)$ of the CUSUM statistic. This completes the proof of (8.147). □

We now address the problem of computing the constants in these approximations.

A special interest represents the case of standard CUSUM that starts from zero, $z = 0$. Write $\bar{\xi}(\theta) = \bar{\xi}(\theta, z = 0)$. Using equality (2.97), we obtain that as long as $J_\theta > 0$ and $\mathsf{E}_\theta Z_1^2 < \infty$ the limiting average overshoot $\varkappa(\theta)$ and the expected value $\bar{\xi}(\theta) = -\mathsf{E}_\theta[\min_{n\ge 0} \lambda_n]$ are related as follows

$$\bar{\xi}(\theta) = \mathsf{E}_\theta[Z_1^2]/2J_\theta - \varkappa(\theta). \tag{8.152}$$

Hence, using (8.146) and (8.152), we obtain the following convenient formula for the ARL function of the standard CUSUM

$$\mathsf{ARL}_0(\theta) = \frac{1}{J_\theta}\left\{h + 2\varkappa(\theta) - \frac{\mathsf{E}_\theta[Z_1^2]}{2J_\theta}\right\} + o(1) \quad \text{as } h \to \infty, \ \theta \in \tilde{\Theta}. \tag{8.153}$$

To compute the stationary distribution note that for CUSUM

$$1 - \mathsf{Q}_{\mathrm{st}}(x) = \mathsf{P}_\infty\left(\max_{n\ge 0} \lambda_n \ge x\right) = \mathsf{P}_\infty(\tau_x < \infty), \quad x \ge 0.$$

This yields the upper bound $1 - \mathsf{Q}_{\mathrm{st}}(x) \le e^{-x}$ and the approximation $1 - \mathsf{Q}_{\mathrm{st}}(x) \sim \zeta(\theta_1)e^{-x}$ for sufficiently large x.

Example 8.2.1 (Change in the parameter of the exponential distribution). In order to verify the accuracy of asymptotic approximations for reasonable values of the threshold h, we performed Monte Carlo simulations for the exponential example where observations are independent, originally having an $\mathsf{Expon}(1)$ distribution, changing at an unknown time to $\mathsf{Expon}(1/(1+\theta))$,

$$f_\theta(y) = (1+\theta)^{-1} e^{-y/(1+\theta)} \mathbb{1}_{\{y \ge 0\}}, \quad \theta \ge 0, \tag{8.154}$$

where the pre-change parameter value $\theta_0 = 0$. Let $\theta_1 > 0$ denote the putative value of the post-change parameter. Then the LLR is

$$Z_n = \frac{\theta_1}{1+\theta_1} X_n - \log(1+\theta_1), \quad X_n \ge 0$$

and the K–L information numbers are $I = \theta_1 - \log(1+\theta_1)$ and $I_0 = \log(1+\theta_1) - \theta_1/(1+\theta_1)$.

We iterate that the most important issue is precision of the approximations for the ARL to false alarm. In this example, the distribution of the LLR Z_1 has the exponential right tail,

$$\mathsf{P}_\theta(Z_1 > z) \propto \exp\left\{-\frac{1+\theta_1}{\theta_1(1+\theta)} z\right\}, \quad \theta \ge 0,$$

so that the distribution of the overshoot over the upper boundary in the SPRT is exactly exponential with the parameter $(1+\theta_1)/\theta_1(1+\theta)$; therefore, the quantities necessary for computing the approximations to the ARL2FA (8.144) and (8.145) are:

$$e_1(\theta_1) = \zeta = 1/(1+\theta_1), \quad \delta_{10} = \theta_1/(1+\theta_1).$$

Applying FO approximation (8.145), we obtain

$$\mathsf{ARL2FA}(h) \approx \frac{(1+\theta_1)^2}{\theta_1 - \log(1+\theta_1)} e^h. \tag{8.155}$$

A higher-order (HO) approximation can be obtained using (8.144):

$$\begin{aligned} \mathsf{ARL2FA}(h) \approx {} & \frac{(1+\theta_1)^2}{\theta_1 - \log(1+\theta_1)} e^h - \frac{1}{\log(1+\theta_1) - \theta_1/(1+\theta_1)} h \\ & - \frac{1+\theta_1}{\theta_1 - \log(1+\theta_1)} - \frac{\theta_1}{(1+\theta_1)\log(1+\theta_1) - \theta_1}. \end{aligned} \tag{8.156}$$

We simulated the CUSUM procedure under the assumption of no change 100,000 times. The results are reported in Table 8.1 for the post-change parameter $\theta_1 = 3$, which is a reasonable value in certain applications such as detection of a randomly appearing target in noisy measurements, in which case θ is the SNR; see, e.g., [451, 452]. It is seen that the approximation (8.155) given in the row FO ARL2FA is not especially accurate. This happens primarily because the first-order approximation takes into account only the first term of expansion and ignores the second term $O(h)$ as well as constants. The row HO ARL2FA corresponds to the higher-order approximation (8.156), which perfectly fits the MC estimates denoted by MC $\widehat{\text{ARL2FA}}$ for all tested threshold values $A = e^h \geq 1.2$. In this table we also present the MC estimates of the standard deviation MC St. Dev. Since the distribution of the CUSUM stopping time is approximately exponential at least for a sufficiently large h [369], the standard deviation is approximately the same as the mean, and the similarity grows as h increases.

Table 8.1: *The ARL2FA vs. threshold $A = e^h$ for $\theta_1 = 3$.*

A	1.2	1.7	2.5	4.6	9.2	13.0	17.1	21.0	41.0
FO ARL2FA	11.90	16.86	24.79	45.61	91.22	128.90	169.55	208.22	406.52
HO ARL2FA	7.96	12.36	19.69	39.56	84.07	121.21	161.43	199.77	397.02
MC $\widehat{\text{ARL2FA}}$	8.04	12.45	19.79	39.57	84.33	121.23	161.88	200.44	397.16
MC St.Dev.	7.49	11.88	19.18	38.61	83.21	119.73	159.91	198.97	396.84

Figure 8.8 shows the logarithm of the empirical (MC estimate) survival function $\log \mathsf{P}_\infty(\tau_A > y)$, where $\tau_A = T_{\rm CS}/\widehat{\text{ARL2FA}}$ is the corresponding standardized stopping time, along with the logarithm of the exponential probability plot $\log e^{-y} = -y$. The figure shows that the exponential distribution approximates the distribution of the stopping time very well already for $A = e^h = 13$ (ARL2FA $\approx$ 120). When considering that in practical applications the values of the ARL to false alarm usually range from 300 and upward, the exponential distribution seems to be a perfect fit.

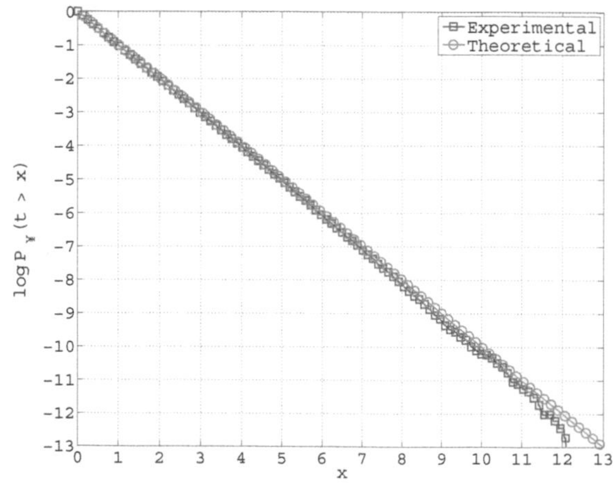

Figure 8.8: *Empirical estimate of* $\log[\mathsf{P}_\infty(\tau_A > y)]$ *for* $A = e^h = 13$.

To compute the ARL function from (8.134), we note that $\omega_0(\theta)$ is the root of the transcendental equation

$$(1+\theta_1)^{-\omega_0}\left[1+\omega_0\frac{\theta_1(1+\theta)}{1+\theta_1}\right] = 1,$$

that

$$J_\theta = (1+\theta)\frac{\theta_1}{1+\theta_1} - \log(1+\theta_1)$$

and that due to exponentiality of the distribution of the upper overshoot

$$e_1(\theta) = \frac{1}{1+\omega_0(\theta)(1+\theta)\theta_1/(1+\theta_1)}, \qquad \delta_{11}(\theta) = \varkappa(\theta) = (1+\theta)\theta_1/(1+\theta_1).$$

Since finding the rest of values in (8.134) is problematic, we have to resort to renewal-theoretic approximations (8.146) and (8.147).

Using (8.153), we obtain that

$$\mathsf{ARL}_0(\theta) \approx \frac{1}{(1+\theta)\frac{\theta_1}{1+\theta_1} - \log(1+\theta_1)} \left\{ h + (1+\theta)\frac{\theta_1}{1+\theta_1} - \frac{[\log(1+\theta_1)]^2}{2\left[(1+\theta)\frac{\theta_1}{1+\theta_1} - \log(1+\theta_1)\right]} \right\}$$

for $\theta > \frac{1+\theta_1}{\theta_1}\log(1+\theta_1) - 1$.

Also, the stationary distribution in this example is (exactly) exponential for $x > 0$ having an atom $\theta_1/(1+\theta_1)$ at zero:

$$\mathsf{Q}_{\mathrm{st}}(x) = \begin{cases} 1-(1+\theta_1)^{-1}e^{-x} & \text{for } x > 0, \\ \theta_1/(1+\theta_1) & \text{for } x = 0 \\ 0 & \text{for } x < 0. \end{cases}$$

8.2.6.5 *Bounds for the ARL Function*

In the previous subsections, we derived several approximations to the ARL function. Bounds for the ARL function are also desirable for two reasons. First, in practice, it is necessary to have reliable estimates of the statistical properties of change detection algorithms, and it turns out that the precision of the previous approximations may be not sufficient in some cases. Second, sometimes it is of interest to prove that one algorithm is better than another in a particular change detection problem. In many cases, the use of bounds is sufficient for this purpose. Relevant bounds are an upper bound for the "worst-worst-case" average delay to detection $\mathsf{ESADD}(T_{\mathrm{cs}})$ and a lower bound for the mean time to false alarm $\mathsf{ARL2FA}(T_{\mathrm{cs}})$ [47, 222, 319, 320].

Consider an arbitrary value of θ and not necessarily the putative values θ_0 and θ_1. Let us begin with an upper bound for the average delay to detection $\mathsf{ESADD}(T_{\mathrm{cs}}) = \mathsf{ARL}_0(\theta)$ where $\theta \in \Theta_1 = \{\theta : J_\theta = \mathsf{E}_\theta(Z_1) > 0\}$. Recalling that the CUSUM algorithm (8.66) can be derived using a sequence of repeated SPRTs with zero lower bound, as explained in Subsection 8.2.2, let us consider the SPRT $\delta^* = (d^*, T^*)$ given by (8.64) with the lower threshold $-\varepsilon = 0$ and the upper threshold $h > 0$ [346]. We obtain

$$\mathsf{ARL}_0(\theta) = \frac{\mathsf{E}_\theta(\lambda_{T^*} | \lambda_{T^*} \le 0)}{J_\theta} \frac{\mathsf{P}_\theta(\lambda_{T^*} \le 0)}{1-\mathsf{P}_\theta(\lambda_{T^*} \le 0)} + \frac{\mathsf{E}_\theta(\lambda_{T^*} | \lambda_{T^*} \ge h)}{J_\theta}. \tag{8.157}$$

Consider the first term on the right-hand side of this equation. Because $\mathsf{E}_\theta(\lambda_{T^*} | \lambda_{T^*} \le 0) \le 0$, the first term is non-positive. Let us now derive an upper bound for the second term in (8.157). Recall the notion of overshoot of the cumulative sum λ_{T^*} over the boundaries introduced in Subsections 3.1.1 and 3.1.3. Clearly, the average overshoot can be bounded as follows [494]:

$$\mathsf{E}_\theta(\lambda_{T^*} | \lambda_{T^*} \ge h) \le h + \varkappa_{\max}(\theta) \tag{8.158}$$

where $\varkappa_{\max}(\theta) = \sup_{t>0} \mathsf{E}_\theta(Z_1 - t | Z_1 \ge t > 0)$ is an upper bound for the average overshoot; see the details in Section 2.5. Therefore, we get the following upper bound

$$\mathsf{ESADD}(T_{\mathrm{cs}}) = \mathsf{ARL}_0(\theta) \le \frac{h + \varkappa_{\max}(\theta)}{J_\theta}, \quad \theta \in \Theta_1. \tag{8.159}$$

Let us add some comments about the computation of $\varkappa_{\max}(\theta)$. If the distribution of the LLR is log-concave (i.e., $\log(p_\theta(x))$, where $p_\theta(x)$ is the pdf of Z_1, is a continuous and convex function of x for some $\theta \in \Theta$, then $\varkappa_{\max}(\theta) = \mathsf{E}_\theta(Z_1|Z_1 > 0)$ [494]. For example, this holds in the case of detecting a change in the mean of a Gaussian random sequence, in the parameter of exponential distribution and many other distributions. In general, the computation of this value is a difficult problem. In [270] a very simple upper bound when $J_\theta > 0$ is suggested under general conditions. See details in Section 2.5.

Let us now describe how to obtain a lower bound for $\mathsf{ARL}_0(\theta)$, where $\theta \in \Theta_0 = \{\theta : J_\theta = \mathsf{E}_\theta(Z_1) < 0\}$. Let us again start from the equation (8.157). We can derive a lower bound of the first term on the right-hand side of (8.157) which we denote by a. Using $\mathsf{P}_\theta(\lambda_{T^*} \le -\varepsilon)$ given in (8.133) with $\varepsilon = 0$, we get that

$$a = \frac{\mathsf{E}_\theta(\lambda_{T^*}|\lambda_{T^*} \le 0)}{J_\theta}\left[\frac{\mathsf{E}_\theta(e^{-\omega_0\lambda_{T^*}}|\lambda_{T^*} \ge h) - 1}{1 - \mathsf{E}_\theta(e^{-\omega_0\lambda_{T^*}}|\lambda_{T^*} \le 0)}\right]. \tag{8.160}$$

From Jensen inequality, it results that

$$1 - \mathsf{E}_\theta\left(e^{-\omega_0\lambda_{T^*}}|\lambda_{T^*} \le 0\right) \le 1 - e^{\mathsf{E}_\theta(-\omega_0\lambda_{T^*}|\lambda_{T^*} \le 0)}. \tag{8.161}$$

Therefore, a lower bound for a is given by

$$a \ge \frac{\mathsf{E}_\theta(\lambda_{T^*}|\lambda_{T^*} \le 0)}{J_\theta}\left[\frac{\mathsf{E}_\theta(e^{-\omega_0\lambda_{T^*}}|\lambda_{T^*} \ge h) - 1}{1 - e^{\mathsf{E}_\theta(-\omega_0\lambda_{T^*}|\lambda_{T^*} \le 0)}}\right]. \tag{8.162}$$

Now, using the inequality $-x \ge 1 - e^x$, and the fact that the ratio in the squared brackets of the previous formula is positive, we obtain

$$a \ge \frac{\mathsf{E}_\theta(\lambda_{T^*}|\lambda_{T^*} \le 0)}{J_\theta}\left[\frac{\mathsf{E}_\theta(e^{-\omega_0\lambda_{T^*}}|\lambda_{T^*} \ge h) - 1}{\omega_0 \mathsf{E}_\theta(\lambda_{T^*}|\lambda_{T^*} \le 0)}\right]. \tag{8.163}$$

And finally, recalling that $\omega_0 < 0$ implies the inequality $\mathsf{E}_\theta(e^{-\omega_0\lambda_{T^*}}|\lambda_{T^*} \ge h) \ge e^{-\omega_0 h}$ and we obtain $a \ge \dfrac{e^{-\omega_0 h} - 1}{J_\theta \omega_0}$. On the other hand, a lower bound for the (negative) second term of (8.157) is

$$\frac{\mathsf{E}_\theta(\lambda_{T^*}|\lambda_{T^*} \ge h)}{J_\theta} \ge \frac{h + \varkappa_{\max}(\theta)}{J_\theta}. \tag{8.164}$$

Therefore,

$$\mathsf{ARL}_0(\theta) \ge \frac{1}{J_\theta}\left[\frac{e^{-\omega_0 h} - 1}{\omega_0} + h + \varkappa_{\max}(\theta)\right]. \tag{8.165}$$

Another possible lower bound for $\mathsf{ARL}_0(\theta)$ when $\mathsf{E}_\theta(Z_1) < 0$ can be obtained by using the following idea. Formula (8.98) can be rewritten in the following manner

$$\mathsf{ARL}_0(\theta) = \frac{\mathsf{E}_\theta(T^*_{0,h})}{\mathsf{P}_\theta(\lambda_{T^*} \ge h)}. \tag{8.166}$$

If $\omega_0 < 0$, the following bound can be used [494]

$$\frac{e^{-\omega_0 h} - 1}{e^{-\omega_0 h} - \eta(\theta)e^{\omega_0 a}} \le Q(\theta) \text{ where } \eta(\theta) = \inf_{\xi > 1} \xi\, \mathsf{E}_\theta\left(e^{-\omega_0\lambda_{T^*}}|e^{-\omega_0\lambda_{T^*}} \le \frac{1}{\xi}\right). \tag{8.167}$$

Hence

$$\mathsf{ARL}_0(\theta) \ge \frac{1}{\mathsf{P}_\theta(\lambda_{T^*} \ge h)} \ge \frac{e^{-\omega_0 h} - \eta(\theta)}{1 - \eta(\theta)} \ge e^{-\omega_0 h}. \tag{8.168}$$

An additional motivation for the derivation of this second bound lies in the following fact. If $\theta = \theta_0$, then $\omega_0(\theta_0) = -1$ and the inequality (8.168) leads to

$$\mathsf{ARL}_0(\theta_0) \geq \frac{1}{\mathsf{P}_{\theta_0}(\lambda_{T^*} \geq h)} \geq e^h, \tag{8.169}$$

which holds under the assumption that $X_n \sim F_{\theta_0}$ and Z_n is the increment of the LLR (and *not* any iid sequence as we assume for the previous bounds). This inequality coincides with the general inequality obtained in Lemma 8.2.1 in a general non-iid case.

This inequality for $\mathsf{ARL}_0(\theta_0)$ is widely used for the investigation of the asymptotic properties of CUSUM-type and GLR algorithms in [271, 272, 367].

It follows from equations (8.159), (8.165), and (8.168) that when θ goes to the value θ^* such that $\mathsf{E}_{\theta^*}(Z_k) = 0$, then the above-mentioned bounds become useless. For this reason, let us compute directly the lower bound for $\mathsf{ARL}_0(\theta^*)$. Putting together the expression for $\mathsf{E}_{\theta^*}(T^*_{0,h})$ and equation (8.98), we get

$$\begin{aligned}
\mathsf{ARL}_0(\theta^*) &= \frac{\mathsf{E}_{\theta^*}(T^*_{0,h})}{1 - \mathsf{P}_{\theta^*}(\lambda_{T^*} \leq 0)} \\
&= \frac{\mathsf{E}_{\theta^*}(\lambda^2_{T^*} | \lambda_{T^*} \leq 0)\mathsf{P}_{\theta^*}(\lambda_{T^*} \leq 0) + \mathsf{E}_{\theta^*}(\lambda^2_{T^*} | \lambda_{T^*} \geq h)(1 - \mathsf{P}_{\theta^*}(\lambda_{T^*} \leq 0))}{\mathsf{E}_{\theta^*}(Z_k^2)(1 - \mathsf{P}_{\theta^*}(\lambda_{T^*} \leq 0))} \\
&= \frac{\mathsf{E}_{\theta^*}(S_k^2 | S_k \leq 0)\mathsf{P}_{\theta^*}(\lambda_{T^*} \leq 0)}{\mathsf{E}_{\theta^*}(Z_k^2)(1 - \mathsf{P}_{\theta^*}(\lambda_{T^*} \leq 0))} + \frac{\mathsf{E}_{\theta^*}(S_k^2 | S_k \geq h)}{\mathsf{E}_{\theta^*}(Z_k^2)} \geq \frac{h^2}{\mathsf{E}_{\theta^*}(Z_k^2)}.
\end{aligned} \tag{8.170}$$

These bounds for $\mathsf{ARL}_0(\theta)$ are very approximate. In some cases, it is possible to obtain more accurate bounds.

8.2.6.6 *Comparing Different ARL Expressions*

Consider the case of a change in the mean μ of an independent Gaussian sequence $(X_n)_{n \geq 1}$ with known variance σ^2. The goal is to compare various approximations and bounds derived above with the ARL obtained solving integral equations numerically which give almost exact performance. We use the CUSUM algorithm with putative values μ_0 and μ_1. In this case, the increment of the cumulative sum (8.66)

$$Z_k = \log \frac{p_{\mu_1}(X_k)}{p_{\mu_0}(X_k)} = \frac{\mu_1 - \mu_0}{\sigma^2} \left(X_k - \frac{\mu_1 + \mu_0}{2} \right) \tag{8.171}$$

is a Gaussian random variable, $Z_k \sim \mathcal{N}(\mu_z, \sigma_z^2)$, with mean $\mu_z = \frac{\mu_1 - \mu_0}{\sigma^2}\left(\mu - \frac{\mu_1 + \mu_0}{2}\right)$ and variance $\sigma_z^2 = \frac{(\mu_1 - \mu_0)^2}{\sigma^2}$. When $\mu = \mu_0$, the mean of the increment is $-I(\mu_0, \mu_1)$ and when $\mu = \mu_1$, it is $+I(\mu_0, \mu_1)$, where $I(\mu_0, \mu_1) = \frac{(\mu_1 - \mu_0)^2}{2\sigma^2}$ is the K–L information.

To get the ARL function $\mathsf{ARL}(\mu)$ we first compute $\varkappa_{\max}(\mu)$,

$$\varkappa_{\max}(\mu) = \mathsf{E}_\mu(Z_1 | Z_1 > 0) = \frac{\int_0^\infty x f_\mu(x)\, \mathrm{d}x}{\int_0^\infty f_\mu(x)\, \mathrm{d}x}, \tag{8.172}$$

where $f_\mu(x)$ is the density of Z_1. Obvious computations yield

$$\varkappa_{\max}(\mu_z) = \frac{\sigma_z \varphi(\mu_z / \sigma_z)}{\Phi(\mu_z / \sigma_z)} + \mu_z, \tag{8.173}$$

where $\varphi(x)$ and $\Phi(x)$ are the Gaussian pdf and cdf.

In the Gaussian case, the non-zero solution of the equation $\mathsf{E}_{\mu_z}(e^{-\omega_0 Z_k}) = 1$ is given by $\omega_0 = \frac{2\mu_z}{\sigma_z^2}$. From (8.159) and (8.165), it follows that

$$\mathsf{ARL}_0(\mu) \leq \frac{h}{\mu_z} + \frac{\sigma_z \varphi(\mu_z/\sigma_z)}{\mu_z \Phi(\mu_z/\sigma_z)} + 1 \text{ for } \mu > \mu^*, \tag{8.174}$$

$$\mathsf{ARL}_0(\mu) \geq \frac{e^{-2\mu_z h/\sigma_z^2} - 1 + 2\mu_z h/\sigma_z^2}{2\mu_z^2/\sigma_z^2} + \frac{\sigma_z \varphi(\mu_z/\sigma_z)}{\mu_z \Phi(\mu_z/\sigma_z)} + 1 \text{ for } \mu < \mu^*, \tag{8.175}$$

where $\mu^* = (\mu_1 + \mu_0)/2$. It follows from (8.121) and (8.124) that Wald's approximation of the ARL function can be written as

$$\mathsf{ARL}_0(\mu) \approx \frac{e^{-2\mu_z h/\sigma_z^2} - 1 + 2\mu_z h/\sigma_z^2}{2\mu_z^2/\sigma_z^2}, \quad \mu \neq \mu^*, \quad \mathsf{ARL}_0(\mu^*) \approx \frac{h^2}{\sigma_z^2}. \tag{8.176}$$

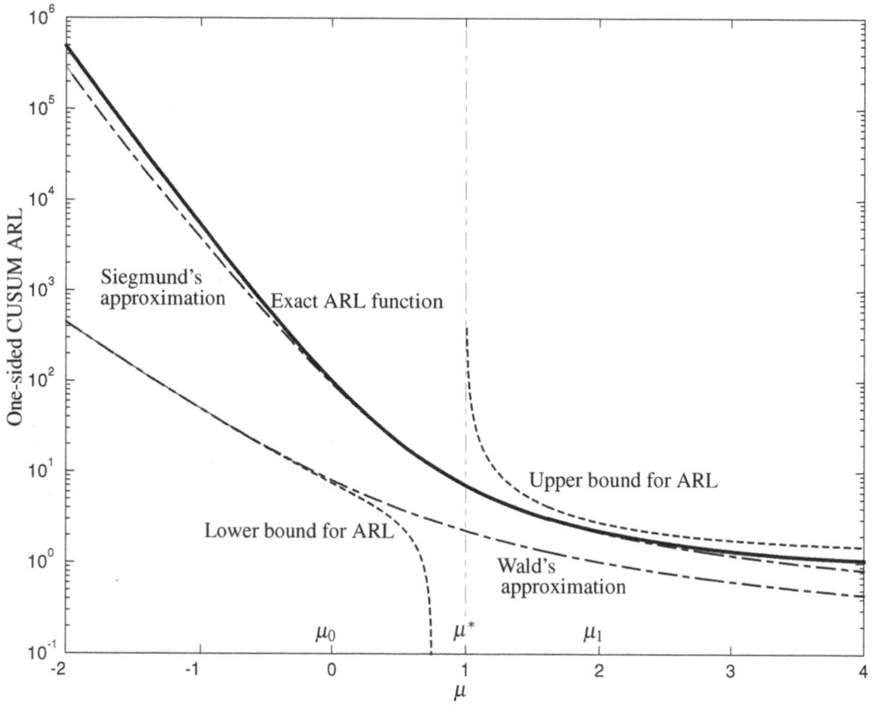

Figure 8.9: *The ARL function* $\mathsf{ARL}_0(\mu)$ *for the Gaussian case — exact function (solid line); Wald's and Siegmund's approximation (dash dotted line); bounds (dashed lines).*

On the other hand, it results from (8.139) and (8.141) that Siegmund's approximation of the ARL function can be written as

$$\mathsf{ARL}_0(\mu) \approx \frac{\exp\left(-2\left(\mu_z h/\sigma_z^2 + 1.166 \cdot \mu_z/\sigma_z\right)\right) - 1 + 2\left(\mu_z h/\sigma_z^2 + 1.166 \cdot \mu_z/\sigma_z\right)}{2\mu_z^2/\sigma_z^2}, \tag{8.177}$$

$$\mathsf{ARL}_0(\mu^*) \approx \left(\frac{h}{\sigma_z} + 1.166\right)^2. \tag{8.178}$$

The typical behavior of these bounds, of the approximations of the ARL, and of the exact ARL obtained by solving the Fredholm integral equations of the second kind is depicted in Figure 8.9.

These results are obtained in the case where $\mu_0 = 0, \mu_1 = 2, \sigma = 1$ and for the threshold $h = 3$. This figure shows that Siegmund's approximation is very close to the exact value of the ARL function, especially for values of μ_z close to 0 due to the small-μ_z asymptotic character of Siegmund's approximation. Wald's approximation is worse, especially for a negative drift of the increment of the decision function, namely for the ARL to false alarm, as we show now.

Let us consider the limit of the difference between Wald's approximation for the ARL and the lower bound for the ARL, i.e.,

$$\frac{\varkappa_{\max}(\mu_z)}{\mu_z} = \frac{\sigma_z \varphi(\mu_z/\sigma_z)}{\mu_z \Phi(\mu_z/\sigma_z)} + 1, \quad \mu_z \to -\infty.$$

For this purpose, we use the asymptotic formula

$$\Phi(-x) \sim \frac{1}{x\sqrt{2\pi}} e^{-\frac{x^2}{2}} \left(1 - \frac{1}{x^2} + \frac{3}{x^4} + \ldots\right) \quad \text{as } x \to +\infty. \tag{8.179}$$

Then

$$\lim_{\mu_z \to -\infty} \frac{\sigma_z \varphi(\mu_z/\sigma_z)}{\mu_z \Phi(\mu_z/\sigma_z)} + 1 = \sigma_z \lim_{\mu_z \to -\infty} \frac{e^{-\frac{\mu_z^2}{2}}}{\mu_z \left(-\frac{1}{\mu_z} e^{-\frac{\mu_z^2}{2}} (1 - \frac{1}{\mu_z^2} + \frac{3}{\mu_z^4} + \ldots)\right)} + 1 = 0. \tag{8.180}$$

In other words, when $\mu_z \to -\infty$, Wald's approximation acts as a bound, as depicted in Figure 8.9.

8.2.6.7 *ARL Function of the Two-sided CUSUM Algorithm*

We now compute the ARL function of the two-sided CUSUM algorithm (8.68)–(8.69) . This problem is considered in [230, 428, 482, 499, 519]. Under quite general conditions, the ARL function $\mathsf{ARL}_{\text{2-CS}}(\theta)$ of the two-sided CUSUM algorithm can be computed from the ARL functions of two one-sided CUSUM algorithms using the following inequality

$$\frac{1}{\mathsf{ARL}_{\text{2-CS}}(\theta)} \geq \frac{1}{\mathsf{ARL}_{\text{CS}}^{\ell}(\theta)} + \frac{1}{\mathsf{ARL}_{\text{CS}}^{u}(\theta)}, \tag{8.181}$$

where $\mathsf{ARL}_{\text{CS}}^{\ell}(\theta)$ is the ARL function of the one-sided CUSUM corresponding to $(\theta_0, \theta_1^{\ell})$ and $\mathsf{ARL}_{\text{CS}}^{u}(\theta)$ is the ARL function of the one-sided CUSUM corresponding to (θ_0, θ_1^{u}) and $\theta_1^{\ell} < \theta_0 < \theta_1^{u}$. In the case (8.69) of a change in the mean μ of a Gaussian sequence, the previous inequality becomes an *equality*

$$\frac{1}{\mathsf{ARL}_{\text{2-CS}}(\mu)} = \frac{1}{\mathsf{ARL}_{\text{CS}}^{\ell}(\mu)} + \frac{1}{\mathsf{ARL}_{\text{CS}}^{u}(\mu)}. \tag{8.182}$$

Let us give a sketch of the proof of this equality. The interested reader is referred to [428, 499] for the details. Note that the stopping time in (8.68) can be written as

$$T_{\text{2-CS}} = \min\left\{T_{\text{CS}}^{u}, T_{\text{CS}}^{\ell}\right\}. \tag{8.183}$$

We now fix μ. The ARL function $\mathsf{ARL}_{\text{CS}}^{\ell}$ can be computed as

$$\begin{aligned} \mathsf{ARL}_{\text{CS}}^{\ell}(\mu) &= \mathsf{E}_{\mu}(T_{\text{CS}}^{\ell}) = \mathsf{E}_{\mu}(T_{\text{2-CS}}) + \mathsf{E}_{\mu}(T_{\text{CS}}^{\ell} - T_{\text{2-CS}}) \\ &= \mathsf{E}_{\mu}(T_{\text{2-CS}}) + \mathsf{E}_{\mu}(T_{\text{CS}}^{\ell} - T_{\text{2-CS}} | T_{\text{CS}}^{\ell} > T_{\text{2-CS}}) \mathsf{P}_{\mu}(T_{\text{CS}}^{\ell} > T_{\text{2-CS}}). \end{aligned} \tag{8.184}$$

It is intuitively obvious and can be formally proven [428] that if $g_k^u \geq h$ for some k, then this implies that $g_k^{\ell} = 0$ for the same k. Hence,

$$\mathsf{ARL}_{\text{CS}}^{\ell}(\mu) = \mathsf{ARL}_{\text{2-CS}}(\mu) + \mathsf{ARL}_{\text{CS}}^{\ell}(\mu) \mathsf{P}(T_{\text{CS}}^{\ell} > T_{\text{2-CS}}) \text{ or } \mathsf{ARL}_{\text{2-CS}}(\mu) = \mathsf{ARL}_{\text{CS}}^{\ell}(\mu) \mathsf{P}(T_{\text{CS}}^{\ell} = T_{\text{2-CS}}). \tag{8.185}$$

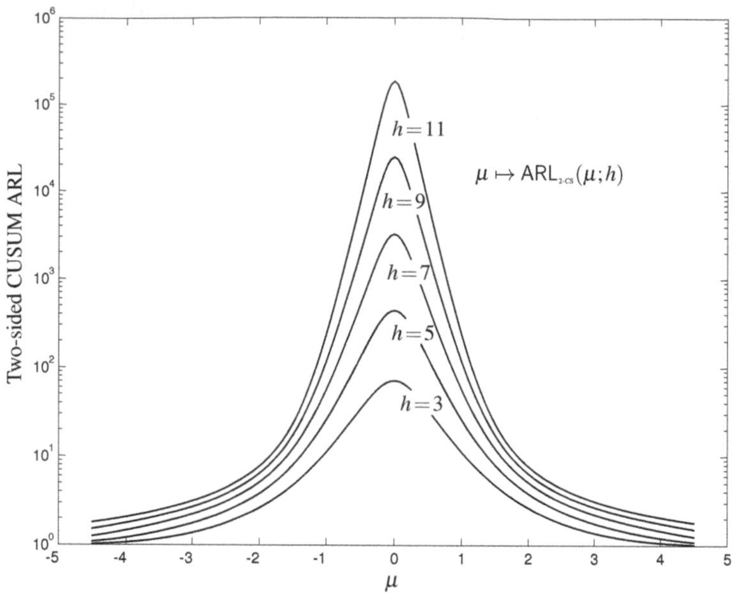

Figure 8.10: *Typical two-sided CUSUM ARL function* $\mathsf{ARL}_{2\text{-}CS}(\mu;h)$ *for different thresholds h.*

A similar result holds for $\mathsf{ARL}^u_{\text{CS}}(\mu)$:

$$\mathsf{ARL}_{2\text{-CS}}(\mu) = \mathsf{ARL}^u_{\text{CS}}(\mu)\mathsf{P}(T^u_{\text{CS}} = T_{2\text{-CS}}). \tag{8.186}$$

Using $\mathsf{P}(T^\ell_{\text{CS}} = T_{2\text{-CS}}) + \mathsf{P}(T^u_{\text{CS}} = T_{2\text{-CS}}) = 1$ and the previous relations, we deduce (8.182).

The ARL function of the two-sided CUSUM algorithm can be computed from (8.182) and Fredholm integral equations, Wald's or Siegmund's approximations to the ARL function of the one-sided CUSUM algorithm. Finally, we give the formula corresponding to the special case of the two-sided CUSUM (8.68) with a zero minimum magnitude $\delta_\mu = 0$ of change in the mean of a Gaussian sequence, for which we use the decision statistic (8.69). Note that in [314] a Brownian motion approximation is used, which leads to the following equation for the ARL

$$\begin{aligned} \mathsf{ARL}_{2\text{-CS}}(\mu_z) &= \left(\frac{h}{\mu_z}\right)\coth\left(\frac{\mu_z h}{\sigma_z^2}\right) - \frac{\sigma_z^2}{2\mu_z^2} - \frac{h^2}{2\sigma_z^2 \sinh^2\left(\frac{\mu_z h}{\sigma_z^2}\right)} \quad \text{when } \mu_z \neq 0, \\ \mathsf{ARL}_{2\text{-CS}}(\mu_z) &= \frac{h^2}{2\sigma_z^2} \quad \text{when } \mu_z = 0, \end{aligned} \tag{8.187}$$

where μ_z and σ_z are the mean and standard deviation of the LLR Z_1.

A typical behavior of the two-sided CUSUM ARL function is shown in Figures 8.10 and 8.11. We wish now to detect a change in the mean μ for the Gaussian example with $\mu_0 = 0$, $\delta_\mu = 1.5$ and $\sigma_p = 3$. The ARL function is calculated by using the Fredholm integral equations (8.99), (8.101), and (8.102). Figure 8.10 corresponds to the case where the true value of standard deviation σ_t is equal to its putative value $\sigma_p = 3$ and h ranges from 3 to 11 and Figure 8.11 to the case when the true value σ_t of standard deviation is unknown. It is assumed that $h = 3$ and the ratio $r = \sigma_t/\sigma_p$ varies from 0.75 to 1.25. It follows from Figure 8.11 that the ARL to false alarm is more sensitive to the deviation of the ratio r from 1 than the "worst-worst-case" ADD $\mathsf{ESADD}(T_{2\text{-CS}})$. Here, the $\mathsf{ESADD}(T_{2\text{-CS}})$ corresponds to the function $\mathsf{ARL}(\mu)$ of two-sided CUSUM, when μ belongs to the interval $|\mu| \in [1.5;5]$.

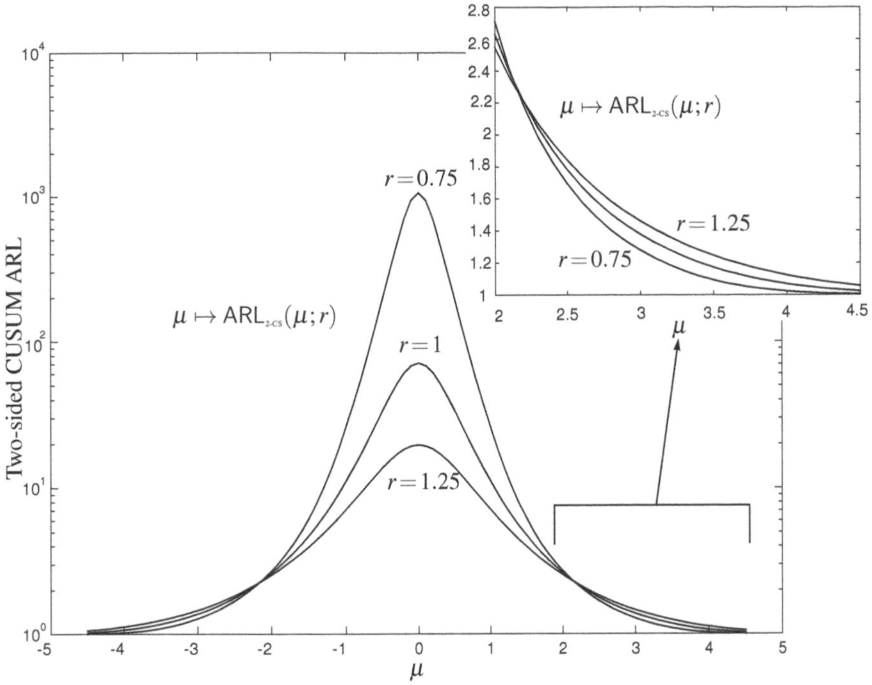

Figure 8.11: *Typical two-sided CUSUM ARL function* $\mathsf{ARL}_{2\text{-}CS}(\mu;r)$ *for different* $r=\sigma_t/\sigma_p$.

8.2.7 A Generalization to a Special Non-iid Case

In this subsection, we provide a generalization of the CUSUM procedure for more sophisticated stochastic models with no assumptions about independence of observations $\{X_n\}_{n\geq 1}$. The generalization of the SPRT for the AR model has been discussed in Subsection 3.2.4 under some special conditions. As in Subsection 3.2.4, assume that the observations $\{X_n\}_{n\geq 1}$ are dependent, but there exists a "whitening" transform $\mathbf{W}(\mathbf{X})=\{W_n(\mathbf{X}_1^n), n\geq 1\}$, not depending on the state of the process (pre- or post-change), such that the transformed random variables $\widetilde{X}_n=W_n(\mathbf{X}_1^n)$, $n\geq 1$ are iid in both situations:

$$p_i(\widetilde{X}_1^n)=\prod_{k=1}^n f_i(\widetilde{X}_k), \quad i=0,1,\ n\geq 1. \tag{8.188}$$

Then the LLR increment of the transformed data is equal to $\widetilde{Z}_k=\log[f_1(\widetilde{X}_k)/f_0(\widetilde{X}_k)]$ and the cumulative LLR $\widetilde{\lambda}_n=\widetilde{Z}_1+\cdots+\widetilde{Z}_n$ is the random walk since $\widetilde{Z}_k$, $k\geq 1$ are iid before and after change. To be more precise, let

$$Z_n=\log\frac{f_1(X_n\mid \mathbf{X}_1^{n-1})}{f_0(X_n\mid \mathbf{X}_1^{n-1})}, \tag{8.189}$$

where $f_i(X_n\mid \mathbf{X}_1^{n-1})$ $(i=0,1)$ are conditional densities of X_n conditioned on the past $\mathbf{X}_1^{n-1}=(X_1,\dots,X_{n-1})$. After whitening the (cumulative) LLR process is $\lambda_n=\sum_{k=1}^n \widetilde{Z}_k$. The CUSUM algorithm (repeated SPRT with a zero-value lower threshold) (8.66) is now written as

$$T_{\mathrm{CS}}=\inf\{n\geq 1: g_n\geq h\}, \quad g_n=\left[g_{n-1}+\widetilde{Z}_n\right]^+, \quad g_0=0. \tag{8.190}$$

For illustration consider the following autoregressive model $\{X_n\}_{n\geq 0}$ with a change in the mean:

$$X_k = \begin{cases} X_0 \sim \mathcal{N}(\mu_1, \sigma_{\rm x}^2) & \text{if } \ 0 = k = \nu \\ X_0 \sim \mathcal{N}(\mu_0, \sigma_{\rm x}^2) & \text{if } \ 0 = k < \nu \\ \rho X_{k-1} + (1-\rho)\mu_0 + \xi_k & \text{if } \ 1 \leq k \leq \nu \\ \rho X_{k-1} + (1-\rho)\mu_1 + \xi_k & \text{if } \quad\ \ k > \nu \end{cases}, \tag{8.191}$$

where $\xi_k \sim \mathcal{N}(0, \sigma^2)$ and $\sigma^2 = (1-\rho^2)\sigma_{\rm x}^2$. The important feature of the above AR model is that the distribution of the initial condition X_0 depends on the changepoint ν. It is assumed that X_0 is available for the CUSUM test. Let us also stress that an abrupt change in the parameter μ leads to an abrupt change in the expectation of the LLR

$$Z_k = \frac{(1-\rho)(\mu_1-\mu_0)}{\sigma^2}\left[X_k - \rho X_{k-1} - \frac{(1-\rho)(\mu_1+\mu_0)}{2}\right], \quad k \geq 1, \tag{8.192}$$

namely,

$$\mathsf{E}Z_k = \begin{cases} -q & \text{if } \ k \leq \nu \\ \ \ q & \text{if } \ k > \nu \end{cases}, \quad q = \frac{(\mu_1-\mu_0)^2(1-\rho)^2}{2\sigma^2} \tag{8.193}$$

but to a transient change in the expectation of X_k,

$$\mathsf{E}X_k = \begin{cases} \mu_0 & \text{if } \ k \leq \nu \\ \mu_1 + (\mu_0-\mu_1)\rho^{k-\nu} & \text{if } \ k > \nu \end{cases} \quad \text{and} \quad \lim_{k\to\infty} \mathsf{E}X_k = \mu_1.$$

Hence, as in Subsection 3.2.4, the cumulative LLR $\lambda_n = \sum_{i=1}^n Z_i$ is the Gaussian random walk with drift $-q$ before change and q after change and with variance $\mathsf{var}[Z_k] = (\mu_1-\mu_0)^2(1-\rho)^2/\sigma^2$. The ARL function can be computed by using $\mathsf{E}Z_k$ and $\mathsf{var}[Z_k]$ exactly as in Subsection 8.2.6.

Let us analyze the impact of the autoregressive coefficient ρ on the ARL. It results from (8.121), (8.176), and (8.193) that Wald's approximation of the ARL function at $\mu = \mu_0$ and $\mu = \mu_1$ is given by [318, 320]

$$\mathsf{ARL}_{z,\rho}(\mu_0) \approx \frac{e^h - e^z - h + z}{q}, \quad \mathsf{ARL}_{z,\rho}(\mu_1) \approx \frac{e^{-h} - e^{-z} + h - z}{q}, \quad q = \frac{(\mu_1-\mu_0)^2}{2\sigma_{\rm x}^2}\cdot\frac{1-\rho}{1+\rho}.$$

Hence, for fixed h, z and $(\mu_1-\mu_0)^2/2\sigma_{\rm x}^2$, Wald's approximation of the ARL is a function of the autoregressive coefficient ρ. For positive correlation ($\rho > 0$), Wald's approximations to $\mathsf{ARL}_{0,\rho}(\mu_i)$, $i = 0, 1$, are approximately $(1+\rho)/(1-\rho)$ times greater than the ARL $\mathsf{ARL}_{0,0}(\mu_i)$ in the iid case ($\rho = 0$). For negative correlation ($\rho < 0$), Wald's approximations to $\mathsf{ARL}_{0,\rho}(\mu_i)$ are approximately $(1+\rho)/(1-\rho)$ times less than the ARL in the iid case.

Let us consider again an exponentially correlated Markov Gaussian process $\{V_t\}_{t\geq 0}$, $\mathsf{E}[V_t] = \mu_i$, $i = 0, 1$, $\mathsf{E}[(V_t-\mu_i)(V_{t+u}-\mu_i)] = \sigma_{\rm v}^2 e^{-\alpha|u|}$ ($\alpha > 0$), as in Remark 3.4.3. Assume that the sampling period Δ can be chosen arbitrarily. By analogy with (3.197)–(3.198), the products $\Delta\cdot\mathsf{ARL}_{z,\Delta}(\mu_i)$, $i = 0, 1$, measured in units of time are given by

$$\begin{aligned} \Delta\cdot\mathsf{ARL}_{z,\Delta}(\mu_0) &\approx \frac{2\sigma_{\rm x}^2 R(\Delta)}{(\mu_1-\mu_0)^2}\left[e^h - e^z - h + z\right], \\ \Delta\cdot\mathsf{ARL}_{z,\Delta}(\mu_1) &\approx \frac{2\sigma_{\rm x}^2 R(\Delta)}{(\mu_1-\mu_0)^2}\left[e^{-h} - e^{-z} + h - z\right]. \end{aligned}$$

As in the case of SPRT (see Remark 3.4.3), the products $\Delta\cdot\mathsf{ARL}_{z,\Delta}(\mu_i)$ are almost linear functions of Δ for large Δ but starting from a certain value of Δ the further reduction of the sampling period has practically no effect on $\Delta\cdot\mathsf{ARL}_{z,\Delta}(\mu_i)$.

Let us now consider a slightly different AR(1) model with an abrupt change in the mean

$$X_k = \begin{cases} \eta_k + \mu_0 & \text{if } \ k \leq \nu \\ \eta_k + \mu_1 & \text{if } \ k > \nu \end{cases}, \quad \eta_0 \sim \mathcal{N}(0, \sigma_{\rm x}^2), \ \ \eta_k = \rho\eta_{k-1} + \xi_k, \ \ k \geq 1. \tag{8.194}$$

It is assumed that the initial condition X_0 is available for the CUSUM test. In contrast to model (8.191), now the expectation $\mathsf{E}X_k$ changes instantaneously but the expectation of the LLR Z_k given by (8.192) changes gradually,

$$\mathsf{E}Z_k = \begin{cases} -q & \text{if} \quad k \le \nu \\ q_1 & \text{if} \quad k = \nu+1 \\ q & \text{if} \quad k > \nu+1 \end{cases}, \quad q_1 = \frac{(\mu_1-\mu_0)^2(1-\rho^2)}{2\sigma^2} = \frac{(\mu_1-\mu_0)^2}{2\sigma_{\mathrm{x}}^2}, \tag{8.195}$$

and q is defined in (8.193). To calculate the ARL function it is necessary to take into account q_1 at the first post-changepoint. Consider the following CUSUM procedure

$$T_{\mathrm{CS}} = \inf\{n \ge 1 : g_n \ge h\}, \; g_n = [g_{n-1} + Z_k]^+, \; g_0 = z, \tag{8.196}$$

where the expectation of the LLR Z_k is given by (8.195). As it follows from the definition of the ARL function (8.3), $\nu = 0$. Denote the ARL function of the CUSUM test (8.196) as $\mathsf{ARL}_{z,\rho}(\mu_1)$. After simple algebra we obtain the following equation [320]

$$\mathsf{ARL}_{z,\rho}(\mu_1) = \mathsf{P}(g_1 > 0) + \int_0^h \mathsf{ARL}_x(\mu_1) f_{z,u}(x)\mathrm{d}x + \mathsf{P}(g_1 = 0)\left(\mathsf{ARL}_0(\mu_1)+1\right), \tag{8.197}$$

where $g_1 = \max\{0, z+Z_1\}$, $\mathsf{ARL}_z(\mu_1)$ denotes the ARL function of the CUSUM procedure with drift q after change, i.e., without taking into account that the drift q_1 of the first post-change term Z_1 is different from the drift q of Z_k, $k > 1$, and $f_{z,u}(x)$ denotes the pdf of the random variable $u = z + Z_1$. For calculation of the ARL function $\mathsf{ARL}_z(\mu_1)$ the equations of Subsection 8.2.6 are applicable with $\mathsf{E}Z_k = q$ and $\mathsf{var}[Z_k]$.

Consider now the autoregressive model with an abrupt change in the variance

$$X_k = \begin{cases} X_0 \sim \mathcal{N}(0, \sigma_{x,1}^2) & \text{if} \quad 0 = k = \nu \\ X_0 \sim \mathcal{N}(0, \sigma_{x,0}^2) & \text{if} \quad 0 = k < \nu \\ \rho X_{k-1} + \xi_k, \; \xi_k \sim \mathcal{N}(0, (1-\rho^2)\sigma_{x,0}^2) & \text{if} \quad 1 \le k \le \nu \\ \rho X_{k-1} + \xi_k, \; \xi_k \sim \mathcal{N}(0, (1-\rho^2)\sigma_{x,1}^2) & \text{if} \quad k > \nu \end{cases}. \tag{8.198}$$

It is easily seen that the observations $\{X_n\}_{n\ge1}$ generated by (8.198) are dependent, but there exists a "whitening" transform $\mathbf{W}(\mathbf{X}) = \{W_n(\mathbf{X}_1^n), n \ge 1\}$ that does not depend on the state of the process (pre- or post-change), such that the transformed random variables $\widetilde{X}_n = W_n(\mathbf{X}_1^n)$, $n \ge 1$ are iid for both situations as previously. An abrupt change in the parameter σ_{x}^2 leads to an abrupt change in the expectation of the LLR given by (3.124) and (3.125). It follows from (3.126) that the LLR process $\{\lambda_n = \sum_{k=1}^n Z_k\}_{n\ge1}$ is a random walk before and after the changepoint. Moreover, because the distribution of Z_k, $k \ge 1$ is independent of ρ, there is no impact of the autocorrelation on the $\mathsf{ARL}_z(\sigma_{\mathrm{x,i}}^2)$, $i = 0, 1$,

$$\mathsf{ARL}_z(\sigma_{\mathrm{x},0}^2) \approx \frac{2\left(e^h - e^z - h + z\right)}{\sigma_{\mathrm{x},0}^2/\sigma_{\mathrm{x},1}^2 - 1 + 2\log(\sigma_{\mathrm{x},1}/\sigma_{\mathrm{x},0})},$$

$$\mathsf{ARL}_z(\sigma_{\mathrm{x},1}^2) \approx \frac{2\left(e^{-h} - e^{-z} + h - z\right)}{\sigma_{\mathrm{x},1}^2/\sigma_{\mathrm{x},0}^2 - 1 + 2\log(\sigma_{\mathrm{x},0}/\sigma_{\mathrm{x},1})}.$$

Assuming that the sampling period Δ can be chosen arbitrarily and considering an exponentially correlated Markov Gaussian process $\{V_t\}_{t\ge0}$, $\mathsf{E}[V_t] = 0$, $\mathsf{E}[V_t V_{t+u}] = \sigma_{\mathrm{V},i}^2 e^{-\alpha|u|}$ $(\alpha > 0)$, $i = 0, 1$, discussed in Subsection 3.2.4, it can be concluded, by analogy with the SPRT, that the products $\Delta \cdot \mathsf{ARL}_z(\sigma_{\mathrm{x,i}}^2)$, $i = 0, 1$ measured in units of time are linear functions of Δ [320]. Hence, the average delay to detection can be made unlimitedly small under the constraint on the ARL2FA when both are measured in units of time.

8.2.8 CUSUM Optimality in the General Non-iid Case

First of all, let us recall that the case of changes in the AR parameters of an AR model leads to the cumulative LLR with dependent increments. Indeed, let us consider the stationary AR(1) model

$$X_1 \sim \mathcal{N}\left(0, \frac{\sigma^2}{1-\rho^2}\right), \quad X_n = \rho X_{n-1} + \xi_n, \quad \xi_n \sim \mathcal{N}(0, \sigma^2), \quad n \geq 2 \tag{8.199}$$

and the hypotheses $H_i : \rho = \rho_i$, $i = 0, 1$, where ρ_0 and ρ_1 are two distinct constants, $|\rho_i| < 1$. It results from (3.200) that the cumulative LLR for the first n observations $\mathbf{X}_1^n$ is

$$\lambda_n = \sum_{k=1}^{n} Z_k = \frac{1}{2}\log\frac{1-\rho_1^2}{1-\rho_0^2} + \frac{X_1^2}{2\sigma^2}(\rho_1^2 - \rho_0^2) + \frac{1}{2\sigma^2}\sum_{k=2}^{n}\left[(X_k - \rho_0 X_{k-1})^2 - (X_k - \rho_1 X_{k-1})^2\right].$$

The cumulative LLR is composed of the increments $Z_k = \frac{1}{2\sigma^2}\left[(X_k - \rho_0 X_{k-1})^2 - (X_k - \rho_1 X_{k-1})^2\right]$, $k \geq 2$. If the hypothesis $H_i : \rho = \rho_i$ is true, then the term $\left\{(X_k - \rho_i X_{k-1})^2\right\}_{k\geq 2}$ produces an iid random sequence but another term $\left\{(X_k - \rho_j X_{k-1})^2\right\}_{k\geq 2}$ with $j \neq i$ produces a dependent random sequence because $\rho_0 \neq \rho_1$. Therefore, the sequence $\{Z_k\}_{k\geq 1}$ is dependent under both hypotheses.

In this subsection, we consider the general non-iid model (6.2) assuming that the observations $\{X_n\}_{n\geq 1}$ are distributed according to conditional pre-change densities $f_{0,n}(X_n|\mathbf{X}_1^{n-1})$ for $n = 1, \dots, \nu$ and according to conditional post-change densities $f_{1,n}(X_n|\mathbf{X}_1^{n-1})$ for $n > \nu$, where post-change densities $f_{1,n}(X_n|\mathbf{X}_1^{n-1}) = f_{1,n}^{(\nu)}(X_n|\mathbf{X}_1^{n-1})$ may generally depend on the changepoint ν. In this general case the LLR process between the hypotheses H_ν and H_∞ is $\lambda_n^\nu = \sum_{i=\nu+1}^n Z_i^{(\nu)}$, where $Z_i^{(\nu)} = \log[f_{1,n}^{(\nu)}(X_n|\mathbf{X}_1^{n-1})/f_{0,n}(X_n|\mathbf{X}_1^{n-1})]$, and the CUSUM statistic W_n is given by $W_n = \max_{0\leq\nu<n}\sum_{i=\nu+1}^n Z_i^{(\nu)}$, $n \geq 1$. The CUSUM stopping time is the first time W_n exceeds a threshold h, $T_{\mathrm{CS}}(h) = \inf\{n : W_n \geq h\}$. Recall definitions of moments $\mathsf{ESM}_m(T)$ and $\mathsf{SM}_m(T)$ given in (8.80). Also, recall that, by Lemma 8.2.1, $\mathsf{ARL2FA}(T_{\mathrm{CS}}(h)) \geq e^h$ for any $h > 0$, so $h_\gamma = \log\gamma$ implies $\mathsf{ARL2FA}(T_{\mathrm{CS}}(\log\gamma)) \geq \gamma$, i.e., $T_{\mathrm{CS}}(h_\gamma) \in \mathbf{C}_\gamma$ for any $\gamma > 1$.

Finally, recall Theorem 8.2.2 that establishes asymptotic lower bounds for infimum of moments in the class $\mathbf{C}_\gamma$ for all $r > 0$,

$$\inf_{T\in\mathbf{C}_\gamma}\mathsf{ESM}_r(T) \geq \inf_{T\in\mathbf{C}_\gamma}\mathsf{SM}_r(T) \geq \left(\frac{\log\gamma}{I}\right)^r (1+o(1)) \quad \text{as } \gamma\to\infty, \tag{8.200}$$

as long as condition (8.84) holds.

Therefore, to establish asymptotic optimality of CUSUM it suffices to show that under certain conditions the lower bounds (8.200) are attained for CUSUM with $h = \log\gamma$ or, more generally, for h selected in such a way that $T_{\mathrm{CS}}(h) \geq \gamma$ and $h \sim \log\gamma$ as $\gamma \to \infty$. This is the subject of the next theorem.

Theorem 8.2.6 (Asymptotic optimality in class $\mathbf{C}_\gamma$). *Consider the general non-iid model* (6.2). *Assume that there exists a finite positive number I such that condition* (8.84) *holds. Further assume that*

$$\sup_{0\leq\nu<k}\operatorname{ess\,sup}\mathsf{P}_\nu\left(n^{-1}\lambda_{k+n}^k < I(1-\varepsilon)|\mathscr{F}_\nu\right) \xrightarrow[n\to\infty]{} 0 \quad \text{for all } 0<\varepsilon<1 \text{ and } k\geq 0. \tag{8.201}$$

Then for all $r > 0$

$$\mathsf{SM}_r(T_{\mathrm{CS}}(h)) \sim \mathsf{ESM}_r(T_{\mathrm{CS}}(h)) \sim \left(\frac{h}{I}\right)^r \quad \text{as } h\to\infty \tag{8.202}$$

and

$$\inf_{T\in\mathbf{C}_\gamma}\mathsf{ESM}_r(T) \sim \inf_{T\in\mathbf{C}_\gamma}\mathsf{SM}_r(T) \sim \left(\frac{\log\gamma}{I}\right)^r \quad \text{as } \gamma\to\infty. \tag{8.203}$$

Therefore, if $h = h_\gamma$ is selected so that $\mathsf{ARL2FA}(T_{\rm CS}(h_\gamma)) \ge \gamma$ and $h_\gamma \sim \log\gamma$, in particular $h_\gamma = \log\gamma$, then the CUSUM procedure is asymptotically minimax in the sense of minimizing expected moments $\mathsf{ESM}_r(T)$ and $\mathsf{SM}_r(T)$ for all $r > 0$ to first order as $\gamma \to \infty$:

$$\inf_{T\in\mathbf{C}_\gamma} \mathsf{ESM}_r(T) \sim \inf_{T\in\mathbf{C}_\gamma} \mathsf{SM}_r(T) \sim \mathsf{ESM}_r(T_{\rm CS}) \sim \mathsf{SM}_r(T_{\rm CS}) \sim \left(\frac{\log\gamma}{I}\right)^r. \tag{8.204}$$

Proof. Note first that asymptotic equalities (8.202) and asymptotic lower bounds (8.200) imply (8.203) and (8.204), so all we need to prove is the validity of asymptotic equalities (8.202).

Replacing $\log\gamma$ with h in the proof of Theorem 8.2.2 yields the lower bounds (under condition (8.84)),

$$\mathsf{ESM}_r(T_{\rm CS}(h)) \ge \mathsf{SM}_r(T_{\rm CS}(h)) \ge \left(\frac{h}{I}\right)^r (1+o(1)) \quad \text{as } h\to\infty$$

for all $r > 0$. Hence, to prove asymptotic equalities (8.202) it suffices to show that under condition (8.201) the following upper bound holds:

$$\mathsf{ESM}_r(T_{\rm CS}(h)) \le \left(\frac{h}{I}\right)^r (1+o(1)) \quad \text{as } h\to\infty. \tag{8.205}$$

Let $N_h = \lfloor h/I(1-\varepsilon)\rfloor$. By condition (8.201), for any $0<\varepsilon<1$, $k\ge 0$, and a sufficiently large h,

$$\sup_{0\le\nu<k} \operatorname{ess\,sup} \mathsf{P}_\nu\left(\lambda_{k+N_h}^k < N_h I(1-\varepsilon)|\mathscr{F}_\nu\right) = \sup_{0\le\nu<k} \operatorname{ess\,sup} \mathsf{P}_\nu\left(\lambda_{k+N_h}^k < h|\mathscr{F}_\nu\right) \le \varepsilon,$$

which implies that for any $\nu \ge 0$ and $\ell \ge 1$,

$$\operatorname{ess\,sup} \mathsf{P}_\nu(T_{\rm CS} - \nu > \ell N_h|\mathscr{F}_\nu) \le \operatorname{ess\,sup} \mathsf{P}_\nu\left(\sum_{i=\nu+(j-1)N_h+1}^{\nu+jN_h} Z_i^{(\nu)} < h, j=1,\dots,\ell|\mathscr{F}_\nu\right) \le \varepsilon^\ell.$$

Therefore, after tedious but straightforward algebra we obtain that for a sufficiently large h and any $0<\varepsilon<1$

$$\begin{aligned}
\mathsf{ESM}_r(T_{\rm CS}) &\le \sum_{j=0}^\infty [(j+1)^r - j^r] \sup_\nu \operatorname{ess\,sup} \mathsf{P}_\nu(T_{\rm CS} > j+\nu|\mathscr{F}_\nu)\\
&\le (1+N_h)^r + \sum_{\ell=0}^\infty \{[(\ell+1)N_h]^r - (\ell N_h)^r\} \sup_\nu \operatorname{ess\,sup} \mathsf{P}_\nu(T_{\rm CS} - \nu > \ell N_h|\mathscr{F}_\nu)\\
&\le (1+N_h)^r + N_h^r \sum_{\ell=0}^\infty [(\ell+1)^r - \ell^r]\varepsilon^\ell = N_h^r(1+o(1)) \quad \text{as } h\to\infty.
\end{aligned}$$

Since ε can be arbitrary small, this implies the asymptotic upper bound (8.205) and the proof is complete. □

Recall that in Subsection 6.3.5 we introduced an alternative false alarm constraint associated with the conditional maximal false alarm probability $\mathsf{MPFA}_m(T) = \sup_{k\ge 0} \mathsf{P}_\infty(T \le k+m|T>k)$ in the fixed window of size $m \ge 1$. Specifically, instead of the class $\mathbf{C}_\gamma$ consider now the class of procedures defined in (6.26) with an upper bound on the maximal PFA, $\mathbf{C}^m(\beta) = \{T : \mathsf{MPFA}_m(T) \le \beta\}$, $0<\beta<1$. The reason for that is twofold. First, in Subsection 6.3.5 we showed that the ARL2FA constraint generally does not guarantee small values of $\mathsf{MPFA}_m(T)$ when γ is large (it only guarantees that there is some k that may depend on γ such that $\mathsf{P}_\infty(T \le k+m|T>k) < m/\gamma$). The class $\mathbf{C}^m(\beta)$ is more stringent than the class $\mathbf{C}_\gamma$ and, as a result, small values of β imply also small values of the ARL2FA (but not *vice versa* in general). Second, in the class $\mathbf{C}^m(\beta)$ one has a chance to find a uniformly optimal procedure that minimizes the conditional average delay to detection

$\mathsf{CADD}_\nu(T) = \mathsf{E}_\nu(T-\nu|T>\nu)$ for all $\nu \ge 0$, at least asymptotically as $\beta \to 0$; see the optimization problem (6.27). More generally, we may ask if the CUSUM procedure minimizes asymptotically (as $\beta \to 0$) higher conditional moments of the detection delay $\mathsf{E}_\nu[(T-\nu)^r|T>\nu]$ for all $\nu \ge 0$ in the class $\mathbf{C}^m(\beta)$. In the following we assume that $\mathsf{MPFA}_m(T) \le \beta$ for some $m = m_\beta$ that depends on β and goes to infinity as β goes to zero but not too fast, so that

$$\liminf_{\beta\to 0}(m_\beta/N_\beta) > 1 \quad \text{but} \quad \lim_{\beta\to 0}[(\log m_\beta)/N_\beta] = 0, \tag{8.206}$$

where $N_\beta = I^{-1}|\log\beta|$. In this case, we simply write $\mathbf{C}(\beta)$ for $\mathbf{C}^{m_\beta}(\beta)$.

The following theorem establishes that under conditions of Theorem 8.2.6 the CUSUM procedure is asymptotically uniformly (for all $\nu \ge 0$) optimal in the class $\mathbf{C}(\beta)$.

Theorem 8.2.7 (Asymptotic optimality in class $\mathbf{C}(\beta)$)**.** *Consider the general non-iid model* (6.2). *Assume that for some finite positive number I condition* (8.84) *for the LLR holds. Assume* (8.206).

(i) *Then as $\beta \to 0$ for all $r > 0$*

$$\inf_{T\in\mathbf{C}(\beta)} \mathsf{E}_\nu[(T-\nu)^r|T>\nu] \ge \left(\frac{|\log\beta|}{I}\right)^r (1+o(1)) \quad \textit{for all } \nu \ge 0. \tag{8.207}$$

(ii) *Assume, in addition, that for some $r \ge 1$ condition* (8.201) *holds. If $h = h_\beta$ is selected so that $\mathsf{MPFA}_{m_\beta}(T_{\mathrm{CS}}(h_\beta)) \le \beta$ and $h_\beta \sim |\log\beta|$, then the lower bound in* (8.207) *is attained for the CUSUM procedure. Thus, it is asymptotically uniformly optimal in the sense of minimizing expected conditional moments $\mathsf{E}_\nu[(T-\nu)^r|T>\nu]$ for all $\nu \ge 0$ to first order as $\beta \to 0$,*

$$\inf_{T\in\mathbf{C}(\beta)} \mathsf{E}_\nu[(T-\nu)^r|T>\nu] \sim \mathsf{E}_\nu[(T_{\mathrm{CS}}-\nu)^r|T_{\mathrm{CS}}>\nu] \sim \left(\frac{|\log\beta|}{I}\right)^r \quad \textit{for all } \nu \ge 0. \tag{8.208}$$

Proof. (i) The method of proof of the asymptotic lower bound (8.207) is similar to the proof of the lower bound (8.86) in Theorem 8.2.2.

For $\gamma > 1$, let $K_\gamma = I^{-1}\log\gamma$ and consider stopping times T such that

$$\sup_{k\ge 0} \mathsf{P}_\infty(T \le k+m|T>k) \le m/\gamma$$

with $m = m_\gamma$ satisfying $\liminf_{\beta\to 0}(m_\gamma/K_\gamma) > 1$ and $\lim_{\gamma\to\infty}[(\log m_\gamma)/K_\gamma] = 0$.

The same reasoning that has led to (8.88) yields (for any $\varepsilon \in (0,1)$):

$$\mathsf{P}_\nu\left\{T \le \nu+(1-\varepsilon)K_\gamma|T>\nu\right\} \le p_{\gamma,\varepsilon}^{(\nu)}(T) + \beta_{\gamma,\varepsilon}^{(\nu)}(T), \tag{8.209}$$

where

$$\beta_{\gamma,\varepsilon}^{(\nu)}(T) = \mathsf{P}_\nu\left\{\frac{1}{(1-\varepsilon)K_\gamma}\max_{1\le n<(1-\varepsilon)K_\gamma} \lambda_{\nu+n}^\nu \ge (1+\varepsilon)I|T>\nu\right\}$$

goes to 0 for any $\nu \ge 0$, any $\varepsilon < 1$ and any T as $\gamma \to \infty$ by condition (8.84) and

$$\begin{aligned} p_{\gamma,\varepsilon}^{(\nu)}(T) &= e^{(1-\varepsilon^2)IK_\gamma}\mathsf{P}_\infty\left\{T \le \nu+(1-\varepsilon)K_\gamma|T>\nu\right\} \le (1-\varepsilon)K_\gamma e^{(1-\varepsilon^2)IK_\gamma}/\gamma \\ &= (1-\varepsilon)I^{-1}(\log\gamma)/\gamma^{\varepsilon^2} \to 0 \end{aligned}$$

also for any $\nu \ge 0$, any $\varepsilon < 1$ and any T. Hence, setting $m_\gamma/\gamma = \beta$ and $N_\beta = I^{-1}|\log\beta|$, we obtain that for all $\varepsilon < 1$ and all $\nu \ge 0$

$$\sup_{T\in\mathbf{C}(\beta)} \mathsf{P}_\nu\left\{T-\nu \le (1-\varepsilon)N_\beta|T>\nu\right\} \to 0 \quad \text{as } \beta \to 0.$$

Now, by the Chebyshev inequality, for any $0 < \varepsilon < 1$ and any $T \in \mathbf{C}(\beta)$

$$\mathsf{E}_\nu\left\{[(T-\nu)^+]^r | T > \nu\right\} \geq [(1-\varepsilon)N_\beta]^r \mathsf{P}_\nu\left\{T - \nu \geq (1-\varepsilon)N_\beta | T > \nu\right\},$$

which along with the former asymptotic relation implies that for all $\nu \geq 0$

$$\inf_{T \in \mathbf{C}(\beta)} \mathsf{E}_\nu\left\{[(T-\nu)^+]^r | T > \nu\right\} \geq [I^{-1}|\log\beta|]^r(1+o(1)) \quad \text{as } \beta \to 0. \tag{8.210}$$

(ii) It follows from (8.202) that

$$\mathsf{E}_\nu\left\{[(T_{\text{CS}}(h)-\nu)^+]^r | T_{\text{CS}}(h) > \nu\right\} \leq \left(\frac{h}{I}\right)^r (1+o(1)) \quad \text{as } h \to \infty. \tag{8.211}$$

Setting $h \sim |\log\beta|$ yields the inequality

$$\mathsf{E}_\nu\left\{[(T_{\text{CS}}(h)-\nu)^+]^r | T_{\text{CS}}(h) > \nu\right\} \leq \left(\frac{|\log\beta|}{I}\right)^r (1+o(1)) \quad \text{as } \beta \to 0,$$

which along with the lower bound (8.207) completes the proof. □

Note, however, that the PFA constraint is not easy to realize in the general non-iid case. In particular, there is no simple formula (and even upper bound) for $\mathsf{MPFA}_{m_\beta}(T_{\text{CS}}(h))$ in general, which is a serious obstacle in practical implementation. This problem is greatly simplified in the iid case (6.4). Indeed, in the iid case, by Pollak and Tartakovsky [369], the CUSUM stopping time $T_{\text{CS}}(h)$ has asymptotically as $h \to \infty$ exponential distribution under P_∞,

$$\lim_{h\to\infty} \mathsf{P}_\infty\left\{T_{\text{CS}}(h)/\mathsf{E}_\infty T_{\text{CS}}(h) > x\right\} = e^{-x}, \quad x > 0.$$

Furthermore, according to computations in Example 8.2.1 and to Figure 8.8, the exponential approximation is very accurate already for small threshold values; therefore, this approximation is expected to work very well not only for a very low false alarm rate but for a moderate one too. Therefore, taking into account that $\mathsf{E}_\infty T_{\text{CS}}(h) \geq e^h$, we obtain that for a sufficiently large h,

$$\mathsf{MPFA}_m(T_{\text{CS}}) \leq 1 - \exp\left\{-m/e^h\right\},$$

i.e., taking $h_\beta(m_\beta) = \log[m_\beta/|\log(1-\beta)|] \approx \log(m_\beta/\beta)$ (for small β) guarantees (at least approximately) that $T_{\text{CS}}(h_\beta)$ belongs to the class $\mathbf{C}(\beta)$. A better precision can be obtained using the approximations to the ARL2FA presented in Subsection 8.2.6. In particular, using first-order approximation (8.145), $\mathsf{E}_\infty T_{\text{CS}}(h) \sim e^h/I\zeta^2$, we obtain that if the threshold h_β is a root of the equation

$$m_\beta I \zeta^2 e^{-h} = \beta, \tag{8.212}$$

then

$$\mathsf{MPFA}_{m_\beta}(T_{\text{CS}}(h_\beta)) \sim \beta \quad \text{as } \beta \to 0. \tag{8.213}$$

If, in particular, m_β is explicitly expressed via β, then $h_\beta = \log(m_\beta I\zeta^2/\beta)$ implies asymptotic equality (8.213). However, m_β may implicitly depend on β via the threshold value. For example, if $m_\beta = O(h_\beta)$, then in order to find h_β one has to solve equation (8.212).

This argument along with Theorem 8.2.7 yields the following theorem for the iid case with I being the K–L number.

Theorem 8.2.8. *Consider the iid model* (6.4). *Assume that the K–L number is positive and finite,* $0 < I < \infty$, *and let* (8.206) *hold.*

(i) *Then for all $r>0$ as $\beta \to 0$*

$$\inf_{T\in \mathbf{C}(\beta)} \mathsf{E}_\nu[(T-\nu)^r | T>\nu] \ge \left(\frac{|\log\beta|}{I}\right)^r (1+o(1)) \quad \text{for all } \nu \ge 0. \tag{8.214}$$

(ii) *If $h=h_\beta$ is selected from equation* (8.212)*, then* $\mathsf{MPFA}_{m_\beta}(T_{\mathrm{CS}}(h_\beta)) \sim \beta$, $h_\beta \sim |\log\beta|$, *and*

$$\inf_{T\in \mathbf{C}(\beta)} \mathsf{E}_\nu[(T-\nu)^r | T>\nu] \sim \mathsf{E}_\nu[(T_{\mathrm{CS}}-\nu)^r | T_{\mathrm{CS}}>\nu] \sim \left(\frac{|\log\beta|}{I}\right)^r \quad \text{for all } \nu \ge 0 \tag{8.215}$$

as $\beta \to 0$.

Proof. (i) Obviously, in the iid case, condition (8.84) gets modified into

$$\lim_{M\to\infty} \mathsf{P}_0\left(M^{-1} \max_{1\le n\le M} \lambda_n^0 > (1+\varepsilon)I\right) = 0,$$

which holds whenever the K–L number, I, is positive and finite. This implies (8.214).

(ii) Condition (8.201) gets modified into

$$\lim_{n\to\infty} \mathsf{P}_0\left\{n^{-1}\lambda_n^0 < I(1-\varepsilon)\right\} = 0,$$

which holds whenever $I<\infty$. Since, obviously, $h_\beta \sim |\log\beta|$ when h_β is selected from equation (8.212) and when conditions (8.206) are satisfied, this completes the proof. □

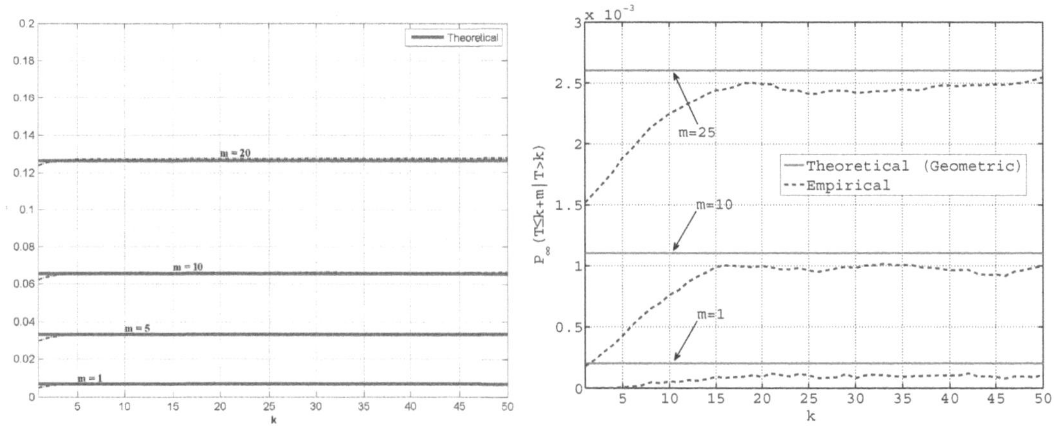

Figure 8.12: *The false alarm probability $\mathsf{P}_\infty(T_{\mathrm{CS}}(h) \le k+m | T_{\mathrm{CS}}(h) > k)$ vs. k for the exponential scenario with $\theta_1 = 2$ (left) and for the Gaussian scenario with $\theta = 1$ (right). Numerical computations are shown by dashed curves and the theoretical approximation by solid lines.*

In Figure 8.12, we show a typical behavior of the conditional false alarm probability $\mathsf{P}_\infty(T_{\mathrm{CS}} \le k+m | T_{\mathrm{CS}} > k)$ as a function of k for several values of m. The computations have been done based on the numerical techniques of Subsection 8.2.6.1, solving numerically the recursive integral equations (8.108), (8.110). The picture on the left corresponds to the exponential example considered in Subsection 8.2.6.4, Example 8.2.1 with $\theta_1 = 2$. The threshold was selected so that the ARL2FA is approximately equal to 148, which is considered a high false alarm rate. The picture on the right corresponds to the Gaussian model when detecting a change from $\mathcal{N}(0,1)$ to $\mathcal{N}(1,1)$ with the threshold chosen so that $\mathsf{ARL2FA}(h) \approx 10^4$ (low false alarm rate). It is seen that in both cases the PFA $\mathsf{P}_\infty(T_{\mathrm{CS}} \le k+m | T_{\mathrm{CS}} > k)$ increases very rapidly to the steady-state value (dashed line). This can be explained by the fact that the stopping time of the CUSUM procedure that starts from the random point distributed according to the quasi-stationary distribution has exactly geometric distribution under P_∞. Hence, when the CUSUM statistic attains the quasi-stationary distribution, the PFA does not change with k. It is also seen that the approximation

$$\mathsf{MPFA}_m(T_{\text{CS}}) \approx 1-\exp\{-m/\mathsf{ARL2FA}(h)\} \approx m/\mathsf{ARL2FA}(h)$$

works extremely well. For the exponential example this approximation is very accurate even for small threshold values (small ARL2FA).

We stress that while in the iid case as well as for weakly dependent models the distribution of the CUSUM stopping time (as well as SR) is approximately exponential, for general non-iid models this is not necessarily the case. Therefore, in general it is difficult to evaluate the conditional probability $\mathsf{P}_\infty(T_{\text{CS}} \le k+m | T_{\text{CS}} > k)$. One way out is to neglect the condition $\{T > k\}$ and instead to consider the unconditional PFA $\mathsf{P}_\infty(k < T_{\text{CS}} \le k+m)$. Correspondingly, introduce the class

$$\mathbf{C}_{\text{un}}(\beta) = \{T : \sup_{k\ge 0} \mathsf{P}_\infty(k < T_{\text{CS}} \le k+m_\beta) \le \beta\}$$

of detection procedures for which this local (maximal) unconditional PFA does not exceed some tolerance level β, where again the size of the interval m_β goes to infinity as $\beta \to 0$ in such a way that conditions (8.206) are satisfied. However, optimization in this class requires considering unconditional average delay to detection $\mathsf{E}_\nu(T-\nu)^+$ instead of $\mathsf{CADD}_\nu(T)$ or more generally unconditional moments $\mathsf{E}_\nu[(T-\nu)^+]^r$, $r \ge 1$, which may cause a certain disadvantage in asymptotic problems, as we discuss below in Remark 8.2.1.

Nevertheless, if we assume that the following relaxed version of condition (8.84) holds,

$$\lim_{n\to\infty} \sup_{\nu\ge 0} \mathsf{P}_\nu\left(M^{-1} \max_{1\le n\le M} \lambda_{\nu+n}^\nu \ge I(1+\varepsilon)\right) = 0 \quad \text{for all } \varepsilon > 0, \tag{8.216}$$

then for all $r > 0$ and any $T \in \mathbf{C}_{\text{un}}(\beta)$ as $\beta \to 0$

$$\mathsf{E}_\nu[(T-\nu)^+]^r \ge \left[\frac{\mathsf{P}_\infty(T>\nu)}{I} + o(1)\right]^r |\log\beta|^r \quad \text{uniformly over } \nu \ge 0, \tag{8.217}$$

i.e., the vanishing term $o(1)$ does not depend on ν. The proof is similar to the proof of the lower estimates (8.200) and (8.207) (for $r=1$, cf. [251, Theorem 2]).

It turns out that the CUSUM procedure is asymptotically optimal attaining the lower bound in (8.217) under condition (8.201). However, the ordinary CUSUM is difficult to implement for general non-iid models, especially when the post-change densities depend on the changepoint, which is often the case. Then computational complexity becomes a serious issue.

An attractive solution is to replace the CUSUM procedure with the window-limited CUSUM

$$T_{\text{WL-CS}}(h) = \inf\left\{n \ge 1 : \max_{n-m_\beta \le j \le n} \sum_{k=j}^n \log \frac{f_1(X_k \mid \mathbf{X}_1^{k-1})}{f_0(X_k \mid \mathbf{X}_1^{k-1})} \ge h\right\}, \tag{8.218}$$

where maximization over all $1 \le j \le n$ is replaced by the maximization in the sliding window of size m_β. Note that this window is large enough but not too large due to conditions (8.206). Roughly speaking this window may be taken on order of $O(I^{-1}|\log\beta|)$. We will use the abbreviation WL-CUSUM for this procedure.

The first observation is that for the WL-CUSUM the unconditional PFA satisfies the inequality

$$\mathsf{P}_\infty(k < T_{\text{WL-CS}} \le k+m_\beta) \le \sum_{j=k-m_\beta+1}^{k+m_\beta} \mathsf{P}_\infty\left(e^{\lambda_n^k} \ge e^h \text{ for some } n \ge k\right) \le 2m_\beta e^{-h},$$

where we used Doob's submartingale inequality

$$\mathsf{P}_\infty\left(e^{\lambda_n^k} \geq e^h \text{ for some } n \geq k\right) \leq e^{-h}$$

applied to the martingale $e^{\lambda_n^k}$ with unit expectation. Thus, if we take $h = \log(2m_\beta/\beta)$, then $T_{\text{WL-CS}} \in \mathbf{C}_{\text{un}}(\beta)$ and also $h_\beta \sim |\log\beta|$ by conditions (8.206). Note that this is critical for asymptotic optimality of WL-CUSUM.

The following theorem establishes uniform first-order asymptotic optimality property of the WL-CUSUM procedure showing that the asymptotic lower bound given in (8.217) is attained by WL-CUSUM with the threshold found from the equation $2m_\beta e^{-h} = \beta$.

Theorem 8.2.9. *Assume that conditions* (8.84), (8.201) *and* (8.206) *hold. Then, as $h \to \infty$*

$$\mathsf{E}_\nu[(T_{\text{WL-CS}} - \nu)^+]^r = \left[\frac{\mathsf{P}_\infty(T_{\text{WL-CS}} > \nu)}{I} + o(1)\right]^r h^r \quad \textit{for all } \nu \geq 0, \tag{8.219}$$

where $o(1) \to 0$ uniformly in $\nu \geq 0$.

If the threshold $h = h_\beta$ is found from the equation $2m_\beta e^{-h} = \beta$, then $T_{\text{WL-CS}} \in \mathbf{C}_{\text{un}}(\beta)$ and as $\beta \to 0$

$$\mathsf{E}_\nu[(T_{\text{WL-CS}} - \nu)^+]^r = \left[\frac{\mathsf{P}_\infty(T_{\text{WL-CS}} > \nu)}{I} + o(1)\right]^r |\log\beta|^r \quad \textit{for all } \nu \geq 0, \tag{8.220}$$

where the vanishing term $o(1)$ is uniform over ν. Therefore, the WL-CUSUM procedure is first-order uniformly optimal.

For $r = 1$, the proof of this theorem can be found in [251]. For $r > 1$, it is similar.

Remark 8.2.1. Despite the fact that formally asymptotic relations (8.217) and (8.220) imply that for any fixed ν

$$\frac{\inf_{T\in\mathbf{C}_{\text{un}}(\beta)} \mathsf{E}_\nu[(T-\nu)^+]^r}{\mathsf{E}_\nu[(T_{\text{WL-CS}} - \nu)^+]^r} \sim 1 \quad \text{as } \beta \to 0$$

(since $\mathsf{P}_\infty(T > \nu) \to 1$ for any $T \in \mathbf{C}_{\text{un}}(\beta)$ as $\beta \to 0$), in practice one may face a difficulty in implementing this result. Indeed, if β is relatively small but not too small, which is typical for many applications, then there will always be the changepoints ν for which the probability $\mathsf{P}_\infty(T > \nu)$ is not close to 1. Moreover, these probabilities may differ substantially for different stopping times. Another issue is that in many cases (in the iid case in the first place) the P_∞-distribution of the CUSUM stopping time (and other stopping times) is approximately exponential, so the unconditional PFA $\mathsf{P}_\infty(k < T_{\text{CS}} \leq k+m) \approx e^{-k} m C e^{-h}$ decays exponentially fast when k increases. While in this case formally $\sup_k \mathsf{P}_\infty(k < T_{\text{CS}} \leq k+m) = \mathsf{P}_\infty(T_{\text{CS}} \leq m) \approx mCe^{-h}$, the exponential decay of the "real" probability $\mathsf{P}_\infty(k < T_{\text{CS}} \leq k+m)$ puts usefulness of the constraint imposed on $\sup_k \mathsf{P}_\infty(k < T_{\text{CS}} \leq k+m)$ under question.

Remark 8.2.2. Theorems 8.2.6 and 8.2.7 hold for the WL-CUSUM procedure when the threshold h can be selected appropriately. Indeed, it can be shown that $\mathsf{E}_\infty T_{\text{WL-CS}}(h) \geq e^h/4(1+o(1))$, so we may take $h = \log(4\gamma)$ and $m_\gamma = O(\log\gamma)$ in the WL-CUSUM and with this selection the WL-CUSUM is asymptotically optimal in the class $\mathbf{C}_\gamma$. Also, $\mathsf{P}_\infty(T_{\text{WL-CS}}(h) \leq k+m | T_{\text{WL-CS}}(h) > k) \leq me^{-h}$ for some k. While in general this does guarantee that the maximal PFA is smaller than me^{-h}, in many particular non-iid cases the value of k for which this inequality holds is essentially finite. In such cases we may take $h = \log(m_\beta/\beta)$ to guarantee $\mathsf{MPFA}(T_{\text{WL-CS}}(h)) \leq \beta$ (at least approximately) and WL-CUSUM is asymptotically optimal in the class $\mathbf{C}(\beta)$.

8.2.9 Local CUSUM

As shown in Section 2.11, the design of optimal tests and their study can be seriously simplified by using the asymptotic local approach. This approach permits to reduce a general hypothesis testing problem to the case of simpler hypothesis testing problem about the mean of a Gaussian law. Now we investigate the use of the local approach for designing change detection algorithms. The asymptotic local approach is introduced in Section 2.11 for FSS hypothesis testing and in Section 3.5 for sequential hypothesis testing. Here we follow the main argument of these sections; see also [47, 60, 317, 319, 320, 376].

We begin with the main idea underlying the design of sequential change detection algorithms using the local expansion. Consider a parametric model $X_n \sim F_\theta(x)$ with the pre-change parameter $\theta = \theta_0$ and the post-change parameter $\theta = \theta_1$, assuming first that the parameter $\theta \in \mathbb{R}$ is scalar. Recall that the key idea of LLR-based CUSUM algorithm (and others too) is that the pre-change mean of the LLR is negative and the post-change mean is positive, $\mathsf{E}_{\theta_0} Z_n < 0$, $\mathsf{E}_{\theta_1} Z_n > 0$. Let us replace the exact LLR with an efficient score. Assume that

$$\theta_i = \theta^* + (-1)^{i+1}\,\frac{\delta\theta}{2}, \quad i = 0, 1, \tag{8.221}$$

where $\delta\theta > 0$ is a small positive number. Let Z_n^* be the contribution of the observation X_n to the efficient score computed at θ^*,

$$Z_n(\theta^*) = Z_n^* = \left.\frac{\partial \log p_\theta(X_n)}{\partial\theta}\right|_{\theta=\theta^*}. \tag{8.222}$$

We first show that, up to second-order terms, the LLR is equivalent to the efficient score. According to the asymptotic expansion (2.369), we have

$$\log\Lambda_n\left(\theta^*, \theta^* + \frac{\delta\theta}{2}\right) \approx \frac{\delta\theta}{2}\left.\frac{\partial \log p_\theta(\mathbf{X}_1^n)}{\partial\theta}\right|_{\theta=\theta^*} - \frac{\delta\theta^2}{8}\,\mathcal{F}_n(\theta^*). \tag{8.223}$$

Therefore, at the points θ_i given in (8.221) the LLR can be written as

$$\log\Lambda_n\left(\theta^* - \frac{\delta\theta}{2}, \theta^* + \frac{\delta\theta}{2}\right) \approx \delta\theta\left.\frac{\partial \log p_\theta(\mathbf{X}_1^n)}{\partial\theta}\right|_{\theta=\theta^*} = \delta\theta\sum_{k=1}^{n} Z_k^*. \tag{8.224}$$

From the last equation, it is obvious that, for small changes, the efficient score has approximately the same property as the LLR, namely :

$$\mathsf{E}_{\theta_0}(Z_n^*) < 0 \ \text{ and } \ \mathsf{E}_{\theta_1}(Z_n^*) > 0, \tag{8.225}$$

In other words, *the change in the parameter θ is reflected into a change in the sign of the expectation of the efficient score*, and the efficient score is used exactly in the same way as the LLR.

Consider now a vector parameter $\theta \in \mathbb{R}^\ell$. Then

$$\theta_i = \theta^* + (-1)^{i+1}\,\frac{\delta\theta}{2}\Upsilon, \quad i = 0, 1,$$

where Υ, $\|\Upsilon\|_2 = 1$, is the *unit* vector of the change direction in the parametric space. The generalization of the previous discussion is straightforward, and the LLR in this case is approximated as

$$\log\Lambda_n\left(\theta^* - \frac{\delta\theta}{2}\Upsilon, \theta^* + \frac{\delta\theta}{2}\Upsilon\right) \approx \delta\theta\,\Upsilon^\top\left.\frac{\partial \log p_\theta(\mathbf{X}_1^n)}{\partial\theta}\right|_{\theta=\theta^*} \approx \delta\theta\,\Upsilon^\top\sum_{k=1}^{n} Z_k^*, \tag{8.226}$$

where $Z_n^* = \nabla_{\theta^*}\left[\log p_\theta(X_n)\right]$ is the ℓ-dimensional efficient score. Therefore, *the change in θ is reflected into a change in the sign of the expectation of the scalar product $\Upsilon^\top Z_n^*$* :

$$\mathsf{E}_{\theta_0}(\Upsilon^\top Z_n^*) < 0 \ \text{ and } \ \mathsf{E}_{\theta_1}(\Upsilon^\top Z_n^*) > 0. \tag{8.227}$$

Assume now that the pre- and post-change hypotheses are composite but local; that is, the true values of $\delta\theta$ before and after the changepoint are unknown but small. The only available information is that the pre-change value of $\delta\theta$ is negative and the post-change value of $\delta\theta$ is positive. Our goal now is to design a local test which is an approximation of the CUSUM test in a small neighborhood of θ^*. In the case of scalar θ, the stopping time of the local CUSUM procedure is defined by analogy with (8.66) as follows

$$T_{\text{LCS}} = \inf\{n \geq 1 : g_n \geq h\}, \quad g_n = [g_{n-1} + Z_n^*]^+, \quad Z_n^* = \left.\frac{\partial \log p_\theta(X_n)}{\partial \theta}\right|_{\theta=\theta^*}, \tag{8.228}$$

and in the vector case as

$$g_n = \left[g_{n-1} + \Upsilon^\top Z_n^*\right]^+, \quad Z_n^* = \nabla_{\theta^*}\left[\log p_\theta(X_n)\right], \tag{8.229}$$

where $g_0 = 0$.

The ARL function $\mathsf{ARL}_z(\theta)$ of the local CUSUM in the independent case can be calculated using the Fredholm integral equations (8.99)–(8.102) given in Subsection 8.2.6.1, where we describe the numerical calculation of the ARL. The kernel $g(x)$ of the integral equations (8.101)–(8.102) is equal to the pdf of the efficient score increment Z_n^* (resp. $\Upsilon^\top Z_n^*$) in the scalar case (resp. vector case).

We now derive approximations for the ARL function, using the same approach as previously, namely (8.99) and Wald's approximation for ESS and OC. We first calculate the non-zero real root of equation $\mathsf{E}_\theta e^{-\omega_0(\theta)Z_k} = 1$ for some $\theta \in \mathbb{R}$ in the neighborhood of θ^*. As in Subsection 3.5.2, using Taylor's expansion of $e^{-\omega_0(\theta)Z_k^*}$ in the neighborhood of θ^*

$$e^{-\omega_0(\theta)Z_k^*} = 1 + \omega_0(\theta)\, Z_k^* + \frac{\omega_0^2(\theta)}{2} Z_k^{*2} - \frac{\omega_0^3(\theta)}{6} Z_k^{*3} e^{-u\omega_0(\theta)\, Z_k^*},$$

where $0 \leq u \leq 1$, we obtain

$$-\omega_0(\theta)\, \mathsf{E}_\theta\left(Z_k^*\right) + \frac{\omega_0^2(\theta)}{2} \mathsf{E}_\theta\left(Z_k^{*2}\right) - \frac{\omega_0^3(\theta)}{6} \mathsf{E}_\theta\left(Z_k^{*3} e^{-u\omega_0(\theta)\, Z_k^*}\right) = 0.$$

It is assumed that $\mathsf{E}_\theta\left(Z_k^*\right) \neq 0$ for $\theta \neq \theta^*$. Let us assume that the regularity conditions defined in [495, Lemma A.1] are fulfilled; see also [104, Sec. 2.3] for the properties of the mgf functions. Hence, if $\mathsf{E}_\theta\left(Z_k^*\right) \neq 0$ then the only real root is non-zero, $\omega_0(\theta) \neq 0$. Assuming that the function $u \mapsto \mathsf{E}_\theta\left(Z_k^{*3} e^{-u\omega_0(\theta)\, Z_k^*}\right)$ is bounded, we get

$$\mathsf{E}_\theta\left(Z_k^*\right) - \frac{\omega_0(\theta)}{2} \mathsf{E}_\theta\left(Z_k^{*2}\right) + o\left(\omega_0(\theta)\right) = 0. \tag{8.230}$$

Since $\mathsf{E}_\theta\left(Z_k^{*2}\right)$ and $\omega_0(\theta)$ are continuous functions of θ in a small neighborhood of θ^* and $\omega_0(\theta^*) = 0$, we obtain the following approximation for the non-zero real root ω_0 of equation $\mathsf{E}_\theta\left(e^{-\omega_0(\theta)Z_k}\right) = 1$:

$$\omega_0(\theta) \approx \frac{2\mathsf{E}_\theta\left(Z_k^*\right)}{\mathsf{E}_{\theta^*}\left(Z_k^{*2}\right)}. \tag{8.231}$$

The previous results can be easily extended to the case of the ℓ-dimensional efficient score Z_k^* where equation (8.231) is rewritten as follows

$$\omega_0(\theta) \approx \frac{2\mathsf{E}_\theta\left(\Upsilon^\top Z_k^*\right)}{\mathsf{E}_{\theta^*}\left[(\Upsilon^\top Z^*)_k^2\right]}. \tag{8.232}$$

To compute the expectations $\mathsf{E}_\theta[\Upsilon^\top Z_k^*]$ and $\mathsf{E}_{\theta^*}[(\Upsilon^\top Z^*)_k^2]$, assume that the regularity conditions defined in [241] are satisfiedd, i.e., the pdf of observations $p_\theta(X)$ can be decomposed in the following form

$$p_\theta(X) = p_{\theta^*}(X)\left[1 + (\theta - \theta^*)^\top Z_k^* + \|\theta - \theta^*\|_2^2 \delta(X;\theta)\right], \tag{8.233}$$

where $\mathsf{E}_\theta\left[\delta(X;\theta)\right]^2 < A$ for $\|\theta-\theta^*\|_2 < \varepsilon$ and $A, \varepsilon \in \mathbb{R}_+$ are positive numbers. Let $\theta - \theta^* = \eta D$, where $\|D\|_2 = 1$, and let the unit vector D define an arbitrary change direction which may differ from the putative change direction Υ. It is also assumed that $\theta = \theta^* + \delta\theta D$ belongs to a small neighborhood of θ^*. Hence,

$$\begin{aligned}\mathsf{E}_\theta\left(\Upsilon^\top Z_k^*\right) &= \int\cdots\int \Upsilon^\top Z_k^*(X) p_\theta(X)\mathrm{d}X \\ &= \int\cdots\int \Upsilon^\top Z_k^*(X) p_{\theta^*}(X)\mathrm{d}X + \eta\int\cdots\int \Upsilon^\top Z_k^*(X) D^\top Z_k^*(X) p_{\theta^*}(X)\mathrm{d}X \\ &\quad +\eta^2\int\cdots\int \Upsilon^\top Z_k^*(X)\delta(X;\theta) p_{\theta^*}(X)\mathrm{d}X. \end{aligned} \tag{8.234}$$

The first integral in (8.234) is equal to zero by definition of θ^* and the third integral is bounded. Therefore, we get

$$\mathsf{E}_\theta\left(\Upsilon^\top Z_k^*\right) = \eta\Upsilon^\top \mathcal{F}(\theta^*) D + o(\eta), \tag{8.235}$$

where $\mathcal{F}(\theta^*)$ is the Fisher information matrix of size $\ell\times\ell$. Also,

$$\mathsf{E}_{\theta^*}\left[\left(\Upsilon^\top Z_k^*\right)^2\right] = \Upsilon^\top \mathcal{F}(\theta^*)\Upsilon. \tag{8.236}$$

The substitution of the expressions for $\mathsf{E}_\theta\left(\Upsilon^\top Z_k^*\right)$ and $\omega_0(\theta)$ into (8.128) yields the following Wald approximation for the ARL in the local case:

$$\mathsf{ARL}_z(\eta) \approx \frac{1}{\eta\Upsilon^\top\mathcal{F}(\theta^*)D}\left(h - z + \frac{e^{-2\eta a h}}{2\eta a} - \frac{e^{-2\eta a z}}{2\eta a}\right),\quad a = \frac{\Upsilon^\top\mathcal{F}(\theta^*)D}{\Upsilon^\top\mathcal{F}(\theta^*)\Upsilon}, \tag{8.237}$$

where $\eta\Upsilon^\top\mathcal{F}(\theta^*)D \neq 0$, and

$$\mathsf{ARL}_z(0) \approx \frac{h^2 - z^2}{\Upsilon^\top\mathcal{F}(\theta^*)\Upsilon}. \tag{8.238}$$

Assume that $z = 0$. Putting together (8.237) and (8.238), we obtain

$$\mathsf{ARL}_0(R) \approx \frac{2R\sqrt{\mathsf{ARL}_0(0)} + e^{-2R\sqrt{\mathsf{ARL}_0(0)}} - 1}{2R^2},\quad R = \eta\frac{\Upsilon^\top\mathcal{F}(\theta^*)D}{\sqrt{\Upsilon^\top\mathcal{F}(\theta^*)\Upsilon}}, \tag{8.239}$$

where $R \neq 0$ plays the role of the SNR.

Let us show now that the Wald's approximation for the ARL function (8.237) of the local CUSUM test (8.228) coincides with the Wald's approximation for the ARL function of the conventional CUSUM test in the case of local hypotheses : $\theta_1 - \theta_0 \to 0$. For the sake of simplicity, consider the scalar case and $z = 0$. It follows from (8.237) that

$$\mathsf{ARL}_0(\eta) \approx \frac{1}{\eta\mathcal{F}(\theta^*)}\left(h_c + \frac{e^{-2\eta h_c}}{2\eta} - \frac{1}{2\eta}\right),$$

where $\eta = \theta - \theta^*$, $\theta^* = \frac{\theta_0+\theta_1}{2}$, and it follows from (8.121), (3.205) and (3.209) that

$$\mathsf{ARL}_0(\theta) \approx \frac{1}{J_\theta}\left(h_\ell + \frac{e^{-\omega_0 h_\ell}}{\omega_0} - \frac{1}{\omega_0}\right),$$

where

$$\omega_0(\theta) \approx \frac{2}{\theta_1 - \theta_0}(\theta - \theta^*)\ \text{ and }\ J_\theta \approx (\theta_1 - \theta_0)(\theta - \theta^*)\ \mathcal{F}(\theta^*).$$

It is easy to see that the above-mentioned right-hand side expressions are equal by setting $h_c = h_\ell/(\theta_1 - \theta_0)$.

8.3 Weighted CUSUM and GLR Algorithms for a Composite Post-Change Hypothesis

In practice, the pre-change model is often completely specified. However, the post-change model is typically not completely known. In particular, in the parametric case the pre-change parameter value $\theta = \theta_0$ is known but the post-change parameter value $\theta = \theta_1$ is rarely known, and the putative value θ_1 may be very different from the true value of θ. In this section, we consider quickest change detection in the case of unknown post-change parameter $\theta \in \Theta_1$. In other words, we assume that the pre-change hypothesis "$H_\infty : \nu = \infty, \theta = \theta_0$" is simple but the post-change hypothesis "$H_k^\vartheta : \nu = k, \theta = \vartheta$", $\vartheta \in \Theta_1$ is composite even when the point of change k is fixed.

As discussed in Section 5.1, there are two popular methods for dealing with the composite hypotheses scenarios. The first one consists in weighting the likelihood ratio with respect to all possible values of the parameter $\theta \in \Theta_1$, using a weighting function $W(\theta)$, which may be interpreted as a prior distribution of the unknown parameter (mixing distribution). The second approach consists in replacing the unknown parameter θ by its maximum likelihood estimate, which results in the generalized likelihood ratio (GLR) algorithm .

More precisely, assuming that the changepoint is fixed, $\nu = k$, the first solution is based upon the weighted likelihood ratio (WLR) for the observations

$$\widetilde{\Lambda}_n^{k+1} = \int_{\Theta_1} \prod_{i=k+1}^{n} \frac{f_\theta(X_i|X_1,\dots,X_{i-1})}{f_{\theta_0}(X_i|X_1,\dots,X_{i-1})} \, \mathrm{d}W(\theta), \tag{8.240}$$

and the second one uses the GLR

$$\widehat{\Lambda}_n^{k+1} = \sup_{\theta \in \Theta_1} \prod_{i=k+1}^{n} \frac{f_\theta(X_i|X_1,\dots,X_{i-1})}{f_{\theta_0}(X_i|X_1,\dots,X_{i-1})}. \tag{8.241}$$

In Chapter 5, we considered the hypothesis testing problems for composite hypotheses in detail; see also Subsection 2.10.3. The changepoint detection problem is similar. The main difference is that now we have to either average the corresponding statistics over the unknown changepoint to obtain a SR-type procedure or to maximize to obtain a CUSUM-type procedure. Therefore, maximizing over k in (8.240), we obtain the weighted CUSUM (WCUSUM) procedure

$$T_{\mathrm{WCS}}(h) = \inf\left\{n \ge 1 : \max_{1 \le k \le n} \log \widetilde{\Lambda}_n^k \ge h\right\}, \tag{8.242}$$

and maximizing over k in (8.241), we obtain the GLR CUSUM algorithm (GCUSUM)

$$T_{\mathrm{GLR}}(h) = \inf\left\{n \ge 1 : \max_{1 \le k \le n} \log \widehat{\Lambda}_n^k \ge h\right\}. \tag{8.243}$$

8.3.1 Asymptotic Optimality of WCUSUM and GLR Algorithms in the iid Case

In this subsection, we study properties of the WCUSUM and GCUSUM procedures for the iid case and the one-parameter exponential family, assuming that $p_\theta(X_{k+1},\dots,X_n) = \prod_{i=k+1}^n f_\theta(X_i)$ and $f_\theta(x)$ has the form (5.47). As usual, without loss of generality, by a linear transformation of coordinates, (5.47) may be transformed so that $\theta_0 = 0$ and $b(0) = \dot{b}(0) = 0$ and hence

$$\frac{f_\theta(x)}{f_{\theta_0}(x)} = \exp\{\theta x - b(\theta)\}, \quad \theta \in \Theta, \tag{8.244}$$

where $\Theta = \{\theta \in \mathbb{R} : \mathsf{E}_0[e^{\theta X_1}] < \infty\}$ is the natural parameter space. Of course we have to assume that the set of post-change parameters does not contain 0, the pre-change parameter value, i.e., $\Theta_1 = \Theta \setminus \{0\}$.

A generalization to the vector, multiparameter case will also be given.

8.3.1.1 Asymptotic Optimality Properties of the Weighted CUSUM Algorithm

Note first that for the exponential family (8.244) the statistic $\widetilde{\Lambda}_n^k$ has the form

$$\widetilde{\Lambda}_n^k = \int_{\Theta_1} \exp\{\theta S_n^k - (n-k+1)b(\theta)\}\mathrm{d}W(\theta), \tag{8.245}$$

where $S_n^k = \sum_{i=k}^n X_i$.

Now, observe that, using Lorden's embedding argument, the WCUSUM stopping time (8.242) can be written as $T_{\text{WCS}} = \inf_{k\geq 1}\{\widetilde{T}_k(h)\}$, where

$$\widetilde{T}_k(h) = \inf\{n \geq k : \log\widetilde{\Lambda}_n^k \geq h\}. \tag{8.246}$$

Given a mixing distribution W, define the probability measure $\bar{\mathsf{P}}^W(\mathcal{A}) = \int_\Theta \mathsf{P}_\theta(\mathcal{A})\,\mathrm{d}W(\theta)$, so that $\widetilde{\Lambda}_n^k = (\mathrm{d}\bar{\mathsf{P}}^W/\mathrm{d}\mathsf{P}_\infty)|_{\mathscr{F}_n^k}$, where $\mathscr{F}_n^k = \sigma(X_k,\dots,X_n)$. Applying Wald's likelihood ratio identity (in just the same way as in (5.181)) yields

$$\mathsf{P}_\infty(\widetilde{T}_k(h) < \infty) = \mathsf{E}^W[(\widetilde{\Lambda}_{\widetilde{T}_k}^k)^{-1}\mathbb{1}_{\{\widetilde{T}_k<\infty\}}] \leq e^{-h}, \quad h > 0.$$

Hence, by Theorem 8.2.4,

$$\mathsf{ARL2FA}(T_{\text{WCS}}) = \mathsf{E}_\infty[T_{\text{WCS}}(h)] \geq e^h \quad \text{for any } h > 0. \tag{8.247}$$

Note that inequality (8.247) holds for an arbitrary model, not necessarily for the exponential family, even for general non-iid models. This can be established using an argument similar to that in the proof of Lemma 8.2.1. See Lemma 8.4.1 in Subsection 8.4.3.2 below.

Clearly, $\mathsf{ARL}_{T_{\text{WCS}}}(\theta) = \mathsf{E}_{0,\theta}[T_{\text{WCS}}(h)] \leq \mathsf{E}_{0,\theta}[\widetilde{T}_1(h)]$, where here and in the following we use $\mathsf{E}_{\nu,\theta}$ ($\mathsf{P}_{\nu,\theta}$) to denote expectation (probability) when the post-change parameter is θ. For brevity we will omit the index ν when $\nu = 0$ and write E_θ and P_θ in place of $\mathsf{E}_{0,\theta}$ and $\mathsf{P}_{0,\theta}$. Recall that in Subsection 5.5.1.1 we studied the properties of the one-sided mixture test $\widetilde{T}_1(h)$ in detail. In particular, using nonlinear renewal theory, we established that, as $h \to \infty$,

$$\mathsf{P}_\infty\left\{\widetilde{T}_1(h) < \infty\right\} \sim e^{-h}\int_{\Theta_1} \zeta_\theta\,\mathrm{d}W(\theta),$$

and if $W(\theta)$ has a positive continuous density $w(\theta)$ with respect to the Lebesgue measure on Θ, then for every $\theta \in \Theta_1$

$$\mathsf{E}_\theta[\widetilde{T}_1(h)] = \frac{1}{I_\theta}\left[h + \frac{1}{2}\log h - \frac{1+\log(2\pi)}{2} + \log\left(\frac{\sqrt{\ddot{b}(\theta)/I_\theta}}{w(\theta)}\right) + \varkappa_\theta\right] + o(1), \tag{8.248}$$

where $I_\theta = \theta\dot{b}(\theta) - b(\theta)$ is the K–L number and ζ_θ, $\varkappa_\theta$ are the numbers associated with the overshoot and defined in (5.179). Therefore,

$$\mathsf{ARL2FA}(T_{\text{WCS}}) = \mathsf{E}_\infty[T_{\text{WCS}}(h)] \geq \frac{e^h}{\int_{\Theta_1}\zeta_\theta\,w(\theta)\,\mathrm{d}\theta}(1+o(1)) \quad \text{as } h \to \infty, \tag{8.249}$$

so setting $h = h_\gamma = \log[\gamma\bar{\zeta}(w)]$ implies $\mathsf{E}_\infty[T_{\text{WCS}}(h_\gamma)] \geq \gamma(1+o(1))$, where

$$\bar{\zeta}(w) = \int_{\Theta_1} \zeta_\theta\,w(\theta)\,\mathrm{d}\theta.$$

Let $\mathsf{SADD}_\theta(T) = \sup_\nu \mathsf{E}_{\nu,\theta}(T-\nu|T>\nu)$ and $\mathsf{ESADD}_\theta(T) = \sup_\nu \operatorname{ess\,sup} \mathsf{E}_{\nu,\theta}[(T-\nu)^+|\mathscr{F}_\nu)$. Substituting h_γ into (8.248) and taking into account that

$$\mathsf{SADD}_\theta(T_{\text{WCS}}) = \mathsf{ESADD}_\theta(T_{\text{WCS}}) = \mathsf{E}_\theta T_{\text{WCS}} \leq \mathsf{E}_\theta\widetilde{T}_1$$

yields the asymptotic inequality

$$\begin{aligned}\mathsf{SADD}_\theta(T_{\text{WCS}}) = \mathsf{ESADD}_\theta(T_{\text{WCS}}) \le \frac{1}{I_\theta}\Bigg[\log\gamma + \frac{1}{2}\log\log\gamma - \frac{1}{2} \\ - \log\left(w(\theta)\bar{\zeta}^{-1}(w)\sqrt{2\pi I_\theta/\ddot{b}(\theta)}\right) + \varkappa_\theta\Bigg] + o(1) \quad \text{as } \gamma\to\infty.\end{aligned} \tag{8.250}$$

Write $\mathsf{ESM}_r^\theta(T) = \sup_\nu \operatorname{ess\,sup} \mathsf{E}_{\nu,\theta}[(T-\nu)^r|\mathscr{F}_\nu]$ and $\mathsf{SM}_r^\theta(T) = \sup_\nu \mathsf{E}_{\nu,\theta}[(T-\nu)^r|T>\nu]$. By Corollary 8.2.1, for all $r>0$

$$\inf_{T\in\mathbf{C}_\gamma} \mathsf{ESM}_r^\theta(T) \ge \inf_{T\in\mathbf{C}_\gamma} \mathsf{SM}_r^\theta(T) \ge (I_\theta^{-1}\log\gamma)^r(1+o(1)) \quad \text{as } \gamma\to\infty, \tag{8.251}$$

so it follows from (8.250) that the WCUSUM procedure asymptotically minimizes to first order the expected detection delay in the worst scenario for all post-change parameter values $\theta\in\Theta_1$:

$$\inf_{T\in\mathbf{C}_\gamma} \mathsf{ESADD}_\theta(T) \sim \inf_{T\in\mathbf{C}_\gamma} \mathsf{SADD}_\theta(T) \sim \mathsf{SADD}_\theta(T_{\text{WCS}}) = \mathsf{ESADD}_\theta(T_{\text{WCS}}) \sim I_\theta^{-1}\log\gamma.$$

However, this asymptotic optimality result can be easily generalized for all positive moments of the detection delay by showing that for all $\theta\in\Theta_1$ and all $r>0$

$$\mathsf{E}_\theta[T_{\text{WCS}}(h)]^r \sim (I_\theta^{-1}h)^r \quad \text{as } h\to\infty,$$

which along with the lower bound (8.251) implies that all positive moments are minimized to first order.

The upper bound (8.250) gives the correct rate of increase of the expected delay proportional to $\log\gamma + \log\sqrt{\log\gamma}$, but the constant can be certainly refined in the asymptotically accurate expansion using nonlinear renewal theory. In what follows we assume that the post-change parameter value is positive and the mixing distribution $W(\theta)$ is concentrated on $\widetilde{\Theta} = \Theta_1\cap(0,\infty)$. Clearly, if $\theta<0$ the argument is the same setting $\widetilde{\Theta} = \Theta_1\cap(-\infty,0)$.

An important observation is that the statistic $\max_{1\le k\le n}\log\widetilde{\Lambda}_n^k$ can be written as the sum of the random walk $\theta S_n - b(\theta)n$ and the slowly changing sequence $\xi_n^\theta = -\log\sqrt{n} + \eta_n^\theta$, where

$$\eta_n^\theta = \log\left\{\sqrt{n}\max_{0\le k\le n}\int_{\widetilde{\Theta}} e^{(t-\theta)S_n - [b(t)-b(\theta)]n - tS_k + b(t)k} w(t)\,dt\right\},$$

so the slowly changing term ξ_n has the form (2.127) with $\ell_n = -\frac{1}{2}\log n$ and η_n^θ given above.

Using an argument similar (but slightly more difficult) to that used in Subsection 5.5.1.1 for power-one tests, it can be shown that η_n^θ converges weakly (under P_θ as $n\to\infty$) to a random variable η_∞^θ with expectation

$$\mathsf{E}_\theta\left[\eta_\infty^\theta\right] = \frac{1}{2} + \log\left[w(\theta)\sqrt{2\pi/\ddot{b}(\theta)}\right] - \mathsf{E}_\theta\left[\min_{k\ge 0}(S_k - b(\theta)k)\right],$$

and applying Theorem 2.6.4(i) we finally obtain

$$\mathsf{ARL}(T_{\text{WCS}};\theta) = \mathsf{E}_\theta[T_{\text{WCS}}(h)] = \frac{1}{I_\theta}\left\{h + \frac{1}{2}\log(h/I_\theta) - \mathsf{E}_\theta[\eta_\infty^\theta] + \varkappa_\theta\right\} + o(1) \quad \text{as } h\to\infty.$$

Therefore, we derived the following asymptotically precise approximation to within the negligible term $o(1)$:

$$\begin{aligned}\mathsf{ARL}(T_{\text{WCS}};\theta) = \mathsf{E}_\theta[T_{\text{WCS}}(h)] = \frac{1}{I_\theta}\Bigg\{h + \frac{1}{2}\log h - \frac{1}{2} - \log\left(w(\theta)\sqrt{2\pi I_\theta/\ddot{b}(\theta)}\right) \\ + \mathsf{E}_\theta\left[\min_{k\ge 0}(\theta S_k - kb(\theta))\right] + \varkappa_\theta\Bigg\} + o(1), \quad h\to\infty.\end{aligned} \tag{8.252}$$

Pollak and Siegmund [367] obtained an approximation similar to (8.252) that ignores the average overshoot $\varkappa_\theta$ using different techniques.

The following theorem systemizes all the results spelled out above. Recall that $\mathsf{SADD}_\theta(T_{\text{WCS}}) = \mathsf{ESADD}_\theta(T_{\text{WCS}}) = \mathsf{E}_\theta T_{\text{WCS}}$.

Theorem 8.3.1. *Consider the one-parameter exponential model* (8.244) *with the natural parameter space* Θ. *Let* $\Theta_1 = \Theta \setminus 0$. *Let* $T_{\text{WCS}}(h)$ *be defined as in* (8.242).

(i) *If the threshold* $h = h_\gamma$ *is selected so that* $\mathsf{ARL2FA}(T_{\text{WCS}}) \ge \gamma$ *and* $h_\gamma \sim \log\gamma$ *as* $\gamma \to \infty$, *in particular* $h_\gamma = \log\gamma$, *then the WCUSUM changepoint detection procedure minimizes asymptotically to first order all positive moments of the delay to detection in the worst-case scenario with respect to the changepoint* ν *and uniformly for all* $\theta \in \Theta_1$ *in the sense that as* $\gamma \to \infty$

$$\begin{aligned} &\inf_{T \in \mathbf{C}_\gamma} \mathsf{ESM}_r^\theta(T) \sim \mathsf{ESM}_r^\theta(T_{\text{WCS}}) \sim (I_\theta^{-1} \log\gamma)^r, \\ &\inf_{T \in \mathbf{C}_\gamma} \mathsf{SM}_r^\theta(T) \sim \mathsf{SM}_r^\theta(T_{\text{WCS}}) \sim (I_\theta^{-1} \log\gamma)^r \end{aligned} \tag{8.253}$$

for all $r > 0$ *and all* $\theta \in \Theta_1$.

(ii) *Let* $\theta > 0$. *If the mixing distribution* $W(\theta)$ *has support on* $\tilde{\Theta} = \Theta_1 \cap (0, \infty)$ *and has a positive continuous density with respect to Lebesgue's measure, then the asymptotic approximation* (8.252) *for* $\mathsf{E}_\theta[T_{\text{WCS}}(h)]$, *and hence, for* $\mathsf{SADD}_\theta(T_{\text{WCS}})$ *and* $\mathsf{ESADD}_\theta(T_{\text{WCS}})$ *holds as* $h \to \infty$. *Moreover, if the threshold is selected as*

$$h = \log\left(\gamma \int_{\tilde{\Theta}} \zeta_\theta \, w(\theta) \, \mathrm{d}\theta\right),$$

then $\mathsf{ARL2FA}(T_{\text{WCS}}) \ge \gamma(1 + o(1))$ *and*

$$\begin{aligned} \mathsf{ARL}(T_{\text{WCS}};\theta) = \mathsf{E}_\theta T_{\text{WCS}} = \frac{1}{I_\theta} \Bigg\{ &\log\gamma + \frac{1}{2}\log\log\gamma - \frac{1+\log(2\pi)}{2} + \log\left(\int_{\tilde{\Theta}} \zeta_t \, w(t) \, \mathrm{d}t\right) \\ &- \log\left(w(\theta) e^{-\varkappa_\theta + \mu_\theta} \sqrt{I_\theta / \ddot{b}(\theta)}\right)\Bigg\} + o(1) \quad \text{as } \gamma \to \infty, \end{aligned} \tag{8.254}$$

where $\mu_\theta = -\mathsf{E}_\theta\left[\min_{k \ge 0}(\theta S_k - kb(\theta))\right]$.

Remark 8.3.1. While we do not have a rigorous proof we believe that the lower bound for the ARL2FA given by (8.249) can be improved as

$$\mathsf{ARL2FA}(T_{\text{WCS}}) = \mathsf{E}_\infty[T_{\text{WCS}}(h)] = \frac{e^h}{\int_{\Theta_1} I_\theta \zeta_\theta^2 w(\theta) \, \mathrm{d}\theta} (1 + o(1)) \quad \text{as } h \to \infty. \tag{8.255}$$

Note that

$$\begin{aligned} \mathsf{P}_\theta\left\{\min_{k \ge 0}(\theta S_k - kb(\theta)) < -x\right\} &= \mathsf{P}_\theta\left\{\theta S_k - kb(\theta) < -x \text{ for some } k \ge 1\right\} \\ &= \mathsf{P}_\theta\left\{e^{-\theta S_k + kb(\theta)} \ge e^x \text{ for some } k \ge 1\right\} \le e^{-x}, \end{aligned}$$

where the last inequality follows from Doob's submartingale inequality since $e^{-\theta S_k + kb(\theta)}$ is a P_θ-martingale with unit expectation. Hence $\mathsf{E}_\theta\left[\min_{k \ge 0}(\theta S_k - kb(\theta))\right] \in [-1, 0)$ and it is approximately equal to -1.

Now, recall that CUSUM T_{CS}^θ tuned to the true post-change parameter value θ is strictly optimal minimizing $\mathsf{ESADD}_\theta(T)$ in the class $\mathbf{C}_\gamma$. Using the asymptotic approximation (8.146), we conclude that

$$\inf_{T\in\mathbf{C}_\gamma} \mathsf{ESADD}_\theta(T) = I_\theta^{-1}\log\gamma + O(1) \quad \text{as } \gamma\to\infty.$$

Comparing with (8.254) shows that the difference between $\mathsf{ESADD}_\theta(T_{\text{WCS}})$ and the optimal asymptotic performance is of order $O(\log\log\gamma/2I_\theta)$ for any mixing distribution,

$$\mathsf{ESADD}_\theta(T_{\text{WCS}}) - \inf_{T\in\mathbf{C}_\gamma} \mathsf{ESADD}_\theta(T) = O\left(\frac{1}{2I_\theta}\log\log\gamma\right) \quad \text{for all } \theta\in\tilde{\Theta}.$$

Evidently, the same is true for the SADD performance measure. The latter order of magnitude cannot be improved for all $\theta\in\tilde{\Theta}$ by any change detection procedure, and it is a price we pay for prior uncertainty with respect to the post-change parameter. However, below we consider a minimax approach associated with the average K–L information that allows us to select an "optimal" mixing distribution $w_0(\theta)$.

Typical choices of the weighting function $W(\theta)$ are the following. The most simple choice is the uniform distribution over a specified finite interval which contains all "interesting" values of the parameter θ. For example, when detecting a change in the mean of a normal population, there is always a value of $\theta_{\max}$ for which the expected detection delay is very small, so considering larger values is impractical. We stress that previous results hold as long as the prior distribution is continuous. Another approach is to consider Dirac masses on some specified values, i.e., discrete prior distributions. In this case the asymptotics are different. See Section 9.2.

We now go on to tackle the minimax problem setting where the speed of detection is expressed by the expected post-change K–L information. This kind of problem was proposed in Subsection 5.5.1.1 in the context of power-one tests. Specifically, define

$$\mathcal{R}_\theta(T) = I_\theta \sup_{\nu\ge 0}\mathsf{E}_{\nu,\theta}(T-\nu|T>\nu) = I_\theta\mathsf{SADD}_\theta(T).$$

This quantity can be interpreted as the maximal expected K–L information required to detect a change when raising an alarm at T since $I_\theta\mathsf{E}_{\nu,\theta}(T-\nu|T>\nu) = \mathsf{E}_{\nu,\theta}(\lambda_T^\nu|T>\nu)$, where $\lambda_n^\nu = \sum_{i=\nu+1}^n Z_i^\theta = \theta S_n^{\nu+1} - (n-\nu)b(\theta)$ is the LLR of a changepoint at ν and a post-change parameter θ ($Z_i^\theta = \theta X_i - b(\theta)$). A reasonable optimality criterion is the minimax criterion with respect to $\mathcal{R}_\theta(T)$, i.e., to find a procedure that minimizes $\sup_{\theta\in\tilde{\Theta}}\mathcal{R}_\theta(T)$ in the class $\mathbf{C}_\gamma$. Using Theorem 5.5.1 in Subsection 5.5.1.1, it can be shown that

$$\inf_{T\in\mathbf{C}_\gamma}\sup_{\theta\in\tilde{\Theta}}\mathcal{R}_\theta(T) \ge \log\gamma + \frac{1}{2}\log\log\gamma + O(1) \quad \text{as } \gamma\to\infty, \tag{8.256}$$

and asymptotic approximation (8.254) implies that this lower bound is attained by the WCUSUM procedure T_{WCS} for any mixing continuous density $w(\theta)$. The residual term $O(1)$ in (8.254) depends on θ. Owing to the minimax principle the idea is to make it independent of θ at least approximately, so that the resulting WCUSUM procedure becomes an equalizer rule to within a negligible term $o(1)$. Evidently, choosing

$$w_0(\theta) = \frac{e^{\varkappa_\theta-\mu_\theta}\sqrt{\ddot{b}(\theta)/I_\theta}}{\int_{\Theta_1} e^{\varkappa_t-\mu_t}\sqrt{\ddot{b}(t)/I_t}\,\mathrm{d}t}, \quad \theta\in\tilde{\Theta}, \tag{8.257}$$

guarantees this property. In this case, we have for all $\theta\in\tilde{\Theta}$

$$\mathcal{R}_\theta(T_{\text{WCS}}) = \log\gamma + \frac{1}{2}\log\log\gamma - \frac{1}{2}[1+\log(2\pi)] + C_{\text{WCS}} + o(1) \quad \text{as } \gamma\to\infty, \tag{8.258}$$

where

$$C_{\text{WCS}} = \log\left(\int_{\tilde{\Theta}} \zeta_t e^{\varkappa_t - \mu_t} \sqrt{\ddot{b}(t)/I_t}\, \mathrm{d}t\right). \tag{8.259}$$

Note that since $0 \le \mu_\theta \le 1$ and ≈ 1 for any θ, as we established above, it can be perhaps excluded from the mixing density w_0 without substantial loss of performance.

However, the difference between the optimal (maximal) risk and $\mathcal{R}_\theta(T_{\text{WCS}}) = \mathcal{R}(T_{\text{WCS}})$ given by (8.258) is not a negligible term $o(1)$, but still a constant, so WCUSUM is second-order minimax but not third-order, as the following theorem shows.

Theorem 8.3.2 (Information lower bound). *Consider the one-parameter exponential family* (8.244) *with the natural parameter space* Θ. *Let* $\mathcal{R}_\theta(T) = I_\theta \sup_{\nu\ge 0} \mathsf{E}_{\nu,\theta}(T-\nu | T>\nu)$. *The following asymptotic (as* $\gamma \to \infty$*) lower bound holds for the expected K–L information in the worst-case scenario:*

$$\inf_{T\in \mathbf{C}_\gamma} \sup_{\theta\in\Theta\setminus 0} \mathcal{R}_\theta(T) \ge \log\gamma + \frac{1}{2}\log\log\gamma - \frac{1}{2}[1+\log(2\pi)] + C_{\text{opt}} + o(1), \tag{8.260}$$

where

$$C_{\text{opt}} = \log\left(\int_{\tilde{\Theta}} \zeta_t e^{\varkappa_t - C(t)} \sqrt{\ddot{b}(t)/I_t}\, \mathrm{d}t\right),$$

and the constant $C(\theta)$ *is defined in* (8.392).

Proof. By Lemma 6.3.1, $\mathsf{P}_\infty(T<k+m|T>k) \le m/\gamma$ for any $T\in\mathbf{C}_\gamma$ and some $k\ge 0$. Now (8.260) follows from Theorem 1 of Siegmund and Yakir [431]. □

The lower bound (8.260) is attained by a randomized version of the weighted SR procedure, as established in Theorem 8.5.2 below.

The assertions of Theorem 8.3.1 also hold for the vector case that includes the multiparameter exponential family (5.69), assuming that the dimension of both θ and X is l. Specifically, part (i) holds without any change. Part (ii) is being modified as follows. Let $\Theta_\varepsilon = \Theta \setminus B_\varepsilon$, where B_ε is a closed ball of radius $\varepsilon > 0$ in $\mathbb{R}^l$. Assume that Θ_ε is compact and $\mathbf{0} \notin \Theta_\varepsilon$ and also that the measure of B_ε converges to 0 as $\varepsilon \to 0$. Suppose that the weighting function $W(\theta)$ over Θ_ε has a positive continuous density $w(\theta)$ with respect to Lebesgue's measure. Then for all $\theta \in \Theta\setminus\mathbf{0}$, as $h\to\infty$,

$$\begin{aligned}\mathsf{E}_\theta[T_{\text{WCS}}(h)] = \mathsf{SADD}_\theta(T_{\text{WCS}}) = \mathsf{ESADD}_\theta(T_{\text{WCS}}) = \frac{1}{I_\theta}\Bigg\{ h + \frac{l}{2}\log h - \frac{l}{2}[1+\log(2\pi)] \\ - \log\left(w(\theta)\sqrt{I_\theta^l/\det[\nabla^2 b(\theta)]}\right) - \mu_\theta + \varkappa_\theta \Bigg\} + o(1),\end{aligned} \tag{8.261}$$

where $\nabla^2 b(\theta)$ is the Hessian matrix of second partial derivatives. If $h = \log\left(\gamma \int_{\Theta_\varepsilon} \zeta_\theta\, w(\theta)\,\mathrm{d}\theta\right)$, then $\mathsf{ARL2FA}(T_{\text{WCS}}) \ge \gamma(1+o(1))$ and

$$\begin{aligned}\mathsf{E}_\theta T_{\text{WCS}} = \frac{1}{I_\theta}\Bigg\{ \log\gamma + \frac{l}{2}\log\log\gamma - \frac{l}{2}[1+\log(2\pi)] + \log\left(\int_{\Theta_\varepsilon} \zeta_t\, w(t)\,\mathrm{d}t\right) \\ - \log\left(w(\theta)e^{-\varkappa_\theta+\mu_\theta}\sqrt{I_\theta^l/\det[\nabla^2 b(\theta)]}\right)\Bigg\} + o(1) \quad \text{as } \gamma\to\infty.\end{aligned} \tag{8.262}$$

Analogously to (8.256), for any mixing density $w(\theta)$ over Θ_ε,

$$\inf_{T\in \mathbf{C}_\gamma} \sup_{\theta\in\Theta\setminus \mathbf{0}} \mathcal{R}_\theta(T) \ge \log\gamma + \frac{1}{2}\log\log\gamma + O(1) \quad \text{as } \gamma\to\infty. \tag{8.263}$$

Comparing with (8.262), we see that the WCUSUM procedure attains the asymptotic lower bound (8.263) for an arbitrary $w(\theta)$. Next, selecting $w(\theta)$ as

$$w(\theta) = w_0(\theta) = \frac{e^{\varkappa_\theta - \mu_\theta}\sqrt{\det[\nabla^2 b(t)]/I_\theta^l}}{\int_{\Theta_\varepsilon} e^{\varkappa_t - \mu_t}\sqrt{\det[\nabla^2 b(t)]/I_t^l}\, \mathrm{d}t}, \quad \theta \in \Theta_\varepsilon, \tag{8.264}$$

we obtain

$$\mathcal{R}_\theta(T_{\text{WCS}}) = \log\gamma + \frac{l}{2}\log\log\gamma - \frac{l}{2}[1+\log(2\pi)] + C_{\text{WCS}} + o(1) \quad \text{as } \gamma \to \infty, \tag{8.265}$$

where

$$C_{\text{WCS}} = \log\left(\int_{\Theta_\varepsilon} \zeta_t e^{\varkappa_t - \mu_t}\sqrt{\det[\nabla^2 b(t)]/I_t^l}\, \mathrm{d}t\right). \tag{8.266}$$

Therefore, if the mixing density is given by (8.264), then the WCUSUM procedure is an almost equalizer with respect to θ since the right side of (8.265) does not depend on θ except for perhaps a vanishing term $o(1)$. Finally, by Lemma 6.3.1 and [431, Theorem 1], as $\gamma \to \infty$,

$$\inf_{T \in \mathbf{C}_\gamma} \sup_{\theta \in \Theta \setminus \mathbf{0}} \mathcal{R}_\theta(T) \geq \log\gamma + \frac{l}{2}\log\log\gamma - \frac{l}{2}[1+\log(2\pi)] + C_{\text{opt}} + o(1), \tag{8.267}$$

where

$$C_{\text{opt}} = \log\left(\int_{\tilde{\Theta}} \zeta_t e^{\varkappa_t - C(t)}\sqrt{\det[\nabla^2 b(t)]/I_t^l}\, \mathrm{d}t\right)$$

and $C(\theta)$ is defined in (8.392). Hence,

$$\sup_{\theta \in \Theta \setminus \mathbf{0}} \mathcal{R}_\theta(T_{\text{WCS}}) - \inf_{T \in \mathbf{C}_\gamma} \sup_{\theta \in \Theta \setminus \mathbf{0}} \mathcal{R}_\theta(T) = O(1),$$

and is approximately equal to $C_{\text{WCS}} - C_{\text{opt}}$.

8.3.1.2 Asymptotic Optimality Properties of the GLR Algorithm

Consider now the GLR-based CUSUM procedure (GCUSUM) defined in (8.243). We will write $\hat{g}_n = \max_{1 \leq k \leq n} \log \widehat{\Lambda}_n^k$, so that the stopping time of the GCUSUM procedure is

$$T_{\text{GLR}}(h) = \inf\{n \geq 1 : \hat{g}_n \geq h\}, \quad h > 0. \tag{8.268}$$

Note that

$$\hat{g}_n = \max_{1 \leq k \leq n} \sup_{\theta \in \Theta_1} \sum_{i=k}^n Z_i^\theta, \quad Z_i^\theta = \log[f_\theta(X_i)/f_{\theta_0}(X_i)]. \tag{8.269}$$

Again, we begin with focusing on the one-parameter exponential family with density (8.244), in which case $Z_i^\theta = \theta X_i - b(\theta)$. We also suppose that $\Theta_1 = \Theta_\varepsilon = (-\infty, -\varepsilon) \cup (\varepsilon, \infty)$ for some $\varepsilon > 0$, which may depend on h and go to zero as $h \to \infty$. Thus,

$$\hat{g}_n = \max_{1 \leq k \leq n} \sup_{|\theta| > \varepsilon} \{\theta(S_n - S_k) - (n-k)b(\theta)\}. \tag{8.270}$$

Let $\tau_h = \inf\{n : \theta S_n - nb(n) \geq h\}$. The first observation is that since $\hat{g}_n \geq \theta S_n - nb(\theta)$ and since $\mathsf{E}_\theta \tau_h \sim (h/I_\theta)^r$ for any $r > 0$, it follows that

$$\mathsf{E}_\theta[T_{\text{GLR}}(h)]^r \leq (h/I_\theta)^r(1+o(1)), \quad h \to \infty.$$

Comparing with the lower bound (8.251) and taking into account that $\mathsf{SADD}_\theta(T_{\text{GLR}}) =$

$\mathsf{ESADD}_\theta(T_{\text{GLR}}) = \mathsf{E}_\theta T_{\text{GLR}}$ shows that if h is selected so that $\mathsf{ARL2FA}(T_{\text{GLR}}) \ge \gamma$ and $h \sim \log\gamma$, then as $\gamma \to \infty$

$$\inf_{T \in \mathbf{C}_\gamma} \mathsf{ESM}_r^\theta(T) \sim \mathsf{ESM}_r^\theta(T_{\text{WCS}}) \sim (I_\theta^{-1} \log\gamma)^r \quad \text{for all } \theta \in \Theta_\varepsilon,$$

and the same is true for $\mathsf{SM}_r(T)$, where ε is either a fixed positive number or $\varepsilon = \varepsilon_\gamma \to 0$ as $\gamma \to \infty$.

The most important issue is how to select the threshold in order to guarantee the inequality $\mathsf{ARL2FA}(T_{\text{GLR}}) \ge \gamma$. Unfortunately, in general there is no answer since the GLR $\widehat{\Lambda}_n$ is not the likelihood ratio and, as discussed in Subsection 5.5.1.1, one cannot obtain a simple upper bound for the probability $\mathsf{P}_0(\hat{T}_k(h) < \infty) = \mathsf{P}_\infty(\hat{T}_k(h) < \infty)$ associated with the one-sided "generating" SPRT stopping time

$$\hat{T}_k(h) = \inf\left\{ n \ge k : \sup_{|\theta| > \varepsilon} [\theta(S_n - S_{k-1}) - (n-k+1)b(\theta)] \ge h \right\}. \tag{8.271}$$

However, for large h, it is possible to obtain an asymptotic approximation. Indeed, by (5.205),

$$\mathsf{P}_\infty(\hat{T}_1(h) < \infty) = K_\varepsilon \sqrt{h} e^{-h} (1 + o(1)) \quad \text{as } h \to \infty,$$

where

$$K_\varepsilon = \frac{1}{\sqrt{2\pi}} \int_{\tilde{\Theta}_\varepsilon} \sqrt{\frac{\ddot{b}(\theta)}{I_\theta}}\, \zeta_\theta \, \mathrm{d}\theta, \quad \tilde{\Theta}_\varepsilon = \{\theta : I_{\min}(\varepsilon) < I_\theta < \bar{I}_{\min}\}. \tag{8.272}$$

Here $I_{\min}(\varepsilon) = \min\{I_{-\varepsilon}, I_\varepsilon\}$, $\bar{I}_{\min} = \min\{I_{\theta_1}, I_{\theta_2}\}$, and the interval (θ_1, θ_2) does not have to be the entire natural parameter space Θ, i.e., $(\theta_1, \theta_2) \subseteq \Theta$. Note that θ_1 may be equal to $-\infty$ and θ_2 to $+\infty$.

Therefore, by Theorem 8.2.4,

$$\mathsf{ARL2FA}(T_{\text{GLR}}) \ge \frac{e^h}{K_\varepsilon \sqrt{h}} (1 + o(1)) \quad \text{as } h \to \infty. \tag{8.273}$$

If ε is fixed, then K_ε is also a fixed constant and setting $h = h_\gamma = \log[K_\varepsilon \gamma \sqrt{\log\gamma}\,]$, we obtain

$$\mathsf{ARL2FA}(T_{\text{GLR}}(h_\gamma)) \ge \gamma(1 + o(1)) \quad \text{as } \gamma \to \infty. \tag{8.274}$$

However, if $\varepsilon = \varepsilon_h \to 0$ as $h \to \infty$, then $I_{\min}(\varepsilon) = O(\varepsilon_h^2)$ and $K_\varepsilon = O(1/\varepsilon_h)$. In this case, asymptotic inequality (8.274) holds if $h = h_\gamma = \log[(\log\log\gamma)\gamma\sqrt{\log\gamma}\,]$ and $\varepsilon_\gamma = O(1/\log\log\gamma)$.

We now proceed with obtaining an asymptotic approximation for $\mathsf{E}_\theta[T_{\text{GLR}}(h)]$ presenting only a proof sketch; see Dragalin [124] for all details. Again, nonlinear renewal theory is a key. Define

$$\xi_n^\theta = \max_{0 \le k < n} \sup_{\vartheta \in \Theta_\varepsilon} \{(\vartheta - \theta)S_n - [b(\vartheta) - b(\theta)]n - \vartheta S_k + b(\vartheta)k\}.$$

Obviously, the GCUSUM statistic $\hat{g}_n$ can be written as the sum of the random walk $\theta S_n - b(\theta)n$ and the sequence ξ_n^θ,

$$\hat{g}_n = \theta S_n - b(\theta)n + \xi_n^\theta, \quad n \ge 1.$$

Next, define the sequence

$$\tilde{\xi}_n^\theta = \frac{1}{2}\left(\frac{S_n - n\dot{b}(\theta)}{\sqrt{\ddot{b}(\theta)n}}\right)^2 - \min_{0 \le k < n} [\theta S_k - b(\theta)k].$$

The sequence $\{\tilde{\xi}_n^\theta\}_{n \ge 1}$ is slowly changing since, as $n \to \infty$, the first term converges weakly to $\chi_1^2/2$ and the second term to $\min_{k \ge 0}[\theta S_k - b(\theta)k]$, so $\tilde{\xi}_n^\theta$ converges to $\tilde{\xi}_\infty^\theta = \chi_1^2/2 - \min_{k \ge 0}[\theta S_k - b(\theta)k]$ under P_θ. Here χ_1^2 is a random variable with χ^2-distribution with 1 degree of freedom. Finally, by

[124, Lemma 1], $\xi^{\theta}_{T_{\mathrm{GLR}}} = \tilde{\xi}^{\theta}_{T_{\mathrm{GLR}}}$ on a set $\mathcal{A}_h$ whose P_θ-probability converges to 1 exponentially fast as $h \to \infty$ with the rate $1 - O(e^{-\lambda h})$ for some $\lambda > 0$. Hence,

$$\mathsf{E}_\theta\left[\xi^{\theta}_{T_{\mathrm{GLR}}}; \mathcal{A}_h\right] \to \mathsf{E}_\theta\left[\tilde{\xi}^{\theta}_{T_{\mathrm{GLR}}}\right] = \frac{1}{2}\mathsf{E}\chi_1^2 - \mu_\theta,$$

where $\mu_\theta = \mathsf{E}_\theta\{\min_{k\ge 0}[\theta S_k - b(\theta)k]\}$ and $\frac{1}{2}\mathsf{E}\chi_1^2 = \frac{1}{2}$. Therefore, by Theorem 2.6.4(i), we obtain

$$\mathsf{ARL}(T_{\mathrm{GLR}}; \theta) = \mathsf{E}_\theta[T_{\mathrm{GLR}}(h)] = \frac{1}{I_\theta}\left\{h - \frac{1}{2} + \mu_\theta + \varkappa_\theta\right\} + o(1) \quad \text{as } h \to \infty. \tag{8.275}$$

Substituting K_ε given in (8.272) into the formula for the threshold $h_\gamma = \log[K_\varepsilon \gamma\sqrt{\log\gamma}\,]$ to obtain (8.274) and then this threshold into (8.275) gives two different approximations to the expected delay depending on whether ε is fixed or approaches zero, i.e., whether the size of the indifference interval $(-\varepsilon, \varepsilon)$ is fixed or shrinks as γ increases.

All the obtained results are summarized in the next theorem.

Theorem 8.3.3. *Consider the one-parameter exponential model* (8.244) *with the natural parameter space* Θ. *Let* $\Theta_1 = \Theta_\varepsilon = (-\infty, -\varepsilon) \cup (\varepsilon, \infty)$ *for some* $\varepsilon > 0$ *which may depend on* γ *and approach zero as* $\gamma \to \infty$. *Let* $T_{\mathrm{GLR}}(h)$ *be defined as in* (8.268).

(i) *If the threshold* $h = h_\gamma$ *is selected so that* $\mathsf{ARL2FA}(T_{\mathrm{GLR}}) \ge \gamma$ *and* $h_\gamma \sim \log\gamma$ *as* $\gamma \to \infty$, *then the GCUSUM procedure minimizes asymptotically to first order all positive moments of the delay to detection in the worst-case scenario with respect to the changepoint* ν *and uniformly for all* $\theta \in \Theta_\varepsilon$ *in the sense that as* $\gamma \to \infty$

$$\begin{aligned} &\inf_{T\in\mathbf{C}_\gamma} \mathsf{ESM}_r^\theta(T) \sim \mathsf{ESM}_r^\theta(T_{\mathrm{GLR}}) \sim (I_\theta^{-1}\log\gamma)^r, \\ &\inf_{T\in\mathbf{C}_\gamma} \mathsf{SM}_r^\theta(T) \sim \mathsf{SM}_r^\theta(T_{\mathrm{GLR}}) \sim (I_\theta^{-1}\log\gamma)^r \end{aligned} \tag{8.276}$$

for all $r > 0$ *and all* $\theta \in \Theta_\varepsilon$.

(ii) *Let* ε *be fixed and let* K_ε *be defined as in* (8.272). *If the threshold is selected as* $h = \log[K_\varepsilon\gamma\sqrt{\log\gamma}\,]$, *then* $\mathsf{ARL2FA}(T_{\mathrm{GLR}}) \ge \gamma(1+o(1))$ *and for all* $\theta \in \Theta_\varepsilon$

$$\begin{aligned} \mathsf{ARL}(T_{\mathrm{GLR}}; \theta) = \mathsf{E}_\theta T_{\mathrm{GLR}} = \frac{1}{I_\theta}\Bigg\{ &\log\gamma + \frac{1}{2}\log\log\gamma - \frac{1}{2}[1+\log(2\pi)] \\ &+ \log\left(\int_{\tilde{\Theta}_\varepsilon} \zeta_t \sqrt{\ddot{b}(t)/I_t}\,\mathrm{d}t\right) - \mu_\theta + \varkappa_\theta\Bigg\} + o(1) \quad \text{as } \gamma \to \infty. \end{aligned} \tag{8.277}$$

(iii) *Let* $\varepsilon_\gamma = O(1/\log\log\gamma)$ *as* $\gamma \to \infty$. *If the threshold is selected as* $h = \log[(\log\log\gamma)\gamma\sqrt{\log\gamma}\,]$, *then* $\mathsf{ARL2FA}(T_{\mathrm{GLR}}) \ge \gamma(1+o(1))$ *and for all* $\theta \in \Theta_\varepsilon$

$$\mathsf{ARL}(T_{\mathrm{GLR}}; \theta) = \mathsf{E}_\theta T_{\mathrm{GLR}} = \frac{1}{I_\theta}\left(\log\gamma + \frac{1}{2}\log\log\gamma + \log\log\log\gamma\right) + O(1) \quad \text{as } \gamma \to \infty. \tag{8.278}$$

Remark 8.3.2. While asymptotic lower bound (8.273) is undoubtedly useful, it is of interest to obtain a non-asymptotic inequality that holds for all $h > 0$. Using [271], it can be shown that

$$\mathsf{ARL2FA}(T_{\mathrm{GLR}}) \ge \frac{e^h}{2(1 + hI_{\min}^{-1}(\varepsilon))} > \frac{e^h}{2h(1 + I_{\min}^{-1}(\varepsilon))} \quad \text{for all } h > 0,$$

and the choice $h = \log[6(1 + I_{\min}^{-1}(\varepsilon))^2\gamma\log\gamma]$ suffices to make $\mathsf{ARL2FA}(T_{\mathrm{GLR}})$ greater than γ. With this choice $h \sim \log\gamma$ for large γ, which is required in part (i). However,

$$\mathsf{ARL}(T_{\mathrm{GLR}}; \theta) = \mathsf{E}_\theta T_{\mathrm{GLR}} = I_\theta^{-1}(\log\gamma + \log\log\gamma) + O(1),$$

giving not an "optimal" second term, which according to (8.277) is $\log\sqrt{\log\gamma}$.

Remark 8.3.3. While we have no rigorous proof, we conjecture that the asymptotic lower bound (8.273) can be improved as follows:

$$\mathsf{ARL2FA}(T_{\text{GLR}}) \sim \frac{e^h}{\widetilde{K}_\varepsilon \sqrt{h}} \quad \text{as } h \to \infty, \quad \widetilde{K}_\varepsilon = \frac{1}{\sqrt{2\pi}} \int_{\Theta_\varepsilon} I_\theta \zeta_\theta^2 \sqrt{\frac{\ddot{b}(\theta)}{I_\theta}}\, \mathrm{d}\theta. \tag{8.279}$$

See [430] for a justification of this asymptotic equality for the Gaussian model with unknown mean.

Again, as in the previous subsection, consider the minimax problem of minimizing the expected post-change K–L information $\sup_\theta \mathcal{R}_\theta(T)$ required for detecting a change ($\mathcal{R}_\theta(T) = I_\theta \mathsf{SADD}_\theta(T)$). Formally, by (8.277),

$$\sup_{\theta\in\Theta_\varepsilon} \mathcal{R}_\theta(T_{\text{GLR}}) = \log\gamma + \frac{1}{2}\log\log\gamma + O(1),$$

so by Theorem 8.3.2 GCUSUM is second-order asymptotically minimax. However, practically speaking this result may not be satisfactory since

$$\sup_{\theta\in\Theta_\varepsilon} \mathcal{R}_\theta(T_{\text{GLR}}) \approx \log\gamma + \frac{1}{2}\log\log\gamma + \log\left(\int_{\tilde{\Theta}_\varepsilon} \zeta_t \sqrt{\ddot{b}(t)/I_t}\, \mathrm{d}t\right) + \sup_{\theta\in\Theta_\varepsilon} \varkappa_\theta,$$

where we ignored the terms $[1+\log(2\pi)]/2$ and $\sup_\theta \mu_\theta \le 1$. The value of $\sup_{\theta\in\Theta_\varepsilon} \varkappa_\theta$ can be large, which diminishes the practical significance of the second-order optimality property.

To overcome this drawback, introduce the weighted GLR-based CUSUM (WGCUSUM) given by the stopping time

$$T_{\text{GLR}}^{w}(h) = \inf\left\{n \ge 1 : \max_{1\le k\le n} w(\hat{\theta}_n^k) \prod_{i=k}^{n} \frac{f_{\hat{\theta}_n^k}(X_i)}{f_{\theta_0}(X_i)} \ge e^h\right\} \tag{8.280}$$

where $\hat{\theta}_n^k$ is the MLE maximizing $\prod_{i=k}^n f_{\theta_n^k}(X_i)$ over $\theta \in \Theta_1$ and $w(\theta)$ is a positive and continuous weight (density). Using the same reasoning as for the conventional GCUSUM with $w(\theta) = 1$, we obtain

$$I_\theta \mathsf{E}_\theta[T_{\text{GLR}}(h)] = h - \frac{1}{2} + \mu_\theta + \varkappa_\theta - \log w(\theta) + o(1) \quad \text{as } h \to \infty \tag{8.281}$$

as well as inequality (8.273) for the ALR2FA with K_ε is now given by

$$K_\varepsilon = \frac{1}{\sqrt{2\pi}} \int_{\tilde{\Theta}_\varepsilon} \zeta_\theta w(\theta) \sqrt{\frac{\ddot{b}(\theta)}{I_\theta}}\, \mathrm{d}\theta. \tag{8.282}$$

Combining (8.281) with (8.282), we obtain

$$\begin{aligned} I_\theta \mathsf{E}_\theta T_{\text{GLR}}^{w} &= \log\gamma + \frac{1}{2}\log\log\gamma - \frac{1}{2}[1+\log(2\pi)] + \log\left(\int_{\tilde{\Theta}_\varepsilon} \zeta_t w(t) \sqrt{\ddot{b}(t)/I_t}\, \mathrm{d}t\right) \\ &\quad - \mu_\theta + \varkappa_\theta - \log w(\theta) + o(1) \quad \text{as } \gamma \to \infty, \end{aligned} \tag{8.283}$$

which generalizes (8.277) for WGCUSUM with an arbitrary continuous weighting density $w(\theta)$. Thus, choosing

$$w(\theta) = \frac{e^{\varkappa_\theta - \mu_\theta}}{\int_{\tilde{\Theta}_\varepsilon} e^{\varkappa_t - \mu_t}\, \mathrm{d}t}, \quad \theta \in \tilde{\Theta}_\varepsilon, \tag{8.284}$$

turns WGCUSUM into the (almost) equalizer rule since for all $\theta \in \Theta_\varepsilon$ as $\gamma \to \infty$

$$\mathcal{R}_\theta(T_{\text{GLR}}^{w}) = \log\gamma + \frac{1}{2}\log\log\gamma - \frac{1}{2}[1+\log(2\pi)] + C_{\text{WGCS}} + o(1), \tag{8.285}$$

where

$$C_{\text{WGCS}} = \log\left(\int_{\tilde{\Theta}_\varepsilon} e^{\varkappa_t - \mu_t} \zeta_t \sqrt{\ddot{b}(t)/I_t}\, \mathrm{d}t\right). \tag{8.286}$$

Using (8.285) and (8.260), we obtain that in this case the difference between the optimal (maximal) risk and $\mathcal{R}_\theta(T^w_{\text{GLR}})$ is approximately constant and equal to $C_{\text{WGCS}} - C_{\text{opt}}$. Also, comparing with (8.258) and (8.259), we see that the asymptotic performance of both WCUSUM and WGCUSUM is the same. Note, however, that if the threshold is selected according to the asymptotic approximation

$$\mathsf{ARL2FA}(T_{\text{GLR}}) \sim \frac{h^{-1/2} e^h}{(2\pi)^{1/2} \int_{\Theta_\varepsilon} w(\theta) I_\theta \zeta_\theta^2 \sqrt{\ddot{b}(\theta)/I_\theta}\, \mathrm{d}\theta}$$

which is an extension of (8.279) to the weighted GCUSUM, then

$$C_{\text{WGCS}} = \log\left(\int_{\tilde{\Theta}_\varepsilon} e^{\varkappa_t - \mu_t} I_t \zeta_t^2 \sqrt{\ddot{b}(t)/I_t}\, \mathrm{d}t\right).$$

Since $I_\theta \zeta_\theta < 1$, it follows that this constant is smaller than that defined in (8.259), so we expect that the WGCUSUM procedure with the above-specified weight will perform somewhat better than the WCUSUM procedure.

In just the same way as in Subsection 8.3.1.1, the above results can be extended to the l-dimensional exponential family (5.69). In particular, the assertion (i) of Theorem 8.3.3 holds as is with $\Theta_\varepsilon = \Theta \setminus B_\varepsilon$, where B_ε is a ball with radius $\varepsilon > 0$. Asymptotic approximation (8.275) also remains unchanged. Inequality (8.273) and the constant K_ε given by (5.206) get modified as

$$\mathsf{ARL2FA}(T_{\text{GLR}}) \ge h^{-l/2} e^h / K_\varepsilon \big(1 + o(1)\big) \quad \text{as } h \to \infty$$

and

$$K_\varepsilon = (2\pi)^{-l/2} \int_{\tilde{\Theta}_\varepsilon} \zeta_\theta \sqrt{\det[\nabla^2 b(\theta)]/I_\theta^l}\, \mathrm{d}\theta,$$

or more generally for the WGCUSUM (8.280)

$$K_\varepsilon^w = (2\pi)^{-l/2} \int_{\tilde{\Theta}_\varepsilon} \zeta_\theta w(\theta) \sqrt{\det[\nabla^2 b(\theta)]/I_\theta^l}\, \mathrm{d}\theta.$$

Therefore, setting $h = \log[K_\varepsilon^w \gamma (\log \gamma)^{l/2}]$ yields $\mathsf{ARL2FA}(T^w_{\text{GLR}}) \ge \gamma(1 + o(1))$ and

$$\begin{aligned} I_\theta \mathsf{E}_\theta T^w_{\text{GLR}} &= \log \gamma + \frac{l}{2} \log\log \gamma - \frac{l}{2}[1 + \log(2\pi)] + \log\left(\int_{\tilde{\Theta}_\varepsilon} \zeta_t w(t) \sqrt{\det[\nabla^2 b(t)]/I_t^l}\, \mathrm{d}t\right) \\ &\quad - \mu_\theta + \varkappa_\theta - \log w(\theta) + o(1) \quad \text{as } \gamma \to \infty \end{aligned}$$

(cf. (8.283)). Choosing the weight $w(\theta)$ as in (8.284) leads to the equalizer (up to $o(1)$) with respect to the parameter θ; with this choice the K–L information $\mathcal{R}_\theta(T^w_{\text{GLR}})$ is given by equality

$$\mathcal{R}_\theta(T^w_{\text{GLR}}) = \log \gamma + \frac{l}{2} \log\log \gamma - \frac{l}{2}[1 + \log(2\pi)] + C_{\text{WGCS}} + o(1)$$

with

$$C_{\text{WGCS}} = \log\left(\int_{\tilde{\Theta}_\varepsilon} e^{\varkappa_t - \mu_t} \zeta_t \sqrt{\det[\nabla^2 b(t)]/I_t^l}\, \mathrm{d}t\right).$$

Comparing with (8.265) and (8.266), one sees that the WCUSUM and WGCUSUM procedures have the same asymptotic performance with respect to the worst expected K–L information.

Finally, comparing with the lower bound (8.267) shows that

$$\sup_{\theta \in \tilde{\Theta}_\varepsilon} \mathcal{R}_\theta(T^w_{\text{GLR}}) - \inf_{T \in \mathbf{C}_\gamma} \sup_{\theta \in \Theta \setminus \mathbf{0}} \mathcal{R}_\theta(T) \approx C_{\text{WGCS}} - C_{\text{opt}}.$$

Change in the Mean of a Gaussian Sequence: Unknown Post-Change Mean. Consider the example of detecting a change in the mean of an independent Gaussian sequence assuming that the observations follow the Gaussian distribution $\mathcal{N}(\theta_0, \sigma^2)$ in the pre-change mode with a known mean value θ_0 and a known variance σ^2 and the Gaussian distribution $\mathcal{N}(\theta, \sigma^2)$ in the post-change mode with an unknown mean value θ. Without loss of generality we set $\theta_0 = 0$. In many applications a minimal magnitude ε of the change can be specified. In this case, using (8.270) and the fact that $b(\theta) = \theta^2/2$, we obtain that the GCUSUM statistic $\hat{g}_n$ reduces to

$$\hat{g}_n = \max_{1 \le j \le n} \sup_{|\theta| \ge \varepsilon} \frac{1}{\sigma^2} \sum_{i=j}^{n} \left(\theta X_i - \frac{\theta^2}{2} \right), \tag{8.287}$$

which yields

$$\hat{g}_n = \frac{1}{\sigma^2} \max_{1 \le j \le n} \sum_{i=j}^{n} \left(\hat{\theta}_{j,n} X_i - \frac{\hat{\theta}_{j,n}^2}{2} \right), \tag{8.288}$$

where the absolute value of the *constrained* change magnitude estimate is

$$|\hat{\theta}_{j,n}| = \left(|\overline{\theta}_{j,n}| - \varepsilon \right)^+ + \varepsilon, \;\; \overline{\theta}_{j,n} = \frac{1}{n-j+1} \sum_{i=j}^{n} X_i,$$

and its sign is the same as the sign of the sample mean $\overline{\theta}_{j,n}$. If a maximal possible change magnitude is also known, the GCUSUM detection statistic is modified accordingly in an obvious manner. If $\varepsilon = 0$ the GCUSUM detection statistic becomes

$$\hat{g}_n = \frac{1}{2\sigma^2} \max_{1 \le j \le n} \frac{1}{n-j+1} \left(\sum_{i=j}^{n} X_i \right)^2. \tag{8.289}$$

The GCUSUM procedure has the form (8.268) with $\hat{g}_n$ defined in (8.288).

In the following without loss of generality we set $\sigma = 1$.

Recall that it is not easy to obtain an accurate lower bound on the ARL2FA of the GCUSUM procedure for all threshold values h. Thus, one of the most important issues is to approximate the ARL2FA at least asymptotically for large threshold values. Indeed, by Remark 8.3.2, Lorden's upper bound on the probability of stopping in the one-sided SPRT under the null hypothesis yields the lower bound

$$\mathsf{ARL2FA}(T_{\text{GLR}}) > \frac{e^h}{2h(1+2/\varepsilon^2)} \quad \text{for all } h > 0,$$

so setting $h = \log[6(1+2/\varepsilon^2)^2 \gamma \log \gamma]$ makes $\mathsf{ARL2FA}(T_{\text{GLR}})$ greater than γ for every $\gamma > 1$. But in this case

$$\mathsf{ARL}(T_{\text{GLR}}; \theta) = \frac{2}{\theta^2} (\log \gamma + \log\log \gamma) + O(1),$$

so that the second term $\log\log\gamma$ is larger than an "optimal" second term $\log \sqrt{\log \gamma} = \frac{1}{2} \log\log \gamma$ (see (8.277)).

By (8.273) and (8.272), the asymptotic lower bound is

$$\mathsf{ARL2FA}(T_{\text{GLR}}) \ge \frac{\sqrt{\pi} e^h}{\sqrt{h} \int_{\varepsilon}^{\infty} \vartheta^{-1} \zeta_\vartheta \, d\vartheta} (1 + o(1)) \quad \text{as } h \to \infty, \tag{8.290}$$

where by Theorem 2.5.3 the term $\zeta_\vartheta = \lim_{a \to \infty} \mathsf{E}_\theta e^{-(\lambda_{\tau_a}^{\vartheta} - a)}$ associated with the overshoot in the one-sided SPRT τ_a is given by

$$\zeta_\vartheta = \frac{2}{\vartheta^2} \exp\left[-2 \sum_{n=1}^{\infty} \frac{1}{n} \Phi\left(-\frac{\vartheta}{2} \sqrt{n} \right) \right], \vartheta > 0. \tag{8.291}$$

For numerical evaluation, the following approximation has been proposed for small ϑ : $\zeta_\vartheta = \exp\{-\rho\vartheta\} + o(\vartheta^2)$ as $\vartheta \to 0$ and $\rho \approx 0.5826$, see [428, 430]. Note that the lower bound (8.290) is poor for small ε, since $\lim_{\vartheta\to 0} \vartheta^{-1}\zeta_\vartheta = \infty$.

By conjecture in Remark 8.3.3, the asymptotic lower bound (8.290) can be improved:

$$\mathsf{ARL2FA}(T_{\mathrm{GLR}}) \sim \frac{\sqrt{\pi}e^h}{\sqrt{h}\int_\varepsilon^\infty \vartheta\zeta_\vartheta^2\,\mathrm{d}\vartheta} \quad \text{as } h\to\infty. \tag{8.292}$$

In contrast to the lower bound (8.290) the asymptotic approximation (8.292) is expected to work well even for $\varepsilon = 0$, since $\lim_{\vartheta\to 0}\vartheta\zeta_\vartheta^2 = 0$. In fact, in the case of $\varepsilon = 0$ (i.e., for the GCUSUM (8.288)) the asymptotic approximation (8.292) along with asymptotic exponentiality of the GCUSUM stopping time $T_{\mathrm{GLR}}(h)$ has been established by Siegmund and Venkatraman [430, Theorem 1]. In other words, in the case of $\varepsilon = 0$,

$$\mathsf{ARL2FA}(T_{\mathrm{GLR}}) \sim \frac{\sqrt{\pi}e^h}{\sqrt{h}\int_0^\infty \vartheta\zeta_\vartheta^2\,\mathrm{d}\vartheta} \quad \text{as } h\to\infty. \tag{8.293}$$

Moreover, a more accurate asymptotically equivalent approximation has been proposed in [430, Theorem 1] :

$$\mathsf{ARL2FA}(T_{\mathrm{GLR}}) \sim \frac{\sqrt{\pi}e^h}{\sqrt{h}\int_0^{\sqrt{2h}} \vartheta\zeta_\vartheta^2\,\mathrm{d}\vartheta} \quad \text{as } h\to\infty. \tag{8.294}$$

Using (8.275), we obtain the asymptotic approximation for the ARL function as $h\to\infty$

$$\mathsf{ARL}(T_{\mathrm{GLR}};\theta) = \frac{2}{\theta^2}\left\{h - \frac{1}{2} + \mathsf{E}_\theta\left[\min_{k\ge 0}\left(\theta S_k - \frac{\theta^2}{2}k\right)\right] + \varkappa_\theta\right\} + o(1)$$

where by Theorem 2.5.3

$$\varkappa_\theta = 1 + \frac{\theta^2}{4} + \theta\sum_{n=1}^\infty \frac{1}{\sqrt{n}}\left[\frac{\theta}{2}\sqrt{n}\,\Phi\left(-\frac{\theta}{2}\sqrt{n}\right) - \varphi\left(\frac{\theta}{2}\sqrt{n}\right)\right]. \tag{8.295}$$

Now, by (2.97),

$$\mathsf{E}_\theta\left[\min_{k\ge 0}\left(\theta S_k - \frac{\theta^2}{2}k\right)\right] = \varkappa_\theta - 1 - \theta^2/4,$$

which yields

$$\mathsf{ARL}(T_{\mathrm{GLR}};\theta) = \frac{2}{\theta^2}\left(h + \varkappa_\theta - 3/2 - \theta^2/4\right) + o(1) \quad \text{as } h\to\infty. \tag{8.296}$$

Alternatively, the quantities ζ_θ and $\varkappa_\theta$ can be computed with the help of Siegmund's corrected Brownian motion approximations as

$$\zeta_\theta \approx \exp\{-\rho\theta\}, \quad \varkappa_\theta \approx 2\rho|\theta| + \theta^2/4.$$

where

$$\rho = -\frac{1}{\pi}\int_0^\infty \frac{1}{t^2}\log\left\{\frac{2\left(1 - e^{-t^2/2}\right)}{t^2}\right\}\mathrm{d}t \approx 0.5826.$$

Substitution into (8.296) yields

$$\mathsf{ARL}(T_{\mathrm{GLR}};\theta) = \frac{2}{\theta^2}(h + 2\rho|\theta| - 3/2) + o(1) \text{ as } h\to\infty. \tag{8.297}$$

These approximations give reasonably good accuracy as long as θ is relatively small.

The accuracy of the approximations for the ARL2FA $\mathsf{E}_\infty[T_{\mathrm{GLR}}]$ and the ARL to detection $\mathsf{E}_\theta[T_{\mathrm{GLR}}]$

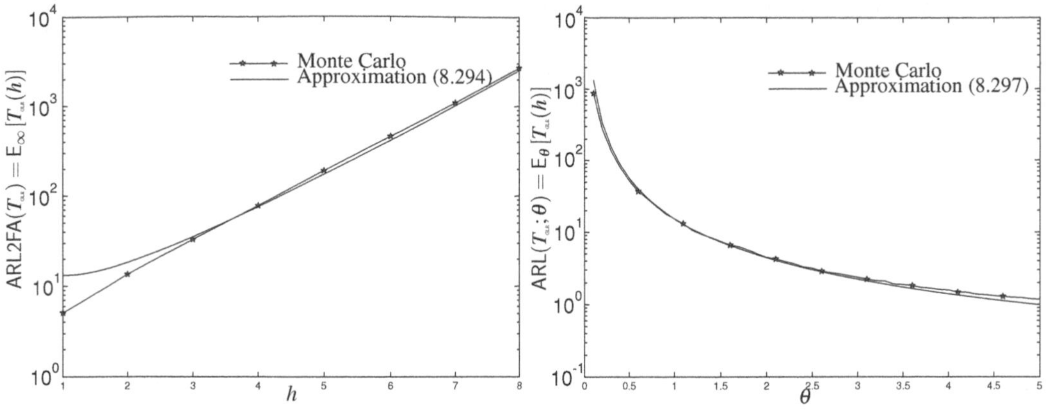

Figure 8.13: *Comparison of the approximations* (8.294) *and* (8.297) *with the results of Monte Carlo simulation.* $h \in [1,8]$ *(left);* $\theta \in [0.1,5]$ *and* $h = 8$ *(right).*

is illustrated in Figure 8.13, showing the comparison of the approximations (8.293)–(8.297) (solid lines) with the results of a 10^5-repetition Monte Carlo simulation of the GCUSUM stopping time given by (8.268) with (8.289). The approximation for the ARL2FA is drawn as a function of h in the left plot of Figure 8.13 for $h \in [1,8]$, and the approximation for the ARL to detection $\mathsf{E}_\theta[T_{\text{GLR}}]$ is drawn as a function of θ for $\theta \in [0.1,5]$ and $h = 8$ ($\mathsf{ARL2}FA = 2533$) on the right. The plots show that the precision of the approximation (8.294) is quite good as long as the threshold h is not too small. As it follows from (8.297), the approximation (8.297) has the same specific feature as an upper bound for the ARL function of the one-sided CUSUM given by equation (8.174). Namely, it does not provide a reasonable approximation for the values of θ close to 0. This conclusion is confirmed by the comparison of the upper bound for the ARL function of the one-sided CUSUM presented in Figure 8.9 with the approximation of $\mathsf{E}_\theta[T_{\text{GLR}}]$ shown in Figure 8.13. For this reason, it is proposed in [250] to use the following modification of equation (8.297):

$$\mathsf{ARL}(T_{\text{GLR}};\theta) = \min\{\mathsf{ARL2FA}(T_{\text{GLR}}), \mathsf{E}_\theta[T_{\text{GLR}}]\}. \tag{8.298}$$

Let us describe the geometric interpretation of the GCUSUM algorithm in the same manner as for the ordinary CUSUM algorithm before, using the reverse time interpretation of the decision function. We begin with a one-sided GCUSUM algorithm, and we will use a symmetry with respect to the horizontal line for the two-sided case as before.

The one-sided version of the GCUSUM procedure (8.287) is

$$T^1_{\text{GLR}}(h) = \inf\left\{n \geq 1 : \sup_{\theta\in[\varepsilon,\overline{\theta})} \sum_{k=1}^{n} \left(\theta X_k - \frac{\theta^2}{2}\right) \geq h\sigma^2\right\},$$

and this stopping rule can be rewritten in reverse time as follows. Since for some time instant n,

$$\sup_{\theta\in[\varepsilon,\overline{\theta})} \sum_{\ell=1}^{n} \left(\theta X_\ell - \frac{\theta^2}{2}\right) \geq h\sigma^2, \tag{8.299}$$

we obtain

$$S_n = \frac{1}{\sigma}\sum_{\ell=1}^{n} X_\ell \geq \inf_{\theta\in[\varepsilon,\overline{\theta})} \left(\frac{h\sigma}{\theta} + \frac{\theta n}{2\sigma}\right). \tag{8.300}$$

Let

$$C(\ell) = \inf_{\theta\in[\varepsilon,\overline{\theta})} \left(\frac{h\sigma}{\theta} + \frac{\theta \ell}{2\sigma}\right) \tag{8.301}$$

denote the stopping boundary for the cumulative sum S_ℓ. It is easy to see that the function $x \mapsto f(x) = h\sigma/x + x\ell/2\sigma$ defined on the positive semi-axis $\mathbb{R}_+$ has a unique minimum $x_0 = \sigma\sqrt{2h/\ell}$. Hence, taking into account the constraints on possible values of $\theta \in [\varepsilon, \overline{\theta})$, three possible situations can be distinguished depending on the value of ℓ:

$$C(\ell) = \begin{cases} (h\sigma)/\overline{\theta} + (\overline{\theta}\ell)/(2\sigma) & \text{if} \quad 1 \le \ell < 2h\left(\sigma/\overline{\theta}\right)^2 \\ \sqrt{2h\ell} & \text{if} \quad 2h\left(\sigma/\overline{\theta}\right)^2 \le \ell \le 2h(\sigma/\varepsilon)^2 \\ (h\sigma)/\varepsilon + (\varepsilon\ell)/(2\sigma) & \text{if} \quad \ell > 2h(\sigma/\varepsilon)^2 \end{cases}. \tag{8.302}$$

This boundary is illustrated in Figure 8.14(left) for $\sigma = 1$, $\varepsilon = 0.3$, $\overline{\theta} = 1$ and $h = 3$. For small ℓ, the boundary in (8.302) is the straight line with maximal angle with respect to the horizontal line. This corresponds to the situation when $\theta = \overline{\theta}$. Recall that the stopping boundary in the form of the straight line has occurred when we considered the geometric interpretation of the CUSUM stopping rule in terms of the V-mask in Section 8.2 (see Figure 8.6). For medium ℓ the boundary is a parabolic curve. This corresponds to the situation when $\varepsilon < \theta < \overline{\theta}$. Finally, for large ℓ, the boundary in (8.302) is the straight line with minimal angle with respect to the horizontal line. This corresponds to the case when $\theta = \varepsilon$. In the case of the two-sided GCUSUM algorithm, we use a symmetry with respect to the horizontal line passing through the point (n, X_n) in reverse time. This leads to the so-called U-mask depicted in Figure 8.14(right). This parabola is inscribed in two V-masks discussed before, because the points of tangency between the straight lines and the parabola have the abscissas $\ell = 2h\left(\sigma/\overline{\theta}\right)^2$ and $\ell = 2h(\sigma/\varepsilon)^2$, respectively, as shown with vertical segments in Figure 8.14(left).

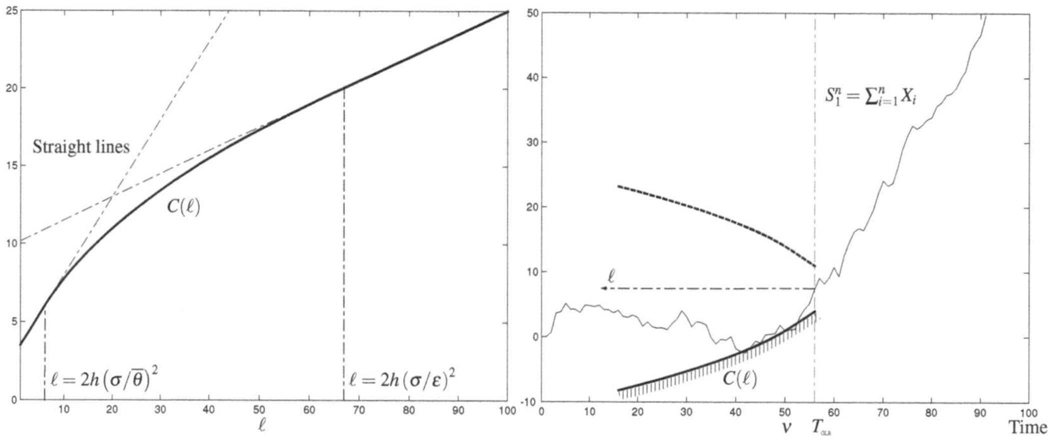

Figure 8.14: *The lower boundary $C(\ell)$ given by* (8.302) *for the cumulative sum S_ℓ in reverse time ℓ (left) and the U-mask for the GCUSUM algorithm (right).*

Implementation of the GCUSUM procedure (8.268) and (8.289) at every time n requires n maximizations of the LLR over $\theta \in \Theta_1$. It is easy to see that the number of maximizations grows to infinity with n. Hence, the computational complexity of GCUSUM is of order $O(\gamma)$ (on average) before a false alarm and renewal. Approximations of this algorithm with a reduced computational cost are thus of interest. Lorden [277] proposed a 2-CUSUM algorithm where two ordinary CUSUM statistics run simultaneously, each tuned to a fixed parameter value. Typically these two CUSUM statistics are designed for detecting changes with large and small magnitudes, respectively. The geometric interpretation of this approximation is that a U-mask can be approximated by the intersection of two V-masks. This point is further discussed in Chapter 9.

8.3.2 *Asymptotic Optimality of WCUSUM and GCUSUM Algorithms in the Non-iid Case*

We now extend the iid model with an unknown post-change parameter considered in Subsection 8.3.1 to the general non-iid case. This section also represents an extension of Subsection 8.2.8, where the CUSUM algorithm has been considered in the case of a simple post-change hypothesis and non-iid observations.

Let $\theta = (\theta_1, \dots, \theta_\ell) \in \Theta \subset \mathbb{R}^\ell$ be a vector parameter. Consider the general non-iid model assuming that the observations $\{X_n\}_{n\geq 1}$ are distributed according to conditional densities $f_{\theta,n}(X_n|\mathbf{X}_1^{n-1})$ with a known pre-change parameter $\theta = \theta_0$ (for $n = 1, \dots, \nu$) and with an unknown post-change parameter $\theta \in \Theta_1 = \Theta \setminus \theta_0$ (for $n > \nu$). As before, we have to assume that in general post-change densities $f_{\theta,n}(X_n|\mathbf{X}_1^{n-1}) = f_{\theta,n}^{(\nu)}(X_n|\mathbf{X}_1^{n-1})$ depend on the changepoint ν. In this general case the LLR process between the hypotheses $H_{\nu,\theta}$ and H_∞ is

$$\lambda_n^\nu(\theta) = \sum_{i=\nu+1}^{n} Z_i^{(\nu)}(\theta), \quad Z_i^{(\nu)}(\theta) = \log \frac{f_{\theta,i}^{(\nu)}(X_i|\mathbf{X}_1^{i-1})}{f_{\theta_0,i}(X_i|\mathbf{X}_1^{i-1})}, \quad \theta \in \Theta_1.$$

We begin with considering the WCUSUM procedure $T_{\text{WCS}}(h)$ defined in (8.242) with

$$\widetilde{\Lambda}_n^k = \int_{\Theta_1} \exp\left\{\lambda_n^{k-1}(\theta)\right\} \mathrm{d}W(\theta),$$

i.e.,

$$T_{\text{WCS}}(h) = \inf\left\{n \geq 1 : \log\left[\max_{1\leq k\leq n} \int_{\Theta_1} \exp\left\{\lambda_n^{k-1}(\theta)\right\} \mathrm{d}W(\theta)\right] \geq h\right\}. \tag{8.303}$$

Similarly to (8.83) let us assume that the LLR $\lambda_{\nu+n}^\nu(\theta) = \sum_{k=\nu+1}^{\nu+n} Z_k^{(\nu)}(\theta)$ normalized to n converges in $\mathsf{P}_{\nu,\theta}$-probability to a positive finite number I_θ:

$$\frac{1}{n}\lambda_{\nu+n}^\nu(\theta) \xrightarrow[n\to\infty]{\mathsf{P}_{\nu,\theta}} I_\theta \quad \text{for all } \theta \in \Theta_1. \tag{8.304}$$

As we discussed before, this number plays the role of the local K–L information. Assume also that

$$\sup_{0\leq\nu<\infty} \operatorname{ess\,sup} \mathsf{P}_{\nu,\theta}\left\{M^{-1} \max_{0\leq n<M} \lambda_{\nu+n}^\nu(\theta) \geq (1+\varepsilon)I_\theta | \mathscr{F}_\nu\right\} \xrightarrow[M\to\infty]{} 0 \tag{8.305}$$

for all $\varepsilon > 0$ and all $\theta \in \Theta_1$ (cf. (8.84)). Then, clearly, assertions of Theorem 8.2.2 hold for all $\theta \in \Theta_1$. In particular, for all $r > 0$ and all $\theta \in \Theta_1$

$$\inf_{T\in\mathbf{C}_\gamma} \mathsf{ESM}_r^\theta(T) \geq \inf_{T\in\mathbf{C}_\gamma} \mathsf{SM}_r^\theta(T) \geq \left(\frac{\log\gamma}{I_\theta}\right)^r (1+o(1)) \quad \text{as } \gamma\to\infty. \tag{8.306}$$

These asymptotic lower bounds are attained by the WCUSUM procedure under certain additional conditions on the left tail distributions similarly to Theorem 8.2.6. In particular, we have to require the following condition

$$\sup_{0\leq\nu<k} \operatorname{ess\,sup} \mathsf{P}_{\nu,\theta}\left(n^{-1}\lambda_{k+n}^k(\theta) < I_\theta(1-\varepsilon)|\mathscr{F}_\nu\right) < \infty \tag{8.307}$$

to hold for some $k \geq 1$, all $0 < \varepsilon < 1$ and all $\theta \in \Theta_1$ (cf. (8.201)). Furthermore, this condition should be strengthened to avoid complications related to too small separation of the pre-change and post-change parameters. The following theorem spells out details.

Theorem 8.3.4 (WCUSUM asymptotic optimality in $\mathbf{C}_\gamma$). *Assume there exist finite positive numbers I_θ, $\theta \in \Theta_1$ such that condition* (8.305) *holds. Further let, for every (sufficiently small) $\varepsilon > 0$,* *there exist a set* $\Theta_\varepsilon \subset \Theta$ *and a number* $n_\varepsilon \ge 1$ *such that* $\theta \in \Theta_\varepsilon$*, the weight* $W(\theta)$ *is strictly positive on* Θ_ε*, and for some* $k > 0$*,*

$$\sup_{n \ge n_\varepsilon} \sup_{0 \le \nu < k} \operatorname{ess\,sup} \mathsf{P}_{\nu,\theta}\left(n^{-1} \inf_{\vartheta \in \Theta_\varepsilon} \lambda_{k+n}^k(\vartheta) < I_\theta(1-\varepsilon) | \mathscr{F}_\nu\right) \le \varepsilon. \tag{8.308}$$

Then, for all $\theta \in \Theta_\varepsilon$ *and all* $r > 0$ *asymptotic lower bounds* (8.306) *hold and*

$$\mathsf{SM}_r^\theta(T_{\text{WCS}}(h)) \sim \mathsf{ESM}_r^\theta(T_{\text{WCS}}(h)) \sim \left(\frac{h}{I_\theta}\right)^r \quad \text{as } h \to \infty \quad \text{for all } \theta \in \Theta_\varepsilon. \tag{8.309}$$

Thus, if $h = h_\gamma$ *is selected so that* $\mathsf{ARL2FA}(T_{\text{WCS}}(h_\gamma)) \ge \gamma$ *and* $h_\gamma \sim \log \gamma$*, in particular* $h = \log \gamma$*, the lower bounds* (8.306) *are attained and the WCUSUM procedure is asymptotically minimax in the sense of minimizing moments* $\mathsf{ESM}_r^\theta(T)$ *and* $\mathsf{SM}_r^\theta(T)$ *to first order as* $\gamma \to \infty$ *uniformly for all* $\theta \in \Theta_\varepsilon$*:*

$$\inf_{T \in \mathbf{C}_\gamma} \mathsf{ESM}_r^\theta(T) \sim \inf_{T \in \mathbf{C}_\gamma} \mathsf{SM}_r^\theta(T) \sim \mathsf{ESM}_r^\theta(T_{\text{WCS}}) \sim \mathsf{SM}_r^\theta(T_{\text{WCS}}) \sim \left(\frac{\log \gamma}{I_\theta}\right)^r. \tag{8.310}$$

Proof. Evidently, (8.306) and (8.309) imply assertion (8.310). Hence, it suffices only to prove assertion (8.309) as well as the fact that setting $h = \log \gamma$ implies the lower bound $\mathsf{ARL2FA}(T_{\text{WCS}}(h_\gamma)) \ge \gamma$. The latter implication follows easily from Lemma 8.4.1 in Section 8.4 by noticing that for any $h > 0$

$$T_{\text{WCS}}(h) \ge T_{\text{WSR}}(h) = \inf\left\{n \ge 1 : \sum_{k=1}^n \widetilde{\Lambda}_n^k \ge e^h\right\}.$$

Condition (8.305) implies the lower bounds

$$\mathsf{ESM}_r^\theta(T_{\text{WCS}}(h)) \ge \mathsf{SM}_r^\theta(T_{\text{WCS}}(h)) \ge \left(\frac{h}{I_\theta}\right)^r (1 + o(1)) \quad \text{as } h \to \infty$$

for all $r > 0$, which can be established as in the proof of Theorem 8.2.2 (replacing $\log \gamma$ with h) or alternatively noticing that $\mathsf{ESM}_r^\theta(T_{\text{WCS}}(h)) \ge \mathsf{ESM}_r^\theta(T_{\text{CS}}^\theta(h/w(\theta)))$, where $T_{\text{CS}}^\theta(h)$ denotes CUSUM tuned to θ. Hence, it suffices to show that under condition (8.308) the following upper bound holds:

$$\mathsf{ESM}_r^\theta(T_{\text{WCS}}(h)) \le \left(\frac{h}{I_\theta}\right)^r (1 + o(1)) \quad \text{as } h \to \infty. \tag{8.311}$$

Since

$$\widetilde{\Lambda}_n^k \ge \left[\inf_{\theta \in \Theta_\varepsilon} e^{\lambda_n^{k-1}(\theta)}\right] W(\Theta_\varepsilon),$$

the proof of inequality (8.311) is analogous to that of (8.205). □

Analogous results also hold in the class $\mathbf{C}(\beta) = \{T : \mathsf{MPFA}(T) \le \beta\}$ that upper-bounds the maximal conditional false alarm probability $\mathsf{MPFA}(T) = \sup_k \mathsf{P}_\infty(T \le k + m_\beta | T > k)$, where hereafter we assume that the size of the interval m_β goes to infinity as $\beta \to 0$ in such a way that

$$m_\beta / |\log \beta| \to \infty \ \text{ but } \ (\log m_\beta)/|\log \beta| \to 0 \quad \text{as } \beta \to 0 \tag{8.312}$$

(cf. (8.206) and Theorem 8.2.7). But it is not obvious how to select the thresholds to guarantee this upper bound, and this is complicated by the fact that the limit distributions of the stopping times under the no-change scenario are not guaranteed to be exponential in the non-iid case. Since it is difficult to evaluate the conditional probability $\mathsf{P}_\infty(T_{\text{WCS}} \le k + m | T_{\text{WCS}} > k)$, as in Subsection 8.2.8 we again consider the unconditional PFA $\mathsf{P}_\infty(k < T_{\text{WCS}} \le k + m)$ and the corresponding class of procedures $\mathbf{C}_{\text{un}}(\beta) = \{T : \sup_{k \ge 0} \mathsf{P}_\infty(k < T \le k + m_\beta) \le \beta\}$. The WCUSUM procedure is asymptotically optimal in the class $\mathbf{C}_{\mathrm{un}}(\beta)$ as well. However, computational complexity is a serious obstacle for implementing this procedure. Indeed, generally, the WLR statistic $\widetilde{\Lambda}_n^k$ cannot be written in a recursive manner. Hence, the number of WLRs at time n that should be maximized over $k \le n$ grows to infinity with n. Then the average number of WLRs that should be calculated at time n before the renewal of the detection statistic that occurs at the time of alarm, assuming that the changepoint is larger than γ, is $O(\gamma)$. Since in practice γ is typically 500 and higher, to reduce the computational cost the window-limited versions ideologically similar to the WL-CUSUM (8.218) are in order.

Consider the window-limited version of WCUSUM (WL-WCUSUM)

$$T_{\text{WL-WCS}}(h) = \inf\left\{n \ge 1 : \max_{\max\{0,n-m_\beta\}+1 \le k \le n} \log \widetilde{\Lambda}_k^n \ge h\right\}, \tag{8.313}$$

where maximization over all $1 \le k \le n$ is replaced by the maximization in the sliding window of size m_β, which is large but not too large due to conditions (8.312). Specifically, $|\log\beta| \ll m_\beta \ll \beta^{-1}$. For $n < m_\beta$, the maximization is performed over $1 \le k \le n$.

Obviously,

$$\mathsf{P}_\infty(k < T_{\text{WL-WCS}} \le k + m_\beta) \le \sum_{j=k-m_\beta+1}^{k+m_\beta} \mathsf{P}_\infty\left(\widetilde{\Lambda}_n^j \ge e^h \text{ for some } n \ge jk\right),$$

where $\{\widetilde{\Lambda}_n^j\}_{n\ge j}$ is a nonnegative martingale under P_∞ with unit expectation, $\mathsf{E}_\infty\widetilde{\Lambda}_n^j = 1$. By the Doob submartingale inequality, $\mathsf{P}_\infty(\widetilde{\Lambda}_n^j \ge e^h \text{ for some } n \ge j) \le e^{-h}$, which implies

$$\sup_{k\ge 0} \mathsf{P}_\infty(k < T_{\text{WL-WCS}} \le k + m_\beta) \le 2m_\beta e^{-h}. \tag{8.314}$$

Therefore, setting $h_\beta = \log(2m_\beta/\beta)$ guarantees $T_{\text{WL-WCS}} \in \mathbf{C}_{\mathrm{un}}(\beta)$ and $h_\beta \sim |\log\beta|$ by conditions (8.312).

The following theorem establishes minimax and uniform first-order asymptotic optimality property of the WL-WCUSUM procedure.

Theorem 8.3.5. *Assume that conditions* (8.305) *and* (8.308) *hold.*

(i) *Let the size of the window m_h in WL-WCUSUM increase with h so that*

$$m_h/h \to \infty \quad \text{and} \quad \log m_h = o(h). \tag{8.315}$$

If the threshold $h = h_\gamma$ is found from the equation

$$\left(\frac{1}{2} - 2m_h e^{-h}\right)\cdot\left(\frac{e^h}{4} - 1\right) = \gamma, \quad \gamma > 1, \tag{8.316}$$

then $T_{\text{WL-WCS}}(h_\gamma) \in \mathbf{C}_\gamma$ and the WL-WCUSUM procedure is asymptotically minimax to first order as $\gamma \to \infty$ uniformly for all $\theta \in \Theta_\varepsilon$:

$$\inf_{T\in\mathbf{C}_\gamma} \mathsf{ESM}_r^\theta(T) \sim \inf_{T\in\mathbf{C}_\gamma} \mathsf{SM}_r^\theta(T) \sim \mathsf{ESM}_r^\theta(T_{\text{WL-WCS}}) \sim \mathsf{SM}_r^\theta(T_{\text{WL-WCS}}) \sim \left(\frac{\log\gamma}{I_\theta}\right)^r. \tag{8.317}$$

(ii) *Let condition* (8.312) *hold. If the threshold $h = h_\beta$ is found from the equation $2m_\beta e^{-h} = \beta$, then $T_{\text{WL-WCS}}(h_\beta) \in \mathbf{C}_{\mathrm{un}}(\beta)$ and the WL-WCUSUM procedure is asymptotically uniformly optimal as $\beta \to 0$ in the sense that for all $\nu \ge 0$, all $\theta \in \Theta_\varepsilon$, and all $r > 0$*

$$\begin{aligned}
\mathsf{E}_{\nu,\theta}[(T_{\text{WL-WCS}} - \nu)^+]^r &= \left[\frac{\mathsf{P}_\infty(T_{\text{WL-WCS}} > \nu)}{I_\theta} + o(1)\right]^r |\log\beta|^r, \\
\mathsf{E}_{\nu,\theta}[(T - \nu)^+]^r &\ge \left[\frac{\mathsf{P}_\infty(T > \nu)}{I_\theta} + o(1)\right]^r |\log\beta|^r \quad \text{for all } T \in \mathbf{C}_{\mathrm{un}}(\beta),
\end{aligned} \tag{8.318}$$

where the residual terms $o(1)$ are uniform over $\nu \ge 0$.

Proof. (i) Let $(j-1)m < k \le jm$. Since for $j = 1, 2, \ldots$,

$$\begin{aligned}\mathsf{P}_\infty(T_{\text{WL-WCS}} < k) &\le \sum_{i=1}^{j} \mathsf{P}_\infty\{(j-1)m \le T_{\text{WL-WCS}} < jm\} \le j\mathsf{P}_\infty(m \le T_{\text{WL-WCS}} < k+m) \\ &\le (1+k/m)\mathsf{P}_\infty(k \le T_{\text{WL-WCS}} < k+m),\end{aligned} \tag{8.319}$$

and $\mathsf{P}_\infty(k \le T_{\text{WL-WCS}} < k+m) \le 2me^{-h}$ by (8.314), it follows that

$$\mathsf{P}_\infty(T_{\text{WL-WCS}} \ge k) \ge 1 - 2m_h e^{-h}(1+k/m_h) \ge \left(\frac{1}{2} - 2m_h e^{-h}\right) \quad \text{for } k \le m_h/(2m_h e^{-h}) = e^h/4.$$

Hence,

$$\mathsf{E}_\infty[T_{\text{WL-WCS}}(h)] \ge \left(\frac{1}{2} - 2m_h e^{-h}\right) \cdot \left(\frac{e^h}{4} - 1\right),$$

implying that $T_{\text{WL-WCS}}(h) \in \mathbf{C}_\gamma$ when h is found from equation (8.316).

Next, note that by condition (8.315), $h_\gamma \sim \log\gamma$ and that

$$T_{\text{WL-WCS}}(h) \le \tau(h) = \inf\left\{n : \left[\max_{n-m \le k \le n} \inf_{\theta \in \Theta_\varepsilon} \lambda_n^k(\theta)\right] \ge h/W(\Theta_\varepsilon)\right\}.$$

Thus, applying the same line of argument to the stopping time $\tau(h)$ as in the proof of inequality (8.205), we obtain the inequality

$$\mathsf{ESM}_r^\theta(T_{\text{WCS}}(h_\gamma)) \le \mathsf{ESM}_r^\theta(\tau(h_\gamma)) \le \left(\frac{h_\gamma}{I_\theta}\right)^r (1+o(1)) = \left(\frac{\log\gamma}{I_\theta}\right)^r (1+o(1)) \quad \text{as } \gamma \to \infty,$$

which along with the lower bound (8.306) proves (8.317).

(ii) The fact that $2m_\beta e^{-h} = \beta$ implies $T_{\text{WL-WCS}}(h_\beta) \in \mathbf{C}_{\text{un}}(\beta)$ follows from (8.314).

For $r = 1$, the proof of assertions (8.318) can be found in [251] (Theorem 2 and Theorem 4(i)). For $r > 1$, the proof is technically more difficult but similar. □

Another approach to the problem of change detection with a composite post-change hypothesis is the GLR approach based on the double maximization. The corresponding GCUSUM scheme is given by the stopping time $T_{\text{GLR}}(h)$ defined in (8.268) with

$$\hat{g}_n = \max_{1 \le k \le n-\widetilde{m}} \sup_{\theta \in \Theta_1} \sum_{i=k}^{n} Z_i^{(k-1)}(\theta),$$

i.e.,

$$T_{\text{GCS}}(h) = \inf\left\{n > \widetilde{m} : \max_{1 \le k \le n-\widetilde{m}} \sup_{\theta \in \Theta_1} \sum_{i=k}^{n} Z_i^{(k-1)}(\theta) \ge h\right\} \tag{8.320}$$

Remark 8.3.4. In practice, to calculate $\hat{g}_n$ it is necessary to find the ML estimates of the parameter $\theta \in \Theta_1$ for each $k : 1 \le k \le n - \widetilde{m}$,

$$\widehat{\theta} = \arg\sup_{\theta \in \Theta_1} \sum_{i=k}^{n} \log f_{\theta,i}^{(k-1)}(X_i | \mathbf{X}_1^{i-1})$$

and, next, to use this sequence of estimates to form the detection statistic

$$\sup_{\theta \in \Theta_1} \sum_{i=k}^{n} Z_i^{(k-1)}(\theta) = \sum_{i=k}^{n} Z_i^{(k-1)}(\widehat{\theta}).$$

The number $\widetilde{m} \ge 0$ defines the smallest sample size $\widetilde{m}+1$ required to warrant existence of the ML estimate of θ.

The GCUSUM procedure has similar optimality properties as WCUSUM. Specifically, under conditions of Theorem 8.3.4 asymptotic approximations (8.309) hold for GCUSUM. However, analogously to sequential testing problems, where the usual difficulty with the GLR method is obtaining the bounds on error probabilities, in the change detection problem this difficulty transfers into obtaining the bounds for the ARL2FA and PFA in the given interval. In particular, in order to guarantee the inequality $\mathsf{ARL2FA}(T_{\text{WCS}}(h_\gamma)) \ge \gamma$ in the WCUSUM procedure it is sufficient to set $h_\gamma = \log \gamma$ due to the nice martingale property of the mixtures $\{\widetilde{\Lambda}_n^k\}$, but in general it is not clear how to select the threshold in the GCUSUM procedure. Certain asymptotic approximations may be obtained based on large-moderate deviation theory under additional conditions on the structure of the parameter space and further smoothness conditions on the distributions; see below.

Similarly to WCUSUM an obvious disadvantage of the GCUSUM procedure is that the number of maximizations at time n grows to infinity with n. As previously, the computational complexity is of order $O(\gamma)$ (on average) before a false alarm and renewal. To reduce the computational cost of the GLR scheme Willsky and Jones [507] introduced the *window-limited* GLR (WL-GLR) scheme

$$T_{\text{WL-GCS}}(h) = \inf\left\{ n > \widetilde{m} : \max_{\max\{0, n-m_\beta\}+1 \le k \le n-\widetilde{m}} \sup_{\theta \in \Theta_1} \sum_{i=k}^{n} Z_i^{(k-1)}(\theta) \ge h \right\}, \tag{8.321}$$

where $0 \le \widetilde{m} < m_\beta$. We will refer to this scheme as the *window-limited GCUSUM* (WL-GCUSUM). Hence, the WL-CUSUM scheme involves $(m_\beta - \widetilde{m})$ LR maximizations at every stage n. In some situations this fact considerably reduces the computational burden (and also memory requirements) and makes this detection scheme manageable in real-time implementations.

Note that $T_{\text{WL-GCS}}(h) \le T_{\text{WL-CS}}^{\theta}(h)$, where $T_{\text{WL-CS}}^{\theta}(h)$ is the WL-CUSUM procedure tuned to $\theta \in \Theta_1$; see (8.218) with $f_1 = f_\theta$. Hence, using (8.219), we obtain that under conditions of Theorem 8.2.9 the following asymptotic inequality holds

$$\mathsf{E}_{\nu,\theta}[(T_{\text{WL-GCS}} - \nu)^+]^r \le \left[\frac{\mathsf{P}_\infty(T_{\text{WL-GCS}} > \nu)}{I_\theta} + o(1)\right]^r h^r \quad \text{for all } \nu \ge 0,$$

where $o(1) \to 0$ uniformly in $\nu \ge 0$. Thus, if $\widetilde{m} = o(|\log \beta|)$ and $h \sim |\log \beta|$, the WL-GCUSUM procedure $T_{\text{WL-GCS}}$ attains asymptotic lower bound (8.318). It therefore remains to obtain either an upper bound or at least an asymptotic approximation for the PFA $\sup_k \mathsf{P}_\infty(k < T_{\text{WL-GCS}} \le k + m_\beta)$. By the same reason as for GCUSUM, this a challenging and tedious task. This task is indeed much more difficult than for the WL-WCUSUM procedure. While for the latter procedure due to the martingale property of the statistic $\widetilde{\Lambda}_n^k$ the false alarm probability can be easily upper-bounded as in (8.314), for WL-GCUSUM there is no simple upper bound. The formal mathematical proof, presented by Lai [251], is very involved, and we give only a sketch of Lai's approach (a detailed proof can be found in [251, 253]).

To obtain an approximation for $\sup_k \mathsf{P}_\infty(k < T_{\text{WL-GCS}} \le k + m_\beta)$ one needs several additional assumptions. The first assumption is related to the structure of the parameter space Θ, which in the case of WCUSUM can be almost arbitrary. Now we have to assume that Θ is a compact l-dimensional submanifold of $\mathbb{R}^l$. This is a crucial assumption that cannot be relaxed since we need Lebesgue measure of Θ, denoted as $|\Theta|$, to be finite. Next, assume that the LLR $Z_n(\theta)$ is of class $\mathbb{C}^2$, i.e., twice continuously differentiable. Here and in the rest of this subsection we omit the superscript $(n-1)$ for brevity. Let $Z_n^*(\theta) = \nabla_\theta [Z_n(\theta)]$ denote the gradient vector and let

$$[H_{n;i,j}(\theta)] = \left[\frac{\partial^2 Z_n(\theta)}{\partial \theta_i \partial \theta_j}\right], \quad i,j = 1,\dots,l$$

denote the Hessian $l \times l$ matrix. Denote by $\widehat{\theta}_j^n$ the ML estimator of θ based on the observations $\mathbf{X}_j^n$ for $j \le n$. Next, assume that $\widehat{\theta}_j^n$ is an interior point of Θ. Then $\sum_{k=j}^n Z_k^*(\widehat{\theta}_j^n) = 0$ and the Hessian matrix $H_{i,j}(\widehat{\theta}_j^n) = \sum_{k=j}^n H_{k;i,j}(\widehat{\theta}_j^n)$ is negative definite. Hence, the LLR can be approximated by the following quadratic function

$$\sum_{k=j}^{n} Z_k(\theta) \approx \sum_{k=j}^{n} Z_k(\widehat{\theta}_j^n) + \frac{1}{2}(\theta - \widehat{\theta}_j^n)^\top H(\widehat{\theta}_j^n)(\theta - \widehat{\theta}_j^n)$$

when $\|\theta - \widehat{\theta}_j^n\|_2$ is small. Finally, assuming that $\widehat{\theta}_k^n$ is an interior point of Θ and the largest eigenvalue $\lambda_{\max}(-H(\widehat{\theta}_k^n))$ of the matrix $-H(\widehat{\theta}_k^n)$ is not too large when $\|\theta - \widehat{\theta}_k^n\|_2$ is small, introduce the following stopping time:

$$\widehat{T}_k = \inf\left\{ n > k + \widetilde{m} : \sup_{\|\theta - \widehat{\theta}_k^n\|_2 \le h^{-\frac{1}{2}}} \lambda_{\max}(-H(\widehat{\theta}_k^n)) \le h^{1+\varepsilon} \text{ and } \sum_{i=k}^{n} Z_i(\widehat{\theta}_k^n) \ge h \right\},$$

where $\varepsilon > 0$. An approximation for the PFA of WL-GCUSUM can be obtained based on the analysis of behavior of the probabilities $\mathsf{P}_\infty(\widehat{T}_k \le k + m_\beta)$, $k \ge 1$ since

$$\mathsf{P}_\infty(k < T_{\text{WL-GCS}} \le k + m_\beta) \le \sum_{i=k-m_\beta+1}^{k+m_\beta} \mathsf{P}_\infty(\widehat{T}_i \le i + m_\beta),$$

and hence,

$$\sup_{k\ge 0} \mathsf{P}_\infty(k < T_{\text{WL-GCS}} \le k + m_\beta) \le 2m_\beta \sup_{k\ge 0} \mathsf{P}_\infty(\widehat{T}_k \le k + m_\beta). \tag{8.322}$$

Introduce the measure $\mathsf{Q}_\nu = \int_\Theta \mathsf{P}_{\nu,\theta}\, d\theta$, which is finite since Θ is a compact set by assumption. Let

$$\bar{\mathcal{L}}_n = \frac{d\mathsf{Q}_\nu^{(n)}}{d\mathsf{P}_\infty^{(n)}} = \int_\Theta e^{\lambda_n^\nu(\theta)}\, d\theta,$$

where $\mathsf{Q}_\nu^{(n)}$ and $\mathsf{P}_\infty^{(n)}$ are restrictions of Q_ν and P_∞ to the sigma algebra $\mathscr{F}_n$. By the Wald likelihood ratio identity,

$$\mathsf{P}_\infty(\widehat{T}_i \le i + m_\beta) = \int_{\{\widehat{T}_i \le i+m_\beta\}} \bar{\mathcal{L}}_{\widehat{T}_i}^{-1}\, d\mathsf{Q}_i = \int_\Theta \left[\int_{\{\widehat{T}_i \le i+m_\beta\}} \bar{\mathcal{L}}_{\widehat{T}_i}^{-1}\, d\mathsf{P}_{i,\theta} \right] d\theta.$$

This representation is a key, and further manipulations show that the right side can be upper-bounded by $(2\pi)^{-l/2} h^{(1+\varepsilon)l/2} e^{-h}(|\Theta| + o(1))$ as $h \to \infty$ (cf. [251, Lemma 2]). Therefore, (8.322) yields

$$\sup_{k\ge 0} \mathsf{P}_\infty(k < T_{\text{WL-GCS}} \le k + m_\beta) \le 2m_\beta (2\pi)^{-\frac{l}{2}} h^{\frac{l(1+\varepsilon)}{2}} e^{-h}\big(|\Theta| + o(1)\big) \quad \text{as } h \to \infty.$$

For small β, the choice $\widetilde{m} = o(|\log \beta|)$ and

$$h = \log \beta^{-1} + \frac{l(1+\varepsilon)}{2} \log\log \beta^{-1} + \log\left[2m_\beta (2\pi)^{-\frac{l}{2}} |\Theta|\right] + o(1)$$

leads to the bound

$$\sup_{k\ge 0} \mathsf{P}_\infty(k < T_{\text{WL-GCS}} \le k + m_\beta) \le \beta(1 + o(1)).$$

Now, if instead of the PFA constraint we are interested in the ARL2FA constraint $\mathsf{ARL2FA}(T_{\text{WL-GCS}}) \ge \gamma$, similar to (8.319)

$$\mathsf{P}_\infty(T_{\text{WL-GCS}} < k) \le (1 + k/m) \mathsf{P}_\infty(k \le T_{\text{WL-GCS}} < k + m),$$

so it follows that

$$\begin{aligned}\mathsf{E}_\infty\left[T_{\text{WL-GCS}}(h)\right] &\geq \left(\frac{1}{2} - 2m_h(2\pi)^{-\frac{l}{2}}h^{\frac{l(1+\varepsilon)}{2}}e^{-h}\left(|\Theta|+o(1)\right)\right)\left(\frac{(2\pi)^{l/2}e^h}{4h^{l(1+\varepsilon)/2}|\Theta|}(1+o(1))-1\right)\\ &= \frac{(2\pi)^{l/2}e^h}{8h^{l(1+\varepsilon)/2}|\Theta|}\left(1+o(1)\right) \quad \text{as } h\to\infty.\end{aligned}$$

We iterate that the GCUSUM procedure in the scalar case has been first considered by Lorden [271, 272], and it is discussed in Subsection 8.3.1.2. The WL-GCUSUM procedure has been studied by Willsky and Jones [507], Lai [251], and Lai and Shan [253].

Finally, we recall Remark 8.2.1 where we questioned usefulness of the unconditional PFA constraint. The same conclusion is certainly true for the composite hypothesis case. However, the more reasonable conditional PFA constraint is often difficult to implement since in general the P_∞-distributions of T_{WCS}, T_{GCS} and their window restricted versions are not exponential.

8.4 The Shiryaev–Roberts Procedure and Its Modifications

In this section, we continue studying the SR procedure that was defined in (7.148)–(7.150) in the iid case. In Theorem 7.6.1 of Section 7.6, we established exact optimality of this procedure for any given ARL2FA $\gamma > 1$ in the generalized Bayesian setting (6.16). In this section, we extend this result to the multicyclic setting and stationary average delay to detection. We also introduce two modifications of this procedure and show that they are almost optimal in the minimax sense when γ is large.

The SR detection procedure is defined by the stopping time

$$T_{\text{SR}}(A) = \inf\{n \geq 1 : R_n \geq A\} \tag{8.323}$$

with the SR statistic R_n given by

$$R_n = \sum_{\nu=0}^{n-1}\prod_{i=\nu+1}^{n}\mathcal{L}_i, \tag{8.324}$$

where in the general non-iid case the LR is $\mathcal{L}_i = f_1^{(\nu)}(X_i|\mathbf{X}_1^{i-1})/f_0(X_i|\mathbf{X}_1^{i-1})$. Note that in general $\mathcal{L}_i = \mathcal{L}_i^{(\nu)}$ depends on the changepoint ν. In a particular case where the LR does not depend on ν the SR statistic R_n can be computed recursively as

$$R_n = (1+R_{n-1})\mathcal{L}_n, \quad n \geq 1, \; R_0 = 0. \tag{8.325}$$

Note that this is always true for the iid model, in which case $\mathcal{L}_n = f_1(X_n)/f_0(X_n)$.

8.4.1 *Optimality of the SR Procedure for a Change Appearing after Many Reruns*

Consider the iid case and the multicyclic detection problem of Subsection 6.3.4 when it is of utmost importance to detect a real change as quickly as possible after its occurrence, even at the price of raising many false alarms (using a repeated application of the same stopping rule) before the change occurs. This essentially means that the changepoint ν is very large compared to the ARL2FA γ which, in this case, represents the mean time between consecutive false alarms. In other words, the change occurs at a distant time horizon and is preceded by a *stationary flow of false alarms*. This scenario is schematically shown in Figure 6.4, Section 6.3.

Recall the definition of the generic multicyclic detection procedure as well as the stationary average delay to detection given in Subsection 6.3.4. Let T be a stopping time and let $T^{(1)}, T^{(2)}, \dots$ denote its sequential independent repetitions. Next, let $\mathcal{T}^{(j)} = T^{(1)} + T^{(2)} + \cdots + T^{(j)}$ be the time of the j-th alarm, and define $N_\nu = \min\{j \geq 1 : \mathcal{T}^{(j)} \geq \nu+1\}$, the number of alarms before detection of the true change. Hence, $\mathcal{T}^{(N_\nu)}$ is the time of detection of the true change after $N_\nu - 1$ false alarms have been raised. The multicyclic detection procedure is then defined as $\mathcal{T} = \{\mathcal{T}^{(j)}\}$, and the associated average delay to detection is $\mathsf{E}_\nu[\mathcal{T}^{(N_\nu)} - \nu]$. The stationary delay to detection (STADD) is the limit $\mathsf{STADD}(\mathcal{T}) := \lim_{\nu\to\infty} \mathsf{E}_\nu[\mathcal{T}^{(N_\nu)} - \nu]$ (cf. (6.21)). As we show below, this limit always exists.

The multicyclic SR procedure is defined similarly. Specifically, let $T_{\rm SR}^{(1)}, T_{\rm SR}^{(2)}, \dots$ be sequential independent repetitions of the SR stopping time $T_{\rm SR}(A)$ defined in (8.323), and let $\mathcal{T}_{\rm SR}{}^{(j)} = T_{\rm SR}^{(1)} + T_{\rm SR}^{(2)} + \dots + T_{\rm SR}^{(j)}$ be the time of the j-th alarm in the multicyclic SR procedure,

$$T_{\rm SR}^{(i)}(A) = \inf\left\{n \ge \mathcal{T}_{\rm SR}{}^{(i-1)} + 1 : R_n^{(i)} \ge A\right\} - \mathcal{T}_{\rm SR}{}^{(i-1)},$$

where $T_{\rm SR}^{(0)} = \mathcal{T}_{\rm SR}{}^{(0)} = 0$ and

$$R_n^{(i)} = \left(1 + R_{n-1}^{(i)}\right)\frac{f_1(X_n)}{f_0(X_n)} \quad \text{for } \mathcal{T}_{\rm SR}{}^{(i-1)} + 1 \le n \le \mathcal{T}_{\rm SR}{}^{(i)}, \quad R_{\mathcal{T}_{\rm SR}{}^{(i-1)}} = 0.$$

Thus, $R_n^{(i)}$, $n \ge \mathcal{T}_{\rm SR}{}^{(i-1)} + 1$ is nothing but the SR statistic that is renewed from scratch after the $(i-1)$-th false alarm (under P_∞) and is applied to the segment of data $X_{\mathcal{T}_{\rm SR}{}^{(i-1)}+1}, X_{\mathcal{T}_{\rm SR}{}^{(i-1)}+2}, \dots$. Let $J_\nu = \min\{j \ge 1 : \mathcal{T}_{\rm SR}{}^{(j)} > \nu\}$, i.e., $\mathcal{T}_{\rm SR}{}^{(J_\nu)}$ is the time of detection of a true change after $J_\nu - 1$ false alarms have been raised. So the STADD of the multicyclic SR procedure $\mathcal{T}_{\rm SR}(A)$ is the limit

$$\mathsf{STADD}(\mathcal{T}_{\rm SR}) = \lim_{\nu\to\infty} \mathsf{E}_\nu\left[\mathcal{T}_{\rm SR}{}^{(J_\nu)} - \nu\right]. \tag{8.326}$$

The next theorem states that the multicyclic SR procedure is strictly optimal for every $\gamma > 1$ with respect to the STADD in the class of multicyclic detection procedures $\mathbf{C}_\gamma$ for which the mean time between false alarms is not less than γ, i.e., it solves exactly the optimization problem (6.22). Recall that by $\mathsf{IADD}(T)$ we denote the integral average delay to detection,

$$\mathsf{IADD}(T) = \frac{\sum_{\nu=0}^\infty \mathsf{E}_\nu(T-\nu)^+}{\mathsf{E}_\infty T}.$$

Theorem 8.4.1. *Let $T_{\rm SR}(A_\gamma)$ be the stopping time defined in* (8.323) *with the threshold $A = A_\gamma$ selected so that $\mathsf{E}_\infty[T_{\rm SR}(A_\gamma)] = \gamma$, $\gamma > 1$. Let $\mathcal{T}_{\rm SR}(A_\gamma)$ be the corresponding multicyclic SR procedure for which the mean time between false alarms is exactly equal to γ, $\mathsf{E}_\infty[T_{\rm SR}^{(i)}(A_\gamma)] = \gamma$ for $i \ge 1$. Suppose a detection procedure T with $\mathsf{E}_\infty T \ge \gamma$ is applied repeatedly. Let $\mathsf{STADD}(\mathcal{T})$ be the stationary average delay to detection of this multicyclic procedure. Then $\mathsf{STADD}(\mathcal{T})$ exists, is equal to $\mathsf{IADD}(T)$ and $\mathsf{STADD}(\mathcal{T}_{\rm SR}) \le \mathsf{STADD}(\mathcal{T})$. That is, the multicyclic SR procedure is optimal for every $\gamma > 1$ in the class $\mathbf{C}_\gamma = \{T : \mathsf{E}_\infty T \ge \gamma\}$.*

Proof. Note that $\mathcal{T}^{(N_\nu-1)}$ is the time interval until the last false alarm. By renewal theory, the distribution of $\nu - \mathcal{T}^{(N_\nu-1)}$ has a limit

$$\lim_{\nu\to\infty} \mathsf{P}_\nu\left(\nu - \mathcal{T}^{(N_\nu-1)} = k\right) = \frac{\mathsf{P}_\infty(T > k)}{\sum_{j=1}^\infty \mathsf{P}_\infty(T \ge j)} \quad \text{for } k = 0, 1, 2, \dots \tag{8.327}$$

See, e.g., Feller [140, page 356]. Clearly, when conditioning on $\nu - \mathcal{T}^{(N_\nu-1)} = k$, the observations $X_{\mathcal{T}^{(N_\nu-1)}+1}, X_{\mathcal{T}^{(N_\nu-1)}+2}, \dots, X_\nu, X_{\nu+1}, \dots$ behave exactly like $X_1, X_2, \dots, X_\nu, X_{\nu+1}, \dots$ when $\nu = k$. Therefore, by conditioning on $\nu - \mathcal{T}^{(N_\nu-1)}$, using (8.327) and letting T be independent of

$T^{(1)}, T^{(2)}, \dots$, we obtain

$$
\begin{aligned}
\mathsf{E}_\nu\left[\mathcal{T}^{(N_\nu)} - \nu\right] &= \mathsf{E}_\nu\left[\mathsf{E}_\nu\left(\mathcal{T}^{(N_\nu)} - \nu | \nu - \mathcal{T}^{(N_\nu - 1)}\right)\right] \\
&= \sum_{k=0}^{\nu} \mathsf{E}_k\left(T - k | \nu - \mathcal{T}^{(N_\nu - 1)} = k, T > k\right) \mathsf{P}_\infty\left(\nu - \mathcal{T}^{(N_\nu - 1)} = k\right) \\
&= \sum_{k=0}^{\nu} \mathsf{E}_k\left(T - k | T > k\right) \mathsf{P}_\infty\left(\nu - \mathcal{T}^{(N_\nu - 1)} = k\right) \\
&\xrightarrow[\nu \to \infty]{} \frac{\sum_{k=0}^{\infty} \mathsf{E}_k\left(T - k | T > k\right) \mathsf{P}_\infty\left(T > k\right)}{\sum_{j=1}^{\infty} \mathsf{P}_\infty\left(T \ge j\right)} \\
&= \frac{\sum_{k=0}^{\infty} \mathsf{E}_k (T-k)^+}{\mathsf{E}_\infty T} = \mathsf{IADD}(T),
\end{aligned}
$$

which proves that $\mathsf{STADD}(\mathcal{T})$ exists and equals $\mathsf{IADD}(T)$.

The same argument yields $\mathsf{STADD}(\mathcal{T}_{\text{SR}}(A_\gamma)) = \mathsf{IADD}(T_{\text{SR}}(A_\gamma))$. By Theorem 7.6.1, $\mathsf{IADD}(T_{\text{SR}}(A_\gamma)) \le \mathsf{IADD}(T)$ for any $T \in \mathbf{C}_\gamma$, which completes the proof. □

In Subsection 8.4.3.2, we will prove that if Z_1 is non-arithmetic and $\mathsf{E}_0|Z_1|^2 < \infty$, then for large threshold values ($A \to \infty$)

$$
\mathsf{IADD}(T_{\text{SR}}(A)) = I^{-1}(\log A + \varkappa - C) + o(1), \quad \mathsf{ARL2FA}(T_{\text{SR}}(A)) \sim A/\zeta,
$$

where the constants $\zeta, \varkappa$ associated with the overshoot over A are defined in (8.355) and the constant C is defined in (8.364); see Theorems 8.4.4 and 8.4.5. Therefore, asymptotically as $\gamma \to \infty$, $A_\gamma = \zeta\gamma$ implies that $\mathsf{ARL2FA}(T_{\text{SR}}(A_\gamma)) \sim \gamma$ and the minimal stationary average delay to detection is approximated as

$$
\mathsf{STADD}(T_{\text{SR}}(A_\gamma)) = I^{-1}(\log\gamma + \log\zeta + \varkappa - C) + o(1).
$$

Completing this subsection we remark that the low cost of false alarms may occur only if there is an additional algorithm that allows for false alarm filtering. This is often the case in applications. An example related to rapid detection of intrusions in computer networks when frequent false alarms can be tolerated is considered in Section 11.3.

8.4.2 *The Shiryaev–Roberts–Pollak Procedure*

Consider now Pollak's minimax problem setting (6.20). Note right away that in general an optimal solution T_{opt} to this problem is not known. In this and the next subsections we are interested in finding detection procedures that are nearly minimax (to third order) for a low false alarm rate, specifically

$$
\inf_{T \in \mathbf{C}_\gamma} \mathsf{SADD}(T) = \mathsf{SADD}(T_{\text{o}}) + o(1) \quad \text{as } \gamma \to \infty.
$$

Recall that the conventional SR statistic given by the recursion (8.325) starts from zero. The conventional SR procedure is only second-order asymptotically optimal,

$$
\inf_{T \in \mathbf{C}_\gamma} \mathsf{SADD}(T) = \mathsf{SADD}(T_{\text{SR}}) + O(1) \quad \text{as } \gamma \to \infty,
$$

which is established in (8.373) below. The same is true for the CUSUM procedure.

It turns out that by tweaking the SR procedure with the head start, i.e., starting it off at a specially designed point, it may be made almost optimal in the minimax sense. In this subsection, we consider a random initialization proposed by Pollak [365] and in the next subsection a deterministic initialization proposed by Moustakides *et al.* [309, 310].

Let Q_A denote the *quasi-stationary distribution* of the SR statistic R_n, defined by

$$\mathsf{Q}_A(x) := \lim_{n\to\infty} \mathsf{P}_\infty\{R_n \le x | R_1 < A, \dots, R_n < A\} = \lim_{n\to\infty} \mathsf{P}_\infty\{R_n \le x | T_{\text{SR}}(A) > n\}. \tag{8.328}$$

Introduce the "randomized" SR statistic

$$R_n^{\mathsf{Q}_A} = (1 + R_{n-1}^{\mathsf{Q}_A})\mathcal{L}_n, \quad n \ge 1, \quad R_0^{\mathsf{Q}_A} \sim \mathsf{Q}_A \tag{8.329}$$

that starts off at a random point $R_0^{\mathsf{Q}_A}$ sampled from this distribution and the corresponding stopping time

$$T_{\text{SRP}}(A) = \inf\{n \ge 1 : R_n^{\mathsf{Q}_A} \ge A\}. \tag{8.330}$$

We will refer to this changepoint detection procedure as the *Shiryaev–Roberts–Pollak* (SRP) procedure.

This procedure has an important property: it is an equalizer rule in the sense that the conditional average delay to detection is constant in the changepoint,

$$\mathsf{E}_\nu(T_{\text{SRP}} - \nu | T_{\text{SRP}} > \nu) = \mathsf{E}_0[T_{\text{SRP}}] \quad \text{for all } \nu = 0, 1, 2, \dots$$

While this fact can be proved rigorously, to understand why it suffices to look at Figure 8.15, which shows a typical behavior of the conditional ADD of the SR procedure *versus* the changepoint. When the changepoint increases the CADD stabilizes at a fixed level $\mathsf{CADD}_\infty(T_{\text{SR}}) = \lim_{\nu\to\infty} \mathsf{CADD}_\nu(T_{\text{SR}})$, and this is happening because the conditional distribution $\mathsf{P}_\infty\{R_\nu \le x | T_{\text{SR}}(A) > \nu\}$ attains its limiting value $\mathsf{Q}_A(x)$ for relatively large but essentially finite ν. Therefore, if we start the SR procedure with a point sampled from the quasi-stationary distribution, the CADD will be constant for all ν. The curves in Figure 8.15 are obtained using integral equations and numerical techniques presented in Subsection 8.4.3.

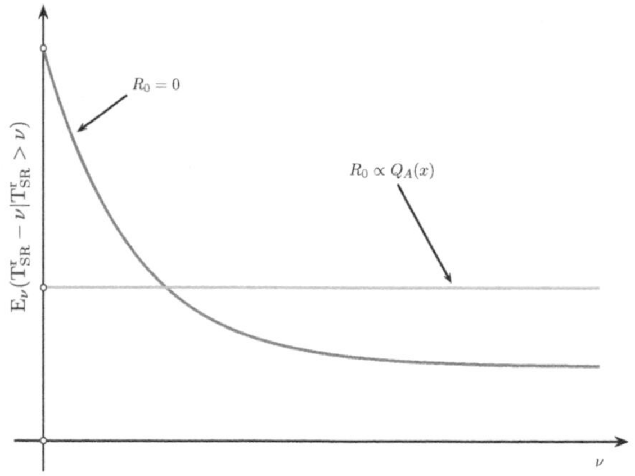

Figure 8.15: *Typical behavior of the conditional average delay to detection as a function of the changepoint* ν *for the SR and SRP procedures. Plots were obtained numerically using methods of Subsection 8.4.3.1.*

Since by the minimax theorem a minimax rule is an equalizer rule, one may conjecture that the SRP procedure is strictly minimax for every $\gamma > 1$ in the class $\mathbf{C}_\gamma$. However, as we will see, this is not true—it is almost minimax but not strictly minimax.

The following theorem, whose proof can be found in the seminal paper by Pollak [365, Theorem 2], establishes third-order asymptotic minimaxity of the SRP procedure.

Theorem 8.4.2. *Let* $T_{\text{SRP}}(A)$ *be defined as in* (8.329)–(8.330). *Assume that the* P_∞*-distribution of the LLR* $Z_1 = \log \mathcal{L}_1$ *is non-arithmetic and that* $\mathsf{E}_0(Z_1)^+ < \infty$. *Then for every* $\gamma > 1$ *there exists* *a positive and finite value of* $A = A_\gamma$ *such that* $\mathsf{E}_\infty[T_{\text{SRP}}(A_\gamma)] = \gamma$*, and with this selection of the threshold* A_γ *the SRP procedure is asymptoticaly minimax to third order,*

$$\inf_{T \in \mathbf{C}_\gamma} \mathsf{SADD}(T) = \mathsf{SADD}(T_{\text{SRP}}(A_\gamma)) + o(1) \quad \textit{as } \gamma \to \infty.$$

Note that this result is extremely strong since $\mathsf{SADD}(T_{\text{SRP}}(A_\gamma))$ is of order $O(\log \gamma)$ while the residual term $o(1)$ vanishes.

The assertions of Theorem 8.4.2 also hold in the arithmetic case. However, in order to guarantee the existence of the threshold A_γ such that the ARL2FA constraint $\mathsf{E}_\infty[T_{\text{SRP}}(A_\gamma)] = \gamma$ would satisfy with equality we need to add the possibility of randomizing at the stopping boundary, i.e., whether to stop or continue sampling when $R_n^{\mathsf{Q}_A} = A$.

The above optimality result requires minimal conditions on the models, e.g., the first moment condition $\mathsf{E}_0|Z_1| < \infty$ (natural for any asymptotic problem) is sufficient. Imposing a stronger second moment condition $\mathsf{E}_0|Z_1|^2 < \infty$ allows us to obtain accurate asymptotic approximations for operating characteristics. The results are given in Subsection 8.4.3.2; see (8.357) in Theorem 8.4.4 and (8.366) in Theorem 8.4.5.

8.4.3 *The Shiryaev–Roberts-r Procedure*

Consider a modification of the SR procedure by initializing the SR statistic not from zero but from any deterministic value $R_0 = r \geq 0$:

$$R_n^r = \left(1 + R_{n-1}^r\right) \mathcal{L}_n, \;\; R_0^r = r, \quad T_{\text{SR}}^r = \inf\{n \geq 1 : R_n^r \geq A\}. \tag{8.331}$$

We refer to this procedure as the *SR-r changepoint detection procedure*. Obviously, to guarantee the given ARL2FA γ the head start r and the threshold A are related via the equation $\mathsf{E}_\infty[T_{\text{SR}}^r(A)] = \gamma$. The goal is to find a head start r that minimizes the SADD (ideally) or at least makes it reasonably small, so that the SR-r procedure would compete with the almost optimal SRP procedure.

8.4.3.1 *SR-r Design*

The idea of the head start optimization is to compare the performance of the SR-r procedure with the lower bound and minimize the discrepancy. The following theorem gives this lower bound.

Theorem 8.4.3. *Let* $T_{\text{SR}}^r(A)$ *be defined as in* (8.331) *and let* $A = A_\gamma$ *be selected so that* $\mathsf{E}_\infty[T_{\text{SR}}^r(A_\gamma)] = \gamma$*. Then for every* $r \geq 0$

$$\inf_{T \in \mathbf{C}_\gamma} \mathsf{SADD}(T) \geq \frac{r\mathsf{E}_0[T_{\text{SR}}^r(A_\gamma)] + \sum_{\nu=0}^\infty \mathsf{E}_\nu[T_{\text{SR}}^r(A_\gamma) - \nu]^+}{r + \mathsf{E}_\infty[T_{\text{SR}}^r(A_\gamma)]}. \tag{8.332}$$

Proof. For any stopping time T

$$\begin{aligned}\sum_{\nu=0}^\infty \mathsf{E}_\nu(T-\nu)^+ &= \sum_{\nu=0}^\infty \mathsf{P}_\nu(T>\nu)\mathsf{E}_\nu(T-\nu|T>\nu) \\ &= \sum_{\nu=0}^\infty \mathsf{P}_\infty(T>\nu)\mathsf{E}_\nu(T-\nu|T>\nu),\end{aligned}$$

where we used the fact that $\mathsf{P}_\nu(T>\nu) = \mathsf{P}_\infty(T>\nu)$. Since $\mathsf{SADD}(T) \geq \mathsf{E}_\nu(T-\nu|T>\nu)$ for any $\nu \geq 0$ and

$$\begin{aligned}\mathsf{SADD}(T) &= \frac{\mathsf{SADD}(T)[r + \sum_{\nu=0}^\infty \mathsf{P}_\infty(T>\nu)]}{r + \sum_{\nu=0}^\infty \mathsf{P}_\infty(T>\nu)} \\ &= \frac{r\mathsf{SADD}(T) + \sum_{\nu=0}^\infty \mathsf{SADD}(T)\mathsf{P}_\infty(T>\nu)}{r + \sum_{\nu=0}^\infty \mathsf{P}_\infty(T>\nu)},\end{aligned}$$

where $\sum_{\nu=0}^{\infty} \mathsf{P}_\infty(T > \nu) = \mathsf{E}_\infty T$, we obtain that for any stopping time T with finite ARL to false alarm

$$\begin{aligned}\mathsf{SADD}(T) &\geq \frac{r\mathsf{E}_0 T + \sum_{\nu=0}^{\infty} \mathsf{E}_\nu(T-\nu|T>\nu)\mathsf{P}_\infty(T>\nu)}{r+\mathsf{E}_\infty T} \\ &= \frac{r\mathsf{E}_0 T + \sum_{\nu=0}^{\infty} \mathsf{E}_\nu(T-\nu)^+}{r+\mathsf{E}_\infty T} := \mathsf{IADD}_r(T).\end{aligned}$$

Therefore,

$$\inf_{T\in\mathbf{C}_\gamma} \mathsf{SADD}(T) \geq \inf_{T\in\mathbf{C}_\gamma} \mathsf{IADD}_r(T).$$

Finally, a slight generalization of the proof of Theorem 7.6.1 for an arbitrary r shows that $\inf_{T\in\mathbf{C}_\gamma} \mathsf{IADD}_r(T) = \mathsf{IADD}_r(T_{\mathrm{SR}}^r(A_\gamma))$, which completes the proof. □

An important observation is that if for a certain r the SR-r procedure is an equalizer rule, i.e., $\mathsf{E}_\nu(T_{\mathrm{SR}}^r - \nu | T_{\mathrm{SR}}^r > \nu)$ is constant for all $\nu \geq 0$, then it is optimal since the right-hand side in (8.332) is equal to $\mathsf{E}_0 T_{\mathrm{SR}}^r$ which in turn is equal to $\sup_{\nu\geq 0} \mathsf{E}_\nu(T_{\mathrm{SR}}^r - \nu | T_{\mathrm{SR}}^r > \nu) = \mathsf{SADD}(T_{\mathrm{SR}}^r)$. This observation is used below in Example 8.4.1 for proving that the SR-r procedure with a specially designed $r = r_A$ is strictly optimal for an exponential model, but the SRP procedure is not optimal.

Recall that in Subsection 8.2.6.1 we derived a system of Fredholm integral equations for various characteristics of a quite general change detection procedure (8.103) in the parametric case when the pre-change parameter is θ_0 and a post-change parameter θ can be arbitrary, not necessarily equal to the putative value θ_1. The corresponding equations (8.105)–(8.109) hold for the SR-r procedure if we set $\Phi(r) = 1 + r$. For the purposes of this subsection, let us set $\theta = \theta_1$ and omit both θ_1 and θ_0 making the results more general (covering not necessarily parametric models), which also simplifies the notation.

Specifically, we write $F_\infty(x) = \mathsf{P}_\infty(\mathcal{L}_1 \leq x)$, $F_0(x) = \mathsf{P}_0(\mathcal{L}_1 \leq x)$ for the corresponding distributions of the LR, assuming that they are continuous;

$$\begin{gathered}\delta_{\nu,r} = \mathsf{E}_\nu(T_{\mathrm{SR}}^r - \nu)^+; \; \rho_{\nu,r} = \mathsf{P}_\infty(T_{\mathrm{SR}}^r > \nu), \quad \nu \geq 0 \; (\rho_{0,r} = 1); \\ \mathsf{ARL}_r = \mathsf{E}_\infty[T_{\mathrm{SR}}^r]; \; \psi_r = \sum_{\nu=0}^{\infty} \mathsf{E}_\nu(T_{\mathrm{SR}}^r - \nu)^+, \\ \mathcal{K}_\infty(x,r) = \frac{\partial}{\partial x} F_\infty\left(\frac{x}{1+r}\right), \quad \mathcal{K}_0(x,r) = \frac{\partial}{\partial x} F_0\left(\frac{x}{1+r}\right), \quad x, r \in [0, A).\end{gathered}$$

Then the equations (8.105)–(8.109) become

$$\mathsf{ARL}_r = 1 + \int_0^A \mathsf{ARL}_x \mathcal{K}_\infty(x,r)\,\mathrm{d}x, \tag{8.333}$$

$$\delta_{0,r} = 1 + \int_0^A \delta_{0,x} \mathcal{K}_0(x,r)\,\mathrm{d}x, \tag{8.334}$$

$$\delta_{\nu,r} = \int_0^A \delta_{\nu-1,x} \mathcal{K}_\infty(x,r)\,\mathrm{d}x, \quad \nu = 1,2,\ldots, \tag{8.335}$$

$$\rho_{\nu,r} = \int_0^A \rho_{\nu-1,x} \mathcal{K}_\infty(x,r)\,\mathrm{d}x, \quad \nu = 1,2,\ldots, \tag{8.336}$$

$$\psi_r = \mathsf{ARL}_r + \int_0^A \psi_x \mathcal{K}_\infty(x,r)\,\mathrm{d}x. \tag{8.337}$$

The CADD and IADD are computed as

$$\mathsf{E}_\nu(T_{\mathrm{SR}}^r - \nu | T_{\mathrm{SR}}^r > \nu) = \frac{\delta_{\nu,r}}{\rho_{\nu,r}}, \quad \nu \geq 0, \quad \mathsf{IADD}_r(T_{\mathrm{SR}}^r) = \frac{r\delta_{0,r} + \psi_r}{r + \mathsf{ARL}_r}. \tag{8.338}$$

The CADD at infinity $\mathsf{CADD}_\infty(T_{\rm SR}^r) = \lim_{\nu\to\infty} \mathsf{E}_\nu(T_{\rm SR}^r - \nu | T_{\rm SR}^r > \nu)$ is computed as

$$\mathsf{CADD}_\infty(T_{\rm SR}^r) = \int_0^A \delta_{0,x}\, q_A(x)\, \mathrm{d}x,$$

where density $q_A(x)$ of the quasi-stationary distribution $\mathsf{Q}_A(x)$ satisfies the integral equation

$$\lambda_A\, q_A(x) = \int_0^A q_A(r)\, \mathcal{K}_\infty(x,r)\, \mathrm{d}r \tag{8.339}$$

(λ_A is the leading eigenvalue of the linear operator associated with the kernel $\mathcal{K}_\infty(x,y)$).

The ARL2FA and SADD of the SRP procedure are computed as

$$\mathsf{E}_\infty\left[T_{\rm SRP}(A)\right] = \int_0^A \mathsf{ARL}_r\, q_A(r)\, \mathrm{d}r, \quad \mathsf{SADD}(T_{\rm SRP}) = \int_0^A \delta_{0,r}\, q_A(r)\, \mathrm{d}r. \tag{8.340}$$

Example 8.4.1 (Exponential model). Consider the exponential model with the pre-change mean 1 and the post-change mean 1/2, i.e., $f_0(x) = e^{-x}\mathbb{1}_{\{x\ge 0\}}$ and $f_1(x) = 2e^{-2x}\mathbb{1}_{\{x\ge 0\}}$. In the sequel we assume that the thresholds in the SR-r and SRP procedures do not exceed 2.

We now show that if in the SR-r procedure $T_{\rm SR}^r$ the initializing value is chosen as $r = r_A = \sqrt{1+A}-1$ and the threshold $A = A_\gamma$ is selected from the transcendental equation

$$A + (\gamma-1)\sqrt{1+A}\log(1+A) - 2(\gamma-1)\sqrt{1+A} = 0, \tag{8.341}$$

then, for every $1 < \gamma < \gamma_0 = (1-0.5\log 3)^{-1} \approx 2.2188$, $\mathsf{ARL2FA}(T_{\rm SR}^{r_A}(A_\gamma)) = \gamma$, and the SR-$r$ procedure is strictly minimax,

$$\mathsf{SADD}(T_{\rm SR}^r) = \inf_{T\in\mathbf{C}_\gamma} \mathsf{SADD}(T). \tag{8.342}$$

However, the SRP procedure is suboptimal, i.e., $\mathsf{SADD}(T_{\rm SRP}(A_*)) > \mathsf{SADD}(T_{\rm SR}^{r_A}(A))$ for all $1 < \gamma < \gamma_0$.

The distribution $F_\infty(x)$ of the LR $\mathcal{L}_1 = 2e^{-X_1}$ is uniform on $[0,2]$ and $F_0(x) = x^2/4$ for $x \in [0,2]$, so the kernels $\mathcal{K}_\infty$ and $\mathcal{K}_0$ are

$$\mathcal{K}_\infty(x,r) = \frac{1}{2(1+r)}\mathbb{1}_{\{0\le x\le 2\}}, \quad \mathcal{K}_0(x,r) = \frac{x}{2(1+r)^2}\mathbb{1}_{\{0\le x\le 2\}}. \tag{8.343}$$

Consider first the SRP procedure $T_{\rm SRP}(A_*)$. As it will become apparent later, the threshold $A_* < 2$ when $\gamma < \gamma_0$. By (8.339) and (8.343), for $A_* < 2$ the quasi-stationary density $q_{A_*}(x) = \mathrm{d}\mathsf{Q}_{A_*}(x)/\mathrm{d}x$ satisfies the integral equation

$$\lambda_{A_*}\, q_{A_*}(x) = \frac{1}{2}\int_0^{A_*} q_{A_*}(r)\, \frac{1}{1+r}\, \mathrm{d}r,$$

which due to the constraint $\int_0^{A_*} q_{A_*}(x)\,\mathrm{d}x = 1$ yields

$$\lambda_{A_*} = \frac{1}{2}\log(1+A_*) \quad \text{and} \quad q_{A_*}(x) = A_*^{-1}\mathbb{1}_{\{x\in[0,A_*)\}}.$$

Thus, for $A_* < 2$ the quasi-stationary distribution $\mathsf{Q}_{A_*}(x) = x/A_*$ is uniform and, moreover, it is attained already for $n = 1$ when the very first observation becomes available.

Clearly, the P_∞-distribution of the SRP stopping time $T_{\rm SRP}$ is geometric with parameter $1-\lambda_{A_*}$, so

$$\mathsf{E}_\infty[T_{\rm SRP}(A_*)] = \frac{1}{1-\lambda_{A_*}} = \frac{1}{1-\frac{1}{2}\log(1+A_*)}. \tag{8.344}$$

It follows that $\mathsf{E}_\infty[T_{\mathrm{SRP}}(A_*)] = \gamma$ when the threshold A_* is chosen as

$$A_* = \exp\left\{\frac{2(\gamma-1)}{\gamma}\right\} - 1 \tag{8.345}$$

and that $A_* < 2$ whenever $\gamma < \gamma_0$.

By (8.340), the maximal average detection delay of the SRP procedure is equal to

$$\mathsf{SADD}(T_{\mathrm{SRP}}(A_*)) = \frac{1}{A_*}\int_0^{A_*} \delta_{0,r}\,\mathrm{d}r, \tag{8.346}$$

so we need to compute the ARL to detection $\delta_{0,r} = \mathsf{E}_0 T^r_{\mathrm{SR}}$ of the SR-r procedure, which also has to be computed for the evaluation of the performance of the SR-r procedure itself.

Assume that $A < 2$. By (8.334) and (8.343), we have

$$\delta_{0,r} = 1 + \frac{1}{2(1+r)^2}\int_0^A \delta_0(x)\,x\,\mathrm{d}x,$$

so that

$$\begin{aligned}\int_0^A \delta_0(r)\,r\,\mathrm{d}r &= \int_0^A r\,\mathrm{d}r + \frac{1}{2}\left[\int_0^A \frac{x\,\mathrm{d}x}{(1+x)^2}\right]\left[\int_0^A \delta_0(r)\,r\,\mathrm{d}r\right] \\ &= \frac{A^2}{2} + \frac{1}{2}\left[\log(1+A) - \frac{A}{1+A}\right]\left[\int_0^A \delta_0(r)\,r\,\mathrm{d}r\right],\end{aligned}$$

which implies that

$$\int_0^A r\delta_{0,r}\,\mathrm{d}r = A^2\left[\frac{A}{1+A} + 2\left(1 - \frac{1}{2}\log(1+A)\right)\right]^{-1}.$$

Consequently,

$$\delta_{0,r} = 1 + \frac{A^2}{2(1+r)^2}\left[\frac{A}{1+A} + 2\left(1 - \frac{1}{2}\log(1+A)\right)\right]^{-1}. \tag{8.347}$$

Using (8.346) and (8.347), we find

$$\mathsf{SADD}(T_{\mathrm{SRP}}(A_*)) = \bar{\delta}_0(A_*), \tag{8.348}$$

where

$$\bar{\delta}_0(y) = 1 + \frac{y^2}{2(1+y)}\left[\frac{y}{1+y} + 2\left(1 - \frac{1}{2}\log(1+y)\right)\right]^{-1}. \tag{8.349}$$

Consider now the SR-r procedure. By (8.333) and (8.343), the ARL to false alarm $\mathsf{ARL}_r = \mathsf{E}_\infty[T^r_{\mathrm{SR}}(A)]$ satisfies the integral equation

$$\mathsf{ARL}_r = 1 + \frac{1}{2(1+r)}\int_0^A \mathsf{ARL}_x\,\mathrm{d}x,$$

which yields

$$\mathsf{ARL}_r = 1 + \frac{A}{2(1+r)}\left[1 - \frac{1}{2}\log(1+A)\right]^{-1}. \tag{8.350}$$

Recall that for $A < 2$ the statistic R^r_n already kicks in the uniform quasi-stationary distribution for $n = 1$ and any $0 \le r < A$, so that T^r_{SR} is an equalizer for $\nu \ge 1$ and any $r \in [0, A)$, i.e., $\delta_{\nu,r} = \bar{\delta}_0(A)$ for all $\nu \ge 1$ and $r < A$ with $\bar{\delta}_0(A)$ given by (8.349). This implies that

$$\mathsf{SADD}(T^r_{\mathrm{SR}}) = \max\left\{\bar{\delta}_0(A), \delta_{0,r}\right\}. \tag{8.351}$$

Let $r = r_A = \sqrt{1+A} - 1$, in which case $\bar{\delta}_0(A) = \delta_{0,r_A}$, i.e., for this value of the head start the SR-r procedure is an equalizer rule for all $\nu \geq 0$. Therefore, by Theorem 8.4.3 the procedure T^r_{SR} that starts from the deterministic point $r = \sqrt{1+A} - 1$ is strictly minimax, and (8.342) holds if the threshold $A = A_\gamma$ is selected so that $\mathsf{E}_\infty T^r_{\text{SR}} = \gamma$. Substituting $r = \sqrt{1+A} - 1$ in (8.350) and equalizing the result to γ, yields transcendental equation (8.341). It is easily verified that $A_\gamma < 2$ for $\gamma < \gamma_0$. Thus, the SR-r procedure is minimax for all $1 < \gamma < \gamma_0$.

In order to show that for every given $\gamma \in (1, \gamma_0)$ the SRP procedure is inferior it suffices to show that $\mathsf{E}_\infty[T^{r_A}_{\text{SR}}(A)] > \mathsf{E}_\infty[T_{\text{SRP}}(A)]$. By (8.350),

$$\mathsf{E}_\infty[T^{r_A}_{\text{SR}}(A)] = \mathsf{ARL}_{r_A} = 1 + \frac{A}{2\sqrt{A+1}}\left[1 - \frac{1}{2}\log(1+A)\right]^{-1}. \tag{8.352}$$

Comparing (8.352) with (8.344), we obtain that we have only to show that

$$1 + \frac{A}{2\sqrt{A+1}}\left[1 - \frac{1}{2}\log(1+A)\right]^{-1} > \left[1 - \frac{1}{2}\log(1+A)\right]^{-1},$$

i.e., that $A/\sqrt{A+1} > \log(A+1)$, which holds for any $A > 0$. Thus, it follows that the SRP procedure is suboptimal.

Let, for example, $\gamma = 2$. Then, by (8.345) and (8.348), $A_* = e - 1 \approx 1.71828$ and $\mathsf{SADD}(T_{\text{SRP}}(A_*)) \approx 1.33275$. For $\gamma = 2$, solving the transcendental equation (8.341) yields $A \approx 1.66485$ and the initialization point $r_A \approx 0.63244$. By (8.351), $\mathsf{SADD}(T^{r_A}_{\text{SR}}(A)) \approx 1.31622$.

We also note that the results analogous to that in this example hold whenever the distributions $F_i(y)$, $i = \infty, 0$ of the LR are such that $F_i(x/(1+r)) = \phi_i(x)\tilde{\phi}(r)$ for some functions ϕ_i and $\tilde{\phi}_i$. In this case kernels $\mathcal{K}_i(x,r)$ are also separable, $\mathcal{K}_i(x,r) = \phi_i'(x)\tilde{\phi}_i(r)$.

We conclude this example with a remark concerning exact optimality of the SR–r procedure in the class $\mathbf{C}^m_\alpha = \{T\colon \sup_{k\geq 0} \mathsf{P}_\infty(k < T \leq k+m | T > k) \leq \alpha\}$, where $\alpha \in (0,1)$ and $m \geq 1$. We first discussed this class in Subsection 6.3.3, where we mentioned that in general it is more stringent than the class $\mathbf{C}_\gamma$. It can be easily verified that the P_∞-distribution of the SR-r stopping time $T^r_{\text{SR}}(A)$ is zero-modified geometric:

$$\mathsf{P}_\infty\{k < T^r_{\text{SR}}(A) \leq k+m | T^r_{\text{SR}}(A) > k\} = 1 - \begin{cases} \left[\frac{1}{2}\log(1+A)\right]^m & \text{for } k \geq 1; \\ \frac{A}{2(1+r)}\left[\frac{1}{2}\log(1+A)\right]^{m-1} & \text{for } k = 0. \end{cases}$$

Thus, in this case, there is a one-to-one correspondence between the classes $\mathbf{C}^m_\alpha$ and $\mathbf{C}_\gamma$. As a result, the SR-r procedure is minimax in the class $\mathbf{C}^m_\alpha$ as well.

In general, however, the SR-r procedure is not strictly optimal for all $\gamma > 1$. As we show below, it is third-order asymptotically minimax (as the SRP procedure). Furthermore, a detailed numerical study performed in the works of Moustakides *et al.* [309, 310] shows that for certain values of the starting point, $R^r_0 = r$, $\mathsf{CADD}_\nu(T^r_{\text{SR}}(A))$ is strictly less than $\mathsf{CADD}(T_{\text{SRP}}(A_*)$ for all $\nu \geq 0$, where A and A_* are such that $\mathsf{E}_\infty[T^r_{\text{SR}}(A)] = \mathsf{E}_\infty[T_{\text{SRP}}(A_*)]$, although the maximal expected delay is only slightly smaller for the SR-r procedure. Figure 8.16 illustrates this point. The plots in this figure are obtained solving the above integral equations for operating characteristics numerically.

Therefore, we should be able to design the initialization point $r = r(\gamma)$ in the SR-r procedure so that this procedure is also third-order asymptotically optimal as the SRP procedure. In this respect, the average delay to detection at infinity $\mathsf{CADD}_\infty(T^r_{\text{SR}}) = \lim_{\nu\to\infty} \mathsf{E}_\nu[T^r_{\text{SR}} - \nu | T^r_{\text{SR}} > \nu]$ plays the critical role. To understand why, let us look at Figure 8.16, which shows the conditional average delay to detection of the SR-r procedure *versus* ν for several initialization values $R^r_0 = r$. For $r = 0$, this is the classical SR procedure (with $R_0 = 0$) whose CADD is monotonically decreasing to its minimum (steady-state value) that is attained at infinity. Note that this steady state is attained for essentially finite values of the changepoint ν. It is seen that there exist values of $r = r_A$ that depend on the threshold A for which the worst point ν is at infinity, $\mathsf{SADD}(T^r_{\text{SR}}) = \mathsf{CADD}_\infty(T^r_{\text{SR}})$. We intend

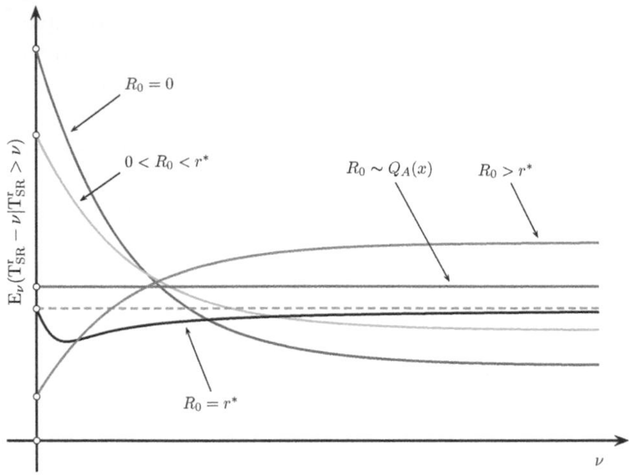

Figure 8.16: *Typical behavior of the conditional expected detection delay* $\mathsf{CADD}_\nu(T^r_{\mathrm{SR}})$ *of the SR-r procedure as a function of the changepoint* ν *for various initialization strategies.*

to find a minimal value of r_A for which this happens. We also note that for the SR-r procedure with initialization $r = r^*$ $(= r^*_A)$ the average detection delay at the beginning and at infinity are approximately equal, $\mathsf{E}_0[T^{r^*}_{\mathrm{SR}}] \approx \mathsf{CADD}_\infty(T^{r^*}_{\mathrm{SR}})$. This allows us to conjecture that an "optimal" SR-r is an equalizer at the beginning ($\nu = 0$) and at sufficiently large values of ν, so that initialization r_A should be selected to achieve this property. It is also important that this value delivers the minimum to the difference between $\mathsf{SADD}(T^r_{\mathrm{SR}})$ and the lower bound $\mathsf{IADD}_r(T^r_{\mathrm{SR}})$ for $\inf_{T \in \mathbf{C}_\gamma} \mathsf{SADD}(T)$ obtained in Theorem 8.4.3,

$$r_A = \arg \inf_{0 \le r < A} \left[\mathsf{SADD}(T^r_{\mathrm{SR}}(A)) - \mathsf{IADD}_r(T^r_{\mathrm{SR}}(A))\right].$$

8.4.3.2 *Asymptotic Operating Characteristics*

Note first that solving integral equations (8.333)–(8.338) numerically allows us to compute (almost precisely) various performance measures of the SR-r procedure, such as CADD, SADD, ARL2FA, PFA. Solving equation (8.339) yields the quasi-stationary distribution, which is used in (8.340) to compute operating characteristics of the SRP procedure. However, precision of numerical techniques depends on the number of grid points, which is large for large threshold values, making numerical methods infeasible for a low FAR.

In this subsection, we provide asymptotic approximations that simplify computations for large threshold values (low FAR).

We begin with the evaluation of the ARL2FA. The following lemma provides a simple lower bound for the ARL2FA of the SR-r procedure, which holds not only in the iid case but for a general non-iid model as well.

Lemma 8.4.1. *Consider the general non-iid model* (6.2). *For any* $A > 0$, *the ARL to false alarm of the SR-r procedure* $T^r_{\mathrm{SR}}(A)$ *satisfies the inequality*

$$\mathsf{ARL2FA}(T^r_{\mathrm{SR}}(A)) \ge A - r. \tag{8.353}$$

Therefore, if $A = A_\gamma$ *is found from the equation*

$$A - r_A = \gamma, \tag{8.354}$$

then $T^r_{\mathrm{SR}}(A_\gamma) \in \mathbf{C}_\gamma$.

Proof. Since $\mathsf{E}_\infty[\mathcal{L}_n^{(k)} \mid \mathscr{F}_{n-1}] = 1$, it is easy to see that $\mathsf{E}_\infty(R_n^r \mid \mathscr{F}_{n-1}) = 1 + R_{n-1}^r$, $n \geq 1$. Thus, the statistic $\{R_n^r - r - n\}_{n\geq 1}$ is a zero-mean $(\mathsf{P}_\infty, \mathscr{F}_n)$–martingale. To avoid triviality assume that $\mathsf{E}_\infty[T_{\text{SR}}^r(A)] < \infty$. Then $\mathsf{E}_\infty(R_{T_{\text{SR}}^r}^r - T_{\text{SR}}^r - r)$ exists and since $0 \leq R_n^r < A$ on the event $\{T_{\text{SR}}^r > n\}$, it follows that

$$\liminf_{n\to\infty} \int_{\{T_{\text{SR}}^r > n\}} |R_n^r - n - r| \, \mathrm{d}\mathsf{P}_\infty = 0.$$

Hence the optional sampling theorem applies to yield $\mathsf{E}_\infty R_{T_{\text{SR}}^r}^r - r = \mathsf{E}_\infty T_{\text{SR}}^r$, and since $R_{T_{\text{SR}}^r}^r \geq A$, we obtain inequality (8.353). □

We now improve this result for the iid model. Recall that $Z_i = \log \mathcal{L}_i$ denotes the LLR for the i-th observation and let $S_n = Z_1 + \cdots + Z_n$. Introduce a one-sided stopping time

$$\tau_a = \inf\{n \geq 1 \colon S_n \geq a\}, \quad a > 0.$$

Let $\kappa_a = S_{\tau_a} - a$ be an overshoot (excess over the level a at stopping), and let

$$\zeta = \lim_{a\to\infty} \mathsf{E}_0[e^{-\kappa_a}], \quad \varkappa = \lim_{a\to\infty} \mathsf{E}_0 \kappa_a. \tag{8.355}$$

The constants $0 < \zeta \leq 1$ and $\varkappa \geq 0$ depend on the model and are the subject of renewal theory.

Below without special emphasis we always assume that the LLR Z_1 is non-arithmetic under both P_0 and P_∞ and that the K–L number $I = \mathsf{E}_0 Z_1$ is positive and finite.

Theorem 8.4.4. *Assume that $r = r^*$ where r^* is either fixed or, more generally, $r^* = r_A^* \to \infty$ in such a way that $r_A^*/A \to 0$ as $A \to \infty$. Then for the SR-r procedure, uniformly in $0 \leq r \leq r_A^*$,*

$$\mathsf{ARL2FA}(T_{\text{SR}}^r(A)) = \zeta^{-1} A(1 + o(1)) \quad \text{as } A \to \infty, \tag{8.356}$$

where the constant ζ is defined in (8.355).

For the SRP procedure

$$\mathsf{ARL2FA}(T_{\text{SRP}}(A)) = \zeta^{-1} A(1 + o(1)) \quad \text{as } A \to \infty. \tag{8.357}$$

Proof. We give only a proof sketch, referring for details to Tartakovsky *et al.* [467, Theorem 3.1]. It follows from Pollak [366, Theorem 1] (see also Yakir [517, Theorem 1]) that for the SR procedure (i.e., when $r = 0$) $\mathsf{E}_\infty[T_{\text{SR}}(A)] \sim A/\zeta$ as $A \to \infty$. Since $R_n^r = re^{S_n} + R_n \geq R_n$, we obtain the upper bound

$$\mathsf{E}_\infty[T_{\text{SR}}^r(A)] \leq \zeta^{-1} A(1 + o(1)) \quad \text{for any } r \geq 0. \tag{8.358}$$

For some positive m, define $M = \inf\{n \colon re^{S_n} \geq m\}$. A quite tedious argument [467] shows that

$$\mathsf{E}_\infty[T_{\text{SR}}^r(A)] \geq \mathsf{E}_\infty[T_{\text{SR}}(A - m)] \left[1 - \mathsf{P}_\infty(M < \infty)\right].$$

Since e^{S_n} is a nonnegative P_∞-martingale with mean 1,

$$\mathsf{P}_\infty(M < \infty) = \mathsf{P}_\infty\left(\inf\{n \colon e^{S_n} \geq m/r\} < \infty\right) < r/m, \tag{8.359}$$

and we obtain

$$\begin{aligned} \mathsf{E}_\infty[T_{\text{SR}}^r(A)] &\geq \mathsf{E}_\infty[T_{\text{SR}}(A - m)](1 - r/m) = \frac{A - m}{\zeta}(1 - r/m)(1 + o(1)) \\ &= (A/\zeta)(1 - m/A)(1 - r/m)(1 + o(1)). \end{aligned} \tag{8.360}$$

Let $r_A^* \to \infty$ and $m = m_A \to \infty$ so that $r_A^*/m_A \to 0$ and $m_A/A \to 0$ (which can always be arranged). Then, uniformly in $0 \leq r \leq r_A^*$,

$$\mathsf{E}_\infty[T_{\text{SR}}^r(A)] \geq \zeta^{-1} A(1 + o(1)), \quad A \to \infty,$$

which along with the reverse inequality (8.358) proves asymptotic equality (8.356) whenever $r_A^* = o(A)$ as $A \to \infty$, and if r^* does not depend on A the result obviously holds.

Consider the SRP procedure, in which case $M = \inf\{n\colon R_0^{\mathsf{Q}_A} e^{S_n} \ge m\}$. Similar to (8.359), $\mathsf{P}_\infty(M < \infty | R_0^{\mathsf{Q}_A} = x) < x/m_A$. Hence, by conditioning on $R_0^{\mathsf{Q}_A}$, we obtain

$$\mathsf{P}_\infty(M < \infty) = \int_0^A \mathsf{P}_\infty(M < \infty | R_0^{\mathsf{Q}_A} = x)\,\mathrm{d}\mathsf{Q}_A(x) \le \frac{1}{m_A}\int_0^A x\,\mathrm{d}\mathsf{Q}_A(x) = \frac{\mu_A}{m_A},$$

where $\mu_A = \mathsf{E}R_0^{\mathsf{Q}_A} = \int_0^A x\,\mathrm{d}\mathsf{Q}_A(x)$ is the mean of the quasi-stationary distribution. Let $\mathsf{Q}_{\mathrm{st}}(x) = \lim_{n\to\infty} \mathsf{P}_\infty(R_n \le x)$ denote stationary distribution of the Markov process R_n. By Kesten [228, Theorem 5],

$$1 - \mathsf{Q}_{\mathrm{st}}(x) \sim 1/x \quad \text{as } x \to \infty, \tag{8.361}$$

which along with the fact that $\mathsf{Q}_A(x) \ge \mathsf{Q}_{\mathrm{st}}(x)$ yields

$$\mu_A = \int_0^A [1 - \mathsf{Q}_A(x)]\,\mathrm{d}x \le \int_0^A [1 - \mathsf{Q}_{\mathrm{st}}(x)]\,\mathrm{d}x = O(\log A). \tag{8.362}$$

Hence, $\mu_A/m_A \le O(m_A^{-1}\log A)$ and to obtain (8.357) it suffices to take $m_A = A^{1/2}$ (say). □

Using (8.356)–(8.357) along with the fact that $\mathsf{E}_\infty[T_{\mathrm{SR}}^r(A)] = \mathsf{E}_\infty R_{T_{\mathrm{SR}}^r(A)} - r$ and $\mathsf{E}_\infty[T_{\mathrm{SRP}}(A)] = \mathsf{E}_\infty R_{T_{\mathrm{SRP}}(A)} - \mu_A$, for practical purposes we suggest the following approximations

$$\mathsf{ARL2FA}(T_{\mathrm{SR}}^r(A)) \approx \zeta^{-1}A - r, \quad \mathsf{ARL2FA}(T_{\mathrm{SRP}}(A)) \approx \zeta^{-1}A - \mu_A. \tag{8.363}$$

To continue with asymptotic approximations for ADD, we need the following additional notation: $V_\infty = \sum_{n=1}^\infty e^{-S_n}$, $\widetilde{\mathsf{Q}}(y) = \mathsf{P}_0(V_\infty \le y)$,

$$C = \mathsf{E}_\infty\{\mathsf{E}_0[\log(1 + R_\infty + V_\infty)|R_\infty]\} = \int_0^\infty\int_0^\infty \log(1 + x + y)\,\mathrm{d}\mathsf{Q}_{\mathrm{st}}(x)\,\mathrm{d}\widetilde{\mathsf{Q}}(y), \tag{8.364}$$

$$C_r = \mathsf{E}_0[\log(1 + r + V_\infty)] = \int_0^\infty \log(1 + r + y)\,\mathrm{d}\widetilde{\mathsf{Q}}(y). \tag{8.365}$$

The following theorem, whose proof is based on nonlinear renewal theory and can be found in [467] (Theorems 3.2 and 3.3), provides asymptotic approximations for the average delay to detection of the SR-r procedure (for large ν and $\nu = 0$), for the maximal average delay to detection $\mathsf{SADD}(T_{\mathrm{SRP}}(A)) = \mathsf{E}_0[T_{\mathrm{SRP}}(A)]$ of the SRP procedure, and for the integral average delay to detection $\mathsf{IADD}(T_{\mathrm{SR}}(A))$ of the SR procedure (within vanishing terms $o(1)$). Recall that, by Theorem 8.4.3, $\mathsf{IADD}(T_{\mathrm{SR}}(A_\gamma))$ gives the lower bound for the SADD in the class $\mathbf{C}_\gamma$, which is used in the next subsection for proving asymptotic optimality of the SR-r procedure.

Theorem 8.4.5. *If $\mathsf{E}_0|Z_1|^2 < \infty$ and Z_1 is non-arithmetic, then for any $r \ge 0$ as $A \to \infty$*

$$\mathsf{CADD}_\infty(T_{\mathrm{SR}}^r(A)) = \mathsf{E}_0[T_{\mathrm{SRP}}(A))] = I^{-1}(\log A + \varkappa - C) + o(1), \tag{8.366}$$

$$\mathsf{E}_0[T_{\mathrm{SR}}^r(A)] = I^{-1}(\log A + \varkappa - C_r) + o(1), \tag{8.367}$$

and

$$\mathsf{IADD}(T_{\mathrm{SR}}(A)) = I^{-1}(\log A + \varkappa - C) + o(1) \quad \text{as } A \to \infty. \tag{8.368}$$

8.4.3.3 Near Optimality

Recall that Figure 8.16 implies that for certain values of the head start $\mathsf{SADD}(T_{\mathrm{SR}}^r) = \mathsf{CADD}_\infty(T_{\mathrm{SR}}^r)$. This is a very important observation, since it allows us to build a proof of asymptotic optimality based on an estimate of $\mathsf{CADD}_\infty(T_{\mathrm{SR}}^r)$ given in Theorem 8.4.5.

Theorem 8.4.6. *Let* $\mathsf{E}_0|Z_1|^2 < \infty$ *and let* Z_1 *be non-arithmetic.*

(i) *Then*

$$\inf_{T \in \mathbf{C}_\gamma} \mathsf{SADD}(T) \ge I^{-1}(\log\gamma + \log\zeta + \varkappa - C) + o(1) \quad \text{as } \gamma \to \infty. \tag{8.369}$$

(ii) *If in the SR-r procedure* $A = A_\gamma = \gamma\zeta$, *and the initialization point r is either fixed or tends to infinity with the rate* $o(\gamma)$ *and is selected so that* $\mathsf{SADD}(T_{\text{SR}}^r(A_\gamma)) = \mathsf{CADD}_\infty(T_{\text{SR}}^r(A_\gamma))$, *then* $\mathsf{ARL2FA}(T_{\text{SR}}^r(A_\gamma)) \sim \gamma$ *and*

$$\mathsf{SADD}(T_{\text{SR}}^r(A_\gamma)) = I^{-1}(\log\gamma + \log\zeta + \varkappa - C) + o(1) \quad \text{as } \gamma \to \infty. \tag{8.370}$$

Therefore, the SR-r procedure is asymptotically optimal to third-order:

$$\inf_{T \in \mathbf{C}_\gamma} \mathsf{SADD}(T) = \mathsf{SADD}(T_{\text{SR}}^r(A_\gamma)) + o(1) \quad \text{as } \gamma \to \infty. \tag{8.371}$$

Proof. (i) By Theorem 8.4.3, $\inf_{T \in \mathbf{C}_\gamma} \mathsf{SADD}(T) \ge \mathsf{IADD}(T_{\text{SR}}(A_\gamma))$. By Theorem 8.4.4, if we set $A = A_\gamma = \zeta\gamma$, then $\mathsf{ARL2FA}(T_{\text{SR}}(A_\gamma)) \sim \gamma$ as $\gamma \to \infty$ and by (8.368)

$$\mathsf{IADD}(T_{\text{SR}}(A_\gamma)) = I^{-1}(\log\gamma + \log\zeta + \varkappa - C) + o(1),$$

which yields (8.369).

(ii) Asymptotic approximation (8.370) follows from (8.366) of Theorem 8.4.5 and assertion (8.371) follows from the asymptotic lower bound (8.369). □

Since for the SR procedure $\mathsf{SADD}(T_{\text{SR}}(A)) = \mathsf{E}_0[T_{\text{SR}}(A)]$, setting $r = 0$ in (8.367) we obtain

$$\mathsf{SADD}(T_{\text{SR}}(A)) = I^{-1}(\log A + \varkappa - C_0) + o(1) \quad \text{as } A \to \infty,$$

where $C_0 = \mathsf{E}_0[\log(1 + V_\infty)]$. Since $A_\gamma = \zeta\gamma$ implies $\mathsf{E}_\infty[T_{\text{SR}}(A)] \sim \gamma$, it follows that with this choice of threshold

$$\mathsf{SADD}(T_{\text{SR}}(A_\gamma)) = I^{-1}(\log\gamma + \log\zeta + \varkappa - C_0) + o(1) \quad \text{as } \gamma \to \infty. \tag{8.372}$$

Comparing (8.372) with the lower bound (8.369) shows that

$$\inf_{T \in \mathbf{C}} \mathsf{SADD}(T) = \mathsf{SADD}(T_{\text{SR}}(A_\gamma)) + O(1) \quad \text{as } \gamma \to \infty. \tag{8.373}$$

Thus, the SR procedure is only second-order asymptotically optimal and the difference is approximately equal to $(C - C_0)/I$. This difference can be quite large when detecting small changes (when I is small).

It is interesting to ask how the conditional average detection delays at infinity $\mathsf{CADD}_\infty(T_{\text{SR}})$, $\mathsf{CADD}_\infty(T_{\text{SR}}^r)$, and $\mathsf{CADD}_\infty(T_{\text{SRP}})$ are related when all three procedures have the same ARL2FA γ. Intuitively, the CADD_∞ should be the smallest for the original SR procedure T_{SR}. This result can be proven in two steps: 1) To show that the ARL2FA of the SRP procedure is increasing in the threshold A (the fact that the ARL2FA of the SR-r procedure is increasing in A for a fixed r is obvious); and 2) To show that the average delay $\mathsf{CADD}_\infty(T_{\text{SRP}})$ of the SRP procedure is increasing in A (obviously, the CADD's at infinity are the same for all three procedures when the same threshold is being used). Since the SR procedure requires the lowest threshold to attain the same false alarm rate, this implies that the SR procedure has the lowest CADD_∞. We believe that $\mathsf{E}_\infty[T_{\text{SRP}}(A)]$ and $\mathsf{CADD}_\infty(T_{\text{SRP}}(A))$ are both increasing in A in the general case. However, we are able to prove this fact only when distribution of $\log\Lambda_1$ is log-concave,[1] both pre-change and post-change, something that guarantees monotonicity properties of the Markov detection statistics. This is restrictive, but it does hold, for example, in detection of a shift of a normal mean and in detection of a change of the parameter of an exponential distribution. It also holds in "Beta" Example 8.4.2 considered in Subsection 8.4.3.5.

The next theorem proves this statement. See [467, Theorem 3.5] for a detailed proof.

[1]The distribution function $F(y)$ of a random variable Y is said to be log-concave if $\log F(y)$ is a concave function.

Theorem 8.4.7. *Suppose that the distribution function of* $\log \Lambda_1$ *is log-concave both pre-change and post-change. Let, for* $1 < \gamma < \infty$, *the threshold* A_γ^r *be such that the ARL2FA of the SR-r procedure* $T_{\mathrm{SR}}^r(A_\gamma^r)$ *is* γ. *Then* $\mathsf{CADD}_\infty(T_{\mathrm{SR}}^r(A_\gamma^r))$ *is an increasing function of r and*

$$\min_{0 \le r < \infty} \mathsf{CADD}_\infty(T_{\mathrm{SR}}^r(A_\gamma^r)) = \mathsf{CADD}_\infty(T_{\mathrm{SR}}(A_\gamma^0)) < \mathsf{CADD}_\infty(T_{\mathrm{SRP}}(A_*)),$$

where A_* *is such that* $\mathsf{ARL2FA}(T_{\mathrm{SRP}}(A_*)) = \gamma$.

Note also that by Theorems 8.4.5 and 8.4.6 the difference between ADDs of all three procedures is $o(1)$ as $\gamma \to \infty$.

8.4.3.4 Selecting the Head Start

Recall that the main idea of designing the SR-r procedure is to select the head start $r = r_A$ in such a way that the difference between the maximal average delay $\mathsf{SADD}(T_{\mathrm{SR}}^r(A))$ and the lower bound

$$\mathsf{IADD}_r(T_{\mathrm{SR}}^r(A)) = \frac{r\mathsf{E}_0[T_{\mathrm{SR}}^r(A_\gamma)] + \sum_{\nu=0}^\infty \mathsf{E}_\nu[T_{\mathrm{SR}}^r(A_\gamma) - \nu]^+}{r + \mathsf{E}_\infty[T_{\mathrm{SR}}^r(A_\gamma)]}$$

is minimized, i.e.,

$$r_A = \arg \inf_{0 \le r < A} \left[\mathsf{SADD}(T_{\mathrm{SR}}^r(A)) - \mathsf{IADD}_r(T_{\mathrm{SR}}^r(A))\right].$$

In the conditions of Theorem 8.4.6, this is equivalent to minimizing the difference

$$\mathsf{CADD}_\infty(T_{\mathrm{SR}}^r(A)) - \mathsf{IADD}_r(T_{\mathrm{SR}}^r(A)).$$

Theorem 8.4.5 suggests that if the head start $r = r^*$ is selected from the equation

$$C_r = C, \tag{8.374}$$

then for the large ARL2FA γ the values of the conditional average delays to detection at zero and infinity are approximately equal, $\mathsf{E}_0[T_{\mathrm{SR}}^{r^*}(A)] \approx \mathsf{CADD}_\infty(T_{\mathrm{SR}}^{r^*}(A))$ (to within small terms $o(1)$). This choice of the head start is intuitively appealing since we intend to make the SR-r procedure look like an equalizer as much as possible. Obviously, the limiting value $\lim_{A\to\infty} r_A^* = r^*$ equating $\mathsf{E}_0[T_{\mathrm{SR}}^{r^*}(A)]$ and $\mathsf{CADD}_\infty(T_{\mathrm{SR}}^{r^*}(A))$ to within $o(1)$ is a fixed number that depends on the model but not on γ. By Theorem 8.4.5, with this selection $\mathsf{CADD}_\infty(T_{\mathrm{SR}}^{r^*}(A)) - \mathsf{IADD}_{r^*}(T_{\mathrm{SR}}^{r^*}(A)) = o(1)$ for large γ, i.e., the SR-r procedure with $r = r^*$ provides almost optimal performance. Evidently, starting the SR-r procedure from the point that equates the conditional average detection delays at zero and at infinity is practically more convenient, as it does not require one to know the lower bound or the quasi-stationary distribution.

We would also like to stress that although starting at r^* causes a faster initial response than starting at $r = 0$, the resemblance to Lucas and Crosier's [281] FIR (fast initial response) scheme is secondary: their method is designed to give a really fast initial response, whereas our goal is to attain asymptotic third-order minimaxity.

In order to select the initializing point r^* from equation (8.374) as well as implement the asymptotic approximations we have to be able to compute the constants C_r and C defined in (8.364) and (8.365), where $\mathsf{Q}_{\mathrm{st}}(x) = \lim_{n\to\infty} \mathsf{P}_\infty(R_n \le x)$ is the stationary distribution of the SR statistic R_n under P_∞ and $\widetilde{\mathsf{Q}}(y) = \lim_{n\to\infty} \mathsf{P}_0(V_n \le y)$ is the limiting distribution of $V_n = \sum_{i=1}^n e^{-S_i}$ under P_0.

Assume that the distribution of Λ_1 is continuous. Then for computing the constants we need to evaluate the two densities $q_{\mathrm{st}}(x) = d\mathsf{Q}_{\mathrm{st}}(x)/dx$ and $\widetilde{q}(x) = d\widetilde{\mathsf{Q}}(x)/dx$. Let R_∞ and V_∞ be random variables that are the limit (in distribution, as $n \to \infty$) of R_n and V_n, respectively, which have densities $q_{\mathrm{st}}(x)$ and $\widetilde{q}(x)$. To find the desired densities, observe that, by recursion (8.325), R_∞ and $(1 + R_\infty)\Lambda_1$ have the same density $q_{\mathrm{st}}(x)$ under P_∞. Similarly V_∞ and $(1 + V_\infty)\Lambda_1^{-1}$ have the same density $\widetilde{q}(x)$ under P_0. To see this note that, by the iid property of the data, V_n has the same P_0-distribution as the random variable $\widetilde{V}_n = \sum_{i=1}^n \prod_{j=i}^n \Lambda_j^{-1}$, which follows the recursion $\widetilde{V}_n = (1+\widetilde{V}_{n-1})\Lambda_n^{-1}$. Therefore, we have the following integral equations for these densities

$$q_{\mathrm{st}}(x) = \int_0^\infty q_{\mathrm{st}}(y)\mathcal{K}_\infty\left(\frac{x}{1+y}\right)\mathrm{d}y; \;\; \widetilde{q}(x) = -\int_0^\infty \widetilde{q}(y)\mathcal{K}_0\left(\frac{1+y}{x}\right)\mathrm{d}y.$$

Thus, $q_{\mathrm{st}}(x)$ and $\widetilde{q}(x)$ are the eigenfunctions corresponding to the unit eigenvalues of the linear operators defined, respectively, with the kernels $\mathcal{K}_\infty(x,y)$ and $-\mathcal{K}_0(x,y)$. The constants C_r and C are then obtained (usually) by numerical integration.

Another reasonable option is to initialize from the mean of the quasi-stationary distribution, $r = \mu_A$. This initialization satisfies at least one of the conditions of Theorem 8.4.6, namely $\mu_A = o(A)$ (since $\mu_A \le O(\log A)$ and in fact $\mu_A = \log A + O(1)$). Computations for several examples show that typically $\mathsf{SADD} = \mathsf{CADD}_\infty$ with this initialization, so the second condition is also satisfied.

8.4.3.5 *Examples: Comparison of Detection Procedures*

Example 8.4.2 (Beta-to-Beta model). To verify the accuracy of the approximations and compare procedures consider the following example. Suppose in the pre-change mode $X_1, X_2, \dots, X_\nu$ are distributed as $\mathsf{beta}(2,1)$ and in the post-change mode $X_{\nu+1}, X_{\nu+2}, \dots$ are distributed as $\mathsf{beta}(1,2)$, i.e., $f_0(x) = 2x\mathbb{1}_{\{0\le x\le 1\}}$ and $f_1(x) = 2(1-x)\mathbb{1}_{\{0\le x\le 1\}}$, in which case there is a sudden shift in the mean from $2/3$ to $1/3$. One of the goals is to verify the accuracy of the asymptotic approximations given in Theorem 8.4.5 and Theorem 8.4.4,

$$\begin{aligned} &\mathsf{SADD}(T_{\mathrm{SR}}^r(A)) \approx \mathsf{SADD}(T_{\mathrm{SRP}}(A)) \approx I^{-1}(\log A + \varkappa - C), \\ &\mathsf{SADD}(T_{\mathrm{SR}}(A)) \approx I^{-1}(\log A + \varkappa - C_0), \end{aligned} \tag{8.375}$$

and the approximations for the ARL2FA

$$\mathsf{ARL2FA}(T_{\mathrm{SR}}^r(A)) \approx A/\zeta - r \quad \text{and} \quad \mathsf{ARL2FA}(T_{\mathrm{SRP}}(A)) \approx A/\zeta - \mu_A. \tag{8.376}$$

To undertake this task, it is necessary to be able to calculate the constants C_r, C, ζ, and $\varkappa$ and also to compute the initialization point r and the mean μ_A of the quasi-stationary distribution Q_A. While usually the constants C_r and C can be evaluated only numerically or by Monte Carlo, for the beta-model these constants are computable analytically.

Indeed, $\mathcal{L}_n = 1/X_n - 1$ and

$$F_\infty(t) = \mathsf{P}_\infty(\mathcal{L}_1 \le t) = 1 - \mathsf{P}_\infty\left(X_n \le \frac{1}{1+t}\right) = 1-(1+t)^{-2}, \tag{8.377}$$

so the quasi-stationary distribution satisfies the integral equation

$$\lambda_A \mathsf{Q}_A(x) = 1 - \int_0^A \frac{(1+y)^2}{(1+x+y)^2}\mathrm{d}\mathsf{Q}_A(y), \tag{8.378}$$

where

$$\lambda_A = 1 - \int_0^A \frac{(1+y)^2}{(1+A+y)^2}\mathrm{d}\mathsf{Q}_A(y)$$

(see (2.16)). By Theorem 2.1.2, the quasi-stationary distribution converges to the stationary distribution of R_n^r, $\mathsf{Q}_A(x) \to \mathsf{Q}_{\mathrm{st}}(x)$ as $A \to \infty$. Thus, by (8.378), the stationary distribution satisfies the following equation

$$\mathsf{Q}_{\mathrm{st}}(x) = 1 - \int_0^\infty \frac{(1+y)^2}{(1+x+y)^2}\mathrm{d}\mathsf{Q}_{\mathrm{st}}(y),$$

and the solution is $\mathsf{Q}_{\rm st}(x) = [x/(1+x)]\,\mathbb{1}_{\{x\ge 0\}}$.

To derive the equation for $\widetilde{\mathsf{Q}}(x)$, observe first that for $t \ge 0$

$$\mathsf{P}_0(\mathcal{L}_1 \ge 1/t) = \mathsf{P}_0\left(X_n \le \frac{t}{1+t}\right) = 1-(1+t)^{-2},$$

which is identical to $\mathsf{P}_\infty(\mathcal{L}_1 \le t)$. As a result, the distribution $\widetilde{\mathsf{Q}}(x)$ satisfies precisely the same equation as $\mathsf{Q}_{\rm st}(x)$ and, therefore,

$$\widetilde{\mathsf{Q}}(x) = \mathsf{Q}_{\rm st}(x) = \frac{x}{1+x}\mathbb{1}_{\{x\ge 0\}}. \tag{8.379}$$

Using (8.364), (8.365), and (8.379), one is able to calculate the constants C_r and C *exactly* as

$$C_r = \frac{1+r}{r}\log(1+r) \quad \text{and} \quad C = \frac{\pi^2}{6} \approx 1.6449. \tag{8.380}$$

In particular, $C_0 = 1$.

Note that the K–L information number $I = 1$, so that

$$\mathsf{SADD}(T^r_{\rm SR}(A)) \approx \mathsf{SADD}(T_{\rm SRP}(A)) \approx \log A + \varkappa - 1.6449, \quad \mathsf{SADD}(T_{\rm SR}(A)) \approx \log A + \varkappa - 1.$$

Unfortunately, neither $\varkappa$ nor ζ are computable exactly. Monte Carlo simulations with 10^6 trials have been used to obtain $\varkappa \approx 1.255$ and $\zeta \approx 0.426$ with the standard error less than 10^{-3}.

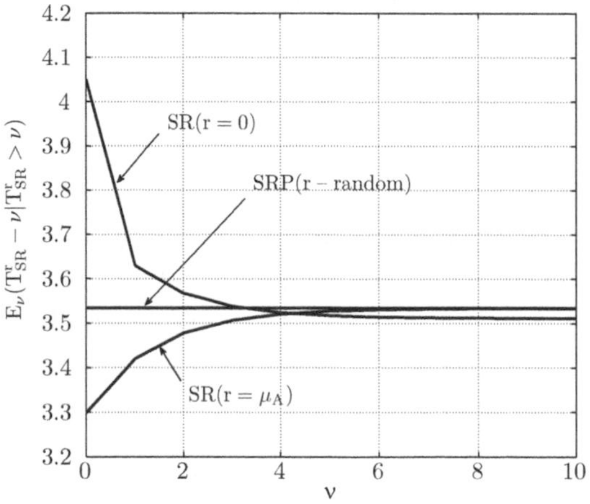

Figure 8.17: *Conditional average detection delay vs. changepoint ν of the SR, SRP, and SR-r ($r = \mu_A$) procedures. The ARL to false alarm* $\mathsf{ARL2FA}(T) \approx 100$.

Operating characteristics (CADD, SADD, IADD, and ARL2FA) were computed solving the corresponding integral equations of Subsection 8.4.3.1 numerically with the number of breakpoints set at 5×10^4, high enough to ensure a relative error in the order of a fraction of a percent. In computations we used two options for choosing the head start r. First, we set $r = \mu_A$. The condition $\mathsf{SADD}(T^{\mu_A}_{\rm SR}) = \mathsf{CADD}_\infty(T^{\mu_A}_{\rm SR})$ of Theorem 8.4.6 is satisfied even for small values of the ARL2FA, as can be seen from Figure 8.17, which shows how $\mathsf{E}_\nu(T-\nu|T>\nu)$ evolves as ν runs from 0 to 10 for the SRP procedure and for the SR-r procedure with $r = \mu_A$. The ARL2FA to false alarm is about 100 for both procedures. Observe that the stationary regime kicks in as early as at $\nu = 6$. In addition, Figure 8.17 illustrates Theorem 8.4.7 — $\mathsf{CADD}_\infty(T)$ is indeed the smallest for the SR procedure, while the difference is small. It is easily shown that the log-concavity conditions of Theorem 8.4.7 hold, i.e., $\log\mathsf{P}_\infty(\log\mathcal{L}_1 \le x)$ and $\log\mathsf{P}_0(\log\mathcal{L}_1 \le x)$ are concave functions. Table 8.2 provides values of the SADD and the lower bound $\mathsf{IADD}(T_{\rm SR})$ *versus* the ARL2FA. Also presented in parentheses are the corresponding theoretical predictions made based on the asymptotic approximations (8.375) and (8.376). The approximations for the ARL2FA are fairly accurate even for small values of the ARL such as 50, while the approximations for the SADD and the lower bound become accurate for the moderate false alarm rate (ARL2FA $= 500$ and higher). Performances of SRP and SR-r with $r = \mu_A$ are indistinguishable.

Table 8.2: *Numerical evaluation of operating characteristics of the SR, SRP, and SR-r procedures. Numbers in parentheses are computed using the asymptotic approximations.*

Procedure	γ	50	500	1000	10000
SR	A	21.0	212.0	424.5	4256.0
	ARL2FA	50.4 (49.3)	499.9 (498.1)	999.8 (997.4)	9999.7 (10000.0)
	SADD	3.41 (3.31)	5.62 (5.62)	6.31 (6.31)	8.61 (8.61)
SRP	A	21.5	213.5	426.5	4259.0
	ARL2FA	49.6 (48.5)	499.4 (497.6)	999.9 (997.4)	9999.8 (10000.1)
	SADD	2.94 (2.67)	5.02 (4.97)	5.69 (5.66)	7.97 (7.97)
SR-r	A	21.5	213.5	426.5	4259.0
	$r = \mu_A$	2.04	4.05	4.71	6.98
	ARL2FA	49.6 (48.5)	500.5 (497.6)	999.8 (997.4)	9999.7 (10000.1)
	SADD	2.94 (2.67)	5.02 (4.97)	5.69 (5.66)	7.97 (7.97)
	Lower Bound	2.94 (2.67)	5.02 (4.97)	5.69 (5.66)	7.97 (7.97)

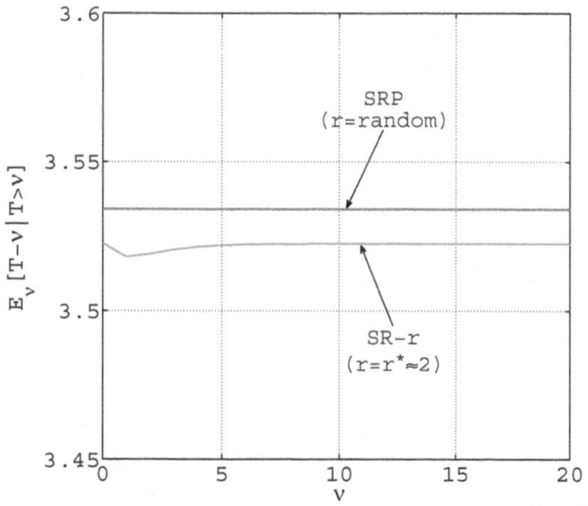

Figure 8.18: *Conditional average detection delay vs. changepoint* ν *for the SRP procedure and for the SR-r procedure with* $r = r^* = 2$*. The ARL to false alarm* ARL2FA$(T) \approx 100$.

The second initialization option is to start from the value of r for which the conditional average detection delay at the point $\nu = 0$ is equal (at least approximately) to the CADD$_\infty$ (in the steady-state mode), as proposed in Subsection 8.4.3.4. In the asymptotic setting, this is equivalent to finding a point $r = r^*$ for which C is equal to C_r, i.e., r^* satisfies equation (8.374). Clearly, r^* is a *fixed* number that does not depend on A since C does not depend on r and A. Using (8.380), we obtain the transcendental equation

$$\frac{1+r^*}{r^*}\log(1+r^*) = \frac{\pi^2}{6},$$

and the solution is $r^* \approx 2$. Figure 8.18 shows the conditional average delay to detection $\mathsf{E}_\nu[T-\nu|T>\nu]$ *versus* the changepoint ν for the SR-r procedure with $r=r^*=2$ and for the SRP procedure. Observe that for the SR-r procedure the average delay at $\nu=0$ is equal to that at infinity, as was planned. More importantly, the point $\nu=0$ is the worst (supremum) point (along with large ν). Also, it can be seen that the SR-r procedure is *uniformly* (i.e., for all $\nu \ge 0$) better than the SRP procedure, although in this example the difference is practically negligible. We also note that this initialization is better than starting off at the mean of the quasi-stationary distribution, while the difference in performance is very small—the SADD is equal to 3.53 for $r=\mu_A$ and 3.52 for $r=r^*$. This allows us to conclude that the SR-r procedure is robust with respect to the initialization point in a certain range.

Example 8.4.3 (Exponential model). Consider now the case where observations are independent, having an exponential distribution with the unit parameter pre-change and the exponential distribution with the parameter $1+\theta$, $\theta>0$, post-change, i.e., $f_0(x)=\exp\{-x\}\,\mathbb{1}_{\{x\ge 0\}}$ and $f_1(x)=\frac{1}{1+\theta}\exp\left\{-\frac{x}{1+\theta}\right\}\mathbb{1}_{\{x\ge 0\}}$. We performed extensive numerical computations for various values of the parameter. Below we present sample results for $\theta=0.1$ corresponding to a relatively small, not easily detectable change. For all of the procedures (SR, SRP, CUSUM), the ARL2FA is let to go up to 10^4. The integration interval $[0,A]$ is sampled from $N=10^5$ equidistant points. We are confident that such sampling is sufficiently fine and provides a very high numerical precision.

Figure 8.19(a) shows operating characteristics in terms of $\mathsf{SADD}(T)$ *versus* the ARL2FA, plus the lower bound $\mathsf{lADD}(T_{\rm SR})$. CUSUM outperforms the classical SR procedure, but SRP and SR-r are more efficient. Figure 8.19(b) is a magnified version of the $\mathsf{SADD}(T)$-vs-$\mathsf{ARL2FA}(T)$ curve for the SR procedure, the SR-r procedure and the lower bound for relatively low FAR, $\mathsf{ARL2FA}(T) \in [5\cdot 10^3, 10^4]$. The best minimax performance is offered by the SR-r procedure. This is expected since by design the SR-r procedure is the closest to the lower bound. In this example, the difference with the lower bound is negligible. This suggests that an unknown optimal procedure can offer only a practically insignificant improvement over the SR-r procedure. The difference in performance between the SRP and SR-r procedures is also very small.

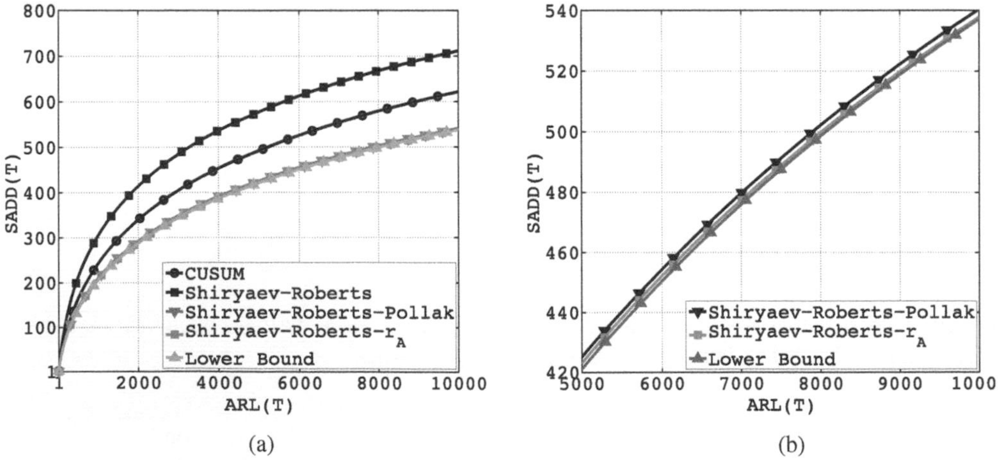

Figure 8.19: *Lower bound* $\mathsf{lADD}(T_{\rm SR})$ *and maximal average detection delay* $\mathsf{SADD}(T)$ *for the CUSUM, SRP, and SR-r procedures vs.* $\mathsf{ARL2FA}$, $\theta=0.1$.

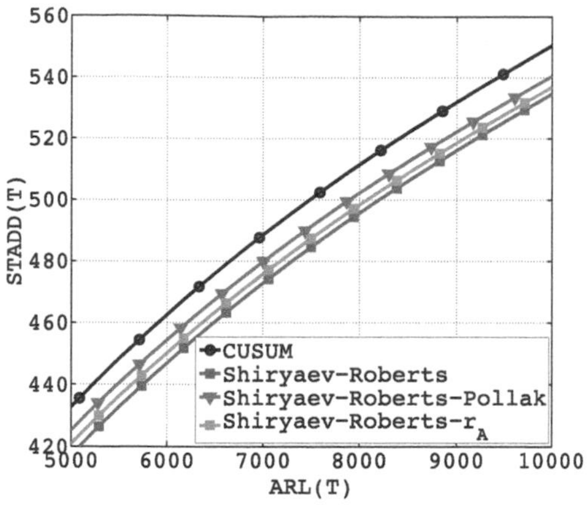

Figure 8.20: *The stationary average detection delay* $\mathsf{STADD}(T)$ *for the CUSUM, SRP, and SR-r procedures vs.* $\mathsf{ARL2FA}$, $\theta = 0.1$.

Figure 8.20 shows the behavior of the stationary average detection delay $\mathsf{STADD}(T)$ against the ARL to false alarm. Since the SR procedure is exactly optimal with respect to STADD its performance is the best among the three procedures, but the difference is relatively small. Note also that for the SRP procedure $\mathsf{STADD}(T_{\text{SRP}})$ is the same as $\mathsf{SADD}(T_{\text{SRP}})$, since the SRP procedure is an equalizer.

8.5 The Weighted Shiryaev–Roberts Procedure for Composite Post-Change Hypothesis

As pointed out in Section 8.3, in most applications the pre-change distribution is known but the post-change distribution is only partially known, often up to an unknown parameter $\theta \in \Theta$, so the post-change hypothesis "$H_k^\vartheta : \nu = k, \theta = \vartheta$", $\vartheta \in \Theta_1$ is composite. In Section 8.3, we established certain asymptotic optimality properties of the Weighted CUSUM and Generalized CUSUM algorithms (8.242) and (8.243). In this section, we consider a natural competitor—the *Weighted Shiryaev–Roberts* (WSR) detection procedure (or mixture SR) given by the stopping time

$$T_{\text{WSR}}(A) = \inf\left\{ n \geq 1 : \sum_{k=1}^{n} \widetilde{\Lambda}_n^k(w) \geq A \right\}, \tag{8.381}$$

where $\widetilde{\Lambda}_n^k(w)$ is the average LR defined in (8.240) and $A > 0$ is the threshold controlling the FAR.

Write

$$R_n^w = \sum_{k=1}^{n} \widetilde{\Lambda}_n^k(w) \equiv \int_{\Theta} R_n(\theta)\, \mathrm{d}W(\theta), \quad n \geq 1, \;\; R_0^w = 0 \tag{8.382}$$

for the weighted SR statistic, where

$$R_n(\theta) = \sum_{k=1}^{n} \prod_{i=k}^{n} \frac{f_\theta(X_i|\mathbf{X}_1^{i-1})}{f_{\theta_0}(X_i|\mathbf{X}_1^{i-1})} \tag{8.383}$$

is the SR statistic tuned to $\theta \in \Theta_1$. Then the stopping time (8.381) is

$$T_{\text{WSR}}(A) = \inf\{n \geq 1 : R_n^w \geq A\}. \tag{8.384}$$

Since $R_n^w - n$ is a zero-mean P_∞-martingale, by Lemma 8.4.1, $\mathsf{ARL2FA}(T_{\text{WSR}}(A)) \geq A$ for any $A > 0$. Note also that the approximation $\mathsf{ARL2FA}(T_{\text{WSR}}(A)) \approx A$ is far more accurate for WSR than for WCUSUM, since the difference is mostly due to the overshoot.

8.5.1 Asymptotic Properties of the Weighted SR Procedure in the iid Case

In the iid case, the weighted SR statistic takes the form

$$R_n^w = \int_\Theta \sum_{k=1}^n \prod_{i=k}^n \frac{f_\theta(X_i)}{f_{\theta_0}(X_i)} \mathrm{d}W(\theta).$$

It is not surprising that the results similar to those obtained in Subsection 8.3.1.1 for WCUSUM also hold for WSR. Consider, in particular, the multiparameter exponential family (5.69), assuming that the dimension of both θ and X_n is l. Then,

$$R_n^w = \int_\Theta \sum_{k=1}^n \prod_{i=k}^n \exp\left\{\theta^\top \sum_{i=k}^n X_i - (n-k+1)b(\theta)\right\} \mathrm{d}W(\theta).$$

A generalization of the argument in [366, 517] (for a single parameter exponential family) yields the following approximation for the ARL2FA,

$$\mathsf{ARL2FA}(T_{\text{WSR}}(A)) \sim \frac{A}{\int_\Theta \zeta_\theta w(\theta)\,\mathrm{d}\theta} \quad \text{as } A \to \infty, \tag{8.385}$$

assuming that $Z_1^\theta = \theta^\top X_1 - b(\theta)$ is non-arithmetic and W has a positive continuous density $w(\theta)$ on Θ.

Let

$$V_n^\theta = \sum_{k=1}^n \exp\left\{-\sum_{i=1}^k Z_i^\theta\right\}, \quad \tilde{\mathsf{Q}}^\theta(x) = \lim_{n\to\infty} \mathsf{P}_{0,\theta}(V_n^\theta \le x)$$

and

$$C_0(\theta) = \mathsf{E}_{0,\theta}\left[\log(1+V_\infty^\theta)\right] = \int_0^\infty \log(1+x)\,\mathrm{d}\tilde{\mathsf{Q}}^\theta(x)$$

(cf. (8.364)).

The following theorem is the analog of Theorem 8.3.1. Note that $\mathsf{SADD}_\theta(T_{\text{WSR}}) = \mathsf{ESADD}_\theta(T_{\text{WSR}}) = \mathsf{E}_{0,\theta} T_{\text{WSR}}$. Recall that by $\nabla^2 b(\theta)$ we denote the Hessian matrix of second partial derivatives.

Theorem 8.5.1. *Consider the l-dimensional exponential model* (5.69) *with natural parameter space Θ. Let $\Theta_1 = \Theta \setminus \mathbf{0}$ and let $T_{\text{WSR}}(A)$ be defined as in* (8.381)*.*

(i) *If the threshold $A = A_\gamma$ is selected so that $\mathsf{ARL2FA}(T_{\text{WSR}}) \ge \gamma$ and $\log A_\gamma \sim \log\gamma$ as $\gamma \to \infty$, in particular $A_\gamma = \gamma$, then the WSR procedure minimizes asymptotically to first order all positive moments of the delay to detection in the worst-case scenario with respect to the changepoint ν and uniformly for all $\theta \in \Theta_1$ in the sense that as $\gamma \to \infty$*

$$\begin{aligned} &\inf_{T \in \mathbf{C}_\gamma} \mathsf{ESM}_r^\theta(T) \sim \mathsf{ESM}_r^\theta(T_{\text{WSR}}) \sim (I_\theta^{-1}\log\gamma)^r, \\ &\inf_{T \in \mathbf{C}_\gamma} \mathsf{SM}_r^\theta(T) \sim \mathsf{SM}_r^\theta(T_{\text{WSR}}) \sim (I_\theta^{-1}\log\gamma)^r \end{aligned} \tag{8.386}$$

for all $r > 0$ and all $\theta \in \Theta_1$.

(ii) *Let, for some $\varepsilon > 0$, $\theta \in \Theta_\varepsilon = \Theta - B_\varepsilon$, where B_ε is a closed ball of radius $\varepsilon > 0$ in $\mathbb{R}^l$. Let $\mathbf{0} \notin \Theta_\varepsilon$ and let Θ_ε be compact. If the mixing distribution $W(\theta)$ with support on Θ_ε has a positive continuous density $w(\theta)$ with respect to Lebesgue's measure, then as $A \to \infty$*

$$\begin{aligned} \mathsf{E}_{0,\theta}[T_{\text{WSR}}(A)] = \frac{1}{I_\theta}\Bigg\{ &\log A + \frac{l}{2}\log\log A - \frac{l}{2}[1+\log(2\pi)] \\ &- \log\left(w(\theta)\sqrt{I_\theta^l / \det[\nabla^2 b(\theta)]}\right) - C_0(\theta) + \varkappa_\theta \Bigg\} + o(1). \end{aligned} \tag{8.387}$$

Moreover, if the threshold is selected as

$$A = \gamma \int_{\Theta_\varepsilon} \zeta_\theta \, w(\theta) \, d\theta,$$

then $\mathsf{ARL2FA}(T_{\text{WSR}}) = \gamma(1+o(1))$ *and*

$$\begin{aligned} \mathsf{E}_{0,\theta} T_{\text{WCS}} = \frac{1}{I_\theta} \Bigg\{ & \log\gamma + \frac{l}{2}\log\log\gamma - \frac{l}{2}[1+\log(2\pi)] + \log\left(\int_{\Theta_\varepsilon} \zeta_t \, w(t) \, dt\right) \\ & - \log\left(w(\theta) e^{-\varkappa_\theta + C_0(\theta)} \sqrt{I_\theta^l / \det[\nabla^2 b(\theta)]} \right) \Bigg\} + o(1) \quad \text{as } \gamma \to \infty. \end{aligned} \tag{8.388}$$

The proof of (ii) is similar to that for the one-sided WSPRT in Subsection 5.5.1. It is based on the observation that, on one hand, the logarithm of the stopped WSR statistic $\log R^w_{T_{\text{WSR}}}$ equals $\log A$ plus an overshoot, and on the other hand, $\log R^w_{T_{\text{WSR}}} = \sum_{i=\nu+1}^{T_{\text{WSR}}} Z_i^\theta + \ell_\nu^\theta(T_{\text{WSR}})$, where $\{\ell_\nu^\theta(n)\}_{n>\nu}$ is a slowly changing sequence. Applying nonlinear renewal theory, we then obtain

$$\begin{aligned} &\mathsf{E}_{\nu,\theta}[\log R^w_{T_{\text{WSR}}} | T_{\text{WSR}} > \nu] = \log A + \varkappa_\theta + o(1), \\ &\mathsf{E}_{\nu,\theta}[\log R^w_{T_{\text{WSR}}} | T_{\text{WSR}} > \nu] = I_\theta \mathsf{E}_{\nu,\theta}(T_{\text{WSR}} - \nu | T_{\text{WSR}} > \nu) + \mathsf{E}_{\nu,\theta}[\ell_\nu^\theta(T_{\text{SR}}) | T_{\text{SR}} > \nu]. \end{aligned}$$

Using a Laplace integration method it can be shown that

$$\begin{aligned} \mathsf{E}_{\nu,\theta}[\ell_\nu^\theta(T_{\text{SR}}) | T_{\text{SR}} > \nu] = -\frac{l}{2}\log\log A + \frac{l}{2}[1+\log(2\pi)] - \frac{1}{2}\log\left(\frac{\det[\nabla^2 b(\theta)]}{I_\theta^l}\right) \\ + \log w(\theta) + C_0(\theta) + o(1). \end{aligned}$$

Combining these three approximations yields (8.387). Technical details are omitted. Note, however, that there is only a minor difference compared to the approximations for the one-sided WSPRT (5.203)–(5.204)—an additional term $C_0(\theta)$ appears in the constant due to an additional term V_n^θ in the slowly changing sequence.

Since by Theorem 8.4.6

$$\inf_{T \in \mathbf{C}_\gamma} \mathsf{SADD}_\theta(T) = I_\theta^{-1} \log\gamma + O(1) \quad \text{as } \gamma \to \infty,$$

it follows from (8.388) that for any mixing distribution

$$\mathsf{SADD}_\theta(T_{\text{WSR}}) - \inf_{T \in \mathbf{C}_\gamma} \mathsf{SADD}_\theta(T) = O\left(\frac{1}{2I_\theta}\log\log\gamma\right) \quad \text{for all } \theta \in \Theta_\varepsilon,$$

and the same is true for the ESADD_θ.

Again, let the risk associated with the detection delay be expressed by the expected post-change K–L information, $\mathcal{R}_\theta(T) = I_\theta \mathsf{SADD}_\theta(T)$, and consider the minimax criterion with respect to $\mathcal{R}_\theta(T)$, i.e., we wish to minimize $\sup_{\theta\in\Theta} \mathcal{R}_\theta(T)$ in the class $\mathbf{C}_\gamma$. For any $l \ge 1$, the asymptotic lower bound (8.256) for $\inf_{T\in\mathbf{C}_\gamma} \sup_{\theta\in\Theta} \mathcal{R}_\theta(T)$ gets modified as

$$\inf_{T \in \mathbf{C}_\gamma} \sup_{\theta\in\Theta} \mathcal{R}_\theta(T) \ge \log\gamma + \frac{l}{2}\log\log\gamma + O(1) \quad \text{as } \gamma \to \infty,$$

and, by (8.388), it is attained by the WSR procedure $T_{\text{WSR}} = T_{\text{WSR}}(w)$ for any mixing continuous density w. Next, if we select mixing density as

$$w_0(\theta) = \frac{e^{\varkappa_\theta - C_0(\theta)} \sqrt{\det[\nabla^2 b(\theta)] / I_\theta^l}}{\int_{\Theta_\varepsilon} e^{\varkappa_t - C_0(t)} \sqrt{\det[\nabla^2 b(t)] / I_t^l} \, dt}, \quad \theta \in \Theta_\varepsilon, \tag{8.389}$$

then the residual term $O(1)$ in (8.388) does not depend on θ. In this case, for all $\theta \in \Theta$

$$\mathcal{R}_\theta(T_{\text{WSR}}) = \log\gamma + \frac{l}{2}\log\log\gamma - \frac{l}{2}[1+\log(2\pi)] + \bar{C}_{\text{WSR}} + o(1) \quad \text{as } \gamma \to \infty, \tag{8.390}$$

where

$$\bar{C}_{\text{WSR}} = \log\left(\int_{\Theta_\varepsilon} \zeta_t e^{\varkappa_t - C_0(t)} \sqrt{\det[\nabla^2 b(t)]/I_t^l}\, \mathrm{d}t\right). \tag{8.391}$$

In other words, this WSR scheme is an almost equalizer rule to within a vanishing term $o(1)$ with respect to the parameter θ. However, it is not an equalizer with respect to the changepoint ν. For this reason, it is a second-order minimax equalizer procedure.

One way to make it third-order minimax is to randomize the initialization point as in the SRP procedure, making the resulting WSR procedure an equalizer in the changepoint ν.

Introduce the following randomized version of this rule obtained by starting off the WSR statistic at a random point

$$R_0^w(Y) = \int_\Theta R_m^\theta(Y) w(\theta)\,\mathrm{d}\theta,$$

where

$$R_m^\theta(Y) = \sum_{k=1}^m \prod_{i=k}^m \frac{f_\theta(Y_i)}{f_{\theta_0}(Y_i)} = \sum_{k=1}^m \exp\left\{\theta^\top \sum_{i=k}^m Y_i - (m-k+1)b(\theta)\right\} w(\theta)\mathrm{d}\theta$$

and where $\{Y_j\}_{1\le j\le m}$ are iid random variables independent of X and sampled from the pre-change distribution f_{θ_0}. This random point, however, is obtained conditioned on the event $\{R_1^w(Y) < A, \dots, R_m^w(Y) < A\}$, i.e., assuming that there is no threshold exceedance prior to starting monitoring the process of interest X. This idea mimics the quasi-stationary concept of the SRP procedure for large enough m. Put otherwise, we attach a large enough sample $\mathbf{Y}_m = (Y_1, \dots, Y_m)$ to $X_1, X_2, \dots$, independent of the observed data X, before starting monitoring and form the statistic $\{R_0^w = R_m^w(Y) : R_1^w(Y) < A, \dots, R_m^w(Y) < A\}$ as a head start for the new WSR procedure.

Denoting this randomized WSR statistic as $\widetilde{R}_n^w$, the modified WSR procedure is defined as $T^*_{\text{WSR}}(w, A) = \inf\left\{n : \widetilde{R}_n^w \ge A\right\}$. Define also the random variable $R_\infty^\theta(Y)$ (independent of V_∞^θ defined above) that has a limiting (stationary) distribution $\mathsf{Q}_{\text{st}}^\theta(x) = \lim_{m\to\infty} \mathsf{P}_\infty(R_m^\theta(Y) \le x)$ and a constant

$$C(\theta) = \mathsf{E}_\infty\left\{\mathsf{E}_{0,\theta}\left[\log(1+V_\infty^\theta + R_0^\theta) | R_0^\theta\right]\right\} = \int_0^\infty \int_0^\infty \log(1+x+y)\,\mathrm{d}\tilde{\mathsf{Q}}^\theta(x)\,\mathrm{d}\mathsf{Q}_{\text{st}}^\theta(y) \tag{8.392}$$

(cf. (8.365)). Recall that the constants $C_0(\theta)$ and $C(\theta)$ can be computed numerically using the method of Subsection 8.4.3.2 or by Monte Carlo.

The following theorem shows that the randomized WSR procedure $T^*_{\text{WSR}}(w_*)$ with a specially designed weight w_* is third-order minimax (to within $o(1)$).

Theorem 8.5.2. *Consider the multiparameter exponential family* (5.69) *with the natural parameter space* Θ. *Let* $\mathcal{R}_\theta(T) = I_\theta \sup_{\nu\ge 0} \mathsf{E}_{\nu,\theta}(T-\nu | T>\nu)$. *Let the LLR* $Z_1^\theta = \theta^\top X_1 - b(\theta)$ *be* P_θ-*nonarithmetic. Let* $\varepsilon = (\log\gamma)^{-1/4}$.

(i) *If the weighting density* $w = w_*$ *is*

$$w_*(\theta) = \frac{e^{\varkappa_\theta - C(\theta)}\sqrt{\det[\nabla^2 b(t)]/I_\theta^l}}{\int_{\Theta_\varepsilon} e^{\varkappa_t - C(t)}\sqrt{\det[\nabla^2 b(t)]/I_t^l}\,\mathrm{d}t}, \qquad \theta \in \Theta_\varepsilon, \tag{8.393}$$

and

$$A_\gamma = \gamma \int_{\Theta_\varepsilon} \zeta_\theta\, w_*(\theta)\,\mathrm{d}\theta, \tag{8.394}$$

then $\mathsf{ARL2FA}(T_{\text{WSR}}(w_*, A_\gamma)) = \gamma(1+o(1))$ *as* $\gamma \to \infty$.

(ii) *If in addition $m = m_\gamma$ satisfies*

$$\lim_{\gamma\to\infty}\left[m_\gamma(\log\gamma)^{-(1+\delta)}\right]=\infty \quad \textit{but} \quad \lim_{\gamma\to\infty}\left[m_\gamma \exp\left\{-(\log\gamma)^{\delta}\right\}\right]=0 \tag{8.395}$$

for some $\delta > 0$, then, as $\gamma\to\infty$,

$$\sup_{\theta\in\Theta}\mathcal{R}_\theta(T^*_{\rm WSR}(w_*,A_\gamma)) = \log\gamma+\frac{1}{2}\log\log\gamma-\frac{1}{2}[1+\log(2\pi)]+C_{\rm opt}+o(1) \tag{8.396}$$

and

$$\inf_{T\in\mathbf{C}_\gamma}\sup_{\theta\in\Theta}\mathcal{R}_\theta(T)=\mathcal{R}_\theta(T^*_{\rm WSR}(w_*,A_\gamma))+o(1), \tag{8.397}$$

where

$$C_{\rm opt}=\log\left(\int_\Theta \zeta_t e^{\varkappa_t - C(t)}\sqrt{\det[\nabla^2 b(t)]/I_t^l}\,\mathrm{d}t\right).$$

Proof. (i) Obviously, $\mathsf{E}_\infty[T^*_{\rm WSR}(w_*,A)]\le\mathsf{E}_\infty[T_{\rm WSR}(w_*,A)]$, so by (8.385)

$$\mathsf{ARL2FA}(T^*_{\rm WSR}(w_*,A))\le\frac{A}{\int_\Theta\zeta_\theta w_*(\theta)\,\mathrm{d}\theta}(1+o(1)) \quad \text{as } A\to\infty. \tag{8.398}$$

Using essentially the same argument as in the proof of Theorem 8.4.4 that has led to the asymptotic approximation (8.357) for the ARL2FA of the SRP procedure yields the lower bound

$$\mathsf{ARL2FA}(T^*_{\rm WSR}(w_*,A))\ge\frac{A}{\int_\Theta\zeta_\theta w_*(\theta)\,\mathrm{d}\theta}(1+o(1)) \quad \text{as } A\to\infty,$$

which along with the upper bound (8.398) yields

$$\mathsf{ARL2FA}(T^*_{\rm WSR}(w_*,A))=\frac{A}{\int_\Theta\zeta_\theta w_*(\theta)\,\mathrm{d}\theta}(1+o(1)) \quad \text{as } A\to\infty,$$

and therefore assertion (i).

(ii) The assertion (ii) follows from Theorems 1 and 2 of Siegmund and Yakir [431]. Indeed, by Theorem 1 in [431] for any stopping rule $T=T_\gamma$ such that for some $k\ge 0$, possibly depending on γ,

$$\mathsf{P}_\infty(T<k+m_\gamma|T>k)\le\frac{m_\gamma}{\gamma}(1+o(1)) \quad \text{as } \gamma\to\infty,$$

where m_γ satisfies conditions (8.395), the following asymptotic lower bound holds for the expected K–L information in the worst-case scenario:

$$\sup_{\theta\in\Theta}\mathcal{R}_\theta(T)\ge\log\gamma+\frac{l}{2}\log\log\gamma-\frac{l}{2}[1+\log(2\pi)]+C_{\rm opt}+o(1), \quad \gamma\to\infty. \tag{8.399}$$

By Lemma 6.3.1, $\mathsf{P}_\infty(T<k+m|T>k)\le m/\gamma$ for any $T\in\mathbf{C}_\gamma$ and some $k\ge 0$. Therefore,

$$\inf_{T\in\mathbf{C}_\gamma}\sup_{\theta\in\Theta}\mathcal{R}_\theta(T)\ge\log\gamma+\frac{l}{2}\log\log\gamma-\frac{1}{2}[1+\log(2\pi)]+C_{\rm opt}+o(1). \tag{8.400}$$

On the other hand, by Theorem 2 in [431], as $A\to\infty$,

$$\mathcal{R}_\theta(T^*_{\rm WSR}(w_*,A))=\log A+\frac{l}{2}\log\log A-\frac{l}{2}[1+\log(2\pi)]+C_{\rm opt}+o(1)$$

whenever the size m of the randomizing sample $(Y_1,\dots,Y_m)$ is large enough, $m/\log A\to\infty$. Hence, substituting $A=A_\gamma$ given in (8.394) in this equality, we obtain asymptotic approximation (8.396). Finally, comparing approximation (8.396) with the lower bound (8.400) yields (8.397), and the proof is complete. □

As in the case of WCUSUM, the implementation of the WSR procedure may be time-consuming because of a large number of integrations that should be performed to compute the WSR statistic R_n^w when n is large. To avoid this excessive computational complexity the following window-limited WSR procedure

$$T_{\text{WL-WSR}}(A) = \inf\left\{n \geq 1 : \sum_{k=n-m+1}^{n} \widetilde{\Lambda}_n^k(w) \geq A\right\} \tag{8.401}$$

can be used. The non-randomized version starts from zero for $n = 0$ and uses summation over $k = 1$ to n for $n \leq m$. It has the same asymptotic properties as WSR so long as the size of the window $m \gg \log\gamma$, i.e., the assertions of Theorem 8.5.1 hold.

The randomized version is obtained as before by generating a sample $(Y_1, \dots, Y_m)$ from the pre-change distribution f_{θ_0} and attaching this sample to the main sample on the left. This randomized window-limited WSR with the mixing distribution w^* defined in (8.393) is also an asymptotic equalizer rule with respect to both the changepoint ν and the parameter θ and is asymptotically minimax relative to the expected K–L information. Specifically, the assertions of Theorem 8.5.2 hold (see [431]). However, even the WL-WSR procedure is difficult, if not impossible, to implement since typically the optimal mixing density $w^*(\theta)$ is impossible to compute analytically (only numerically), the issue we already discussed in Subsection 5.5.1 for a hypothesis testing problem. A discretization of the parameter space and implementing a discrete version of the WSR procedure resolves the computational complexity issue. This is, obviously, equivalent to assuming that the mixing distribution is discrete, not continuous.

In other words, as in Subsection 5.5.1.2, the set Θ_1 is approximated by a discrete set $\{\theta_1, \dots, \theta_N\} \subset \Theta_1$, in which case the discrete mixture SR statistic is easily computable, $R_n^w = \sum_{i=1}^N W_i R_n(\theta_i)$, $W_i = W(\theta_i)$. In the discrete case, the asymptotic approximation for the ADD is substantially different than in the continuous case (especially when N is not too large). For the sake of generality, consider not necessarily the exponential family but an arbitrary family of distributions with density $f_0(x)$ (pre-change) and densities $f_i(x)$, $i = 1, \dots, N$ (post-change), assuming that the LLR $Z_1^i = \log[f_i(X_1)/f_0(X_1)]$ is non-arithmetic and $\mathsf{E}_{0,i}|Z_1^i|^2 < \infty$, which also implies that $I_i = \mathsf{E}_{0,i} Z_1^i < \infty$.

Let each partial SR statistic $R_n(\theta_i)$ start from a positive point R_0^i, which may be deterministic or random (for the conventional WSR $R_0^i = 0$). The log-WSR statistic can be written as $\log R_n^w = \sum_{k=1}^n Z_k^i + Y_n^i$, where

$$Y_n^i = \log\left\{W_i(1 + V_n^i + R_0^i) + \sum_{j \neq i}\left[W_j \exp\left(-\sum_{k=1}^{n}(Z_k^i - Z_k^j)\right)(1 + V_n^j + R_0^j)\right]\right\},$$

$$V_n^i = \sum_{k=1}^{n} \exp\left\{-\sum_{s=1}^{k} Z_s^i\right\}.$$

The last term under the logarithm converges (as $n \to \infty$) to 0 $\mathsf{P}_{0,i}$-a.s., so Y_n^i converges to the random variable $\log[W_i(1 + V_\infty^i + R_0^i)]$ as $n \to \infty$ that has expectation $\log W_i + C(i)$, where $C(i) = \mathsf{E}_{0,i}[\log(1 + V_\infty^i + R_0^i)]$. In particular, $C(i) = C_0(i) = \mathsf{E}_{0,i}[\log(1 + V_\infty^i)]$ for the conventional WSR procedure with the zero head start. Applying the second Nonlinear Renewal Theorem yields

$$\mathsf{E}_{0,i}[T_{\text{WSR}}(A)] = I_i^{-1}\left[\log A + \varkappa_i - C_0(i) - \log W_i\right] + o(1) \quad \text{as } A \to \infty. \tag{8.402}$$

Also,

$$\mathsf{ARL2FA}(T_{\text{WSR}}(A)) \sim \frac{A}{\sum_{i=1}^N \zeta_i W_i} \quad \text{as } A \to \infty,$$

so taking $A = A_\gamma = \gamma \sum_{i=1}^N \zeta_i W_i$, we obtain $\mathsf{ARL2FA}(T_{\text{WSR}}(A)) \sim \gamma$ and

$$\mathsf{E}_{0,i}[T_{\text{WSR}}(A_\gamma)] = I_i^{-1}\left[\log\gamma + \log\left(\sum_{i=1}^{N} \zeta_i W_i\right) + \varkappa_i - C_0(i) - \log W_i\right] + o(1). \tag{8.403}$$

Since

$$\inf_{T \in \mathbf{C}_\gamma} \mathsf{SADD}_i(T) = I_i^{-1} \log \gamma + O(1) \quad \text{as } \gamma \to \infty,$$

it follows from (8.403) that for any discrete mixing distribution

$$\mathsf{SADD}_i(T_{\text{WSR}}) - \inf_{T \in \mathbf{C}_\gamma} \mathsf{SADD}_i(T) = O(1) \quad \text{for all } i = 1, \dots, N,$$

as long as N is fixed, and the same is true for the ESADD_i. Therefore, in the discrete case the WSR procedure is second-order asymptotically minimax with respect to the expected delay to detection for any mixing distribution and all $i = 1, \dots, N$.

Let $\mathcal{R}_i(T) = I_i \mathsf{SADD}_i(T)$. Again, consider the randomized variant $T^*_{\text{WSR}}(W, A)$, which guarantees an asymptotic (up to $o(1)$) equalizer property of the expected K–L information $I_i \mathsf{E}_{\nu,i}(T^*_{\text{WSR}} - \nu | T^*_{\text{WSR}} > \nu)$ with respect to ν for any discrete weights $W = \{W_i\}$:

$$I_i \mathsf{E}_{\nu,i}(T^*_{\text{WSR}} - \nu | T^*_{\text{WSR}} > \nu) = \log \gamma + \log \left(\sum_{i=1}^N \zeta_i W_i \right) + \varkappa_i - C(i) - \log W_i + o(1) \tag{8.404}$$

for all $\nu \geq 1$ as $\gamma \to \infty$, where $C(i) = \mathsf{E}_\infty \left\{ \mathsf{E}_{0,i} \left[\log(1 + V^i_\infty + R^i_0) | R^i_0 \right] \right\}$ (cf. (8.392) with $\theta = i$). Finally, it follows from (8.404) that to make the discrete WSR procedure an equalizer with respect to $i = 1, \dots, N$ the weights $W_i = W_i^*$ should be selected as

$$W_i^* = \frac{e^{\varkappa_i} - C(i)}{\sum_{j=1}^N e^{\varkappa_j - C(j)}}, \quad i = 1, \dots, N. \tag{8.405}$$

In this case,

$$I_i \mathsf{E}_{\nu,i}(T^*_{\text{WSR}} - \nu | T^*_{\text{WSR}} > \nu) = \log \gamma + \log \left(\sum_{j=1}^N \zeta_j e^{\varkappa_i - C(i)} \right) + o(1) \quad \text{as } \gamma \to \infty. \tag{8.406}$$

Note that this asymptotic approximation also holds for the randomized WSR algorithm with the randomization from the quasi-stationary distributions $\mathsf{Q}^i_A(x) = \lim_{n \to \infty} \mathsf{P}_\infty \left\{ R_n(i) \leq x | T^i_{\text{SR}} > n \right\}$ of the SR statistics $R_n(i)$ $(i = 1, \dots, N)$ and setting $R^w_0 = \sum_{i=1}^N W_i^* R^i_0$. In the discrete case, this procedure can be realized in two ways. The first is to pre-compute the quasi-stationary distributions Q^i_A, $i = 1, \dots, N$ using numerical techniques of Subsection 8.4.3.1 and then sample R^i_0 from Q^i_A. Another way is to compute $R^i_m(Y_i)$ based on the samples $(Y^i_1, \dots, Y^i_m)$ from f_0 independent of X prior to monitoring X. The size m should be large enough satisfying $\log \gamma \ll m \ll \gamma$.

The following theorem summarizes the above reasoning. Its rigorous and detailed proof will be presented elsewhere.

Theorem 8.5.3. *Suppose that $\mathsf{E}_{0,i}|Z^i_1|^2 < \infty$ and that Z^i_1 are P_i-nonarithmetic for $i = 1, \dots, N$. Let the mixing distribution in the randomized WSR procedure $T^*_{\text{WSR}}(W^*, A_\gamma)$ be chosen as in (8.405) and the threshold as $A_\gamma = \gamma \sum_{i=1}^N \zeta_i W_i^*$. Then, as $\gamma \to \infty$, $\mathsf{ARL2FA}(T^*_{\text{WSR}}(W^*, A_\gamma)) = \gamma(1 + o(1))$,*

$$\inf_{T \in \mathbf{C}_\gamma} \max_{i=1,\dots,N} \mathcal{R}_i(T) = \max_{i=1,\dots,N} \mathcal{R}_i(T^*_{\text{WSR}}(W^*, A_\gamma)) + o(1),$$

$$\max_{i=1,\dots,N} \mathcal{R}_i(T^*_{\text{WSR}}(W^*, A_\gamma)) = \log \gamma + \log \left(\sum_{j=1}^N \zeta_j e^{\varkappa_j - C(j)} \right) + o(1).$$

Therefore, the randomized WSR procedure is third-order asymptotically minimax with respect to the expected K–L information.

Since the values of $\varkappa_i$, ζ_i, and $C(i)$ can be precomputed (at least numerically) off-line, this discrete version of the WSR procedure is easily implementable on-line.

8.5.2 *Asymptotic Properties of the SR and Weighted SR Procedures in the Non-iid Case*

Consider now the general non-iid model (6.2). The SR procedure is again defined by the stopping time (8.323), where R_n is defined by (8.324) with $\mathcal{L}_i = f_1^{(\nu)}(X_i|\mathbf{X}_1^{i-1})/f_0(X_i|\mathbf{X}_1^{i-1})$. Recall that by Lemma 8.4.1, $\mathsf{ARL2FA}(T_{\text{SR}}(A)) \geq A$ for any $A > 0$, so $A = \gamma$ implies that $\mathsf{ARL2FA}(T_{\text{SR}}(\gamma)) \geq \gamma$ for any $\gamma > 1$.

In Subsection 8.2.8, we established asymptotic optimality properties of the CUSUM algorithm as $\gamma \to \infty$ in the class $\mathbf{C}_\gamma = \{T : \mathsf{E}_\infty T \geq \gamma\}$ as well as in the class $\mathbf{C}(\beta) = \{\mathsf{MPFA}_{m_\beta}(T) \leq \beta\}$ as $\beta \to 0$. Absolutely analogous results hold for the SR procedure when the threshold h (in the CUSUM procedure) is replaced with $\log A$.

More specifically, the assertions of Theorem 8.2.6 hold for the SR procedure: set $h = \log A$ in (8.202) and (8.204) is satisfied if we set $A = \gamma$. Theorem 8.2.7 also holds if we set $h_\beta = \log A_\beta$. Therefore, under appropriate conditions the SR procedure minimizes asymptotically as $\gamma \to \infty$ and $\beta \to 0$ to first-order moments of the detection delay in the classes $\mathbf{C}_\gamma$ and $\mathbf{C}(\beta)$.

For the same computational reason as for CUSUM it is of interest to introduce a window-limited SR procedure (WL-SR)

$$T_{\text{WL-SR}}(h) = \inf\left\{ n \geq 1 : \sum_{j=n-m_\beta+1}^{n} \sum_{k=j}^{n} \frac{f_1(X_k \mid \mathbf{X}_1^{k-1})}{f_0(X_k \mid \mathbf{X}_1^{k-1})} \geq A \right\},$$

where summation over all $1 \leq j \leq n$ in the SR statistic is replaced by summation in the sliding window of size m_β (cf. (8.218)). This window is large enough but not too large due to conditions (8.206). Assertions of Theorem 8.2.9 hold for the WL-SR procedure, again if we replace h with $\log A$.

In the non-iid composite-hypothesis case, the results similar to those obtained in Subsection 8.3.2 for the WCUSUM procedure also hold for the WSR procedure and its window-limited version. Recall that in the general non-iid case the WSR procedure is defined by (8.382)–(8.384). Theorem 8.3.4 holds for WSR with $h = \log A$, and the results analogous to Theorem 8.3.5 are true for the WL-WSR procedure.

The proofs are similar and we omit further details.

Chapter 9

Multichart Changepoint Detection Procedures for Composite Hypotheses and Multipopulation Models

In this chapter, we discuss in detail changepoint detection by using multichart procedures for iid and general non-iid models, which are outlined in Chapter 6, Section 6.3.

9.1 Motivation for Applying Multichart Detection Procedures

As follows from Section 8.3, there are no recursive or other computationally feasible implementations of the weighted CUSUM and GLR algorithms in general. An obvious disadvantage of such schemes is that the number of maximizations grows to infinity with n. For example, if sampling is performed with the rate of a given ARL to false alarm γ (which is too optimistic and usually even not realistic since the rate is expected to be much higher), then the mean number of maximizations of the LR over $\theta_1 \in \Theta_1$ that should be performed at time $n \in [1,\gamma]$ is $O(\gamma)$. In the case of weighted CUSUM, the mean number of WLR calculations (related to the integration over possible values of the parameter θ) also grows to infinity with n with the same rate.

In Section 8.3, we discussed several methods that had been proposed to reduce the computational cost of the weighted CUSUM, SR, and GLR schemes, in particular the window-limited GLR (WL GLR) scheme; see (8.313) and (8.321). As has been discussed in Section 8.3, the "optimal" choice of the moving window length m_γ is roughly speaking $m_\gamma = O(\log \gamma)$, where θ_0 and θ_1 are the pre- and post-change values of the parameter vector; see [250, 251, 253] for further details. The statistical interpretation of this result is very simple: the optimal mean detection delay for the alternative θ_1 is asymptotically equal to $\log \gamma / I(\theta_1, \theta_0)$, hence, to extract almost all useful information from the observations to make a decision on the existence or not of an abrupt change it is necessary to carry out at least $O(\log \gamma / I(\theta_1, \theta_0))$ last observations. Consequently, the WL GLR scheme involves at least $\log \gamma / I(\theta_1, \theta_0)$ LR maximizations at every stage n. In some situations this fact considerably reduces the computational burden (and also memory requirements) of the GLR scheme and makes this detection scheme manageable in real-time implementations.

Let us discuss now the most interesting practical case when $I(\theta_0, \theta_1) \in [I_0, I_1]$ and $0 < I_0 \ll I_1 < \infty$. This case can be considered without any loss of generality because in technical systems I_0 is a fixed positive constant (obtained from standard requirements). From the statistical point of view, this value establishes the indifference zone between the hypotheses H_0 and H_1. On the other hand, the parameter I_1 is chosen so that the mean detection delay is small. In this case the WL GLR scheme involves at least $\log \gamma / I_0$ LR maximizations at every stage n (we assume that $\widetilde{m}$ defined in Remark 8.3.4 is negligible). Hence, if the ratio $\log \gamma / I_0$ is large, then the WL GLR scheme is still time consuming. Another problem with the WL GLR is an adequate choice of the window size $m_\gamma = O(\log \gamma)$ for a finite value of γ, which is vague. Moreover, as the Monte Carlo simulation shows, the choice of m_γ as the smallest integer greater than $\log \gamma / I_0$ (according to the above proposal) leads to a substantial degradation of the WL GLR scheme performance compared to the optimal one.

In this chapter we argue that discretizing a parameter space and exploiting the corresponding multichart procedures is an efficient alternative to the "continuous" WL change detection procedures.

9.2 Multichart CUSUM and Shiryaev–Roberts Procedures

As in Subsection 8.5.1, consider the iid case and assume that the pre-change hypothesis $H_\infty : f = f_0$ is simple but the post-change hypothesis $H_k : \nu = k, f \in \{f_1, \dots, f_N\}$ is composite and discrete (not necessarily parametric), where $f_0, f_1, \dots, f_N$ are arbitrary densities with respect to some non-degenerate sigma-finite measure. Let $Z_n^i = \log[f_i(X_n)/f_0(X_n)]$ be the LLR for the n-th observation associated with post-change density f_i and $\lambda_n^k(i) = \sum_{t=k+1}^n Z_t^i$.

In this section, we are interested in the properties of the following multichart extensions of the CUSUM and SR procedures . For $i = 1, \dots, N$, define the "local" CUSUM and SR statistics

$$V_n^i = \max_{1 \le k \le n} \exp\left\{\lambda_n^{k-1}(i)\right\}, \quad R_n^i = \sum_{k=1}^n \exp\left\{\lambda_n^{k-1}(i)\right\}$$

and the corresponding local Markov times

$$T_{\mathrm{CS}}^i(B_i) = \inf\left\{n \ge 1 : V_n^i \ge B_i\right\}, \quad T_{\mathrm{SR}}^i(A_i) = \inf\left\{n \ge 1 : R_n^i \ge A_i\right\}, \tag{9.1}$$

where $B_i > 0, A_i > 0$ are thresholds. The stopping times of the multichart CUSUM and SR procedures, to which we will refer as MCUSUM and MSR, are

$$\begin{aligned} T_{\mathrm{MCS}}(B_1, \dots, B_N) &= \min\left\{T_{\mathrm{CS}}^1(B_1), \dots, T_{\mathrm{CS}}^N(B_N)\right\}, \\ T_{\mathrm{MSR}}(A_1, \dots, A_N) &= \min\left\{T_{\mathrm{SR}}^1(A_1), \dots, T_{\mathrm{SR}}^N(A_N)\right\}. \end{aligned} \tag{9.2}$$

Let $I_i = \mathsf{E}_{0,i}[Z_1^i]$ and $I_{0,i} = \mathsf{E}_\infty[-Z_1^i]$ be the K–L numbers ($i = 1, \dots, N$). As usual, we use $\kappa_a^i = \lambda_{T_a^i}^0(i) - a$ (on $\{T_a^i < \infty\}$) to denote the overshoot over $a > 0$ in the one-sided SPRT $T_a^i = \inf\left\{n : \lambda_n^0(i) \ge a\right\}$, and by $\mathsf{H}_i(x) = \lim_{a\to\infty} \mathsf{P}_{0,i}(\kappa_a^i \le x)$ we denote its limiting distribution. Let

$$\zeta_i = \int_0^\infty e^{-x}\, \mathsf{H}_i(\mathrm{d}x), \quad \varkappa_i = \int_0^\infty x\, \mathsf{H}_i(\mathrm{d}x), \quad i = 1, \dots, N.$$

In the following for the sake of simplicity we suppose that the LLRs Z_1^i, $i = 1, \dots, N$ are nonarithmetic with respect to P_∞ and $\mathsf{P}_{0,i}$.

While practically all the results hold for the general case, we will focus on a special multi-population case (a multisample slippage problem) similar to that treated in Subsection 5.5.1.3 for hypothesis testing. The multisample slippage scenario is most natural for considering the discrete post-change hypothesis, in addition to being useful as an approximation to a continuous-parameter testing problem discussed above, and it has a wide range of applications associated with multichannel or multisensor systems [465, 472, 475]. Suppose there are N populations some of which may become anomalous at an unknown point in time ν. For example, N sensors monitor different areas and a signal may appear in one or more of these areas. We wish to detect signal appearance as quickly as possible without identifying its location.

Specifically, we observe mutually independent populations $X_n^1, \dots, X_n^N$, $n = 1, 2, \dots$, which, for $\nu \ge n$, are distributed according to densities $p_0^i(X_n^i)$, $i = 1, \dots, N$. At an unknown time ν a change occurs, and one of the populations (and only one, which is the hardest case) changes its statistical properties.[1] If the change occurs in the j-th population, then for $n > \nu$ the pdf of X_n^j is $p_1^j(X_n^j)$. This

[1] Results similar to that obtained below can be also obtained in the case of a distributed structured change when the signal signature is known; say it is known that L nearest channels are affected when the change occurs.

scenario is a special case of the problem with $X_n = (X_n^1, \dots, X_n^N)$ being the N-dimensional vector with independent components,

$$f_0(X_n) = \prod_{j=1}^N p_0^j(X_n^j), \quad f_i(X_n) = p_1^i(X_n^i) \prod_{\substack{j=1 \\ j\neq i}}^N p_0^j(X_n^j), \quad i = 1, \dots, N.$$

Evidently, the joint pdf of the vector $\mathbf{X}_1^n = (X_1, \dots, X_n)^\top$ conditioned on the hypothesis H_ν^j that the change happens in the j-th population at time ν has the form

$$p(\mathbf{X}_1^n|H_\nu^j) = \left[\prod_{s=1}^{\nu}\prod_{i=1}^{N} p_0^i(X_s^i)\right] \times \left[\prod_{s=\nu+1}^{n} p_1^j(X_s^j) \cdot \prod_{\substack{i=1 \\ i\neq j}}^N p_0^i(X_s^i)\right], \quad \nu < n,$$

$$p(\mathbf{X}_1^n|H_\nu^j) = p(\mathbf{X}_1^n|H_\infty) = \prod_{s=1}^{n}\prod_{i=1}^{N} p_0^i(X_s^i), \quad \nu \geq n.$$

Thus, the LLRs take the form

$$\lambda_n^\nu(i) =: \log \frac{p(\mathbf{X}_1^n|H_\nu^i)}{p(\mathbf{X}_1^n|H_\infty)} = \sum_{t=\nu+1}^{n} Z_t^i, \quad Z_t^i = \log \frac{p_1^i(X_t^i)}{p_0^i(X_t^i)}, \quad i = 1, \dots, N.$$

The following lemma shows that asymptotically as $B_{\min} = \min_{1\leq i\leq N} B_i \to \infty$ and $A_{\min} = \min_{1\leq i\leq N} A_i \to \infty$ the P_∞-distributions of the stopping times T_{MCS} and T_{MSR} are exponential and provides useful asymptotic approximations for the ARL2FA.

Lemma 9.2.1. *Let the K–L numbers I_i and $I_{0,i}$ be positive and finite and let Z_1^i, $i = 1, \dots, N$ be nonarithmetic with respect to P_∞ and $\mathsf{P}_{0,i}$. Then as $B_{\min}, A_{\min} \to \infty$ the stopping times T_{MCS} and T_{MSR} are asymptotically exponentially distributed under P_∞ with expectations*

$$\mathsf{E}_\infty T_{\text{MCS}} \sim \frac{1}{\sum_{i=1}^N (I_i\zeta_i^2/B_i)}, \quad \mathsf{E}_\infty T_{\text{MSR}} \sim \frac{1}{\sum_{i=1}^N (\zeta_i/A_i)}. \tag{9.3}$$

Proof. By (8.145) and (8.356), $\mathsf{E}_\infty T_{\text{CS}}^i \sim B_i/I_i\zeta_i^2$ and $\mathsf{E}_i T_{\text{SR}}^i \sim A_i/\zeta_i$ as $B_i, A_i \to \infty$. Hence, by [369, Theorem 1], the P_∞-distributions of the stopping times $T_{\text{CS}}^i/(B_i/I_i\zeta_i^2)$ and $T_{\text{SR}}^i/(A_i/\zeta_i)$ are asymptotically $\mathsf{Expon}(1)$ as $B_{\min}, A_{\min} \to \infty$. Since $T_{\text{CS}}^1, \dots, T_{\text{CS}}^N$ as well as $T_{\text{SR}}^1, \dots, T_{\text{SR}}^N$ are independent random variables, the result follows from elementary calculation. □

The asymptotic equalities (9.3) imply that if the thresholds B_i and A_i satisfy equations

$$\sum_{i=1}^N (I_i\zeta_i^2/B_i) = 1/\gamma, \quad \sum_{i=1}^N (\zeta_i/A_i) = 1/\gamma, \tag{9.4}$$

then $\mathsf{ARL2FA}(T_{\text{MCS}}) \sim \gamma$ and $\mathsf{ARL2FA}(T_{\text{MSR}}) \sim \gamma$ as $\gamma \to \infty$.

Note also that if the thresholds are found from the equations $\sum_{i=1}^N B_i^{-1} = 1/\gamma$ and $\sum_{i=1}^N A_i^{-1} = 1/\gamma$, then $\mathsf{ARL2FA}(T_{\text{MCS}}) \geq \gamma$ and $\mathsf{ARL2FA}(T_{\text{MSR}}) \geq \gamma$ for all $\gamma > 1$.

Since equations (9.4) (each) include N unknowns, there are many ways to select the thresholds. One way is to use equal thresholds $B_i = B$ and $A_i = A$, which has been explored in [459]. Another way is to equalize the ARLs to false alarms in each channel, i.e., to add the constraints $B_1/I_1\zeta_1^2 = \cdots = B_N/I_N\zeta_N^2$ and similarly for the MSR procedure. The third reasonable option is to equalize the expected K–L information to detection in all channels

$$I_1\mathsf{SADD}_1(T_{\text{CS}}^1) = \cdots = I_N\mathsf{SADD}_N(T_{\text{CS}}^N), \quad I_1\mathsf{SADD}_1(T_{\text{SR}}^1) = \cdots = I_N\mathsf{SADD}_N(T_{\text{SR}}^N), \tag{9.5}$$

which is the most reasonable option, as will become apparent later on. Hereafter (as in the previous chapter) $\mathsf{SADD}_i(T) = \sup_\nu \mathsf{E}^i_\nu(T-\nu|T>\nu)$ and $I_i\mathsf{SADD}_i(T) = \sup_\nu \mathsf{E}^i_\nu(\lambda^i_T - \lambda^i_\nu|T>\nu)$.

It is worth noting that Lemma 9.2.1 allows us also to evaluate the maximal false alarm probability $\mathsf{MPFA}_m(T) = \sup_{k\geq 0}\mathsf{P}_\infty(T\leq k+m|T>k)$. Indeed, it follows from the lemma that as $B_{\min}, A_{\min}\to\infty$

$$\mathsf{MPFA}_m(T_{\text{MCS}}) \sim 1-\exp\left\{-m\sum_{i=1}^N (I_i\zeta_i^2/B_i)\right\}, \quad \mathsf{MPFA}_m(T_{\text{MSR}}) \sim 1-\exp\left\{-m\sum_{i=1}^N (\zeta_i/A_i)\right\}.$$

Therefore, if

$$\sum_{i=1}^N (I_i\zeta_i^2/B_i) = \alpha/m, \quad \sum_{i=1}^N (\zeta_i/A_i) = \alpha/m, \tag{9.6}$$

then $\mathsf{MPFA}_m(T_{\text{MCS}})\sim\alpha$ and $\mathsf{MPFA}_m(T_{\text{MSR}})\sim\alpha$ as $\alpha\to 0$.

Recall that $\mathbf{C}_\gamma = \{T : \mathsf{ARL2FA}(T)\geq\gamma\}$ and $\mathbf{C}^{m_\alpha}(\alpha) = \mathbf{C}(\alpha) = \{T : \mathsf{MPFA}_{m_\alpha}(T)\leq\alpha\}$, where m_α is allowed to depend on the PFA constraint α. Let $K_\alpha = |\log\alpha|/\min_i I_i$. As before (see (8.206)), we assume that the time window m_α depends on α in such a way that $m_\alpha/|\log\alpha|\gg 1$ but $(\log m_\alpha)/|\log\alpha|\ll 1$; specifically,

$$\liminf_{\alpha\to 0}(m_\alpha/K_\alpha) > 1 \quad \text{but} \quad \lim_{\alpha\to 0}[(\log m_\alpha)/K_\alpha] = 0. \tag{9.7}$$

The following theorem establishes the first-order asymptotic minimax optimality of the MCUSUM and MSR procedures in the classes $\mathbf{C}_\gamma$ and $\mathbf{C}(\alpha)$ with respect to positive moments of the detection delay $\mathsf{SM}^i_r(T) = \sup_\nu \mathsf{E}_{\nu,i}[(T-\nu)^r|T>\nu]$ and $\mathsf{ESM}^i_r(T) = \sup_\nu \operatorname{ess\,sup}\mathsf{E}_{\nu,i}\{[(T-\nu)+]^r|\mathscr{F}_\nu]$, $r>0$, $i=1,\dots,N$.

Theorem 9.2.1 (Asymptotic optimality). *Let the K–L numbers be positive and finite, $0<I_i<\infty$, $0<I_{0,i}<\infty$, $i=1,\dots,N$.*

(i) *Then for any $r>0$ and all $i=1,\dots,N$*

$$\begin{aligned}
\mathsf{SM}^i_r(T_{\text{MCS}}) \sim \mathsf{ESM}^i_r(T_{\text{MCS}}) &\sim \left(\frac{\log B_i}{I_i}\right)^r \quad \text{as } B_{\min}\to\infty,\\
\mathsf{SM}^i_r(T_{\text{MSR}}) \sim \mathsf{ESM}^i_r(T_{\text{MSR}}) &\sim \left(\frac{\log A_i}{I_i}\right)^r \quad \text{as } A_{\min}\to\infty
\end{aligned} \tag{9.8}$$

as long as $\lim_{B_{\min}\to\infty}(B_{\min}/B_i)$ and $\lim_{A_{\min}\to\infty}(A_{\min}/A_i)$ are bounded away from zero.

(ii) *If the thresholds $B_i = B_i(\gamma)$ are selected so that $\mathsf{ARL2FA}(T_{\text{MCS}})\geq\gamma$ and $\log B_i(\gamma)\sim\log\gamma$ and likewise for the MSR procedure, then for all $r>0$ and $i=1,\dots,N$, as $\gamma\to\infty$,*

$$\begin{aligned}
\inf_{T\in\mathbf{C}_\gamma}\mathsf{ESM}^i_r(T) \sim \inf_{T\in\mathbf{C}_\gamma}\mathsf{SM}^i_r(T) \sim \mathsf{ESM}^i_r(T_{\text{MCS}}) \sim \mathsf{SM}^i_r(T_{\text{MCS}}) \sim \left(\frac{\log\gamma}{I_i}\right)^r,\\
\inf_{T\in\mathbf{C}_\gamma}\mathsf{ESM}^i_r(T) \sim \inf_{T\in\mathbf{C}_\gamma}\mathsf{SM}^i_r(T) \sim \mathsf{ESM}^i_r(T_{\text{MSR}}) \sim \mathsf{SM}^i_r(T_{\text{MSR}}) \sim \left(\frac{\log\gamma}{I_i}\right)^r.
\end{aligned} \tag{9.9}$$

(iii) *Let (9.7) hold. If the thresholds $B_i = B_i(\alpha)$ are selected so that $\mathsf{MPFA}_{m_\alpha}(T_{\text{MCS}})\leq\alpha$ and $\log B_i(\alpha)\sim|\log\alpha|$ and likewise for the MSR procedure, then for all $r>0$ and $i=1,\dots,N$, as $\alpha\to 0$,*

$$\begin{aligned}
\inf_{T\in\mathbf{C}(\alpha)}\mathsf{ESM}^i_r(T) \sim \inf_{T\in\mathbf{C}(\alpha)}\mathsf{SM}^i_r(T) \sim \mathsf{ESM}^i_r(T_{\text{MCS}}) \sim \mathsf{SM}^i_r(T_{\text{MCS}}) \sim \left(\frac{|\log\alpha|}{I_i}\right)^r,\\
\inf_{T\in\mathbf{C}(\alpha)}\mathsf{ESM}^i_r(T) \sim \inf_{T\in\mathbf{C}(\alpha)}\mathsf{SM}^i_r(T) \sim \mathsf{ESM}^i_r(T_{\text{MSR}}) \sim \mathsf{SM}^i_r(T_{\text{MSR}}) \sim \left(\frac{|\log\alpha|}{I_i}\right)^r.
\end{aligned} \tag{9.10}$$

Moreover, the asymptotic equalities (9.10) also hold for the expectations $\mathsf{E}_{\nu,i}[(T-\nu)^r|T>\nu]$ uniformly over all $\nu\geq 0$.

Proof. We give the proofs of the three assertions only for MCUSUM. For MSR the proofs are essentially the same.

(i) Since obviously $T_{\text{MCS}} \le T^i_{\text{CS}}$ (for any i), it follows from Theorem 8.2.3(i) that

$$\mathsf{SM}^i_r(T_{\text{MCS}}) \le \mathsf{ESM}^i_r(T_{\text{MCS}}) \le \left(\frac{\log B_i}{I_i}\right)^r (1+o(1)) \quad \text{as } B_{\min} \to \infty.$$

In the iid case,

$$\sup_{\nu \ge 0} \operatorname{ess\,sup} \mathsf{P}_{\nu,i}\left\{M^{-1} \max_{0\le n<M} \lambda^{\nu}_{\nu+n}(i) \ge (1+\varepsilon)I_i | \mathscr{F}_\nu\right\} = \mathsf{P}_{0,i}\left\{M^{-1} \max_{0\le n<M} \lambda^0_n(i) \ge (1+\varepsilon)I_i\right\},$$

and by the SLLN,

$$\lim_{M\to\infty} \mathsf{P}_{0,i}\left\{M^{-1} \max_{0\le n<M} \lambda^0_n(i) \ge (1+\varepsilon)I_i\right\} = 0 \quad \text{for all } \varepsilon > 0.$$

Hence, by Theorem 8.2.2, the lower bound

$$\mathsf{ESM}^i_r(T) \ge \left(\frac{\log B_i}{I_i}\right)^r (1+o(1)) \quad \text{as } B_i \to \infty$$

holds for any stopping time T with $\mathsf{E}_\infty T \ge CB_i$ and $\log(CB_i) \sim \log B_i$ as $B_i \to \infty$. Since $C = 1 + \sum_{k\neq i}(B_i/B_k) < \infty$ whenever $B_k/B_{\min}$, $k = 1,\dots,N$ are bounded away from infinity, it follows that

$$\mathsf{ESM}^i_r(T_{\text{MCS}}) \ge \mathsf{SM}^i_r(T_{\text{MCS}}) \ge \left(\frac{\log B_i}{I_i}\right)^r (1+o(1)) \quad \text{as } B_{\min} \to \infty,$$

which along with the previous reverse inequality implies (9.8).

(ii) By Theorem 8.2.2,

$$\inf_{T\in\mathbf{C}_\gamma} \mathsf{ESM}^i_r(T) \ge \inf_{T\in\mathbf{C}_\gamma} \mathsf{SM}^i_r(T) \ge \left(\frac{\log \gamma}{I_i}\right)^r (1+o(1)) \quad \text{as } \gamma \to \infty,$$

which is attained by the MCUSUM procedure if $\log B_i(\gamma) \sim \log\gamma$ by (9.8).

(iii) By Theorem 8.2.8(i),

$$\inf_{T\in\mathbf{C}(\alpha)} \mathsf{E}_{\nu,i}[(T-\nu)^r | T > \nu] \ge \left(\frac{|\log\alpha|}{I_i}\right)^r (1+o(1)) \quad \text{for all } \nu \ge 0.$$

By (9.8), this bound is attained by MCUSUM if $\log B_i(\alpha) \sim |\log\alpha|$. □

As discussed above, there are several ways to select thresholds that satisfy equations (9.4) and (9.6). If we add the constraints (9.5) that equalize the expected K–L information for all $i = 1,\dots,N$ (approximately up to asymptotically small terms), then the first-order approximations (9.8) yield $\log B_i = \mathtt{constant}$, i.e., the first-order equalizer prescribes constant thresholds $B_i = B$ and $A_i = A$ that do not depend on the population number. If

$$B = \gamma \sum_{i=1}^N \zeta_i^2 I_i, \quad A = \gamma \sum_{i=1}^N \zeta_i,$$

then $\mathsf{ARL2FA}(T_{\text{MCS}}) \sim \mathsf{ARL2FA}(T_{\text{MSR}}) \sim \gamma$ as $\gamma \to \infty$ and if

$$B = \frac{m_\alpha}{\alpha} \sum_{i=1}^N \zeta_i^2 I_i, \quad A = \frac{m_\alpha}{\alpha} \sum_{i=1}^N \zeta_i,$$

then $\mathsf{MPFA}_{m\alpha}(T_{\text{MCS}}) \sim \mathsf{MPFA}_{m\alpha}(T_{\text{MSR}}) \sim \alpha$ as $\alpha \to 0$. Therefore, by Theorem 9.2.1, the MCUSUM and MSR procedures with constant thresholds are asymptotically optimal in the classes with $\mathsf{ARL2FA}(T) \geq \gamma(1+o(1))$ and $\mathsf{MPFA}_{m\alpha}(T) \leq \alpha(1+o(1))$.[2] Note that in the case of thresholds independent of the population number, the MCUSUM and MSR stopping times are

$$T_{\text{MCS}}(B) = \inf\left\{n \geq 1 : \max_{1\leq i\leq M} V_n^i \geq B\right\}, \quad T_{\text{MSR}}(A) = \inf\left\{n \geq 1 : \max_{1\leq i\leq M} R_n^i \geq A\right\}.$$

That is, MCUSUM is nothing but GCUSUM (GLR CUSUM) in this particular case.

We now proceed with higher-order expansions for the expected detection delay for the general case of different thresholds. To this end, we rewrite the stopping times T_{MCS} and T_{MSR} in the form of a random walk crossing a threshold plus a nonlinear slowly changing term. This allows us to apply nonlinear renewal theory. The argument is only slightly more complicated that in the case of a single population; see Theorem 8.2.5. For this reason, we omit technical details and give only some heuristics.

For large threshold values, the probability $\mathsf{P}_{\nu,i}\left\{T_{\text{MCS}} = T_{\text{CS}}^i | T_{\text{MCS}} > \nu\right\}$ is close to 1, so as $B_{\min} \to \infty$ and $A_{\min} \to \infty$,

$$\begin{aligned} \mathsf{CADD}_{\nu,i}(T_{\text{MCS}}) &= \mathsf{CADD}_{\nu,i}(T_{\text{CS}}^i) + o(1), \\ \mathsf{CADD}_{\nu,i}(T_{\text{MSR}}) &= \mathsf{CADD}_{\nu,i}(T_{\text{MSR}}^i) + o(1). \end{aligned} \tag{9.11}$$

Indeed, it can be shown that $\mathsf{P}_{0,i}(T_{\text{MCS}} \neq T_{\text{CS}}^i) \leq O(1/B_i)$, so taking into account that by (9.8) $\mathsf{E}_{i,0} T_{\text{MCS}}^2 = O((\log B_i)^2)$, we obtain

$$\mathsf{E}_{0,i}(T_{\text{CS}}^i - T_{\text{MCS}}) \leq \mathsf{E}_{0,i}\left[T_{\text{CS}}^i \mathbb{1}_{\{T_{\text{MCS}} \neq T_{\text{CS}}^i\}}\right] \leq \sqrt{\mathsf{E}_{0,i}[(T_{\text{CS}}^i)^2] \mathsf{P}_{0,i}(T_{\text{MCS}} \neq T_{\text{CS}}^i)} = O\big((\log B_i)/B_i^{1/2}\big),$$

where we used the Cauchy–Schwarz–Bunyakovsky inequality. Therefore, it suffices to evaluate $\mathsf{CADD}_{\nu,i}(T_{\text{CS}}^i) = \mathsf{E}_{\nu,i}(T_{\text{CS}}^i - \nu | T_{\text{CS}}^i > \nu)$ for large B_i. As mentioned in Subsection 8.2.6.4, this task in manageable only for $\nu = 0$ and large ν, i.e., for $\mathsf{CADD}_{0,i}(T) = \mathsf{E}_{0,i} T$ and $\mathsf{CADD}_{\infty,i}(T) = \lim_{\nu\to\infty} \mathsf{E}_{\nu,i}(T - \nu | T > \nu)$.

Write $g_n^i = \max(0, \log V_n^i)$. Let $\mathsf{Q}_{\text{cs}}^i(y) = \lim_{n\to\infty} \mathsf{P}_\infty(g_n^i \leq y)$ and $\mathsf{Q}_{\text{sr}}^i(y) = \lim_{n\to\infty} \mathsf{P}_\infty(R_n^i \leq y)$ be the stationary distributions of the CUSUM and SR statistics g_n^i and R_n^i, respectively. Define

$$\begin{aligned}
\widetilde{\mathsf{Q}}_{\text{cs}}^i(y) &= \mathsf{P}_{0,i}\left(\min_{n\geq 0} \lambda_n^0(i) \leq y\right), \quad \widetilde{\mathsf{Q}}_{\text{sr}}^i(y) = \mathsf{P}_{0,i}\left(\sum_{n=1}^{\infty} e^{-\lambda_n^0(i)} \leq y\right), \\
C_{\text{cs}}^i &= -\mathsf{E}_{0,i}\left[\min_{n\geq 0} \lambda_n^0(i)\right] = -\int_{-\infty}^{0} y \, \mathrm{d}\widetilde{\mathsf{Q}}_{\text{cs}}^i(y), \\
\widetilde{C}_{\text{cs}}^i &= -\mathsf{E}_\infty\left\{\mathsf{E}_{0,i}\left[\min\left(-g_\infty^i, \min_{n\geq 0} \lambda_n^0(i)\right)\right] \Big| g_\infty^i\right\} = -\int_0^\infty \int_{-\infty}^{-z} y \, \mathrm{d}\widetilde{\mathsf{Q}}_{\text{cs}}^i(y) \mathrm{d}\mathsf{Q}_{\text{cs}}^i(z), \\
C_{\text{sr}}^i &= \mathsf{E}_{0,i}\left[\log\left(1 + \sum_{n=1}^{\infty} e^{-\lambda_n^0(i)}\right)\right] = \int_0^\infty \log(1+y) \, \mathrm{d}\widetilde{\mathsf{Q}}_{\text{sr}}^i(y), \\
\widetilde{C}_{\text{sr}}^i &= \mathsf{E}_\infty\left\{\mathsf{E}_{0,i}\left[\log\left(1 + R_\infty^i + \sum_{n=1}^{\infty} e^{-\lambda_n^0(i)}\right)\right] \Big| R_\infty^i\right\} = \int_0^\infty \int_0^\infty \log(1+r+y) \, \mathrm{d}\widetilde{\mathsf{Q}}_{\text{sr}}^i(y) \, \mathrm{d}\mathsf{Q}_{\text{sr}}^i(r).
\end{aligned}$$

Using Theorem 8.2.5 and Theorem 8.4.5, we obtain the following result.

Theorem 9.2.2. *Assume that $\mathsf{E}_{0,i}|Z_1^i|^2 < \infty$ and that Z_1^i are $\mathsf{P}_{0,i}$-nonarithmetic, $i = 1, \dots, N$. Then*

[2]Note that there is no guarantee that the inequalities $\mathsf{ARL2FA}(T_{\text{MCS}}) \geq \gamma$ and $\mathsf{MPFA}_{m\alpha}(T) \leq \alpha$ are satisfied for all $\gamma > 1$ and $\alpha < 1$ with these selections of thresholds.

for MCUSUM, as $B_{\min} \to \infty$,

$$\mathsf{CADD}_{0,i}(T_{\text{MCS}}) = I_i^{-1}\left(\log B_i + \varkappa_i - C_{\text{cs}}^i\right) + o(1), \tag{9.12}$$

$$\mathsf{CADD}_{\infty,i}(T_{\text{MCS}}) = I_i^{-1}\left(\log B_i + \varkappa_i - \widetilde{C}_{\text{cs}}^i\right) + o(1), \tag{9.13}$$

and for MSR, as $A_{\min} \to \infty$,

$$\mathsf{CADD}_{0,i}(T_{\text{MSR}}) = I_i^{-1}\left(\log A_i + \varkappa_i - C_{\text{sr}}^i\right) + o(1), \tag{9.14}$$

$$\mathsf{CADD}_{\infty,i}(T_{\text{MSR}}) = I_i^{-1}\left(\log A_i + \varkappa_i - \widetilde{C}_{\text{sr}}^i\right) + o(1). \tag{9.15}$$

As for the single-chart CUSUM and SR procedures, the supremum of the CADD for the MCUSUM and MSR procedures is attained at $\nu = 0$. Hence, after omission of the vanishing terms $o(1)$ in (9.12) and (9.14) the equalizer property (9.5) leads to the equations

$$\log B_i + \varkappa_i - C_{\text{cs}}^i = c, \quad \log A_i + \varkappa_i - C_{\text{sr}}^i = c^*, \quad i = 1,\dots,N,$$

which imply $B_i = e^{c-(\varkappa_i - C_{\text{cs}}^i)}$ and $A_i = e^{c^*-(\varkappa_i - C_{\text{sr}}^i)}$. Using these equations together with (9.4) yields

$$B_i = \gamma\left(\sum_{j=1}^N \zeta_j^2 I_j e^{\varkappa_j - C_{\text{cs}}^j}\right) e^{-(\varkappa_i - C_{\text{cs}}^i)}, \quad A_i = \gamma\left(\sum_{j=1}^N \zeta_j e^{\varkappa_j - C_{\text{sr}}^j}\right) e^{-(\varkappa_i - C_{\text{sr}}^i)}. \tag{9.16}$$

With this choice of thresholds the MCUSUM and MSR procedures are equalizers with respect to the post-change hypothesis up to an asymptotically negligible term $o(1)$: for all $i = 1,\dots,N$ as $\gamma \to \infty$

$$\begin{aligned} I_i \mathsf{SADD}_i(T_{\text{MCS}}) &= \log\gamma + \log\left(\sum_{j=1}^N \zeta_j^2 I_j e^{\varkappa_j - C_{\text{cs}}^j}\right) + o(1), \\ I_i \mathsf{SADD}_i(T_{\text{MSR}}) &= \log\gamma + \log\left(\sum_{j=1}^N \zeta_j e^{\varkappa_j - C_{\text{sr}}^j}\right) + o(1). \end{aligned} \tag{9.17}$$

According to Theorem 8.5.3 both procedures are second-order asymptotically minimax up to a constant. Indeed, by this theorem,

$$\inf_{T \in \mathbf{C}_\gamma} \max_{1 \le i \le N} [I_i \mathsf{SADD}_i(T)] = \log\gamma + \log\left(\sum_{j=1}^N \zeta_j e^{\varkappa_j - \widetilde{C}_{\text{sr}}^j}\right) + o(1),$$

where $\widetilde{C}_{\text{sr}}^j > C_{\text{cs}}^j > C_{\text{sr}}^j$.

To make the MSR procedure almost optimal up to an $o(1)$ term, one may initialize each local SR statistic at random according to the quasi-stationary distributions $\mathsf{Q}_{A_i}(x) = \lim_{n\to\infty} \mathsf{P}_\infty(R_n^i \le x | T_{\text{CS}}^i > n)$, $i = 1,\dots,N$. Denote this randomized MSR procedure as T_{MSR}^Q. For this procedure,

$$I_i \mathsf{SADD}_i(T_{\text{MSR}}^Q) = I_i \mathsf{CADD}_{\infty,i}(T_{\text{SR}}^i) + o(1) = \log\gamma + \log\left(\sum_{j=1}^N \zeta_j e^{\varkappa_j - \widetilde{C}_{\text{sr}}^j}\right) + o(1).$$

Similar results hold in the case of the PFA constraint if one replaces γ with m_α/α.

To summarize, the MCUSUM and MSR procedures with the thresholds given by (9.16) have two optimality properties: (1) they are second-order asymptotically optimal (minimax) with respect to the maximal average delay to detection $\mathsf{SADD}_i(T)$ (and with respect to $\mathsf{ESADD}_i(T)$ too) uniformly for all $i = 1,\dots,N$, as well as minimize asymptotically all moments of the detection delay to first-order, and (2) they are second-order (double) asymptotically minimax with respect to the K–L expected information $\max_i[I_i \mathsf{SADD}_i(T)] = \max_{i,\nu}[I_i \mathsf{CADD}_{\nu,i}(T)]$ required for detecting a change. Furthermore, by randomization at the beginning the MSR procedure can be made third-order asymptotically minimax.

Example 9.2.1. Consider a multipopulation Gaussian model where both the pre- and post-change observations are iid unit-variance normal random variables, having expected values zero and $\theta_i > 0$, respectively, $p_0(X_n^i) \sim \mathcal{N}(0,1)$, $p_i(X_n^i) \sim \mathcal{N}(\theta_i, 1)$, $i = 1,\dots,N$. In this example, $Z_n^i(X_n^i) = \theta_i X_n^i - \theta_i^2/2$, $I_i = \theta_i^2/2$, $C_{\rm cs}^i = \mathsf{E}_i(Z_1^i)^2/2I_i - \varkappa_i = 1 + \theta_i^2/4 - \varkappa_i$ (cf. (8.152)),

$$\varkappa_i = 1 + \frac{\theta_i^2}{4} - \theta_i \sum_{k=1}^{\infty} \left[\frac{1}{\sqrt{k}} \varphi\left(\frac{\theta_i}{2}\sqrt{k}\right) - \frac{\theta_i}{2} \Phi\left(-\frac{\theta_i}{2}\sqrt{k}\right) \right],$$

$$\zeta_i = \frac{2}{\theta_i^2} \exp\left\{ -2 \sum_{k=1}^{\infty} \frac{1}{k} \Phi\left(-\frac{\theta_i}{2}\sqrt{k}\right) \right\}.$$

These two latter expressions do not simplify, but are amenable to numerical calculation with any given precision. The constant $C_{\rm sr}^i$ can be computed numerically either by Monte Carlo simulation or by numerical integration; see Subsection 8.4.3.4.

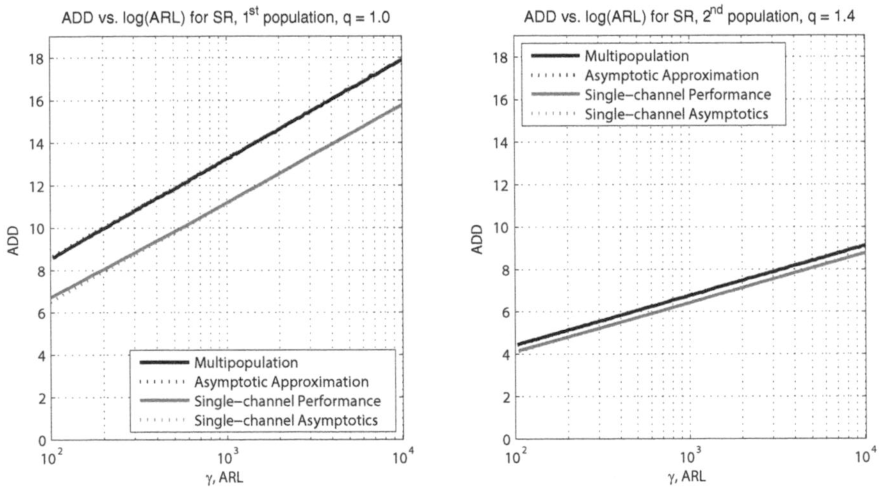

(a) The MSR procedure.

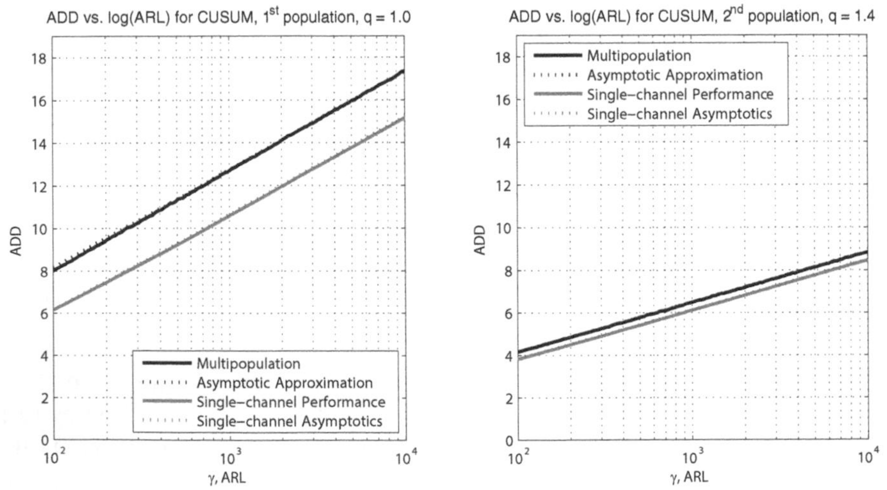

(b) The MCUSUM procedure.

Figure 9.1: *Operating characteristics of the MSR and MCUSUM procedures.*

Almost exact operating characteristics of the MCUSUM and MSR procedures can be computed solving numerically integral equations that are generalizations of the integral equations (8.105)–(8.109) to the vector (multipopulation) case [436].

We now present sample results for a two-population case ($N = 2$) with post-change parameters $\theta_1 = 1.0$ and $\theta_2 = 1.4$. The single-population model serves as a benchmark. Figure 9.1 shows the maximal expected delays to detection $\mathsf{SADD}_i(T)$, $i = 1,2$ *versus* the ARL to false alarm in the range from 10^2 to 10^4, i.e., from high FAR to low FAR. The solid lines correspond to the exact characteristics computed by numerical integration and the dashed lines to the asymptotic approximations (9.17). The asymptotic approximations work very well, with error less than 5% achieved at ARL2FA as low as 100. This means that one can use the proposed expansions for choosing the design parameters and resort to the numerical framework, which can be costly for high levels of threshold, only at low values of ARL2FA to obtain more accurate results. The single-population performance is of course always better since there is a price to pay for prior uncertainty, but the difference is not dramatic. Also, MCUSUM is more efficient than MSR, as expected, but the difference is very small.

9.3 Quickest Detection of Unstructured Changes in Multiple Populations

We now consider a more difficult and challenging multipopulation scenario where there are multiple streams of data and the number and indexes of the affected streams are unknown. We use the notation of the previous section. In other words, we assume that at an unknown time ν there are changes in distributions of data streams X_n^i for $i \in \mathcal{N} \subset \{1,\dots,N\}$, where the subset $\mathcal{N}$ and its size $|\mathcal{N}|$ ($1 \le |\mathcal{N}| \le N$) are not known. Furthermore, we are mostly interested in the *unstructured* changepoint problem when the geometric structure (spatial signature) of a change across populations is not known. This problem is often of interest in applications. For example, it is a generalization of the target detection problem in multisensor (or multichannel) target detection systems considered in the previous section in a single-target context, to the practically important multitarget case with an *a priori* unknown number of targets. Yet another challenging application is detection of intrusions in computer networks using multichannel intrusion detection systems where a certain unknown number of channels can be affected; see Section 11.3.

Specifically, given N populations (sensors, channels), the observations represent the N-component vector $X_n = (X_n^1,\dots,X_n^N)^\top$, $n \ge 1$ with components independent within and across populations. If the j-th component changes its statistical properties, the pre-change density $p_0^j(X_n^j)$ changes to the post-change density $p_1^j(X_n^j)$, so that the joint density of vector $\mathbf{X}_1^n = (X_1,\dots,X_n)^\top$ conditioned on the hypothesis $H_\nu^{\mathcal{N}}$ that the change occurs in the subset of streams $\mathcal{N} \subset \{1,\dots,N\}$ at time $\nu \in \{0,1,\dots\}$ has the form

$$p(\mathbf{X}_1^n|H_\nu^{\mathcal{N}}) = \left[\prod_{s=1}^{\nu}\prod_{i=1}^{N} p_0^i(X_s^i)\right] \times \left[\prod_{s=\nu+1}^{n}\prod_{j\in\mathcal{N}} p_1^j(X_s^j)\cdot\prod_{i\notin\mathcal{N}} p_0^i(X_s^i)\right], \quad \nu < n,$$

$$p(\mathbf{X}_1^n|H_\nu^{\mathcal{N}}) \equiv p(\mathbf{X}_1^n|H_\infty) = \prod_{s=1}^{n}\prod_{i=1}^{N} p_0^i(X_s^i), \quad \nu \ge n.$$

The subset of affected components $\mathcal{N} \subset \{1,\dots,N\}$ is not known in advance, and the goal is to design a detection procedure that (asymptotically as $\gamma \to \infty$) minimizes the maximal average delay to detection $\mathsf{SADD}^{\mathcal{N}}(T) = \sup_{\nu\ge 0} \mathsf{E}_\nu^{\mathcal{N}}(T-\nu|N>\nu)$ in the class $\mathbf{C}_\gamma$ for every possible subset $\mathcal{N}$, where $\mathsf{E}_\nu^{\mathcal{N}}$ is the expectation when the change occurs in the subset $\mathcal{N}$ at time ν.

In the extreme case where the change occurs in a single unknown component, $|\mathcal{N}| = 1$, the multichart CUSUM and SR procedures of the previous section are asymptotically optimal as $\gamma \to \infty$. These procedures are expected to work very well if the proportion of affected components is small compared to the total number of components [472]. In another extreme case where $\mathcal{N} = \{1,\dots,N\}$, nearly optimal changepoint detection procedures are either SRP or SR–r with the cumulative LR $\mathcal{L}_n = \prod_{i=1}^N [p_1^i(X_n^i)/p_0^i(X_n^i)]$.

However, the real challenge is when $\mathcal{N}$ is an arbitrary subset of $\{1,\dots,N\}$ with an arbitrary, unknown size. Let $W_n^i = \log V_n^i$ be the CUSUM statistic for the i-th component, $W_n^i = (W_{n-1}^i)^+ + Z_n^i$, $W_0^i = 0$. Mei [294] proposed the detection procedure based on the sum of local CUSUM statistics, $M_a = \inf\{n : \sum_{i=1}^N W_n^i \ge a\}$. We believe that Mei's procedure can also be improved by introducing a "pilot" thresholding of the component CUSUM statistics $W_n^i \ge h_i$ and then accounting in the sum only those that exceed preliminary thresholds h_j, making it more adaptive and reducing "noise." The same method applies to the SR statistics. That is, one may add $\log R_n^i$ for different populations and use the stopping time $K_a = \inf\{n : \sum_{i=1}^N \log R_n^i \ge a\}$. It can be relatively easily shown that for both detection procedures the maximal ADD is asymptotically given by

$$\mathsf{SADD}^{\mathcal{N}}(M_a) \sim \mathsf{SADD}^{\mathcal{N}}(K_a) \sim \frac{a}{\sum_{i\in\mathcal{N}} I_i} \quad \text{for any } \mathcal{N} \in \{1,\dots,N\} \ \text{ as } a \to \infty.$$

Therefore, if $a = a(N,\gamma)$ is selected so that $\mathsf{ARL2FA}(a) \sim \gamma$ and $a(N,\gamma) \sim \log\gamma$, then these procedures are asymptotically first-order minimax as $\gamma \to \infty$,

$$\inf_{T\in\mathbf{C}_\gamma} \mathsf{SADD}^{\mathcal{N}}(T) \sim \mathsf{SADD}^{\mathcal{N}}(M_a) \sim \mathsf{SADD}^{\mathcal{N}}(K_a) \sim \frac{\log\gamma}{\sum_{i\in\mathcal{N}} I_i} \quad \text{for any } \mathcal{N} \in \{1,\dots,N\}.$$

In many applications the maximal proportion L/N of the components that can be affected is known in advance. Moreover, this proportion can be relatively small. One application area where this is the case is considered by Siegmund [429, Section 3.5]; another interesting application related to detection and tracking of an unknown number of targets is discussed in the end of this section. In this case, a good algorithm should necessarily take this information into account restricting a number of component statistics in the cumulative decision statistic by this proportion. One way of doing so is to rank the local CUSUM statistics in Mei's algorithm and apply the stopping time $M_a^* = \inf\{n : \sum_{i=1}^L W_n^{(i)} \ge a\}$, where $W_n^{(i)}$, $i = 1,2,\dots,L$ is the space-ordered sequence of statistics, $W_n^{(1)} \ge W_n^{(2)} \ge \cdots \ge W_n^{(L)}$. Another way is to use the assumed fraction $|\mathcal{N}|/N$ to compute a prior probability of populations being affected by a change.

To understand the structure of a reasonable detection procedure it is instructive to consider a Bayesian problem with geometric prior distribution for the changepoint, $\mathsf{P}(\nu = k) = \rho(1-\rho)^k$, $k \ge 0$, assuming also that, conditioned on the event that the change occurs at a certain time $\nu = k < \infty$, the populations are affected independently of each other with probabilities $\mathsf{P}(\theta_i = 1) = \pi_i$, where θ_i is a Bernoulli random variable that takes the value 1 if the i-th population is affected and 0 otherwise. Then the posterior probability $\mathsf{P}(\nu < n|\mathbf{X}_1^n)$ of the change being in effect before time n can be shown to be proportional to the statistic

$$R_n^\rho(N) = \sum_{k=1}^n (1-\rho)^{k-1} \prod_{i=1}^N \left[1 - \pi_i + \pi_i \Lambda_n^{k-1}(i)\right],$$

where

$$\Lambda_n^\nu(i) = \exp\{\lambda_n^\nu(i)\} = \frac{p(X_1^i,\dots,X_n^i|H_\nu^i)}{p(X_1^i,\dots,X_n^i|H_\infty)} = \prod_{k=\nu+1}^n \frac{p_1^i(X_k^i)}{p_0^i(X_k^i)}$$

is the local LR of the hypothesis H_ν^i that the change occurs in the i-th population at time $\nu < \infty$ *versus* H_∞ that there is never a change. In the hypothesis testing context, the statistic

$$\Lambda_n^\nu = \prod_{i=1}^N [1 - \pi_i + \pi_i \Lambda_n^\nu(i)] \tag{9.18}$$

can be interpreted as the LR of the hypothesis that at least one population is affected against none. Letting ρ go to 0, namely taking the uniform improper prior for the changepoint, we obtain that the statistic $R_n^\rho(N)$ converges to the statistic

$$R_n(N) = \sum_{k=1}^n \prod_{i=1}^N \left[1 - \pi_i + \pi_i \Lambda_n^{k-1}(i)\right]. \tag{9.19}$$

Hence, it stands to reason that the stopping rule $T_{\mathrm{SR}}^N(A) = \inf\{n : R_n(N) \ge A\}$ should be a "good" detection procedure at least asymptotically for low FAR. Note that the statistic $R_n(N)$ is a multipopulation prototype of the SR statistic. Another reasonable option is to replace summation with maximization over $k = 1,\dots,n$, which leads to a multipopulation version of the CUSUM detection procedure (after logarithmic transformation),

$$W_n(N) = \max_{1\le k\le n} \sum_{i=1}^{N} \log\left[1-\pi_i+\pi_i\Lambda_n^{k-1}(i)\right], \quad T_{\mathrm{CS}}^N(a) = \inf\{n\ge 1 : W_n(N) \ge a\}. \tag{9.20}$$

As usual, the thresholds A and a are selected so that the constraint imposed on the FAR is met. The LRs $\Lambda_n^{k-1}(i)$ in these formulas can be further thresholded to neglect the populations that are not affected, e.g., one may replace $\Lambda_n^{k-1}(i)$ with $\max\{\Lambda_n^{k-1}(i), 1\}$. Also, to simplify the problem one may assume equal values of the probabilities $\pi_i = \pi$ for all $i = 1,\dots,N$. Since $\theta_1,\dots,\theta_N$ are independent Bernoulli random variables, the probability π is found from the equation

$$\frac{N!}{L!(N-L)!}\pi^L(1-\pi)^{N-L} = \frac{L}{N}.$$

Xie and Siegmund [516] considered a Gaussian model with a change in the mean assuming in addition that the post-change mean values $\mu_1,\dots,\mu_N$ are unknown and being replaced with maximum likelihood estimators, which results in the multipopulation GLR CUSUM-type algorithm of the form

$$T^N(a) = \inf\left\{n : \max_{1\le k\le n}\sum_{i=1}^{N}\log\left[1-\pi+\pi\exp\left[(n-k+1)^{-1}\left(\left(\sum_{j=k}^{n} X_j^i\right)^+\right)^2\right]\right] \ge a\right\}.$$

Moreover, for practical purposes they study a window limited version of this algorithm where the maximization is restricted to a sliding window of size m such that $m/a \to \infty$ as $a \to \infty$. Accurate approximations for the ARL2FA and the ADD were obtained in [516], but asymptotic optimality properties were not discussed.

Observe further that for large values of $\lambda_n^{k-1}(i)$,

$$1-\pi+\pi\Lambda_n^{k-1}(i) = 1-\pi+\pi\exp\left\{\lambda_n^{k-1}(i)\right\} \approx \pi\exp\left\{\lambda_n^{k-1}(i)\right\},$$

which suggests the detection algorithms

$$T_*^N(A) = \inf\left\{n\ge 1 : \sum_{k=1}^{n}\exp\left[\sum_{i=1}^{N}\left(\lambda_n^{k-1}(i)\right)^+\right] \ge A\right\},$$

$$T_\star^N(a) = \inf\left\{n\ge 1 : \max_{1\le k\le n}\sum_{i=1}^{N}\left(\lambda_n^{k-1}(i)\right)^+ \ge a\right\},$$

where summation $\sum_{i=1}^N \left(\lambda_n^{k-1}(i)\right)^+$ is replaced with summation of the maximal L values of $\lambda_n^{k-1}(i)$ if the number of possible affected populations is *a priori* bounded by $L < N$.

Establishing asymptotic optimality properties (in the minimax sense with respect to ν and uniform in $\mathcal{N}$) of the above introduced change detection procedures is an open question. See, however, [294] regarding asymptotic optimality of the rule M_a. To establish asymptotic optimality of other procedures introduced above one may use nonlinear renewal theory along with obtaining asymptotic approximations for the ARL2FA. Consider, for example, the procedure $T_{\mathrm{CS}}^N(a)$ defined in (9.20). Observe that

$$\sum_{i=1}^{N}\log\left(1-\pi+\pi e^{\lambda_n^0(i)}\right) = \sum_{i\in\mathcal{N}}\lambda_n^0(i) + \xi_n,$$

where the first term is a random walk with mean $n\sum_{i\in\mathcal{N}} I_i$ and

$$\xi_n = |\mathcal{N}|\log\pi + \sum_{i\in\mathcal{N}}\log\left(1+\frac{1-\pi}{\pi}e^{-\lambda_n^0(i)}\right) + \sum_{i\notin\mathcal{N}}\log\left(1-\pi+\pi e^{\lambda_n^0(i)}\right)$$

is a slowly changing sequence, since by the SLLN it converges $\mathsf{P}_0^{\mathcal{N}}$-a.s. to the constant

$$C(\pi,N,\mathcal{N}) = |\mathcal{N}|\log\pi + (N-|\mathcal{N}|)\log(1-\pi).$$

Hence, assuming that $\mathsf{E}_0^i|\lambda_1^0(i))|^2<\infty$, by the nonlinear renewal theorem

$$\mathsf{E}_0^{\mathcal{N}}[T_{\rm CS}^N(a)] = \frac{1}{\sum_{i\in\mathcal{N}} I_i}\left[a+\varkappa_{\mathcal{N}} - C(\pi,N,\mathcal{N})\right]+o(1) \quad \text{as } a\to\infty, \tag{9.21}$$

where $\varkappa_{\mathcal{N}} = \lim_{a\to\infty}\mathsf{E}_0^{\mathcal{N}}[\sum_{i\in\mathcal{N}}\lambda_{\tau_a^{\mathcal{N}}}^0(i)-a]$ in the limiting average overshoot in the one-sided SPRT $\tau_a^{\mathcal{N}} = \inf\{n:\sum_{i\in\mathcal{N}}\lambda_n^0(i)\ge a\}$.

Next, let

$$T_a(\pi,N) = \inf\left\{n\ge 1: \sum_{i=1}^N \log\left[1-\pi+\pi e^{\lambda_n^0(i)}\right]\ge a\right\}.$$

Observing that $T_a(\pi,N)\ge \inf\{n:\sum_{i=1}^N \max_{1\le j\le n}\lambda_j^0(i)\ge a\}$ and

$$\mathsf{P}_\infty\left(\inf\left\{n: \max_{1\le j\le n}\lambda_j^0(i)\ge a\right\}<\infty\right)\le e^{-a},$$

we obtain

$$\mathsf{P}_\infty\{T_a(\pi,N)<\infty\}\le e^{-a}\sum_{j=0}^{N-1}\frac{a^j}{j!}.$$

Note that the CUSUM procedure $T_{\rm CS}^N(a)$ is nothing but a repeated implementation of the one-sided test $T_a(\pi,N)$, so $T_a(\pi,N)$ is a generating time for the procedure $T_{\rm CS}^N(a)$ and by Theorem 8.2.4

$$\mathsf{E}_\infty[T_{\rm CS}^N(a)]\ge\frac{1}{\mathsf{P}_\infty\{T_a(\pi,N)<\infty\}} = (N-1)!a^{-(N-1)}e^a(1+o(1)), \quad a\to\infty.$$

Therefore, setting

$$a = \log\gamma + (N-1)\log\log\gamma - \log(N-1)! \tag{9.22}$$

yields $\mathsf{E}_\infty[T_{\rm CS}^N(a)] = \gamma(1+o(1))$ as $\gamma\to\infty$. Substituting (9.22) in (9.21) we obtain that for any $\mathcal{N}\in\{1,\dots,N\}$

$$\begin{aligned}\mathsf{E}_0^{\mathcal{N}}[T_{\rm CS}^N(a)] = \frac{1}{\sum_{i\in\mathcal{N}} I_i}\Big[&\log\gamma+(N-1)\log\log\gamma+\varkappa_{\mathcal{N}}-\log(N-1)!\\ &-|\mathcal{N}|\log\pi-(N-|\mathcal{N}|)\log(1-\pi)\Big]+o(1) \quad \text{as } \gamma\to\infty.\end{aligned}$$

Since $\mathsf{SADD}^{\mathcal{N}}(T_{\rm CS}^N(a)) = \mathsf{E}_0^{\mathcal{N}}[T_{\rm CS}^N(a)]$ and

$$\inf_{T\in\mathbf{C}_\gamma}\mathsf{SADD}^{\mathcal{N}}(T)\ge\frac{\log\gamma}{\sum_{i\in\mathcal{N}} I_i}(1+o(1)) \quad \text{for any } \mathcal{N}\in\{1,\dots,N\} \text{ as } \gamma\to\infty,$$

it follows that the procedure $T_{\rm CS}^N(a)$ is first-order asymptotically minimax for all possible subsets $\mathcal{N}\in\{1,\dots,N\}$.

Finally, we consider practical applications where the above results are of importance. In a number of applications such as computer vision, security, and remote sensing one deals with detection and tracking of objects observed in a cluttered sequence of images obtained by video or infrared sensors placed on still or moving platforms; see, e.g., [464]. Formally, it is assumed that we observe a sequence of two-dimensional $N_x \times N_y$ images (at the output of a sensor):

$$X_n(\mathbf{r}_{ij}) = \sum_{k=1}^{L} A_n(k) S(\mathbf{r}_{ij} - \mathbf{r}_n(k) - \boldsymbol{\delta}_n) \mathbb{1}_{\{\nu<n\}} + b_n(\mathbf{r}_{ij} - \boldsymbol{\delta}_n) + \xi_n(\mathbf{r}_{ij}), \; n \geq 1, \tag{9.23}$$

where $\xi_n(\mathbf{r}_{ij})$ is sensor noise which is uncorrelated in time and space and often can be well modeled by a white Gaussian process; $b(\mathbf{r}_{ij})$ is clutter (background), usually highly correlated in time and also correlated (structured) in space; $A_n(k)S(\mathbf{r}_{ij} - \mathbf{r}_n(k))$ is a signal from the k-th object with spatial coordinates $\mathbf{r}_n(k) = (x_n(k), y_n(k))$ and maximal intensity $A_n(k)$; $S(\mathbf{r}_{ij})$ is the object signature related to the sensor's point spread function; L is an unknown number of objects; $\boldsymbol{\delta}_n = (\delta_x(n), \delta_y(n))$ is an unknown 2D shift due to jitter (vibrations); $\mathbf{r}_{ij} = (x_i, y_j)$ is the pixel in the plain image with coordinates (x_i, y_j), $i = 1, \ldots, N_x$, $j = 1, \ldots, N_y$; ν is an unknown moment of object appearance. The value of L can be between 0 and $L_{\max}$. If $L = 0$, then there are no objects at all. Depending on the application, $L_{\max}$ may vary between 1 and several dozens. Also, depending on the application, the effective size of the function $S(\mathbf{r})$ may be sub-pixel (point objects) or slightly larger than a pixel size (slightly extended objects). Typically, a spatiotemporal clutter rejection algorithm is first applied to the sequence of observed images (9.23), which estimates and compensates jitter (registers images), estimates clutter, and subtracts the resulting estimate from the observed image [464]. This results in data whitening, and clutter-suppressed (whitened) frames have the form

$$\widetilde{X}_n(\mathbf{r}_{ij}) = \sum_{\nu=1}^{L} A_n(\nu) \widetilde{S}(\mathbf{r}_{ij} - \mathbf{r}_n(\nu)) \mathbb{1}_{\{\nu<n\}} + \tilde{\xi}_n(\mathbf{r}_{ij}), \;\; n \geq 1,$$

where $\widetilde{S}(\mathbf{r})$ is a target signature at the output of the clutter rejection filter and $\tilde{\xi}_n(\mathbf{r})$ are residuals that behave like random (sensor) noise ξ_n if clutter is being estimated accurately. Typical raw (highly cluttered) and clutter-suppressed (whitened) images are shown in Figure 9.2. There are several point dim objects in the scene not visible to the naked eye, and sophisticated techniques are needed to detect and track these objects.

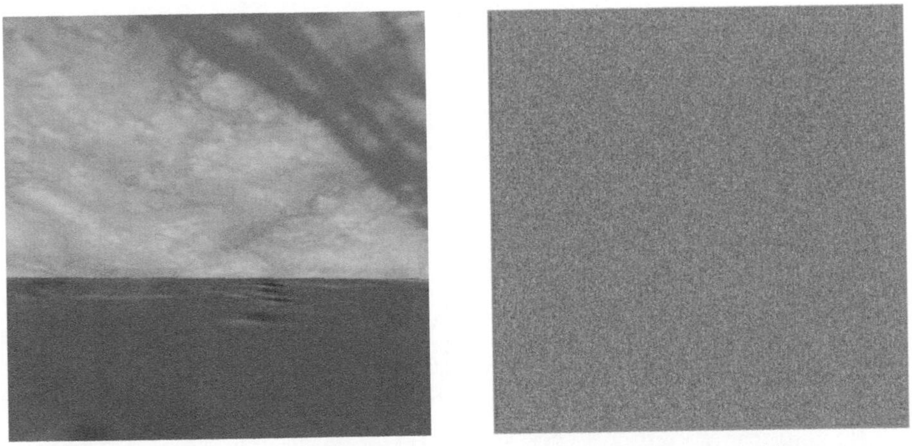

(a) Original raw infrared image (b) Clutter-suppressed image

Figure 9.2: *Typical infrared raw and whitened images with point objects that are not visible to the naked eye.*

Since neither the velocities nor the locations of the objects are known, the detection system should monitor a number N_s of spatial channels and a number N_v of velocity channels. Obviously, this problem has all the features of the general multistream changepoint detection problem with a known upper bound $L_{\max}/N_s$ for the proportion of streams which can be affected, and typically

$L_{\max}/N_s \ll 1$. For point objects, for example, the number of spatial channels may be as large as the image size, e.g., $N_s = 512 \times 512$, plus a few dozen velocity channels. Thus, the dimensionality of this problem becomes tremendous, and special techniques are needed. Generally, objects may appear and disappear at different times.

9.4 Composite Hypothesis: Linear Gaussian Model, ε-Optimality

9.4.1 The Concept of ε-Optimality

Let us discuss the ε-optimality concept [329, 330]. Recall the maximal conditional average delay to detection $\mathsf{SADD}(T) = \sup_{0 \le \nu < \infty} \mathsf{E}_\nu(T - \nu | T > \nu)$ defined in (6.19). Let $\mathsf{LB}(\gamma) = \inf_{T \in \mathbf{C}_\gamma} \mathsf{SADD}(T)$ denote the lower bound for the SADD in the class $\mathbf{C}_\gamma$ for every $\gamma > 1$. It follows from Chapter 8 that the asymptotic lower bound is given by

$$\mathsf{LB}(\gamma) \sim \frac{\log \gamma}{I(\theta_1, \theta_0)} \quad \text{as } \gamma \to \infty.$$

In the sequel, we write $I(\theta_1) = I(\theta_0, \theta_1)$, since θ_0 is known. Because the K–L information $I(\theta_1)$, $\theta_1 \in \Theta_1$, can vary, it is convenient to consider the $\mathsf{SADD}(T)$ and its lower bound LB as functions of I and γ: $(I, \gamma) \mapsto \mathsf{SADD}(T; I, \gamma)$ and $(I, \gamma) \mapsto \mathsf{LB}(I, \gamma)$. Typically the uniformly (asymptotically) optimal algorithm, which attains the lower bound for all I simultaneously, is very time/memory consuming (as the GLR scheme). For example, the GLR scheme is equivalent to running an *infinite number* of CUSUM procedures in parallel, each designed to detect a particular value θ_1. The idea of the ε-optimality concept is to reduce the number of CUSUM procedures such that this new (finite) set of recursive procedures will be almost optimal with respect to the quickest detection criterion. We propose to decompose a given parameter set Θ_1 defining the post-change model into several subsets chosen so that in each subset the detection problem can be solved with a small relative loss of optimality ε by a recursive change detection algorithm. The set of such parallel recursive algorithms establishes the ε-optimal detection scheme which reduces the computational cost of the GLR scheme. By choosing an acceptable value ε of *non-optimality*, the statistician can easily find a tradeoff between the complexity of this ε-optimal change detection algorithm and its efficiency.

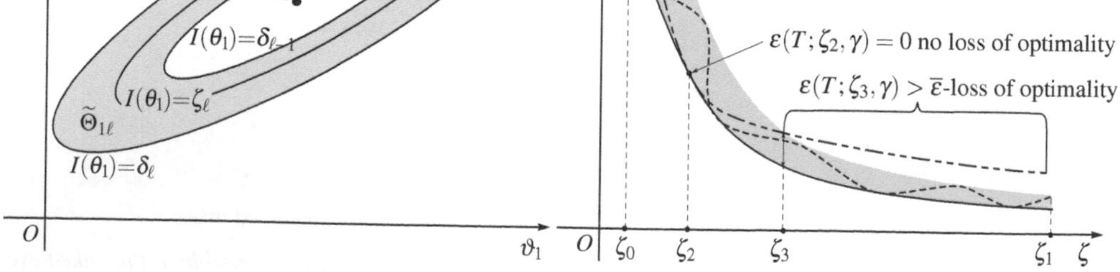

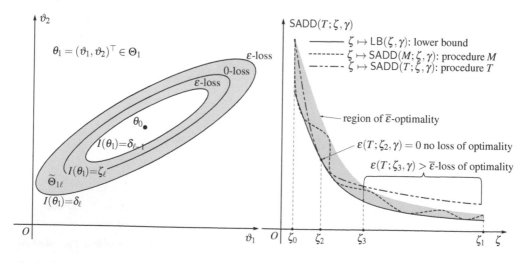

Figure 9.3: *The concept of ε-optimality. A parametrized family of level surfaces $\mathcal{S} : I(\theta_1) = \zeta$ (left). The region of $\bar{\varepsilon}$-optimality (right) (Reprinted from Signal Processing, Vol. 81(1), I. Nikiforov, "A simple change detection scheme", 149-172, Copyright (2001), with kind permission from Elsevier).*

The ε-optimality concept is shown in Figure 9.3. Let T be the stopping time from the class $\mathbf{C}_\gamma$.

Assume that this detection procedure is (or can be approximated without loss of optimality by) a recursive procedure which is able to detect any change with a constant value of $I(\theta_1) = \zeta$. It is assumed that a parametrized family of level surfaces $\mathcal{S} : I(\theta_1) = \zeta_\ell$, $\ell = 1,\ldots,L$, is defined on Θ_1 as displayed in Figure 9.3 (left).

For the given values of ζ and γ, define the loss of optimality of the procedure T compared to the optimal procedure (or relative efficiency) as

$$\varepsilon(T;\zeta,\gamma) = 1 - \frac{\mathsf{LB}(\zeta,\gamma)}{\mathsf{SADD}(T;\zeta,\gamma)}.$$

The closer the value of ε is to zero, the better the procedure T is. If $\varepsilon = 0$, then the procedure T is optimal. Typical graphs of the functions $\zeta \mapsto \mathsf{LB}(\zeta,\gamma)$ and $\zeta \mapsto \mathsf{SADD}(T;\zeta,\gamma)$ are shown in Figure 9.3 (right), where the function $\zeta \mapsto \mathsf{LB}(\zeta,\gamma)$ is shown by a solid line and the function $\zeta \mapsto \mathsf{SADD}(T;\zeta,\gamma)$ is shown by a dash-dotted line. If the procedure T is optimal for a given value of ζ, say ζ_2 (see Figure 9.3 (right)), then $\varepsilon(T;\zeta_2,\gamma) = 0$, i.e., there is no loss of optimality. But for another value of the K–L information, say $\zeta_3 \neq \zeta_2$, the procedure T is not optimal and this leads to an ε-loss of optimality $\varepsilon(T;\zeta_3,\gamma) > 0$; see Figure 9.3 (right).

Let us fix the maximum acceptable loss of optimality $\overline{\varepsilon}$ and define the following function f : $\zeta \mapsto f(\zeta,\gamma) = \mathsf{LB}(\zeta,\gamma)/(1-\overline{\varepsilon})$. It is easy to see that the region between the graphs of $\mathsf{LB}(\zeta,\gamma)$ and $f(\zeta,\gamma)$ represents the zone where the loss of optimality is bounded by $\overline{\varepsilon}$: $\varepsilon(T;\zeta,\gamma) \le \overline{\varepsilon}$. This region is shown by a shaded area in Figure 9.3 (right). Let us characterize the asymptotic loss of optimality $\varepsilon(T)$ of the procedure T as ζ ranges over the interval $[I_0;I_1]$, where $I_0 = \inf_{\theta_1 \in \Theta_1} I(\theta_1)$ and $I_1 = \sup_{\theta_1 \in \Theta_1} I(\theta_1)$. Because all theoretical results on optimal detection have an asymptotic character, we suggest to define an asymptotic loss of efficiency $\varepsilon(T)$ as the maximal over $\zeta \in [I_0;I_1]$ value of the limit $\varepsilon(T;\zeta)$ of the function $\varepsilon(T;\zeta,\gamma)$ as $\gamma \to \infty$, i.e.,

$$\varepsilon(T) \triangleq \sup_{\zeta \in [I_0;I_1]} \varepsilon(T;\zeta); \;\; \varepsilon(T;\zeta) \triangleq \lim_{\gamma\to\infty} \left(1 - \frac{\mathsf{LB}(\zeta,\gamma)}{\mathsf{SADD}(T;\zeta,\gamma)}\right). \tag{9.24}$$

Let $0 < \overline{\varepsilon} < 1$ be a given constant. We say that the detection procedure T is $\overline{\varepsilon}$-optimal if $\varepsilon(T) = \overline{\varepsilon}$. The concept of $\overline{\varepsilon}$-optimality is illustrated in Figure 9.3 (right). Let us analyze two procedures N and M which operating characteristics are shown by dash-dot and dotted lines, respectively. The shaded area consists of all points for which the loss of optimality is bounded by $\overline{\varepsilon}$. As follows from Figure 9.3 (right), $\overline{\varepsilon}$-optimality is satisfied for the procedure M, i.e., the procedure M is $\overline{\varepsilon}$-optimal, and it is not satisfied for N. Hence, to reduce the computational cost of the GLR scheme, the following scheme is proposed:

(i) The first step is to cover a given domain Θ_1 by a set of L subsets (zones of responsibility) $\widetilde{\Theta}_{11},\ldots,\widetilde{\Theta}_{1L}$, as shown in Figure 9.3 (left). These subsets are defined as follows

$$\widetilde{\Theta}_{1\ell} = \{\theta_1 : \delta_{\ell-1} \le I(\theta_1) \le \delta_\ell\}, \;\; \ell = 1,\ldots,L. \tag{9.25}$$

Define now a subdivision $\sigma = \{\zeta_1,\ldots\zeta_L\}$ of the closed interval $[I_0;I_1]$, where the values ζ_ℓ are chosen so that $I_0 = \delta_0 < \zeta_1 < \delta_1,\ldots,\delta_{L-1} < \zeta_L < \delta_L = I_1$.

(ii) The second step is to design L parallel recursive procedures, each of them is asymptotically optimal for detection of changes for $I(\theta_1) = \zeta_\ell$, $\ell = 1,\ldots,L$ (see Figure 9.3 (left)). Moreover, these recursive schemes should hold their SADD stability (with ε variations) against small changes in the K–L information number. Therefore, if L zones of responsibility (9.25) and L putative values $\zeta_1,\ldots,\zeta_L$ are chosen so that the variation of the actual value ζ of the K–L information number $I(\theta_1)$, where $\theta_1 \in \widetilde{\Theta}_{1\ell}$, around the putative value ζ_ℓ is limited for each zone, then we can expect that the set of such recursive procedures will be ε-optimal.

To simplify the notation, from now on, we consider an ℓ-th zone of responsibility and we omit the index ℓ. As follows from Section 8.3, to design an asymptotically optimal procedure, two different approaches can be used: the GLR procedure and the WLR procedure.

9.4.2 *Detection of Changes in the Mean of a Gaussian Vector*

Consider the following Gaussian model, which plays an important role in change detection because frequently the detection of *additive* changes in state-space, regression, and ARMA models can be reduced (sometimes asymptotically) to this *basic Gaussian* model by using a residual generation mechanism. See [36, 47, 250, 329, 330] for more detailed discussions.

Let $(X_k)_{k\geq 1}$ $(X_k \in \mathbb{R}^r, r > 1)$ be the following independent sequence of Gaussian random vectors:

$$\mathcal{L}(X_k) = \begin{cases} \mathcal{N}(\theta_0, \Sigma) & \text{if} \quad k \leq \nu \\ \mathcal{N}(\theta_1, \Sigma) & \text{if} \quad k > \nu, \end{cases} \tag{9.26}$$

where the mean vector θ_0 and variance-covariance matrix Σ are known *a priori*. The K–L information number is $I(\theta_1) = d^2/2$, where $d^2 = (\theta_1 - \theta_0)^\top \Sigma^{-1}(\theta_1 - \theta_0)$ is the SNR. Thus, the lower bound for the SADD is given by

$$\mathsf{LB}(d, \gamma) \sim \frac{2 \log \gamma}{d^2} \quad \text{as } \gamma \to \infty. \tag{9.27}$$

9.4.2.1 *WL GLR Scheme*

For the detection of a change in the mean of a Gaussian vector (9.26), the WL GLR scheme has been studied in [250, 251, 253, 507]. It follows from equation (8.321) that the WL GLR stopping time is

$$\widehat{N}_m = \inf\left\{ n \geq 1 : \max_{\max\{0, n-m\}+1 \leq k \leq n} \sup_{\theta_1} \left\{ \sum_{j=k}^{n} \log \frac{f_{\theta_1}(X_j)}{f_{\theta_0}(X_j)} \right\} \geq h \right\}, \tag{9.28}$$

where $f_\theta(X) = 1/((2\pi)^{n/2}(\det \Sigma)^{1/2}) \exp\{-\frac{1}{2}(X-\theta)^\top \Sigma^{-1}(X-\theta)\}$ is the pdf of the observation vector X. After simple transformations, we get

$$\begin{aligned} S_{n,k}(\theta_1) = \sum_{j=k}^{n} \log \frac{f_{\theta_1}(X_j)}{f_{\theta_0}(X_j)} &= \frac{n-k+1}{2} \Big\{ -\left(\overline{X}_{n,k} - \theta_1\right)^\top \Sigma^{-1} \left(\overline{X}_{n,k} - \theta_1\right) \\ &+ \left(\overline{X}_{n,k} - \theta_0\right)^\top \Sigma^{-1} \left(\overline{X}_{n,k} - \theta_0\right) \Big\}, \end{aligned} \tag{9.29}$$

where $\overline{X}_{n,k} = \frac{1}{n-k+1} \sum_{j=k}^{n} X_j$. Hence, the stopping time of the GLR scheme can be written as

$$\widehat{N}_m = \inf\left\{ n \geq 1 : \max_{\max\{0, n-m\}+1 \leq k \leq n} \sup_{\theta_1} S_{n,k}(\theta_1) \geq h \right\}, \tag{9.30}$$

where

$$\begin{aligned} \sup_{\theta_1} S_{n,k}(\theta_1) &= \frac{n-k+1}{2} \left(\overline{X}_{n,k} - \theta_0\right)^\top \Sigma^{-1} \left(\overline{X}_{n,k} - \theta_0\right) \\ &= \frac{1}{2(n-k+1)^2} \chi^2_{n,k}, \quad \chi^2_{n,k} = \Big[\sum_{j=k}^{n} (X_j - \theta_0)\Big]^\top \Sigma^{-1} \Big[\sum_{j=k}^{n} (X_j - \theta_0)\Big]. \end{aligned} \tag{9.31}$$

9.4.2.2 χ^2 *Detection Schemes: the K–L Information Is Known*

In contrast to the case of unknown θ_1, the CUSUM and GLR schemes can be rewritten recursively if the K–L information is known. Assume now that the unknown vector θ_1 is constrained to lie on the level surface $\theta_1 : (\theta_1 - \theta_0)^\top \Sigma^{-1}(\theta_1 - \theta_0) = d^2$.

There are two recursive χ^2 detection schemes adapted to the case of known K–L information:

- the recursive χ^2-CUSUM stopping rule;

- the recursive χ^2-GLR stopping rule.

Consider the χ^2-SPRT defined by (5.141)–(5.142) in Subsection 5.4.3. Since the CUSUM procedure can be represented via a repeated SPRT (see (8.64)), we will use the χ^2-SPRT to distinguish between the following two hypotheses

$$H_0 = \{\theta = \theta_0\} \text{ and } H_1 = \{\theta_1 : (\theta_1 - \theta_0)^\top \Sigma^{-1} (\theta_1 - \theta_0) = d^2\}. \tag{9.32}$$

It follows from Example 5.4.2 in Subsection 5.4.3 that the problem (9.32) can be replaced with the following hypothesis testing problem concerning the noncentrality parameter c of the χ^2-distribution:

$$H_0 : c = 0 \text{ and } H_1 : c = d^2. \tag{9.33}$$

Hence, the χ^2-SPRT $(\widetilde{d}, \widetilde{T})$ applied to the observations $X_k, X_{k+1}, \dots$ is given by

$$\widetilde{T} = \inf\left\{n \ge k : \widetilde{\lambda}_{n,k} \notin (0,h)\right\}, \quad \widetilde{d} = \begin{cases} 1 & \text{if } \widetilde{T} \ge h \\ 0 & \text{if } \widetilde{T} \le 0, \end{cases} \tag{9.34}$$

where

$$\widetilde{\lambda}_{n,k} = -(n-k+1)\frac{d^2}{2} + \log{}_0F_1\left(\frac{r}{2}, \frac{d^2\chi^2_{n,k}}{4}\right). \tag{9.35}$$

It is worth noting that, in contrast to the conventional CUSUM procedure described in Section 8.2, the χ^2 detection schemes represented as an open-ended χ^2-SPRT is not equivalent to the repeated χ^2-SPRT. Analogously to (9.30), the open-ended χ^2-SPRT leads to the following non-recursive χ^2-CUSUM stopping rule

$$\widetilde{N} = \inf\{n \ge 1 : \max_{\max\{0, n-m\}+1 \le k \le n} \widetilde{\lambda}_{n,k} \ge h\}. \tag{9.36}$$

The recursive χ^2-CUSUM stopping rule as a repeated χ^2-SPRT has been introduced in [319] and studied in [47, 321, 329, 330]:

$$\widetilde{N}_r = \inf\{n \ge 1 : \widetilde{S}_n \ge h\}, \tag{9.37}$$

$$\widetilde{S}_n = -n_n\frac{d^2}{2} + \log{}_0F_1\left(\frac{r}{2}, \frac{d^2\chi^2_n}{4}\right), \quad \chi^2_n = V_n^\top \Sigma^{-1} V_n, \tag{9.38}$$

$$V_n = \mathbb{1}_{\{\widetilde{S}_{n-1} > 0\}} V_{n-1} + (X_n - \theta_0), \quad n_n = \mathbb{1}_{\{\widetilde{S}_{n-1} > 0\}} n_{n-1} + 1, \tag{9.39}$$

where h is a threshold and n_n is the counter of the observations in each successive cycle of the repeated χ^2-SPRT. The initial condition is $\widetilde{S}_0 = 0$. In practice, the function $z \mapsto \log{}_0F_1\left(\frac{r}{2}, \frac{z^2}{4}\right)$ can be easily computed by the following method. Let us fix a large positive constant $\overline{z}$ and decompose the working interval $[0, \overline{z}]$ by inserting $(j-1)$ points of subdivision, say, $z_1, z_2, \dots, z_{j-1}$. There are two possible situations:

- if $z \le \overline{z}$ then a polynomial interpolation is applied to

$$(0;0), \left(z_1; \log{}_0F_1\left(\frac{r}{2}, \frac{z_1^2}{4}\right)\right), \dots, \left(z_{j-1}; \log{}_0F_1\left(\frac{r}{2}, \frac{z_{j-1}^2}{4}\right)\right), \left(\overline{z}; \log{}_0F_1\left(\frac{r}{2}, \frac{\overline{z}^2}{4}\right)\right);$$

- if $z > \overline{z}$ then the following asymptotic approximation [1, Ch. 13] is used

$$\log{}_0F_1\left(\frac{r}{2}, \frac{z^2}{4}\right) = z + \log\frac{\Gamma(r-1)}{\Gamma\left(\frac{r-1}{2}\right)} + (1-r)\log z - \left(\frac{1-r}{2}\right)\log 4 + \log\left(1 + O\left(\frac{1}{z}\right)\right).$$

We now discuss the χ^2-GLR procedure. The detailed analysis of the GLR procedure, including the proof of its first-order optimality, can be found in [47, Chap. 4 and 7] and [321, 329, 330]. It follows from (9.29) that in the case of known K–L information number $I(\theta_1) = d^2/2$, the χ^2-GLR procedure is based on the statistic

$$\begin{aligned}
\widehat{S}_{n,k} &= \sup_{\theta_1 : I(\theta_1)=d^2/2} S_{n,k}(\theta_1) = \sup_{\theta_1 : I(\theta_1)=d^2/2} \left\{ \sum_{j=k}^{n} \log \frac{f_{\theta_1}(X_j)}{f_{\theta_0}(X_j)} \right\} \\
&= \frac{n-k+1}{2} \sup_{\theta_1 : I(\theta_1)=d^2/2} \left\{ -\left(\overline{X}_{n,k} - \theta_1\right)^\top \Sigma^{-1} \left(\overline{X}_{n,k} - \theta_1\right) + \left(\overline{X}_{n,k} - \theta_0\right)^\top \Sigma^{-1} \left(\overline{X}_{n,k} - \theta_0\right) \right\} \\
&= (-n+k-1)\frac{d^2}{2} + d|\chi_{n,k}|.
\end{aligned}$$

The definition of the χ^2-GLR stopping rules is analogous to the χ^2-CUSUM rules. The non-recursive χ^2-GLR procedure is given by

$$\widehat{N} = \inf\{n \geq 1 : \max_{\max\{0,n-m\}+1 \leq k \leq n} \widehat{S}_{n,k} \geq h\}. \tag{9.40}$$

The recursive χ^2-GLR stopping rule was first introduced in [362]:

$$\widehat{N}_r = \inf\{n \geq 1 : \widehat{S}_n \geq h\}, \; \widehat{S}_n = -n_n \frac{d^2}{2} + d|\chi_n|, \; \chi_n^2 = V_n^\top \Sigma^{-1} V_n, \tag{9.41}$$

$$V_n = \mathbb{1}_{\{\widehat{S}_{n-1}>0\}} V_{n-1} + (X_n - \theta_0), \; n_n = \mathbb{1}_{\{\widehat{S}_{n-1}>0\}} n_{n-1} + 1. \tag{9.42}$$

9.4.2.3 Statistical Properties of the χ^2 Detection Schemes: the K–L Information Is Known

Asymptotic optimality of the non-recursive χ^2 change detection schemes has been investigated in [47, 321, 327] using Lorden's "worst-worst-case" (essential supremum) ADD defined by (6.17).

Theorem 9.4.1. *Let $\widetilde{N}$ be the stopping time of the non-recursive χ^2-CUSUM procedure* (9.36) *and $\widehat{N}$ be the stopping time of the non-recursive χ^2-GLR procedure* (9.40)*. Assume that the pre-change parameter θ_0 and the K–L information number $I(\theta_1) = d^2/2$, where $d^2 = (\theta_1 - \theta_0)^\top \Sigma^{-1}(\theta_1 - \theta_0)$, are known. Then these change detection procedures are asymptotically optimal in the class $\mathbf{C}_\gamma$:*

$$\mathsf{ESADD}(\widetilde{N}; d, \gamma) \sim \mathsf{ESADD}(\widehat{N}; d, \gamma) \sim \mathsf{LB}(d, \gamma) \sim \frac{2\log\gamma}{d^2} \quad \text{as } \gamma \to \infty.$$

Proof. See the proof of Theorem 1 in [321] and the proof of Theorem 1 in [327]. □

Next, we discuss the asymptotic optimality property of the *recursive* χ^2 change detection schemes [329, 330]. Our goal is to show that the recursive χ^2-CUSUM procedure (9.37)–(9.39) and the recursive χ^2-GLR procedure (9.41)–(9.42) attain the asymptotic lower bound given by (9.27), i.e., $\mathsf{LB}(d,\gamma) = \inf_{T \in \mathbf{C}_\gamma} \mathsf{SADD}(T) \sim 2\log\gamma/d^2$ as $\gamma \to \infty$. First, we have to show that the maximal expected detection delay satisfies the following asymptotic relations $\mathsf{SADD}(\widetilde{N}_r; d, \gamma) \sim \mathsf{SADD}(\widehat{N}_r; d, \gamma) \sim 2h/d^2$ as $h \to \infty$, where h is the threshold. Lemma 3 in [329] establishes this result. To explain the complexity of this problem let us recall that the change time ν is unknown. Due to this fact, the last cycle of the repeated SPRT can contain a "tail" of pre-change observations. To show that the impact of this tail on the asymptotic detection delay is negligible, we also use another technical result established in [329, Lemma 4]. Next, we have to show that the ARL to false alarm of the recursive χ^2-CUSUM procedure satisfies the inequality $\log \mathsf{ARL2FA}(\widetilde{N}_r) = \log \mathsf{E}_\infty(\widetilde{N}_r) \geq h$ and the stopping time $\widehat{N}_r$ satisfies the inequality $\log \mathsf{ARL2FA}(\widehat{N}_r) = \log \mathsf{E}_\infty(\widehat{N}_r) \geq h(1+o(1))$ as $h \to \infty$. These results are given in [329, Lemma 5]. Finally, by putting the above results together, the relation between the SADD and the ARL2FA is established and this proves that the recursive χ^2-procedures attain the asymptotic lower bound $\mathsf{LB}(d,\gamma)$ in the class $\mathbf{C}_\gamma$.

Theorem 9.4.2. *Let $\widetilde{N}_r$ be the stopping time of the recursive χ^2-CUSUM procedure* (9.37)–(9.39) *and $\widehat{N}_r$ be the stopping time of the recursive χ^2-GLR procedure* (9.41)–(9.42)*. Assume that the pre-change parameter θ_0 and the K–L information number $I(\theta_1) = d^2/2$, where $d^2 = (\theta_1-\theta_0)^\top\Sigma^{-1}(\theta_1-\theta_0)$, are known. Then these procedures are asymptotically optimal in the class $\mathbf{C}_\gamma$:*

$$\mathsf{SADD}(\widetilde{N}_r;d,\gamma) \sim \mathsf{SADD}(\widehat{N}_r;d,\gamma) \sim \mathsf{LB}(d,\gamma) \sim \frac{2\log\gamma}{d^2}, \quad \text{as } \gamma\to\infty. \tag{9.43}$$

Proof. See the proof of Theorem 1 in [329]. □

9.4.2.4 *Statistical Properties of the χ^2 Detection Schemes: the K–L Information Is Unknown*

Let compute the SADD of the recursive χ^2-CUSUM and χ^2-GLR detection procedures. It is supposed that these procedures are designed to detect a change when the K–L information number is fixed $I(\theta_1) = d^2/2$ but the post-change observations $X_{\nu+1}, X_{\nu+2},\ldots$ are distributed as $\mathcal{N}(\theta,\Sigma)$, where

$$\theta\in\Theta_b = \left\{\theta : (\theta-\theta_0)^\top\Sigma^{-1}(\theta-\theta_0) = b^2 > \frac{1}{4}d^2\right\}.$$

Hence, the true value of the K–L information number $b^2/2$ is different from its putative value $d^2/2$. It can be shown that the asymptotic relation between the SADD and the threshold h for the recursive χ^2-CUSUM and χ^2-GLR procedures is given by

$$\mathsf{SADD}(\widetilde{N}_r) \sim \mathsf{SADD}(\widehat{N}_r) \sim \frac{2h}{b^2-(b-d)^2} \quad \text{as } h\to\infty;$$

and the asymptotic relation between the SADD and the ARL2FA is

$$\mathsf{SADD}(\widetilde{N}_r;b,\gamma) \sim \mathsf{SADD}(\widehat{N}_r;b,\gamma) \le \min\left\{\frac{2\log\gamma}{b^2-(b-d)^2},\gamma\right\}(1+o(1)) \quad \text{as } \gamma\to\infty. \tag{9.44}$$

9.4.2.5 *ε-Optimal Multichart Tests*

Let us continue discussing the ε-optimality concept started in Subsection 9.4.1. The ε-optimal multichart test is based on L parallel recursive χ^2-CUSUM (or χ^2-GLR) procedures, each of which is optimal only for the putative value of the post-change SNR a_ℓ, $\ell = 1,\ldots,L$. Hence, it is necessary to show that the optimal recursive χ^2-CUSUM (or GLR) procedure has a certain stability when the true value of the post-change SNR b differs from the putative one a_ℓ. It is also necessary to find a convenient zone of responsibility $[\underline{b}_\ell, \overline{b}_\ell]$ for each of the L recursive χ^2-procedures.

Consider L parallel recursive χ^2-CUSUM (or GLR) procedures. The stopping time of the ε-optimal scheme is given by:

$$N_{\varepsilon r} = \min\left\{N_r(a_1), N_r(a_2),\ldots,N_r(a_L)\right\}, \tag{9.45}$$

where $N_r(a_\ell)$ is the stopping time (9.37) (or (9.41)) of a recursive χ^2-procedure designed to detect a change with the putative SNR a_ℓ. The goal is to find the number L of parallel χ^2-procedures and the subdivision $\sigma = \{a_1 < a_2 < \cdots < a_L\}$ of the interval $[d_0, d_1]$ such that $d_0 \le a_1 < 2d_0$, $a_L \le d_1$ and

$$\inf_{b\in[d_0,d_1]}\lim_{\gamma\to\infty}\left\{\frac{n(b,\gamma)}{\mathsf{SADD}(N_{\varepsilon r};b,\gamma)}\right\} \ge \overline{e} = 1-\overline{\varepsilon},$$

where $\overline{e}$ is the efficiency of the ε-optimal scheme and $\mathsf{SADD}(N_{\varepsilon r};b,\gamma)$ is the maximal conditional average delay to detection of the ε-optimal rule (9.45).

9.4.2.6 *Statistical Properties of the ε-Optimal Multichart Tests*

We first discuss the relation between the SADD and the ARL2FA of the ε-optimal rule, and we next compute a suitable subdivision σ of the interval $[d_0, d_1]$. The following lemma establishes the ARL2FA of the recursive rules $\widetilde{N}_{\varepsilon r}$ and $\widehat{N}_{\varepsilon r}$.

Lemma 9.4.1. *Let $\widetilde{N}_{\varepsilon r}$ and $\widehat{N}_{\varepsilon r}$ be the stopping times of the recursive ε-optimal rules. Then*

$$\begin{cases} \log \mathsf{ARL2FA}(\widetilde{N}_{\varepsilon r}) = \log \mathsf{E}_\infty(\widetilde{N}_{\varepsilon r}) & \geq \; h - \log L, \\ \log \mathsf{ARL2FA}(\widehat{N}_{\varepsilon r}) = \log \mathsf{E}_\infty(\widehat{N}_{\varepsilon r}) & \geq \; (h - \frac{r-1}{2} \log 2h - \log L)(1 + o(1)) \end{cases} \quad \text{as } h \to \infty, \quad (9.46)$$

where the notation $\widetilde{N}_{\varepsilon r}$ (or $\widehat{N}_{\varepsilon r}$) means that the ε-optimal rule is designed by using the χ^2-CUSUM (or χ^2-GLR) procedure and h is the threshold of the recursive χ^2-procedures.

Proof. See the proof of Lemma 5 in [329]. □

It follows from the definition (9.45) of the ε-optimal multichart test that

$$\mathsf{SADD}(N_{\varepsilon r}; b) \leq \min\{\mathsf{SADD}(N_r(a_1; b)), \dots, \mathsf{SADD}(N_r(a_L; b))\}.$$

On the right-hand side of this inequality the minimum is attained for $\ell_0 = \arg\min_{\ell=1,\dots,L} |b - a_\ell|$ and is given by

$$\mathsf{SADD}(N_{\varepsilon r}; b) \leq \frac{2h}{b^2 - (b - a_{\ell_0})^2}(1 + o(1)) \text{ as } h \to \infty,$$

where $b \in [d_0, d_1]$. This leads to the following asymptotic relation between the SADD and the ARL2FA for the stopping time $N_{\varepsilon r}$:

$$\mathsf{SADD}(N_{\varepsilon r}; b, \gamma) \leq \frac{2 \log \gamma}{b^2 - (b - a_{\ell_0})^2}(1 + o(1)) \text{ as } \gamma \to \infty, \quad (9.47)$$

when $b \in [d_0, d_1]$. To design the ε-optimal scheme we have to define a zone of responsibility for each of the L χ^2-procedures. Putting the lower bound $\mathsf{LB}(b, \gamma) \sim 2 \log \gamma / b^2$ together with (9.47) into (9.24), we get $\varepsilon(b) = (b - a_\ell)^2 / b^2$, $1 \leq \ell \leq L$; see Figure 9.3. This yields an equation for the bounds $\underline{b}_\ell$ and $\overline{b}_\ell$ of the ℓ-th zone of responsibility. Namely, to find $\underline{b}_\ell$ or $\overline{b}_\ell$, the following equation should be solved with respect to b : $b^2 \overline{\varepsilon} = (b - a_\ell)^2$. Two solutions of this equation are

$$\underline{b}_\ell = \frac{a_\ell}{1 + \sqrt{\overline{\varepsilon}}} \text{ and } \overline{b}_\ell = \frac{a_\ell}{1 - \sqrt{\overline{\varepsilon}}}.$$

To cover the interval $[d_0, d_1]$ with a set of L sub-intervals $[\underline{b}_\ell, \overline{b}_\ell]$, where $\overline{b}_\ell = \underline{b}_{\ell+1}$ and $\underline{b}_1 = d_0$, the values a_ℓ is chosen as follows:

$$a_\ell = d_0 \frac{(1 + \sqrt{\overline{\varepsilon}})^\ell}{(1 - \sqrt{\overline{\varepsilon}})^{\ell - 1}}, \quad \ell = 1, 2, \dots, L. \quad (9.48)$$

Finally, to get $\overline{b}_L \geq d_1$, the number L of χ^2-procedures is given by

$$L = \left\lceil \log \frac{d_1}{d_0} \left(\log \frac{1 + \sqrt{\overline{\varepsilon}}}{1 - \sqrt{\overline{\varepsilon}}} \right)^{-1} \right\rceil, \quad (9.49)$$

where the function $\lceil x \rceil$ gives the smallest integer $\geq x$. The above results are summarized in Table 9.1. By using this table a comparison of the efficiency *versus* complexity of the recursive ε-optimal procedure ($N_{\varepsilon r}$) and the WL GLR procedure ($\widehat{N}_{m, \widetilde{m}}$) can be easily done.

Table 9.1: *Efficiency vs. complexity for the ε-optimal multichart $N_{\varepsilon r}$ and WL GLR $\widehat{N}_m$ procedures as $\gamma \to \infty$ (Reprinted from Signal Processing, Vol. 81(1), I. Nikiforov, "A simple change detection scheme", 149-172, Copyright (2001), with kind permission from Elsevier).*

Test	Loss of Optimality	Number of LR Maximizations
ε-optimal	ε	$\left\lceil \log \frac{d_1}{d_0} \left(\log \frac{1+\sqrt{\varepsilon}}{1-\sqrt{\varepsilon}} \right)^{-1} \right\rceil$
WL GLR (optimal)	0	$O(2\log\gamma/d_0^2)$

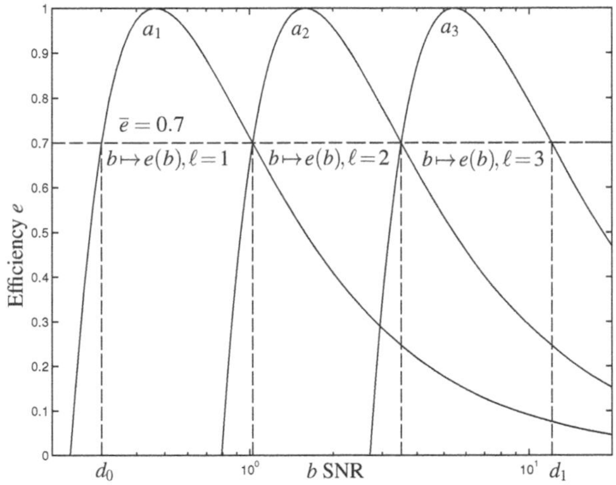

Figure 9.4: *The functions $b \mapsto e(b)$ for three χ^2-CUSUM (or χ^2-GLR) recursive procedures and their zones of responsibility.*

9.4.2.7 *Comparison of the ε-Optimal Procedure with the WL GLR Procedure*

Let us compare the statistical properties and the computational complexity of the recursive ε-optimal multichart and WL GLR procedures by using the asymptotic approximations and the Monte-Carlo simulations. It is assumed that the SNR b varies from $d_0 = 0.3$ to $d_1 = 12$, $X_k \in \mathbb{R}^2$ and $\overline{e} = 1 - \overline{\varepsilon} = 0.7$. It follows from equation (9.49) that it is enough to run three parallel χ^2-procedures to get this level of efficiency (see the horizontal dashed line $\overline{e} = 0.7$ in Figure 9.4). The putative values of the SNR to which the procedures are tuned are $a_1 = 0.464$, $a_2 = 1.589$ and $a_3 = 5.437$. The curves $b \mapsto e(b) = 1 - \varepsilon(b)$ for these three χ^2-CUSUM (or χ^2-GLR) procedures are shown in Figure 9.4. The zones of responsibility $[0.3,\ 1.027]$, $[1.027,\ 3.513]$ and $[3.513,\ 12.022]$ are shown by vertical dashed lines. Hence, the WL GLR procedure $\widehat{N}_m$ involves m computations of $V_{n,k}$ and $\chi^2_{n,k}$ at every stage and the ε-optimal procedure ($N_{\varepsilon r}$) involves L computations of $V_{n,k}$ and $\chi^2_{n,k}$. The relative complexity of $\widehat{N}_m$ with respect to $N_{\varepsilon r}$ is given in Table 9.2. The ARL2FA γ varies from 10^4 to 10^{10}. Because the WL GLR procedure is asymptotically optimal, we use (9.27) as a theoretic approximation for the mean detection delay:

$$\mathsf{SADD}(\widehat{N}_m; b, \gamma) \sim \mathsf{LB}(b, \gamma) \sim \frac{2\log\gamma}{b^2} \quad \text{as } \gamma \to \infty. \tag{9.50}$$

Figure 9.5 presents the comparison of 10^3-repetition Monte Carlo simulations with asymptotic approximations for the above three χ^2-GLR charts $\widehat{N}_{0.3,r}$ and WL GLR procedure $\widehat{N}_m$ when $\gamma = 10^5$. The zones of responsibility for the three parallel χ^2-GLR procedures and the interval $[d_0, d_1]$ are

Table 9.2: *The relative complexity of $\widehat{N}_m$ with respect to $N_{\varepsilon r}$ when $d_0 = 0.3$, $d_1 = 12$, $\overline{e} = 0.7$ and $\gamma = 10^4$ to 10^{10}.*

Relative Complexity	ARL2FA						
	$\gamma=10^4$	$\gamma=10^5$	$\gamma=10^6$	$\gamma=10^7$	$\gamma=10^8$	$\gamma=10^9$	$\gamma=10^{10}$
$\frac{\text{Complexity of } \widehat{N}_m}{\text{Complexity of } N_{\varepsilon r}}$	68	85	102	119	136	153	170

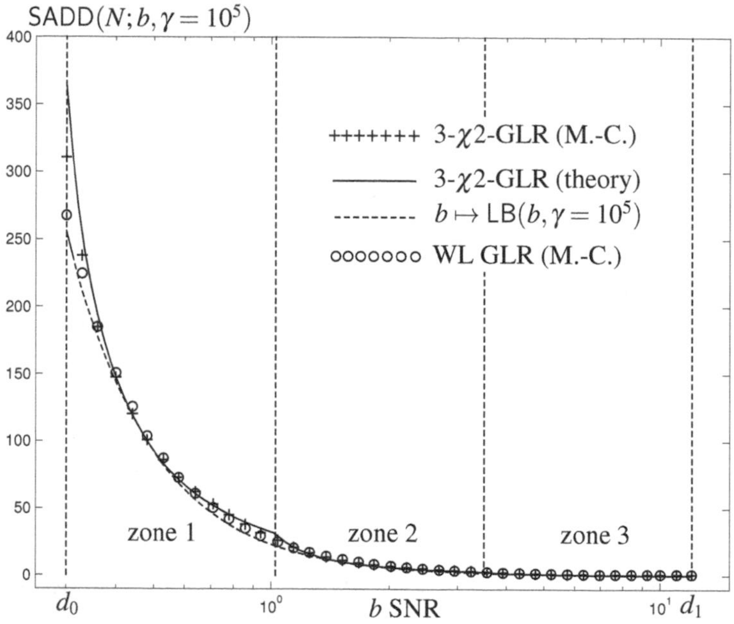

Figure 9.5: *The function $d \mapsto \mathsf{SADD}(N; d, \gamma)$ for the procedures $\widehat{N}_{0.3,r}$ and $\widehat{N}_{800,0}$ when $d_0 = 0.3$, $d_1 = 12$, $r = 2$ and $\gamma = 10^5$: comparison between the asymptotic approximations and Monte Carlo simulations.*

shown by vertical dashed lines in Figure 9.5. Both the simulated and theoretic $\mathsf{SADD}(N; b, \gamma = 10^5)$ as functions of the SNR b for these procedures are presented here. The lower bound $\mathsf{LB}(b, \gamma = 10^5)$ given by (9.50) is used as a theoretic approximation for the SADD of the WL GLR procedure $\widehat{N}_m$. Equation (9.47) is used as a theoretic SADD for $\widehat{N}_{0.3,r}$. Two versions of the ε-optimal 3-χ^2-chart have been studied : three parallel χ^2-CUSUM procedures $\widetilde{N}_{0.3,r}$ and three parallel χ^2-GLR procedures $\widehat{N}_{0.3,r}$ (see details in [329, Table III]).

As it follows from extensive Monte Carlo simulations, the asymptotic choice $m \gtrsim \log \gamma / I_0$ but $\log m = o(\log \gamma)$ as $\gamma \to \infty$ is not very practicable. To be sure that the comparison is correct, we first make the simulation of the WL GLR procedure when $m = \lceil 2 \log \gamma / d_0^2 \rceil$ and next we repeat the simulation with a bigger value of m. For the WL GLR $\widehat{N}_m$ this parameter has been chosen as $m = 256$ and, next, $m = 800$. As it follows from [329, Table III], the choice $m = 256$ leads to some degradation of the SADD for small values of $b \simeq 0.3$. The thresholds for the procedures have been chosen so that $\mathsf{E}_\infty(\widetilde{N}_{0.3,r}) = \mathsf{E}_\infty(\widehat{N}_{0.3,r}) = \mathsf{E}_\infty(\widehat{N}_m) = \gamma = 10^5$ by using a 10^3-repetition Monte Carlo simulation in each case. Next, the maximal conditional average delays to detection $\mathsf{SADD}(\widetilde{N}_{0.3,r}; b, \gamma = 10^5)$, $\mathsf{SADD}(\widehat{N}_{0.3,r}; b, \gamma = 10^5)$ and $\mathsf{SADD}(\widehat{N}_{800}; b, \gamma = 10^5)$ have been evaluated for b varying from 0.3 to 12. The simulation shows that the variation of the estimated maximal conditional average delays to detection is negligible with respect to its standard deviation when $\nu > 100$. Hence, to estimate

$\mathsf{SADD}(N;b,\gamma=10^5)$ the value $\mathsf{SADD}(N)=\sup_{0\le\nu<\infty}\mathsf{E}_\nu(N-\nu|N>\nu)$ has been replaced with its Monte-Carlo estimate

$$\mathsf{SADD}(N)\simeq\max_{1\le\nu\le 100}\left\{\frac{1}{n_d}\sum_{i=1}^{10^3}(N_i-\nu)\mathbb{1}_{\{N_i>\nu\}}\right\},\tag{9.51}$$

where N_i is the time of detection in the i-th statistical experiment and $n_d=\sum_{i=1}^{10^3}\mathbb{1}_{\{N_i>\nu\}}$.

The simulation shows that for large values of the SNR b the approximations $\mathsf{SADD}(N;b,\gamma)$ for all the procedures differ from the simulated ones. This is due to an asymptotic character of equations (9.47) and (9.50). The following heuristic modification for (9.47) is proposed to fix this gap:

$$\overline{\mathsf{E}}_b(N_{\varepsilon r};\gamma)\approx\max\left\{1,\frac{2\log\gamma}{b^2-(b-a_{\ell_0})^2}\right\}.$$

Figure 9.5 completely confirms (except the asymptotic character of approximations) the theoretic performances of the ε-optimal rule (in our case the efficiency level is $\overline{e}=0.7$). The extensive Monte Carlo simulations show that $\widehat{N}_{0.3r}$ and $\widetilde{N}_{0.3r}$ perform even better than expected from asymptotic theory. As it follows from [329, Table III], in detecting changes with $b\simeq a_\ell$ the performance of $\widehat{N}_{0.3r}$ is better than the performance of $\widehat{N}_{256,0}$ and even than the performance of $\widehat{N}_{800,0}$. The asymptotic approximation of the parameter m of the WL GLR scheme should be multiplied by 2 or 3 when $\gamma=10^5$. This means that the asymptotic approximation of the WL GLR scheme complexity should be also multiplied by 2 or 3.

9.4.3 Detection of Changes in the Linear Regression Model

The linear regression model is widely used in control applications and also in statistical signal processing (trajectography, adaptive arrays, adaptive filters, *etc.*). Here we follow the main argument of the previous section; see also [330, 518] for further details on the detection of changes in regression parameters.

In contrast to the additive change model (9.26) where the change affects the mean value of the observations, the following regression changepoint model is non-additive in that the changes occur in the regressive parameter vector θ:

$$y_n=\begin{cases}X_n^\top\theta_0+\xi_n & \text{if}\quad n\le\nu\\ X_n^\top\theta_1+\xi_n & \text{if}\quad n>\nu\end{cases},\tag{9.52}$$

In this model, y_n is the scalar output, ξ_n is an independent scalar Gaussian sequence with zero mean, $\xi_n\sim\mathcal{N}(0,1)$, X_n is the measured input, $X\in\mathbb{R}^r$, and X_n and ξ_n are independent.

Remark 9.4.1. Signal processing algorithms often carry out narrow-band signals. The complex representation is common for such signals, which contain information in the carrier phase and the amplitude. Hence, the scalar $y\in\mathbb{C}$ and the vectors $X,\theta\in\mathbb{C}^r$ are complex in this case. The case of complex signals will not be developed here, but it is worth noting that the extension of the proposed change detection scheme to the case of complex signal can be easily done by using the technique of a complex gradient operator [84].

First of all, let us show that there exists a set Θ_1 of post-change parameters θ_1 such that the K–L information number $I(\theta_1)$ is lower and upper bounded for the regression model (9.52). In fact this set is given (at least approximately) by the following equation:

$$\Theta_1=\left\{\theta_1\ :\ \frac{d_0^2}{2}\le I(\theta_1)=\frac{b^2}{2}=\frac{1}{2}(\theta_1-\theta_0)^\top R(\theta_1-\theta_0)\le\frac{d_1^2}{2}\right\},\tag{9.53}$$

where $R\in\mathbb{R}^{r\times r}$ is a positive definite matrix. If X_n is a stationary random sequence, R is the matrix of second moments, given by $R \triangleq \mathsf{E}(X_n X_n^\top)$. If X_n is a deterministic sequence, we assume that [518]

$$\frac{1}{n}\sum_{i=k}^{k+n-1} X_i\, X_i^\top \to R \ \text{ as } \ n\to\infty \ \text{ uniformly in } \ k\geq 1. \tag{9.54}$$

Let us compute the K–L information number $I(\theta_1)$ by using the equation (2.234). The expectation of the normalized LLR $\frac{1}{n}S_{k+n-1,k}(\theta_1)$ is

$$I_{n,k}(\theta_1) = \mathsf{E}_{\theta_1}\left(n^{-1}S_{k+n-1,k}(\theta_1)\right) = \frac{1}{2}\,(\theta_1-\theta_0)^\top\,\frac{1}{n}\sum_{i=k}^{k+n-1} X_i\, X_i^\top (\theta_1-\theta_0). \tag{9.55}$$

Assuming that condition (9.54) is satisfied, putting (9.54) and (9.55) together, and taking the limit as $n\to\infty$, we get the K–L information number

$$I_{n,k}(\theta_1) \to I(\theta_1) = \frac{1}{2}\,(\theta_1-\theta_0)^\top R(\theta_1-\theta_0) \ \text{ as } \ n\to\infty \ \text{ uniformly in } \ k\geq 1.$$

Therefore, analogously to the previous case, the domain Θ_1 in (9.53) can be expressed in terms of $b^2 = 2I(\theta_1)$.

9.4.3.1 The WL GLR Procedure

To simplify the notations we apply the transformation $e_n = y_n - X_n^\top\theta_0$ to the output y_n. This leads to the following representation of the regression model (9.52):

$$e_n = \begin{cases} \xi_n & \text{if } \ n\leq\nu \\ X_n^\top\beta+\xi_n & \text{if } \ n>\nu \end{cases},$$

where $\beta = \theta_1-\theta_0$. Let us briefly discuss how to derive the WL GLR procedure for the regression model (9.52). The details can be found in [253]. It follows from (8.321) that

$$\sup_\beta \sum_{j=k}^n \log\frac{f_\beta(e_j)}{f_0(e_j)} = \sup_\beta \sum_{j=k}^n \log\frac{\exp\left\{-\frac{1}{2}(e_j-X_j^\top\beta)^2\right\}}{\exp\left\{-\frac{1}{2}e_j^2\right\}},$$

where $f_\beta(y) = \frac{1}{\sqrt{2\pi}}\exp\left\{-\frac{1}{2}(y-X^\top\beta)^2\right\}$ is density of the observation vector X. After simple algebra we obtain

$$\sup_\beta \sum_{j=k}^n \log\frac{f_\beta(e_j)}{f_0(e_j)} = \frac{1}{2}\sum_{j=k}^n\left[2e_jX_j^\top\widehat{\beta}-\widehat{\beta}^\top X_jX_j^\top\widehat{\beta}\right] = \frac{1}{2}\left[\sum_{j=k}^n 2e_jX_j^\top\right]\widehat{\beta}-\frac{1}{2}\widehat{\beta}^\top\left[\sum_{j=k}^n X_jX_j^\top\right]\widehat{\beta}$$

where $\widehat{\beta} = \left(\sum_{j=k}^n X_jX_j^\top\right)^{-1}\sum_{j=k}^n e_jX_j$. Hence, the stopping time of the WL GLR procedure is given by

$$\widehat{N}_{m,r} = \inf\left\{n>r: \max_{\max\{0,n-m\}+1\leq k\leq n-r}\frac{1}{2}\sum_{j=k}^n X_j^\top e_j\cdot\left(\sum_{j=k}^n X_jX_j^\top\right)^{-1}\sum_{j=k}^n e_jX_j\geq h\right\}, \tag{9.56}$$

where $1\leq r<m$, $e_n = y_n - X_n^\top\theta_0$ and the inverse matrix $P_{n,k} \triangleq \left(\sum_{j=k}^n X_jX_j^\top\right)^{-1}$ is recursively calculated for $k = n, n-1, n-2, \dots$ as

$$P_{n,k} = \left(\mathbb{I} - P_{n,k+1}\frac{X_nX_n^\top}{1+X_n^\top P_{n,k+1}X_n}\right)P_{n,k+1} \ \text{with} \ P_{n,n+1} = R^{-1} \ \text{ if } \ k=n.$$

9.4.3.2 Recursive Constrained GLR Procedure

Let us now discuss the constrained GLR procedure for the regression model (9.52). It is assumed that the vector $\theta_1 = (\vartheta_{1,1},\ldots,\vartheta_{1,r})^T$ is unknown but the K–L information is known: $2I(\theta_1) = d^2$. The solution consists in maximizing the LLR $S_{n,k}(\theta_1)$ when θ_1 is constrained to lie on the level surface $\mathcal{S} : 2I(\theta_1) = d^2$:

$$\widehat{S}_{n,k} = \sup_{\theta_1 : 2I(\theta_1)=d^2} S_{n,k}(\theta_1), \tag{9.57}$$

which results in the constrained GLR procedure. To solve this maximization problem, let us apply the method of Lagrange multipliers :

$$\begin{cases} \nabla\left[S_{n,k}(\theta_1)\right] + \lambda \nabla\left[I(\theta_1)\right] = 0 \\ 2\,I(\theta_1) = d^2 \end{cases}, \tag{9.58}$$

where $\nabla\left[S_{n,k}(\theta_1)\right] = \left(\frac{\partial S_{n,k}(\theta_1)}{\partial \vartheta_{1,1}},\ldots,\frac{\partial S_{n,k}(\theta_1)}{\partial \vartheta_{1,r}}\right)^\top$ is the gradient of $S_{n,k}(\theta_1)$ and $\nabla\left[I(\theta_1)\right]$ is the gradient of $I(\theta_1)$. This system of $r+1$ simultaneous equations need to be solved (if possible) for $r+1$ variables which are $\vartheta_{1,1},\ldots,\vartheta_{1,r}$ and λ. It follows from equation (9.57) that

$$\widehat{S}_{n,k} = \max_{\beta : 2I(\beta)=d^2} \left\{ \sum_{i=k}^{n} \beta^\top X_i e_i - 1/2\, \beta^\top \sum_{i=k}^{n} X_i X_i^\top \beta \right\},$$

where $2I(\beta) = \beta^\top \sum_{i=k}^{n} X_i X_i^\top \beta/(n-k+1)$. The method of Lagrange multipliers (9.58) leads to the following results:

$$\begin{cases} \nabla\left[S_{n,k}(\beta)\right] + 2\lambda\, \nabla\left[I(\beta)\right] = \sum_{i=k}^{n} X_i e_i - \left(1-\frac{2\lambda}{n-k+1}\right)\sum_{i=k}^{n} X_i X_i^\top \beta = 0 \\ 2I(\beta) = \beta^\top \sum_{i=k}^{n} X_i X_i^\top \beta/(n-k+1) = d^2 \end{cases}.$$

If the matrix $\sum_{i=k}^{n} X_i X_i^\top$ is nonsingular, then there is a unique solution for the above system. The solution leads to the following expression for the GLR statistic

$$\widehat{S}_{n,k} = -(n-k+1)\frac{d^2}{2} + d|\chi_{n,k}|,\ \chi_{n,k}^2 = \sum_{i=k}^{n} X_i^\top e_i \left(\frac{1}{n-k+1}\sum_{i=k}^{n} X_i X_i^\top\right)^{-1} \sum_{i=k}^{n} X_i e_i.$$

We now discuss how to select the parameter $\tilde{m}$. The LLR $S_{n,k}(\theta_1)$ in (9.58) is based on $n-k+1 \ge \tilde{m}+1$ last observations. Hence, the choice of the parameter $\tilde{m}$ defines the minimum number of samples to compute the LLR $S_{n,k}(\theta_1)$. To get a unique solution of the above system the parameter $\tilde{m}$ should be chosen as $\tilde{m} = r$. Taking into account this fact and the recursive method of computing the inverse of $\sum_{i=k}^{n} X_i X_i^\top$ knowing $\sum_{i=k}^{n-1} X_i X_i^\top$, we get the recursive constrained GLR procedure:

$$\widehat{N}_r(d) = \inf\{n \ge 1 : \hat{S}_n \ge h\}, \tag{9.59}$$

$$\widehat{S}_n = \begin{cases} 0 & \text{if } n_n < r+1 \\ -n_n\frac{d^2}{2} + d\sqrt{V_n^\top n_n P_n V_n} & \text{if } n_n \ge r+1 \end{cases},$$

$$P_n = \left(\mathbb{I} - P_{n-1}\frac{X_n X_n^\top}{1+X_n^\top P_{n-1} X_n}\right) P_{n-1} \text{ with } P_{n-1} = P_0 \text{ if } n_n = 1, \tag{9.60}$$

where

$$P_n \triangleq \left(\sum_{i=n-n_n+1}^{n} X_i\, X_i^\top\right)^{-1},\quad V_n = \mathbb{1}_{\{(n_{n-1}<r+1)\cup(\tilde{S}_{n-1}>0)\}} V_{n-1} + X_n e_n,$$

$$n_n = \mathbb{1}_{\{(n_{n-1}<r+1)\cup(\tilde{S}_{n-1}>0)\}} n_{n-1} + 1.$$

The initial conditions are $\widehat{S}_0 = 0$, $n_0 = r+1$ and $P_0 = R^{-1}$. In practice, the matrix P_0 can be also chosen as $\omega\mathbb{I}$, where ω is a large constant.

9.4.3.3 Statistical Properties of the ε-Optimal Multichart Procedures

In the case of the regression model for Lorden's "worst-worst-case" ADD defined in (6.17), an asymptotic optimality has been discussed in [26, 250, 251, 518]. It has been shown that

$$\mathsf{ESADD}(\widehat{N}_{m,r};b,\gamma) \sim \mathsf{LB}(b,\gamma) \sim \frac{2\log\gamma}{b^2} \text{ as } \gamma\to\infty. \tag{9.61}$$

Most certainly asymptotic relations (9.61) also hold for $\mathsf{SADD}(\widehat{N}_{m,r};b,\gamma)$.

To define the ε-optimal multichart procedure, we consider, as previously, a set of L parallel recursive constrained GLR procedures:

$$N_{\varepsilon r} = \min\left\{\widehat{N}_r(a_1),\widehat{N}_r(a_2),\ldots,\widehat{N}_r(a_L)\right\}. \tag{9.62}$$

We assume that the asymptotic relation between the SADD and the ARL to false alarm for the stopping time $N_{\varepsilon r}$ given by (9.47) is valid also for the regression model,

$$\mathsf{SADD}(N_{\varepsilon r};b,\gamma) \le \max\left\{r+1, \frac{2\log\gamma}{b^2-(b-a_{\ell_0})^2}\right\}(1+o(1)), \tag{9.63}$$

where $b\in[d_0,d_1]$ and $\ell_0 = \arg\min_{\ell=1,\ldots,L}|b-a_\ell|$. To find the number L and the subdivision σ we apply again the equations (9.48)–(9.49). The results of the Monte-Carlo simulation show that this approach can be used at least as an asymptotic approximation.

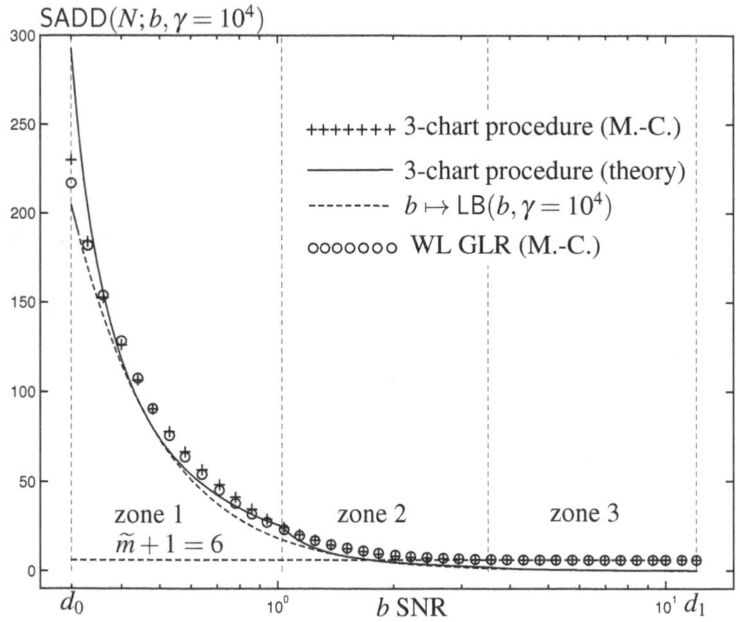

Figure 9.6: *The function $d\mapsto \mathsf{SADD}(N;d,\gamma)$ for the procedures $\widehat{N}_{0.3,r}$ and $\widehat{N}_{800,0}$ when $d_0=0.3$, $d_1=12$, $r=5$ and $\gamma=10^4$: comparison between the asymptotic approximations and Monte Carlo simulations.*

Let us compare the statistical performance and the efficiency of the recursive ε-optimal multichart and WL GLR procedures in the case of regression model (9.52). It is assumed that the inputs X_n are zero mean Gaussian random vectors with covariance matrix R. The parameters have been chosen as $r=5$, $\theta_0^\top=(0,\ldots,0)$, $\xi_n\sim\mathcal{N}(0,1)$. The detailed description of the model can be found in [330]. It is also assumed that the parameter θ_1 is given by the equation $\theta_1^\top R_5\theta_1 = b^2$, where the SNR b varies from $d_0=0.3$ to $d_1=12$. The efficiency has been fixed as $\overline{e}=1-\overline{\varepsilon}=0.7$ and the

ARL2FA as $\gamma = 10^4$. It follows from equation (9.49) that it is enough to run three parallel recursive constrained GLR procedures to get the non-optimality level $\overline{\varepsilon} = 0.3$ and that the values of the SNR to which the procedures should be tuned are $a_1 = 0.464$, $a_2 = 1.589$ and $a_3 = 5.437$.

Extensive Monte Carlo simulations have been performed for different values of b; see also [330]. Figure 9.6 presents the comparison of 10^3-repetition Monte Carlo simulations with asymptotic approximations for the above ε-optimal 3-chart procedure $\widehat{N}_{0.3,r}$ given by (9.59) and (9.62) and for the WL GLR procedure $\widehat{N}_{m,r}$ given by (9.56) when $\gamma = 10^4$. The zones of responsibility for the three parallel recursive constrained GLR procedures and the interval $[d_0, d_1]$ are shown by vertical dashed lines. The thresholds for the procedures have been chosen so that $\mathsf{E}_\infty(\widehat{N}_{0.3,r}) = \mathsf{E}_\infty(\widehat{N}_{m,r}) = \gamma = 10^4$ by using a 10^3-repetition Monte Carlo simulation in each case. As before, the asymptotic choice of the parameter $m = \lceil \log \gamma / I_0 \rceil = 206$ leads to a certain degradation of the SADD for small values of $b \simeq 0.3$. For this reason the size of the sliding window has been chosen as $m = 400$. The estimated SADD has been calculated by using (9.51). Both the simulated and theoretic SADD as functions of the SNR b for these procedures are presented in Figure 9.5. As previously, because the WL GLR procedure $\widehat{N}_{m,r}$ is asymptotically optimal, the lower bound $\mathsf{LB}(b, \gamma = 10^4)$ is used as a theoretic expression for the SADD of this procedure. On the other hand, (9.63) is used as a theoretic SADD for the ε-optimal 3-chart procedure $\widehat{N}_{0.3,r}$. Figure 9.5 confirms (except the asymptotic character of approximations) the theoretic performances of the ε-optimal rule. The extensive Monte Carlo simulations show that $\widehat{N}_{0.3r}$ performs even better than expected from asymptotic theory. Analogously to the previous case, the asymptotic approximation of the parameter m of the WL GLR scheme should be multiplied by 2 when $\gamma = 10^4$.

Chapter 10

Sequential Change Detection and Isolation

In this chapter, we generalize the quickest changepoint detection problem to the following multi-decision detection–isolation problem. There are M "isolated" points/hypotheses associated with a change that has to be detected and identified (isolated) as soon as possible after the change occurs. Alternatively, there may be M populations that are either statistically identical or the change occurs in one of them at an unknown point in time. It is necessary to detect the change in distribution as soon as possible and indicate which hypothesis is true after a change occurs or which population is "corrupted." Both the rate of false alarms and the misidentification (misisolation) rate should be controlled by given (usually low) levels. We propose certain multihypothesis detection–isolation procedures that asymptotically minimize the tradeoff between expected value of the detection lag and the false alarm/misisolation rates in the worst-case scenario. At the same time the corresponding sequential detection–isolation procedures are computationally simple and can be easily implemented on-line.

This multidecision quickest change detection problem is of importance for a variety of applications. For example, efficient statistical decision tools are needed for detecting and isolating abrupt changes in the properties of stochastic signals and dynamical systems, ranging from on-line fault diagnosis in complex technical systems to detection/classification in radar, infrared, and sonar signal processing. Specifically, the early on-line fault diagnosis (detection and isolation) in industrial processes helps in preventing these processes from more catastrophic failures. The problem of on-line target detection and classification is important when detecting and classifying a type of a target that appears at an unknown point in time or detecting a randomly appearing target in a multichannel system.

Below we discuss in detail the changepoint detection and isolation problem for iid models outlined in Section 6.3. We restrict ourselves to considering the non-Bayesian detection–isolation problem.

10.1 Problem Formulation

Consider the iid case, assuming that there is a finite family $\mathcal{P} = \{\mathsf{P}_\ell,\ \ell = 0,\dots,M\}$ of $M+1$ distinct distributions $F_\ell(x)$ with densities $f_\ell(x)$, $\ell = 0,\dots,M$ such that $f_i(x) \not\equiv f_j(x)$ for $i \neq j$. Suppose that the observations $\{X_n\}_{n\geq 1}$ are sequentially observed, *independent* and such that $X_1,\dots,X_\nu$ are each distributed according to a common distribution $F_0(x)$ (density $f_0(x)$), while $X_{\nu+1}, X_{\nu+2},\dots$ each follow a common distribution $F_\ell(x)$

$$X_n \sim \begin{cases} F_0(x) & \text{if} \quad 1 \leq n \leq \nu \\ F_\ell(x) & \text{if} \quad n \geq \nu+1 \end{cases}, \tag{10.1}$$

where the changepoint ν and the number $\ell = 1,\dots,M$ are unknown and deterministic (not random). The problem is to detect a change as quickly as possible and to identify which post-change hypothesis H_ℓ, $\ell = 1,2,\dots,M$ is true. Let P_k^ℓ and E_k^ℓ denote the probability measure and expectation when $\nu = k$ and H_ℓ is the true post-change hypothesis, and let P_∞ and E_∞ denote the same when $\nu = \infty$ (i.e., there never is a change). As has been mentioned in Section 6.3, a sequential change detection–

isolation procedure is a *pair* $\delta = (T,d)$ depending on the observations, where T is a stopping time and d is a $\{\mathscr{F}_T\}$- measurable final decision with values in the set $\{1,2,\dots,M\}$. In other words, if $T < \infty$ and $d = \ell$, then at time T the post-change hypothesis H_ℓ is accepted.

10.2 Fixed Sample Size Change Detection–Isolation Algorithms

In this section, we consider fixed sample size (FSS) change detection–isolation algorithms. We also derive some analytical properties of these algorithms and discuss numerical approximations. These algorithms are extensions of the FSS change detection algorithm discussed in Subsection 8.1.1 to the case of multiple hypotheses.

The FSS change detection–isolation algorithms are based on the following rule: samples with fixed size m are taken, and at the end of each sample a decision function is computed to test between the null hypothesis $H_0 : \mathsf{P} = \mathsf{P}_\infty$ that there is no change and the hypotheses H_ℓ, $\ell = 1,\dots,M$ that the change has occurred and the post-change distribution is F_ℓ. Sampling is stopped after the first sample of K observations for which the decision $\overline{d}_K$ is made in favor of $H_{\overline{d}} : \overline{d} > 0$.

10.2.1 A Multisample Sequential Slippage Problem

Let us continue discussing the slippage problem started in Subsection 2.9.6.2. Suppose that the independent M-dimensional vectors $\mathbf{X}_n = (X_{1,n},\dots,X_{M,n})^\top$, $\{\mathbf{X}_n\}_{n\geq 1}$, are sequentially observed. It is assumed that before change ($n \leq \nu$) the elements of the vector $\mathbf{X}_n$ are iid with common distribution $F_0(x)$ and density $f_0(x)$ but after change ($n \geq \nu + 1$) under post-change hypothesis H_ℓ, the elements $X_{1,n},\dots,X_{M,n}$ are independent, $X_{1,n},\dots,X_{\ell-1,n},X_{\ell+1,n},\dots,X_{M,n}$ are identically distributed with common distribution $F_0(x)$ and density $f_0(x)$ and $X_{\ell,n}$ has distribution $F_\ell(x)$ and density $f_\ell(x)$,[1]

$$\mathbf{X}_n \sim \begin{cases} \prod_{i=1}^M F_0(X_{i,n}) & \text{if} \quad 1 \leq n \leq \nu \\ F_\ell(X_{\ell,n}) \cdot \prod_{i=1, i\neq\ell}^M F_0(X_{i,n}) & \text{if} \quad n \geq \nu + 1 \end{cases}, \quad \ell = 1,\dots,M. \tag{10.2}$$

For the sake of simplicity consider a symmetric case where the post-change distributions $F_\ell(X_{\ell,n})$, $\ell = 1,\dots,M$ do not depend on ℓ. As discussed in Subsection 2.9.6.2, due to the invariant property, the Bayesian slippage test d is equivalent to the MP slippage test (2.316) which maximizes the common power $\beta = \mathsf{P}_\ell(d = \ell), \forall \ell = 1,\dots,M$ in the class $\mathbf{C}(\alpha) = \{d : \mathsf{P}_0(d \neq 0) \leq \alpha\}$. By applying the slippage test (2.316) at the end of each sample of size m, we get the following pair (stopping time and final decision) $\delta_{\text{FSS}} = (T_{\text{FSS}}, d_{\text{FSS}})$ of the FSS change detection–isolation algorithm

$$T_{\text{FSS}} = m \cdot \inf\{K \geq 1 : d_K \geq 1\},\; d_K = d(\mathbf{X}_{(K-1)m+1},\dots,\mathbf{X}_{Km}), \tag{10.3}$$

$$d_{\text{FSS}} = d(\mathbf{X}_{T_{\text{FSS}}-m+1},\dots,\mathbf{X}_{T_{\text{FSS}}}), \tag{10.4}$$

where

$$d\,(\mathbf{X}_1,\dots,\mathbf{X}_m) = \begin{cases} \widehat{\ell}(\mathbf{X}_1,\dots,\mathbf{X}_m) & \text{when} \quad \widehat{\lambda}\,(\mathbf{X}_1,\dots,\mathbf{X}_m) \geq h \\ 0 & \text{when} \quad \widehat{\lambda}\,(\mathbf{X}_1,\dots,\mathbf{X}_m) < h \end{cases}, \tag{10.5}$$

and

$$\widehat{\ell} = \arg\max_{1\leq j\leq M} \lambda_j, \quad \widehat{\lambda} = \max_{1\leq j\leq M} \lambda_j, \quad \lambda_j = \log\Lambda_j = \sum_{n=1}^m \log\frac{f_1(X_{j,n})}{f_0(X_{j,n})}. \tag{10.6}$$

Let us compute the ARL for this FSS change detection–isolation algorithm. The number of samples T_{FSS} has a geometric distribution

$$\mathsf{P}_\ell(T_{\text{FSS}} = k) = (1 - \mathsf{P}_\ell(d\,(\mathbf{X}_1,\dots,\mathbf{X}_m) \neq 0))^{k-1}\mathsf{P}_\ell(d\,(\mathbf{X}_1,\dots,\mathbf{X}_m) \neq 0), \;\; \ell = 0,1,\dots,M.$$

[1] The results can be easily generalized to the case where the pre-change distributions $F_0(x) = F_0^\ell(x)$ are different for different populations.

Hence

$$\mathsf{ARL} = \mathsf{E}_0^{\ell}(T_{\text{FSS}}) = \frac{m}{\mathsf{P}_{\ell}(d(\mathbf{X}_1,\ldots,\mathbf{X}_m) \neq 0)}, \quad \ell = 0,1,\ldots,M. \tag{10.7}$$

In particular, the ARL2FA is given by

$$\mathsf{ARL2FA} = \mathsf{E}_0^{0}(T_{\text{FSS}}) = \mathsf{E}_{\infty}(T_{\text{FSS}}) = \frac{m}{\alpha_0} \tag{10.8}$$

and due to the symmetry the ARL to detection is

$$\mathsf{ARL} = \mathsf{E}_0^{\ell}(T_{\text{FSS}}) = \mathsf{E}_0^{1}(T_{\text{FSS}}) = \frac{m}{1-\zeta}, \quad \ell = 1,\ldots,M, \tag{10.9}$$

where $\alpha_0 = \mathsf{P}_0(d(\mathbf{X}_1,\ldots,\mathbf{X}_m) \neq 0)$ and $\zeta = \mathsf{P}_1(d(\mathbf{X}_1,\ldots,\mathbf{X}_m) = 0)$.

Let us now discuss the statistical performance indices of this algorithm by using the minimax criterion introduced in Subsection 6.3.6. Let $\mathbf{C}_{\gamma}$ be the class of detection–isolation procedures defined in (6.31) for which the minimum of ARL2FA and ARL2FI (ARL to false isolation) is at least $\gamma > 1$. The risk associated with detection delay is defined here analogously to Lorden's "worst-worst-case" (essential supremum) ADD in (6.32)

$$\mathsf{ESADD}(\delta) = \max_{1\le j\le M}\sup_{0\le \nu<\infty}\left\{\operatorname{ess\,sup}\mathsf{E}_{\nu}^{j}[(T-\nu)^{+}|\mathscr{F}_{\nu}]\right\}. \tag{10.10}$$

It follows from [326] that the ESADD for the FSS change detection–isolation algorithm (10.3)–(10.6) is given by the following expression

$$\mathsf{ESADD}(\delta_{\text{FSS}}) = \max\left\{\frac{m}{1-\zeta}, \max_{1<k\le m}\left[(m-k+1)+\frac{m}{1-\zeta}\right]\right\} = m-1+\frac{m}{1-\zeta} \tag{10.11}$$

(cf. (10.9)).

Consider the following sequences of stopping times $T_0 = 0 < T_1 < \cdots < T_r < \cdots$ and final decisions $d_1,\ldots,d_r,\ldots$, where $T_{r\ge 1} = T_{\text{FSS}}$ and $d_{r\ge 1} = d_{\text{FSS}}$ are the stopping time and final decision of the test δ_{FSS} applied to $X_{T_{r-1}+1}, X_{T_{r-1}+2},\ldots$. It is assumed that after an alarm the observation process immediately restarts from scratch. It follows from (10.7) that the ARL2FA of j-type is

$$\mathsf{ARL2FA}_j = \mathsf{E}_0^{0}\left(\inf_{r\ge 1}\{T_{\text{FSS},r} : d_{\text{FSS},r}=j\}\right) = \mathsf{E}_{\infty}\left(\inf_{r\ge 1}\{T_{\text{FSS},r} : d_{\text{FSS},r}=j\}\right) = \frac{m}{\alpha_{0,j}}, \tag{10.12}$$

where $\alpha_{0,j} = \mathsf{P}_0[d(\mathbf{X}_1,\ldots,\mathbf{X}_m) = j > 0]$ is the probability of false alarm of j-type for the M-slippage problem. Note that due to the symmetry $\alpha_{0,j} = \alpha_{0,1}$ for $j = 1,\ldots,M$. It is easy to see that

$$\sum_{j=1}^{M}\mathsf{P}_0[d(\mathbf{X}_1,\ldots,\mathbf{X}_m) = j > 0] + \mathsf{P}_0[d(\mathbf{X}_1,\ldots,\mathbf{X}_m) = 0] = 1.$$

Hence,

$$\alpha_{0,j} = \frac{1-\prod_{j=1}^{M}\mathsf{P}_0(\lambda_j < h)}{M} = \frac{1-[\mathsf{P}_0(\lambda_1 < h)]^M}{M}.$$

The ARL2FI of a j-type, when the hypothesis H_{ℓ} is true, is given by exactly the same equation as in the case of ARL2FA

$$\mathsf{ARL2FI}_{\ell,j} = \mathsf{E}_0^{\ell}\left(\inf_{r\ge 1}\{T_{\text{FSS},r} : d_{\text{FSS},r}=j\}\right) = \mathsf{E}_{\ell}\left(\inf_{r\ge 1}\{T_{\text{FSS},r} : d_{\text{FSS},r}=j\}\right) = \frac{m}{\alpha_{\ell,j}}, \tag{10.13}$$

where $\alpha_{\ell,j} = \mathsf{P}_{\ell}[d(\mathbf{X}_{n-m+1},\ldots,\mathbf{X}_n) = j > 0]$ is the probability of a j-th type false isolation for the

M-slippage problem when the hypothesis H_ℓ is true. It is easy to see that $\alpha_{\ell,j} = \alpha_{\ell,k}$ for $\ell, j, k = 1,\ldots,M$ and $\ell \neq j \neq k$, so it suffices to compute the probability $\alpha_{2,1}$. First, it can be shown that

$$\alpha_{2,1} = \mathsf{P}_2\left[\{\lambda_1 \geq h\}\bigcap\left\{\bigcap_{j=2}^{M}\{\lambda_1 > \lambda_j\}\right\}\right] \leq \mathsf{P}_0\left[\{\lambda_1 \geq h\}\bigcap\left\{\bigcap_{j=2}^{M}\{\lambda_1 > \lambda_j\}\right\}\right].$$

Hence, $\alpha_{\ell,j} \leq \alpha_{0,j}$, when $\ell > 0$. Second, from the definition of the decision function d it follows that

$$\alpha_{2,1} = \mathsf{P}_2\left[\{\lambda_1 \geq h\}\bigcap\left\{\bigcap_{j=2}^{M}\{\lambda_1 > \lambda_j\}\right\}\right] \geq \underline{\alpha}_{2,1} = \mathsf{P}_0(\lambda_1 \geq h)\cdot \mathsf{P}_2(\lambda_2 < h)\cdot \prod_{j=3}^{M}\mathsf{P}_0(\lambda_j < h),$$

where $\prod_{j=k}^{\ell} = 1$ if $k > \ell$. Finally, we get the following inequality for the probability of false isolation

$$\underline{\alpha}_{\ell,j} \leq \alpha_{\ell,j} \leq \alpha_{0,j}$$

for $\ell, j = 1,\ldots,M$ and $\ell \neq j$ and it can be concluded that

$$\mathsf{ARL2FA}_j \leq \mathsf{ARL2FI}_{\ell,j} \leq \frac{m}{\underline{\alpha}_{\ell,j}} \quad \text{for } \ell, j = 1,\ldots,M \text{ and } \ell \neq j. \tag{10.14}$$

There are two tuning parameters of this FSS algorithm: the sample size m and the threshold h. To optimize the FSS algorithm it is necessary to find the probabilities ζ and $\alpha_{0,j}$ as functions of m and h and to minimize the ESADD subject to $\mathsf{ARL2FA}_j = \gamma$, as it follows from (10.14) and the definition of the class $\mathbf{C}_\gamma$:

$$\mathbf{C}_\gamma = \left\{\delta = (T,d) : \min_{0\leq \ell\leq M}\min_{1\leq j\neq \ell\leq M} \mathsf{E}_0^\ell\left(\inf_{r\geq 1}\{T_r \,:\, d_r = j\}\right) \geq \gamma\right\}. \tag{10.15}$$

10.2.1.1 Detection and Isolation of a Change in the Mean of a Gaussian Vector

To continue the study and optimization of the FSS algorithm, consider the following independent sequence $\{\mathbf{X}_n\}_{n\geq 1}$ of Gaussian vectors

$$\mathbf{X}_n \sim \begin{cases} \mathcal{N}(\theta_0, \sigma^2\mathbb{I}_M) & \text{if } \; 1 \leq n \leq \nu \\ \mathcal{N}(\theta_\ell, \sigma^2\mathbb{I}_M) & \text{if } \; n \geq \nu+1, \end{cases}, \quad \ell = 1,\ldots,M, \tag{10.16}$$

where $\mathbf{X}_n \in \mathbb{R}^M$, $\theta \in \mathbb{R}^M$, $M \geq 2$, $\theta_0 = (0,\ldots,0)^\top$, $\theta_\ell = (0,\ldots,0,\vartheta_\ell,0,\ldots,0)^\top$, $\vartheta_1 = \cdots = \vartheta_M = \vartheta > 0$ and σ^2 are known constants. In other words, we have M independent channels and the "target" ϑ can appear at an unknown moment $\nu+1$ and in an unknown channel ℓ. This situation can be called the case of orthogonal alternatives, i.e., $\theta_j \perp \theta_\ell$ for $j \neq \ell$.

The FSS change detection–isolation algorithm $\delta_{\text{FSS}} = (T_{\text{FSS}}, d_{\text{FSS}})$ is defined as previously (10.3)–(10.6) with the LLR λ_ℓ given by

$$\lambda_\ell = \sum_{n=1}^{m} \log\frac{f_\ell(\mathbf{X}_n)}{f_0(\mathbf{X}_n)} = \frac{\vartheta}{\sigma^2}\sum_{n=1}^{m} X_{\ell,n} - \frac{\vartheta^2 m}{2\sigma^2}. \tag{10.17}$$

The optimal relations between the ESADD, ARL2FA, and ARL2FI for this FSS algorithm are defined by the following theorem.

Theorem 10.2.1. *For the symmetric Gaussian model* (10.16)*, the ESADD of the FSS change detection–isolation algorithm* (10.3)–(10.6) *is given by*

$$\mathsf{ESADD}(\delta_{\text{FSS}}) = \frac{\sigma^2(x-y)^2}{\vartheta^2\{1-\Phi(y)\Phi^{M-1}(x)\}} + \frac{\sigma^2(x-y)^2}{\vartheta^2} - 1, \tag{10.18}$$

where $x = (h + m_\rho)/\sqrt{2m_\rho}$, $y = (h - m_\rho)/\sqrt{2m_\rho}$, $m_\rho = m\vartheta^2/(2\sigma^2)$*; the ARL2FA is given by*

$$\mathsf{ARL2FA}_j(\delta_{\scriptscriptstyle FSS}) = \frac{\sigma^2(x-y)^2 M}{\vartheta^2\{1-\Phi^M(x)\}}, \quad j = 1,\ldots,M; \tag{10.19}$$

the bounds for the ARL2FI are given by

$$\mathsf{ARL2FA}_j(\delta_{\scriptscriptstyle FSS}) \le \mathsf{ARL2FI}_{\ell,j}(\delta_{\scriptscriptstyle FSS}) \le \frac{\sigma^2(x-y)^2}{\vartheta^2[1-\Phi(x)]\Phi(y)\Phi^{M-2}(x)}, \quad \ell, j = 1,\ldots,M, \ \ell \ne j; \tag{10.20}$$

and the bounds for the probability of false isolation are given by

$$\frac{[1-\Phi(x)]\Phi(y)\Phi^{M-2}(x)}{1-\Phi(y)\Phi^{M-1}(x)} \le \alpha_{\ell,j}(\delta_{\scriptscriptstyle FSS}) \le \frac{1-\Phi^M(x)}{M\{1-\Phi(y)\Phi^{M-1}(x)\}} \quad \ell, j = 1,\ldots,M, \ \ell \ne j. \tag{10.21}$$

The detailed proof of Theorem 10.2.1 may be found in [326].

Remark 10.2.1. It is easy to see that the K–L information number between two alternative (post-change) hypotheses H_ℓ and H_j ($\ell \ne j$) is greater than the K–L information number between the pre-change hypothesis H_0 and a post-change alternative H_ℓ

$$I(\theta_\ell, \theta_0) = \frac{\vartheta^2}{2\sigma^2} \text{ and } I(\theta_\ell, \theta_j) = \frac{\vartheta^2}{\sigma^2}, \quad 1 \le \ell \ne j \le M.$$

Hence, it seems intuitively plausible that the detection step has to be more difficult than the isolation step and that the ARL2FI is greater than the ARL2FA, which is confirmed by (10.20).

We have discussed the optimal FSS algorithm when $M = 1$ in Subsection 8.1.1. It was shown that

$$\mathsf{ESADD}(\delta_{\scriptscriptstyle FSS})_{M=1} \sim 2\frac{\log\gamma}{I(\vartheta)} \quad \text{as } \gamma \to \infty,$$

where $I(\vartheta) = \vartheta^2/(2\sigma^2)$. Now, this result can be generalized as follows. It immediately follows from equations (10.18)–(10.20) that

$$\begin{aligned}\mathsf{ESADD}(\delta_{\scriptscriptstyle FSS})_{M\ge 1} &\le \frac{\sigma^2(x-y)^2}{\vartheta^2[1-\Phi(y)]} + \frac{\sigma^2(x-y)^2}{\vartheta^2} - 1,\\ \mathsf{ARL2FI}_{\ell,j}(\delta_{\scriptscriptstyle FSS}) &\ge \mathsf{ARL2FA}_j(\delta_{\scriptscriptstyle FSS}) \ge \frac{\sigma^2(x-y)^2}{\vartheta^2[1-\Phi(x)]}.\end{aligned}$$

The right sides of the above inequalities are equal to the right sides of the equations for $\mathsf{ESADD}(\delta_{\scriptscriptstyle FSS})$ and γ in the case of $M = 1$; for a detailed proof see [326, Theorems 1 and 2]. Therefore, $\mathsf{ESADD}(\delta)_{M=1}$ plays the role of an upper bound for $\mathsf{ESADD}(\delta_{\scriptscriptstyle FSS})_{M\ge 2}$ and the following asymptotic result can be established. Let $\mathsf{ARL2FI}_{\ell,j}(\delta_{\scriptscriptstyle FSS}) \ge \mathsf{ARL2FA}_j(\delta_{\scriptscriptstyle FSS}) = \gamma$, then

$$\mathsf{ESADD}(\delta_{\scriptscriptstyle FSS})_{M\ge 2} \le \mathsf{ESADD}(\delta_{\scriptscriptstyle FSS})_{M=1} \sim 2\frac{\log\gamma}{I(\vartheta)} \quad \text{as } \gamma \to \infty. \tag{10.22}$$

Remark 10.2.2. Let us pursue the discussion of Remark 8.1.1 on the relation between the ARL and the ESADD for the FSS change detection-isolation algorithm. It can be shown that the asymptotically optimal choice of the tuning parameters m^* and h^* is given by

$$m^* \sim \frac{\log\gamma}{I(\vartheta)} \text{ and } h^* \sim \log\gamma \text{ as } \gamma \to \infty. \tag{10.23}$$

As follows from [326, Theorems 1 and 2], the ARL, or the mean detection delay when $\nu = 0$, is asymptotically equal to

$$\mathsf{ARL}(\theta_\ell) = \mathsf{E}_0^\ell(T_{\scriptscriptstyle FSS}) \sim \frac{\log\gamma}{I(\vartheta)} \quad \text{as } \gamma \to \infty, \ \ell = 1,\ldots,M.$$

Hence, analogously to the case of pure change detection ($M = 1$), the $\mathsf{ESADD}(\delta_{\text{FSS}})$ is asymptotically twice as great as the ARL for the FSS change detection–isolation algorithm (10.3)–(10.6).

The non-asymptotically optimal choice of the parameters $x = x(m,h)$ and $y = y(m,h)$ of the FSS algorithm can be reduced to the following optimization problem:

$$\begin{cases} \arg\min_{x,y} \mathsf{ESADD}(x,y\,;\delta_{\text{FSS}}) & = & (x^*,y^*) \\ \text{subject to } \mathsf{ARL2FA}_j((x,y\,;\delta_{\text{FSS}}) & = & \gamma \end{cases}. \tag{10.24}$$

Hence, we fix a prescribed value γ and deduce the optimal values x^* and y^* by minimizing the function $(x,y) \mapsto \mathsf{ESADD}(x,y;\delta_{\text{FSS}})$ (see (10.18)) with respect to x and y under the constraint $\mathsf{ARL2FA}_j((x,y\,;\delta_{\text{FSS}}) = \gamma$ (see (10.19)). The minimum of $\mathsf{ESADD}(x,y\,;\delta_{\text{FSS}})$ is given by $\mathsf{ESADD}(x^*,y^*;\delta_{\text{FSS}})$. The relation between $\gamma = \mathsf{ARL2FA}_j$ and $\mathsf{ESADD}(x^*,y^*;\delta_{\text{FSS}})$, obtained by nu-

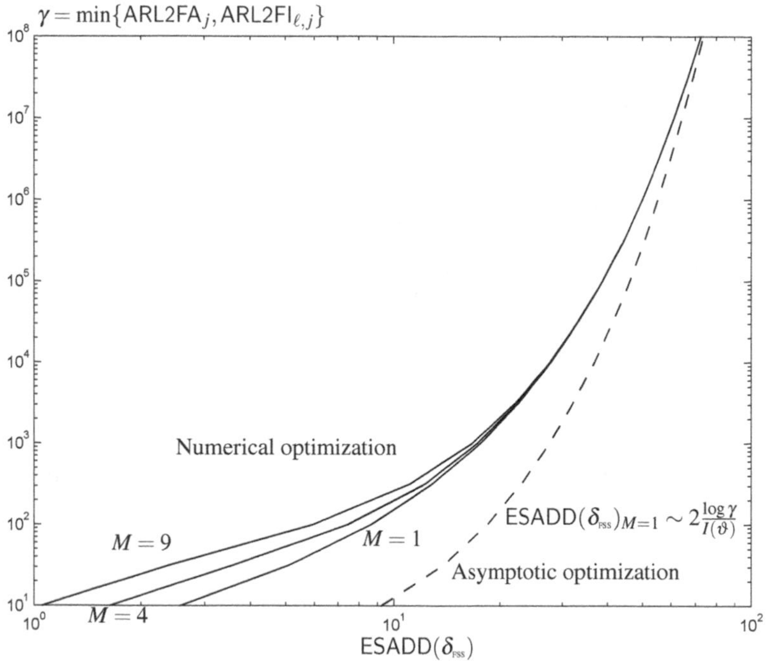

Figure 10.1: *Numerical optimization of the FSS algorithm.*

merical optimization for the SNR $\frac{\vartheta}{\sigma} = 1$ and $M = 1,4,9$, and their asymptotic relation (10.22) for the case $M = 1$ are shown in Figure 10.1. This figure shows that the non-asymptotic "exact" curve of $\mathsf{ESADD}(\delta_{\text{FSS}})$ obtained by numerical optimization is close to the asymptotic upper bound $(4\sigma^2 \log\gamma)/\vartheta^2$ when γ is large.

The further extension of the scalar independent Gaussian channels to the vector independent channels has been proposed in [252]. Let each independent channel be represented by the observation vector $\mathbf{X}_{nj} \in \mathbb{R}^{p_j}$ (not necessarily of the same dimension). Before the change, i.e., for $n \le \nu$, all channels contain only noise, $\mathsf{E}_{n\le\nu}(\mathbf{X}_n) = 0$. After the change, i.e., for $n \ge \nu+1$, the "target" appears in an unknown channel ℓ, i.e., $\mathsf{E}_{n\ge\nu+1}(\mathbf{X}_{n\ell}) = \theta^\ell$ and $\mathsf{E}_{n\ge\nu+1}(\mathbf{X}_{nj}) = 0$ for $j \ne \ell$. The concatenation of these vectors produces the total observation vector $\mathbf{X}_n = (\mathbf{X}_{n1}^\top \cdots \mathbf{X}_{nM}^\top)^\top$.

Let $\{\mathbf{X}_n\}_{n\ge1}$ be the following independent sequence of Gaussian vectors

$$\mathbf{X}_n \sim \begin{cases} \mathcal{N}(\theta_0,\Sigma) & \text{if} \quad 1 \le n \le \nu \\ \mathcal{N}(\theta_\ell,\Sigma) & \text{if} \quad n \ge \nu+1, \end{cases}, \quad \ell = 1,\ldots,M, \tag{10.25}$$

where $\Sigma = \mathsf{diag}\{\Sigma_1,\cdots,\Sigma_M\}$ is a covariance block diagonal matrix of size $p\times p$, $p = \sum_{j=1}^M p_j$, $\theta_0 = (0,\ldots,0)^\top$, $\theta_\ell = (0,\ldots,0,\theta^{\ell T},0,\ldots,0)^\top$, $\theta \in \mathbb{R}^p$.

Without assuming θ_ℓ to be specified in advance, the FSS detection–isolation algorithm based on the GLR statistics is proposed in [252]. Nevertheless, it is assumed that the K–L information numbers $I(\theta_\ell,\theta_0)$ are known in advance and they are the same $I(\theta_\ell,\theta_0)=I$ for $1\le\ell\le M$. The independent GLR statistics are χ^2-distributed in the Gaussian case.

The FSS change detection–isolation procedure $\delta_{\text{FSS}}=(T_{\text{FSS}},d_{\text{FSS}})$ is defined as previously by (10.3)–(10.6), but now the GLR statistic is used instead of the LLR statistics in (10.6),

$$\lambda_\ell=\frac{1}{2m}\left[\sum_{n=1}^{m}\mathbf{X}_{n\ell}\right]^\top\Sigma_\ell^{-1}\left[\sum_{n=1}^{m}\mathbf{X}_{n\ell}\right]. \tag{10.26}$$

As in the previous case, taking into account the block-diagonal covariance matrix of $\mathbf{X}_n$ and the definition of the vector θ, see (10.25), it can be shown that the K–L information number between two alternative hypotheses H_ℓ and H_j when $\ell\neq j$ is greater than the K–L information number between the pre-change hypothesis and a post-change alternative

$$I(\theta_\ell,\theta_0)=\frac{1}{2}\theta_\ell^\top\Sigma_\ell^{-1}\theta_\ell \text{ and } I(\theta_\ell,\theta_j)=\frac{1}{2}\left[\theta_\ell^\top\Sigma_\ell^{-1}\theta_\ell+\theta_j^\top\Sigma_j^{-1}\theta_j\right],\quad 1\le\ell\neq j\le M.$$

An alternative to the class $\mathbf{C}_\gamma$ defined in (6.31) has been proposed by Lai [252] who suggested to define a time window of length m_α and to impose a constraint on the probability of false alarm in any such window or the probability of false isolation within a time window of length m_α starting from the changepoint ν to be upper bounded by αm_α. In other words, define the following class:

$$\mathbf{C}_\alpha=\left\{\delta=(T,d):\ \sup_{\nu\ge1}\mathsf{P}_\infty(\nu\le T<\nu+m_\alpha)\le\alpha m_\alpha,\ \sup_{\nu\ge0}\mathsf{P}_\nu^\ell(\nu<T\le\nu+m_\alpha)\le\alpha m_\alpha\right\} \tag{10.27}$$

Let $\delta_{\text{FSS}}=(T_{\text{FSS}},d_{\text{FSS}})$ be the FSS change detection–isolation algorithm (10.3)–(10.6) with (10.26). It has been established that if $m_\alpha>m$ and the threshold h in (10.5) is chosen so that (10.27) is satisfied, then the mean detection delay of the FSS algorithm $\delta_{\text{FSS}}\in\mathbf{C}_\alpha$ when $\nu=0$ can attain the asymptotic lower bound for the mean detection delay in this class

$$\mathsf{E}_\nu^\ell(T-\nu)^+\ge\frac{\mathsf{P}_\infty(T>\nu)}{\min\limits_{0\le j\neq\le M}I(\theta_\ell,\theta_j)+o(1)}|\log\alpha|. \tag{10.28}$$

For a detailed proof see [252, Theorem 6]. Hence, $\mathsf{E}_0^\ell(T_{\text{FSS}})\sim\left(I^{-1}(\theta_\ell,\theta_0)+o(1)\right)|\log\alpha|$ as $\alpha\to0$. To attain this lower bound, the sample size m should be chosen as

$$m\sim\frac{|\log\alpha|}{I}\quad\text{as }\alpha\to0. \tag{10.29}$$

Recall that we consider a symmetric case where $I(\theta_\ell,\theta_0)=I$. This leads to the following asymptotic equation for the maximal conditional average delay to detection

$$\mathsf{SADD}(\delta_{\text{FSS}})=\sup_{0\le\nu<\infty}\mathsf{E}_\nu^\ell(T_{\text{FSS}}-\nu|T_{\text{FSS}}>\nu)\sim\frac{2|\log\alpha|}{I}\quad\text{as }\alpha\to0. \tag{10.30}$$

Remark 10.2.3. We have discussed the two FSS change detection–isolation algorithms. The first procedure is based on the LR because the alternative hypotheses are simple whereas the second one is based on the GLR because the alternative hypotheses are composite. Nevertheless, both procedures necessitate that the K–L information numbers should be considered known in advance. Indeed, as follows from (10.23) and (10.29), to attain the smallest value of ESADD or SADD, it is necessary to know in advance the K–L information numbers $I(\theta_\ell,\theta_0)$ for $1\le\ell\le M$ to find an asymptotically optimal parameter m. In the second case the vectors θ^ℓ can be unknown in advance but the K–L information numbers $I(\theta_\ell,\theta_0)$ have to be known in advance and moreover they should be the same $I(\theta_\ell,\theta_0)=I$ for $1\le\ell\le M$ if we wish to get an optimal detection for any channel ℓ.

10.2.2 A General Changepoint Model: Constrained Minimax FSS Algorithm

We now apply the constrained minimax multihypothesis FSS test introduced and discussed in Subsection 2.9.6.3 to the general changepoint model defined by (10.1) with the family $\mathcal{P} = \{\mathsf{P}_\ell,\ \ell = 0,\dots,M\}$ of $M+1$ distinct distributions $F_\ell(x)$ with densities $f_\ell(x)$, $\ell = 0,\dots,M$ such that $f_i(x) \not\equiv f_j(x)$ for $i \neq j$. In contrast to the case of independent channels discussed in Subsection 10.2.1, the general change detection and isolation problem does not possess any natural symmetry or invariance under the group of permutations of the set $\{H_1,\dots,H_M\}$. For this reason, the only solution is to use the "equalizer test" (2.321) which maximizes the common power in the class $\mathbf{C}(\alpha)$. The FSS change detection–isolation procedure $\delta_{\text{FSS}} = (T_{\text{FSS}}, d_{\text{FSS}})$ is defined as previously by (10.3)–(10.6), but now the LR statistics Λ_j, $j = 1,\dots,M$ are used with additional weights q_j,

$$\lambda_j = \log \Lambda_j = \log q_j + \sum_{n=1}^{m} \log \frac{f_j(X_n)}{f_0(X_n)}, \tag{10.31}$$

and the weight coefficients $q_1,\dots,q_M$ are selected so that the probability of false classification is constant over the set of alternative hypotheses $H_1,\dots,H_M$

$$\alpha_\ell(d) = \mathsf{P}_\ell\left[d\,(X_1,\dots,X_m) \neq \ell\right] = \alpha_j(d), \quad \forall \ell, j \neq 0.$$

This "equalizer test" maximizes the common power

$$\beta = \mathsf{P}_\ell\left[d\,(X_1,\dots,X_m) = \ell\right] = \mathsf{P}_j\left[d\,(X_1,\dots,X_m) = j\right], \quad \forall \ell, j \neq 0$$

in the class $\mathbf{C}(\alpha)$. The equations previously obtained for the ESADD, ARL2FA, and ARL2FI, see (10.11), (10.13), and (10.12), are applicable in the case of the constrained minimax FSS algorithm.

Let us consider again the Gaussian model but now with non-orthogonal alternatives. In contrast to the case of orthogonal alternatives discussed in Subsection 10.2.1.1, this more general model cannot be reduced to independent channels. The motivation of non-orthogonal alternatives is the following. Often, the observation $\mathbf{Y}_n$ can be seen as an output of a linear model including non-random nuisance parameters. Some examples of such situations can be found in Section 2.10. It follows from the theory of invariance that the rejection of nuisance parameters implies the projection of the vector $\mathbf{Y}_n$ onto a subspace containing the maximal invariant statistics (see Subsections 2.9.5 and 2.10.3.1). Let us continue the discussion of the linear Gaussian model $\mathbf{Y}_n = H\mathbf{X}_n + \mu + \xi_n$ with nuisance parameters $\mathbf{X}_n$ and several alternative hypotheses $H_j : \{\mu = \rho\,\theta_j,\ \rho \geq \rho_j\}$ considered in Subsection 2.10.3.2. As it follows from equation (2.360), the dimension of the nuisance-free subspace $\mathbf{Z}_n = W\mathbf{Y}_n = W[H\mathbf{X}_n + \mu + \xi_n]$ is necessarily less than M. Moreover, in the general changepoint model the number of non-orthogonal alternative hypotheses can be much greater than the dimension p of the nuisance-free subspace.

For the sake of simplicity, consider the nuisance-free sequence $\{\mathbf{Z}_n\}_{n\geq 1}$ of iid Gaussian vectors observed sequentially, $\mathbf{Z}_n \sim \mathcal{N}(\theta, \Sigma)$, where Σ is a (known) covariance matrix. By using the invariance properties of the Gaussian family $\mathcal{N}(\theta, \Sigma)$ (see Subsections 2.9.5), without loss of generality the change detection and isolation problem can be reduced to the following problem

$$\mathbf{Z}_n \sim \begin{cases} \mathcal{N}(\theta_0, \sigma^2 \mathbb{I}_p) & \text{if} \quad 1 \leq n \leq \nu \\ \mathcal{N}(\theta_\ell, \sigma^2 \mathbb{I}_p) & \text{if} \quad n \geq \nu + 1, \end{cases} \quad \ell = 1,\dots,M, \tag{10.32}$$

where $\mathbf{Z}_n \in \mathbb{R}^p$, $\theta_\ell \in \mathbb{R}^p$, $1 \leq \ell \leq M$, $M \geq 2$, $\theta_0 = (0,\dots,0)^\top$ and $\mathbb{I}_p$ is the identity matrix of size $p \times p$. The vectors θ_ℓ are known and have the same norm $\|\theta_\ell\|^2 = c^2$ for all ℓ. It is assumed that $\theta_\ell \neq \pm\theta_j$ for all $\ell \neq j$.

As we have previously mentioned, typically Lorden's "worst-worst-case" ADD (10.10) is too conservative and another measure of the detection speed, namely the maximal conditional average delay to detection $\mathsf{SADD}(T)$ defined in (6.35), is better suited for practical purposes. The goal is to minimize the SADD of the FSS algorithm in the class $\mathbf{C}_{\gamma,\beta}$ defined in (6.34). As it follows from

Definition 2.10.3 of the constrained asymptotically uniformly minimax test and Theorem 2.10.4, the negative impact of constant weight coefficients $q_1 = \cdots = q_M$ becomes negligible if the SNR $\omega^2 = c^2/\sigma^2$ tends to infinity. For the FSS change detection and isolation problem this is equivalent to the asymptotic case when $\gamma \to \infty$ and $\beta \to 0$ in the class $\mathbf{C}_{\gamma,\beta}$.

The FSS change detection–isolation algorithm $\delta_{\text{FSS}} = (T_{\text{FSS}}, d_{\text{FSS}})$ is defined by (10.3)–(10.6) with the following LLR

$$\lambda_\ell = \sum_{n=1}^{m} \log \frac{f_\ell(\mathbf{Z}_n)}{f_0(\mathbf{Z}_n)} = \frac{1}{\sigma^2} \sum_{n=1}^{m} \theta_\ell^\top \mathbf{Z}_n - m \frac{\|\theta_\ell\|_2^2}{2\sigma^2}. \tag{10.33}$$

The statistical characteristics of the FSS algorithm depend on the mutual "geometry" of the alternative hypotheses expressed in terms of the K–L information (distance) $I_{\ell,j} = \frac{\|\theta_\ell - \theta_j\|^2}{2\sigma^2}$, $0 \le \ell \ne j \le M$ when the vectors $\theta_1, \dots, \theta_M$ are not orthogonal. Let $\delta_{\ell,j} = \frac{1}{2}\,\|e_\ell - e_j\|^2$ be the K–L distance between two unit vectors $e_\ell = \theta_\ell/c$ and $e_j = \theta_j/c$, where $e_\ell \ne \pm e_j$ for all $1 \le \ell \ne j \le M$. The numbers $\{\delta_{\ell,j}\}_{1 \le \ell \ne j \le M}$ describe the mutual "geometry" of the hypotheses. Assuming that $e_0 = 0$, we get the K–L distances δ_{det} and δ_{isol} characterizing the detection and isolation steps, respectively, $\delta_{\text{det}} = \min_{1 \le j \le M} \delta_{0,j} = 1/2$, $\delta_{\text{isol}} = \min_{1 \le i \ne j \le M} \delta_{i,j}$. Clearly that $0 < \delta_{\text{isol}} < 2$. It follows that $I_{\text{det}} = \omega^2 \delta_{\text{det}}$ and $I_{\text{isol}} = \omega^2 \delta_{\text{isol}}$, where $\omega^2 = c^2/\sigma^2$ is the SNR. The main result is given by the following theorem whose proof may be found in [146].

Theorem 10.2.2. *Consider the model* (10.32). *Let* $\delta_{\mathit{FSS}} = (T_{\mathit{FSS}}, d_{\mathit{FSS}})$ *be the FSS change detection–isolation algorithm given by* (10.3)–(10.6) *with the LLR* (10.33). *Define a subclass* $\tilde{\mathbf{C}}_{\gamma,\beta}$ *of the class* $\mathbf{C}_{\gamma,\beta}$ *in the following manner* $\min\{\delta^2_{\mathit{isol}}, \delta_{\mathit{isol}}/2\} \log \gamma \ge \log \beta^{-1} (1 + |o(1)|)$ *as* $\beta \to 0$. *If the tuning parameters* h^* *and* m^* *are selected as*

$$h^* \sim \log \gamma \text{ and } m^* \sim \frac{\log \gamma}{I_{\mathit{det}}} \quad \text{as } \gamma \to \infty,$$

then

$$\mathsf{SADD}(\delta_{\mathit{FSS}}) \sim \frac{2 \log \gamma}{I_{\mathit{det}}} \quad \text{as } \gamma \to \infty. \tag{10.34}$$

10.3 The Generalized CUSUM Change Detection–Isolation Algorithms

In this section we extend the CUSUM procedure introduced in Section 8.2 to the case of multiple hypotheses. We consider the general changepoint model defined by (10.1) with the family $\mathcal{P} = \{\mathsf{P}_\ell,\ \ell = 0, \dots, M\}$ of $M+1$ distinct distributions $F_\ell(x)$ with densities $f_\ell(x)$, $\ell = 0, \dots, M$ such that $f_i(x) \not\equiv f_j(x)$ for $i \ne j$. The generalized CUSUM change detection–isolation algorithm (GCS) $\delta_{\text{GCS}} = (T_{\text{GCS}}, d_{\text{GCS}})$ is given by

$$T_{\text{GCS}} = \min\{T_1, \dots, T_M\},\ \ d_{\text{GCS}} = \arg\min\{T_1, \dots, T_M\}, \tag{10.35}$$

where the stopping time T_ℓ is responsible for the detection of post-change hypothesis H_ℓ and is defined as follows

$$T_\ell = \inf_{k \ge 1} T_\ell(k),\ \ T_\ell(k) = \inf\left\{ n \ge k : \min_{0 \le j \ne \ell \le M} \lambda_{\ell,j}(k,n) \ge h \right\}, \tag{10.36}$$

where

$$\lambda_{\ell,j}(k,n) = \lambda_\ell(k,n) - \lambda_j(k,n) = \sum_{i=k}^{n} \log \frac{f_\ell(X_i)}{f_j(X_i)}$$

is the LLR between the hypotheses H_ℓ and H_j. It is easy to see that this "generalized" CUSUM algorithm is based on the sequential test for multiple simple hypotheses defined in Chapter 4. The key statistical properties of the LLR are as follows

$$\mathsf{E}_j\left[\lambda_{\ell,j}(k,n)\right] < 0 \text{ and } \mathsf{E}_\ell\left[\lambda_{\ell,j}(k,n)\right] > 0.$$

In other words, a change in the statistical model (10.1) is reflected as a change in the sign of the LLR mean. The CUSUM algorithm (8.60) which has to *detect* a change in distribution is based on the comparison of the difference between the value of the LLR and its current minimum value with a given threshold h

$$g_n = \lambda(1,n) - \min_{0 \le k < n} \lambda(1,k) = \max_{1 \le k \le n} \lambda(k,n) \gtrless h, \;\; \lambda(1,0) = 0.$$

To *detect and isolate* a change in the model we may exploit the idea of CUSUM with some modifications. Now we have a composite but finite null hypothesis $\mathbf{H}_0 = \bigcup_{0 \le j \ne \ell \le M} H_j$ and a simple alternative H_ℓ. The GLR for testing $\mathbf{H}_0$ against H_ℓ is given by

$$\log \frac{\prod_{i=k}^{n} f_\ell(X_i)}{\max\limits_{0 \le j \ne \ell \le M} \prod_{i=k}^{n} f_j(X_i)} = \log \min_{0 \le j \ne \ell \le M} \prod_{i=k}^{n} \frac{f_\ell(X_i)}{f_j(X_i)} = \min_{0 \le j \ne \ell \le M} \lambda_{\ell,j}(k,n). \tag{10.37}$$

For this reason T_ℓ stops if, for some $k < T_\ell$, the observations $X_k, \dots, X_{T_\ell}$ are *significant* for accepting the hypothesis H_ℓ with respect to this set of alternatives,

$$T_\ell = \inf\left\{ n \ge 1 : \max_{1 \le k \le n} \min_{0 \le j \ne \ell \le M} \lambda_{\ell,j}(k,n) \ge h \right\}. \tag{10.38}$$

Now we investigate the statistical properties of the GCS (10.35)–(10.36). Let $\mathbf{C}_\gamma$ be the class of detection–isolation procedures defined in (10.15) for which the minimum of ARL2FA and ARL2FI is at least $\gamma > 1$. The asymptotic relation between the ESADD (10.10) and γ is stated in the following theorem.

Theorem 10.3.1. *Let $\delta_{\text{GCS}} = (T_{\text{GCS}}, d_{\text{GCS}})$ be the detection–isolation algorithm* (10.35)–(10.36) *from the class $\mathbf{C}_\gamma$ and $h \sim \log \gamma$ as $\gamma \to \infty$, in particular $h = \log \gamma$. Then*

$$\mathsf{ESADD}(\delta_{\text{GCS}}) \sim \max_{1 \le \ell \le M} \mathsf{E}_\ell(T_{\text{GCS}}) \sim \frac{\log \gamma}{I^*} \;\; \text{as} \;\; \gamma \to \infty \tag{10.39}$$

where

$$I^* = \min_{1 \le \ell \le M} \min_{0 \le j \ne \ell \le M} I_{\ell,j}, \;\; 0 < I_{\ell,j} < \infty \;\; \text{for all} \;\; 0 \le \ell \ne j \le M$$

is the minimal K–L information number between two closest hypotheses.

The detailed proof may be found in [322]. It is essentially based on the following lemmas.

Lemma 10.3.1. *Let T_ℓ be stopping variables* (10.38) *with respect to $X_1, X_2, \dots \sim \mathsf{P}_\ell$. Then*

$$\mathsf{E}_\ell(T_j) \ge e^h \;\; \text{for} \;\; \ell = 0, \dots, M, \;\; 1 \le j \ne \ell \le M. \tag{10.40}$$

Corollary 10.3.1. *Consider the sequences of stopping times $\{T_k\}_{k \ge 0}$ and final decisions $\{d_k\}_{k \ge 1}$, where $T_{k \ge 1} = T_{\text{GCS}}$ and $d_{k \ge 1} = d_{\text{GCS}}$ are the stopping time and the final decision of the procedure δ_{GCS} applied to $X_{T_{k-1}+1}, X_{T_{k-1}+2}, \dots$. Assume that after an alarm the observation process immediately restarts from scratch. Then for any $h > 0$, the minimum of the mean times before a false alarm or a false isolation satisfies the inequality*

$$\min_{0 \le \ell \le M} \min_{1 \le j \ne \ell \le M} \mathsf{E}_0^\ell \left(\inf_{k \ge 1} \{ T_k : d_k = j \} \right) \ge e^h. \tag{10.41}$$

Note that (10.41) implies that if $h = \log \gamma$, then $\delta_{\text{GCS}} \in \mathbf{C}_\gamma$ for every $\gamma > 1$.

Lemma 10.3.2. *Let T_j be stopping variables* (10.38) *with respect to $X_1, \dots, X_k, X_{k+1}, \dots \sim \mathsf{P}_k^j$. Then*

$$\sup_{0 \le \nu < \infty} \left\{ \operatorname{ess\,sup} \mathsf{E}_\nu^j [(T_{\text{GCS}} - \nu)^+ | \mathscr{F}_\nu] \right\} \le \mathsf{E}_j[T_j(1)] \sim \frac{h}{\min_{0 \le \ell \ne j \le K-1} I_{j,\ell}} \;\; \text{as} \;\; h \to \infty. \tag{10.42}$$

Corollary 10.3.2. *Let $\delta_{GCS} = (T_{GCS}, d_{GCS})$ be the detection–isolation algorithm* (10.35)–(10.36). *Then*

$$\mathsf{ESADD}(\delta_{GCS}) \le \max_{1\le j\le M} \mathsf{E}_j[T_j(1)] \sim \frac{h}{I^*} \quad \text{as } h\to\infty \tag{10.43}$$

and taking $h = \log\gamma$ yields (10.39).

Remark 10.3.1. As follows from (10.37)–(10.38), the stopping time T_ℓ is computed by using the GLR test between a composite null hypothesis and a simple alternative. We have seen in Section 8.3 that the GLR change detection tests cannot be written in a recursive manner. The stopping time T_ℓ is also nonrecursive. An obvious disadvantage of such schemes is that the number of maximizations at time n grows to infinity with n. Hence, for practical purposes it is of interest to find *another* appropriate recursive computational scheme in order to reduce the amount of numerical operations, which should be performed for every new observation without losing optimality.

Easily implementable recursive change detection–isolation algorithms based on CUSUM and SR statistics have been proposed in [343, 461]. The idea of the CUSUM-based algorithm is to replace *max min* in (10.38) with *min max*. This permutation leads to a recursive *matrix CUSUM test.* The pair $\delta_{\text{MCS}} = (T_{\text{MCS}}, d_{\text{MCS}})$ is again defined by (10.35)–(10.36) but the stopping time T_ℓ is replaced by

$$T_{\text{MCS},\ell} = \inf\left\{n\ge 1 : \min_{0\le j\neq\ell\le M}\max_{1\le k\le n} \lambda_{\ell,j}(k,n)\ge h\right\} = \inf\left\{n\ge 1 : \min_{0\le j\neq\ell\le M} g_{\ell,j}(n)\ge h\right\} \tag{10.44}$$

where $g_{\ell,j}(n) = \left[g_{\ell,j}(n-1) + \lambda_{\ell,j}(n,n)\right]^+$. A more general case of the matrix CUSUM test is considered in [461]

$$T_{\text{MCS},\ell} = \inf\left\{n\ge 1 : \min_{0\le j\neq\ell\le M}\left[g_{\ell,j}(n) - h_j\right]\ge 0\right\}, \tag{10.45}$$

where different thresholds may be used in order to handle different constraints for the false alarm and false isolation rates. This generalization will be discussed later on in more detail.

The matrix CUSUM algorithm is fully recursive and its statistical properties asymptotically coincide with the generalized CUSUM algorithm (10.35)–(10.36).

More specifically, the following results hold.

Lemma 10.3.3. *Let $\delta_{MCS} = (T_{MCS}, d_{MCS})$ be the detection–isolation algorithm defined in* (10.35), (10.36), *and* (10.44). *Consider the sequences of stopping times $\{T_k\}_{k\ge 0}$ and final decisions $\{d_k\}_{k\ge 1}$, where $T_{k\ge 1} = T_{MCS}$ and $d_{k\ge 1} = d_{MCS}$ are the stopping time and the final decision of the test δ_{MCS} applied to $X_{T_{k-1}+1}, X_{T_{k-1}+2}, \ldots$. Then*

$$\min_{0\le\ell\le M}\min_{1\le j\neq\ell\le M} \mathsf{E}_0^\ell\left(\inf_{k\ge 1}\{T_k : d_k = j\}\right)\ge e^h \tag{10.46}$$

and

$$\mathsf{ESADD}(\delta_{MCS}) \le \max_{1\le j\le M}\mathsf{E}_j[T_{MCS,j}(1)] \sim \frac{h}{I^*} \quad \text{as } h\to\infty. \tag{10.47}$$

The following theorem establishes an asymptotic lower bound for the ESADD over the class $\mathbf{C}_\gamma$ of sequential change detection–isolation procedures $\delta = (T,d)$, which along with the previous results allows us to conclude that the properly designed algorithms δ_{GCS} and δ_{MCS} are first-order asymptotically optimal.

Theorem 10.3.2. *Let $\mathbf{C}_\gamma$ be the class of detection–isolation procedures defined in* (10.15) *for which the minimum of ARL2FA and ARL2FI is at least $\gamma > 1$. Suppose that the K–L numbers $I_{\ell,j}$ are positive and finite, $0 < I_{\ell,j} < \infty$ for all $1\le\ell\neq j\le M$. Then*

$$\inf_{\delta\in\mathbf{C}_\gamma}\mathsf{ESADD}(\delta)\ge\frac{\log\gamma}{I^*}(1+o(1)) \quad \text{as } \gamma\to\infty. \tag{10.48}$$

The detailed proof may be found in [322].

Corollary 10.3.3. *Let the thresholds h in the detection–isolation algorithms δ_{GCS} and δ_{MCS} be selected so that they belong to the class $\mathbf{C}_\gamma$ and $h \sim \log\gamma$, in particular $h = \log\gamma$. Then both algorithms are asymptotically optimal in the class $\mathbf{C}_\gamma$,*

$$\inf_{\delta\in\mathbf{C}_\gamma} \mathsf{ESADD}(\delta) \sim \mathsf{ESADD}(\delta_{GCS}) \sim \mathsf{ESADD}(\delta_{MCS}) \sim \frac{\log\gamma}{I^*} \quad as\ \gamma\to\infty. \tag{10.49}$$

Example 10.3.1. Let us consider again the detection and isolation of a target appearing in one of M independent Gaussian channels defined in Subsection 10.2.1.1 by equation (10.16). The goal of this example is to compare the FSS and GCS detection–isolation algorithms using asymptotic and non-asymptotic optimization. It is assumed that the SNR is $\vartheta/\sigma = 1$ and $M = 5$. The relation between $\gamma = \min\{\mathsf{ARL2FA}_j, \mathsf{ARL2FI}_{\ell,j}\}$ and $\mathsf{ESADD}(\delta_{\mathrm{FSS}})$ for the FSS detection–isolation algorithm (10.3)–(10.6) with the LLR (10.17) is calculated using asymptotic equation (10.22) and non-asymptotic numerical optimization problem (10.24). The same relation for the GCS detection–isolation algorithm is calculated by asymptotic equation (10.49). Hence,

$$\mathsf{ESADD}(\delta_{\mathrm{GCS}}) \sim \frac{2\sigma^2\log\gamma}{\vartheta^2} \text{ and } \mathsf{ESADD}(\delta_{\mathrm{FSS}}) \le \frac{4\sigma^2\log\gamma}{\vartheta^2}(1+o(1)) \text{ as } \gamma\to\infty.$$

The results of comparison are shown in Figure 10.2. This figure shows that the curve of $\mathsf{ESADD}(\delta_{\mathrm{FSS}})$ does not reach the lower bound $\inf_{\delta\in\mathbf{C}_\gamma}\mathsf{ESADD}(\delta)$ when γ is large. The non-asymptotic "exact" curve of $\mathsf{ESADD}(\delta_{\mathrm{FSS}})$ obtained by numerical optimization is close to its asymptotic upper bound $(4\sigma^2\log\gamma)/\vartheta^2$ when γ is large, exactly as in the case of pure detection, $M = 1$. The optimal sequential change detection–isolation algorithm is asymptotically twice as good as the FSS competitor.

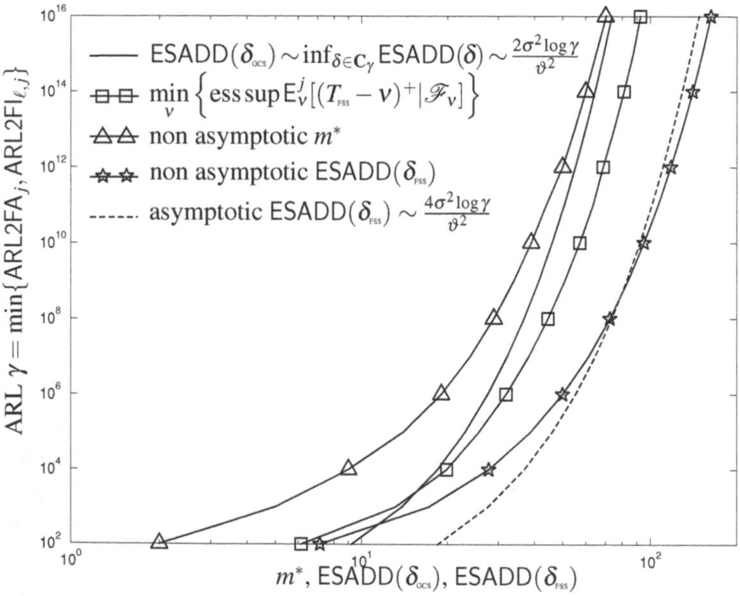

Figure 10.2: *Numerical and asymptotic comparisons of the GCS and FSS detection–isolation algorithms.*

Remark 10.3.2. Let us add one comment on the Lorden's "worst-worst-case" ADD for the FSS change detection–isolation tests given by (10.3)–(10.6) with the LLR (10.17). As it follows from

(10.11), the essential supremum in (10.10) can be rewritten as

$$\begin{aligned}\mathsf{ESADD}(\delta_{\text{FSS}}) &= \sup_{0\le\nu<\infty}\left\{\operatorname*{ess\,sup}\mathsf{E}_\nu^j[(T_{\text{FSS}}-\nu)^+|\mathscr{F}_\nu]\right\}\\ &= \max\left\{\frac{m^*}{1-\zeta},\ \max_{1<k\le m^*}\left[(m^*-k+1)+\frac{m^*}{1-\zeta}\right]\right\},\end{aligned}$$

where $\zeta_j = \mathsf{P}_j(d(\mathbf{X}_1,\ldots,\mathbf{X}_m)=0)=\zeta$ (due to the symmetry) and m^* is the optimal fixed sample size. Let us assume that the changepoint ν belongs to the following subset of natural numbers : $\mathbb{N}_m = \{0, m^*, 2m^*,\ldots\}$. It follows from the above equation that the essential supremum $\min_\nu\left\{\operatorname*{ess\,sup}\mathsf{E}_\nu^j[(T_{\text{FSS}}-\nu)^+|\mathscr{F}_\nu]\right\}$ attains its minimum $m^*/(1-\zeta)$ for such values of $\nu\in\mathbb{N}_m$. It can be shown that $\zeta\to 0$ as $\gamma\to\infty$ (see details in [326]) and the optimal fixed sample size m^* attains the asymptotic lower bounds $\inf_{\delta\in\mathbf{C}_\gamma}\mathsf{ESADD}(\delta)\sim(2\sigma^2\log\gamma)/\vartheta^2$ for ESADD in the class $\mathbf{C}_\gamma$ (see (10.23)). Figure 10.2 shows that the curve $\min_\nu\left\{\operatorname*{ess\,sup}\mathsf{E}_\nu^j[(T_{\text{FSS}}-\nu)^+|\mathscr{F}_\nu]\right\} = m^*/(1-\zeta)$ very slowly converges to the optimal fixed sample size m^* obtained by numerical optimization and to the asymptotic lower bound $m^*\sim\inf_{\delta\in\mathbf{C}_\gamma}\mathsf{ESADD}(\delta)\sim(2\sigma^2\log\gamma)/\vartheta^2$ because the probability $\zeta(\gamma)\sim 1/(\sqrt{2\pi\log\gamma\cdot\log\log\gamma})$ is a slowly decreasing function, which tends to 0 as $\gamma\to\infty$. The behavior of the functions ESADD and m^* is quite similar to the behavior of the same functions in the case of pure detection illustrated in Figure 8.2.

The same conclusion is true for the FSS change detection–isolation test (10.3)–(10.6) with the GLR (10.26) with respect to the minimum value of $\mathsf{E}_\nu^\ell(T_{\text{FSS}}-\nu)^+$, which attains its lower bound (10.28) in the class $\mathbf{C}_\alpha$ for $\nu\in\mathbb{N}_m$.

Another possible, perhaps even more practical approach is to express the false isolation rate via the maximal probabilities of false isolation

$$\sup_{0\le\nu<\infty}\mathsf{P}_\nu^\ell(d=j|T>\nu),\quad \ell,j=1,\ldots,M,\ \ell\ne j, \tag{10.50}$$

which are upper-bounded by given numbers. However, there are technical difficulties in proving upper bounds for the maximal probabilities (10.50) of the GCS and MCS procedures. For this reason, Nikiforov [323] suggested to consider the maximal error probabilities over ℓ, j at the onset time $\nu=0$, i.e., $\max_{0\le\ell\le M}\max_{1\le j\ne\ell\le M}\mathsf{P}_0^\ell(d=j)$. Therefore, the class of detection–isolation procedures $\mathbf{C}_\gamma$ defined in (10.15) is now modified as

$$\mathbf{C}_{\gamma,\beta_0} = \left\{\delta=(T,d):\ \min_{1\le j\le M}\mathsf{E}_\infty\left(\inf_{r\ge1}\{T_r\ :\ d_r=j\}\right)\ge\gamma,\ \max_{0\le\ell\le M}\max_{1\le j\ne\ell\le M}\mathsf{P}_0^\ell(d=j)\le\beta_0\right\}, \tag{10.51}$$

where $\gamma>1$ and $0<\beta_0<1$ are given numbers. Hence, the minimax optimization problem seeks to

$$\begin{gathered}\text{find } \delta_{\text{opt}}\in\mathbf{C}_{\gamma,\beta_0} \text{ such that } \mathsf{ESADD}(\delta_{\text{opt}}) = \inf_{\delta\in\mathbf{C}_{\gamma,\beta_0}}\mathsf{ESADD}(\delta)\\ \text{for every } \gamma>1 \text{ and } \beta_0\in(0,1),\end{gathered} \tag{10.52}$$

where $\mathbf{C}_{\gamma,\beta_0}$ is the class of detection–isolation procedures defined in (10.51). The asymptotic relation between the ESADD (10.10), γ and β_0 and optimality of the test δ_{GCS} in the class $\mathbf{C}_{\gamma,\beta_0}$ is stated in the following theorem [323].

Theorem 10.3.3. *Let $0<I_{\ell,j}<\infty$ for all $1\le\ell\ne j\le M$. We assume that the GCS algorithm $\delta_{GCS}=(T_{GCS},d_{GCS})$ defined by (10.35) with the slightly modified stopping time T_ℓ*

$$T_\ell = \inf_{k\ge1}T_\ell(k),\ \ T_\ell(k)=\inf\left\{n\ge k:\ \min_{0\le j\ne\ell\le M}\lambda_{\ell,j}(k,n)-h_{\ell,j}\ge0\right\}, \tag{10.53}$$

belongs to the class $\mathbf{C}_{\gamma,\beta_0}$ with the thresholds $h_{\ell,j}$ chosen by the following equation:

$$h_{\ell,j} = \begin{cases} h_{det} & \text{if} \quad 1 \le \ell \le M \quad \text{and} \quad j = 0 \\ h_{isol} & \text{if} \quad 1 \le j, \ell \le M \quad \text{and} \quad j \ne \ell \end{cases}, \tag{10.54}$$

where $h_{det} \sim \log\gamma$ as $\gamma \to \infty$ and $h_{isol} \sim \log\beta_0^{-1}$ as $\beta_0 \to 0$. Then

$$\begin{aligned} \inf_{\delta \in \mathbf{C}_{\gamma,\beta_0}} \mathsf{ESADD}(\delta) &\sim \mathsf{ESADD}(\delta_{GCS}) \\ &\sim \max\left\{\frac{\log\gamma}{I^*_{det}}, \frac{\log\beta_0^{-1}}{I^*_{isol}}\right\} \quad \text{as } \gamma \to \infty,\ \beta_0 = c\gamma^{-1} \to 0, \end{aligned} \tag{10.55}$$

where $c > 0$ is a positive constant, $I^*_{det} = \min_{1 \le j \le M} I_{j,0}$ and $I^*_{isol} = \min_{1 \le \ell \le M} \min_{1 \le j \ne \ell \le M} I_{\ell,j}$. Hence, the detection–isolation algorithm $\delta_{GCS} = (T_{GCS}, d_{GCS})$ is asymptotically optimal in the class $\mathbf{C}_{\gamma,\beta_0}$.

A similar result, which can be derived from [461], holds for the matrix CUSUM algorithm $\delta_{\rm MCS}$. In fact, the results can be generalized to the case of the class $\mathbf{C}(\gamma, \{\alpha_j\})$ which confines the ARL2FA and all error probabilities (misidentification),

$$\mathbf{C}(\gamma, \{\alpha_i\}) = \left\{\delta = (T,d) \colon \mathsf{E}_\infty T \ge \gamma,\ \mathsf{P}_0^i(d \ne i) \le \alpha_i,\ i = 1, \dots, M\right\}, \tag{10.56}$$

where $\gamma > 1$ and $\alpha_i \in (0,1)$ are given numbers. The following theorem establishes the operating characteristics and the asymptotic optimality properties of the matrix CUSUM procedure $\delta_{\rm MCS} = (T_{\rm MCS}, d_{\rm MCS})$ given by $T_{\rm MCS} = \min_{1 \le \ell \le M} T_{{\rm MCS},\ell}$, $d_{\rm MCS} = \arg\min_{1 \le \ell \le M} T_{{\rm MCS},\ell}$, where the stopping time $T_{{\rm MCS},\ell}$ is defined in (10.45). Define

$$\zeta_i = \lim_{a \to \infty} \mathsf{E}_i\{\exp[-(\lambda_{i,0}(1,\tau_i) - a)]\}, \quad \tau_i = \inf\{n \ge 1 : \lambda_{i,0}(1,n) \ge a\}.$$

Also, write

$$\mathsf{ESADD}_i(\delta) = \sup_{0 \le \nu < \infty} \operatorname{ess\,sup} \mathsf{E}_\nu^i[(T-\nu)^+ | \mathscr{F}_\nu], \quad i = 1, \dots, M.$$

Recall that in our previous notation $\mathsf{ESADD}(\delta) = \max_{1 \le i \le M} \mathsf{ESADD}_i(\delta)$. Write $h_{\min} = \min(h_0, h_1, \dots, h_M)$, $\alpha_{\max} = \max(\alpha_1, \dots, \alpha_M)$.

Theorem 10.3.4. *Assume that the K–L numbers $I_{i,j}$, $i, j = 1, \dots, M$, $i \ne j$ are positive and finite.*
(i) *For all $h_j > 0$, $j = 0, 1, \dots, M$ the ARL2FA satisfies the inequality*

$$\mathsf{E}_\infty T_{MCS} \ge M^{-1} e^{h_0}.$$

If in addition $\mathsf{E}_i|\lambda_{i,j}(1,1)|^2 < \infty$, $i \ne j$ and $\lambda_{i,j}(1,1)$ are non-arithmetic, then

$$\mathsf{E}_\infty T_{MCS} \ge \frac{e^{h_0}}{\sum_{i=1}^M \zeta_i^2 I_{i,0}} (1 + o(1)) \quad \text{as } h_0 \to \infty.$$

(ii) *For all $h_j > 0$, $j = 0, 1, \dots, M$, the error probabilities satisfy the inequalities*

$$\mathsf{P}_0^i(d_{MCS} \ne i) \le (M-1)\, e^{-h_i}\, \mathsf{E}_i T_{MCS}.$$

(iii) *For all $i = 1, \dots, M$,*

$$\mathsf{ESADD}_i(\delta_{MCS}) \sim \max\left[\frac{h_0}{I(i,0)}, \max_{\substack{1 \le k \le M \\ k \ne i}} \left(\frac{h_k}{I(i,k)}\right)\right] \quad \text{as } h_{\min} \to \infty, \tag{10.57}$$

and therefore,

$$\mathsf{P}_0^i(d_{MCS} \ne i) \le (M-1)\, e^{-h_i} \max\left[\frac{h_0}{I(i,0)}, \max_{\substack{1 \le k \le M \\ k \ne i}} \left(\frac{h_k}{I(i,k)}\right)\right] (1 + o(1)) \quad \text{as } h_{\min} \to \infty.$$

(iv) *If $h_0 = \log(\gamma \sum_{i=1}^{M} \zeta_i^2 I_{i,0})$ and the thresholds h_ℓ, $\ell = 1, \dots, M$ are selected from the equations*

$$e^{h_\ell} = \max\left[\frac{h_0}{I(i,0)}, \max_{\substack{1 \le k \le M \\ k \ne i}} \left(\frac{h_k}{I(i,k)}\right)\right] \frac{M-1}{\alpha_\ell},$$

then $\mathsf{E}_\infty T_{MCS} \ge \gamma(1+o(1))$, $\mathsf{P}_0^i(d_{MCS} \ne i) \le \alpha_i(1+o(1))$ as $\gamma \to \infty$ and $\alpha_{\max} \to 0$, and asymptotically as $\gamma \to \infty$ and $\alpha_{\max} \to 0$ the matrix CUSUM procedure is first-order optimal for all post-change hypotheses

$$\inf_{\delta \in \mathbf{C}(\gamma, \{\alpha_i\})} \mathsf{ESADD}_i(\delta) \sim \mathsf{ESADD}_i(\delta_{MCS}) \sim \max\left[\frac{\log \gamma}{I(i,0)}, \max_{\substack{1 \le k \le M \\ k \ne i}} \left(\frac{|\log \alpha_k|}{I(i,k)}\right)\right], \quad i = 1, \dots, M.$$

The characteristic feature of this minimax approach based on Lorden's "worst-worst-case" ESADD is a pessimistic estimation of the detection and isolation delay and an optimistic estimation of the ARL2FI or the probability of false isolation. Recall that when defining the classes $\mathbf{C}_{\gamma,\beta_0}$ and $\mathbf{C}(\gamma, \{\alpha_i\})$ we assumed that the change occurs at the onset time $\nu = 0$ instead of taking the worst point to avoid theoretical difficulties with obtaining approximations to the maximal over ν probabilities of false isolations. For some special cases, like slippage problems with independent channels, this last assumption is not too restrictive [461]. But for certain asymmetric geometries of the alternative hypotheses $\{H_1, \dots, H_M\}$ the rate of false isolations can dramatically increase when $n \to \infty$. It occurs due to an uncontrolled growth of cumulative sums $\lambda_{\ell,j}(1,n)$ when $X_1, \dots, X_n \sim \mathsf{P}_0$.

Example 10.3.2. Let $\{\mathbf{X}_n\}_{n \ge 1}$ be the following independent sequence of Gaussian vectors

$$\mathbf{X}_n \sim \begin{cases} \mathcal{N}(\theta_0, \mathbb{I}_2) & \text{if} \quad 1 \le n \le \nu \\ \mathcal{N}(\theta_\ell, \mathbb{I}_2) & \text{if} \quad n \ge \nu + 1, \end{cases}, \quad \ell = 1, 2, \tag{10.58}$$

where $\theta_0 = (0,0)^\top$, $\theta_1 = (1,0)^\top$, $\theta_2 = (2,1)^\top$. Consider the change detection–isolation tests δ_{GCS} and δ_{MCS} with the threshold $h = 10$. A typical behavior of the LLR $\lambda_{\ell,j}(1,n)$ is shown in Figure 10.3. It is assumed that $\nu = 100$ and the true post-change hypothesis is H_2. The post-change hypotheses θ_1 and θ_2 are not orthogonal. Now, the mutual geometry of θ_0, θ_1, and θ_2 is such that the LLR $\lambda_{1,2}(1,n)$ has a positive pre-change drift $\mathsf{E}_\infty(\lambda_{1,2}(1,n)) > 0$. On the other hand, the LLR $\lambda_{1,0}(1,n)$ has a positive post-change drift $\mathsf{E}_\nu^2(\lambda_{1,0}(\nu+1,n)) > 0$ when the true post-change hypothesis is H_2. Recall that the stopping time and the final decision of the GCS test are defined as $T_{\text{GCS}} = \min\{T_1, T_2\}$ and $d_{\text{GCS}} = \arg\min\{T_1, T_2\}$. Clearly, both stopping times T_1 and T_2 have similar distributions when $\nu \to \infty$. For this reason the event $\{T_1 < T_2\}$ is not rare when $\nu > 0$. Hence, the probability of false isolation grows when $\nu \to \infty$. A typical behavior of the CUSUM statistics $g_1(n) = \min\{g_{1,0}(n), g_{1,2}(n)\}$ and $g_2(n) = \min\{g_{2,0}(n), g_{2,1}(n)\}$ in the case of the test δ_{MCS} is similar to the behavior of the test δ_{GCS}.

To avoid this uncontrolled behavior of cumulative sums, another recursive vector CUSUM change detection–isolation algorithm (VCS) has been proposed in [328, 331]. The pair $\delta_{\text{VCS}} = (T_{\text{VCS}}, d_{\text{VCS}})$ is again defined by (10.35)–(10.36) but the stopping time T_ℓ is replaced with

$$T_{\text{VCS},\ell} = \inf\left\{n \ge 1 : \min_{0 \le j \ne \ell \le M} \left[g_\ell(n) - g_j(n) - h_{\ell,j}\right] \ge 0\right\} \tag{10.59}$$

where the recursive decision function $g_\ell(n)$ is defined by

$$g_\ell(n) = \left[g_\ell(n-1) + \lambda_{\ell,0}(n,n)\right]^+ \tag{10.60}$$

with $g_\ell(0) = 0$ for every $1 \le \ell \le M$ and $g_0(n) \equiv 0$. The thresholds $h_{\ell,j}$ are chosen by the following formula:

$$h_{\ell,j} = \begin{cases} h_{\text{det}} & \text{if} \quad 1 \le \ell \le M \quad \text{and} \quad j = 0 \\ h_{\text{isol}} & \text{if} \quad 1 \le j, \ell \le M \quad \text{and} \quad j \ne \ell \end{cases}, \tag{10.61}$$

where h_{det} is the detection threshold and h_{isol} is the isolation threshold.

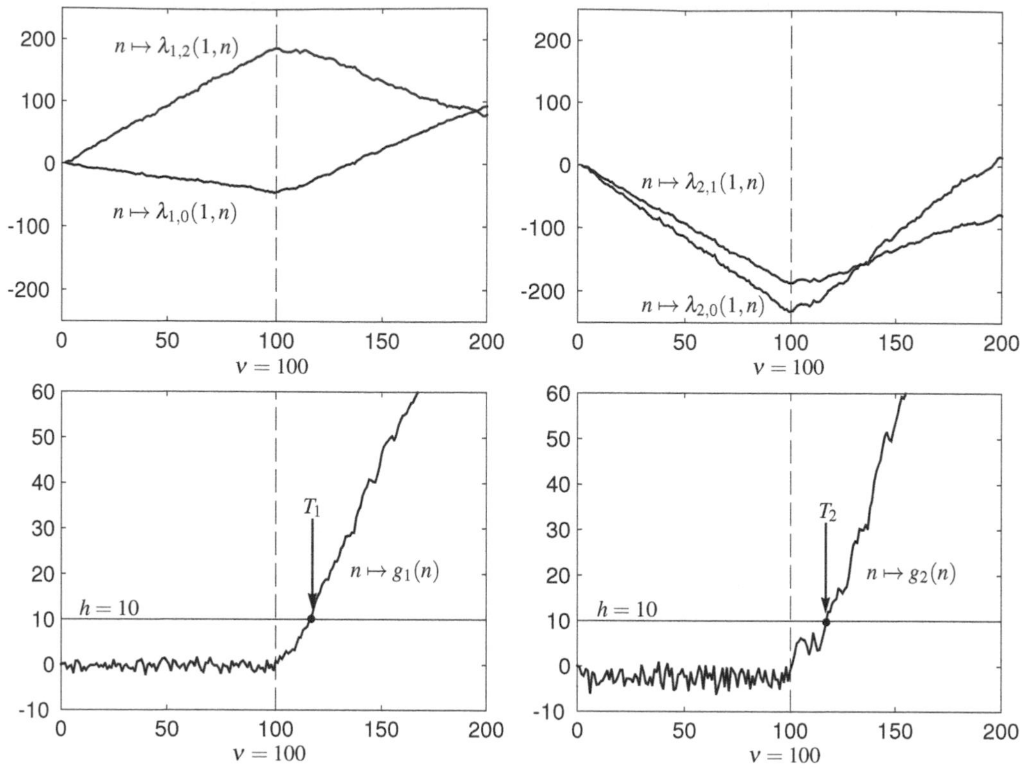

Figure 10.3: *The behavior of the LLR (top):* $\lambda_{1,0}(1,n)$, *and* $\lambda_{1,2}(1,n)$ *(left),* $\lambda_{2,0}(1,n)$ *and* $\lambda_{2,1}(1,n)$ *(right). The behavior of the vector CUSUM for change detection-isolation (bottom):* $g_1(n) = \max_{1\le k\le n}\min\{\lambda_{1,0}(k,n),\lambda_{1,2}(k,n)\}$ *(left)* $g_2(n) = \max_{1\le k\le n}\min\{\lambda_{2,0}(k,n),\lambda_{2,1}(k,n)\}$ *(right).*

Remark 10.3.3. Let us comment on some specific features of the test δ_{VCS} with respect to the test δ_{GCS}. It is easy to see that the test δ_{VCS} is nothing but M parallel scalar CUSUM tests (see the decision functions $g_1(n),\dots,g_M(n)$) and a simple logical rule which compares the quantity $g_\ell(n) - g_j(n)$ with the thresholds $h_{\ell,j}$. The ℓ-th CUSUM test is designed to detect a change from P_0 to P_ℓ. During the pre-change period the nonnegative functions $g_1(n),\dots,g_M(n)$ are stochastically small (because $\mathsf{E}_0(\lambda_{\ell,0}(n,n)) < 0$) and, hence, only an insignificant growth of the probability of false isolation takes place when $\nu > 0$. Let us also note that $\lambda_{\ell,j}(\nu+1,n) \simeq g_\ell(n) - g_j(n)$ when $n \gg \nu+1$ and $\mathsf{E}_\ell(g_j(n)) \ge 0$. Therefore, both algorithms extract approximately the same information from the observations $X_{\nu+1},\dots,X_n$. Nevertheless, if $\mathsf{E}_\ell(g_j(n)) < 0$ then the recursive algorithm partly loses the information from these observations. Hence, it seems plausible that asymptotic optimality of the test δ_{VCS} can be achieved only under the constraint $h_{\text{det}} \ge h_{\text{isol}}$.

The statistical properties of the test $\delta_{\text{VCS}} = (T_{\text{VCS}}, d_{\text{VCS}})$ have been investigated in [328] with respect to the criterion (6.34)–(6.36). The maximal conditional average delay to detection $\mathsf{SADD}(T)$ is used now instead of too conservative Lorden's "worst-worst-case" ADD (10.10),

$$\mathsf{SADD}(\delta) = \max_{1\le j\le M}\sup_{0\le\nu<\infty} \mathsf{E}_\nu^j(T-\nu \mid T>\nu). \tag{10.62}$$

The main result concerning the SADD of the test $\delta_{\text{VCS}} = (T_{\text{VCS}}, d_{\text{VCS}})$ is given in the following theorem.

Theorem 10.3.5. *Suppose that* $0 < I_{\ell,j} < \infty$ *for all* $0 \le \ell \ne j \le M$ *and that the mgf* $\varphi(\zeta) = \mathsf{E}_\ell\left(e^{\zeta\lambda_{\ell,j}(n,n)}\right) < \infty$ *exists for all real* $\zeta \in]-\eta;\eta[$, *where* $\eta > 0$, *and for all* $1 \le \ell \le M$ *and*

$0 \le j \ne \ell \le M$. *Let* $h_{\ell,j}$ *be given by* (10.61) *and* $h_{det} \ge h_{isol}$. *Then, as* $h_{isol} \to \infty$,

$$\sup_{0 \le \nu < \infty} \mathsf{E}_\nu^\ell(T_{VCS} - \nu | T_{VCS} > \nu) \le \max\left(\frac{h_{det}}{I_{\ell,0}}, \frac{h_{isol}}{\min\limits_{1 \le j \ne \ell \le M} I_{\ell,j}}\right)(1+o(1)), \tag{10.63}$$

$$\mathsf{SADD}(\delta_{VCS}) \le \max\left(\frac{h_{det}}{I^*_{det}}, \frac{h_{isol}}{I^*_{isol}}\right)(1+o(1)). \tag{10.64}$$

The detailed proof may be found in [328]. Let us discuss now the probability of false isolation. From Theorem 10.3.5 it follows that $\sup_{0 \le \nu < \infty} \mathsf{E}_\nu^\ell(T_{\mathrm{VCS}} - \nu | T_{\mathrm{VCS}} > \nu)$ is mainly defined by the stopping time $T_{\mathrm{VCS},\ell}$ when the hypothesis H_ℓ is true. The false isolation, i.e., the event $\{d_{\mathrm{VCS}} = j\}$ when the true hypothesis is H_ℓ means that due to noise $T_{\mathrm{VCS},j} < \min_{i \ne j}\{T_{\mathrm{VCS},i}\}$ given that $T_{\mathrm{VCS}} > \nu$. Naturally, this event is rare. Roughly speaking, to estimate the probability $\mathsf{P}_\nu^\ell(d = j | T_{\mathrm{VCS}} > \nu)$ we have to compute the conditional probability to stop the observation process by the "false" stopping time $T_{\mathrm{VCS},j}$ before it will be stopped by the "true" stopping time $T_{\mathrm{VCS},\ell}$ given $T_{\mathrm{VCS}} > \nu$. The asymptotic upper bound for the maximal probability $\max_{\ell,j} \sup_{\nu \ge 0} \mathsf{P}_\nu^\ell(d = j | T_{\mathrm{VCS}} > \nu)$ is given by the following theorem whose detailed proof may be found in [328].

Theorem 10.3.6. *Let the conditions of Theorem 10.3.5 be satisfied. Then, as* $h_{isol} \to \infty$,

$$\max_{1 \le \ell \le M} \max_{1 \le j \ne \ell \le M} \sup_{\nu \ge 0} \mathsf{P}_\nu^\ell(d = j | T_{VCS} > \nu) \le e^{-h_{isol}}\left\{\max\left(\frac{h_{det}}{I^*_{det}}, \frac{h_{isol}}{I^*_{isol}}\right) + h_{isol}\right\}(1+o(1)). \tag{10.65}$$

Consider the sequences of stopping times $\{T_k\}_{k \ge 0}$ and final decisions $\{d_k\}_{k \ge 1}$, where $T_{k \ge 1} = T_{\mathrm{VCS}}$ and $d_{k \ge 1} = d_{\mathrm{VCS}}$ are the stopping time and final decision of the test δ_{VCS} applied to $X_{T_{k-1}+1}, X_{T_{k-1}+2}, \dots$. It is assumed that after an alarm the observation process immediately restarts from scratch. It results from [322, 328] that the ARL2FA of the type ℓ for the test $\delta_{\mathrm{VCS}} = (T_{\mathrm{VCS}}, d_{\mathrm{VCS}})$ is given by

$$\mathsf{ARL2FA}_\ell = \mathsf{E}_\infty\left(\inf_{k \ge 1}\{T_k \,:\, d_k = \ell\}\right) \ge e^h, \text{ for } \ell = 1, \dots, M. \tag{10.66}$$

Corollary 10.3.4. *Let* $\gamma \to \infty$, $\beta \to 0$ *and* $\log\gamma \ge \log\beta^{-1}(1+o(1))$. *Consider the VCS procedure* (10.35) *and* (10.59)–(10.61). *If thresholds are selected as* $h_{det} \sim \log\gamma$ *as* $\gamma \to \infty$ *and* $h_{isol} \sim \log\beta^{-1}$ *as* $\beta \to 0$, *then*

$$\max_{1 \le \ell \le M} \mathsf{ARL2FA}_\ell \ge \gamma, \quad \max_{1 \le \ell \le M} \max_{1 \le j \ne \ell \le M} \sup_{\nu \ge 0} \mathsf{P}_\nu^\ell(d = j | T_{VCS} > \nu) \le \beta(1+o(1))$$

and

$$\mathsf{SADD}(\delta_{VCS}) \le \max\left\{\frac{\log\gamma}{I^*_{det}}, \frac{\log\beta^{-1}}{I^*_{isol}}\right\}(1+o(1)). \tag{10.67}$$

A detailed comparison of the asymptotic formula (10.67) with the results of Monte Carlo simulation may be found in [328].

Corollary 10.3.5. *The change detection–isolation procedure* δ_{VCS} *is asymptotically equivalent to the tests* δ_{GCS} *and* δ_{MCS} *in a subclass of the class* $\mathbf{C}_{\gamma,\beta_0}$ *such that* $\log\gamma \ge \log\beta_0^{-1}(1+o(1))$:

$$\mathsf{SADD}(\delta_{GCS}) \sim \mathsf{SADD}(\delta_{MCS}) \sim \mathsf{SADD}(\delta_{VCS}) \sim \inf_{\delta \in \mathbf{C}_{\gamma,\beta_0}} \mathsf{SADD}(\delta) \sim \max\left\{\frac{\log\gamma}{I^*_{det}}, \frac{\log\beta_0^{-1}}{I^*_{isol}}\right\} \tag{10.68}$$

as $\beta_0 \to 0$.

In contrast to the above definition of the ARL2FA, we use a slightly different ARL2FA in the definition of the class $\mathbf{C}_{\gamma,\beta}$ in (6.34), namely $\mathsf{E}_\infty T \geq \gamma$. To show that this difference does not change the asymptotic relation between $\mathsf{SADD}(\delta_{\text{VCS}})$, γ and β in equation (10.67) we need to establish that $\log \mathsf{E}_\infty(T_{\text{VCS}}) \geq h_{\text{det}} - \log(M)$. The detailed proof may be found in [331].

Finally, it can be shown that the following asymptotic lower bound

$$\inf_{\delta\in\mathbf{C}_{\gamma,\beta_0}} \mathsf{SADD}(\delta) \geq \max\left\{\frac{\log\gamma}{I^*_{\text{det}}}, \frac{\log\beta^{-1}}{I^*_{\text{isol}}}\right\}(1+o(1)) \text{ as } \min\{\gamma,\beta^{-1}\}\to\infty$$

holds for the SADD over the class

$$\mathbf{C}_{\gamma,\beta} = \left\{\delta : \mathsf{E}_\infty T \geq \gamma, \max_{1\leq\ell\leq M}\max_{1\leq j\neq\ell\leq M}\sup_{\nu\geq 0}\mathsf{P}^\ell_\nu(d=j|T>\nu)\leq\beta\right\},$$

where $\gamma > 1$ and $\beta\in(0,1)$ are given numbers. This bound along with the previous results immediately implies first-order asymptotic optimality of the VCS procedure.

Theorem 10.3.7. *Let* $0 < I_{\ell,j} < \infty$ *for all* $1\leq\ell\neq j\leq M$. *Let* $\mathbf{C}_{\gamma,\beta}$ *be the class of detection–isolation procedures for which the ARL2FA is at least* $\gamma > 1$ *and the maximal probability of false isolation is upper bounded by* β. *If the thresholds in the detection–isolation procedure* δ_{VCS} *are selected as* $h_{det}\sim\log\gamma$ *as* $\gamma\to\infty$ *and* $h_{isol}\sim\log\beta^{-1}$ *as* $\beta\to 0$, *then*

$$\mathsf{E}_\infty T_{VCS} \geq \gamma, \quad \max_{1\leq\ell\leq M}\max_{1\leq j\neq\ell\leq M}\sup_{\nu\geq 0}\mathsf{P}^\ell_\nu(d=j|T_{VCS}>\nu)\leq\beta(1+o(1))$$

and

$$\mathsf{SADD}(\delta_{VCS}) \sim \inf_{\delta\in\mathbf{C}_{\gamma,\beta}}\mathsf{SADD}(\delta) \sim \max\left\{\frac{\log\gamma}{I^*_{det}}, \frac{\log\beta^{-1}}{I^*_{isol}}\right\} \tag{10.69}$$

as $\gamma\to\infty$, $\beta\to 0$ *and* $\log\gamma\geq\log\beta^{-1}(1+o(1))$.

Remark 10.3.4. The above results can be generalized to higher moments of the detection delay

$$\max_{1\leq j\leq M}\sup_{0\leq\nu<\infty}\mathsf{E}^j_\nu[(T-\nu)^m|T>\nu)$$

using techniques developed in [461]. Specifically, under conditions of Theorem 10.3.7, for every $m > 0$,

$$\begin{aligned}\inf_{\delta\in\mathbf{C}_{\gamma,\beta}}\max_{1\leq j\leq M}\sup_{0\leq\nu<\infty}\mathsf{E}^j_\nu[(T-\nu)^m|T>\nu) &\sim \max_{1\leq j\leq M}\sup_{0\leq\nu<\infty}\mathsf{E}^j_\nu[(T_{\text{VCS}}-\nu)^m|T_{\text{VCS}}>\nu)\\ &\sim\left[\max\left(\frac{\log\gamma}{I^*_{\text{det}}}, \frac{\log\beta^{-1}}{I^*_{\text{isol}}}\right)\right]^m.\end{aligned}$$

Remark 10.3.5. Similar results also hold for the SR-based procedure where the CUSUM statistics (10.60) are replaced by the SR statistics

$$\log R_\ell(n) = \log(1+R_\ell(n-1)) + \lambda_{\ell,0}(n,n).$$

See [461] for the matrix SR detection–isolation procedure.

Part III

Applications

Chapter 11

Selected Applications

11.1 Navigation System Integrity Monitoring

11.1.1 Introduction

We address the problem of *fault detection, isolation, and reconfiguration* (FDIR) in dynamic systems with random disturbances [505]. This problem has received extensive research attention in the aerospace domain, for navigation and flight control systems integrity monitoring. To be specific, we consider the problem of fault tolerant navigation. In this case the FDIR algorithms are designed to *detect, isolate (identify)* a faulty sensor (or subsystem), and to automatically remove faulty sensor (or subsystem) from the navigation solution [112, 226, 290, 505, 507].

Navigation systems are standard equipment for vehicles such as aircrafts, boats, and missiles. Conventional navigation systems use the measurements of sensors (and the equations of the vehicle motion) to estimate all desired parameters such as the locations and velocities of the vehicle, *etc.* The problem of optimal estimating the desired navigation parameters $\{X_k\}_{k\geq 1}$, where $k \in \{0,1,2,\dots\}$ is the discrete time, is mainly considered for two models [47, 112, 336].

Regression model.

$$Y_k = HX_k + V_k + \Upsilon(k,\nu), \tag{11.1}$$

where $X_k \in \mathbb{R}^n$ is an unknown and non-random vector containing the desired navigation parameters, H is a known $r \times n$ full rank matrix. The profile of abrupt change $\Upsilon(k,\nu)$ is defined as follows

$$\Upsilon(k,\nu) = \begin{cases} 0 & \text{if} \quad k \leq \nu \\ \theta(k-\nu) & \text{if} \quad k \geq \nu+1 \end{cases}. \tag{11.2}$$

We assume that there exists measurement redundancy, namely that $r > n$. The known covariance matrix R is diagonal of size r.

State-space model.

$$\begin{aligned} X_{k+1} &= F(k+1,k)X_k + W_k + \Upsilon_1(k,\nu), \\ Y_k &= H(k)X_k + V_k + \Upsilon_2(k,\nu), \end{aligned} \tag{11.3}$$

where $X_k \in \mathbb{R}^n$ is the state vector containing the desired navigation parameters and physical errors, $Y_k \in \mathbb{R}^r$ is the measurement vector (namely the measured outputs of the inertial navigation system, which are in fact transformed sensor outputs), $W_k \in \mathbb{R}^n$ and $V_k \in \mathbb{R}^r$ are nonstationary zero mean Gaussian white noises having covariance matrices $Q(k) \geq 0$ and $R(k) > 0$, respectively. The initial state X_0 is a Gaussian zero mean vector with covariance matrix $P_0 > 0$. The vectors of abrupt change $\Upsilon_i(k,\nu)$ are defined as in (11.2). The matrices F, H, Q, R, P_0 are known, the change time ν and change vectors $\theta_i(k-\nu)$ are unknown.

The regression model with measurement redundancy given by (11.1) adequately describes the output signals of strapdown inertial reference units (SIRU) [108, 156, 175, 181, 182] and global navigation satellite systems (GNSS) [137, 169]. The state-space model given by (11.3) is usually used to describe the output signals of inertial navigation systems (INS) [137, 169]. The state-space model is also used for output signals of hybrid INS-GPS navigation systems [137, 138, 169]. Under normal operating conditions (i.e., in the pre-change mode when $k \leq \nu$) $\Upsilon_i(k,\nu) = 0$, the output signals of a navigation system contain useful information and normal operation errors. Consider the first two moments of the estimator $\widehat{X}_k$:

$$b_k = \mathsf{E}\left(\widehat{X}_k - X_k\right), \quad P_k = \mathrm{cov}\left(\widehat{X}_k - X_k\right). \tag{11.4}$$

It is well-known that under the assumption of linearity, the *optimal unbiased estimate* of the unknown vector X_k in the model (11.1) is given by the least squares (LS) algorithm [237]. The estimator of the state vector X_k in the model (11.3) that minimizes any scalar-valued monotonically nondecreasing function of P_k is the conditional expectation:

$$\widehat{X}_k = \mathsf{E}(X_k|Y_1,\ldots,Y_k). \tag{11.5}$$

A numerical method to compute this expectation is the *Kalman filter* [138]. Therefore, in both cases

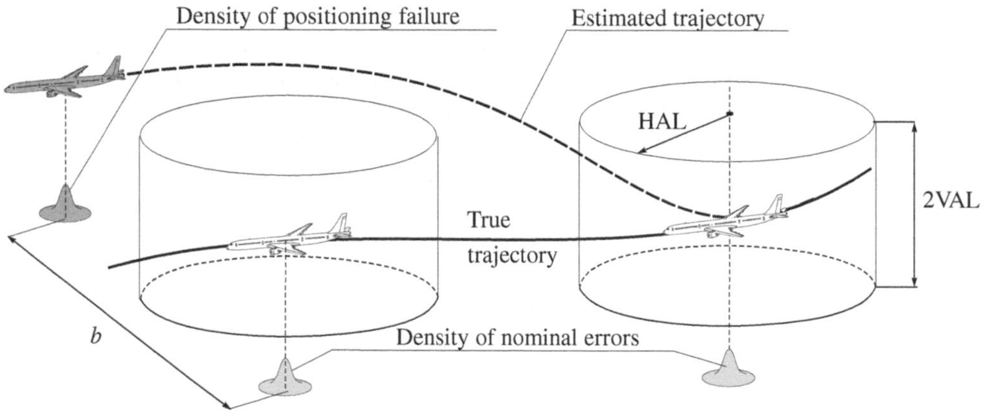

Figure 11.1: *Navigation system integrity monitoring.*

$b_k = 0$ and the covariance matrix P_k attains its lower bound in the class of unbiased estimators. If a *sensor fault* (or a channel/subsystem degradation) occurs at an unknown change time ν then the vector $\Upsilon_i(k,\nu)$ changes from 0 to $\theta_i(k-\nu)$. The LS estimator and the conditional expectation $\mathsf{E}(X_k|Y_1,\ldots,Y_k)$ are not *optimal solutions* when $k \geq \nu+1$. A sensor fault leads to a biased estimate with a bias $b_k \neq 0$ which is clearly *very undesirable*.

The risk of integrity failure is usually defined per unit of time or per approach (non-precision and precision approaches are phases of flight before landing). The protected zone is a cylinder with radius HAL (Horizontal Alert Limit) and height 2VAL (Vertical Alert Limit) with its center being at the true aircraft position. This scenario is shown in Figure 11.1. The protected zone is required to contain the indicated horizontal (resp. vertical) position with the probability of $1-p_r$ per unit of time or per approach. To define the smallest magnitude $\theta(k-\nu)$ of a fault $\Upsilon(k,\nu)$ which leads to a positioning failure, let us consider the case of horizontal positioning failure. A simplified definition of such a fault is the following. Let us assume that $\nu = 0$. A fault $\Upsilon(k,0) = \theta$, $k = 1,2,\ldots$, is considered as a horizontal positioning failure if its impact violates an acceptable risk of integrity p_r, i.e.,

$$(1-p_f)\mathsf{P}_\infty\left(\|X_h-\widehat{X}_h\|_2 > \mathrm{HAL}\right) + p_f\mathsf{P}_{1,\theta}\left(\|X_h-\widehat{X}_h\|_2 > \mathrm{HAL}\right) > p_r, \tag{11.6}$$

where $X_h = (x,y)^\top$ is the true horizontal position of the aircraft, $\widehat{X}_h = (\widehat{x},\widehat{y})^\top$ is the estimated horizontal position of the aircraft, p_f is an *a priori* probability that a fault occurs per unit of time,

$\mathsf{P}_\infty\left(\|X_h - \widehat{X}_h\|_2 > \mathrm{HAL}\right)$ is the probability that a disk in the horizontal plane, with its center being at the true position X_h and the radius HAL, does not contain the indicated horizontal position $\widehat{X}_h$ given that the fault is absent, $\mathsf{P}_{1,\theta}\left(\|X_h - \widehat{X}_h\|_2 > \mathrm{HAL}\right)$ is the probability of the event $\{\|X_h - \widehat{X}_h\|_2 > \mathrm{HAL}\}$ given that the fault is $\Upsilon(k,0) = \theta$.

The navigation system degradation should be detected as soon as possible when it leads to an unacceptable growth of the output errors (see (11.6)). Fast detection and isolation are necessary because between the fault onset time ν and its detection and isolation users operate with abnormal measurements, $b_k \neq 0$. On the other hand, false alarms or false isolations also result in lower accuracy of the estimate $\widehat{X}_k$ because some correct information is not used and, conversely, some incorrect information is used. The optimal solution involves a tradeoff between these contradictory requirements.

11.1.2 Inertial Navigation Integrity Monitoring: A Toy Example

Consider the following example [47, 234, 315, 335, 336]. Modern civil airplanes are usually equipped with a triplicate strapdown INS. This system is made of two types of sensors: gyros and accelerometers [137, 138, 218].

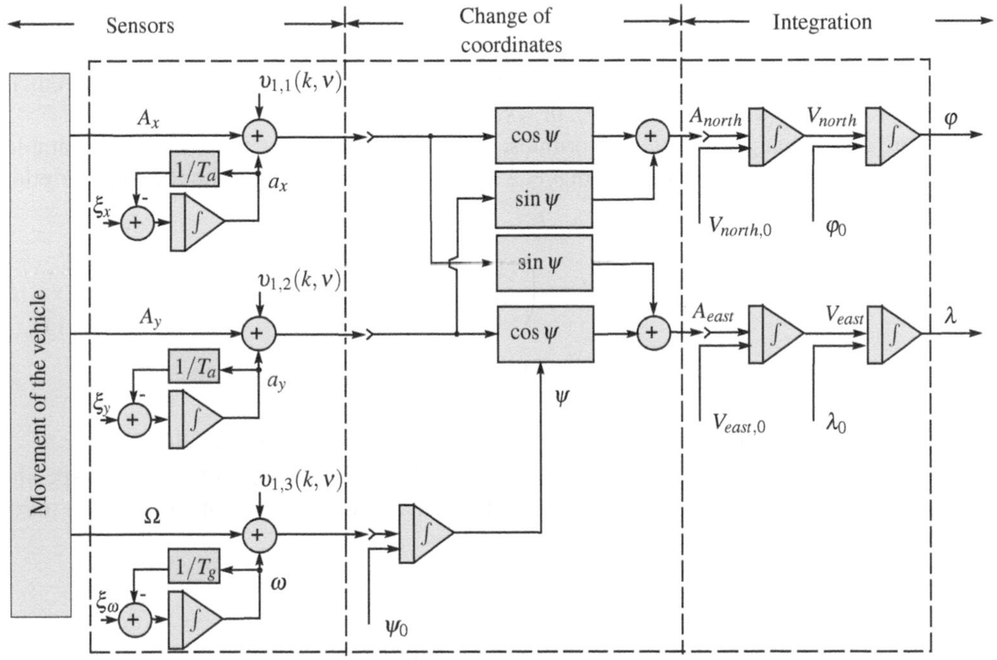

Figure 11.2: *Simplified horizontal channels of the INS.*

A very simplified scheme of INS horizontal channels is shown in Figure 11.2. It is assumed that altitude, pitch, and roll are equal to zero. The vehicle accelerations A_x and A_y are measured by accelerometers in the body frame (denoted by (x,y) and rigidly attached to the vehicle). The yaw rate Ω in the body-frame is measured by gyro. By integrating Ω, the acceleration vector (A_x, A_y) is transformed to the navigation frame denoted by (north,east). Finally, the current north-east coordinates (φ, λ) are obtained by double integration of two accelerations (A_{north}, A_{east}). The simplest accelerometer (or gyro) stochastic drift model, discussed in [131, 291, 315], consists of a low-pass filter fed by white Gaussian noise ξ_x (resp. ξ_x or ξ_ω); see Figure 11.2. Hence, the gyros and accelerometers errors a_x, a_y, and ω are given by the following Ornstein–Uhlenbeck processes [291]

$$\begin{aligned} da_x(t) &= -\frac{1}{T_a} a_x(t)\,dt + \sigma_x dW_x(t), \quad da_y(t) = -\frac{1}{T_a} a_y(t)dt + \sigma_y dW_y(t), \\ d\omega(t) &= -\frac{1}{T_g}\,\omega(t)dt + \sigma_\omega dW_\omega(t), \end{aligned} \tag{11.7}$$

where $W_x(t)$, $W_y(t)$, $W_\omega(t)$ denote the Wiener processes, T_a is the accelerometer error time constant, T_g is the gyro error time constant, and $\sigma_x^2, \sigma_y^2, \sigma_\omega^2$ denote variances. The soft drifting-type fault of an accelerometer or gyro is modeled by $\upsilon_{1,j}(k,\nu)$, $j = 1,2,3$. See Figure 11.2.

The detection and isolation of soft drifting-type faults in one of these sensors using the output signals of the INS is of great interest. A sensor fault should be detected as soon as possible when it leads to unacceptable growing of output errors. Therefore, it is reasonable to consider fault detection as the quickest change detection problem.

It is known [197, 291] that the INS error models can be reduced to (11.3). As follows from (11.7), errors in each sensor can be described as a stationary AR(1) model with a large time constant. This information is captured by the matrix F in (11.3). For the detection of soft drifting-type faults, it is proposed to observe the differences between the state vectors of pairs of INS [225, 315] :

$$\begin{aligned} \Delta X_{k+1} &= F(k+1,k)\Delta X_k + \sqrt{2} W_k + \Upsilon(k,\nu), \\ \Delta Y_k &= H(k)\Delta X_k + \sqrt{2} V_k, \end{aligned} \tag{11.8}$$

where $\Delta X_k = X_k^1 - X_k^2$, $\Delta Y_k = Y_k^1 - Y_k^2$, X_k^i is the state vector of INS$_i$ and Y_k^i is the measurement vector of INS$_i$, $i = 1,2$, $\Upsilon(k,\nu) = (\upsilon_{1,1}(k,\nu), \upsilon_{1,2}(k,\nu), \upsilon_{1,3}(k,\nu))^\top$.

To compare three change detection algorithms, consider a simple but representative example. We assume the state-space model (11.3) with $n = 2$ and $r = 1$, representing a simplified description of the INS heading gyro error with

$$\begin{aligned} &Y = y_1, \quad X = \begin{pmatrix} x_1 \\ x_2 \end{pmatrix}, \quad F = \begin{pmatrix} 1 & \delta \\ 0 & 1 - \frac{\delta}{T_g} \end{pmatrix}, \quad H = \begin{pmatrix} 1 & 0 \end{pmatrix}, \\ &R = \sigma_V^2, \quad Q = \begin{pmatrix} 0 & 0 \\ 0 & \sigma_W^2 \end{pmatrix}, \quad \Upsilon(k,\nu) = \begin{pmatrix} \upsilon(k,\nu) \\ 0 \end{pmatrix}, \end{aligned} \tag{11.9}$$

where δ is the sampling period, $\delta \ll T_g$; see [47, 181, 197, 315, 336] for the details.

We now compare the GLR (GCUSUM) and double-sided CUSUM algorithms with a specific fault detection algorithm, based upon the Kalman filter state estimate $\widehat{X}_{k|k}$, which has been introduced in [225]. This estimate can be computed recursively

$$\widehat{X}_{k|k} = F\widehat{X}_{k-1|k-1} + K_k\,\varepsilon_k,$$

where K_k is the Kalman gain and $\{\varepsilon_k\}_{k\geq 0}$ is the innovation process. Moreover, it turns out that it is relevant to use the second component $(\widehat{x}_2)_{k|k}$:

$$(\widehat{x}_2)_{k|k} = (1-\alpha)\,(\widehat{x}_2)_{k-1|k-1} + k_2\,\varepsilon_k, \tag{11.10}$$

where $\alpha = \delta/T_g$ and k_2 is the second component of the Kalman gain K_k. Now assume that the Kalman gain is constant, which is the case in the steady-state mode. For the sake of simplicity, assume that the signature of the gyro fault $\upsilon(k,\nu)$ on the Kalman filter innovation standardized by its variance is a step function, i.e.,

$$\varepsilon_k \sim \begin{cases} \mathcal{N}(0,1) & \text{for } k \leq \nu \\ \mathcal{N}(\mu,1) & \text{for } k \geq \nu+1 \end{cases},$$

where $|\mu|$ is the magnitude of the change.

With these assumptions, the two-sided CUSUM INS integrity monitoring algorithm defined in (8.68)–(8.69) has the form

$$T_{2-\mathrm{CS}} = \inf\{k \geq 1 : (g_k^u \geq h_1) \cup (g_k^\ell \geq h_1)\},$$
$$g_k^u = \left[g_{k-1}^u + \varepsilon_k - \frac{\delta_\mu}{2}\right]^+, \quad g_k^\ell = \left[g_{k-1}^\ell - \varepsilon_k - \frac{\delta_\mu}{2}\right]^+, \tag{11.11}$$

and the GLR-based CUSUM algorithm defined in (8.268)–(8.269) becomes

$$T_{\mathrm{GLR}} = \inf\{n \geq 1 : g_n \geq h_2\}, \quad g_n = \max_{1 \leq j \leq n} \frac{1}{n-j+1} \left(\sum_{i=j}^{n} \varepsilon_i\right)^2. \tag{11.12}$$

Finally, the third integrity monitoring algorithm based on the Kalman filter is defined in the following manner:

$$T_{\mathrm{EWMA}} = \inf\{k \geq 1 \ : |(\widehat{x}_2)_{k|k}| \geq h_3\}, \quad (\widehat{x}_2)_{k|k} = (1-\alpha)\,(\widehat{x}_2)_{k-1|k-1} + k_2\,\varepsilon_k. \tag{11.13}$$

It is easy to see that this last algorithm is nothing but the two-sided EWMA procedure (up to a change in the scale of ε) given by (8.44)–(8.46).

Analytical Comparison. In Chapter 6, we discussed several relevant criteria for performance evaluation of change detection algorithms associated with the average delay to detection (ADD) *vs.* the FAR. In this subsection, we use ESADD and the ARL function, i.e., the FAR is measured via the $\mathsf{ARL2FA}(T)$.

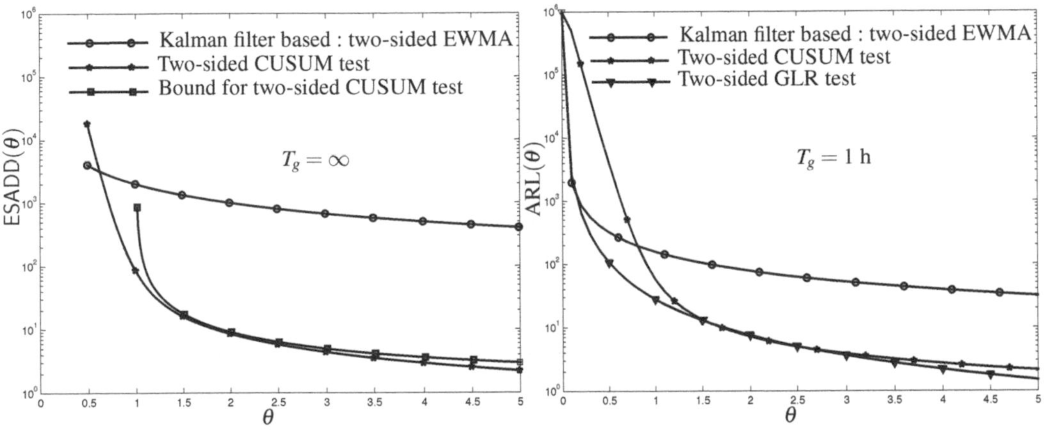

Figure 11.3: *Comparison of three INS fault detection algorithms.*

The comparison of the fault detection algorithms (11.11)–(11.13) is presented in Figure 11.3. Let us first compare the two-sided CUSUM and EWMA tests using the analytical bounds for the ESADD and the ARL2FA [47, 336]. The results of this comparison are presented in the left plot of Figure 11.3. For the two-sided CUSUM test, it is assumed that the putative change magnitude is known $\delta_\mu = 2$. The thresholds h_1 and h_2 have been chosen to get for each algorithm the level of $\mathsf{ARL2FA} = 10^6$. Because we want to show an advantage of the two-sided CUSUM algorithm, we need to know the upper bound for $\mathsf{ESADD}(T_{\mathrm{CS}})$ and the lower bound for $\mathsf{ARL2FA}(T_{\mathrm{CS}})$ given in Subsection 8.2.6.5 by equations (8.174) and (8.175). These bounds are valid for the one-sided CUSUM test, but an upper bound $\mathsf{ESADD}(T_{\mathrm{CS}})$ will be also valid for the two-sided one, and $\mathsf{ARL2FA}(T_{\mathrm{CS}})/2$ gives the ARL2FA for the two-sided CUSUM test. Nevertheless, these bounds are too conservative. For this reason, we mainly use Siegmund's corrected Brownian motion approximation given in Subsection 8.2.6.3 and we apply a bound for $\mathsf{ESADD}(T_{\mathrm{CS}})$ just to confirm the conclusion.

The operating characteristics of the EWMA procedure are calculated as follows. It is well known that for INS systems where the gyro (resp. accelerometer) error time constant T_g (resp. T_a) is very large, i.e., the values of the constant $\alpha = \delta/T_g$ in (11.10) are close to 0. In order to study the case of small values of α, let us consider the limit case $\alpha = 0$. In this case, the filter equation (11.10) defines a cumulative sum of SPRT. The bounds for the ASN in sequential analysis are discussed in Subsection 3.1.2. To compare the two-sided EWMA test with the two-sided CUSUM algorithm, we need to know the lower bound for $\mathsf{ESADD}(T_{\text{EWMA}})$ and the upper bound for $\mathsf{ARL2FA}(T_{\text{EWMA}})$. It follows from (3.39) and (3.43) that

$$\mathsf{ESADD}(T_{\text{EWMA}}) \geq \max_{0\leq\varepsilon\leq h_3} \left\{ \frac{h_3+\varepsilon}{|\mu|} - \left(\frac{2h_3}{|\mu|} + \frac{\varphi(|\mu|)}{|\mu|\,\Phi(-|\mu|)} - 1 \right) \bar{Q}(|\mu|) \right\},$$

$$\mathsf{ARL2FA}(T_{\text{EWMA}}) \leq h_3^2 + 1 + \frac{4h_3}{\sqrt{2\pi}},$$

where

$$\bar{Q}(|\mu|) = \frac{\Phi(-|\mu|)\, e^{-2(h_3+\varepsilon)|\mu|} - \Phi(|\mu|)}{\phi(-|\mu|)\, e^{-2(h_3+\varepsilon)|\mu|} - \Phi(|\mu|)\, e^{2(h_3-\varepsilon)|\mu|}}.$$

As it follows from the left plot of Figure 11.3, the two-sided CUSUM test performs much better than the two-sided EWMA test even for this comparison which is favorable to the Kalman filter based EWMA test because the curve of the EWMA test shown in the left plot corresponds to a lower bound of $\mathsf{ESADD}(T_{\text{EWMA}})$. The two-sided CUSUM test is only inferior to the Kalman filter based EWMA test for detecting changes of magnitude less than $\theta = 0.62$. The explanation of such behavior of the two-sided EWMA test is that the choice of the constant $\alpha = \delta/T_g$ is based on the physical model of gyros (resp. accelerometers), and this choice is not optimal from the abrupt change detection point of view.

Numerical Comparison. The numerical comparison of the three fault detection algorithms in terms of the ARL-functions is presented in the right plot of Figure 11.3. It is assumed that the gyro error time constant is $T_g = 1$ hour and the sampling period is $\delta = 1$ sec. The thresholds h_1, h_2, and h_3 have been chosen to get for each algorithm the level of $\mathsf{ARL2FA} = 10^6$. For the two-sided CUSUM test, it is assumed that the putative change magnitude is known $\delta_\mu = 2$. The ARL-functions of the fault detection algorithms $\mathsf{ARL}(\theta)$ are drawn for $\theta \in [0.1, 5]$. The ARL functions of the two-sided EWMA and CUSUM tests have been computed by using the Fredholm integral equations (8.43) and (8.99) with (8.101) and (8.102). The ARL function of the two-sided GLR test has been computed using the approximations (8.294) and (8.297).

First of all, it can be concluded that the GLR test performs almost uniformly better than two other competitors. The CUSUM test is only slightly more efficient than the GLR one around the putative change magnitude $\delta_\mu = 2$, namely when $|\mu| \in [1.6; 2.6]$, and it is less efficient otherwise. But, as we discuss in Chapter 9, the GLR algorithm can be approximated by a multichart changepoint detection procedure, i.e., by several parallel CUSUM algorithms. In order to attain a tradeoff between complexity and efficiency of the algorithms, it is useful, when the range of the possible change magnitude is wide, to use this approximation. The performance of the Kalman filter based EWMA procedure is far worse compared to both 2−CUSUM and GCUSUM.

Let us pursue our discussion of the triple strapdown INS integrity monitoring. Let us denote the output Y_k^i of INS_i, $i = 1, 2, 3$, and assume that only *one INS* can fail at a time. The difference $\Delta Y_k^{ij} = Y_k^i - Y_k^j$ is given by (11.8). To detect a change $\Upsilon(k, \nu)$ we have to compute three differences ΔY_k^{ij} and Kalman filter innovations ε_k^{ij} and then we need to detect a change in the mean of each innovation sequence ε_k^{ij} standardized by its variance by using the algorithms described above. The detection is declared when at *least one* of the three change detection algorithms stops and declares the presence of a gyro (resp. accelerometer) fault: $T = \min_{1\leq i\neq j\leq 3}\{T^{ij}\}$, where T^{ij} is the stopping time associated with the innovation ε_k^{ij}. When the second alarm is declared, the *isolation* of the failed INS is realized by *voting*.[1] The proposed methodology (a set of two-sided CUSUM with the voting rule) has been successfully implemented to experimental data obtained from test flights of the IL-96-300 airplane; see [336] for details.

11.1.3 Strapdown Inertial Reference Unit Integrity Monitoring

Conventional INSs, discussed in Subsection 11.1.2, are designed with three gyros and three accelerometers mounted on orthogonal axes. The integrity monitoring can be achieved externally only by comparing the output signals Y_k^i of two or more such INSs. Skewed axis strapdown inertial reference units (SIRU) which contain several sensors, such as six single degrees of freedom gyros and six accelerometers, can provide internal integrity monitoring. The benefits of the INS with one redundant sensor (or SIRU) *vs.* two or more conventional INSs can be significant [446]. Fault detection and isolation in SIRU is of primary interest for reliability reasons [108, 156, 181, 182, 212, 290, 446].

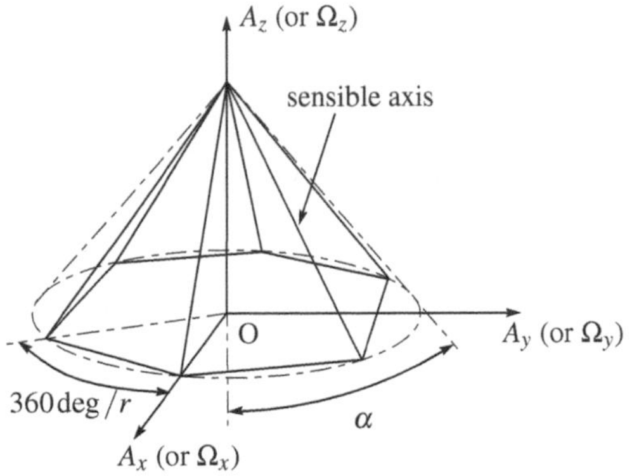

Figure 11.4: *Typical orientations of inertial sensors equally spaced on a cone.*

Conventional redundant SIRU incorporates $r \geq 5$ single degree-of-freedom sensors (laser giros or accelerometers) [156, 175, 182, 446]. We assume that r skewed axis inertial sensors are equally spaced on a cone with half-angle $\alpha = 54.736\,\text{deg}$. The geometric illustration of such sensor configuration is shown in Figure 11.4. A simplified SIRU measurement model is defined by the following regression model with redundancy:

$$\mathbf{Y}_k = H\mathbf{X}_k + \zeta_k, \quad \zeta_k = a\zeta_{k-1} + \xi_k + \Upsilon(k,\nu), \tag{11.14}$$

where $\mathbf{X}_k \in \mathbb{R}^3$ is a non-random unknown state vector (of accelerations, $\mathbf{X} = (A_x, A_y, A_z)^\top$, or angular velocities, $\mathbf{X} = (\Omega_x, \Omega_y, \Omega_z)^\top$), $\mathbf{Y}_k \in \mathbb{R}^r$ is a vector of measurements, $\xi_k \in \mathbb{R}^r$ is Gaussian white noise with zero mean and covariance $R = \sigma^2 \mathbb{I}_r$, $\sigma^2 > 0$, a is the AR coefficient of random noise (practically, $a \simeq 1$), $H = (h_{ij})$ is a full column rank matrix of size $r \times 3$, $h_{i1} = \cos\beta_i \sin\alpha$, $h_{i2} = \sin\beta_i \sin\alpha$, $h_{i3} = -\cos\alpha$, $\beta_i = 360\,\text{deg}(i-1)/r$, $i = 1, \ldots, r$ and $\Upsilon(k,\nu)$ is a sensor fault signature (i.e., the additional drift(s) of sensor(s)) occurring at time $\nu+1$. It is easy to see that equation (11.14) can be reduced to the following regression model

$$\mathbf{G}_k = H\mathbf{U}_k + \xi_k + \Upsilon(k,\nu), \quad \mathbf{G}_k \triangleq \mathbf{Y}_k - a\mathbf{Y}_{k-1}, \quad \mathbf{U}_k \triangleq \mathbf{X}_k - a\mathbf{X}_{k-1}. \tag{11.15}$$

The physical quantities $\mathbf{U}_k$ should be considered as a nuisance parameter. Hence, the detection

[1] A combination of the CUSUM procedures with voting was first used in [181] for two-degrees-of-freedom inertial sensors.

and isolation of faults should be done by using the statistic which is invariant under the group of translations $\mathbf{G} = \{g : g(Y) = \mathbf{Y} + H\mathbf{C}\}$, $\mathbf{C} \in \mathbb{R}^3$; see details in Subsection 2.10.3.1. To get this invariant statistic $\mathbf{Z}$, we have to project the vector $\mathbf{G}$ on the orthogonal complement $R(H)^\perp$ of the column space $R(H)$ (left null space of the matrix H), i.e., $\mathbf{Z} = W\mathbf{G}$. The matrix $W^\top = (w_1, \dots, w_{r-3})$ of size $r \times (r-3)$ is composed of the eigenvectors $w_1, \dots, w_{r-3}$ of the projection matrix $P_H = \mathbb{I}_r - H(H^\top H)^{-1} H^\top$.[2]

The FDI of sensor faults in SIRUs has been studied using both FSS and sequential approaches. In particular, in [108, 156, 182, 446] the FDI problem has been addressed using the FSS approach, and in [181, 234, 335, 507], it has been shown that the SIRU FDI problem can be effectively solved using a sequential approach (CUSUM and/or GLR procedures).

11.1.3.1 SIRU Fault Detection

Consider the invariant statistic $\mathbf{Z} = W\mathbf{G}$. Then the changepoint model can be rewritten via the independent sequence of Gaussian vectors $\{\mathbf{Z}_n\}_{n \ge 1}$:

$$\mathbf{Z}_n \sim \mathcal{N}(\Upsilon(n,\nu), \sigma^2 \mathbb{I}_{r-3}), \quad \Upsilon(n,\nu) = \begin{cases} 0 & \text{if} \quad n \le \nu \\ W\theta & \text{if} \quad n \ge \nu + 1 \end{cases}. \tag{11.16}$$

It is assumed that $r = 5$ and the SNR $b = \sqrt{\theta^\top P_H \theta}/\sigma$ belongs to an interval $[b_0, b_1]$. We compare the χ^2-FSS change detection test (8.34), or χ^2-Shewhart chart, with the recursive ε-optimal multichart test based on three recursive χ^2-tests described in Subsection 9.4.2. The obvious advantage of the χ^2 statistic is its ability to detect any additive SIRU fault (additional drift(s) of sensor(s)), even if more than one of the SIRU sensors is affected by this fault. The stopping time of the χ^2-FSS change detection test is given by

$$T_{\text{FSS}} = m \cdot \inf\{K \ge 1 : d_K = 1\}, \; d_K = \begin{cases} 0 & \text{if} \quad \left\| \overline{\mathbf{Z}}_K \right\|_2 < mh \\ 1 & \text{if} \quad \left\| \overline{\mathbf{Z}}_K \right\|_2 \ge mh \end{cases}, \quad \overline{\mathbf{Z}}_K = \sum_{i=(K-1)m+1}^{Km} \mathbf{Z}_i. \tag{11.17}$$

The stopping time of the ε-optimal scheme is given by

$$N_{\varepsilon r} = \min\left\{ \widehat{N}_r(a_1), \widehat{N}_r(a_2), \widehat{N}_r(a_3) \right\}, \tag{11.18}$$

where $\widehat{N}_r(a_\ell)$ is the stopping time of a recursive χ^2-test designed to detect a change with the SNR a_ℓ,

$$\widehat{N}_r(b) = \inf\{n \ge 1 : \widehat{S}_n \ge h\}, \; \widehat{S}_n = -n_n \frac{b^2}{2} + b|\chi_n|, \; \chi_n^2 = \mathbf{V}_n^\top \Sigma^{-1} \mathbf{V}_n, \tag{11.19}$$

$$\mathbf{V}_n = \mathbb{1}_{\{\widehat{S}_{n-1} > 0\}} \mathbf{V}_{n-1} + \mathbf{Z}_n, \; n_n = \mathbb{1}_{\{\widehat{S}_{n-1} > 0\}} n_{n-1} + 1, \; \widehat{S}_0 = 0. \tag{11.20}$$

As previously discussed, a criterion appropriate for performance evaluation of fault detection algorithms is $\mathsf{SADD}(T)$ *vs.* $\mathsf{ARL2FA} = \gamma$ and a magnitude of a change. To obtain an expression of SADD, let us rewrite the expression for ESADD (8.35) for the χ^2-FSS change detection test. Putting together (8.35) and the definition of the SADD (6.19), we get

$$\mathsf{SADD}(T_{\text{FSS}}; m, h) = \max\left\{ \frac{m}{1-\beta}, \max_{1 \le \nu \le m-1} \left[m - \nu + \frac{m}{1-\beta} \mathsf{P}\left(\chi^2_{r-3,\lambda_\nu} < \frac{m^2 h^2}{m-\nu} \right) \right] \right\}, \tag{11.21}$$

where $\lambda_\nu = (m-\nu) b^2$, $\beta = \mathsf{P}(\chi^2_{r-3, mb^2} < mh^2)$ and $\chi^2_{p,\lambda}$ is distributed according to a noncentral χ^2 law with p degrees of freedom and noncentrality parameter λ. To minimize the SADD of the χ^2-FSS change detection test, the optimization problem (8.37) is solved now with $\mathsf{SADD}(T_{\text{FSS}}; m, h)$ instead of $\mathsf{ESADD}(T_{\text{FSS}}; m, h)$. The relation between the SADD and the ARL2FA for the ε-optimal

[2]The very first application of the invariant test theory to the SIRU FDIR can be found in [182].

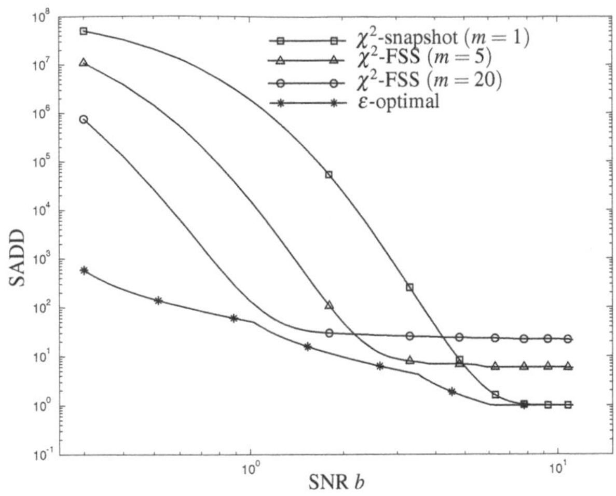

Figure 11.5: *Comparison of the χ^2-FSS change detection test and the recursive ε-optimal χ^2 multichart test;* ARL2FA $= 10^8$.

scheme is given by (9.47)–(9.50). The relative loss of optimality has been chosen $\varepsilon = 0.3$. As follows from (9.48)–(9.49), the subdivision $\{a_1 < a_2 < a_3\}$ of the interval $[b_0, b_1] = [0.3, 11.2]$ is defined as follows: $a_1 = 0.4643$, $a_2 = 1.5889$ and $a_3 = 5.4374$. The results of the comparison between the χ^2-FSS change detection test and the recursive ε-optimal χ^2 multichart test are presented in Figure 11.5 for $\gamma = 10^8$. This figure shows operating characteristics in terms of $\mathsf{SADD}(T)$ *vs.* the magnitude of a change for the ε-optimal χ^2 multichart test and the χ^2-FSS change detection test with $m = 1, 5, 20$. Two conclusions can be made. First, the sequential procedure performs significantly better than the classical FSS test even for the optimal choice of m. The only exception is the case of SNR $b = 10$. For very large SNRs both strategies, the sequential and FSS procedures (which is called "snapshot" when $m = 1$), are equivalent because one post-change observation $\mathbf{Z}_n$ is sufficient to make a decision on the state of SIRU. Second, the performance of the χ^2-FSS change detection test is heavily based on the SNR; see Subsection 8.1.1. This fact is explained by the strong relation between the optimal value of the sample size m^* and the SNR for the χ^2-FSS change detection test.

11.1.3.2 SIRU Fault Detection and Isolation

Let us again consider the independent sequence of Gaussian vectors (invariant statistics) $\{\mathbf{Z}_n = W\mathbf{G}_n\}_{n\geq 1}$:

$$\mathbf{Z}_n \sim \mathcal{N}(\Upsilon_\ell(n,\nu), \sigma^2 \mathbb{I}_{r-3}),\ \Upsilon_\ell(n,\nu) = \begin{cases} 0 & \text{if } \ n \leq \nu \\ W\theta_\ell & \text{if } \ n \geq \nu + 1 \end{cases},\ \ell = 1, \dots, r, \tag{11.22}$$

where the vector $\theta_\ell = (0, \dots, \vartheta, \dots, 0)^\top$ has only one non-zero component $\vartheta > 0$. Hence, it is assumed that a fault can appear only in one sensor at a time. A fault occurring in the ℓ-th sensor is projected by the matrix W on the subspace of the invariant statistic (parity space) $\delta_\ell = W\theta_\ell$. Therefore, each sensor fault generates a specific fault direction in the parity space. This situation is illustrated in Figure 11.6. It is assumed that $r = 5$ and the SNR is $b = \vartheta/\sigma = 4$.

The pair (stopping time and final decision) $\delta_{\text{FSS}} = (T_{\text{FSS}}, d_{\text{FSS}})$ of the FSS change detection–isolation algorithm is defined by (10.3)–(10.6) with the following LLR

$$\lambda_\ell = \frac{1}{\sigma^2} \sum_{n=1}^{m} \theta_\ell^\top \mathbf{Z}_n - m \frac{\|\theta_\ell\|_2^2}{2\sigma^2}. \tag{11.23}$$

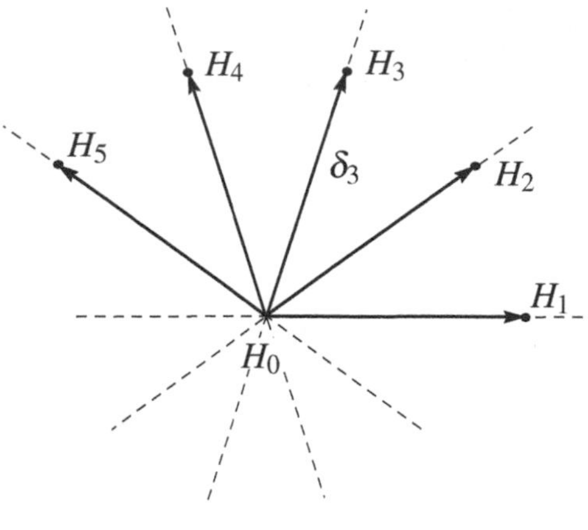

Figure 11.6: *The fault directions in the parity space (the space of invariant statistics).*

The pair $\delta_{\text{VCS}} = (T_{\text{VCS}}, d_{\text{VCS}})$ of the VCS detection–isolation procedure is defined by (10.35)–(10.36) and (10.59) with the recursive CUSUM decision function $g_\ell(n)$

$$g_\ell(n) = \left[g_\ell(n-1) + \lambda_{\ell,0}(n,n)\right]^+, \; \lambda_{\ell,0}(n,n) = \frac{1}{\sigma^2}\theta_\ell^\top \mathbf{Z}_n - \frac{\|\theta_\ell\|_2^2}{2\sigma^2} \tag{11.24}$$

with $g_\ell(0) = 0$ for every $1 \le \ell \le r$ and $g_0(n) \equiv 0$.

The criterion appropriate for performance evaluation of fault detection–isolation algorithms is $\mathsf{SADD}(T)$ *vs.* $\mathsf{ARL2FA} = \gamma$ and $\max_{1\le\ell\le M} \max_{1\le j\ne\ell\le M} \sup_{\nu\ge 0} \mathsf{P}_\nu^\ell(d = j | T > \nu) = \beta$ for various SNRs. The above-mentioned FSS change detection–isolation algorithm has two tuning parameters m and h. Hence, the SADD can be minimized for the given value γ but the worst-case probability of false isolation β cannot be minimized and is completely defined by the statistical model (11.22). The VCS detection–isolation algorithm is first-order asymptotically optimal in a subclass of the class $\mathbf{C}_{\gamma,\beta}$ such that $\log\gamma \ge \log\beta^{-1}(1+o(1))$ as $\beta \to 0$. The detection threshold h_{det} (resp. the isolation threshold h_{isol}) is chosen as a function of γ (resp. as a function of β) in the above subclass. As follows from (10.34) and (10.69),

$$\mathsf{SADD}(\delta_{\text{FSS}}) \sim \frac{2\log\gamma}{I_{\text{det}}}, \quad \mathsf{SADD}(\delta_{\text{VCS}}) \sim \max\left\{\frac{\log\gamma}{I^*_{\text{det}}}, \frac{\log\beta^{-1}}{I^*_{\text{isol}}}\right\}, \tag{11.25}$$

where $\log\gamma \ge \log\beta^{-1}(1+o(1))$ as $\beta \to 0$. Therefore, the VCS sequential change detection–isolation algorithm is asymptotically twice as good as the FSS competitor if the following condition is satisfied $\log\gamma/I^*_{\text{det}} \gtrsim \log\beta^{-1}/I^*_{\text{isol}}$. The following relation between γ and β^{-1} has been used for the comparison: $\log\gamma/I^*_{\text{det}} \sim 2\log\beta^{-1}/I^*_{\text{isol}}$.

The results of the comparison of the FSS change detection–isolation algorithm and the VCS change detection–isolation algorithm are presented in Figure 11.7. The picture on the left displays operating characteristics in terms of $\mathsf{SADD}(T)$ *vs.* the ARL2FA for two competitors, computed using asymptotic equations (in the case of VCS, with Siegmund's corrected Brownian motion approximation) and 10^6-repetition Monte Carlo simulation. The picture on the right displays the worst-case probability of false isolation *vs.* the ARL2FA. Two conclusions can be made. First, the asymptotic approximations (11.25) are relatively close to true values of SADD for both algorithms. Second, the bounds for the worst-case probability of false isolation are very conservative for both algorithms; see also [146, 328, 461]. It can be also concluded that the VCS change detection–isolation algorithm is not only asymptotically twice as good as the FSS algorithm for the SADD *vs.* the ARL2FA but it also performs much better in terms of the worst-case probability of false isolation *vs.* the ARL2FA.

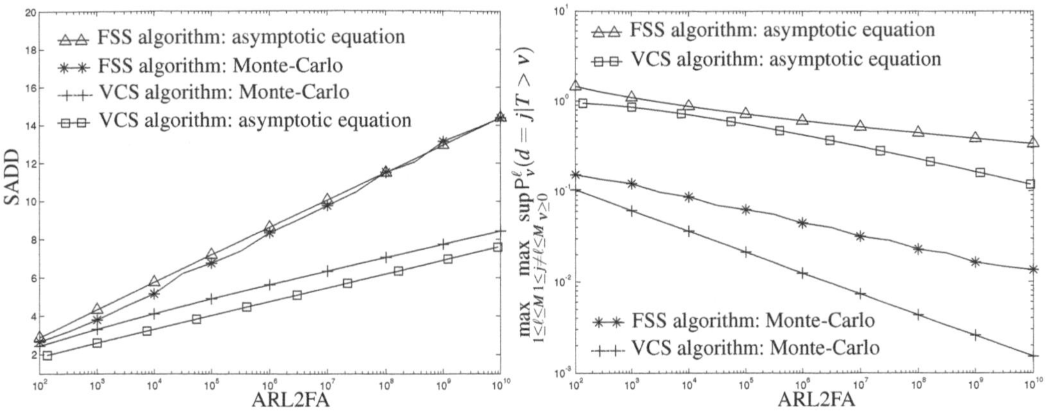

Figure 11.7: *Comparison of the FSS and VCS fault detection–isolation algorithms.*

11.1.4 Radio-Navigation Integrity Monitoring

For many safety-critical applications, the main issue with existing GNSS consists in integrity monitoring. Integrity monitoring requires that a GNSS detects, isolates, and removes faulty satellite channel from the navigation solution before it sufficiently contaminates the output. There are two main methods of GNSS integrity monitoring:

- On-board integrity monitoring: the receiver autonomous integrity monitoring (RAIM) is a method of GNSS integrity monitoring that uses redundant GNSS (or differential GNSS) measurements at the user's receiver. A special RAIM module is integrated in the GPS or/and Glonass or Galileo navigation receiver in order to detect, isolate, and remove the contaminated GNSS channel.
- Ground station-based integrity monitoring: a ground monitoring station at an exactly known position is used to detect and isolate the contaminated channel. When a fault is detected and isolated, the corresponding information is transmitted to the user via the integrity channel.

As has been mentioned previously, GNSS integrity monitoring consists of two operations: *detection* of the fact that a satellite channel produced wrong data and *isolation* or identification of which channel(s) is/are malfunctioning. The GNSS degradation, when it leads to an unacceptable growth of the position errors, should be detected as soon as possible subject to the constraints on the false alarms and false isolations.

11.1.4.1 On-Board Integrity Monitoring

The navigation solution is based upon accurately measuring the distance (range) from several satellites with known locations to a user (vehicle). Let us assume that there are n satellites located at the known positions $X_i = (x_i, y_i, z_i)^\top, i = 1, \dots, n$, and a user at $X_u = (x, y, z)^\top$. The pseudorange r_i from the i-th satellite to the user can be written as

$$r_i = d_i(x,y,z) + c\,t_r + \xi_i, \;\; i = 1, \dots, n,$$

where $d_i(x,y,z) = \|X_i - X_u\|_2$ is the true distance from the i-th satellite to the user, t_r is the user clock bias, $c \simeq 2.9979 \cdot 10^8 \mathrm{m/s}$ is the speed of light and ξ_i is an additive pseudorange error at the user's position.

Introducing the vectors $R = (r_1, \dots, r_n)^\top$ and $X = (X_u^\top, t_r)^\top$ and linearizing the pseudorange equation with respect to the state vector X_u around the working point X_{u0}, we get the measurement equation

$$Y = R - D_0 \simeq H(X - X_0) + \xi, \tag{11.26}$$

where $D_0 = (d_{1_0}, \ldots, d_{n_0})^\top$, $d_{i_0} = \|X_i - X_{u_0}\|_2$, $\xi = (\xi_1, \ldots, \xi_n)^\top$, $H = \frac{\partial R}{\partial X}\Big|_{X=X_0}$ is the Jacobian matrix of size $n \times 4$. If we assume that $n \geq 5$, $\mathsf{E}(\xi) = 0$ and $\mathsf{cov}(\xi) = \Sigma$, where $\Sigma = \mathsf{diag}\{\sigma_1^2, \ldots, \sigma_n^2\}$ is a known diagonal matrix of order n, then the *iterative least squares* (LS) algorithm provides us with an optimal solution.

The pseudorange errors $\{\xi_k\}_{k\geq 1}$ are usually strongly autocorrelated due to the tropospheric and ionospheric refraction. This autocorrelation should be taken into account for the integrity calculation [332, 333]. The exponential character of the sampling ACF shows that the following simple autoregressive model can be proposed as an approximation:

$$\xi_k = a\xi_{k-1} + \sqrt{1-a^2}\zeta_k, \quad \zeta_k \sim \text{i.i.d.}\, \mathcal{N}(0, \Sigma). \tag{11.27}$$

Hence, a simplified measurement model of GNSS is defined by the following regression model with redundancy:

$$\widetilde{\mathbf{Y}}_k = \widetilde{H}\mathbf{X}_k + \widetilde{\xi}_k + \Upsilon(k, \nu), \;\; \widetilde{\xi}_k = a\widetilde{\xi}_{k-1} + \sqrt{1-a^2}\widetilde{\zeta}_k, \quad \widetilde{\zeta}_k \sim \text{i.i.d.}\, \mathcal{N}(0, \mathbb{I}_n), \tag{11.28}$$

where $\widetilde{\mathbf{Y}}_k = \Sigma^{-1/2}\mathbf{Y}_k$, $\widetilde{H} = \Sigma^{-1/2}H$, $\widetilde{\xi}_k = \Sigma^{-1/2}\xi_k$, $\mathbf{X}_k \in \mathbb{R}^4$ is a non-random unknown state vector (of positions $X - X_0$ and user clock bias t_r). Hence, as in Subsection 11.1.3, the detection and isolation of faults should be performed using the statistic invariant under the group of translations $\mathbf{G} = \{g : g(Y) = \mathbf{Y} + \widetilde{H}\mathbf{C}\}$, $\mathbf{C} \in \mathbb{R}^3$; see details in Subsections 11.1.3 and 2.10.3.1.

In the case when only one individual satellite fault is assumed at a time, any RAIM fault detection and exclusion (FDE) algorithm is defined by a pair (T, d) based on the measured observations (pseudo-ranges) $\mathbf{Y}_k$, where T is the alarm time (i.e., the time when a positioning failure is detected) and $d \in \{1, \ldots, n\}$ is the final decision. Let us extend the set of states of d by adding an additional state 0. If the RAIM FDE algorithm decides that the exclusion is not available (failed exclusion), say due to the poor geometry of GNSS constellation, then it is flagged by $d = 0$. If $d > 0$, then the RAIM FDE algorithm decides that the exclusion is possible and d is equal to the number of contaminated GNSS channels. The following statistical criteria are usually used [401]: a probability of a false alert (exclusion is impossible or failed exclusion),

$$\alpha(T, k) = \mathsf{P}_\infty\left(\{k \leq T \leq m_\alpha + k - 1\} \cap \{d = 0\}\right), \tag{11.29}$$

where m_α is a typical period to compute the false alert probability, $\mathsf{P}_\infty(\ldots)$ means that there is no positioning failure; a probability of a missed alert,

$$\begin{aligned}\beta(T, k_0) = \max_{1\leq \ell \leq n(k_0)} &\Big\{\mathsf{P}_{k_0}^{\ell}(T - k_0 + 1 > m_\tau) \\ &+ \mathsf{P}_{k_0}^{\ell}\left(\{k_0 \leq T \leq m_\tau + k_0 - 1\} \cap \{d \neq \ell\} \cap \{d > 0\}\right)\Big\},\end{aligned} \tag{11.30}$$

where m_τ is the time to alert, $\mathsf{P}_{k_0}^{\ell}(\ldots)$ means that there is a positioning failure, the onset of this failure is k_0 and the ℓ-th channel is contaminated, $n(k_0)$ is the number of visible satellites at the instant k_0; and, finally, a probability of a failed exclusion,

$$\omega(T, k_0) = \max_{1\leq \ell \leq n(k_0)} \mathsf{P}_{k_0}^{\ell}\left(\{k_0 \leq T \leq m_\tau + k_0 - 1\} \cap \{d = 0\}\right). \tag{11.31}$$

As is mentioned in [401] "The detection function is defined to be available when the constellation of satellites provides a geometry for which the missed alert and false alert requirement can be met on all satellites being used for the applicable alert limit and time-to-alert[...]." Let the missed alert and false alert probabilities be bounded from above as $\beta(T, k_0) \leq 10^{-3}/$ failure and $\alpha(T, k) \leq 10^{-5}/\text{h}$

or $\alpha(T,k) \leq 2 \cdot 10^{-5}$/approach. If these inequalities are satisfied for the HAL, VAL, integrity risk p_r and m_τ which are specified for a given mode of flight, then the detection function of the RAIM FDE algorithm (T,d) is defined to be available [401]. The availability of a RAIM FDE function is computed as a fraction of the typical time interval L of the GNSS constellation(s) during which this RAIM FDE function is available for a certain geographic position (latitude and longitude). For example, $L = 24$ hours for a single GPS or Galileo constellation and $L = 72$ hours for a combined GPS/Galileo constellation.

The exclusion function of the RAIM FDE algorithm (T,d) is defined to be available if $\omega(T,\nu,k_0) \leq 10^{-3}$/failure for the HAL, VAL, p_r, m_τ specified for a given mode of flight and the detection function of the algorithm (T,d) still available after exclusion, given the ℓ-th satellite failed (i.e., with $n-1$ visible satellites).

Five particular RAIM FDE schemes have been compared in [332, 333]. The quality of these RAIM FDE schemes has been characterized by the availability of the detection and exclusion functions.

We discuss now only two algorithms: the snapshot LS-based RAIM FDE algorithm (i.e., the FSS change detection–isolation algorithm with $m = 1$) and the sequential RAIM FDE algorithm based on the constrained GLR test. The detailed description of these RAIM FDE algorithms can be found in [332, 333].

Table 11.1: *Comparison of snapshot and sequential RAIM FDE algorithms in the case of Galileo E1/E5.*

RAIM	Detection Availability			Exclusion Availability		
	Min	Max	Mean	Min	Max	Mean
Snapshot	0.87	1.00	0.97	0.61	1.00	0.82
Sequential	0.89	1.00	0.97	0.65	1.00	0.89

Stepwise faults in a single satellite channel are assumed. The amplitudes of these faults are chosen to satisfy the condition of a minimal positioning failure (vertical or horizontal) as defined in (11.6). The availability of RAIM FDE functions has been calculated for both competitors in each of the grid points in latitude and longitude spread over the Earth's surface. The results of comparison of the snapshot and sequential RAIM FDE algorithms in the case of Galileo E1/E5 signals are summarized in Table 11.1. Here the minimal, maximal, and average availabilities of detection and exclusion functions are calculated over the set of designated grid points (in latitude and longitude) for snapshot and sequential RAIM FDE algorithms. The proposed sequential RAIM FDE algorithms perform better than the standard existing RAIM FDE algorithms. The improvement is in the range of a few percent of availability gain in most of the cases tested.

11.1.4.2 Ground Station-Based Integrity Monitoring

Let us consider a ground station at the known position $X_s = (x,y,z)^\top$. The difference between the pseudorange r_i from the i-th satellite to the station and the true distance d_i from the i-th satellite to the station can be written as

$$y_i = r_i - d_i = c\,t_s + \xi_i, \quad i = 1,2,\ldots,n$$

(see [324]), where r_i is the *pseudorange* from the i-th satellite to the station and $d_i = \|X_i - X_s\|_2$ is the known distance from the i-th satellite to the station, t_s is the user clock bias, c is the speed of light, $\xi = (\xi_1,\ldots,\xi_n)^\top$ is the vector of additive pseudorange errors at the ground station position, $\xi \sim \mathcal{N}(0,\Sigma_s)$ and $\Sigma_s = \mathrm{diag}\{\sigma_{s,1}^2,\ldots,\sigma_{s,n}^2\}$ is a known diagonal matrix of order n.

A fault is modeled by the vector θ of additional pseudorange biases:

$$\widetilde{Y}_{s,k} = \widetilde{\mathbf{1}}_n\mu + \widetilde{\xi}_k + \Upsilon_s(k,\nu), \quad \widetilde{\xi}_k = a\widetilde{\xi}_{k-1} + \sqrt{1-a^2}\,\widetilde{\zeta}_k, \quad \Upsilon_s(n,\nu) = \begin{cases} 0 & \text{if} \quad n \leq \nu \\ \theta & \text{if} \quad n \geq \nu+1 \end{cases},$$

where $\widetilde{Y}_{s,k} = \Sigma_s^{-1/2} Y_{s,k}$, $,\widetilde{\mathbf{1}}_n = \Sigma_s^{-1/2}\mathbf{1}_n$, $\widetilde{\xi}_k = \Sigma_s^{-1/2}\xi_k$, $\widetilde{\zeta} = (\zeta_1,\ldots,\zeta_n)^\top \sim \mathcal{N}(0,\mathbb{I}_n)$, Σ_s is a known diagonal matrix of order n, $Y_s = (y_1, y_2, \ldots, y_n)^\top$ is the vector of station measurements, $\mathbf{1}_n = (1,\ldots,1)^\top$ and $\mu = c\,t_s$, $\mu \in \mathbb{R}$, is the station clock bias multiplied by the speed of light, measured in meters and considered as a nuisance parameter.

A detailed discussion of GPS fault detection and isolation algorithms using a ground station-based integrity monitoring and the comparison of sequential and FSS algorithms can be found in [324].

Remark 11.1.1. In principle, several GNSS channels can be contaminated simultaneously. Due to the fact that the station position $X_s = (x,y,z)^\top$ is perfectly known, the number of nuisance parameters is reduced to the minimum. In this case, the only unknown parameter is the station clock bias. For this reason, the station-based integrity monitoring capacity is higher than the capacity of the RAIM algorithm. Nevertheless, some simultaneous faults can be masked by the unknown station clock bias. Let us represent the additional biases vector θ in the following manner $\theta = \widetilde{\mathbf{1}}_n\theta_\mu + W^\top\theta_W$, $\theta_\mu \in \mathbb{R}$, $\theta_W \in \mathbb{R}^{n-1}$, where the nuisance rejection $((n-1)\times n)$ matrix $W\ : W\widetilde{\mathbf{1}}_n = 0$ is composed of $n-1$ basis vectors which span the orthogonal complement of $R(\widetilde{\mathbf{1}}_n)$ and it satisfies the conditions defined by (2.359). It follows from the above definition of θ that the sub-vector θ_μ is undetectable: the subspace of θ_μ coincides with the subspace of the nuisance parameter μ and, hence, θ_μ is masked by μ. The detectable part of θ is represented by $W^\top\theta_W$.

Let us analyze the impact of this undetectable part $\widetilde{\mathbf{1}}_n\theta_\mu$ of the vector fault θ on the user's positioning. Consider a user (aircraft) at the position $X_u = (x_u, y_u, z_u)^\top$. The linearized measurement equation of the user is given by (11.28). Let us additionally assume that $\Upsilon_s(k,\nu) = \Upsilon(k,\nu)$ and $\Sigma_s = \Sigma$, which is realistic in the case of a dense ground station network. The additional bias vector θ affecting the GNSS channels implies an additional error $b = \mathsf{E}(\widehat{X} - X) = (\widetilde{H}^\top\widetilde{H})^{-1}\widetilde{H}^\top\theta$ in the vector $\widehat{X}$. Fortunately, the impact $b = \theta_\mu(\widetilde{H}^\top\widetilde{H})^{-1}\widetilde{H}^\top\widetilde{\mathbf{1}}_n$ of such an undetectable bias $\widetilde{\mathbf{1}}_n\theta_\mu$ on the first three components $\widehat{x}_u, \widehat{y}_u, \widehat{z}_u$ representing the aircraft position is equal to zero, i.e., $b_x = b_y = b_z = 0$. Therefore, undetectable (by a ground monitoring station) pseudorange biases are not dangerous for the navigation.

11.2 Vibration-Based Structural Health Monitoring

11.2.1 Introduction

In many applications the problem of fault detection and isolation (FDI) is a crucial issue which has been investigated with different types of approaches; see, e.g., the overviews [35, 151, 158, 205, 505] and books [47, 72, 76, 119, 159, 206, 349]. An increasing interest in condition-based maintenance has appeared in industrial applications. The key idea is to replace systematic inspections by condition-based inspections, i.e., inspections decided upon the continuous monitoring of the considered system (machine, structure, process or plant), based on the sensors data, in order to prevent a possible malfunction or damage before it happens. A solution consists in the early detection of slight deviations with respect to a characterization of the system in its usual working conditions. Indeed, if such an early detection can be performed while preserving robustness with respect to changes in normal operating conditions, one can reasonably expect being able to prevent larger deviations resulting from malfunction, fatigue, faults, or damage before they happen, and consequently to increase the availability of the system. It should be clear that the local approach already discussed in Chapters 2, 3, and 8 provides us with tools which perform this early detection task.

Mechanical systems integrity monitoring, introduced in Subsection 1.3.5, is a typical example of this kind of monitoring problem. Structural Health Monitoring (SHM) is the whole process of the design, development, and implementation of techniques for the detection, localization, and estimation of damages, for monitoring the integrity of structures and machines within the aerospace, civil and mechanical engineering infrastructures [95, 136, 481, 528]. In addition to these key driving application areas, SHM is also spreading over most transportation infrastructures and vehicles, within the naval, railway, and automobile domains. Examples of structures or machines to be monitored include aircraft, spacecraft, buildings, bridges, dams, ships, offshore platforms, on-shore and off-shore wind energy systems, turbo-alternators, and other heavy machinery.

SHM is a topic of growing interest, due to the aging of many engineering constructions and machines and to increased safety norms. Many current approaches still rely on visual inspections or local non-destructive evaluations performed manually, e.g., acoustic, ultrasonic, radiographic, or eddy-current methods. These experimental approaches assume an *a priori* knowledge and the accessibility of a neighborhood of the damage location. Nevertheless it is now recognized that useful alternatives to those local evaluations consist in automatic global vibration-based monitoring techniques [23, 121, 136, 528].

Many structures to be monitored, e.g., civil engineering structures subject to wind and earthquakes, aircrafts subject to turbulence, are subject to both fast and unmeasured variations in their environment and small slow variations in their modal or vibrating properties. While any change in the excitation is meaningless, damages or fatigues on the structure are of interest. But the available measurements do not separate the effects of the external forces from the effect of the structure. Moreover, the changes of interest, that may be as small as 1% in the eigenfrequencies, are visible neither on the signals nor on their spectra.

Most classical modal analysis and vibration monitoring methods basically process data registered either on test beds or under specific excitation or rotation speed conditions. However, a need has been recognized for vibration monitoring algorithms devoted to the processing of data recorded in-operation, namely during the actual functioning of the considered structure or machine, without artificial excitation, slowing down or stopping [40, 355].

It is the purpose of this section to describe a solution to this problem. This solution builds on the following facts.

- The vibration-based structural health monitoring problem translates into the problem of detecting small changes in the eigenstructure of the state transition matrix F of a linear dynamical state-space system. It can also be stated as the problem of detecting small changes in the autoregressive (AR) part of a multivariable autoregressive moving average (ARMA) model having nonstationary MA coefficients.
- The likelihood function is not the relevant estimating function in such a situation. The reason is typically that the Fisher information matrix of an ARMA process is not block-diagonal w.r.t. the AR and MA matrix coefficients. However, a more appropriate parameter estimating function results from a class of estimation algorithms known under the name of subspace-based identification.
- The statistical local approach, a relevant tool for detecting small changes, can be applied to estimating functions different from the likelihood [31, 32, 33, 36, 60, 62, 166, 176, 190] and in particular to the parameter estimating function associated with subspace-based identification [38, 45].
- Both a batch-wise change detection algorithm and a sample-wise recursive CUSUM detection algorithm are of interest in the context of SHM.

11.2.2 Subspace-Based Identification and Parameter Estimating Function

It is well established [135, 191, 214, 354] that the vibration-based structural analysis and health monitoring problems translate into the identification and monitoring of the eigenstructure of the state transition matrix F of a linear dynamical state-space system excited by a zero mean Gaussian white noise sequence $(V_n)_{n\geq 1}$:

$$X_{n+1} = F\, X_n + V_{n+1}, \quad Y_n = H\, X_n, \tag{11.32}$$

namely the $(\lambda, \varphi_\lambda)$ defined by

$$\det(F - \lambda I) = 0, \quad (F - \lambda I)\,\phi_\lambda = 0, \quad \varphi_\lambda \triangleq H\,\phi_\lambda. \tag{11.33}$$

The associated parameter vector is

$$\theta \triangleq \begin{pmatrix} \Lambda \\ \text{vec}\Phi \end{pmatrix} \tag{11.34}$$

where Λ is the vector whose elements are the eigenvalues λ, Φ is the matrix whose columns are the φ_λ's, and vec is the column stacking operator. This parameter is canonical, that is invariant w.r.t. a change in the state-space basis.

Subspace-based methods is the generic name for linear systems identification algorithms based on either time domain measurements or output covariance matrices, in which different subspaces of Gaussian random vectors play a key role. Subspace fitting estimates take advantage of the orthogonality between the range (or left kernel) spaces of certain matrix-valued statistics. There has been a growing interest in these methods [483, 485, 486], their connection to instrumental variables [487] and maximum likelihood [344] approaches, and their invariance properties [92]. They are actually well suited for identifying the system eigenstructure.

Dealing with long samples of multisensor measurements can be mandatory for in-operation structural analysis under non-stationary natural or ambient excitation. Processing output covariance matrices is of interest in such cases. The difference between the covariance-driven form of subspace algorithms which is described here and the usual data-driven form [483] is minor, at least for eigenstructure identification [354].

Covariance-driven Subspace Identification. Let $R_i \triangleq \mathsf{E}\left(Y_k\, Y_{k-i}^\top\right)$ and

$$\mathcal{H}_{p+1,q} \triangleq \begin{pmatrix} R_0 & R_1 & \vdots & R_{q-1} \\ R_1 & R_2 & \vdots & R_q \\ \vdots & \vdots & \vdots & \vdots \\ R_p & R_{p+1} & \vdots & R_{p+q-1} \end{pmatrix} \triangleq \text{Hank}(R_i) \tag{11.35}$$

be the output covariance and Hankel matrices, respectively; and $G \triangleq \mathsf{E}t(X_k\, Y_k^\top)$. Direct computations of the R_i's from equations (11.32) lead to the well-known key factorizations [442]

$$R_i = H\,F^i\,G, \tag{11.36}$$

$$\mathcal{H}_{p+1,q} = \mathcal{O}_{p+1}(H,F)\,\mathcal{C}_q(F,G), \tag{11.37}$$

where

$$\mathcal{O}_{p+1}^\top(H,F) \triangleq \left(H^\top\ (HF)^\top\ \dots\ (HF^p)^\top\right) \text{ and } \mathcal{C}_q(F,G) \triangleq (G\ FG\ \dots\ F^{q-1}G) \tag{11.38}$$

are the observability and controllability matrices, respectively. In factorization (11.37), the left factor $\mathcal{O}$ only depends on the pair (H,F), and thus on the system eigenstructure in (11.32), whereas the excitation V_k only affects the right factor $\mathcal{C}$ through the cross-covariance matrix G.

The observation matrix H is then found in the first block-row of the observability matrix $\mathcal{O}$. The state-transition matrix F is obtained from the shift invariance property of $\mathcal{O}$

$$\mathcal{O}_p^\uparrow(H,F) = \mathcal{O}_p(H,F)\,F, \quad \text{where } \mathcal{O}_p^{\uparrow\top}(H,F) \triangleq \left((HF)^\top\ (HF^2)^\top\ \dots\ (HF^p)^\top\right).$$

Recovering F requires to assume that $\text{rank}(\mathcal{O}_p) = \dim F$, and thus that the number of block-rows in $\mathcal{H}_{p+1,q}$ is large enough. The eigenstructure (λ, ϕ_λ) then results from (11.33).

The actual implementation of this subspace algorithm, known under the name of balanced realization (BR) [5] has the empirical covariances

$$\widehat{R}_i = 1/(N-i) \sum_{k=i+1}^{N} Y_k \, Y_{k-i}^{\top} \tag{11.39}$$

substituted for R_i in $\mathcal{H}_{p+1,q}$, yielding the empirical Hankel matrix

$$\widehat{\mathcal{H}}_{p+1,q} \stackrel{\Delta}{=} \text{Hank}\left(\widehat{R}_i\right). \tag{11.40}$$

Since the actual model order is generally not known, this procedure is run with increasing model orders [40, 297]. The singular value decomposition (SVD) of $\widehat{\mathcal{H}}_{p+1,q}$ and its truncation at the desired model order yield, in the left factor, an estimate $\widehat{\mathcal{O}}$ for the observability matrix $\mathcal{O}$:

$$\widehat{\mathcal{H}} = U\,\Delta\,V^{\top} = U \begin{pmatrix} \Delta_1 & 0 \\ 0 & \Delta_0 \end{pmatrix} V^{\top},$$
$$\widehat{\mathcal{O}} = U\,\Delta_1^{1/2}, \quad \widehat{\mathcal{C}} = \Delta_1^{1/2}\,V^{\top}.$$

From $\widehat{\mathcal{O}}$, estimates $(\widehat{H},\widehat{F})$ and $(\widehat{\lambda},\widehat{\phi}_{\lambda})$ are recovered as sketched above. The CVA algorithm basically applies the same procedure to a Hankel matrix pre- and post-multiplied by the covariance matrix of future and past data, respectively [6, 38].

A minor but extremely fruitful remark is that it is possible to write the covariance-driven subspace identification algorithm under a form which involves a parameter estimating function. This is explained next.

Associated Parameter Estimating Function. Choosing the eigenvectors of matrix F as a basis for the state space of model (11.32) yields the following representation of the observability matrix

$$\mathcal{O}_{p+1}^{\top}(\theta) = \left(\Phi^{\top}\ (\Phi\Delta)^{\top}\ \dots\ (\Phi\Delta^{p})^{\top}\right), \tag{11.41}$$

where $\Delta \stackrel{\Delta}{=} \mathsf{diag}(\Lambda)$, and Λ and Φ are as in (11.34). Whether a nominal parameter θ_0 fits a given output covariance sequence $(R_j)_j$ is characterized by [38, 487]

$$\mathcal{O}_{p+1}(\theta_0) \text{ and } \mathcal{H}_{p+1,q} \text{ have the same left kernel space.} \tag{11.42}$$

This property can be checked as follows. From the nominal θ_0, compute $\mathcal{O}_{p+1}(\theta_0)$ using (11.41), and perform a SVD of $\mathcal{O}_{p+1}(\theta_0)$ for extracting a matrix S such that

$$S^{\top}\,S = I_s \text{ and } S^{\top}\,\mathcal{O}_{p+1}(\theta_0) = 0. \tag{11.43}$$

Matrix S is not unique (two such matrices relate through a post-multiplication with an orthonormal matrix), but can be regarded as a function of θ_0 for reasons which will become clear in Subsection 11.2.3. Then the characterization writes

$$S(\theta_0)^{\top}\,\mathcal{H}_{p+1,q} = 0. \tag{11.44}$$

For a multivariable random process $(Y_k)_k$ whose distribution is parametrized by a vector θ, a parameter estimating function [166, 190] is a vector function $\mathcal{K}$ of the parameter θ and a finite size sample of observations[3] $\mathcal{Y}_{k,\rho}^{\top} = (Y_k^{\top}\ \dots\ Y_{k-\rho+1}^{\top})$, such that

$$\mathsf{E}_{\theta}\,\mathcal{K}(\theta_0,\mathcal{Y}_{k,\rho}) = 0 \text{ iff } \theta = \theta_0,$$

[3] More sophisticated functions of the observations may be necessary for complex dynamical processes [62, 166, 190, 292].

of which the empirical counterpart defines an estimate $\widehat{\theta}$ as a root of the estimating equation

$$1/N \sum_k \mathcal{K}(\theta, \mathcal{Y}_{k,\rho}) = 0. \tag{11.45}$$

Since subspace algorithms exploit the orthogonality between the range (or left kernel) spaces of matrix-valued statistics, the estimating equations associated with subspace fitting have the following particular product form [92, 486]

$$1/N \sum_k \mathcal{K}(\theta, \mathcal{Y}_{k,\rho}) \triangleq \text{vec}\left(\mathcal{S}^\top(\theta)\, \widehat{\mathcal{N}}_N\right) = 0, \tag{11.46}$$

where $\mathcal{S}(\theta)$ is a matrix-valued function of the parameter and $\widehat{\mathcal{N}}_N$ is a matrix-valued statistic based on an N-size data sample. Choosing the Hankel matrix $\mathcal{H}$ as the statistics $\mathcal{N}$ provides us with the estimating function associated with the covariance-driven subspace identification algorithm

$$\text{vec}\left(\mathcal{S}^\top(\theta)\, \widehat{\mathcal{H}}_N\right) = 0 \tag{11.47}$$

which of course is coherent with (11.44).

The reasoning above holds in the case of known system order. However, in most practical cases the data are generated by a system of higher order than the model. The nominal model characterization and parameter estimating function relevant for that case are investigated in [38].

Other Uses of the Key Factorizations. Factorization (11.37) is the key for the characterization (11.44) of the canonical parameter vector θ in (11.34), and for deriving a residual adapted to detection purposes. This is explained in Subsections 11.2.3 and 11.2.4. Factorization (11.36) is also the key for

- Designing various *input–output* covariance-driven subspace identification algorithms adapted to the presence of both known controlled inputs and unknown ambient excitations [299];
- Designing an extension of covariance-driven subspace identification algorithm adapted to the presence and fusion of non-simultaneously recorded multiple sensors setups [297];
- Proving consistency and robustness results [61, 63], including for that extension [298].

11.2.3 Batch-Wise Change Detection Algorithm

Change detection is a natural approach to fault/damage detection. Indeed the damage detection problem can be stated as the problem of detecting a change in the modal parameter vector θ defined in (11.34). In this subsection we describe a batch-wise change detection algorithm built on the covariance-driven subspace-based estimating function and the statistical local approach to the design of change/fault/damage detection algorithms. It is assumed that a reference value θ_0 is available, generally identified using data recorded on the undamaged system. Based on a new data sample $Y_1, \ldots, Y_N$, the damage detection problem is to decide whether the new data are still well described by this parameter value or not. The modal diagnosis problem is to decide which components of the modal parameter vector θ have changed. The damage localization problem is to decide which parts of the structure have been damaged, or equivalently to decide which elements of the structural parameter matrices have changed.

We concentrate here on the damage detection problem for which we describe a χ^2-test based on a residual associated with the subspace identification algorithm described in Subsection 11.2.2. The modal diagnosis problem can be solved with similar χ^2-tests focused onto the modal subspaces of interest, using selected sensitivities of the residual w.r.t. the modal parameters and the damage localization problem can be solved with similar χ^2-tests focused onto the structural subspaces of interest, plugging sensitivities of the modal parameters w.r.t. the structural parameters of a finite elements model in the above setting [45].

Subspace-based Residual. For checking whether the new data $Y_1, \dots, Y_N$ are well described by the reference parameter vector θ_0, the idea is to use the parameter estimating function in (11.47), namely to compute the empirical Hankel matrix $\widehat{\mathcal{H}}_{p+1,q}$ in (11.40)-(11.39) and to define the vector

$$\zeta_N(\theta_0) \triangleq \sqrt{N}\, \mathrm{vec}\left(S(\theta_0)^\top\, \widehat{\mathcal{H}}_{p+1,q}\right). \tag{11.48}$$

Let θ be the actual parameter value for the system which generated the new data sample, and E_θ be the expectation when the actual system parameter is θ. From (11.44), we know that

$$\mathsf{E}_\theta\left(\zeta_N(\theta_0)\right) = 0 \quad \text{iff} \quad \theta = \theta_0, \tag{11.49}$$

namely vector $\zeta_N(\theta_0)$ in (11.48) has zero mean when θ does not change, and non-zero mean in the presence of a change (damage). Consequently $\zeta_N(\theta_0)$ plays the role of a residual.

It turns out that this residual has highly interesting properties in practice, both for damage detection [38] and localization [45], and for flutter monitoring [296]. Even when the eigenvectors are not monitored, they are explicitly involved in the computation of the residual. It is our experience [45] that this fact may be of crucial importance in structural health monitoring, especially when detecting small deviations in the eigenstructure.

The Residual is Gaussian. To decide whether $\theta = \theta_0$ holds true or not, or equivalently whether the residual ζ_n is significantly different from zero, requires the knowledge of the probability distribution of $\zeta_N(\theta_0)$, which unfortunately is generally unknown. One manner to circumvent this difficulty is to assume close hypotheses:

$$(\textit{Safe})\; H_0 : \; \theta = \theta_0 \quad \text{and} \quad (\textit{Damaged})\; H_1 : \; \theta = \theta_0 + \delta\theta/\sqrt{N} \tag{11.50}$$

where vector $\delta\theta$ is unknown, but fixed. For large N hypothesis H_1 corresponds to small deviations in θ, as in the statistical local approach, of which the main result is the following [36, 47, 60, 525].

Let $\Sigma(\theta_0) \triangleq \lim_{N\to\infty} \mathsf{E}_{\theta_0}\left(\zeta_N\, \zeta_N^\top\right)$ be the residual covariance matrix (it is assumed that the limit exists). The estimation of Σ may be somewhat tricky [525, 524]. Provided that $\Sigma(\theta_0)$ is positive definite, the residual ζ_N in (11.48) is asymptotically Gaussian distributed with the same covariance matrix $\Sigma(\theta_0)$ under both H_0 and H_1, that is [38]

$$\zeta_N(\theta_0) \xrightarrow[N\to\infty]{} \begin{cases} \mathcal{N}(\,0,\, \Sigma(\theta_0)\,) & \text{under} \quad H_0 \\ \mathcal{N}(\,\mathcal{J}(\theta_0)\,\delta\theta,\, \Sigma(\theta_0)\,) & \text{under} \quad H_1 \end{cases} \tag{11.51}$$

where $\mathcal{J}(\theta_0)$ is the Jacobian matrix containing the sensitivities of the residual w.r.t. the modal parameters

$$\mathcal{J}(\theta_0) \triangleq 1/\sqrt{N}\; \partial/\partial\theta\; \mathsf{E}_\theta\, \zeta_N(\theta_0)|_{\theta=\theta_0}. \tag{11.52}$$

As seen in (11.51), a deviation $\delta\theta$ in the system parameter θ is reflected into a change in the mean value of residual ζ_N, which switches from zero in the undamaged case to $\mathcal{J}(\theta_0)\;\delta\theta$ in case of small damage. Note that matrices $\mathcal{J}(\theta_0)$ and $\Sigma(\theta_0)$ depend on neither the sample size N nor the fault vector $\delta\theta$ in hypothesis H_1. Thus they can be estimated prior to testing, using data on the safe system exactly as the reference parameter θ_0. In case of non-stationary excitation, a similar result has been proven, for scalar output signals, and with matrix Σ estimated on newly collected data [308].

χ^2-test for Damage Detection. Let $\widehat{\mathcal{J}}$ and $\widehat{\Sigma}$ be consistent estimates of $\mathcal{J}(\theta_0)$ and $\Sigma(\theta_0)$, and assume additionally that $\mathcal{J}(\theta_0)$ is full column rank. Then, thanks to (11.51), testing between the hypotheses H_0 and H_1 in (11.50) can be achieved with the aid of the following χ^2-test

$$\chi_N^2 \triangleq \zeta_N^\top\, \widehat{\Sigma}^{-1}\, \widehat{\mathcal{J}}\left(\widehat{\mathcal{J}}^\top\, \widehat{\Sigma}^{-1}\, \widehat{\mathcal{J}}\right)^{-1} \widehat{\mathcal{J}}^\top\, \widehat{\Sigma}^{-1}\, \zeta_N \tag{11.53}$$

which should be compared to a threshold. Note that the IV-based test proposed in [42] can be seen as a particular case of (11.53) [38].

In (11.53), the dependence on θ_0 has been removed for simplicity. The only term which should be computed after data collection is residual ζ_N in (11.48). Thus the test can be computed on-board. Test statistics χ_N^2 is asymptotically distributed as a χ^2-variable, with $\mathsf{rank}(\mathcal{J})$ degrees of freedom. From this, a threshold for χ_N^2 can be deduced, for a given false alarm probability. The noncentrality parameter of this χ^2-variable under H_1 is $\delta\theta^\top \mathcal{J}^\top \Sigma^{-1} \mathcal{J}\, \delta\theta$. How to select a threshold for χ_N^2 from histograms of empirical values obtained on data for undamaged cases is explained in [300].

From the expressions in (11.48) and (11.53), it is easy to show that this test enjoys some invariance property: any pre-multiplication of the left kernel S by an invertible matrix factors out in χ_N^2 [37]. This is why S defined in (11.43) can be considered as a function of θ_0, as announced in Subsection 11.2.2.

The asymptotic properties of the test (11.53) have been investigated in [399] for the IV-based version, and in [242] in the case of more general estimating functions not limited to subspace.

11.2.4 Sample-Wise Recursive CUSUM Detection Algorithm

Another asymptotic for this estimating function is used for designing a sample-wise recursive CUSUM detection algorithm. Indeed, in some applications, it is necessary to design detection algorithms working sample point-wise rather than batch-wise. For example, as explained in Subsection 11.2.5, the early warning of deviations in specific modal parameters is required for new aircraft qualification and exploitation, and especially for handling the flutter monitoring problem.

A simplified although well-sounded version of the flutter monitoring problem consists in monitoring a specific damping coefficient. It is known, e.g., from Cramer–Rao bounds, that damping factors are difficult to estimate accurately [157]. However, detection algorithms usually have a much shorter response time than identification algorithms. Thus, for improving the estimation of damping factors and achieving this in real-time, the idea is to design an on-line detection algorithm able to detect whether a specified damping coefficient ρ decreases below some critical value ρ_c [296]

$$H_0:\ \rho \geq \rho_c \quad \text{and} \quad H_1:\ \rho < \rho_c. \tag{11.54}$$

A good candidate for designing this test is the residual associated with subspace-based covariance-driven identification defined in (11.48), which can be computed recursively as

$$\zeta_N(\theta_0) = \sum_{k=q}^{N-p} Z_k(\theta_0)/\sqrt{N}, \tag{11.55}$$

where

$$Z_k(\theta_0) \triangleq \operatorname{vec}(S(\theta_0)^\top \mathcal{Y}_{k,p+1}^{+} \mathcal{Y}_{k,q}^{-\top}) \tag{11.56}$$

and $\mathcal{Y}_{k,p+1}^{+\top} \triangleq (Y_k^\top \ \dots \ Y_{k+p}^\top)$, $\mathcal{Y}_{k,q}^{-\top} \triangleq (Y_k^\top \ \dots \ Y_{k-q+1}^\top)$.

Since the hypothesis (11.54) regarding the damping coefficient is non local any more (compare with (11.50)), the asymptotic local approach used in Subsection 11.2.3 can no longer be used for that residual, and another asymptotic should be used instead. From (11.51) and (11.55), we know that $\sum_{k=q}^{N-p} Z_k(\theta_0)/\sqrt{N}$ is asymptotically Gaussian distributed, with mean zero under $\theta = \theta_0$ and $\mathcal{J}(\theta_0)\delta\theta$ under $\theta = \theta_0 + \delta\theta/\sqrt{N}$. Now, the arguments in Subsection 5.4.1 of [62] lead to the following approximation: for k large enough, $Z_k(\theta_0)$ can itself be regarded as asymptotically Gaussian distributed with *zero mean* under $\theta = \theta_0$, and the $Z_k(\theta_0)$'s are *independent*. Furthermore, a change in θ is reflected into a change in the mean vector ν of $Z_k(\theta_0)$. This paves the road for the use of CUSUM tests for detecting such changes, according to the type and amount of *a priori* information available for the parameters to be monitored [47].

For monitoring a damping coefficient (scalar parameter θ_a), the CUSUM test writes

$$S_n(\theta_a) \triangleq \sum_{k=q}^{n-p} Z_k(\theta_a) \tag{11.57}$$

$$T_n(\theta_a) \triangleq \max_{q \le k \le n-p} S_k(\theta_a)$$

$$g_n(\theta_a) \triangleq T_n(\theta_a) - S_n(\theta_a) \tag{11.58}$$

and an alarm is raised when $g_n(\theta_a) \geq \gamma$ for some threshold γ. Since it is known neither what is the actual hypothesis when this processing starts, nor what are the actual sign and magnitude of the change in θ_a that will occur, a relevant procedure consists in introducing a *minimum magnitude of change* $\nu_m > 0$, running two tests in parallel for a decreasing and an increasing parameter, respectively, making a decision from the first test which fires, and resetting all sums and extrema to zero and switching to the other one afterwards. This is investigated in [296].

For addressing the more realistic problem of monitoring two pairs of frequencies and damping coefficients possibly subject to specific time variations, multiple CUSUM tests for single parameters can be run in parallel [44]. The approach was extended to monitor other stability criteria [44, 296, 527, 529, 530]. Experiments showed good performances in detecting the deviation of the system with respect to the reference toward instability. However, in all those algorithms the reference corresponds in practice to normal flight conditions quite far from flutter. It results that the flutter onset time estimate is conservative, namely alarms are raised too early with respect to the true flutter airspeed, even considering the 15% safety margin between flight envelope and flutter considered in air worthiness regulations. An adaptive subspace-based detection approach has been thus proposed in [531] to overcome this limitation. This approach involves the updating of the reference left kernel matrix and of two calibration matrices during on-line testing. A first algorithm performs batch computations within a moving window with fixed size. A second algorithm achieves recursive updating within a growing window with increasing size.

11.2.5 Typical Application Examples

11.2.5.1 Monitoring the Integrity of the Civil Infrastructure

The change detection algorithm described in Subsection 11.2.3 has been applied [300] to the Swiss Z24 bridge, a benchmark of the BRITE/EURAM project SIMCES on identification and monitoring of civil engineering structures, for which EMPA (the Swiss Federal Laboratory for Materials Testing and Research) has carried out tests and data recording. The response of the bridge to traffic excitation under the bridge has been measured over one year in 139 points, mainly in the vertical and transverse directions, and sampled at 100 Hz. The global χ^2-test has been applied to data of the four reference stations. Thus the test has been evaluated for several data sets, for both the safe and damaged structures.

Two damage scenarios are considered: pier settlement of 20 mm and 80 mm, respectively, further referred to as DS1 and DS2. Even though the effect of the damages on the natural frequencies is really small (no more than 1% for DS1), the χ^2-test is very sensitive: for DS1, 1000 times larger than for the safe case.

The implementation and tuning of an on-line monitoring system for automated damage detection have also been achieved. Monitoring results based on three sensors have been analyzed, from which the following conclusions have been drawn. The overall increase in the test value is slightly hidden by its daily fluctuations. These fluctuations are due to changes in the modal parameters themselves, due to variations in environmental variables such as temperature, precise hour of measurements, speed of wind, ... and can be higher than the changes of the modal characteristics due to damage. However, modal variations due to damage imply greater variations of the test than those due to environmental changes.

Another major issue is to take care of the fluctuations of the excitation, due for example to changes in the traffic or neighboring activities (a new bridge was in construction a few hundred meters away), and to avoid running the test when the excitation is clearly different from the excitation of the reference model. A good way to avoid interference between these changes and the test result is to calibrate several reference data sets corresponding to different values of the environmental variables, including excitation and temperature, and to run the test upon matching the environmental characteristics of both the reference and the fresh data sets. Another approach is to include these variables into the model and consider them as nuisance information. This has been investigated in [24, 25, 43].

11.2.5.2 Flight Flutter Monitoring

The improved safety and performance of aerospace structures and reduced aircraft development and operating costs are major concerns. One of the critical objectives is to ensure that the newly designed aircraft is stable throughout its operating range. A critical aircraft instability phenomenon, known as flutter, results from an unfavorable interaction of aerodynamic, elastic, and inertial forces, and may cause major failures. A careful exploration of the dynamical behavior of the structure subject to vibration and aeroservoelastic forces is thus required. A major challenge is the in-flight use of flight test data. The flight flutter monitoring problem can be addressed on-line as the problem of detecting that some instability indicators decrease below some critical value. CUSUM-type change detection algorithms, as described in Subsection 11.2.4, are useful solutions to these problems [41, 46, 296, 531].

A toolbox for in-operation modal analysis and damage detection and localization can be downloaded from http://www.irisa.fr/i4s/constructif/modal.htm.

11.3 Rapid Detection of Intrusions in Computer Networks

11.3.1 Introduction

Cybersecurity has evolved into a critical 21st century problem that affects governments, businesses, and individuals. Recently, cyber threats have become more diffuse, more complex, and harder to detect. Malicious activities and intrusion attempts such as spam campaigns, phishing, personal data theft, worms, distributed denial-of-service (DDoS) attacks, address resolution protocol man-in-the-middle attacks, fast flux, *etc.*, occur every day, have become commonplace in contemporary computer networks, and pose enormous risks to users for a multitude of reasons. Malicious events usually produce (relatively) abrupt changes in network traffic profiles, which have to be detected and isolated rapidly while keeping a low FAR so as to respond appropriately and eliminate the negative consequences for the users.

Detection of traffic anomalies in computer networks is performed by employing *Intrusion Detection Systems* (IDS). Such systems in one way or another capitalize on the fact that *maltraffic* is noticeably different from *legitimate* traffic. There are two categories of IDSs: signature-based and anomaly-based. For an overview, see [113, 133, 224]. A signature-based IDS (SbIDS) inspects passing traffic to find matches against already known malicious patterns. Examples of SbIDSs are Snort [396] and Bro [353]. An anomaly-based IDS (AbIDS) is first trained to recognize normal network behavior and then seeks deviations from the normal profile, which are classified as potential attacks [224, 471, 472, 473].

As an example, consider DDoS attacks [280, 302], which typically involve many traffic streams resulting in a large number of packets aimed at congesting the target's server or network. Such attacks can be detected by noticing a change in the average number of packets sent through the victim's link per unit time. Intuitively, it is appealing to formulate the problem of detecting DDoS as a *quickest changepoint detection* problem. That is, to detect changes in statistical models as rapidly as possible maintaining the FAR at a given low level.

Changepoint detection theory allows us to develop solutions that are easily implemented, have certain optimality properties and, therefore, can be effectively used for designing AbIDSs for the early detection of intrusions in high-speed computer networks.

AbIDSs and SbIDSs are complementary. Neither alone is sufficient to detect and isolate the anomalies generated by attacks. The reason is that both these types of IDSs, when working independently, are plagued by a high rate of false positives and are susceptible to carefully crafted attacks that "blend" themselves into normal traffic. The ability of changepoint detection techniques to run at high speeds and with small detection delays presents an interesting opportunity. What if one could combine these techniques with signature-type methods that offer very low FAR but are too computationaly heavy to use at line speeds? Do such synergistic IDSs exist, and if so, how can they be integrated?

In this section, we describe a "hybrid" approach that integrates two substantially different detection techniques — anomaly changepoint detection methods with signature–spectral detection techniques. We demonstrate that the resulting two-stage *hybrid anomaly–signature IDS* performs better than any individual system. This two-stage hybrid approach allows augmenting hard detection decisions with profiles that can be used for further analysis, in particular for filtering false positives and confirming real attacks both at single-sensor and network levels. The hybrid IDS is tested on real attacks, and the results presented in Subsection 11.3.3 demonstrate its effectiveness.

11.3.2 Anomaly-Based Intrusion Detection System

11.3.2.1 Score-Based CUSUM and SR Procedures

Network anomalies (malicious or legitimate) occur at unknown points in time and produce abrupt changes in statistical properties of traffic data. Hence, we consider the problem of anomaly detection in computer networks a quickest changepoint detection problem: to detect changes in network traffic as rapidly as possible while maintaining a tolerable level of false alarms.

In network monitoring, one can observe various useful features from packet headers, e.g., packet size, source IP address, destination IP address, source port, destination port, types of protocols (ICMP, UDP, TCP), *etc*. In the case of UDP (User Datagram Protocol) flooding attacks, potentially useful observables include packet sizes, source ports, destination ports, and destination prefix. In the case of TCP (Transmission Control Protocol) flooding attacks, conceivably, we could have multiple channels that record counts of different flags (SYN, ACK, PUSH, RST, FIN, URG) from TCP header. Another plausible observable is a number of half-open connections for the detection of SYN flooding attacks. We could also have channels that keep track of the discrepancies in TCP SYN-FIN or TCP SYN-RST pairs. Furthermore, in order to detect file-sharing, we could monitor arrival (packet, byte or flow) counts, port numbers, and source-destination prefixes.

Recall that there are two competitive changepoint detection procedures, CUSUM and SR, which are given by

$$\begin{aligned} T_{\text{CS}} &= \inf\{n \ge 1 : W_n \ge h\}, \quad W_n = (W_{n-1} + Z_n)^+, \\ T_{\text{SR}} &= \inf\{n \ge 1 : R_n \ge A\}, \quad R_n = (1 + R_{n-1})\mathcal{L}_n, \end{aligned} \tag{11.59}$$

respectively, where $W_0 = R_0 = 0$, $h, A > 0$, $Z_n = \log \mathcal{L}_n$, and $\mathcal{L}_n = f_1(X_n|\mathbf{X}_1^{n_1})/f_0(X_n|\mathbf{X}_1^{n-1})$ is the LR in the general non-iid case (assuming it does not depend on the changepoint), which simplifies to $\mathcal{L}_n = f_1(X_n)/f_0(X_n)$ in the iid case. Both procedures have certain optimality properties discussed in Chapters 7 and 8 in detail.

In network security, however, typically neither the pre- nor post-change distributions are known, since the behavior of traffic is poorly understood. Consequently, one can no longer rely on the LR $\mathcal{L}_n$, demanding an alternative approach. One way is to replace the LLR with some reasonable statistic $S_n(\mathbf{X}_1^n)$ that is sensitive to the expected change and will be referred to as a *score*. The corresponding score-based modification of the CUSUM and SR procedures are

$$W_n^{\text{sc}} = \max\{0, W_{n-1}^{\text{sc}} + S_n\}, \quad T_{\text{CS}}^{\text{sc}} = \inf\{n \ge 1 : W_n^{\text{sc}} \ge h\} \tag{11.60}$$

and

$$R_n^{\mathrm{sc}} = (1+R_{n-1}^{\mathrm{sc}})e^{S_n}, \quad T_{\mathrm{SR}}^{\mathrm{sc}} = \inf\{n \geq 1 : R_n^{\mathrm{sc}} \geq A\}, \tag{11.61}$$

where $W_0^{\mathrm{sc}} = 0 = R_0^{\mathrm{sc}}$ and $h, A > 0$ are detection thresholds which determine the FAR. Recall Figure 6.3 in Chapter 6 that illustrates a typical behavior of the detection statistics. As long as the observed process $\{X_n\}_{n\geq 1}$ is "in-control," the statistics W_n and R_n fluctuate not far from the zero barrier. But as soon as $X_{\nu+1}$, the first "out-of-control" measurement, is observed, the behavior of the statistics makes a complete 180° turn — now they start rapidly drifting up eagerly trying to hit the threshold h. This behavior is guaranteed by the fact that the score statistics have a negative drift in the normal regime, $\mathsf{E}_\infty S_n < 0$, but a positive drift in the abnormal regime, $\mathsf{E}_\nu S_n > 0$ for $\nu < n$. Note that the resulting score-based (semiparametric or nonparametric) CUSUM and SR procedures are no longer guaranteed to be optimal.

Assume that the score S_n obeys the SLLN

$$\frac{1}{n}\sum_{i=\nu+1}^{\nu+n} S_i \xrightarrow[n\to\infty]{\mathsf{P}_\nu-\text{a.s.}} Q \quad \text{for all } \nu \geq 0$$

with a positive and finite number $Q = \lim_{n\to\infty} n^{-1}\mathsf{E}_0\left[\sum_{i=1}^n S_i\right]$. If, in addition, we postulate a certain rate of convergence in the SLLN, analogously to Theorem 8.3.4 it can be shown that

$$\mathsf{SADD}(T_{\mathrm{CS}}^{\mathrm{sc}}) \sim \mathsf{STADD}(T_{\mathrm{CS}}^{\mathrm{sc}}) \sim h/Q \quad \text{as } h \to \infty, \tag{11.62}$$

and similar asymptotic approximations hold for the score-based SR procedure with h replaced by $\log A$; see also Theorem 3 in [472]. In general, however, it is impossible to approximate ARL2FA unless S_n is connected to the LLR. So it is unclear how to select the thresholds h and A to guarantee the given FAR level. In general, Monte Carlo simulations seem to be the only way.

The score function S_n can be chosen in a number of ways, each particular choice depending crucially on the expected type of change. For example, detecting a shift in the mean value and a change in the variance requires different score functions. In the applications of interest, the problem can be usually reduced to detecting changes in mean values or in variance or in both mean and variance. In [472, 473], a linear memoryless score was proposed for detecting changes in the mean. In [463, 471], this score was generalized to linear-quadratic to simultaneously handle changes in both mean and variance.

Specifically, let $\mu_\infty = \mathsf{E}_\infty X_n$, $\sigma_\infty^2 = \mathsf{var}_\infty[X_n]$ and $\mu = \mathsf{E}_0 X_n$, $\sigma^2 = \mathsf{var}_0[X_n]$ denote the pre- and post-anomaly mean values and variances, respectively. Write $Y_n = (X_n - \mu_\infty)/\sigma_\infty$ for the centered and scaled observation at time n. In the real-world applications, the pre-change parameters μ_∞ and σ_∞^2 are estimated from the training data and periodically re-estimated due to the non-stationarity of network traffic; they can therefore be assumed known. Introduce the following memoryless linear-quadratic score

$$S_n(Y_n) = a_1 Y_n + a_2 Y_n^2 - a_0, \tag{11.63}$$

where a_0, a_1, a_2 are nonnegative design numbers. In the case where the variance either does not change or changes insignificantly compared to the change in the mean, the coefficient a_2 may be set to zero. In the opposite case, where the mean changes only slightly compared to the variance, one may take $a_1 = 0$. The first linear case is typical for many cybersecurity applications such as ICMP and UDP DDoS attacks. However, in certain cases, such as the TCP SYN attacks considered in [375, 471], both the mean and the variance change significantly.

Further improvement may be achieved by using either mixtures or adaptive versions with generalized likelihood ratio-type statistics similar to that discussed in Section 8.3. Also, in certain cases an improvement can be obtained by running several CUSUM (or SR) algorithms in parallel. These multichart CUSUM and SR procedures are robust and very efficient; see Section 9.2 and [468, 469].

In certain conditions splitting packets in "bins" and considering multichannel detectors helps localize and detect attacks more rapidly [463, 472, 473]. Consider the multichannel scenario where the vector observations $X_n^1, \dots, X_n^N$, $n \geq 1$ are used to decide on the presence of anomalies. Here X_n^i is a sample obtained at time n in the i-th channel. For example, in the case of UDP flooding attacks the channels correspond to packet sizes (size bins), while for TCP SYN attacks they correspond to IP addresses (IP bins).

Similarly to the single-channel case, for $i = 1, \dots, N$, introduce the score functions $S_n^i = a_1^i Y_n^i + a_2^i (Y_n^i)^2 - a_0^i$ and the corresponding score-based CUSUM and SR statistics

$$W_n^i = \max\left\{0, W_{n-1}^i + S_n^i\right\}, \quad R_n^i = (1 + R_n^i) \exp\left\{S_n^i\right\}, \quad R_0^i = 0 = W_0^i. \tag{11.64}$$

Typically, the statistics W_n^i and $\log R_n^i$ ($i = 1, \dots, N$) remain close to zero in normal conditions; when the change occurs in the j-th channel, the j-th statistics W_n^j and $\log R_n^j$ start rapidly drifting upward; see Figure 6.3.

The "MAX" algorithm previously proposed in [472, 473] is based on the maximal statistic $W_{\max}(n) = \max_{1 \leq i \leq N} W_n^i$, which is compared to a threshold h that controls the FAR, i.e., the algorithm stops and declares the attack at

$$T_{\max}(h_N) = \inf\{n \geq 1 : W_{\max}(n) \geq h_N\}. \tag{11.65}$$

Clearly, the MAX algorithm is a particular case of the multichart CUSUM algorithm (9.1)–(9.2) studied in Section 9.2 in detail for the LLR-based scores. This method shows very high performance and is the best one can do when attacks are visible in one or very few channels. In the iid case with known models, the latter conclusion follows from the near optimality results presented in Section 9.2 (see Theorem 9.2.1).

However, the most general case is where the number of affected channels is *a priori* unknown and may vary from small to large. This challenging problem was considered in Section 9.3 where several asymptotically optimal LR-based detection procedures were suggested for known pre- and post-change models. When models are unknown similar procedures can be used with the LLRs Z_n^i replaced with the scores S_n^i, $i = 1, \dots, N$. In particular, the reasonable detection statistic is $W_n^c = \sum_{i=1}^N W_n^i$, or if the maximal percentage, p, of the affected channels is *a priori* known, then $W_n^{c,p} = \sum_{i=1}^{pN} W_n^{(i)}$, where $W_n^{(i)}$, $i = 1, \dots, N$ are ordered versions, $W_n^{(1)} \geq W_n^{(2)} \geq \cdots \geq W_n^{(N)}$. A similar approach can be used to form the SR-type multichannel detection procedure which is based on the thresholding of the statistic $\sum_{i=1}^{pN} \log R_n^{(i)}$, i.e.,

$$T_\star^{\mathrm{SRsc}} = \inf\left\{n \geq 1 : \sum_{i=1}^{pN} \log R_n^{(i)} \geq h\right\}, \quad R_n^{(1)} \geq R_n^{(2)} \geq \cdots \geq R_n^{(N)}.$$

Efficient SR and CUSUM-type unstructured multichannel procedures can also be constructed based on the statistics given in (9.19) and (9.20) where the LRs $\Lambda_n^{k-1}(i)$ are replaced with scores $\exp\{\sum_{j=k}^n S_j^i\}$. Finally, the following algorithm is worth trying

$$T_*^{\mathrm{sc}} = \inf\left\{n \geq 1 : \max_{1 \leq k \leq n} \sum_{i=1}^N \left(\sum_{j=k}^n S_j^i\right)^+ \geq h\right\}.$$

Yet another approach is to exploit a nonparametric algorithm with binary quantization and optimization of the quantization threshold. In this case, it is possible to implement optimal binary quantized CUSUM and SR algorithms that are based on true likelihood ratios for Bernoulli sequences at the output of quantizers. See [472, Section 4] for details.

11.3.2.2 Detection of DDoS Attacks

Detection of the TCP SYN DoS Attack (CAIDA). To validate usefulness of a multichannel AbIDS as opposed to the single-channel system, we used the eight-hour real backbone data captured by

CAIDA (The Cooperative Association for Internet Data Analysis, San Diego). A bidirectional link from San Jose, CA to Seattle, WA belonging to the US backbone Internet Service Provider (ISP) was monitored. This data set contains a TCP SYN flooding attack. The attack's aim is to congest the victim's link with a series of SYN requests.

We compare the single-channel CUSUM (SC-CUSUM) to the multichannel CUSUM (MC-CUSUM) algorithm in terms of the maximal ADD given by (6.19) and the FAR expressed via the average run length to false alarm $\mathsf{ARL2FA}(T) = \mathsf{E}_\infty T$.

In the case of SYN attack detection, one can observe the number of SYNs that would cause denial of service. Since we are interested in the SYN arrival rate at a destination, we divide the channels based on their destination IP addresses. There are many ways to divide the IP address space. We take the following approach to set up the multichannel detection problem. One specific group of IP addresses that belong to the same 8-bit prefix is considered. This group of IP addresses (/8) is further subdivided into 256 channels, each containing all the IP addresses that have the same first 16 bits (/16), so we have $N = 256$ channels. In each channel, we monitor the number of SYN packets sent per second for the entire 8-hour duration. In the single channel case, the time series is formed by monitoring the total number of SYNs per second for all the IP addresses that have the same 8-bit prefix; the pre-change mean value $\mu_\infty = 3$ SYNs/sec and the post-change (attack) mean $\mu = 19$ SYNs/sec with the same (approximately) variance. Thus, a linear score has been used ($a_2 = 0$ in (11.63)). In the multichannel case, the attack occurs in the channel $i = 113$ with mean values $\mu_\infty^{113} = 0.0063$ SYNs/sec and $\mu^{113} = 15.3$ SYNs/sec. It is therefore obvious that localizing the attack with the multichannel MAX algorithm (11.65) based on the thresholding of the statistic $W_{\max}(n) = \max_{1 \le i \le 256} W_n^{(i)}$ enhances the detection capability.

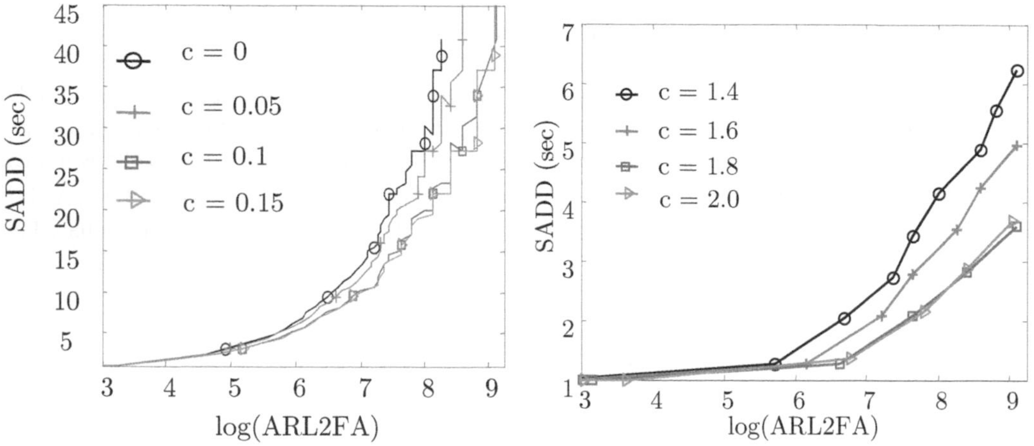

(a) The SC-CUSUM algorithm

(b) The MAX MC-CUSUM algorithm

Figure 11.8: *SADD versus* $\log \mathsf{ARL2FA}$ *for the SC-CUSUM and MAX MC-CUSUM algorithms.*

Figures 11.8(a) and 11.8(b) illustrate the relation between the SADD and $\log(\mathsf{ARL2FA})$ for the SC-CUSUM and MAX MC-CUSUM detection algorithms, respectively. The optimal value of the design parameter $a_0 = c$ is $c_{\text{opt}} = 0.1$ for the single channel case and $c_{\text{opt}} = 1.8$ for the multichannel case. In the extreme right of the plot, we achieve the ARL2FA of 2.25 hours ($\log(\mathsf{ARL2FA}) = 9$). For this FAR, the SADD for the MAX MC-CUSUM is 3.5 sec, while the SADD for the SC-CUSUM is 45 sec, about thirteen times higher. Since the average delay to detection dramatically increases as the FAR decreases, for the lower FAR the SC-CUSUM algorithm may be unable to detect short attacks.

We therefore conclude that in certain scenarios the use of multichannel intrusion detection systems may be very important.

Detection of the TCP SYN DDoS Attack (LANDER Project). Next, we present the results of testing the single-channel (score-based) CUSUM and SR detection algorithms with respect to the stationary average delay to detection (STADD) defined in (6.21) and illustrated in Figure 6.4, i.e., when the attack occurs long after surveillance begins and is preceded by multiple false alarms. This testing is performed for a real data set containing a DDoS SYN flood attack. The data is courtesy of the Los Angeles Network Data Exchange and Repository (LANDER) project (see `http://www.isi.edu/ant/lander`). This is a research-oriented traffic capture, storage, and analysis infrastructure that operates under Los Nettos, a regional ISP in the Los Angeles area. The aggregate traffic volume through Los Nettos measures over 1 Gigabit/s each way, and the ISP's backbone is 10 Gigabit. Leveraged by a Global Positioning System (GPS) clock, LANDER's capture equipment is able to collect data at line speed and with a down-to-the-nanosecond time precision.

Specifically, the trace is flow data captured by Merit Network Inc. (see `http://www.merit.edu`) and the attack is on a University of Michigan IRC server. It starts at approximately 550 seconds into the trace and lasts for 10 minutes. Figure 11.9 shows the number of attempted connections (the connections birth rate) as a function of time. While the attack can be seen with the naked eye, it is not completely clear when it starts. In fact, there is fluctuation (a spike) in the data before the attack.

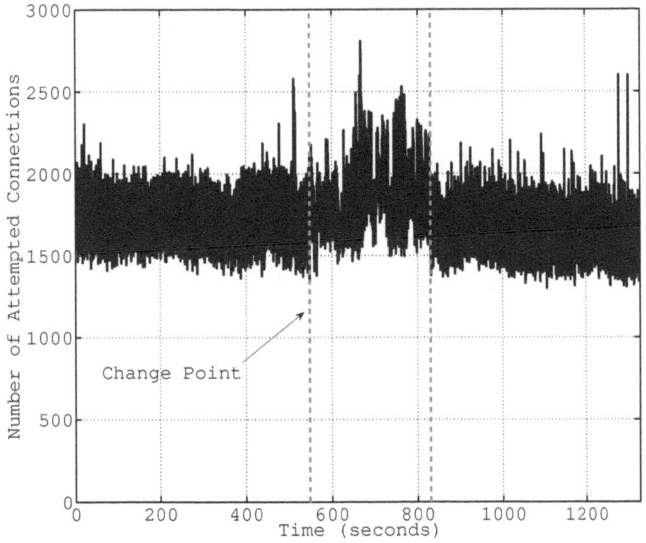

Figure 11.9: *The connections birth rate for LANDER data.*

The observations $\{X_n\}_{n\geq 1}$ represent the number of connections during 20 msec batches. The estimated values of the connections birth rate mean and standard deviation for legitimate and attack traffic are: $\mu_\infty \approx 1669, \sigma_\infty \approx 114$ and $\mu \approx 1888, \sigma \approx 218$ (connections per 20 msec). Therefore, this attack leads to a considerable increase in both the mean and the standard deviation of the connections birth rate.

Statistical analysis of this data set shows that the distribution of the number of attempted connections for legitimate traffic is very close to Gaussian, but for attack traffic it is not; see Tartakovsky *et al.* [471] for details. We implement the score-based multicyclic SR and CUSUM procedures with the linear-quadratic memoryless score (11.63). When choosing the design parameters a_0, a_1, a_2 we assume the Gaussian pre-attack model. We set the detection thresholds $A = 1900$ and $h = 6.68$ so as to ensure approximately the same level of ARL2FA at approximately 500 samples (10 sec) for both procedures.

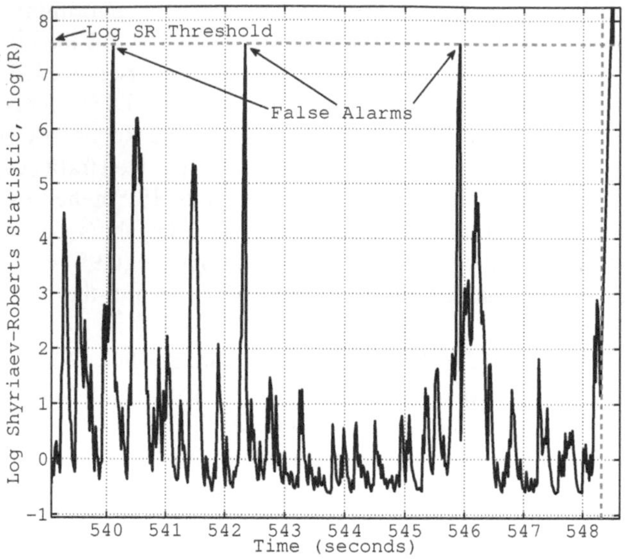

Figure 11.10: *Long run of the SR procedure (logarithm of the SR statistic versus time) for SYN flood attack.*

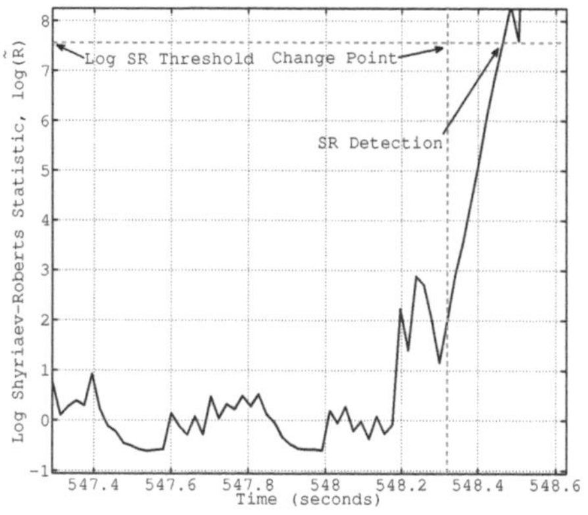

(a) The multicyclic SR procedure

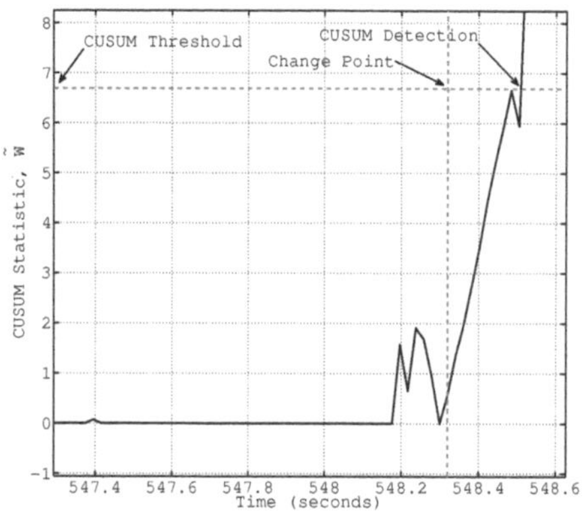

(b) The multicyclic CUSUM procedure

Figure 11.11: *Detection of the SYN flood attack by the multicyclic SR and CUSUM procedures.*

The results are illustrated in Figures 11.10 and 11.11. Figure 11.10 shows a relatively long run of the SR statistic with several false alarms and then the true detection of the attack with a very small detection delay (at the expense of raising many false alarms prior to the correct detection). Recall that the idea of minimizing the STADD is to set the detection thresholds low enough in order to detect attacks very quickly, which unavoidably leads to multiple false alarms prior to the attack starts. These false alarms should be filtered by a specially designed algorithm, as has been suggested by Pollak and Tartakovsky [370] and will be further discussed in Subsection 11.3.3. Figure 11.11(a) shows the behavior of $\log R_n^{\rm sc}$ shortly prior to the attack and right after the attack starts until detection. Figure 11.11(b) shows the CUSUM score-based statistic $W_n^{\rm sc}$. Both procedures successfully detect the attack with very small delays and raise about 7 false alarms per 1000 samples.

The detection delay is approximately 0.14 seconds (7 samples) for the repeated SR procedure, and about 0.21 seconds (10 samples) for the CUSUM procedure. As expected, the SR procedure is slightly better.

Detection of the ICMP DDoS Attack (LANDER Project). Yet another test data, used for studying the behavior of the SR and CUSUM algorithms, contains the Internet Control Message Protocol (ICMP) reflector attack. This data set is again courtesy of the LANDER project. The essence of this DDoS attack is to congest the victim's link with echo reply (ping) requests sent by a large number of separate compromised machines (reflectors) so as to have the victim's machine exhaust all of its resources. The attack starts at roughly 102 seconds into the trace and lasts for about 240 seconds. The data trace (packet rate) is shown in Figure 11.12.

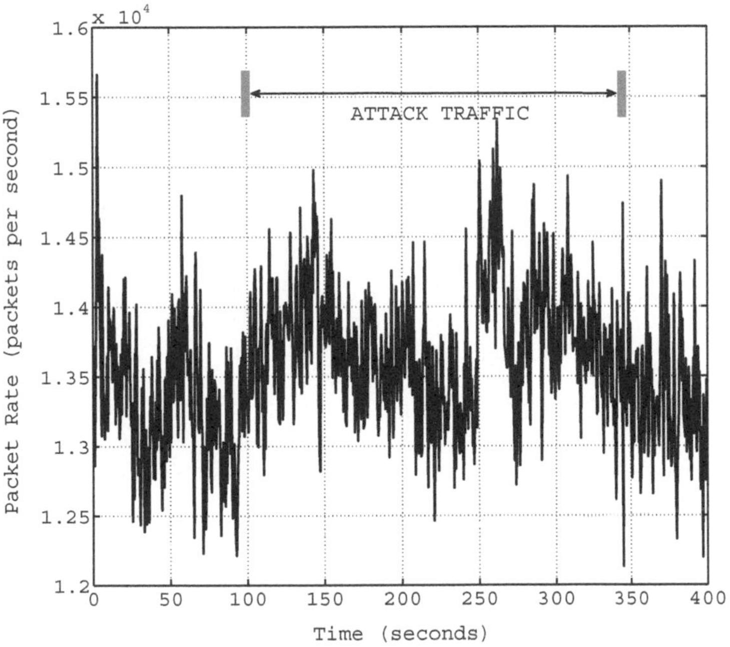

Figure 11.12: *Packet rate for the ICMP reflector attack.*

A detailed statistical analysis proves that distributions of the packet rate are Gaussian for both legitimate and attack traffic [375]. Furthermore, the model is not just Gaussian but a special case where the ratios of pre-change and post-change mean-to-variance values are the same, $\mu_\infty/\sigma_\infty^2 = \mu/\sigma^2 = a$, $a > 0$. In other words, the data in the trace behave according to the $\mathcal{N}(\mu_\infty, a\mu_\infty)$-to-$\mathcal{N}(\mu, a\mu)$ model, and accurate estimation shows that $\mu_\infty \approx 13330$, $\mu \approx 13600$ and $a \approx 20$. Thus, we can apply the LR-based CUSUM and SR procedures (11.59) with the LR given by

$$\mathcal{L}_n = \sqrt{\frac{\mu_\infty}{\mu}} \exp\left\{-\frac{\mu - \mu_\infty}{2a}\right\} \exp\left\{\frac{\mu - \mu_\infty}{2a\mu\mu_\infty} X_n^2\right\}.$$

To make sure that the ARL2FA for each procedure is at the same desired level $\gamma > 1$, we recall that for sufficiently large γ,

$$\mathsf{ARL2FA}(T_{\text{CS}}) \approx e^h/(I\zeta^2) - h/I_0 - 1/(I\zeta), \quad \mathsf{ARL2FA}(T_{\text{SR}}) \approx A/\zeta$$

(see (8.144) and (8.356)). In our case, $I \approx 0.137$, $I_0 \approx 0.134$ and $\zeta \approx 0.731$. If, for example, $\gamma = 1000$, the threshold for the CUSUM procedure should be set to $h = \log 76.32 \approx 4.34$, and for the SR procedure to 731.3. We confirmed both thresholds numerically using the numerical methodology of Subsection 8.4.3: the actual ARL2FA for the CUSUM procedure is 998.4 and that for the SR procedure is 1000.1. The result of employing the detection procedures is shown in Figure 11.13. Figure 11.13(a) shows the behavior of the SR detection statistic for the first 120 seconds of the trace (recall that the attack starts 102 seconds into the trace), and Figure 11.13(b) shows the same for the CUSUM detection statistic $\widetilde{W}_n = e^{W_n}$ to be on the same scale as the SR statistic. Both procedures successfully detect the attack, though at the expense of raising three false alarms. The detection delay for the SR procedure is approximately 15 seconds (or 30 samples), and for the CUSUM procedure it is about 18 seconds (or 36 samples). Thus, the SR procedure is slightly better. The difference is small simply because the trace is too short: if the attack took place at a point farther into the trace, and we could repeat the SR procedure a few more times, the detection delay would be much smaller than that of the CUSUM procedure.

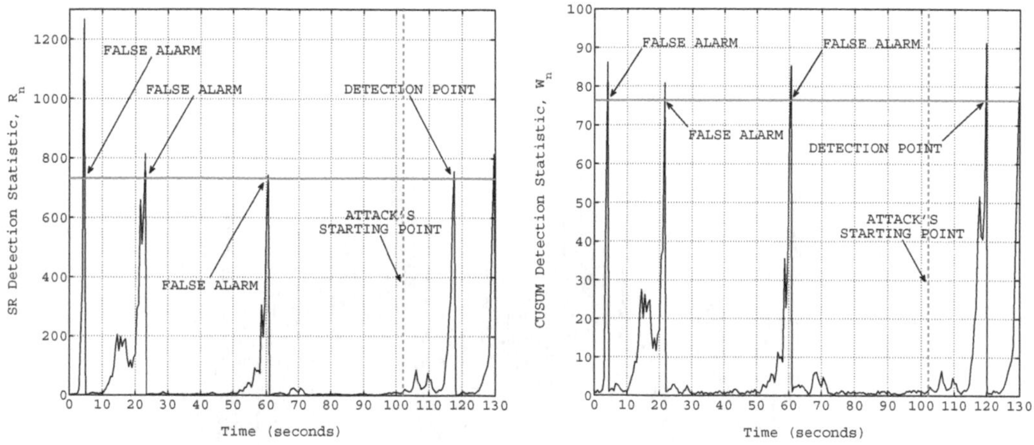

(a) The SR procedure

(b) The CUSUM procedure

Figure 11.13: *Results of detection of the ICMP reflector attack.*

11.3.2.3 Rapid Spam Detection

Most organizations run spam filters at their local networks, such as Bayesian filters or a block list. These filters examine the content of each message as well as the IP address. If the message matches known spam signatures, it is marked as spam. These techniques work well but have high operational costs.

Another approach to dealing with spam is to monitor traffic at the network level looking for spam behavior. Detecting spammers at the network level has several advantages such as no privacy issues related to message content examination, near real-time detection based on network behavior, and minimizing collateral damage because dynamic addresses released by spammers can be removed from block lists quickly.

Prior work has shown that there is a number of useful features for detecting spammers such as the autonomous system the IP address belongs to, message size, blocked connections, and message length. All these features can be determined from network traffic and used in conjunction with changepoint detection to detect when traffic patterns from a particular host match known spammer patterns.

We now illustrate rapid spam detection for a particular real-world data set, obtained from a regional ISP, to prove the feasibility of change detection methods. The trace contains email flows to a mail server from a number of hosts. The records are sorted by the source IP address. Our objective is to isolate suspicious hosts and extract the typical behavioral pattern. Examining how the email size changes with time shows that it is very stable with some occasional bursts. The individual producing such bursts is very likely to be a spammer. Figure 11.14 shows the detection of a real-world spammer using the AbIDS. The email (SMTP) traffic generated by a certain host is under surveillance. Ordinarily, SMTP traffic produced by a user sending legitimate messages is characterized by a relatively steady intensity, i.e., the number of messages sent per unit time remains more or less constant, with no major bursts or drops. However, the behavior changes abruptly once a spam attack begins: the number of sent messages explodes, possibly for a very short period of time. The topmost plot in Figure 11.14 illustrates this typical behavior. The spike in the traffic intensity that appears in the far right of the plot is easily detected by our score-based changepoint detection algorithms. The behavior of the linear score-based CUSUM and SR statistics ($a_2 = 0$ in (11.63)) is plotted in the middle and bottom pictures, respectively. Both statistics behave similarly — an alarm is raised as soon as the traffic intensity blunder caused by the spam attack is encountered. The spammer is detected immediately after he/she starts activity. The difference is mainly prior to the attack — there are two false detections produced by CUSUM, while none by SR.

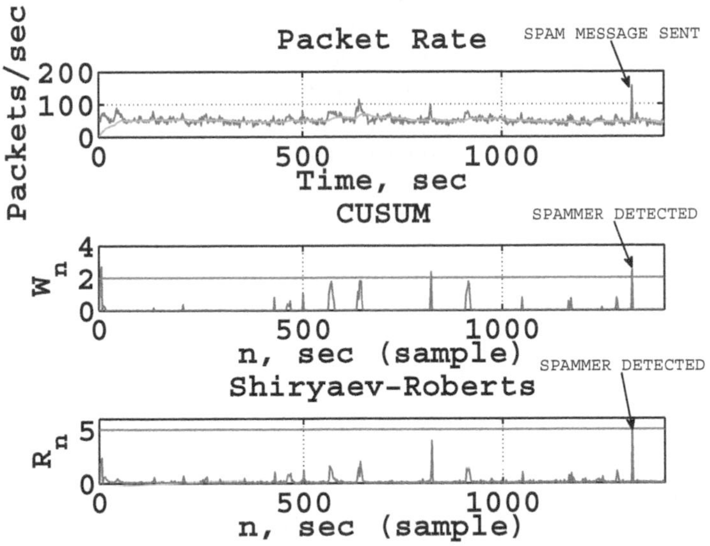

Figure 11.14: *Spam detection. Raw data (top); CUSUM statistic (middle); SR statistic (bottom).*

11.3.3 Hybrid Anomaly–Signature Intrusion Detection System

11.3.3.1 IDS Structure

Since in real life legitimate traffic dominates, comparing various AbIDSs using the multicyclic approach and the stationary average detection delay $\mathsf{STADD}(T)$ is the most appropriate method for cybersecurity applications. However, even an optimal changepoint detection method is subject to large detection delays if the FAR is maintained at a low level. Hence, as already mentioned, employing one such scheme alone will lead to multiple false detections, or if detection thresholds are increased to lower the FAR, the delays will be too large, and attacks may proceed undetected.

In this subsection, we show how changepoint detection techniques can be combined with other methods that offer very low FAR but are too time-consuming to use at on-line speeds. This novel approach leads to a two-stage hybrid anomaly–signature IDS (HASIDS) with profiling, false alarm filtering, and true attack confirmation capabilities. Specifically, consider complementing a changepoint detection-based AbIDS with a flow-based signature IDS that examines the traffic's spectral profile and reacts to changes in spectral characteristics of the data. The main idea is to integrate anomaly- and spectral-signature-based detection methods so that the resulting HASIDS overcomes the shortcomings of current IDSs. We propose using "flow-based" signatures in conjunction with anomaly-based detection algorithms. In particular, Fourier and wavelet spectral signatures and related *spectral analysis techniques* can be exploited, as shown in [187, 199, 200]. This approach is drastically different from traditional signature-based systems because it depends not on packet content but on communication patterns alone.

At the first stage, we use either CUSUM or SR multicyclic (repeated) changepoint detection algorithm to detect traffic anomalies. Recall that in network security applications it is of utmost importance to detect attacks that may occur in a distant future very rapidly (using a repeated application of the same anomaly-based detection algorithm), in which case the true detection of a real change is preceded by a long interval with frequent false alarms that should be filtered by a separate algorithm. This latter algorithm is based on spectral signatures, so at the second stage we exploit a spectral-based IDS that filters false detections and confirms true attacks.

In other words, the methodology is based on using the changepoint detection method for preliminary detection of attacks with low threshold values and a discrete Fourier (or wavelet) transform to reveal periodic patterns in network traffic to confirm the presence of attacks and reject false detections produced by the anomaly IDS. When detection thresholds are low, the AbIDS produces an intense flow of false alarms. However, these false alarms can be tolerated at the level of minutes or even seconds, since they do not lead to real false alarms in the whole system. An alarm in the AbIDS triggers a spectral analyzer. This alarm will either be rejected or confirmed, in which case a final alarm will be raised. Schematically, the system is shown in Figure 11.15.

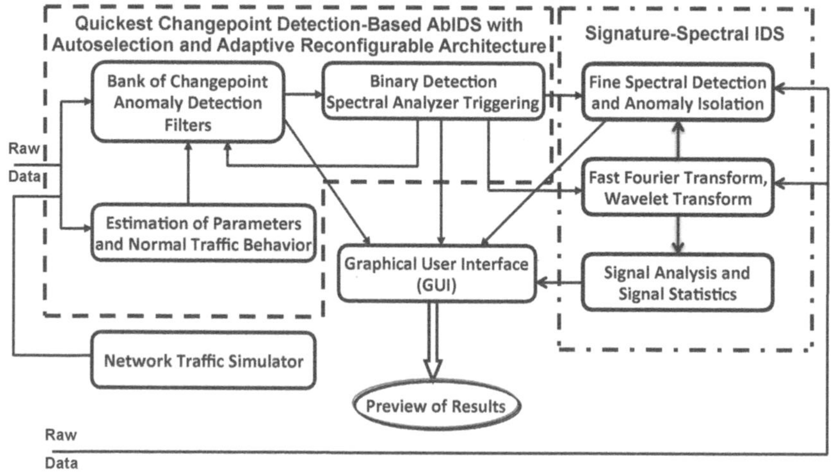

Figure 11.15: *Block diagram of the hybrid anomaly–signature intrusion detection system.*

To summarize, the HASIDS combines the changepoint detection-based AbIDS and the signature (spectral) IDS in one unit. The AbIDS allows for detecting attacks quickly due to low threshold setting, which leads to frequent false alarms that are filtered by a separate algorithm. This separate algorithm is based on the spectral IDS which rejects false detections and confirms true attacks. Further details may be found in Tartakovsky [463, Section 2.4]. This hybrid approach allows us not only to detect attacks with small delays and a low FAR but also to isolate/localize anomalies precisely, e.g., low-rate pulsing attacks. See Figure 11.17 in the next subsection for further details.

11.3.3.2 Experimental Study

We now present sample testing results that illustrate the efficiency of the HASIDS for LANDER data sets.

Detection of the ICMP DoS Flood Attack (LANDER). The first data set is a tcpdump trace file containing a fragment of real-world malicious network activity, identified as an ICMP attack. The trace was captured on one of the Los Nettos private networks (a regional ISP in Los Angeles).

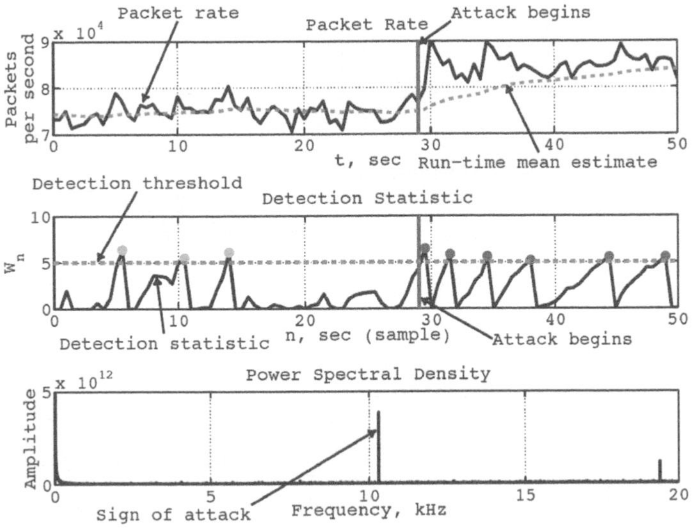

Figure 11.16: *Detection of the ICMP DDoS attack with HASIDS.*

Figure 11.16 demonstrates the attack detection. It shows raw data (top), the behavior of the multicyclic CUSUM statistic (middle), and the power spectral density of the data (bottom). The hybrid IDS filtered all false alarms (shown by circles prior to the attack) and detected the attack very rapidly after its occurrence. Note that the spectral analyzer is triggered only when a threshold exceedance occurs. None of the false alarms passed the hybrid system since the peak in spectrum appeared only after the attack began. This allowed us to set a very low threshold in the anomaly IDS, resulting in a very small delay to detection of the attack.

Detection of the UDP DoS Flood Attack (LANDER). The next experiment demonstrates the supremacy of HASIDS over AbIDS by applying both systems to detect and isolate a real-world "double-strike" UDP DDoS attack. The attack is composed of two consecutive "pulses" in the traffic intensity, shown in Figure 11.17(a). Each pulse is a sequence of UDP packets sent at a rate of about 180 Kbps, which is approximately thrice the intensity of the background traffic (about 53 Kbps). One would think that because the attack is so strong, any IDS will be able to detect it quickly. However, the challenge is that each pulse is rather short, the gap between the two pulses is very short, and the source of the attack packets for each pulse is different. Therefore, if the detection speed is comparable to the short duration of the pulses, the attack will get through undetected. Furthermore, if the renewal time (the interval between the most recent detection and the time the IDS is ready for a new detection) is longer than the distance between the pulses, then even though the first pulse may be detected, the second one is likely to be missed. Hence, this scenario is challenging and illustrates the efficiency of the proposed HASIDS not only for detecting attacks quickly but also for isolating closely located anomalies.

Figure 11.17 illustrates the detection by the AbIDS (Figure 11.17(b)) and by the HASIDS (Figure 11.17(c)). Figure 11.17(b) shows that the AbIDS detects the first pulse but misses the second one because the threshold in the AbIDS is high to guarantee the given low FAR. With the selected threshold, this trace shows no false alarms. If the threshold is lowered, as it is in Figure 11.17(c), both segments of the attack are perfectly detected and localized. However, this brings many false alarms at the output of the changepoint detection based AbIDS. What can be done? The hybrid IDS offers an answer. The FFT (fast Fourier transform) spectral module is triggered by the AbIDS when

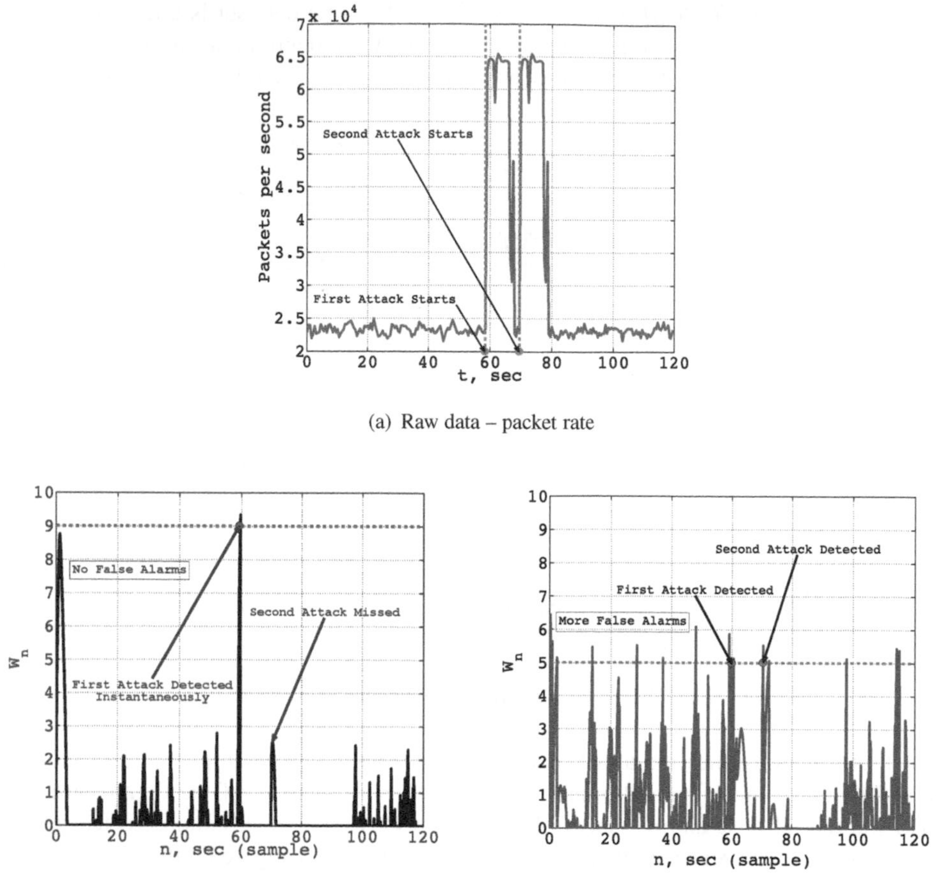

(a) Raw data – packet rate

(b) Detection by AbIDS

(c) Detection by HASIDS

Figure 11.17: *Detection of a short UDP DoS attack with AbIDS and HASIDS. The second "pulse" is missed by the AbIDS but not by the HASIDS.*

detections (false or true) occur. In this particular experiment, all false alarms were filtered by the FFT spectral analyzer. This allowed us to lower the detection threshold in the AbIDS and, as a result, detect both pulses with a very small detection delay and no false alarms, since in the HASIDS false alarms were filtered by the spectral module. See Figure 11.17(c).

11.3.3.3 Conclusion

The experimental study shows that the proposed AbIDS, which exploits score-based CUSUM and SR changepoint detection methods, is robust and efficient for detecting a multitude of computer intrusions such as UDP, ICMP, and TCP SYN DDoS attacks as well as spammers. More importantly, devising the hybrid anomaly–signature IDS that fuses quickest change detection techniques with spectral signal processing methods solves both aspects of the intrusion detection problem. It achieves unprecedented speeds of detection and simultaneously has a very low FAR in ultra-high-speed networks. In addition to achieving high performance in terms of the tradeoff between delays to detection, correct detections, and false alarms, the hybrid system allows one to estimate anomaly length and distinguish between anomalies, i.e., efficient isolation-localization of anomalies.

Bibliography

[1] M. Abramowitz and I. A. Stegun. *Handbook of Mathematical Functions (5th ed.)*. Dover Publications, Inc., New York, USA, 1964.

[2] E. Ahmed. *Monitoring and Analysis of Internet Traffic Targeting Unused Address Spaces*. Ph.D. Thesis, Information Security Institute, Queensland University of Technology, Brisbane, AU, May 2010. Online.

[3] E. Ahmed, A. Clark, and G. Mohay. Effective change detection in large repositories of unsolicited traffic. In S. Georgescu, S. Heikkinen, and M. Popescu, editors, *Proceedings of the 4th International Conference on Internet Monitoring and Protection (ICIMP'09), Venice/Mestre, IT*, pages 1–6, May 2009.

[4] H. Akaïke. A new look at the statistical model identification. *IEEE Transactions on Automatic Control*, 19(6):716–723, Dec. 1974.

[5] H. Akaïke. Stochastic theory of minimal realization. *IEEE Transactions on Automatic Control*, 19(6):667–674, Dec. 1974.

[6] H. Akaïke. Markovian representation of stochastic processes by canonical variables. *SIAM Journal of Control*, 13(1):162–173, 1975.

[7] E. Al Solami, C. Boyd, A. Clark, and A. K. Islam. Continuous biometric authentication: Can it be more practical? In *Proceedings of the 12th IEEE International Conference on High Performance Computing and Communications, Melbourne, AU*, pages 647–652, Los Alamitos, CA, USA, Sept. 2010. IEEE Computer Society.

[8] G. E. Albert. Accurate sequential tests on the mean of an exponential distribution. *Annals of Mathematical Statistics*, 27(2):460–470, June 1956.

[9] T. W. Anderson. A modification of the sequential probability ratio test to reduce the sample size. *Annals of Mathematical Statistics*, 31(1):165–197, Jan. 1960.

[10] R. André-Obrecht. A new statistical approach for the automatic segmentation of continuous speech signals. *IEEE Transactions on Acoustics, Speech, and Signal Processing*, 36(1):29–40, Jan. 1988.

[11] U. Appel and A. von Brandt. Adaptive sequential segmentation of piecewise stationary time series. *Information Sciences*, 29(1):27–56, Feb. 1983.

[12] P. Armitage. Sequential analysis with more than two alternative hypotheses, and its relation to discriminant function analysis. *Journal of the Royal Statistical Society - Series B Methodology*, 12(1):137–144, 1950.

[13] P. Armitage, C. K. McPherson, and B. C. Rowe. Repeated significance tests on accumulating data. *Journal of the Royal Statistical Society - Series A General*, 132(2):235–244, 1969.

[14] L. A. Aroian and H. Levene. The effectiveness of quality control procedures. *Journal of the American Statistical Association*, 45(252):520–529, Dec. 1950.

[15] K. J. Arrow, D. Blackwell, and M. A. Girshick. Bayes and minimax solutions of sequential decision problems. *Econometrica*, 17(3–4):213–244, July 1949.

[16] D. Assaf, M. Pollak, Y. Ritov, and B. Yakir. Detecting a change of a normal mean by dynamic sampling with a probability bound on a false alarm. *Annals of Statistics*, 21(3):1155–1165,

Sept. 1993.

[17] K. E. Atkinson. *A Survey of Numerical Methods for the Solution of Fredholm Integral Equations of the Second Kind.* SIAM, Philadelphia, PA, USA, 1976.

[18] K. E. Atkinson and W. Han. *Theoretical Numerical Analysis: A Functional Analysis Framework (3rd ed.)*, volume 39 of *Texts in Applied Mathematics.* Springer-Verlag, 2009.

[19] F. Baggiani and S. Marsili-Libelli. Real-time fault detection and isolation in biological wastewater treatment plants. *Water Science & Technology*, 60(11):2949–2961, 2009.

[20] R. R. Bahadur. Sufficiency and statistical decision functions. *Annals of Mathematical Statistics*, 25(3):423–462, 1954.

[21] B. Bakhache and I. Nikiforov. Reliable detection of faults in measurement systems. *International Journal of Adaptive Control Signal Processing*, 14(7):683–700, Nov. 2000.

[22] P. A. Bakut, I. A. Bolshakov, B. M. Gerasimov, A. A. Kuriksha, V. G. Repin, G. P. Tartakovsky, and V. V. Shirokov. *Statistical Radar Theory*, volume 1 (G. P. Tartakovsky, Editor). Sovetskoe Radio, Moscow, USSR, 1963. In Russian.

[23] D. Balageas, C.-P. Fritzen, and A. Güemes, editors. *Structural Health Monitoring*. Lavoisier, Paris, F., 2006.

[24] E. Balmès, M. Basseville, L. Mevel, and H. Nasser. Handling the temperature effect in vibration-based monitoring of civil structures: A combined subspace-based and nuisance rejection approach. *Control Engineering Practice*, 17(1):80–87, Jan. 2009.

[25] E. Balmès, M. Basseville, L. Mevel, H. Nasser, and W. Zhou. Statistical model-based damage localization: A combined subspace-based and substructuring approach. *Structural Control and Health Monitoring*, 15(6):857–875, Oct. 2008.

[26] R. K. Bansal and P. Papantoni-Kazakos. An algorithm for detecting a change in a stochastic process. *IEEE Transactions on Information Theory*, 32(2):227–235, Mar. 1986.

[27] Y. Bar-Shalom and X. R. Li. *Estimation and Tracking: Principles, Techniques and Software.* Artech House Radar Library. Artech House, Boston-London, 1993.

[28] M. Baron and A. G. Tartakovsky. Asymptotic optimality of change-point detection schemes in general continuous-time models. *Sequential Analysis*, 25(3):257–296, Oct. 2006. Invited Paper in Memory of Milton Sobel.

[29] E. D. Barraclough and E. S. Page. Tables for Wald tests for the mean of a normal distribution. *Biometrika*, 46(1–2):169–177, June 1959.

[30] J. Bartroff and T. L. Lai. Multistage tests of multiple hypotheses. *Communications in Statistics - Theory and Methods*, 39(8–9):1597–1607, 2010. Special Issue: Recent Advances in Statistical Inference – In Honor of Professor Masafumi Akahira.

[31] I. V. Basawa. Neyman–Le Cam tests based on estimating functions. In L. Le Cam and R. A. Olsen, editors, *Proceedings of the Berkeley Conference in Honor of Jerzy Neyman and Jack Kiefer*, volume 2, pages 811–825. University of California Press, Berkeley, CA, USA, 1985.

[32] I. V. Basawa. Generalized score tests for composite hypotheses. In V. V. Godambe, editor, *Estimating Functions*, pages 121–132. Clarendon Press, Oxford, 1991.

[33] I. V. Basawa, V. P. Godambe, and R. L. Taylor, editors. *Selected Proceedings of the Symposium on Estimating Functions held at the University of Georgia, Athens, GA, March 21–23, 1996*, volume 32 of *Lecture Notes - Monograph Series.* Institute of Mathematical Statistics, Hayward, CA, USA, 1997.

[34] A. E. Basharinov and B. S. Fleishman. *Methods of the Statistical Sequential Analysis and Their Radiotechnical Applications.* Soviet Radio Publishing House, Moscow, RU, 1962. In Russian.

[35] M. Basseville. Detecting changes in signals and systems - A survey. *Automatica*, 24(3):309–

326, May 1988.

[36] M. Basseville. On-board component fault detection and isolation using the statistical local approach. *Automatica*, 34(11):1391–1416, Nov. 1998.

[37] M. Basseville. An invariance property of some subspace-based detection algorithms. *IEEE Transactions on Signal Processing*, 47(12):3398–3401, Dec. 1999.

[38] M. Basseville, M. Abdelghani, and A. Benveniste. Subspace-based fault detection algorithms for vibration monitoring. *Automatica*, 36(1):101–109, Jan. 2000.

[39] M. Basseville and A. Benveniste. Design and comparative study of some sequential jump detection algorithms for digital signals. *IEEE Transactions on Acoustics, Speech, and Signal Processing*, 31(3):521–535, June 1983.

[40] M. Basseville, A. Benveniste, M. Goursat, L. Hermans, L. Mevel, and H. Van der Auweraer. Output-only subspace-based structural identification: From theory to industrial testing practice. *ASME Journal of Dynamic Systems, Measurement, and Control*, 123(4 - Special issue on Identification of Mechanical Systems):668–676, Dec. 2001.

[41] M. Basseville, A. Benveniste, M. Goursat, and L. Mevel. In-flight monitoring of aeronautic structures: Vibration-based on-line automated identification versus detection. *IEEE Control Systems Magazine*, 27(5):27–42, Oct. 2007.

[42] M. Basseville, A. Benveniste, G. V. Moustakides, and A. Rougée. Detection and diagnosis of changes in the eigenstructure of nonstationary multivariable systems. *Automatica*, 23(4):479–489, July 1987.

[43] M. Basseville, F. Bourquin, L. Mevel, H. Nasser, and F. Treyssède. Handling the temperature effect in vibration monitoring: Two subspace-based analytical approaches. *ASCE Journal of Engineering Mechanics*, 136(3):367–378, Mar. 2010.

[44] M. Basseville, M. Goursat, and L. Mevel. Multiple CUSUM tests for flutter monitoring. In *Proceedings of the 14th Symposium on System Identification (SYSID'06), Newcastle, AU*. IFAC/IFORS, Mar. 2006.

[45] M. Basseville, L. Mevel, and M. Goursat. Statistical model-based damage detection and localization: Subspace-based residuals and damage-to-noise sensitivity ratios. *Journal of Sound and Vibration*, 275(3–5):769–794, Aug. 2004.

[46] M. Basseville, L. Mevel, and R. Zouari. Variations on CUSUM tests for flutter monitoring. In *Proceedings of the 2nd International Workshop in Sequential Methodologies (IWSM), Troyes, FR*, June 2009.

[47] M. Basseville and I. V. Nikiforov. *Detection of Abrupt Changes - Theory and Application*. Information and System Sciences Series. Prentice-Hall, Inc., Englewood Cliffs, NJ, USA, 1993. Online.

[48] M. Basseville and I. V. Nikiforov. Fault isolation for diagnosis: Nuisance rejection and multiple hypotheses testing. *Annual Reviews in Control*, 26(2):189–202, Dec. 2002.

[49] M. Basseville and I. V. Nikiforov. Handling nuisance parameters in systems monitoring. In *Proceedings of the 44th IEEE Conference on Decision and Control and European Control Conference (CDC-ECC'05), Seville, SP*, pages 3832–3837. IEEE, Dec. 2005.

[50] C. W. Baum and V. V. Veeravalli. A sequential procedure for multihypothesis testing. *IEEE Transactions on Information Theory*, 40(6):1994–2007, Nov. 1994.

[51] L. E. Baum and M. Katz. Convergence rates in the law of large numbers. *Transactions of the American Mathematical Society*, 120(1):108–123, Oct. 1965.

[52] E. J. Baurdoux and A. E. Kyprianou. The Shepp–Shiryaev stochastic game driven by spectrally negative Levy process. *Theory of Probability and its Applications*, 53(3):588–609, 2009.

[53] P. Baxendale. T. E. Harris's contributions to recurrent Markov processes and stochastic flows. *Annals of Probability*, 39(2):417–428, Mar. 2011.

[54] B. Baygün and A. O. Hero. Optimal simultaneous detection and estimation under a false alarm constraint. *IEEE Transactions on Information Theory*, 41(3):688–703, May 1995.

[55] R. Bechhofer. A note on the limiting relative efficiency of the Wald sequential probability ratio test. *Journal of the American Statistical Association*, 55(292):660–663, Dec. 1960.

[56] R. E. Bechhofer, J. Kiefer, and M. Sobel. *Sequential Identification and Ranking Procedures (with Special Reference to Koopman–Darmois Populations)*, volume 3 of *Statistical Research Monographs*. The University of Chicago Press, Chicago, IL, USA, 1968.

[57] M. Becker, M. Karpytchev, M. Davy, and K. Doekes. Impact of a shift in mean on the sea level rise: Application to the tide gauges in the Southern Netherlands. *Continental Shelf Research*, 29(4):741–749, Mar. 2009.

[58] N. Bedjaoui and E. Weyer. Algorithms for leak detection, estimation, isolation and localization in open water channels. *Control Engineering Practice*, 19(6):564–573, June 2011.

[59] J. P. Bello, L. Daudet, S. Abdallah, C. Duxbury, M. Davies, and M. B. Sandler. A tutorial on onset detection in music signals. *IEEE Transactions on Speech and Audio Processing*, 13(5):1035–1047, Sept. 2005.

[60] A. Benveniste, M. Basseville, and G. Moustakides. The asymptotic local approach to change detection and model validation. *IEEE Transactions on Automatic Control*, 32(7):583–592, July 1987.

[61] A. Benveniste and J.-J. Fuchs. Single sample modal identification of a non-stationary stochastic process. *IEEE Transactions on Automatic Control*, 30(1):66–74, Jan. 1985.

[62] A. Benveniste, M. Métivier, and P. Priouret. *Adaptive Algorithms and Stochastic Approximations*, volume 22 of *Applications of Mathematics*. Springer-Verlag, New York, USA, 1990.

[63] A. Benveniste and L. Mevel. Nonstationary consistency of subspace methods. *IEEE Transactions on Automatic Control*, 52(6):974–984, June 2007.

[64] R. H. Berk. Locally most powerful sequential test. *Annals of Statistics*, 3(2):373–381, Mar. 1975.

[65] S. Bersimis, S. Psarakis, and J. Panaretos. Multivariate statistical process control charts: An overview. *Quality and Reliability Engineering International*, 23(5):517–543, Aug. 2007.

[66] R. N. Bhattacharya and R. R. Rao. *Normal Approximations and Asymptotic Expansions*, volume 64 of *Classics in Applied Mathematics*. SIAM, Philadelphia, PA, USA, 2010.

[67] A. Bissell. CUSUM techniques for quality control. *Journal of the Royal Statistical Society - Series C Applied Statistics*, 18(1):1–30, 1969.

[68] S. S. Blackman. *Multiple-Target Tracking with Radar Applications*. Artech House Radar Library. Artech House, Dedham, UK, 1986.

[69] S. S. Blackman, R. J. Dempster, and T. J. Broida. Multiple hypothesis track confirmation for infrared surveillance systems. *IEEE Transactions on Aerospace and Electronic Systems*, 29(3):810–823, July 1993.

[70] D. Blackwell and M. A. Girshick. *Theory of Games and Statistical Decisions*. Wiley Publications in Statistics. John Wiley & Sons, Inc., New York, USA, 1954.

[71] R. E. Blahut. *Principles and Practice of Information Theory*. Series in Electrical and Computer Engineering. Addison Wesley Publ. Co., 1987.

[72] M. Blanke, M. Kinnaert, J. Lunze, J. Schröder, and M. Staroswiecki. *Diagnosis and Fault-Tolerant Control (2nd ed.)*. Springer-Verlag, Berlin, DE, 2006.

[73] G. Bodenstein and H. M. Praetorius. Feature extraction from the encephalogram by adaptive segmentation. *Proceedings of the IEEE*, 65(5):642–652, May 1977.

[74] S. Bohacek, K. Shah, G. R. Arce, and M. Davis. Signal processing challenges in active queue management. *IEEE Signal Processing Magazine*, 21(5):69–79, Sept. 2004.

[75] W. Bohm and P. Hackl. Improved bounds for the average run length of control charts based on finite weighted sums. *Annals of Statistics*, 18(4):1895–1899, Dec. 1990.

[76] C. Boller, F.-K. Chang, and Y. Fujino, editors. *Encyclopedia of Structural Health Monitoring*. John Wiley & Sons, Inc., 2009.

[77] E. Borel. La théorie du jeu et les équations intégrales à noyau symétrique. *Comptes Rendus Hebdomadaires des Séances de l'Académie des Sciences*, 173:1304–1308, Dec. 1921. In French.

[78] L. I. Borodkin and V. V. Mottl'. Algorithm for finding the jump times of random process equation parameters. *Automation and Remote Control*, 39(6):23–32, 1976.

[79] A. A. Borovkov. *Mathematical Statistics*. Gordon and Breach Sciences Publishers, Amsterdam, NL, 1998.

[80] G. E. Box and A. Luceno. *Statistical Control by Monitoring and Feedback Adjustment*. John Wiley & Sons, Inc., New York, USA, 1997.

[81] G. E. Box, A. Luceno, and M. del Carmen Paniagua-Quinones. *Statistical Control by Monitoring and Adjustment (2nd ed.)*. John Wiley & Sons, Inc., New York, USA, 2009.

[82] P. Braca, S. Marano, V. Matta, and P. Willett. Asymptotic optimality of running consensus in testing binary hypotheses. *IEEE Transactions on Signal Processing*, 58(2):814–825, Feb. 2010.

[83] P. Braca, S. Marano, V. Matta, and P. Willett. Consensus-based Page's test in sensor networks. *Signal Processing*, 91(4):919–930, Apr. 2011.

[84] D. H. Brandwood. A complex gradient operator and its application in adaptive array theory. *IEE Proceedings F, Communications, Radar and Signal Processing*, 130(1):11–16, Feb. 1983.

[85] A. Briassouli, V. Tsiminaki, and I. Kompatsiaris. Human motion analysis via statistical motion processing and sequential change detection. *EURASIP Journal on Image and Video Processing*, 2009, 2009. Special issue on video-based modeling, analysis, and recognition of human motion, Article ID 652050.

[86] B. E. Brodsky and B. S. Darkhovsky. *Nonparametric Methods in Change-point Problems*. Series on Mathematics and Its Applications. Kluwer Academic Publishers, Dordrecht, NL, 1993.

[87] S. Bruni, R. Goodall, T. X. Mei, and H. Tsunashima. Control and monitoring for railway vehicle dynamics. *Vehicle System Dynamics*, 45(7–8):743–779, 2007.

[88] D. L. Burkholder. Distribution function inequalities for martingales. *Annals of Probability*, 1(1):19–42, Feb. 1973.

[89] D. L. Burkholder, B. J. Davis, and R. F. Gundy. Integral inequalities for convex functions of operators on martingales. In L. M. Le Cam, J. Neyman, and E. L. Scott, editors, *Proceedings of the Sixth Berkeley Symposium on Mathematical Statistics and Probability, June 21–July 18, 1970*, volume 2: Probability Theory, pages 223–240. University of California Press, Berkeley, CA, USA, 1972.

[90] D. L. Burkholder and R. A. Wijsman. Optimum properties and admissibility of sequential tests. *Annals of Mathematical Statistics*, 34(1):1–17, Mar. 1963.

[91] J. Bussgang and D. Middleton. Optimum sequential detection of signals in noise. *IRE Transactions on Information Theory*, 1(3):5–18, Dec. 1955.

[92] J.-F. Cardoso and É. Moulines. Invariance of subspace based estimators. *IEEE Transactions on Signal Processing*, 48(9):2495–2505, Sept. 2000.

[93] J. Carroll. Vulnerability assessment of the transportation infrastructure relying on the global positioning system. Technical report, John A. Volpe National Transportation Systems Center, Cambridge, MA, USA, Aug. 2001. Online.

[94] H. P. Chan and T. L. Lai. Asymptotic approximations for error probabilities of sequential or fixed sample size tests in exponential families. *Annals of Statistics*, 28(6):1638–1669, Dec. 2000.

[95] F.-K. Chang. Structural health monitoring: Promises and challenges. In *Proceedings of the 30th Annual Review of Progress in Quantitative NDE (QNDE), Green Bay, WI, USA*. American Institute of Physics, July 2003.

[96] A. Chen, T. Wittman, A. G. Tartakovsky, and A. L. Bertozzi. Efficient boundary tracking through sampling. *Applied Mathematics Research Express*, 2(2):182–214, 2011.

[97] H. Chernoff. Sequential design of experiments. *Annals of Mathematical Statistics*, 30(3):755–770, Sept. 1959.

[98] H. Chernoff. Sequential tests for the mean of a normal distribution III (small t). *Annals of Mathematical Statistics*, 36(1):28–54, Feb. 1965.

[99] Y. S. Chow and T. L. Lai. Some one-sided theorems on the tail distribution of sample sums with applications to the last time and largest excess of boundary crossings. *Transactions of the American Mathematical Society*, 208(1):51–72, July 1975.

[100] Y. S. Chow, H. Robbins, and D. Siegmund. *Great Expectations: The Theory of Optimal Stopping*. Houghton Mifflin Co., Boston, MA, USA, 1971.

[101] Y. S. Chow, H. Robbins, and D. Siegmund. *The Theory of Optimal Stopping*. Dover Publications, Inc., New York, USA, 1991. Second and corrected edition of [100].

[102] E. Côme, M. Cottrell, M. Verleysen, and J. Lacaille. Aircraft engine health monitoring using self-organizing maps. In P. Perner, editor, *Advances in Data Mining - Applications and Theoretical Aspects. Proceedings of the 10th Industrial Conference on Data Mining (ICDM'10), Berlin, DE*, volume 6171 of *Lecture Notes in Computer Science*, pages 405–417. Springer-Verlag, July 2010.

[103] E. T. Copson. *Asymptotic Expansions*. Cambridge Tracts in Mathematics. Cambridge University Press, Cambridge, UK, 1965.

[104] D. R. Cox and H. D. Miller. *The Theory of Stochastic Processes*. John Wiley & Sons, Inc., New York, USA, 1965.

[105] S. V. Crowder. A simple method for studying run-length distributions of exponentially weighted moving average charts. *Technometrics*, 29(4):401–407, Nov. 1987.

[106] S. V. Crowder, D. M. Hawkins, M. R. Reynolds Jr., and E. Yashchin. Process control and statistical inference. *Journal of Quality Technology*, 29(2):134–139, Apr. 1997.

[107] C. Curry, R. L. Grossman, D. Locke, S. Vejcik, and J. Bugajski. Detecting changes in large data sets of payment card data: A case study. In *Proceedings of the 13th ACM SIGKDD International Conference on Knowledge Discovery and Data Mining (KDD'07), San Jose, CA, USA*, pages 1018–1022, New York, USA, 2007. ACM.

[108] K. C. Daly, E. Gai, and J. V. Harrison. Generalized likelihood test for FDI in redundant sensor configurations. *AIAA Journal of Guidance, Control, and Dynamic*, 2(1):9–17, Jan. 1979.

[109] R. B. Davies. Asymptotic inference in stationary Gaussian time series. *Advances in Applied Probability*, 5(3):469–497, 1973.

[110] S. Dayanik, W. Powell, and K. Yamazaki. An asymptotically optimal strategy in sequential change detection and identification applied to problems in biosurveillance. In J. Li, D. Aleman, and R. Sikora, editors, *Proceedings of the 3rd Workshop on Data Mining and Health Informatics (DM-HI 2008), Washington D.C., USA*. INFORMS, Oct. 2008.

[111] S. K. De and M. Baron. Step-up and step-down methods for testing multiple hypotheses in sequential experiments. *Journal of Statistical Planning and Inference*, 142(7):2059–2070, July 2012.

[112] P. Debanne and I. Nikiforov. Contrôle d'intégrité dans les systèmes de navigation aéronautiques. In *Colloque ATMA94, Sécurité et Fiabilité dans les domaines Maritime et Aéronautique, Paris*, pages 75–88, 27-28 Avril 1994.

[113] H. Debar, M. Dacier, and A. Wespi. Toward a taxonomy of intrusion detection systems. *Computer Networks*, 31(8):805–822, Apr. 1999.

[114] E. del Castillo. *Statistical Process Adjustment for Quality Control.* Series in Probability and Statistics. John Wiley & Sons, Inc., 2002.

[115] J. DeLucia and H. V. Poor. Performance analysis of sequential tests between Poisson processes. *IEEE Transactions on Information Theory*, 43(1):221–238, Jan. 1997.

[116] L. M. Delves and J. Walsh, editors. *Numerical Solution of Integral Equations.* Oxford University Press, Oxford, UK, Oct. 1974.

[117] R. J. Di Francesco. Real-time speech segmentation using pitch and convexity jump models: Application to variable rate speech coding. *IEEE Transactions on Acoustics, Speech, and Signal Processing*, 38(5):741–748, May 1990.

[118] P. Diaconis and D. Ylvisaker. Conjugate priors for exponential families. *Annals of Statistics*, 7(2):269–281, Mar. 1979.

[119] S. X. Ding. *Model-based Fault Diagnosis Techniques: Design Schemes, Algorithms, and Tools.* Springer-Verlag, Berlin, DE, 2008.

[120] N. Dobigeon and J.-Y. Tourneret. Joint segmentation of wind speed and direction by using a hierarchical model. *Computational Statistics and Data Analysis*, 51(12):5603–5621, Aug. 2007.

[121] S. W. Doebling, C. R. Farrar, and M. B. Prime. Summary review of vibration-based damage identification methods. *The Shock and Vibration Digest*, 30(2):91–105, Mar. 1998.

[122] A. Doucet, N. De Freitas, and N. J. Gordon, editors. *Sequential Monte Carlo Methods in Practice.* Statistics for Engineering and Information Science. Springer-Verlag, Berlin, New York, 2001.

[123] V. P. Dragalin. Asymptotic solution of a problem of detecting a signal from k channels. *Russian Mathematical Surveys*, 42(3):213–214, 1987.

[124] V. P. Dragalin. Optimality of the generalized CUSUM procedure in the quickest detection problem. In E. F. Mishchenko, editor, *Statistics and Control of Random Processes: Proceedings of the Steklov Institute of Mathematics*, volume 202, pages 107–119. American Mathematical Society, Providence, RI, USA, Nov. 1994.

[125] V. P. Dragalin. A multi-channel change point problem. In *Proceedings of the 3rd Umea-Wuzburg Conference in Statistics, Umea, SE*, pages 97–108. Umea University, 1995.

[126] V. P. Dragalin and A. A. Novikov. Asymptotic solution of the Kiefer–Weiss problem for processes with independent increments. *Theory of Probability and its Applications*, 32(4):617–627, Oct. 1987.

[127] V. P. Dragalin and A. A. Novikov. Adaptive sequential tests for composite hypotheses. In *Statistics and Control of Random Processes: Proceedings of the Steklov Institute of Mathematics*, volume 4, pages 12–23. TVP Science Publisher, Moscow, RU, 1995.

[128] V. P. Dragalin, A. G. Tartakovsky, and V. V. Veeravalli. Multihypothesis sequential probability ratio tests - Part I: Asymptotic optimality. *IEEE Transactions on Information Theory*, 45(11):2448–2461, Nov. 1999.

[129] V. P. Dragalin, A. G. Tartakovsky, and V. V. Veeravalli. Multihypothesis sequential probabil-

ity ratio tests - Part II: Accurate asymptotic expansions for the expected sample size. *IEEE Transactions on Information Theory*, 46(4):1366–1383, Apr. 2000.

[130] A. J. Duncan. *Quality Control and Industrial Statistics (5th ed.).* Richard D. Irwin Professional Publishing, Inc., 1986.

[131] A. Dushman. On gyro drift models and their evaluation. *IRE Transactions on Aerospace and Navigational Electronics*, 9(4):230–234, 1962.

[132] B. Eisenberg and B. K. Ghosh. The sequential probability ratio test. In B. K. Ghosh and P. K. Sen, editors, *Handbook of Sequential Analysis*, volume 118 of *Statistics Textbooks and Monographs*, pages 47–65. Marcel Dekker, Inc., New York, Basel, Hong Kong, 1991.

[133] J. Ellis and T. Speed. *The Internet Security Guidebook: From Planning to Deployment.* Academic Press, 2001.

[134] J. D. Esary, F. Proschan, and D. W. Walkup. Association of random variables, with applications. *Annals of Mathematical Statistics*, 38(5):1466–1474, Oct. 1967.

[135] D. J. Ewins. *Modal Testing: Theory, Practice, and Applications (2nd ed.).* Research Studies Press, Letchworth, Hertfordshire, 2000.

[136] C. R. Farrar, S. W. Doebling, and D. A. Nix. Vibration-based structural damage identification. *Philosophical Transactions A - Philosophical Transactions of the Royal Society: Mathematical, Physical and Engineering Sciences*, 359(1778):323–345, Jan. 2001.

[137] J. Farrell and M. Barth. *The Global Positioning System and Inertial Navigation.* McGraw Hill, 1999.

[138] P. Faurre. Kalman filtering and the advancement of navigation and guidance. In A. C. Antoulas, editor, *Mathematical Systems Theory: The Influence of R. E. Kalman*, pages 89–134. Springer-Verlag, 1991.

[139] E. A. Feinberg and A. N. Shiryaev. Quickest detection of drift change for Brownian motion in generalized Bayesian and minimax settings. *Statistics and Decisions*, 24(4):445–470, Oct. 2006.

[140] W. Feller. *An Introduction to Probability Theory and Its Applications (2nd ed.)*, volume 2 of *Series in Probability and Mathematical Statistics*. John Wiley & Sons, Inc., 1966.

[141] G. Fellouris and A. G. Tartakovsky. Nearly minimax one-sided mixture-based sequential tests. *Sequential Analysis*, 31(3):297–325, 2012.

[142] G. Fellouris and A. G. Tartakovsky. Almost optimal sequential tests of discrete composite hypotheses. *Statistica Sinica*, 23(4):1717–1741, 2013.

[143] T. S. Ferguson. *Mathematical Statistics: A Decision Theoretic Approach.* Probability and Mathematical Statistics. Academic Press, 1967.

[144] T. S. Ferguson. Who solved the secretary problem? *Statistical Science*, 4(3):282–289, Aug. 1989.

[145] L. Fillatre. Constrained epsilon-minimax test for simultaneous detection and classification. *IEEE Transactions on Information Theory*, 57(12):8055–8071, Dec. 2011.

[146] L. Fillatre and I. Nikiforov. A fixed-size sample strategy for the sequential detection and isolation of non-orthogonal alternatives. *Sequential Analysis*, 29(2):176–192, Apr. 2010.

[147] L. Fillatre and I. Nikiforov. Asymptotically uniformly minimax detection and isolation in network monitoring. *IEEE Transactions on Signal Processing*, 60(7):3357–3371, July 2012.

[148] R. A. Fisher. Theory of statistical estimation. *Mathematical Proceedings of the Cambridge Philosophical Society*, 22(5):700–725, July 1925.

[149] M. F. Fishman. Optimization of the algorithm for the detection of a disorder based on the statistic of exponential smoothing. *Statistical Problems of Control*, 83:146–151, 1988. In Russian.

[150] M. Fouladirad and I. Nikiforov. Optimal statistical fault detection with nuisance parameters. *Automatica*, 41(7):1157–1171, July 2005.

[151] P. M. Frank. Fault diagnosis in dynamic systems using analytical and knowledge based redundancy - A survey and some new results. *Automatica*, 26(3):459–474, May 1990.

[152] P. A. Fridman. A method of detecting radio transients. *Monthly Notices of the Royal Astronomical Society*, 409(2):808–820, Dec. 2010.

[153] C. Fuchs and R. Kennet. *Multivariate Quality Control – Theory and Application*, volume 54 of *Series on Quality and Reliability*. Marcel Dekker, Inc., New York, USA, 1998.

[154] C.-D. Fuh. SPRT and CUSUM in hidden Markov models. *Annals of Statistics*, 31(3):942–977, June 2003.

[155] C.-D. Fuh. Asymptotic operating characteristics of an optimal change point detection in hidden Markov models. *Annals of Statistics*, 32(5):2305–2339, Oct. 2004.

[156] E. Gai, J. V. Harrison, and K. C. Daly. FDI performance of two redundant sensor configurations. *IEEE Transactions on Aerospace and Electronic Systems*, 15(3):405–413, May 1979.

[157] W. Gersch. On the achievable accuracy of structural parameter estimates. *Journal of Sound and Vibration*, 34(1):63–79, 1974.

[158] J. J. Gertler. Survey of model-based failure detection and isolation in complex plants. *IEEE Control Systems Magazine*, 8(6):3–11, Dec. 1988.

[159] J. J. Gertler. *Fault Detection and Diagnosis in Engineering Systems*. Marcel Dekker, Inc., New York, USA, 1998.

[160] B. K. Ghosh, editor. *Sequential Tests of Statistical Hypotheses*. Addison Wesley Publ. Co., Reading, MA, USA, 1970.

[161] B. K. Ghosh. A brief history of sequential analysis. In B. K. Ghosh and P. K. Sen, editors, *Handbook of Sequential Analysis*, volume 118 of *Statistics Textbooks and Monographs*, pages 1–19. Marcel Dekker, Inc., New York, Basel, Hong Kong, 1991.

[162] B. K. Ghosh and P. K. Sen, editors. *Handbook of Sequential Analysis*, volume 118 of *Statistics Textbooks and Monographs*. Marcel Dekker, Inc., New York, Basel, Hong Kong, 1991.

[163] M. Ghosh, N. Mukhopadhyay, and P. K. Sen, editors. *Sequential Estimation*. Wiley Series in Probability and Statistics. John Wiley & Sons, Inc., New York, USA, Mar. 1997.

[164] I. I. Gikhman and A. V. Skorokhod. *Stochastic Differential Equations*. Springer-Verlag, Berlin, New York, 1972.

[165] M. A. Girshick and H. Rubin. A Bayes approach to a quality control model. *Annals of Mathematical Statistics*, 23(1):114–125, Mar. 1952.

[166] V. P. Godambe, editor. *Estimating Functions*. Clarendon Press, Oxford, UK, 1991.

[167] A. L. Goel and S. M. Wu. Determination of the ARL and a contour nomogram for CUSUM charts to control normal mean. *Technometrics*, 13(2):221–230, May 1971.

[168] G. K. Golubev and R. Z. Khas'minskii. Sequential testing for several signals in Gaussian white noise. *Theory of Probability and its Applications*, 28(3):573–584, 1984.

[169] M. S. Grewal, L. R. Weill, and A. P. Andrews. *Global Positioning Systems, Inertial Navigation, and Integration*. John Wiley & Sons, Inc., 2001.

[170] R. Grossman, M. Sabala, A. Aanand, S. Eick, L. Wilkinson, P. Zhang, J. Chaves, S. Vejcik, J. Dillenburg, P. Nelson, D. Rorem, J. Alimohideen, J. Leigh, M. Papka, and R. Stevens. Real time change detection and alerts from highway traffic data. In *Proceedings of the ACM/IEEE Supercomputing Conference (SC'05), Seattle, WA, USA*, pages 69–74, Nov. 2005.

[171] L. Grunske and P. Zhang. Monitoring probabilistic properties. In H. van Vliet and V. Issarny, editors, *Proceedings of the 7th Joint Meeting of the European Software Engineering*

Conference and the ACM SIGSOFT Symposium on The Foundations of Software Engineering (ESEC/FSE'09), Amsterdam, NL, pages 183–192, Aug. 2009.

[172] D. E. Gustafson, A. S. Willsky, J.-Y. Wang, M. C. Lancaster, and J. H. Triebwasser. ECG/VCG rhythm diagnosis using statistical signal snalysis - I. Identification of persistent rhythms. *IEEE Transactions on Biomedical Engineering*, 25(4):344–353, July 1978.

[173] A. Gut. *Stopped Random Walks: Limit Theorems and Applications*, volume 5 of *Series in Applied Probability*. Springer-Verlag, New York, USA, 1988.

[174] C. Hagwood and M. Woodroofe. On the expansion for expected sample size in non-linear renewal theory. *Annals of Probability*, 10(3):844–848, Aug. 1982.

[175] S. R. Hall, P. Motyka, E. Gai, and J. J. Deyst Jr. In-flight parity vector compensation for FDI. *IEEE Transactions on Aerospace and Electronic Systems*, 19(5):668–676, Sept. 1983.

[176] W. J. Hall and D. J. Mathiason. On large sample estimation and testing in parametric models. *International Statistical Review*, 58(1):77–97, 1990.

[177] W. J. Hall, R. A. Wijsman, and J. K. Ghosh. The relationship between sufficiency and invariance with applications in sequential analysis. *Annals of Mathematical Statistics*, 36(2):575–614, Apr. 1965.

[178] Z. Hameed, Y. S. Hong, Y. M. Cho, S. H. Ahn, and C. K. Song. Condition monitoring and fault detection of wind turbines and related algorithms: A review. *Renewable and Sustainable Energy Reviews*, 13(1):1–39, 2009.

[179] T. E. Harris. The existence of stationary measures for certain Markov processes. In *Proceedings of the Third Berkeley Symposium on Mathematical Statistics and Probability, December 1954 and July–August 1955*, volume 2: Contributions to Probability Theory, pages 113–124. University of California Press, Berkeley, CA, USA, 1956.

[180] T. E. Harris. *The Theory of Branching Processes*. Springer-Verlag, Berlin, DE, 1963.

[181] J. V. Harrison and T. T. Chien. Failure isolation for a minimally redundant inertial sensor system. *IEEE Transactions on Aerospace and Electronic Systems*, 11(3):349–357, May 1975.

[182] J. V. Harrison and E. G. Gai. Evaluating sensor orientations for navigation performance and failure detection. *IEEE Transactions on Aerospace and Electronic Systems*, 13(6):631–643, Nov. 1977.

[183] E. Hartikainen, S. Ekelin, and J. M. Karlsson. Change detection and estimation for network-measurement applications. In *Proceedings of the 2nd ACM Workshop on Performance Monitoring and Measurement of Heterogeneous Wireless and Wired Networks (PM2HW2N'07), Chania, Crete Island, GR*, pages 1–10, New York, USA, Oct. 2007. ACM.

[184] D. M. Hawkins and D. H. Olwell. *Cumulative Sum Charts and Charting for Quality Improvement*. Series in Statistics for Engineering and Physical Sciences. Springer-Verlag, USA, 1998.

[185] D. M. Hawkins, P. Qiu, and C. W. Kang. The changepoint model for statistical process control. *Journal of Quality Technology*, 35(4):355–366, Oct. 2003.

[186] A. Hazan, M. Verleysen, M. Cottrell, and J. Lacaille. Trajectory clustering for vibration detection in aircraft engines. In P. Perner, editor, *Advances in Data Mining - Applications and Theoretical Aspects. Proceedings of the 10th Industrial Conference on Data Mining (ICDM'10), Berlin, DE*, volume 6171 of *Lecture Notes in Computer Science*, pages 362–375. Springer-Verlag, July 2010.

[187] X. He, C. Papadopoulos, J. Heidemann, U. Mitra, and U. Riaz. Remote detection of bottleneck links using spectral and statistical methods. *Computer Networks*, 53(3):279–298, Feb. 2009.

[188] D. Henry, S. Simani, and R. J. Patton. Fault detection and diagnosis for aeronautic and

aerospace missions. In C. Edwards, T. Lombaerts, and H. Smaili, editors, *Fault Tolerant Flight Control - A Benchmark Challenge*, volume 399 of *Lecture Notes in Control and Information Sciences*, pages 91–128. Springer-Verlag, Berlin, Heidelberg, DE, Apr. 2010.

[189] N. Herrmann and T. H. Szatrowski. Sample size savings for curtailed one-sample nonparametric tests for location shift. *Annals of Statistics*, 15(1):296–313, Mar. 1987.

[190] C. C. Heyde. *Quasi-Likelihood and Its Application*. Series in Statistics. Springer-Verlag, New York, USA, 1997.

[191] W. Heylen, S. Lammens, and P. Sas. *Modal Analysis Theory and Testing*. Mechanical Engineering Dept., Katholieke Universiteit Leuven, Leuven, B., 1995.

[192] W. Hoeffding. A lower bound for the average sample number of a sequential test. *Annals of Mathematical Statistics*, 24(1):127–130, Mar. 1953.

[193] W. Hoeffding. Lower bounds for the expected sample size and the average risk of a sequential procedure. *Annals of Mathematical Statistics*, 31(2):352–368, June 1960.

[194] P. G. Hoel and R. P. Peterson. A solution to the problem of optimum classification. *Annals of Mathematical Statistics*, 20(3):433–438, Sept. 1949.

[195] P. L. Hsu and H. Robbins. Complete convergence and the law of large numbers. *Proceedings of the National Academy of Sciences of the United States of America*, 33(2):25–31, Feb. 1947.

[196] B. Huang. Detection of abrupt changes of total least squares models and application in fault detection. *IEEE Transactions on Control Systems Technology*, 9(2):357–367, Mar. 2001.

[197] J. R. Huddle. Inertial navigation system error-model considerations in Kalman filtering applications. In C. T. Leondes, editor, *Nonlinear and Kalman Filtering Techniques, Part 2 of 3*, volume 20 of *Control and Dynamic Systems*, pages 293–339. Academic Press, 1983.

[198] M. D. Huffman. An efficient approximate solution to the Kiefer–Weiss problem. *Annals of Statistics*, 11(1):306–316, Mar. 1983.

[199] A. Hussain, J. Heidemann, and C. Papadopoulos. A framework for classifying denial of service attacks. In *Proceedings of the Conference on Applications, Technologies, Architectures, and Protocols for Computer Communications, Karlsruhe, DE*, pages 99–110. ACM SIGCOMM, Aug. 2003.

[200] A. Hussain, J. Heidemann, and C. Papadopoulos. Identification of repeated denial of service attacks. In *Proceedings of the 25th IEEE International Conference on Computer Communications (INFOCOM'06), Barcelona, SP*, pages 1–15. IEEE, Apr. 2006.

[201] C.-H. Hwang, G.-L. Lai, and S.-C. Chen. Spectrum sensing in wideband OFDM cognitive radios. *IEEE Transactions on Signal Processing*, 58(2):709–719, Feb. 2010.

[202] I. A. Ibragimov and R. Z. Khasminskii. *Statistical Estimation - Asymptotic Theory*, volume 16 of *Applications of Mathematics Series*. Springer-Verlag, New York, USA, 1981.

[203] A. Irle. Extended optimality of sequential probability ratio tests. *Annals of Statistics*, 12(1):380–386, Mar. 1984.

[204] A. Irle and N. Schmitz. On the optimality of the SPRT for processes with continuous time parameter. *Statistics - A Journal of Theoretical and Applied Statistics*, 15(1):91–104, 1984.

[205] R. Isermann. Process fault detection based on modeling and estimation methods - A survey. *Automatica*, 20(4):387–404, July 1984.

[206] R. Isermann. *Fault Diagnosis Systems: An Introduction From Fault Detection To Fault Tolerance*. Springer-Verlag, Berlin, DE, 2005.

[207] N. Ishii, A. Iwata, and N. Suzumura. Segmentation of nonstationary time series. *International Journal of Systems Science*, 10(8):883–894, Aug. 1979.

[208] A. K. Iyengar, M. S. Squillante, and L. Zhang. Analysis and characterization of large-scale web server access patterns and performance. *World Wide Web*, 2(1–2):85–100, June 1999.

[209] J. E. Jackson and R. A. Bradley. Sequential χ^2 and t^2 tests. *Annals of Mathematical Statistics*, 32(4):1063–1077, Dec. 1961.

[210] J. Jacod. *Calcul Stochastique et Problèmes de Martingales*, volume 714 of *Lecture Notes in Mathematics*. Springer-Verlag, Berlin, 1979. In French.

[211] R. Jana and S. Dey. Change detection in teletraffic models. *IEEE Transactions on Signal Processing*, 48(3):846–853, Mar. 2000.

[212] M. K. Jeerage. Reliability analysis of fault-tolerant IMU architectures with redundant inertial sensors. *IEEE Aerospace and Electronic Systems Magazine*, 5(7):23–28, 1990.

[213] R. H. Jones, D. H. Crowell, and L. E. Kapuniai. Change detection model for serially correlated multivariate data. *Biometrics*, 26(2):269–280, June 1970.

[214] J. N. Juang. *Applied System Identification*. Prentice-Hall, Inc., Englewood Cliffs, NJ, USA, 1994.

[215] J. Jung, V. Paxson, A. W. Berger, and H. Balakrishnan. Fast portscan detection using sequential hypothesis testing. In *Proceedings of the 2004 IEEE Symposium on Security and Privacy, Oakland, CA, USA*, pages 211–225, May 2004.

[216] S. Kanev and T. van Engelen. Wind turbine extreme gust control. *Wind Energy*, 13(1):18–35, Jan. 2010.

[217] L. V. Kantorovich and V. I. Krylov. *Approximate Methods of Higher Analysis*. Interscience Publishers, New York, USA, 1958. Translated by Curtis D. Benster.

[218] M. Kayton and W. R. Fried, editors. *Avionics Navigation Systems*. John Wiley & Sons, Inc., 1996.

[219] K. W. Kemp. Formula for calculating the operating characteristic and average sample number of some sequential tests. *Journal of the Royal Statistical Society - Series B Methodology*, 20(2):379–386, 1958.

[220] K. W. Kemp. The average run length of the cumulative sum chart when a V-mask is used. *Journal of the Royal Statistical Society - Series B Methodology*, 23(1):149–153, 1961.

[221] K. W. Kemp. Formal expressions which can be used for the determination of the operating characteristic and average sample number of a simple sequential test. *Journal of the Royal Statistical Society - Series B Methodology*, 29(2):248–262, 1967.

[222] K. W. Kemp. A simple procedure for determining upper and lower limits for the average sample run length of a cumulative sum scheme. *Journal of the Royal Statistical Society - Series B Methodology*, 29(2):263–265, 1967.

[223] J. H. B. Kemperman. *The General One-dimensional Random Walk with Absorbing Barriers: with Applications to Sequential Analysis*. 's-Gravenhage, Amsterdam, NL, 1951.

[224] S. Kent. On the trail of intrusions into information systems. *IEEE Spectrum*, 37(12):52–56, Dec. 2000.

[225] T. H. Kerr. False alarm and correct detection probabilities over a time interval for restricted classes of failure detection algorithms. *IEEE Transactions on Information Theory*, 28(4):619–631, July 1982.

[226] T. H. Kerr. Decentralized filtering and redundancy management for multisensor navigation. *IEEE Transactions on Aerospace and Electronic Systems*, 23(1):83–119, Jan. 1987.

[227] T. H. Kerr. A critique of several failure detection approaches for navigation systems. *IEEE Transactions on Automatic Control*, 34(7):791–792, July 1989.

[228] H. Kesten. Random difference equations and renewal theory for products of random matrices. *Acta Mathematica*, 131(1):207–248, 1973.

[229] R. A. Khan. Wald's approximations to the average run length in cusum procedures. *Journal of Statistical Planning and Inference*, 2(1):63–77, 1978.

[230] R. A. Khan. A note on Page's two-sided cumulative sum procedures. *Biometrika*, 68(3):717–719, Dec. 1981.

[231] J. Kiefer and J. Sacks. Asymptotically optimal sequential inference and design. *Annals of Mathematical Statistics*, 34(3):705–750, Sept. 1963.

[232] J. Kiefer and L. Weiss. Properties of generalized sequential probability ratio tests. *Annals of Mathematical Statistics*, 28(1):57–74, Jan. 1957.

[233] S. J. Kim, G. Doretto, J. Rittscher, P. Tu, N. Krahnstoever, and M. Pollefeys. A model change detection approach to dynamic scene modeling. In *Proceedings of the 6th IEEE International Conference on Advanced Video and Signal Based Surveillance (AVSS'09), Genoa, IT*, pages 490–495, Washington, DC, USA, 2009. IEEE Computer Society.

[234] V. Kireichikov, V. Mangushev, and I. Nikiforov. Investigation and application of CUSUM algorithms to monitoring of sensors. *Statistical Problems of Control*, 89:124–130, 1990. In Russian.

[235] G. Kitagawa and W. Gersch. A smoothness prior time-varying AR coefficient modeling of nonstationary covariance time series. *IEEE Transactions on Automatic Control*, 30(1):48–56, Jan. 1984.

[236] S. Knoth. Computation of the ARL for CUSUM-S^2 schemes. *Computational Statistics and Data Analysis*, 51(2):499–512, Nov. 2006.

[237] K.-R. Koch. *Parameter Estimation and Hypothesis Testing in Linear Models (2nd ed.)*. Springer-Verlag, Berlin, Heidelberg, DE, 1999.

[238] D. Kohlruss. Exact formulas for the OC and ASN functions of the SPRT for Erlang distributions. *Sequential Analysis*, 13(1):53–62, Jan. 1994.

[239] R. Kress. *Numerical Analysis*, volume 181 of *Graduate Texts in Mathematics*. Springer-Verlag, New York, USA, 1998.

[240] S. Kullback. *Information Theory and Statistics (2nd ed.)*. Dover Books on Mathematics. Dover Publications, Inc., 1997.

[241] A. F. Kushnir and A. I. Pinskii. Asymptotically optimal tests for testing hypotheses for interdependent samples of observations. *Theory of Probability and its Applications*, 16(2):281–292, 1971.

[242] L. Ladelli. Diffusion approximation for a pseudo-likelihood test process with application to detection of change in a stochastic system. *Stochastics and Stochastics Reports*, 32(1–2):1–25, 1990.

[243] T. L. Lai. Optimal stopping and sequential tests which minimize the maximum expected sample size. *Annals of Statistics*, 1(4):659–673, July 1973.

[244] T. L. Lai. Control charts based on weighted sums. *Annals of Statistics*, 2(1):134–147, Jan. 1974.

[245] T. L. Lai. A note on first exit times with applications to sequential analysis. *Annals of Statistics*, 3(4):999–1005, July 1975.

[246] T. L. Lai. On uniform integrability in renewal theory. *Bulletin of the Institute of Mathematics Academia Sinica*, 3(1):99–105, Jan. 1975.

[247] T. L. Lai. On *r*-quick convergence and a conjecture of Strassen. *Annals of Probability*, 4(4):612–627, Aug. 1976.

[248] T. L. Lai. Asymptotic optimality of invariant sequential probability ratio tests. *Annals of Statistics*, 9(2):318–333, Mar. 1981.

[249] T. L. Lai. Nearly optimal sequential tests of composite hypotheses. *Annals of Statistics*, 16(2):856–886, June 1988.

[250] T. L. Lai. Sequential changepoint detection in quality control and dynamical systems (with

discussion). *Journal of the Royal Statistical Society - Series B Methodology*, 57(4):613–658, 1995.

[251] T. L. Lai. Information bounds and quick detection of parameter changes in stochastic systems. *IEEE Transactions on Information Theory*, 44(7):2917–2929, Nov. 1998.

[252] T. L. Lai. Sequential multiple hypothesis testing and efficient fault detection-isolation in stochastic systems. *IEEE Transactions on Information Theory*, 46(2):595–608, Mar. 2000.

[253] T. L. Lai and J. Z. Shan. Efficient recursive algorithms for detection of abrupt changes in signals and control systems. *IEEE Transactions on Automatic Control*, 44(5):952–966, May 1999.

[254] T. L. Lai and D. Siegmund. A nonlinear renewal theory with applications to sequential analysis I. *Annals of Statistics*, 5(5):946–954, Sept. 1977.

[255] T. L. Lai and D. Siegmund. A nonlinear renewal theory with applications to sequential analysis II. *Annals of Statistics*, 7(1):60–76, Jan. 1979.

[256] T. L. Lai and H. Xing. Sequential change-point detection when the pre- and post-change parameters are unknown. *Sequential Analysis*, 29(2):162–175, Apr. 2010.

[257] T. L. Lai and L. Zhang. A modification of Schwarz's sequential likelihood ratio tests in mulitvariate sequential analysis. *Sequential Analysis*, 13(1):79–96, Feb. 1994.

[258] S. Lalley. Non-linear renewal theory for lattice random walks. *Sequential Analysis*, 1(3):193–205, July 1982.

[259] S. M. LaValle. *Planning Algorithms*. Cambridge University Press, Cambridge, UK, May 2006. Online.

[260] L. Le Cam. *Asymptotic Methods in Statistical Decision Theory*. Series in Statistics. Springer-Verlag, New York, Berlin, Heidelberg, 1986.

[261] E. L. Lehmann. *Testing Statistical Hypotheses*. John Wiley & Sons, Inc., New York, USA, 1968.

[262] D. Lelescu and D. Schonfeld. Statistical sequential analysis for real-time video scene change detection on compressed multimedia bitstream. *IEEE Transactions on Multimedia*, 5(1):106–117, Mar. 2003.

[263] H. Li, H. Dai, and C. Li. Collaborative quickest spectrum sensing via random broadcast in cognitive radio systems. *IEEE Transactions on Wireless Communications*, 9(7):2338–2348, July 2010.

[264] Y. V. Linnik. *Statistical Problems with Nuisance Parameters*, volume 20 of *Translations of Mathematical Monographs*. American Mathematical Society, Providence, RI, USA, 1968.

[265] R. S. Liptser and A. N. Shiryaev. *Theory of Martingales*. Kluwer Academic Publishers, Dordrecht, NL, 1989.

[266] R. S. Liptser and A. N. Shiryaev. *Statistics of Random Processes - Part I: General Theory*, volume 5 of *Stochastic Modelling and Applied Probability*. Springer-Verlag, Berlin, New York, 2001.

[267] R. S. Liptser and A. N. Shiryaev. *Statistics of Random Processes - Part II: Applications*, volume 6 of *Stochastic Modelling and Applied Probability*. Springer-Verlag, Berlin, New York, 2001.

[268] M. Loève. *Probability Theory (4th ed.)*. Springer-Verlag, New York, USA, 1977.

[269] G. Lorden. Integrated risk of asymptotically Bayes sequential tests. *Annals of Mathematical Statistics*, 38(5):1399–1422, Oct. 1967.

[270] G. Lorden. On excess over the boundary. *Annals of Mathematical Statistics*, 41(2):520–527, Apr. 1970.

[271] G. Lorden. Procedures for reacting to a change in distribution. *Annals of Mathematical Statistics*, 42(6):1897–1908, Dec. 1971.

[272] G. Lorden. Open-ended tests for Koopman–Darmois families. *Annals of Statistics*, 1(4):633–643, July 1973.

[273] G. Lorden. 2-SPRT's and the modified Kiefer–Weiss problem of minimizing an expected sample size. *Annals of Statistics*, 4(2):281–291, Mar. 1976.

[274] G. Lorden. Nearly optimal sequential tests for exponential families. 1977. Unpublished Manuscript.

[275] G. Lorden. Nearly-optimal sequential tests for finitely many parameter values. *Annals of Statistics*, 5(1):1–21, Jan. 1977.

[276] G. Lorden. Structure of sequential tests minimizing an expected sample size. *Probability Theory and Related Fields*, 51(2):291–302, Jan. 1980.

[277] G. Lorden and I. Eisenberger. Detection of failure rates increases. *Technometrics*, 15(1):167–175, Feb. 1973.

[278] V. I. Lotov. On the asymptotic behavior of characteristics of the sequential likelihood ratio test. *Theory of Probability and its Applications*, 30(1):181–187, 1986.

[279] V. I. Lotov. Asymptotic expansions in a sequential likelihood ratio test. *Theory of Probability and its Applications*, 32(1):57–67, 1987.

[280] G. Loukas and G. Öke. Protection against denial of service attacks: A survey. *The Computer Journal*, 53(7):1020–1037, Sept. 2010.

[281] J. M. Lucas and R. B. Crosier. Fast initial response for CUSUM quality control schemes: Give your CUSUM a head start. *Technometrics*, 24(3):199–205, Aug. 1982.

[282] A. Luceño and J. Puig-Pey. An accurate algorithm to compute the run length probability distribution, and its convolutions, for a Cusum chart to control normal mean. *Computational Statistics and Data Analysis*, 38(3):249–261, Jan. 2002.

[283] Y. L. Luke. *Mathematical Functions and Their Approximations*. Academic Press, New York, San Francisco, USA, 1975.

[284] R. Lund and J. Reeves. Detection of undocumented changepoints: A revision of the two-phase regression model. *Journal of Climatology*, 15(17):2547–2554, Sept. 2002.

[285] J. Luo, M. Namburu, K. R. Pattipati, L. Qiao, and S. Chigusa. Integrated model-based and data-driven diagnosis of automotive antilock braking systems. *IEEE Transactions on Systems, Man and Cybernetics, Part A: Systems and Humans*, 40(2):321–336, Mar. 2010.

[286] I. B. MacNeill. Detection of interventions at unknown times. In A. H. El-Shaarawi and S. R. Esterby, editors, *Time Series Methods in Hydrosciences - Proceedings of an International Conference Held at Canada Centre for Inland Waters*, volume 17 of *Developments in Water Science*, pages 16–26. Elsevier, 1982.

[287] D. P. Malladi and J. L. Speyer. A generalized Shiryayev sequential probability ratio test for change detection and isolation. *IEEE Transactions on Automatic Control*, 44(8):1522–1534, Aug. 1999.

[288] R. L. Mason and J. C. Young. *Multivariate Statistical Process Control with Industrial Application*. SIAM, Philadelphia, PA, USA, 2001.

[289] T. K. Matthes. On the optimality of sequential probability ratio tests. *Annals of Mathematical Statistics*, 34(1):18–21, Mar. 1963.

[290] P. S. Maybeck. Failure detection without excessive hardware redundancy. In *Proceedings of the National Aerospace and Electronics Conference (NAECON'76), Dayton, OH, USA*, pages 315–322. IEEE, May 1976.

[291] P. S. Maybeck. *Stochastic Models, Estimation, and Control*, volume 141 of *Mathematics in*

Science and Engineering. Academic Press, 1979.

[292] D. L. McLeish and C. G. Small. *The Theory and Application of Statistical Inference Functions*, volume 44 of *Lecture Notes in Statistics*. Springer-Verlag, Berlin, 1988.

[293] Y. Mei. Is average run length to false alarm always an informative criterion? *Sequential Analysis*, 27(4):354–376, Oct. 2008.

[294] Y. Mei. Efficient scalable schemes for monitoring a large number of data streams. *Biometrika*, 97(2):419–433, Apr. 2010.

[295] S. P. Mertikas. Automatic and online detection of small but persistent shifts in GPS station coordinates by statistical process control. *GPS Solutions*, 5(1):39–50, July 2001.

[296] L. Mevel, M. Basseville, and A. Benveniste. Fast in-flight detection of flutter onset: A statistical approach. *AIAA Journal of Guidance, Control, and Dynamic*, 28(3):431–438, May 2005.

[297] L. Mevel, M. Basseville, A. Benveniste, and M. Goursat. Merging sensor data from multiple measurement setups for nonstationary subspace-based modal analysis. *Journal of Sound and Vibration*, 249(4):719–741, Jan. 2002.

[298] L. Mevel, A. Benveniste, M. Basseville, and M. Goursat. Blind subspace-based eigenstructure identification under nonstationary excitation using moving sensors. *IEEE Transactions on Signal Processing*, 50(1):41–48, Jan. 2002.

[299] L. Mevel, A. Benveniste, M. Basseville, M. Goursat, B. Peeters, H. Van der Auweraer, and A. Vecchio. Input/output versus output-only data processing for structural identification - Application to in-flight data analysis. *Journal of Sound and Vibration*, 295(3):531–552, Aug. 2006.

[300] L. Mevel, M. Goursat, and M. Basseville. Stochastic subspace-based structural identification and damage detection and localization - Application to the Z24 bridge benchmark. *Mechanical Systems and Signal Processing*, 17(1):143–151, Jan. 2003.

[301] T. G. Mikhailova, I. N. Tikhonov, V. A. Mangushev, and I. V. Nikiforov. Automatic identification of overlapping of signals produced by two subsequent shocks on a seismogram. *Volcanology and Seismology*, (5):94–102, 1990. In Russian.

[302] J. Mirkovic, S. Dietrich, D. Dittrich, and P. Reiher. *Internet Denial of Service Attack and Defense Mechanisms*. The Radia Perlman Series in Computer Networking and Security. Prentice Hall - PTR, New Jersey, USA, 2004.

[303] D. C. Montgomery. *Introduction to Statistical Quality Control (6th ed.)*. John Wiley & Sons, Inc., 2008.

[304] F. Mosteller. A *k*-sample slippage test for an extreme population. *Annals of Mathematical Statistics*, 19(1):58–65, Jan. 1948.

[305] G. V. Moustakides. Optimal stopping times for detecting changes in distributions. *Annals of Statistics*, 14(4):1379–1387, Dec. 1986.

[306] G. V. Moustakides. Optimality of the CUSUM procedure in continuous time. *Annals of Statistics*, 32(1):302–315, Feb. 2004.

[307] G. V. Moustakides. Sequential change detection revisited. *Annals of Statistics*, 36(2):787–807, Mar. 2008.

[308] G. V. Moustakides and A. Benveniste. Detecting changes in the AR parameters of a nonstationary ARMA process. *Stochastics*, 16(1):137–155, 1986.

[309] G. V. Moustakides, A. S. Polunchenko, and A. G. Tartakovsky. Numerical comparison of CUSUM and Shiryaev–Roberts procedures for detecting changes in distributions. *Communications in Statistics - Theory and Methods*, 38(16–17):3225–3239, 2009.

[310] G. V. Moustakides, A. S. Polunchenko, and A. G. Tartakovsky. A numerical approach

to performance analysis of quickest change-point detection procedures. *Statistica Sinica*, 21(2):571–596, Apr. 2011.

[311] N. Mukhopadhyay, S. Datta, and S. Chattopadhyay. *Applied Sequential Methodologies: Real-World Examples with Data Analysis*, volume 173 of *Statistics Textbooks and Monographs*. Marcel Dekker, Inc., New York, USA, 2004.

[312] N. Mukhopadhyay and T. K. S. Solanky. *Multistage Selection and Ranking Procedures: Second Order Asymptotics*, volume 142 of *Statistics Textbooks and Monographs*. Marcel Dekker, Inc., New York, USA, 1994.

[313] S. Muthukrishnan, E. van den Berg, and Y. Wu. Sequential change detection on data streams. In *Proceedings of the 2nd International Workshop on Data Stream Mining and Management (DSMM'07), Omaha, NE, USA*, Oct. 2007.

[314] J. Nadler and N. B. Robbins. Some characteristics of Page's two-sided procedure for detecting a change in location parameter. *Annals of Mathematical Statistics*, 42(2):538–551, 1971.

[315] P. M. Newbold and Y. C. Ho. Detection of changes in the characteristics of a Gauss–Markov process. *IEEE Transactions on Aerospace and Electronic Systems*, 4(5):707–718, Sept. 1968.

[316] I. V. Nikiforov. Sequential analysis applied to autoregression processes. *Automation and Remote Control*, 36(8):174–178, 1975.

[317] I. V. Nikiforov. A statistical method for detecting the time at which the sensor properties change. In *Proceedings of the IMEKO Symposium on Applications of Statistical Methods in Measurement, Leningrad, RU*, pages 1–7. TC7 Proceedings, May 1978.

[318] I. V. Nikiforov. Cumulative sums for detection of changes in random process characteristics. *Automation and Remote Control*, 40(2):48–58, 1979.

[319] I. V. Nikiforov. Modification and analysis of the cumulative sum procedure. *Automation and Remote Control*, 41(9):74–80, 1980.

[320] I. V. Nikiforov. *Sequential Detection of Abrupt Changes in Time Series Properties*. Nauka, Moscow, RU, Sept. 1983. In Russian.

[321] I. V. Nikiforov. On the first-order optimality of an algorithm for detection of a fault in the vector case. *Automation and Remote Control*, 55(1):68–82, 1994.

[322] I. V. Nikiforov. A generalized change detection problem. *IEEE Transactions on Information Theory*, 41(1):171–187, Jan. 1995.

[323] I. V. Nikiforov. On two new criteria of optimality for the problem of sequential change diagnosis. In *Proceedings of the 1995 American Control Conference, Seattle, WA, USA*, volume 1, pages 97–101, June 1995.

[324] I. V. Nikiforov. New optimal approach to global positioning system/differential global positioning system integrity monitoring. *AIAA Journal of Guidance, Control, and Dynamic*, 19(5):1023–1033, Sept. 1996.

[325] I. V. Nikiforov. Sequential, FSS and snapshot approaches to GPS/DGPS integrity monitoring. In *Proceedings of the 10th International Meeting of the Satellite Division of the Institute of Navigation, Kansas City, MO, USA*, pages 449–458. ION, Sept. 1997.

[326] I. V. Nikiforov. Two strategies in the problem of change detection and isolation. *IEEE Transactions on Information Theory*, 43(2):770–776, Mar. 1997.

[327] I. V. Nikiforov. Quadratic tests for detection of abrupt changes in multivariate signals. *IEEE Transactions on Signal Processing*, 47(9):2534–2538, Sept. 1999.

[328] I. V. Nikiforov. A simple recursive algorithm for diagnosis of abrupt changes in random signals. *IEEE Transactions on Information Theory*, 46(7):2740–2746, July 2000.

[329] I. V. Nikiforov. A suboptimal quadratic change detection scheme. *IEEE Transactions on*

Information Theory, 46(6):2095–2107, Sept. 2000.

[330] I. V. Nikiforov. A simple change detection scheme. *Signal Processing*, 81(1):149–172, Jan. 2001.

[331] I. V. Nikiforov. A lower bound for the detection/isolation delay in a class of sequential tests. *IEEE Transactions on Information Theory*, 49(11):3037–3046, Nov. 2003.

[332] I. V. Nikiforov and B. Roturier. Statistical analysis of different RAIM schemes. In *Proceedings of the 15th International Meeting of the Satellite Division of the Institute of Navigation, Portland, OR, USA*, pages 1881–1892. ION GPS, 2002.

[333] I. V. Nikiforov and B. Roturier. Advanced RAIM algorithms: First results. In *Proceedings of the 18th International Technical Meeting of the Satellite Division of the Institute of Navigation (GNSS 2005), Long Beach, CA, USA*, pages 1789–1800. ION, Sept. 2005.

[334] I. V. Nikiforov and I. N. Tikhonov. Application of change detection theory to seismic signal processing. In M. Basseville and A. Benveniste, editors, *Detection of Abrupt Changes in Signals and Dynamical Systems*, volume 77 of *Lecture Notes in Control and Information Sciences*, pages 355–373. Springer-Verlag, Berlin, DE, 1985.

[335] I. V. Nikiforov, V. Varavva, and V. Kireichikov. Application of statistical fault detection algorithm for navigation system monitoring. In R. Isermann, editor, *Proceedings of the 1st Symposium on Fault Detection, Supervision and Safety for Technical Processes (SAFEPROCESS'91), Baden-Baden, DE*, pages 351–356. IFAC, 1991.

[336] I. V. Nikiforov, V. Varavva, and V. Kireichikov. Application of statistical fault detection algorithms to navigation systems monitoring. *Automatica*, 29(5):1275–1290, Sept. 1993.

[337] A. Novikov and B. Ergashev. Analytical approach to the calculation of moving average characteristics. *Statistical Problems of Control*, 83:110–114, 1988. In Russian.

[338] A. A. Novikov. On discontinuous martingales. *Theory of Probability and its Applications*, 20(1):11–26, 1975.

[339] A. A. Novikov. Martingale identities and inequalities and their applications in nonlinear boundary-value problems for random processes. *Mathematical Notes*, 35(3):241–249, 1984.

[340] J. S. Oakland. *Statistical Process Control (6th ed.)*. Butterworth and Heinemann, Oxford, UK, 2007.

[341] A. R. Offringa, A. G. de Bruyn, M. Biehl, S. Zaroubi, G. Bernardi, and V. N. Pandey. Post-correlation radio frequency interference classification methods. *Monthly Notices of the Royal Astronomical Society*, 405(1):155–167, June 2010.

[342] M. Ortner and A. Nehorai. A sequential detector for biochemical release in realistic environments. *IEEE Transactions on Signal Processing*, 55(7):4173–4182, July 2007.

[343] T. Oskiper and H. V. Poor. Online activity detection in a multiuser environment using the matrix CUSUM algorithm. *IEEE Transactions on Information Theory*, 48(2):477–493, Feb. 2002.

[344] B. Ottersten, M. Viberg, and T. Kailath. Analysis of subspace fitting and ML techniques for parameter estimation. *IEEE Transactions on Signal Processing*, 40(3):590–599, Mar. 1992.

[345] E. S. Page. An improvement to Wald's approximation for some properties of sequential tests. *Journal of the Royal Statistical Society - Series B Methodology*, 16(1):136–139, 1954.

[346] E. S. Page. Continuous inspection schemes. *Biometrika*, 41(1–2):100–114, June 1954.

[347] E. S. Page. Control charts for the mean of a normal population. *Journal of the Royal Statistical Society - Series B Methodology*, 16(1):131–135, 1954.

[348] S. H. Park and G. G. Vining. *Statistical Process Monitoring and Optimization*, volume 160 of *Statistics Textbooks and Monographs*. Marcel Dekker, Inc., 1999.

[349] R. J. Patton, P. M. Frank, and R. N. Clarke, editors. *Issues of Fault Diagnosis for Dynamic*

Systems. Springer-Verlag, London, UK, 2000.

[350] E. Paulson. A sequential decision procedure for choosing one of k hypotheses concerning the unknown mean of a normal distribution. *Annals of Mathematical Statistics*, 34(2):549–554, June 1963.

[351] I. V. Pavlov. A sequential decision rule for the case of many composite hypotheses. *Engineering Cybernetics*, 22:19–23, 1984.

[352] I. V. Pavlov. Sequential procedure of testing composite hypotheses with applications to the Kiefer–Weiss problem. *Theory of Probability and its Applications*, 35(2):280–292, 1990.

[353] V. Paxson. Bro: A system for detecting network intruders in real-time. *Computer Networks*, 31(23–24):2435–2463, Dec. 1999.

[354] B. Peeters and G. De Roeck. Reference-based stochastic subspace identification for output-only modal analysis. *Mechanical Systems and Signal Processing*, 13(6):855–878, Feb. 1999.

[355] B. Peeters and G. De Roeck. Stochastic system identification for operational modal analysis: A review. *ASME Journal of Dynamic Systems, Measurement, and Control*, 123(4 - Special issue on Identification of Mechanical Systems):659–667, Dec. 2001.

[356] L. Pelkowitz and S. C. Schwartz. Asymptotically optimum sample size for quickest detection. *IEEE Transactions on Aerospace and Electronic Systems*, 23(2):263–272, Mar. 1987.

[357] T. Peng, C. Leckie, and K. Ramamohanarao. Proactively detecting distributed denial of service attacks using source IP address monitoring. In N. Mitrou, K. Kontovasilis, G. N. Rouskas, I. Iliadis, and L. Merakos, editors, *Networking 2004*, volume 3042 of *Lecture Notes in Computer Science*, pages 771–782. Springer-Verlag, Berlin, DE, 2004.

[358] G. Peskir. Optimal stopping games and Nash equilibrium. *Theory of Probability and its Applications*, 53(3):558–571, 2009.

[359] G. Peskir and A. N. Shiryaev. Sequential testing problems for Poisson processes. *Annals of Statistics*, 28(3):837–859, June 2000.

[360] G. Peskir and A. N. Shiryaev. *Optimal Stopping and Free Boundary Problems*. Lectures in Mathematics, ETH Zürich. Birkhäuser, 2006.

[361] M. Pettersson. Monitoring a freshwater fish population: Statistical surveillance of biodiversity. *Environmetrics*, 9(2):139–150, Mar. 1998.

[362] J. J. Pignatiello and G. C. Runger. Comparisons of multivariate CUSUM charts. *Journal of Quality Technology*, 22(3):173–186, July 1990.

[363] V. F. Pisarenko, A. F. Kushnir, and I. V. Savin. Statistical adaptive algorithms for estimation of onset moments of seismic phases. *Physics of the Earth and Planetary Interiors*, 47:4–10, Aug. 1987.

[364] M. Pollak. Optimality and almost optimality of mixture stopping rules. *Annals of Statistics*, 6(4):910–916, July 1978.

[365] M. Pollak. Optimal detection of a change in distribution. *Annals of Statistics*, 13(1):206–227, Mar. 1985.

[366] M. Pollak. Average run lengths of an optimal method of detecting a change in distribution. *Annals of Statistics*, 15(2):749–779, June 1987.

[367] M. Pollak and D. Siegmund. Approximations to the expected sample size of certain sequential tests. *Annals of Statistics*, 3(6):1267–1282, Dec. 1975.

[368] M. Pollak and D. Siegmund. Convergence of quasi-stationary to stationary distributions for stochastically monotone Markov processes. *Journal of Applied Probability*, 23(1):215–220, Mar. 1986.

[369] M. Pollak and A. G. Tartakovsky. Asymptotic exponentiality of the distribution of first exit times for a class of Markov processes with applications to quickest change detection. *Theory*

of Probability and its Applications, 53(3):430–442, 2009.

[370] M. Pollak and A. G. Tartakovsky. Optimality properties of the Shiryaev–Roberts procedure. *Statistica Sinica*, 19(4):1729–1739, Oct. 2009.

[371] P. K. Pollet. Quasi-stationary distributions: A bibliography. Technical report, School of Mathematics and Physics, The University of Queensland, Australia, July 2012. Online.

[372] A. S. Polunchenko, G. Sokolov, and A. G. Tartakovsky. Optimal design and analysis of the exponentially weighted moving average chart for exponential data. *Sri Lankan Journal of Applied Statistics, Special Issue: Contemporary Statistical Science*, 2014. To appear.

[373] A. S. Polunchenko and A. G. Tartakovsky. On optimality of the Shiryaev–Roberts procedure for detecting a change in distribution. *Annals of Statistics*, 38(6):3445–3457, Dec. 2010.

[374] A. S. Polunchenko and A. G. Tartakovsky. State-of-the-art in sequential change-point detection. *Methodology and Computing in Applied Probability*, 14(3):649–684, Sept. 2012.

[375] A. S. Polunchenko, A. G. Tartakovsky, and N. Mukhopadhyay. Nearly optimal change-point detection with an application to cybersecurity. *Sequential Analysis*, 31(3):409–435, July 2012.

[376] H. V. Poor and O. Hadjiliadis. *Quickest Detection*. Cambridge University Press, Cambridge, UK, 2009.

[377] T. D. Popescu. Detection and diagnosis of model parameter and noise variance changes with application in seismic signal processing. *Mechanical Systems and Signal Processing*, 25(5):1598–1616, July 2011.

[378] S. C. Port. Escape probability for a half-line. *Annals of Mathematical Statistics*, 35(3):1351–1355, Sept. 1964.

[379] D. Potin, P. Vanheeghe, E. Duflos, and M. Davy. An abrupt change detection algorithm for buried landmines localization. *IEEE Transactions on GeoSciences and Remote Sensing*, 44(2):260–272, Feb. 2006.

[380] P. E. Protter. *Stochastic Integration and Differential Equations (2nd ed.)*. Springer-Verlag, Berlin, New York, 2004.

[381] S. J. Qin. Statistical process monitoring: Basics and beyond. *Journal of Chemometrics*, 17(8–9):480–502, Aug. 2003.

[382] R. Rajagopal, X. Nguyen, S. C. Ergen, and P. Varaiya. Simultaneous sequential detection of multiple interacting faults, Dec. 2010. arXiv.

[383] J. F. Ramil. Algorithmic cost estimation for software evolution. In *Proceedings of the 22nd International Conference on Software Engineering (ICSE'00), Limerick, IE*, pages 701–703, 2000.

[384] J. Ramírez and J. Jesús. A Contour nomogram for designing CUSUM charts for variance. Technical Report 33, University of Wisconsin–Madison, Center for Quality and Productivity Improvement, Madison, WI, USA, Feb. 1989.

[385] A. R. Rao, K. H. Hamed, and H.-L. Chen. *Nonstationarities in Hydrologic and Environmental Time Series*. Series on Water Science and Technology Library. Kluwer Academic Publishers, Dordrecht, NL, 2003.

[386] B. Reynolds and D. Ghosal. Secure IP telephony using multi-layered protection. In *Proceedings of the Network and Distributed System Security Symposium (NDSS'03), San Diego, CA, USA*, 2003.

[387] M. R. Reynolds. Approximations to the average run length in cumulative sum control charts. *Technometrics*, 17(1):65–71, Feb. 1975.

[388] J. A. Rice. *Mathematical Statistics and Data Analysis*. Duxbury Press, Cengage Learning Ltd., Andover, Hampshire, UK, 1995.

[389] Y. Ritov. Decision theoretic optimality of the CUSUM procedure. *Annals of Statistics*, 18(3):1464–1469, Sept. 1990.

[390] H. Robbins. Statistical methods related to the law of the iterated logarithm. *Annals of Mathematical Statistics*, 41(5):1397–1409, Oct. 1970.

[391] H. Robbins and D. Siegmund. A class of stopping rules for testing parameter hypotheses. In L. M. Le Cam, J. Neyman, and E. L. Scott, editors, *Proceedings of the Sixth Berkeley Symposium on Mathematical Statistics and Probability, June 21–July 18, 1970*, volume 4: *Biology and Health*, pages 37–41. University of California Press, Berkeley, CA, USA, 1972.

[392] H. Robbins and D. Siegmund. The expected sample size of some tests of power one. *Annals of Statistics*, 2(3):415–436, Sept. 1974.

[393] M. W. Robbins, R. B. Lund, C. M. Gallagher, and Q. Q. Lu. Changepoints in the North Atlantic tropical cyclone record. *Journal of the American Statistical Association*, 106(493):89–99, Mar. 2011.

[394] S. W. Roberts. A comparison of some control chart procedures. *Technometrics*, 8(3):411–430, Aug. 1966.

[395] P. B. Robinson and T. Y. Ho. Average run lengths of geometric moving average charts by numerical methods. *Technometrics*, 20(1):85–93, Feb. 1978.

[396] M. Roesch. Snort – lightweight intrusion detection for networks. In *Proceedings of the 13th Systems Administration Conference (LISA), Seattle, WA, USA*, pages 229–238. USENIX, Nov. 1999.

[397] K. Rohloff and T. Başar. The detection of RCS worm epidemics. In *Proceedings of the ACM Workshop on Rapid Malcode (WORM'05), Fairfax, VA, USA*, pages 81–86, New York, USA, 2005. ACM.

[398] D. Rosenfield, E. Zhou, F. H. Wilhelm, A. Conrad, W. T. Roth, and A. E. Meuret. Change point analysis for longitudinal physiological data: Detection of cardio-respiratory changes preceding panic attacks. *Biological Psychology*, 84(1):112–120, Apr. 2010.

[399] A. Rougée, M. Basseville, A. Benveniste, and G. V. Moustakides. Optimum robust detection of changes in the AR part of a multivariable ARMA process. *IEEE Transactions on Automatic Control*, 32(12):1116–1120, Dec. 1987.

[400] G. G. Roussas. *Contiguity of Probability Measures: Some Applications in Statistics*, volume 63 of *Cambridge Tracts in Mathematics and Mathematical Physics*. Cambridge University Press, Cambridge, UK, 1972.

[401] RTCA. *Minimum Operational Performance Standards for Global Positioning Systems/ Wide Area Augmentation System Airborne Equipment*. RTCA, Inc., Washington, D.C., USA, May 1998.

[402] D. Rudoy, S. G. Yuen, R. D. Howe, and P. J. Wolfe. Bayesian change-point analysis for atomic force microscopy and soft material indentation. *Journal of the Royal Statistical Society - Series C Applied Statistics*, 59(4):573–593, June 2010.

[403] I. Samy, I. Postlethwaite, and D.-W. Gu. Survey and application of sensor fault detection and isolation schemes. *Control Engineering Practice*, 19(7):658–674, 2011.

[404] A. C. Sanderson and J. Segen. Hierarchical modeling of eeg signals. *IEEE Transactions on Pattern Analysis and Machine Intelligence*, 2(4):405–414, July 1980.

[405] I. R. Savage and J. Sethuraman. Stopping time of a rank-order sequential probability ratio test based on Lehmann alternatives. *Annals of Mathematical Statistics*, 37(5):1154–1160, Sept. 1966.

[406] S. E. Schechter, J. Jung, and A. W. Berger. Fast detection of scanning worm infections. In *Proceedings of the 7th International Symposium on Recent Advances in Intrusion Detection*

(RAID'04), Sophia Antipolis, FR, pages 59–81, Sept. 2004.

[407] G. Schwarz. Asymptotic shapes of Bayes sequential testing regions. *Annals of Mathematical Statistics*, 33(1):224–236, Mar. 1962.

[408] G. Seco-Granados, J. A. Lopez-Salcedo, D. Jimenez-Banos, and G. Lopez-Risueno. Challenges in indoor global navigation satellite systems: Unveiling its core features in signal processing. *IEEE Signal Processing Magazine*, 29(2):108–131, Mar. 2012.

[409] O. Seidou and T. B. M. J. Ouarda. Recursion-based multiple changepoint detection in multiple linear regression and application to river streamflows. *Water Resources Research*, 43(7), July 2007. W07404.

[410] L. Shepp and A. N. Shiryaev. The Russian option: Reduced regret. *Annals of Applied Probability*, 3(3):631–640, Aug. 1993.

[411] W. A. Shewhart. *Economic Control of Quality of Manufactured Products*. D. Van Nostrand Co., New York, USA, 1931.

[412] W. A. Shewhart. *Statistical Method from the Viewpoint of Quality Control*. Washington D.C.: Graduate School of the Department of Agriculture, Washington D.C., USA, 1939.

[413] A. N. Shiryaev. The detection of spontaneous effects. *Soviet Mathematics – Doklady*, 2:740–743, 1961. Translation from Doklady Akademii Nauk SSSR, 138:799–801, 1961.

[414] A. N. Shiryaev. The problem of the most rapid detection of a disturbance in a stationary process. *Soviet Mathematics – Doklady*, 2:795–799, 1961. Translation from Doklady Akademii Nauk SSSR, 138:1039–1042, 1961.

[415] A. N. Shiryaev. On optimum methods in quickest detection problems. *Theory of Probability and its Applications*, 8(1):22–46, 1963.

[416] A. N. Shiryaev. On the detection of disorder in a manufacturing process - I. *Theory of Probability and its Applications*, 8(3):247–265, 1963.

[417] A. N. Shiryaev. On the detection of disorder in a manufacturing process - II. *Theory of Probability and its Applications*, 8(4):402–413, 1963.

[418] A. N. Shiryaev. Some exact formulas in a "disorder" problem. *Theory of Probability and its Applications*, 10(2):348–354, 1965.

[419] A. N. Shiryaev. *Statistical Sequential Analysis: Optimal Stopping Rules*. Nauka, Moscow, RU, 1969. In Russian.

[420] A. N. Shiryaev. *Optimal Stopping Rules*, volume 8 of *Series on Stochastic Modelling and Applied Probability*. Springer-Verlag, New York, USA, 1978.

[421] A. N. Shiryaev. Quickest detection problems in the technical analysis of the financial data. In H. Geman, D. Madan, S. R. Pliska, and T. Vorst, editors, *Mathematical Finance–Bachelier Congress 2000*, Selected Papers from the First World Congress of the Bachelier Finance Society, Paris, June 29-July 1, 2000, pages 487–521. Springer Finance, Berlin, DE, 2002.

[422] A. N. Shiryaev. From disorder to nonlinear filtering and martingale theory. In A. A. Bolibruch, Y. S. Osipov, and Y. G. Sinai, editors, *Mathematical Events of the Twentieth Century*, pages 371–397. Springer-Verlag, Berlin, Heidelberg, DE, 2006.

[423] A. N. Shiryaev. Quickest detection problems: Fifty years later. *Sequential Analysis*, 29(4):345–385, Oct. 2010.

[424] G. Shmueli and H. Burkom. Statistical challenges facing early outbreak detection in biosurveillance. *Technometrics*, 52(1):39–51, Feb. 2010.

[425] D. Siegmund. Error probabilities and average sample number of the sequential probability ratio test. *Journal of the Royal Statistical Society - Series B Methodology*, 37(3):394–401, 1975.

[426] D. Siegmund. Corrected diffusion approximations in certain random walk problems. *Ad-*

vances in Applied Probability, 11(4):701–719, Dec. 1979.

[427] D. Siegmund. Corrected diffusion approximations and their applications. In L. Le Cam and R. A. Olsen, editors, *Proceedings of the Berkeley Conference in Honor of Jerzy Neyman and Jack Kiefer*, volume 2, pages 599–617. University of California Press, Berkeley, CA, USA, 1985.

[428] D. Siegmund. *Sequential Analysis: Tests and Confidence Intervals*. Series in Statistics. Springer-Verlag, New York, USA, 1985.

[429] D. Siegmund. Change-points: From sequential detection to biology and back. *Sequential Analysis*, 32(1):2–14, Jan. 2013.

[430] D. Siegmund and E. S. Venkatraman. Using the generalized likelihood ratio statistic for sequential detection of a change-point. *Annals of Statistics*, 23(1):255–271, Feb. 1995.

[431] D. O. Siegmund and B. Yakir. Minimax optimality of the Shiryayev–Roberts change-point detection rule. *Journal of Statistical Planning and Inference*, 138(9):2815–2825, Sept. 2008.

[432] G. Simons. Lower bounds for average sample number of sequential multihypothesis tests. *Annals of Mathematical Statistics*, 38(5):1343–1364, Oct. 1967.

[433] V. A. Siris and F. Papagalou. Application of anomaly detection algorithms for detecting SYN flooding attacks. *Computer Communications*, 29(9):1433–1442, May 2006.

[434] G. M. Smith. *Statistical Process Control and Quality Improvement (3rd ed.)*. Prentice-Hall, Inc., 1997.

[435] M. Sobel and A. Wald. A sequential decision procedure for choosing one of three hypotheses concerning the unknown mean of a normal distribution. *Annals of Mathematical Statistics*, 20(4):502–522, Dec. 1949.

[436] G. Sokolov. *Multi-population Optimal Change-Point Detection*. Ph.D. Thesis, Department of Mathematics, University of Southern California, Los Angeles, CA, USA, May 2014.

[437] Y. G. Sosulin, A. G. Tartakovsky, and M. M. Fishman. Sequential detection of a correlated Gaussian signal in white noise. *Radio Engineering and Electronic Physics*, 24(4):720–732, 1979.

[438] G. Soudlenkov and V. V. Kitaev. Computationally efficient algorithm for fast transients detection, Apr. 2011. arXiv.

[439] R. Sparks, T. Keighley, and D. Muscatello. Early warning CUSUM plans for surveillance of negative binomial daily disease counts. *Journal of Applied Statistics*, 37(11):1911–1930, Nov. 2010.

[440] M. S. Srivastava and Y. Wu. Comparison of EWMA, CUSUM and Shiryayev–Roberts procedures for detecting a shift in the mean. *Annals of Statistics*, 21(2):645–670, June 1993.

[441] W. Stadje. On the SPRT for the mean of an exponential distribution. *Statistics & Probability Letters*, 5(6):389–395, Oct. 1987.

[442] P. Stoïca and R. L. Moses. *Introduction to Spectral Analysis*. Prentice-Hall, Inc., Upper Saddle River, NJ, USA, 1997.

[443] Z. G. Stoumbos, M. R. Reynolds Jr., T. P. Ryan, and W. H. Woodall. The state of statistical process control as we proceed into the 21st century. *Journal of the American Statistical Association*, 95(451):992–997, Sept. 2000.

[444] V. Strassen. Almost sure behavior of sums of independent random variables and martingales. In L. M. Le Cam and J. Neyman, editors, *Proceedings of the Fifth Berkeley Symposium on Mathematical Statistics and Probability, June 21–July 18, 1965 and December 27, 1965–January 7, 1966*, volume 2: Contributions to Probability Theory. Part 1, pages 315–343. University of California Press, Berkeley, CA, USA, 1967.

[445] R. L. Stratonovich. A new representation for stochastic integrals and equations. *SIAM Jour-*

nal of Control, 4(2):362–371, 1966.

[446] M. A. Sturza. Navigation system integrity monitoring using redundant measurements. *Journal of The Institute of Navigation*, 35(4):483–501, 1988.

[447] G. Tagaras. A survey of recent developments in the design of adaptive control charts. *Journal of Quality Technology*, 30(3):212–231, July 1998.

[448] D.-I. Tang and N. L. Geller. Closed testing procedures for group sequential clinical trials with multiple endpoints. *Biometrics*, 55(4):1188–1192, Dec. 1999.

[449] A. G. Tartakovskii. Sequential composite hypothesis testing with dependent non-stationary observations. *Problems of Information Transmission*, 17(1):29–42, 1981.

[450] A. G. Tartakovskii and I. A. Ivanova. Approximations in sequential rules for discrimination of composite hypotheses and their precision in the problem of signal detection from post-detector data. *Problems of Information Transmission*, 28(1):55–66, Jan.–Mar. 1992.

[451] A. G. Tartakovskii and I. A. Ivanova. Comparison of some sequential rules for detecting changes in distributions. *Problems of Information Transmission*, 28(2):117–124, Apr.–Jun. 1992.

[452] A. G. Tartakovsky. *Sequential Methods in the Theory of Information Systems*. Radio i Svyaz', Moscow, RU, 1991. In Russian.

[453] A. G. Tartakovsky. Asymptotically minimax multialternative sequential rule for disorder detection. In *Statistics and Control of Random Processes: Proceedings of the Steklov Institute of Mathematics*, volume 202, pages 229–236. American Mathematical Society, Providence, RI, USA, 1994.

[454] A. G. Tartakovsky. Minimax-invariant regret solution to the N-sample slippage problem. *Mathematical Methods in Statistics*, 6(4):491–508, 1997.

[455] A. G. Tartakovsky. Asymptotic optimality of certain multihypothesis sequential tests: Non-i.i.d. case. *Statistical Inference for Stochastic Processes*, 1(3):265–295, Oct. 1998.

[456] A. G. Tartakovsky. Asymptotic solution to a multi-decision change-point problem. Unpublished manuscript, 1998.

[457] A. G. Tartakovsky. Asymptotically optimal sequential tests for nonhomogeneous processes. *Sequential Analysis*, 17(1):33–62, Jan. 1998.

[458] A. G. Tartakovsky. An efficient adaptive sequential procedure for detecting targets. In D. A. Williamson, editor, *Proceedings of the IEEE Aerospace Conference, Big Sky, MT, USA*, volume 4, pages 1581–1596. IEEE, Mar. 2002.

[459] A. G. Tartakovsky. Asymptotic performance of a multichart CUSUM test under false alarm probability constraint. In *Proceedings of the 44th IEEE Conference Decision and Control and European Control Conference (CDC-ECC'05), Seville, SP*, pages 320–325. IEEE, Omnipress CD-ROM, 2005.

[460] A. G. Tartakovsky. Discussion on "Is average run length to false alarm always an informative criterion?" by Yajun Mei. *Sequential Analysis*, 27(4):396–405, Oct. 2008.

[461] A. G. Tartakovsky. Multidecision quickest change-point detection: Previous achievements and open problems. *Sequential Analysis*, 27(2):201–231, Apr. 2008.

[462] A. G. Tartakovsky. Asymptotic optimality in Bayesian changepoint detection problems under global false alarm probability constraint. *Theory of Probability and its Applications*, 53(3):443–466, 2009.

[463] A. G. Tartakovsky. Rapid detection of attacks in computer networks by quickest changepoint detection methods. In N. Adams and N. Heard, editors, *Data Analysis for Network Cyber-Security*, pages 33–70. Imperial College Press, London, UK, 2014.

[464] A. G. Tartakovsky and J. Brown. Adaptive spatial-temporal filtering methods for clut-

ter removal and target tracking. *IEEE Transactions on Aerospace and Electronic Systems*, 44(4):1522–1537, Oct. 2008.

[465] A. G. Tartakovsky, X. R. Li, and G. Yaralov. Sequential detection of targets in multichannel systems. *IEEE Transactions on Information Theory*, 49(2):425–445, Feb. 2003.

[466] A. G. Tartakovsky and G. V. Moustakides. State-of-the-art in Bayesian changepoint detection. *Sequential Analysis*, 29(2):125–145, Apr. 2010.

[467] A. G. Tartakovsky, M. Pollak, and A. S. Polunchenko. Third-order asymptotic optimality of the generalized Shiryaev–Roberts changepoint detection procedures. *Theory of Probability and its Applications*, 56(3):457–484, 2012.

[468] A. G. Tartakovsky and A. S. Polunchenko. Decentralized quickest change detection in distributed sensor systems with applications to information assurance and counter terrorism. In *Proceedings of the 13th Annual Army Conference on Applied Statistics, Rice University, Houston, TX, USA*, Oct. 2007.

[469] A. G. Tartakovsky and A. S. Polunchenko. Quickest changepoint detection in distributed multisensor systems under unknown parameters. In *Proceedings of the 11th IEEE International Conference on Information Fusion, Cologne, DE*, July 2008.

[470] A. G. Tartakovsky and A. S. Polunchenko. Minimax optimality of the Shiryaev–Roberts procedure. In *Proceedings of the 5th International Workshop on Applied Probability (IWAP'10), Madrid, SP*, Universidad Carlos III de Madrid, Colmenarejo Campus, July 2010.

[471] A. G. Tartakovsky, A. S. Polunchenko, and G. Sokolov. Efficient computer network anomaly detection by changepoint detection methods. *IEEE Journal of Selected Topics in Signal Processing*, 7(1):4–11, Feb. 2013.

[472] A. G. Tartakovsky, B. L. Rozovskii, R. B. Blažek, and H. Kim. Detection of intrusions in information systems by sequential change-point methods. *Statistical Methodology*, 3(3):252–293, July 2006.

[473] A. G. Tartakovsky, B. L. Rozovskii, R. B. Blažek, and H. Kim. A novel approach to detection of intrusions in computer networks via adaptive sequential and batch-sequential change-point detection methods. *IEEE Transactions on Signal Processing*, 54(9):3372–3382, Sept. 2006.

[474] A. G. Tartakovsky, B. L. Rozovskii, and K. Shah. A nonparametric multichart CUSUM test for rapid intrusion detection. In *Proceedings of the Joint Statistical Meetings, Minneapolis, MN, USA*, Aug. 2005.

[475] A. G. Tartakovsky and V. V. Veeravalli. Change-point detection in multichannel and distributed systems. In N. Mukhopadhyay, S. Datta, and S. Chattopadhyay, editors, *Applied Sequential Methodologies: Real-World Examples with Data Analysis*, volume 173 of *Statistics: a Series of Textbooks and Monographs*, pages 339–370. Marcel Dekker, Inc., New York, USA, 2004.

[476] A. G. Tartakovsky and V. V. Veeravalli. General asymptotic Bayesian theory of quickest change detection. *Theory of Probability and its Applications*, 49(3):458–497, 2005.

[477] M. Thottan and C. Ji. Anomaly detection in IP networks. *IEEE Transactions on Signal Processing*, 51(8):2191–2204, Aug. 2003. Special Issue on Signal Processing in Networking.

[478] D. Tjostheim. Autoregressive representation of seismic *P*-wave signals with an application to the problem of short period discriminant. *Geophysical Journal of the Royal Astronomical Society*, 43(2):269–291, 1975.

[479] S. Trivedi and R. Chandramouli. Active steganalysis of sequential steganography. In E. J. Delp III and P. W. Wong, editors, *Proceedings of SPIE*, volume 5020 - Security and Watermarking of Multimedia Contents V, pages 123–130, Santa Clara, CA, USA, Jan. 2003.

[480] S. Trivedi and R. Chandramouli. Secret key estimation in sequential steganography. *IEEE Transactions on Signal Processing*, 53(2):746–757, Feb. 2005.

[481] H. Van der Auweraer and B. Peeters. International research projects on structural health monitoring: An overview. *Structural Health Monitoring*, 2(4):341–358, Dec. 2003.

[482] C. S. Van Dobben de Bruyn. *Cumulative Sum Tests: Theory and Practice*, volume 24 of *Statistics Monograph*. Charles Griffin and Co. Ltd., London, UK, 1968.

[483] P. Van Overschee and B. De Moor. *Subspace Identification for Linear Systems*. Kluwer Academic Publishers, Boston, MA, USA, 1996.

[484] N. V. Verdenskaya and A. G. Tartakovskii. Asymptotically optimal sequential testing of multiple hypotheses for nonhomogeneous Gaussian processes in an asymmetric situation. *Theory of Probability and its Applications*, 36(3):536–547, 1991.

[485] M. Viberg. Subspace-based methods for the identification of linear time-invariant systems. *Automatica*, 31(12):1835–1853, Dec. 1995.

[486] M. Viberg and B. Ottersten. Sensor array processing based on subspace fitting. *IEEE Transactions on Signal Processing*, 39(5):1110–1121, May 1991.

[487] M. Viberg, B. Wahlberg, and B. Ottersten. Analysis of state space system identification methods based on instrumental variables and subspace fitting. *Automatica*, 33(9):1603–1616, Sept. 1997.

[488] J. von Neumann. Zur theorie der gessellshaftspiele. *Mathematische Annalen*, 100:295–320, 1928. In German.

[489] J. von Neumann and O. Morgenstern. *Theory of Games and Economic Behavior*. Princeton University Press, Princeton, NJ, USA, 1944.

[490] A. Wald. Tests of statistical hypotheses concerning several parameters when the number of observations is large. *Transactions of the American Mathematical Society*, 54(3):426–482, Nov. 1943.

[491] A. Wald. On cumulative sums of random variables. *Annals of Mathematical Statistics*, 15(3):283–296, Sept. 1944.

[492] A. Wald. Sequential tests of statistical hypotheses. *Annals of Mathematical Statistics*, 16(2):117–186, June 1945.

[493] A. Wald. Differentiation under the expectation sign in the fundamental identity of sequential analysis. *Annals of Mathematical Statistics*, 17(4):493–497, Dec. 1946.

[494] A. Wald. *Sequential Analysis*. John Wiley & Sons, Inc., New York, USA, 1947.

[495] A. Wald. *Statistical Decision Functions*. John Wiley & Sons, Inc., New York, USA, 1950.

[496] A. Wald and J. Wolfowitz. Optimum character of the sequential probability ratio test. *Annals of Mathematical Statistics*, 19(3):326–339, Sept. 1948.

[497] A. Wald and J. Wolfowitz. Bayes solutions of sequential decision problems. *Annals of Mathematical Statistics*, 21(1):82–99, Mar. 1950.

[498] L. Weiss. On sequential tests which minimize the maximum expected sample size. *Journal of the American Statistical Association*, 57(299):551–557, Sept. 1962.

[499] G. B. Wetherill and D. W. Brown. *Statistical Process Control: Theory and Practice (3rd ed.)*. Texts in Statistical Science. Chapman and Hall, London, UK, 1991.

[500] D. J. Wheeler. *Advanced Topics in Statistical Process Control: The Power of Shewhart's Charts (2nd ed.)*. SPC Press, Inc., 2004.

[501] D. J. Wheeler and D. S. Chambers. *Understanding Statistical Process Control (2nd ed.)*. SPC Press, Inc., 1992.

[502] J. Whitehead. *The Design and Analysis of Sequential Clinical Trials*. Statistics in Practice. John Wiley & Sons, Inc., Chichester, UK, 1997.

[503] R. A. Wijsman. Exponentially bounded stopping time of sequential probability ratio tests for

composite hypotheses. *Annals of Mathematical Statistics*, 42(6):1859–1869, Dec. 1971.

[504] R. A. Wijsman. Examples of exponentially bounded stopping time of invariant sequential probability ratio tests when the model may be false. In L. M. Le Cam, J. Neyman, and E. L. Scott, editors, *Proceedings of the Sixth Berkeley Symposium on Mathematical Statistics and Probability, June 21–July 18, 1970*, volume 1: Theory of Statistics, pages 109–128. University of California Press, Berkeley, CA, USA, 1972.

[505] A. S. Willsky. A survey of design methods for failure detection in dynamic systems. *Automatica*, 12(6):601–611, Nov. 1976.

[506] A. S. Willsky, J. J. Deyst, and B. S. Crawford. Two self-test methods applied to an inertial system problem. *Journal of Spacecrafts and Rockets*, 12(7):434–437, July 1975.

[507] A. S. Willsky and H. L. Jones. A generalized likelihood ratio approach to the detection and estimation of jumps in linear systems. *IEEE Transactions on Automatic Control*, 21(1):108–112, Feb. 1976.

[508] S. P. Wong. Asymptotically optimum properties of certain sequential tests. *Annals of Mathematical Statistics*, 39(4):1244–1263, Aug. 1968.

[509] W. H. Woodall. Control charts based on attribute data: Bibliography and review. *Journal of Quality Technology*, 29(2):172–183, Apr. 1997.

[510] W. H. Woodall. Controversies and contradictions in statistical process control. *Journal of Quality Technology*, 32(4):341–350, Oct. 2000.

[511] W. H. Woodall and D. C. Montgomery. Research issues and ideas in statistical process control. *Journal of Quality Technology*, 31(4):376–386, Oct. 1999.

[512] M. Woodroofe. A renewal theorem for curved boundaries and moments of first passage times. *Annals of Probability*, 4(1):67–80, Feb. 1976.

[513] M. Woodroofe. *Nonlinear Renewal Theory in Sequential Analysis*, volume 39 of *CBMS-NSF Regional Conference Series in Applied Mathematics*. SIAM, Philadelphia, PA, USA, 1982.

[514] M. Woodroofe and H. Takahashi. Asymptotic expansions for the error probabilities of some repeated significance tests. *Annals of Statistics*, 10(3):895–908, Sept. 1982.

[515] R. H. Woodward and P. L. Goldsmith. *Cumulative Sum Techniques*, volume 3 of *Mathematical and Statistical Techniques for Industry*. Oliver and Boyd for Imperial Chemical Industries, Ltd., Edinburgh, UK, 1964.

[516] Y. Xie and D. Siegmund. Sequential multi-sensor change-point detection. *Annals of Statistics*, 41(2):670–692, Mar. 2013.

[517] B. Yakir. A note on the run length to false alarm of a change-point detection policy. *Annals of Statistics*, 23(1):272–281, Feb. 1995.

[518] Q. Yao. Asymptotically optimal detection of a change in a linear model. *Sequential Analysis*, 12(3–4):201–210, Feb. 1993.

[519] E. Yashchin. On a unified approach to the analysis of two-sided cumulative sum schemes with headstarts. *Advances in Applied Probability*, 17(3):562–593, Sept. 1985.

[520] A. Yazidi, O.-C. Granmo, and B. Oommen. Tracking the preferences of users using weak estimators. In D. Wang and M. Reynolds, editors, *AI 2011: Advances in Artificial Intelligence*, volume 7106 of *Lecture Notes in Computer Science*, pages 799–808. Springer-Verlag, 2011.

[521] L. Yu, L. Yu, Y. Qi, J. Wang, and H. Wen. Traffic incident detection algorithm for urban expressways based on probe vehicle data. *Journal of Transportation Systems Engineering and Information Technology*, 8(4):36–41, 2008.

[522] S. Zacks. *Theory of Statistical Inference (Probability & Mathematical Statistics)*. John Wiley & Sons, Inc., 1971.

[523] C. H. Zhang. A nonlinear renewal theory. *Annals of Probability*, 16(2):793–825, Apr. 1988.

[524] Q. Zhang and M. Basseville. Advanced numerical computation of χ^2-tests for fault detection and isolation. In *Proceedings of the 5th Symposium on Fault Detection, Supervision and Safety for Technical Processes (SAFEPROCESS'03), Washington D.C., USA*, pages 211–216. IFAC/IMACS, June 2003.

[525] Q. Zhang, M. Basseville, and A. Benveniste. Early warning of slight changes in systems. *Automatica*, 30(1):95–113, Jan. 1994.

[526] X. Zhang, F. W. Zwiers, and P. A. Stott. Multi-model multi-signal climate change detection at regional scale. *Journal of Climatology*, 19(17):4294–4307, Sept. 2006.

[527] W. Zhou, S. Maalej, R. Zouari, and L. Mevel. Flutter monitoring with the statistical subspace method for the 2-D structure. In *Proceedings of the 25th International Modal Analysis Conference (IMAC-XXV), Orlando, FL, USA*, Feb. 2007.

[528] D. C. Zimmerman, M. Kaouk, and T. Simmermacher. Structural health monitoring using vibration measurements and engineering insight. *ASME Journal of Mechanical Design*, 117(3):214–221, June 1995.

[529] R. Zouari, L. Mevel, and M. Basseville. CUSUM test for flutter monitoring of modal dynamics. In *Proceedings of ISMA 2006 - Noise and Vibration Engineering Conference, Leuven, BE*, Sept. 2006.

[530] R. Zouari, L. Mevel, and M. Basseville. Mode-shapes correlation and CUSUM test for online flutter monitoring. In *Proceedings of the 17th IFAC Symposium on Automatic Control in Aerospace (ACA'07), Toulouse, FR*, June 2007.

[531] R. Zouari, L. Mevel, and M. Basseville. An adaptive statistical approach to flutter monitoring. *AIAA Journal of Aircraft*, 49(3):735–748, May 2012.

Index

D

E

F

G

O

P

Q

R

S